D0762168

McGraw-Hill Encyclopedia of Astronomy

McGraw-Hill Encyclopedia of Astronomy

SECOND EDITION

EDITORS IN CHIEF

Sybil P. Parker
McGraw-Hill, Inc.

Jay M. Pasachoff
Director, Hopkins Observatory
Williams College

McGraw-Hill, Inc.
New York San Francisco Washington, D.C. Auckland Bogotá
Caracas Lisbon London Madrid Mexico City Milan Montreal
New Delhi San Juan Singapore Sydney Tokyo Toronto

1 2 3 4 5 6 7 8 9 DOW/DOW 9 8 7 6 5 4 3

Library of Congress Cataloging in Publication data

McGraw-Hill encyclopedia of astronomy / editors in chief, Sybil P. Parker, Jay M. Pasachoff.—2nd ed.
p. cm.
"Most of the material in this volume has been published previously in the McGraw-Hill encyclopedia of science & technology, seventh edition" —T.p. verso.
Includes bibliographical references and index.
ISBN 0–07–045314–4
1. Astronomy—Encyclopedias. I. Parker, Sybil P. II. Pasachoff, Jay M. III. Title: McGraw-Hill encyclopedia of science & technology.
QB14.M3725 1993
520′.3—dc20 92-40523
CIP

ISBN 0-07-045314-4

McGraw-Hill Staff

Jonathan Weil
Editor

Patricia W. Albers
Editorial Administrator

Ron Lane
Art Director

Vincent Piazza
Assistant Art Director

Joe Faulk
Editing Manager

Frank Kotowski, Jr.
Senior Editing Supervisor

Ruth W. Mannino
Editing Supervisor

Suppliers

Typeset and composed by the Clarinda Company, Clarinda, Iowa.

Printed and bound by R. R. Donnelley & Sons Company, the Lakeside Press at Willard, Ohio.

Color plates printed by the Lehigh Press Inc., Pennsauken, New Jersey.

First Edition Preface

Astronomy is the most ancient science. The constellations were recognized in Mesopotamia as early as 3000 B.C.; the Egyptians were using calendars based on solar events by 2000 B.C.; and Ptolemy provided a geometrical description of the motions of celestial bodies in the second century. By the fifteenth century we had entered the age of modern astronomy, and the seventeenth century witnessed major contributions from Sir Isaac Newton (interpreted the motions of the planets); Galileo (invented the telescope); C. Huygens (discovered a satellite of Saturn); and G. D. Cassini (observed Saturn's rings). While there were many noteworthy contributions to dynamical and observational astronomy during the eighteenth and nineteenth centuries, it was not until well into the twentieth century—1957—that we entered the Space Age with the launch of *Sputnik I*.

The science of astronomy is concerned with the study of radiation received from all parts of the universe. Such observation was limited to the visual portion of the electromagnetic spectrum during the greater part of human history. This limitation was eliminated by development of sophisticated instruments with sensitivities extending into other portions of the spectrum. The development of the photographic plate in the last half of the nineteenth century made possible the study of astrophysics, one of the most active branches of present-day astronomy. Over the past few decades, data have been accumulated and discoveries made at an unprecedented rate, and there are few signs of slowing down in the near future.

Today, astronomers are exploring the universe with satellites, space probes, rockets, and high-altitude balloons, as well as optical, radio, infrared, ultraviolet, x-ray, gamma-ray, cosmic-ray, and neutrino telescopes. The discovery of quasars, pulsars, complex interstellar molecules, cosmic microwave radiation, radio galaxies, supernovae, and peculiar objects, such as SS 443, opened new vistas in observational astronomy. Similarly, theoretical astronomy and astrophysics have undergone revolutionary changes as scientists now focus on problems of cosmology and cosmogony—the expanding universe, stellar structure and evolution, the composition and age of the solar system, the interpretation of black holes—trying to answer the most fundamental questions about the nature of the universe.

The *Encyclopedia of Astronomy* provides authoritative and up-to-date coverage of this dynamic field of science. This a reference work *of,* not *about,* astronomy. As such, the articles are arranged alphabetically, rather than topically, and have been prepared by more than one hundred of the world's leading scientists, each writing about his or her area of specialization. The Encyclopedia includes entries on the theoretical, observational, and experimental aspects of astronomy as well as on the instruments used to obtain the data.

The 230 articles were taken from the *McGraw-Hill Encyclopedia of Science and Technology* (5th ed., 1982). In addition to some 400 photographs, charts, graphs, and drawings, there are 13 pages in full color. Most articles include a bibliography, and cross references are used extensively to guide the reader to related topics. There is also a detailed index for rapid access to information.

Professor George O. Abell, who served as Consulting Editor in astronomy on the parent project mentioned above, has contributed invaluable guidance in creating this specialized volume. We thank him for his interest, and thank too the many Contributors whose expertise is reflected in the articles.

This Encyclopedia fills a need for a truly comprehensive and scholarly treatment of astronomy. It will be an indispensable tools for scientists, students, librarians, hobbyists, and all others who need accurate, detailed information in a readily accessible format.

Sybil P. Parker
Editor in Chief

Second Edition Preface

The decade since the publication of the first edition of this Encyclopedia has witnessed major advances in both observational and theoretical astronomy, which are reflected in this second edition. The *Voyager* probes have completed their exploration of the solar system, providing the first close-up views of the Uranus and Neptune systems, and the *Magellan* spacecraft has carried out a remarkably detailed radar mapping of cloud-shrouded Venus. A new generation of space-based astronomical observatories has been inaugurated, while there has been a strong resurgence in the design and construction of large optical and infrared telescopes on the ground, and the laser guide-star technique promises to eventually provide ground based telescopes with resolution equaling that of telescopes in space. Meanwhile, the first naked-eye supernova in nearly 400 years has provided astronomers with a new object to test many of their theories for the first time.

The gains in observational power have been matched by advances in understanding. The observation of such objects as molecular clouds, Herbig-Haro objects, blue straggler stars, and starburst galaxies has led to greater comprehension of the evolution of both the stars and the galaxies they make up. Competing theories of the evolution of the universe itself have been subjected to increasingly stringent observational tests. The rapid pace of discovery is reflected in the fact that 21 new articles have been written for this Encyclopedia and 20 articles have been revised or rewritten to incorporate important new developments, including the discovery of fluctuations in the cosmic microwave radiation.

The Editors

McGraw-Hill Encyclopedia of Astronomy

A-Z

Aberration

The apparent change in direction of a source of light caused by an observer's component of motion perpendicular to the impinging rays; also, a departure of an optical image forming system from ideal behavior.

Aberration of Light

This effect can be visualized by first imagining a stationary telescope (**illus.** *a*) aimed at a luminous source such as a star, with photons traveling concentrically down the tube to an image at the center of the focal plane. Next the telescope is given a component of motion perpendicular to the incoming rays (illus. *b*). Photons passing the objective require a finite time to travel the length of the tube. During this time the telescope has moved a short distance, causing the photons to reach a spot on the focal plane displaced from the former image position. To return the image to the center, the telescope must be tilted in the direction of motion by an amount sufficient to ensure that the photons once again come concentrically down the tube in its frame of reference (illus. *c*). The necessary tilt angle α is given by $\tan \alpha = v/c$, where v is the component of velocity perpendicular to the incoming light and c is the velocity of light. (An analogy illustrating aberration is the experience that, in order for the feet to remain dry while walking through vertically falling rain, it is necessary to tilt an umbrella substantially forward.)

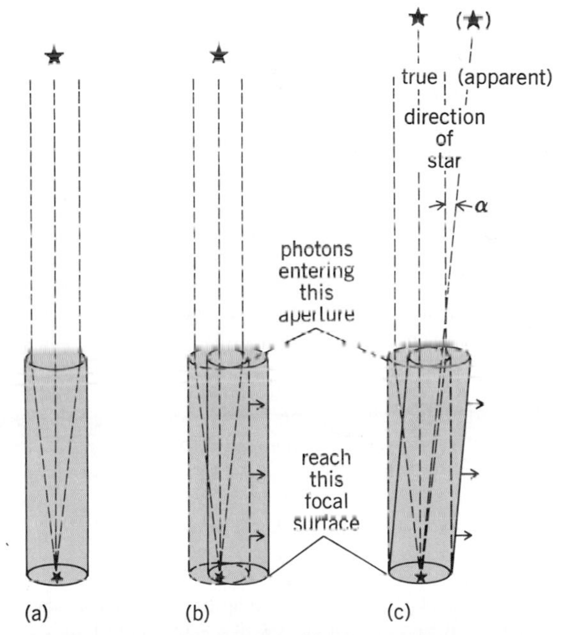

Demonstration of aberration. (*a*) Fixed telescope; photons form image at center of focal plane. (*b*) Moving telescope; image is displaced from center. (*c*) Tilted moving telescope is required to restore image to center.

Aberration of light was discovered by the English astronomer James Bradley in 1725 while searching unsuccessfully for parallax of nearby stars, using the Earth's orbital diameter as baseline. He found instead the necessity to compensate continuously for the Earth's velocity in its elliptical orbit, an effect about a hundred times greater than the parallax of typical nearby stars. This discovery provided the first direct physical confirmation of the Copernican theory.

Annual aberration causes stars to appear to describe ellipses whose major axes are the same length, the semimajor axis (usually expressed in arc-seconds) being the constant of aberration $\alpha = 20''.496$. The minor axis of the ellipse depends on the star's ecliptic latitude β, and is given by $2\alpha \sin \beta$. The small diurnal aberration arising from the Earth's axial rotation depends on geographic latitude ϕ, and is given by $0''.31 \cos \phi$.

A second important application of aberration has been its clear-cut demonstration that, as is axiomatic to special relativity, light reaching the Earth has a velocity unaffected by the relative motion of the source toward or away from the Earth. This is shown by the fact that the aberration effect is the same for all celestial objects, including some quasars with apparent recessional velocities approaching the speed of light. *See* Earth rotation and orbital motion; Parallax; Quasar.

Harlan J. Smith

Optical Aberration

The standard aberrations that are often found in optical systems include spherical aberration, chromatic aberration, coma, and astigmatism. These aberrations can appear in systems involving both lenses and mirrors.

Spherical aberration. This is a deviation from perfect focus because spherical mirrors do not focus parallel light rays to a point. While a parabolic surface focuses the parallel rays that come from distant objects to a single point, a spherical surface focuses to a point at its center only the rays that come from the same point. Thus, systems that use spherical optics do not bring parallel rays to a perfect focus. Nonetheless, for many purposes the spherical aberration is sufficiently small that the ease in making spherical optics overwhelms the disadvantages. Optical systems, both reflective and refractive, with different focal lengths for rays hitting the optics at different distances from the central axis have spherical aberration.

The Hubble Space Telescope, launched in 1990, is the best-known example of spherical aberration. Its 2.4-m (94-in.) mirror was tested with an improperly adjusted null corrector, so images are not as compact as planned. Only 15% of the light is focused into the central 0.15 arc-second, instead of the approximately 70% planned. *See Satellite astronomy.*

Chromatic aberration. This results because lenses have different focal lengths for different wavelengths. Thus, different colors focus at different points. For example, if the image is recorded on black-and-white film or a CCD that is sensitive over a wide range of colors, the image appears blurred by the chromatic aberration. Use of a monochromatic filter, which passes only one color, allows lenses with chromatic aberration to be used to give excellent images, though the system must be focused differently for images at different wavelengths. Nonetheless, lenses with known chromatic aberration are often useful. In solar coronagraphs, for example, single lenses are used in spite of their chromatic aberration because they have lower scattering properties than the compound lenses that would have to be used to compensate for the chromatic aberration. *See Coronagraph.*

Chromatic aberration arises because glasses and other transparent substances have different indices of refraction at different wavelengths. Chromatic aberration can be controlled by using compound lenses in which each element is made of a different material. The materials can be chosen so that specific wavelengths are focused at the same position. It is not possible to find combinations that focus all wavelengths at the same place, so some chromatic aberration is always present. Achromatic lenses have two wavelengths that focus at the same position. The maximum deviation from focus at wavelengths in between is, at least, more controlled than in single lenses. Apochromatic lenses have three wavelengths that focus at the same position, and the maximum deviation in between is usually less than in achromats. New technologies that control the index of refraction of materials have the potential of making lenses with lower chromatic aberration than before.

Coma. This optical aberration arises when rays incident on an axis far from the central axis of a lens form an asymmetric image. The images of points appear elongated and spread out, like tiny comets, giving this aberration its name. Wide-field photographs often show coma near their edges. Coma shows readily on the outer parts of photographs of star fields taken with ordinary cameras.

Astigmatism. This optical aberration results from an optical system focusing asymmetrically. Thus, astigmatism makes light from an object that crosses the lens along one line focus on a different surface than light that crosses the lens along other lines. Cylindrical correction compensates for astigmatism. The human eye is often astigmatic, and such cylindrical correction commonly appears in eyeglasses.

Field curvature. This aberration results in extended objects being in focus over a curved surface rather than a plane. Sometimes field curvature is compensated for by curving the detector surface. In Schmidt cameras, for example, the plates or film are curved to match the field curvature. This curvature is often temporary, caused and controlled by pressure or by vacuum. *See Schmidt camera.*

Distortion. In this aberration, the shape of the optical element does not transform straight lines into straight lines. Thus the lens or mirror transforms straight lines into curves. Pincushion distortion and barrel distortion curve off-axis lines in opposite directions so that a square centered at the axis maps into a quadrilateral with concave or convex sides, respectively. *See Optical telescope.*

Jay M. Pasachoff

Bibliography. G. O. Abell, D. Morrison, and S. Wolff, *Exploration of the Universe*, 6th ed., 1991; M. Cagnet, *Atlas of Optical Phenomena*, 1971; E. Hecht and A. Zajac, *Optics*, 2d ed., 1987; J. M. Pasachoff, *Astronomy: From the Earth to the Universe*, 4th ed. rev., 1993; W. H. Price, The photographic lens, *Sci. Amer.*, 235(2):72–83, August 1976; R. Wolfson and J. M. Pasachoff, *Physics*, 2d ed., 1993; M. Zeilik, S. A. Gregory, and E. v. P. Smith, *Introductory Astronomy and Astrophysics*, 3d ed., 1992.

Adaptive optics

The science of optical systems in which a controllable optical element, commonly a deformable mirror, is used to optimize the performance of the system, for example, to maintain a sharply focused image in the presence of wavefront aberrations due to distortion of a telescope mirror or air turbulence in the viewing path. A distinction is made between active optics, which covers any optical component that can be modified or adjusted by external control, and adaptive optics, which applies to closed-loop feedback systems employing sensors and data processors.

In a typical adaptive optical system (**Fig. 1**), the distorted light beam to be compensated is reflected from the deformable mirror and is sampled by a beam splitter. The light sample is analyzed in a wavefront sensor that determines the error in each part of the beam. The required corrections are computed and applied to the deformable mirror whose surface forms the shape necessary to flatten the reflected wavefront. The result is to remove the optical error at the sampling point so that the light passing through the beam splitter may be focused to a sharp image. Nonlinear optical devices are also capable of performing some adaptive optics functions; these devices operate at high optical power levels.

Since the development of adaptive optics in the late 1960s, its main applications have been in astronomical instrumentation and laser systems. Only astronomical applications will be discussed in this article. In the current generation of ground-based astronomical telescopes, active optical systems are used to control the figure of the primary mirror, while adaptive optical systems provide real-time compensation for atmospheric turbulence.

Primary-mirror control. It is recognized that the conventional method of casting mirror blanks is impractical for large primary mirrors over about 5 m (16 ft) in diameter (the size of the Palomar telescope) because of their excessive weight and thermal inertia. Three different approaches are being pursued for new mirrors of 8–10-m (26–33-ft) diameter: lightweighted or honeycomb mirror structures, thin meniscus mirrors, and segmented mirrors. Lightweighted mirrors have a stiffness similar to that of a solid blank but with only about one-quarter of the mass; they thus reach temperature equilibrium quickly. Thin meniscus mirrors employ a thin, flexible facesheet on an array of adjustable supports (actuators). Segmented mirrors (like the Keck 10-m or 33-ft telescope opened in 1992) employ an array of individual panels, each supported by at least three actuators.

In each case, to obtain the full angular resolving power of the optical aperture, the whole mirror surface must be kept in phase to better than one-quarter wavelength of light in the presence of temperature variations and changes in the direction of the gravity vector due to telescope pointing. This phasing is achieved by active control of the mirror supports. In many cases the control is purely mechanical, using pressure and position sensors on the mirror structure. True adaptive optical control of such mirrors requires measurement of the optical quality of the mirror by using a wavefront sensor with an external reference such as a star. An example is the European Southern Observatory 3.5-m (11.5-ft) New Technology Telescope, installed at La Silla, Chile, in 1990.

Atmospheric turbulence compensation. The major application of adaptive optics in astronomy is compensation of atmospheric turbulence, which causes the effect known as seeing, in which the image of a star, essentially a point source, is expanded into a blurred disk. The size of this seeing disk depends on the turbulence strength and is typically about 1 second of arc, independent of the size of the telescope. Because of atmospheric turbulence, large ground-based telescopes cannot achieve their inherent angular resolving power, the angular resolution being reduced to approximately that of a 10-cm (4-in.) aperture. Adaptive optics can in principle compensate the optical effects of turbulence and restore the full resolution of the telescope.

For turbulence compensation at visible wavelengths, a fast adaptive optics system with a bandwidth of about 300 Hz is required. Large telescope mirrors cannot be driven at this rate. The solution is to employ a small high-bandwidth deformable mirror in the imaging path. For average turbulence, one actuator is required for each 10 × 10-cm (4 × 4-in.) area of the primary mirror. Thus to compensate a telescope of 1-m (39-in.) aperture about 80 actuators are required, while a 4-m (158-in.) telescope would require over 1200 actuators.

To provide the data to drive the deformable mirror, a very sensitive high-speed wavefront sensor is required that will operate by using the radiation from a star or similar object. There are two fundamental limitations to the use of adaptive optics for turbulence compensation: the photon limit and the size of the isoplanatic patch. To make useful wavefront measurements, it is necessary to collect at least 100 photons from the reference source within the atmospheric change time of 1 or 2 milliseconds, which necessitates a minimum brightness of about the 10th visual magnitude for the reference star, greatly restricting the number of objects that can be compensated and viewed directly. The isoplanatic patch is the field of view around the reference star over which the wavefront measurement is valid. Because of the presence of turbulent layers high in the atmosphere, this angle is extremely small, often only a few seconds of arc at visible wavelengths. The result of these limitations is that full turbulence compensation at visible wavelengths is possible only in the immediate vicinity of a few bright stars or similar objects.

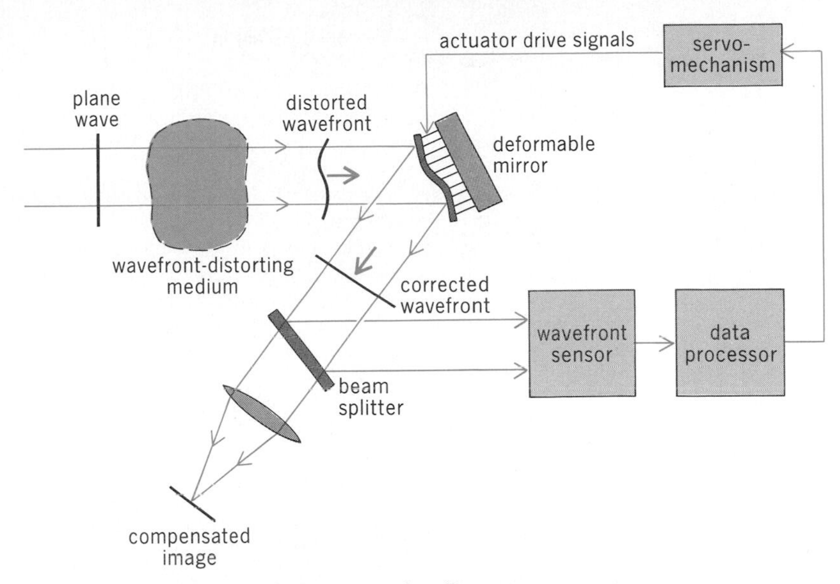

Fig. 1. Typical adaptive optical system using discrete components.

Several approaches are being pursued to overcome these limitations to allow the more general use of adaptive optics in astronomy.

Use of partial compensation. Subapertures can be used that are larger than the coherence length of the atmospheric turbulence. The larger subapertures enable more photons to be collected per measurement from the reference source, and also increase the isoplanatic patch size, at the expense of less perfect compensation. This approach is particularly useful at the long-wavelength end of the visible spectrum (0.6–0.8 micrometer), where the effects of turbulence are less severe.

Observation at infrared wavelengths. Several atmospheric windows exist in the infrared band between 2 and 5 μm, through which celestial radiation passes through the Earth's atmosphere. Because of wavelength scaling effects, a dramatic improvement in adaptive optical performance capability is obtained at these wavelengths: fewer actuators are required, the time available for each measurement is longer, and the isoplanatic patch is larger so that the field compensated by each guide star is expanded. At a wavelength of 2.2 μm, the sky coverage with natural guide stars approaches 30%, while at 5 μm it approaches 100%. SEE INFRARED ASTRONOMY.

Laser guide stars. Neither of the above approaches

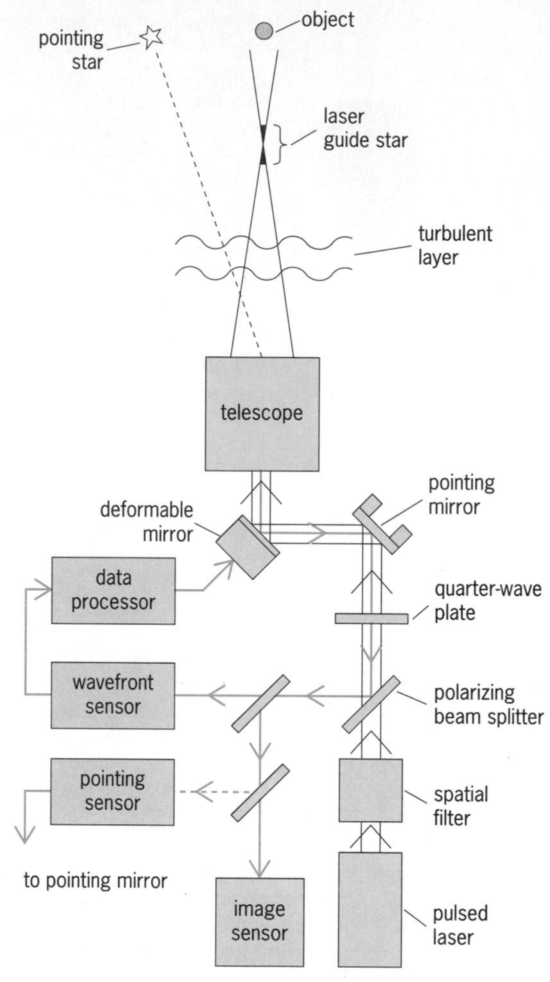

Fig. 2. Adaptive optical system using a laser guide star.

solves the problem of full compensation at shorter wavelengths, down to the atmospheric ultraviolet cut-off at about 0.35 μm. The only known solution to full image compensation with adaptive optics over the entire band is the use of laser guide stars. Artificial stars may be created in the upper atmosphere either by backscatter from air molecules (Rayleigh scattering) at 10–40-km (6–25-mi) altitude or by stimulating fluorescence in the mesospheric sodium layer at around 90 km (55 mi). The principle of adaptive optics using laser guide stars is shown in **Fig. 2**. The operating sequence is generally as follows. The deformable mirror is first flattened, and the telescope is aimed at the object by using offset pointing from a natural star. A laser pulse of typically 5-microsecond duration is then projected through the telescope to produce a laser guide star of length about 1.5 km (0.9 m) and diameter about 1 m (3 ft). The backscattered light passes through the turbulent atmosphere, is collected by the telescope, and is directed to the wavefront sensor, which is optically gated in time to select the required altitude. The static focus error of the received wavefront (due to the finite altitude of the laser guide star) is removed, and the residual wavefront errors are measured. The drive signals required to null the wavefront error are then computed and applied to the deformable mirror. The imaging sensor then records a compensated image of the object. The time between measurement and correction must be less than the change time of the atmospheric turbulence. The repetition rate of the correction cycle is limited mainly by the pulse repetition time of the laser. This time is usually longer than the relaxation time of the atmosphere, so that successive measurements are uncorrelated. The adaptive optics components then operate in an open-loop mode and must be accurately calibrated.

A single laser guide star samples a partial (conical) volume of the complete optical path, resulting in an additional error termed focal anisoplanatism. The size of this error depends on the altitude of the laser guide star and is most serious with lower-altitude Rayleigh scattering. The solution is to use multiple laser guide stars and to stitch together the wavefront measurements so as to cover the complete optical field of view. The feasibility of this process has been demonstrated.

System design and performance. The performance of an adaptive optical system for astronomical telescopes depends on external phenomena such as turbulence strength, isoplanatic angle, and wind velocity, which vary over wide ranges, and on internal factors that are designed into the system such as actuator spacing, mirror influence function, wavefront sensor detector efficiency, and response time. To determine the overall system performance, the error induced by these parameters on each major component is evaluated; these errors are summed to find the total system error. In this way an adaptive optical system can be optimized for any specificed set of conditions, and the image quality can be predicted.

The major errors remaining after the adaptive optical system has compensated the incoming wavefront are fitting error, isoplanatic error, temporal error, and photon error.

Fitting error. This error is due to the inability of the wavefront corrector exactly to match the incoming wavefront. It is a function of the ratio d/r_0, where d is the actuator spacing and r_0 is the turbulence coherence length, which is typically 5–10 cm (2–4 in.) at visible wavelengths. Fitting error also depends on the influence function of the corrector, which in the case of a deformable mirror is the shape of the deformation produced by each actuator. Continuous faceplate mirrors have a smooth gaussian-type influence function that gives a smaller fitting error per actuator than segmented mirrors with piston-type actuators. In all cases, reducing d gives better compensation at the expense of a larger number of actuators.

Isoplanatic error. This error occurs when the reference beam used for wavefront measurement does not exactly coincide with the optical path used for imaging. In most cases, angular displacements produce the most serious effect, but errors can also be caused by lateral displacement (using different sections of the telescope aperture) and by large differences in range (focal anisoplanatism), which is a serious problem with laser guide stars. The isoplanatic angle θ_0 is defined as the angle between two lines of sight for which the root-mean-square difference in the wavefront error is 1 radian. The field of view over which an adaptive optics system will produce useful compensation is about twice the isoplanatic angle.

Temporal error. Atmospheric turbulence is a dynamic process. Temporal errors are caused by the time delay between the measurement and correction of a wavefront disturbance. The system response time is composed of the integration time of the wavefront sensor detectors, the time required to compute the necessary corrections, and the response time of the deformable mirror actuators. The temporal error depends on the ratio of this response time to the characteristic change time of the atmospheric turbulence, which is roughly equal to the coherence length r_0 divided by the effective wind velocity.

Photon error. This error is caused by the quantum nature of light. Because of dim reference sources and short integration times, wavefront sensors for astronomical applications typically operate under photon-starved conditions in which the number of photons detected per measurement may be on the order of only 100. The signal-to-noise ratio of the wavefront measurement cannot exceed the square root of the number of detected photons, resulting in significant photon error. The photon count may be increased by using larger subapertures or by increasing the detector integration time, but as noted above, both of these strategies increase other errors. Thus, for each set of external conditions, there are optimum values of subaperture size and integration time that minimize the total system error.

Wavelength dependence. The atmospheric parameters r_0 and θ_0 have a powerful effect on the overall performance of adaptive optical systems, and both of these scale with the 6/5 power of wavelength. Operating at longer wavelengths allows larger actuator spacing and integration times, thereby reducing all the basic errors and allowing larger sky coverage with fainter reference sources. At wavelengths approaching 10 μm, the value of r_0 increases to several meters so wavefront compensation is no longer required for most telescopes.

Strehl ratio. A useful parameter for comparing optical performance is the Strehl ratio, which is defined as the ratio of the measured peak intensity of the image of a point source to the peak intensity of an unaberrated system. An optically perfect system has a Strehl ratio of unity. For small aberrations, the Strehl ratio can be approximated as $\exp(-\sigma^2)$, where σ^2 is the wavefront-error variance over the aperture in radians squared. In laser systems, where energy is very expensive and must be concentrated in the smallest spot, high Strehl ratios (greater than 0.9) are desired, requiring residual root-mean-square wavefront errors of 1/20 wave or less. For imaging systems, however, the required Strehl ratio depends on many factors such as the image structure and signal-to-noise ratio. For many imaging tasks, a Strehl ratio of 0.3 is acceptable, relaxing the residual root-mean-square error requirement to 1/6 wave. This relaxation greatly simplifies the design of adaptive optical systems. For comparison, the uncompensated Strehl ratio is generally less than 0.05, although it may be considerably higher for brief periods.

Technology. Adaptive optical systems may be broadly classified into conventional systems employing discrete optical and electronic components, and unconventional systems using integrated devices in which the basic functions are performed by physical processes within a nonlinear medium. For astronomical use, photodetectors having the ultimate sensitivity are required for the wavefront sensor, which at present mandates the use of discrete components. Integrated devices using nonlinear optics have been developed for laser systems operating at relatively high power levels, but these devices do not currently have sufficient sensitivity for astronomical instrumentation.

Conventional systems employ three main components: a wavefront sensor, a data processor, and a wavefront compensator.

Wavefront sensors. Most practical wavefront sensors measure the gradient of the wavefront over an array of subapertures in the incoming beam. These local gradient measurements are then reconstructed into a map of the wavefront error over the whole aperture. The two main techniques for gradient sensing use the imaging Hartmann sensor and the shearing interferometer. The imaging Hartmann sensor is a modification of a classic method for measuring the optical figure of large telescope mirrors. The optical beam to be measured is divided into an array of contiguous subapertures, each containing a lens that brings the light from the reference source to a separate focus (**Fig. 3*a***). A wavefront tilt in any subaperture displaces the focused spot. The displacement in each subaperture is measured by a photodetector array. With a plane-wave input, the focused spot in each subaperture is centered on its detector. Displacement of the spots reveals the direction and magnitude of the two-dimensional wavefront gradient, and the output of each detector channel is an electrical signal representing this gradient.

The principle of shearing interferometry is to split the wavefront to be measured into two replicas that are mutually displaced by a small distance and then recombined. By equalizing the optical paths, an interference pattern is formed with incoherent light. Lateral displacement or shear is most often used, in which case the intensity of the interference pattern is proportional to the wavefront gradient in the direction of shear. By using a diffraction grating to produce the replicas (Fig. 3*b*), the shear distance becomes proportional to wavelength, in which case the interference fringes coincide at all wavelengths so that the sensor operates with white light. An array of photodetectors is used to measure the fringe pattern. A further advantage of using a grating to produce the shear is that it enables the detector signal to be modulated or chopped simply by moving the grating, eliminating errors due to calibration and drift. Two such interferometers are normally used, making wavefront gradient measurements in two orthogonal directions. *See Interferometry*.

Other approaches to wavefront sensing are image sharpening and curvature sensing. The principle of image sharpening is to make trial phase changes in

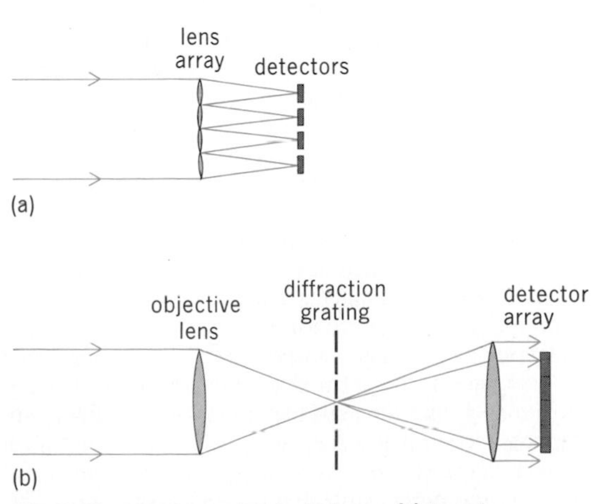

Fig. 3. Wavefront-sensing techniques. (*a*) Imaging Hartmann sensor. (*b*) Shearing interferometer that uses diffraction grating.

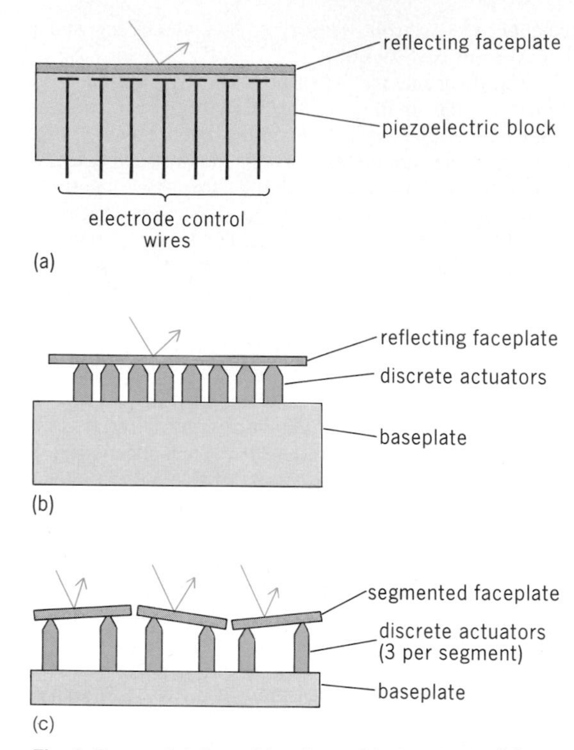

Fig. 4. Types of deformable mirror. (*a*) Monolithic. (*b*) Thin-plate. (*c*) Segmented.

the optical aperture—by using a deformable mirror, for example—and to evaluate the effect of each change on the image quality. For a given reference-star brightness and subaperture size, the wavefront measurement error of an image-sharpening sensor increases with the number of subapertures (and therefore with the area of the telescope aperture), whereas for wavefront gradient sensors (Hartmann and shearing interferometer), for the same conditions, the error is independent of the number of subapertures. In spite of its optical simplicity, therefore, image sharpening has limited value as a wavefront sensing technique for astronomical applications, even for small apertures.

Curvature sensing is a technique that promises to reduce considerably the computation needed in discrete adaptive optics systems. Instead of measuring wavefront slope (a vector quantity), the local wavefront curvature (a scalar quantity) of the input beam is measured. The curvature measurements can be applied directly to a deformable mirror by employing bimorph actuators that control the local curvature of the wavefront.

Data processors. The gradient measurements of wavefront sensors must be reconstructed into a coherent error map covering the whole aperture before they can be used to control a deformable mirror. There are two main approaches: the parallel analog network and digital computation. In the analog network, the wavefront slope measurements are converted into electric currents that are injected into a resistor network representing the actuator array. The voltages appearing at the nodes of the network then represent the actuator positions giving the least overall error. The digital approach accomplishes the same result by solving a set of simultaneous equations. For a system with N actuators, the number of multiply-add operations is $2N \times N$. For large apertures this computational load is considerable.

Wavefront compensators. In principle, wavefront compensators may either shift the optical phase of a light beam or change the optical path length. Phase-shift devices such as electrooptical crystals have a limited range of correction (one wave) and are often spectrally limited. Deformable mirrors employ mechanical displacement of the reflecting surface to control the optical path length. A major problem in the design of these mirrors is to provide the necessary surface control without compromising overall mechanical stability. Advantages of deformable mirrors include high reflection efficiency, wide spectral bandwidth, and large correction range.

Three main types of deformable mirror have been developed: the monolithic deformable mirror, the thin-plate mirror using discrete actuators, and the segmented mirror. The monolithic mirror employs a solid block of piezoelectric ceramic such as lead zirconium titanate, with an array of electrodes at the top surface, to which is bonded a thin sheet of glass forming the reflecting surface (**Fig. 4***a*). Control voltages applied to the electrodes produce local deformation of the surface. These devices have good stability and frequency response but require high voltages (typically 2000 V for a 1-μm deflection).

The thin-plate mirror employs an array of discrete multilayer piezoelectric actuators mounted on a rigid baseplate (Fig. 4*b*). The use of multilayer actuators considerably reduces the control voltages required, 300 V typically producing a stroke of 7 μm.

In contrast to the devices discussed above, which have continuous facesheets, segmented mirrors are composed of many separate panels (Fig. 4*c*). Each segment is normally supported on three actuators that provide tip, tilt, and piston adjustments. One advantage of segmented mirrors is that the individual panels are relatively small and easier to fabricate than one large mirror. However, the alignment of the segments to maintain a properly phased surface is quite difficult.

Prospects. Adaptive optical systems have been employed in the past only for a few specialized applications; the general use of this technology in ground-based astronomy has been slowed by the small sky coverage offered by conventional systems. The development of laser guide stars removes this constraint, and is expected to stimulate wider use of adaptive optics for compensation of atmospheric turbulence in the next generation of Earth-bound telescopes.

John W. Hardy

Bibliography. M. A. Ealey (ed.), Active and adaptive optical systems, *Proc. SPIE*, vol. 1542, 1991; M. A. Ealey (ed.), Active and adaptive optical components, *Proc. SPIE*, vol. 1543, 1991; J. W. Hardy, Active optics: A new technology for the control of light, *Proc. IEEE*, 66:651–697, 1978; R. A. Humphries et al., Atmospheric-turbulence measurements using a synthetic beacon in the mesospheric sodium layer, *Opt. Lett.*, 16:1367–1369, 1991; D. V. Murphy et al., Experimental demonstration of atmospheric compensation using multiple synthetic beacons, *Opt. Lett.*, 16:1797–1799, 1991; R. K. Tyson, *Principles of Adaptive Optics*, 1991.

Albedo

A term referring to the reflecting properties of surfaces. White surfaces have albedos close to 1; black surfaces have albedos close to 0.

Several types of albedos are in common use. The Bond albedo (A_B) determines the energy balance of a planet or satellite and is defined as the fraction of the total incident solar energy that the planet or satellite reflects to space. The "normal albedo" of a surface, more properly called the normal reflectance (r_n), is a measure of the relative brightness of the surface when viewed and illuminated vertically. Such measurements are referred to as a perfectly white Lambert surface—a surface which absorbs no light and scatters the incident energy isotropically—usually approximated by magnesium oxide (MgO), magnesium carbonate ($MgCO_3$), or some other bright material. *See* Planet.

Bond albedos for solar system objects range from 0.9 for Saturn's icy satellite Enceladus and Neptune's Triton to values as low as 0.01–0.02 for dark objects such as the satellites of Mars (**Table 1**). Cloud-shrouded Venus has the highest Bond albedo of any planet (0.76). The value for Earth is 0.35. The Bond albedo is defined over all wavelengths, and its value therefore depends on the spectrum of the incident radiation. For objects in the outer solar system not yet visited by spacecraft (such as Pluto), the values of A_B in Table 1 are estimates derived indirectly, since for these bodies it is impossible to measure the scattered radiation in all directions from Earth.

Normal reflectances of some common materials are listed in **Table 2**. The normal reflectances of many materials are strongly dependent on wavelength, a fact that is commonly used in planetary science to infer the composition of surfaces remotely. While the Bond albedo cannot exceed unity, the normal reflectance of a surface can if the material is more backscattering at opposition than the reference surface.

In the case of solar system objects, a third type of albedo, the geometric albedo (p), is commonly defined. It is the ratio of incident sunlight reflected in the backscattering direction (zero phase angle or opposition) by the object, to that which would be reflected by a circular disk of the same size but covered with a perfectly white Lambert surface. Objects like the Moon, covered with dark, highly textured surfaces, show uniformly bright disks at opposition (no limb darkening), for these, $p = r_n$. At the other extreme, a planet covered with a Lambert-like visible surface (frost or bright cloud) will be limb-darkened as the cosine of the incidence angle at opposition, and $p = \frac{2}{3} r_n$. Table 1 lists the visual geometric albedo p_v, which is the geometric albedo at a wavelength of 550 nanometers.

For solar system objects, the ratio A_B/p is called the phase integral and is denoted by q. Here p is the value of the geometric albedo averaged over the incident spectrum, and does not generally equal p_v given in Table 1. Values range from near 1.5 for cloud-covered objects (1.4 for Saturn and 1.3 for Venus) to 0.2 for the very dark and rugged satellites of Mars: Phobos and Deimos. The value for the Moon is about 0.6.

It is possible to define other types of albedos based on the geometry and nature (diffuse or collimated) of the incident beam and scattered beams. For such bidirectional albedos or reflectances, the angles of incidence and scattering as well as the angle between these two directions (phase angle) must be specified. Finally, in determining the energy balance of surfaces, one is often interested in the analog for a flat surface of the bond albedo, that is, the fraction of the incident energy that the surface reflects. For a plane parallel beam incident at an angle i, this quantity, $A(i)$, is generally close to the value A_B, but can be significantly larger than A_B for some surfaces, for large values of i.

Joseph Veverka

Table 1. Bond albedos and visual geometric albedos of selected solar system objects

Object	Bond albedo (A_B)	Visual geometric albedo (p_v)
Mercury	0.12	0.14
Venus	0.76	0.59
Earth	0.35	0.37
Mars	0.24	0.15
Jupiter	0.34	0.45
Saturn	0.34	0.46
Uranus	0.34	0.48
Neptune	0.28	0.50
Pluto	0.6(?)	0.76
Moon	0.12	0.14
Phobos (M1)	0.02	0.05
Deimos (M2)	0.02	0.06
Io (J1)	0.56	0.63
Europa (J2)	0.58	0.68
Ganymede (J3)	0.38	0.43
Callisto (J4)	0.13	0.17
Amalthea (J5)	0.02	0.06
Mimas (S1)	0.60	0.75
Enceladus (S2)	0.90	1.00
Tethys (S3)	0.60	0.80
Dione (S4)	0.45	0.55
Rhea (S5)	0.45	0.65
Titan (S6)	0.20	0.20
Ariel (U1)	0.21	0.39
Umbriel (U2)	0.10	0.21
Titania (U3)	0.15	0.27
Oberon (U4)	0.12	0.23
Miranda (U5)	0.18	0.32
Triton (N1)	0.90	0.75
Nereid (N2)	0.07	0.14
1989N1	0.03	0.06
Ceres	0.03	0.06
Vesta	0.12	0.23
Comet Halley (nucleus)	0.01	0.03

Table 2. Normal reflectances of materials*

Material	Albedo	Material	Albedo
Lampblack	0.02	Granite	0.35
Charcoal	0.04	Olivine	0.40
Carbonaceous meteorites	0.05	Quartz	0.54
Volcanic cinders	0.06	Pumice	0.57
Basalt	0.10	Snow	0.70
Iron meteorites	0.18	Sulfur	0.85
Chondritic meteorites	0.29	Magnesium oxide	1.00

*Powders; for wavelengths near 0.5 micrometer.

Alfvén waves

Propagating oscillations in electrically conducting fluids or gases in which a magnetic field is present.

Magnetohydrodynamics deals with the effects of

magnetic fields on fluids and gases which are efficient conductors of electricity. Molten metals are generally good conductors of electricity, and they exhibit magnetohydrodynamic phenomena. Gases can be efficient conductors of electricity if they become ionized. Ionization can occur at high temperatures or through the ionizing effects of high-energy (usually ultraviolet) photons. A gas which consists of free electrons and ions is called a plasma. Most gases in space are plasmas, and magnetohydrodynamic phenomena are expected to play a fundamental role in the behavior of matter in the cosmos. *See Cosmic electrodynamics.*

Waves are a particularly important aspect of magnetohydrodynamics. They transport energy and momentum from place to place and may, therefore, play essential roles in the heating and acceleration of cosmical and laboratory plasmas. Waves occur whenever there is a temporal change somewhere in the fluid. In effect, the waves tell the rest of the fluid how to adjust to the change. Since nothing in the universe is temporally steady, waves are expected to occur ubiquitously.

A wave is a propagating oscillation. If waves are present, a given parcel of the fluid undergoes oscillations about an equilibrium position. The parcel oscillates because there are restoring forces which tend to return it to its equilibrium position. In an ordinary gas, the only restoring force comes from the thermal pressure of the gas. This leads to one wave mode: the sound wave. If a magnetic field is present, there are two additional restoring forces: the tension associated with magnetic field lines, and the pressure associated with the energy density of the magnetic field. These two restoring forces lead to two additional wave modes. Thus there are three magnetohydrodynamic wave modes. However, each restoring force does not necessarily have a unique wave mode associated with it. Put another way, each wave mode can involve more than one restoring force. Thus the usual sound wave, which involves only the thermal pressure, does not appear as a mode in magnetohydrodynamics.

The three modes have different propagation speeds, and are named fast mode (F), slow mode (S), and intermediate mode (I). (The intermediate mode is sometimes called the Alfvén wave, but some scientists refer to all three magnetohydrodynamic modes as Alfvén waves. The intermediate mode is also called the shear wave. Some scientists give the name magnetosonic mode to the fast mode.)

Basic equations. The magnetohydrodynamic wave modes are analyzed by using the magnetohydrodynamic equations for the motion of a conducting fluid in a magnetic field, combined with Maxwell's equations and Ohm's law.

In studies of waves, the most important of the magnetohydrodynamic equations is called the momentum equation. It expresses how fluid parcels accelerate in response to the various forces acting on the fluid. The essential forces come from the thermal pressure in the fluid, and from the tension and pressure forces associated with the magnetic field. The magnetic field can exert forces on the fluid only when an electric current flows in the fluid; thus magnetohydrodynamic waves can exist only in an electrically conducting fluid. In analyzing magnetohydrodynamic waves, it is usually assumed that the electrons and ions move together so that the fluid is electrically quasineutral. In that case, the forces due to electric fields are negligible.

Two other equations express the conservation of mass of the fluid and the conservation of entropy. However, entropy is conserved only when viscosity, heat conduction, and electrical resistivity are ignored.

The fourth equation is Ohm's law, which expresses the relationship between the electric current and the electric field in the fluid. If the fluid has no electrical resistivity, any nonzero electric field would lead to an infinite current, which is physically unacceptable. Thus the electric field "felt" by any fluid parcel must be zero. However, detailed analysis of the motions of electrons and ions reveals that this statement is valid only if the wave frequency is much less than the cyclotron frequency of the plasma ions; thus magnetohydrodynamic waves are very low-frequency waves.

It is possible to combine Ohm's law with Faraday's law of induction. The resultant equation is called the magnetohydrodynamic induction equation, which is the mathematical statement of the "frozen-in" theorem. This theorem states that magnetic field lines can be thought of as being frozen into the fluid, with the proviso that the fluid is always allowed to slip freely along the field lines. The coupling between the fluid and the magnetic field is somewhat like the motion of a phonograph needle in the grooves of a record: the needle is free to slip along the grooves (the field lines) while otherwise being constrained from crossing the grooves. It is this coupling between the fluid and the magnetic field which makes magnetohydrodynamic waves possible. The oscillating magnetic field lines cause oscillations of the fluid parcels, while the fluid provides a mass loading on the magnetic field lines. This mass loading has the effect of slowing down the waves, so that they propagate at speeds much less than the speed of light (which is the propagation speed of waves in a vacuum).

The final equation is Ampère's law, which states that an electric current always produces a magnetic field. (In magnetohydrodynamics, it is usually possible to neglect the displacement current in Maxwell's equations. This is valid as long as the wave propagation speeds are small compared to the speed of light.)

Thus there are six equations (mass and entropy conservation, the momentum equation, Ohm's law, Faraday's law, and Ampère's law) for six unknowns (the pressure, density, and velocity of the fluid; the electric current; and the electric and magnetic fields). Six equations are in principle sufficient to determine six unknowns.

Linearization of equations. Unfortunately, the six equations are too difficult to be of much use. The essence of the difficulty is that some of the equations are nonlinear; that is, they contain products of the quantities for which a solution is sought. Nonlinear magnetohydrodynamics is still only in its infancy, and only a few specialized solutions are known. In order to get solvable equations, scientists accept the limitation of dealing with small-amplitude waves and linearize the equations, so that products of the unknowns are removed. Fortunately, much can still be learned from this procedure. The linearization consists of four steps: (1) All quantities are split into two parts: a background part (denoted by the subscript 0) which is constant in space and time, and a wave part (denoted by the prefix δ) which oscillates in space and time; for example, $\mathbf{B} = \mathbf{B}_0 + \delta\mathbf{B}$ and $\mathbf{v} = \mathbf{v}_0 + \delta\mathbf{v}$, where $\mathbf{B}$ denotes the vector magnetic field and $\mathbf{v}$ denotes fluid flow velocity. (2) These definitions are inserted into the six fundamental equations. (3) It is assumed that the waves are of such small amplitude that products of the wave parts can be ignored. (4) This step is for convenience only; it is assumed that there is no background flow. This simplifies the mathematics, but it does so without loss of generality, since the calculation without flow is equivalent to the case with flow as viewed in a frame of reference moving with the flow.

The resulting equations have solutions which are harmonic in time and space, that is, which vary as exp $(i\mathbf{k} \cdot \mathbf{r} - i\omega t)$. Here $\mathbf{r}$ is spatial position, ω is (angular) frequency, and $\mathbf{k}$ is called the wave vector. The magnitude of the wave vector represents how rapidly the wave oscillates in space, and its direction represents the direction of propagation. The magnitude of the angular frequency represents how rapidly the wave oscillates in time. Since the equations are linear, more general solutions can be constructed by adding together an arbitrary number of harmonic solutions.

Dispersion relation. The linearized equations have solutions only when ω and $\mathbf{k}$ are related to one another in a special way. The equation which expresses this relationship between ω and $\mathbf{k}$ is called the dispersion relation. It is a fundamental relationship in the theory of waves.

The dispersion relation for magnetohydrodynamic waves is given by Eq. (1). The quantity M_0^2 is given

$$(\omega^2 - k_0^2 v_A^2)(k_\perp^2 + M_0^2) = 0 \qquad (1)$$

by Eq. (2). The quantity $k_\parallel$ is the component of $\mathbf{k}$

$$M_0^2 = \frac{(k_\parallel^2 c_s^2 - \omega^2)(k_\parallel^2 v_A^2 - \omega^2)}{(c_s^2 + v_A^2)(k_\parallel^2 c_T^2 - \omega^2)} \qquad (2)$$

parallel to $\mathbf{B}_0$ while $k_\perp$ is the component perpendicular to $\mathbf{B}_0$. Equation (2) contains several velocities which are fundamental velocities in magnetohydrodynamics. The quantity v_A is called the Alfvén speed. The Alfvén speed is equivalent to the speed at which a wave travels along a taut string, except that the tension of the string is replaced by the tension in the magnetic field lines. The quantity c_s is the usual speed of sound in a gas, and c_T is called the tube speed, given by Eq. (3).

$$c_T^2 = c_s^2 v_A^2 / c_s^2 + v_A^2 \qquad (3)$$

Equation (1) is a cubic equation for ω^2. The three roots represent the three magnetohydrodynamic wave modes. For each of the three values of ω^2, there are two values of ω (one positive and one negative) which correspond to waves propagating in opposite directions.

A particularly useful quantity is the propagation speed, or phase speed, of the wave, given by Eq. (4).

$$v_{\text{ph}} = \omega / k \qquad (4)$$

This quantity can be obtained by solving Eq. (1). **Figure 1** shows the phase speeds of the three modes, for the two illustrative cases $c_s/v_A = 0.75$ and $c_s/v_A = 1.5$. The three modes are designated fast, intermediate, and slow, according to their phase speeds. The ordinate in Fig. 1 is v_{ph}, normalized to the Alfvén speed, while the abscissa is the angle θ between the wave propagation direction and the direction of $\mathbf{B}_0$. Thus θ satisfies Eq. (5). The graphs change character

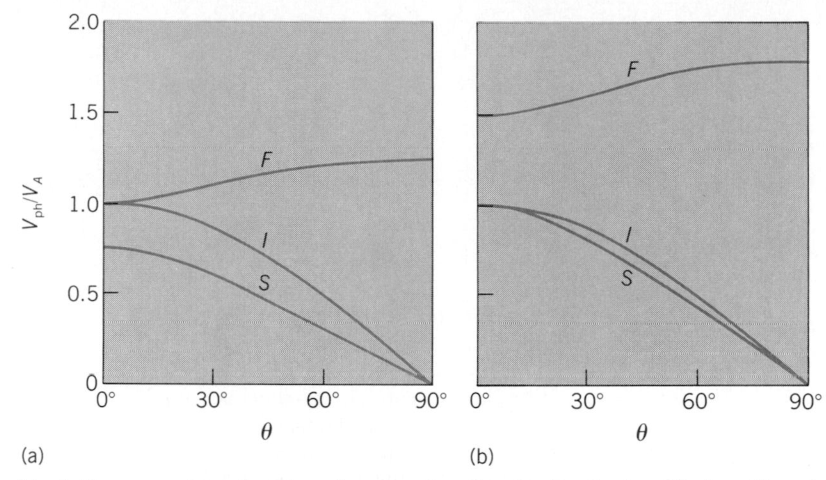

Fig. 1. Phase speed as a function of propagation direction for the fast (F), slow (S), and intermediate (I) modes. (a) $c_s/v_A = 0.75$. (b) $c_s/v_A = 1.5$.

$$\tan \theta = k_\perp / k_\parallel \qquad (5)$$

depending on whether c_s/v_A is less than or greater than 1.

All three modes can propagate along the magnetic field, while only the fast mode can propagate across the magnetic field. For propagation along the field ($\theta = 0$), two of the modes have $v_{\text{ph}} = v_A$, while the third has $v_{\text{ph}} = c_s$; the latter mode is essentially a sound wave unaffected by the magnetic field, while the other two modes are intrinsically magnetic. The intermediate mode has $v_{\text{ph}} = v_A \cos \theta$; this mode is always magnetic in character, and thus the sound speed does not affect its phase speed. In all other cases the fast and slow modes have a mixed character, and all three restoring forces combine in a complicated way. The essential properties of the modes are summarized below.

Intermediate mode. The velocity and magnetic field fluctuations in this mode are found to be perpendicular to the plane containing $\mathbf{k}$ and $\mathbf{B}_0$. This means that the motions are pure shears. There is no compression of the plasma. This is why the sound speed does not appear in the dispersion relation. The tension in the magnetic field lines is the only restoring force involved in the propagation of the wave. This

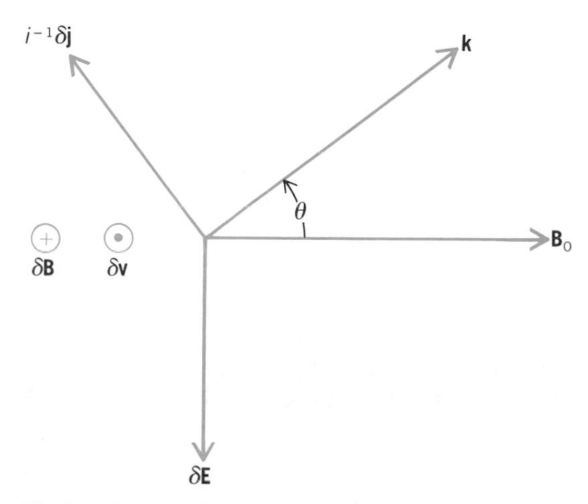

Fig. 2. Vector relationships between the various fluctuating wave quantities for the intermediate mode.

mode is therefore closely analogous to the propagation of waves on a string: the tension in the magnetic field lines plays the same role as the tension in the string.

The dispersion relation is given by Eq. (6). The

$$\omega^2 = k_\parallel^2 v_A^2 \quad (6)$$

quantity $k_\perp$ does not appear because the motions on neighboring field lines do not communicate with one another. This is a consequence of the fact that the motions are shears, so that neighboring field lines never bump together.

Figure 2 shows how the various fluctuating quantities are related. The symbols next to $\delta\mathbf{B}$ and $\delta\mathbf{v}$ indicate that these quantities are pointing into and out of the paper, respectively. The electric current $\delta\mathbf{j}$ must be multiplied by i^{-1} (or $1/\sqrt{-1}$) because the current is 90° out of phase with the other quantities. The quantities $\delta\mathbf{v}$ and $\delta\mathbf{B}$ are closely related, and in fact satisfy Eq. (7). (The symbol sqn means the al-

$$\delta\mathbf{v}/v_A = -\operatorname{sqn}(\mathbf{k}\cdot\mathbf{B}_0)\,\delta\mathbf{B}/B_0 \quad (7)$$

gebraic sign of the quantity in parentheses.) Equation (7) has been used to identify the presence of the intermediate mode in the solar wind. There, spacecraft-borne instruments have been used to measure the magnetic field and plasma velocity directly. The fluctuations have been found to obey Eq. (7) fairly closely. It appears that intermediate waves are rather copious in the solar wind. Further investigations have revealed that the waves usually propagate away from the Sun. The Sun may thus be an emitter of these waves. *SEE SOLAR WIND; SUN.*

Waves propagate energy. Since the intermediate mode does not involve the thermal properties of the plasma, its energy flux density is given by the time average of the electromagnetic Poynting flux. For the intermediate mode, the wave energy flux density is always parallel or antiparallel to $\mathbf{B}_0$. Thus the intermediate mode always channels energy along the magnetic field. In fact, the energy flux density of the wave consists of the wave's kinetic energy density plus the wave's magnetic energy density, propagating along the magnetic field at speed v_A.

Because these waves channel energy along magnetic fields, they may be responsible for the observed fact that cosmical plasmas are strongly heated in the presence of magnetic fields. For example, they may heat the solar atmosphere and the solar wind. They may also accelerate the solar wind through an effect which is closely analogous to the radiation pressure exerted by light waves. Intermediate waves have also been proposed as a means of heating plasmas in controlled thermonuclear fusion devices.

Fast mode. The fast mode is difficult to analyze. However, in many cosmical and laboratory plasmas, v_A^2 is much greater than c_s^2. This is the strong-magnetic-field case. In that case the fast mode is more easily understood.

From Eq. (1), the fast-mode dispersion relation is approximately given by Eq. (8). The direction of $\mathbf{k}$

$$\omega^2/k^2 \approx v_A^2 \quad (8)$$

does not appear in Eq. (8). Thus the waves propagate isotropically.

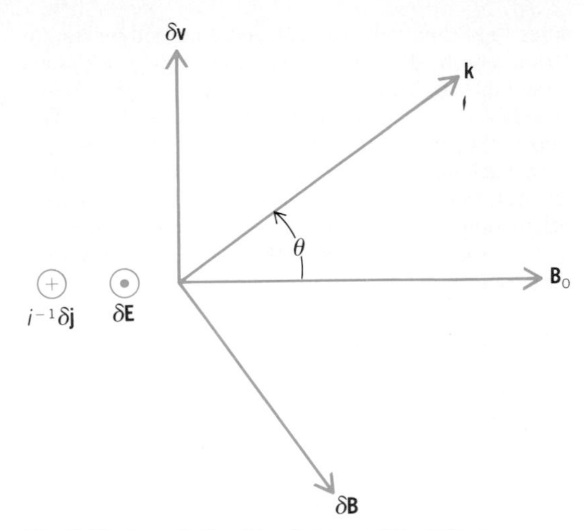

Fig. 3. Vector relationships between the various fluctuating wave quantities for the fast mode in a strong magnetic field, where v_A^2 is much greater than c_s^2.

Figure 3 shows how the various fluctuating quantities are related. Fast waves are compressive, and the magnetic field strength fluctuates as well. Thus fast waves are governed by the two restoring forces associated with the tension and pressure in the magnetic field.

Because the magnetic field is strong, the wave energy flux density is mainly due to the time average of the Poynting flux. For this mode, it is evident that the energy flux is along $\mathbf{k}$. Thus the fast mode can propagate energy across the magnetic field.

Slow mode. Like the fast mode, the slow mode is difficult to study in general, and the discussion will again be confined to strong magnetic fields, so that v_A^2 is much greater than c_s^2. The slow mode in a strong field is equivalent to sound waves which are guided along the strong magnetic field lines. The strong magnetic field lines can be thought of as a set of rigid pipes which allow free fluid motion along the pipes, but which restrict motion in the other two directions. The motions on the individual pipes are not coupled together, and thus the slow mode is analogous to the sound waves on a set of independent organ pipes. The slow mode channels energy along the magnetic field. Because the sound speed is small, by assumption, the slow mode transmits energy less effectively than the fast or intermediate modes.

Nonlinear effects. Only small-amplitude waves have been considered. Real waves have finite amplitude, and nonlinear effects can sometimes be important. One such effect is the tendency of waves to steepen, ultimately forming magnetohydrodynamic shock waves and magnetohydrodynamic discontinuities. There is an abundance of magnetohydrodynamic discontinuities in the solar wind. It is presumed that these are the result of the nonlinear steepening of the intermediate waves.

It is also possible that waves can degenerate into turbulence. There are indications that this too happens in thc solar wind.

Dissipation. Viscosity, heat conduction, and electrical resistivity have all been ignored in the above discussion. These effects will all tend to dissipate the wave energy and heat the plasma.

In some plasmas, the electrons and ions almost never collide with one another. In such nearly collisionless plasmas, the fast and slow modes can dissipate by the process of Landau damping.

Shock formation and turbulence also lead to wave dissipation.

Surface waves. Only waves in a spatially uniform background have been considered. Real plasmas are never perfectly uniform. The analysis of magnetohydrodynamic waves in a nonuniform background is complicated, and a subject of active investigation. However, it is possible to consider an extreme limit, in which the background is uniform except at certain surfaces where it changes discontinuously. It is then found that the surfaces can support magnetohydrodynamic surface waves, which are in some respects similar to waves on the surface of a lake. These waves can propagate energy along the surface, and they may play important roles in heating cosmical and laboratory plasmas. SEE MAGNETOHYDRODYNAMICS.

Joseph V. Hollweg

Bibliography. R. L. Carovillano and J. M. Forbes (eds.), *Solar-Terrestrial Physics,* 1983; J. V. Hollweg, Hydromagnetic waves in interplanetary space, *Publ. Astron. Soc. Pacific,* 86:561–593, 1974; E. R. Priest, *Solar Magnetohydrodynamics,* 1982; G. Schmidt, *Physics of High Temperature Plasmas,* 2d ed., 1979.

Almanac

A book that contains astronomical or meteorological data arranged according to days, weeks, and months of a given year and may also include diverse information of a nonastronomical character. This article is restricted to astronomical and navigational almanacs.

Development. The earliest known almanac material was computed by the Egyptians in A.D. 467, and astronomical ephemerides appeared irregularly thereafter. The first regular almanac appears to have been produced in Germany from 1475 to 1531. The first almanac to be published regularly by a national government was the *Connaissance des Temps* (1679) by France. In 1767 the British began publishing the *Nautical Almanac* to improve the art of navigation. The *Berliner Astronomisches Jahrbuch* was first published in Germany for 1776, and the *Efemerides Astronomicas* was published in Spain in 1791. In the United States *The American Ephemeris and Nautical Almanac* has been published since 1855. Beginning with the issue for 1981, the two series of publications titled *The Astronomical Ephemeris,* which had previously replaced *The Nautical Almanac and Astronomical Ephemeris,* and *The American Ephemeris and Nautical Almanac* were continued with the title of *The Astronomical Almanac*.

Over the years cooperation has developed between the organizations responsible for the preparation and publication of the almanacs. Under the auspices of the International Astronomical Union (IAU), agreements are reached concerning the bases and constants to be used in the publications and the exchange of data in different forms.

Astronomical Almanac. *The Astronomical Almanac* contains ephemerides, which are tabulations, at regular time intervals, of the orbital positions and rotational orientation of the Sun, Moon, planets, satellites, and some minor planets. It also contains mean places of stars, quasars, pulsars, galaxies, and radio sources, and the times for astronomical phenomena such as eclipses, conjunctions, occultations, sunrise, sunset, twilight, moonrise, and moonset. This volume contains the fundamental astronomical data needed by astronomers, geodesists, navigators, surveyors, and space scientists. SEE ASTRONOMICAL COORDINATE SYSTEMS; EPHEMERIS.

Navigational almanacs. While *The Astronomical Almanac* is basically designed for the determination of positions of astronomical objects as observed from the Earth, *The Nautical Almanac* and *The Air Almanac* are designed to determine the navigator's position from the tabulated position of the celestial object. The ground point of the celestial object is used with the North and South poles and the observer's assumed position to establish a spherical triangle known as the navigational triangle. By means of spherical trigonometry or navigational tables, the navigator determines a computed altitude and the direction of the ground point (azimuth) of the celestial object with respect to his or her position. The computed altitude and azimuth enable the navigator to plot a line of position that goes through or near this assumed position. The computed altitude is compared to the observed altitude of the celestial object. From the combination of two or more observations the navigator is able to determine position. SEE CELESTIAL NAVIGATION.

The Nautical Almanac contains hourly values of the Greenwich hour angle and declination of the Sun, Moon, Venus, Mars, Jupiter, and Saturn and the sidereal hour angle and declination of 57 stars for every third day. Monthly apparent positions are tabulated for an additional 173 navigational stars. The positions are tabulated to an angular accuracy of 0.1 minute of arc, which is equivalent to 0.1 nautical mile (0.2 km). Since tabular quantities must be combined to derive the navigational fix, this tabular accuracy is sufficient to produce a computed position with an error no greater than 0.3 to 0.4 nmi (0.6 to 0.7 km). *The Nautical Almanac* also contains the times of sunrise, sunset, moonrise, moonset, and twilight for various latitudes. A diary of astronomical phenomena, predictions of lunar and solar eclipses, visibility of planets, and other information of interest to the navigator are also included.

The Air Almanac, published semiannually, is arranged with two pages of data back to back on a single sheet for each day; the positions of the Sun, first point of Aries, three planets, and the Moon are given at 10-min intervals. As necessary, information is adjusted so that the tabulated data at any given time can be used during the interval to the next entry, without interpolation, to an accuracy sufficient for practical air navigation. The times of sunrise, sunset, twilight, moonrise, and moonset are also given daily. Also provided are star recognition charts; a sky diagram; rising, setting, and depression graphs; standard times; and correction tables necessary for air navigation. While designed for air navigators, *The Air Almanac* is used by mariners who accept the reduced accuracy in exchange for its greater convenience compared with *The Nautical Almanac*. A number of countries publish in their languages air and nautical almanacs that are based on, or similar to, the English language versions.

Other almanacs. The *Almanac for Computers,* published annually by the U.S. Naval Observatory since 1977, provides the coefficients for power series and Chebyshev polynomials, which permit the computation for any specific time of the data contained in the nautical, air, and astronomical almanacs. A hand calculator or small computer is necessary for the computations, but the data can be calculated directly for the desired time, thus avoiding interpolation. Also, algorithms are included for performing astronomical and navigational calculations. *See* CALCULATORS.

The Astronomical Phenomena, published annually, contains the times for phenomena such as eclipses, conjunctions, occultations, sunrise, sunset, moonrise, moonset, and moon phases, seasons, and visibility of planets. Also, the dates of civil and religious holidays are included.

The positions of the Sun, Moon, and planets in various coordinate systems at reduced accuracy for an extended period of time are tabulated in *Planetary and Lunar Coordinates*. This publication continues the "Planetary Coordinates" series, which gave the planetary positions for 1800 to 1980 in three volumes. The first edition of the new series covered the period 1980 to 1984; the second edition covers the period 1984 to 2000. The new series includes lunar coordinates and some phenomena, such as dates of eclipses and lunar phases.

For surveyors, the U.S. Naval Observatory prepares and the Bureau of Land Management of the Department of Interior publishes *The Ephemeris*. The Royal Greenwich Observatory publishes *The Star Almanac*. Both of these publications are designed to permit the surveyor to determine a geographical position from celestial observations.

The *Apparent Places of Fundamental Stars* is prepared annually by the Astronomisches Rechen Institut of Heidelberg, Germany. The mean and apparent positions of 1535 stars are tabulated. *See* FUNDAMENTAL STARS.

Ephemerides of Minor Planets is prepared annually by the Institute of Theoretical Astronomy, St. Petersburg, Russia, and published by the Academy of Sciences of the former Soviet Union. This volume contains the elements, opposition dates, and opposition ephemerides of all numbered minor planets. *See* ASTEROID.

Machine-readable data. The data from the publications as well as additional data, such as long-term ephemerides and many star catalogs, are available in machine-readable form. However, in spite of the increased use of computers, there is a continuing need for the printed volumes.

Publication and distribution. *The Astronomical Almanac, The Nautical Almanac, The Air Almanac, Astronomical Phenomena,* and *Planetary and Lunar Coordinates* are cooperative publications of the U.S. Naval Observatory and the Royal Greenwich Observatory, and are available from the U.S. Government Printing Office and Her Majesty's Stationery Office. These publications are used in many countries, and similar references, generally based on the same basic data, are published in various languages by the Spanish, Chinese, C.I.S. (formerly, Soviet), Japanese, German, French, Indian, Argentinian, Brazilian, Danish, Greek, Indonesian, Italian, Korean, Mexican, Norwegian, Peruvian, Philippine, and Swedish governments. Examples of publications comparable to *The Astronomical Almanac* are the *Connaissance des Temps* by France, the *Astronomical Almanac of the USSR* (now C.I.S.), the *Efemerides Astronomicas* by Spain, the *Chinese Astronomical Ephemeris,* the *Indian Ephemeris,* and the *Japanese Ephemeris.*

P. K. Seidelmann

Altitude

In astronomical, navigational, and surveying practice, an angle at the observer, measured positively from 0° to 90° above the observer's horizon along the vertical circle through the object being observed and the observer's zenith. It is also the complement of the angle at the observer along the vertical circle from the observer's zenith to the object. Practitioners often refer to this angle as elevation. *See* ASTRONOMICAL COORDINATE SYSTEMS; CELESTIAL NAVIGATION.

In aeronautics and astronautics, an altitude is the vertical distance (height) of an object such as an aircraft or spacecraft above a reference surface. Aircraft altitude is usually referenced to sea level; spacecraft altitude may be referenced to some arbitrary fiducial surface on the body around which the spacecraft is orbiting.

Raynor L. Duncombe

Bibliography. C. H. Cotter, *The Elements of Navigation and Nautical Astronomy,* 1981; A. Frost, *The Principles and Practice of Navigation,* 1981.

Andromeda Galaxy

The spiral galaxy of type Sb nearest to the Milky Way system. This galaxy is a member of a small cluster of galaxies known as the Local Group. This group contains also the Milky Way system, the Triangulum Nebula (M33), the Large and Small Magellanic Clouds, NGC 6822, and several faint dwarf elliptical galaxies.

The Andromeda Galaxy M31 or NGC 224 is particularly important because it is close enough to the Earth for its stellar and other content to be studied in great detail. Observations show that the Great Galaxy M31 contains bright blue stars of absolute magnitude $M_B \approx -9.5$, cepheid variable stars, bright supergiant red stars, clusters of stars (both globular and open clusters), planetary nebulae, normal novae, gas, and dust.

The approximate distance to M31 is known from the apparent luminosities of the cepheid variable stars. The period-luminosity relation for cepheids, combined with the apparent brightness of these stars, gives a distance of about 2,500,000 light-years (1.5×10^{19} mi or 2.4×10^{19} km). The accuracy of this value depends on the calibration of the period-luminosity relation for cepheids, which now appears to be well known. *See* CEPHEIDS; VARIABLE STAR.

The major axis of the photographic image of M31 is 200 minutes of arc to an isophotal level of 25 magnitudes per square second of arc. This corresponds to a linear diameter of 150,000 light-years (9×10^{17} mi or 1.4×10^{18} km) if the distance of 2,500,000 light-years is adopted.

The nebula is rotating about its center with a period of about 2×10^8 years. The evidence comes from radial velocities of selected objects in its disk. This rotational period, combined with the diameter, gives the mass of M31 as 10^{11} solar masses. The total number of stars is greater than 2×10^{11}. The luminosity in the wavelength range from $\lambda = 390$ nanometers to $\lambda = 500$ nm is $M_B = -20.3$, which corresponds to 2×10^{11} equivalent suns. All studies show that M31 is typical of other regular spiral galaxies of the Sb class. *See Galaxy, external; Nebula.*

Allan Sandage

Aphelion

In astronomy, that point at one extremity of the major axis of an elliptical orbit about the Sun where the orbiting body is farthest from the Sun. The Earth passes aphelion on or near July 4, referred to as the time of aphelion passage. Because the orbit of the Earth is nearly a circle (eccentricity 0.017), the Earth is then only some 1.55×10^6 mi (2.5×10^6 km) farther from the Sun than at its mean distance of 9.3×10^7 mi (1.496×10^8 km). *See Celestial mechanics; Orbital motion.*

Raynor L. Duncombe

Apogee

The position most distant from Earth in the orbit of a satellite, as in the orbit of the Moon or of an artificial satellite. The Moon at apogee is 5.5% further from Earth than at its mean distance; that is, its orbital eccentricity is 0.055. *See Aphelion; Celestial mechanics; Orbital motion; Perigee.*

Gerard P. Kuiper

Apsides

In astronomy, the two points in an elliptical orbit that are closest to, and farthest from, the primary body about which the secondary revolves. In the orbit of a planet or comet about the Sun, the apsides are, respectively, perihelion and aphelion. In the orbit of the Moon, the apsides are called perigee and apogee, while in the orbit of a satellite of Jupiter, these points are referred to as perijove and apojove. The major axis of an elliptic orbit is referred to as the line of apsides. *See Celestial mechanics; Orbital motion; Perigee; Perihelion.*

Raynor L. Duncombe

Aquarius

The Water Bearer, in astronomy, a large zodiacal constellation visible in both summer and autumn. Aquarius is the eleventh sign of the zodiac. To ancients, the constellation resembled a man pouring a stream of water from a jar. Four stars, η, ζ, π, and γ, arranged like a Y form the head of the Water Bearer (see **illus.**). The stream of water flows into the Fish's Mouth (Fomalhaut) in the constellation Pisces Austrinus (Southern Fish). Fomalhaut, bright and solitary in this part of the sky, is one of the relatively important navigational stars. From earliest time this constellation has been associated with water, probably because the Sun is seen in Aquarius during the rainy season of February. *See Constellation; Zodiac.*

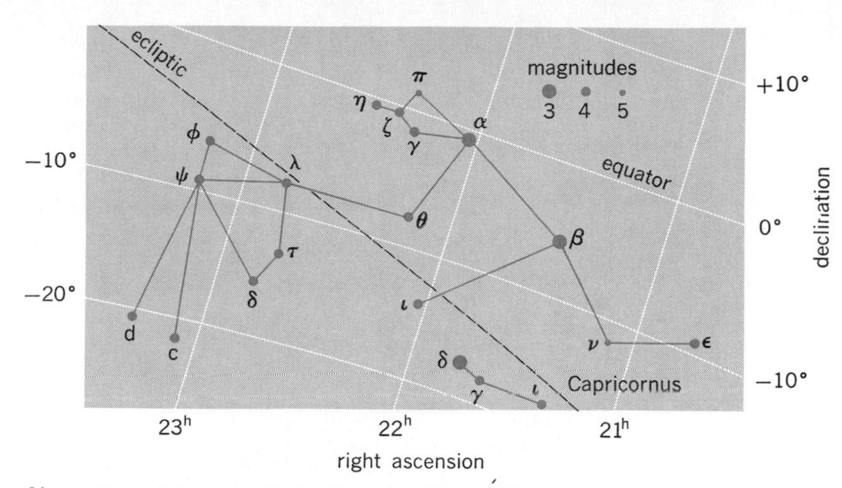

Line pattern of the constellation Aquarius. The grid lines represent the coordinates of the sky. The apparent brightness, or magnitude, of the stars is shown by the sizes of the dots, which are graded by appropriate numbers as indicated.

Ching-Sung Yu

Archeoastronomy

The field of science that attempts to determine how much astronomy prehistoric people knew and how it influenced their lives. It involves multiple disciplines: astronomy to chart the heavens, archeology to probe the cultural context, engineering to survey sites, and ethnology to provide clues to the cultural past. Archeoastronomy has prompted valuable insights into the astronomy of the past, even to revolutionizing some of the models of prehistoric cultures.

Archeoastronomers work throughout the world. All prehistoric and preliterate peoples have observed the sky. These observations are usually related to keeping a calendar (both solar and lunar) and to regulating sacred time, the sequence of rituals. This article highlights some fruitful work in archeoastronomy, touching on both the Old World and New World, especially the southwestern part of North America.

The work of archeoastronomers at any site must always be placed in a cultural context. Without that context, finding an astronomical alignment—some celestial event like the northernmost azimuth of the Sun aligned with ground objects an seen by an observer—may be no more than coincidence biased by wanting to find it and knowing where to look for "significant" results. It is usually possible to find an astronomical alignment at a site; the real question is whether or not the builders intended it as such. The great danger is the imposition of modern knowledge of astronomy upon an alien culture of the past.

While the accuracy of archeoastronomers' ideas about the past can never be known for certain, researchers stand on firmer ground if ethnographic information is available to bolster the archeological evidence. In this regard, the New World has a distinct advantage over the Old: the culture of people who were observing the sky before the arrival of Europeans still survives among their descendants. This culture, although suffering from pressures to change, serves as a caution and guide to interpreting the past.

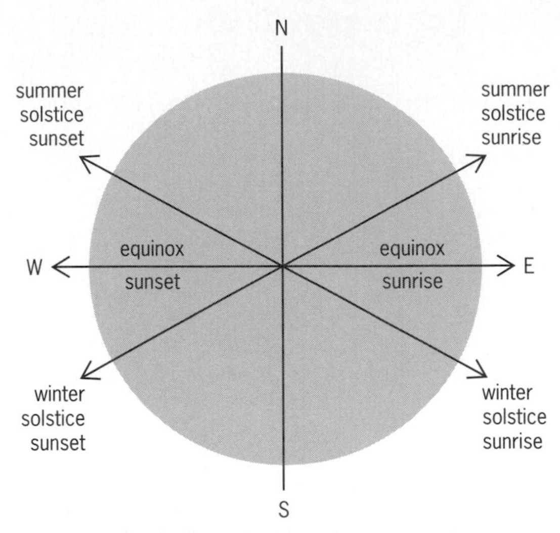

Fig. 1. Angular swing of the Sun along the horizon at rising and setting from solstice to solstice. The angle is for a latitude of 36°, that of the North American Southwest.

Naked-Eye Astronomy

The kinds of observations of key celestial cycles that can be made without a telescope are reviewed in this section. Only the Sun and the Moon will be considered.

Sun. Most people are aware that the height of the Sun in the sky at noon changes with the seasons. The Sun reaches its highest noon point on the summer solstice (around June 22); drops to its lowest on the winter solstice (December 22); and is at the middle at the equinoxes (March 21 and September 23). (These dates refer to the Northern Hemisphere; in the Southern Hemisphere, they are reversed.) *See Equinox; Seasons; Solstice.*

There is, however, less familiarity with the Sun's seasonal motion along the horizon. On the day of the summer solstice, for example, the Sun rises the farthest north of east that it will get for the year (**Fig. 1**). On the equinoxes, it rises due east. And on the winter solstice, it reaches its farthest point south of east. (The same occurs, mirror-reflected, at sunset.)

Thus, from summer to winter, the sunrise point moves to the south; from winter to summer, to the north. The rate at which the sunrise point moves from day to day varies during the year. At the solstices, the sunrise points do not noticeably move for a few days. The Sun appears to ''stand still'' (which is the meaning of the word solstice). In contrast, at the equinoxes, the sunrise points move at their fastest rate, by almost the Sun's own diameter in a day at midlatitudes.

This seasonal voyage of the Sun along the horizon differs with latitude with respect to the size of the solstice-to-solstice swing along the horizon. At 36°, the latitude of the North American Southwest, the swing amounts to 60°, one-sixth of the total horizon circle (Fig. 1). Farther north, the arc is greater; at the latitude of Stonehenge, it is about 80°. At more southerly latitudes, the arc is less; at 20° (middle of Mexico), it varies a total of 50°.

Moon. The Moon's most obvious change is that of its phases, referring to how much of the illuminated side of the Moon is visible from the Earth. The month of phases (synodic month) is simply the time from one phase of the Moon to the repetition of that phase, say from full moon to full again. It averages 29.5 days. If the interval is counted from the first visible crescent to the last visible crescent, it averages 28 days. *See Phase.*

Suppose the point of moonrise is observed for a month. It would be seen that the moonrise point varies from a point farthest south to one farthest north during the month. In other words, the moonrise motion mimics the sunrise motion but occurs about 12 times as fast. Depending on when the observations are made, the moonrise arc may be larger than, the same as, or smaller than the sunrise arc. This difference results because the Moon's path in the sky with respect to the stars is not the same as the Sun's, but is inclined at about 5°, crossing the Sun's path at two points. Thus, the Moon can appear as much as 5° below the Sun's path, 5° above it, or right on it. *See Moon.*

Complication. The matter is complicated in that the two points where the Sun's and Moon's paths cross (the nodes) move with respect to the stars, taking 18.6 years to circle the sky once. The result is that when the Moon's path reaches its highest point above the Sun's, the Moon's horizon swing is greater than the Sun's. When the two line up, the swings are the same. When the Moon's path falls below the Sun's, the total arc is less.

How much greater or less can be considerable. At a latitude of 36°, the Moon moves through a maximum arc of 70° and a minimum arc of 45° during the 18.6-year cycle (**Fig. 2**). In analogy to the Sun standing still at the solstices, the two extremes of the Moon's positions are also called standstills: major standstill for the maximum angle and minor standstill for the minimum, with 9.3 years between. Again in analogy to the Sun, these angular changes are more pronounced at more northern latitudes.

When the Moon moves from minor to major standstill, its monthly arc stays within that of the Sun for about half the time. During this time, an alignment that works for the Moon will work for the Sun at some time during the year. Hence, it may be that an orientation is not specifically for the Moon. However,

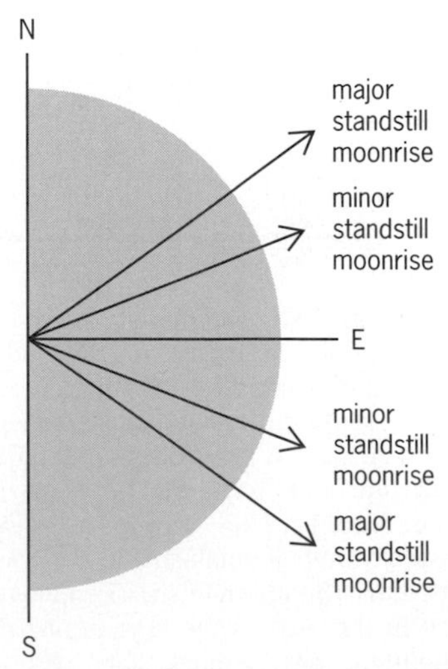

Fig. 2. Monthly angular swing of the Moon along the horizon at moonrise, for times of maximum angle (major standstill) and minimum angle (minor standstill). The angles are shown for a latitude of 36°.

during the time when the Moon's swing lies outside the angle of the Sun, an orientation can apply to the Moon only, and not at all to the Sun.

Horizon-marking system. It can now be seen how a simple horizon-marking system is set up. First, a location with a clear view of the horizon must be found, with at least a few prominent features over the angular range of the sunrise (or sunset). Then it is necessary to return to this spot daily and note the rising positions of the Sun throughout the year at significant times: the solstices, the equinoxes, and, perhaps, important times to plant crops. Thus, a basic solar calendar is established. Since the Sun's positions at various dates along the arc remain fixed for a long time, once established the calendar will be good for many years.

A solar-horizon calendar keeps reliable track of the seasonal year. To subdivide that interval, people usually rely upon a lunar calendar that follows the phases of the Moon. The essential astronomical problem is that a seasonal cycle and a lunar-phase cycle are incommensurate. The solution can involve complicated, long-term strategies, such as that used in the Gregorian calendar, or more casual, practical ones, such as counting a ''short'' month of a few days at the end of each year. *See Calendar*.

The Old World: Stonehenge and Other Megalithic Sites

In a direct sense, Stonehenge created the interest in archeoastronomy, and the site exemplifies the problems and potential of the archeoastronomical enterprise.

Stonehenge is popularly known for the massive upright stones that form a central horseshoe and circle (some 65 ft or 25 m in diameter) in the center of the site (**Fig. 3**). Such large stones are commonly called megaliths in Great Britain; this term has come to be applied to all sites where stones, even fairly small ones, are arranged in some pattern. The horseshoe opens out on the main axis of Stonehenge, called the Avenue. Some 260 ft (80 m) from the center, within but not in the center of the Avenue, sits the tilted Heel Stone. This main axis of Stonehenge aligns roughly with the summer solstice sunrise. In fact, at summer solstice the Sun rises somewhat to the left of the Heel Stone as seen by an observer at the center of the structure.

Astronomical alignments at Stonehenge. Despite earlier interpretations of the astronomical use of Stonehenge, the modern controversy developed in the 1960s. G. Hawkins, an astronomer, searched for astronomical alignments to the Sun, Moon, stars, and planets for the main features of the site. He found them for the Sun and the Moon, including moonrise and moonset during major and minor standstills. Later, he proposed that the site could even have been used to warn of the times of possible eclipses.

Radiocarbon dates indicate that Stonehenge was built over a span from 3100 to 1000 B.C. in three separate stages. The muddle over the astronomical use of Stonehenge comes, in large part, from the fact that it is a mosaic of structures, most likely built by different people, perhaps for different reasons. The great stones were erected between 2000 and 1500 B.C.; it is not clear what their cultural connection was to the earlier structure.

It is the earliest parts of Stonehenge, constructed between 3100 and 2100 B.C., that have the most astronomical promise (**Fig. 4**). These comprise the outer earthwork ring and ditch (about 330 ft or 100 m in diameter and 7 ft or 2 m high) broken only in the direction of the Heel Stone; a ring of 56 holes (the Aubrey Holes) that were dug and then quickly filled with chalk; an array of postholes near the opening to the Heel Stone; and the four Station Stones that lie along the circle of the Aubrey Holes.

Lunar and solar observing can be done with these

Fig. 3. Inner great trilithons of Stonehenge. (*Courtesy of O. Gingerich*)

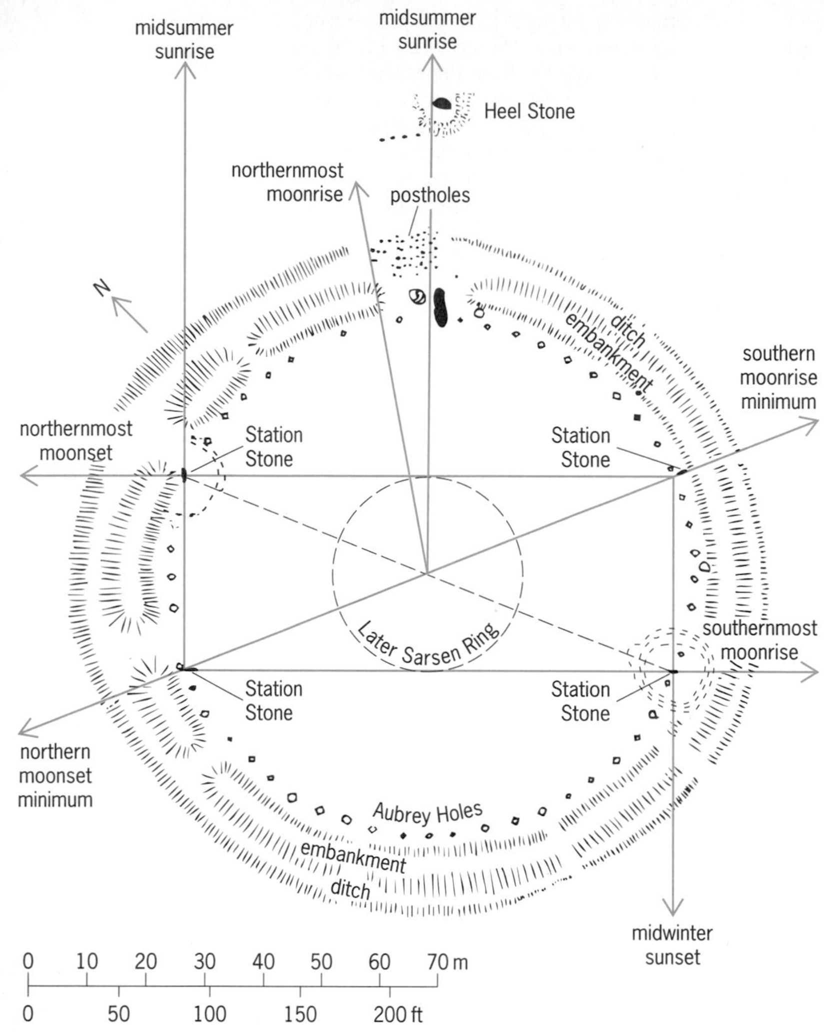

Fig. 4. Diagram of the major features of Stonehenge that may have been used astronomically. *(After O. Gingerich, in K. Brecher and M. Feirtag, eds., Astronomy of the Ancients, MIT Press, 1979)*

elements. The four Station Stones form a fairly good rectangle. From its center, the summer solstice Sun rises along the opening to the Heel Stone. The short sides of the rectangle are parallel to this line, so they point to the summer solstice sunrise and winter solstice sunset. The long sides of the rectangle and its diagonals line up the moonrises and moonsets at the major and minor standstills.

Hawkins also contended that the inner megaliths of the horseshoe sighting outward through the ring around them also aligned to important settings and risings of the Sun and Moon. Hawkins then argued that the Aubrey Holes were used to indicate ''danger seasons'' when eclipses might occur. One lunar eclipse cycle (it is not the only one) takes 56 years (three times the 18.6-year standstill cycle). In this picture, the Aubrey Holes were used as a counter to keep track of the years within these cycles.

Evaluation of claims. Both astronomical and archeological criticism can be applied to these arguments. First, Stonehenge can be criticized as a lunar eclipse anticipator. A 56-year cycle does exist, but once worked out, it fails to apply after a few cycles. Also, to work out the cycle in the first place requires hundreds of years of careful observing and a preserved record (probably oral) from which to infer the cycle. Given problems with bad weather hiding eclipses and the difficulty of preserving a nonwritten record for such a long time, the establishment seems highly improbable. Second, the purported alignments with the inner megaliths are also questionable; the gaps are rather wide and, depending on where one stands, can cover a large angle on the horizon. Their crudeness suggests that alignments attributed to them are probably accidents of the layout.

The inner rectangle seems much better astronomically; because of its fairly large size (112 by 260 ft or 34 by 79 m), it results in fairly accurate sightlines. These also contain nice symmetries, which increase their appeal. However, the archeologist R. J. C. Atkinson noted that only two of the four stones actually survive; of these two (91 and 93), one has fallen and one seems to be a later replacement. Thus the original positioning of the stones is not known with accuracy.

All told, the older parts of Stonehenge make a reasonable solar and lunar observatory. The extensive symmetries are very appealing and more important than the accuracies of the sightlines. Finally, even archeologists admit that the summer solstice alignment, more than any other, appears as an intention of the original construction.

Other megalithic sites. The basic problem in all of this is that even if the astronomy works, the cultural context is unclear. Horizon watching can be used simply to tell the time of year or more forcefully to set a ritual calendar. The only hope of guessing about the importance of astronomy in megalithic societies is to examine other sites along with Stonehenge.

Two large burial sites, constructed about 3000–2500 B.C., also display winter solstice orientations. At Newgrange in Ireland, a long passage directs sunlight to reach the center of the tomb every morning for about a week around the winter solstice. At Maes Howe in the Scottish Orkney Islands, the setting Sun's rays illuminate the central chamber around the time of the winter solstice.

Hundreds of prehistoric sites in Great Britain and France have been carefully surveyed by A. Thom, an engineer. He first found indications of alignments for the solstices and equinoxes, then for the lunar standstills. He promoted the idea that megalithic astronomers made extremely accurate observations of the Moon (using very distant foresights, tens of kilometers long) so as to pick out very small, long-term variations of the Moon's motions.

The precise lunar observations have been questioned, and different analyses lead to the conclusion that the Moon was observed, but not with the precision inferred by Thom. From the view of cultural necessity, it is unclear how such precision would benefit megalithic people in terms of simple survival value. However, more so than Hawkins's efforts, Thom's work forced archeologists to account for the astronomy in megalithic cultures.

Lacking rich cultural evidence, investigations must turn to the statistical evidence of a large number of sites, selected in an unbiased way, from ones of a similar cultural background. That work is still in progress, but it has revealed so far (mostly for sites in Scotland) strong hints of rough (within a degree or so) orientations to the position of the Moon at major and minor standstills. Hence, general trends show up, even if embedded in a background of noise (that is, orientations that have no relation to astronomy).

Despite disputes over many points, archeoastronomy has forced a reexamination of the standard picture of megalithic life. Certainly astronomy was important, even if it is not known exactly how or why.

Skywatching in the New World

Compared to the Old World, the New World archeoastronomer has the advantage of the survival of remnants of the cultures from pre-Columbian times. Even the great destruction wielded by the Spanish in Mesoamerica, especially their burning of Mayan books that contained much astronomy, could not wipe out completely the astronomy inherent in that culture.

In the Southwest, the Spanish encountered adobe villages, which they called pueblos, of the native peoples who had lived in them at least 1000 years. Many of these pueblos disappeared in historic times (from 1540 onward); those that survived are the cultural connection to the people called the Anasazi, who occupied a vast area in the Southwest, centered on the Four Corners area (where New Mexico, Arizona, Utah, and Colorado now meet). Here stand ruins deserted from A.D. 1000 to 1400, stone and adobe constructions that provide some insight into the life of the Anasazi.

Hopi and Zuñi astronomy. The Hopi (in Arizona) and Zuñi (in New Mexico) pueblos provide the best clues to the past because these villages were touched only lightly by the Spanish. Ethnographers gathered cultural information here around 1900, before the severe pressures on the part of Anglos occurred. It is inferred that the Hopi and Zuñi are cultural descendants of the Anasazi (although it is not known from which specific Anasazi sites). Thus, these pueblos preserve a remarkable cultural connection to prehistory.

Among the Hopi and Zuñi, astronomy played a central role in the agricultural and ceremonial life. The seasonal cycle of the Sun set the ritual calendar and determined the times of specific crop plantings and harvestings. The dry Southwest demands an observant farming, for raising crops is a marginal activity; in the past, failed crops could mean death. So solar astronomy carried a practical weight as well as a religious one. The counting of months by lunar phases played a secondary role in tracking the ritual calendar. Notched calendar sticks were used to assist in the lunar count.

The observing was invested in a religious office, usually called the Sun Priest. He watched daily from a special spot within the pueblo or not far outside it and carefully observed sunrise (or sunset) relative to the horizon features. From past experience, he knew the points that marked the summer and winter solstices and the times to plant crops.

A crucial aspect of the Sun Priest's work was the ability to forecast ceremonial dates; he did so by making anticipatory observations about 2 weeks ahead of time. At that time, the rising (or setting) points of the Sun showed a daily change that could be reliably discerned against a horizon profile. By counting down a certain number of days, the Sun Priest could announce ahead of time the day for the ritual, allowing the people of the pueblo enough time to prepare for the ceremony. These forecasting procedures enabled the historic pueblo Sun Priest to predict the actual dates of the solstices with good accuracy, mostly within 1 day of the actual astronomical dates.

The proper choice of ceremonial dates was the major responsibility of the Sun Priest, and it is likely that a prehistoric Sun Priest had the same responsibilities for his work. The winter solstice marked the heart of the Hopi and Zuñi ritual year. For the Hopi, each month was named, and the passing of a month sometimes was used to set the time for a ceremony.

Along with horizon features, the Zuñi Sun Watcher, called Pekwin, used a natural pillar to chart the seasons. When the shadow cast by the pillar lined up in a special fashion, Pekwin knew that the summer solstice would soon occur. Also, within the pueblo, special windows and portholes allowed sunlight to hit special plates or markings on the walls at significant times of the year. Thus light and shadows, along with horizon features, made up the basis of the puebloan solar astronomy. The ancestors are believed to have done much the same as an adaptive survival strategy in a similar cultural context and environment.

Anasazi astronomy. Around A.D. 1000, the Anasazi prospered in the San Juan Basin and other regions of the Colorado Plateau. They built community houses that were four or five stories high, contained hundreds of rooms, and many large and small kivas—round, underground rooms used for ritual purposes.

Chaco Canyon in northwestern New Mexico grew to be a center of Anasazi culture. By 1130, eight

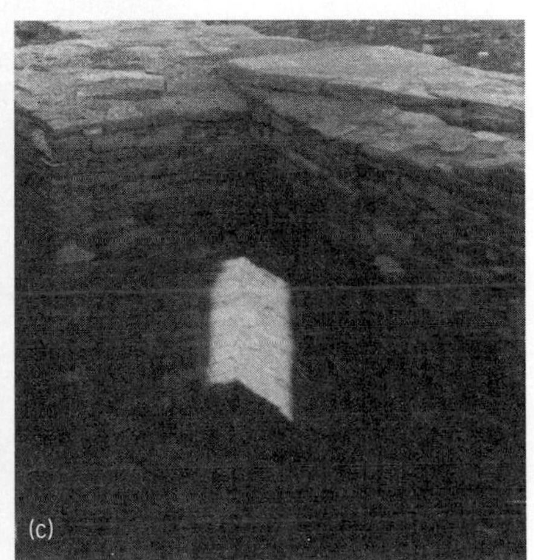

Fig. 5. Pueblo Bonito. (*a*) View of the southeast area. The two corner windows may have been used to anticipate and confirm the winter solstice. Room 228 is at the left. (*b*) Light at sunrise entering room 228 at the end of October. (*c*) Light from the winter solstice sunrise streaming into the corner of room 228. (*Photos by M. Zeilik*)

large villages were located within 9 mi (15 km) of the central canyon. An extensive road system within the San Juan Basin, perhaps a trade network, connected Chaco to many outlying villages. Perhaps a few thousand people lived here in the large and small villages. These Chacoans faced climatic conditions similar to those of today; like the historic pueblos, they probably also had Sun Priests and seasonal solar calendars (and perhaps lunar ones).

Three problems must be kept in mind when evaluating sites in Chaco Canyon. Many Chacoan buildings have been reconstructed, perhaps in ways quite different from the original way, so building orientations, unless based on general alignments of the original foundations, must be taken with caution. Second, few locations within the canyon offer good horizon profiles for a seasonal watch. Finally, it is not known whether each large village had its own, independent Sun Priest (as is true in the historic pueblos) or whether a central religious figure had the authority and responsibility for sun watching for the all the villages (including those at the outlier sites networked by the road system).

Pueblo Bonito. This D-shaped apartment house of over 800 rooms was built close to the north wall of the canyon. Within it are a number of corner doorways and windows, which are rather unusual in Anasazi architecture: Pueblo Bonito contains over half of the known examples. The archeologist J. Reyman noted that two of the windows in rooms in the southeast part of the ruin have a clear view of the winter solstice sunrise (**Fig. 5***a*)—if, in fact, an outer wall did not obstruct the view.

A Sun Priest at Pueblo Bonito could, in fact, keep a horizon calendar from anywhere on the east side of the pueblo, perhaps from the highest point on the roof. However, the horizon profile lacks clear features for sunrise during the span from the end of October up to the winter solstice. The solstice would be forecast from a rather long countdown from the last sun-

Fig. 6. Fajada Butte in Chaco Canyon. (*a*) View of entire butte (*photo by M. Zeilik*). (*b*) Rock slabs near the top of the butte. They rest against the rock surface on which two spirals are pecked. Their upper edges cause sunlight late in the morning to play upon the rock face and around the spirals (*photo by R. Elston*). (*c*) Summer solstice sunlight cutting through the large spiral at about 11:13 a.m. in 1980. Because of a shift of the middle slab, the current view is somewhat different (*photo by W. Wampler*).

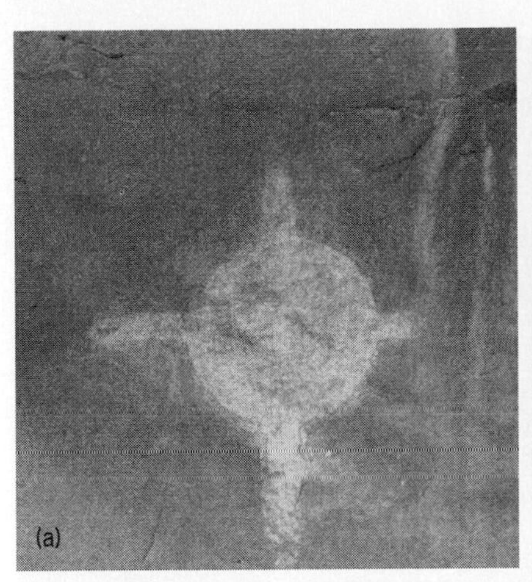

Fig. 7. Wijiji site in Chaco Canyon. (*a*) Painted white sun symbol above the ledge near the ruin. The emblem resembles that of the Zia pueblo sun symbol and a sun symbol on a war shield from Jemez pueblo. (*b*) Winter solstice sunrise observed from the ledge at a position near the boulders. The angular width of the rock pillar is a little smaller than the angular diameter of the Sun. (*Photos by M. Zeilik*)

rise observed behind a dominant horizon feature. The use of one of the southeast openings could have served as a complementary technique. The end of October is the first time that sunlight enters at sunrise. It does so at an oblique angle, so that the illumination on the wall appears as a shaft of light (Fig. 5*b*). As the sunrise points move southward during the approach to the winter solstice, the light shaft marches across the wall at an average rate of about 3 cm (roughly 1 in.) per day (Fig. 5*c*). By observing this motion relative to markers in the wall, the Sun Priest could easily forecast accurately the day of the winter solstice.

Fajada Butte. This butte (**Fig. 6***a*) thrusts upward at the eastern end of the canyon, the lone dramatic break in the landscape. Atop the butte lies a sun marker that tracks the seasons. Within 30 ft (10 m) of the summit, three rock slabs lie against the butte's southeast face (Fig. 6*b*). The slabs are a few meters long with about 4-in. (10-cm) gaps between them. They shield the rock face on which they rest from the Sun except at times before local solar noon. Then the edges of the slabs allow sunlight to strike the rock face, on which are carved two spirals: a large one (almost 1.5 ft or 0.5 m wide) right behind the slabs, and a smaller one below and to the left of the larger.

The spirals mark the Sun's yearly cycle by light patterns visible late in the morning. On the summer solstice, a shaft of light materializes above the large spiral at about 11 a.m. In about 20 min, it descends and slices through the heart of the spiral design (Fig. 6*c*). On the winter solstice, two shafts of light appear on the outside of the large spiral and pass through its outer edges at about 10 a.m. At both equinoxes, two shafts appear, one shorter and to the left of the other. The shorter shaft cuts through the center of the small spiral, while the larger one drops through one side of the large spiral.

In addition to the solar markings, A. Sofaer and her colleagues argue that it was also used to mark the 18.6-year lunar standstill cycle at moonrise for phases between full and waning crescent.

It is possible that the play of light and shadow from a natural rock fall was noted by the Anasazi, and that they then made the spirals on the rock face. Certainly the equinoxes and solstices have distinctive patterns of light and shadow. However, it is unlikely that these patterns were used to forecast, say, the solstices with the precision achieved by the historic pueblo Sun Priests, since the horizontal motion of the main shaft of light is much too small on an average daily basis to predict the solstices any better than within a week or so. The site might well have served as a sun shrine, a place to which religious officials journeyed at important times of the year to place offerings to the Sun. In the historic pueblos, sun shrines can be natural rock formations located some distance from the pueblo; a priest would visit them before sunrise.

Unfortunately, it may never be possible to validate these interpretations about the site. The middle rock slab moved in the mid-1980s, and the pattern of the light shafts has changed considerably. Hence, what is now seen is not what was seen before, and may not have been the pattern viewed by the Anazasi if erosion has shifted the slabs in the past 1000 years.

Wijiji. This is a pueblo built about A.D. 1100, late in Chaco's history. About 0.6 km (1 mi) to the east of it, a large rincon (valley) opens up in the mesa. On the northwest side of the rincon runs a narrow ledge, which can be reached by climbing a prehistoric staircase. Here on the wall is painted a large four-pointed symbol (**Fig. 7***a*) that resembles the Zia Pueblo sun sign.

North of the symbol, three boulders rest on the ledge; the largest has a double spiral carved in its surface. The design and technique are clearly Anasazi. Eastward from the ledge, a large rock pillar, across the rincon, rises above the horizon. From a spot a few meters south of the boulder with the double spiral, the winter solstice Sun rises behind the pillar (Fig. 7*b*). Because the angular width of the pillar against the sky is somewhat smaller than the angular diameter of the Sun, the winter solstice sunrise behind the pillar is a clear event when viewed from the location near the boulders. If an observer moves a few meters (yards) to either side, the shift moves the Sun to either side of the pillar. Near the sun symbol the Sun rises behind the pillar 16 days prior to the winter solstice, making possible an anticipatory observation that works extremely well to forecast the day of the winter solstice.

Was this the intention of the Anasazi? The area of

the ledge contains much rock art, some of it Navajo. R. Williamson has argued that the white "sun" symbol is Navajo in origin and that the site was used by Navajo (perhaps in the late seventeenth century) for sun watching. That may be the case, but there are certainly Anasazi relics here, too. Thus the site may have been first used by the Anasazi and then adopted by the Navajo for ceremonial purposes.

By 1250, Chaco was deserted. No one knows why. Some of these people settled the pueblos that are known today, taking with them the astronomy that was part of their culture.

Other Places and Cultures

Indigenous astronomy did not develop only in Great Britain and North America. In fact, every prehistoric or preliterate culture appears to have developed its own astronomy. The traditional navigators of the Pacific, both Micronesians and Polynesians, used naked-eye observations of the stars to sail thousands of miles successfully. The Carib people of northern South America developed a calendar that relied on the positions of the stars relative to each other and to the Sun at times of rising and setting. Incised pieces of bone hint that Ice Age people kept track of the phases of the Moon. The Ngas of Nigeria, Africa, predict the position of the new moon so that they can "shoot the moon" with a spear or arrow propelled by song. In the Eddas, the primary works of Nordic mythology, the motions of the Sun, Moon, and planets are presented in a mythic guise.

The understanding of the use of astronomy in the cultures in Mesoamerica and South American prior to the arrival of the Spanish has also expanded considerably. The practices go well beyond complex systems of keeping astronomical cycles and calendars. For instance, the ruling class at some Maya cities (such as Bonampak) used their astronomical knowledge of the cycles of Venus in order to wage war. In other places, such as Palenque and Yaxchilan, astronomy also had social meaning in the context of political power and transfer of power within the ruling class; political events took place at the time of the summer solstice and, perhaps, with conjunctions of Jupiter and Saturn. Ancient Inca planning and politics also incorporated astronomy, such as in the ceque system of radial lines from the Temple of the Sun in the valley of Cuzco. These lines mark the directions to sacred places as well as to astronomical phenomena. The ceque system had a calendric manifestation in knotted cords that could tally the days of the agricultural year.

Michael Zeilik

Bibliography. A. F. Aveni (ed.), *Archaeoastronomy in the New World,* 1982; A. F. Aveni (ed.), *World Archeoastronomy,* 1989; A. F. Aveni and G. Urton (eds.), Ethnoastronomy and archaeoastronomy in the American tropics, *Ann. N.Y. Acad. Sci.,* vol. 385, 1982; D. C. Heggie (ed.), *Archaeoastronomy in the Old World,* 1982.

Areal velocity

The rate at which a line that joins a fixed point and a moving particle sweeps out a surface area is called the areal velocity with respect to the fixed point.

In polar coordinates, where ϕ is the central angle and ρ is the distance between the fixed point and the moving particle, the areal velocity dA/dt equals $(1/2)\rho^2 d\phi/dt$. In elliptical motion, if the origin is at one focus, $\rho^2 d\phi/dt$ is a constant. In astronomy, Kepler's law of areas expresses this characteristic. SEE CELESTIAL MECHANICS.

Raynor L. Duncombe

Aries

The Ram, in astronomy, a zodiacal and autumnal constellation. In Aries, there are a few stars brighter than fourth magnitude, three of which, α, β, and γ, form an obtuse triangle (see **illus.**). The bright star,

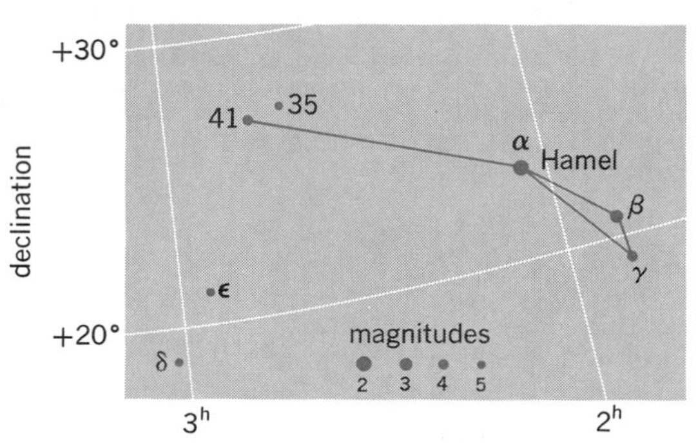

Line pattern of constellation Aries. Grid lines represent the coordinates of the sky. Apparent brightness, or magnitudes, of stars are shown by sizes of the dots, which are graded by appropriate numbers as indicated.

α-Hamel (called the Ram), an eye of the Ram, is a navigational star. Among the 12 zodiacal constellations, Aries was considered as the first, because about 2000 years ago when the zodiacal constellations were organized, the Sun was in Aries where it crossed the Equator at vernal equinox. Today, because of the precession of the equinoxes, this reference point has moved into the constellation Pisces. However, Aries remains the first sign of the zodiac. SEE CONSTELLATION; PISCES; PRECESSION OF EQUINOXES; ZODIAC.

Ching-Sung Yu

Asteroid

One of the many thousands of small planets (minor planets) revolving around the Sun, mainly between the orbits of Mars and Jupiter. The presence of a gap in J. A. Bode's empirical law of planetary spacings motivated a search for the missing planet. The Italian astronomer G. Piazzi discovered Ceres on January 1, 1801. Three other small planets were discovered in the next few years, leading H. Olbers to suggest that they were all fragments of a disrupted planet. Visual and, later, photographic searches for additional asteroids have continued to the present day. Newly discovered ones are assigned a catalog number and name (such as 433 Eros) only after they are observed often enough to compute an accurate orbit. The Harvard-Smithsonian Center for Astrophysics maintains a complete file of all measurements of asteroid positions. The Institute for Theoretical Astronomy in St. Petersburg publishes an annual ephemeris of predicted

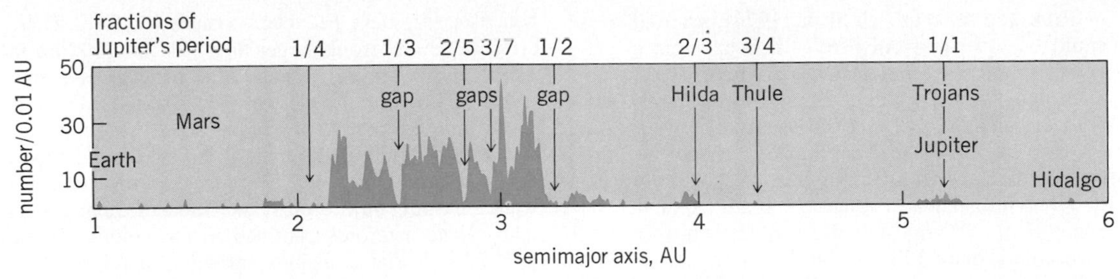

Fig. 1. Distribution of numbered asteroids, with distance from the Sun, between the Earth and Jupiter. Major fractions of asteroid orbital periods to Jupiter's period are shown. The arrows point to associated clusters or Kirkwood gaps.

asteroid positions for the nearly 4000 cataloged asteroids. *SEE PLANET.*

Orbits. The vast majority of asteroids have semimajor axes (mean distances to the Sun; symbolized a) between 2.2 and 3.2 astronomical units (1 AU = distance from Earth to the Sun = 1.496×10^8 km = 9.3×10^7 mi). However, dozens of small asteroids orbit between Venus and Mars, and two groups, the Trojan asteroids, orbit at Jupiter's distance from the Sun. The farthest asteroid discovered so far is 2060 Chiron, which is located between Saturn and Uranus at a = 13.6 AU. The innermost has a = 0.83 AU. *SEE TROJAN ASTEROIDS.*

Most asteroid orbits are more elliptical and inclined to the plane of the ecliptic than the orbits of major planets. Eccentricities (e) average about 0.15, and inclinations (i) about 10°; occasionally they exceed 0.5 and 30°, approaching the characteristics of short-period comet orbits. A number of small asteroids (Amor objects) cross, but do not intersect, the orbit of Mars, and a few even cross the Earth's orbit (Apollo objects).

Asteroids are not uniformly distributed in a, e, and i. **Figure 1** shows the vacant lanes in the main asteroid belt, known as Kirkwood gaps, which occur at distances where the periods of revolution would be a simple fraction (such as ⅓ or ⅖) of Jupiter's period of 11.86 years. Asteroids in these gaps may have been preferentially removed when Jupiter's powerful gravity sent them into wild, chaotic orbits; in these orbits, they may have collided with Mars, the Earth, or another planet. Most asteroids originally beyond the ½ resonance were directly ejected from such orbits by nearby Jupiter early in the history of the solar system, except those grouped near the stable ⅔ (Hilda group) and ¾ (Thule) resonances, which may survive because there are few nearby asteroids with which to collide.

Clusterings of asteroids with similar a, e, and i are known as Hirayama families, named for the Japanese astronomer K. Hirayama who discovered some of the major ones in 1918. About 40% of the numbered asteroids are members of over 100 families tabulated by J. Williams. Some of the larger families are composed of asteroids that have the same colors, and are presumably made of the same minerals. Such a family probably consists of fragments of a larger, precursor asteroid that was broken apart by a catastrophic collision with another asteroid or comet. Some of these families are immersed in huge, torus-shaped dust belts, discovered by the *Infrared Astronomical Satellite*. The colliding asteroids are gradually grinding each other down to dust, which then spirals into the Sun or is blown away by light pressure from the Sun. *SEE ORBITAL MOTION.*

Shapes, spins, and satellites. When the brightnesses of most asteroids are measured, they vary in a periodic manner, occasionally by more than one magnitude (a factor of 2.5), but more commonly 0.3 magnitude or less. The light curves usually are double-peaked (**Fig. 2**), characteristic of an irregularly shaped body spinning in space rather than the more complex curves that would result from albedo differences (spots) on a spherical body. Light curves have been measured for many hundreds of asteroids. The rotation periods are typically about 8 to 11 h, but range from 2½ h to several weeks. The fact that most asteroids rotate with periods similar to the major planets suggests the spins may have a primordial origin. But there is some evidence that at least the smaller asteroids' spins may result from collisons they have suffered. Such collisions also produce the irregular, nonspherical shapes of asteroids. The larger asteroids, especially those composed of weak materials, cannot maintain highly irregular shapes against the compression of their own gravitational fields, which is why only smaller asteroids and some of the larger ones probably made of a strong material like iron have large light-curve amplitudes. Asteroids could not rotate much faster than 2½ h, otherwise the centripetal acceleration at their equator would exceed their gravity and they would fly apart.

Some large, rapidly spinning asteroids have large amplitudes; they may be weak "rubble piles," previously fractured by collisions and now distorted into elongated equilibrium ellipsoids by their rapid spins.

Several kinds of observations have suggested that at least a few asteroids have "minor satellites" orbiting about them. The observations are controversial and need to be checked in the future.

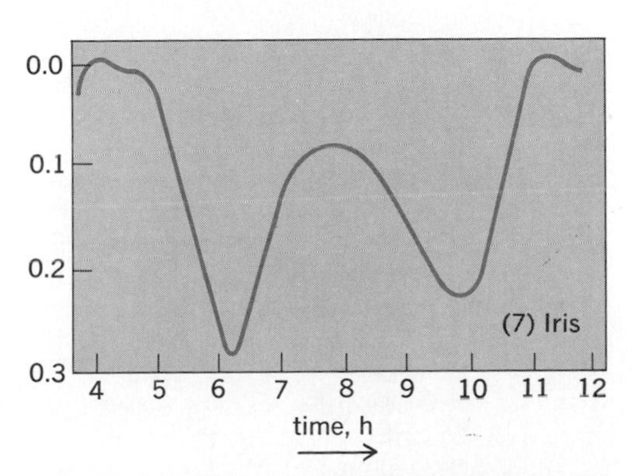

Fig. 2. Variation with time in the brightness of 7 Iris due to its irregular shape and 7.1-h rotation. Ordinate is in tenths of a stellar magnitude. ***(After G. Kuiper)***

Sizes and masses. Until the 1970s, asteroid sizes could be estimated only from their apparent brightnesses in the sky. But asteroids are faint, starlike objects; only Vesta is bright enough to be faintly seen with the unaided eye, and no asteroid shows a disk, even in the largest telescopes, large enough to measure accurately. It was impossible to tell, for instance, if 324 Bamberga is a relatively bright-colored body about 43 mi (70 km) in diameter or (as is now known) a very dark body 150 mi (240 km) in diameter. Two techniques were used in the 1970s to measure the diameters and albedos of over 130 asteroids: the polarimetric technique, based on an empirical correlation between the albedos of materials and how they polarize light; and the radiometric technique, which measures diameters by comparing the brightness of reflected visible sunlight from an asteroid with the brightness of the asteroid's emitted thermal radiation in the infrared. In the mid-1980s, the radiometric technique was used by the *Infrared Astronomical Satellite* to measure the diameters and albedos of over 2000 asteroids. The two techniques have been checked by timings of the disappearances of stars when asteroids chance to pass between the Earth and a star. *See Albedo; Infrared astronomy.*

It is now known that asteroids are much darker, hence larger, than had been assumed before. There are about 30 asteroids larger than 124 mi (200 km) in diameter (see **table**); about 75% of them are soot black (geometric albedos of 3–5%). Asteroids are much more numerous at smaller sizes, generally following a size distribution characteristic of fragmentation processes, as would be expected if the asteroids were smashing into each other. Indeed, there are so many large asteroids confined in the volume of the asteroid belt that collisions sufficient to fragment all but the larger asteroids occur every few billion years, and much more often for smaller ones. Thus the asteroids are mainly collisional fragments.

Masses of asteroids are not easily measured since they are so small as to exert a negligible effect on the orbits of other planets. But a few large asteroids regularly pass very close to some other asteroids and affect their orbits measurably. Measurements of these effects have yielded the masses of the three largest ones: Ceres (2.6×10^{21} lb or 1.2×10^{24} g), Pallas (4.9×10^{20} lb or 2.2×10^{23} g), and Vesta (6.0×10^{20} lb or 2.7×10^{23} g). From the diameters and masses together, densities of 2.3, 2.6, and 3.3 times that of water are implied, respectively. Vesta's density is similar to that of ordinary rocks, but the other two seem as underdense as carbonaceous chondritic

Noteworthy asteroids

Size rank	Number and name	Spectral type	Diameter, mi (km)	Spin period, h	Orbital elements		
					a, AU	e	i
1	1 Ceres	G (C-like)	578 (930)	9.1	2.768	0.08	10.6°
2	2 Pallas	B (C-like)	343 (552)	7.8	2.773	0.23	34.8
3	4 Vesta	Achondrite	324 (521)	5.3	2.362	0.09	7.1
4	10 Hygiea	C	260 (419)	17.5	3.138	0.12	3.8
5	704 Interamnia	F (C-like)	203 (327)	8.7	3.060	0.15	17.3
6	511 Davida	C	200 (322)	5.2	3.181	0.17	15.9
7	52 Europa	C	183 (295)	11.3	3.095	0.11	7.5
8	87 Sylvia	P	172 (277)	5.2	3.483	0.09	10.9
9	65 Cybele	P	167 (269)	6.1	3.428	0.11	3.6
10	15 Eunomia	S	161 (259)	6.1	2.642	0.19	11.8
11	16 Psyche	M	155 (249)	4.2	2.922	0.14	3.1
12	31 Euphrosyne	B (C-like)	154 (248)	5.5	3.148	0.23	26.3
13	451 Patienta	B (C-like)	153 (247)	9.7	3.065	0.07	15.2
14	3 Juno	S	150 (242)	7.2	2.671	0.25	13.0
15	324 Bamberga	C	149 (240)	29.4	2.685	0.34	11.2
16	13 Egeria	G (C-like)	139 (224)	7.0	2.576	0.09	16.5
17	45 Eugenia	F (C-like)	139 (223)	5.7	2.720	0.08	6.6
18	624 Hektor	D	186 × 93 (300 × 150)	6.9	5.153	0.03	18.3
19	532 Herculina	S	137 (220)	9.4	2.774	0.17	16.3
20	107 Camilla	C	137 (220)	4.9	3.487	0.07	10.0
21	423 Diotima	P?	135 (217)	4.6	3.069	0.03	11.2
22	121 Hermione	C	135 (217)	8.9	3.460	0.14	7.6
23	19 Fortuna	C	130 (210)	7.5	2.442	0.16	1.6
24	24 Themis	C	129 (207)	8.4	3.129	0.13	0.8
25	7 Iris	S	127 (204)	7.1	2.386	0.23	5.5
26	6 Hebe	S	126 (202)	7.3	2.424	0.20	14.8
27	702 Alauda	C	126 (202)	8.4	3.195	0.03	20.5
28	88 Thisbe	C	124 (200)	6.0	2.769	0.16	5.2
			Other interesting asteroids				
	41 Daphne	C	116 (187)	6.0	2.767	0.27	15.8
	44 Nysa	Aubrite?	42 (68)	6.4	2.423	0.15	3.7
	165 Loreley	C	99 (160)	6.?	3.140	0.07	11.2
	216 Kleopatra	M	87 (140)	5.4	2.790	0.25	13.2
	250 Bettina	M	53 (86)	5.1	3.140	0.14	12.9
	349 Dembowska	Achondrite?	90 (145)	4.7	2.926	0.09	8.3
	433 Eros	Chon./S?	22 × 7 (36 × 12)	5.3	1.458	0.22	10.8
	747 Winchester	P	116 (186)	8.?	3.004	0.34	18.2
	1566 Icarus	Chon.?	1 (2)	2.3	1.078	0.83	22.9

meteorites, which contain a large fraction of volatiles. The total mass of all asteroids is only three times that of Ceres alone, or about 5% that of the Moon. *See Ceres.*

Surface compositions. Spectra of sunlight reflected from asteroids have shapes, including absorption bands, characteristic of different rock-forming minerals (**Fig. 3**). Combined with the albedo data from polarimetry and radiometry, the spectral colors of surfaces of over 1000 asteroids show that more than three-quarters of them have very low albedos and are composed of carbon-rich material (often with hydrated, or water-rich, minerals). The black asteroids located in the middle and outer parts of the belt (called C type) resemble carbonaceous meteorites, which are believed to be among the most primitive materials in the solar system, unaltered since the planets were forming. The black asteroids near the outer edge of the main belt, and most of the Hildas, have a reddish tinge and are not represented by known meteorites on the Earth; they are called P types, and may be even richer in organic components. Still farther out, many of the Trojans are even redder and more mysterious; they are termed D types.

Closer to the inner edge of the belt, most asteroids are so-called S types, characterized by moderately high albedos and by absorption bands due to the common silicate minerals pyroxene and olivine. They also contain considerable metal, and probably are akin to either the stony-iron meteorites or the ordinary chondritic meteorites. Calculations show that fragments of S types located near the ⅓ Kirkwood gap may be converted into chaotic orbits that reach to the Earth; thus those S types may be the parent bodies for many of the meteorites. Although S types constitute only about 15% of the total asteroid population, most of the brighter asteroids visible from the Earth are S's, because they are both brighter and closer to the Earth. *See Meteorite.*

The general progression of asteroid compositions, from S types in the inner belt, to C's, to P's, and then to D's at Jupiter's distance, is thought to reflect the variation with distance from the Sun in the composition of the original nebular dust from which the planets were formed. *See Cosmochemistry; Solar system.*

There are also some less common asteroid compositions, which are nevertheless interesting because they imply that some of the asteroids were somehow heated to the point that they melted. The asteroid Vesta is apparently covered with solidifed lava flows, made of the same minerals as the basaltic achondritic meteorites. When a planetary body is heated to the point that volcanic lavas flow onto its surface, metal sinks to the center, forming a core. A number of asteroids, notably 16 Psyche, have so-called M-type spectra and radar echoes suggesting they are made of solid metal. Astronomers believe that Psyche's parent body was melted into a body that looked like Vesta does today, with a basaltic surface, an olivine-rich mantle, and a metallic core. Then enormous asteroid collisions fragmented and stripped away the rocky crust and mantle, exposing the core, which is Psyche as it is seen today. It is difficult to understand why Psyche and other M types were so thoroughly fragmented by collisions, while Vesta's crust is preserved intact. Also, there are only about eight small asteroids known (olivine-rich A types) which are made of the rocks expected to occur in great abundance in the mantles of once-melted asteroids. Where all the olivine went when the numerous M types were collisionally stripped to their cores remains a mystery.

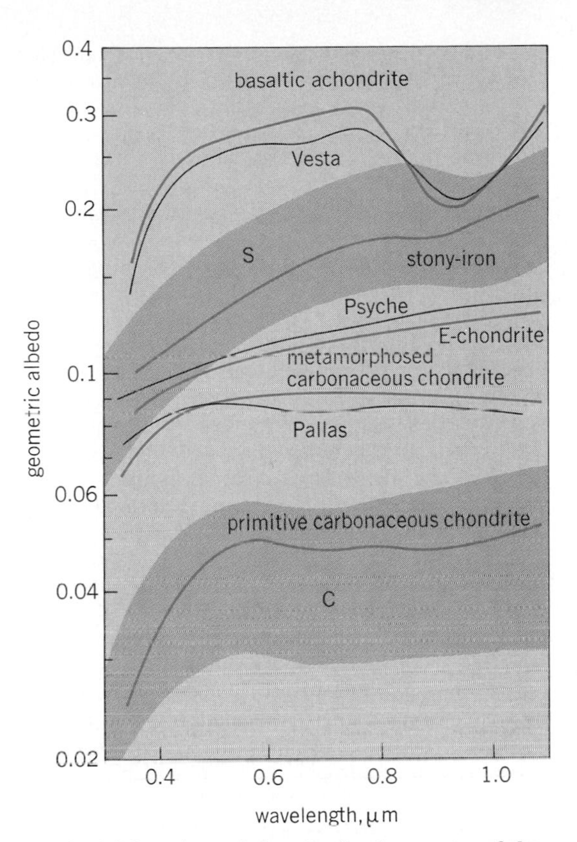

Fig. 3. Visible and near-infrared reflection spectra of the major S and C compositional classes of asteroids, plus several unusual ones. Laboratory spectra for some similar meteorites are shown with colored lines.

Surface conditions. Surfaces of belt asteroids are cratered repeatedly. Larger asteroids may be fragmented and pulverized to great depths. But most excavated ejecta escape the weak gravity fields of smaller asteroids, so thick deposits of fragmental soil (regoliths) do not develop. Radar echoes from Eros, an elongated (22 × 7 mi or 36 × 12 km) Mars-crosser, show that it lacks a thick regolith on its probably crater-scarred surface. But polarimetry of asteroids suggests that even the smallest are dusty. Subsolar surface temperatures mostly range from −80 to +10°F (210 to 260 K). Asteroids lack even thin atmospheres because of their low gravity.

Earth-approaching asteroids. Apollo and Amor asteroids are of special interest, particularly because they stand a chance of striking the Earth. Indeed, Meteor Crater (Arizona), and other craters on the Earth and the Moon, testify to the potential for collisions with Earth approachers. Many scientists believe that just such a collision 6.5×10^7 years ago rendered many species of life, including the dinosaurs, extinct. It has been estimated that even today the chances of an impact occurring that could destroy civilization may be as high as 1 in 100,000 each year. Although dozens of Earth approachers have been discovered, it is estimated that many hundreds larger than 1 mi (1.6 km) across remain to be found.

These small asteroids have not always been in Earth-approaching orbits. After some tens of millions of years, most of the current crop will have struck the Earth, the Moon, or one of the other inner planets, or will have been ejected from the solar system. Most are probably fragments of main-belt asteroids, traveling in chaotic orbits, just like their smaller cousins, the meteorites. (In fact, some meteorites are probably chips broken off Earth approachers.) Other Apollo and Amor asteroids are dead comet nuclei, which have lost their ices after thousands of years spent close to the Sun in the inner solar system. *See Comet.*

Origin and evolution. Olbers's idea that the asteroids are fragments of an exploded planet is no longer attractive. It is hard to imagine what could produce an explosion sufficient to disperse a full-sized planet against its own gravity. Moreover, meteorites show evidence of having formed in bodies the size of present-day asteroids. Since current cosmogonical models for the origin of planets involve accretion from myriads of asteroidlike planetesimals, it is likely that asteroids are a remnant of the planetesimals that failed to accrete into a planet between Mars and Jupiter. Perhaps bombardment of the asteroid zone by large planetesimals scattered from massive, nearby Jupiter increased the relative velocities of asteroids to the present value of 3 mi/s (5 km/s) so that asteroids fragment rather than accrete when they meet each other. Instead of forming a planet, the asteroids have been smashing each other to bits, and those seen now are mostly fragments, except for the "lucky few" that have escaped catastrophic collisions so far.

Evidently some asteroids of primitive, nonvolatile solar composition were heated within the first few hundred million years after the origin of the solar system, perhaps by the solar wind or extinct radionuclides, and they melted. While the unmelted, weak, C-type asteroids may have been depleted by a large factor by collisions, most of the strong stony-iron cores of the melted proto-asteroids have survived, perhaps as the S-type asteroids observed today. The asteroids still collide and fragment, occasionally spraying the inner solar system with chips that produce craters or fall as meteorites.

This picture of the evolution of the asteroids is incomplete and tentative. As new data are collected all the time, research continues on alternative interpretations. A spacecraft mission to several asteroids could reveal many further clues that the asteroids contain about the formative period of solar system history. It is hoped that it will be possible to fly a multiasteroid rendezvous mission by the year 2000.

Clark R. Chapman

Bibliography. J. K. Beatty, *The New Solar System*, 3d ed., 1989; R. Binzel, T. Gehrels, and M. Mathews (eds.), *Asteroids II*, 1989; C. J. Cunningham, *Introduction to Asteroids*, 1988; H.Y. McSween, Jr., *Meteorites and Their Parent Planets*, 1987.

Astrometry

Measurement of the angular positions of celestial objects on the celestial sphere and the study of the changes of those positions with time.

Fundamental measurements. The positions of celestial objects are measured visually or photoelectrically with small, stable telescopes designed to measure meridian altitudes and meridian crossing times with the utmost precision. These transit (meridian-circle) instruments provide a fundamental reference frame for the positions of objects on the celestial sphere. *See Astronomical coordinate systems; Celestial sphere; Optical telescope; Telescope.*

The positions are in the form of two coordinates, right ascension and declination, oriented with respect to the celestial equator and referred to the vernal equinox as the origin of the former. Differences in positions of objects at different epochs yield values of the precession of the Earth's polar axis along with the individual proper motions of the stars. Results of these observations are compiled in catalogs of fundamental star positions and proper motions, such as the FK4, which contains values for 3522 stars. The FK4 defines a fundamental reference system against which all other star positions may be measured. The improved FK5 survey will extend the fundamental reference frame to fainter stars with improved positional accuracy. *See Astronomical catalogs; Astronomical coordinate systems; Precession of equinoxes.*

Photographic astrometry. Photographic plates taken with large telescopes can yield data on larger numbers of stars to fainter limiting magnitudes than can transit measurements. The photographic plates are measured with precision measuring machines, the most advanced of which are automated and can reach positional accuracies across the plates of 0.2–0.3 micrometer. *See Astronomical photography.*

Wide-field astrometric studies are conducted by using telescopes with focal lengths of 2–7 m (6–23 ft) and fields of view of 2–10°. These wide-field studies are used to produce comprehensive catalogs of star positions referred to a fundamental reference frame, and catalogs of stellar proper motions. Wide-field astrometry can also be used to detect stars of high proper motion, thus selecting, from the large number of faint stars in the sky, those likely to be nearest the Sun. *See Star.*

Long-focus astrometric photography uses telescopes with focal lengths of 10–20 m (33–66 ft) and fields of view of less than a degree. These studies are used to measure (1) the annual parallaxes of stars, which provide the underpinning of the cosmic distance scale; (2) the relative motions of binary stars, which can provide the masses of stars and can reveal the presence of unseen companions; and (3) the distributions of proper motions in both open and globular star clusters, which help determine membership in the clusters, the total masses of clusters (by applying the virial theorem), the dynamical states of the clusters, and, in conjunction with radial-velocity measurements, the distances to the clusters. *See Binary star; Parallax; Star clusters.*

Advanced techniques and instrumentation. The precision and productivity of astrometric studies continue to improve owing to new techniques and better instrumentation. The technique of speckle interferometry, which eliminates much of the blurring of star images caused by atmospheric turbulence, is routinely used to measure the relative positions of binary stars too close to each other for their separations to be measured visually. Arrays of solid-state detectors (charge-coupled devices), more sensitive and dimensionally stable than photographic plates, promise to improve positional accuracy over small fields of the sky to better than 0.001 arc-second. Photoelectric scanners, which directly measure the separation of images in the focal plane of the telescope, can provide differential measurements of faint star positions to comparable accuracy. *See Charge-coupled devices; Speckle.*

Observations made from space promise improvements. The Hubble Space Telescope is able to conduct astrometry free of the atmospheric and gravitational limitations of Earth-based instruments. The European Space Agency's dedicated astrometric satellite, *Hipparchos*, is designed to determine the positions of about 100,000 stars down to the eleventh magnitude, along with parallaxes and proper motions of high precision. SEE SATELLITE ASTRONOMY.

Laurence A. Marschall

Radio astrometry. The field of radio astrometry has achieved very precise positions (0.0002 arcsecond) of celestial radio sources. The technique utilizes two radio telescopes configured as an interferometer. One method employs two radio telescopes, situated a few miles apart, linked by cables to a receiving system that instantly measures the difference in phase between the radio signals arriving at the telescope. Another technique, very long-baseline interferometry (VLBI), utilizes a very accurate timing system, usually atomic clocks, to record the signals received at the end of a baseline several thousand miles apart. With computers, a vast amount of radio data may be recorded, stored, and processed. SEE ATOMIC CLOCK.

The National Radio Observatory's Very Large Array (VLA) in New Mexico contains antennas in a Y-shaped configuration, with arms 13, 13, and 11.8 mi (21, 21, and 19 km) long. It gives radio astrometric positions to high accuracy. A dedicated continental-scale interferometric array, the Very Long Baseline Array (VLBA), is under construction. SEE RADIO TELESCOPE.

The U.S. Naval Observatory Green Bank 22-mi (35-km) interferometer system has a second element, located orthogonal to, and 20 mi (32 km) distant from, the previous baseline. This interferometer is being developed as part of an all-Navy VLBI system, primarily to determine Earth orientation parameters. SEE EARTH ROTATION AND ORBITAL MOTION.

Benny L. Klock

Bibliography. H. K. Eichhorn, *Astronomy of Star Positions*, 1974; W. A. Hiltner (ed.), *Stars and Stellar Systems*, vol. 2: *Astronomical Techniques*, 1962; K. J. Johnston, *Radio Interferometry*, International Astronomical Union, 20th General Assembly, 1988; D. Monet, Recent advances in optical astrometry, *Annu. Rev. Astron. Astrophys.*, 26:413–440, 1988; A. C. Readhead, Radio astronomy by very long baseline interferometry, *Sci. Amer.*, 246(6):38–47, 1982; W. F. van Altena, Astrometry, *Annu. Rev. Astron. Astrophys.*, 21:131–164, 1983; P. van de Kamp, *Principles of Astrometry*, 1967; H. G. Walter, The precision of astrometric surveys of radio sources, *Abh. Hamburger Sternwarte*, 10:145–149, 1982.

Astronomical atlases

Sets of maps of celestial phenomena. Often developed in conjunction with catalogs that list position, brightness, and other features, maps provide a clear picture of the spatial relations between the phenomena. Over many centuries various peoples observed the positions and motions of some celestial bodies, but the most systematic observations and detailed maps are the product of Western culture.

Ptolemaic catalog. The oldest extant star catalog is that compiled by Ptolemy and included in the *Almagest,* written around A.D. 150. The Ptolemaic catalog covers 1028 stars visible from the Mediterranean and bright enough to be distinguished by the naked eye. The historical importance of the Ptolemaic catalog can hardly be overemphasized. For more than 1400 years Islamic and European astronomers limited their observations to those stars listed by Ptolemy. Ptolemy did not make any star maps or atlases. Instead, he plotted the stars onto a globe which, unfortunately, no longer survives. But it was probably the prototype for the numerous globes made for Islamic astronomers, several of which can still be seen in museum collections.

Early maps and atlases. The oldest extant flat maps of the stars were made in Germany, and date from the mid-fifteenth century. The oldest printed star maps are woodcuts, dated 1515, drawn by the Nuremberg artist Albrecht Dürer. These two planispheres—one of the northern ecliptic hemisphere and one of the southern—depict the Ptolemaic stars and constellations. The stellar longitudes were intended to be correct for the year 1500. Since, however, Dürer and his associates used a theory of precession now known to be erroneous, the star positions are actually correct for about 1440. As European astronomers undertook systematic observations and improved their understanding of precession, the star positions in catalogs and on maps gained in accuracy. On Gerhardus Mercator's globe of 1551, the star positions are correct for the year 1550. SEE PRECESSION OF EQUINOXES.

The increased literacy and enthusiasm for science of the early "scientific revolution" is reflected in the proliferation of star maps and atlases in that period. Indicative of this trend is Alessandro Piccolomini's *De Le Stelle Fisse Libro Uno* (Venice, 1540), an atlas of 48 maps, one for each Ptolemaic constellation. This was one of the first scientific texts written in the vernacular (in this case, Italian) rather than in Latin, the language of scholars, and it proved remarkably popular. Within a century there were at least ten more Italian editions, and three French and three Latin translations.

Toward the end of the sixteenth century astronomers began to observe the stars that Ptolemy had neglected or could not see, and as their catalogs expanded, star maps and atlases increased. Johann Bayer's *Uranometria* (Augsburg, 1603) is based on Tycho Brahe's extended catalog of northern stars, and includes a map of southern stars recently observed by Dutch navigators sailing through southern waters. In this atlas Bayer identified the stars by letters—Greek for the brighter, lowercase Roman for the fainter—which, in time, became their standard designations. The finest seventeenth-century atlas was Johannes Hevelius's *Firmamentum Sobiescianum sive Uranographia* (Gdansk, 1687), based on Hevelius's observations of the northern stars and Edmund Halley's telescopic observations of the southern stars.

John Flamsteed, the first Astronomer Royal, compiled the first telescopic catalog of the positions and magnitudes of the northern stars. The *Atlas Coelestis* (London, 1729) is based on these observations. The last great atlas illustrated with constellation figures was Johann Elert Bode's *Uranographia* (Berlin, 1801). Containing 17,240 stars and 99 constellations, it is spendid in its complexity, but difficult to read. SEE CONSTELLATION.

Modern atlases. In the nineteenth century the introduction of larger telescopes, steadier mounts, better graduated circles, and filar micrometers led to more extensive and precise star catalogs and atlases. Premier among these was the *Atlas des nordlichen gestirnten Himmels* (Bonn, 1863), organized by F. W. A. Argelander. These 40 charts showed the positions and magnitudes of 324,198 stars in the northern hemisphere. The charts, along with their companion star catalog, the *Bonner Durchmusterung*, or "B.D.," are still in use today. A supplement showing stars between 1 and 23° south declination appeared in 1887. The third edition, revised, of Argelander's northern atlas appeared in 1954; the second edition, revised, of the southern atlas appeared in 1951. SEE ASTRONOMICAL CATALOGS.

Photographic atlases. By the 1880s photography was sufficiently well developed for astronomers to begin considering a photographic atlas of the heavens. The *Carte du Ciel* was an international project of long-exposure, dry-plate photography, using 17 identical refracting telescopes, corrected for photography and situated around the world. This ambitious atlas was never fully published. In 1949, under sponsorship of the National Geographic Society, was begun the *Palomar Observatory Sky Survey*—the first photographic atlas that showed the sky in two colors. Taken with the 48-in. (1.2-m) Schmidt telescope, it reveals stars brighter than magnitude 20 situated north of −33°.

A useful addition to the *Palomar Sky Survey* is a set of transparent overlays published by Ohio State University Radio Observatory (*Maps of the Palomar Sky Survey Photographs*, Columbus, 1981). These overlays are imprinted with the positions and catalog numbers of stars, nebulae, and so forth. Placing an overlay on top of a Sky Survey photographic print makes it possible to identify many of the cataloged objects on that print.

A new *Palomar Observatory Sky Survey* was begun in 1985. This atlas will use improved photographic plates to reach magnitude 22, in three colors.

The southern hemisphere of the sky has already been photographed to magnitude 22, with the 48-in. (1.2-m) United Kingdom Schmidt telescope in Siding Spring, Australia, on blue-sensitive plates. A 40-in. (1.02-m) Schmidt telescope at the European Southern Observatory in Chile has also photographed the southern sky to a magnitude limit of 20, by using red-sensitive plates. All of these surveys have been distributed in the form of photographic atlases on paper, film, or glass plates. SEE ASTRONOMICAL PHOTOGRAPHY.

Printed atlases. In addition to the photographic atlases, printed atlases continue to be published, usually in conjunction with star catalogs. Among the more popular atlases of this type are the *Smithsonian Astrophysical Observatory Star Catalog* and *Star Atlas of Reference Stars and Nonstellar Objects* (Washington, 1966); Antonin Becvar's *Skalnate Pleso Atlas* (Cambridge, 1949) and its successors; *Norton's Star Atlas* (Cambridge, 1979); and W. Tirion's *Sky Atlas 2000.0* (Cambridge, 1982).

Atlases of nonstellar objects. By the late nineteenth century photography had revealed countless hitherto-inaccessible details of nebulae and galaxies, and astronomers had begun to recognize the significance of these objects. An important pioneer in nebular photography was Edward Emerson Barnard, whose *Photographic Atlas of the Milky Way* appeared in 1927. Another pioneer was Isaac Robets, whose *Atlas of 52 Regions* was published in Paris in 1929. Edwin Hubble, known as the founder of modern extragalactic astronomy, charted the distribution of galaxies in space and their morphology. The *Hubble Atlas of Galaxies* appeared in 1961, edited by Allan Sandage. SEE GALAXY, EXTERNAL; MILKY WAY GALAXY; NEBULA.

Atlases of solar system objects. Galileo, one of the first to observe the Moon through a telescope, produced the first detailed drawings of the lunar surface. These were succeeded by Johannes Hevelius's *Selenographia* (Gdansk, 1647) and Johann Schmidt's *Mondcharte in 25 sectionen* (Berlin, 1878). Lunar photography dates from the 1850s, but not until late in the nineteenth century did photography provide any advantage over visual observations at the telescope. G. P. Kuiper's *Photographic Lunar Atlas* (Chicago, 1960) contains the best set of Moon plates based on terrestrial observations to date. On October 7, 1959, *Lunik III* took the first photographs of the back side of the Moon, and subsequent probes generated countless closeups of the lunar surface. A particularly beautiful photographic atlas of the Moon is the NASA publication *The Moon as Viewed by Lunar Orbiter* (Washington, 1970). SEE MOON.

As spacecraft explore more of the solar system, atlases of more objects are becoming available. The planets Mercury and Mars have been thoroughly photographed. The satellites of Mars, Jupiter, and Saturn have also been mapped. The surface of cloud-shrouded Venus has been mapped by terrestrial and space-borne radar. More atlases of individual solar system objects are sure to appear after each new spacecraft carries out its mission. SEE JUPITER; MARS; MERCURY; SATURN; SOLAR SYSTEM; VENUS.

Deborah Jean Warner; Charles T. Kowal

Astronomical catalogs

Lists or enumerations of astronomical data, generally ordered by increasing right ascension of the objects listed. Astronomical catalogs vary a great deal in form and content depending upon their use, which may be purely astronomical, or for navigation, time determination, geodesy, or space science applications. In some catalogs the essential data are stellar positions and motions, while in others astrophysical data, such as magnitudes, spectra, and radial velocities of stars, are important. There are also catalogs of special stellar and of nonstellar objects. SEE ASTRONOMICAL COORDINATE SYSTEMS.

Catalogs of stellar positions. A catalog regarded as the best representation of the celestial coordinate system at the time of its publication is called a fundamental star catalog.

The *Fourth Fundamental Catalogue*, designated *FK4*, was published by the Astronomisches Rechen-Institut in Heidelberg, West Germany, in 1963. The catalog contains the positions (right ascensions and declinations) and their changes with time (proper motions and precession) of 1535 stars. These data were compiled from transit circle observations made over a span of 110 years. The successor, the *FK5*, appeared in 1985 with data compiled from more than 250 catalogs with a combined number of 3,000,000 observations. SEE OPTICAL TELESCOPE.

The positions and proper motions of the stars in *FK5* provide a fundamental system for measurements of other star positions and proper motions, which may be carried out for a variety of problems arising in stellar astronomy. The reference demands homogeneity over the entire sky, which means that the positions and proper motions of the stars in different parts of the sky represent the celestial coordinate system with the same precision. With few exceptions, the stars in the *FK5* are brighter than ninth magnitude. Positions of the stars fainter than those in the *FK5* on a fundamental system are obtained by a close coordination between visual and photographic programs. *SEE FUNDAMENTAL STARS.*

Moderately bright stars (seventh to ninth magnitude), selected on the basis of one star per square degree of the sky, are related to the fundamental system by meridian circle observations. These stars form a system of sufficient density to serve as position references for photographic observations. Typical of catalogs of such reference stars is the *AGK3R* of some 21,000 stars in the northern celestial hemisphere, which is the result of an international cooperative program involving observations with 12 meridian circles over a period of 6 years, beginning in 1956. This catalog provides approximately one star for each square degree of the sky, which is sufficient to provide the necessary reference stars for the reduction of photographs taken with wide-angle cameras to the system of the catalog. The observations for a similar project for the southern hemisphere were completed with international cooperation, and a final catalog, the *SRS* (*Southern Reference Stars*) of some 20,000 stars, appeared in 1984.

A photographic survey of the northern celestial hemisphere resulted in the *AGK3* (*Dritter Astronomische Gesellschaft Katalog*, 1975), the third in a series of catalogs published by the German Astronomical Society. The catalog contains position and proper motions of 183,000 stars, including all stars brighter than ninth magnitude and some as faint as eleventh magnitude. The *AGK3R* was used for reference stars for this catalog (hence its name) to bring the *AGK3* on the *FK4* system. The photographic observations for a similar catalog for the southern hemisphere, using the *SRS* for reference stars, have been completed and the measurements of the plates essentially finished.

An important photographic catalog covering the entire sky to a limiting magnitude of 13 is the result of an international undertaking involving 19 observatories, with each assigned zones of declination and observing with nearly identical telescopes. The catalog, known as the *Carte du Ciel* (*CdC*) or *Astrographic Catalogue* (*AC*), finally completed for all zones in 1964, provides the positions of the stars in the form of rectangular coordinates, as measured on the plates. By means of auxiliary tables, these coordinates can be translated into right ascension and declination, which has been done for a number of zones. The total work involved the measurements of roughly 1,500,000 stars. *SEE ASTROMETRY.*

A catalog which was initially compiled for the determination of artificial satellite positions, but has since found wide use in various computer-oriented research projects in astronomy and geodesy, is the *Smithsonian Astrophysical Observatory Star Catalogue* (*SAOC*). It is derived from selected visual and photographic catalogs, contains positions and proper motions of 259,000 stars, and covers the entire sky in zones of 10°-wide declination. It was published in book form (1966), but is also available in machine-readable form.

Catalogs of astrophysical data. Among the numerous catalogs in this category is the monumental *Henry Draper Catalog* (*HD*) of spectral classification, which with its extension includes data for 275,000 stars published in 10 volumes (*Annals of the Astronomical Observatory of Harvard College*, vols. 91–100, 1918–1924). The introduction of luminosity classes in spectral classification, such as the prevailing Morgan-Keenan-Kellman (MKK) system (1943), led to classification of some 33,000 stars in this system, compiled into a catalog by the Mount Stromlo Observatory in Australia (1981). This catalog also contains the magnitudes and colors of stars on the U,B,V photometric system. The general acceptance of this photometric system since its inception in the 1950s is shown by the compilation of a general catalog containing data for 53,000 stars (*Astronomy and Astrophysics*, suppl., vol. 29, 1977). The major catalogs of stellar radial velocities are a *General Catalogue* (1953) and a compilation of such data published since, available in machine-readable form at the centers listed below. *SEE ASTRONOMICAL SPECTROSCOPY; MAGNITUDE.*

Catalogs of special stellar objects. Among the many catalogs in this category are a *General Catalog of Variable Stars*, compiled by B. V. KuKarkin and colleagues (1971), an index catalog of some 73,000 known double stars, a revised *Bright Star Catalog* (1982) of all stars brighter than magnitude 6.5, a general catalog of stellar parallaxes, and a catalog of known nearby stars within 22 parsecs of the solar system; 1 parsec = 3.0857×10^{16} m. *SEE BINARY STAR; PARALLAX; VARIABLE STAR.*

Catalogs of nonstellar objects. L. E. Dreyer's *New General Catalogue of Nebulae and Starclusters* (1888) contains the objects originally classified as nonstellar, with galaxies included as nebulae. The catalog contains 7840 objects, and was supplemented by two index catalogs in 1895 and 1908 with an additional 5386 objects. The NGC and IC numbers assigned in these catalogs remain the most commonly used designations. *SEE GALAXY, EXTERNAL; NEBULA; STAR CLUSTERS.*

Since Dreyer's time, the number of known nonstellar optical objects has increased by an order of magnitude, and the data are scattered through the astronomical literature, generally cataloged in special lists of physically similar objects. For some objects there are compendia of all previous catalogs of that type of object, but for most objects no such compendia exist. R. S. Dixon and G. Sonneborn compiled *A Master List of Nonstellar Astronomical Objects* (1980) with approximately 185,000 listings from 270 catalogs, with multiple listings of objects appearing in several catalogs.

With the expansion of astronomical research beyond the visible part of the electromagnetic spectrum to the ultraviolet and x-ray regions at the short wavelengths, and to infrared and radio wavelengths at the long wavelengths, the number of nonstellar objects has increased dramatically. There are 25,000 to 30,000 extragalactic radio sources catalogued for which as many as 75,000 to 100,000 names are in use, since in many cases the same object appears in several survey catalogs with different identification. *SEE RADIO ASTRONOMY.*

Among the more than 50 catalogs of these sources are the Cambridge Observatory discovery catalogs of quasars named *3C* and *4C* (*Memoirs of the Royal Astronomical Society*, vol. 68, 1959, and vol. 69, 1969). An index of extragalactic radio source catalogs, published before 1976, by M. J. L. Kesteven and A. H. Bridle (*Royal Canadian Society Journal*, vol. 71, 1977) gives precise references to the individual catalogs. Another reference is the *Master List* by Dixon, maintained at the Ohio State University Radio Observatory since 1970, and frequently updated. SEE QUASAR.

The number of known infrared sources has rapidly increased, especially since the launch in 1982 of the *Infrared Astronomical Observatory Satellite (IRAS).* Complete listings of all known sources by January 1983 have been published, with literature references in a NASA reference publication (1984). SEE INFRARED ASTRONOMY.

Ultraviolet astronomical observations of nonstellar objects have been obtained from sounding rocket vehicles and from several spacecraft, such as *Copernicus,* the *International Ultraviolet Explorer (IUE),* and *Voyager 1* and *2.* Numerous complications of observations of the interstellar medium and all types of nebulae including galaxies have been published, but no major catalog is available. SEE SATELLITE ASTRONOMY; ULTRAVIOLET ASTRONOMY.

Early discoveries of x-ray sources were from rocket observations, but a rapid increase in known sources occurred after the launch of the *UHURU* satellite in 1970 and the *Ariel 5* in 1973, followed by such satellites as the *High Energy Astronomical Observatory (HEAO),* the *European X-ray Observatory Satellite (EXOSAT),* and the *Einstein Observatory (HEAO 2).* SEE X-RAY ASTRONOMY.

Principal catalogs of x-ray sources are the third and forth UHURU catalogs (*Astrophysical Journal Supplement,* vol. 27, 1974, and vol. 38, 1978), and the *Ariel (3A) Catalogue (Monthly Notices, Royal Astronomical Society, vol. 197, 1981).*

Astronomical data centers. Astronomical data centers have been established from which catalogs previously published in print are now (with a few exceptions) available in machine-readable form or on microfiche. Centers are located at the NASA Goddard Space Flight Center, Greenbelt, Maryland; at the Centre de Données Stellairs, Strasbourg, France; and in the Commonwealth of Independent States, Japan, and Germany. SEE STAR.

Kaj Aa. Strand

Astronomical coordinate systems

Schemes for locating astronomical objects on the celestial sphere. To an observer on the Earth's surface, the stars of the night sky appear to be placed upon a spherical shell of infinite radius with the observer at the center. Celestial objects appear to move with respect to the stars, and at any given time their position on this imaginary sphere, called the celestial sphere, can be specified by two angles, called celestial coordinates, whose values depend upon what coordinate system is used. SEE CELESTIAL SPHERE.

Each coordinate system is defined by a fundamental plane and a principal axis. For example, on the Earth's surface, longitude and latitude coordinates are used to determine positions. In this system, which is analogous to astronomical coordinate systems, the fundamental plane is that of the Earth's Equator, and the principal axis is defined by a line running from the Earth's center to a point on the Equator at the longitude of Greenwich, England. Longitude is measured east or west of Greenwich. Angular positions along the fundamental plane, like longitude, can be expressed in terms of either degrees or hours of time. One hour equals 15°. Angular measurements north or south of the fundamental plane, like latitude, are measured from 0 to 90° at the poles. In astronomical coordinate systems, the degrees north are denoted with a plus sign (+), and degrees south are with a minus sign (−). SEE LATITUDE AND LONGITUDE.

Horizon system. The boundary between the hemisphere of the sky that is visible and the hemisphere which is hidden from view by the Earth is called the horizon (**Fig. 1**). The observer is located at the center of the system (*O*), the pole directly overhead is termed the zenith, and the opposite pole, the nadir. These pole directions are aligned with a plumb line, which is determined by the observer's local gravity. The fundamental horizon plane is 90° from the poles, and for astronomical applications the principal axis is most often taken to pass through the north point. Great circles that pass through the zenith and nadir are termed vertical circles; the one passing through the east and west points is termed the prime vertical, and that passing through the north and south points is called the celestial meridian. SEE HORIZON; ZENITH.

The longitudinal coordinate of a celestial object is termed its azimuth and is most often measured eastward from the north point to the object's vertical circle; and the latitudinal coordinate, termed its altitude, is measured along the object's vertical circle, north or south from the horizon plane to the object. For example, the position of the star *R* in Fig. 1 would have an approximate azimuth and altitude of 210° and +60°. The object's zenith distance, measured along the object's vertical circle, is the angle measured from the zenith point to the object. SEE AZIMUTH.

Because the horizon coordinate system is fixed with respect to the observer on the Earth's surface, it must rotate with the Earth. Hence the azimuth and altitude of a celestial object are constantly changing with time. For many astronomical applications, such as the listing of stellar and planetary positions, a nonrotating, coordinate system is required. The most common of these nonrotating, or inertial, coordinate systems is the equatorial system.

Equatorial system The fundamental plane of the equatorial coordinate system can be visualized by imagining that the Earth's equatorial plane is extended to intersect the celestial sphere. An alternate fundamental plane, the ecliptic plane, is the extension of the Earth's orbital plane onto the celestial sphere (**Fig. 2**). These planes intersect at two points, called equinoxes, with the angle between them ϵ being termed the obliquity of the ecliptic. This angle is about 23.5°. SEE ECLIPTIC; EQUINOX.

Due to the Earth's motion about the Sun, observers on Earth see an apparent motion of the Sun along the

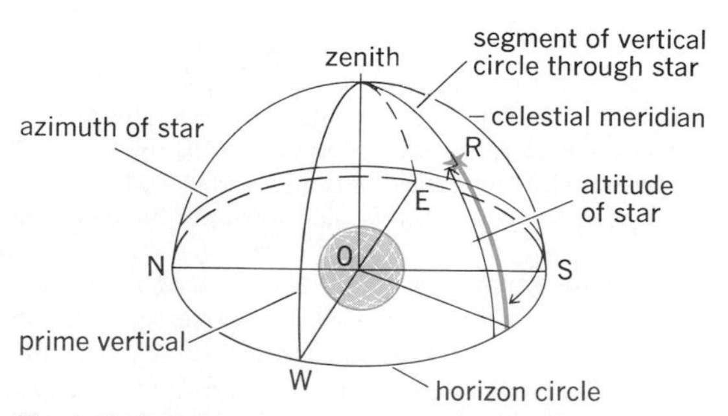

Fig. 1. Horizon system of astronomical coordinates.

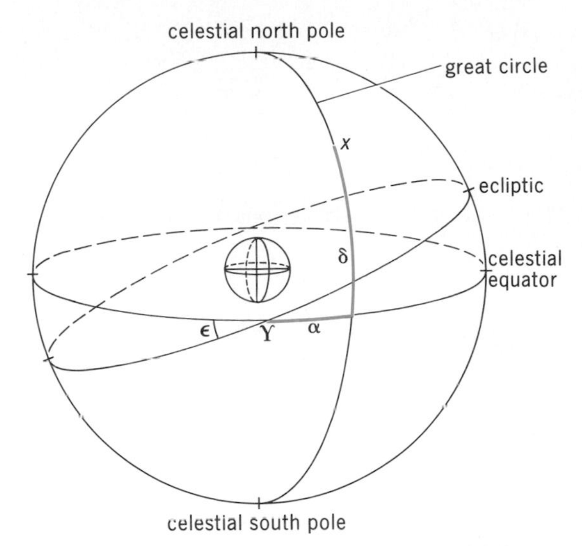

Fig. 2. Equatorial system of astronomical coordinates.

ecliptic plane. The point where the Sun's annual apparent motion takes it northward across the equatorial plane is called the vernal equinox ϒ, and the line between the Earth's center and this point defines the principal axis for both the equatorial and ecliptic coordinate systems. The apparent passage of the Sun through the vernal equinox, on about March 21, marks the beginning of spring in the Northern Hemisphere. Because of disturbing effects of the Sun and Moon on the Earth's figure, the Earth's rotation axis precesses, causing the celestial pole to describe an approximate circular motion about the ecliptic pole once every 26,000 years. This causes the location of the vernal equinox to drift westward along the ecliptic about 50 arc-seconds each year. Hence for an inertial coordinate system, where the principal axis is not moving, an epoch must be specified at which time the coordinate system is held fixed. In practice, the beginnings of the years 1950 or 2000 are most often used as epochs. *See Earth rotation and orbital motion; Precession of equinoxes.*

The north and south celestial poles represent the extension of the Earth's North and South poles onto the celestial sphere. For a celestial object (for example, object *X* in Fig. 2), the longitudinal coordinate is termed the right ascension α and is measured eastward along the celestial equator from the vernal equinox ϒ to the great circle passing through the object and the north and south celestial poles. The latitudinal coordinate, called the declination δ, is then measured along the object's great circle, north or south from the Equator to the object.

Ecliptic system. The ecliptic coordinate system is often used when representing the orbital motions of the planets, asteroids, and comets. The fundamental plane is that of the ecliptic, and as in the equatorial system, the principal axis is the line extending from the Earth's center to the vernal equinox. The position of a celestial object is defined, in the ecliptic system, by the ecliptic longitude and latitude. The longitude is measured eastward along the ecliptic plane from the vernal equinox to the great circle that passes through the object as well as the north and south ecliptic poles. The ecliptic latitude is then measured along this great circle either north or south from the ecliptic plane to the object.

Galactic system. Astronomers working with stars and other objects within the Milky Way Galaxy often find it convenient to use the galactic disk as the fundamental plane of their coordinate system, and the line extending from the galactic center to the Sun's location as the principal axis. With the Sun as the origin, galactic longitude is measured from the principal axis along the galactic equator eastward to the great circle that passes through the object and the north and south galactic poles. Galactic latitude is measured along the object's great circle, either north or south from the galactic plane to the object. In equatorial coordinates, the position of the north galactic pole, at the 1950 epoch, is approximately $\alpha = 12.8$ h (192.2°) and $\delta = +27.4°$. *See Milky Way Galaxy.*

Donald K. Yeomans

Bibliography. P. Duffett-Smith, *Practical Astronomy with Your Calculator*, 1981; A. E. Roy, *Orbital Motion*, 1978; L. G. Taff, *Computational Spherical Astronomy*, 1981.

Astronomical observatory

A building or group of buildings which house optical telescopes or the electronics required for radio telescopes or space telescopes, and the astronomers studying the observations. There are over 300 active astronomical observatories in at least 43 countries around the world, over 50 of them in the United States (see **table**). This omits hundreds of amateur telescopes, five or six Earth-orbiting space observatories controlled by radio and computers operated by astronomers on the ground, and at least one high-flying airplane equipped with telescopes.

Size of telescopes. The optical telescopes used at these observatories vary widely in size, from the huge 200-in. (5-m) mirror reflector (see **illus.**) and 48-in.

The 200-in. (5-m) observatory located on Palomar Mountain, California, at night. The photograph was made by moonlight.

Major astronomical observatories, by country*

Date founded, name, and place	Apertures and types of telescopes	Observing programs
Argentina		
1870, Córdoba (Bosque Alegre)	61-in. (1.5-m) mirror	Spectra, faint objects
1882, La Plata	33-in. (84-cm) mirror	Spectra, colors
	17-in. (43-cm) lens	Positions
	100-ft (30-m) radio dish	Interstellar gas
1964, Yale-Columbia Austral (El Leoncito)	20-in. (51-cm) lenses (2)	Positions, colors
Australia		
1924, Mt. Stromlo (Canberra)	74-, 50-in. (1.9-, 1.3-m) mirrors	Faint objects
	40-, 30-in. (1-, 0.8-m) mirrors	Spectra, colors
	26-in. (66-cm) lens	Positions
	20-in. (51-cm) Schmidt	Spectra, colors
1952, CSIRO (Parks)	210-ft (64-m) radio dish	Interstellar gas
1964, Sydney Univ. (Narrabri)	260-in. (6.6-m) mirrors (2)	Star diameters
	500-ft (150-m) radio array	Radio sources
1975, Anglo-Australian (Siding Springs)	150-in. (3.8-m) mirror	Faint objects, spectra
	48-in. (1.2-m) Schmidt	Faint objects
	40-in. (1-m) mirror	Colors, spectra
Canada		
1917, Dominion Astrophysical (Victoria)	72-, 48-in. (1.8-, 1.2-m) mirrors	Spectra, colors
1932, David Dunlap (Toronto)	74-, 24-in. (1.9-, 0.6-m) mirrors	Spectra, colors
1959, Penticton (B.C.)	84-ft (26-m) radio dish	Interstellar gas
	4000-ft (1220-m) radio reflectors	Interstellar gas
	5-ft (1.5-m) radio dish	Sun
	280-ft (85-m) radio reflector	Cosmic-ray effects
1960, Algonquin (Ontario)	150-ft (46-m) radio dish	Interstellar gas
Chile		
1962, Inter-American (Cerro Tololo)	150-in. (3.8-m) mirror	Faint objects
	60-, 36-in. (1.5-, 0.9-m) mirrors	Faint objects, spectra, planets
	24-in. (61-cm) Schmidt	Colors
1964, European Southern (La Silla)	142-in. (3.6-m) mirror (Cassegrain)	Faint objects, spectra
	142-in. (3.6-m) mirror (new technology)	Faint objects, spectra
	87-in. (2.2-m) mirror (Max Planck)	Faint objects, spectra
	59-in. (1.5-m) mirrow (Danish)	Faint objects, spectra
	55-in. (1.4-m) mirror (coudé)	Spectra
	40-in. (1-m) Schmidt	Galaxy survey, colors
	40-in. (1-m) mirror (very thin)	Faint objects
	49-ft. (15-m) radio dish (submillimeter)	Interstellar clouds
1968, Carnegie (Las Campanes)	100-in. (2.5-m) mirror	Faint objects
China		
(600 B.C.) 1958, Beijing	24-in. (61-cm) Schmidt	Colors
	16-in. (41-cm) lenses (2)	Positions
(Sha-ho)	24-in. (61-cm) mirror	Sun
(Mi-yun)	35-ft (9-m) radio dishes (20)	Radio sources
(Hsing-lung)	78-in. (2-m) mirror	Faint objects, spectra
	24-in. (61-cm) Schmidt	Colors, spectra
	16-in. (40-cm) lens	Variable stars
1872, Shanghai	20-ft (6-m) radio dish	Time, positions
	16-in. (40-cm) lenses (2)	Positions
1934, Purple Mountain (Nanking)	24-in. (60-cm) Schmidt	Satellites, positions
	16-in. (40-cm) lens	Variable stars
	12-in. (30-cm) lens	Sun
1972, Yunnan (Kunming)	39-in. (1-m) mirror	Satellites, positions
	12-in. (30-cm) lens	Sun
Czechoslovakia		
1928, Ondrejov	78-in. (2.0-m) mirror	Spectra, faint objects
	25-in. (64-cm) mirror	Colors
	10-in. (25-cm) lens	Sun
1953, Skalnate Pleso	24-in. (61-cm) mirror	Colors
	12-in. (30-cm) lens	Positions
	8-in. (20-cm) lens	Sun
Denmark		
1963, Copenhagen (Brorfelde)	21-in. (53-cm) Schmidt	Spectra
	20-, 16-in. (50-, 40-cm) mirrors	Colors
Egypt		
1905, Helwan (Cairo)	30-in. (76-cm) mirror	Spectra, colors
(Kottamia)	74-in. (1.9-m) mirror	Faint objects, spectra
France		
1670, Paris (Meudon)	39-, 24-in. (1.0-, 0.6-m) mirrors	Spectra
	33-in. (84-cm) lens	Planets
	16-in. (40-cm) lenses (2)	Sun

*For brevity, this list omits many active observatories. Moreover, new telescopes will be added in future years.

Major astronomical observatories, by country (cont.)

Date founded, name, and place	Apertures and types of telescopes	Observing programs
France (cont.)		
1887, Nice	30-, 16-in. (75-, 40-cm) lenses	Double stars, positions
1930, Pic du Midi	41-in. (1-m) mirror	Spectra, Moon, planets
(Pyrenean Mountains)	24-in. (60-cm) lens	Planets
	16-in. (40-cm) lens	Sun
1938, Haute-Provence (St. Michele)	76-, 60-in. (1.9-, 1.5-m) mirrors	Spectra, faint objects
	24-in. (60-cm) Schmidt	Colors
1955, Univ. of Paris	650-ft (200-m) radio reflector	Interstellar gas
1975, CERGA (Grasse)	98-in. (2.5-m) Schmidt	Positions
	59-, 39-in. (1.5-, 1.0-m) mirrors	Moon and satellite ranging
	39-in. (1.0-m) mirrors (4)	Optical and infrared interferometry
Germany		
1705, Berlin (Babelsberg)	28-, 21-in. (71-, 53-cm) mirrors	Colors
	26-in. (65-cm) lens	Positions
1823, Hamburg	40-, 24-in. (100-, 60-cm) mirrors	Faint objects
(Bergedorf)	24-in. (60-cm) lenses (2)	Positions
	8-in. (20-cm) lens	Sun
	32-in. (80-cm) Schmidt	Spectra, colors
1836, Bonn (Hoher List)	42-, 14-in. (110-, 35-cm) mirrors	Colors, spectra
	13-in. (33-cm) Schmidt	Colors, spectra
	14-in. (35-cm) lens	Positions
	82-ft (25-m) radio dish	Interstellar gas
1960, Karl Schwarzschild (Tautenburg)	54-in. (1.4-m) Schmidt	Faint objects
Great Britain		
1675, Royal Greenwich	36-, 30-in. (91-, 76-cm) mirrors	Colors
(Herstmonceux, 1948)	28-, 26-in. (71-, 66-cm) lenses	Positions
	6-, 4-in. (15-, 10-cm) lenses	Sun
(Canary Islands)	98-in. (2.5-m) mirror "Herschel"	Faint objects, spectra
1818, Edinburgh Royal	36-, 20-in. (91-, 51-cm) mirrors	Spectra, colors
	24-, 16-in. (60-, 40-cm) Schmidts	Positions, colors, spectra
1820, Cambridge	36-in. (91-cm) mirror	Spectra
	25-, 12-in. (63-, 30-cm) lenses	Sun
	17-in. (43-cm) Schmidt	Colors
1945, Mullard (Cambridge)	1-mi (1.6-km) radio array	Radio sources
	1500-ft (460-m) arrays (2)	Radio sources, Sun, planets
1949, Jodrell Bank	250-, 125-ft (76-, 38-m) dishes	Planets, interstellar gas
	218-ft (66-m) radio bowl	Radio sources
	120-ft (37-m) radio reflector	Radio sources
	25-, 10-ft (7.6-, 3.3-m) dishes	Radio sky survey
Iraq		
1984, Iraqi National (Mt. Korek)	99-ft (30-m) radio dish	Interstellar molecules
	138-in. (3.5-m) mirror	Faint objects, spectra
	47-in. (1.2-m) mirror	Faint objects, spectra
Ireland		
1790, Armagh	10-in. (25-cm) lens	Star catalogs
	15-in. (39-cm) mirror	Star clusters
Italy		
1578, Vatican	25-in. (63-cm) Schmidt	Spectra, colors
(Castel Gandolfo)	24-in. (61-cm) mirror	Spectra
	16-in. (40-cm) lenses (2)	Positions
1880, Arcetri Royal	15-, 12-in. (38-, 30-cm) lenses	Sun
1940, Padua (Asiago)	72-, 48-in. (1.8-, 1.2-m) mirrors	Faint objects
	26-, 16-in. (65-, 40-cm) Schmidts	Faint objects, spectra
1980, Italian National Council (Gornergrat)	59-in. (1.5-m) mirror	Infrared brightnesses
Japan		
1920, Tokyo (Mitaka)	36-, 12-in. (90-, 30-cm) mirrors	Colors
	26-in. (65-cm) lens	Spectra
	18-, 8-in. (45-, 20-cm) lenses	Sun
	20-in. (50-cm) Schmidt	Positions
	78-ft (24-m) radio bowl	Radio sources
	33-, 20-ft (10-, 6-m) dishes	Sun, polarization
1960, Okayama	74-, 36-in. (1.9-, 0.9-m) mirrors	Faint objects
	26-in. (65-cm) lens	Sun
Mexico		
1942, Tonantzintla	40-in. (1-m) mirror	Faint objects
	26-in. (65-cm) Schmidt	Colors, spectra
1978, Baja California (San Pedro Martir)	79-in. (2-m) mirror	Faint objects
Netherlands		
1955, Univ. of Leiden	82-ft (25-m) radio dishes (2)	Interstellar gas
(Dwingeloo)	33-, 25-ft (10-, 7.5-m) dishes	Interstellar gas
(Westerbork)	4700-ft (1.4-km) radio array	Radio sources

Major astronomical observatories, by country (cont.)

Date founded, name, and place	Apertures and types of telescopes	Observing programs
South Africa		
1820, Cape of Good Hope Royal	40-, 30-in. (100-, 76-cm) mirrors	Colors
	24-, 8-in. (61-, 20-cm) lenses	Positions, Sun
1928, Boyden (Mazelspoort)	60-in. (1.5-m) mirror	Spectra, colors
	32-in. (81-cm) Schmidt	Colors, positions
(1772) Radcliffe (Sutherland, 1978)	74-in. (1.9-m) mirror	Spectra, colors
Soviet Union (now Commonwealth of Independent States)		
1839, Pulkovo (Leningrad)	28-, 20-in. (70-, 50-cm) mirrors	Planets, spectra
	26-in. (65-cm) lens	Positions
	27-in. (69-cm) Maksutov	Faint objects, colors
	20-, 10-in. (50-, 25-cm) lenses	Sun
	12-in. (30-cm) transit	Positions
	394-ft (120-m) radio reflector	Sun, planets, radio sources
	52-, 39-ft (16-, 12-m) dishes	Radio sources, Sun, planets
	350-ft (105-m) radio array	Polarization
1948, Crimean Astrophysical	104-in. (2.5-m) mirror	Faint objects
	48-, 20-in. (1.2, 0.5-m) mirrors	Planets, Sun
	25-in. (65-cm) Maksutov	Faint objects
	16-, 5-in. (40-, 13-cm) lenses	Sun
	72-ft (22-m) radio dish	Planets, interstellar gas
	23-, 13-ft (7-, 4-m) radio dishes	Sun
1952, Burakan (Armenia)	40-, 21-in. (1-, 0.5-m) Schmidts	Faint objects
	20-in. (50-cm) mirror	Faint objects
1976, Zelenchukskaya	236-in. (6-m) mirror	Faint objects, spectra
	1950-ft (600-m) radio array	Radio sources
Spain		
1979, German-Spanish (Calar Alto)	59-in. (1.5-m) mirror	Faint objects, spectra
	32-in. (0.8-m) Schmidt	Faint objects
	47-in. (1.2-m) mirror	Faint objects, photometry
	86-in. (2.2-m) mirror	Faint objects, spectra
	138-in. (3.5-m) mirror	Faint objects, spectra
Sweden		
1730, Uppsala	40-in. (1-m) Schmidt	Faint objects
1748, Stockholm (Saltsjöbaden)	40-in. (1-m) mirror	Spectra, colors
	24-, 16-in. (60-, 40-cm) lenses	Positions, colors
	20-in. (50-cm) Schmidt	Spectra, colors
Switzerland		
1855, Zurich (Arosa)	10-in. (25-cm) lenses (2)	Sun
1895, Basel (Metzerlen)	24-in. (60-cm) mirror	Colors, faint objects
	16-in. (40-cm) Schmidt	Colors
1961, Geneva (Jungfraujoch)	30-in. (76-cm) mirror	Spectra, colors
United States (East and Midwest)		
1830, U.S. Naval (D.C.)	26-, 15-in. (65-, 38-cm) lenses	Planets, positions
	6-in. (15-cm) transit	Positions
	8-, 26-in. (20-, 65-cm) zenith tubes	Time
1830, Yale (Conn.)	40-, 20-in. (1-, 0.5-m) mirrors	Positions, colors
	15-in. (38-cm) lens	Positions
	200-ft (60-m) radio array	Planets, Sun, polarization
1836, Hopkins (Williams College, Mass.)	7-in. (0.18-m) lens	Teaching
1839, Harvard (Mass.)	61-, 24-in. (1.5-, 0.6-m) mirrors	Faint objects, spectra
	15-in. (38-cm) lens	Colors
	Patrol cameras	Variable stars
1840, Van Vleck (Wesleyan Univ., Conn.)	24-in. (61-cm) mirror	Spectra, colors
	20-in. (51-cm) lens	Positions
1855, Univ. of Michigan (Ann Arbor)	38-in. (96-cm) mirror	Spectra
1866, Princeton (N.J.)	36-in. (91-cm) mirror	Spectra
1878, Univ. of Wisconsin (Madison)	36-in. (91-cm) mirror	Spectra, colors
	16-in. (40-cm) lens	Colors
1883, Leander McCormick (Univ. of Virginia)	32-in. (81-cm) mirror	Colors, spectra
	26-in. (66-cm) lens	Positions
1897, Yerkes (Williams Bay, Wis.)	41-, 24-in. (105-, 61-cm) mirrors	Spectra, colors, polarization
	40-in. (1-m) lens	Spectra, positions
1917, Sproul (Swarthmore, Penn.)	24-in. (61-cm) lens	Positions
1954, Carnegie (D.C.)	120-, 97-, 60-ft (37-, 30-, 18-m) radio dishes	Interstellar gas, sources
	650-ft (200-m) reflector	Radio sources, Sun
	3000-, 1500-ft (900-, 450-m) arrays	Radio sources, Sun
1956, Haystack Hill (Mass.)	120-, 84-ft (36-, 25-m) radio dishes	Planets, radio sources
1958, National Radio Astronomy (W. Va.)	300-, 140-ft (91-, 43-m) dishes	Interstellar gas, planets
	85-, 40-ft (26-, 12-m) dishes	Radio sources, planets
	15-ft (4.6-m) radio horn	Radio flux standard

Major astronomical observatories, by country (cont.)

Date founded, name, and place	Apertures and types of telescopes	Observing programs
United States (East and Midwest) (cont.)		
1962, Univ. of Illinois	600-ft (180-m) radio bowl	Radio sources, Moon
(Danville)	28-ft (8.5-m) radio dish	Sun
1963, Arecibo (Puerto Rico)	1000-ft (300-m) radio bowl	Planets, interstellar gas
United States (West)		
1875, Lick	120-, 36-in. (3-, 0.9-m) mirrors	Faint objects, spectra
(Mt. Hamilton, Calif.)	36-, 20-in. (91-, 51-cm) lenses	Positions
1894, Lowell	42-, 24-in. (1-, 0.6-m) mirrors	Faint objects, spectra
(Flagstaff, Ariz.)	24-, 13-in. (61-, 33-cm) lenses	Planets, positions
Mt. Wilson and Los Campanas	100-, 60-in. (2.5-, 1.5-m) mirrors	Spectra, faint objects
(Mt. Wilson, Calif.)	12-in. (30-cm) mirrors (2)	Sun
1916, Steward (Kitt Peak)	85-in. (2.2-m) mirror	Faint objects
1937, McDonald	107-in. (2.7-m) mirror	Faint objects, planets
(Mt. Locke, Tex.)	82-, 36-in. (2.1-, 0.9-m) mirrors	Faint objects, spectra
	16-ft (5-m) radio dish	Interstellar gas
(Marfa, Tex.)	2-mi (3.3-km) radio array	Radio source survey
1940, High Altitude	16-, 10-in. (41-, 25-cm) mirrors	Sun
(Climax, Colo.)		
1948, California Institute of	200-, 60-in. (5.1-, 1.5-m) mirrors	Faint objects, spectra
Technology/Palomar		
(Palomar Mt., Calif.)	48-, 18-in. (120-, 46-cm) Schmidts	Faint objects, colors
1949, Sacramento Peak (N. Mex.)	16-, 12-in. (40-, 30-cm) mirrors	Sun
1955, U.S. Naval	61-, 40-in. (1.5-, 1-m) mirrors	Positions, colors
(Flagstaff, Ariz.)	23-in. (58-cm) lens	Colors
	12-in. (30-cm) mirror transit	Positions
1955, Univ. of Colorado (Boulder)	80-ft (25-m) radio dishes (2)	Sun, planets, radio sources
1958, Jet Propulsion Lab	210-ft (65-m) radio dish	Satellites, planets
(Goldstone, Calif.)	85-ft (26-m) radio dishes (2)	Planets, satellites
1959, Kitt Peak National	150-, 84-in. (3.8-, 2.1-m) mirrors	Faint objects, spectra
(Tucson, Ariz.)	36-, 16-in. (91-, 40 cm) mirrors	Colors, polarization
	60-, 36-in. (1.5-, 0.9-m) mirrors	Sun
1959, Stanford Univ. (Calif.)	150-ft (46-m) radio dish	Moon, planets, source sizes
	60-ft (18-m) radio dishes (3)	Radio sources, Moon
	1150-ft (350-m) radio array	Planets, Sun, Moon
1961, Perkins (Flagstaff, Ariz.)	69-in. (1.7-m) mirror	Spectra, colors
1967, Mt. Hopkins (Ariz.)	6 × 72-in. (0.15 × 1.8 m) multiple mirror	Faint objects, positions, spectra
	60-, 12-in. (1.5-, 0.3-m) mirrors	Spectra, planets
	20-in. (51 cm) Schmidt	Positions
	34-ft (10-m) Cerenkov	Gamma-ray effects
1967, Hawaii (Mauna Kea)	84-, 60-in. (2.1-, 1.5-m) mirrors	Faint objects
Univ. of Michigan (Mauna Kea)	48-in. (1.2-m) mirrors (2)	Spectra, colors
1979, NASA Hawaii (Mauna Kea)	120-in. (3-m) mirror	Infrared observations
1979, United Kingdom (Mauna Kea)	150-in. (3.8-m) mirror	Infrared observations
1979, France-Canada-Hawaii	140-in. (3.6-m) mirror	Faint objects, infrared
(Mauna Kea)		
1981, Very Large Array	85-ft (26-m) radio dishes (27)	Interstellar clouds, galaxies,
(Socorro, N. Mex.)		quasars

(1.2-m) Schmidt at Palomar Mountain in California, used mostly to study faint stars, nebulae, and galaxies, to the 10-in. (0.25-m) lens telescopes at Zurich, Switzerland, used to study sunspots. Radio telescopes vary from the 1000-ft (305-m) bowl at Arecibo, Puerto Rico, through the 300-ft (91-m) dish at the National Radio Astronomy Observatory in West Virginia, to arrays of much smaller dishes in the United States and 25 other countries. The most powerful such array is the Very Large Array (VLA) near Socorro, New Mexico, consisting of 27 radio dishes, each 85 ft (26 m) in diameter, which can be moved along the railroad tracks in the shape of a Y with 12-mi-long (20-km) arms. The telescopes in orbit outside the Earth's atmosphere are smaller, and designed for special purposes, although the Hubble Space Telescope has an aperture of 95 in. (2.4 m). *See Optical telescope; Radio astronomy; Radio telescope; Satellite astronomy; Schmidt camera; Telescope.*

Large-aperture telescopes have the advantages of detecting fainter objects and yielding sharper images. The Very Large Array radio telescope has an effective aperture of 22 mi (35 km). The largest optical telescope in operation is the Palomar 200-in. (5-m), but several larger ones are under construction, including the 394-in. (10-m) Keck for Mauna Kea, Hawaii, and the Very Large Telescope (VLT) array of four 315-in. (8-m) telescopes for the European Southern Observatory at La Silla, Chile, which will have an effective aperture of 630 in. (16 m).

Affiliation. Most of these observatories are linked with universities or research institutes, some of them far from the telescope, which is best located on a remote mountain far from city lights. For instance, the Palomar Mountain Observatory is linked with the California Institute of Technology in Pasadena, some 125 mi (200 km) away. Universities provided the rapid increase in numbers of observatories in western Europe and the United States during the nineteenth century, when small lens telescopes were used by the faculty to teach astronomy students. In 1839 Harvard College Observatory was founded in Cambridge, Massachusetts, and has increased in size over the years to become the Center for Astrophysics, with over 200 employees. Earlier, some governments founded observatories for measuring time and helping navigators. For instance, the Royal Greenwich Observatory was founded in 1675 near London, England,

and in 1830 the U.S. Naval Observatory was founded in Washington, D.C. Even earlier, monarchs often supported small groups of astronomers or individuals such as Tycho Brahe, who had his own observatory built by the king's money at Uraniborg on a Danish island, before telescopes were invented. Tycho made many accurate measurements of the positions of planets by using a 19-ft (5.8-m) quadrant, a quarter circle marked from 0 to 90°. Earlier observatories in Samarkand, north of Afghanistan in central Asia, and Beijing, China, had elaborate pillars and bowls where the astronomers walked around to sight the stars and measure their positions. Stonehenge, in southern England, is thought to have been used in the same way as an observatory over 3000 years ago.

Equipment. The typical optical observatory has a telescope in a building with a hemispherical, rotatable dome in which there is a slot up one side with a cover that can be pulled aside from within, so that the telescope can "see out." Large optical telescopes are housed in domes 60 ft (18 m) or more in diameter (although the Multiple Mirror Telescope is housed in a rectangular shed that rotates as a whole), while most radio telescopes are out in the open. Starting about 1890, astronomers took photographs, using the telescope as a large camera. Observatories are equipped with darkrooms where photographic plates can be developed, plate files where they are stored, and measuring engines which are used to measure star images on the plates accurately. After about 1910, spectrographs were attached to the telescopes to spread the light of a star into a spectrum from red to blue. Photographs of these spectra (spectrograms) are measured on a one-dimensional measuring engine—a microscope mounted on a long, accurate screw. *SEE ASTRONOMICAL PHOTOGRAPHY; ASTRONOMICAL SPECTROSCOPY.*

Electronics. Since the mid-1930s, electronic detectors have been used on optical telescopes to measure starlight accurately. The electronics in large observatories have become increasingly complex. In the Multiple Mirror Telescope Observatory, opened in 1979 near Tucson, Arizona, the astronomer need not peer through an eyepiece at the telescope, but can sit in a heated room watching a television screen and controlling the telescope from a computer terminal. Since radio and space telescopes are all electronic, modern observatories require electronic technicians and computers. There are also many electric motors required to rotate the dome of an optical observatory and to point the telescope.

Neutrino observatories. Neutrinos from the Sun and other sources have been detected since the mid-1970s. These small, nearly massless particles are released during nuclear reactions in the cores of stars, and they pass through the matter in stars and planets. A neutrino detector contains a large volume of water in which a very small percentage of the millions of neutrinos passing through result in brief flashes of light that are counted by sensitive photocells. At least two neutrino observatories detected the neutrino burst emitted by Supernova 1987A, and astronomers now use several more in the United States, Japan, and Europe. *SEE SOLAR NEUTRINOS; SUPERNOVA.*

Observations in space. Space telescopes are in orbit many miles above the Earth's surface. The *International Ultraviolet Explorer* (*IUE*) is in geosynchrononous orbit about 22,000 mi (35,500 km) from the Earth's center, but the NASA Hubble Space Telescope is in a much lower orbit, circling the Earth every 90 min about 270 mi (430 km) above the surface. These two space telescopes are linked by radio to observatory control rooms in the Goddard Space Flight Center near Washington, D.C. Other satellite telescopes include the NASA *Infrared Astronomical Satellite* (*IRAS*), cooled by liquid helium (to detect infrared wavelengths of 12, 25, 60, and 100 micrometers), which operated for almost a full year in an orbit 600 mi (1000 km) above the Earth's surface. X-ray and gamma-ray observations of astronomical objects have been made from satellite observatories launched by European countries as well as NASA. (Far-infrared radiation, far-ultraviolet radiation, x-rays, and gamma rays cannot penetrate the Earth's atmosphere.) *SEE GAMMA-RAY ASTRONOMY; INFRARED ASTRONOMY; ULTRAVIOLET ASTRONOMY; X-RAY ASTRONOMY.*

In April 1972, a far-ultraviolet camera was set up on the Moon by *Apollo 16* astronauts and obtained 184 photos and spectra of astronomical objects over 3 days' time. This first observatory on the Moon will soon be followed by others now being planned. The advantages offered by the Moon are little or no atmosphere, a solid base, and low gravity (one-sixth of the Earth's); the disadvantage is the high cost of shipping telescope parts (and astronomers and their habitats) from Earth to Moon. Radio telescopes on the Moon's far side have the advantage of being shielded from Earth's radio noise. Infrared telescopes may be able to operate at very low temperatures in craters near the Moon's poles. Thus, there are active plans for a very low-frequency radio telescope, a small optical telescope with a highly accurate spectrograph, a far-infrared telescope, and an ultraviolet interferometer.

Thornton Page

Bibliography. G. P. Kuiper and B. Middlehurst, *Telescopes*, 1960; National Aeronautics and Space Administration, *Future Astronomical Observatories on the Moon*, NASA CP2489, 1988; T. Page, *Observatories of the World*, Smithsonian Astrophysical Observatory, 1970; T. Page and L. Page, *Telescopes*, 1966.

Astronomical photography

The application of the photographic process to astronomy. Over the past century, photographic observations have made profound contributions to astronomy and astrophysics; and for much of this period, photographs were the best light dectectors available for a wide range of applications. Since the mid-1970s, a variety of image detectors utilizing photocathodes or silicon diode arrays have offered detective performance several times better than the best photographic materials. However, it is clear that many astronomers will continue to use photographic techniques for some time to come and for a number of reasons, including low initial cost, relative ease of use, and tradition. In particular, for applications where coverage of a wide field is required, photography is still the detector of choice, because emulsions can be manufactured in much larger formats than electronic detectors. Moreover, the emulsion itself is an effective and efficient storage device, requiring much less space than the equivalent digital image on a magnetic tape and presenting the information in a direct visual format. Indeed, many of the most important advances in astronomy and astrophysics have resulted from visual evaluation of photographs. Emphasis has been shifting from visual evaluation to pure machine measurements, partly because precision microphotometers are more readily available and partly because of advances in computer processing of digital images.

Monochrome Photography

Large photographic plates are still the best detectors for survey work, especially when used with Schmidt telescopes for direct imagery (**Figs. 1** and **2**) or with thin objective prisms for deep spectroscopic surveys. In modern astronomy the other chief uses of photography are photometry of stars and galaxies at the prime focus of large reflectors, astrometry with long-focal-length telescopes, and spectroscopy. For some of these applications, photography is often used as the readout for image intensifiers. *See Astrometry; Astronomical spectroscopy; Schmidt camera.*

Detective performance. Since the invention of the daguerreotype in 1839, dramatic improvements have been made in the sensitivity of photography (**Fig. 3**). Major revolutions occurred around 1851 and 1875 with the introduction of the wet plate and dry plate. In the twentieth century, progress has largely been achieved by improvements in silver halide emulsions. Under optimum conditions modern emulsions can deliver a detective quantum efficiency of about 3% of the ideal, and in principle it should be possible to improve photographic performance by factors of 2 and more if some of the degrading mechanisms, such as recombination losses, could be eliminated.

Photographic photometry can be defined as any procedure that allows one to determine the amount of light that exposes each picture element (pixel). In general, the blackening (density) that results on a photograph depends on the details of the chemical

Fig. 1. The 48-in. (1.2-m) Schmidt telescope, Palomar Mountain.

Fig. 2. A 48-in. (1.2-m) Schmidt plate of a field in the Milky Way showing bright and dark nebulosities. (*California Institute of Technology/Palomar Observatory*)

processing as well as on the exposure. For photometry of spectra or diffuse images it is customary to calibrate the scale of relative intensity versus density on each photograph, typically by exposing an unused portion of the same photograph with a series of spots with known relative intensities. For stellar photometry it is better to have a sequence of stars with known magnitudes on the same photograph. The ratio of exposing signal to the uncertainty with which the signal was inferred (using photographic photometry) is often called signal-to-noise. For most photometric applications signal-to-noise is the most important characteristic of a photograph, because it offers a quantitative evaluation of the accuracy to expect from a measurement. *See Magnitude.*

Nearly all astronomical photography is done with special spectroscopic plates. These materials come in several classes of grain sizes. The ultimate accuracy with which photographic photometry can be carried out is set by the uncertainties due to graininess. In general, this uncertainty depends strongly on the density as well as the grain size of the emulsion, with optimum signal-to-noise for small pixel sizes occurring near density 1. For a pixel with area of 1000 square micrometers, the optimum signal-to-noise ranges from approximately 15 to 52 for various types of commonly used plates. The signal-to-noise varies with the square root of pixel area [just as it should (**Fig. 4**)]. However, when the linear scale of a pixel begins to exceed 0.4 in. (1 mm) or so, large-scale nonuniformities can limit the signal-to-noise to about 100, or an accuracy of about 1%. For large plates very careful agitation is needed during development in order to keep the large-scale nonuniformities below 5 or 10% on an equivalent exposure basis.

Spectroscopic plates are available in several spectral sensitizations (**Fig. 5**). Types with the suffix -O are blue-sensitive, -J extends well into the green, -E and -F are panchromatic with good response out to

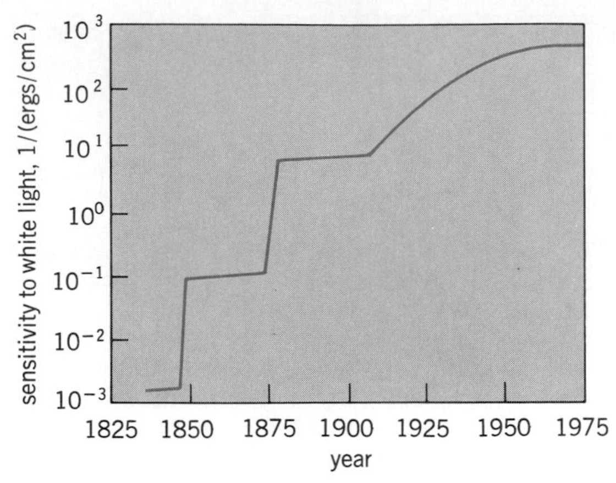

Fig. 3. History of photographic sensitivity. The vertical scale is the reciprocal of the energy, expressed in ergs/cm², that is required to create an image. 1 erg/cm² = 10^{-3} J/m². (*After A. G. Smith and A. A. Hoag, Advances in astronomical photography at low light levels, Annu. Rev. Astron. Astrophys., 17:43–71, 1979*)

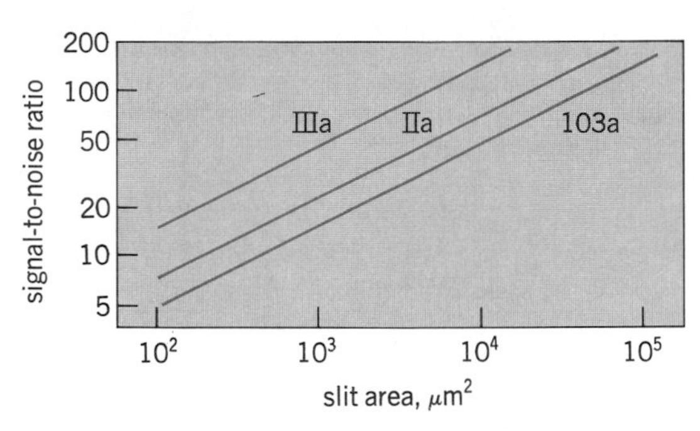

Fig. 4. Signal-to-noise as a function of microphotometer slit area for various photographic plates. (*After R. M. West and J. L. Heudier, eds., Modern Techniques in Astronomical Photography, European Southern Observatory, Geneva, 1978*)

about 700 nanometers, and -N is the best available in the near infrared out to 900 nanometers. The absolute values of the various curves in Fig. 5 are not reliable because of uncertainties in processing and emulsion type and batch, but the variation of sensitivity with wavelength is more accurate.

For long exposure times there is a serious breakdown in the reciprocity law that would predict that the photographic sensitivity should depend only on the product of the exposing intensity I and the exposure time t. The loss of sensitivity at low light levels and long exposure times is often called low-intensity reciprocity failure. Fortunately, emulsion makers have discovered methods for reducing reciprocity failure. Spectroscopic plates that have been treated specially for low-intensity reciprocity failure usually are distinguished by an "a" in their designation. For example, the two plates compared in **Fig. 6** have similar sensitivities for short exposure times, but the plate designated IIa-O is nearly 10 times less degraded than II-O for 1-h exposures. However, even the specially treated plates will have substantial reciprocity failure for long exposures. There have been major advances in understanding the sources of low-intensity reciprocity failure and in developing various treatments for its reduction. Some of the techniques are described below.

Hypersensitization techniques. Most hypersensitization techniques involve preexposure treatments such as evacuation, baking, bathing, or hypering with hydrogen gas. Many of the procedures work best if the exposures are made in a controlled atmosphere by using special cassettes. Since at least five different mechanisms are involved, it is often beneficial to combine treatments. However, since so many different combinations are possible, astronomers often disagree on what is the best recipe. Several of the most successful protocols are outlined below.

Chemical treatments. Infrared plates are so slow that they are rarely used without some sort of preexposure chemical treatment. Simply bathing the plates in water and then drying before exposure can routinely gain a factor of 10 in speed, while factors of 100 are more common if the plates are exposed wet or in an oxygen-free 100% relative-humidity atmosphere. Another treatment that has been used for decades is bathing in ammonia. This can also provide gains as large as a factor of 100, but it is trickier to get good uniformity and low background fog. A treatment of bathing in dilute solutions of silver nitrate has had widespread success, and this is probably the best procedure for infrared plates. Bathing increases the Ag^+ concentration in the emulsion and may also increase the effectiveness of the dye sensitization. However, bathing is rarely used for plates other than the infrared, because more effective procedures than bathing are available for the other materials.

Removal of oxygen and water. Oxygen, especially in the presence of water, is a powerful desensitizer, with the effect being most detrimental at low light levels and long exposures. Careful removal of oxygen and water before and during exposure can all but elimi-

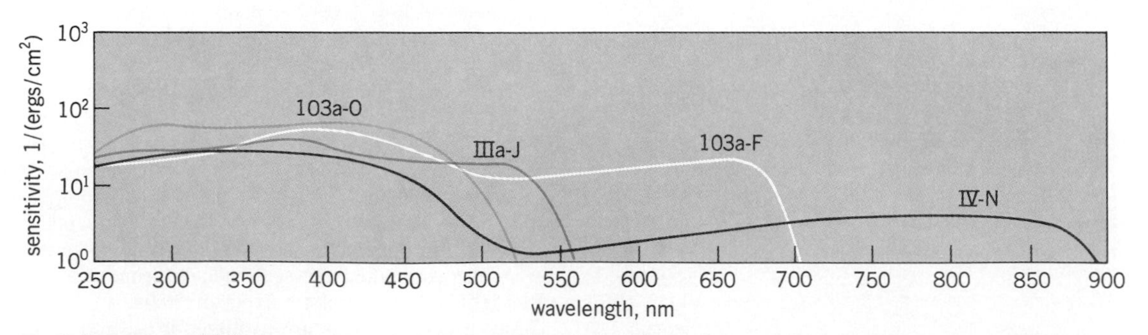

Fig. 5. Spectral sensitivity curves for various photographic plates. The vertical scale is the reciprocal of the exposure, expressed in ergs/cm², required to produce a density of 0.6 above fog. The IV-N curve is with hypersensitization. 1 erg/cm² = 10^{-3} J/m². (*After Kodak Plates and Films for Scientific Photography, Kodak Publ. P-315, Eastman Kodak Co., 1973*)

nate low-intensity reciprocity failure for several types of astronomical plates. Two strategies for removal of oxygen and water are in common use: evacuation with a mechanical pump and trap for several hours is effective, and flushing with dry nitrogen for extended periods has also been used. Typical speed gains are factors of 2 to 4 for long exposures.

Baking. In order to provide a reasonable shelf life, the chemical sensitization of astronomical emulsions is not pushed to the optimum during manufacture. However, users can finish off the ''ripening'' process shortly before use by baking in vacuum or dry nitrogen. This technique is especially effective if oxygen and water have first been removed, and storage in a dry atmosphere often allows the baked plates to be useful days or weeks after treatment. The gains in speed from baking are typically factors of 2 to 5.

Hydrogen. Substantial speed gains can be achieved by soaking plates in hydrogen gas. The mechanism involved is probably reduction sensitization, and therefore can be used in conjunction with removal of oxygen and water and with baking to get the full benefit of all three techniques. A widely followed practice is to do the baking in a dilute mixture of hydrogen in nitrogen, known commercially as forming gas. This avoids the risks of a hydrogen explosion. The background fog comes up very quickly with excessive hydrogen treatment, and speed gains of more than a factor of 2 beyond that achieved with drying and baking are rarely practical. Moreover, drying and baking can actually improve the detective quantum efficiency, while hydrogen treatment only increases the speed.

Preflashing. Occasionally a better picture will result if a slight amount of uniform preexposure is impressed on the plate, but the signal-to-noise can only be degraded if the density of the preflash exposure exceeds about 0.2. In effect, the preflash moves faint features from a domain of low contrast near the photographic threshold into a domain of higher contrast. Thus the improvement is often one of perception, not of photographic detection.

Push development. Extended development times can improve the speed and detective quantum efficiency of some materials. However, the newer fine-grained monodisperse emulsions have shown little or no gain from longer development.

Cooling. If emulsions are cooled during exposure, some of the low-intensity reciprocity failure can be eliminated. However, since removal of oxygen and water is a more effective way to reduce low-intensity reciprocity failure, cooling is rarely used with astronomical emulsions. It is occasionally justified when special materials with extreme low-intensity reciprocity failure, such as color film, are being used.

Intensified photography. Even when they are optimally hypersensitized, astronomical plates have a detective quantum efficiency of only about 2 or 3%. Photocathodes, on the other hand, can have quantum efficiencies as high as 25% or more. Even the image tubes produced in large numbers for military night-vision applications routinely achieve quantum efficiencies in excess of 10%. A stack of image tubes is often used at the focus of a spectrograph or direct camera, with the image from the final phosphor screen transferred optically onto a film or plate for recording a long exposure. The disadvantages of intensified photography are the distortions, nonuniformities, blemishes, and ion flashes that plague many

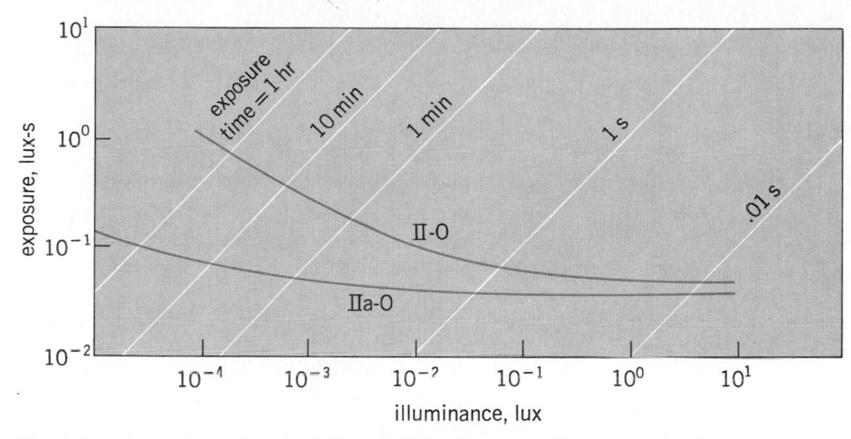

Fig. 6. Low-intensity reciprocity failure (LIRF) of a "normal" emulsion (II-O) compared with a similar material (IIa-O) that has been treated during manufacture to reduce reciprocity failure. Curves show exposure required to produce a density of 0.6 above fog; the rise of each curve at left thus indicates decreasing sensitivity at long exposures. (*After Kodak Films and Plates for Scientific Photography, Kodak Publ. P-315, Eastman Kodak Co., 1973*)

tubes. Moreover, the tubes must be cooled to reduce the dark current for long exposures. Perhaps the biggest disadvantage for survey work is that the largest photocathodes are rarely 4 in. (100 mm) in diameter, and the military tubes are more often 1 in. (25 mm) than 1.5 in. (40 mm).

Astronomers have also developed special image tubes which use a nuclear track emulsion to record directly the electrons produced at the photocathode. Several observatories have used these electronographic cameras successfully for both spectroscopy and direct photography. *SEE IMAGE TUBE.*

Electronic imaging devices. Enormous effort has gone into the development of detector systems which give out the image as an electrical signal which can be digitized either for immediate computer processing or for magnetic storage for later analysis. These systems divide roughly into two classes. In the first class, a photocathode device such as an image intensifier chain is used to detect the incoming light, while an electronic device such as a television tube or diode array is used to read out the intensifier image. In the second class, silicon diode arrays are illuminated directly. The advantage of the intensified detectors is that they can detect individual photons and thus can be used at the very lowest light levels. The silicon arrays have the highest quantum efficiency, reaching nearly 75% in some devices, but the readout noise limits the operation to an integrating mode where a few tens of photons must be detected before looking at the picture. Size is also a problem since these detectors are limited by the dimensions of the chips, which are even smaller than photocathodes. *SEE CHARGE-COUPLED DEVICES.*

Image processing. While the development of electronic imaging hardware has improved the sensitivity of the big telescopes dramatically, perhaps the greatest advances in astronomical imagery, both photographic and electronic, have come in the area of image processing. Not only are modern high-precision microphotometers commonly available for digitizing photographic images, but elaborate computer schemes for processing and analyzing the images have been developed at several centers. Although it is still not common practice, it is feasible to extract nearly all of the information stored on a large direct photograph. For example, procedures have been developed for automatic computer identification and classification of the thousands of images on a single plate taken with a 158-in. (4-m) telescope or large Schmidt telescope. Other groups have software systems which process digitized spectra completely, identifying lines, solving for wavelengths and intensities, and determining radial velocities with the use of cross correlations. In some cases, these systems have been set up with electronic detectors and powerful displays right at the telescope so that much of the data analysis can go on in real time as the detector is integrating.

David W. Latham

Color Photography

The language of astronomy abounds with references to color, and the concept of color is implicit in many of the measurements that astronomers make. Color index, for example, is a quantity related to the temperature of a star, while the redshift of a galaxy is used to indicate its recessional velocity. More directly, stars may be described as red giants or white dwarfs or even blue stragglers. *See Redshift; Star.*

These names reflect the underlying importance of color in astrophysics and cosmology, and though the colors involved are subtle and difficult to distinguish by the eye, in its dark-adapted state, special photographic techniques can be used to display them. A realistic representation of the true colors of celestial bodies can reveal new relationships in familiar objects and add an important third dimension to the morphology and brightness information of the more usual monochrome representations.

Color films. As discussed above, special photographic materials are necessary to accommodate the unusual requirements of photography in astronomy. Not only is the amount of incoming radiation to be detected extremely small, but it is accompanied by unwanted light from the night sky (the airglow). The materials must therefore combine extreme sensitivity at long exposures with high contrast. The ability to detect faint objects is ultimately more dependent on the contrast and resolution of the photographic material than on the light grasp of the telescope or available observing time. On the other hand, color films are designed for general use at levels of illumination where high contrast and low-light-level efficiency are unimportant. In addition, these films are intended primarily to reproduce the broadband colors of everyday life, and for this the rather uneven spectral response of the individual layers shown in **Fig. 7** is specifically intended. Unfortunately, gaseous nebulae emit most of their visible radiation in the form of monochromatic emission lines from the ionized elements present. In the visible region, the strong green forbidden lines of doubly ionized oxygen [O III] near 500 nm and the rich red of a hydrogen recombination line Hα at 656 nm often predominate. The green line falls at a minimum in the blue-green sensitivity of a typical color film and is not well recorded; however, the red line coincides with the peak sensitivity of the red-sensitive layer. Thus, color films always show gaseous nebulae as red, largely irrespective of the contribution from the green oxygen line, whereas yellow (red + green) would be a more realistic representation. The line spectrum of the Orion Nebula is shown superimposed on the spectral sensitivity curves of a typical color film in Fig. 7. *See Nebula.*

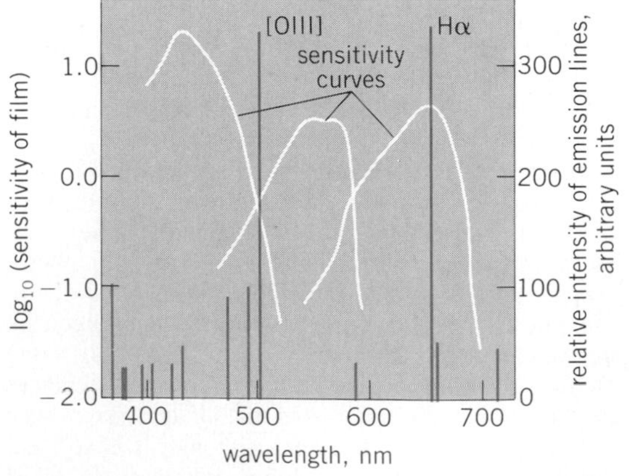

Fig. 7. Comparison of the spectral sensitivity curves of a color film with the emission-line spectrum of the Orion Nebula.

A further problem is the effect of long exposures on the relative sensitivity of the three layers, which are differently affected by low-intensity reciprocity failure. Changes in both sensitivity and contrast of the layers are found, and exposures which are long enough to be astronomically useful often produce severe color-balance distortion. However, in the absence of materials prepared specifically for color astrophotography, attempts have been made to minimize the problems of ordinary color films and exploit their advantages of low cost, ease of use, and ready availability. Both reversal and negative films have been used. Most workers appear to favor the higher contrast and brighter colors of reversal films, but the greater latitude and color-correction possibilities of negative materials are considerable advantages, especially since exposure times are difficult to assess. The long-exposure properties of both types of film can be improved by user-applied preexposure treatments.

Cooled cameras. As discussed above, low-intensity reciprocity failure of both color and monochrome films is reduced if the long exposure is made at a low temperature. Most experiments have been made with cameras designed for fairly small formats and cooled to about −103°F (−75°C) with solid carbon dioxide. Care must be taken to avoid the condensation of water vapor on the film during exposure and to ensure that the cooling is uniform across the frame. This becomes increasingly difficult with the large formats used in professional astronomy, and is impossible in some types of telescopes. Exposure at low temperature reduces the effects of low-intensity reciprocity failure on both the speed and color balance of all types of films.

Hypersensitization of color film. Some of the techniques which are used for spectroscopic plates may also be applied to color films. Baking both in nitrogen and in forming gas, a 2–4% hydrogen-in-nitrogen mixture, is useful. Forming gas is available in special kits specifically for hypersensitizing small quantities of films for astrophotography. Films are baked for several hours in a flow of the gas at 150°F (65°C) just prior to exposure and then (preferably) exposed in a nitrogen atmosphere. Substantial long-exposure speed gains are reported. However, the process affects each of the three sensitive layers differently, and some shift in color balance may be experienced.

These user-applied processes reduce some of the disadvantages of color films for astrophotography,

Emulsion-filter combinations suitable for making three-color separations of astronomical objects

Color	Emulsion*	Passband, nm: Filter cut-on	Passband, nm: Emulsion cut-off	Contrast, γ†	Exposure, min‡
Blue	IIa–O	385	~500	1.7	20
Green	IIa–D	475	~610	1.7	20
Red	098–04	610	~690	1.9	35

*Eastman Kodak spectroscopic plates.
†The 098–04 emulsion is developed for 5 min at 68°F (20°C) with developer D19; the others are developed for 8 min.
‡Exposure time required to reach sky limit (maximum output signal-to-noise) on an f/3 telescope. Contrast of the red image is lowered to match the other two during the contact-copying stage. The exposure times are for plates which have been hypersensitized in nitrogen and hydrogen.

and push development may be used in addition to the above, to increase both speed and contrast. However, the basic problem of uneven spectral response remains. As a result, color films can reproduce only realistic colors of the brighter, continuous-spectrum objects, such as planets, stars, and galaxies. Faint objects and emission nebulae are not well recorded.

Indirect color photography. An alternative approach, which avoids the imperfections of color films, uses the oldest system of color photography, the three-color separation technique. Three separate exposures are made onto plates sensitive to blue, green, and red light. Suitable passbands for the three colors are provided by the well-established emulsion-filter combinations used in astronomy for photographic photometry. These are listed in the **table**, together with typical exposure times. The overlap of the passbands ensures a much more even response across the spectrum than that available from color films. The exposures are made on spectroscopic plates which are responsive to hypersensitization and which are designed for low-light-level photography in the presence of night-sky airglow. Separation negatives can also be produced from image tubes which use photography as a detector. Two distinct methods of combining the images into color pictures have been used, subtractive printing and additive processes.

Subtractive printing. In subtractive printing, the original negatives are copied separately onto three monochrome films which are processed to produce positives. The positive images are then transformed into an equivalent amount of a complementary-colored dye, either by a color-forming redevelopment or imbibition of dye into hardened gelatin. The image from the red-sensitive plate is dyed cyan, the green image is dyed magenta, and the blue image is made yellow. The films containing the dye images are superimposed in register or transferred by imbibition, again in register, to a special receiving layer (dye transfer process). The subtractive processes permit some adjustment of contrast and color balance during assembly, but in practice are complicated to operate and are now little used in astronomy.

Additive processes. From the original separation negatives, monochrome positive copies are made by contact onto any suitable film material. At this stage, a wide range of image-manipulation techniques can be applied to enhance small or faint features and to adjust the contrast of the original images. The positives are subsequently enlarged sequentially in register onto a positive-working color material (such as Cibachrome or Ektachrome) or onto a negative stock for subsequent printing by normal substractive methods. The positives are enlarged through the equivalent of the taking filters, the red positive through a red filter, and the green through a green filter, and so forth. This process is easier to use and more flexible than any of the subtractive systems and can yield color prints of both scientific and esthetic merit.

Color balance. The eye is a poor discriminator of color at low light levels, and it is not possible to check the color balance of astronomical photographs against the original scene or against color memory to verify color fidelity. Color pictures produced on the 154-in. (4-m) Anglo-Australian telescope are balanced by comparison with a neutral gray step-wedge image projected onto the plates during the telescope exposure. The light source of the projector is filtered to a color temperature of 5500 K (9400°F; equivalent to sunlight). Thus, sunlike stars appear white on pictures from this observatory, and stars hotter or cooler than the Sun appear blue and yellow respectively. Likewise, the balance of color in an emission-line nebula is adjusted with reference to the same gray step wedge on the assumption that the spectrum has been recorded equivalently in all three passbands.

David F. Malin

Bibliography. D. F. Malin, Colour photography in astronomy, *Vistas Astron.*, 24:219–238, 1980; D. F. Malin, Photographic image enhancement in astronomy, *J. Photog. Sci.*, 29:199–205, 1981; D. F. Malin and P. G. Murdin, *The Colours of the Stars*, 1984; A. G. Smith and A. A. Hoag, Advances in astronomical photography at low light levels, *Annu. Rev. Astron. Astrophys.*, 17:43–71, 1979; R. M. West and J. L. Heudier (eds.), *Modern Techniques in Astronomical Photography*, European Southern Observatory, Geneva, 1978.

Astronomical spectroscopy

The use of spectroscopy as a technique for obtaining observational data on the chemical compositions, physical conditions, and velocities of astronomical objects. Astronomical applications of optical spectroscopy from ground-based observatories cover the electromagnetic spectrum from the infrared through the visible to the near-ultraviolet. Space-age techniques—high-altitude balloons, rockets, and satellites—have extended spectroscopic observations through the far-infrared to the domain of radio astronomy, and through the far-ultraviolet to the region of x-rays and gamma rays. Work in these longest and shortest wavelengths requires techniques other than those discussed here. SEE GAMMA-RAY ASTRONOMY; IN-

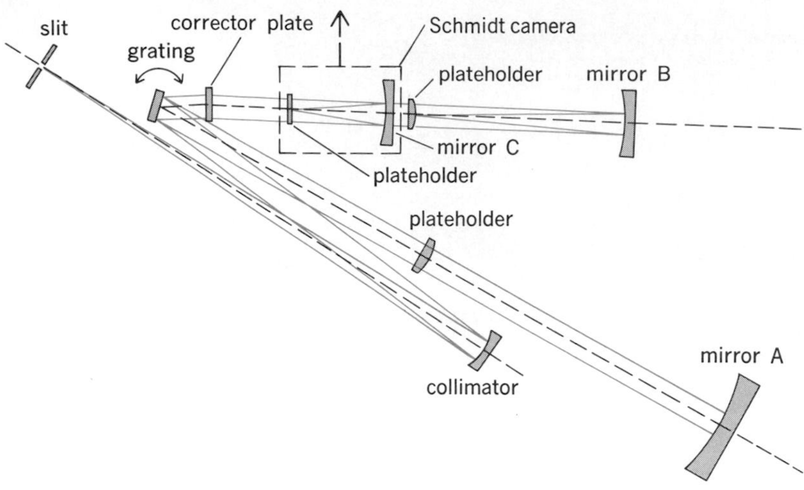

Fig. 1. Optical system of coudé spectrograph of Mount Wilson 2.5-m telescope. The three mirrors A, B, and C, with apertures of 91, 56, and 56 cm and focal lengths *f* of 290, 185, and 81 cm, are the camera objectives which can be chosen to photograph the spectrum; the camera with the focal length of 81 cm must be removed if the one with the focal length of 185 cm is to be used.

FRARED ASTRONOMY; RADIO ASTRONOMY; ROCKET ASTRONOMY; SATELLITE ASTRONOMY; ULTRAVIOLET ASTRONOMY; X-RAY ASTRONOMY.

In optical spectroscopy a spectrograph is attached to a telescope, usually a reflector, which serves as a light collector. Prime-focus, newtonian, Cassegrain, and coudé spectrographs are distinguished by the focus at which they are used. Alternatively, an echelle system may be used. The coudé focus is preferred for most high-dispersion work. The image of the celestial body being studied is focused on the spectrograph slit by the telescope objective. The light from the slit passes through the collimator, which is a lens or mirror. The parallel rays of light are dispersed by glass or quartz prisms or, as in most modern spectrographs, by a diffraction grating. The grating is usually "blazed" by shaping its grooves in such a fashion that most of the dispersed light is cast into a chosen order of the spectrum. *SEE OPTICAL TELESCOPE.*

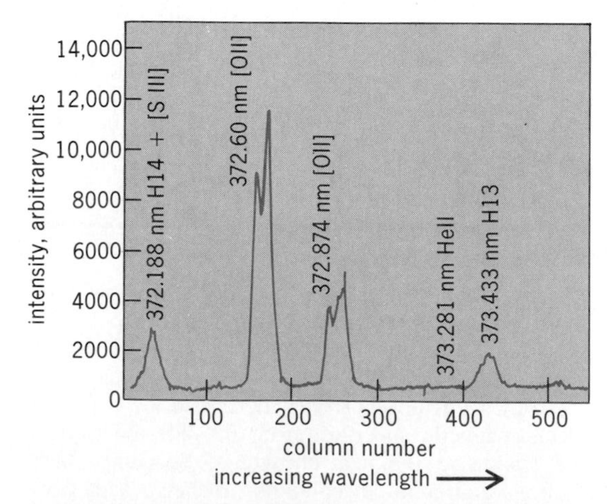

Fig. 2. Raw scan of a small portion (wavelengths ranging from about 372.1 to about 378.8 nm) of the spectrum of the planetary nebula NGC 7027, obtained with the Hamilton spectrograph at Lick Observatory. The column number of the charge-coupled device is very nearly linearly proportional to wavelength.

Dispersion. The reciprocal linear dispersion provided by stellar spectrographs ranges from about 100 nm/mm to about 0.01 nm/mm, the higher dispersions usually being provided by stationary spectrographs at the coudé focus, while the other foci are employed for lower dispersions. The optical system of the coudé spectrograph of the 100-in. (2.5-m) Mount Wilson telescope is shown in **Fig. 1**. An increasingly popular alternative arrangement to a coudé system for obtaining high dispersion is the echelle spectrograph, which achieves comparable results in a more compact instrument and can be mounted at one of the other foci. An echelle differs fundamentally from a diffraction spectrograph in that a spectrum is produced by the interference of light waves reflected from a steplike structure. It yields short-wavelength segments of successive high-order grating spectra, which would normally lie on top of each other. By introducing a prism with dispersion at right angles to that produced by the echelle, the successive spectral strips are separated from one another so that a large range of the spectrum is obtained in parallel strips on a rectangular area, rather than being spread out linearly, as in the coudé arrangement. With a charge-coupled device as a detector in place of a photographic plate, this format is much superior to that of a conventional coudé. **Figure 2** shows one of these spectral segments.

Recording methods. In the simplest form of an astronomical spectrograph, a prism is placed in front of the objective of the telescope; spectra of all stars in the field of the telescope are photographed in one exposure. Objective-prism spectrograms are used primarily for spectral and luminosity classification.

Until the 1970s, spectra were recorded photographically (**Fig. 3**). When high photographic speeds are required, Schmidt cameras are often used, as in Fig. 1. If only a small field (about 1 in. or 2–3 cm) in the focal plane suffices, an image tube can be used. The intensified image thus obtained can be registered on a photographic plate or scanned electronically as in the Robinson-Wampler image tube scanner. If several image tubes are used in tandem, spectral resolution tends to be degraded, but this problem was largely overcome in the image photon counting system devised by A. Boksenberg. Charge-coupled devices, which have high quantum efficiencies, are now generally used. All modern spectroscopic equipment employs high-speed computers so the scanner output can be displayed, processed, and stored. With large telescopes it is thus possible to record spectra of dim stars, remote galaxies, and faint quasars that could not have been attempted previously. *SEE CHARGE-COUPLED DEVICES; IMAGE TUBE.*

Fourier transform spectroscopy. Fourier transform spectroscopy, used particularly in infrared work, employs a concept entirely different from the spectrographs described. Instead of being dispersed in a spectrograph, the light of a wide band of wavelengths is passed through a Michelson interferometer with variable spacing of its two apertures. The resulting interferogram, which is an electronic record of the interference signal produced by the interferometer as the separation of the apertures is varied, is converted into a record of intensity versus wavelength by a high-speed computer. *SEE INTERFEROMETRY.*

Solar spectrum. The visual spectra of the Sun and most other stars consist of a continuous background ranging in color from deep red to violet, which is cut

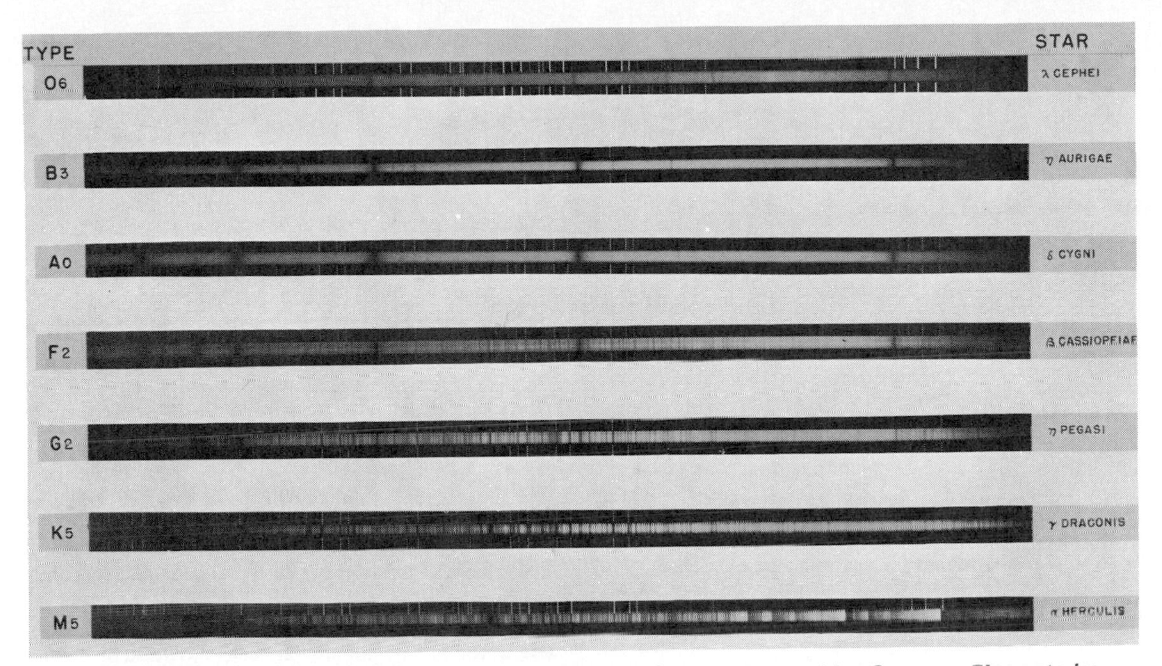

Fig. 3. **Photographic recordings of principal types of stellar spectra. (*Mount Wilson and Los Campanas Observatories, Carnegie Institute of Washington*)**

by numerous dark absorption lines (Fraunhofer lines), first mapped by J. von Fraunhofer in 1814. SEE SUN.

Planetary spectra. In the optical region the spectra of planets and their satellites are essentially reflected sunlight. When planets have atmospheres, absorption lines and bands due to molecules therein contained are also observed. From quantitative studies of these absorption bands, much information on chemical compositions and isotope ratios is found. SEE PLANET.

Emission-line spectra. Attenuated, incandescent, gaseous clouds in the Milky Way Galaxy and in other galaxies produce bright line spectra, which are sometimes superposed on a continuous background. Extended stellar envelopes (**Fig. 4**), quasars, and the outer envelopes of the Sun (chromosphere and corona) also produce emission-line spectra. SEE NEBULA; QUASAR.

Classification of stellar spectra. Although stellar spectra show a great variety of appearances, most can be classified in a two-parameter spectral class and luminosity system. Strengths and appearance of different spectral lines (usually on low-to-moderate-dispersion spectrograms) supply classification criteria. Classification indices are usually temperature-sensitive lines; luminosity criteria are usually density-sensitive lines. Stars of high luminosity have lower atmospheric densities. The spectral sequence is described by letters O, B, A, F, G, K, M (ranging from hot, about 72,000°F or 40,000 K, to cool, about 5000°F or 3000 K) with decimal subdivisions. Luminosity classes range from I (very luminous) to V (main-sequence dwarf stars). The Sun's spectral class is G2V, which means that it lies closer to G0 than to K0 and is a dwarf star. SEE STAR.

Spectrophotometry. For many purposes a plot of true intensity against wavelength is needed. Over a wide wavelength range (several hundred nanometers) the intensity distribution depends mostly on the temperature. With high spectral resolution it is possible to measure the detailed profile of a line (**Fig. 5**), which supplies information on the abundance of the chemical element or compound involved, the temperature and density distribution in the stellar atmosphere, and large-scale mass motions and rotation. Such measurements are usually made with photoelectric scanners or with charge-coupled device detectors.

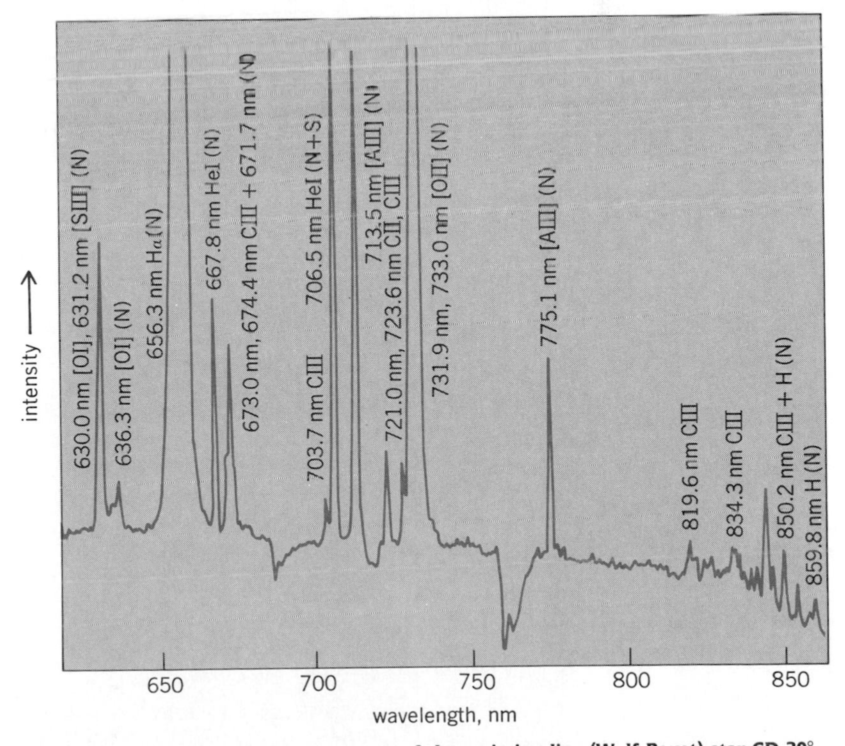

Fig. 4. **Scan of a portion of the spectrum of the emission-line (Wolf-Rayet) star CD-30° 15469 surrounded by a small incandescent nebula; recorded by the Robinson-Wampler scanner at Lick Observatory. Lines of nebular origin are indicated by (N). The other lines come from the stellar envelope.**

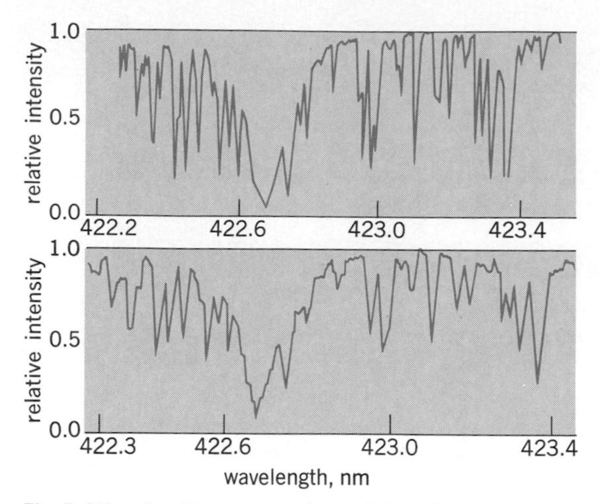

Fig. 5. Microdensitometer tracings of the solar spectrum as a function of wavelength. The same section of spectrum is shown; the upper tracing was made from a spectrogram with 0.03 nm/per millimeter of linear dispersion; the lower tracing was made from one with 0.28 nm/mm.

Radial velocity. Measurements of the wavelength displacement, due to the Doppler effect, between the spectral lines in a celestial object and a laboratory source (such as an iron arc or discharge tube) give the line-of-sight velocity of the source with respect to the observer. Radial velocity measurements enable establishment of orbits of close double stars (spectroscopic binaries, **Fig. 6**), and when combined with photometric data have yielded dimensions of the stars themselves. Also, such measurements have given the motion of the Sun with respect to nearby stars, the rotation of the Galaxy and other galaxies, and recession velocities of external galaxies (up to 125,000 mi/s or 200,000 km/s). Wavelength shifts of lines in spectra of some quasars imply velocities exceeding 90% of the speed of light. Radial velocity measurements revealed the remarkable nature of the object SS 433 with its jets moving at velocities of 30,000 mi/s (50,000 km/s). *See Binary star; Doppler effect; Galaxy, external; SS 433.*

Lawrence H. Aller

Bibliography. D. F. Gray, *Observation and Analysis of Stellar Photospheres*, 1976; W. A. Hiltner (ed.), *Astronomical Techniques: Stars and Stellar Systems*, vol. 2, 1962; G. P. Kuiper and B. Middlehurst (eds.), *Telescopes: Stars and Stellar Systems*, vol. 1, 1960; D. J. Schroeder, *Astronomical Objects*, 1987.

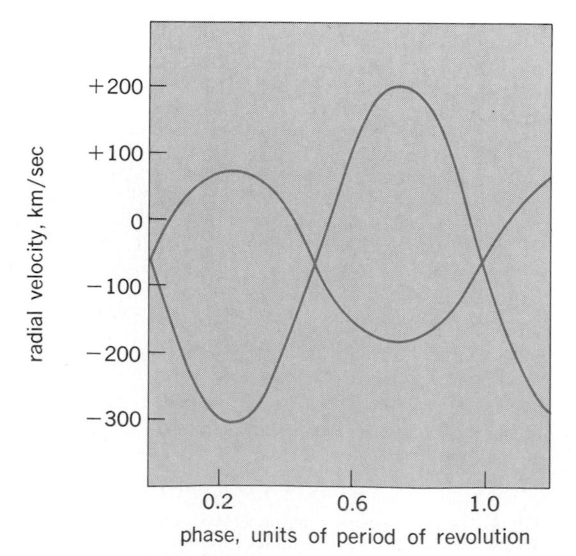

Fig. 6. Radial-velocity curve of a spectroscopic binary (W Ursae Majoris); the spectra of both components have been measured.

Astronomical unit

The basic unit of length in the solar system. The astronomical unit (AU) is also used to a limited extent for interstellar distances through the definition of the parsec (1 pc = 206,265 AU). It is nearly equal to the mean distance a between the center of mass of the Sun and the center of mass of the Earth-Moon system (a = 1.00 000 23 AU), and for that reason it is often convenient to think of it as the mean distance between the Sun and Earth. *See Parsec.*

The most accurate determination of the length of the astronomical unit in physical units, such as meters, is obtained from phase-modulated continuous-wave (CW) radio signals beamed to other planets. The round-trip travel times of the signals are determined by cross-correlating the returned signal from the planet with the transmitted signal, and as a result, planetary distances are measured directly.

Prior to the inception of planetary radar technology around 1961, determinations of the astronomical unit relied on parallax effects in the positions of planets and asteroids as viewed from distant points on the surface of the Earth, or on spectroscopic Doppler shifts in the light of stars near the plane of the ecliptic. These techniques were limited in accuracy to about 43,000 mi (70,000 km). On the other hand, early radar determinations were accurate to 300 mi (500 km), and by accumulating several years of planetary distance measurements, it was possible to refine this accuracy to 0.6 mi (1 km), the limit of passive radar bounce technology. Continuous-wave signals returned from the two Viking orbiters and two landers on the surface of Mars made possible a determination of the astronomical unit to an accuracy of 65 ft (20 m). *See Doppler effect; Earth rotation and orbital motion; Parallax.*

With an adopted value of 299,792,458 m/s (186,282.39705 mi/s) for the speed of electromagnetic propagation in vacuum, the value of the astronomical unit from the Viking data is 92,955,807.25 mi (149,597,870,660 m).

John D. Anderson

Astronomy

The study of the universe and the objects in it. As perhaps the oldest science, dating from early observation of the heavens, astronomy has a long tradition. Modern astronomy uses techniques from many other sciences, including physics, mathematics, chemistry, geology, and biology. The astronomers' point of view, however, is often different from that of scientists in other disciplines, because the goal of astronomical research is to understand celestial objects or groups of objects by using whatever techniques are appropriate. Astronomers' breadth of knowledge of different types of astronomical objects often gives insights into understanding any given object or process.

Scope. The distinction between astronomy and astrophysics has largely evaporated. The older view that astronomers study objects in the sky and astrophysicists seek to explain them has been superseded. The new broadening of astronomy is illustrated by the use of chemical knowledge and methods in studying molecules in interstellar clouds; the use of geological knowledge and methods in analyzing closeup observations from spacecraft of planets and their moons; and the use of biological knowledge and methods in analyzing the origin and evolution of life on the Earth and in the search for extraterrestrial life. *SEE ASTROPHYSICS.*

Astronomy has been transformed by new technology, notably electronics. Computers are integral to most astronomical research, both observational and theoretical. Electronic devices have transformed observation; for example, the use of electronic chips known as charge-coupled devices (CCDs) as detectors and of optical fibers to carry stellar images to spectrograph slits has reduced the time needed to determine the redshift of many galaxies from perhaps 8 h to about 1 min. This factor of 500 has led to a qualitative change, whereby the mapping of space becomes possible to a new degree of completeness. *SEE ASTRONOMICAL SPECTROSCOPY; CHARGE-COUPLED DEVICES.*

Divisions. The divisions of astronomy include stel lar astronomy, interstellar astronomy, galaxies, cosmology, solar physics, solar-system astronomy, the search for extraterrestrial life, astrometry, ground-based telescope building, telescopes in space, computational astronomy, celestial mechanics, and historical astronomy.

Stellar astronomy. This is the study of stars, especially through observations of the intensity of light and its variation (photometry) and stellar spectra (spectroscopy). This work, formerly carried out exclusively in the visible spectrum, has now been extended to the x-ray region, the ultraviolet, and the infrared. The 1987 explosion of the brightest supernova since 1604 has enabled scientists to test and refine theories of stellar evolution. *SEE INFRARED ASTRONOMY; MAGNITUDE; SPECTROGRAPHY; STAR; STELLAR EVOLUTION; SUPERNOVA; ULTRAVIOLET ASTRONOMY; X-RAY ASTRONOMY.*

Interstellar astronomy. This is the study of interstellar space, especially the hydrogen gas and the molecules observed there. The techniques of radio and infrared astronomy are especially important in this field. Giant molecular clouds are now known to be dominant structures in the Milky Way Galaxy. *SEE INTERSTELLAR MATTER; MILKY WAY GALAXY; RADIO ASTRONOMY.*

Galaxies. Millions of galaxies can now be studied, principally by spectroscopy, with new techniques such as those described above. Large telescopes and sensitive electronic detectors have led to the discovery of galaxies so old that they were formed only 1×10^9 years after the origin of the universe. X-ray observations from spacecraft have been used to study not only individual galaxies but also intergalactic gas. *SEE GALAXY, EXTERNAL.*

Cosmology. Mapping of the distance to thousands of galaxies has enabled three-dimensional maps of regions of space to be made; the structure of space must be interpreted in models of galaxy formation. The big bang theories, with their inflationary-universe current versions, now dominate cosmological thought. It is now realized that the extremely high energies available in the first fractions of a second of the universe led to the formation of elementary particles beyond the ability of earthbound particle accelerators to create. The analysis of the early universe has led to a significant merger between cosmology and elementary-particle physics. *SEE BIG BANG THEORY; COSMOLOGY; INFLATIONARY UNIVERSE COSMOLOGY.*

Solar physics. The study of the Sun, traditionally known as solar physics rather than solar astronomy, treats it not only as an individual object close enough to study in detail but also as a prototype of more distant stars. Studies across the spectrum and at special times such as during solar eclipses have enabled the three-dimensional structure of at least the upper layers of the Sun to be analyzed. High resolution is available on the Sun uniquely among the stars. New techniques such as helioseismology bring promise of understanding the deep solar interior. The solar neutrino problem, in which the number of neutrinos received is only about one-fourth the number predicted, has raised fundamental questions about the present understanding either of stellar structure or of neutrino physics, and a new generation of experiments may shortly resolve some of the outstanding problems. *SEE ECLIPSE; HELIOSEISMOLOGY; SOLAR NEUTRINOS; SUN.*

Solar-system astronomy. The planets can be studied with telescopes from the ground, from spacecraft in Earth orbit, and from spacecraft passing near them or landing on them. The detailed closeup views of, for example, Mars from the Soviet *Phobos* mission or Uranus's moon Miranda from the American *Voyager 2* mission have continued to advance the hybrid discipline of astrogeology. Studies of Halley's Comet from a variety of spacecraft and ground-based telescopes during its 1986 apparition are part of a search for the most primitive material in the solar system, whose location is needed in order to improve models for the origin of the solar system. *SEE COMET; HALLEY'S COMET; PLANET; PLANETARY PHYSICS; SOLAR SYSTEM.*

Search for extraterrestrial life. Serious observations are being made or are proposed for searches for signs of celestial signals being sent by extraterrestrial beings. The radio spectrum, in particular, is being searched for signals of types that would not be generated naturally. *SEE EXTRATERRESTRIAL INTELLIGENCE.*

Astrometry. This is the study of the positions and motions of stars and other celestial objects. Traditionally part of astronomy rather than astrophysics, this field should be greatly advanced by leaps in accuracy provided by the European Space Agency's *Hipparchos* spacecraft and NASA's Hubble Space Telescope. *SEE ASTROMETRY.*

Telescopes. New technologies have led to advances in telescopes in all parts of the spectrum. These include a 10-m (395-in.) optical telescope in Hawaii, designed to provide a fourfold increase in collecting area over telescopes built decades ago. Arrays of radio telescopes, including the existing Very Large Array and the Very Long Baseline Array, can provide unprecedented resolution. Some new telescopes, such as those for the submillimeter spectrum, are studying parts of the spectrum for essentially the first time. *SEE OPTICAL TELESCOPE; RADIO TELESCOPE; SUBMILLIMETER ASTRONOMY; TELESCOPE.*

The many spacecraft already launched have been joined or superseded by a new generation of more permanent observatories in space. Besides the Hubble Space Telescope, designed to study the visible and infrared with resolution at least five times superior to that available from the Earth's surface, these include such major observatories as NASA's Advanced X-Ray Facility and Gamma Ray Observatory and the European Space Agency's Infrared Space Observatory. The Soviet Union installed an x-ray telescope aboard its space station. *See* GAMMA-RAY ASTRONOMY; SATELLITE ASTRONOMY.

Computational astronomy. Small- and moderate-sized computers run telescopes and assist with data reduction. Supercomputers allow the solution of classes of problems that could not be previously solved, such as those involving hydrodynamics. Models of matter falling into black holes or of Jupiter's Great Red Spot are among problems that have been treated. *See* BLACK HOLE; JUPITER.

Celestial mechanics. Another topic that remains traditionally part of astronomy rather than astrophysics is modern celestial mechanics, which became prominent in such matters as predicting the path of Halley's Comet with sufficient accuracy to allow spacecraft rendezvous. Computations involving celestial mechanics are also important in arranging spacecraft paths, such as the *Voyager* missions' paths to the outer planets and the *Galileo* mission's intricate orbit among Jupiter's satellites. *See* CELESTIAL MECHANICS.

Historical astronomy. Studying astronomy's rich history not only enables the roots of the science to be better understood but also provides data for comparison with the present. For example, the paths of solar eclipses from hundreds or thousands of years ago can be analyzed together with data about current eclipses to check whether the Sun has changed in size.

Jay M. Pasachoff

Astrophysics

The application of modern physics to the problems of astronomy. Astrophysics embraces much of the activity of present-day astronomy; specifically excluded are the study of the motions of the planets and satellites (celestial mechanics), the measurement of positions and motions (astrometry), and usually also the structure and dynamics of the Milky Way Galaxy and other galaxies, although such topics as the density wave theory of spiral arms are sometimes considered to be part of astrophysics. *See* ASTRONOMY; CELESTIAL MECHANICS; MILKY WAY GALAXY.

Instrumentation. From an operational point of view, astrophysics differs from terrestrial physics in that it is primarily an observational subject, although certain special problems lend themselves to experimental treatment. The basic problem is often the measurement of the quantity and quality, that is, the energy distribution and polarization, of the electromagnetic radiation of the Sun, stars, planets, and nebulae, and the study of their spectra. A variety of telescopes and detectors are employed. Detectors include the photographic plate with suitable filters, the photoelectric cell, lead sulfide and lead telluride cells, thermocouples, charge-coupled devices (CCDs), reticons, image tube scanners (ITS), and image photon counting systems (IPCS). For many problems, the photographic plate is being replaced by electronic detectors, particularly charge-coupled devices. The radio-frequency region requires antennas [large interferometer arrays such as the Very Large Array (VLA) and steerable dishes] and radio-frequency detectors appropriate to the problem at hand. The infrared region has assumed increasing importance. Especially significant is the development of areal detectors that will permit the recording of extended surfaces, such as nebulae and galaxies. *See* ASTRONOMICAL PHOTOGRAPHY; INFRARED ASTRONOMY; RADIO ASTRONOMY; TELESCOPE.

Spectrographs employ linear dispersions ranging from several tens of millimeters per nanometer of wavelength to several tens of nanometers per millimeter. In solar work a host of special instruments are used to secure monochromatic photographs of the Sun, its faint outer envelopes, chromosphere, corona, and transient phenomena such as flares. Special instruments have been devised to measure the magnetic field of the Sun and to study radio-frequency radiation produced at times of high solar activity. Particle, x-ray, ultraviolet, and cosmic rays of solar origin are also observed. *See* ASTRONOMICAL SPECTROSCOPY.

Important advances have been made with the aid of telescopes equipped with infrared, x-ray, and gamma-ray detectors flown in balloons (stratoscopes), rockets, and satellites. Particularly important have been the *Copernicus* and *International Ultraviolet Explorer* (*IUE*) satellites, which cover the range between the Lyman limit (91.2 nanometers) and the optical region (300 nm). The *Einstein* satellite was equipped with a grazing incidence telescope that actually forms images in soft x-rays and produces high-resolution pictures of x-ray sources in the Milky Way Galaxy and other galaxies. Some are optically identified. The *IRAS* satellite covers an infrared range not observable from the surface of the Earth. It has revealed a large number of new infrared sources, particularly among galaxies, cool dust clouds where star formation may be occurring, and a cloud of cool particles around the star Vega. The most important instrument designed for launch into orbit is the Hubble Space Telescope. Solar x-rays and ultraviolet radiation can be monitored continuously, and monochromatic pictures of the Sun in this radiation provide a synoptic record of solar activity. Several improved designs for infrared, ultraviolet, x-ray, and gamma-ray telescopes have been put into various stages of development and implementation. *See* GAMMA-RAY ASTRONOMY; ROCKET ASTRONOMY; SATELLITE ASTRONOMY; ULTRAVIOLET ASTRONOMY; X-RAY ASTRONOMY; X-RAY TELESCOPE.

The field of experimental astrophysics involves measurement of shapes and strengths of spectral lines emitted under controlled conditions of temperature and pressure. Techniques involve the use of shock tubes, atomic beams, controlled arcs, beam foils, laser excitation of discrete levels, and so forth. Also involved are measurements of cross sections for atomic and molecular collisional excitation and ionization, low-energy nuclear transformations, and even synthesis of molecular fragments that may be found in interstellar space.

Neutrino astronomy has become an important facet of research. To the solar neutrinos detected by Raymond Davis have now been added the neutrinos from Supernova 1987A. This supernova played an important role in verifying theories and improving concepts of the supernova phenomenon. In addition to radio-frequency, infrared, optical, and ultraviolet radiation, and neutrinos, it emitted gamma rays. *See* SOLAR NEUTRINOS; SUPERNOVA.

Divisions. Theoretical astrophysics constitutes a branch of theoretical physics and embraces many subjects that usually are considered the province of the latter. For purposes of discussion, astrophysics may be divided somewhat as follows, according to the types of celestial bodies that are involved.

Solar physics. Solar physics includes all phenomena connected with the Sun and overlaps with geophysics in the consideration of solar-terrestrial relationships, for example, the connection between solar activity and auroras, magnetic storms, and sudden ionospheric disturbances. Some aspects of solar physics are concerned with the quiet or undisturbed Sun, for example, the granulation observed in the white light photosphere, problems of the dark-line or Fraunhofer spectrum, the chromosphere, and the corona as observed at sunspot minimum. Study of minute oscillations of the solar surface (solar seismology) makes it possible to probe the solar interior and to assess models of its structure and scenarios of its evolution. A study of the active Sun involves the 22.5-year magnetic cycle, sunspots and centers of activity, and flares, which often include ejection of streams of energetic particles, x-ray emission, and nonthermal radio-frequency emission. *See Sun.*

Solar system. The physics of the solar system includes the nature of planetary atmospheres and interiors and the chemical and physical constitution of comets, meteors, and small particles such as those that constitute the zodiacal cloud. In addition, much attention is paid to interplanetary plasmas (the solar wind) and magnetic fields, and to the magnetosphere of the Earth and Jupiter. The study of meteorites offers great opportunities in cosmochronology and cosmochemistry, since they supply essential information on the initial chemical composition and age of the solar system. The techniques involved are those of nuclear and microanalytical chemistry. *See Comet; Cosmochemistry; Interplanetary matter; Magnetosphere; Meteor; Meteorite; Planetary physics; Solar wind.*

Stellar atmospheres. The study of stellar atmospheres, including certain aspects of the solar atmosphere, constitutes an important and active field of astrophysics. Modern work is concerned not only with quiet, static structures but also with turbulent atmospheres, buffeted by powerful winds, with the envelopes of close binaries, variable stars, and stars of unusual chemical composition. *See Star.*

Nebulae and interstellar matter. Much effort is devoted to gaseous nebulae and the interstellar medium. The former includes examples of both thermal excitation, as in Orion and planetary nebulae, and nonthermal excitation, as in the Crab Nebula. They may be excited by stellar radiation, by dissipation of shock waves, or by synchrotron radiation. The interstellar medium pervades the galactic plasma and extends beyond it as the galactic halo. It is very inhomogeneous, including cool (a few kelvins) clouds with solid grains and often complex molecules all shielded from ultraviolet starlight, relatively attentuated domains at temperatures of a few thousand kelvins, regions where hydrogen is ionized, and extremely hot (100,000 K or 180,000°F) volumes presumably produced by supernova detonations. It is the milieu from which stars are being formed and to which they return their outer envelopes when they perish. *See Interstellar matter; Nebula.*

Stellar structure and evolution. Stellar structure and evolution trace the development of a star from its condensation from the interstellar medium, through the initial gravitational contraction phase, to the state in which it converts hydrogen to helium in its core. It is then called a main-sequence star. When the hydrogen in the central region has all been transmuted into helium, the conversion takes place in a thin shell around the inert core and the star becomes a giant or supergiant. The evolution can now become very complex. Eventually, the outer stellar layers are lost to space, and the core normally settles down to become a white dwarf. The escaping shell may form a planetary nebula. If the star is massive, a quiet demise may not occur—the star may explode as a supernova, and the remnant may become a neutron star or even a black hole. Evolution in close binaries may be very complicated as material from one star may be dumped upon the other, sometimes inducing outbursts, especially if the second star has evolved into a white dwarf. *See Binary star; Black hole; Nova; Stellar evolution; Supernova; White dwarf star.*

High-energy astrophysics. Radio astronomy data and observations secured from above the Earth's atmosphere have greatly stimulated the development of high-energy astrophysics, particularly the study of the acceleration of charged particles to high energies in supernova remnants, pulsars, radio galaxies, certain binaries, and quasistellar sources.

It is clear that some binary systems, such as Cygnus X-3, accelerate particles to very high energies, producing copious amounts of gamma rays, which are observed directly, and cosmic rays, whose presence is inferred.

Much attention is paid to galaxies, both normal and exotic. Studies of normal galaxies are concerned with questions such as the relation between a galaxy's mass and angular momentum and its evolution, both dynamical and as reflected in the changing character of the stars that are involved and the chemical composition and physical state of the interstellar medium. Exotic galaxies such as strong radio emitters, Seyfert galaxies, and N galaxies usually contain active nuclei radiating large amounts of energy. The source of activity in many of these systems is often attributed to black holes. *See Astrophysics, high-energy; Galaxy, external; Pulsar; Quasar.*

General relativity. Verifications of general relativity are astronomical or astrophysical. In addition to classical tests such as deflection of light by the Sun or advance of the perihelion of Mercury, predictions of dissipation of energy by gravity waves have been verified by observations of a pulsar binary. General relativity is applied in interpretation of black holes, difficult to observe but believed to occur in certain binary star systems and nuclei of galaxies. It is also used in constructing large-scale models of the physical universe, in cosmology, and in describing the first moments of the universe following the big bang. *See Big bang theory.*

Relation to physics. The fields or problems of astrophysics may also be classified in terms of the underlying or supporting fields in physics. Classical mechanics is the basis of the mechanics of the solar and galactic systems; it supplies important data on the masses and density concentrations in the stars. Thermodynamics and statistical mechanics underlie the derivation of the basic equations of L. Boltzmann and M. N. Saha and the equations of molecular dissociation, which are necessary in the calculation of equilibrium conditions in stellar atmospheres and interiors. Geometrical optics and physical optics underlie both radio and optical astronomical instrumentation. Physical optics provides fundamental formulas for the dispersion and absorption of spectral lines and the scattering of light by small particles in space; it lies behind the theory of operation of all radio telescopes. Molecular and atomic structure provide the basis for an interpretation of spectra of the Sun, planets, comets, stars, and nebulae. Solid-state physics provides the background for various detection devices, for the investigation of the formation of grains in space, of meteorites, and of the internal structure of planets.

Electricity and magnetism and plasma physics have become very important in astrophysics, particularly in connection with solar physics and synchrotron radiation. The origin and interpretation of the solar cycle, the shapes of gaseous nebulae, and the acceleration of high-energy particles in flares, the Crab Nebula, and other nonthermal sources all require techniques of magnetohydrodynamics. Nuclear physics furnishes information on problems of the generation of energy in stars, on stellar structure and evolution, and on the origin of elements. The importance of general relativity has been noted above. High-energy and fundamental particle physics are connected to astrophysics in GUTS (grand unification theories) cosmology—an effort to understand the first instants down to 10^{-37} s of the inflationary universe, and relationships between weak and strong nuclear forces, and the electromagnetic forces in nature. A further goal is to relate the present inhomogeneities in the large-scale structure of the universe to events transpiring in the very earliest interval of time. *SEE MAGNETOHYDRODYNAMICS.*

It is now recognized that the origin of cosmic rays is an astrophysical problem concerned with the acceleration of charged particles to very high energies. Nearly every branch of physics has some application to astronomy. *SEE COSMIC RAYS.*

Unsolved problems. Some outstanding problems in theoretical and basic experimental astrophysics include the computation of basic atomic parameters such as *f* values, damping constants, target areas for the collisional excitation of various levels, and molecular dissociation energies, together with experimental checks whenever these can be obtained. Difficult problems are associated with stellar structure, even for normal stars during the quiet phases of their lives. The stability of compressible fluids in gravitational fields where there is energy loss by radiation as well as by convection currents will be an important area of study. The active regions of the Sun, where magnetic fields play an important role, as well as sunspots, the chromosphere, and the corona pose impressive challenges to both observer and theoretician.

The basic physics and chemistry of the interstellar medium are incompletely understood. It is not exactly known how grains are formed, or how complex molecules are built up on the surfaces of grains, or how stars form out of the interstellar medium.

Although the broad principles of stellar structure and evolution seem well enough understood (except for rapid and catastrophic changes), the discordance between measured and predicted solar neutrino fluxes indicate uncertainties either in stellar structure theory or in basic physics. The initial formation of a star is not well understood, and neither are the details of the transition from giant to white dwarf phase. *SEE SOLAR NEUTRINOS.*

Immense amounts of energy are involved in x-rays, gamma rays, and cosmic rays. Thus, among the most striking problems of all are those associated with high-energy astrophysics: the radio galaxies, with their giant lobes of nonthermal radiation; violent events in galactic nuclei; pulsars; and the quasistellar sources whose exact nature is not yet understood. It is often supposed that vastly extended clouds of relativistic electrons observed in quasars and radio galaxies are powered by central black holes.

The most fundamental questions pertain to cosmology—the details of the big bang, and whether the universe will go on expanding forever or collapse upon itself. It is not yet known if the density of matter in the universe is sufficient to slow down and eventually stop the expansion. One of the most tantalizing problems is the search for so-called dark matter. The density of the visible material in the universe fails by a factor of 10 to ensure it will not expand forever. As early as the 1930s, observations of galaxies had suggested the existance of large amounts of material of very low luminosity, but this elusive material has not yet been identified. A better understanding of galaxies, clusters of galaxies, and the intergalactic medium is needed. *SEE COSMOLOGY.*

Lawrence H. Aller

Bibliography. C. W. Allen, *Astrophysical Quantities*, 1973; K. G. Burbidge (ed.), *Annual Reviews of Astronomy and Astrophysics*; M. Harwit, *Astrophysical Concepts*, 1973; *International Astronomical Union Symposia*; R. Lang, *Astrophysical Formulae*, 1978.

Astrophysics, high-energy

The study of astronomical phenomena which generate particles or radiation with energies which range from the order of 10^3 to 10^{21} eV, the highest observed.

Historical development. Shortly after the discovery of radioactivity and x-rays, scientists found that air was ionized by these radiations, leading to electrical conduction. But a puzzling natural background of ionization, whose origin was unknown, also existed. Measurements made from balloons showed that the background ionization increased with height, proving the existence of cosmic rays. Many experiments before World War II showed that cosmic rays are very energetic particles striking the atmosphere, where they produce secondary particles that reach the ground. These cosmic rays were the first indication that high-energy processes are important in astrophysics. *SEE COSMIC RAYS.*

Radio astronomy began when Karl Jansky discovered that the Milky Way is a major source of radio noise. Grote Reber mapped this radiation just before World War II, but the many advances in radio and radar equipment during the war led to a tremendous increase in radio astronomy after 1945. The first radio source identifications were made: Cas A, a supernova remnant in the Milky Way; and Cyg A, a distant peculiar galaxy with enormous radio luminosity. These sources radiate by synchrotron radiation, produced by high-energy electrons spiraling in a magnetic field. While the observed radio photons have very low energy, as low as 1 microelectronvolt, the electrons producing the radio waves must have energies greater than 1 gigaelectronvolt, quite comparable to the energies of cosmic rays, which range from 10^9 to 10^{21} eV. *SEE RADIO ASTRONOMY.*

The first astronomical x-ray observations were pictures of the Sun made by using pinhole cameras flown on rockets. By 1962, x-ray detectors had been improved to the point that scientists expected to be able to see the x-ray fluorescence from the Moon, which would help determine the chemical composition of the lunar surface. A rocket flight by R. Giacconi and colleagues to measure this effect discovered instead a source in the constellation Scorpius, now known as Sco X-1, the brightest nonsolar x-ray source in the sky. This discovery led to the development of the totally new field of x-ray astronomy, which now dominates high-energy astrophysics. *SEE X-RAY ASTRONOMY.*

Astrophysical sources. Astrophysical sources of high-energy particles or radiation include supernovas, neutron stars, binary systems containing a compact object, supermassive black holes, gamma-ray burst sources, x-ray bursters, and the x-ray background.

Supernovas. When a star explodes in a supernova, it becomes 10^{10} times brighter than the Sun for several weeks. This explosion drives most of the star out into space, where it forms an expanding blast wave. The gas behind this blast wave is heated to temperatures of 10^8 K (2×10^8 °F), so typical electrons have energies of 10 keV. These electrons radiate by the bremsstrahlung process, producing 10-keV x-rays. The cosmic-ray electrons hit by this blast wave are accelerated to energies of 1 GeV and more, leading to emission of radio waves by the synchrotron process. Thus supernova remnants are strong sources of both x-rays and radio waves. Examples are the Crab Nebula, which is the remnant of a supernova of A.D. 1054, and the remnants of the Tycho (1572) and Kepler (1604) supernovae. The radio source Cas A is believed to come from a supernova that exploded in 1667, but was unnoticed at the time. *SEE BREMSSTRAHLUNG; STELLAR EVOLUTION; SUPERNOVA.*

Neutron stars. When a star explodes, a neutron star, an incredibly dense sphere with a radius of 6 mi (10 km) and a mass of at least 1.4 solar masses, may be left behind. Neutron stars have magnetic fields 10^{12} times stronger than the magnetic field of the Earth, and particles are easily trapped and channeled by this field. When a neutron star rotates, radiation is produced, and for proper orientations periodically sweeps across the Earth with a period set by the speed of rotation. Pulsed radio radiation is observed; these rotating neutron stars are thus known as pulsars. The fastest known pulsar spins 642 times per second. The first pulsars to be discovered were radio pulsars, and about 400 examples are known. Almost all (99%) are isolated neutron stars, and only one, the Crab pulsar, is also a strong x-ray source. A few radio pulsars are weak x-ray sources. Only a handful are observable in visible light. *SEE NEUTRON STAR; PULSAR.*

Binary star systems. A second class of pulsar, the x-ray pulsar, consists of a rotating, magnetic neutron star in a binary star system. The strongest pointlike x-ray sources in the Milky Way are all binaries consisting of a normal star orbiting around a compact companion, that is, a white dwarf, neutron star, or black hole. Material from the normal star falls into the gravitational potential well of the compact object, releasing a large amount of energy in the process, just as water falling over a waterfall can be used to produce energy (**Fig. 1**). When this material falls, or accretes, onto a white dwarf, a cataclysmic variable, or dwarf nova, is the result. These stars are moderate x-ray sources. When the material falls onto a neutron star or black hole, a powerful x-ray source can result. *SEE NOVA; WHITE DWARF STAR; X-RAY STAR.*

Because the orbit of the compact object around the normal star can be measured, it is possible to determine the mass of neutron stars and black holes in compact binaries. The best measurements are for neutron stars, which have masses around 1.4 solar masses. Black-hole masses are much harder to measure, because black holes do not produce pulses. However, the compact object in the source Cyg X-1 seems to be heavier than 6 solar masses. Because theoretical predictions do not allow neutron stars to be this heavy, Cyg X-1 almost certainly contains a black hole. *SEE BINARY STAR; BLACK HOLE.*

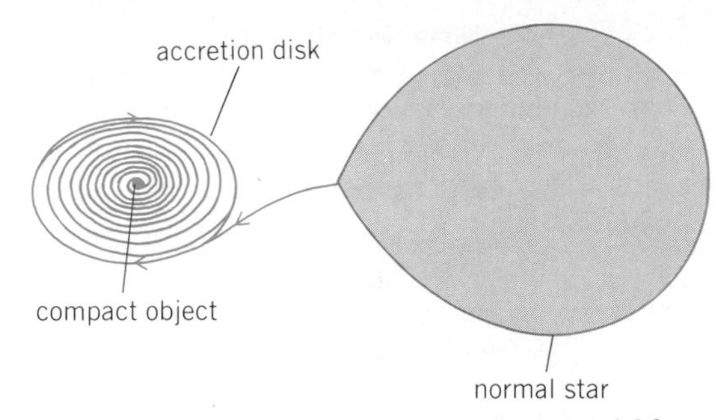

Fig. 1. A binary star x-ray source, showing material from the normal star falling onto the accretion disk surrounding a compact companion: a white dwarf, a neutron star, or a black hole.

Supermassive black holes. Supermassive black holes in the nuclei of galaxies and quasars produce large x-ray luminosities. These black holes must have masses of 10^8–10^{10} solar masses. All of the stars in a galaxy orbit around the nucleus, and those that get too close to the black hole will suffer tidal stripping, star–star collisions, and lose part or all of their matter into the black hole. The energy released as this material falls into the gravity well produces very powerful x-ray, optical, and radio emission from quasars and active galactic nuclei. There is a small point radio source in the nucleus of the Milky Way that could be a miniature version of this process, with a mass of 10^6 solar masses. *SEE MILKY WAY GALAXY; QUASAR.*

Gamma-ray burst sources. A puzzling class of high-energy sources is the gamma-ray burst sources. These are bursts of low-energy (about 100 keV) gamma rays that last for about 10 s. They were discovered by the Vela satellites, whose principal purpose is to watch for nuclear bomb tests in space. Very precise positions for several of these sources have been found by measuring the time of flight of the gamma rays across the solar system by using detectors on several planetary probes, but no certain identifications have been made. One unusually strong burst came from the direction of a supernova remnant in the Large Magellanic Cloud: If this is a correct identification, then the peak burst luminosity was 10^{11} solar luminosities. The burst positions are distributed isotropically all over the sky, so the sources must be either nearby, within 100 parsecs (1 parsec = 3.26 light-years = 1.92×10^{13} mi = 3.09×10^{13} km), or in very distant galaxies over 200 megaparsecs away. *SEE GAMMA-RAY ASTRONOMY.*

X-ray bursters. X-ray bursters are also observed, but these are well understood. When material accretes onto a neutron star at a slow rate, the surface temperature of the star is not high enough to burn the nuclear fuel. After several hours or days the layer of nuclear fuel is thick enough to reach densities and temperatures suitable for nuclear ignition at the bottom of the fuel layer. Once ignited, the entire fuel supply is

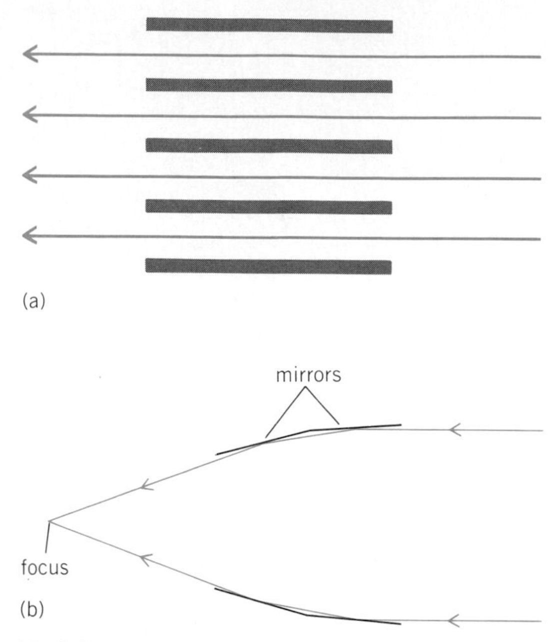

Fig. 2. **Two types of x-ray instruments: (*a*) stacked x-ray collimators, and (*b*) a grazing-incidence x-ray telescope.**

burned in milliseconds. The surface of the neutron star is heated to 10^8 K (2×10^8 °F) and cools in 10 s by radiating x-rays.

X-ray background. In addition to discrete sources, the x-ray sky is remarkable in being quite bright: there is an isotropic x-ray background that is much brighter than the Milky Way at energies of 10–100 keV. By contrast, any optical background is at least 100 times fainter than the Milky Way. The source of this background is controversial: at least some of the background is due to x-rays from millions of faint quasars, but some flux may come from a very hot, thin intergalactic medium with temperatures greater than 4×10^8 K (7×10^8 °F).

Instrumentation. High-energy astrophysics is a field whose evolution has been driven by the very rapid improvement of instruments and telescopes. The x-ray telescopes used to detect Sco X-1 and other bright sources in the Milky Way were not really telescopes, but collimators. A collimator is like a cardboard mailing tube: it limits the field of view to a particular direction, but it does not concentrate light (**Fig. 2***a*). However, polished metal surfaces will reflect x-rays which arrive at grazing incidence, allowing the construction of true x-ray telescopes (Fig. 2*b*). The first satellite using such a detector for objects other than the Sun was the *Einstein Observatory*, flown by NASA, which lasted from 1978 to 1981. This telescope was 1000 times more sensitive than any previous x-ray telescope. *See* X-RAY TELESCOPE.

ROSAT (Roentgen Satellite), a project of the German Federal Republic, contains a large x-ray telescope. It was designed to spend 6 months creating an all-sky survey 100 times more sensitive than any previous survey (the *Einstein Observatory* did not do an all-sky survey), and then to observe selected areas with a sensitivity a few times better than *Einstein*.

AXAF (Advanced X-ray Astrophysics Facility) is the next x-ray observatory planned by NASA. It will have a telescope with several times greater collecting area than the telescope on the *Einstein Observatory*, and the angular resolution will be 10 times better. Thus *AXAF* will be many times more sensitive than *Einstein*, and designed for decades of use as the x-ray equivalent of the Hubble Space Telescope.

GRO (*Gamma Ray Observatory*) is a satellite designed to carry several experiments to measure gamma rays with energies from 1 MeV to 3 GeV. These rays are so penetrating that no satellite can carry enough lead to make a collimator, and grazing incidence telescopes will not work. Thus the gamma-ray sources seen in the Milky Way by the European *COS-B* satellite generally have not been identified with optical counterparts. *GRO* is to be much larger than any previous gamma-ray satellite, with a total weight of 15 tons (13,500 kg). The gamma-ray collecting power of *GRO* is to be 20 times greater than that of *COS-B*. The large number of gamma rays seen by *GRO* should lead to improved positions and more reliable identifications. One type of source that *GRO* is designed to observe is type 1 supernovae. Current models assert that radioactive decay provides the energy source for these objects; if so, the gamma-ray lines from the decays should be detected by a spectrometer on *GRO*. *See* SATELLITE ASTRONOMY.

Edward L. Wright

Bibliography. E. L. Chupp, *Gamma Ray Astronomy*, 1976; M. S. Longair, *High Energy Astrophysics*, 1981; P. W. Sanford, P. Laskarides, and J. Salton (eds.), *Galactic X-Ray Sources*, 1982.

Atomic clock

A device that uses an internal resonance frequency of atoms (or molecules) to measure the passage of time. The terms atomic clock and atomic frequency standard are often used interchangeably. A frequency standard generates pulses at regular intervals. A frequency standard can be made into a clock by the addition of an electronic counter, which records the number of pulses.

Basic principles. Most methods of timekeeping rely on counting some periodic event, such as the rotation of the Earth, the motion of a pendulum in a grandfather clock, or the vibrations of a quartz crystal in a watch. An atomic clock relies on counting periodic events determined by the difference of two different energy states of an atom. According to quantum mechanics, the internal energy of an atom can assume only certain discrete values. A transition between two energy states with energies E_1 and E_2 may be accompanied by the absorption or emission of a photon (particle of electromagnetic radiation). The frequency ν of this radiation is given by the equation below, where h is Planck's constant. A basic advan-

$$h\nu = |E_2 - E_1|$$

tage of atomic clocks is that the frequency-determining elements, atoms of a particular isotope, are the same everywhere. Thus, atomic clocks constructed and operated independently will measure the same time interval, that is, the length of time between two events. In order for the two clocks to agree on the time, they must be synchronized at some earlier time.

An atomic frequency standard can be either active or passive. An active standard uses as a reference the electromagnetic radiation emitted by atoms as they decay from a higher energy state to a lower energy state. An example is a self-oscillating maser. A passive standard attempts to match the frequency of an electronic oscillator or laser to the resonant frequency of the atoms by means of a feedback circuit. The cesium atomic beam and the rubidium gas cell are examples of passive standards. Either kind of standard requires some kind of frequency synthesis to produce an output near a convenient frequency, such as 5 MHz, that is proportional to the atomic resonance frequency.

Two different gages of the quality of a clock are accuracy and stability. The accuracy of a frequency standard is defined in terms of the deviation of its frequency from an ideal standard. In practice, it might be defined in terms of the frequency differences measured between independently constructed and operated standards of the same type. Improving the accuracy depends on understanding and controlling all the parameters that might cause the frequency to shift. The stability of a frequency standard is defined in terms of the constancy of its average frequency from one interval of time to the next. For many frequency standards, the stability initially improves with increasing measurement time but eventually gets worse. That is, a more precise measurement of the frequency can be made by averaging together successive measurements, until some imperfection in the apparatus causes the frequency to change. The stability increases with increased Q (resonance frequency divided by the width of the resonance) and with increased measurement signal-to-noise ratio.

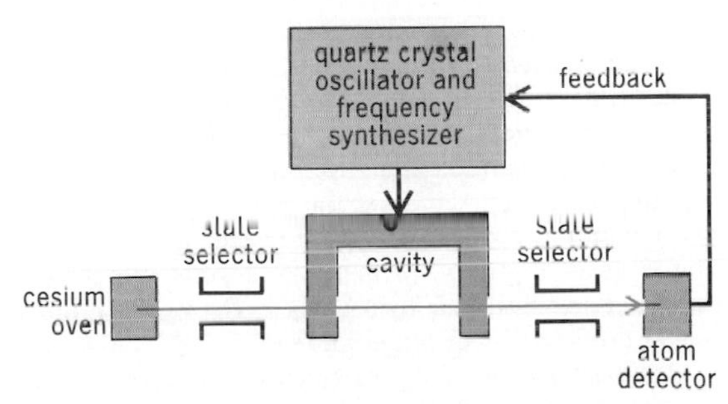

Fig. 1. Cesium atomic-beam clock. *(After G. Kamas, ed., Time and Frequency Users' Manual, NBS Tech. Note 695, 1977)*

Common types. The three most commonly used types of atomic clock are the cesium atomic beam, the hydrogen maser, and the rubidium gas cell. The cesium clock has high accuracy and good long-term stability. The hydrogen maser has the best stability for periods of up to a few hours. The rubidium cell is the least expensive and most compact and also has good short-term stability.

Cesium atomic-beam clock. This clock (**Fig. 1**) uses a 9193-MHz transition between two hyperfine energy states of the cesium-133 atom. Both the atomic nucleus and the outermost electron have magnetic moments; that is, they are like small magnets, with a north and a south pole. The two hyperfine energy states differ in the relative orientations of these magnetic moments. The cesium atoms travel in a collimated beam through an evacuated region. Atoms in the different hyperfine states are deflected into different trajectories by a nonuniform magnetic field. Atoms in one of the two states are made to pass through a microwave cavity, where they are exposed to radiation near their resonance frequency. The resonant radiation may cause the atom to make a transition from one state to the other; if that happens, the atom is deflected by a second, nonuniform magnetic field onto a detector.

The Q of the resonance is over 10^8 for some laboratory standards and somewhat less for the smaller standards that are commercially available. The cesium atomic beam is the most accurate of all atomic clocks. The best models have an error of only about 2 parts in 10^{14}, or about 1 s in 10^6 years. For this reason, cesium has become the basis of the international definition of the second: the duration of 9,192,631,770 periods of the radiation corresponding to the transition between the two hyperfine states of the ground state of the cesium-133 atom. The cesium clock is especially well suited for applications such as timekeeping, where absolute accuracy without recalibration is necessary. Measurements from many cesium clocks throughout the world are averaged together to define an international time scale that is uniform to parts in 10^{14}, or about 1 microsecond in a year. SEE ATOMIC TIME; DYNAMICAL TIME.

Hydrogen maser. This instrument (**Fig. 2**) is based on the hyperfine transition of atomic hydrogen, which has a frequency of 1420 MHz. Atoms in the higher hyperfine energy state are selected by a focusing magnetic field, so that they enter an evacuated storage bulb inside a microwave cavity. The atoms bounce off the Teflon-coated walls for about 1 s before they are induced to make a transition to the lower hyperfine state by a process called stimulated emission. The stimulated emission from many atoms creates a self-sustaining microwave oscillation.

The resonance Q is about 10^9. The best hydrogen masers have a stability of about 1 part in 10^{15} for averaging periods of 10^4 s. Over longer periods of time, the frequency drifts, primarily because of changes of the cavity tuning. Collisions with the walls cause the frequency to be shifted by about 1 part in 10^{11} relative to that of a free atom, but the magnitude of the shift varies from one device to another. This shift limits the accuracy of the hydrogen maser to about 1 part in 10^{12}.

The hydrogen maser can also be operated as a passive device, with improved long-term stability, due to the addition of automatic cavity tuning. The short-term stability is worse than that for an active maser.

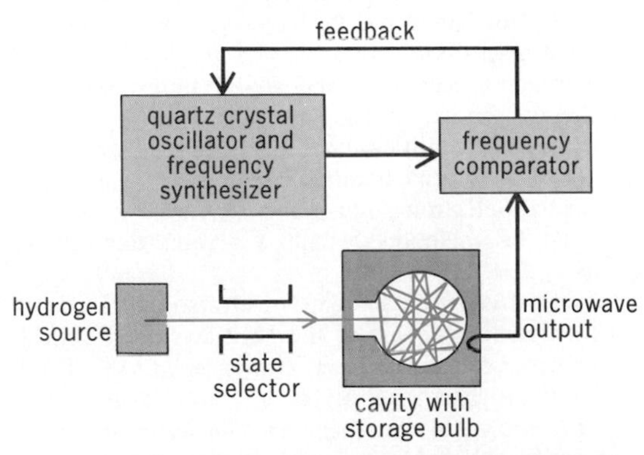

Fig. 2. Hydrogen maser. *(After G. Kamas, ed., Time and Frequency Users' Manual, NBS Tech. Note 695, 1977)*

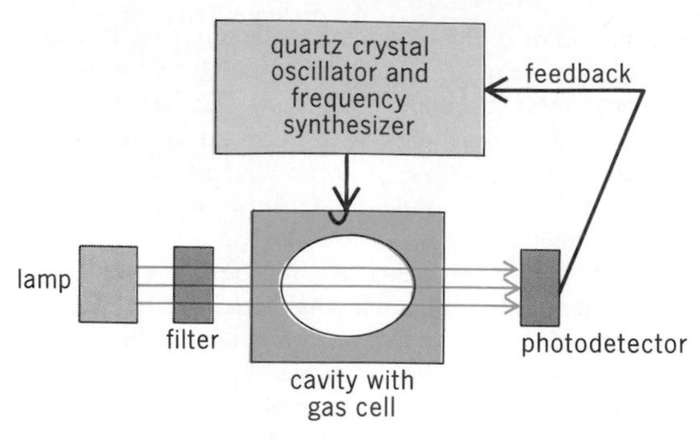

Fig. 3. Rubidium gas cell. *(After G. Kamas, ed., Time and Frequency Users' Manual, NBS Tech. Note 695, 1977)*

Rubidium gas cell. This device (**Fig. 3**) is based on the 6835-MHz hyperfine transition of rubidium-87. The rubidium atoms are contained in a glass cell together with a buffer gas, such as argon, that prevents them from migrating to the cell walls. A method called optical pumping is used to prepare the atoms in one hyperfine state. Filtered light from a rubidium resonance lamp is absorbed by atoms in one of the two hyperfine states, causing them to be excited to a higher state, from which they quickly decay to the other hyperfine state. If the atoms are then subjected to microwave radiation at the hyperfine transition frequency, they are induced to make transitions back to the other hyperfine state. They can then absorb light again from the lamp; this results in a detectable decrease in the light transmitted through the cell.

The Q is only about 10^7, but the short-term stability is quite good, reaching 1 part in 10^{13} for averaging times of 1 day. After longer periods, changes in the buffer gas pressure and the lamp cause the frequency to drift. The accuracy is not better than 1 part in 10^{10}. Rubidium standards are used in applications that do not require the accuracy of a cesium standard.

Experimental types. Many other kinds of atomic clocks, such as thallium atomic beams and ammonia and rubidium masers, have been demonstrated in the laboratory. The first atomic clock, constructed at the National Bureau of Standards in 1949, was based on a 24-GHz transition in the ammonia molecule.

Some laboratories have tried to improve the cesium atomic-beam clock by replacing the magnetic state selection with laser optical pumping and fluorescence detection. Improved performance is expected because of increased signal-to-noise ratio and a more uniform magnetic field. Other laboratories have studied atomic-beam standards based on magnesium, calcium, or methane, which have frequencies higher than that of cesium.

Hydrogen masers operated at low temperatures (a few kelvins above absolute zero) may be capable of much better stabilities than conventional hydrogen masers. The improvement is expected to come from the reduced electronic noise and the fact that oscillation can be maintained with a greater number of atoms.

Atomic frequency standards can also be based on optical transitions. One of the best-developed optical frequency standards is the 3.39-micrometer (88-THz) helium-neon laser, stabilized to a transition in the methane molecule. Frequency synthesis chains have been built to link the optical frequency to radio frequencies.

Ion traps, which confine ions in a vacuum by electric and magnetic fields (**Fig. 4**), have been studied for use in atomic clocks. They provide a benign environment for the ions while still allowing a long measurement time. Clocks based on optically pumped mercury-199 ions have been built and show good stability. Other trapped ion standards make use of laser cooling to reduce frequency errors due to Doppler shifts. Laser cooling is a method by which resonant light pressure is used to damp the motion of atoms. A Q of 3×10^{11} has been observed on a hyperfine transition of laser-cooled beryllium-9 ions. Even higher Q's may be observable on certain optical transitions. An optical transition has been observed in a single, trapped mercury ion with a Q of about 3×10^{11}, and it should be possible to increase this Q by a factor of 10^3. An optical frequency standard based on such an ion might be capable of an accuracy of 1 part in 10^{18}.

Applications. Atomic clocks are used in applications for which less expensive alternatives, such as quartz oscillators, do not provide adequate performance. The use of atomic clocks in maintaining a uniform international time scale has already been mentioned; other applications are described below. *SEE QUARTZ CLOCK.*

Navigation. The Global Positioning System is a satellite-based system that enables a user with a suitable radio receiver to determine position within about 10 m (33 ft). The satellites send out accurately timed radio pulses, from which the user's receiver can calculate its location and time. The satellites and the ground stations, but not the users, need atomic clocks (usually cesium clocks).

Communications. Various digital communications systems require precise synchronization of transmitters and receivers in a network. Some systems use time-division multiplexing, in which many channels of information are sent over the same line by sequentially allotting a small time slot to each channel. Timing is very critical when there are several sources of information with their own clocks. The primary timing is provided by cesium clocks.

Radio astronomy. Very long-baseline interferometry is a technique that allows two or more widely separated radio telescopes to achieve very high angular resolution by correlation of their signals. The system

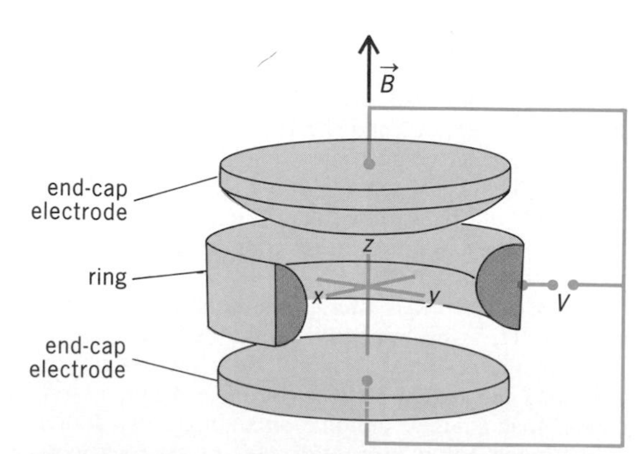

Fig. 4. Electrodes used to create the confining electric potential for a Penning ion trap or a Paul ion trap. An electric potential *V*, which is static for a Penning trap and oscillating for a Paul trap, is applied between the ring electrode and the end-cap electrodes. The Penning trap requires a uniform magnetic field B *(After F. J. Rogers and H. E. Dewitt, eds., Strongly Coupled Plasma Physics, Plenum, 1987)*

has the resolution that a single telescope would have if its aperture were equal to the distance between the telescopes. This can be thousands of miles. The accurate timing needed to correlate the signals is provided by hydrogen masers. *SEE RADIO ASTRONOMY; RADIO TELESCOPE.*

Space exploration. Navigation of space probes by Doppler tracking requires very stable local oscillators, derived from atomic frequency standards. Doppler tracking relies on determining the velocity of the spacecraft by measuring the frequency shift of a signal after it has been echoed to the Earth by a transponder on the spacecraft. Stable local oscillators are also needed for studies of planetary atmospheres and rings by fluctuations of the radio signals transmitted through them. *SEE SPACE NAVIGATION AND GUIDANCE.*

Fundamental science. According to A. Einstein's special and general theories of relativity, a moving clock runs slower than a stationary one, and a clock on the Earth's surface runs slower than one far from the Earth. These predictions were verified to high accuracy by an experiment in which a hydrogen maser was launched in a rocket to an altitude of 10,000 km (6000 mi). *SEE TIME.*

Wayne M. Itano

Bibliography. H. Hellwig, Atomic frequency standards: A survey, *Proc. IEEE*, 63(2):212–229, 1975; H. Hellwig, K. M. Evenson, and D. J. Wineland, Time, frequency and physical measurement, *Phys. Today*, 31(12):23–30, 1978; J. Jespersen and J. Fitz-Randolph, *From Sundials to Atomic Clocks*, 1982; N. F. Ramsey, History of atomic clocks, *J. Res. Nat. Bur. Stand.*, 88(5):301–320, 1983; N. F. Ramsey, Precise measurement of time, *Amer. Sci.*, 76(1):42–49, 1988; D. J. Wineland, Trapped ions, laser cooling, and better clocks, *Science*, 226:395–400, 1984.

Atomic time

Time based on quantum transitions. The practical scale of atomic time is obtained from the running of atomic clocks since mid-1955, corrected for height above sea level but for no other relativistic effect. *SEE ATOMIC CLOCK.*

An atom which drops from an energy level, E_2, to a lower one, E_1, emits radiation of frequency $f = (E_2 - E_1)/h$, where h is Planck's constant. For a selected transition of cesium-133, $f = 9{,}192{,}631{,}770$ cycles per second of ephemeris time was determined in 1958. The second in the International System of Units (SI) was defined in 1967 as this number of periods of the radiation.

International Atomic Time (French abbreviation, TAI) is formed by the Bureau International de l'Heure (BIH), in Paris, from the operation of about 150 commercial atomic clocks and about 7 laboratory-built, primary standards, located about the world. The clocks are related chiefly by Loran C, clock transport, and satellites. The primary standards, accurate to about 5 parts in 10^{14}, provide the long-time stability of TAI, which reproduces the SI second as defined to about 1 part in 10^{13}.

TAI is a highly precise atomic time scale used, for example, in determining variations in the Earth's speed of rotation, computing orbits, and tracking celestial objects, including spacecraft. Several navigational systems not dependent upon any time scale, such as Loran C, nevertheless use atomic clocks tied to TAI for maximum accuracy.

Civil clock time. Universal Time One (UT1), based on the Earth's rotation, is now losing about 1 s per year relative to TAI, because of tidal friction and irregular variations in the Earth's rotational speed. Coordinated Universal Time (UTC) was formed to provide civil time related to atomic time. Legal time based on a standard time zone differs from UTC by an integral number of hours.

The Bureau International de l'Heure keeps UTC an exact number of seconds from TAI and within 0.9 s of UT1 by adding a leap second to UTC as needed, usually to the last minute of December or June. A leap second is announced at least 8 weeks in advance. About 1 leap second a year is inserted. *SEE EARTH ROTATION AND ORBITAL MOTION.*

Relativistic effects. Let C_F be a fictitious clock, fixed in a distant inertial system and at constant gravitational potential. Then by relativity theory a clock C_E fixed on the Earth will have periodic variations relative to C_F, totalling P_R, due to the Earth's motions relative to the Sun, Moon, and planets, which cause variations in speed and potential. The largest effect is due to the Earth's eccentric orbit about the Sun. As would be judged by C_F, the frequency of C_E would be lower at perihelion (around January 1) than at aphelion by 6.6 parts in 10^{10}, and C_E would be less advanced by 3.3 ms on April 1 than on October 1.

TAI is not corrected for P_R. A simpler time system results for general use by correcting only for height above sea level of C_E. Frequency increases by 1.1 parts in 10^{16} per meter (3.4 parts in 10^{17} per foot) above sea level. Corrections for P_R must be made for some observations, however—for example, the timing of pulsars.

The dynamical scale, Barycentric Dynamical Time (TDB), does include P_R. An auxiliary atomic scale, Terrestrial Dynamical Time (TDT), is used, however, as the time entry in apparent geocentric ephemerides to simplify the comparison of theory and observation.

Cosmology. A question of cosmological importance is whether dynamical and atomic times differ (apart from P_R). According to some theories, the gravitational constant G decreases with time and, depending upon the theory, dynamical time either accelerates or decelerates secularly relative to atomic time. If confirmed, then the times are intrinsically different. *SEE DYNAMICAL TIME; GRAVITATION; TIME.*

William Markowitz

Bibliography. Bureau International de l'Heure, *Annual Reports;* G. M. Clemence and V. Szebehely, Annual variation of an atomic clock, *Astron. J.*, 72:1234–1236, 1967; W. Markowitz et al., Frequency of cesium in terms of ephemeris time, *Phys. Rev. Lett.*, 1:105–106, 1958; T. C. Van Flandern, Is the gravitational constant changing?, *Astrophys. J.*, 248:813–816, 1981.

Azimuth

An angle measured in the plane of the observer's horizon from one of several arbitrary departure points; also the angle at the observer's zenith between the local meridian and the vertical circle passing through the object being observed. *SEE ASTRONOMICAL COORDINATE SYSTEMS; CELESTIAL NAVIGATION.*

In astronomical and surveying practice azimuth has been referenced to each of the four cardinal points of the horizon; north, east, south, and west. In navigation practice, azimuth is measured from the north point of the horizon eastward from 0 to 360°. The horizon angle measured from 0° at the north or south horizon point eastward or westward through 90 or 180° is connoted azimuth angle in contrast to azimuth.

Raynor L.Duncombe

Bibliography. C. H. Cotter, *The Elements of Navigation and Nautical Astronomy*, 1981; A. Frost, *The Principles and Practice of Navigation*, 1981.

Big bang theory

The theory that the universe began in a state of extremely high density and has been expanding since some particular instant that marked the origin of the universe. A widely accepted version of big bang theory incorporates developments in elementary particle theory, and is known as the inflationary universe theory. The predictions of the inflationary universe and standard big bang theories are the same after the first 10^{-35} s.

Three observations are at the base of observational big bang cosmology. First, the universe is expanding uniformly, with objects at greater distances receding at a greater velocity. Second, the Earth is bathed in an isotropic glow of radiation that has the characteristics expected from the remnant of a hot primeval fireball. Third, the abundances of the light elements and isotopes (chiefly, deuterium, helium, and lithium) are explained as a result of the hot, dense state between 1 and 1000 s after the big bang.

Cosmological theory in general and the big bang theory in particular are currently based on the theory of gravitation advanced by Albert Einstein in 1916 and known as the general theory of relativity. Though the predictions of the general theory have little effect in the limited sphere of the Earth, they dominate on as large a scale as the universe.

Expansion of the universe. Only in the 1920s did it become clear that the "spiral nebulae," clouds of gas with arms spiraling outward from a core, were galaxies on the scale of the Milky Way Galaxy. This fact was established in 1925, when Edwin Hubble reported that his observations of variable stars in several galaxies enabled the distance to these galaxies to be determined with some accuracy.

Starting in 1912, V. M. Slipher observed the spectra of several of these spiral nebulae, and discovered that they almost all had large redshifts. According to the Doppler effect, these large redshifts correspond to large velocities of recession from the Earth. Hubble estimated distances to the objects measured by Slipher, and in 1929 showed that there appeared to be a direct correlation between the distance to a galaxy and its velocity of recession. Hubble and M. L. Humason extended this work and by 1931 had clearly established the redshift–distance relation. SEE DOPPLER EFFECT; REDSHIFT.

The relation is known as Hubble's law: $v = H_0 d$, where v is the velocity of recession, d is the distance to the galaxy, and H_0 is Hubble's constant. Determining Hubble's constant requires the independent measurement of the distances to galaxies; the redshift can easily be measured on spectra.

The data from which Hubble's law is derived may be plotted on the Hubble diagram (**Fig. 1**), in which the horizontal axis is the apparent magnitude, corrected for various instrumental and observational effects, and the vertical axis is the redshift $z = \Delta\lambda/\lambda$,

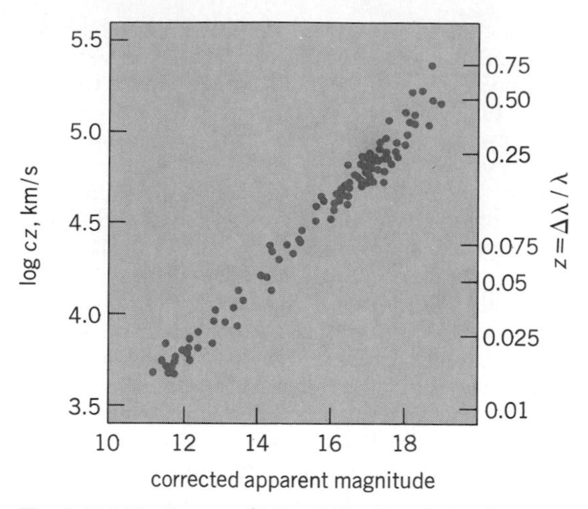

Fig. 1. Hubble diagram. (*After J. Kristian, A. Sandage, and J. A. Westphal, The extension of the Hubble diagram, II. New redshifts and photometry of very distant galaxy clusters: First indication of a deviation of the Hubble diagram from a straight line, Astrophys. J., 221:383–394, 1978*)

where $\Delta\lambda$ is the shift in radiation of wavelength λ. The quantity cz (where $c \cong 3 \times 10^5$ km · s^{-1} or 1.86×10^5 mi/s is the speed of light), also plotted on the vertical axis, is approximately equal to the recession velocity for velocities much smaller than c, and the distance to a galaxy of given luminosity is directly related to its apparent magnitude. Thus the Hubble law is expressed by the fact that the data points in Fig. 1 lie near a single straight line.

If velocity is expressed in kilometers per second, and distance is expressed in megaparsecs (where 1 parsec is the distance from which the radius of the Earth's orbit would subtend 1 second of arc, and is equivalent to 3.26 light-years or 3.09×10^{13} km or 1.92×10^{13} mi), then Hubble's constant H_0 is given in km · s^{-1} · Mpc^{-1}. The measurements of A. Sandage and G. Tammann with the 200-in. (5-m) Hale telescope on Palomar Mountain in California have given the standard value now in wide use for Hubble's constant, 50 km · s^{-1} · Mpc^{-1}. Other astronomers such as G. de Vaucouleurs have derived somewhat higher values of Hubble's constant, in the neighborhood of 75–100 km · s^{-1} · Mpc^{-1}. Several new investigations are giving values closer to 75. It is widely hoped that observations with the Hubble Space Telescope will resolve this major controversy. SEE GALAXY, EXTERNAL; HUBBLE CONSTANT.

Tracing the expansion of the universe back in time shows that the universe would have been compressed to infinite density approximately 13–20 $\times 10^9$ years ago (for Hubble's constant = 50). In the big bang theory, the universe began at that time as a big bang began the expansion. The big bang was the origin of space and time.

In 1917 Einstein found a solution to his own set of equations from his general theory of relativity that predicted the nature of the universe. His universe, though, was unstable: it could only be expanding or contracting. This seemed unsatisfactory at the time, for the expansion had not yet been discovered, so Einstein arbitrarily introduced a special term—the cosmological constant—into his equations to make the universe static. The need for the cosmological constant disappeared with Hubble's discovery of the expansion.

W. de Sitter worked out his own solution to Einstein's equations (with the cosmological constant) later in 1917. De Sitter's solutions, though, were valid only for a universe that did not contain any matter. This is not an impossible approximation to the current universe, however, because the cosmic density of matter is very low. Nonetheless, de Sitter's own work in 1930 found the density of the universe too high for his solution to be valid.

In 1922 A. Friedmann worked out solutions that are at the basis of the cosmological models that are now generally accepted. G. Lemaître in 1927 independently found similar solutions. Lemaître's solutions indicated that the original "cosmic egg" from which the universe was expanding was hot and dense. This was the origin of the current view that the universe was indeed very hot in its early stages. Friedmann's solution did not involve the cosmological constant and Lemaître's did, but otherwise the solutions were similar.

Early universe. Modern theoretical work has been able to trace the universe back to the first instants in time. In the big bang theory and in related theories that also propose a hot, dense early universe, the universe may have been filled in the earliest instants with exotic elementary particles, some of which are now being studied by physicists with large accelerators but others of which are yet to be discovered. Individual quarks may also have been present. By 1 microsecond after the universe's origin, many of the exotic particles and the quarks had been incorporated in other fundamental particles. Most cosmologists now think that more than 90% of the matter in the universe is in nonbaryonic form and has thus been undetectable; a leading theory has this additional material in a form known as cold dark matter. *See* ELEMENTARY PARTICLE; QUARKS.

Work in the early 1980s incorporated the effect of elementary particles in cosmological models. The research seems to indicate that the universe underwent a period of extremely rapid expansion in which it inflated by a factor of billions in a very short time. This inflationary universe model provides an explanation for why the universe is so homogeneous: Before the expansion, regions that now seem too separated to have been in contact were close enough to interact. After the inflationary stage, the universe is in a hot stage and is still dense; the models match the big bang models thereafter.

In the inflationary universe models, the universe need not have arisen from a single big bang; matter could have appeared as fluctuations in the vacuum.

It is not definitely known why there is an apparent excess of matter over antimatter, though attempts in elementary particle physics to unify the electromagnetic, the weak, and the strong forces show promise in explaining the origin of the matter-antimatter asymmetry. The asymmetry seems to have arisen before the first millisecond.

By 5 s after the origin of the universe, the temperature had cooled to 10^9 K (2×10^9°F), and only electrons, positrons, neutrinos, antineutrinos, and photons were important types of ordinary matter. A few protons and neutrons were mixed in, and they grew relatively more important as the temperature continued to drop. The universe was so dense that photons traveled only a short way before being reabsorbed. By the time 4 min had gone by, elements had started to form.

After about a million years, when the universe cooled to 3000 K (5000°F) and the density dropped sufficiently, the protons and electrons suddenly combined to make hydrogen atoms, a process called recombination. Since hydrogen's spectrum absorbs preferentially at the wavelengths of sets of spectral lines rather than continuously across the spectrum, and since there were no longer free electrons to interact with photons, the universe became transparent at that instant. The average path traveled by a photon—its mean free path—became very large. The blackbody spectrum of the gas at the time of recombination was thus released and has been traveling through space ever since. As the universe expands, this spectrum retains its blackbody shape though its characteristic temperature drops.

Background radiation. Between 1963 and 1965 A. A. Penzias and R. W. Wilson, working with a well-calibrated horn antenna, discovered an isotropic source of radio noise whose strength was independent of time of day and of season, and whose intensity at the observing wavelength of 7 cm (2.8 in.) was equivalent to that which would be emitted by a blackbody—an idealized radiating substance—at a temperature of about 3 K (−454°F). P. J. E. Peebles in R. H. Dicke's group had carried out calculations that indicated that the remnant radiation from the big bang might be detectable, and this led to the announcement of both the discovery of the radiation and its interpretation in terms of fossil radiation from the big bang. A prior theoretical prediction, made in the 1940s by G. Gamow, R. Alpher, and G. Hermann, was then recalled. Later, other apparatus observed the radiation at other frequencies in the microwave region, but final confirmation of the blackbody nature of the radiation came only in 1975, when balloon observations succeeded in measuring the spectrum of the background radiation in a frequency range that included the peak of the blackbody curve and extended into the infrared. The best current observations of the background radiation are from NASA's *Cosmic Background Explorer (COBE)* spacecraft. The spectrum is in excellent agreement with a blackbody curve at 2.735 ± 0.060 K, 2.735°C (4.92°F) above absolute zero. Once a smoothly varying dipole anisotropy caused by the 600-km/s (370-mi/s) motion of the Milky Way Galaxy is removed, ripples on the order of 15–30 μK are detected. Though most of the apparent ripples seen in all-sky maps are noise, calculations show that some are real signals, showing inhomogeneities in the early universe. The fluctuations seem to be the same on all observed scales, matching predictions of the inflationary-universe version of big bang theory. The background radiation continues to be strong evidence for the big bang theory that the universe started with a hot, dense phase. *See* COSMIC BACKGROUND RADIATION.

Nucleosynthesis. As the early universe cooled, the temperatures became sufficiently low for element formation to begin. By about 100 s, deuterium (one proton plus one neutron) formed. When joined by another neutron to form tritium, the amalgam soon decayed to form an isotope of helium. Ordinary helium, with still another neutron, also resulted.

Some of the first work on nucleosynthesis was carried out in 1948 by Alpher and Gamow. The current formulation of nucleosynthesis in the early universe is that of R. V. Wagoner, W. A. Fowler, and F. Hoyle, updated by D. N. Schramm, K. A. Olive, T. P. Walker, G. Steigman, and H.-S. Kang. The relative abundances of isotopes of the light elements in the first few minutes, according to this model, are shown

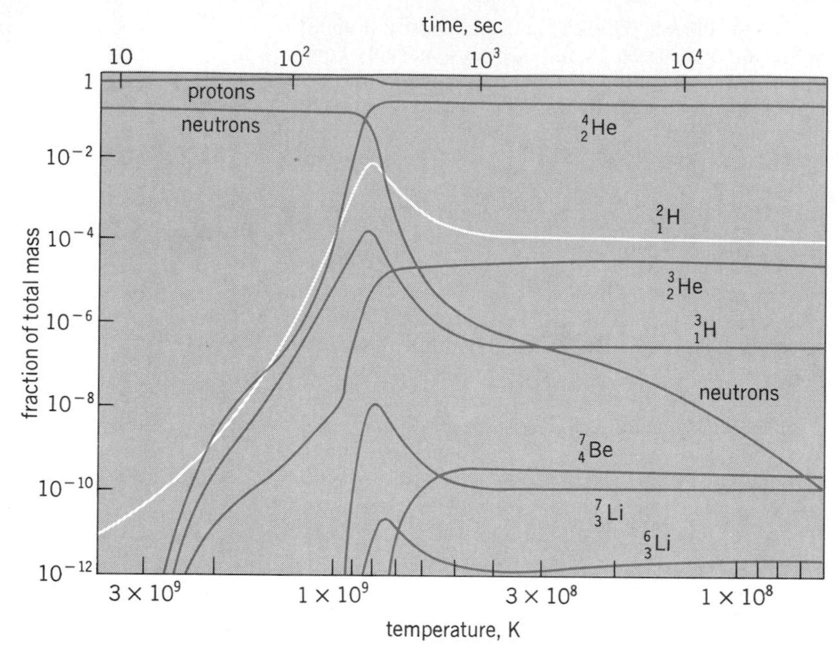

Fig. 2. Relative abundance of isotopes of the light elements in the first few minutes after the big bang according to the model of R. V. Wagoner, for a certain assumed density at some early time. (*After J. M. Pasachoff, Astronomy: From the Earth to the Universe, 4th ed., Saunders College Publishing, 1993*)

in **Fig. 2**. Time is shown on the top axis, and the corresponding temperature is shown on the bottom axis. The calculations shown are for a certain assumed density at some early time; if a different density were present, then the curves would be different.

The calculations show that within minutes the temperature drops to 10^9 K (2 × 10^9°F), too low for most nuclear reactions to continue. Most models give a resulting abundance of about 25% of the mass in the form of helium, regardless of the density of the universe. This abundance is hard to determine observationally. Current results, about 25% of the mass, are in rough agreement with the theoretical value.

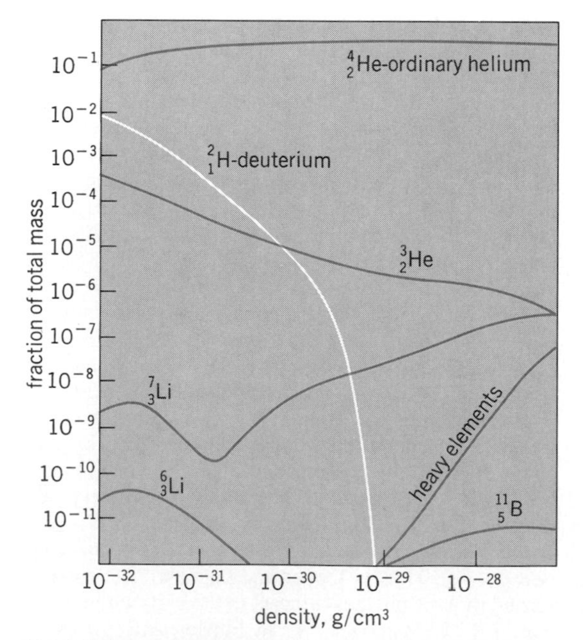

Fig. 3. Relative abundances of isotopes and elements as a function of present-day density of matter, according to the model of R. V. Wagoner. (*After J. M. Pasachoff, Astronomy: From the Earth to the Universe, 4th ed., Saunders College Publishing, 1993*)

The abundances of others of the light elements are more sensitive to parameters of matter in the early universe. **Figure 3** shows abundances as a function of the present-day density of matter. From knowledge of the approximate rate of expansion of the universe, the density of matter in the early universe can be deduced from the current density, and abundances can then be calculated. In particular, the deuterium abundance is especially sensitive to the cosmic density at the time of deuterium formation, because the rate at which deuterium is "cooked" into tritium increases rapidly with increasing density.

Big bang nucleosynthesis, although at first thought to be a method of forming all the elements, foundered for the heavy elements at mass numbers 5 and 8. Isotopes of these mass numbers are too unstable to form heavier elements quickly enough. The gap is bridged only in stars, through processes worked out by E. M. Burbidge, G. Burbidge, Fowler, and Hoyle in 1957. Thus the lightest elements were formed as a direct result of the big bang (Fig. 3), while the heavier elements as well as additional quantities of most of the lighter elements were formed later in stars or supernovae. *See Nucleosynthesis*.

Open versus closed universe. The two basic possibilities for the future of the universe are that the universe will continue to expand forever, or that it will cease its expansion and begin to contract. It can be shown that the case where the universe will expand forever corresponds to an infinite universe. The term applied is the open universe. The case where the universe will begin to contract corresponds to a finite universe. The term applied is the closed universe.

Deceleration parameter. One basic way to test whether the universe is open or closed is to determine the rate at which the expansion of the universe is slowing, that is, the rate at which it is deviating from Hubble's law. The slowing is measured with a term called q_0, the deceleration parameter. The dividing line is marked by $q_0 = ½$, with smaller values corresponding to an open universe and larger values corresponding to a closed universe.

The most obvious way to measure q_0 is by looking at the most distant galaxies or clusters of galaxies, measuring their distances independently from Hubble's law, and plotting Hubble's law in a search for deviations. Deviations have been reported, for example, in the 1978 work by J. Kristian, Sandage, and J. A. Westphal. But all such observations are subject to the considerable uncertainty introduced by the fact that observations deep into space also see far back into time. The galaxies then were surely different than they are now, so evolutionary effects must be taken into account. It is not even known for certain whether galaxies were brighter or dimmer in the distant past, and thus the method may tell more about the evolution of galaxies than about cosmology.

A. Yahil, Sandage, and Tammann analyzed the deceleration parameter through studies of nearby galaxies. They were investigating the differential motions that have been generated since the beginning of the universe by the gravitational accelerations due to nearby fluctuations in the cosmic density. In particular, they studied motions in a complex of galaxies surrounding the Virgo Cluster out to a radius of 1000 km · s^{-1} (600 mi/s), about 20 Mpc using a value of 50 for Hubble's constant. In 1979 they reported $q_0 = 0.06$, corresponding to an open universe.

From studies of elliptical and spiral galaxies, a group of scientists has found a bulk streaming motion toward a region it calls the Great Attractor. The region would be a concentration in Centaurus, 20 times more populous than the Virgo Cluster. Its distortion of the Hubble flow may have led to misevaluation of the Hubble constant.

Cosmic deuterium abundance. Because deuterium is a sensitive indicator of the cosmic density at the time of its formation soon after the big bang, and because deuterium can only be destroyed and not formed in stars, the study of the cosmic deuterium abundance is one of the best methods for determining the future of the universe. Basically, it involves assessing whether there is enough gravity in the universe to halt the expansion.

Counts can be made of all the stars, galaxies, quasars, interstellar matter, and so forth, and the sum gives a density much too low to close the universe. But this method does not assess the amount of invisible matter, which could be in the form of intergalactic gas, black holes, and so forth. Indeed, studies of the motions of galaxies in clusters of galaxies often indicate that much more mass is present inside galaxy clusters than is visible. The amount of this missing mass may be 50 times the amount of visible mass.

Assessing the density of the universe through studies of the deuterium abundance is independent of whether the matter is visible or invisible. Although deuterium is present even on Earth as a trace isotope, with an abundance 1/6600 that of normal hydrogen in seawater, it is difficult to detect in interstellar space. Finally, in 1972, deuterium was first detected in the interstellar medium through radio astronomical investigations. A number of determinations have followed, including studies in the ultraviolet made with telescopes in space. A Hubble Space Telescope result published in 1993 by J. Linsky and colleagues for deuterium between the Earth and the star Capella is 1.5×10^{-5}. Determinations of deuterium in the atmospheres of the planets have also been carried out. Though there are some discrepancies remaining between abundances determined in different locations and in different fashions, the deuterium observations seem to indicate clearly that the density of the universe in the form of baryons is very low and hence that the universe is open. If, however, the prediction of the inflationary universe is accepted that the universe is on the boundary between open or closed, then the rest of the matter must be nonbaryonic. The question of whether this matter is in the form of neutrinos, which are rapidly moving and thus termed hot, or slowly moving particles of unknown type known as cold dark matter, is a current field of research.

X-ray and neutrino background. A diffuse background of x-rays had been observed over the decade of the 1970s. The first NASA High-Energy Astronomy Observatory, *HEAO 1*, measured its spectrum. These observations led to the suspicion that a lot of hot material, previously undiscovered, may have been present between the galaxies. *See* X-RAY ASTRONOMY.

But the second High-Energy Astronomy Observatory, known as the *Einstein Observatory*, changed the picture in 1979. The observatory had the capability of forming x-ray images, and discovered that at least some of the x-ray background came from faint quasars, which appeared on long exposures of even fields that had been thought to be blank. Continued analysis of the data from the *Einsten Observatory* and the subsequent *ROSAT* is showing that the quasars can account for much of the x-ray background. If so, then the x-ray background would not be revealing enough matter in the form of hot intergalactic gas to close the universe.

Evidence suggesting that neutrinos may have a small rest mass would change this picture. Since there are about 100 neutrinos in each cubic centimeter (1500 in each cubic inch) of the universe, much mass would be in that form if neutrinos had even a small rest mass. But the experimental suggestion in the early 1980s that a neutrino rest mass had been discovered seems not to have been confirmed.

Inflationary scenarios. Although several lines of evidence, including studies of differential velocities of nearby galaxies caused by density perturbations and of the cosmic abundance of deuterium, indicate that the universe is open, it was not clear why the universe was so close to the dividing line between being open or closed that the subject was so much in doubt. The inflationary universe model, which is based on ideas of Alan Guth, Andrei Linde, Andreas Albrecht, and Paul Steinhardt, provides a natural explanation for the universe being on this dividing line. After expansion slows down at the close of the inflationary stage (thus causing a phase change, much like water boiling into steam), the universe necessarily approaches this line. Thus whether the universe was open or closed was perhaps the wrong question to ask. Further work on inflationary scenarios is necessary to see whether certain problems, such as the inflationary model's predictions of the density fluctuations that lead to the coalescence of galaxies, can be accounted for. *See* COSMOLOGY; INFLATIONARY UNIVERSE COSMOLOGY; UNIVERSE.

Jay M. Pasachoff

Bibliography. J. D. Barrow and M. S. Turner, The inflationary universe: Birth, death and transfiguration, *Nature*, 298:801–805, 1983; R. A. Carrigan, Jr., and W. P. Trower (eds.), *Particle Physics in the Cosmos*, 1988; A. H. Guth and P. J. Steinhardt, The inflationary universe, *Sci. Amer.*, 250(5):116–128, 1984; S. W. Hawking, *A Brief History of Time*, 1987; J. C. Mather et al., A preliminary measurement of the cosmic microwave background spectrum by the *Cosmic Background Explorer* (*COBE*) satellite, *Astrophys. J. Lett.*, 354:L37–L40, 1990; N. Mandolesi and N. Vittorio (eds.), *The Cosmic Microwave Background: 25 Years Later*, 1989; J. M. Pasachoff, *Astronomy: From the Earth to the Universe*, 4th ed., 1993; J. Silk, *The Big Bang*, 2d ed., 1989; G. Smoot et al., Structure in the COBE DMR first year maps, *Astrophys. J. Lett.*, 396:L1, September 1, 1992; S. Weinberg, *The First Three Minutes*, 1977.

Binary star

Two stars held together by their mutual gravitational attraction in a permanent (or at least long-term) association. The two components of this binary system revolve in close elliptical (or circular) orbits around their common center of gravity, which carries them through space so that they also have a common proper motion.

Binary stars play an important role in astrophysics for three reasons:

1. They are a very common phenomenon. It is more likely for a given star to be a member of a binary or multiple system than to be single. Out of the 10 nearest stars visible to the naked eye, 6 are double or multiple objects and only 4 are single. Among the

4 is the Sun, which is not, strictly speaking, single because of its planetary system.

2. Binary systems reveal much more information about stellar properties than do single stars. In particular, masses of stars can be measured only by their gravitational attraction on nearby bodies. The radii, luminosities, and effective temperatures of stars in binary systems can also be obtained, and the mutual relations of these quantities can then be studied. *SEE HERTZSPRUNG-RUSSELL DIAGRAM; MASS-LUMINOSITY RELATION.*

3. In binary stars where the components interact strongly, new phenomena and new types of stars occur which otherwise would not exist: novae, cataclysmic variables, and binary x-ray sources are examples.

Dimensions, orbits, and periods. The dimension of a binary system is determined by the separation of the component stars, that is, the distance between their centers. If the separation is denoted by A, then the period of the orbital motion, P, is given by the relation $P^2 = A^3/(M_1 + M_2)$. In this formula, component masses M_1 and M_2 are expressed in units of the mass of the Sun (2×10^{30} kg or 4.4×10^{30} lbm), and A is in astronomical units (1 AU $= 0.930 \times 10^8$ mi $= 1.496 \times 10^8$ km); then the period is in years. According to circumstances, it is possible to measure and observe the absolute motion of either component with respect to the center of gravity, or the relative motion of one component with respect to the other. For a circular orbit, the separation A is simply the radius of the relative orbit. Many orbits are ellipses; then the separation is understood to be the semimajor axis of the relative ellipse. For absolute orbits, referred to the common center of gravity, the period remains the same, the individual orbits are similar ellipses with the same eccentricities, and the semimajor axes are inversely proportional to masses.

Separations vary over an enormous range from one binary system to another. The upper limit is dictated by the general gravitational field of the neighboring stars, which in random encounters would disrupt a binary system in which the gravitational bond is too weak because of a large separation. This upper limit depends critically on the density of stars in space. In the broader vicinity of the Sun, it has been estimated to be between 5×10^3 and 4×10^4 AU. The lower limit on separations is given by the physical dimensions of the component stars. Two stars like the Sun will be in contact (at one point on the surface) when their separation is 1.1×10^6 mi (1.8×10^6 km). Small compact stars like white degenerate dwarfs or neutron stars could be even closer together.

For a pair of Sun-like stars, the period corresponding to the upper limit of separations is $10^5 - 10^7$ years. Two Suns in contact would revolve about each other in one-third of a day. Very close pairs of dwarf stars in contact are very common (the W UMa stars). Ultrashort-period binaries are also known; one component is usually a degenerate dwarf. The recurrent nova WZ Sge has a period of 81.6 min. Pairs of white dwarfs, if they exist, could have periods shorter than 1 min. *SEE CELESTIAL MECHANICS.*

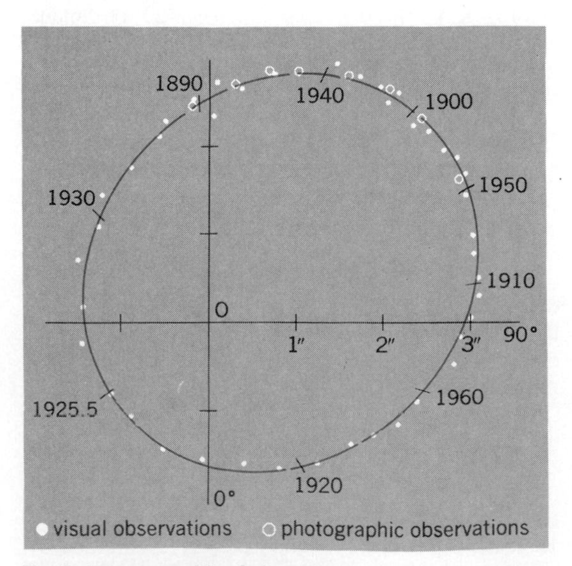

Fig. 1. Apparent orbit of visual binary Krüger 60.

Main categories. Systems of large separations are often observable directly through a telescope as two separate stars. They are then called visual binary stars. For a closer pair, the orbital velocities may be sufficiently large to be detectable by a periodic shift of spectral lines caused by the Doppler effect. These objects are called spectroscopic binaries. When the two stars are so close to each other that their dimensions are not negligible compared to their separation, they can eclipse each other, provided that the orbital plane is suitably oriented. These objects are then easily detected photometrically from the periodic changes of the light received from the system, and are called eclipsing binaries or photometric binary stars. *SEE DOPPLER EFFECT.*

If the separation of the components is so large as to make the period longer than about 10^3 years, the orbital motion will be practically imperceptible, but the binary system is nevertheless often recognizable as a common proper-motion pair. The nearest star to the Sun, Proxima Centauri, was discovered as late as 1915 just because of its common proper motion with the much brighter star Alpha Centauri (itself a visual binary). Another famous common proper-motion pair is Mizar and Alcor in the ''handle'' of the Big Dipper, often used as an informal test of good sight.

Visual binaries. About 75,000 visual binary stars have been discovered and cataloged, but orbits are known for only about 700 of them. Each orbit determination requires many years of painstaking measurements. The relative position of the fainter component (the secondary) with respect to the primary is usually measured with a filar micrometer attached to a long-focus refractor. (This is one of the very few areas in astronomy where the human eye is still directly used in measurements.) As **Fig. 1** shows, the orbit is plotted point by point by determining two polar coordinates, the angular separation and the position angle, as functions of time. This apparent orbit is the projection, onto the plane of the sky, of the actual orbital ellipse, which must be reconstructed mathematically.

A telescope with an aperture of D centimeters or D' inches can, under very good conditions, resolve a double star with an angular separation $a = 12''/D = 5''/D'$, but the atmosphere sets a limit for even the largest telescopes at about 0.″1. A reliable orbit requires separations of at least 0.″5. The relation between the angular separation a'' and the actual linear separation A in astronomical units is $a'' = A/d$, where d is the distance to the system in parsecs. Thus a binary at a distance of 20 parsecs must have a true separation at least 10 AU, or a period of 32 years for Sunlike stars. Larger separations, better for accurate measurements, quickly lead to long periods. Visual binaries can be studied only among the nearby stars.

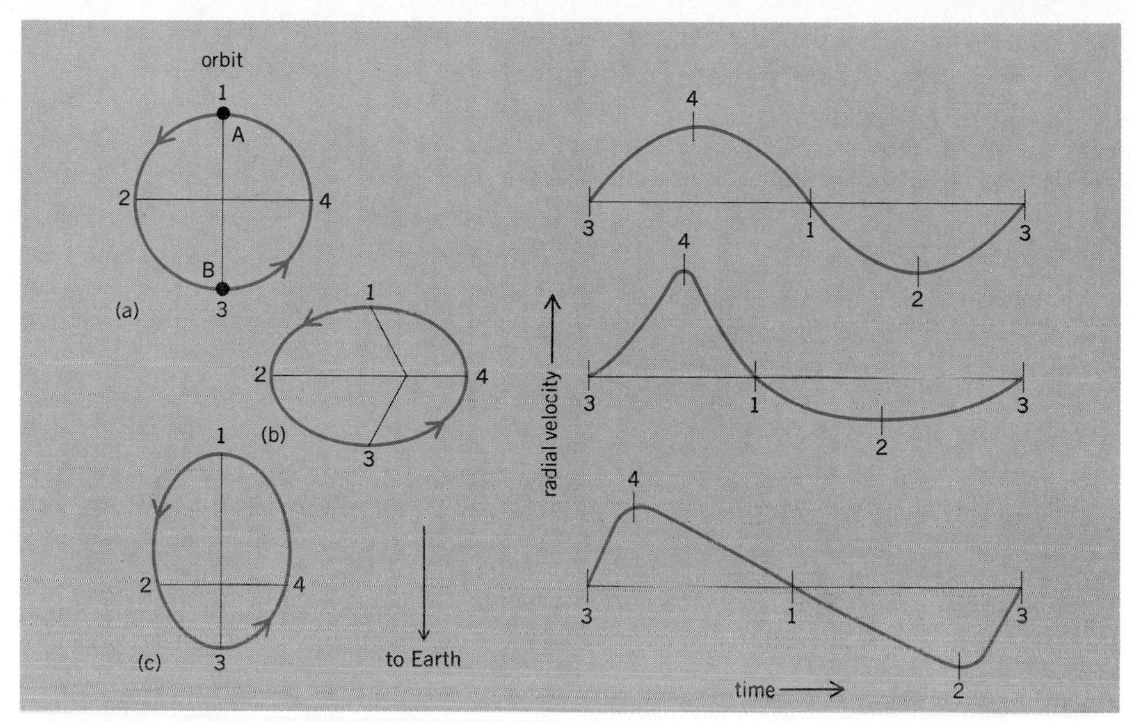

Fig. 2. Diagrams showing possible radial velocity curves (graphs of radial velocity versus time) of spectroscopic binaries. (*a*) Circular orbit. (*b*) Elliptical orbit with eccentricity $e = 0.5$ and major axis perpendicular to direction to Earth. (*c*) Elliptical orbit with $e = 0.5$ and major axis along direction to Earth.

These in turn are almost exclusively dwarf stars, smaller, less massive, and cooler than the Sun.

Within these limitations, visual binary stars furnish valuable fundamental data on stars. If the parallax of the primary star is reliably determined, the distance to the system is known, hence also the true linear separation of the components A. Knowing also the period P, the sum of the masses of the components, (M_1+M_2), is obtained from Kepler's third law formulated above. Individual masses can be obtained if the motion of each component is measured with respect to the center of mass of the system (that is, absolute orbits are determined). About 30 visual binary pairs have yielded reliable masses in this way. Their number is growing very slowly, since typical well-determined orbits correspond to periods between 20 and 350 years.

Spectroscopic binaries. A binary component moving about the center of gravity of the system in a circular orbit with radius A_1 with a period P has an orbital speed $V_1 = 2\pi A_1/P$. By measuring the periodic shifts of spectral lines caused by the Doppler effect, the radial component of this velocity (along the line of sight) can be determined.

The manner in which the radial velocity changes during the orbital cycle is shown in **Fig. 2**, which is drawn for the case when the observer is in the orbital plane of the system (here the plane of the paper). Figure 2*a* shows two identical stars A and B moving in circular orbits about their common center of gravity. When either star is in position 2, it is approaching the Earth with its full orbital speed. Since the Doppler shift in this case diminishes the wavelengths of the spectral lines, the radial velocity is taken as negative; thus position 2 corresponds to the bottom of the radial velocity curve (the graph of radial velocity versus time) plotted on the right. Half a period later, the same star recedes from the Earth at maximum speed: position 4 corresponds to the maximum point (4) on the radial velocity curve. When the stars are at points 1 and 3 (conjunctions), they move perpendicularly to the line of sight, and no radial velocity is observed for the components, except that the center of mass has its own radial velocity, which both components share. The radial velocity of only one component is plotted in Fig 2. The other curve is its mirror image. At conjunctions, both stars have the same radial velocity, and their spectral lines coincide (**Fig. 3***a*). At the quadratures, points 2 and 4, they are most widely split and accessible to observations (Fig. 3*b*).

If the orbits are ellipses, the radial velocity curves become skew-symmetrical. Also, the shape of the radial velocity curve depends on the orientation of the orbital ellipse with respect to the observer (Fig. 2*b* and *c*). In each case, the (unplotted) radial velocity curve of the other component is a mirror image of the curve shown. If the mass of the other component is smaller by the ratio M_B/M_A, the amplitude of its curve will be larger at each point by a factor M_A/M_B. Also, in general, the orbit will not be seen exactly edge-on as plotted here. Rather, the orbital plane will be tilted to the plane of the sky (which is perpendicular to the

Fig. 3. Spectra of the close binary β Aurigae (period = 3.960 days) at: (*a*) conjunctions, with lines coincident; (*b*) quadratures, with the lines of the two components well separated.

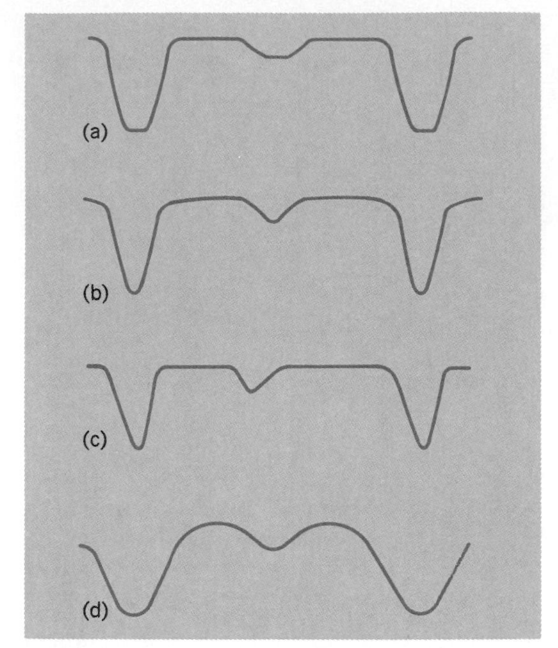

Fig. 4. Light variations from eclipsing binary systems. (*a*) Alternations of total and annular eclipses. (*b*) Partial eclipses; reflection effect present. (*c*) Eccentric orbit; partial eclipses. (*d*) Tidal deformation, showing convex light curve.

line of sight) at an angle i (the inclination). Inclination $i = 90°$ means that the system is being observed edge-on, as in Fig. 2. Generally for inclinations $i < 90°$, the radial velocities are reduced by a factor $\sin i$. If a system is seen pole-on ($i = 0°$), no radial velocity variations can be observed. Such systems can be recognized only if the two components differ greatly in spectral types but not in brightness; they are then called spectrum binaries.

About 2000 spectroscopic binaries are known, but they are mostly single-spectrum systems, that is, the secondary star is too faint to be seen. More information is obtained from double-spectrum binaries, since the ratio of masses then follows from the ratio of the amplitudes of the radial velocity curves. However, masses can be obtained unambiguously only if the inclination can be determined, and this is possible when the binary system is also eclipsing. *See Astronomical spectroscopy.*

Eclipsing binaries. At conjunctions (points 1 and 3 in Fig. 2), the two stars eclipse each other, since in the case represented by Fig. 2 they are exactly aligned on the line of sight. Binary systems with inclinations 90° are of course rare, but because of the finite size of the component stars, eclipses may occur at inclinations less than 90°, provided the separation between the components is sufficiently small.

The effect of the eclipses on the total light received from the binary system is seen from **Fig. 4**. Figure 4*a* shows the light changes in a typical system displaying alternate total and annular eclipses. The system consists of a smaller but hotter star and a larger but cooler secondary. Maximum light is obtained outside eclipses. Since it is assumed here that the components are spherical, there is no variation in the out-of-eclipse light (as indicated by the flat maxima of the light curve). When the secondary begins to eclipse the brighter primary, light drops rapidly during the phases of the partial eclipse, until the primary component is totally eclipsed. Then only the light of the cooler, fainter star is seen (corresponding to the flat bottom of the light curve). Half a period later, the secondary star is eclipsed, but the secondary minimum of light is less deep: first, the secondary contributes less light, and second, the primary star is smaller and causes only an annular eclipse of the primary. After another half period, the primary eclipse occurs again.

Figure 4*b* shows the light curve of a similar system seen at an angle significantly different from 90°. The eclipses are only partial, and their depths diminish as one proceeds to inclinations more and more different from 90°. Figure 4*c* shows again the same system, but with an eccentric orbit. Since the orbital velocity in an elliptical orbit is variable, the intervals between the primary and secondary eclipses are unequal, depending on the orientation of the orbital elipse, as can be figured out by means of Fig. 2. Figure 4*b* and *d* indicates what kind of other complications may be encountered. When the two stars are close together, their mutual illumination invokes an additional light variation between the eclipses known as the reflection effect (Fig. 4*b*). If the proximity becomes such that the tides which the stars raise on each other appreciably deform their shapes, these tidally distorted ellipsoids will expose varying cross sections to the observer and produce an ellipticity effect (Fig. 4*d*). In systems of low inclination, all the observed light variation may be due to proximity effects only; such systems are known as ellipsoidal variables.

More than 1000 eclipsing binaries are fairly well known, although only about 200 have well-determined orbits. They yield the value of inclination needed to complete the determination of the spectroscopic elements and obtain masses; moreover, they yield radii from the relative durations of the eclipses. Most eclipsing binaries have periods of a few days, so that the accumulation of observational material proceeds much faster than with the visual binaries. Thus, eclipsing binaries are the most valuable source of data on stars. Such data represent normal single stars only if the presence of the other component does not alter significantly the structure and evolution of the star. Systems in which the component stars do affect each other's evolution are called close binaries. *See Eclipsing variable stars.*

Evolution of close binaries. Stars spend most of their lifetime on the main sequence, generating energy by conversion of hydrogen into helium in their cores. When the hydrogen in the core is exhausted, the stars expand. Single stars encounter no problem in this expansion, but the gravitational field of a nearby companion sets a definite upper limit to the volume available for expansion to a binary star component; it also considerably distorts the shape of the

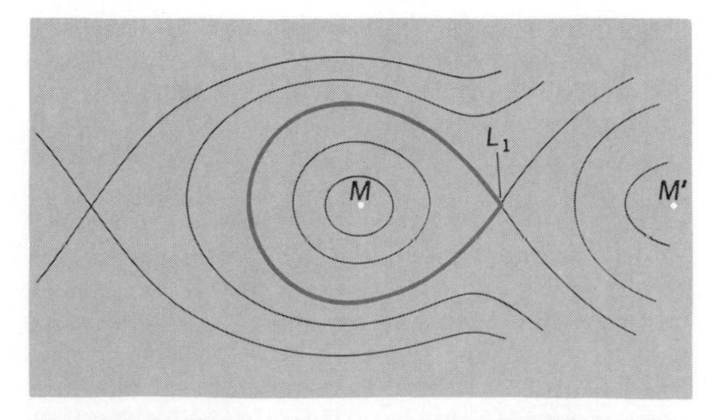

Fig. 5. Nature of equipotential surfaces. The Roche limit is marked by a colored line.

expanded star. When the expanding component reaches the limiting surface, the evolution of both stars is completely changed. More massive stars evolve faster; the more massive component thus always reaches the limit first. *See* STELLAR EVOLUTION.

The sequence of shapes through which an expanding component passes is shown in **Fig. 5**. The curves drawn are cross sections of three-dimensional equipotential surfaces, for a system, with stars of masses M and M', viewed edge-on. In the so-called Roche model, three forces are assumed to affect the structure of the expanding star's outer layers: its own gravitational attraction, the gravitational attraction of the other star, and centrifugal force due to the orbital evolution of the system. When a star is still small with respect to the dimensions of the system, it is little affected by the external force field, and is very nearly spherical. As it expands, its shape becomes that of a distorted ellipsoid more and more extended toward the companion, until the star reaches the critical surface marked with a heavy line in Fig. 5. The next equipotential surface already surrounds both stars. A star that has reached its critical surface will lose mass in the vicinity of the point L_1 (the first Lagrangian point) where the effective gravity is zero, and a gas stream will flow from there toward the other star. The stream either impacts directly on the companion or leads to the formation of an accretion disk surrounding it. In either case, the star gains (accretes) mass either directly from the stream or through the disk, in which viscosity makes the particles fall on the star. An as yet unknown part of the mass lost by the "loser" may, however, escape from the system. *See* ROCHE LIMIT.

Classification of close binaries. According to Z. Kopal (1955), three main categories are distinguished: (1) In detached systems, both components are smaller than their respective critical lobes. Most unevolved binaries, with components on the main sequence, fall in this category; they are suitable for determining characteristics well representing ordinary single stars. (2) In semidetached systems, one component fills its critical Roche lobe. Many eclipsing binaries, often called Algols after their prototype, are of this type. (3) In contact systems, both components fill their critical lobes, or even overflow them and possess a common envelope. The W UMa systems, dwarf stars with periods less than 1 day, are of this type.

Evolution with mass transfer. When the more massive star reaches its Roche lobe, it will begin to lose mass very rapidly, and in an astronomically short time it becomes the less massive component. Then the mass loss slows down considerably (so that about 10^{-8} solar mass per year is transferred), but the loser still fills its critical lobe. Mass loss stops when the loser either ignites helium (for more massive stars) or loses almost all its hydrogen-rich envelope. In both cases, it shrinks rapidly and becomes either a helium dwarf or a degenerate white dwarf. The rapid mass-transfer process (with rates of 10^{-5} solar mass per year or higher) is too short to be often observed.

Ultraviolet spectrograms, obtained with the *International Ultraviolet Explorer* (*IUE*) satellite, show that the accreting star is embedded in a hot plasma and that a strong stellar wind may be induced, which possibly carries a large fraction of the circumstellar material away from the system. Stellar wind (a rapid radial outflow of gas from the atmosphere of a star), as an alternative to the Roche lobe overflow, may play an important role in many interacting binary systems. Best-known examples are the supergiant stars Zeta Aurigae and VV Cephei, and symbiotic stars. *See* SYMBIOTIC STAR; ULTRAVIOLET ASTRONOMY.

X-ray binaries. Since 1970 x-ray satellites have discovered a number of strong sources of x-rays clearly associated with binary systems; the x-ray source undergoes eclipses or the optical counterpart is recognized as a single-spectrum spectroscopic binary, a binary, or both. Some sources emit x-rays in regular pulses (Cen X-3 with a period of 4.8 s, Her X-1 at 1.2 s), and the periodic modulation of this average pulsed period over the orbital period (which is 2.1 days for Cen X-3, 1.7 days for Her X-1), interpreted as Doppler shift, permits the orbital elements to be determined. Her X-1 was identified with a peculiar variable star, HZ Her, whose light variations are caused by one hemisphere being heated by x-rays. The source of the x-rays itself must be a compact star, since only such stars can pulse with so short a period; in most cases it is a neutron star, but a black hole is also a possible candidate.

The optically visible star is unstable and loses mass, either by stellar wind or by Roche lobe overflow. The stream flows to the compact star and falls into a very deep "potential well," since the compact star is extremely small. Usually the accretion occurs through a viscous disk, but eventually always a large fraction of the potential energy of the particles is released as radiation (either abruptly in a series of shocks or more slowly in the viscous heating of the accretion disk). The hot gas reaches temperatures of order $50–500 \times 10^{6}$°C ($90–900 \times 10^{6}$°F) and radiates mostly in x-rays. Some of the x-ray binary sources are among the most powerful energy emitters in the Galaxy, radiating up to 10^{38} ergs/s or 10^{31} W (the Sun's total energy output is 4×10^{33} ergs/s or 4×10^{26} W). The regular rapid pulses of x-rays are believed to be generated by a rapid rotation of the magnetized neutron star. The neutron stars found in binary x-ray sources are thought to be remants of supernova explosions, but the remnant can also be a black hole. A black hole can generate x-rays equally well as a neutron star. The decision often hinges on the mass of the invisible, x-ray-producing compact object in an x-ray binary. *See* ASTROPHYSICS, HIGH-ENERGY; BLACK HOLE; NEUTRON STAR; PULSAR; X-RAY STAR.

Cataclysmic variables. Degenerate dwarfs, the end products of evolution of stars of moderate and small masses, when paired with a mass-losing star in a close binary system, produce a cataclysmic variable. These dwarfs include the novae (conspicuous by a very high increase in light in a short, one-event outburst), recurrent novae (which repeat the outbursts, usually on a diminished scale, over time intervals of decades), and dwarf novae (which display even smaller outbursts on time scales of months). All these objects appear to be binaries of a basically similar nature: a cooler object, usually a red dwarf (but occasionally, as in the recurrent nova T CrB, a giant), is unstable and transfers mass to a hotter and much more compact object, a degenerate white dwarf or a similar body. A viscous disk surrounding the accreting star plays an important role. Often it is so dense that a hot spot forms where the stream impacts. A hot spot or a hot region can naturally also form at the surface of the star proper. Variable rates of mass transfer, variable viscosity, and continual interaction between the star, the disk, and the stream then lead to instabilities and eruptive events. In the more violent cases, thermonuclear reactions at the surface of

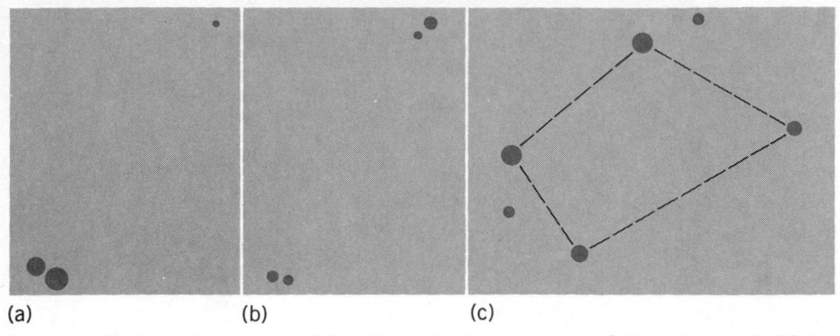

Fig. 6. Multiple stellar systems. (*a*) A hierarchical triple system (Alpha Centauri). (*b*) A hierarchical quadruple system (Epsilon Lyrae). (*c*) A system of the Trapezium type (Theta Orionis).

the accreting star are believed to be the source of the energy for a nova outburst. *See Nova; Variable star.*

Origin. Just what fraction of all stellar objects are binary stars is very difficult to determine because of the large number of selection effects in their discoveries. Estimates run between 30 and 60%. In any case, it is clear that the formation of binary stars must be a very common process related to stellar formation itself. Capture of one star by another requires the presence of a third body to carry away the excess of energy, and is therefore an inefficient process. Fission of a rapidly rotating star into two is hampered by the large density concentration in stars; perhaps it could occur at the very early stages of stellar evolution, and would then probably account for some binaries with rather unequal masses. According to L. Lucy, the bulk of binary systems are probably formed by successive fragmentations of protostars.

Triple and multiple systems. If fragmentation of the protostar is the main process producing binary systems, systems with more than two components should be anticipated. Indeed, it is estimated that about one-third of all binary stars are actually triple stars, and probably again one-third of all triple systems will eventually be found to be actually quadruple. A triple system consists typically of a close binary pair accompanied by a much more distant third component (**Fig. 6*a***). A good example is the pair Alpha Centauri accompanied by Proxima Centauri. Quadruple systems usually consist of two close pairs separated by a large distance (for example, Epsilon Lyrae; Fig. 6*b*). The famous Trapezium (Theta Orionis; Fig. 6*c*) at the center of the Orion Nebula is an unstable system, but it (and similar multiple stars) consists of very young, very hot stars which are short-lived by themselves.

Astrometric binaries. Very careful measurements of the proper motions of the nearest stars disclose that some of them show deviations from a rectilinear motion. Some of these deviations, although very minute, appear periodic, and may mean that these stars are attended by companions of low mass. Systems discovered by these astrometric measurements are called astrometric binaries. The new observing technique of speckle interferometry suggests that cold bodies may indeed frequently accompany many of the nearby stars. *See Astrometry; Speckle; Star.*

Mirek J. Plavec

Bibliography. A. H. Batten, *Binary and Multiple Systems of Stars*, 1973; P. P. Eggleton and J. E. Pringle, *Interacting Binaries*, 1985; W. Heintz, *Double Stars*, 1978; M. J. Plavec, D. M. Popper, and R. K. Ulrich (eds.), *Close Binary Stars: Observation and Interpretation*, 1980; J. Sahade and F. B. Wood, *Interacting Binary Stars*, 1978.

Black hole

One of the end points of gravitational collapse, in which the collapsing matter fades from view, leaving only a center of gravitational attraction behind. General relativity predicts that, if a star of more than about 3 solar masses has completely burned its nuclear fuel, it should collapse to a configuration known as a black hole. The resulting object is independent of the properties of the matter that produced it and can be completely described by stating its mass, spin, and charge. The most striking feature of this object is the existence of a surface, called the horizon, which completely encloses the collapsed matter. The horizon is an ideal one-way membrane: that is, particles and light can go inward through the surface, but none can go outward. As a result, the object is dark, that is, black, and hides from view a finite region of space (a hole). Arguments concerning the existence of black holes originally centered on the fact that there are many stars of over 3 solar masses and that there seemed to be no other outcome of collapse than the formation of a black hole. In 1971, however, some direct observational evidence was obtained for a black hole in the binary x-ray system Cygnus X-1. Since that time, black holes have been identified by similar evidence in two other x-ray binaries, LMC X-3 and A0620-00. In addition, supermassive black holes may be responsible for the large energy output of quasars and other active galactic nuclei, and there is growing evidence that black holes exist also at the center of many other galaxies, including the Milky Way Galaxy. *See Gravitational collapse.*

Theory. Shortly after Albert Einstein formulated the general theory of relativity in 1916, the solution of the field equations corresponding to a nonrotating black hole was found. For many years this solution, called the Schwarzschild solution, was used to describe the gravitational attraction outside a spherical star. However, the interpretation of the Schwarzschild solution as a solution for a black hole was not made at the time. More than 20 years elapsed before it was shown that such a black hole could, and probably would, be formed in the gravitational collapse of a nonrotating star of sufficient mass. It was not until 1963 that the solution for a spinning black hole, the Kerr solution, was found. This was particularly important, since most stars are rotating, and the rotation rate is expected to increase when such stars collapse. Although some collapsing, rotating stars might avoid becoming black holes by ejecting matter, thereby reducing their mass, many stars will evolve to a stage of catastrophic collapse in which the formation of a black hole is the only conceivable outcome. However, unlike the case of nonrotating black holes, no one has shown that a collapsing, rotating star of sufficient mass must form a Kerr black hole. On the other hand, it has been shown that if the collapse of a star proceeds past a certain point, the star must evolve to a singularity, that is, an infinitely dense state of matter beyond which no further evolution is possible. Such singularities are found inside black holes in all known black hole solutions, but it has only been conjectured that the singularity produced in a collapse must be inside a black hole. *See Stellar evolution.*

Black hole solutions have also been found for the case in which the black holes have a charge, that is, an electrical as well as a gravitational influence. However, since matter on the large scale is electrically neutral, black holes with any significant charge are not expected in astronomy. Similarly, black hole solutions allow black holes to possess magnetic charge, that is, a magnetic single-pole interaction. Although some elementary-particle theories predict that there should exist particles with magnetic charge, called magnetic monopoles, sufficient experimental evidence is not yet available to confirm their existence. Even if monopoles did exist, they would play little part in the formation of black holes, and so astronomical black holes are expected to be both electrically and magnetically neutral.

Uniqueness theorems about black holes make it likely that at least some Kerr black holes would be formed. Uniqueness theorems address the question of how many kinds of black holes could exist and how complicated their structure could be. These theorems show that black holes must have a simple structure. In fact, the mass, spin, charge, and magnetic charge are all that are needed to specify completely a black hole. Further, any distortion of a black hole, such as is caused by a chunk of matter falling inside, is removed by a burst of radiation. Therefore, although the collapse of a rotating star would be quite complicated, it appears that the final system, the Kerr black hole, would be relatively simple and independent of the details of collapse.

The possible formation of black holes depends critically on what other end points of stellar evolution are possible. There can always be chunks of cold matter which are stable, but their mass must be considerably less than that of the Sun. For masses on the order of a solar mass, only two stable configurations are known for cold, evolved matter. The first, the white dwarf, is supported against gravitational collapse by the same quantum forces that keep atoms from collapsing. However, these forces cannot support a star which has a mass in excess of about 1.2 solar masses. (A limiting value of 1.4 solar masses was first found by S. Chandrasekhar and is known as the Chandrasekhar limit. More realistic models of white dwarfs, taking into account nuclear reactions, lower this number somewhat, but the actual value depends on the composition of the white dwarf.) The second stable configuration, the neutron star, is supported against gravitational collapse by the same forces that keep the nucleus of an atom from collapsing. There is also a maximum mass for a neutron star, estimated to be between 1 and 3 solar masses, the uncertainty being due to the poor knowledge of nuclear forces at high densities. Both white dwarfs and neutron stars have been observed, the former for many years and the latter more recently in the studies of pulsars and binary x-ray sources. *See Neutron star; Pulsar; White dwarf star.*

It would appear from the theory that if a collapsing star of over 3 solar masses does not eject matter, it has no choice but to become a black hole. There are, of course, many stars with mass larger than 3 solar masses, and it is expected that a significant number of them will reach the collapse stage without having ejected sufficient matter to take them below the 3-solar-mass limit. Further, more massive stars evolve more rapidly, enhancing the rate of formation of black holes. It seems reasonable to conclude that a considerable number of black holes should exist in the universe. One major problem is that, since the black hole is dark, it is itself essentially unobservable. Fortunately, some black holes may be observable in the sense that the black hole maintains its gravitational influence on other matter, and thus it can make its presence known. Otherwise, the detection of a black hole would be limited to observations of the collapse of a star to a black hole, a rare occurrence in the Milky Way Galaxy and one which happens very quickly.

Structure. For a nonrotating black hole, the radius of the horizon (Schwarzschild radius) is determined entirely by the mass. Defining R so that the surface area of the spherical horizon is $4\pi R^2$, the equation relating R to the mass M is $R = 2GM/c^2$, where G is the constant of gravity and c is the speed of light. Classical general relativity would allow M to take on all possible values, but quantum effects suggest that the lowest possible value of M is about 10^{-8} kg (2×10^{-8} lbm). However, the lower-mass black holes may not be astronomically relevant, since collapsing stars with masses less than about a solar mass (2×10^{30} kg or 4×10^{30} lbm) would become white dwarfs or neutron stars. It is thought that low-mass black holes could exist only if they were created at the time of the origin of the universe.

An astronomical black hole of 3 solar masses would have a radius of about 20 km (12 mi). This size is comparable to that expected for neutron stars. The density to which matter would have to be compressed in order to form such a black hole is comparable to that found in neutron stars or in the nuclei of atoms. Black holes with a mass larger than 1000 solar masses may be formed from the collapse of supermassive stars. The existence of such supermassive stars has not been proven, but supermassive stars and supermassive black holes have been proposed to be at the cores of quasars and active galactic nuclei. A supermassive black hole with a mass of a few thousand solar masses would have a radius comparable to the radius of the Earth. Black holes of 1,000,000 to 100,000,000 solar masses are likely to exist in the center of some galaxies, including the Milky Way Galaxy. These could be formed either in the collapse of a supermassive star with a possible subsequent accretion of matter, or in the coalescing of a large number of black holes of more modest mass. The density required to form these very massive black holes is low, approaching that of ordinary terrestrial densities. *See Galaxy, external; Milky Way Galaxy; Quasar; Supermassive stars.*

For nonrotating black holes, the horizon is also a surface of infinite redshift; that is, light emitted from just above the surface reaches a distant observer with a much lower frequency and energy than it had when it was emitted. As a result, an object which falls into a black hole appears to an observer to take an infinite time to reach the Schwarzschild radius, with the observed light coming from the object redshifting quickly to darkness as the approach is made. The picture would be quite different for a person riding on the falling object. The "rider" would reach the Schwarzschild radius in a finite time, feeling nothing more unusual than gravitational tidal forces. However, once inside, the person would be trapped since even light which moves outward cannot escape. The

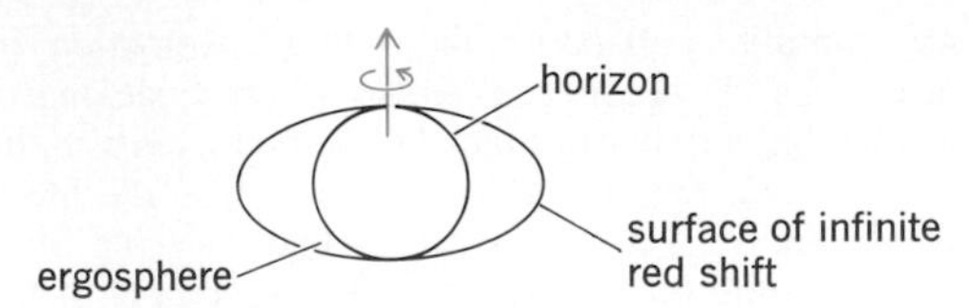

Fig. 1. The Kerr, or rotating, black hole. (*After P. C. Peters, Black holes: New horizons in gravitational theory, Amer. Sci., 62(5):575–583, 1974*)

theory predicts that this individual, as well as everything else within the horizon, would be crushed to infinite density within a short time.

For rotating black holes, the surface of infinite redshift lies outside the horizon except at the poles, as illustrated in **Fig. 1**. The region between the two surfaces is called the ergosphere. This region is important because it contains particle trajectories which have negative energy relative to an outside observer. Roger Penrose showed that it is possible, though perhaps unlikely astronomically, to use these trajectories to recover even more than the rest mass energy of matter sent into a rotating black hole, the extra energy coming from the slowing down of the rotation of the black hole. Others have proposed that radiation incident on a rotating black hole could be similarly amplified.

The black hole solutions of general relativity, ignoring quantum-mechanical effects as described below, are completely stable. Once massive black holes form, they will remain forever; and subsequent processes, for example, the accumulation of matter, only increase their size. Two black holes could coalesce to form a single, larger black hole, but a single black hole could not split up into two smaller ones. This irreversibility in time led researchers to consider analogies between black holes and thermal properties of ordinary matter, in which there is a similar irreversibility as matter becomes more disordered as time goes on. Finally Steven Hawking showed that when quantum effects are properly taken into account, a black hole should emit thermal radiation, composed of all particles and quanta of radiation which exist. This established the black hole as a thermal system, having a temperature inversely proportional to its mass. Since a radiating system loses energy and therefore loses mass, a black hole can shrink and decay if it is radiating faster than it is accumulating matter. For black holes formed from the collapse of stars, the temperature is about 10^{-7} K (2×10^{-7} °F above absolute zero, -459.67°F). Regardless of where such black holes are located, the ambient radiation incident on the black hole from other stars, and from the big bang itself, is much larger than the thermal radiation emitted by the black hole, implying that the black hole would not shrink. Even if the ambient radiation is shielded from the black hole, the time for the black hole to decay is much longer than the age of the universe, so that, in practice, black holes formed from collapse of a star are essentially as stable as they were thought to be before the Hawking radiation was predicted.

Theoretically, black holes of any mass could have been created at the beginning of the universe in the big bang. For smaller-mass black holes, the Hawking radiation process would be quite important, since the temperatures would be very high. For example, a black hole created with mass of 10^{12} kg (2×10^{12} lbm), about the mass of a mountain, would have just radiated away all of its mass, assuming that no mass had been accreted in the meantime. Black holes created with a mass smaller than 10^{12} kg would have disappeared earlier, and those with a larger mass would still exist. The final stage of evaporation would be quite violent and would take place quickly. As a black hole radiates, it loses mass and its temperature rises. But a higher temperature means that it radiates and loses mass at a faster rate, raising its temperature even further. The final burst, as the black hole gives up the remainder of its mass, would be a spectacular event. The final emission would contain all radiation and particles which could exist, even those not generated by existing accelerators. At present, there is no evidence which points to the existence of black holes with small mass or to their evaporation by the Hawking radiation process.

Observation. Because black holes themselves are unobservable, their existence must be inferred from their effect on other matter. Thus a theoretical model must be constructed to explain the observational data and to show that the most reasonable explanation is that a black hole is responsible for the observed effects. Such is the case with the binary x-ray star system Cygnus X-1. There are a number of binary x-ray systems known. The model which best explains the data is one in which a fairly normal star is in mutual orbit about a very compact object. Because these two are so close, mass flows from the star onto an accreting disk about the compact object, as shown in **Fig. 2**. As the mass in the disk spirals inward, it heats up by frictional forces. Because the central body is so compact, the matter heats to a temperature at which thermal x-rays are produced. The only compact objects known that could accomplish this are neutron stars and black holes. The existence of very short-time bursts of radiation also points to an object of small diameter, that is, compact. In some of these binary x-ray systems, there is also a regular pulsed component to the x-rays, indicating a rotating neutron star (by reasoning similar to that given for pulsars). In these systems, the compact object could not be a black hole because that would imply a more complicated structure than a black hole would allow. In other systems, however, there are only irregular pulsations or fluctuations; they are candidates for possible black holes.

The crucial evidence comes from the mass determination of the compact object. Because the inclination of the orbit is not known, a range of masses is found; however, there will be a typical mass obtained by assuming that the orbit is not in an extreme orientation. For three x-ray binaries, Cygnus X-1, LMC X-3, and A0620-00, the typical mass of the compact

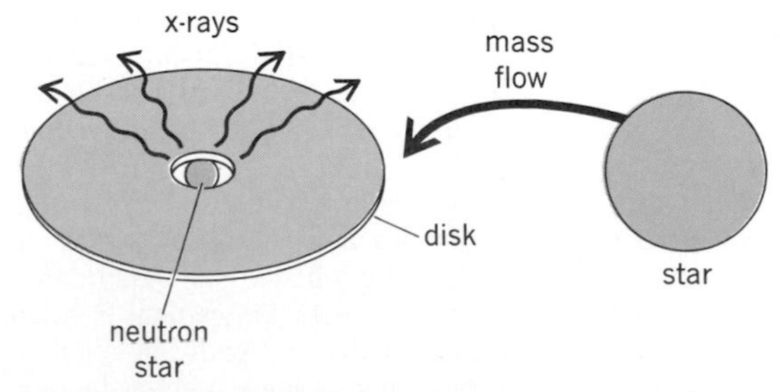

Fig. 2. Schematic diagram of a binary x-ray source. (*After P. C. Peters, Black holes: New horizons in gravitational theory, Amer. Sci., 62(5):575–583, 1974*)

Andromeda Nebula, M31 or NGC 224, the spiral galaxy which is nearest the Milky Way system. Also called the Great Nebula, it rotates about its center with a period of approximately 2×10^8 years and is believed to be some 2.5×10^6 light-years distant. It contains bright blue stars, bright supergiant red stars, Cepheid variable stars, star clusters, planetary nebulae, novae, gas, and dust. Photograph was made with 48-in. (1.2-m) Schmidt telescope on Palomar Mountain. (*California Institute of Technology/Palomar Observatory*)

body is about 10 solar masses, much larger than the maximum mass of a neutron star. In fact, the compact objects in the first and third binary systems are more massive than the maximum mass of a neutron star, no matter what orientation the orbit is assumed to have. Assuming that general relativity is the correct theory of gravitation (and this assumption is now supported very well experimentally), there can be no compact objects of such a mass other than a black hole. In this sense it can now be said that black holes exist.

While the evidence is less direct and more model-dependent, there is growing acceptance of the idea that supermassive, rotating black holes exist at the cores of nuclei of active galaxies, including quasars and radio galaxies. Here, the black hole is assumed to interact with accreting matter in such a way as to provide a source of energy to power these ultraluminous objects. The arguments used to explain the acceleration of particles and the production of radiation are quite involved and technical. However, the black hole model seems to be the most plausible explanation, but that does not establish the existence of black holes in as direct a way as for the x-ray binaries.

Black holes are thought to exist in the nuclei of other galaxies as well, their presence not giving rise to amounts of radiation as spectacular as for active galactic nuclei only because of differing conditions near the black hole. In the Milky Way Galaxy, there is strong evidence for the existence of a massive black hole of the order of 1–3 × 10^6 solar masses at the galactic center. The mass distribution of matter in the core of the Galaxy, inferred from velocities of stars and gas in the vicinity, suggests a pointlike source of gravitation. There is also a very compact, nonthermal radio source at that location, the radiation supposedly arising from accretion of matter onto a black hole. This source exhibits small variations on short time scales, further limiting its size. However, despite this evidence, the existence of such a massive black hole is by no means certain. The binary x-ray sources still provide the best observational evidence for the existence of black holes. *See* Astrophysics, high-energy; Binary star; X-ray astronomy; X-ray star.

Philip C. Peters

Bibliography. J. D. Bekenstein, Black-hole thermodynamics, *Phys. Today*, 33(1):24–31, 1980; S. Detweiler, Resource letter BH-1: Black holes, *Amer. J. Phys.*, 49(5):394–400, 1981; S. W. Hawking, The quantum mechanics of black holes, *Sci. Amer.*, 236(1):34–40, 1977; J. V. Narlikar, *From Black Clouds to Black Holes*, 1985; R. Price and K. S. Thorne, The membrane paradigm for black holes, *Sci. Amer.*, 258(4):69–77, April 1988; S. L. Shapiro and S. A. Teukolsky, *Black Holes, White Dwarfs, and Neutron Stars: The Physics of Compact Objects*, 1983.

Blue straggler star

A star that is a member of a stellar association and is located at an unusual position on the association's color–magnitude diagram, above the turnoff from the main sequence.

A stellar cluster is observationally characterized by its color–magnitude diagram, which represents each member's luminosity plotted against its color, which in turn is related to its effective temperature. The location and structure of the various regions of a cluster's color–magnitude diagram can be made to yield crucial information on its age and chemical composition by exploiting the knowledge of the evolutionary tracks of stars with a specified range of initial masses and abundances. In order to make this method work properly, however, it is assumed that all the stars in a cluster are coeval (born at the same time), single, and isolated objects and that they are internally stable, nonrotating, nonmagnetic, and unmixed. The remarkable overall consistency between the observed color–magnitude diagrams and stellar evolutionary theory provides the most convincing evidence for the basic soundness of both the theory and the assumptions. *See* Color index; Hertzsprung-Russell diagram; Magnitude.

The existence of a special group of stars known as blue stragglers represents a challenge to this tidy scenario. They were discovered by A. Sandage in 1953 in the galactic globular cluster M3, whose most up-to-date color–magnitude diagram is shown in **Fig. 1**. The blue stragglers are the objects located below

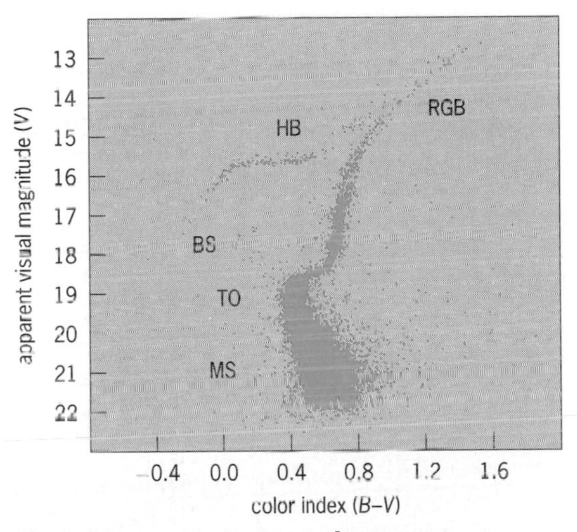

Fig. 1. Color–magnitude diagram [apparent visual magnitude (*V*) versus color index (*B–V*)] of 10,637 stars in the galactic globular cluster M3 (NGC 5272). The important evolutionary stages are marked by MS (main sequence), TO (turn-off), RGB (red giant branch), HB (horizontal branch), and BS (blue stragglers). (*After R. Buonanno et al., High precision photometry of 10,000 stars in M3, Mem. Soc. Astron. It, 57:391–393, 1986*)

the horizontal branch and above the turnoff from the main sequence. They form a new sequence extending from the extrapolation of the main sequence to higher luminosities on the left to the red giant branch on the right. While the majority of the cluster members fall rather precisely on the expected isochrones for the approximately 15 × 10^9-year age of the cluster, the 50 or so blue stragglers in M3 are located on much younger isochrones. In the case of the globular cluster NGC 5053, for example, the blue straggler sequence of approximately 25 objects can be well fit by isochrones giving an age of 2–4 × 10^9 years, while the rest of the cluster is consistent with an age of approximately 18 × 10^9 years. It seems, in other words, as if the blue stragglers somehow never made it across the gap between the main sequence and the red giant branch as expected from stellar evolutionary theory and are, therefore, true stragglers with respect to the main body of stars in the cluster.

Occurrence. Approximately 600 objects of this type have been found so far in every known type of stellar association, including open and globular clusters and dwarf spheroidal galaxies. They represent, therefore, a ubiquitous phenomenon. Blue stragglers are not found in all clusters, implying that their existence is not a universal phenomenon and may depend on environmental parameters. It is not yet clear how complete a sample the observed blue stragglers represent, however, since many blue stragglers may escape observation in very dense cores of clusters, and others may masquerade as normal yellow stars when observed at visible wavelengths and can be detected only through ultraviolet observations. In the case of the globular clusters, the oldest known stellar associations in the universe, where the blue straggler dilemma is most pronounced, all indications point to objects having masses consistent with their position on the color–magnitude diagram. This fact is well established both indirectly by their tendency to settle to the cluster core because of mass segregation, and directly by the recent discovery of a few pulsating dwarf cepheid blue stragglers from which a mass can be derived from their pulsation mode and light curve. Blue stragglers in globular clusters can be clearly defined as objects having the following general range of physical parameters: temperatures between 6000 and 10,000 K, masses between 1 and 2.5 solar masses, and luminosities between 3 and 30 solar luminosities. SEE CEPHEIDS; LOCAL GROUP.

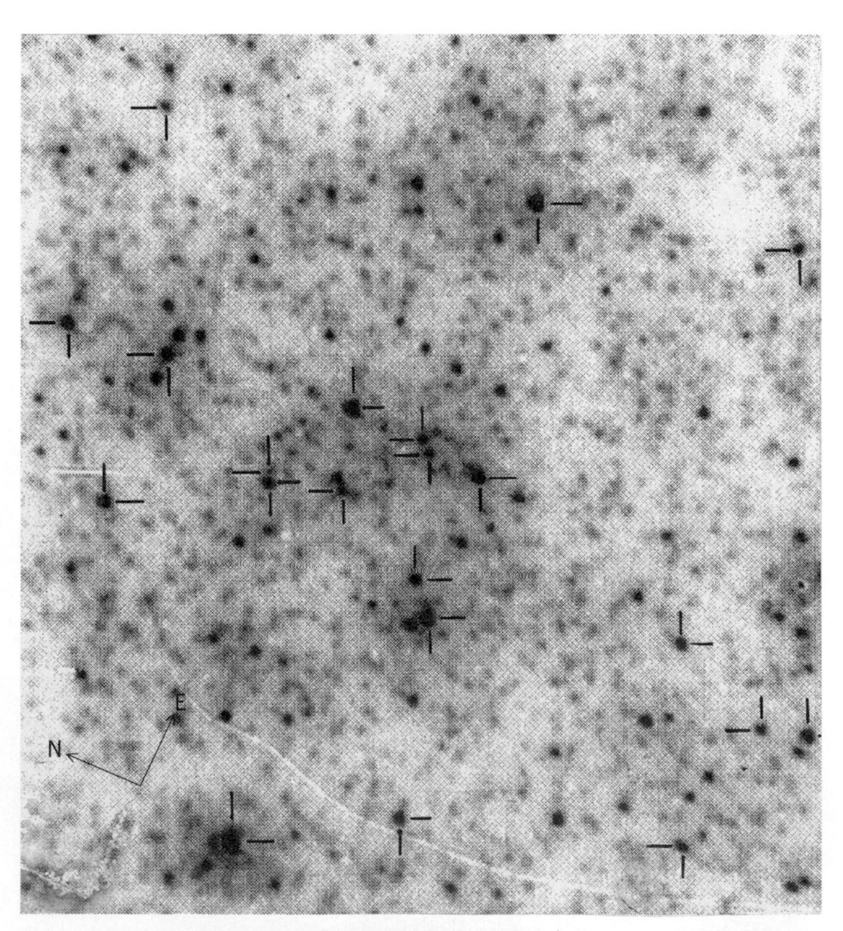

Fig. 2. Hubble Space Telescope Faint Object Camera image of the core of 47 Tucanae. The field is 22 × 22 arc-seconds squared in size. The 21 blue stragglers found within 0.5 core radii of the center are indicated by horizontal and vertical bars. (*NASA/ESA*).

Origin. Since such stars cannot exist in the cluster under the assumptions made above, some or all of the assumptions must be wrong. The simplest explanation of the phenomenon, of course, is violation of the coeval hypothesis; that is, the blue stragglers really are younger than the rest of the cluster members, perhaps because of a recent burst of star formation. Since there is independent evidence for such events having occurred in open clusters and some luminous dwarf spheroidal galaxies, this delayed-formation scenario is the most likely in these particular cases. On the other hand, since there is no evidence at all for recent star formation episodes in the galactic globular clusters, some other assumption must be flawed if the blue stragglers in these systems are to be explained. Several possibilities related to delayed evolution or extended lifetime in a coeval ensemble have been analyzed. The most plausible are internal mixing that supplies new fuel to the core, mass transfer in binary systems in which one component has increased its mass at the expense of the other, and the merging of two low-mass stars to form a more massive star via direct collision. In some environments, all of these processes may be occurring simultaneously, giving rise to different types of blue stragglers. Some, but not all, blue stragglers are known to be specific types of variable stars, for example, contact, semidetached, and eclipsing binaries, implying that the mass-transfer scenario must be occurring in at least some clusters. SEE BINARY STAR; ECLIPSING VARIABLE STARS; VARIABLE STAR.

Considerable effort is presently being expended in locating and classifying blue stragglers. The Hubble Space Telescope, launched in April 1990, is particularly helpful in this regard because of its high spatial resolution and ultraviolet sensitivity. It has been able to resolve, for the first time, the cores of very dense clusters where blue stragglers tend to aggregate. The first discovery of this type was that of 21 blue stragglers in the core of the globular cluster 47 Tucanae. Their position in the cluster is indicated in **Fig. 2**, which shows an image of the core taken by the Faint Object Camera on the Hubble Space Telescope through an ultraviolet filter centered at a wavelength of 220 nm. The field of view of this image is 22 × 22 arc-seconds squared in size corresponding to approximately 0.5 of the core radius. A few blue stragglers have been found outside the core. SEE SATELLITE ASTRONOMY.

The likely association of blue stragglers with binaries of all types, including x-ray sources and millisecond pulsars, should help clarify their critical role in the dynamics and evolution of globular clusters before, during, and after core collapse. By their very existence, blue stragglers sharpen the understanding of the nature of the stellar populations and their interactions in all clusters. They represent, therefore, a crucial tool in deciphering the origin and history of these fascinating objects. SEE PULSAR; STAR CLUSTERS; STELLAR EVOLUTION; X-RAY ASTRONOMY.

Francesco Paresce

Bibliography. R. Buonanno et al., High precision photometry of 10,000 stars in M3, *Mem. Soc. Astron. It.*, 57:391–393, 1986; F. Fusi-Pecci et al., On the blue stragglers and horizontal branch morphology in galactic globular clusters, *Astron. J.*, 104:1831–1849, 1992; P. J. T. Leonard, Stellar collisions in globular clusters and the blue straggler problem, *Astron. J.*, 98:217–226, 1989; J. M. Nemec, Anomalous cepheids and population II blue stragglers, in E. G.

Schmidt (ed.), *The Use of Pulsating Stars in Fundamental Problems of Astronomy*, IAU Colloq. 111, 1989; F. Paresce et al., Blue stragglers in the core of the globular cluster of 47 Tucanae, *Nature*, 352:297–301; 1991.

Boötes

The Bear Driver, in astronomy, a northern and summer constellation. Boötes is one of the earliest recorded constellations. Arcturus, an orange, first-magnitude navigational star, dominates the constellation. It has been known and admired for ages and is one of the few stars mentioned in the Bible. Five prominent stars in this constellation, β, γ, ρ, ϵ, and δ, form a pentagon, shaped much like an elongated kite, with Arcturus at the junction of the tail. Boötes is conventionally pictured as a driver of the Bear (Ursa Major) nearby. In a more recent version he is seen seated, smoking a pipe, his feet dangling and with one hand holding the leash of the Hunting Dogs (Canes Venatici). *See* Constellation; Ursa Major.

Ching-Sung Yu

Bremsstrahlung

Electromagnetic radiation emitted when a charged particle is accelerated by the electric field of another charged particle to which it is not bound. Bremsstrahlung is also called free-free emission. Encounters between two identical particles (two electrons, two protons, and so forth) yield almost no radiation and will not be discussed.

Physical process. Any plasma (ionized gas) emits bremsstrahlung radiation. The brightness of this radiation can be calculated by considering the force exerted on an electron as it passes close to a proton (or other positively charged nucleus), as a function of its speed and distance of closest approach. The emission from a volume of gas is then just the sum of the contributions of the individual electrons. The result is given by Eqs. (1) and (2), where N_e and N_i are the

$$\epsilon = 1.43 \times 10^{-27} N_e N_i Z^2 T^{1/2} g_{\text{III}} \text{ erg} \cdot \text{cm}^{-3} \cdot \text{s}^{-1} \text{ (cgs units)} \quad (1)$$

$$\epsilon = 1.43 \times 10^{-40} N_e N_i Z^2 T^{1/2} g_{\text{III}} \text{ J} \cdot \text{m}^{-3} \cdot \text{s}^{-1} \text{ (SI units)} \quad (2)$$

number densities of electrons and ions (number per cubic centimeter in cgs units, number per cubic meter in SI units), Z is the charge on the ions (one for protons), and T is the temperature of the gas (in kelvins). The coefficient g_{III}, called the Gaunt factor, is a quantum-mechanical correction factor not very different from 1; a value of 1.2 is correct to within 20% for most astronomical applications.

The radiation is distributed almost uniformly across a very wide range of wavelengths (see **illus.**). The long-wavelength (low-frequency) cutoff occurs when the gas begins to absorb its own radiation, yielding emission indistinguishable from blackbody radiation at the same temperature. This cutoff falls within the radio band for most astronomical plasmas. The short-wavelength (high-frequency) cutoff happens when an electron does not have enough energy to emit even one photon at that frequency. It is in the infrared for gas at 10^4 K, in the extreme ultraviolet for 10^5–10^6 K, and at x-ray wavelengths for all hotter gases. The very flat (wavelength-independent) spectrum with a short-wavelength exponential cutoff is a signature of this type of radiation, often called thermal bremsstrahlung. Another signature is the simultaneous presence of emission or absorption lines due to incompletely ionized atoms of iron and other metals. These together distinguish bremsstrahlung from synchrotron and inverse-Compton emission.

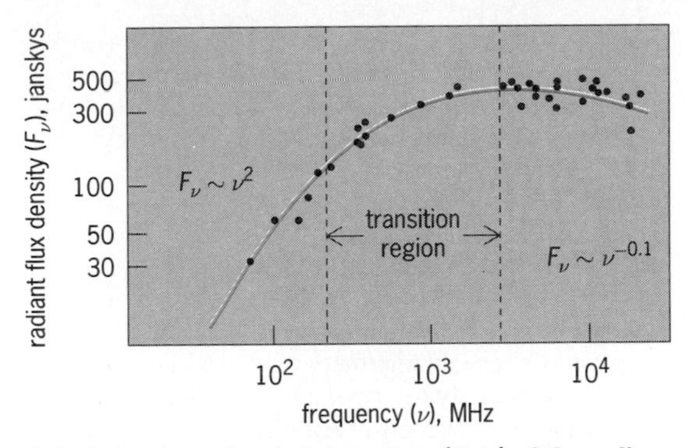

Calculation (curve) and observations (dots) of the radio bremsstrahlung from the hot, ionized gas in the Orion Nebula. 1 jansky = 10^{-26} W · m^{-2} · Hz^{-1}. At low frequencies, the nebula is optically thick (opaque to its own radiation), and the emission is identical to a blackbody spectrum. At higher frequencies, flux depends only logarithmically on wavelength of observation. At still higher frequencies (above about 10^8 MHz, in the infrared), the spectrum cuts off exponentially. (*After F. H. Shu, The Physics of Astrophysics, vol. 1: Radiation Processes, University Science Books, 1991*)

Astrophysical applications. Most of the material in the universe, except for the terrestrial planets and the coldest clouds of interstellar molecular gas, is in the form of a plasma capable of bremsstrahlung emission. This emission is, however, apparent only in cases where the gas does not absorb its own radiation (is optically thin) and no other process emits more radiation at the same wavelengths.

Situations in which much of the observed radiation is bremsstrahlung include radio emission from H II regions, planetary nebulae, and the coronae of the Sun and stars; and the x-ray emission from some supernova remnants, elliptical galaxies and clusters of galaxies, and the coronae of the Sun and other stars. *See* Interstellar matter; Planetary nebula; Radioastronomy; Supernova; X-ray astronomy.

Bremsstrahlung x-rays from the solar corona were observed by the first x-ray telescope successfully carried by a rocket above the Earth's atmosphere in 1947. The detection demonstrated conclusively that the corona is hot (about 10^6 K or hotter) and that material is, therefore, flowing out of it into a solar wind, not down onto the Sun. Other stars show similar emission from coronae or winds, sometimes much brighter than the 3×10^{26} ergs/s (3×10^{19} W) coming from the Sun. Binary stars where the winds from the two stars collide and heat each other (RS Canum Venaticorum stars) are among the brightest stellar bremsstrahlung sources at 10^{30}–10^{31} ergs/s (10^{23}–10^{24} W). Radio bremsstrahlung from coronae and winds is also common among stars brighter than the Sun. Flares, bursts, and other sorts of variability are observed in both radio and x-ray emission, some of it (particularly in the radio) attributable to other emission mechanisms. *See* Binary star; Solar corona; Sun.

The illustration shows the radio bremsstrahlung of the Orion Nebula. The low-frequency cutoff implies a temperature near 8000 K and a particle density of about 2000 cm^{-3}, consistent with what is found for the nebula by other methods. *SEE ORION NEBULA.*

Both clusters of galaxies and single elliptical galaxies are often sources of bremsstrahlung x-rays. The emission comes from hot gas confined by the gravitational field of the galaxy or cluster. There can be as much as 10^{10} solar masses of hot gas in a bright galaxy and 10^{14} solar masses in a cluster. The temperature, amount, and extent of the gas can be used to determine the total amount of mass (visible and invisible) that is producing the gravitational field. That amount, especially in rich clusters, is considerably larger than the mass observed directly in the galaxies and gas, indicating that a large part of the universe consists of dark matter. Surprisingly, the intracluster gas is not the pure hydrogen and helium that should be left over from galaxy formation. It shows emission lines of iron, indicating that at least some of it has been either blown out of the galaxies or polluted by a generation of very massive stars that lived and died before the galaxies formed. *SEE COSMOLOGY; GALAXY, EXTERNAL; UNIVERSE.*

Virginia Trimble

Bibliography. G. B. Rybicki and A. P. Lightman, *Radiation Processes in Astrophysics*, 1979; F. H. Shu, *The Physics of Astrophysics*, vol. 1: *Radiation Processes*, 1991.

Brown dwarf

One of the least massive self-gravitating objects that are formed in the fragmentation of an interstellar cloud. Brown dwarfs are less massive than the least massive true stars, the red dwarf stars, and are comparable to or heavier than the most massive planets, that is, the gas-giant planets such as Jupiter.

Brown dwarfs are distinguished from stars in that they are insufficiently massive to sustain long-term nuclear burning of hydrogen, although they undergo a brief period of deuterium (heavy-hydrogen) burning. Accordingly, brown dwarfs cool and fade away over periods of time that are short compared to the age of the Sun (5×10^9 years), while red dwarf stars burn hydrogen continuously at a relatively low rate and should continue to shine over times longer than the present age of the universe (about 20×10^9 years). The current search for brown dwarfs is important because studying them may elucidate the processes by which interstellar clouds break up and condense, and should demonstrate how to distinguish brown dwarfs from extrasolar planets (the planets of stars other than the Sun).

Mass range. Because the origin of a given object in space is rarely observed and the nature of the nuclear burning within an object is also not observed, astronomers use the working definition that objects in the mass range from 10 to 80 jupiters are brown dwarfs. One jupiter, the mass of the planet Jupiter, is equal to about one-thousandth the mass of the Sun. However, the definition of a planet, namely an object that does not sustain nuclear energy generation and that is formed by accretion in a viscous disk surrounding a star or protostar, does not necessarily exclude objects that may be as massive as 20 jupiters. Further, the theory of the internal constitution and nuclear-energy generation of red dwarf stars and brown dwarfs is sufficiently imprecise that the possibility exists that objects as light as 80 jupiters are red dwarf stars or that objects as heavy as 90 jupiters are brown dwarfs. These definitions can be improved only when brown dwarfs are identified with certainty and are subjected to detailed study.

Search for brown dwarfs. Several objects were discovered during the late 1980s that may be brown dwarfs, although no case is proven. Methods used to search for brown dwarfs are astrometry, infrared photometry, infrared speckle interferometry, and radial-velocity monitoring. Astrometry seeks to detect brown dwarfs that are companies of known stars through the slight wobbling that the presence of a brown dwarf must induce in the motion of its accompanying star. Infrared photometry seeks to detect the infrared radiation of a brown dwarf that is the companion of a small hot star such as a white dwarf star. Infrared speckle interferometry is used to attempt to obtain images of infrared radiation sources, such as brown dwarfs, that are close companions of known stars. Radial-velocity monitoring is conducted to detect the orbital motion of a known star around the center of mass of the binary system that it forms with a brown dwarf. *SEE ASTROMETRY; ASTRONOMICAL SPECTROSCOPY; SPECKLE.*

It appears from radial-velocity monitoring that although extrasolar planets may exist (which remains to be proven), sunlike stars are not likely to have brown dwarf companions. A striking result from infrared speckle interferometry is that red dwarf stars generally lack brown dwarf companions. Astrometry and infrared speckle interferometry have detected a few objects which may be brown dwarf companions of low-mass stars but which could well be very light red dwarf stars.

The strongest case for the existence of brown dwarfs comes from infrared photometry, which has revealed the existence of faint companions, glowing in infrared light, to the white dwarf stars Giclas 29-38 and GD165. The best case, GD165B (the companion of GD165), is clearly seen on infrared images, although its existence was first inferred from photometry of the infrared light from the vicinity of GD165. Although these observations may be beyond dispute, the interpretation that GD165B is a brown dwarf is not iron-clad, because its estimated mass is about 60 to 80 jupiters, and, given the uncertainties in the theory of red dwarf stars and brown dwarfs, it is possible that the star is an extremely low-mass red dwarf. *SEE BINARY STAR; INFRARED ASTRONOMY; PLANET; RED DWARF STAR; STAR; STELLAR EVOLUTION.*

Stephen P. Maran

Bibliography. E. E. Becklin and B. Zuckerman, A low-temperature companion to a white dwarf star, *Nature*, 336:656–658, 1988; M. C. Kafatos, R. S. Harrington, and S. P. Maran (eds.), *Astrophysics of Brown Dwarfs*, 1986; S. P. Maran, Very-low-mass companions—extrasolar planets?, *Physics News in 1987*, p. S-14, 1988, and *Phys. Today*, 40(1):S-14, January 1988; D. W. McCarthy, Jr., and T. J. Henry, Direct infrared observations of the very low mass object Gliese 623B, *Astrophys. J. (Lett.)*, 319:L93–L98, 1987.

Calendar

A list, usually in the form of a table, showing the correspondence between days of the week and days of the month; also, a list of special observances with

Perpetual calendar*

Day of the week for any known date from the beginning of the Christian Era to the year 2400

					Julian calendar								Gregorian calendar			
	Century:			0†	0‡ 700 1400	100 800 1500¶	200 900	300 1000	400 1100	500 1200	600 1300	1600 2000	1700 2100	1800 2200	1500§ 1900 2300
Year															
0				. . .	DC	ED	FE	GF	AG	BA	CB	BA	C	E	G
1	29	57	85	A	B	C	D	E	F	G	A	G	B	D	F
2	30	58	86	G	A	B	C	D	E	F	G	F	A	C	E
3	31	59	87	F	G	A	B	C	D	E	F	E	G	B	D
4	32	60	88	E	FE	GF	AG	BA	CB	DC	ED	DC	FE	AG	CB
5	33	61	89	. . .	D	E	F	G	A	B	C	B	D	F	A
6	34	62	90	. . .	C	D	E	F	G	A	B	A	C	E	G
7	35	63	91	. . .	B	C	D	E	F	G	A	G	B	D	F
8	36	64	92	. . .	AG	BA	CB	DC	ED	FE	GF	FE	AG	CB	ED
9	37	65	93	. . .	F	G	A	B	C	D	E	D	F	A	C
10	38	66	94	. . .	E	F	G	A	B	C	D	C	E	G	B
11	39	67	95	. . .	D	E	F	G	A	B	C	B	D	F	A
12	40	68	96	. . .	CB	DC	ED	FE	GF	AG	BA	AG	CB	ED	GF
13	41	69	97	. . .	A	B	C	D	E	F	G	F	A	C	E
14	42	70	98	. . .	G	A	B	C	D	E	F	E	G	B	D
15	43	71	99	. . .	F	G	A	B	C	D	E	D	F	A	C
16	44	72		. . .	ED	FE	GF	AG	BA	CB	DC	CB	ED	GF	BA
17	45	73		. . .	C	D	E	F	G	A	B	A	C	E	G
18	46	74		. . .	B	C	D	E	F	G	A	G	B	D	F
19	47	75		. . .	A	B	C	D	E	F	G	F	A	C	E
20	48	76		. . .	GF	AG	BA	CB	DC	ED	FE	ED	GF	BA	DC
21	49	77		. . .	E	F	G	A	B	C	D	C	E	G	B
22	50	78		. . .	D	E	F	G	A	B	C	B	D	F	A
23	51	79		. . .	C	D	E	F	G	A	B	A	C	E	G
24	52	80		. . .	BA	CB	DC	ED	FE	GF	AG	GF	BA	DC	FE
25	53	81		. . .	G	A	B	C	D	E	F	E	G	B	D
26	54	82		. . .	F	G	A	B	C	D	E	D	F	A	C
27	55	83		. . .	E	F	G	A	B	C	D	C	E	G	B
28	56	84		. . .	DC	ED	FE	GF	AG	BA	CB	BA	DC	FE	AG

Month					Dominical letter						
Jan., Oct.					A	B	C	D	E	F	G
Feb., Mar., Nov.					D	E	F	G	A	B	C
Apr., July					G	A	B	C	D	E	F
May					B	C	D	E	F	G	A
June					E	F	G	A	B	C	D
Aug.					C	D	E	F	G	A	B
Sept., Dec.					F	G	A	B	C	D	E
1	8	15	22	29	Sun.	Sat.	Fri.	Thurs.	Wed.	Tues.	Mon.
2	9	16	23	30	Mon.	Sun.	Sat.	Fri.	Thurs.	Wed.	Tues.
3	10	17	24	31	Tues.	Mon.	Sun.	Sat.	Fri.	Thurs.	Wed.
4	11	18	25		Wed.	Tues.	Mon.	Sun.	Sat.	Fri.	Thurs.
5	12	19	26		Thurs.	Wed.	Tues.	Mon.	Sun.	Sat.	Fri.
6	13	20	27		Fri	Thurs.	Wed.	Tues.	Mon.	Sun.	Sat.
7	14	21	28		Sat.	Fri.	Thurs.	Wed.	Tues.	Mon.	Sun.
											Mon.

To find the calendar for any year of the Christian Era, first find the Dominical letter for the year in the upper section of the table. Two letters are given for leap years; the first is to be used for January and February, the second for the other months. In the lower section of the table, find the column in which the Dominical letter for the year is in the same line with the month for which the calendar is desired; this column gives the days of the week that are to be used with the month. For example, in the table of Dominical letters the letter for 1962 is G; in the line with July, this letter occurs in the first column; hence July 4, 1962, is Wednesday.

*After *Smithsonian Physical Tables*, 9th ed., Washington, D.C., 1954.
†A.D. 1 through A.D. 4 only.
‡A.D. 5 through A.D. 99 only.
§On and after 1582, October 15 only.
¶On and before 1582, October 4 only.

their dates, especially those that fall on different dates in different years, as a church calendar; also, a set of rules that serves to attach to any day a specific number or name or combination of number and name. The calendars used at various times and places are numerous and diverse. Most represent attempts to divide the synodic month or the tropical year into numbered days. According to whether the emphasis is on month or year, calendars are called lunar or solar.

Gregorian calendar. The calendar used for civil purposes throughout the world, known in Western countries as the Gregorian calendar, was established by Pope Gregory XIII, who decreed that the day following Thursday, October 4, 1582, should be Friday, October 15, 1582, and that thereafter centennial years (1600, 1700, and so on) should be leap years only when divisible by 400 (1600, 2000, and so on), other years being leap years when divisible by four, as previously. The effect of the reform was to restore the vernal equinox (beginning of spring, which governs the observance of Easter) to March 21, and to reduce the number of days in 400 calendar years by 3, making the average number of days in the calendar year 365.2425, instead of 365.24220 days. The new calendar was adopted by Great Britain in 1752, but did not become universal until well into the twentieth century.

Although the Gregorian calendar is a solar calendar, lunar vestiges remain in it, as is shown by the division of the year into months. The week is independent of any astronomical phenomenon; it was established before the beginning of the Christian Era, and the cyclical succession of the days of the week has not since been disturbed. The connection between the week and the ancient Jewish cycle of seven days has not been established; for example, it is not known whether the modern Saturday is identical with the ancient seventh day.

The calendar generally used before the adoption of the Gregorian calendar is called the Julian calendar; their relationship is shown in the **table**. Dates recorded for civil and commercial purposes before October 4, 1582, may be assumed to be in the Julian calendar, but dates recorded since then may require considerable care in interpretation.

Calendar reform. The Gregorian calendar is not perfectly adjusted to the tropical year, the error accumulating to 1 day after about 3300 years. Numerous proposals for calendar reform have been made with the object of reducing the error, and there have also been many proposals for altering the succession of days for the purpose of bringing greater regularity into the calendar. For example, one proposal would divide the year into 13 months of 4 weeks each, which would make 364 days. The 365th day, and the 366th in leap years, would have a special name and would not belong to any month or to any week. Thus, every month of every year would begin on Sunday and end with Saturday, and annual calendars would become unnecessary. The 13-month calendar has the disadvantage, serious from the standpoint of business, of not being divisible into quarters. Another proposal, which overcomes this difficulty, retains 12 months but would give them successively 31, 30, 30, 31, 30, 30, 31, 30, 30, 31, 30, 30 days, the extra day or days again being outside the system. No attempt to regularize the calendar can succeed if the succession of week days and the average length of the calendar year are both preserved. *See* Day; Month; Time; Year.

Gerald M. Clemence

Bibliography. E. Achelis, *Of Time and the Calendar*, 1955; H. J. Cowan, *Time and Its Measurement: From the Stone Age to the Nuclear Age*, 1958; W. M. O'Neil, *Time and the Calendars*, rev. ed., 1978.

Cancer

The Crab, in astronomy, a winter constellation and the faintest of the zodiacal groups. Cancer, the fourth sign of the zodiac, is important because during early times it marked the northernmost limit of the ecliptic, when the zodiacal system was adopted. The Tropic of Cancer takes its name from this constellation. The four faint stars, α, β, δ, and ι, form a rough Y outline, which is suggestive of a crab (see **illus.**). In the center of Cancer is a hazy object. This is a magnificent cluster of faint stars called Praesepe (the Beehive) or the Manger. *See* Constellation; Zodiac.

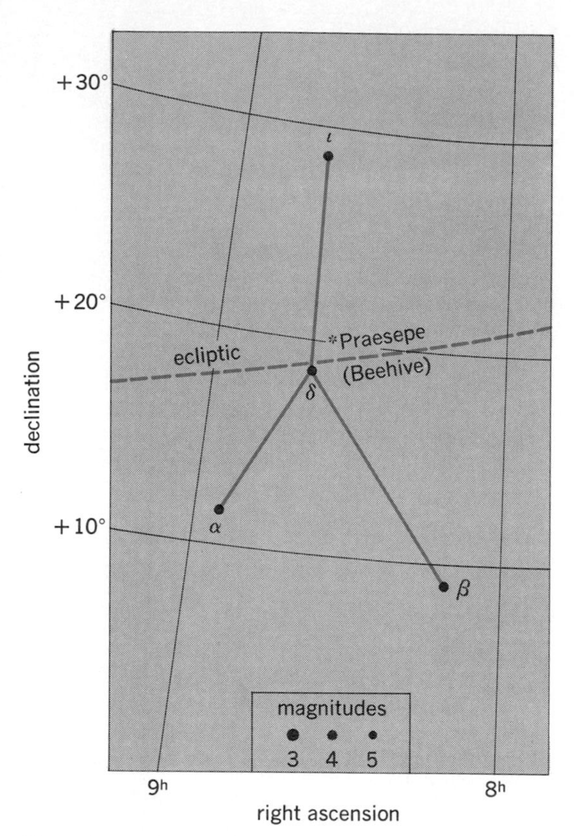

Line pattern of constellation Cancer. Grid lines represent the coordinates of the sky. Apparent brightness, or magnitudes, of stars are shown by sizes of dots, which are graded by appropriate numbers as indicated.

Ching-Sung Yu

Capricornus

The Sea Goat, in astronomy, an inconspicuous zodiacal constellation in the southern sky lying between Aquarius and Sagittarius. Capricornus is the tenth sign of the zodiac. It has two third-magnitude stars, β and δ, the remainder being of fourth magnitude or

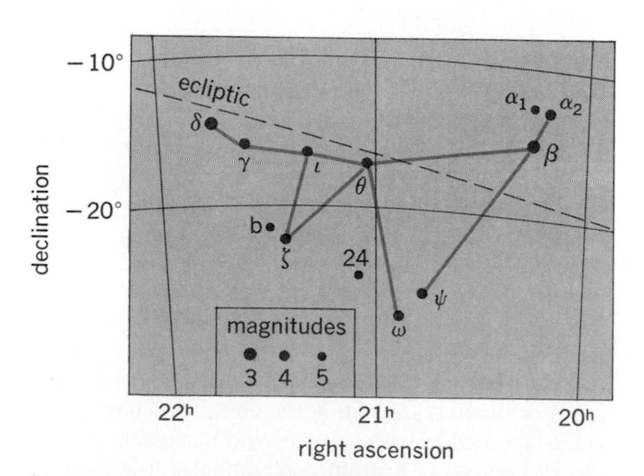

Line pattern of constellation Capricornus. Grid lines represent coordinates of the sky. Apparent brightness, or magnitudes, of stars are shown by size of dots, which are graded by appropriate numbers as indicated.

fainter. The constellation has been described from the earliest times as a goat, or as a figure that is part goat with the tail of a fish (see **illus.**). Together with the neighboring zodiacal constellations Aquarius and Pisces, Capricornus forms the great heavenly sea; all the names are related to water. The Tropic of Capricorn originates from this constellation, which marked the southern limit of the ecliptic in ancient times. *See Aquarius; Constellation; Pisces; Zodiac.*

Ching-Sung Yu

Carbon-nitrogen-oxygen cycles

A group of nuclear reactions that involve the interaction of protons (nuclei of hydrogen atoms, designated by ^{1}H) with carbon, nitrogen, and oxygen nuclei. The cycle involving only isotopes of carbon and nitrogen is well known as the carbon-nitrogen (CN) cycle. These cycles are thought to be the main source of energy in main-sequence stars with mass 20% or more in excess of that of the Sun. Completion of any one of the cycles results in consumption of four protons (4 ^{1}H) and the production of a helium (^{4}He) nucleus plus two positrons (e^{+}) and two neutrinos (ν). The two positrons are annihilated with two electrons (e^{-}), and the total energy release is 26.73 MeV. Ap-

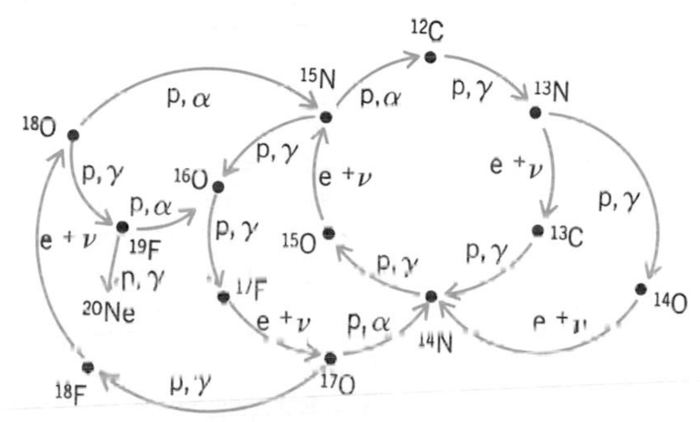

Carbon-nitrogen-oxygen cycles. (*After J. Audouze, ed., CNO Isotopes in Astrophysics, D. Reidel, 1977*)

proximately 1.7 MeV is released as neutrino energy and is not available as thermal energy in the star. The energy E = 26.73 MeV arises from the mass difference between four hydrogen atoms and the helium atom, and is calculated from the Einstein mass-energy equation $E = \Delta mc^2$, where Δm is the mass difference and c^2 is the square of the velocity of light. Completion of a chain can be thought of as conversion of four hydrogen atoms into a helium atom. Because the nuclear fuel that is consumed in these processes is hydrogen, they are referred to as hydrogen-burning processes by means of the carbon-nitrogen-oxygen (CNO) cycles. *See Solar neutrinos.*

Carbon-nitrogen cycle. The original carbon-nitrogen cycle was suggested independently by H. A. Bethe and C. F. von Weizsäcker in 1938 as the source of energy in stars. In the first reaction of the carbon-nitrogen cycle, a carbon nucleus of mass 12 captures a proton, forming a nitrogen nucleus of mass 13 and releasing a photon of energy, 1.943 MeV. This may be written: $^{12}\mathrm{C} + {}^{1}\mathrm{H} \rightarrow {}^{13}\mathrm{N} + \gamma$, or $^{12}\mathrm{C}(p,\gamma)^{13}\mathrm{N}$. The nucleus ^{13}N is unstable and decays by emitting a positron (e^{+}) and a neutrino (ν). In reaction form, $^{13}\mathrm{N} \rightarrow {}^{13}\mathrm{C} + e^{+} + \nu$, or $^{13}\mathrm{N}(e^{+}\nu)^{13}\mathrm{C}$. The cycle continues through ^{14}N, ^{15}O, and ^{15}N by the reactions and decays shown in $^{12}\mathrm{C}(p,\gamma)^{13}\mathrm{N}(e^{+}\nu)$-$^{13}\mathrm{C}(p,\gamma)^{14}\mathrm{N}(p,\gamma)^{15}\mathrm{O}(e^{+}\nu)^{15}\mathrm{N}(p,\alpha)^{12}\mathrm{C}$. These reactions form the second cycle from the right in the **illustration**. The reaction $^{15}\mathrm{N}(p,\alpha)^{12}\mathrm{C}$ represents emission of an alpha particle (^{4}He nucleus) when ^{15}N captures a proton and this cycle returns to ^{12}C. Because of the cycling, the total number of carbon, nitrogen, and oxygen nuclei remains constant, so these nuclei act as catalysts in the production of a helium nucleus plus two positrons and two neutrinos with the release of energy. The positrons that are created annihilate with free electrons rapidly after creation, so the energy used in their creation is returned to the energy fund. Synthesis of some of the rare carbon, nitrogen, and oxygen isotopes is accomplished through hydrogen burning by means of the carbon-nitrogen-oxygen cycles.

Table 1 shows the energy release, Q, and the average release of thermal energy, Q (thermal), in each reaction of the cycle, as well as the maximum neutrino energy $E_{\nu}^{\max}$ in the two reactions in which a neutrino is produced.

A standard model for the Sun has been developed by J. N. Bahcall and colleagues to determine the flux of solar neutrinos that should be coming to the Earth. In this model, the site of maximum hydrogen burning is just outside the central core of the Sun. At a distance of 0.0511 solar radius from the center, chosen to lie in the most active part of that site for hydrogen burning in the proton-proton chains, the temperature is 14×10^{6} K (25×10^{6} °F), the density is ρ = 112 g/cm^{3} (112 times the density of water), and the mass fraction of hydrogen is $X(^{1}\mathrm{H})$ = 0.483. The reaction rates, R, calculated for the carbon-nitrogen cycle reactions at that temperature, are shown in **Table 2**. Small corrections for screening of the Coulomb field between nuclei by electrons have been neglected. Taking into account the mass fraction of hydrogen and the density at the site, the number of reactions per nucleus per second is given by $\lambda = \rho R X(^{1}\mathrm{H})/$

Table 1. Energies involved in the carbon-nitrogen cycle

Reaction	Q, MeV	Q (thermal), MeV	$E_{\nu}^{\max}$, MeV
$^{12}\mathrm{C} + {}^{1}\mathrm{H} \rightarrow {}^{13}\mathrm{N} + \gamma$	1.943	1.943	
$^{13}\mathrm{N} \rightarrow {}^{13}\mathrm{C} + e^{+} + \nu$	2.221	1.51	1.199
$^{13}\mathrm{C} + {}^{1}\mathrm{H} \rightarrow {}^{14}\mathrm{N} + \gamma$	7.551	7.551	
$^{14}\mathrm{N} + {}^{1}\mathrm{H} \rightarrow {}^{15}\mathrm{O} + \gamma$	7.297	7.297	
$^{15}\mathrm{O} \rightarrow {}^{15}\mathrm{N} + e^{+} + \nu$	2.753	1.75	1.731
$^{15}\mathrm{N} + {}^{1}\mathrm{H} \rightarrow {}^{12}\mathrm{C} + {}^{4}\mathrm{He}$	4.966	4.966	

Table 2. Reaction rates and mean lifetimes for carbon-nitrogen cycle in the Sun

Reaction	R, $s^{-1}/(mole/cm^3)$*	λ, s^{-1}	τ, years
$^{12}C + {}^1H \rightarrow {}^{13}N + \gamma$	8.19×10^{-17}	4.40×10^{-15}	7.20×10^6
$^{13}N \rightarrow {}^{13}C + e^+ + \nu$		1.16×10^{-3}	2.73×10^{-5}
$^{13}C + {}^1H \rightarrow {}^{14}N + \gamma$	2.83×10^{-16}	1.52×10^{-14}	2.08×10^6
$^{14}N + {}^1H \rightarrow {}^{15}O + \gamma$	2.94×10^{-19}	1.58×10^{-17}	2.01×10^9
$^{15}O \rightarrow {}^{15}N + e^+ + \nu$		5.68×10^{-3}	5.58×10^{-6}
$^{15}N + {}^1H \rightarrow {}^{12}C + {}^4He$	7.27×10^{-15}	3.91×10^{-13}	8.11×10^4

*From G. R. Caughlan and W. A. Fowler, Thermonuclear reaction rates, V, *Atom. Nucl. Data Tables*, vol. 40, 1988.

1.0078. By inverting λ, the mean lifetimes, τ, of the carbon, nitrogen, and oxygen nuclei are obtained; these are shown in the last column of Table 2 in years. The fraction of energy from CNO cycles in the Sun is calculated to be only 1.5%; the remaining 98.5% of the energy is from hydrogen burning in the proton-proton chains.

CNO bicycle. Nuclear research in the 1950s led to the addition of a second cycle to the processes of hydrogen burning by the CNO cycles. Laboratory research has shown that in a very small number of proton captures by ^{15}N a photon is emitted with the formation of an oxygen nucleus of mass 16, leading to reactions which may cycle back to ^{14}N by $^{15}N(p,\gamma)^{16}O(p,\gamma)^{17}F(e^+\nu)^{17}O(p,\alpha)^{14}N$. The pair of cycles consisting of this cycle and the carbon-nitrogen cycle, that forms the main cycle and the first cycle to the left of it in the illustration, is called the carbon-nitrogen-oxygen bicycle. The rate of the $^{12}N(p,\alpha)^{12}C$ reaction is of the order of 10^3 times that of the $^{15}N(p,\gamma)^{16}O$ reaction at most temperatures expected in stellar interiors.

Other CNO cycles. In research since 1960, it has become apparent that many possible branches among the nuclei must be included in any analysis of hydrogen burning by carbon, nitrogen, and oxygen nuclei. For example, if the unstable nucleus ^{13}N manages to capture a proton before it decays in its mean lifetime of 862 s, a third cycle can occur through $^{13}N(p,\gamma)^{14}O(e^+\nu)^{14}N$. This cycle is displayed in the illustration by the branch on the right-hand side of the main carbon-nitrogen cycle and is known as the fast or the hot carbon-nitrogen cycle. The other possible branches leading to additional cycles shown in the diagram are due to competition between (p,γ) and (p,α) reactions. The added reactions are $^{17}O(p,\alpha)^{18}F(e^+\nu)^{18}O(p,\alpha)^{15}N$ and $^{18}O(p,\alpha)^{19}F(p,\alpha)^{16}O$. The reaction $^{19}F(p,\gamma)^{20}Ne$ shown in the illustration leads out of the carbon, nitrogen, and oxygen nuclei and hence away from CNO cycles.

There are two additional branches that may occur if the unstable fluorine nuclei (^{17}F and ^{18}F) capture protons before they can decay. SEE NUCLEOSYNTHESIS; PROTON-PROTON CHAIN; STELLAR EVOLUTION.

Georgeanne R. Caughlan

Bibliography. J. Audouze (ed.), *CNO Isotopes in Astrophysics*, 1977; J. N. Bahcall et al., Standard solar models and the uncertainties in predicted capture rates of solar neutrinos, *Rev. Mod. Phys.*, 54:767–799, 1982; H. A. Bethe, Energy production in stars, *Phys. Rev.*, 55:103, 434–456, 1939; E. M. Burbidge et al., Synthesis of elements in stars, *Rev. Mod. Phys.*, 29:547–650, 1957; G. R. Caughlan and W. A. Fowler, Thermonuclear reaction rates, V, *Atom. Nucl. Data Tables*, vol. 40, 1988; W. A. Fowler, *Nuclear Astrophysics,* 1967; M. J. Harris et al., Thermonuclear reaction rates, III, *Annu. Rev. Astron. Astrophys.*, 21:165–176, 1983; C. F. von Weizsäcker, Uber Elementwandlungen im Innern der Sterne, II, *Phys. Z.*, 39:633–646, 1938.

Carbon star

Any of a class of stars with an apparently high abundance ratio of carbon to hydrogen. In normal stars, oxygen is more abundant than carbon, and carbon and nitrogen are about equal. In carbon stars, however, carbon is more abundant than oxygen. The majority of the carbon-rich stars are low-temperature red giants of the C class, showing bands of C_2, CN, and CH (the older designation of this is R or N). A few hot carbon-rich stars are also known. An excess abundance of carbon is also found in the S (zirconium oxide) stars and the barium (heavy-element) stars. The C-class stars have surface temperatures in the range 5500–2500 K (9500–4000°F). Heavy absorption produced by bands ascribed to C_3 and SiC_2 exists in cooler objects. An unexpected peculiarity is the strength of the resonance lines of neutral lithium; this element is easily destroyed by nuclear reactions with hydrogen. Its synthesis by helium burning may occur in late stages of red giant evolution.

The degree of enhancement of carbon varies from star to star. In a related phenomenon found in global clusters, variations in the C/N/O ratios are found. An outstanding peculiarity is the presence of the isotope ^{13}C in some of these stars, as shown by the presence of $^{12}C^{13}C$ and $^{13}C^{13}C$ isotope bands, with an abnormally high $^{13}C/^{12}C$ ratio, between 1 to 4 and 1 to 10. In some other C-class stars the $^{12}C^{13}C$ bands are absent, indicating a low $^{13}C/^{12}C$ ratio, possibly near the terrestrial value of 1 to 90. An important discovery was that nearly all red giants, not only carbon stars, have systematically higher values of the $^{13}C/^{12}C$ ratio than the Sun.

The difference between normal red giant spectra and those of the carbon-rich group arises apparently because of the strong binding of the compound CO. In stars for which there is more oxygen than carbon, at temperatures below 5000 K (8500°F), almost all the carbon is bound in the form of CO and therefore unavailable to form further carbon compounds. In the carbon-rich stars the CO consumes all the oxygen, but leaves some carbon in free atomic form, available for carbon compounds. The importance of the C/O ratio lies in the theory of the origin of chemical elements, in which it is shown that carbon is easily synthesized at temperatures found in cores of evolving red giants. *See Nucleosynthesis; Star.*

Jesse L. Greenstein

Cardinal points

The four intersections of the horizon with the meridian and with the prime vertical circle, or simply prime vertical, the intersections with the meridian being designated north and south, and the intersections with the prime vertical being designated east and west (see **illus.**). The cardinal points are 90° apart; they lie in a plane with each other and correspond to the cardinal regions of the heavens. The four intermediate points, northeast, southeast, northwest, and southwest, are the collateral points. *See Astronomical coordinate systems.*

Frank H. Rockett

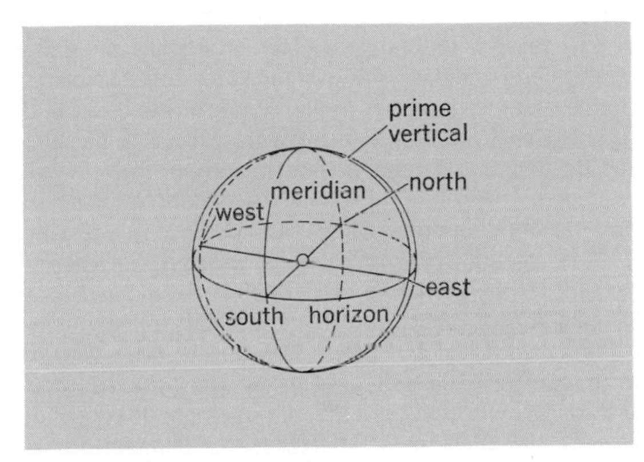

Cardinal points around horizon.

Cassiopeia

In astronomy, a prominent northern circumpolar constellation as seen from the middle latitudes. The five main bright second- and third-magnitude stars of Cassiopeia form the rather distorted W or M by which the constellation is usually identified (see **illus.**). The W is also called Cassiopeia's Chair, because on its side the W suggests the shape of a chair. Most old sky maps show Cassiopeia as a queen seated upon a throne. Lying as it does in the Milky Way, the entire region of Cassiopeia is rich in star fields containing several beautiful star clusters. Cassiopeia and the Big Dipper lie across the opposite sides of the North Celestial Pole. *See Constellation; Ursa Major.*

Ching-Sung Yu

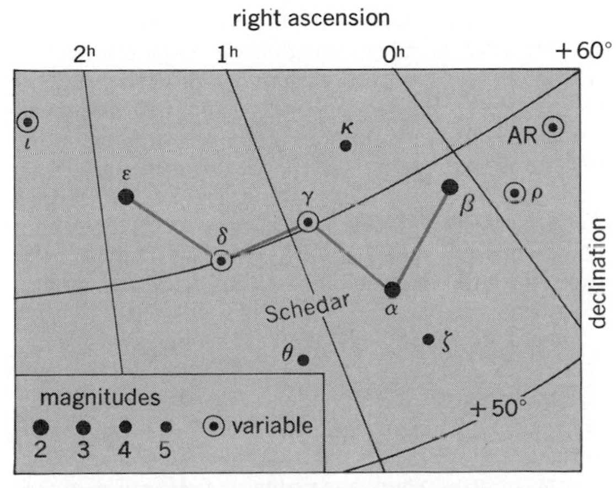

Line pattern of the constellation Cassiopeia. The grid lines represent the coordinates of the sky. Magnitudes of the stars are shown by the sizes of the "dots."

Celestial mechanics

The field of dynamics as applied to celestial bodies moving under their mutual gravitational influence in systems with few bodies. It usually describes and predicts motions in the solar system, both of natural bodies such as planets, satellites, asteroids, and comets, and of artificial bodies such as space probes. It can also be applied to small stellar systems.

Newton's laws. Isaac Netwon's law of universal gravitation is the foundation of most of the field. It states that the force produced by one particle upon another is attractive along the line connecting the bodies, is proportional to the product of the masses of the bodies, and is inversely proportional to the square of the distance between the bodies. The constant of proportionality is G, the universal constant of gravitation. Newton's second law of motion then says that the acceleration experienced by a body is equal to the force on that body divided by its mass.

Two-body problem. The simplest and only exactly solvable problem in celestial mechanics is that of one particle moving about another. Since any body with spherical symmetry looks gravitationally like a point mass from the outside, the results from this problem may be used to describe approximately the relative motion of two finite bodies, such as a planet around the Sun or a satellite around a planet. The principal results from this problem had already been recognized empirically by Johannes Kepler and are embodied in his three laws of planetary motion. Usually the motion of the smaller body (the secondary) is described relative to the larger one (the primary). This relative motion is confined to a plane, and the path traced is a conic section such that the primary occupies one focus. If the bodies are gravitationally bound, the conic is an ellipse. The longest segment connecting opposite points on the ellipse is called the major axis, and half this length is called the semimajor axis a (see **illus.**). The departure of the ellipse from a circle is called the eccentricity e, which is usually quite small for planetary orbits. The tilt of the plane from some reference plane is called the inclination, and for the solar system that reference plane is the plane of the Earth's orbit, known as the ecliptic plane. Planetary inclinations are also usually quite small. The line of intersection of the plane of motion with the reference plane is called the line of nodes. The point on the orbit closest to the primary, which is at one end of the major axis, is called the pericenter (specifically for planetary orbits, the perihelion), and its angular distance from the node is called the argument of pericenter. The time at which the secondary passes through the pericenter is called the epoch of pericenter. A seventh parameter is the period of revolution, and the cube of the semimajor axis divided by the

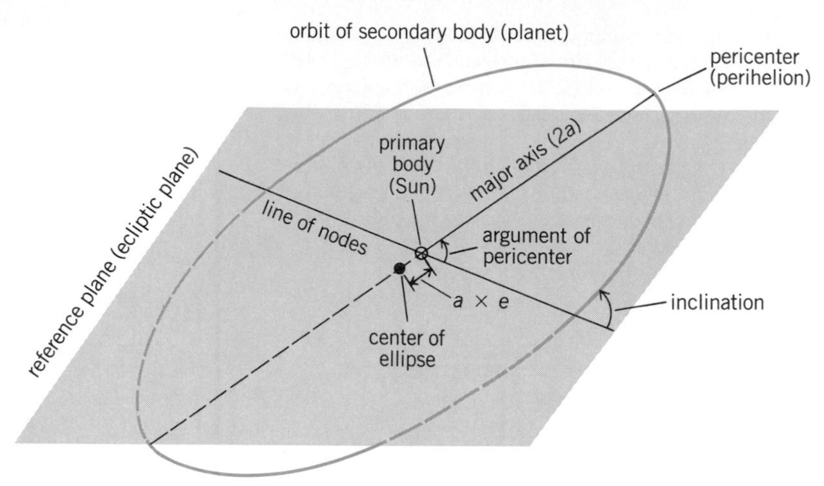

Relative motion of one body about another when the bodies are gravitationally bound. Parameters used to describe the motion are shown. (Terms used to describe the motion of a planet about the Sun are given in parentheses.) e = eccentricity.

square of the period is proportional to the sum of the two masses. Since a planetary mass is small compared to that of the Sun, this ratio is essentially constant for the planets; this is Kepler's third law, also known as the harmonic law.

A second result applies whenever the forces are directed along the line connecting the two bodies. Angular momentum is conserved, which causes the line connecting the two bodies to sweep out equal areas in equal times, a result stated in Kepler's second law. This results in the relative velocity in the orbit being inversely proportional to the square root of the separation. Ellipses are not the only type of relative motion permitted, and the type of conic depends on the total energy in the orbit. If there is just enough energy for the bodies to escape from each other, the relative orbit is a parabola. If there is more than enough energy for escape, such that some relative velocity would still remain, the orbit is a hyperbola. A hyperbola would also describe the relative motion of two independent bodies encountering each other, as in the case of two stars within the galaxy. SEE ORBITAL MOTION; PLANET.

Orbit determination. One of the major operational problems of celestial mechanics is that of determining the orbit of a body in the solar system from observations of its position, or distance plus line-of-sight velocity, at various times. The objective is to determine the numerical values of the parameters characterizing the orbit, known as orbital elements. It is usually first assumed that the motion is that of a two-body system and that the mass of the body is negligible. Thus, six elements must be determined. Each observation consists of two values at each time. Since six numbers are needed to solve for six other ones, a minimum of three observations is required. Usually there are more than three observations, which means a unique solution is not possible, but best values must be estimated in some statistical sense.

The problem is made difficult because observations are being made from the surface of the Earth, and hence parameters easily described by the two-body problem are not observed. The location of the point of observation with respect to the Sun must be accurately known, and this is often the limiting factor in orbit determination. In many cases, it is the improvement in knowledge of the orbit of the Earth that is the most useful result of a set of observations. SEE EARTH ROTATION AND ORBITAL MOTION.

Ephemeris generation. Once the orbit is known, the future locations of the object can be predicted. A table of predicted positions is called an ephemeris, and the generation of such a table is relatively straightforward. Usually, further observations are then obtained and checked for discrepancies, which in turn leads to improved values for the elements in a continually repeated cycle. SEE EPHEMERIS.

Space flight and ballistics. Another important problem is determining the proper orbit to get from one point at one time to another point at another time. This may involve getting from one body to another (space flight) or from one point to another on the same body (ballistics). For space flight, the approach is to consider several two-body problems and then patch them together. For instance, for interplanetary flight, there is an initial hyperbolic orbit with respect to one planet, an elliptic orbit with respect to the Sun for the most part, and a final hyperbolic orbit with respect to the other planet. For economy, an orbit with as little energy change as possible is desired. This dictates an elliptic heliocentric orbit that is just tangent to one planetary orbit at one extreme and just tangent to the other orbit at the other extreme. Such an orbit is known as a Hohmann transfer orbit, and it is unique for each pair of planets. The period of the orbit is known, and the tangent points are at opposite ends of the major axis. Therefore, the launch time is dictated by the relative locations of the planets. Since in practice there is a little room for adjustment, there usually is a period of time, known as the launch window, during which a launch is possible.

The ballistic problem requires determining an Earth-centered elliptic orbit that will connect launch and target points. The actual path will be that part of the ellipse above the Earth's surface. Usually the two points are at least approximately at the same distance from the center of the Earth, but the Earth rotates. This requires a cyclic procedure of estimating travel time, allowing for the shift in target location during that time, improving the orbit, and so on.

Binary stars. An astronomical application of the two-body problem is the relative motion of components of a binary star. It was realized in the eighteenth century that this motion obeys the same laws as the solar system, which established the case for the universality of the law of gravitation. In this application, there is the advantage of being on the outside looking in, but there is no information on separation in the line of sight. The same techniques of orbit determination can be applied in usually simplified form, but since the masses are not known, period must be treated as a seventh independent parameter. This leads to the primary astronomical interest in these objects, because if the orbit is known, the masses can be determined, and this is the only presently known way to determine directly these important stellar data. SEE BINARY STAR.

Restricted three-body problem. Only slightly increased in complexity is this problem of the motion of a massless particle moving in the gravitational field of two bodies moving around each other in two-body motion. The simplest version of this problem is called planar circular, which means the particle moves in the plane of revolution of the two bodies which are themselves in circular orbits. This can be applied to objects like a spacecraft going from the Earth to the Moon or asteroids under the influence of the Sun and Jupiter. This problem has no general solution; the analytic and numerical study of the problem is concerned with stability, periodic orbits, and topology of solutions. There are five specific solutions—the fixed points or libration points. If the massless particle is placed at any of these points with zero velocity in the coordinate system rotating with the primaries, it will remain at that point in the rotating system. Three of these points are located along the line connecting the primaries, one between them and one outside of each; these are known as the linear points. Their exact distances from the primaries depend on the ratio of the masses of the primaries.

The other two points form equilateral triangles with the primaries, one ahead and one behind as they revolve. Unlike the linear points, these triangular points can be stable, in that a slight displacement of the massless particle away from the point will not produce unbounded motion but rather an oscillation (called a libration) about the point. The requirement for this is that the primaries' mass ratio must be below about 0.04. This is indeed true for the Sun-Jupiter system, and there are asteroids, known as the Trojans, librating about both triangular points in this system. It is also true for the Earth-Moon systems but the presence of the Sun makes matters more complicated. However, it is possible to keep the particle reasonably close to one of these lagrangian or *L* points for extended periods of time if initial conditions are chosen properly. *SEE TROJAN ASTEROIDS.*

Perturbations. If there are three or more bodies, all of which have mass and therefore all of which influence each other, the problem becomes almost hopeless. The degree of complexity is essentially independent of the number of bodies, so the problem is called the n-body problem. This is usually studied by purely numerical means, but in two extreme cases some analytical progress can be made. One is when the number of bodies, n, becomes so large that statistical approaches are possible; this leads into the dynamics of star clusters and galaxies and out of the field of celestial mechanics. The other is when relative geometries or masses are such that the situation becomes a series of two-body problems with small coupling influences, or perturbations. These perturbations can be treated in some approximate way, such as series expansions or iterative solutions. *SEE GALAXY, EXTERNAL; MILKY WAY GALAXY; STAR CLUSTERS.*

The major efforts in these perturbation studies have been the development of general theories of motion. These are elaborate mathematical approximate solutions which almost invariably involve series expansions in some small parameter of the problem. These can be solutions for slow variations of the elements that are fixed in the two-body problem, or they can be for slow divergence from the coordinates that would result from the two-body problem, or a mixture of the two. The numerical parameters are carried as literal expressions as far as possible, but before application, actual numerical values must be estimated. This usually involves a first estimate from two-body results, then similar ephemeris generation, comparison with observations, improved estimates of the parameters, and so on.

There are two classical areas of general perturbation theory. One is the development of lunar theory, the representation of the motion of the Moon about the Earth, under the influence of rather strong perturbations from the Sun. The small parameter is the period of the Moon (a month) compared to the period of the Earth (a year), and the motion of the Earth-Moon system about the Sun is considered known. This parameter is not really very small, producing slow rates of convergence and thus expressions with thousands of terms. *SEE MOON.*

The other major development has been planetary theory, the description of the motion of planets (either major or minor) about the Sun, under the influence of (other) major planets. The small parameter is the mass of the perturbing planet or planets compared with the Sun, but series expansions involving powers of ratios of mutual distances are also required. These ratios can become quite large (for Neptune and Pluto, it can become unity), again producing slow convergence. Furthermore, there are many planets, which add more terms to the resulting expressions. However, the biggest complication is resonances, a situation realized when the revolution periods of two planets are related by small integers. (The most famous of these is the great inequality of Jupiter-Saturn, produced by five of Jupiter's almost-12-year periods being almost exactly the same as two of Saturn's almost-30-year periods). Fortunately, these perturbations are long-period, large-amplitude, and sensitive to the perturbing mass, making them useful for planetary mass determinations.

The use of the general theory is now giving way to that of the special or numerical theory. The full equations of motion are solved on computers in approximated numerical form to simulate the motions of the bodies and produce tables of coordinates as functions of time. Again, ephemerides must be generated and compared to observations to establish the best numerical values of the starting conditions. In addition, the extrapolatory value of these theories is more uncertain than that of analytic ones, because of the approximations inherent in the numerical solutions. *SEE PERTURBATION.*

Oblateness and tides. Another set of perturbations from two-body theory arises from the finite sizes of the actual bodies. Because of rotation during the time of formation, no body is completely spherically symmetric, but rather possesses at least a small bulge around its equator. This bulge will cause slow changes in the orientation of the orbit of any object circling the oblate body. It will also cause the orbiter to pull asymmetrically on the oblate body and change its orientation in space. These slow changes in orientation are collectively referred to as precession. *SEE PRECESSION OF EQUINOXES.*

If a body of finite size is under the influence of another one, the parts closer to the perturber will feel a stronger pull and the parts farther away will feel a weaker pull. If the body is not infinitely rigid, this will result in a deformation known as tides. If it has a liquid or gaseous surface and is rotating, the tidal bulges will be carried at least part way around the body, but will reach a point at which the horizontal tidal forces compensate frictional drag and an equilibrium is reached. This drag will change the rotation rate of the body, at the same time releasing energy from the system, and the asymmetric bulge will perturb the orbit of the circling body to change its period and distance. The end result in one extreme is that the system will evolve until rotation and revolution rates are equal. Many planetary satellites are thus locked into synchronous rotation with their planets, and the Earth-Moon system is evolving to a time when both the day and the month will be equal to about 47 of our present days. However, solar tides will carry Earth rotation past the synchronous state, and the lunar distance will start to decrease again. Eventually the opposite extreme situation will arise, when the tidal forces on the Moon will exceed the cohesive ones and the Moon will break up. The limit on how closely a body can approach another one without breaking up is known as Roche's limit. Its exact value depends on the nature of the bodies involved, but it is usually around 2½ times the radius of the larger body. SEE ROCHE LIMIT; SATURN; TIDE.

Unseen companions. The most complex situation is where there is a perturbation but no visible perturber. Unexplained variations can exist in observed orbits that can be due only to bodies not yet seen and therefore not yet modeled. Such was the case for Uranus after its discovery, and ultimately Neptune was discovered. Such is the case for both Uranus and Neptune, and while Pluto has been discovered, its mass is much too low to be responsible for the observed perturbations. A comparable situation can arise for a star, in which a wobble is seen in its motion across the sky, due to an unseen companion. Such detections have led to the discovery of many faint, substellar objects and may even result in the location of extrasolar planets. SEE ASTROMETRY; NEPTUNE.

Post-newtonian theories. The newtonian law of universal gravitation has been remarkably successful in explaining most astronomical dynamical phenomena. However, there have been some discrepancies, the most glaring being a small unexplainable motion in the perihelion of Mercury. For a time, a planet interior to Mercury ("Vulcan") was suspected, but the problem was resolved by Einstein's theory of general relativity. Philosophically, gravitation is quite different in the two theories, but the mathematical description of motion in general relativity shows that Newton's simple relationship is "almost" correct. The errors depend on such quantities as the square of the object's velocity compared to the square of the speed of light, which make them virtually undetectable for astronomical bodies (with the exception of Mercury). These effects are, however, easily detectable in spacecraft trajectories, and thus now have to be routinely considered. Post-Einstein theories of gravitation have also been proposed, but there is no observational support for them and thus no observational need to adopt a more complex theory. SEE GRAVITATION.

Robert S. Harrington

Bibliography. R. R. Bates, D. D. Mueller, and J. E. White, *Fundamentals of Astrodynamics*, 1971; D. Brouwer and G. M. Clemence, *Methods of Celestial Mechanics*, 1961; Y. Hagihara, *Celestial Mechanics*, 1970-1976; W. G. Hoyt, *Planets X and Pluto*, 1980; V. Szebehely, *Theory of Orbits*, 1967.

Celestial navigation

Navigation with the aid of celestial bodies, primarily for determination of position when landmarks are not available. In celestial navigation, position is not determined relative to the objects observed, as in navigation by piloting, but in relation to the points on the Earth having certain celestial bodies directly overhead.

Celestial bodies are also used for determination of horizontal direction on the Earth, and for regulating time, which is of primary importance in celestial navigation because of the changing positions of celestial bodies in the sky as the Earth rotates daily on its axis.

The navigator is concerned less with the actual motions of celestial bodies than with their apparent motions as viewed from the Earth. The heavens are pictured as a hollow celestial sphere of infinite radius, with the Earth as its center and the various celestial bodies on its inner surface. The navigator visualizes this sphere as rotating on its axis once in about 23 h 56 min—one sidereal day. The stars are then back where they were when the period started.

Bodies closer to the Earth appear to change position at a different rate than do those at a distance. The Sun appears to make a complete revolution among the stars once a year, as the Earth makes one revolution in its orbit. The apparent motion is along a great circle called the ecliptic, which is inclined nearly 23.5° to the plane of the Equator of the Earth. All of the planets stay within 8° of the ecliptic, in a band called the zodiac. Within this band they appear to move among the stars. The Moon, too, stays within the zodiac as it revolves around the Earth—or more properly as the Earth and Moon revolve around their common center of mass—once each lunar month. SEE EARTH ROTATION AND ORBITAL MOTION.

Body selection and identification. The navigator uses a limited number of celestial bodies—the Sun, Moon, 4 planets, and perhaps 20–30 stars. Although 173 stars are listed in the *Nautical Almanac*, and 57 of these are listed in the *Air Almanac* and on the daily pages of the *Nautical Almanac*, the majority of them are normally not used unless the navigator's favorite ones are unavailable. Some of the navigational stars are not seen at the latitudes traveled by many navigators. SEE ALMANAC.

With relatively few bodies in use and little change in star positions from one evening to the next, identification is seldom a problem. A tentative selection is often made in advance by means of some form of star finder. When set for the latitude of the observer and the time and date of observation, this device provides a graphical indication of the approximate altitude and azimuth of each star shown. The relative positions of

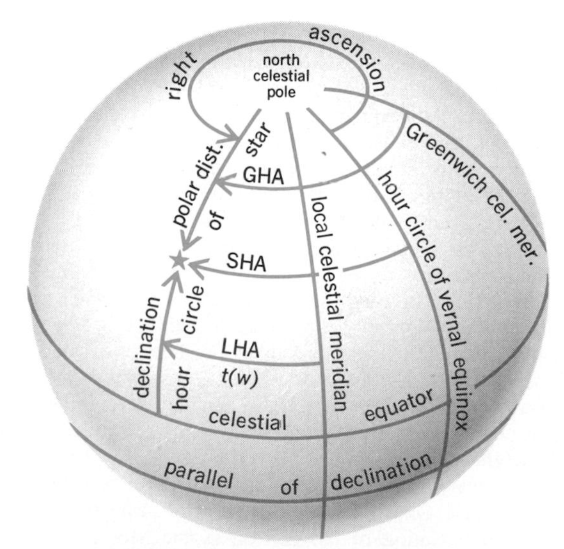

Fig. 1. Celestial equator system of coordinates.

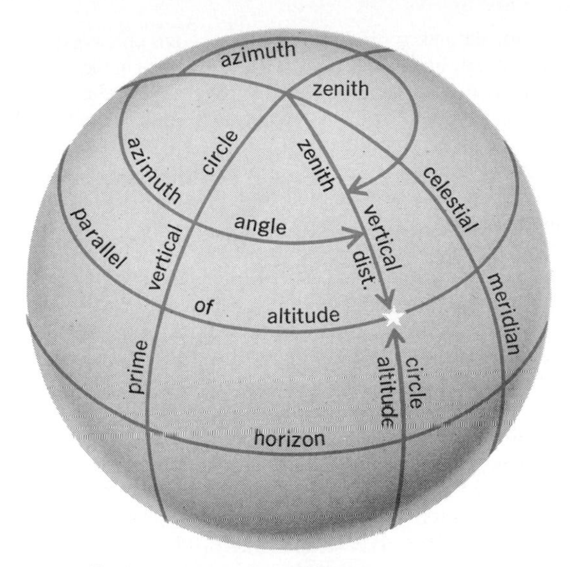

Fig. 2. Horizon system of coordinates.

other bodies can be plotted by the user. Star finders, tables, and star charts may be used to identify bodies observed before identification.

Celestial equator coordinate system. Several systems of coordinates are available to identify points on the celestial sphere. The celestial equator system of coordinates is an extension of the equatorial system commonly used on the Earth. The intersection of the plane of the terrestrial Equator, extended, with the celestial sphere is a great circle called the celestial equator. The Earth's axis, extended, intersects the celestial sphere at the north and south celestial poles. Small circles parallel to the celestial equator, similar to parallels of latitude on the Earth, are called parallels of declination. Each of these connects points of equal declination, the celestial coordinate similar to latitude on the Earth.

Great circles through the celestial poles, similar to meridians on the Earth, are called celestial meridians if they are considered to remain fixed in relation to terrestrial meridians, and hour circles if considered to remain fixed on the rotating celestial sphere.

Several different quantities on the celestial sphere are analogous to those of longitude on the Earth. Greenwich hour angle is measured westward from the Greenwich celestial meridian, through 360°. Local hour angle is similarly measured from the celestial meridian of the observer. Meridian angle is measured eastward and westward from the local meridian, through 180°. Sidereal hour angle is measured westward from the hour circle of the veneral equinox, the point at which the Sun crosses the celestial equator on its northward travel in spring, through 360°. Right ascension is measured eastward from the hour circle of the vernal equinox, usually in hours, minutes, and seconds, from 0 through 24 h. The various relationships of the celestial equator system are shown in **Fig. 1**, where GHA is Greenwich hour angle, SHA is sidereal hour angle, LHA is local hour angle, and t is meridian angle (shown here as westerly).

With the exception of right ascension, all of these quantities are customarily stated to a precision of one minute of arc by the air navigator, and to one-tenth of a minute of arc by the marine navigator. The celestial equator system is used in the almanacs for indicating positions of celestial bodies at various times. SEE ASTRONOMICAL COORDINATE SYSTEMS.

Horizon system of coordinates. The navigator also uses the horizon system of coordinates, which is similar to the celestial equator system. The primary great circle is the horizon of the observer. The pole vertically overhead is the zenith, and the opposite pole is the nadir. Small circles parallel to the horizon are called parallels of altitude, each connecting all points having the same altitude. Angular distance downward from the zenith is called zenith distance. Great circles through the zenith and nadir are vertical circles. The prime vertical circle passes through the east and west points of the horizon. Azimuth is measured clockwise around the horizon, from 000° (it is generally expressed in three figures) at the north point, to 360°. Azimuth angle is measured eastward and westward to 180°, starting from north in the Northern Hemisphere, and from south in the Southern Hemisphere. Thus, it starts directly below the elevated pole, the celestial pole above the horizon. The relationships of the horizon system are shown in **Fig. 2**. The navigator uses the horizon system because it offers the most practical references for the origin of his or her measurements.

Altitude and zenith distance are customarily stated by the air navigator to a precision of one minute of arc, and azimuth and azimuth angle to a precision of one degree; the marine navigator states them to one-tenth of a minute and one-tenth of a degree, respectively.

Two similar systems based upon the ecliptic and the galactic equator are used by astronomers, but not by navigators.

Position determination. Position determination in celestial navigation is primarily a matter of converting one set of coordinates to the other. This is done by solution of a spherical triangle called the navigational triangle.

The concept of the spherical navigational triangle is graphically shown in **Fig. 3**, a diagram on the plane of the celestial meridian. The celestial meridian passes through the zenith of the observer, and is therefore a vertical circle of the horizon system. Elements of both systems are shown in Fig. 3, indicating that an approximate solution can be made graphically.

The vertices of the navigational triangle are the elevated pole (P_n), the zenith (Z), and the celestial body

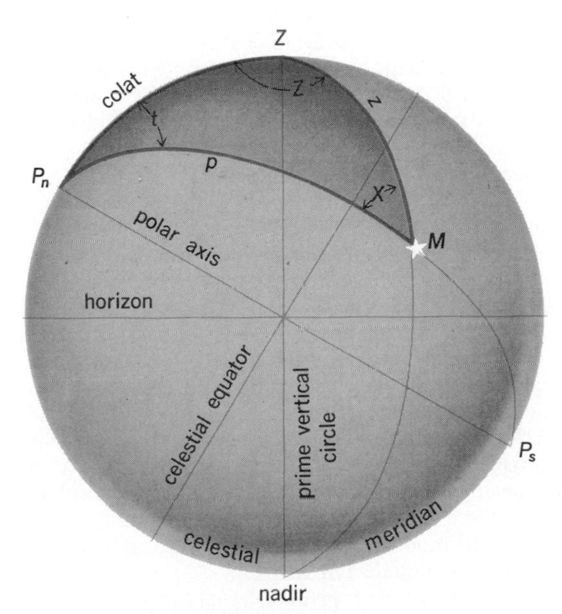

Fig. 3. Navigational triangle.

(M). The angles at the vertices are, respectively, the meridian angle (t), the azimuth angle (Z), and the parallactic angle (X). The sides of the triangle are the codeclination of the zenith or the colatitude (colat) of the observer, the coaltitude or zenith distance (z) of the body, and the codeclination or polar distance (p) of the body.

A navigational triangle is solved, usually by computation, and compared with an observed altitude to obtain a line of position by a procedure known as sight reduction.

Observed altitude. To establish a celestial line of position, the navigator observes the altitude of a celestial body, noting the time of observation. Observation is made by a sextant, so named because early instruments had an arc of one-sixth of a circle. By means of the double reflecting principle, the altitude of the body is double the amount of arc used. Similar instruments were called octants, quintants, and quadrants, depending upon the length of the arc. Today, all such instruments, regardless of length of the arc, are generally called sextants.

The marine sextant uses the visible horizon as the horizontal reference. An air sextant has an artificial, built-in horizontal reference based upon a bubble or occasionally a pendulum or gyroscope. The sextant altitude, however measured, is subject to certain errors, for which corrections are applied.

When a marine sextant is used, observations can be made only when both the horizon and one or more celestial bodies are visible. This requirement generally eliminates the period between the end of evening twilight and the beginning of morning twilight. The navigational stars and planets are therefore usually observable only during twilight. Air navigation is not subject to this limitation, star and planet observations being available all night.

In selecting bodies for observation, the navigator considers difference in azimuth, magnitude (brightness of the body), altitude (avoiding both extremes), and sometimes other factors. If speed is of particular concern, the navigator selects a body nearly ahead or astern to provide a speed line. A body near the ship's beam provides a course line. One north or south provides a latitude line, while one east or west provides a longitude line. One perpendicular to a shoreline provides an indication of distance offshore.

Many navigators prefer to observe three bodies differing in azimuth by about 120°, or four bodies differing by 90°, and preferably at about the same altitude. In this way any constant error in the altitudes is eliminated.

Sight reduction. The process of deriving from an observation the information needed for establishing a line of position is called sight reduction. A great variety of methods has been devised. That now in general use is called the Marcq St.-Hilaire method, after the French naval officer who proposed it in 1875, and is described as follows.

Having obtained the body's corrected sextant altitude, called the observed altitude, the navigator uses the almanac to obtain the Greenwich hour angle and declination of the body. The navigator converts the former to local hour angle or meridian angle for an assumed position in the vicinity of actual position. With these two quantities and the latitude of the assumed position, the navigator solves the navigational triangle for altitude and azimuth.

Each altitude of a given celestial body at any one instant defines a circle having the geographical position of the body (the point on the Earth at which the body is momentarily vertically overhead) as the center, and the zenith distance as the radius. The difference between the computed altitude for the assumed position and the observed altitude at the actual position of the observer, called the altitude difference or altitude intercept, is the difference in radii of the circles of equal altitude through the two positions.

This difference is measured along an azimuth line through the assumed position, each minute of altitude difference being considered one nautical mile. A perpendicular through the point so located is considered a part of the circle of equal altitude through the position of the observer. This is the line of position sought. The intersection of two or more nonparallel lines of position adjusted to a common time defines a fix (or running fix if the elapsed time between observations is more than a few minutes) locating the position of the observer at the time of observation. Although two lines of position are sufficient for a fix, most navigators prefer to observe three or more to provide a check and to decrease somewhat the probable error of the fix.

Sometimes, however, the navigational triangle is solved in reverse, starting with altitude, assumed latitude, and declination, and solving for meridian angle. This is then compared with Greenwich hour angle at the time of observation to determine the longitude at which the line of position crosses the assumed latitude. An American, Capt. Charles H. Sumner, used this method in his discovery in 1837 of the celestial line of position. Others used one point, found by this so-called time sight method, and the azimuth of the celestial body to establish a line of position.

Special short-cut methods have been used for a body observed on or near the celestial meridian and for Polaris. Since simple methods of sight reduction have become widely used, the popularity of such special methods has decreased.

A large number of methods have been devised for solution of the navigational triangle. The most widely accepted have been mathematical, but a great many graphical and mechanical solutions have been proposed. Many of these methods are discussed in U.S. Defense Mapping Agency Hydrographic Topographic Center Publication 9, *American Practical Navigator*, originally by N. Bowditch. Most American navigators and many of other nationalities now use one of two methods, both published by the U.S. Defense Mapping Agency Hydrographic Topographic Center. Publication 229, *Sight Reduction Tables for Marine Navigation*, is intended for use with the *Nautical Almanac* for marine navigation. Publication 249, *Sight Reduction Tables for Air Navigation*, is intended for use with the *Air Almanac* for air navigation. Editions of both of these sets of tables are also published in certain other countries.

With the emergence of electronic computers and hand-held calculators, sight reduction has been performed increasingly with limited use or elimination of tables. The extent to which a computer replaces tables depends upon the capability of the available computer and the preference of the navigator. The principal use of a nonprogrammable hand-held calculator is to perform simple arithmetical calculations such as addition, subtraction, and interpolation, with respect to both the almanac and slight reduction tables. If a programmable calculator with stored programs is available, the complete sight reduction process can be performed without the aid of tables, and if suitable data are available, even the ephemerides of celestial bodies can be computed, eliminating the need for an almanac in the usual form. An *Almanac for Computers*, providing the essential data, is published by the U.S. Naval Observatory. In the most sophisticated systems, a central computer combines sight reduction data with outputs of electronic positioning systems, perhaps including those associated with navigational satellites, and dead-reckoning data to provide continuous readout of the current most probable position.

Time relationships. Time is repeatedly mentioned as an important element of a celestial observation because the Earth rotates at the approximate rate of 1 minute of arc each 4 s of time. An error of 1 s in the timing of an observation might introduce an error in the line of position of as much as one-quarter of a mile. Time directly affects longitude determination, but not latitude. The long search for a method of ascertaining longitude at sea was finally solved two centuries ago by the invention of the marine chronometer, a timepiece with a nearly steady rate.

Several different kinds of time are used by the navigator. A timepiece, which keeps watch time, usually has a small watch error. When this is applied to watch time, the result is usually zone time. This is familiar to most people as standard time (such as Pacific Standard Time) or, when clocks are set an hour ahead, as daylight saving time (such as Central Daylight Saving Time). At sea the zones may be set by each vessel or aircraft, but they are generally 15° wide, centered on the meridians exactly divisible by 15°.

When zone time is increased or decreased by 1 h per 15° longitude (the amount of the zone description, for example, +7 for Mountain Standard Time), Greenwich mean time is obtained. This is the time used in the almanacs.

Local mean time differs from zone time by the difference in longitude between the meridian of the observer and the zone meridian, at the rate of 4 min of time for each degree of longitude, the meridian to the eastward having the later time. Local mean time is used in tables indicating time of sunrise, sunset, moonrise, moonset, and beginning and ending of twilight.

All forms of mean time are based upon apparent motions of a fictitious mean sun which provides an essentially uniform time. Apparent time, based upon the apparent (visible) Sun, may differ from mean time by a maximum of nearly 16½ min. Apparent time, plus or minus 12 h, indicates the actual position of the Sun with respect to the celestial meridian. At local apparent noon the Sun is on the celestial meridian, and the local apparent time is 1200. (Navigators customarily state time in four digits without punctuation, from 0000 at the start of a day to 2400 at the end of a day.)

Sidereal time, based upon motion of the stars, is used (indirectly in many cases) with a star finder or a star chart.

The custom of setting navigational timepieces to Greenwich mean time is growing, particularly among air navigators. This time is used almost invariably in polar regions.

Time signals are broadcast from a number of stations throughout the world to permit checking of standard timepieces. Marine chronometers are not reset by the user, an accurate record being kept of chronometer time, chronometer error, and chronometer rate. SEE TIME.

Day's work. A typical day's work of a marine navigator at sea, when using celestial navigation, is as follows:

1. Plot of dead reckoning.
2. Morning twilight observations for a fix.
3. Report of 0800 position to the commanding officer.
4. Morning sun line and compass check.
5. Winding of chronometers and determination of error and rate.
6. Noon sun line, advance of morning sun line for running fix, and report of 1200 position to the commanding officer.
7. Afternoon sun line and compass check.
8. Determination of time of sunset and preparation of a list of bodies available for observation during evening twilight.
9. Evening twilight observations for a fix.
10. Report of 2000 position to the commanding officer.
11. Determination of time of beginning of morning twilight, time of sunrise, and preparation of a list of bodies available for observation during morning twilight.
12. Time of moonrise and moonset.

Electronic applications, however, have gradually changed the pattern of celestial navigation in at least three important respects: (1) by providing noncelestial position information at sea, (2) by providing devices for automatic observation and sight reduction, and (3) by extending use of celestial navigation to all weather conditions at all times of day or night, by the use of electronic star trackers and of radio astronomy. SEE RADIO ASTRONOMY.

Alton B Moody

Bibliography. E. S. Maloney, *Dutton's Navigation and Piloting*, 1978; U.S. Defense Mapping Agency Hydrographic Topographic Center, *Air Navigation* (Publ. 216, 1966), *American Practical Navigator* (Publ. 9, 1977), *Sight Reduction Tables for Air Navigation* (3 vols., Publ. 249), *Sight Reduction Tables for Marine Navigation* (6 vols., Publ. 229); U.S. Naval Observatory, *Air Almanac, Almanac for Computers*, and *Nautical Almanac*.

Celestial sphere

The imaginary sphere, on the inside surface of which the astronomical objects appear to be located. Its center is the center of the Earth. The sphere is so large in proportion to the size of the Earth that its center can be considered as the same point as the observer, wherever he or she may be on the Earth.

A small section of the celestial sphere has been removed in the **illustration** to show the Earth at the center. The Earth's axis has been extended from the North Pole to intersect the celestial sphere in a point called the north celestial pole, which is only about 1° from Polaris, the North Star. Since the northern part of the Earth is tilted outward from the page in the illustration, the south poles of the Earth and the celestial sphere are not shown.

Halfway between the north and south celestial poles is the celestial equator. Parallel to it are circles of declination. Declination is the angular distance

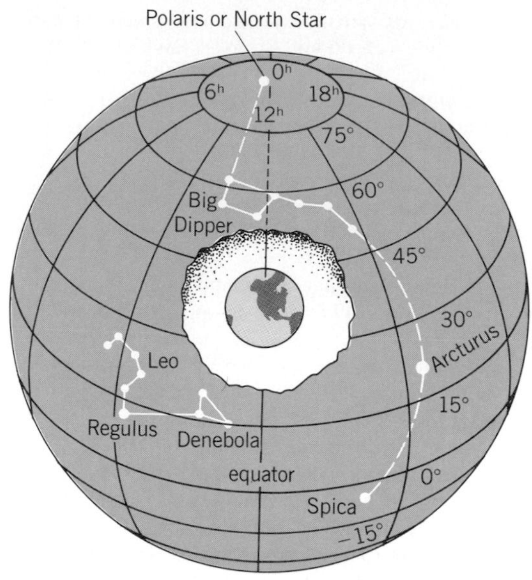

The celestial sphere and the Earth. (*After C. H. Cleminshaw, The Beginner's Guide to the Skies, T. Y. Crowell, 1977*)

north or south of the celestial equator, corresponding to latitude on the Earth.

Corresponding to the Earth's meridians, which run from pole to pole, are the hour circles on the celestial sphere. Similar to the way in which longitude is measured on Earth, right ascension is measured along the celestial equator in hours of time. *See Latitude and longitude.*

The celestial sphere is always viewed from the inside, but the illustration and a celestial globe represent it as it would appear from the outside. Hence the constellations of the Big Dipper and Leo, which are shown in the illustration, appear backward from the way in which they look to an observer on the Earth. *See Astronomical coordinate systems.*

Clarence H. Cleminshaw

Centaurus

The Centaur, in astronomy, one of the most magnificent of the southern constellations. Two first-magnitude navigational stars, Alpha and Beta Centauri, mark the right and left front feet, respectively, of the centaur. The former is Rigil Kentaurus, or simply Rigil Kent, and the latter Hadar. Rigil Kent is the third brightest star in the whole sky, ranking after Sirius and Canopus. The line joining Rigil Kent and Hadar points to the constellation Crux (Southern Cross). Thus they are also called the Southern Pointers, in contradistinction to the northern pointers of the Big Dipper. *See Constellation; Crux; Ursa Major.*

Ching-Sung Yu

Central force

A force whose line of action is always directed toward a fixed point. The central force may attract or repel. The point toward or from which the force acts is called the center of force. If the central force attracts a material particle, the path of the particle is a curve concave toward the center of force; if the central force repels the particle, its orbit is convex to the center of force. Undisturbed orbital motion under the influence of a central force satisfies Kepler's law of areas. *See Celestial mechanics.*

Raynor L. Duncombe

Cepheids

A class of brightness-variable stars whose prototype is the star Delta Cephei in the constellation Cepheus. Hundreds are known in the Milky Way Galaxy and other nearby galaxies. While both bluer and redder stars also vary in their intrinsic light, the properties of these β Cephei, ZZ Ceti, RV Tauri, and Mira variables are much less understood than the yellow-color Cepheids. These yellow stars are known to be pulsating in radius by as much as 10% or more. Their light variations are due to their changing surface area and, more importantly, their changing surface temperature. Larger yellow stars are intrinsically brighter because they have more surface area, and they have larger pulsation periods because they have a larger radius.

The interest in these stars is twofold. If their intrinsic brightnesses can be inferred from their pulsation period, the brightnesses can be used as indicators of their distance from the Earth. The observed period and a calibrated period-luminosity relation is used to give an intrinsic brightness. The observed distance-dependent apparent brightness then gives the actual distance. The second, and more current, interest in Cepheids is that their pulsation properties reveal their masses and internal structure, which help in understanding how stars age. Thus, Cepheids and the related classes of yellow pulsating stars have been extremely useful in mapping the scale of the universe and in probing the details of stellar interiors.

Types of Cepheids. Classical Cepheids are high-luminosity variables with masses between 3 and 18 times that of the Sun. Their pulsation periods range from 2 to over 100 days. They are nearing the end of their thermonuclear energy production lives, having exhausted most of their hydrogen fuel by converting it to helium in their central regions. In most of these variable stars, helium at the very center is being further converted to carbon and oxygen by even higher-temperature thermonuclear burning.

The luminosity of more slowly evolving stars with lower mass, radius, and pulsation period is still due to hydrogen converting to helium. Those stars with masses between 1 and 3 solar masses and periods between about 0.1 and 1 day are in the related class of stars called Delta Scuti variables. Both the Cepheids and the Delta Scuti variables have original hydrogen, helium, and heavier-element compositions similar to that of the Sun, and all are called population I stars.

As stars of about 1 solar mass age, they become Delta Scuti variables and later evolve again to pulsation conditions after some loss of mass. They then are called RR Lyrae variables and are seen mostly in the very old globular star clusters, with their most massive stars already evolved to their death as white dwarfs or other compact objects. These RR Lyrae variables are also seen in the general galactic field. The name population II has been given to these very old stars which have masses like the Sun but ages two or more times longer. Since they were born before some of the elements such as carbon, nitrogen, oxygen, and iron were made in earlier generations of stars, they are very deficient in these elements. As these RR Lyrae variables age and grow even larger in radius, converting their central helium to carbon and oxygen, they become population II Cepheids with periods about the same as the classical, massive, and younger population I Cepheids. *SEE NUCLEOSYNTHESIS*.

Thus the population I stars become variables when their surface temperatures are 7000–8000 K (12,100–13,900°F) as Delta Scuti variables, and, for more massive and luminous stars, when their surface temperatures are 5000–7000 K (8500–12,100°F) as Cepheids. The population II variable-star history with mass much like that of the Sun is as a Delta Scuti variable (7000–8000 K or 12,100–13,900°F), an RR Lyrae variable (6500–7500 K or 11,200–13,000°F), and finally as a population II Cepheid (5000–7000 K or 8500–12,100°F). The radii of both population types range from just over the solar radius for the shortest-period Delta Scuti variable to over 100 times that value for the longest-period Cepheids. The mean luminosity of these variables is from less than 10 to over 100,000 times that of the Sun.

Observation of pulsations. The pulsations of the Cepheids and the related classes can be detected not only by their varying luminosity but also by their outward and inward motions as detected by the periodic Doppler shifts of their spectral lines. The light range of the Cepheid variation is about one magnitude, and the peak-to-peak velocity range is between 6 and 40 mi/s (10 and 70 km/s). These variations, which are mirror images of one another with maximum light at minimum radial velocity, are frequently smooth. The interpretation of this relative variation is that the surface temperature is largest when the star is passing through its mean position in its outward expansion. However, for the Cepheids between 5- and 15-day periods, bumps occur at a phase that is late in declining light for a period of 5 days, at the peak of the luminosity for a period of 10 days, and at the very start of rising light at a period of 15 days. All these variable stars pulsate in radial modes, that is, they retain their spherical shapes even as they pulsate.

Period-luminosity-color relation. The use of the Cepheids as distance indicators requires calibration of the period-luminosity-color relation. Observations in 1912 of the Cepheids in the Small Magellanic Cloud showed that, for these stars all at the same distance from the Earth, the mean luminosity increased with period. This period-luminosity relationship can also be derived now from stellar evolution and pulsation theories. A further small effect is also known from observation and theory, that is, the period depends also on color or, physically, on the surface temperature. Two stars with the same mean luminosity can have two different mean radii, that is, different temperatures and colors, and therefore have two different periods. This period-luminosity-color relation is calibrated from stars in nearby star clusters or associations with distances known by other considerations. Then the intrinsic mean luminosity and color are directly related to an observed period. Care must be taken that the color is properly corrected for the absorption of light by interstellar matter. Since Cepheids are seen in several other nearby galaxies, they serve to measure these galaxy distances with fair accuracy.

The 1969 period-luminosity-color relation developed by A. Sandage and G. Tammann has been replaced by one based on the infrared colors J, H, and K for bands, respectively, at 1.25, 1.65, and 2.2 micrometers. This new relation is not very dependent on the pulsation phase and temperature, that is, on the color, making it unnecessary to observe the distance-indicating Cepheid around its entire pulsation cycle. It also is very insensitive to the abundance of metals in the atmosphere, the position in the instability strip (discussed below), or the reddening of the light due to interstellar absorption (see **illus.**).

Structure and evolution. Cepheids have also given insight into stellar evolution theory. These yellow variables pulsate because of an instability in the near-surface layers which periodically inhibits the flow of radiation to the surface. The detailed properties of hydrogen and helium as these elements become ionized give the instability, which grows until internal radiative damping exactly balances the ionization driving. Many linear (assuming an infinitesimal amplitude) and nonlinear (allowing for an observable amplitude) calculations have clarified the pulsation details in the several longest-period modes with only a few internal nodes.

There is a definite surface temperature, which is only slightly hotter or cooler for lesser or greater luminosity, above which the Cepheids are not observed to pulsate. The theoretical explanation for this cutoff is that when the ionization mechanisms get too shallow they do not involve much mass and do not give enough driving to overcome the deep damping. This observed blue edge of the pulsation instability strip in the log L − log T_e (where L is the luminosity and T_e is the surface effective temperature), or Hertzsprung-Russell, diagram can be used to determine a minimum helium content, because the ionization of the element helium causes most of the driving. *SEE HERTZSPRUNG-RUSSELL DIAGRAM*.

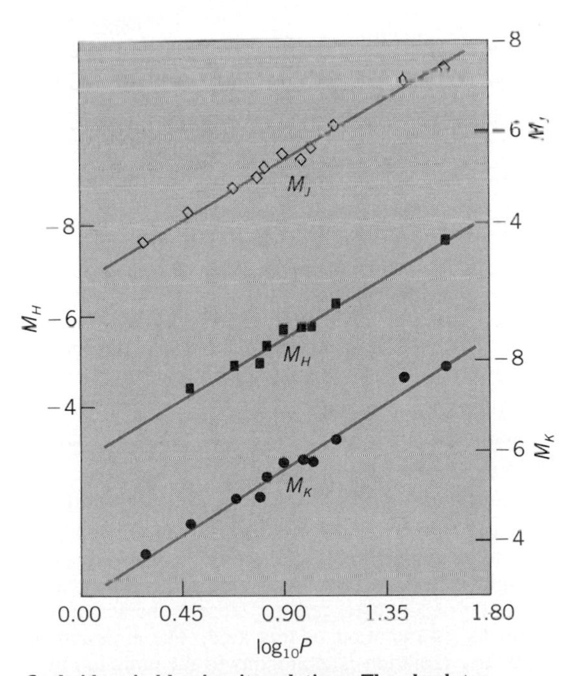

Cepheid period-luminosity relations. The absolute magnitudes M_J, M_H, and M_K, based on infrared colors *J*, *H*, and *K* for bands respectively at 1.25, 1.65, and 2.2 micrometers, are plotted as a function of the logarithm of the period *P*, in days. Data points indicate measured values; lines indicate best fits to data. (*After R. McGonegal et al., The near infrared Cepheid distance scale. I. Preliminary galactic calibration, Astrophys. J., 269:641–644, 1983*)

An observed red edge of the instability strip seems to be caused by the occurrence of convection in the surface layers. While the detailed pulsation damping mechanisms of convection are not so well understood, it is clear that there is a red (or cool) limit to the ionization instability mechanism, even though redder, often irregular variables exist.

Mass anomalies are displayed by those Cepheids which have bumps in their light variation cycles and those that exhibit two periods. The light-curve bumps and the ratios of the two periods indicate through pulsation calculations that these two classes have masses that are only one-half or even only one-third those expected from evolution calculations or from more conventional mass-determination methods. It is thought that the internal opacity of the material to radiation flow or, alternatively, unconventional composition structures might be the explanation of these smaller masses based on period ratios. Another area of research involves the consideration that these stars actually lose significant mass as they age.

The yellow pulsating stars of all classes mentioned here can reveal their mean radii, which can be compared to those predicted from stellar evolution theory. At two different phases of the pulsation cycle, when the surface temperatures are equal, the differing luminosity of the stars is due only to the differing surface areas. Use of simultaneous observations of the light and radial velocity variations can give the actual motion of the stellar surface during this time between equal-temperature phases. Analysis then gives the mean (or Wesselink) radius, which even for the anomalous-mass Cepheids agrees with evolution theory. Further, the use of the theoretical relation between the period and the mean density of the star gives a mass consistent with evolution theory. SEE STAR; STELLAR EVOLUTION; VARIABLE STAR.

Arthur N. Cox

Bibliography. S. A. Becker, L. Iben, and R. S. Tuggle, On the frequency-period distribution of Cepheid variables in galaxies in the Local Group, *Astrophys. J.*, 218:633–653, 1977; M. Breger, Delta Scuti and related stars, *Publ. Astron. Soc. Pacific,* 91:5–26, 1979; A. N. Cox, Cepheid masses from observations and pulsation theory, *Astrophys. J.*, 229:212–222, 1979; R. P. Kraft, The absolute magnitudes of classical Cepheids, in K. A. Strand (ed.), *Stars and Stellar Systems,* vol. 3, pp. 421–447, 1963; R. McGonegal et al., The intermediate Cepheid distance scale. I. Preliminary galactic calibration, *Astrophys. J.*, 269:641–644, 1983.

Cerenkov radiation

Light emitted by a high-speed charged particle when the particle passes through a transparent, nonconducting, solid material at a speed greater than the speed of light in the material. The blue glow observed in the water of a nuclear reactor, close to the active fuel elements, is radiation of this kind. The emission of Cerenkov radiation is analogous to the emission of a shock wave by a projectile moving faster than sound, since in both cases the velocity of the object passing through the medium exceeds the velocity of the resulting wave disturbance in the medium. This radiation, first predicted by P. A. Cerenkov in 1934 and later substantiated theoretically by I. Frank and I. Tamm, is used as a signal for the indication of high-speed particles and as a means for measuring their energy in devices known as Cerenkov counters.

Direction of emission. Cerenkov radiation is emitted at a fixed angle θ to the direction of motion of the particle, such that $\cos\theta = c/nv$, where v is the speed of the particle, c is the speed of light in vacuum, and n is the index of refraction of the medium. The light forms a cone of angle θ around the direction of motion. If this angle can be measured, and n is known for the medium, the speed of the particle can be determined. The light consists of all frequencies for which n is large enough to give a real value of $\cos\theta$ in the preceding equation.

Cerenkov counters. Particle detectors which utilize Cerenkov radiation are called Cerenkov counters. They are important in the detection of particles with speeds approaching that of light, such as those produced in large accelerators and in cosmic rays, and are used with photomultiplier tubes to amplify the Cerenkov radiation. These counters can emit pulses with widths of about 10^{-10} s, and are therefore useful in time-of-flight measurements when very short times must be measured. They can also give direct information on the velocity of the passing particle.

Dielectrics such as glass, water, or clear plastic may be used in Cerenkov counters. Choice of the material depends on the velocity of the particles to be measured, since the values of n are different for the materials cited. By using two Cerenkov counters in coincidence, one after the other, with proper choice of dielectric, the combination will be sensitive to a given velocity range of particles.

The counters may be classified as nonfocusing or focusing. In the former type, the dielectric is surrounded by a light-reflecting substance except at the point where the photomultiplier is attached, and no use is made of the directional properties of the light emitted. In a focusing counter, lenses and mirrors may be used to select light emitted at a given angle

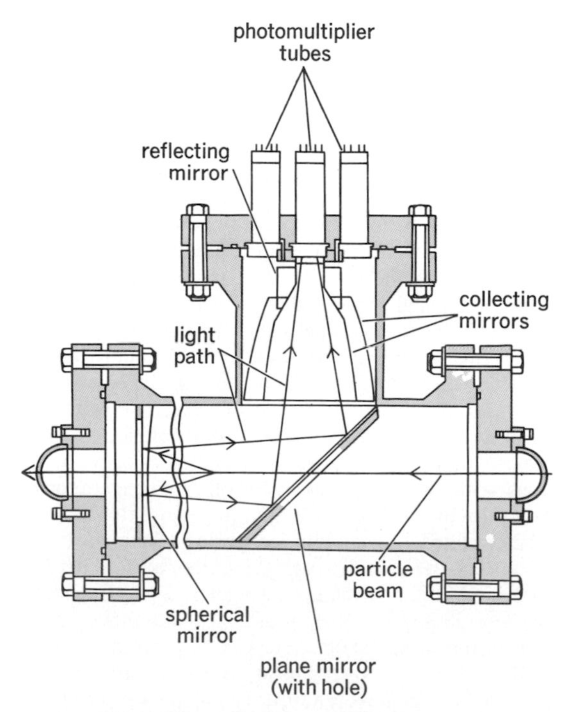

Fig. 1. Differential gas Cerenkov counter.

and thus to give information on the velocity of the particle.

Cerenkov counters may be used as proportional counters, since the number of photons emitted in the light beam can be calculated as a function of the properties of the material, the frequency interval of the light measured, and the angle θ. Thus the number of photons which make up a certain size pulse gives information on the velocity of the particle.

Gas, notably carbon dioxide (CO_2; see **Fig. 1**), may also be used as the dielectric in Cerenkov counters. In such counters the intensity of light emitted is much smaller than in solid or liquid dielectric counters, but the velocity required to produce a count is much higher because of the low index of refraction of gas.

William B. Fretter

Cerenkov astronomy. Faint flashes of Cerenkov light are emitted when energetic cosmic-ray particles penetrate the Earth's atmosphere. Observatories that are especially designed to record such radiation give information on the origin of cosmic rays and on some of the highest-energy phenomena to occur in nature.

In **Fig. 2**, the initiating radiation for the cosmic-ray air shower is a very high-energy gamma ray. For such radiation the atmosphere is practically opaque; the probability of its reaching the ground without interacting is approximately 1 in 10^{12}. The cascade of electrons and positrons shown is generated from the interplay of two processes: electron-positron pair production from gamma rays and gamma-ray emission as the electrons and positrons are accelerated by the electric fields of nuclei in the atmosphere (bremsstrahlung). For a primary gamma ray having an energy of 10^{12} eV (1 teraelectronvolt), as many as a thousand or more electrons and positrons will contribute to the cascade. The combined Cerenkov light of the cascade is beamed to the ground over an area of a few hundred meters in diameter and marks the arrival direction of the initiating gamma ray to about 1°. On a clear, dark night this radiation may be detected as a very short pulse of light lasting a few nanoseconds, by using an optical reflector (**Fig. 3**). *See* *Bremsstrahlung*.

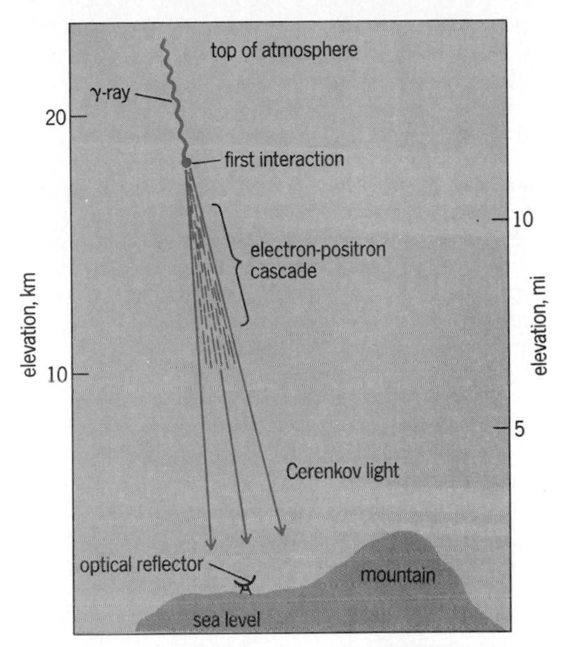

Fig. 2. Schematic view of air shower initiated by a very high-energy gamma ray.

Most cosmic-ray air showers are initiated by electrically charged particles. The magnetic fields of the Milky Way Galaxy prevent such particles from traveling in a straight line. For example, a 1-TeV proton travels along a twisted path with a radius of curvature of approximately 10^{-3} of a light-year. Therefore, the directional information regarding the source of the energetic cosmic ray is hopelessly scrambled. If the apparent arrival directions of such cosmic rays are plot-

Fig. 3. The 34-ft (10-m) optical reflector at the Smithsonian Institution's Whipple Observatory near Tucson, Arizona.

ted on a map of the sky, a uniform distribution is seen. However, high-energy gamma rays preserve their source direction; thus possible cosmic sources of high-energy gamma rays are seen as positional enhancements in the shower rate. One of the first clear source detections using this technique was the observations of Cygnus X-3 at TeV energies in 1972. *SEE MILKY WAY GALAXY*.

Since this discovery, more than a dozen other sources of very high-energy radiation have been reported. In some cases the detection technique has not depended on observing a directional enhancement but on observing a periodicity that is characteristic of a suspected source. The Crab Pulsar, the collapsed neutron star remnant of a supernova explosion in 1054 in the Milky Way Galaxy, spins with a period of 33 milliseconds. Radiation produced from beams generated near this neutron star is modulated with this same period, just as the light from a rotating beacon flashes on and off with the rotation period of the beacon. Very energetic gamma rays with this characteristic period have been observed, as well as unmodulated gamma rays coming from the diffuse nebula that surrounds the pulsar. *SEE CRAB NEBULA*.

The observation of very energetic radiation from the Crab Pulsar was, to some extent, anticipated. However, subsequent observations using the modulation technique have shown that some neutron stars in binary systems also emit pulsed radiation at TeV energies. This is surprising since the two physical situations are quite different. The Crab's neutron star is isloated. The power source is the gradual decay of the neutron star's rotational kinetic energy as its strong magnetic field interacts with the plasma environment. On the other hand, the energy source for the neutron star binaries is the strong gravitational field of the neutron star. Matter from the companion star falls onto the neutron star and, in the process, gains kinetic energy, some of which is liberated as radiation. *SEE NEUTRON STAR; PULSAR*.

There is no firm theoretical understanding of the mechanism (or mechanisms) by which the high-energy radiation from neutron star binaries is produced. The progenitors of the gamma rays must be electrically charged particles with individual particle energies in excess of the energies of the gamma rays themselves.

Perhaps the most powerful of the binary sources is Cygnus X-3. Observations have shown that it may be producing radiation at energies of 10^4 TeV. Calculations indicate that the very high-energy beams responsible for such radiation would be sufficient to account for a substantial fraction of the Milky Way Galaxy's store of very high-energy cosmic rays. An implication of these observations is that a significant portion of the more energetic cosmic rays observed on Earth may arise from the action of similar sources within the Milky Way Galaxy during the past few million years. *SEE COSMIC RAYS; GAMMA-RAY ASTRONOMY*.

Richard C. Lamb

Bibliography. J. V. Jelley, Cerenkov radiation from extensive air showers, *Prog. Elem. Part. Cosmic Ray Phys.*, 9:41–159, 1967; R. C. Lamb and T. C. Weekes, Very high energy gamma-ray binary stars, *Science*, 238:1528–1534, 1987; J. Litt and R. Meunier, Cerenkov counter technique in high-energy physics, *Annu. Rev. Nucl. Sci.*, 23:1–43, 1973; T. C. Weekes, Very high energy gamma-ray astronomy, *Phys. Rep.*, 160:1–121, 1988; L. C. L. Yuan and C. S. Wu (eds.), *Nuclear Physics*, pp. 162–194, vol. 5A of L. Marton (ed.), *Methods of Experimental Physics*, 1961.

Ceres

The first asteroid discovered. It was found serendipitously by G. Piazzi on January 1, 1801. Although it is located at the heliocentric distance predicted by the Titius-Bode law for the "missing" planet between Mars and Jupiter, additional asteroids were not located thus, and so the significance, if any, of this "law" is unknown.

With an effective diameter of 580 mi (933 km), Ceres is the largest asteroid but not the brightest since it reflects only 10% of the visual light it receives. (Vesta, though only about half as large as Ceres, is the brightest asteroid since it is closer to the Sun, 2.4 versus 2.8 astronomical units, or 2.2 versus 2.6 × 10^8 mi, or 3.6 versus 4.2 × 10^8 km, and reflects 38% of its incident light.) Ceres' mass of 2.19 × 10^{21} lbm (9.95 × 10^{20} kg, that is, 1.7 × 10^{-4} that of the Earth) contains approximately 40% of the asteroid belt's total mass. Its relatively low density, approximately 2.3 that of water (144 lbm/ft^3 or 2.3 g/cm^3), may, at least in part, be due to the fact that its surface material, and hence possibly its interior, contains water of hydration.

Ceres is a spheroidal object with an equatorial diameter of 596 mi (959 km) and a polar diameter of 564 mi (907 km), displays no large-scale color or brightness variations over its surface, and has a rotation period of 9^h03^m. *SEE ASTEROID; ASTRONOMICAL UNIT; PLANET*.

Edward F. Tedesco

Bibliography. R. Millis et al., The size, shape, density, and albedo of Ceres from its occultation of BD + 8°471, *Icarus*, 72:507–518, 1987; M. Standish and H. Hoffmann, Mass determinations of asteroids, in R. Binzel, T. Gehrels, and M. Matthes (eds.), *Asteroids II*, 1989; E. Tedesco et al., Worldwide photometry and light curve observations of 1 Ceres during the 1975–1976 apparition, *Icarus*, 54:23–29, 1983.

Charge-coupled devices

Semiconductor devices wherein minority charge is stored in a spatially defined depletion region (potential well) at the surface of a semiconductor, and is moved about the surface by transferring this charge to similar adjacent wells. The formation of the potential well is controlled by the manipulation of voltage applied to surface electrodes. Such devices can be made into astronomical detectors.

The charge-coupled-device (CCD) camera has largely replaced the photomultiplier tube and the photographic plate as the detector of choice in optical astronomy. It has light sensitivity throughout the wavelength range visible to the human eye and is nearly 100% efficient at detecting extremely faint astronomical sources. Because its response to light is linear, it is easy to use in quantitative measurements of brightness (photometry); and its output is normally digital, so charge-coupled-device images are compatible with the latest generation of computer image-processing systems.

Operation. Most astronomical charge-coupled-device cameras involve imaging onto the detector array, which has typically 1024 × 1024 picture elements called pixels (see **illus.**). Each pixel responds to a light photon by creating an electron-hole pair, and the separated charge is stored in a portion of the pixel where atoms have been specially implanted. In practical use the camera is first exposed to light by a mechanical shutter until it is time to read the camera into a digital memory. The readout cycle is done with the shutter closed because the device is still light-sensitive during readout.

Readout is accomplished by an elaborate sequence of electrical clocking pulses applied to the detector by a control electronics package that is separate from the charge-coupled-device detector itself. The clocking pulses first cause each pixel to deposit its accumulated charge in the pixel above it and to receive the charge package from the pixel below. The charge packages that were in the top row are shifted during this process into a special readout register that is adjacent to the top row. In the readout register, a separate system of clocking pulses shifts the charge packages from left to right, one pixel at a time. At the right side of the readout register, an amplifier and converter change the charge packets into digital numbers that can be stored in a digital computer memory. When the readout register has sequentially processed all of its charge packets, the top row of the image has been read into digital computer memory, and another clocking pulse causes each pixel to again dump into the pixel above it, with the top row again dumping into the readout register. In this way the picture is read out a row at a time until the complete image is digitally stored in computer memory, where it is available to be reformatted into a picture to be displayed. This process takes several seconds.

For astronomical applications, charge-coupled devices need to be operated at temperatures near −110°C (−166°F) to prevent their becoming saturated from thermally induced electron-hole pairs. Because this temperature is well below dry-ice temperature, liquid nitrogen coolant or a thermoelectric cooler must be used.

Advantages. Compared to chemical photography, the principal advantages of charge-coupled devices are stability, linearity, and sensitivity. Photographic emulsions have somewhat variable thickness across a single plate, and from plate to plate, at the level of a few percent; these thickness variations cause small variations in sensitivity. Once the plate has been exposed and chemically processed, it is not possible to calibrate these variations. The charge-coupled-device camera has small pixel-to-pixel sensitivity variations that are stable and that can be mapped and corrected for by simply imaging a uniformly illuminated screen (or the daytime or twilight sky) in a process called flat-fielding. *See Astronomical photography.*

Whereas photographic materials have a complex nonlinear response to light, with thresholds and saturations that depend on the history of plate storage and use as well as chemical processing, charge-coupled-device detectors have a very linear light response so that the recorded digital signal is proportional to stellar brightness. If saturation occurs because the storage register of an individual pixel fills, charge usually spreads to an adjacent pixel.

Because charge-coupled-device cameras are more efficient than chemical photographic materials, exposures can be shorter by a factor of about 25. This shortened exposure time can be an overwhelming advantage for astronomy, with its tradition of all-night or even all-week exposures. The maximum sensitivity is in the red spectral region, and the dectector has the further advantage of responding to wavelengths up to 1000 nanometers in the near-infrared. To obtain sensitivity in the deep violet and ultraviolet, special coatings can be applied. The charge-coupled-device can also be used as an x-ray detector.

Compared to photomultipliers, which have dominated brightness measurements since 1950, the principal advantage of the charge-coupled-device is its ability to measure brightnesses of all the stars in a field at once. Because the terrestrial atmosphere constantly changes its transmission properties at the 1% level, simultaneous measurement with charge-coupled devices has increased photometric accuracies in many programs by factors of 10 or 100. The multiple-detector advantage has made the charge-coupled device the detector of choice in many spectrometers, polarimeters, and other astronomical instruments. *See Polarimetry; Spectrograph.*

Disadvantages. Probably the principal disadvantage of charge-coupled devices for some time will be their size. The largest practical charge-coupled devices available presently have 2000 × 2000 pixels, and their information content is about that of an average 35-mm slide. It is extremely difficult to make large detectors because extremely pure silicon substrates and processing are required. Moreover, a single malfunctioning pixel tends to cause blockage of all upstream charge packets, resulting in a line defect extending across all images. Cosmic rays leave small bright spots where they strike, and at high-altitude observatories exposures are therefore usually limited to 20 min. For deeper images, several frames are recorded, individually cleaned of cosmic-ray events, and coadded. An extremely bright star in the field can mar an image when the detector keeps creating electron-hole pairs that keep filling adjacent pixels until the star image is a long line the length of the detector. Moreover, start-up costs are likely to be higher than for film because of the complex electronics, computer, and cooling systems needed.

Prospects. With such successful implementation of charge-coupled devices at all major observatories, developments of all kinds are in progress. Rectangular devices optimized for spectroscopy are being developed. So-called skipper amplifiers with multiple read

Modern charge-coupled-device detector. This 2048 × 2048-pixel detector was custom-manufactured for astronomical image applications and incorporates a skipper amplifier. The large central area is the light-sensitive surface, and the pins around the perimeter allow the electrical connections that shift the charge packets.

cycles to reduce read noise are available. More complex electronics packages with options to read only a restricted part of the detector, or to accomplish on-chip binning, are experimentally available. Drift scan techniques for sky surveys, in which the clocking rate of charge-coupled-device rows is matched to the sidereal rate for a fixed telescope, are being perfected.

Rudolph E. Schild

Bibliography. S. Howell, *Astronomical CCD Observing and Reduction Techniques*, Astronomical Society of the Pacific Conference Series, vol. 23, 1992; G. Jacoby (ed.), *CCDs in Astronomy*, Astronomical Society of the Pacific Conference Series, vol. 8, 1990.

Chromosphere

A transparent, tenuous layer of gas occurring in the atmosphere of the Sun and resting on the opaque photosphere. It is a thermal buffer zone between the low temperature at the top of the photosphere (4400 K or 7500°F) and the high temperature at the bottom of the corona ($0.5–1 \times 10^6$ K or $1–2 \times 10^6$ °F). Its temperature of about 8000 to 20,000 K (14,000 to 36,000°F) is due to the input of mechanical energy. The chromosphere consists of a homogeneous atmosphere between 1500 and 2000 km (900 and 1200 mi) thick, and an embedded fur of hairy spikes known as spicules, some 2000 to 10,000 km (1200 to 6000 mi) high. The interface between the chromosphere and the corona is a thin layer, the transition zone, in which the temperature rises and the density of the gas falls very abruptly to coronal levels. SEE SUN.

John W. Evans

Chronometer

A large, strongly built watch especially designed for precise timekeeping on ships at sea. The name is sometimes loosely applied to any fine watch.

The features that distinguish a chronometer from a watch are (1) a heavy balance wheel, the axis of which is kept always vertical by mounting the entire instrument within two concentric rings, so pivoted as to permit the chronometer to remain undisturbed despite considerable tilting of the box containing it, as the ship rolls and pitches; (2) a balance spring wound in cylindrical shape, instead of a nearly flat helix; (3) a special escapement; and (4) a fusee, by means of which the power of the mainspring is made to work through a lever arm of continuously changing length, being shortest when the spring is tightly wound and longest when it has run down, thus regulating the transmitted power so that it is approximately constant at all times.

These mechanical chronometers are being replaced by quartz digital chronometers, which operate on the same principle as a quartz watch. SEE QUARTZ CLOCK.

Oceangoing ships formerly relied completely on chronometers keeping Greenwich mean time to determine longitude. The broadcast of radio time signals that became widespread in the 1920s, and subsequently the loran, Omega, and satellite navigation systems, have made Greenwich mean time available to mariners at almost any instant, and chronometers are no longer indispensable for determining longitude at sea. However, they are still a necessary backup, and naval vessels will not leave port without one or more chronometers aboard. SEE HOROLOGY; TIME.

Steven J. Dick

Bibliography. M. E. Whitney, *The Ship's Chronometer*, 1985.

Coal Sack

An area in one of the brighter regions of the Southern Milky Way which to the naked eye appears entirely devoid of stars and hence dark with respect to the surrounding Milky Way region. Telescopic and spectroscopic observations reveal that the Coal Sack is a cloud of small, solid particles approximately 120 parsecs (400 light-years or 2.3×10^{15} mi or 3.7×10^{15} km) from Earth and 10–15 parsecs ($2–3 \times 10^4$ mi or $3–4.5 \times 10^{14}$ km) in diameter. The cloud not only absorbs about two-thirds of the visible light passing through it but also strongly reddens the transmitted light. The Coal Sack is located just to the southeast of the Southern Cross. SEE CRUX; INTERSTELLAR MATTER.

William Liller

Color index

A quantitative measure of a star's color. Even a casual look at the night sky reveals the stars to be colored. From spectral analysis and the laws of radiation, it has been known since the 1920s that such color is closely related to temperature. Stars are reddish at around 3000 K (5000°F), orange about 4500 K (7500°F), yellowish about 6000 K (10,300°F), white about 10,000 K (17,500°F), and bluish above 10,000 K. With proper calibration, color provides a means by which the temperature and spectral class can be evaluated. Visual examination is too crude,

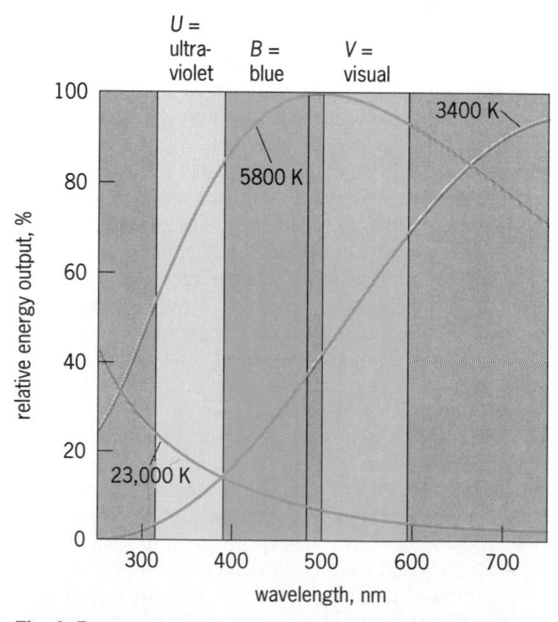

Fig. 1. Responses of the *U*, *B* (close to photographic), and *V* (close to visual) filters are superimposed onto black-body curves for 3400, 5800, and 23,000 K (5700, 10,000, and 41,000°F). The higher-temperature body is brighter in the blue and has a lower magnitude (*After S. P. Wyatt and J. B. Kaler, Principles of Astronomy: A Short Version, Allyn and Bacon, 1981*)

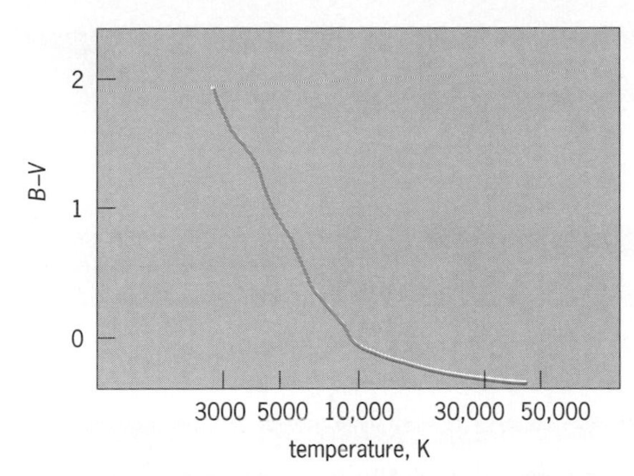

Fig. 2. Graph of the color index *B–V* versus temperature. As temperature climbs, the *B–V* color index decreases.

however. Instead color must be properly defined and quantified. SEE SPECTRAL TYPE; STAR.

Color is defined through the star's magnitude m, which indicates apparent brightness. The naked eye stars are historically grouped into six magnitude classes, 1st magnitude being the brightest and 6th magnitude the faintest. The system was quantified in the nineteenth century so that five magnitude divisions were made to correspond to a factor of 100 in brightness. One magnitude division is then a factor of 2.512. The star Vega, with $m = 0.03$, now provides a fundamental calibration point. SEE MAGNITUDE.

The original magnitudes were visual, being estimated with the unaided eye, and depended on the eye's response, which is maximized for yellow light. Photography provided a way of recording starlight and of measuring magnitudes. Early photographic emulsions, however, were sensitive primarily to blue light. As a result, blue stars look relatively brighter to the photographic plate than they do the eye, and red stars considerably fainter (**Fig. 1**). It was then necessary to establish a separate magnitude system for photography, called m_{ptg}. The original visual magnitudes were distinguished by calling them m_v. The two magnitudes were set equal for white stars with temperatures of 9300 K (16,280°F). Color can then be quantified by taking the difference between the two magnitudes, the color index becoming $m_{ptg} - m_v$. Blue stars have slightly negative color indices, red stars positive ones (**Fig. 2**). SEE ASTRONOMICAL PHOTOGRAPHY.

Photoelectric photometry allowed the establishment of the more precise *UBV* system, wherein *V* (yellow) and *B* (blue) filters respectively mimic the response of the human eye and the untreated photographic plate, and a *U* filter is added in the ultraviolet (Fig. 1). The traditional color index is then replaced by *B–V*, and the addition of *U* allows a second color index, *U–B*. Main-sequence stars trace a locus on the color-color diagram in which *U–B* is plotted against *B–V*. Deviations from the plot allow the evaluation of the degree of interstellar extinction. SEE HERTZSPRUNG-RUSSELL DIAGRAM; INTERSTELLAR EXTINCTION.

The standard system has been expanded into the red with an *R* filter, into the near-infrared with an *I* filter, and even deeper into the infrared with other filters. Numerous color indices are then available to examine stars that radiate primarily in the infrared. Other photometric schemes, such as the Strömgren four-color system (uvby, for ultraviolet, violet, blue, and yellow), allow more sophisticated color indices that are responsive not only to temperature but to a variety of other parameters. SEE INFRARED ASTRONOMY.

James B. Kaler

Bibliography. J. Dufay, *Introduction to Astrophysics: The Stars*, 1961; J. B. Kaler, *Stars and Their Spectra: An Introduction to the Spectral Sequence*, 1989; K. Aa. Strand (ed.), *Basic Astronomical Data*, 1963.

Comet

One of the major types of objects that move in closed orbits around the Sun. Compared to the orbits of planets and asteroids, comet orbits are more eccentric and have a much greater range of inclinations to the ecliptic (the plane of the Earth's orbit). Physically, a comet is a small, solid body which is roughly a half mile in diameter, contains a high fraction of icy substances, and shows a complex morphology, often including the production of an extensive atmosphere and tail, as it approaches the Sun. SEE ASTEROID; PLANET.

About 10 comets are discovered or rediscovered each year. On the average, one per year is a bright comet, visible to the unaided eye and generating much interest among the public as well as among comet workers.

Astronomers consider comets to be worthy of detailed study for several reasons: (1) They are intrinsically interesting, involving a large range of physical and chemical processes. (2) They are valuable tools for probing the solar wind. (3) They are considered to be remnants of the solar system's original material and, hence, prime objects to be studied for clues about the nature of the solar system in the distant past. (4) Comets may be required to explain other solar system phenomena.

Fig. 1. Comet West as photographed on March 9, 1976, showing the general appearance of a bright comet. The fan-shaped structure emanating from the head or coma is the dust tail, while the single straight structure is the plasma tail. (*S. M. Larson, Lunar and Planetary Laboratory, University of Arizona*)

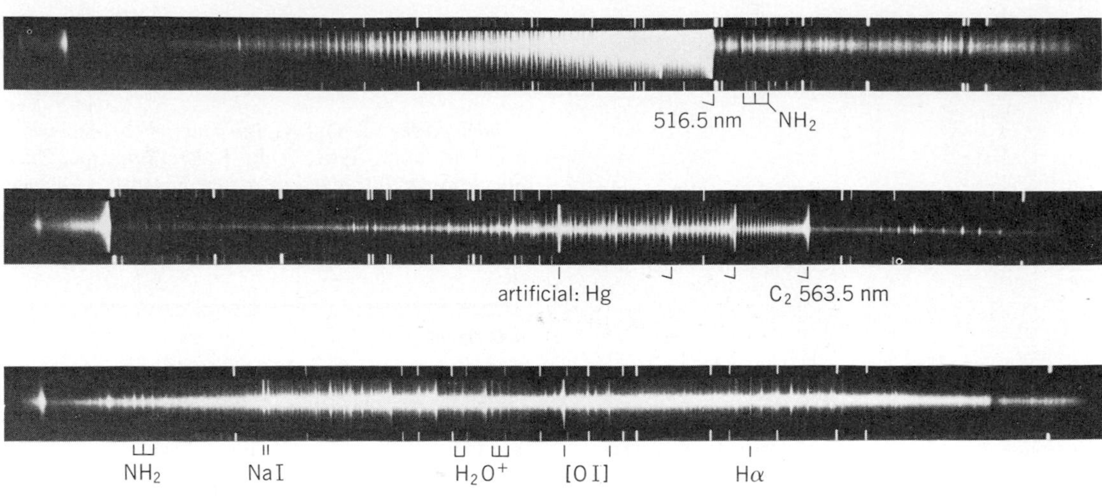

Fig. 2. High-dispersion spectrograms of Comet Kohoutek on January 9 and 11, 1974, showing the typical bright-band structure in the visible region of the spectrum. The bands are produced by molecular species; the amine radical (NH_2), diatomic carbon (C_2), and ionized water (H_2O^+) are marked. The lines of the atomic species sodium (NaI), oxygen ([OI]), and hydrogen (Hα) are also from the comet, but the mercury line (Hg) is from street lights. (*Lick Observatory, University of California*)

Appearance. As seen from Earth, comets are nebulous in appearance, and the tail is usually the most visually striking feature. This tail can in some cases stretch along a substantial arc in the sky. An example is given in **Fig. 1**, which shows Comet West dominating the eastern sky on March 9, 1976, with tails some 30° in length. Some fainter comets, however, have little or no tail.

The coma or head of a comet is seen as the ball of light from which the tail or tails emanate. Within the coma is the nucleus, the origin of the material in the tail and coma.

Discovery and designation. Comets are discovered by both amateur and professional astronomers. The fainter ones are often discovered by professionals on wide-field photographic plates taken for other purposes. Amateurs usually carry out systematic searches of the sky using wide-field binoculars or telescopes. Discoveries are communicated to the Bureau for Astronomical Telegrams, Smithsonian Astrophysical Observatory, Cambridge, Massachusetts, and are then announced by the International Astronomical Union.

Normally, comets are named after their discoverers, and up to three independent codiscoverers are allowed. An example is Comet Kobayashi-Berger-Milon. The need for such a rule is easily understood when one realizes that some bright comets have been discovered almost simultaneously by dozens of individuals. Comets are numbered in two ways. The first comet discovered in 1990 would be 1990a, the second 1990b, and so on. After the orbits have been calculated, they are assigned a roman numeral in order of perihelion (point of closest approach to the Sun) passage. The third comet to pass perihelion in 1990 would be 1990 III. Halley's Comet at its last appearance was first comet 1982i and then comet 1986 III. Halley's Comet is also designated P/Halley, the P indicating a periodic comet (a comet with period less than 200 years).

Occasionally, a comet has been named after the person who computed its orbit. Examples are Halley's Comet and Encke's Comet.

Oort Cloud and comet evolution. A major step in understanding the origin of comets was taken in 1950 by J. Oort. He developed the idea that comets are in effect "stored" in a Sun-centered spherical cloud of radius approximately 50,000 AU from the Sun (1 AU is the distance from the Sun to the Earth, 9.3×10^7 mi or 1.50×10^8 km). Occasionally, gravitational perturbations by passing stars send comets into the inner solar system, where they are discovered and their phenomena are observed. This process, of course, leads to the eventual destruction of the comet, because passages near the Sun cause a loss of material which is not recovered by the comet. Very rough estimates indicate that a comet loses 1% of its mass at each perihelion passage. Thus, a comet with radius of 0.6 mi (1 km) would lose a layer approximately 10 ft (3 m) thick on each passage. The origin and evolution of comets are discussed in more detail below.

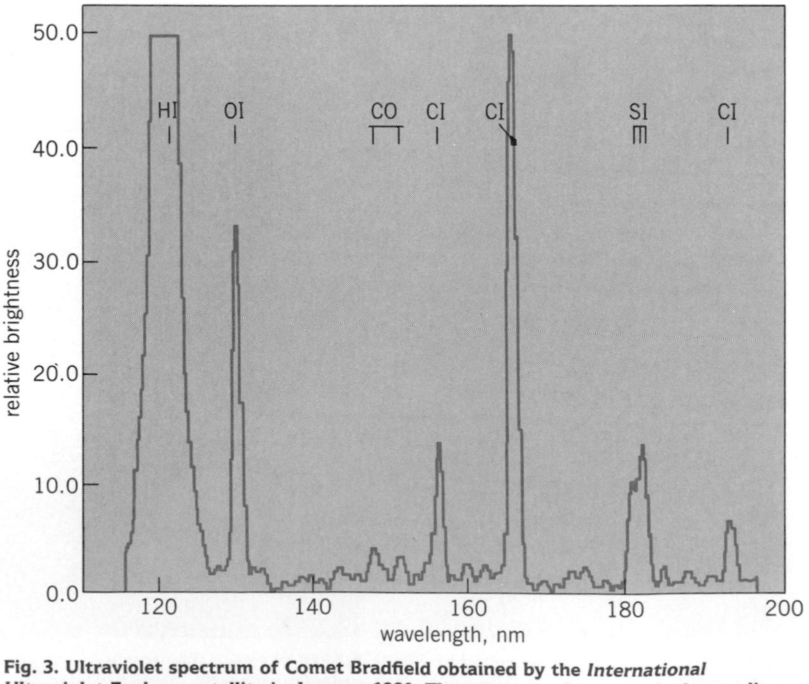

Fig. 3. Ultraviolet spectrum of Comet Bradfield obtained by the *International Ultraviolet Explorer* satellite in January 1980. The spectrum shows the hydrogen line (H I) from the hydrogen cloud as well as lines of oxygen (O I), carbon monoxide (CO), carbon (C I), and sulfur (S I). (*After P. D. Feldman et al., IUE observations of the UV spectrum of Comet Bradfield, Nature, 236, 132–135, 1980*)

A multiple exposure of the annular eclipse of May 30, 1984, photographed from Picayune, Mississippi, through a neutral density filter. An exposure was made every 10 minutes. The foreground was filled in on a separate exposure without the filter and with the camera slightly displaced. (*Courtesy of William P. Sterne, Jr.*)

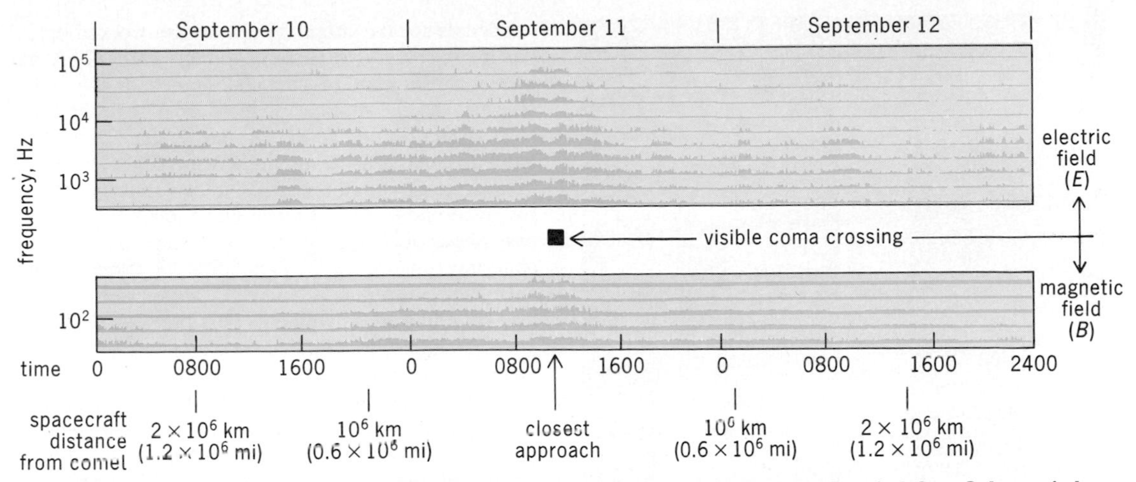

Fig. 4. Observations of plasma waves, produced when cometary ions interact with the solar wind, for a 3-day period centered on the time of closest approach of the *International Cometary Explorer* spacecraft on September 11, 1985. The peak amplitudes for the different frequency ranges are shown on a logarithmic scale. The region of intense plasma wave activity is much larger than the visible coma. (*F. L. Scarf, TRW Space and Technology Group*)

The current view considers the Oort Cloud as a steady-state reservoir which loses comets as just described and gains them from an inner cloud of comets located between the orbit of Neptune and the traditional Oort Cloud. Ideas that periodic comet showers could result from a tenth planet in the inner cloud or a faint solar companion, called Nemesis, are novel but not generally accepted.

Observations. Observations of comets run the gamut of modern observing techniques. These include photographs in visual and ultraviolet wavelengths, photometry (accurate brightness measurements) in visual and infrared wavelengths, spectral scans in many wavelength regions, radio observations, and observations in extreme ultraviolet wavelengths from rockets and orbiting spacecraft above the Earth's atmosphere. A sample spectrum in the visual wavelength range with some constituents identified is shown in **Fig. 2**. Visual wavelength spectra of comets show considerable variation. Ultraviolet spectra, obtained regularly with the *International Ultraviolet Explorer* satellite, have been found to be rather similar. A sample spectrum in ultraviolet wavelengths is shown in **Fig. 3**. Observations of meteors also contribute to the understanding of comets. *SEE ASTRONOMICAL SPECTROSCOPY*.

Measurements. The advent of space missions to comets has added an entirely new dimension to obtaining cometary information. A wide variety of direct measurement techniques have already been applied. Mass spectrometers have determined the properties of neutral gases, plasmas, and dust particles (via impact ionization). Dust counters have recorded the fluxes of dust and determined the mass distribution. Magnetometers have measured the strength and orientation of the magnetic field. Further plasma properties have been determined by plasma wave detectors.

As illustrations of this vast data source, **Fig. 4** shows the measurements of plasma wave activity around Comet Giacobini-Zinner, and **Fig. 5** shows the measurements of the densities of water (H_2O) and carbon dioxide (CO_2) in Halley's coma.

In addition, detailed imaging of the nucleus was made possible by the space missions. *SEE HALLEY'S COMET*.

Near-Sun and near-Earth comets. Most comets are observed under so-called normal circumstances; that is, the comet is between 0.5 and 1.5 AU from the Sun and a few tenths of an AU from Earth. Two specific exceptions are discussed here.

While comets have been observed at distances from the Sun of less than 0.1 AU for decades, only in 1979 was it learned that comets occasionally impact the Sun (**Fig. 6**). This fact was determined from a satellite (*SOLRAD*) used to monitor the Sun's corona; hence, the cometary discovery was not planned. The instrument carries an occulting disk which blocks out the bright glare of the Sun's surface, producing the dark circle with an apparent diameter of 5 solar radii in Fig. 6. Comet Howard-Koomen-Michels hit the Sun on August 30, 1979; the estimated time of impact was 22^h29^m Greenwich mean time. Figure 6 clearly shows the comet approaching the Sun, but its reappearance, which would have produced a cometary image on the frames shown, did not occur. Instead, a brightening of a major portion of the corona was ob-

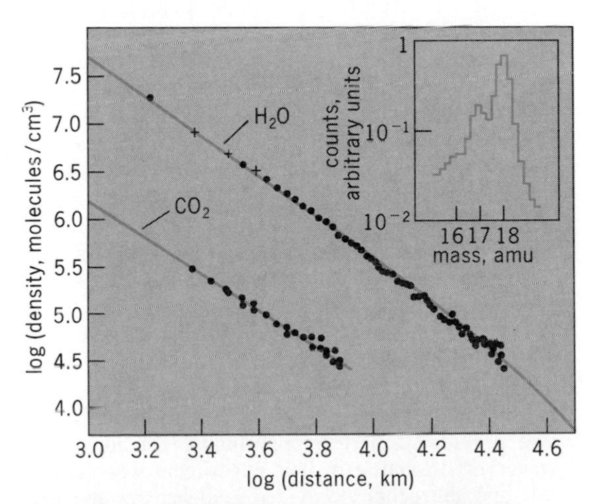

Fig. 5. *Giotto* measurements of number densities of water vapor (H_2O) and carbon dioxide (CO_2) molecules as a function of distance from the nucleus of Halley's Comet. The inset shows measurements in the mass range including 16, 17, and 18 amu and clearly indicates that H_2O (mass = 18 amu) is the dominant species. 1 km = 0.6 mi. 1 molecule/cm^3 = 16.4 molecules/in.^{3} (*After D. Krankowsky et al., In situ gas and ion measurements at Comet Halley, Nature, 321:326–329, 1986*)

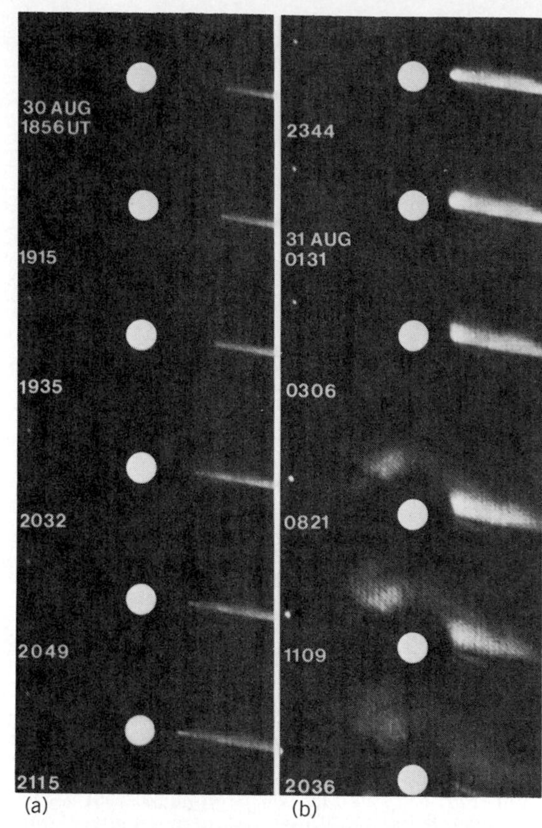

Fig. 6. Sequence of photographs obtained from a *SOLRAD* satellite showing Comet Howard-Koomen-Michels hitting the Sun in August 1979. Captions give date and universal time (Greenwich mean time) of each photograph. (*a*) Before impact. (*b*) After impact. (*Naval Research Laboratory, Washington, D.C.*)

served which then slowly faded away. The brightening was presumably caused by cometary debris ultimately blown away by the Sun's radiation pressure following the disintegration of the nucleus.

Analysis of the data from this solar monitoring satellite has turned up several more cometary impacts with the Sun. These results imply that solar impacts by relatively small comets are quite frequent.

Comets very near the Earth are rare, but provide unique opportunities for new discoveries. Comet IRAS-Araki-Alcock was another example of a comet being first discovered by a satellite, in this case the *Infrared Astronomical Satellite* (*IRAS*), which also discovered additional comets. This comet passed within 0.031 AU of Earth on May 11, 1983. This distance was only about 12 times the average distance from the Earth to the Moon. Photographs of the comet showed a faint tail on May 9 (**Fig. 7***a*); but by May 12 the tail was not visible (Fig. 7*b*) and the coma clearly displayed a distinct dual structure with bright inner and faint outer components.

The small distance to Comet IRAS-Araki-Alcock also provided a perfect opportunity to bounce radar waves off the nucleus. The experiment was successful, and the echoes provided evidence for the existence of a solid cometary nucleus with diameter of 0.6 mi (1 km).

The discovery of small comets hitting the Sun and the comets discovered by *IRAS* (including Comet IRAS-Araki-Alcock) imply that small comets are more numerous than had been generally believed. As the details for the origin of comets are worked out, it will be necessary to account for this fact. There has been increasing interest in small comets or cometlike-objects, and the number of these objects in the inner solar system may be much larger than previously thought. Such objects might be responsible for brightness decreases in the Earth's upper atmosphere and for a component of Lyman-alpha emission. In the past, these objects may have been important in lunar cratering, and the total population of comets might have been an important source of atmospheres for the terrestrial planets as well as a possible source of moderately complex organic molecules necessary for the initial development of life on Earth.

Orbits. The first closed comet orbit to be calculated was Edmond Halley's elliptical orbit for the comet of 1680. This work indicated that comet orbits were ellipses with the Sun at one focus. In subsequent work, Halley noticed the striking similarity of the orbits of what were thought to be three different comets observed in 1531, 1607, and 1682. In 1705 he concluded that these were the same comet with a period of 75 or 76 years, and predicted its return in 1758.

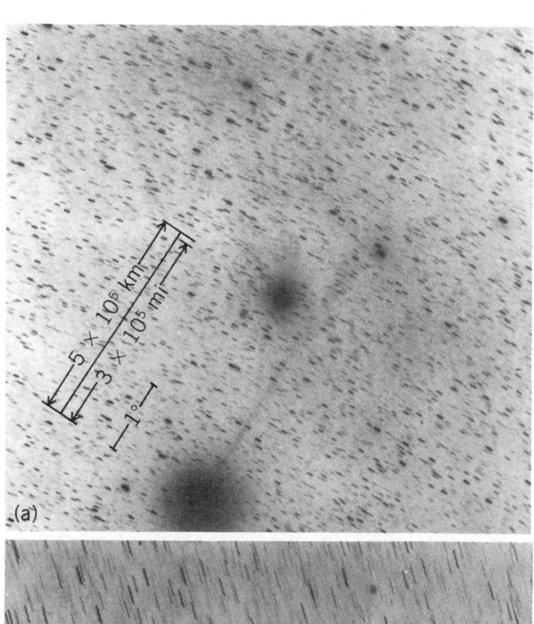

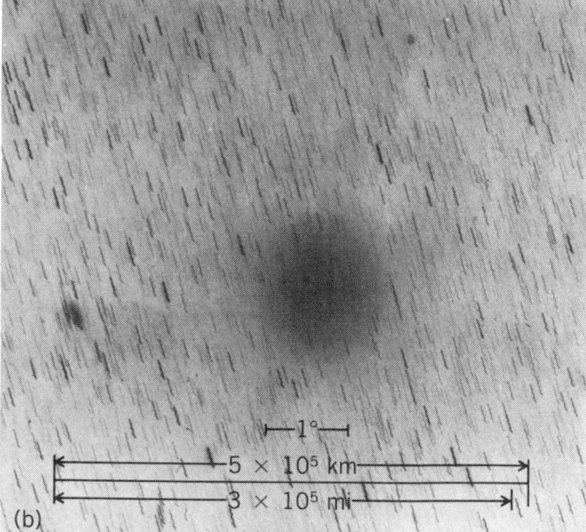

Fig. 7. Comet IRAS-Araki-Alcock. (*a*) May 9, 1983. (*b*) May 12, 1983. Scales are marked on the photograph. (*Joint Observatory for Cometary Research, operated by NASA–Goddard Space Flight Center and New Mexico Institute of Mining and Technology*)

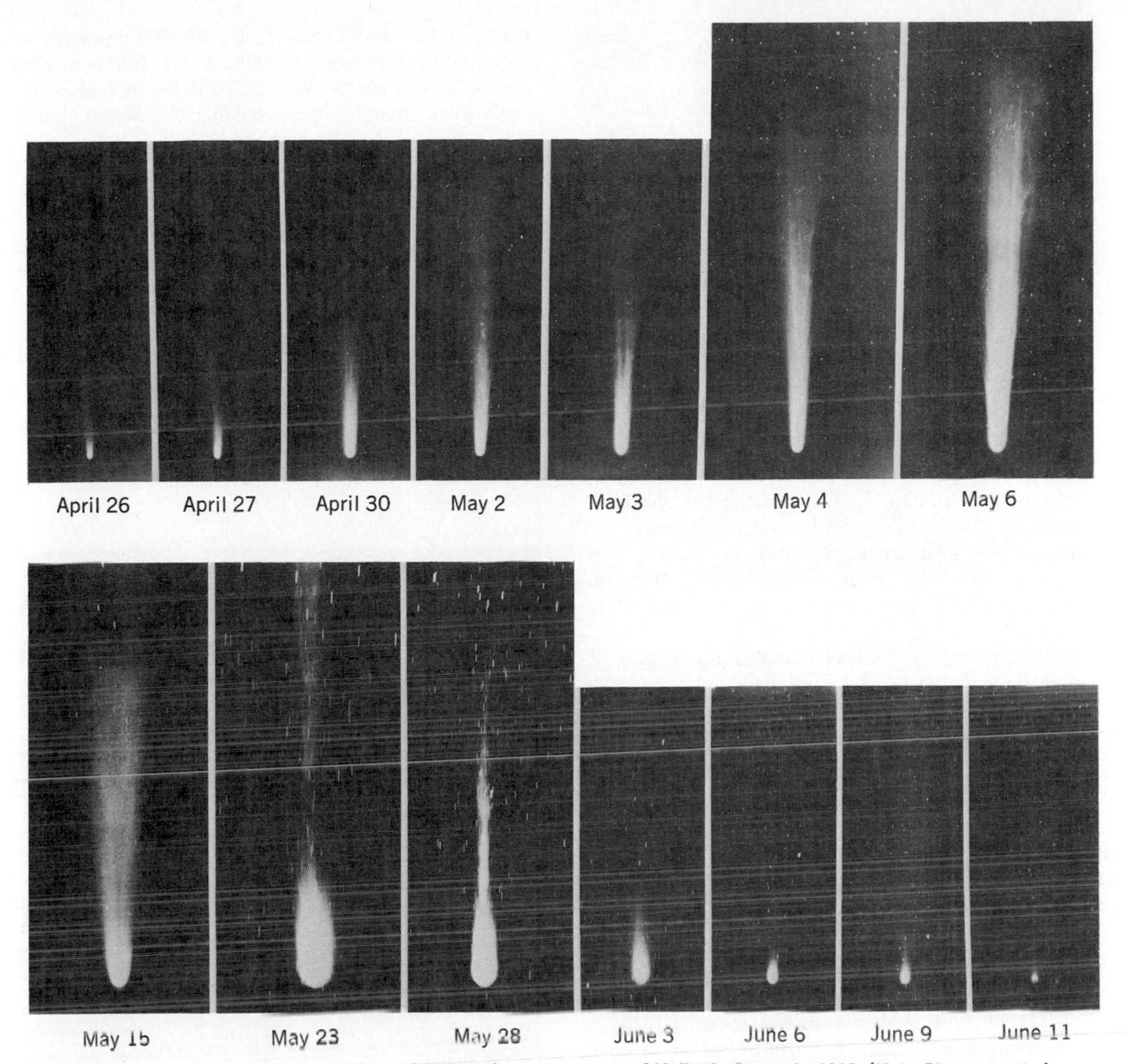

Fig. 8. Sequence of photographs showing the changing appearance of Halley's Comet in 1910. (*Hale Observatories*)

This comet is the one bearing Halley's name. It made an appearance in 1910 (**Fig. 8**), and again in 1985 and 1986.

The second comet to have its return successfully predicted was named after J. F. Encke. This comet has the shortest known period, 3.3 years. At its 1838 return, Encke's Comet showed another common property of comets, that is, a steadily changing period. This phenomenon is now known to result from the so-called nongravitational forces which must be explained by any successful comet model.

Six parameters are necessary to describe completely the orbit of a comet. They specify the orientation of the orbital plane in space, the orientation of the orbit in this plane, the size and shape of the ellipse, and the position of the comet along the orbit. In principle, three observations of position on the celestial sphere are sufficient, because each observation consists of independent measures of two coordinates. In practice, definitive orbits are derived from many observations, often hundreds.

While comet orbits are represented by ellipses to a good approximation, there are departures caused by the nongravitational forces and by the gravitational perturbations of the planets. When orbital parameters are listed for a comet, they refer to an ellipse which exactly matches the comet's position and motion at a specific time. Such an osculating orbit, as it is called, forms the starting point for studies of orbital evolution and for accurate predictions of the time and location of a comet's appearance in the sky. *See Celestial mechanics*.

Orbital parameters have been determined for over 750 individual comets. Over 600 are classified as long-period comets, that is, comets with orbital periods greater than 200 years. The orbital planes of the long-period comets have approximately random inclinations with respect to the ecliptic. This means that there are as many comets with direct orbits (revolving around the Sun in the same sense as the planets) as with retrograde orbits (revolving in the sense opposed to the planets' motion). Careful examination of the original (that is, inbound) orbits of the long-period comets shows none that are hyperbolas; that is, no interstellar comets have yet been observed. This fact strongly implies that the cloud of comets is gravitationally bound to the Sun and therefore is a part of the solar system.

About 150 short-period comets are mostly in direct orbits with inclinations of less than 30°. The distribution of periods shows a peak between 7 and 8 years, and the majority have an aphelion (point of

Atomic and molecular species observed in comets*

Neutrals	Ions
H, OH, O, S, H_2O, H_2CO C, C_2, C_3, CH, CN, CO, CS, S_2 HCN, CH_3CN, NH NH_2, Na, Fe, K Ca, V, Cr, Mn, Co, Ni, Cu $(H_2CO)_n$	CO^+, CO_2^+, H_2O^+, OH^+, H_3O^+ CH^+, N_2^+, Ca^+, C^+, CN^+

*Species observed in coma and tail in spectroscopic studies and direct spacecraft measurements.

greatest distance from the Sun) near the orbit of Jupiter.

Composition. The results of many spectroscopic studies and direct spacecraft measurements on the composition of comets are summarized in the **table**, where the atoms, molecules, ions, and classes of substances that have been observed are listed. Fairly complex molecules such as polymerized formaldehyde $[(H_2CO)_2]$ and methyl cyanide (CH_3CN) are present.

Observations of the gas composition in Halley's coma have found these values by number of molecules: water, approximately 80%; carbon monoxide, roughly 10%; carbon dioxide, approximately 3.5%; complex organic compounds such as polymerized formaldehyde, a few percent; and trace substances, the remainder. These values were obtained from direct measurements, except for that of carbon monoxide, which was determined from a rocket experiment. These values are for the outer layers of a comet which has passed through the inner solar system many times. The values of a new comet or the interior of Halley may be different with a higher fraction of carbon dioxide likely.

Measurements of dust composition were made by instruments on the spacecraft sent to Halley's Comet. Some of the particles contain essentially only the atoms hydrogen, carbon, nitrogen, and oxygen; these are the ''CHON'' particles. Other particles have a silicate composition, that is, they resemble rocks found throughout the solar system. The dominant dust composition is a mixture of these two and resembles carbonaceous chondrites enriched in the light elements hydrogen, carbon, nitrogen, and oxygen.

Structure. The main components of a comet are the nucleus, coma, hydrogen cloud, and tails.

Nucleus. Strong evidence points to the existence of a central nuclear body or nucleus for all comets from which all cometary material, both gas and dust, originates. In the early 1950s, F. L. Whipple proposed an icy conglomerate model of the nucleus. In this model, the nucleus is not a cloud of particles as had previously been thought (the ''sandbank'' model), but a single mass of ice with embedded dust particles, commonly called a dirty snowball. Such a nucleus could supply an adequate amount of gas to explain cometary phenomena and last through many apparitions because only a relatively thin surface layer would be eroded away (by sublimation) during each passage near the Sun.

The nucleus of Halley's Comet has been directly observed. Its shape was highly irregular, the surface exhibited features and structure, and the albedo (fraction of reflected light) was very low. While the confirmation of the existence of a single, nuclear body is important, Halley is just one comet and generalizations from it may not be valid. *See Albedo.*

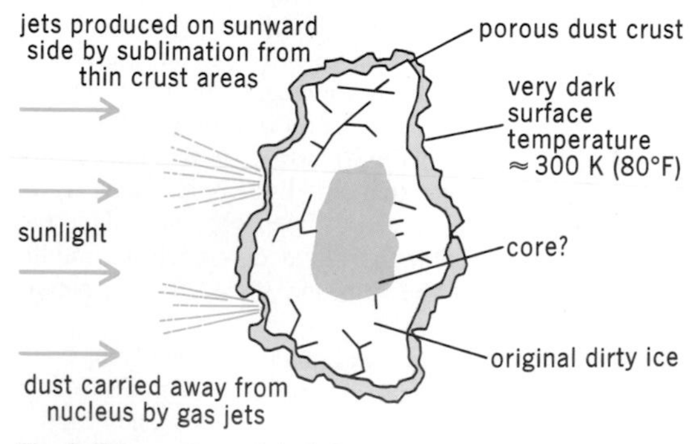

Fig. 9. Schematic model of the cometary nucleus.

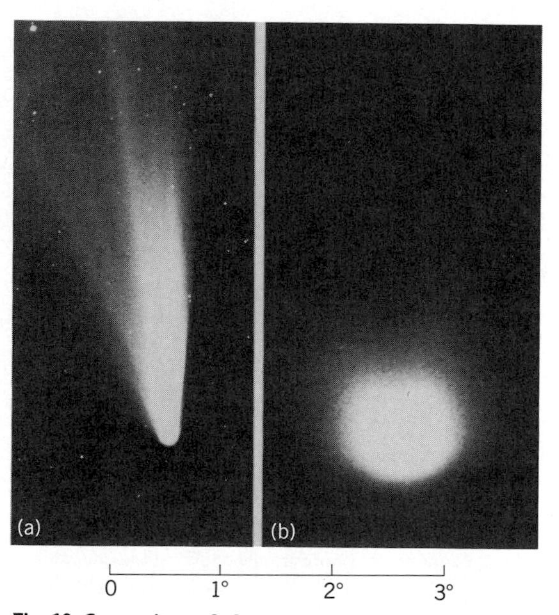

Fig. 10. Comparison of photographs of Comet West, obtained from a rocket on March 5, 1976, and printed to same scale. (*a*) Visual photograph (*P. D. Feldman, Johns Hopkins University*). (*b*) Ultraviolet photograph showing hydrogen cloud (*C. B. Opal and G. R. Carruthers, Naval Research Laboratory*).

A schematic drawing of a cometary nucleus, embodying the best available information, is given in **Fig. 9**. The generic body is probably irregular, with radius ranging from around 1000 ft (300 m) to 12 mi (20 km). Masses would range in order of magnitude from 10^{11} to 10^{16} kg (10^{11} to 10^{16} lbm), with roughly equal parts of ice and dust. The average density would be in the range 0.2 to 2 g/cm^3 (0.2 to 2 times that of water). Nuclei of comets sometimes split into two or more pieces; Comet West (1976 VI) is an example.

Coma. The coma is observed as an essentially spherical cloud of gas and dust surrounding the nucleus. The principal gaseous constituents are the neutral molecules listed in the table, and the dust composition has been described above. The coma can extend as far as 10^5 to 10^6 km (10^5 to 10^6 mi) from the nucleus, and the material is flowing away from the nucleus at a typical speed of 0.6 mi/s (1.0 km/s). As the gas flows away from the nucleus, the dust particles are dragged along. For comets at heliocentric distances greater than 2.5 to 3 AU, the coma is not normally visible and is presumed not to be present.

Hydrogen cloud. In 1970, observations of Comet Tago-Sato-Kosaka (1969g) and Comet Bennett (1969i) from orbiting spacecraft showed that these comets were surrounded by a giant hydrogen cloud that extends to distances on the order of 10^7 km (10^7 mi), or a size larger than the Sun. The observations were made in the resonance line of atomic hydrogen at 121.6 nm. Hydrogen clouds have since been observed for many other comets. A photograph of Comet West's hydrogen cloud is given in **Fig. 10**, along with a visual photograph to the same scale for comparison. Fairly bright comets (such as Bennett) have a hydrogen production rate by sublimation from the nucleus at heliocentric distance of 1 AU in the range 3 to 8 × 10^{29} atoms/s. The size of the hydrogen cloud depends on the velocity of the outflowing hydrogen atoms, and 5 mi/s (8 km/s) has been derived. This velocity would arise (from energy balance considerations) if most of the hydrogen in the cloud were produced by the photodissociation of the hydroxyl radical (OH).

Tails. Photographs of bright comets generally show two distinct types of tails (Fig. 1): the dust tails and the plasma tails. They can exist separately or together in the same comet. In a color photograph, the dust tails appear yellow because the light is reflected sunlight, and the plasma tails appear blue from emission due to ionized carbon monoxide, CO^+.

Studies of features in plasma tails led L. Biermann in 1951 to postulate the existence of a continuous outflow of ionized material from the Sun, which he called the solar corpuscular radiation, now called the solar wind. The interaction of the solar wind and its magnetic field, as suggested by H. Alfvén in 1957, plays an important role in cometary physics and in the formation of plasma tails. The direct measurements on the sunward side by the spacecraft sent to Halley's Comet and on the tailward side by the *International Cometary Explorer* mission to Comet Giacobini-Zinner in 1985 have confirmed these views. SEE SOLAR WIND.

Fig. 11. Halley's Comet as recorded by the United Kingdom Schmidt Telescope in Australia on February 22, 1986. The image shows dust tail structures (above, right), plasma tail (right, below), and an antitail (left, above). (*Royal Observatory, Edinburgh*)

The plasma tails are generally straight and have lengths that range in order of magnitude from 10^7 to 10^8 km (10^7 to 10^8 mi). The Great Comet of 1843 had a plasma tail extending over 2 AU in length. The plasma in these tails is composed of electrons and molecular ions. The dominant visible ion is CO^+, and the other ions known to be present are listed in the table. The zone of production for the molecular plasma appears to be in the coma near the sunward side of the nucleus. The material in the plasma tails is concentrated into thin bundles or streamers, and ad-

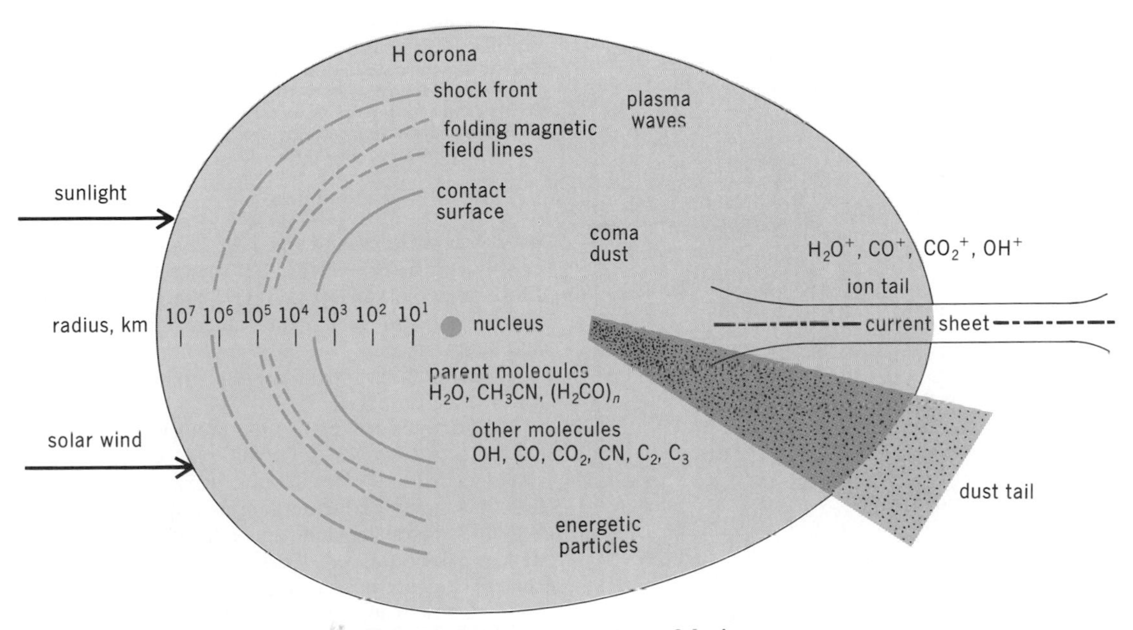

Fig. 12. Cometary features diagrammed on a logarithmic scale. 1 km = 0.6 mi.

ditional structure is found in the tail in the form of knots and kinks. These features appear to move along the tail away from the head at speeds of 6 mi/s (10 km/s) to 120 mi/s (200 km/s). Plasma tails are generally not observed beyond heliocentric distances of 1.5 to 2 AU; an exception is Comet Humason, which showed a spectacular, disturbed plasma tail well beyond normal distances.

The dust tails are usually curved and have lengths that range in order of magnitude from 10^6 to 10^7 km (10^6 to 10^7 mi). Normally, the dust tails are relatively homogeneous; an exception, Comet West, is shown in Fig. 1. Observations indicate that the dust particles are typically 1 micrometer in diameter and are probably silicate in composition. Occasionally dust tails are seen which appear to point in the sunward direction, the so-called antitails. Examples are Comet Arend-Roland in 1957 and Halley's Comet in 1986 (**Fig. 11**). These are not truly sunward appendages but are the result of projection effects.

The structure and dimensions of the constituent parts of comets—nucleus, coma, hydrogen cloud, and tails—are summarized schematically in **Fig. 12**.

Modern theory. The goal of modern comet theory is to explain the facts about comets outlined above. The broad, theoretical approach appears to be in reasonably good shape and has survived the tests of direct exploration by spacecraft.

Sublimation from nucleus. The cornerstone of the current best ideas is F. L. Whipple's icy conglomerate model of the nucleus as further developed by A. Delsemme. As a comet approaches the Sun on its orbit, sunlight supplies radiant energy to the surface of the nucleus. The energy received heats the nucleus when it is far from the Sun. As the comet continues toward the Sun, the temperature of the surface layers increases to a value, determined by the thermodynamic properties of the ice, where sublimation (passage from the solid state directly to the gaseous state) occurs. Then, most of the incident energy goes to the sublimation of ices. The situation has at least one additional complexity. Sublimation of ice from a dust–ice mixture probably leaves a dust crust which is heated to temperatures higher than the sublimation temperature of water ice. The heat is conducted inward to the ice layers, and the sublimated gases then pass outward through the dust crust. The jets of material seen on Halley's Comet originate in areas of minimal dust crust.

The onset of activity in comets at 2.5 to 3 AU is entirely consistent with water ice as the dominant ice constituent of the nucleus. But a problem still exists. Most other possible constituents are predicted from theory to begin sublimation much farther from the Sun than water ice, but, for example, molecular emissions from CN are visible essentially at the onset of activity.

The accepted solution to the problem is to assume that comets are made up of the type of ice called a clathrate hydrate. Very simply, ice in the crystalline form has cavities which are formed by the bonds that hold the crystal together and into which other substances can be trapped. Thus, the release of most minor constituents is controlled by the thermodynamic properties of water ice and not by the properties of the minor constituent. If all the available cavities are

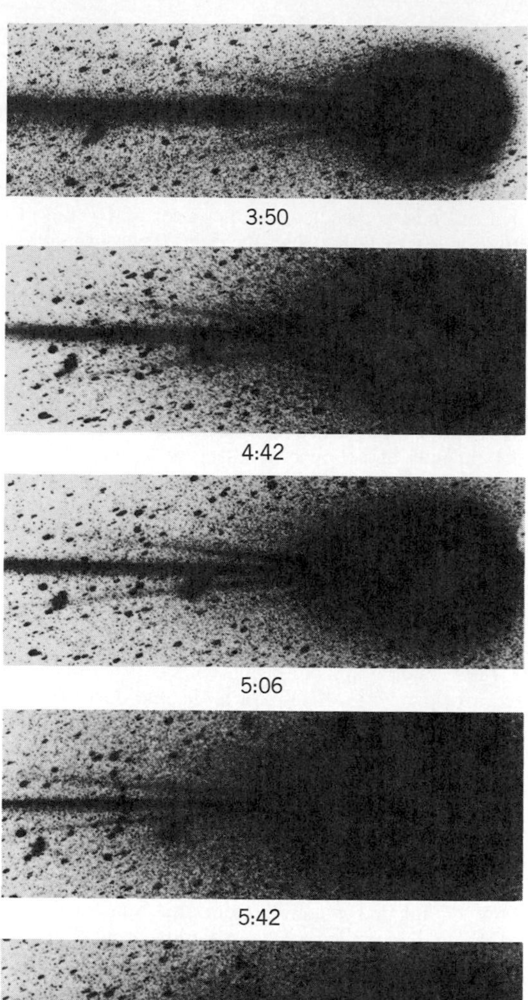

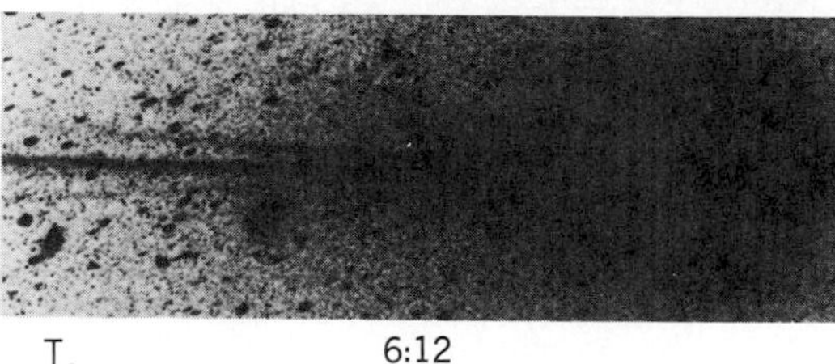

Fig. 14. Sequence of photographs of Comet Kobayashi-Berger-Milon on July 31, 1975, showing the capture of magnetic field from the solar wind. The dominant pair of tail streamers visible on either side of the main tail lengthen and turn toward the tail axis in this sequence. (*Joint Observatory for Cometary Research, operated by NASA–Goddard Space Flight Center and New Mexico Institute of Mining and Technology*)

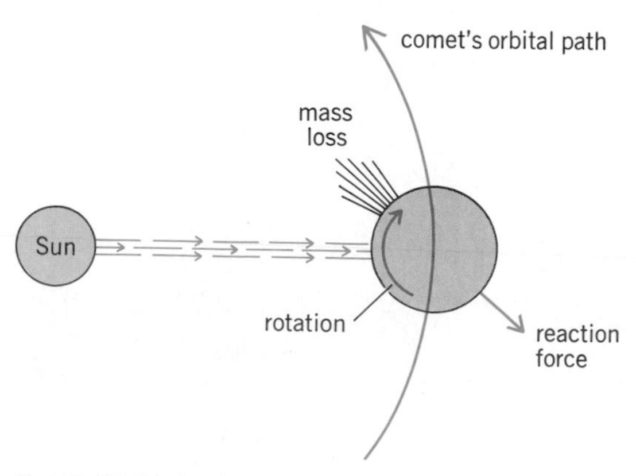

Fig. 13. Mechanism by which mass loss can produce a reaction force on the nucleus by the rocket effect.

filled, the minor constituents can amount to 17% of the icy lattice material by number of atoms. If more minor constituents, such as carbon dioxide (CO_2), are present in the nucleus, they would be vaporized well beyond 3 AU on the comet's first approach to the Sun. This phenomenon could explain comets found to be anomalously bright at large distances from the Sun and the dimming of some comets after their first perihelion passage.

Origin of nongravitational forces. The nongravitational forces have a ready explanation if the cometary nucleus is rotating. The result of radiant energy producing sublimation of ices in an ice-dust mixture is to leave a crust of dusty material. Thus, there is a time lag while the heat traverses the dust layer between maximum solar energy received and maximum loss through sublimation. If the nucleus were not rotating, the maximum mass loss would be directly toward the Sun. For a rotating nucleus, the mass loss occurs away from the sunward direction, toward the afternoon side of the comet. The analogous situation on Earth (that is, the time lag between cause and effect) produces the warmest time of day in the afternoon, not at noon. The mass loss under these circumstances produces a force on the nucleus via the rocket effect (reaction force), and this force can accelerate or retard the motion of the comet in its orbit, as shown schematically in **Fig. 13**. Detailed studies of the nongravitational forces in comets show that they are entirely consistent with water ice as the controlling substance.

Formation of coma. The sublimated gases, mostly neutral molecules, flow away from the nucleus, dragging some of the dust particles with them to form the coma. Close to the nucleus, the densities are high enough that chemical reactions can occur between molecular species. Photodissociation is also important. Thus, the molecules observed spectroscopically far from the nucleus often are not the same as the initial composition.

Formation of dust tails. The dust particles carried away from the nucleus by the flow of coma gases are blown in the antisolar direction by the Sun's radiation pressure to form the dust tails. The general theory has been developed by M. L. Finson and R. F. Probstein, and good agreement can be obtained with the observed shapes and sizes of the tails if the emission of dust from the nucleus has a peak before perihelion. The larger particles liberated from the nucleus can orbit the Sun and reflect sunlight to produce the zodiacal light.

Formation of plasma tails. The gas flowing away from the nucleus has a more involved fate. Under normal circumstances, when a comet's heliocentric distance is about 1.5 to 2 AU, significant ionization of the coma molecules occurs (probably by solar radiation), and this triggers a reaction with the solar wind. At the Earth, the solar wind, a fully ionized proton-electron gas, flows away from the Sun at 250–310 mi/s (400–500 km/s) and has an embedded magnetic field. Because of the magnetic field, the ionized cometary molecules cause the solar-wind field lines to slow down in the vicinity of the comet while proceeding at the full solar-wind speed away from the comet. This situation causes the field lines and the trapped plasma to wrap around the nucleus like a folding umbrella, to form the plasma tail. This picture has been completely confirmed by the spacecraft measurements. In addition, the folding can be seen and photographed because of the emission from the trapped ions (such as CO^+) which serve as tracers of the field lines. A photographic sequence showing this phenomenon is given in **Fig. 14**. Thus, while the ionized molecules are indeed swept in the antisolar direction by the solar wind, the plasma tail should be thought of as a part of the comet attached to the near-nuclear region by the magnetic field captured from the solar wind.

Exceptions occur at times apparently when the polarity of the solar-wind magnetic field changes. This can disrupt the magnetic connection to the near-nuclear region and literally causes the old plasma tail to disconnect while the new tail is forming. This process is quite common. **Figure 15** shows an example from Comet Morehouse in 1908, and **Fig. 16** from Comet Halley in 1986. The physics of this phenomenon is complex and not generally understood.

Fate of comets. When the process of sublimation has been carried out over an extensive period of time, as would be the case for the short-period comets, the ices will be exhausted and the inactive or "dead" comet should consist of dust particles and larger-sized rocky material. These remnants are dispersed along

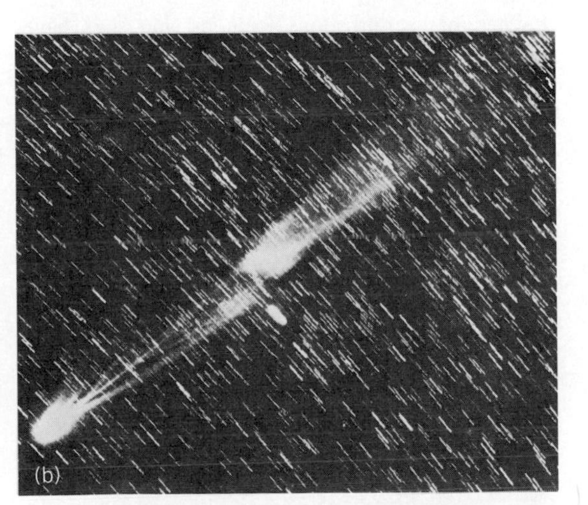

Fig. 15. Comet Morehouse. (*a*) Beginning of separation of tail from comet's head on September 30, 1908. (*b*) Tail widely separated from the head on October 1, 1908. (*Yerkes Observatory, University of Chicago*)

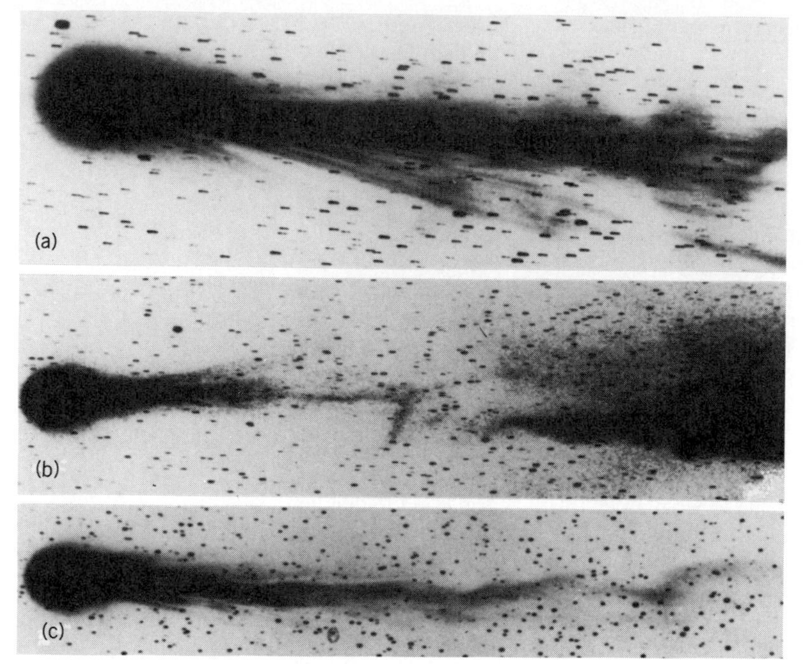

Fig. 16. Images of Halley's Comet clearly showing a disconnection event, taken on *(a)* January 9 and *(b)* January 10, 1986, at the Calar Alto Observatory of the Max-Planck Institute für Astronomy, and *(c)* January 11, 1986, at the Haute-Provence Observatory.

the comet's orbit by perturbations, and are the particles responsible for producing meteors when they enter the Earth's atmosphere. Small particles, very probably of cometary origin, have been collected by high-flying aircraft. *See* Meteor.

A detailed summary of the physical processes in comets is given in **Fig. 17**.

Origin. For years, it seemed likely that the population of short-period comets could be produced from the population of the long-period comets—those in Oort's Cloud—by their gravitational interaction with Jupiter. It is now understood that they originate in a ring of comets in the vicinity of the orbits of Neptune and Uranus.

The Oort Cloud is the source of the long-period comets. The evidence from the statistics of cometary orbits indicates an essentially spherical cloud of comets with dimensions in the range 10^4 to 10^5 AU. Gravitational perturbations from passing stars have several effects on the cloud. They limit its size and tend to make the orbits random (as observed). Most importantly, the perturbations continually send new comets from the cloud into the inner solar system, where they are observed. Thus, the Oort Cloud can be considered as a steady-state reservoir for new comets. Evidence is mounting for the view that the Oort Cloud is supplied by an inner cloud of comets.

The current consensus on the origin of comets holds that they condensed from the solar nebula at the same time as the formation of the Sun and planets. In other words, although the details are sketchy, comets are probably a natural by-product of the solar system's origin. They may be remnants of the formation process, and their material may be little altered from the era of condensation to the present time. *See* Solar system.

The consensus scenario for the origin is as follows. Generally accepted models of the solar nebula have temperature and density conditions suitable for the condensation of cometary materials at solar distances around the orbits of Uranus and Neptune. The condensate could coalesce to produce a ring of comets or cometesimals. Gravitational perturbations by the major planets would disperse the ring of comets, sending some into the inner cloud or Oort's Cloud and sending some inward to become short-period comets.

Some scenarios for formation and evolution assign additional roles to the comets. They may have been a major source of the atmospheres of the terrestrial planets and might even have provided the organic molecules necessary for the evolution of life on Earth. While some of these ideas are speculative, they help show why comets are objects of keen interest and active study, and why several space agencies have launched deep-space probes to comets during the 1980s.

Space missions to comets. Interest in sending a spacecraft to a comet was heightened by the appearance of Halley's Comet in 1985 and 1986. The first mission was the diversion of the third *International Sun-Earth Explorer* (*ISEE 3*) from Earth orbit to pass through the tail of Comet Giacobini-Zinner in September 1985. The spacecraft, renamed *International Cometary Explorer* (*ICE*), measured the properties of the tail, including the magnetic field, and the general plasma environment.

For a discussion of the missions to Halley's Comet *see* Halley's Comet. Some of these spacecraft may be diverted to new targets.

Even as the data from the first space missions are under detailed analysis, plans are being made for the next generation of missions. All of the first missions were flybys, and the next step should be rendezvous missions carried out hopefully by the year 2000. Beyond this time period, missions could involve the return to Earth of cometary material.

Comets and the public. There is no scientific evidence for the belief that comets are harbingers, omens, or actual producers of evil, disasters, or natural calamities. Some facts about comets can be distorted to provide an apparent basis for some fears. There are toxic gases in comets, but the amount is very small and, moreover, the cometary gas would not penetrate the Earth's atmosphere. The only real problem would be an impact of a comet on the Earth's surface. The odds of an impact are very small, and studies indicate that the somewhat fragile cometary nucleus would break up high in the Earth's atmosphere. Thus, the Tunguska event, which took place in central Siberia on June 30, 1908, and which produced substantial devastation, is no longer believed to have a cometary origin.

The opportunities for much of the public to view bright comets are becoming increasingly rare due to the combination of atmospheric pollution and bright city lights. In January 1974 there was general public disappointment over the poor visibility of Comet Kohoutek, particularly as viewed from cities in the populous northeastern United States, although it was a conspicuous object to astronomers at mountaintop observatories in the southwestern United States. Halley's Comet was not a spectacular object for public viewing in 1985 and 1986. In part, this was due to the circumstances of the comet's orbit, which led to the comet's times of greatest intrinsic brightness occurring well away from the Earth in contrast to the situation in 1910. The march of light pollution is

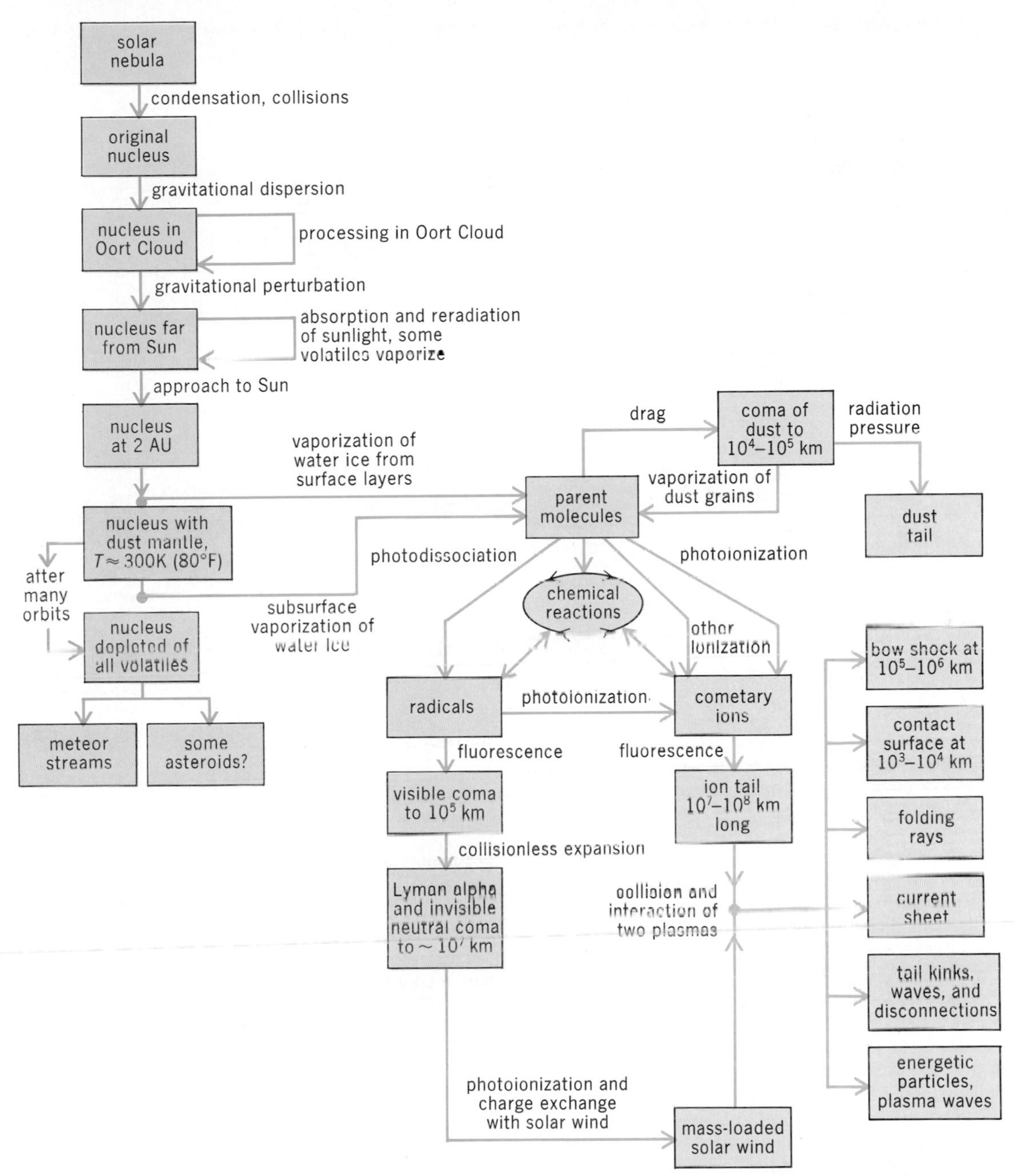

Fig. 17. Summary of processes involved in the formation of comets and the interaction with sunlight and the solar wind. 1 km = 0.6 mi.

inexorable, and the interested viewer of comets will require an observing site well away from urban population centers.

John C. Brandt

Bibliography. J. C. Brandt, *Comets: A Scientific American Reader,* 1981; J. C. Brandt and R. D. Chapman, *Introduction to Comets,* 1981; R. D. Chapman and J. C. Brandt, *The Comet Book,* 1984; M. Grewing, F. Praderie, and R. Reinhard (eds.), *Exploration of Halley's Comet*, 1988; The *International Cometary Explorer* Mission to Comet Giacobini-Zinner, *Science*, 232:353–385, 1986; Voyages to Comet Halley, *Nature*, 321:259–365, 1986; F. L. Whipple, *The Mystery of Comets*, 1985; L. L. Wilkening, *Comets*, 1982.

Constellation

One of the 88 areas into which the sky is divided. Each constellation has a name that reflects its earliest recognition. Though pictures are associated with the constellations, they have no official status, and constellations have been depicted differently by different artists.

The identifications of the constellations are lost in antiquity. No doubt ancient peoples associated myths with the heavens and imagined pictures connecting or surrounding the bright stars. The names of some constellations have been handed down by the Chaldeans or Egyptians, but most of those that can be seen from midnorthern latitudes have Greek or Roman origins.

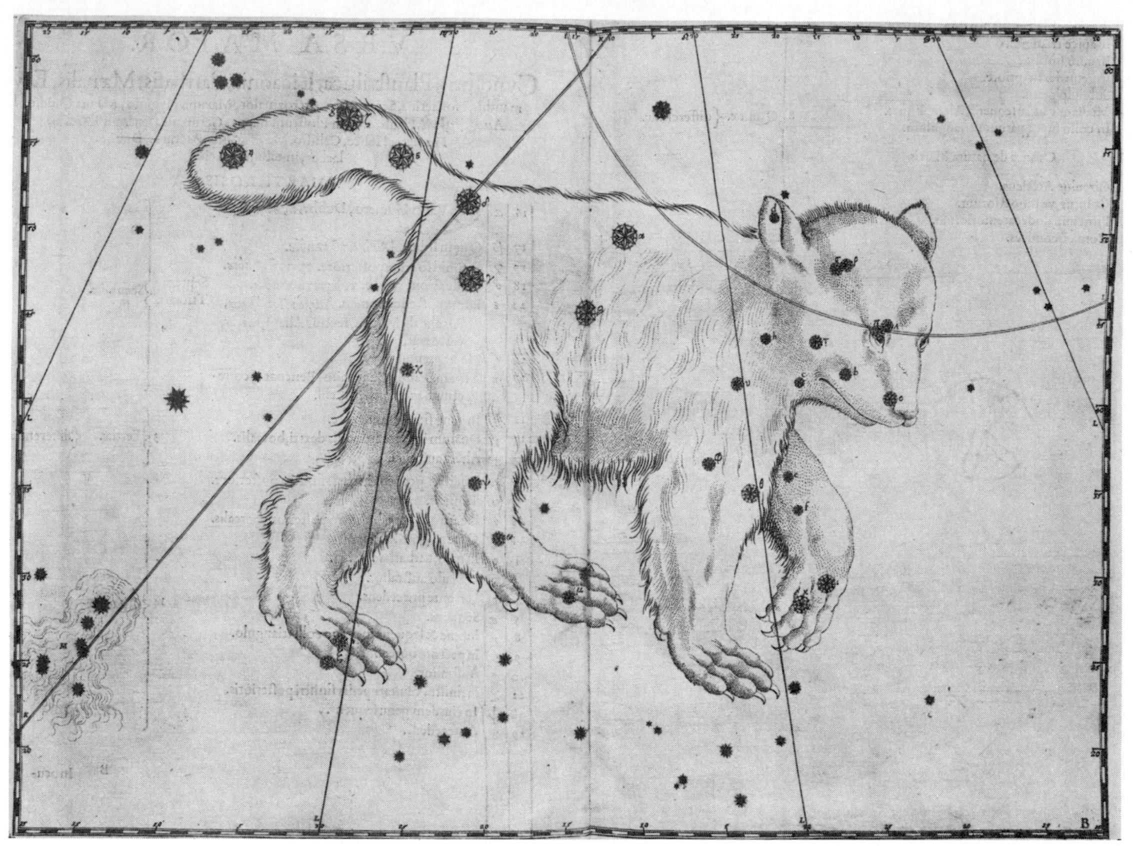

Fig. 1. The constellation Ursa Major from Bayer's star atlas.

Star maps are found in various cultures, and constellation figures and myths usually differ from those found in Greek and Roman sources.

Star catalogs and atlases. The catalog of Ptolemy, in Hellenic Alexandria in the second century of the Christian Era, included over 1000 stars grouped into 48 constellations. Many of the images of constellations are derived from the beautiful engravings by Alexander Mair in Johann Bayer's *Uranometria* (1603) **[Fig. 1]**. Bayer included the constellations listed by Ptolemy and also named 12 new ones containing stars observed on expeditions to the Southern Hemisphere. Bayer originated the scheme of labeling individual stars in constellations with Greek and other letters, roughly in order of brightness, and the genitive form of the constellation name. For example, the bright star Betelgeuse (the second brightest in the constellation Orion) is alpha Orionis (alpha of Orion), and Sirius, the brightest star in the sky, is alpha Canis Majoris (alpha of the Big Dog). In some cases, Bayer labeled stars in order around figures in the sky, as for the Big Dipper. Bayer also used some uppercase (as in the star P Cygni) and lowercase Latin letters. In 1624, Jakob Bartsch placed three new constellations in gaps in Bayer's atlas and separated the southern constellation Crux from Centaurus. Coma Berenices (Berenice's Hair) was also added at about this time, reportedly by Tycho Brahe.

Johannes Hevelius added nine more southern constellations in his 1690 star atlas, *Firmamentum Sobiescianum sive Uranographia*. His figures represent the view of a celestial globe from the outside and thus are reversed from Bayer's. Nicolas Louis de Lacaille added 14 constellations in 1763 from his expedition to the Cape of Good Hope, using names reflecting the mechanical age, such as air pump, microscope, and telescope.

John Flamsteed (1729) and Johan Elert Bode (1801) produced other elegant star atlases, with engravings of constellations. Since the mid-1800s, Ptolemy's largest constellation, Argo Navis, has been divided into Carina, the keel; Puppis, the stern; and Vela, the sails.

Modern star atlases, from *Uranometria 2000.0* to the *Guide Star Catalogue* compiled in the late 1980s for the use of observers with the Hubble Space Telescope, do not usually show the figures of tradition. *SEE* ASTRONOMICAL ATLASES; ASTRONOMICAL CATALOGS.

International agreement. In 1928, the International Astronomical Union formally accepted the division of the sky into 88 constellations (see **table**), with the final list provided 2 years later; each star now falls in only one constellation (**Fig. 2**). The boundaries follow north-south or east-west celestial coordinates (right ascension and declination lines) from the year 1875; because of precession, the current boundaries do not match rounded values of celestial coordinates. *SEE* ASTRONOMICAL COORDINATE SYSTEMS; PRECESSION OF EQUINOXES.

Asterisms. Some of the most familiar patterns in the sky are asterisms rather than constellations. For example, the asterism known as the Big Dipper is part of the constellation Ursa Major. The asterism known as the Great Square of Pegasus has three of its corners in Pegasus but the fourth in Andromeda. The Northern Cross is made of stars in Cygnus.

Zodiac. The Sun, the Moon, and the planets move through a band in the sky known as the ecliptic. The constellations that fall close to the ecliptic through which the Sun traditionally moves are known as the

The constellations*

Latin name	Genitive	Abbreviation	English translation
Andromeda	Andromedae	And	Andromeda[†]
Antlia	Antliae	Ant	Pump
Apus	Apodis	Aps	Bird of Paradise
Aquarius	Aquarii	Aqr	Water Bearer
Aquila	Aquilae	Aql	Eagle
Ara	Arae	Ara	Altar
Aries	Arietis	Ari	Ram
Auriga	Aurigae	Aur	Charioteer
Boötes	Boötis	Boo	Herdsman
Caelum	Caeli	Cae	Chisel
Camelopardalis	Camelopardalis	Cam	Giraffe
Cancer	Cancri	Cnc	Crab
Canes Venatici	Canum Venaticorum	CVn	Hunting Dogs
Canis Major	Canis Majoris	CMa	Big Dog
Canis Minor	Canis Minoris	CMi	Little Dog
Capricornus	Capricorni	Cap	Goat
Carina	Carinae	Car	Ship's Keel[‡]
Cassiopeia	Cassiopeiae	Cas	Cassiopeia[†]
Centaurus	Centauri	Cen	Centaur[†]
Cepheus	Cephei	Cep	Cepheus[†]
Cetus	Ceti	Cet	Whale
Chamaeleon	Chamaeleonis	Cha	Chameleon
Circinus	Circini	Cir	Compass
Columba	Columbae	Col	Dove
Coma Berenices	Comae Berenices	Com	Berenice's Hair[†]
Corona Australis	Coronae Australis	CrA	Southern Crown
Corona Borealis	Coronae Borealis	CrB	Northern Crown
Corvus	Corvi	Crv	Crow
Crater	Crateris	Crt	Cup
Crux	Crucis	Cru	Southern Cross
Cygnus	Cygni	Gyg	Swan
Delphinus	Delphini	Del	Dolphin
Dorado	Doradus	Dor	Swordfish
Draco	Draconis	Dra	Dragon
Equuleus	Equulei	Equ	Little Horse
Eridanus	Eridani	Eri	River Eridanus[†]
Fornax	Fornacis	For	Furnace
Gemini	Geminorum	Gem	Twins
Grus	Gruis	Gru	Crane
Hercules	Herculis	Her	Hercules[†]
Horologium	Horologii	Hor	Clock
Hydra	Hydrae	Hya	Hydra[†] (water monster)
Hydrus	Hydri	Hyi	Sea Serpent
Indus	Indi	Ind	Indian
Lacerta	Lacertae	Lac	Lizard
Leo	Leonis	Leo	Lion
Leo Minor	Leonis Minoris	LMi	Little Lion
Lepus	Leporis	Lep	Hare
Libra	Librae	Lib	Scales
Lupus	Lupi	Lup	Wolf
Lynx	Lyncis	Lyn	Lynx
Lyra	Lyrae	Lyr	Harp
Mensa	Mensae	Men	Table (mountain)
Microscopium	Microscopii	Mic	Microscope
Monoceros	Monocerotis	Mon	Unicorn
Musca	Muscae	Mus	Fly
Norma	Normae	Nor	Level (square)
Octans	Octantis	Oct	Octant
Ophiuchus	Ophiuchi	Oph	Ophiuchus[†] (serpent bearer)
Orion	Orionis	Ori	Orion[†]
Pavo	Pavonis	Pav	Peacock
Pegasus	Pegasi	Peg	Pegasus[†] (winged horse)
Perseus	Persei	Per	Perseus[†]
Phoenix	Phoenicis	Phe	Phoenix
Pictor	Pictoris	Pic	Easel
Pisces	Piscium	Psc	Fish
Piscis Austrinus	Piscis Austrini	PsA	Southern Fish
Puppis	Puppis	Pup	Ship's Stern[‡]
Pyxis	Pyxidis	Pyx	Ship's Compass[‡]
Reticulum	Reticuli	Ret	Net
Sagitta	Sagittae	Sge	Arrow
Sagittarius	Sagittarii	Sgr	Archer
Scorpius	Scorpii	Sco	Scorpion
Sculptor	Sculptoris	Scl	Sculptor
Scutum	Scuti	Sct	Shield
Serpens	Serpentis	Ser	Serpent
Sextans	Sextantis	Sex	Sextant
Taurus	Tauri	Tau	Bull
Telescopium	Telescopii	Tel	Telescope
Triangulum	Trianguli	Tri	Triangle
Triangulum Australe	Trianguli Australis	TrA	Southern Triangle
Tucana	Tucanae	Tuc	Toucan

The constellations* (continued)

Latin name	Genitive	Abbreviation	English translation
Ursa Major	Ursae Majoris	UMa	Big Bear
Ursa Minor	Ursae Minoris	UMi	Little Bear
Vela	Velorum	Vel	Ship's Sails‡
Virgo	Virginis	Vir	Virgin
Volans	Volantis	Vol	Flying Fish
Vulpecula	Vulpeculae	Vul	Little Fox

*After J. M. Pasachoff, *Contemporary Astronomy*, 4th ed., 1989.
†Proper names.
‡Formerly formed the consellation Argo Navis, the Argonauts' Ship.

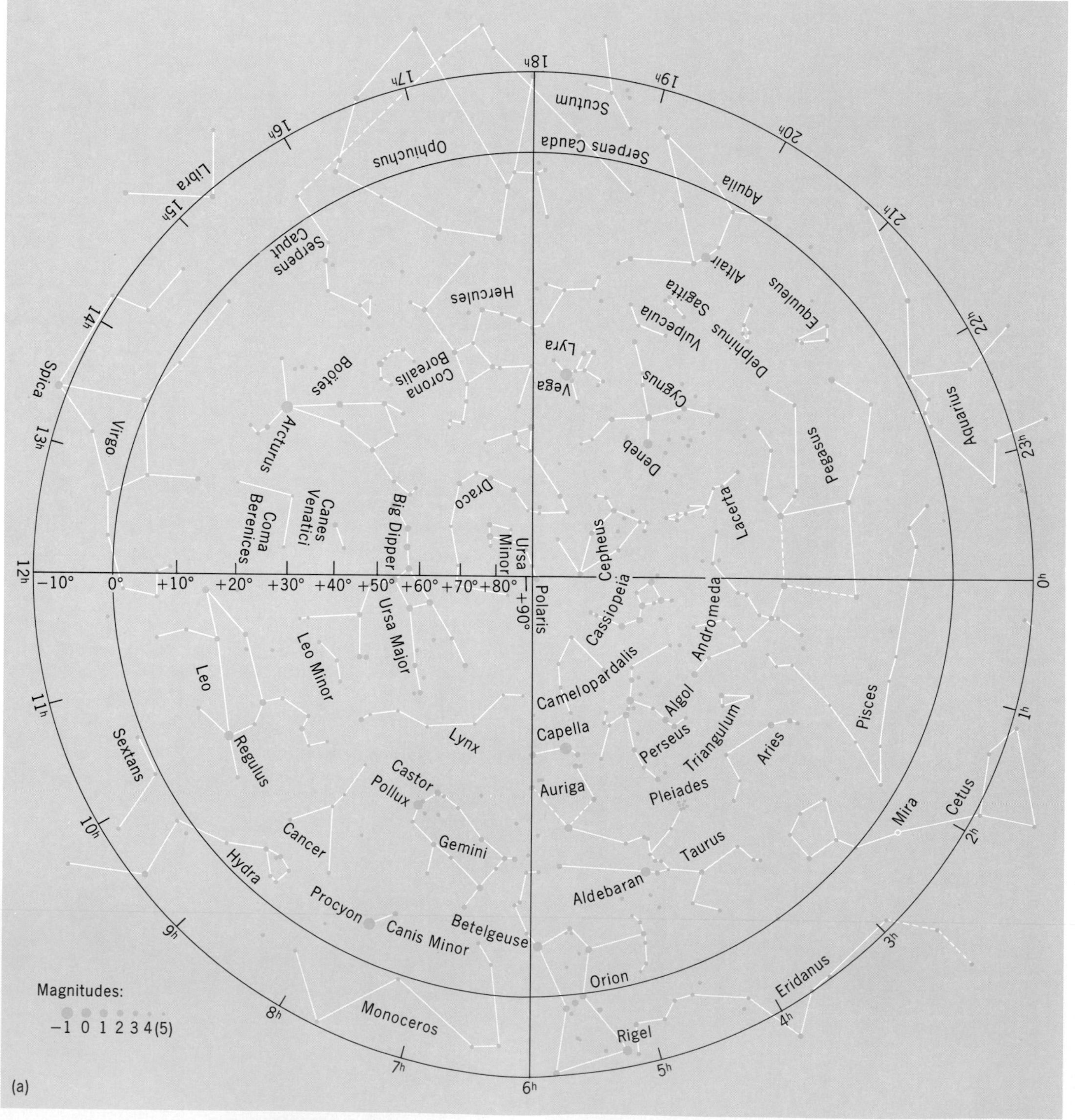

Fig. 2. Universally accepted constellations; brighter stars are shown by larger dots. (*a*) Northern hemisphere.

zodiac. Actually, the Sun moves through 13 constellations, and precession has changed the dates at which the Sun passes through the zodiacal constellations. The intersection of the ecliptic and the celestial equator (the extension of the Earth's Equator into the sky) is known as the vernal equinox, or the first point of Aries. It has been used to mark the zero point of the celestial coordinate system. Westward from Aries, the zodiacal constellations are Taurus, the Bull; Gemini, the Twins; Cancer, the Crab; Leo, the Lion; Virgo, the Virgin; Libra, the Scales; Scorpius, the Scorpion; Sagittarius, the Archer; Capricornus, the Seat Goat; Aquarius, the Water Bearer; and Pisces, the Fish. *See* Ecliptic; Equinox; Zodiac.

With the advent of computer control of telescopes, astronomers no longer use the constellations to locate

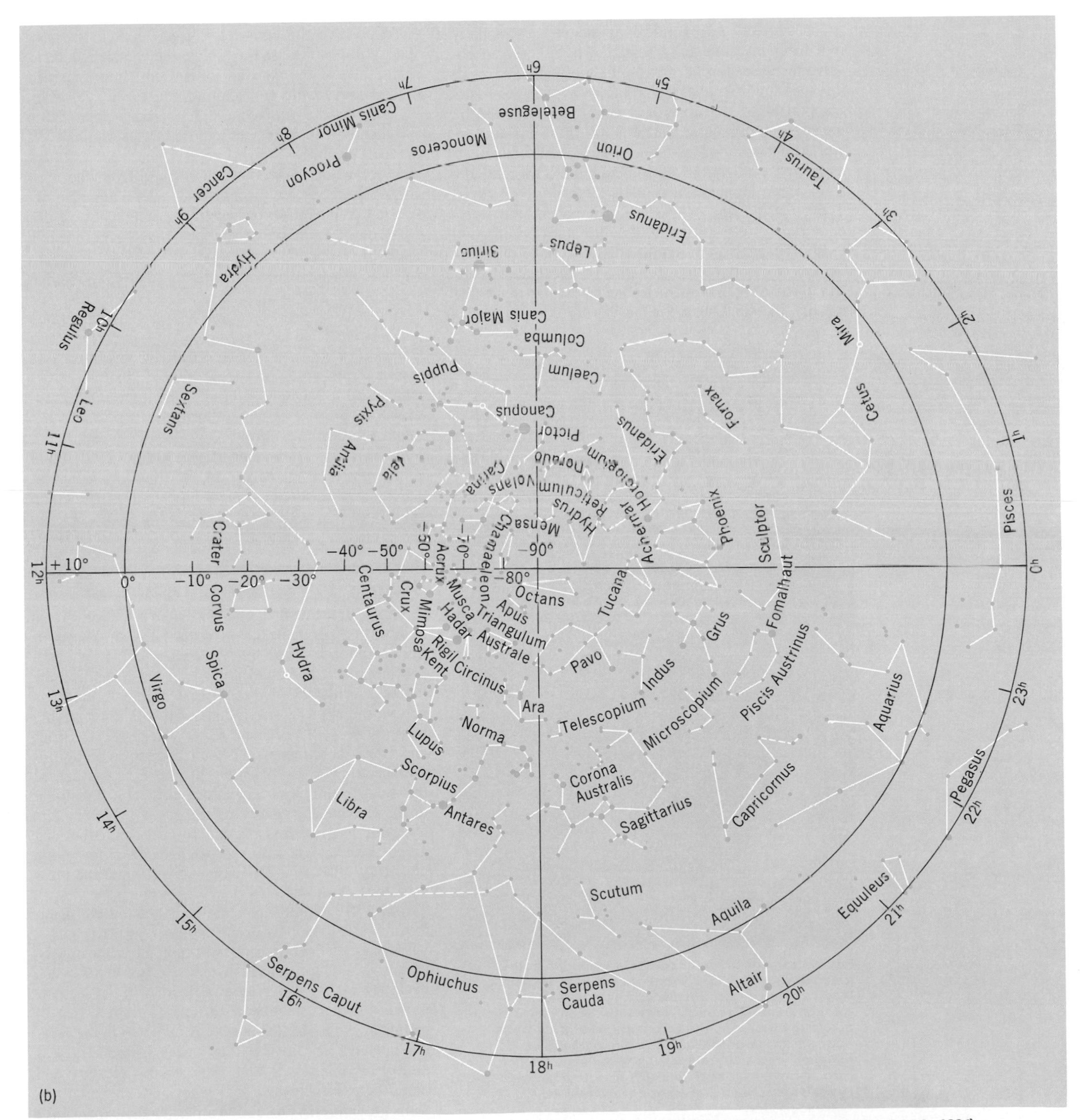

(b) Southern hemisphere. (*After J. M. Pasachoff and D. H. Menzel, A Field Guide to the Stars and Planets, Houghton Mifflin, 2d ed., 1984*)

objects in the sky. Even amateur telescopes are often equipped with computer locators.

Jay M. Pasachoff

Bibliography. J. M. Pasachoff, *Peterson's First Guide to Astronomy,* 1988; J. M. Pasachoff and D. H. Menzel, *A Field Guide to the Stars and Planets,* 2d ed., 1983; H. A. Rey, *The Stars*, rev. ed., 1988; W. Tirion, B. Rappaport, and G. Lovi, *Uranometria 2000.0*, 1987; D. J. Warner, *The Sky Explored: Celestial Cartography 1500–1800*, 1979.

Coronagraph

A specialized astronomical telescope substantially free from instrumentally scattered light, used to observe the solar corona, the faint atmosphere surrounding the Sun. Coronagraphs can record the emission component of the corona (spectral lines emitted by high-temperature ions surrounding the Sun) and the white-light component (solar photospheric light scattered by free electrons surrounding the Sun) routinely from high mountain sites under clear sky conditions. The emission and white-light components are typically only a few millionths the brightness of the Sun itself. Hence, the corona is difficult to observe unless the direct solar light is completely rejected, and unless instrumentally diffracted and scattered light that reaches the final image plane of the coronagraph is small relative to the coronal light. SEE SOLAR CORONA; SUN.

The basic design **(Fig. 1)**, as invented by B. Lyot, has an occulting disk in the primary image plane of the telescope to block the image of the Sun itself. In addition, the primary objective (a lens or superpolished mirror) is specially fabricated to minimize scattered light. Also, light diffracted by the objective rim must be suppressed. For this, an aperture (Lyot stop) is placed at an image of the objective as produced by a field lens, with the aperture diameter slightly smaller than that of the objective image. A camera lens behind the Lyot stop forms the coronal image at the final image plane. SEE OPTICAL TELESCOPE.

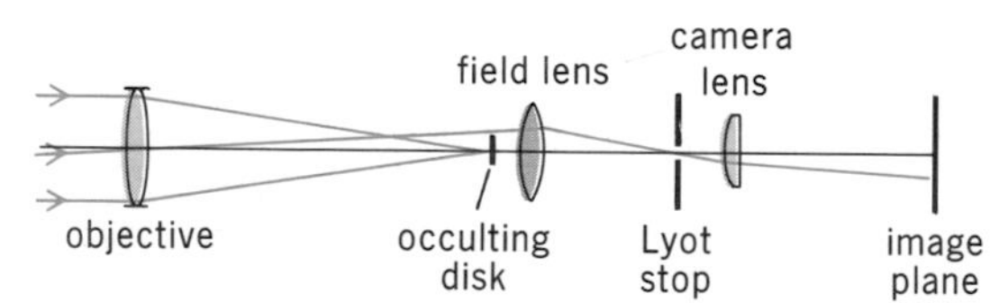

Fig. 1. Diagram showing the key features of a coronagraph used for observing the solar corona.

To observe the emission corona, optical filters or spectrographs are used to isolate the wavelengths corresponding to the various coronal emission lines and thus minimize the contribution due to the sky **(Fig. 2)**. To observe the white-light corona, polarization subtraction techniques are required to discriminate the coronal electron-scattered light (polarized with the electric vector tangential to the limb) from the variably polarized sky background. Coronagraphs designed for operation on satellites usually have a circular mask in front of the objective to completely shade it. This external occulting is more efficient, when no sky background is present, than if an internal occulting disk is used.

Fig. 2. Image of the solar corona as recorded in the green spectral line (530.3 nm) of Fe(XIV). The complex loop structures trace corresponding magnetic field configurations. The black central disk is an image of the occulting disk inside the coronagraph, slightly larger than the image of the Sun. The photograph was recorded with a 20-cm-aperture (8-in.) emission-line coronagraph. (*National Solar Observatory/Sacremento Peak Association of Universities for Research in Astronomy, Inc.*)

Raymond Smartt

Bibliography. D. E. Billings, *A Guide to the Solar Corona*, 1966; R. R. Fisher et al., New Mauna Loa coronagraph systems, *Appl. Opt.*, 20:1094–1101, 1981; R. M. MacQueen et al., The high altitude observatory coronagraph polarimeter on the Solar Maximum Mission, *Solar Phys.*, 65:91–107, 1980; R. N. Smartt, R. B. Dunn, and R. R. Fisher, Design and performance of a new emission-line coronograph, *S.P.I.E.* (Society of Photo-optical Instrumentation Engineers), 288:395–402, 1981.

Cosmic background radiation

A nearly uniform flux of microwave radiation that is believed to permeate all of space. The discovery of this radiation in 1965 by A. Penzias and R. W. Wilson has had a profound impact on understanding the nature and history of the universe. The interpretation of this radiation as the remnant fireball from the big bang by P. J. E. Peebles, R. H. Dicke, R. G. Roll, and D. T. Wilkinson, and their correct prediction of the spectrum to be that of a blackbody, was one of the great triumphs of cosmological theory. The flight of the *Cosmic Background Explorer* (*COBE*) satellite has both verified the basic nature of the radiation and given deep insight into the mechanism of formation of galaxy clusters in the early universe and the presence of dark matter.

Origin of radiation. In the theory of the big bang, the universe began with an explosion $10–15 \times 10^9$ years ago. This big bang was not an explosion of matter into empty space but an explosion of space itself. The early universe was filled with dense, hot, glowing matter; there was no region of space free of matter or radiation. (This state is reflected in the present universe by the fact that space is more or less uniformly filled with galaxies; galaxy clusters and holes between clusters are believed to have grown from gravitational instabilities during the expansion.)

The explosion of space increased the volume of the matter and radiation and thus reduced the density and temperature. The initial temperature was so high that even for several hundred thousand years after the initial explosion the universe was still as hot as the surface of the present-day Sun. At this temperature the matter of the universe was in the form of a plasma of electrons, protons, alpha particles (helium nuclei), and photons. The photons were strongly absorbed and reemitted by the electrons, and their spectrum was similar to that of the Sun. About 500,000 years after the initial explosion, the expansion caused the temperature to drop enough that the electrons and protons recombined to form hydrogen atoms. Unlike the previous plasma, which was opaque to light, neutral hydrogen is transparent. From that time (called the time of the decoupling or of recombination) until now, the cosmic photons have been traveling virtually unscattered, carrying information about the nature of the universe at the time of the decoupling (**Fig. 1**). *See Big bang theory.*

To an observer moving with the plasma, the photons have a blackbody spectrum with a characteristic temperature of a few thousand kelvins. (A blackbody spectrum is the characteristic emission from a perfectly absorbing object heated to the characteristic temperature. The orange glow emitted from a heated pan is approximately blackbody, as is the light emitted from the filament of a light bulb or from the surface of the Sun.) Although the glow from the plasma is in the visible region, as a result of the recessional velocity of the plasma from the Earth the radiation is redshifted from the visible into the microwave region, with a characteristic temperature of 3 K (5°F above absolute zero, −459.67°F). Detection of the radiation is really the observation of the shell of matter that last scattered the radiation. *See Redshift.*

The microwave radiation is coming from the most distant region of space ever observed, and was emitted earlier in time than any other cosmological signal. The radiation was originally termed cosmic background radiation because the discoverers foresaw that it would cause a background interference with satellite communications, but the term has taken on a vivid new meaning: the radiating shell of matter forms the spatial background in front of which all other astrophysical objects, such as quasars, lie. Until methods are devised to detect the neutrinos or gravity waves that were decoupled earlier, there will be no direct means of viewing beyond this background. *See Gravitational radiation detectors.*

Measurements. Subsequent to its discovery, experimental work on the microwave background has been concerned primarily with the measurement of the radiation's color (its spectrum of intensity at different frequencies) and with its isotropy (its intensity as a function of direction in the sky). Measurements were initially made from mountaintops, balloons, and airplanes in order to minimize interference from atmospheric microwaves. In 1989, the National Aeronautics and Space Administration (NASA) launched the *COBE* satellite (**Fig. 2**) into a nearly polar orbit 900 km (560 mi) high. The *COBE* spectrophotometer measures the spectrum in the range 30–3000 GHz (wavelengths of 1 cm to 0.1 mm), and the radiometers measure the isotropy (uniformity in different directions) at frequencies of 31, 53, and 90 GHz.

Spectrum. The big bang theory predicted that the spectrum would be similar to that of a blackbody. Confirmation of this prediction by balloon and mountaintop experiments led to the rejection of other cosmological models (such as the steady-state universe). Thus, by the late 1970s the big bang theory had become the standard model for the early universe.

However, small but statistically significant deviations had been repeatedly reported, and many theories had been developed to account for slight deviations. For example, the universe is not completely transparent to the radiation, and scattering of background photons from free electrons can increase the photon's energy, reducing the intensity of the radiation at low frequencies and increasing it at high frequencies, a

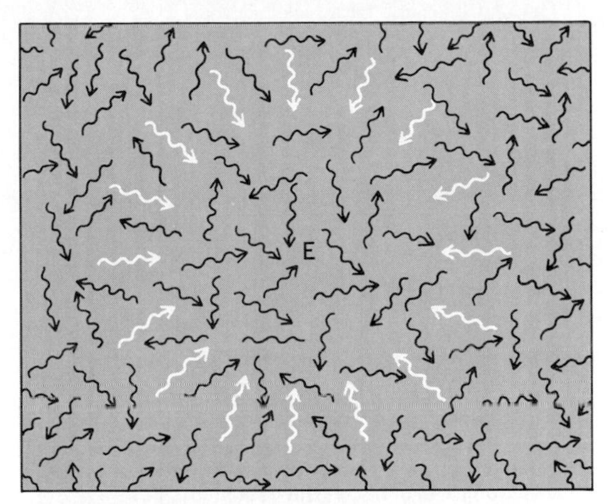

Fig. 1. Origin of cosmic background radiation. Arrows represent photons at a time when the universe became transparent, about 500,000 years after the big bang. Those photons that are reaching the Earth now are indicated by the white arrows. The position of the Earth is indicated by the letter E.

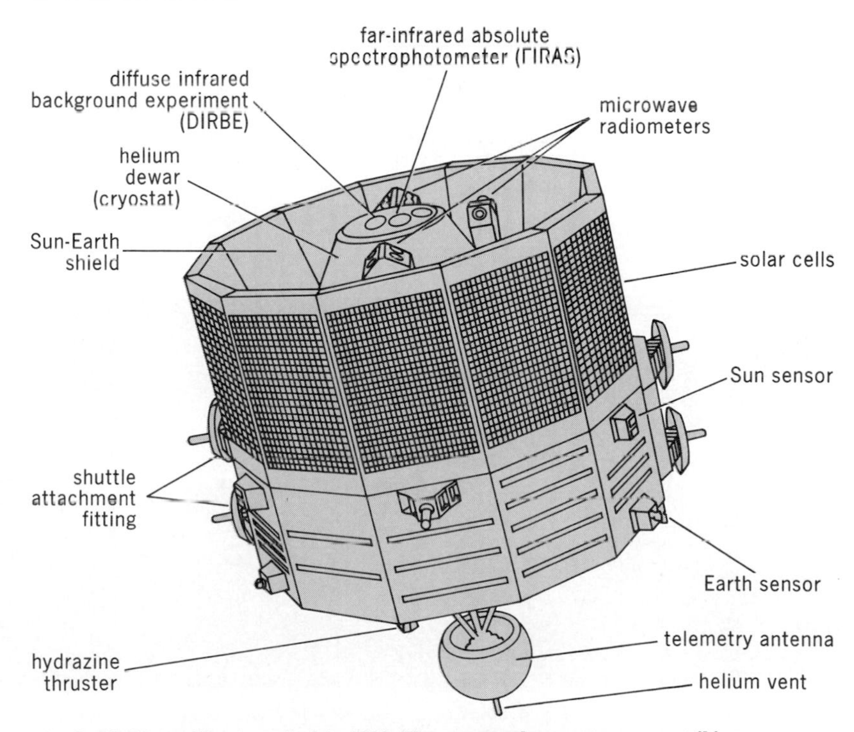

Fig. 2. *COBE* satellite, launched in 1989. The spectrophotometer gave striking confirmation of the predicted blackbody spectrum of the cosmic microwave radiation, and the radiometers showed ripples in the radiation that are believed to be early indications of the formation of galaxy clusters.

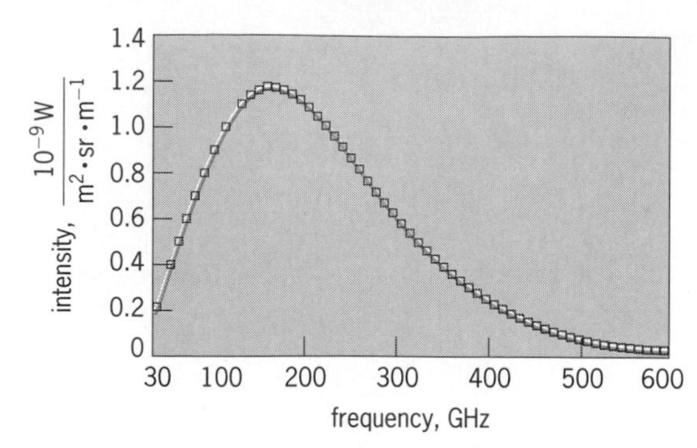

Fig. 3. Spectrum of the cosmic background radiation measured by the *COBE* satellite. The squares are the data, and the smooth curve is the best-fit curve for blackbody emission. The fact that the theoretical curve passes through virtually all the data points is a remarkable confirmation of the simple big bang theory.

process known as the Sunyaev-Zel'dovich effect. In addition, dust in the early universe, and early energy release from star formation or the decay of unknown (but theoretically possible) particles such as massive neutrinos, can also distort the spectrum.

Given the prior hints of distortion, it came as a surprise when the initial results from the *COBE* satellite showed that the spectrum was indeed that of a blackbody to better than 1% accuracy. **Figure 3** shows these measurements, taken in the direction of the sky called the North Galactic Pole, which has minimal interference from the Milky Way Galaxy. The small boxes indicate 1% uncertainties, and the theoretical blackbody curve passes through all of them. *COBE* may evenually produce a plot of similar or greater accuracy for nearly every direction in the sky, with a precision approaching 0.2%, and perhaps at this increased sensitivity some distortion will be seen. The spectrum is identical to that of a blackbody at a temperature above absolute zero of 2.735 ± 0.060 K.

The lack of deviations is just as exciting and puzzling to cosmologists as the presence of such structure would be, for it places strong limits on the nature of material present at the time of decoupling, just a half million years after the creation of the universe.

Isotropy. In describing their discovery, Penzias and Wilson said that the intensity of the radiation in different directions was isotropic (uniform in all directions) to better than 10%. Within a few years, measurements showed that the radiation was isotropic to better than 1%. This uniformity posed a difficult problem for cosmological theory, since according to the simple big bang theory the different parts of the sky that gave rise to the microwave background had not been close enough together to reach an equilibrium temperature. Thus there was no way to understand how the intensity could be uniform to even 10%. This theoretical problem was solved by a variation on the big bang theory called the inflationary universe model. In this picture the early universe was much smaller than had previously been supposed, giving it time to attain a uniform temperature; the rapid expansion of the universe occurred at a later time, the period of inflation. *SEE INFLATIONARY UNIVERSE COSMOLOGY.*

Anisotropy was first observed in the 1970s from airplane and balloon flights. **Figure 4*a*** shows a more recent map of intensity produced by the *COBE* satellite. The smooth yin-yang pattern seen in the plot is the cosine variation from the 600-km/s (370-mi/s) motion of the Milky Way Galaxy. The variation is only about 3.4 mK, or about 0.1% as large as the constant 2.7-K past. The small horizontal spur on the plot (near the middle) is interfering microwave radiation from the Milky Way Galaxy itself. *SEE MILKY WAY GALAXY.*

The smoothness and large angular scale of the variation observed suggest that its cause is not cosmological but local: the motion of the Earth relative to the radiation. If the Earth is moving relative to the shell of matter that emitted the radiation, radiation coming from the direction in which the Earth is moving will be more intense, and radiation coming from the opposite direction will be weaker because of the Doppler effect. The intensity of the radiation due to the motion should be proportional to the cosine of the angle between the direction of motion and the direction of observation, and this angular dependence is exactly what is observed. *SEE DOPPLER EFFECT.*

The 600-km/s (370-mi/s) velocity of the Milky Way Galaxy was much higher than had been anticipated. It is believed to result from the gravitational acceleration of the Milky Way Galaxy toward the

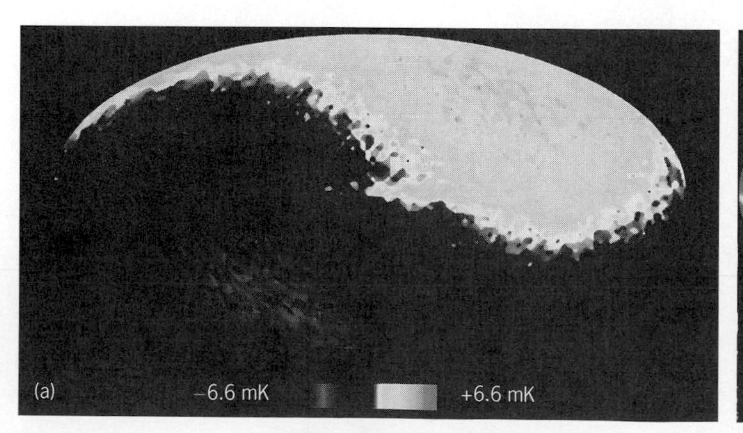

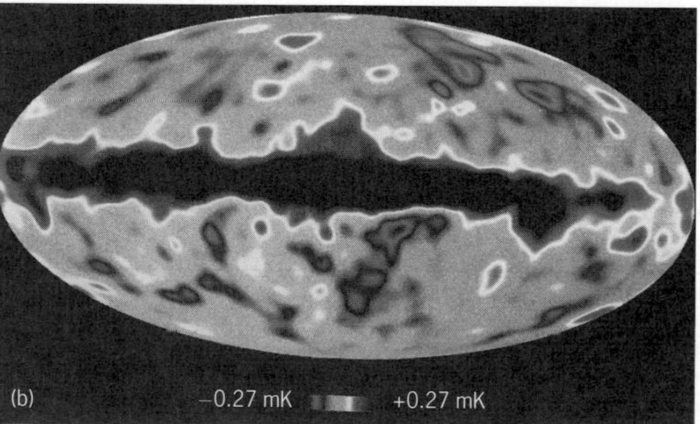

Fig. 4. Maps of the cosmic microwave background measured at the frequency 53 GHz (0.57 mm wavelength). The constant term (2.7 K) has been removed from both maps. Galactic coordinates are used, so that the region of the Milky Way is a horizontal line across the middle of each map. (*a*) Yin-yang, or cosine, pattern that results from the motion of the Milky Way Galaxy with respect to the cosmological shell of matter that emitted the radiation. (*b*) Intensity variation that remains when the cosine pattern is removed.

center of a large local supercluster of galaxies. Theoretical analysis of this motion, taking into account determinations of the mass distribution of nearby galaxies, suggests that the mass density of the local supercluster may be at the critical density, that is, the density which, if it were representative of all space, would cause the geometry of the universe to be finite in extent, according to general relativity theory. SEE GALAXY, EXTERNAL; UNIVERSE.

In 1992 the *COBE* team announced that they had found small variations in the microwave map that appear to be cosmological in origin. Figure 4*b* shows the map of intensity with the cosine term removed. The bright horizontal band comes from synchrotron emission in the Milky Way Galaxy (due to free electrons accelerated in galactic magnetic fields); it is the same as the small spurs seen in Fig. 4*a*. Galactic emission is also considered local, since it comes from within 10,000 light-years.

The most exciting regions of the plot are the marbled areas above and below the synchrotron stripe, which the *COBE* team called ripples in the radiation. The small blobs seen in this region of the map are believed to be the first indications that the early universe was not completely uniform. The structure has a magnitude of approximately 15–30 μK, a factor of 200,000 times smaller than the 2.7-K radiation itself.

Interpretation. The inflationary universe model accounts for the uniformity of the radiation but still leaves the puzzle that the present universe is highly nonuniform, with most of the mass clumped into stars, galaxies, and clusters of galaxies. Most astrophysicists believe that the clumping came about from the mutual gravitational attraction of the matter created in the big bang. The cosmological anisotropy observed by the *COBE* satellite is interpreted as the early sign of the clumping, taking place a half million years after the creation of the universe.

Theories of galaxy formation must postulate large amounts of dark, unseen matter in order to provide the strong gravitational fields necessary for sufficient clumping. The nature of this matter is unknown, but the matter must be more massive than all that observed in stars and galaxies. The radiation coming from the otherwise unseen clumps undergoes a gravitational redshift, according to the general theory of relativity, and it is this effect that *COBE* is believed to be observing. These residual lumps are consistent in magnitude and form with that predicted by the inflationary universe version of the big bang theory. Thus the ripples observed in the radiation are in fact a map of the distribution of dark matter in the very early universe. SEE COSMOLOGY; GRAVITATIONAL REDSHIFT.

Advanced experiments. *COBE* will continue to produce important results for several years, as it accumulates additional data on the isotropy and as systematic effects in the spectrum data are analyzed (for example, effects from microwave emissions of the Earth, the Sun, and the planets, which can be subtracted once understood). New technologies will mean that terrestrial measurements will once again contribute significantly to the data. Measurements at the South Pole (where interfering water vapor is very low) will give precise measurements of small-angular-scale anisotropy, and of the spectrum in the long-wavelength bands (greater than 1 cm), where *COBE* is insensitive. The development of sensitive microwave detectors and amplifiers will allow new balloon measurements to contribute to knowledge of the spectrum.

Richard A. Muller

Bibliography. R. A. Muller, The cosmic background radiation and the new aether drift, *Sci. Amer.*, 238(5):64–74, 1978; G. F. Smoot et al., Structure in the *COBE* differential microwave radiometer first year maps, *Astrophys. J. Lett.*, 396:L1–L5, 1992; R. A. Sunyaev and Ya. B. Zel'dovich, Cosmic background radiation as a probe of the contemporary structure and history of the universe, *Annu. Rev. Astron. Astrophys.*, 18:537–560, 1988; S. Weinberg, *The First Three Minutes*, 1977.

Cosmic electrodynamics

The science concerned with electromagnetic phenomena in ionized media encountered in interstellar space, in stars, and above the atmosphere. Because these ionized materials are excellent electrical conductors, they are strongly linked to magnetic-field lines; they can travel freely along but not across the field lines. Statistically this linkage tends to equalize the energies in the magnetic field and in the turbulent motion of the ionized material. Phenomena treated under cosmic electrodynamics include acceleration of charged particles to cosmic-ray energies, both in the galaxy and on the surface of the Sun; collisions between galaxies; the correlation of magnetic fields with galactic structure; sunspots and prominences; magnetic storms; aurora; and Van Allen radiation belts.

Alfvén waves—transverse waves which travel along the magnetic-field lines in a manner similar to the way waves travel along a stretched string—are often important. SEE MAGNETOHYDRODYNAMICS.

Rolf K. M. Landshoff

Bibliography. H. Alfvén, *Cosmic Plasma*, 1981; J. H. Piddington, *Cosmic Electrodynamics*, 2d ed., 1981.

Cosmic rays

Electrons and the nuclei of atoms—largely hydrogen—that impinge upon Earth from all directions of space with nearly the speed of light. These nuclei with relativistic speeds are often referred to as primary cosmic rays, to distinguish them from the cascade of secondary particles generated by their impact against air nuclei at the top of the terrestrial atmosphere. The secondary particles shower down through the atmosphere and are found all the way to the ground and below.

Cosmic rays are studied for a variety of reasons, not the least of which is a general curiosity over the process by which nature can produce such energetic nuclei. Apart from this, the primary cosmic rays provide the only direct sample of matter from outside the solar system. Measurement of their composition can aid in understanding which aspects of the matter making up the solar system are typical of the Milky Way Galaxy as a whole and which may be so atypical as to yield specific clues to the origin of the solar system. Cosmic rays are electrically charged; hence they are deflected by the magnetic fields which are thought to exist throughout the Milky Way Galaxy, and may be used as probes to determine the nature of these fields far from Earth. Outside the solar system the energy contained in the cosmic rays is comparable to that of the magnetic field, so the cosmic rays probably play a major role in determining the structure of the field. Collisions between the cosmic rays and the nuclei of the atoms in the tenuous gas which permeates the Milky Way Galaxy change the cosmic-ray composition in a measurable way and produce gamma rays which can be detected at Earth, giving information on the distribution of this gas.

This modern understanding of cosmic rays has arisen by a process of discovery which at many times produced seemingly contradictory results, the ultimate resolution of which led to fundamental discoveries in other fields of physics, most notably high-energy particle physics. At the turn of the century several different types of radiation were being studied, and the different properties of each were being determined with precision. One result of many precise experiments was that an unknown source of radiation existed with properties that were difficult to characterize. In 1912 Viktor Hess made a definitive series of balloon flights which showed that this background radiation increased with altitude in a dramatic fashion. Far more penetrating then any other known at that time, this radiation had many other unusual properties and became known as cosmic radiation, because it clearly did not originate in the Earth or from any known properties of the atmosphere.

Unlike the properties of alpha-, beta-, gamma-, and x-radiation, the properties of cosmic radiation are not of any one type of particle, but are due to the interactions of a whole series of unstable particles, none of which was known at that time. The initial identification of the positron, the muon, the π meson or pion, and certain of the K mesons and hyperons were made from studies of cosmic rays.

Thus the term cosmic ray does not refer to a particular type of energetic particle, but to any energetic particle being considered in its astrophysical context.

Cosmic-ray detection. Cosmic rays are usually detected by instruments which classify each incident particle as to type, energy, and in some cases time and direction of arrival. A convenient unit for measuring cosmic-ray energy is the electronvolt which is the energy gained by a unit charge (such as an electron) accelerating freely across a potential of 1 V. One electronvolt equals about 1.6×10^{-19} joule. For nuclei it is usual to express the energy in terms of electronvolts per nucleon, since as a function of this variable the relative abundances of the different elements are nearly constant. Two nuclei with the same energy per nucleon have the same velocity.

Flux. The intensity of cosmic radiation is generally expressed as a flux by dividing the average number seen per second by the effective size or ''geometry factor'' of the measuring instrument. Calculation of the geometry factor requires knowledge of both the sensitive area (in square centimeters) and the angular acceptance (in steradians) of the detector, as the arrival directions of the cosmic rays are randomly distributed to within 1% in most cases. A flat detector of any shape but with area of 1 cm^2 has a geometry factor of π cm^2·sr if it is sensitive to cosmic rays entering from one side only. The total flux of cosmic rays in the vicinity of the Earth but outside the atmosphere is about 0.3 nucleus/(cm^2·s·sr) [2 nuclei/(in.^{2}·s·sr)]. Thus a quarter dollar, with a surface area of 4.5 cm^2 (0.7 in.^{2}), lying flat on the surface of the Moon will be struck by $0.3 \times 4.5 \times 3.14 = 4.2$ cosmic rays per second.

Energy spectrum. The flux of cosmic rays varies as a function of energy. This type of function is called an energy spectrum, and may refer to all cosmic rays or to only a selected element or group of elements. Since cosmic rays are continuously distributed in energy, it is meaningless to attempt to specify the flux at any one exact energy. Normally an integral spectrum is used, in which the function gives the total flux of particles with energy greater than the specified energy [in particles/(cm^2·s·sr)], or a differential spectrum, in which the function provides the flux of particles in some energy interval (typically 1 MeV/nucleon wide) centered on the specified energy, in particles/[cm^2·s·sr·(MeV/nucleon)]. The basic problem of cosmic-ray research is to measure the spectra of the different components of cosmic radiation and to deduce from them and other observations the nature of the cosmic-ray sources and the details of where the particles travel on their way to Earth and what they encounter on their journey.

Types of detectors. All cosmic-ray detectors are sensitive only to moving electrical charges. Neutral cosmic rays (neutrons, gamma rays, and neutrinos) are studied by observing the charged particles produced in the collision of the neutral primary with some type of target. At low energies the ionization of the matter through which they pass is the principal means of detection. Such detectors include cloud chambers, ion chambers, spark chambers, Geiger counters, proportional counters, scintillation counters, solid-state detectors, photographic emulsions, and chemical etching of certain mineral crystals or plastics in which ionization damage is revealed. The amount of ionization produced by a particle is given by the square of its charge multiplied by a universal function of its velocity. A single measurement of the ionization produced by a particle is therefore usually not sufficient both to identify the particle and to determine its energy. However, since the ionization itself represents a significant energy loss to a low-energy particle, it is possible to design systems of detectors which trace the rate at which the particle slows down and thus to obtain unique identification and energy measurement.

At energies above about 500 MeV/nucleon, almost all cosmic rays will suffer a catastrophic nuclear interaction before they slow appreciably. Some measurements are made using massive calorimeters which are designed to trap all of the energy from the cascade of particles which results from such an interaction. More commonly an ionization measurement is combined with measurements of physical effects which vary in a different way with mass, charge, and energy. Cerenkov detectors and the deflection of the particles in the field of large superconducting magnets or the magnetic field of the Earth itself provide the best means of studying energies up to a few hundred gigaelectronvolts per nucleon. Detectors employing the phenomenon of x-ray transition radiation promise to be useful for measuring composition at energies up to a few thousand GeV per nucleon. Transition radiation detectors have already been used to study electrons having energies of 10–200 GeV which, because of their lower rest mass, are already much more relativistic than protons of the same energies.

Above about 10^{14} eV, direct detection of individual particles is no longer possible since they are so rare. Such particles are studied by observing the large showers of secondaries they produce in Earth's atmosphere. These showers are detected either by counting the particles which survive to strike ground-level detectors or by looking at the flashes of light the showers produce in the atmosphere with special telescopes and photomultiplier tubes. It is not possible to directly determine what kind of particle produces any given shower. However, because of the extreme energies involved, which can be measured with fair accuracy and have been seen as high as 10^{20} eV (16 J), most of the collision products travel in the same direction as the primary and at essentially the speed of light. This center of intense activity has typical dimensions of only a few tens of meters, allowing it to be tracked (with sensitive instruments) like a miniature meteor across the sky before it hits the Earth at a well-defined location. In addition to allowing determination of the direction from which each particle came, the development of many such showers through the atmosphere may be studied statistically to gain an idea of whether the primaries are protons or heavier nuclei. Basically the idea behind these studies is that a heavy nucleus, in which the energy is initially shared among several neutrons and protons, will cause a shower that starts higher in the atmosphere and develops more regularly than a shower which has the same total energy but is caused by a single proton.

Atmospheric cosmic rays. The primary cosmic-ray particles coming into the top of the terrestrial atmosphere make inelastic collisions with nuclei in the atmosphere. The collision cross section is essentially the geometrical cross section of the nucleus, of the order of 10^{-26} cm^2 (10^{-27} in.^{2}). The mean free path for primary penetration into the atmosphere is given in **Table 1**. (Division by the atmospheric density in g/cm^3 gives the value of the mean free path in centimeters.)

When a high-energy nucleus collides with the nucleus of an air atom, a number of things usually occur. Rapid deceleration of the incoming nucleus leads to production of pions with positive, negative, or neutral charge; this meson production is closely analogous to the generation of x-rays, or bremsstrahlung, produced when a fast electron is deflected by impact with the atoms in a metal target. The mesons, like the bremsstrahlung, come off from the impact in a narrow cone in the forward direction. Anywhere from 0 to 30 or more pions may be produced, depending upon the energy of the incident nucleus. The ratio of neutral to charged pions is about 0.75. *See Bremsstrahlung.*

A few protons and neutrons (in about equal proportions) may be knocked out with energies up to a few GeV. They are called knock-on protons and neutrons.

A nucleus struck by a proton or neutron of the nucleonic component with an energy greater than approximately 300 MeV may have its internal forces momentarily disrupted so that some of its nucleons are free to leave with their original nuclear kinetic energies of about 10 MeV. The nucleons freed in this fashion appear as protons, deuterons, tritons, alpha particles, and even somewhat heavier clumps, radiating outward from the struck nucleus. In photographic emulsions the result is a number of short prongs radiating from the point of collision, and for this reason is called a nuclear star.

All these protons, neutrons, and pions generated by collision of the primary cosmic-ray nuclei with the nuclei of air atoms are the first stage in the development of the secondary cosmic-ray particles observed inside the atmosphere. Since several secondary particles are produced by each collision, the total number of energetic particles of cosmic-ray origin will at first increase with depth, even while the primary density is decreasing. Since electric charge must be conserved and the primaries are positively charged, the positive particles outnumber the negative particles in the secondary radiation by a factor of about 1.2. This factor is called the positive excess.

Table 1. Mean free paths for primary cosmic rays in the atmosphere

Charge of primary nucleus	Mean free path in air, g/cm^2*
$Z = 1$	60
$Z = 2$	44
$3 \leq Z \leq 5$	32
$6 \leq Z \leq 9$	27
$10 \leq Z \leq 29$	21

*1 g/cm^2 = 1.42×10^{-2} lb/in.^{2}

Electromagnetic cascade. The uncharged π^0 mesons decay into two gamma rays with a lifetime of about 9×10^{-17} s. The decay is so rapid that π^0 mesons are not directly observed among the secondary particles in the atmosphere. The two gamma rays, which together have the rest energy of the π^0, about 140 MeV, plus the π^0 kinetic energy, each produce a positron-electron pair. Upon passing sufficiently close to the nucleus of an air atom deeper in the atmosphere, the electrons and positrons convert their energy into bremsstrahlung. The bremsstrahlung in turn creates new positron-electron pairs, and so on. This cascade process continues until the energy of the initial π^0 has been dispersed into a shower of positrons, electrons, and photons with insufficient individual energies ($\lesssim 1$ MeV) to continue the pair production. The shower, then being unable to reproduce its numbers, is dissipated by ionization of the air atoms. The electrons and photons of such showers are referred to as the soft component of the atmospheric (secondary) cosmic rays, reaching a maximum intensity at an atmospheric depth of 150–200 g/cm^2 and then declining by a factor of about 10^2 down to sea level. *See Electron-positron pair production.*

Muons. The $\pi^{\pm}$ mesons produced by the primary collisions have a lifetime about 2.6×10^{-8} s before they decay into muons: $\pi^{\pm} \rightarrow \mu^{\pm}$ + neutrino. With a lifetime of this order a $\pi^{\pm}$ possessing enough energy (greater than 10 GeV) to experience significant relativistic time dilatation may exist long enough to interact with the nuclei of the air atoms. The cross section for $\pi^{\pm}$ nuclear interactions is approximately the geometrical cross section of the nucleus, and the result of such an interaction is essentially the same as for the primary cosmic-ray protons. Most low-energy $\pi^{\pm}$ decay into muons before they have time to undergo nuclear interactions.

Except at very high energy (above 500 GeV), muons interact relatively weakly with nuclei, and are too massive (207 electron masses) to produce bremsstrahlung. They lose energy mainly by the comparatively feeble process of ionizing an occasional air atom as they progress downward through the atmosphere. Because of this ability to penetrate matter, they are called the hard component. At rest their lifetime is 2×10^{-6} s before they decay into an electron or positron and two neutrinos, but with the relativistic time dilatation of their high energy, 5% of the muons reach the ground. Their interaction with matter is so weak that they penetrate deep into the ground, where

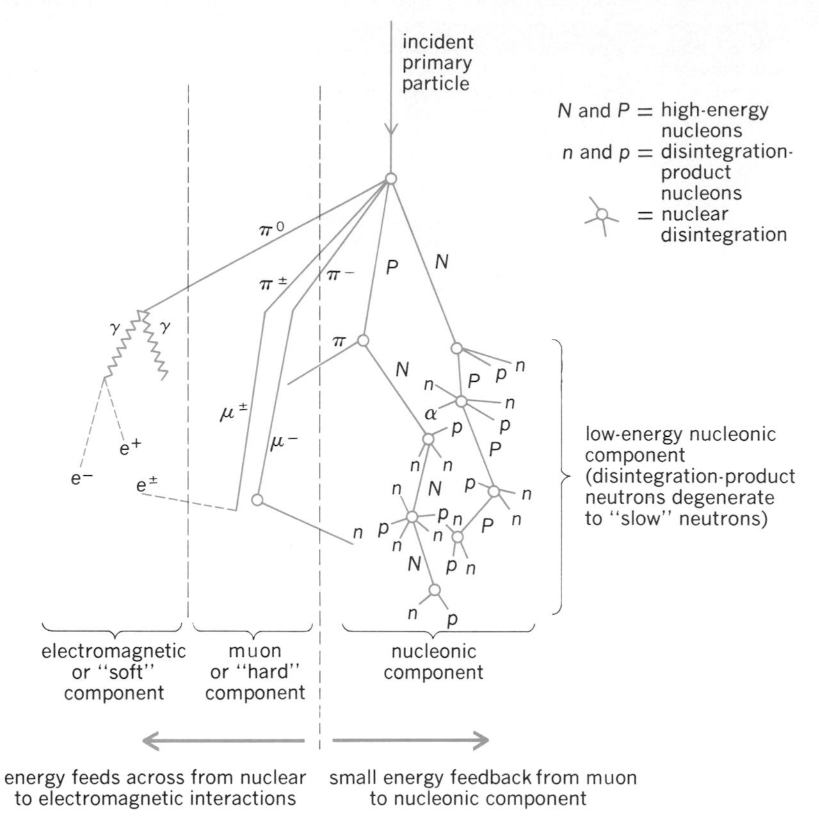

Fig. 1. Cascade of secondary cosmic-ray particles in the terrestrial atmosphere.

they are the only charged particles of cosmic-ray origin to be found. At a depth equivalent of 300 m (990 ft) of water the muon intensity has decreased from that at ground level only by a factor of 20; at 1400 m (4620 ft) it has decreased by a factor of 10^3.

Nucleonic component. The high-energy nucleons—the knock-on protons and neutrons—produced by the primary-particle collisions and a few pion collisions proceed on down into the atmosphere. They produce nuclear interactions of the same kind as the primary nuclei, though of course with diminished energies. This cascade process constitutes the nucleonic component of the secondary cosmic rays.

When the nucleon energy falls below about 100 MeV, stars and further knock-ons can no longer be produced. At the same time the protons are rapidly disappearing from the cascade because their ionization losses in the air slow them down before they can make a nuclear interaction. The neutrons are already dominant at 3500 m (11,550 ft), about 300 g/cm^2 (4.3 lb/in.^{2}) above sea level, where they outnumber the protons four to one. Thus the final stages in the lower atmosphere are given over almost entirely to neutrons in a sequence of low-energy interactions which convert them to thermal neutrons (neutrons of kinetic energy of about 0.025 eV) in a path of about 90 g/cm^2 (1.3 lb/in.^{2}). These thermal neutrons are readily detected in boron trifluoride (BF_3) counters. The nucleonic component increases in intensity down to a depth of about 120 g/cm^2, and thereafter declines in intensity, with a mean absorption length of about 200 g/cm^2 (2.8 lb/in.^{2}).

The various cascades of secondary particles in the atmosphere are shown schematically in **Fig. 1**. Note that about 48% of the initial primary cosmic-ray energy goes into charged pions, 25% into neutral pions, 7% into the nucleonic component, and 20% into stars. The nucleonic component is produced principally by the lower-energy (about 5 GeV) primaries. Higher-energy primaries put their energy more into meson production. Hence in the lower atmosphere, a Geiger counter responds mainly to the higher-energy primaries (about 5 GeV) because it counts the muons and electrons, whereas a BF_3 counter detecting thermal neutrons responds more to the low-energy primaries.

Neutrinos. Cosmic neutrinos, detected for the first time from the explosion of the supernova 1987A, provide confirmation of theoretical calculations regarding the collapse of the cores of massive stars. Although neutrinos are produced in huge numbers (over 10^{15} passed through a typical human body from this supernova), they interact with matter only very weakly, necessitating a very large detector. Detectors consisting of huge tanks containing hundreds of tons of pure water located deep underground to reduce the background produced by other cosmic rays recorded less than two dozen neutrino events. Still larger detectors will permit observation of more distant supernovae and allow sensitive searches for point sources of high-energy neutrinos. Also, by measuring the fraction of non-neutrino-induced events containing multiple muons, these new detectors should determine the composition of cosmic rays at energies above 10^{15} eV. Some preliminary measurements indicate that these high-energy cosmic rays may consist primarily of iron nuclei rather than the protons that dominate at lower energies. Measurement of the flux of solar neutrinos, which is really quite a different problem, has begun to cause fundamental changes in thought about the physics of the Sun. *SEE SOLAR NEUTRINOS; SUPERNOVA.*

Relation to particle physics. Investigations of cosmic rays continue to make fundamental contributions to particle physics. Neutrino detectors have set the best limit yet (about 10^{32} years) on the lifetime of the proton. These detectors are also able to study the physics of the neutrino and specifically to search for

Table 2. Properties of particles when all have a rigidity of 1 GV

Particle	Charge	Nucleons	Kinetic energy, MeV	Kinetic energy, MeV/nucleon	Momentum, MeV/c
Electron	1	—	1000	—	1000
Proton	1	1	430	430	1000
^{3}He	2	3	640	213	2000
^{4}He	2	4	500	125	2000
^{16}O	8	16	2000	125	8000

oscillations, or spontaneous conversions of one type of neutrino into another. Cosmic rays remain the only source of particles with energies above 1000 GeV. With the continued increase in the size and sensitivity of detectors, study of cosmic rays should continue to provide the first indications of new physics at ultrahigh energies.

Geomagnetic effects. The magnetic field of Earth is described approximately as that of a magnetic dipole of strength 8.1×10^{15} weber-meters (8.1×10^{25} gauss · cm^3) located near the geometric center of Earth. Near the Equator the field intensity is 3×10^{-5} tesla (0.3 gauss), falling off in space as the inverse cube of the distance to the Earth's center. In a magnetic field which does not vary in time, the path of a particle is determined entirely by its rigidity, or momentum per unit charge; the velocity simply determines how fast the particle will move along this path. Momentum is usually expressed in units of eV/c, where c is the velocity of light, because at high energies, energy and momentum are then numerically almost equal. By definition, momentum and rigidity are numerically equal for singly charged particles. The unit so defined is normally called the volt, but should not be confused with the standard and nonequivalent unit of the same name. **Table 2** gives examples of these units as applied to different particles with rigidity of 1 GV. This corresponds to an orbital radius in a typical interplanetary (10^{-9} tesla or 10^{-5} gauss) magnetic field of approximately 10 times the distance from the Earth to the Moon.

The minimum rigidity of a particle able to reach the top of the atmosphere at a particular geomagnetic latitude is called the geomagnetic cutoff rigidity at that latitude, and its calculation is a complex numerical problem. Fortunately, for an observer near the ground, obliquely arriving secondary particles, produced by the oblique primaries, are so heavily attenuated by their longer path to the ground that it is usually sufficient to consider only the geomagnetic cutoff for vertically incident primaries, which is given in **Table 3**. Around the Equator, where a particle must come in perpendicular to the geomagnetic lines of force to reach Earth, particles with rigidity less than 10 GV are entirely excluded, though at higher latitudes where entry can be made more nearly along the lines of force, lower energies can reach Earth. Thus, the cosmic-ray intensity is a minimum at the Equator, and increases to its full value at either pole—this is the cosmic-ray latitude effect. Even deep in the atmosphere the variation with latitude is easily detected with BF_3 counters, as shown in **Fig. 2**. North of 45° the effect is slight because the additional primaries admitted are so low in energy that they produce few secondaries.

Accurate calculations of the geomagnetic cutoff must consider the deviations of the true field from that of a perfect dipole and the change with time of these deviations. Additionally the distortion of the field by the pressure of the solar wind must often be accounted for, particularly at high latitude. Such corrections vary rapidly with time because of sudden bursts of solar activity and because of the rotation of the Earth. Areas with cutoffs of 400 MV during the day may have no cutoff at all during the night. This day-night effect is confined to particles with energies so low that neither they nor their secondaries reach the ground, and is thus observed only on high-altitude balloons or satellites.

Table 3. Geomagnetic cutoff

Geomagnetic lat.	Vertical cutoff, GV
0°	15
±20°	11.5
+40°	5
±60°	1
±70°	0.2
±90°	0

Since the geomagnetic field is directed from south to north above the surface of Earth, the incoming cosmic-ray nuclei are deflected toward the east. Hence an observer finds some 20% more particles incident from the west. This is known as the east-west effect. *See* *Geomagnetism*.

Solar modulation. Figure 3 presents portions of the proton and alpha-particle spectra observed near the Earth but outside of the magnetosphere in 1973. Below 20 GeV/nucleon the cosmic-ray intensity varies markedly with time. S. Forbush was the first to show that the cosmic-ray intensity was low during the years of high solar activity and sunspot number, which follow an 11-year cycle. This effect is clearly seen in the data of Fig. 2 and has been extensively studied with ground-based and spacecraft instruments. While this so-called solar modulation is now understood in general terms, it has not been calculated in detail, in large part because of the lack of any direct measurements of conditions in the solar system out of the ecliptic plane, to which all present spacecraft are confined because of limitations on the power of rockets.

There is indirect evidence that the solar system is not at all spherically symmetric and that conditions near the ecliptic plane are quite special.

The primary cause of solar modulation is the solar wind, a highly ionized gas (plasma) which boils off the solar corona and propagates radially from the Sun at a velocity of about 400 km/s (250 mi/s). The wind is mostly hydrogen, with typical density of 5 protons/cm^3 (80 protons/in.^{3}). This density is too low for collisions with cosmic rays to be important. Rather, the high conductivity of the medium traps part of the solar magnetic field and carries it outward. The rotation of the Sun and the radial motion of the plasma com-

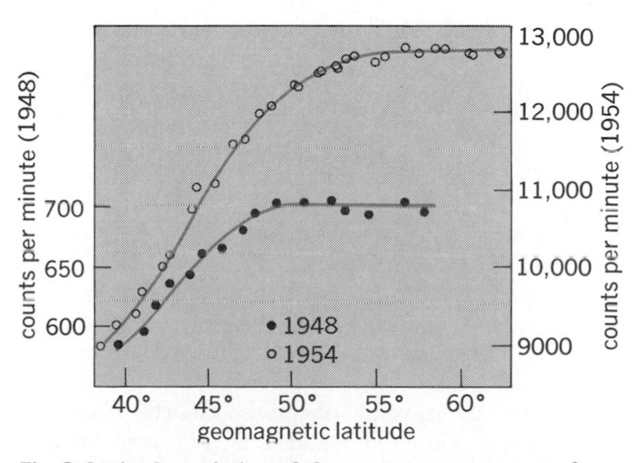

Fig. 2. Latitude variation of the neutron component of cosmic rays in 80°W longitude and at a height corresponding to an atmospheric pressure of 30 kPa (22.5 cm of mercury) in 1948, when the Sun was active, and 1954, when the Sun was deep in a sunspot minimum.

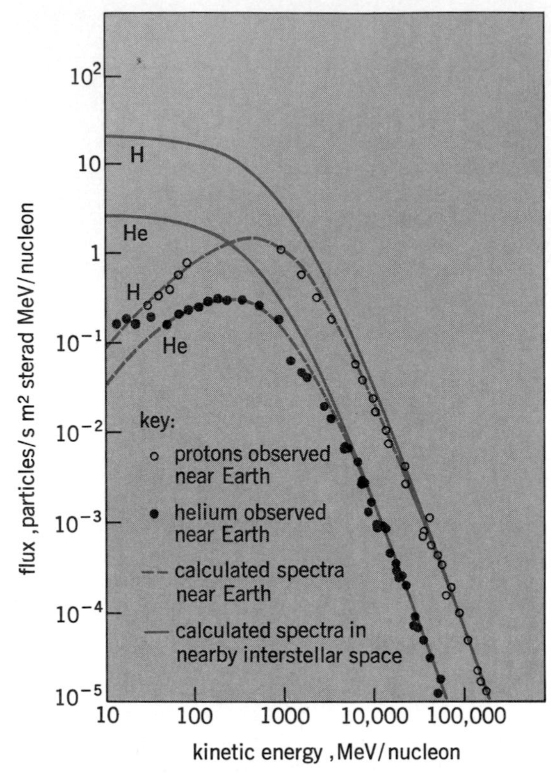

Fig. 3. Spectra of cosmic-ray protons and helium at Earth and in nearby interstellar space, showing the effect of solar modulation. Observations were made in 1973, when the Sun was quiet.

bine to create the observed archimedean spiral pattern of the average interplanetary magnetic field. Turbulence in the solar wind creates fluctuations in the field which often locally obscure the average direction and intensity. This complex system of magnetic irregularities propagating outward from the Sun deflects and sweeps the low-rigidity cosmic rays out of the solar system.

In addition to the bulk sweeping action, another effect of great importance occurs in the solar wind, adiabatic deceleration. Because the wind is blowing out, only those particles which chance to move upstream fast enough are able to reach Earth. However, because of the expansion of the wind, particles interacting with it lose energy. Thus, particles observed at Earth with 10 MeV/nucleon energy actually started out with several hundred megaelectronvolts per nucleon in nearby interstellar space, and those with only 100–200 MeV/nucleon initial energy probably never reach Earth at all. This is particularly unfortunate because at these lower energies the variation with energy of nuclear reaction probabilities would allow much more detailed investigation of cosmic-ray history. Changes in the modulation with solar activity are caused by the changes in the pattern of magnetic irregularities rather than by changes in the wind velocity, which are quite small. *SEE MAGNETOHYDRODYNAMICS.*

There are several phenomenological classes of cosmic-ray variation besides the 11-year variation which are associated with the short-term variations of the solar wind and generally affect only low-energy particles.

A sudden outburst at the Sun, at the time of a large flare, yields a cosmic-ray decrease in space which extends to very high energies, 50 GeV or so. The magnetic fields carried in the blast wave from the flare sweep back the cosmic rays, causing a decrease in their intensity. This is termed the Forbush-type decrease, where the primary intensity around the world may drop in an irregular way as much as 20% in 15 h or 8% in 3 h, slowly recovering in the days or weeks that follow. Often, but not always, the Forbush decrease and geomagnetic storms accompany each other. Striking geographical variation is to be seen in the sharper fluctuations during the onset of a Forbush decrease.

The region in space where solar modulation is important is probably a sphere with a radius of 50–80 AU, although this is not at all certain. (1 AU, or astronomical unit, is the mean Earth-Sun separation, 1.49×10^8 km or 9.26×10^7 mi.) The *Pioneer 11* spacecraft did not find abrupt changes in intensity indicative of crossing any type of boundary out to a distance of 40 AU, although the fluxes detected clearly increased as it traveled outward. A distinct change in interplanetary phenomena seems to take place in the 10–15 AU range to a quieter, more azimuthally uniform, and less temporally varying situation. *SEE SOLAR MAGNETIC FIELD; SOLAR WIND; SUN.*

Composition of cosmic rays. Nuclei ranging from protons to lead have been identified in the cosmic radiation. The relative abundances of the elements ranging up to nickel are shown in **Fig. 4**, together with the best estimate of the ''universal abundances'' obtained by combining measurements of solar spectra, lunar and terrestrial rocks, meteorites, and so forth. Most obvious is the similarity between these two distributions. However, a systematic deviation is quickly apparent: the elements lithium-boron and scandium-manganese as well as most of the odd-charged nuclei are vastly overabundant in the cosmic radiation. This effect has a simple explanation: the cosmic rays travel great distances in the Milky Way Galaxy and occasionally collide with atoms of interstellar gas—mostly hydrogen and helium—and fragment. This fragmentation, or spallation as it is called, produces lighter nuclei from heavier ones but does not change the energy/nucleon very much. Thus the energy spectra of the secondary elements are similar to those of the primaries.

Calculations involving reaction probabilities determined by nuclear physicists show that the overabundances of the secondary elements can be explained by assuming that cosmic rays pass through an average of about 5 g/cm² (0.07 lb/in.²) of material on their way to Earth. Although an average path length can be obtained, it is not possible to fit the data by saying that all particles of a given energy have exactly the same path length; furthermore, results indicate that higher-energy particles traverse less matter in reaching the solar system, although their original composition seems energy independent. *SEE ELEMENTS, COSMIC ABUNDANCE OF.*

When spallation has been corrected for, differences between cosmic-ray abundances and solar-system or universal abundances still remain. The most important question is whether these differences are due to the cosmic rays having come from a special kind of material (such as would be produced in a supernova explosion), or simply to the fact that some atoms might be more easily accelerated than others. It is possible to rank almost all of the overabundances by considering the first ionization potential of the atom and the rigidity of the resulting ion, although this calculation gives no way of predicting the magnitude of the enhancement expected. It is also observed that the relative abundances of particles accelerated in solar flares are far from constant from one flare to the next.

Isotopes. The possibility of such preferential acceleration is one of the reasons why much cosmic-ray study is concentrated on determining the isotopic composition of each element, as this is much less likely to be changed by acceleration. It is apparent that the low-energy helium data in Fig. 3 do not fit the calculated values. Since it is known that this low-energy helium is nearly all ^{4}He, whereas the higher-energy helium contains 10% ^{3}He, one can be fairly certain that the deviation is due to a local source of energetic ^{4}He within the solar system rather than a lack of understanding of the process of solar modulation. Similarly, a low-energy enhancement of nitrogen is pure ^{14}N, whereas the higher-energy nitrogen is almost 50% ^{15}N.

High-accuracy measurements of isotopic abundance ratios became available only in the late 1970s, but rapidly one puzzle emerged. The ratio of ^{22}Ne to ^{20}Ne in the cosmic-ray sources is estimated to be 0.37, while the accepted solar system value for this number is 0.12, which agrees well with the abundances measured in solar-flare particles. However, another direct sample of solar material—the solar wind—has a ratio of 0.08, indicating clearly that the isotopic composition of energetic particles need not reflect that of their source. Conclusions drawn from the observed difference in the solar and cosmic-ray values must be viewed as somewhat tentative until the cause of the variation in the solar material is well understood.

Electron abundance. Cosmic-ray electron measurements pose other problems of interpretation, partly because electrons are nearly 2000 times lighter than protons, the next lightest cosmic-ray component. Protons with kinetic energy above 1 GeV are about 100 times as numerous as electrons above the same energy, with the relative number of electrons decreasing slowly at higher energies. But it takes about 2000 GeV to give a proton the same velocity as a 1-GeV electron. Viewed in this way electrons are several thousand times more abundant than protons. (Electrical neutrality of the Milky Way Galaxy is maintained by lower-energy ions which are more numerous than cosmic rays although they do not carry much energy.) It is thus quite possible that cosmic electrons have a different source entirely from the nuclei. It is generally accepted that there must be direct acceleration of electrons, because calculations show that more positrons than negatrons should be produced in collisions of cosmic-ray nuclei with interstellar gas. Measurements show, however, that only 10% of the electrons are positrons. As the number of positrons seen agrees with the calculated secondary production, added confidence is gained in the result that there is indeed an excess of negatrons.

Electrons are light enough to emit a significant amount of synchrotron radiation as they are deflected by the 10^{-10}-tesla (10^{-6}-gauss) galactic magnetic field. Measurement of this radiation by radio telescopes provides sufficient data for an approximate calculation of the average energy spectrum of electrons in interstellar space and other galaxies. Comparison of spectra of electrons and positrons measured at Earth with those calculated to exist in interstellar space provides the most direct measurement of the absolute amount of solar modulation. *SEE RADIO ASTRONOMY.*

Properties of energy spectrum. At energies above 10^{10} eV, the energy spectra of almost all cosmic rays are approximated over many decades by functions in which the flux decreases as the energy raised to some negative, nonintegral power referred to as the spectral index. Such a power-law relationship is of course a

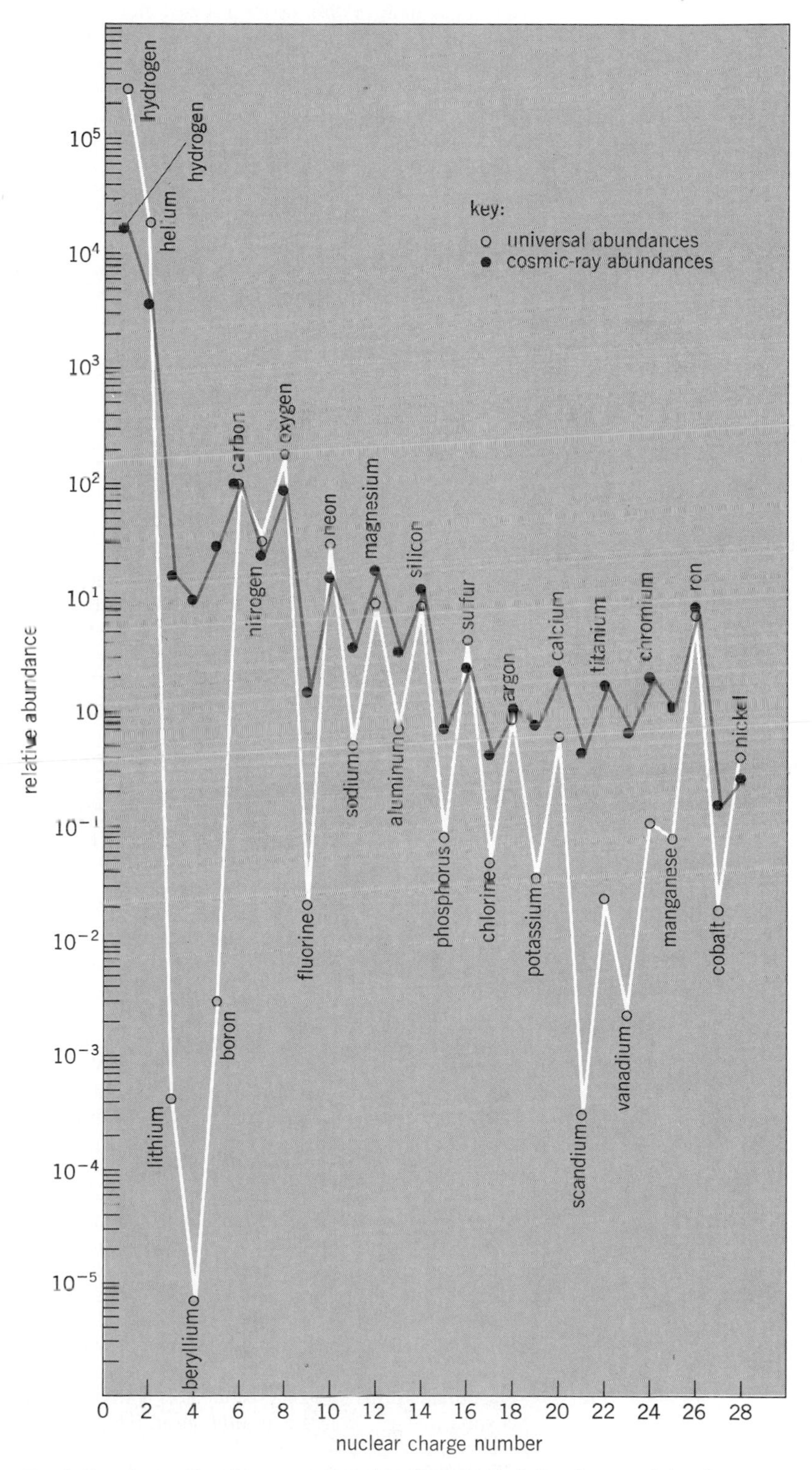

Fig. 4. Cosmic-ray abundances compared to the universal abundances of the elements. Carbon is set arbitrarily to an abundance of 100 in both cases.

straight line when plotted using logarithmic axes. A steep (that is, more rapidly falling with increasing energy) spectrum thus has a higher spectral index than a flat spectrum. The straight-line regions of the spectra in Fig. 3 correspond to a variation of flux with a spectral index of -2.7. A spectral index of -2.7 provides a good fit with the data up to 10^{15} eV total energy. Between 10^{15} and 10^{19} eV a steeper spectrum, with an index around -3.0, seems to be well established. Above 10^{19} eV the spectrum surprisingly flattens once more, returning to an index of about -2.7. The spectral index above 10^{20} eV has not been determined, because particles are so rare that they are almost never seen, even in detectors which cover several square kilometers and operate for many years. At such high energies, the individual particles are not identified, and changes in the measured-energy spectrum could be the result of composition changes. However, the evidence available indicates that the composition is essentially unchanged.

Age. Another important result which can be derived from detailed knowledge of cosmic-ray isotopic composition is the "age" of cosmic radiation. Certain isotopes are radioactive, such as beryllium-10 (^{10}Be) with a half-life of 1.6×10^6 years. Since beryllium is produced entirely by spallation, study of the relative abundance of ^{10}Be to the other beryllium isotopes, particularly as a function of energy to utilize the relativistic increase in this lifetime, will yield a number related to the average time since the last nuclear collision. Measurements show that ^{10}Be is nearly absent at low energies and yield an estimate of the age of the cosmic rays of approximately 10^7 years. An implication of this result is that the cosmic rays propagate in a region in space which has an average density of 0.1–0.2 atom/cm^3 (1.5–3 atoms/in.^{3}). This is consistent with some astronomical observations of the immediate solar neighborhood.

Very high-energy particles cannot travel long distances in the 2.7 K blackbody-radiation field which permeates the universe. Electrons of 15 GeV energy lose a good portion of their energy in 10^8 years by colliding with photons via the (inverse) Compton process, yet electrons are observed to energies of 100 GeV and over. A similar loss mechanism becomes effective at approximately 10^{20} eV for protons. These observations are of course not conclusive, but a safe statement is that a cosmic-ray age of 10^7 years is consistent with all currently available data. *See* COSMIC BACKGROUND RADIATION.

Several attempts have been made to measure the constancy of the cosmic-ray flux in time. Variations in ^{14}C production, deduced from apparent deviations of the archeological carbon-dating scale from that derived from studies of tree rings, cover a period of about 10^3 years. Radioactive ^{10}Be in deep-sea sediments allows studies over 10^6 years, whereas etching of tracks left by cosmic rays in lunar minerals covers a period of 10^9 years. None of these methods has ever indicated a variation of more than a factor of 2 in average intensity. There are big differences in these time scales, and the apparent constancy of the flux could be due to averaging over variations which fall in the gaps as far as the time scales are concerned. Nevertheless, the simplest picture seems to be that the cosmic rays are constant in time at an intensity level which is due to a long-term balance between continuous production and escape from the Galaxy, with an average residence time of 10^7 years.

Origin. Although study of cosmic rays has yielded valuable insight into the structure, operation, and history of the universe, their origin has not been determined. The problem is not so much to devise processes which might produce cosmic rays, but to decide which of many possible processes do in fact produce them.

In general, analysis of the problem of cosmic-ray origin is broken into two major parts: origin in the sense of where the sources are located, whatever they are, and origin in the sense of how the particles are accelerated to such high energies. Of course, these questions can never be separated completely.

Location of sources. It is thought that cosmic rays are produced by mechanisms operating within galaxies and are confined almost entirely to the galaxy of their production, trapped by the galactic magnetic field. The intensity in intergalactic space would only be a few percent of the typical galactic intensity, and would be the result of a slow leakage of the galactic particles out of the magnetic trap. It has not been possible to say much about where the cosmic rays come from by observing their arrival directions at Earth. At lower energies (up to 10^{15} eV) the anisotropies which have been observed can all be traced to the effects of the solar wind and interplanetary magnetic field. The magnetic field of the Milky Way Galaxy seems to be completely effective in scrambling the arrival directions of these particles.

Between 10^{15} and 10^{19} eV a smoothly rising anisotropy is measured, ranging from 0.1 to 10%, but the direction of the maximum intensity varies in a nonsystematic way with energy. At these energies, particles have a radius of curvature which is not negligible compared to galactic structures, and thus their arrival direction could be related to where they came from but in a complex way.

Above 10^{19} eV the radius of curvature in the galactic magnetic field becomes comparable to or larger than galactic dimensions, making containment of such particles in the Milky Way Galaxy impossible. Only a few hundred events greater than 10^{19} eV have been detected, but the directions from which they have come are plainly nonrandom. A clear minimum appears in the direction of the disk of the Milky Way Galaxy and a clear maximum near the north galactic pole, which is also the direction of the majority of the galaxies in the so-called local supercluster. The average distance of these galaxies is such that these particles in fact would be able to propagate to the Earth from them without losing their energy to photon collisions. This clearly defined anisotropy begins at the same energy at which the energy spectrum changes—further indication that these particles may have a different source from the low-energy particles. An alternative explanation, that particles are absorbed by material in the galactic plane, cannot at present be ruled out because no large detectors presently view the sky in the region of the south galactic pole. Thus it is not known whether the presence of the cluster of galaxies in the direction of the maximum flux is only coincidental.

Much effort has been devoted to construction of air-shower detectors with good directional resolution in order to search for point sources of gamma rays with energies greater than 10^{12} eV. A number of claims have been made regarding the detection of such high-energy gamma rays from the directions of well-known x-ray binary sources. It is generally be-

(a)

(b)

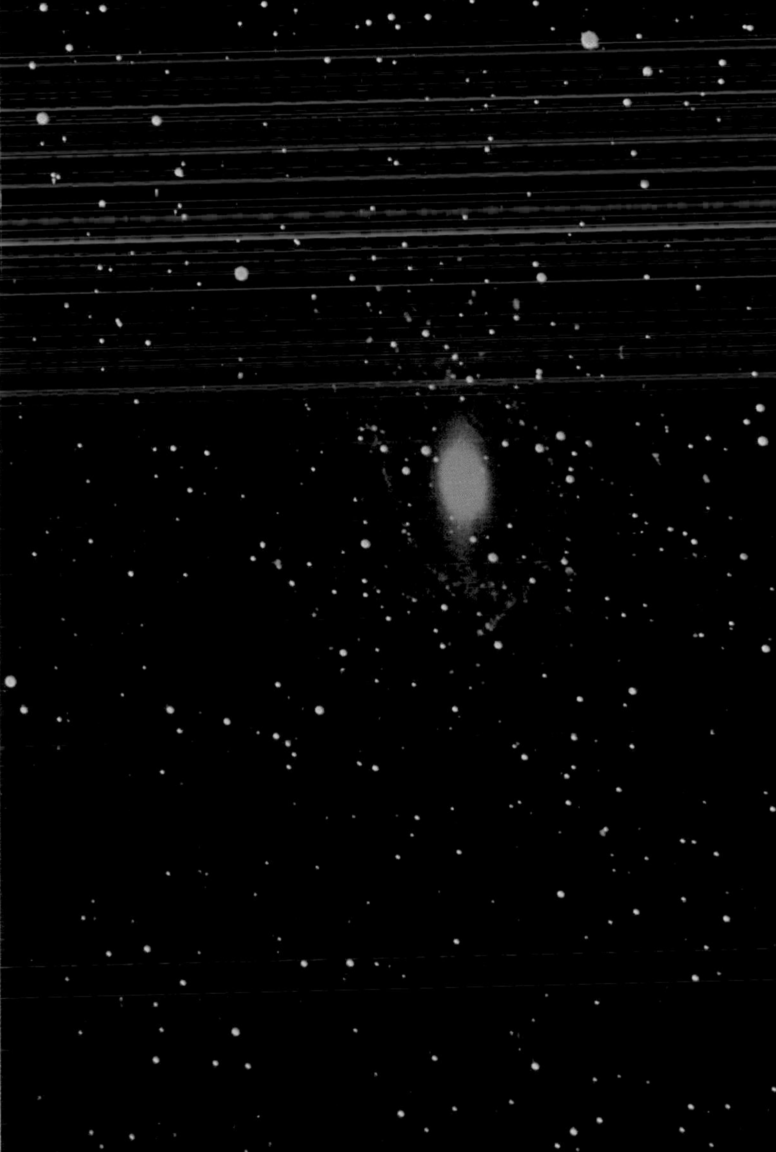
(c)

(*a*) M83, a prominent southern spiral galaxy. Blue patches are associations of very young stars; dark lanes and patches are produced by dust; red blobs are H II regions. (*b*) Galaxy NGC 5128. (*c*) NGC 6744, a barred spiral galaxy. (*Photographs by R. J. Dufour*)

lieved that gamma rays of energy greater than 10^{12} eV can be produced only through interactions of high-energy protons or other hadrons. Unambiguous detection and study of these gamma rays will provide important information on the structure and operation of these exotic objects. *SEE ASTROPHYSICS, HIGH-ENERGY; BINARY STAR; GAMMA-RAY ASTRONOMY; X-RAY STAR.*

Direct detection of cosmic rays propagating in distant regions of the Milky Way Galaxy is possible by observing the electromagnetic radiation produced as they interact with other constituents of the Milky Way Galaxy. Measurement of the average electron spectrum using radio telescopes has already been mentioned. Proton intensities are mapped by studying the arrival directions of gamma rays (at about 50 MeV) produced as they collide with interstellar gas. Unfortunately, the amount of radiation in these processes depends upon both the cosmic-ray flux and the magnetic-field intensity or density of interstellar gas. Areas where cosmic rays are known to exist can be pointed out because the radiation is observed. But where no radiation is seen, it is not known whether its absence results from lack of cosmic rays or lack of anything for them to interact with. In particular, very little radiation is seen from outside the Milky Way Galaxy, but there is also very little gas or magnetic field there. There is therefore no direct evidence either for or against galactic containment.

A major difficulty with the concept of cosmic radiation filling the universe is the large amount of energy needed to maintain the observed intensity in the face of an expanding universe—probably more energy than is observed to be emitted in all other forms put together. *SEE COSMOLOGY.*

Confinement mechanisms. Three possible models of cosmic-ray confinement are under investigation. All assume that cosmic rays are produced in sources, discrete or extended, scattered randomly through the galactic disk. Most popular is the "leaky box" model, which proposes that the particles diffuse about in the magnetic field for a few million years until they chance to get close to the edge of the Milky Way Galaxy and escape. This is a phenomenological model in that no mechanism is given by which either the confinement time or the escape probability as a function of energy can be calculated from independent observations of the galactic structure. Its virtue is that good fits to the observed abundances of spallation products are obtained by using only a few adjustable parameters. Variations of the model which mainly postulate boxes within boxes—ranging from little boxes surrounding sources to a giant box or static halo surrounding the whole Milky Way Galaxy—can be used to explain variations from the simple predictions. However, all attempts to calculate the details of the process have failed by many orders of magnitude, predicting ages which are either far older or far younger than the observed age.

A second model is that of the dynamical halo. Like the earlier static-halo model, it is assumed that cosmic rays propagate not only in the galactic disk but also throughout a larger region of space, possibly corresponding to the halo or roughly spherical but sparse distribution of material which typically surrounds a galaxy. This model is based on the observation that the energy density of the material which is supposed to be contained by the galactic magnetic field is comparable to that of the field itself. This can result in an unstable situation in which large quantities of galactic material stream out in a galactic wind similar in some respects to the solar wind. In this case the outward flow is a natural part of the theory, and calculations have predicted reasonable flow rates. In distinction to the solar wind, in which the cosmic rays contribute almost nothing to the total energy density, they may provide the dominant energy source in driving the galactic wind.

A third model assumes that there is almost no escape; that is, cosmic rays disappear by breaking up into protons which then lose energy by repeated collision with other protons. To accept this picture, one must consider the apparent 5-g/cm^2 (0.07-lb/in.^{2}) mean path length to be caused by a fortuitous combination of old distant sources and one or two close young ones. Basically, the objections to this model stem from the tendency of scientists to accept a simple theory over a more complex (in the sense of having many free parameters) or specific theory when both explain the data. *SEE MILKY WAY GALAXY.*

Acceleration mechanisms. Although the energies attained by cosmic-ray particles are extremely high by laboratory standards, their generation can probably be understood in terms of known astronomical objects and laws of physics. Even on Earth, ordinary thunderstorms generate potentials of millions of volts, which would accelerate particles to respectable cosmic-ray energies (a few gigaelectronvolts) if the atmosphere were less dense. Consequently, there are many theories of how the acceleration could take place, and it is quite possible that more than one type of source exists. Two major classes of theories may be identified—extended-acceleration regions and compact-acceleration regions.

Extended-acceleration regions. Acceleration in extended regions (in fact the Milky Way Galaxy as a whole) was first proposed by E. Fermi, who showed that charged particles could gain energy from repeated deflection by magnetic fields carried by the large clouds of gas which are known to be moving randomly about the Milky Way Galaxy. Many other models based on such statistical acceleration have since been proposed, the most recent of which postulates that particles bounce off shock waves traveling in the interstellar medium. Such shocks, supposed to be generated by supernova explosions, undoubtedly exist to some degree but have an unknown distribution in space and strength, leaving several free parameters which may be adjusted to fit the data.

Compact-acceleration regions. The basic theory in the compact-acceleration class is that particles are accelerated directly in the supernova explosions themselves. One reason for the popularity of this theory is that the energy generated by supernovas is of the same order of magnitude as that required to maintain the cosmic-ray intensity in the leaky box model.

However, present observations indicate that the acceleration could not take place in the initial explosion. Cosmic rays have a composition which is similar to that of ordinary matter and is different from the presumed composition of the matter which is involved in a supernova explosion. At least some mixing with the interstellar medium must take place. Another problem with an explosive origin is an effect which occurs when many fast particles try to move through the interstellar gas in the same direction: the particles interact with the gas through a magnetic field which they generate themselves, dragging the gas along and rapidly losing most of their energy. In more plausible theories of supernova acceleration, the particles are accelerated gradually by energy stored up in the remnant by the explosion or provided by the intense magnetic field of the rapidly rotating neutron star or pulsar which is formed in the explosion.

Such acceleration of high-energy particles is clearly observed in the Crab Nebula, the remnant of a supernova observed by Chinese astronomers in A.D. 1054. This nebula is populated by high-energy electrons which radiate a measurable amount of their energy as they spiral about in the magnetic field of the nebula. So much energy is released that the electrons would lose most of their energy in a century if it were not being continuously replenished. Pulses of gamma rays also show that bursts of high-energy particles are being produced by the neutron star—the gamma rays coming out when the particles interact with the atmosphere of the neutron star. Particles of cosmic-ray energy are certainly produced in this object, but it is not known whether the particles escape from the trapping magnetic fields in the nebula and join the freely propagating cosmic-ray population. *See Crab Nebula; Neutron star; Pulsar; X-ray astronomy.*

Acceleration in the solar system. The study of energetic particle acceleration in the solar system is valuable in itself, and can give insight into the processes which produce galactic cosmic rays. Large solar flares, about one a year, produce particles with energies in the gigaelectronvolt range, which can be detected through their secondaries even at the surface of the Earth. It is not known if such high-energy particles are produced at the flare site itself or are accelerated by bouncing off the shock fronts which propagate from the flare site outward through the solar wind. Nuclei and electrons up to 100 MeV are regularly generated in smaller flares. In many events it is possible to measure gamma rays and neutrons produced as these particles interact with the solar atmosphere. X-ray, optical and radio mapping of these flares are also used to study the details of the acceleration process. By relating the arrival times and energies of these particles at detectors throughout the solar system to the observations of their production, the structure of the solar and interplanetary magnetic fields may be studied in detail.

In addition to the Sun, acceleration of charged particles has been observed in the vicinity of the Earth, Mercury, Jupiter, and Saturn—those planets which have significant magnetic fields. Again, the details of the acceleration mechanism are not understood, but certainly involve both the rotation of the magnetic fields and their interactions with the solar wind. Jupiter is such an intense source of electrons below 30 MeV that it dominates other sources at the Earth when the two planets lie along the same interplanetary magnetic field line of force. Although the origin of the enhanced flux of ^{4}He has not been identified with certainty, it may be generated by the interaction of the solar wind with interstellar gas in the regions of the outer solar system where the wind is dying out and can no longer flow smoothly. *See Jupiter; Mercury; Planetary physics; Saturn.*

Direct observation of conditions throughout most of the solar system will be possible in the next few decades, and with it should come a basic understanding of the production and propagation of energetic particles locally. This understanding will perhaps form the basis of an understanding of the problem of galactic cosmic rays, which will remain for a very long time the only direct sample of material from the objects of the universe outside the solar system.

Paul Evenson

Bibliography. A. M. Hillas, *Cosmic Rays*, 1972; F. K. Lamb (ed.), *High Energy Astrophysics*, 1985; M. S. Longair, *High Energy Astrophysics*, 1981; J. L. Osborne and A. W. Wolfendale (eds.), *Origin of Cosmic Rays: Proceedings of the NATO Advanced Study Institute*, 1975; M. A. Pomerantz, *Cosmic Rays*, 1971; G. Setti and G. Spada (eds.), *Origin of Cosmic Rays*, 1981; M. M. Shapiro (ed.), *Cosmic Radiation in Contemporary Astrophysics*, 1986.

Cosmic spherules

Solidified droplets of extraterrestrial materials that melted either during high-velocity entry into the atmosphere or during hypervelocity impact of large meteoroids onto the Earth's surface. Cosmic spherules are rounded particles that are millimeter to microscopic in size and that can be identified by unique physical properties. Although great quantities of the spheres exist on the Earth, they are ordinarily found only in special environments where they have concentrated and are least diluted by terrestrial particulates. *See Meteor.*

The most common spherules are ablation spheres produced by aerodynamic melting of meteoroids as they enter the atmosphere. Typical ablation spheres are produced by melting of submillimeter asteroidal and cometary fragments that enter the atmosphere at velocities ranging from 6.5 to 43 mi (11 to 72 km) per second. Approximately 10,000 tons of such particles collide with the Earth each year, and cosmic spheres in the size range of 0.004 to 0.04 in. (0.1 to 1.0 mm) are the most abundant form of this material that survives to reach the Earth's surface. The spheres are formed near 48 mi (80 km) altitude, where deceleration, intense frictional heating, melting, partial vaporization, and solidification all occur in only a few seconds time. During formation, the larger particles can be seen as luminous meteors or shooting stars. Impact spheres constitute a second and rarer class of particles that are produced when giant meteoroids impact the Earth's surface with sufficient velocity to produce explosion craters that eject molten droplets of both meteoroid and target materials.

Impact spheres. These are very abundant on the Moon, but they are rare on the Earth, and they have been found in only a few locations. Impacts large enough to produce explosion craters occur on the Earth every few tens of thousands of years, but the spheres and the craters themselves are rapidly degraded by weathering and geological processes. Meteoritic spherules have been found around a number of craters, including Meteor Crater in Arizona, Wabar in Saudi Arabia, Box Hole and Henbury in Australia, Lonar in India, and Morasko in Poland. They have also been found at the Sikhote-Alin meteorite shower site and at the location of the Tunguska explosion in the Soviet Union. Spheres from these craters include ablation spheres as well as true impact spheres produced either by shock melting of target and meteoroid or by condensation from impact-generated vapor. Impact spheres are also produced by the larger cratering events, some of which may have played roles in biological extinctions. Silica-rich glass spheroids (microtektites) are found in thin layers that are contemporaneous with the conventional tektites. Microtektites are believed to be shock-melted sedimentary materials that were ejected from large impact craters. They were ejected as plumes that covered substantial fractions of the surface of the Earth. The cumulative mass of microtektites in the 35-million-year-old North American tektite field is estimated to be equivalent in mass to 36 mi^3 (100 km^3) of solid rock. Microspherules of a different composition have been found in the thin iridium-rich layer associated with the global mass extinctions at the Cretaceous–Tertiary boundary. *See* Tektite.

Ablation spheres. These fall to Earth at a rate of one 0.1-mm-diameter sphere per square meter per year, and every rooftop contains these particles. Unfortunately they are usually mixed in with vast quantities of terrestrial particulates, and they are very difficult to locate. They can, however, be easily found in special environments that do not contain high concentrations of terrestrial particles that could be confused with cosmic spheres larger than 0.004 in. (0.1 mm) in diameter. Such environments include the mid-Pacific ocean floor and certain ice deposits in Greenland and Antarctica. In mid-Pacific sediments the accumulation of terrestrial sediments is only a few meters per million years, and the spherules larger than 0.004 in. (0.1 mm) than are found in concentrations of roughly 10–100 per kilogram of sediment. Cosmic spheres are easily extracted from the sediment because most spherules are ferromagnetic and much larger than typical sediment particles. The highest ablation spherule concentrations on Earth (over one per gram of sediment) are found in the melt zones of the Greenland ice cap, where melting ice leaves dust particles concentrated directly on the ice surface.

Cosmic ablation spherules can be grouped into two major types: type S (stony) and type I (iron). In polished sections it can be seen that the S spheres are composed of olivine, magnetite, and glass, with textures consistent with rapid crystallization from melt. Both of the type S spherules shown in the **illustration** have elemental compositions similar to those of stony meteorites. In general, types S spheres have compositions that are a close match with chondritic meteorites except for depletion of volatile elements such as sulfur and sodium that are lost during atmospheric entry. Many spheres are also depleted in nickel by a process that may also lead to the formation of the iron spheres. Typical type I spheres consist of iron oxide (magnetite and wustite) surrounding a core of either nickel-iron metal or a small nugget of platinum group elements. Nickel and the platinum group elements are concentrated during brief oxidation when the sphere is molten. Some of the iron spheres may be droplets of iron meteorites, but the most common ones appear to be droplets that separated from the stony spheres during atmospheric melting.

Identification. Cosmic spherules are of particular scientific interest because they provide information about the composition of comets and asteroids and also because they can be used as tracers to identify debris resulting from the impact of large extraterrestrial objects. The ablation spherules can be positively identified because of several unique properties, including distinctive oxygen isotope compositions and the presence of isotopes that are produced by cosmic rays such as aluminum-26 and manganese-53. Some spheres did not melt entirely, and they also retain high concentrations of noble gases implanted by the solar wind. In general, cosmic spherules can be confidently identified on the basis of their elemental and mineralogical compositions, which are radically different from nearly all spherical particles of terrestrial origin.

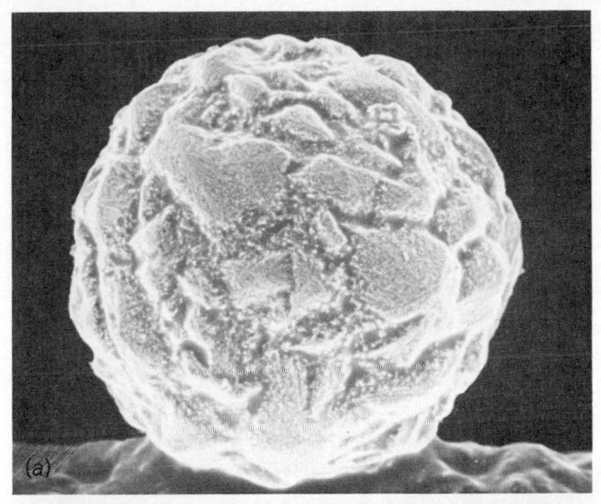

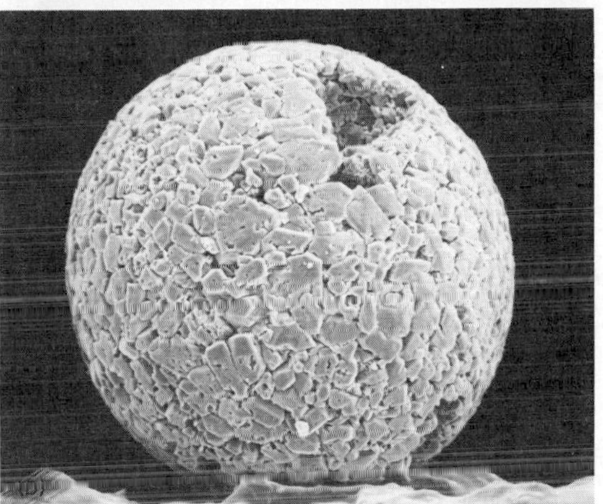

Scanning electron photomicrographs of stony cosmic spherules about 300 μm in diameter collected from the mid-Pacific ocean floor. (*a*) Sphere with so-called turtleback texture, indicative of rapid cooling from a very hot molten droplet. (*b*) Sphere with porphyritic texture, indicative of formation of a particle that was not so strongly heated during hypervelocity entry into the atmosphere as the sphere in *a*.

Don E. Brownlee, II

Bibliography. D. E. Brownlee, Cosmic dust: Collection and research, *Annu. Rev. Earth Planet. Sci.,* 13:147–173, 1985; C. Emiliani (ed.), *The Sea,* 1981; B. P. Glass et al., North American microtekitites from the Caribbean Sea and their fission track ages, *Earth Planet Sci. Lett.,* 19:184–192, 1973; P. W. Hodge, *Interplanetary Dust,* 1981; J. A. M. McDonnel, *Cosmic Dust,* 1978.

Cosmic string

A hypothetical object that may account for large-scale structure in the universe. The problem of the form and origin of such structure is one of the major challenges in science. The hot big bang theory provides a remarkably simple and successful description of the broad features of the universe, but it provides no mechanism through which structure could have formed. If the universe began, as is supposed, in a

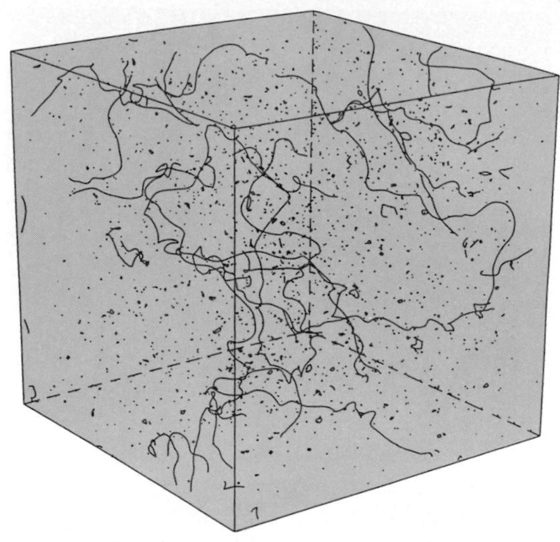

Computer model of a cubical box of cosmic strings. The strings oscillate at speeds close to the speed of light and reconnect if they cross. The length of an edge of the box is one-half the Hubble radius (approximately the distance traveled by light in the time since the big bang). *(After A. Albrecht and N. Turok, Evolution of cosmic string networks, Phys. Rev. D, 40:973–988, 1989)*

very uniform state, then according to the known laws of physics it would have remained smooth on any macroscopic scale right up to the present, and there would be no stars, galaxies, or galaxy superclusters. *SEE BIG BANG THEORY; GALAXY, EXTERNAL.*

Cosmic strings were the earliest theory of the formation of structure in the universe to emerge from high-energy particle theory in the early 1980s. It was realized that certain grand unified theories and superstring theories, which seek to unify the disparate forces of nature within a single theoretical framework, automatically lead to the production of a network of cosmic strings in the very early universe. Eventually, disturbances produced by the strings would lead to the formation of structures like galaxies and galaxy clusters.

Origin. Cosmic strings are analogous to defects that form when water is suddenly frozen to form ice. Freezing is an important example of a phase transition in which symmetry is broken. The water is symmetric under rotations; it looks the same from any angle. However, the crystalline structure of ice is not; its crystalline planes pick out definite directions. As the water freezes, at each point in space the crystalline structure of the solid picks an orientation in which to form, but this orientation varies from place to place. At some points, there is a mismatch between the orientation of neighboring crystalline regions. The resulting topological defects can take the form of sheets, lines, or points.

All unified field theories are based on the concept of symmetry breaking. The idea is that at high energies and temperatures the forces of nature should be indistinguishable and matter should exist in a highly symmetric state. However, at low energies the symmetry is broken and the forces are distinct. But the symmetry breaking that occurred as the very hot early universe cooled would have produced defects, just as symmetry breaking produces defects in ice when water freezes. Cosmic strings are closely analogous to the linelike defects in ice, often called dislocation lines. *SEE INFLATIONARY UNIVERSE COSMOLOGY.*

Properties. Cosmic strings are very thin, approximately 10^{-15} the radius of a proton. They can take the form of closed loops, or infinite strings that wander on forever. When the cosmic strings form, around 10^{-34} s after the big bang, most of the string is in very long strings, which wander right across the universe. In effect, the universe is filled with a random, tangled network of spaghetti. As the universe expands, the string network chops itself up into tiny loops, which lose energy by radiating gravity waves until they shrink and disappear. However, the long strings cannot be eliminated; if a long string intersects itself, it loses some string to a loop but the long string remains. Thus, long strings would survive right up to the present.

Cosmic strings are also very massive; 1 m of grand-unified cosmic string would weigh 10^{20} kg. Their mass presents little obstacle to their motion, however, because they have an enormous tension, precisely equal to their mass per unit length times the speed of light c squared. Thus, wiggles on a cosmic string propagate at the speed of light. The motion of a cosmic string depends only on the speed of light and on the metric of the background universe in which they are moving (and the standard hot big bang theory provides this universe). So it is in principle possible to calculate precisely how the tangled string network would have evolved.

Furthermore, there are reasons to believe that this evolution proceeds in a scaling manner. That is, after initial transients have died down, the network should look the same (in a statistical sense) at all times but with the overall scale changing with time. At any time, a region the size of the horizon (the distance that light has traveled since the big bang) would look somewhat like the **illustration**.

Origin of structure in the universe. Strings disturb matter through their gravitational field. The dimensionless number measuring the strength of the gravitational field they produce is given by Newton's constant G times the mass per unit length of the string, μ, divided by the squared speed of light, c^2. For grand-unified strings, $G\mu/c^2 \sim 10^{-6}$. Because this is much smaller than unity, such strings are quite unlikely to lead to the formation of black holes, despite their extreme density. They would provide a contribution to the density of the universe of order 10^{-4} of the total density. *SEE BLACK HOLE; GRAVITATION.*

This contribution is just about what is required to explain the structure observed in today's universe. Disturbances start to grow under their own self-gravity when the universe becomes dominated by matter, roughly 10,000 years after the big bang. This process can amplify disturbances as small as 1 part in 10,000, such as the strings would provide, into gravitationally bound objects like galaxies and galaxy clusters. Furthermore, the characteristic length scale on the string network is of order the horizon scale at all times. At matter domination, the horizon scale corresponds to a scale of around 60 megaparsecs today (taking account of the subsequent expansion of the universe), the approximate scale of the bubbles revealed by galaxy surveys.

Detection. There are three principal effects through which cosmic strings may be detected. The most distinctive signal of their presence would be linelike discontinuities in the microwave radiation pattern observed on the sky. These discontinuities would be produced by a sort of gravitational slingshot effect.

The fractional temperature difference across the line on the sky would be of order $8\pi G\mu v/c^3 \approx 10^{-5}$, where v is the velocity of the string, of order half the speed of light. The detection in 1992 of temperature fluctuations in the microwave sky by the *Cosmic Background Explorer* (*COBE*) satellite is about at the predicted level. However, the beam width of the *COBE* detectors is about 7°, larger than the expected interstring spacing. More accurate maps of the microwave sky, with better angular resolution, will make it possible to determine whether the detected fluctuations are really due to a network of cosmic strings. SEE COSMIC BACKGROUND RADIATION.

Second, the strings would act as gravitational lenses. If a pointlike source of light were viewed directly behind the string, two images would be observed, separated by an angle of approximately 4 arc-seconds. A long string could produce a line of double galaxies on the sky. SEE GRAVITATIONAL LENS.

Third, if cosmic strings exist, the universe should be filled with gravity waves of all frequencies, emitted by the loops chopped off as the string network straightened out in the very early universe. If a gravity wave were to pass between the Earth and a pulsar, it would change the distance traveled by the radio waves from the pulsar, and those waves would be Doppler shifted in frequency. Current experiments are very close to the sensitivity required to detect the gravity-wave background. SEE DOPPLER EFFECT; PULSAR.

As well as these distinctive tests, the wealth of data being gathered on the distribution and velocities of galaxies, and the discovery of galaxies and quasars at high redshift, provide ever tighter constraints on theories like cosmic strings. Even if all the current theoretical ideas are ruled out by new observations (as they may well be), many useful theoretical tools will be developed, and important insights will be gained as to what a correct theory might look like. SEE COSMOLOGY; UNIVERSE.

Neil Turok

Bibliography. G. W. Gibbons, S. W. Hawking, and T. Vachaspati, *The Formation and Evolution of Cosmic Strings*, 1990; T. W. B. Kibble, Topology of cosmic domains, *J. Phys.*, A9:1387–1398, 1976; D. Spergel and N. Turok, Texture and cosmic structure, *Sci. Amer.*, 266(3):52–59, March 1992; A. Vilenkin, Cosmic strings and domain walls, *Phys. Rep.*, 121:263–315, 1985.

Cosmochemistry

The science of the chemistry of the universe, particularly that beyond the Earth—that is, the abundance and distribution of elements, chemical compounds, and minerals; chemical processes, particularly in the formation of cosmic bodies; isotopic variations; radioactive transformations; and nuclear reactions, including those by which the elementary constituents were formed. Cosmochemistry is often functionally considered as a branch or extension of geochemistry. It is also closely related to astronomy and astrophysics. SEE GEOCHEMISTRY.

Techniques and sources of information. Many techniques of geochemistry can be applied to extraterrestrial samples which can be obtained for analysis in the laboratory. These include chemical analysis by a variety of methods, including neutron activation analysis for trace elements; electron-microprobe and ion-microprobe analysis of very small areas of polished surfaces; petrographic and x-ray-diffraction characterization of minerals and rocks; studies of magnetic properties and remanent magnetism; determination of relative abundances of isotopes of the elements, particularly by mass spectrometry; radiometric age determinations; and measurement of cosmic-ray-induced radioactivities and stable reaction products.

The most important of the available samples are meteorites, which are generally considered to be fragments of minor planets and possibly outgassed comets, and lunar samples that have been returned to Earth by spacecraft. Remote chemical studies have been made on the surfaces and in the atmospheres of planets and in the vicinity of comets. SEE METEORITE.

For most of the objects of the universe, chemical information comes from analysis and theoretical interpretation of the radiations received from them on Earth and by artificial satellites. Spectral analysis of the light (visible, ultraviolet, infrared) received from the Sun, stars, and luminous nebulae give information on their elemental contents and on the few compounds that are stable in their exposed regions. Radio astronomy provides information on the interstellar medium, including a considerable number of molecular constituents. The interstellar medium and very high-temperature regions of stars, nebulae, and galaxies are studied in the short-wavelength ultraviolet and x-ray spectral regions by instruments aboard rockets and artificial satellites. SEE ASTRONOMICAL SPECTROSCOPY; INFRARED ASTRONOMY; RADIO ASTRONOMY; SATELLITE ASTRONOMY; ULTRAVIOLET ASTRONOMY; X-RAY ASTRONOMY.

Nonluminous bodies of the solar system can be characterized chemically to some extent by their reflection spectra—that is, albedo as a function of wavelength. Meteor spectra give information about these small fragments of asteroids and comets. Observations of the solar wind and energetic solar particles give information about the Sun's outer layers. Cosmic rays bring information about more distant regions of space. SEE ALBEDO; COSMIC RAYS; METEOR; SOLAR WIND.

Solar system. Study of the solar system involves the Sun, planets, asteroids, and comets.

Sun. The elemental composition of the Sun, which accounts for 99.9% of the mass of the solar system, is essentially that of the system as a whole, and undoubtedly also of the presumed precursor solar nebula. Lines of 74 elements have been identified in the solar spectrum, and it is probable that all of the 81 stable and 2 long-lived elements are present.

Many of the abundant elements can be determined quantitatively by analysis of their Fraunhofer lines. Helium can be detected, but its quantification in the Sun is difficult; the solar He:H atomic ratio is believed to be similar to that observed in other population I objects (hot stars, luminous nebulae), about 0.098 (ignoring the transformation of hydrogen to helium by thermonuclear reactions in the core, which does not mix with the surface). Carbon, nitrogen, and oxygen are determined relative to hydrogen in the solar photosphere. For most of the other elements, the relative abundances are assumed to be the same as in type I carbonaceous chondrite meteorites, which are believed to have condensed from the solar nebula with little or no fractionation. The relative abundances of the noble gases and mercury are deduced from population I objects or by use of nuclear systematics.

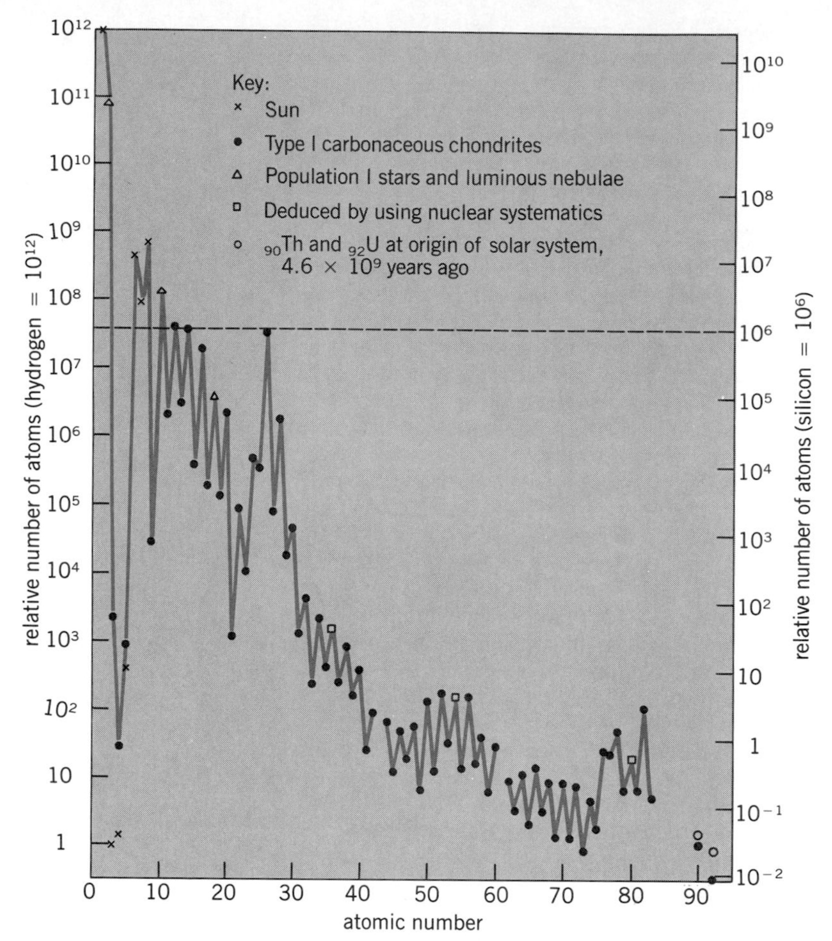

Fig. 1. Solar system elemental abundances. Meteoritic abundances relative to silicon have been normalized to solar abundances relative to hydrogen by using $^{14}Si/^{1}H$ = 3.58×10^{-5}. (*From data of E. Anders and N. Grevesse, Abundances of the elements: Meteoritic and Solar, Geochim. Cosmochim. Acta, 53:197–214, 1988*)

Carbon, nitrogen, and oxygen, like the noble-gas elements, are highly depleted in meteorites because of the volatility of the elements and their compounds. On the other hand, the observed solar values for lithium, beryllium, and boron are distinctly lower than those derived from meteorites, because those elements are being destroyed by thermonuclear reactions at the bottom of the Sun's outer convective layer. **Figure 1** shows the relative atomic abundances of the elements deduced for the solar system. The corresponding mass fractions are approximately 70.7% hydrogen, 27.4% helium, and 1.9% elements 3–92. *SEE ELEMENTS, COSMIC ABUNDANCE OF; SOLAR SYSTEM; SUN.*

Planets and asteroids. The terrestrial planets (Mercury, Venus, Earth, Mars), their satellites, and the asteroids consist largely of solids which have been separated from volatile elements. Their outer parts consist mainly of silicates, and each of these planets has a metallic core, the fraction of metal decreasing with increasing solar distance. The more massive Jovian planets (Jupiter, Saturn, Uranus, and Neptune) have retained large amounts of hydrogen, helium, other noble gases, and volatile compounds such as methane (CH_4), ammonia (NH_3), and water (H_2O). The overall compositions of these planets may be similar to that of the solar nebula, but the metallic and "rocky" elements are evidently concentrated in cores, and helium is probably partially concentrated in the interiors of Jupiter and Saturn since its abundances in these planets' atmospheres are less than the solar value. For a detailed discussion of the compositions of the planets *SEE PLANETARY PHYSICS*. For compositions of the individual planets *SEE EARTH; JUPITER; MARS; MERCURY; NEPTUNE; PLUTO; SATURN; URANUS; VENUS*. For discussions of planetary satellites *SEE* the articles on their planets and *MOON*.

Reflection spectroscopy of asteroids (minor planets) indicates considerable variety in surface compositions, many corresponding to known types of meteorites. The most abundant have low albedos (~2–5%) and spectra resembling carbonaceous chondrites. *SEE ASTERIOD*.

Comets. The fluorescent emission spectra of the nearly straight type I tails of luminous comets show the presence of atoms, molecules, radicals, and ions consisting of hydrogen, carbon, nitrogen, oxygen, and sulfur. In the comas of bright comets, metallic elements are also seen. Shower meteors, which are outgassed fragments of comets, show spectral lines of common metallic elements. Huge atomic-hydrogen clouds have been observed extending from comets in the antisolar directions, and large amounts of hydroxyl radical (OH) are seen. Spacecraft measurements near Comet Halley indicated that, except for hydrogen and noble gases, its overall composition is similar to the Sun's, with minor fractionation of nonvolatile elements. These observations are consistent with the icy-conglomerate model, according to which comets were formed as masses of water, methane, ammonia, and carbon dioxide, in which are embedded organic compounds and metallic, siliceous, and carbonaceous particles. Comets, such as Halley, which have passed close to the Sun many times have dark mantles of devolatilized and partially sintered dust, and emissions are concentrated in localized areas. Dust particles entrained by the sublimating gases form curved type II reflecting tails. Photodecomposition of vaporized water produces the atomic hydrogen and hydroxyl clouds. *SEE COMET; INTERPLANETARY MATTER*.

Stars. Most stars in the Milky Way and other galaxies can be grouped into populations having various distinguishing characteristics, which are fundamentally related to their ages or times of formation. Within each population, most main-sequence stars and many of the evolved stars exhibit similar surface chemical compositions. Contents of heavier elements, such as carbon, nitrogen, oxygen, neon, and metallic elements, increase with decreasing population age. Regardless of age, the hydrogen-helium ratio seems to be roughly constant at ~0.1 on an atomic basis, and the relative amounts of the heavy elements to each other are roughly constant.

These observations can be explained by three sets of circumstances: (1) The interstellar medium, from which stars are formed, has gradually increased its content of heavy elements (thus also dust), rapidly in the early history of the Milky Way Galaxy and slowly subsequently, but the amount of helium resulting from hydrogen fusion and escaping from stars is small relative to the primordial amount. (2) The heavy elements ejected into the interstellar medium are produced by a variety of processes in many stars of a variety of types, and the mix of products is roughly constant. (3) All but the least luminous main-sequence stars and some moderately evolved stars have nonconvecting (radiative-energy-transporting) zones somewhere in their interiors, and so the products of nuclear transformations in their cores are never mixed into their surface layers; and low-luminosity stars have as yet transformed very little of their hydrogen into helium in their lifetimes. Therefore, all of these stars have preserved in their atmospheres the elementary composition of the media from which they were born, which has evolved over time.

Some main-sequence and slightly evolved stars and many highly evolved stars have spectra showing nonstandard elemental contents which can be attributed to the presence in their surfaces of products of nuclear transformations in their atmospheres or interiors. Study of such stars helps scientists to understand the locales and mechanisms of energy generation and nucleosynthesis. *SEE NUCLEOSYNTHESIS; STAR; STELLAR EVOLUTION.*

Interstellar medium. This involves interstellar gas, dust, and molecules, as well as their chemistry.

Interstellar gas. Between the stars of the Milky Way are large quantities of gas and dust, with overall elemental composition similar to that of population I stars including the Sun, though definitely variable in elemental, chemical, and isotopic composition. Everywhere hydrogen dominates, and helium is the second most abundant element.

Neutral atomic hydrogen (H I) is detected and mapped by its characteristic 21-cm fine-structure radio emission (seen also in absorption in some cases). It is found to be concentrated in the spiral arms of the disk, mostly in diffuse clouds. Such clouds are typically ~30 parsecs (1 pc = 3.09×10^{13} km = 1.92×10^{13} mi = 3.26 light-years) in diameter with hydrogen concentration ~10–100 atoms/cm³ and temperature ~50–100 K (−370 to −280°F). They often show filamentary or shell-like structure, and amount to several percent of the mass of the stars.

Perhaps half of the interstellar matter is present as giant dense molecular clouds, of typical mass ~10^5 solar masses and dimension ~40 pc. These are most abundant in the inner galactic disk and central region. In these, molecular hydrogen (H_2) is undoubtedly the dominant constituent, although it cannot be observed there directly.

In less dense regions, molecular hydrogen can be detected by electronic absorption lines superposed on the ultraviolet continua of distant hot stars and extragalactic objects, and in somewhat warmer regions by infrared vibration-rotation emission lines. In the same regions, carbon monoxide can be detected by the 2.6-mm radio emissions of the abundant CO (that is, $^{12}C^{16}O$) and its rarer isotopic forms ^{13}CO and $C^{18}O$. Radio emission from ^{13}CO remains unsaturated in even the densest regions and thus serves as a tracer for molecular hydrogen and the giant molecular clouds. Molecular hydrogen is thereby found to have concentrations typically ~300 molecules/cm³ in the exteriors and up to ~10^7 molecules/cm³ in the densest cores of the giant cloud complexes.

More conspicuous though quantitatively less abundant are various kinds of luminous nebulae, collectively called H II regions, which owe their high temperatures (~10^4 K), ionization, and luminosity (fluorescence) to neighboring hot stars (both young and highly evolved). Typical H^+ concentrations are ~10–10^3/cm³. *SEE NEBULA.*

Interstellar dust. In cooler regions the gas is always accompanied by dust, whose mass is ~1% of that of the gas. The dust reveals itself mainly by absorption and scattering of starlight, this extinction being greater at shorter wavelengths and thus reddening the transmitted light. The spectral extinction indicates dominant particle sizes of ~0.01–0.1 micrometer. It shows maxima at wavelengths attributed to graphite and/or amorphous carbon, amorphous solid water, and silicates. In addition, refractory oxides and metal (predominantly iron) may be present, and water, ammonia, and methane ices may mantle cores of refractory constituents. More complex molecules and organic polymers may also be present. *SEE INTERSTELLAR EXTINCTION.*

The *Infrared Astronomical Satellite* (*IRAS*) observed extended sources of infrared emission called infrared cirrus, often associated with H I clouds and molecular regions and correlated with extinction of starlight. The relative intensities at different wavelengths indicate frequent occurrence of "cold" dust (~30 K or −400°F) and "warm" dust (~300 K or 80°F). The latter contains a major component of very small (≤2 nanometers) grains, possibly carbonaceous, or very large molecules, which become temporarily heated internally by absorption of individual starlight photons.

Dust also produces dark nebulae seen in front of star fields of the Milky Way. Large clouds have a typical dimension of ~4 pc and mass of ~2000 solar masses, and smaller ones (Bok globules) are typically ~1 pc in diameter and have mass of ~60 solar masses. Frequently associated with H II regions are even smaller Bok globules (~0.1 pc or less in diameter and of ~1 solar mass). *SEE GLOBULE.*

In moderately transparent interstellar regions, absorption lines of heavy atoms and ions are observed. These elements are depleted relative to hydrogen as compared to population I stars; this is correlated positively with expected condensation temperature (**Fig. 2**). This strongly suggests that the atoms missing from the gas are bound in solid oxide or silicate grains formed at high temperatures. That is believed to occur in the expanding and cooling ejecta of giant stars with

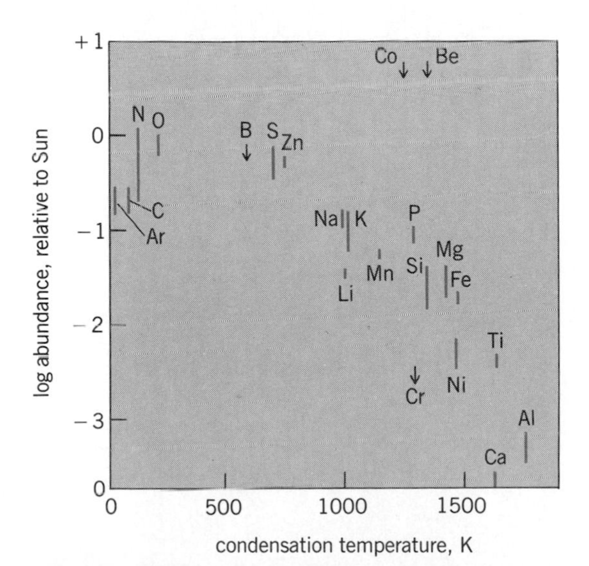

Fig. 2. Relative abundances of elements in a region of the interstellar medium versus temperature of condensation from gas of solar composition cooling in thermal and chemical equilibrium. Abundances are determined by absorption of ultraviolet light from the star Zeta Ophiuchi. Arrows indicate observational upper limits. (*After B. D. Savage and J. S. Mathis, Interstellar dust, Annu. Rev. Astron. Astrophys., 17:73, 1979*)

strong stellar winds, planetary nebulae, novae, and supernovae, and in interstellar shock fronts.

Interstellar molecules. Any neutral or charged aggregate of two or more atoms that exists in space is called a molecule; this includes many species that are known in chemistry as radicals and radical ions. Although very short-lived at the relatively high pressures of terrestrial environments, they can have long lifetimes in even dense interstellar regions because of the infrequency of collisions.

A number of simple molecules are observed in diffuse clouds by visible and ultraviolet absorption and by infrared and radio-line emission. Many more molecules and several molecular ions are observed by radio-line emission in the giant clouds. Over 75 molecules and a half dozen molecular ions have been identified, and dozens of weak lines are not yet identified. Molecules containing up to 13 atoms have been observed. In diffuse clouds, photodissociation allows only relatively simple molecules to be built up to detectable concentrations.

Interstellar chemistry. Two principal modes of formation of interstellar molecules have been proposed: grain-surface reactions (heterogeneous catalysis) and gas-phase reactions. Only reactions which are exoergic or very nearly so can occur by either mode at the low temperatures (below 100 K or −280°F) of molecular clouds.

A solid grain can catalyze the combination of two (or more) atoms or molecules by adsorbing both (or all) on its surface until they migrate into contact and combine, following which the product, generally a saturated molecule, escapes. Reactive centers on grain surfaces are probably rapidly covered by permanently bound atoms and molecules, and so the additional binding, effective in the process just described, is mainly by van der Waals forces.

When a particular reaction can occur by both modes, the grain-surface mode should be more effective because of the much greater frequency of collisions involving the larger particles in spite of their low number abundance. A large fraction of low-temperature atom- and molecular-grain collisions results in sticking. Since hydrogen atoms are quite mobile on adsorbing surfaces and the adsorption energy for molecular hydrogen is small, the grain-catalyzed mechanism is undoubtedly the major one for the formation of molecular hydrogen. A few other small molecules may have appreciable formation rates on grains. However, the van der Waals binding energy increases with the number of atoms of a molecule in contact with the surface, inhibiting its release. Most heavy molecules formed on grain surfaces probably stay there. Thus it appears that most heavy molecules found in the gas phase must have been formed in that phase.

Most two-body reactions involving neutral molecules have activation energies, resulting in strong temperature dependence of reaction rates and effectively preventing low-temperature collisions from leading to reaction. However, exoergic reactions between positive ions and neutral molecules generally have zero activation energy, and thus are nearly insensitive to temperature and can take place at near-zero temperatures. The same is true for some reactions of free atoms or radicals with each other or molecules. Such gas-phase reactions account for most of the interstellar molecules heavier than molecular hydrogen in the cold molecular clouds.

The necessary ionization is provided by ultraviolet photons in the diffuse regions and by cosmic rays, which can penetrate even dense interstellar regions. Most of the initial ionization forms H_2^+, H^+, and He^+, which can extract electrons from heavier atoms

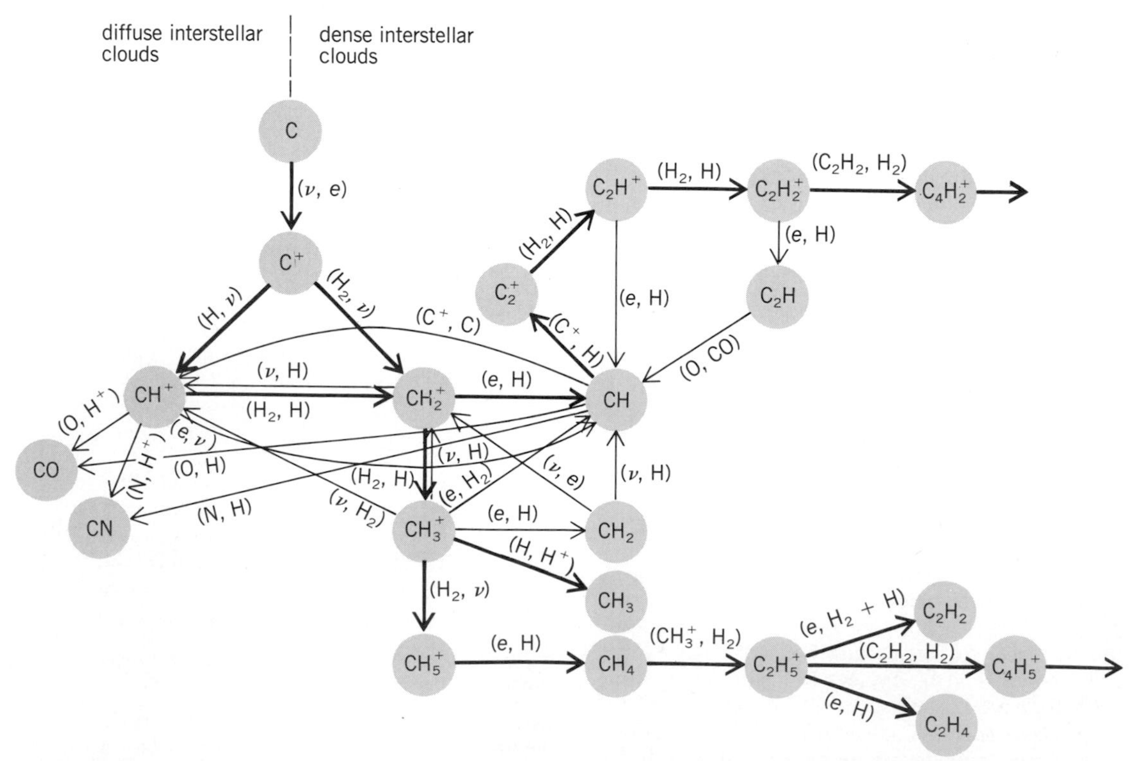

Fig. 3. Gas-phase reactions between carbon and hydrocarbon molecules in cold interstellar media. Within parentheses are the reactant and product not shown. ν = photon (or, as reactant in dense clouds, cosmic-ray particle), e = negative electron. Heavy arrows show principal reaction paths leading to higher-order hydrocarbons. (*After G. Winnewisser, The chemistry of interstellar molecules, Top. Curr. Chem., 99:39, 1981*)

and molecules to form principally C^+, O^+, N^+, and CO^+. Secondary reactions, especially chain reactions involving cations, then lead to a great diversity of products. Reactions involving the abundant molecular hydrogen and carbon monoxide are prominent.

The interaction of various reactions in cold interstellar regions is illustrated in **Fig. 3** for the simpler compounds of carbon. Similar schemes exist for other elements, especially oxygen and nitrogen.

In high-temperature regions, additional reactions, including many which have activation energies and some which are endoergic, can take place by ordinary thermal mechanisms. Elevated temperatures can be produced by interstellar shocks and by radiations from newly formed stars in collapsing clouds. *SEE INTERSTELLAR MATTER*.

Cosmic rays. Galactic cosmic rays (those not originating from the Sun) are highly energetic atomic nuclei, whose relative abundances reflect the composition of their source regions, but with considerable modification because of elemental differences in the acceleration processes and nuclear spallation-fragmentation reactions by collisions between heavy nuclei and hydrogen and helium nuclei in space. Modern balloon- and spacecraft-borne detectors can resolve individual masses up to the iron region, and the mass spectra are similar to the solar system abundances (Fig. 1). However, spallation and fragmentation greatly increase the relative abundances of lithium, beryllium, boron, and fluorine and the rarer isotopes of the other elements. Most of the solar system and stellar-atmosphere ^{6}Li, ^{9}Be, ^{10}B, and ^{11}B is believed to have been produced in this way. *SEE COSMIC RAYS*.

Galaxies. The mass fraction of heavy elements is enhanced in the central regions of the Milky Way and in similar spiral galaxies such as the Andromeda Galaxy (Messier 31), presumably because of the above-average rates of star formation and death of massive stars, which are chiefly responsible for dispersing products of nucleosynthesis into the interstellar medium.

The overall chemical composition of galaxies depends chiefly on the different stellar population distributions. Elliptical galaxies consist almost exclusively of population II stars, with very little interstellar gas and dust, and so virtually no formation of new stars occurs. Therefore they have relatively low contents of heavy elements. Spiral and undifferentiated irregular galaxies have high contents of gas and dust, resulting in high rates of star formation and nucleosynthesis, and thus considerable amounts of population I stars and relatively high contents of heavy elements. In some tidally interacting galaxies, induced star formation, called starburst, occurs, resulting in enhanced heavy-element abundances. *SEE GALAXY, EXTERNAL; MILKY WAY GALAXY*.

Cosmochronology. The radioactive dating methods based on primary natural radionuclides which are used in geochemistry are also applicable to cosmochemical samples in terrestrial laboratories. Methods based on short-lived extinct natural radionuclides are applicable to times very early in the solar system. Methods based on cosmic-ray-induced nuclides are especially useful for meteorites and lunar-surface materials, which have not been shielded by atmospheres.

Long-lived radionuclides. A number of meteoritic uranium,thorium-lead (U,Th-Pb), rubidium-strontium (Rb-Sr), and samarium-neodymium (Sm-Nd) ages have been obtained which are close to 4.56×10^9 years, and no reliable ages are older. This is regarded as the age of solid objects in the inner solar system. A few younger dates indicate subsequent disturbances due to igneous fractionation, metamorphism, or impact-induced shock, which "reset the clocks." The potassium-argon (K-Ar) and uranium, thorium-helium (U,Th-He) methods, which date retention of gaseous ^{40}Ar and ^{4}He, provide information about heating and shock events.

Extinct natural radionuclides. Excess ^{129}Xe from the decay of now-extinct ^{129}I has been observed in a number of meteorites, especially chondrites. This indicates that they were formed as cool solids and incorporated live ^{129}I no later than about 10 times its half-life (1.59×10^7 years) following its last prior nucleosynthesis in a supernova. Differences in the predecay $^{129}I/^{127}$ ratio in different meteorites indicate a spread of at least 16 million years in their gas-retention-onset times.

Xenon isotopes attributable to spontaneous fission of ^{244}Pu (half-life 8.1×10^7 years) are also widely observed, especially in achondrites. Excess of fission-fragment radiation-damage tracks in some meteoritic minerals is also ascribed to this extinct radionuclide.

Aluminum-correlated ^{26}Mg excesses have been found in high-temperature condensates occurring as inclusions in carbonaceous chondrites, and can definitely be ascribed to the decay of extinct ^{26}Al (half-life 7.1×10^5 years). Similarly, ^{107}Ag excesses strongly correlated with Pd/Ag ratios in iron meteorites, which were differentiated in planetary bodies, require that the parent ^{107}Pd (half-life $\sim 6.5 \times 10^5$ years) was alive in the early solar system.

Decay products of other now extinct radionuclides have been detected in meteorites as well, while traces of some nuclides with suitable half-lives have not been detected so far. A special case is ^{22}Na (2.6 years) $\rightarrow$ ^{22}Ne (Ne E), discussed below.

Nucleosynthesis of solar nebula material. The hydrogen and the bulk of the helium of the solar system undoubtedly originated in the cooling and expanding big bang fireball. The nuclides ^{129}I, ^{244}Pu, ^{235}U, ^{238}U, and ^{232}Th all result from the *r*-process of nucleosynthesis and were contributed by many supernovae to the interstellar medium from which the solar nebula formed. Analysis of the relative abundances of these radionuclides in meteorites at the time of their formation suggests that their nucleosynthesis began 8–15×10^9 years ago and terminated about 1–2×10^8 years before meteorite formation, with several percent of the *r*-products coming from the last contributing supernova. The ^{26}Al observations are interpreted by some as indicating a minor nucleosynthetic contribution much later. It is probable that the bulk of the stable elementary matter was synthesized on a similar time scale. *SEE SUPERNOVA*.

Ages of bodies in solar system. Dating of lunar rocks and soil indicates an age of the Moon as a body of about 4.4–4.6×10^9 years, about the same as that of the meteorites, and it is believed that all of the planets were formed at close to the same time. Dating of lunar lava basalts indicates that igneous processes were active on the Moon as recently as 3.1×10^9 years ago. It seems likely that most meteorites are fragments of asteroids, some of which must have been strongly heated and then cooled rather rapidly following their formation. However, a few rare basaltic-type meteorites, the SNC (Shergotty-Nakhla-Chassigny) group, have crystallization ages of about 1.3×10^9 years, and it is widely postulated that they are ejecta from Mars, which must have been volcanically active then, although it is no longer so.

Measurements of stable and radioactive cosmogenic (cosmic-ray-produced) nuclides in meteorites have shown that most iron meteorites have existed as small bodies in space, produced by collisional fragmentation from larger bodies, for less than 1×10^9 years, and most stony meteorites, which are more fragile, for less than 2×10^7 years.

Isotopes. Limited information is available on the isotopic composition of matter outside the solar system. In some cool giant stars, carbon and oxygen isotopes can be detected in optical and infrared spectra, and $^{13}C/^{12}C$, $^{17}O/^{16}O$, $^{18}O/^{16}O$, and $^{17}O/^{18}O$ ratios are generally greater than in the terrestrial elements. This can be attributed in part to variable participation of the carbon-nitrogen-oxygen thermonuclear-energy cycle in different stars. *SEE* CARBON-NITROGEN-OXYGEN CYCLES.

In the interstellar medium, isotope ratios of several elements can be determined from ultraviolet absorption and microwave molecular-emission spectra. The $D/^1H$ ratio in atomic hydrogen, $\sim 1.8 \times 10^{-5}$, is considerably lower than the terrestrial value ($\sim 1.5 \times 10^{-4}$, which is elevated because of selective escape of ^1H from the atmosphere), but it is greatly enhanced in molecules because of mass-dependent chemical fractionation in ion–molecule reactions. The $^{13}C/^{12}C$, $^{18}O/^{16}O$, $^{17}O/^{16}O$, and $^{15}N/^{14}N$ ratios also differ markedly from their solar system values and vary significantly from the galactic center to the disk. These observations are clues to past stellar activity in the Milky Way Galaxy.

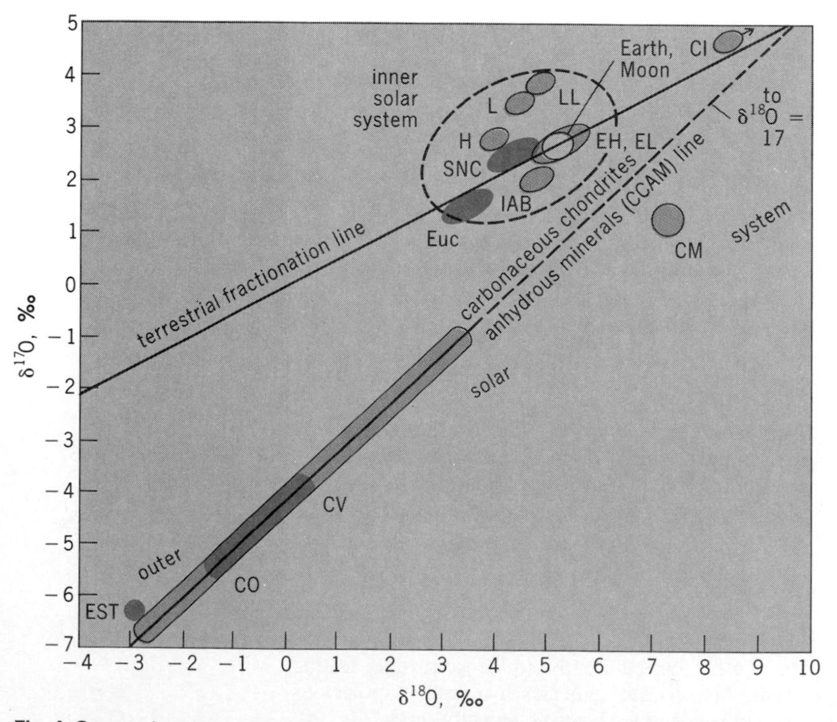

Fig. 4. Oxygen-isotope composition of solar system objects. $\delta^{17}O$ and $\delta^{18}O$ are the relative changes in the ratios $^{17}O/^{16}O$ and $^{18}O/^{16}O$ from their values in standard mean ocean water. CI, CM, CV, CO: types of carbonaceous chondrites. H, L, LL: types of ordinary chondrites. EH, EL: types of enstatite chondrites. Euc: eucrite, howardite, and diogenite achondrites, mesosiderites, and pallasites. SNC: Shergotty-Nakhla-Chassigny group of achondrites. IAB: iron meteorites of groups IA and IB. EST: Eagle Station trio of pallasites. (*After J. T. Wasson, Meteorites: Their Record of Early Solar-System History, W. H. Freeman, 1985*)

Studies of meteorites and the Moon have shown that for the most part there is a close similarity between the isotopic composition of meteoritic and lunar elements and their terrestrial counterparts. This is evidence that the bulk of the matter of the solar nebula, which was presumably derived from a number of different nucleosynthetic sources, was well mixed both physically and chemically before the formation of planetary bodies.

However, careful examination of many meteorite types and particularly of selected small phases of meteoritic matter has shown that there are numerous cases of differences. Those which cannot be accounted for by radioactive and cosmic-ray-induced nuclear transformations within the samples are referred to collectively as isotopic anomalies, and can be attributed to isotopic fractionation and to variable nuclear processes acting on matter which was not thoroughly mixed in the solar nebula.

The $D/^1H$ ratio is variable in meteorites, greater than 30–fold enhancements being found in some organic polymer fractions, which are presumed to be derived from relict interstellar molecules. The $^{13}C/^{12}C$ ratio is close to the terrestrial mean (~ 0.011) in major meteorite phases, but is greatly enhanced in some minor carbonaceous particles in some primitive chondrites. Considerable enhancements of $^{15}N/^{14}N$ have been observed in minor nonmetallic phases of some iron meteorites. These are probably derived from surviving interstellar phases of anomalous composition.

Oxygen is the principal major meteoritic element to exhibit significant isotopic variations. Because it has three stable isotopes, fractionation and mixing of isotopically different components can be distinguished. **Figure 4** illustrates schematically many of the observations. (The complete range of variations is considerably greater than shown here, $\delta^{18}O$ ranging from about −41‰ to about +35‰ and $\delta^{17}O$ from about −42‰ to about +17‰) Mass-dependent fractionation yields points on a line of slope 0.5 on this diagram, as illustrated by the terrestrial fractionation line. The Earth-Moon field refers to estimated bulk compositions. Meteorites having similar compositions are inferred to have formed in the inner part of the solar nebula. Objects away from this line must have been derived partly or wholly from matter of different oxygen composition. The array near a line of slope close to 1.0 suggests mixing of two end-member reservoirs, one near the terrestrial fractionation line and one very rich in ^{16}O but with about the same $^{18}O/^{17}O$ ratio ($\delta^{18}O$ about −40‰, $\delta^{17}O$ about −42‰). The array with slope ~ 1.0 above the terrestrial fractionation line implies at least one more ^{16}O-poor reservoir, and combined fractionation and mixing effects are also indicated. It is believed that the ^{16}O-rich and ^{16}O-poor oxygen resided in dust components that were not completely vaporized and mixed in the solar nebula, and the intermediate fraction, which accounted for the majority of planetary-object oxygen, resided in the gas.

The next most abundant meteoritic element showing large anomalies is magnesium. It also has three stable isotopes, but because one of them is partly radiogenic (^{26}Mg from extinct ^{26}Al) distinction between fractionation and nucleosynthetic anomalies is complicated. The larger anomalies imply an incompletely mixed interstellar dust component. Smaller anomalies are due to mass fractionation, as is also the case for silicon.

Calcium, titanium, and chromium exhibit anomalies, principally variable excesses of their heaviest isotopes ^{48}Ca, ^{50}Ti, and ^{54}Cr. These are attributed to an incompletely mixed dust component containing matter nucleosynthesized in a neutron-rich environment.

The noble gases in meteorites, being highly depleted, show strong effects of cosmic-ray irradiation, which enhances principally the lighter isotopes. The most pronounced anomaly is that of ^{22}Ne, which is strongly enhanced in some differential-heating release fractions. Indications are of the presence of a component of possibly pure ^{22}Ne (called NE E), which is attributed to the decay of short-lived ^{22}Na in dust particles which incorporated that radionuclide close to sites of its production, most likely in novae. Krypton and xenon show components with the isotopic signature of the *s*-process of nucleosynthesis, which occurs in red giant stars. These must have been mixed to the surfaces and incorporated into stellar winds, or expelled in planetary nebulae, and trapped or implanted in circumstellar dust particles. Noble-gas anomalies often occur in particles showing anomalies also of carbon, oxygen, and other mineral-forming elements. SEE NOVA; PLANETARY NEBULA.

Chemistry in cosmic evolution. According to current orthodoxy, the matter of the universe was created in an expanding "primordial fireball" (following the big bang) about 1–2 × 10^{10} years ago, mostly in the form of ^{1}H and ^{4}He in close to the present proportions, about 4:1 on a mass basis. Small amounts of ^{2}H, ^{3}He, and ^{7}Li were also formed then. SEE BIG BANG THEORY; COSMOLOGY.

Galactic evolution. Successive condensations in the expanding universe produced galaxy clusters, galaxies, star clusters, and individual stars. The most massive early stars generated heavy elements in their interiors and ejected them into the interstellar medium mainly through stellar winds and supernova eruptions. Later-formed stars in the Milky Way Galaxy, formed before it had collapsed to a disk, thus inherited small amounts of heavy elements. These are the population II stars, including those in globular clusters, which have typically 0.01–0.1% of elements heavier than helium. Subsequent star formation occurred mainly in the disk, with a rather rapid buildup of heavy elements, reflected in the initial composition of successive generations of stars. Most population I stars, of which the Sun is a member, have about 1–3% of heavy elements.

Solar system formation. About 4.6 × 10^{9} years ago a fragment of a giant molecular gas-dust cloud in a spiral arm of the Milky Way collapsed gravitationally to form the solar nebula. The collapse may have been initiated or accelerated by a nearby supernova, some of whose newly formed stable and radioactive nuclides may have been incorporated into the nebula. In the center of the nebula, accelerated contraction produced the Sun, initially much more luminous than now, probably in a T Tauri phase characterized by strong stellar winds. The outer parts collapsed into a disk, and the inner regions at least were heated by gravitational compression, solar radiation, and electromagnetic disturbances to temperatures sufficient to vaporize most of the dust grains.

As the Sun decreased in luminosity and became a stable main-sequence star, the nebula cooled while maintaining a distribution of temperature decreasing with increasing distance from the Sun. As the temperature at a particular radial distance decreased, a series of solids condensed from the gas. The **table**, based on thermochemical calculations, indicates the probable order of condensation of minerals, many of which are now found in meteorites. There is evidence, particularly from isotope fractionations, that some of the matter was subjected to repeated temperature decreases and increases with condensation and partial or complete revolatilization.

Stability fields of equilibrium solar nebula condensates*

Phase	Formula	Temperature limits, K†	
		Upper‡	Lower¶
Corundum	Al_2O_3	1758	1513
Perovskite	$CaTiO_3$	1647	1393
Melilite	$Ca_2Al_2SiO_7$–$Ca_2MgSi_2O_7$	1625	1450
Spinel	$MgAl_2O_4$	1513	1362
Metallic iron	(Fe,Ni)	1473	
Diopside	$CaMgSi_2O_6$	1450	
Forsterite	Mg_2SiO_4	1444	
	Ti_3O_5	1393	1125
Anorthite	$CaAl_2Si_2O_8$	1362	
Enstatite	$MgSiO_3$	1349	
Eskolaite	Cr_2O_3	1294	
Metallic cobalt	Co	1274	
Alabandite	MnS	1139	
Rutile	TiO_2	1125	
Alkali feldspar	$(Na,K)AlSi_3O_8$	~1000	
Troilite	FeS	700	
Magnetite	Fe_3O_4	405	
Water ice	H_2O	≤200	

*For total pressure of 10^{-3} atm (100 pascals).
†°F = (K × 1.8) − 459.67.
‡For condensation with decreasing temperature or disappearance with rising temperature.
¶For conversion to or from forms stable at lower temperatures.
SOURCE: L. Grossman, Condensation in the primitive solar nebula, *Geochim. Cosmochim. Acta*, 36:597–619, 1972.

When the temperature had dropped to somewhat below the condensation temperature, metal, sulfide, and siliceous grains initially of equilibrium composition adhered and accreted into small, loosely bound aggregates. Many of these were flash-heated by an uncertain process or processes so that they were instantly melted and rapidly cooled to form spherical objects. These comprise at least some of the abundant millimeter-sized chondrules now observed in chondritic meteorites. Further aggregation of dust and particles produced larger objects called planetesimals, perhaps meter- and kilometer-sized and even larger. The increased transparency of the nebula allowed solar radiation pressure and a strong solar wind in the inner regions to drive away the remaining gases. The planetesimals coalesced further to form the terrestrial planets, fortuitously large ones sweeping up all of the others at about their solar distances. The proportion of low-temperature condensates, including most oxides and silicates, relative to iron, a high-temperature condensate, increased with increasing distance from the Sun, and this is reflected in the densities and core sizes of the terrestrial planets.

In the outer part of this region, conditions were such that a single large planet did not form; but instead a number of minor planets, the progenitors of the asteroids, resulted. Gravitational perturbations by nearby massive Jupiter may have played a role. Reflection spectroscopy of the asteroids indicates a variety of chemical and mineralogical types, with some correlation with mean solar distance. In particular, in the outer part of the asteroid belt there is a high proportion of apparently carbonaceous bodies, indicating lower temperatures of formation than in the inner part of the belt. For the minor planets represented by the chondritic meteorites, there is evidence based mainly on trace-element analyses that the temperature was decreasing during their accretion, so that a layered structure resulted.

The formation of Earth's Moon is a special case, but whatever process was responsible created a fractionation of refractory elements, which are enhanced in the Moon, from more volatile elements, which are enriched in the Earth.

Still farther out, where enormous amounts of gases and solids were present, gravitational attraction became important, and uncondensed gases and unaccreted grains were also swept up by the giant planets which formed there. Each of these planets may have accreted nearly all of the hydrogen and helium that were associated with the heavier elements in its region of the nebula; the differing proportions of hydrogen, helium, and heavier elements in the planets' visible parts may reflect different fractionation and redistribution of elements in their interiors. Pluto's low density indicates that it is composed largely of ices, probably mostly water, though spectroscopy indicates solid and gaseous methane on its surface. This reflects both its low gravity and the extremely low temperature far out in the solar nebula.

The formation of the large satellites of the Jovian planets, which mostly have direct low-eccentricity orbits nearly in their parents' equatorial planes, presumably followed a pattern similar to the formation of the Sun's satellites, but with the difference that the temperature was much lower. Consequently they have low densities and consist largely of ices, predominantly water, surrounding small rocky cores.

At much greater distances, innumerable comets formed as aggregates of dust grains and ices. Those with the longest periods have randomly distributed orbital orientations and make up the enormous Oort cloud, which formed before the collapse of the solar nebula to a disk. Those of somewhat shorter periods have predominantly prograde orbits and make up the Kuiper belt beyond the orbit of Neptune, evidently formed after the collapse and shrinkage of the disk but before appreciable heating and evaporation of ices.

Planetary evolution. After and even during accretion, evolutionary changes within the planetary bodies and their atmospheres began to take place. Most chondrites show evidence that each was at one time a part of the regolith of a bombarded planetary body, probably still in the accretionary stage, and that it was later subjected to higher temperatures and pressures in the interior, resulting in lithification and varying degrees of metamorphism.

The major heat sources were probably gravitation, radioactive disintegration, and electromagnetic energy propagated in the solar wind. The last might have been important for asteroidal and smaller bodies. If appreciable amounts of intermediate-lived and now-extinct radionuclides (particularly ^{26}Al and ^{60}Fe) were present, they could have had important heating effects in asteroidal bodies. Relatively strong short-lived heat sources in the meteorite parents, which lose heat rapidly by surface radiation, were necessary to account for formation of the achondrite and iron meteorites, which are igneous rocks, followed by rapid cooling. Gravitational energy of accretion becomes important for the larger asteroids and the planets, which retain heat better, and long-lived radionuclides (^{235}U, ^{40}K, ^{238}U, and ^{232}Th) are effective in large bodies on long time scales. Their interiors rise to high temperatures, and when melting occurs, mass redistribution releases still more gravitational energy, especially if an iron or troilite core is formed.

Melting, gravitational redistribution, recrystallization, and outgassing produced stratified interiors and crusts, hydrospheres and atmospheres, volcanism, other geologic phenomena, and igneous rocks, both plutonic and extrusive, in the terrestrial planets. Any original atmospheres of the latter were probably swept away by the solar wind, and the present atmospheres of Venus, Earth, and Mars presumably resulted from interior outgassing. High temperatures and solar wind have stripped Mercury's atmosphere, and thermal upper-atmosphere evaporation has caused loss of much hydrogen from the other planets' atmospheres and most of the other light gases from Mars. Conversion of the presumably carbon dioxide (CO_2)-rich early atmosphere of the Earth to an oxygen (O_2)-rich atmosphere is probably the result of its biosphere's activity. *SEE PLANETARY PHYSICS; SOLAR WIND.*

Truman P. Kohman

Bibliography. E. Anders and N. Grevesse, Abundances of the elements: Meteoritic and solar, *Geochim. Cosmochim. Acta*, 53:197–214, 1989; J. F. Kerridge and M. S. Mathews (eds.), *Meteorites and the Early Solar System*, 1988; J. M. Moran and P. T. P. Ho (eds.), *Interstellar Matter*, 1988; S. K. Runcorn, G. Turner, and M. M. Woolfson (eds.), *The Solar System: Chemistry as a Key to Its Origin*, 1988; J. W. Truran, Nucleosynthesis, *Annu. Rev. Nucl. Part. Phys.*, 34:53–97, 1984; G. Turner and C. T. Pillinger (eds.), *Diffuse Matter in the Solar System: Comet Halley and Other Studies*, 1987.

Cosmology

The study of the structure and the origin of the universe, including the origin of galaxies, the elements, and matter itself.

Structure of the universe. Modern cosmology began in the early twentieth century with theoretical work on the cosmological implications of A. Einstein's theory of general relativity and with the astronomical debate over the nature of spiral nebulae.

Astronomers were uncertain over the nature of the faint spiral nebulae. H. Shapley contended that these faint nebulae were regions of star formation that were part of the Milky Way Galaxy. H. D. Curtis argued that these faint nebulae were distant galaxies much like the Milky Way. This debate, presented in a meeting of the National Academy of Sciences in 1920, was settled in 1923 by E. Hubble's discovery of 12 Cepheid variable stars in M31, the Andromeda Nebula. There is a simple empirical relationship between the period of the Cepheids and their luminosity. Thus, Hubble's observations allowed him to determine that the Andromeda Nebula was at a very large distance and was a galaxy much like the Milky Way. *SEE ANDROMEDA NEBULA; CEPHEIDS; GALAXY, EXTERNAL.*

Hubble continued his study of galaxies and found that most were receding from the Earth. Hubble proposed a simple linear relationship between the distance to a galaxy and its recessional velocity, given by Eq. (1), where v is the recessional velocity of the

$$v = Hr \qquad (1)$$

galaxy, usually measured in kilometers per second, and r is the distance to the galaxy, usually measured in megaparsecs (1 Mpc = 3.26×10^6 light-years = 3.08×10^{22} m = 1.9×10^{19} mi). Astronomers are still trying to accurately measure H, the Hubble constant. *See* HUBBLE CONSTANT.

Size of the universe. Measuring the distance to astronomical objects remains a great scientific challenge. Astronomers are able to measure the distance to the nearest stars by using parallax. They then must rely on empirical properties of stars to extrapolate to more distant objects. Observations of globular clusters and open clusters in the Milky Way Galaxy are needed to calibrate the luminosity-period relation for Cepheids. *See* PARALLAX; STAR.

Primary extragalactic distance indicators are used to measure the distance to nearby galaxies such as the Magellanic Clouds and Andromeda. Variable stars, such as Cepheids and RR Lyrae stars, are still important tools in determining the distances to these objects. Observations of the expansion of the dust shell around supernova 1987A in the Large Magellanic Cloud have provided an alternative method for measuring distances. All of these techniques imply that the Large Magellanic Cloud is at a distance of roughly 50,000 parsecs (150,000 light-years). *See* MAGELLANIC CLOUDS; SUPERNOVA; VARIABLE STAR.

Secondary distance indicators are then used to determine the relative distance to the Virgo Cluster. For example, the brightest red and blue stars in a distant galaxy in the Virgo Cluster are assumed to have the same luminosity as the brightest stars in Andromeda. Other secondary distance indicators include the brightness of typical globular clusters or planetary nebulae. Estimates of the distance to the Virgo Cluster range from 10 to 20 Mpc (30 to 60 $\times$ 10^6 light-years) and are a major source of uncertainty in estimates of the Hubble constant. *See* VIRGO CLUSTER.

Tertiary distance indicators are then used to extrapolate from the Virgo Cluster to more distant clusters. Galaxies in the Virgo Cluster are observed to have a simple relationship, noted by B. Tully and R. Fisher, between their gas velocities and their luminosity. This relation is used to measure the relative distance to distant galaxies. Together with Doppler-shift measurements of recessional velocity, this technique yields estimates of the Hubble constant. *See* DOPPLER EFFECT.

Current best estimates on the Hubble constant range from 42 to 100 (km/s)/Mpc. These estimates imply that the radius of the visible universe, c/H, where c is the speed of light, is between 3000 and 7000 Mpc (10 and 23 $\times$ 10^9 light-years). A fully operational Hubble Space Telescope should be able to detect Cepheids in the Virgo Cluster and reduce the uncertainty in the measurements of the Hubble constant. Several other techniques for measuring distance have been developed since the late 1980s. These techniques, which include using the expansion rates of supernovae, may also help determine more accurately the size of the universe. *See* SATELLITE ASTRONOMY.

Age of the universe. The universe ought to be older than any visible star; thus, the inferred ages of the stars in the Milky Way Galaxy place a lower limit on age of the universe. The oldest known stars are in globular clusters, dense concentrations of roughly 100,000 stars, believed to be 12–20 $\times$ 10^9 years old. These age estimates are based on models of stellar evolution that predict when a star of a given mass becomes a red giant. *See* STAR CLUSTERS; STELLAR EVOLUTION.

Just as radioactivity lifetimes can be used to date archeological artifacts, nuclear dating provides an alternative independent measure of the age of the Milky Way Galaxy. Using these techniques, W. A. Fowler has estimated a minimum age of the Galaxy of 10×10^9 years. The oldest white dwarf stars detected are 9×10^9 years old, and the Sun is believed to be 4.5×10^9 years old. These age measurements place a firm lower limit on the age of the universe. *See* SOLAR SYSTEM.

Homogeneity of the universe. Observations of the distribution of galaxies show significant inhomogeneities. Galaxies show a strong tendency to associate in groups, clusters, and superclusters. Even galaxies that do not lie in the centers of rich clusters appear to lie in large coherent structures such as the so-called Great Wall. Redshift surveys reveal that the many galaxies tend to lie on relatively narrow walls surrounding empty voids that are roughly 30 Mpc (10^8 light-years) across. *See* UNIVERSE.

Some cosmologists have suggested that the large-scale structure is fractal. An example of a fractal is the shoreline of Spain, which shows interesting structure on all scales. If the large-scale structure were fractal, then a map of the galaxy distribution that was 30 Mpc across would appear as a scaled-down version of a map of the galaxy distribution that was 300 Mpc across. This would imply that the basic assumptions of the big bang model were not valid.

Observations of the distribution of galaxies on large scales, however, show that the galaxy distribution appears uniform on very large scales. Deep optical surveys by R. Kirshner and collaborators show no evidence for voids larger than 50 Mpc. Studies based on the *Infrared Astronomy Satellite* (*IRAS*) survey also do not find significant fluctuations in the galaxy counts on scales larger than 60 Mpc (180 $\times$ 10^6 light-years). Thus, while the universe appears clumpy on small scales, when viewed on a large enough scale the universe does appear uniform. *See* INFRARED ASTRONOMY.

Big bang model. When Einstein proposed his theory of general relativity, the universe was believed to be static. Einstein had to modify his equations so that general relativity would allow a static universe by adding a cosmological constant term. Hubble's observations of the expanding universe then implied that these modifications were unnecessary. In the 1920s, G. Lemaître and A. Friedmann independently proposed a general relativistic model of the expanding universe. One of the simplest solutions to Einstein's relativity equation, it assumes that the universe is homogeneous and expanding. When Lemaître and Friedmann made their proposal, there was no real evidence for their simplifying assumptions. Only since the late 1980s have observations become sensitive enough to confirm their assumptions.

The Friedmann-Lemaître model, often called the big bang model, implies that the universe began in an extremely dense state and expanded and cooled. In this model, the Hubble law is predicted as an approximate description of the expansion valid for galaxies within a few hundred megaparsecs of the Milky Way

Galaxy. The model implies that radiation is redshifted as the universe expands. Thus, radiation from distant objects should appear at lower frequencies than those at which it was emitted. Observations of atomic lines from distant quasars confirm that radiation is redshifted just as predicted. SEE QUASAR; REDSHIFT.

The Friedmann-Lemaître model, while fully relativistic, can be described in the language of newtonian physics. The Hubble law implies that a shell of galaxies of radius R and mass m expands with velocity HR. Thus, the kinetic energy of the shell is $m(HR)^2/2$. If M is the mass interior to the shell, then the gravitational binding energy of the shell is GMm/R, where G is the newtonian constant of gravitation. The total energy E of the shell is therefore given by Eq. (2). Since it has been assumed that the universe

$$E = \frac{m(HR)^2}{2} - \frac{GMm}{R} \tag{2}$$

is uniform, the mass M within the shell can be replaced with the quantity $4\pi\rho R^3/3$, where ρ is the density of the universe and $4\pi R^3/3$ is the volume of a shell of radius R. Then Eq. (2) can be rewritten as Eq. (3), where Ω is the ratio of the density of the universe

$$\begin{aligned} \frac{E}{mR^2} &= \frac{H^2}{2}\left(1 - \frac{8\pi G\rho}{3H^2}\right) \\ &= \frac{H^2}{2}(1 - \Omega) \end{aligned} \tag{3}$$

to the critical density of the universe, $3H^2/(8\pi G)$. If $H = 50$ (km/s)/Mpc, the critical density now is 4.7×10^{-27} kg/m^3 (7×10^{10} solar masses/Mpc^3). If $\Omega < 1$, the total energy of the shell is positive, and the universe will continue to expand forever. If $\Omega > 1$, the total energy of the shell is negative, gravity will eventually stop the expansion, and the universe will eventually collapse. Some physicists speculate that this so-called big crunch will be followed by a future big bang. If $\Omega = 1$, the total energy is zero, and the universe stands on the balance between open and closed and corresponds to a special solution called the Einstein–de Sitter model. SEE GRAVITATION.

This simple newtonian model, while accurately describing the dynamics of the expanding universe, can lead to a conceptual error. The newtonian shell has a center, since newtonian theory cannot deal with a uniform mass density. General relativity, however, allows the universe to be isotropic and expanding uniformly without having a special center point. In the Friedmann-Lemaître model, the Milky Way Galaxy is not a special place in the universe.

Geometry of the universe. The density of the universe determines not only the final fate of the universe but also its geometry. If $\Omega > 1$, the universe is closed and its geometry is that of a three-dimensional sphere. (A circle is a one-sphere and the surface of the Earth is a two-sphere.) If two pilots leave New York in opposite directions, one heading east and the other heading west, they will meet somewhere over the Pacific. If the Earth were flat, the two pilots would continue to head directly away from each other. Similarly, if the universe is closed, two light rays sent off in opposite directions will eventually bend toward each other. If $\Omega = 1$, the universe is flat and the two light rays will continue to move away from each other. If $\Omega < 1$, the geometry of the universe is hyperbolic, much like that of a saddle.

Determination of age of universe. The density of the universe, together with the Hubble constant, determines the age of the universe in the big bang model. If Ω is much less than 1, the age of the universe is the inverse of the Hubble constant, $1/H$. If the Hubble constant is 50 (km/s)/Mpc, this implies that the universe is 2×10^{10} years old. If the Hubble constant is 100 (km/s)/Mpc and the universe is open, the big bang happened 10^{10} years ago. If $\Omega = 1$, integrating Eq. (3) would imply a smaller age for the universe, $2/(3H)$. Thus, in an Einstein–de Sitter universe a Hubble constant of 50 (km/s)/Mpc implies that the universe is 13.3×10^9 years old, while a Hubble constant of 100 (km/s)/Mpc implies that the universe is only 6.7×10^9 years old. Such a short age is in conflict with astrophysical estimates of stellar ages. Thus, if future observations find a large Hubble constant and a large value of Ω, the big bang paradigm would have to be reconsidered.

Open versus closed universe. Observations of the dynamics of galaxies can be used to determine the mean density of the universe, and thus to determine Ω. These observations suggest that $\Omega \approx 0.1$–0.2 and thus imply that the universe is open. However, if there exists some nonluminous matter that does not collect in clusters, Ω can be higher. The existence of dark matter in galaxies, discussed below, has led to speculation that dark matter may exist in the voids between galaxies.

Astronomical observations could also potentially measure the geometry of the universe. If the universe is closed, there is less volume associated with a given redshift than if it is open. The implication is that fewer galaxies should be detectable at large distances than in an open universe. The current observational situation is confused; galaxy counts using infrared techniques favor a flat or closed universe, while galaxy counts using optical techniques favor an open universe.

Observations of supernovae are another potential probe of the geometry of the universe. It is suspected that all type Ia supernovae, supernovae produced by the explosive burning of a white dwarf star, have the same luminosity. Thus, observations of distant supernovae can potentially determine the distances to galaxies at high redshift. Since the relation between distance and redshift differs in different cosmologies, this test can be used to measure the geometry of the universe. However, the geometry and density of the universe cannot yet be determined accurately.

Microwave background radiation. One of the most dramatic discoveries of modern physics was the detection of the microwave background radiation by A. Penzias and R. Wilson. Most cosmologists believe that this radiation is the leftover heat from the big bang.

In the hot big bang model, the universe started in an extremely hot dense state. In this state, the universe was composed of electrons, positrons, quarks, neutrinos, and photons. As the universe expanded, most of the matter annihilated with antimatter into photons. These photons then cooled as the universe expanded. Thermal physics predicts that the spectrum of radiation from the big bang would be similar to that emitted by a blackbody, a so-called Planck spectrum. The Friedmann-Lemaître model predicts that this radiation should be uniform since the big bang started in a uniform state.

The observations of the *Cosmic Background Ex-*

plorer (*COBE*) satellite, launched in 1989, provided strong confirmation of the hot big bang model. The observed spectrum of the microwave background radiation agrees closely with the predicted Planck spectrum. The *COBE* experiment also confirmed that the microwave background radiation is uniform to nearly 1 part in 50,000, consistent with the homogeneity assumption of the hot big bang model. *See* Cosmic background radiation.

Nucleosynthesis. Within the context of the hot big bang model, the conditions in the universe now can be extrapolated back to the first moments after the big bang. The temperature of the cosmic background radiation now is 2.73 kelvins above absolute zero. When the universe was half of its present size, the background temperature was twice as high. One second after the big bang, the universe was only 3×10^{-11} of its present size and the temperature of the microwave background was roughly 10^{10} K (1.8×10^{10}°F). At this high temperature, the universe consisted of a thermal sea of photons, electrons, positrons, and neutrinos. In addition, there was a handful of protons and neutrons. There was approximately 1 baryon (proton or neutron) for every 10^{10} photons.

As the universe cooled, the protons and neutrons combined to make deuterium. Most of this deuterium then interacted to make helium, while trace amounts combined to make lithium. Most of the deuterium and helium in the early universe is believed to have been produced in the first minutes of the big bang. One of the successes of the big bang model is its ability to account for the observed abundances of light elements. The universe today consists of roughly 28% helium, 70% hydrogen, and 2% other elements. Nucleosynthesis in stars produces roughly equal amounts of helium and heavier elements from hydrogen. Thus, it is difficult to understand the helium abundances without big bang nucleosynthesis.

The amounts of deuterium, helium, and lithium produced in the big bang depend sensitively on the number of baryons per photon in the early universe and on the number of neutrino flavors. Based on the observed abundances of these elements in old stars, D. N. Schramm, J. E. Gunn, and G. Steigman predicted that there should be only three flavors of neutrinos. Experiments at accelerators at the European Center for Nuclear Research (CERN) and the Stanford Linear Accelerator Center (SLAC) have confirmed this prediction based on the first moments of the early universe.

The observed light-element abundances also lead to a prediction for the density of protons and neutrons. The current best estimates suggest that if the universe is made only of protons, neutrons, and electrons, then Ω is between 0.01 and 0.1. Thus, unless there exists some exotic form of matter, big bang nucleosynthesis suggests that the universe is open. (As is discussed below, some exotic matter may possibly exist.) *See* Big bang theory.

Alternatives to the hot big bang. There have been a variety of models proposed as alternatives to the hot big bang. Historically, the most important alternative model is the steady-state model, which assumes a homogeneous expanding universe that is not evolving because matter is continuously being created out of the vacuum. This model has great difficulty accounting for the observed shape of the microwave background radiation and cannot account for the observed abundance of helium.

Another alternative is the cold big bang model in which the universe began in a big bang that started at absolute zero. E. Wright has argued that in this model the microwave background radiation could be due to reemitted light from iron needles. This model requires special dust grains and also cannot explain the abundances of light elements.

In H. Alfven's plasma universe, the Milky Way Galaxy is part of a finite cloud of material expanding into empty flat space. It is very difficult for this model to explain the observed uniformity of the microwave background spectrum or the light-element abundances.

E. Segal has proposed an alternative to cosmology based on general relativity. In this chronometric theory, there is a quadratic rather than a linear relationship between redshift and distance. Modern measurements of the distances to rich clusters find that these objects obey Hubble's linear relationship rather than Segal's quadratic law. Observations by M. Cohen of the correlation between jet expansion velocity and redshift are a direct test of the chronometric hypothesis, which it fails.

Dark-matter problem. Astronomers are in an embarrassing situation: the amount of mass in galaxies measured dynamically by observing stellar and gas motions is roughly 10 times the mass observed in dust. This discrepancy suggests that either there is a basic flaw in newtonian physics or 90% of the mass in galaxies is in some yet unknown form, usually referred to as the dark matter.

Evidence for dark matter. Some of the strongest evidence for existence of dark matter comes from radio observations of hydrogen gas and optical observations of star motions in spiral galaxies like the Milky Way Galaxy. The neutral gas and stars in the disks of galaxies are moving on nearly circular orbits. The centrifugal acceleration of the gas (and stars), v^2/r outward (where v is the velocity of motion and r is the orbit radius), must balance the gravitational acceleration of the galaxy, yielding Eq. (4), where $M(r)$ is the

$$v^2 = \frac{GM(r)}{r} \tag{4}$$

mass of the galaxy within radius of the gas or stellar orbit. Thus, observations of the gas or stellar velocity are a direct measure of the mass interior to its orbit. Most of the light in a galaxy like the Milky Way Galaxy is located in the inner 15,000 parsecs (50,000 light-years). The Sun is roughly 8000 parsecs (26,000 light-years) from the galactic center. If mass were distributed as light, the gas rotation velocity should begin to fall off outside the optical edge of the galaxy. In all optical and radio observations of spiral galaxies, the gas rotation velocity is not falling but is either flat or slowly rising. This situation implies that there is mass where there is no light.

Observations of hot x-ray-emitting gas in ellipitcal galaxies provide a method for measuring their mass distribution. Since the gas pressure gradients must balance the force of gravity, gas density and temperature profiles can be directly related to the elliptical galaxy mass profiles. Data from x-ray satellites suggest that the galactic dark-matter problem is ubiquitous: all galaxies, whether spiral or elliptical, dwarf or normal, seem to have halos of dark matter.

Observations of groups and clusters of galaxies provide an alternative method of weighing galaxies.

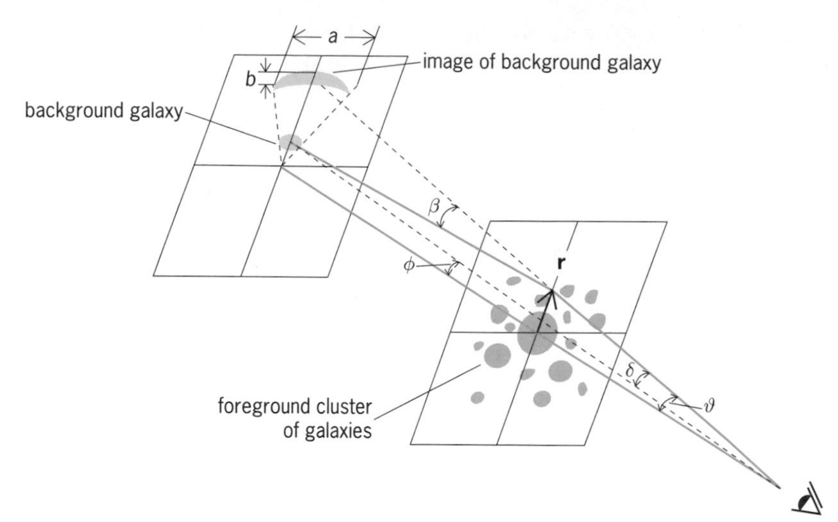

Gravitational displacement and distortion (lensing) of a distant background galaxy by a compact foreground cluster of galaxies. A light ray passing the cluster plane at an impact parameter vector *r* is gravitationally bent through an angle β. Thus, it is seen displaced (through an angle δ) from φ, its true angular distance from the cluster centroid, to the larger angle ϑ. Because of its finite width, the image also is distorted into a circular arc (of length *a* and width *b*) concentric with the cluster. (*After A. Tyson, Mapping dark matter with gravitational lenses, Phys. Today, 45(6):24–32, June 1992*)

The velocities of galaxies in clusters can be used to determine the mass of the entire cluster. This technique for measuring galactic mass, first applied by F. Zwicky in the 1930s, also implies a dynamical galactic mass that vastly exceeds the mass in luminous material.

General relativity implies that mass curves space. Thus, the path of light rays moving through a cluster of galaxies is bent by the mass within the cluster (see **illus.**). Hence, dense clusters of galaxies can act as gravitational lenses that distort the images of galaxies behind the cluster into arcs. Measurements of these distorted arcs by J. A. Tyson and collaborators confirm the evidence for enormous amounts of dark matter in clusters. *See Gravitational lens.*

Candidates for dark matter. Candidates for the dark matter range in mass from microelectronvolt axions to 10^6-solar-mass black holes. Many groups of physicists and astronomers are actively searching for various candidates for the dark matter.

Baryons (ordinary protons and neutrons) make up most of the mass of the Earth and the Sun. They are the most obvious dark-matter candidate. These baryons cannot be bound into luminous stars, nor can they be in either hot or cold gas. Cold gas can be detected through hyperfine and molecular lines. Hot gas is detectable by ultraviolet and x-ray satellites. Current observations constrain the gas mass to be much less than the mass needed to account for the dark matter.

Baryons can escape direct detection only if they are bound together into clumps, either as comets, bound by atomic forces, or as planets or very low-mass stars, bound by gravitational forces. These low-mass stars or planets can, however, be detected indirectly through their gravitational effects. Just as the clusters of galaxies do, these dark-matter objects distort space. If one of the objects passes between the Earth and a distant star, the light from the background star will be focused by the low-mass object and the background star will appear to brighten. Thus, low-mass objects may be detected through gravitational microlensing. Several groups are monitoring roughly 10^6 stars in the nearby Magellanic Clouds, hoping to detect these microlensing events. This technique could detect objects more massive than Jupiter regardless of their composition.

Black holes have been proposed as another possible candidate for the dark matter. If the first generation of star formation consisted of very massive stars, these objects would rapidly burn hydrogen to heavier elements and then collapse to black holes. Observational constraints on element abundance require that these massive stars not lose much of the carbon, oxygen, or iron produced during their stellar evolution to the interstellar medium. Models of very massive star evolution suggest that these stars could swallow most of their mass and form a halo of black holes. This scenario is rather tentative, since very little is understood about pregalactic star formation. These black holes could also be detected in the current lensing searches. *See Black hole.*

Neutrinos are another viable candidate for the dark matter. In the hot big bang model, neutrinos are produced in roughly equal numbers as the photons that make up the microwave background radiation. If the mass of the neutrino is roughly 50 eV, 10^{-4} the mass of the electron, then the neutrino is massive enough to be the dark matter. Experimental limits on the electron neutrino imply that its mass is less than 15 eV. However, current experimental limits on the tau and mu neutrino mass do not rule out cosmologically interesting masses. The only hope for measuring tau and mu neutrino masses in the 30–100-eV range is the detection of time-of-flight delay in the arrival of low-energy neutrinos from a galactic supernova. Neutrinos have fallen from favor as astrophysical dark-matter candidates; computer simulations of galaxy formation in a universe dominated by the neutrinos do not appear consistent with the observations of galaxy clustering.

Yet another possibility is that the dark matter consists of some as yet undetected particle. Particle physicists have suggested several possible candidates for the dark matter. The most popular of these particles is predicted in an extension of standard physics called supersymmetry. This particle, named a WIMP (weakly interacting massive particle), could potentially be detected in deep underground experiments. *See Weakly interacting massive particle (WIMP).*

Origin of large-scale structure. One of the most active areas of research in cosmology has been the attempt to understand the formation of galaxies and the origin of large-scale structure. As discussed above, the distribution of galaxies shows significant inhomogeneities on the scale of 30 Mpc. Astrophysicists have been struggling to understand the origin of these structures.

Most popular models of structure formation are based on gravitational instability. These models assume that there was some initial source of weak density fluctuation. In the expanding universe, these weak density fluctuations grow gravitationally. Regions that are slightly overdense expand more slowly; thus, these overdense regions collapse and form galaxies and clusters. On the other hand, underdense regions expand more rapidly than the surrounding universe. These underdense regions eventually develop into voids. The gravitational instability can successfully explain many of the statistical properties of the observed galaxy clustering pattern.

The details of the gravitational formation scenario depend upon the nature of the dark matter. If the dark matter is in the form of either baryons or neutrinos, small-scale density fluctuations are erased. In these models, the first objects to form are the rich clusters. Galaxies form later through the fragmentation of rich clusters. If the dark matter is in the form of WIMPs, it can cluster more easily. In these cold dark-matter models, the first structures to form are usually subgalactic objects, which later condense to form galaxies. Galaxies can then agglomerate gravitationally and form clusters. Observations of quasars at high redshifts suggest that at least some small objects must have formed in the early universe and appear to favor the cold dark-matter variant of the gravitational formation scenario.

Hydrodynamical processes and radiation physics may also play an important role in the formation of galaxies and large-scale structure. Astronomers are still struggling to understand the formation of stars in the Milky Way Galaxy now. Thus, it is difficult, if not impossible, to correctly model the formation of stars in the early universe. A complete picture for structure formation will likely require these processes in addition to gravitational instability. *See Protostar.*

Microwave background fluctuations. Observations of the microwave background radiation probe conditions in the early universe. In most cosmology models, the microwave photons that are detected on Earth last interacted with electrons when the universe was one-thousandth its present size. Thus, fluctuations in the microwave background temperature reflect the density fluctuations in the early universe.

The gravitational instability picture implies that there should be density fluctuations in the early universe of at least 1 part in 10,000. These density fluctuations would produce temperature fluctuations of roughly a few parts in 100,000. Over 25 years of searches for these fluctuations were rewarded in 1992, when the *COBE* satellite detected fluctuations in the microwave background temperature of about 2 parts in 100,000. The amplitude of these fluctuations is consistent with the predictions of gravitational instability theory.

Source of primordial fluctuations. While the gravitational instability picture can explain how initial weak fluctuations grew to form large-scale structure, it does not explain the initial source of these fluctuations. There are several competing theories.

The inflationary universe model is an attractive scenario for simulanteously explaining the homogeneity of the universe on large scales and the fluctuations in density observed on the smaller scales. In this scenario, the universe in its first few moments underwent a period of exponential expansion. This rapid expansion, which took place before nucleosynthesis, stretched out initial density variations and eliminated any monopoles or other unwanted objects that formed before the inflationary epoch. During this inflationary epoch, quantum fluctuations generated new density fluctuations that are predicted to be the initial source of structure. The inflationary scenarios predict that the density parameter Ω should be extremely close to 1. *See Inflationary universe cosmology.*

Ordering dynamics of a cosmic field provides an alternative scenario for structure formation. Cosmic fields behave much like iron atoms in a solid. They tend to order themselves and align as the universe expands. The energy fluctuations associated with this alignment process produce density fluctuations. In one version of this scenario, a phase transition in the early universe leads to the formation of cosmic strings, which accrete matter and produce density fluctuations. A related version of this scenario has the texture or global monopoles, which are produced at this phase transition, generate the structure. These models do not require that Ω be near 1. However, they lack an explanation for the observed large-scale homogeneity. *See Cosmic string.*

Explosive galaxy formation may provide another mechanism for amplifying and perhaps generating density fluctuations. Galaxies undergoing rapid star formation may produce a powerful wind that will sweep up gas. This swept-up gas may eventually cool to form another generation of galaxies that will in turn produce a strong wind. This process can be a runaway mechanism for generating fluctuations hydrodynamically. This violent amplification process will produce a great deal of hot gas, which may eventually be detectable.

Maps of the microwave background fluctuations may eventually reveal the source of the initial fluctuations. Different models make different predictions for the amplitude of microwave fluctuations on different scales. The inflationary model predicts that there should be fluctuations of roughly the same amplitude on all angular scales on the sky. The model also suggests that the fluctuations should be random and show no correlations. Models for structure formation based on ordering dynamics (cosmic strings, textures, and so forth) predict that the microwave maps should show phase correlations. Observations will definitively test these hypotheses.

David N. Spergel

Bibliography. M. V. Berry, *Principles of Cosmology and Gravitation*, 1989; W. A. Fowler, The age of the observable universe, *Quart. J. Roy. Astron. Soc.*, 28:87–108, 1987; P. J. E. Peebles et al., The case for the relativistic hot big bang cosmology, *Nature*, 352:769–776, 1991; M. Rowan Robinson, *The Cosmological Distance Ladder: Distance and Time in the Universe*, 1985; W. H. Tucker and K. Tucker, *The Dark Matter: Contemporary Science's Quest for the Mass Hidden in Our Universe*, 1988.

Crab Nebula

The Crab Nebula in Taurus (see **illus.**) is the most remarkable known gaseous nebula. Observed optically, it consists of an amorphous mass that radiates a continuous spectrum and is involved in a mesh of

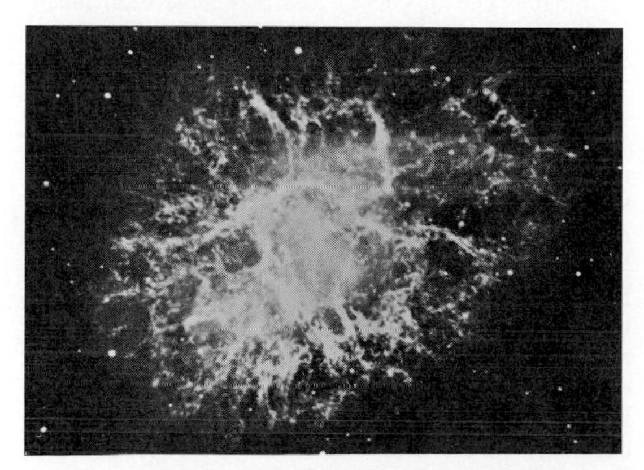

Crab Nebula, in the constellation Taurus, emitter of strong radio waves and of x-rays.

delicate filaments which radiate a bright line spectrum characteristic of typical gaseous nebulae.

This nebula has been identified as the expanding remnant of a supernova that appeared in A.D. 1054, reached an apparent magnitude of -5, corresponding to an absolute magnitude of about -18 (roughly 1.6×10^9 times as bright as the Sun). From a comparison of the angular expansion rate of the elliptical nebular shell with the expansion velocity in kilometers per second, measured by the Doppler shift, distance estimates ranging from 1300 to 2000 parsecs (1 parsec = 1.9×10^{13} mi = 3.1×10^{13} km) have been obtained. If a distance of 1800 parsecs is adopted, the nebular size is 3.1×2.1 parsecs. The total mass of the nebula is comparable to that of the Sun. Presumably it comprises only material from the star not yet mixed with the interstellar medium. *SEE DOPPLER EFFECT; SUPERNOVA.*

The Crab Nebula also radiates strongly in the radio range, and the infrared, ultraviolet, x-ray, and gamma-ray spectral regions. Furthermore, the radiation of the central amorphous mass is strongly polarized, even in the x-ray region—indicating that the source of the radiation must be the synchrotron emission of electrons accelerated in a field on the order of 10^{-8} tesla. Energy must be supplied to these electrons at at a rate equivalent to about 30,000 times the solar power output.

The source of energy for the Crab Nebula appears to be a remarkable, rapidly spinning object known as a pulsar, or neutron star, which is presumed to be the residue of the supernova that created the nebula. Its period of variability (which is associated with its rotation period) is 0.033 s. The period increases uniformly with time except for sudden spin-ups or glytches, when the rotation suddenly speeds up. These glytches are interpreted as readjustments in a crust of a neutron star, that is, as "star-quakes." Energy output variations are observed in the radio, infrared, optical, and x-ray range. The total amount of electromagnetic energy emitted by the star is about 10^{29} W (10^{36} ergs/s), but this is only a small fraction of the amount supplied to high-energy particles. The complex pulses observed both optically and at radio frequencies can be interpreted by an oblique magnetized rotator, but the mechanisms responsible for accelerations of particles to high energies are not fully understood. The ultimate source of the energy of the Crab Nebula is the rotational energy of the neutron star; in the period 1054–1975 the pulsar lost about 10^{43} joules (10^{50} ergs), mostly in particle energy. At present, the pulsar is losing about 10^{31} W (10^{38} ergs/s), but the rate of energy loss may have been much higher at an earlier epoch. The Crab Nebula may be regarded as the Rosetta Stone of high-energy astrophysics. *SEE ASTROPHYSICS, HIGH-ENERGY; NEUTRON STAR; PULSAR.*

One might anticipate that a search of the sky would find some object roughly similar to the Crab Nebula. Supernova remnants (for example, the Veil Nebula in Cygnus) abound in the Milky Way Galaxy, in the Magellanic Clouds, and in the Triangulum Spiral, M33, but in many instances what is observed is material thrown off by the supernova mixed with the neighboring interstellar medium. There is no trace of the residual star. Only the Vela supernova pulsar has been found optically, a twenty-sixth-magnitude star detected with the Anglo-Australian telescope. It is the faintest star ever observed. Remote pulsars cannot be detected, but it would seem that in some instances the star simply shattered completely or the residue disappeared as a black hole. *SEE BLACK HOLE; NEBULA.*

Lawrence H. Aller

Bibliography. *Crab Nebula: 46th International Astronomical Union Symposium*, Manchester, 1971; S. Mitton, *The Crab Nebula*, 1978.

Crux

The Southern Cross, in astronomy, the most celebrated of the constellations of the far south. The four principal bright stars of the group α, γ, β, and δ form the figure of a cross, giving the constellation its name, in contradistinction to the Northern Cross of Cygnus in the north, which is larger and more distinct, having a bright star to mark its center (see **illus.**). The brightest star in Crux is Alpha Crucis at the foot of the cross, which has received the artificial name, Acrux. This is a navigational star. The star at the top of the cross is Gamma Crucis, sometimes called Gacrux. The line joining these two stars points approximately toward the South Celestial Pole. *SEE CONSTELLATION; CYGNUS.*

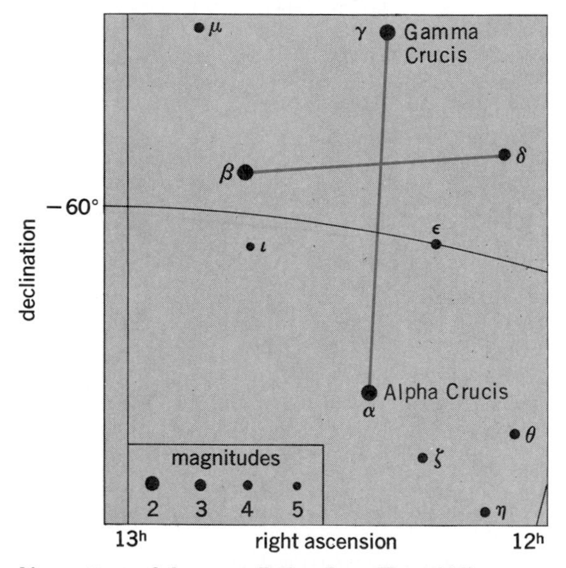

Line pattern of the constellation Crux. The grid lines represent the coordinates of the sky. The apparent brightness, or magnitude, of the stars is shown by the sizes of dots, graded by appropriate numbers as indicated.

Ching-Sung Yu

Cygnus

The Swan, in astronomy, is a conspicuous northern summer constellation. The five major stars of the group, α, γ, β, ϵ, and δ, are arranged in the form of a cross (see **illus.**). Hence Cygnus is often called the Northern Cross, to distinguish it from the Southern Cross of the constellation Crux. The constellation is represented by a swan with widespread wings flying southward. The bright star Deneb, signifying tail, is the tail of the swan; it lies at the head of the cross.

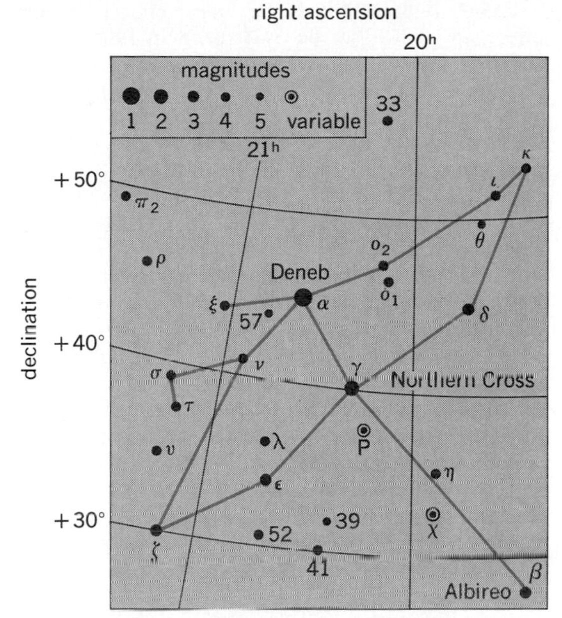

Line pattern of constellation Cygnus. Grid lines represent coordinates of sky. Apparent brightness, or magnitudes, of stars is shown by sizes of "dots," which are graded by appropriate numbers as indicated.

Albireo, a beautiful double star of contrasting orange and blue colors, is the head of the swan. The whole constellation lies in, and parallel to, the path of the Milky Way. It contains several splendid star fields. *See Constellation; Crux.*

Ching-Sung Yu

Day

A unit of time equal to the period of rotation of Earth. Different sorts of day are distinguished, according to how the period of rotation is reckoned with respect to one or another direction in space.

Solar day. The apparent solar day is the interval between any two successive meridian transits of the Sun. It varies through the year, reaching about 24 h 30 s of ordinary clock time in December and about 23 h 59 min 39 s in September.

The mean solar day is the interval between any two successive meridian transits of an imagined point in the sky that moves along the celestial equator with a uniform motion equal to the average rate of motion of the Sun along the ecliptic. Ordinary clocks are regulated to advance 24 h during a mean solar day.

Sidereal day. The sidereal day is the interval between any two successive meridian transits of the vernal equinox. Similarly, as for the solar day, a distinction is made between the apparent sidereal day and the mean sidereal day which, however, differ at most by a small fraction of a second. A mean sidereal day comprises 23 h 56 min 4.09054 s of a mean solar day.

The period of rotation of Earth with respect to a fixed direction in space is 0.0084 s longer than a sidereal day. No special name has been given to this kind of day, and although of theoretical interest, it is not used in practice.

Variations in duration. The mean solar day, the sidereal day, and the day mentioned in the preceding paragraph all vary together in consequence of variations in the speed of rotation of Earth, which are of three sorts: seasonal, irregular, and secular. The seasonal variations are probably caused, at least in part, by the action of winds and tides; the effect is to make the day about 0.001 s longer in March than in July, and is nearly repetitive from year to year. The irregular variations are probably the result of interactions between motions in the core of Earth and the outer layers; the effect is to cause more or less abrupt changes of several thousandths of a second in the length of the day, which persist for some years. The secular variation is the result of tidal friction, mainly in shallow seas, which causes the duration of the day to increase about 0.001 s in a century. *See Earth rotation and orbital motion; Time.*

Gerald M. Clemence

Bibliography. H. Jeffreys, *The Earth: Its Origin, History and Physical Constitution*, 6th ed., 1976; K. Lambeck, *The Earth's Variable Rotation: Geophysical Causes and Consequences*, 1980.

Doppler effect

The change in the frequency of a wave observed at a receiver whenever the source or the receiver is moving relative to each other or to the carrier of the wave (the medium). The effect was predicted in 1842 by C. Doppler, and first verified for sound waves by C. H. D. Buys-Ballot in 1845 from experiments conducted on a moving train.

Acoustical Doppler effect. The Doppler effect for sound waves is now a commonplace experience: If one is passed by a fast car or a plane, the pitch of its noise is considerably higher in approaching than in parting. The same phenomenon is observed if the source is at rest and the receiver is passing it.

Quantitative relations are obtained from a galilean transformation between the system of the receiver, moving with velocity $\vec{v}$, and that of the source, moving with velocity $\vec{w}$ with respect to the medium, which may be air taken at rest. In the rest frame the wave propagates with the proper sound velocity given by Eq. (1), where λ and ν are the wavelength and

$$|\vec{c}| = c = \lambda\nu \approx 333 \text{ m/s} = 1015 \text{ ft/s} \quad (1)$$

frequency measured in that system. Consider propagation of the wave in the direction of the vector $\vec{c}$, forming angles α and β with the velocities $\vec{v}$ and $\vec{w}$ (**Fig. 1**). The frequency ν_R received is by definition

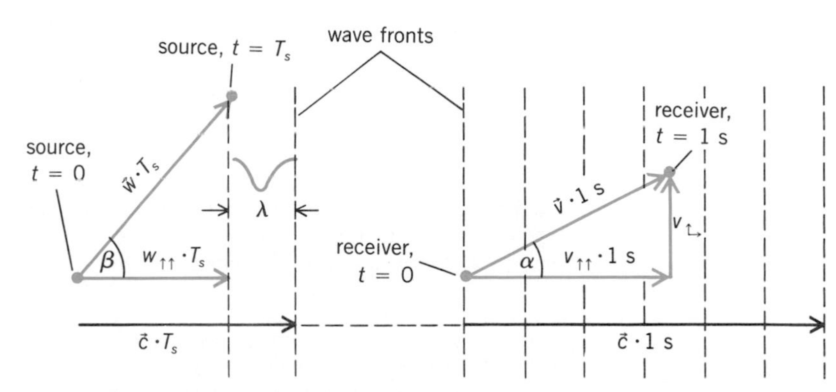

Fig. 1. Motions of source, receiver, and propagating sound that produce the acoustical Doppler effect.

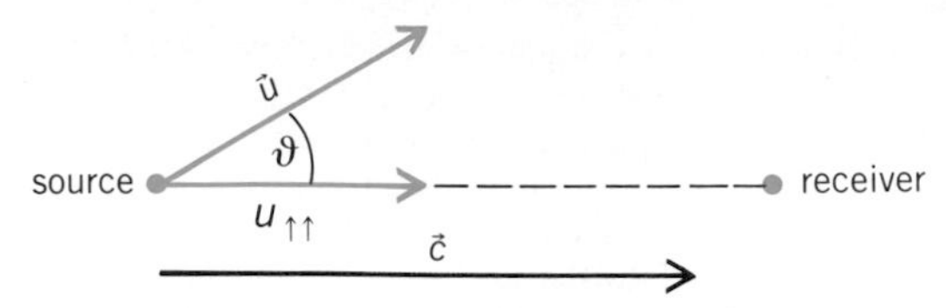

Fig. 2. Motions of source and propagating light, in the system of the receiver, that produce the optical Doppler effect.

the number of wavefronts (wave crests) passing the receiver per unit of time. This is given by calculating the difference between the sound velocity c and the component of the receiver velocity parallel to it $v_{\uparrow\uparrow}$ divided by the wavelength as in Eq. (2).

$$\nu_R = \frac{1}{\lambda}(c - v_{\uparrow\uparrow}) = \nu\left(1 - \frac{v_{\uparrow\uparrow}}{c}\right) = \nu\left(1 - \frac{v\cos\alpha}{c}\right) \quad (2)$$

A moving source acts differently since it effectively changes the wavelength in the medium, as seen from the following argument: Let the source emit a wave crest at time $t=0$, traveling during one period of the source oscillation, $T_s = 1/\nu_s$, a distance $c \cdot T_s$. The source follows the wave crest with its velocity component $w_{\uparrow\uparrow}$, parallel to $\vec{c}$. Therefore it emits the next wave crest at a distance from the first wave crest given by Eq. (3), which is by definition the wave-

$$s = (c - w_{\uparrow\uparrow})\,T_s = \frac{c - w_{\uparrow\uparrow}}{\nu_s} = \lambda = \frac{c}{\nu} \quad (3)$$

length λ in the medium. Hence Eq. (4) is obtained for

$$\nu = \frac{\nu_s}{1 - (w_{\uparrow\uparrow}/c)} = \frac{\nu_s}{1 - [(w\cos\beta)/c]} \quad (4)$$

the frequency ν observed in the frame of the medium.

In both cases the transverse velocity components $v_{\llcorner}$ and $w_{\llcorner}$ have no effect since they run parallel to the wavefront. In the supersonic regime where w is greater than c, Eq. (4) holds only inside the Mach cone.

Combining Eqs. (2) and (4) relates the received frequency to the emitted frequency by the general relation of Eq. (5), which may be expanded up to sec-

$$\nu_R = \nu_s \frac{1 - (v_{\uparrow\uparrow}/c)}{1 - (w_{\uparrow\uparrow}/c)} \quad (5)$$

ond order in the velocities, as in Eq. (6). The first-

$$\nu_R = \nu_s\left(1 + \frac{w_{\uparrow\uparrow} - v_{\uparrow\uparrow}}{c} + \frac{(w_{\uparrow\uparrow} - v_{\uparrow\uparrow})w_{\uparrow\uparrow}}{c^2} + \cdots\right) \quad (6)$$

order term, the so-called linear Doppler effect, depends only on the relative motion between source and receiver, that is, on Eq. (7), where $\vec{u}$ is the relative

$$u_{\uparrow\uparrow} = w_{\uparrow\uparrow} - v_{\uparrow\uparrow} \quad (7)$$

velocity of the source with respect to the receiver, whereas the higher-order terms depend explicitly on the motion of the source relative to the medium.

Optical Doppler effect. The linear optical Doppler effect was first observed by J. Stark in 1905 from a shift of spectral lines emitted by a beam of fast ions (canal rays) emerging from a hole in the cathode of a gas discharge tube run at high voltage. Still, their velocity was several orders of magnitude below that of light in vacuum c_0. This precluded a test of the second-order effect which would be sensitive to any kind of medium carrying light waves, the so-called light ether, if it existed. However, the precise interferometric experiments of A. A. Michelson and E. W. Morley (1887) which are closely related to the above considerations were sensitive to second-order effects. They showed clearly that the velocity of light is not bound to any ether, but is measured to be the same in any moving system. This result was a crucial check for A. Einstein's theory of special relativity (1905), which also makes a clear prediction for the optical Doppler effect with the help of a Lorentz transformation between source and receiver instead of a galilean transformation. This prediction is given by Eq. (8), where $\vec{u}$ is the relative velocity, and ϑ is the angle between $\vec{u}$ and the light propagation velocity $\vec{c}$, both measured in the system of the receiver (**Fig. 2**). Equation (8) agrees to first order with Eq. (6), but

$$\nu_R = \nu_s \frac{\sqrt{1 - u^2/c_0^2}}{1 - [(u\cos\vartheta)/c_0]} \quad (8)$$

differs in higher orders since it does not contain the source velocity with respect to any ether, but depends entirely on the relative velocity $\vec{u}$. *SEE INTERFEROMETRY.*

In contrast to the classical Doppler formula of Eq. (5), the exact relativistic optical Doppler effect of Eq. (8) also exhibits a shift for transverse motion ($\vartheta = 90°$), namely by $\sqrt{1 - u^2/c_0^2}$. This transverse Doppler effect results in a lowering of the received frequency (a redshift), which is equivalent to the relativistic time dilatation in the moving system (a clock runs slower). A verification of the higher-order Doppler effect (and hence of special relativity) to a precision of 4×10^{-5} was obtained recently by two-photon spectroscopy (see below) on a beam of fast neon atoms. Since the first-order Doppler effect cancels in this method, the experiment yielded directly the redshift of the atomic transition in question.

Applications. The Doppler effect has important applications in remote-sensing, high-energy physics, astrophysics, and spectroscopy.

Remote sensing. Let a wave from a sound source or radar source, or from a laser, be reflected from a moving object back to the source, which may itself move as well. Then a frequency shift $\Delta\nu$ is observed by a receiver connected to the source, which is given to first order by Eq. (9), where $u_{\uparrow\uparrow}$ is given by Eq.

$$\Delta\nu = \frac{2\nu_s u_{\uparrow\uparrow}}{c_0} \quad (9)$$

(7) and c_0 is the propagation speed of sound or electromagnetic radiation. The factor 2 is due to reflection. The measurement of $\Delta\nu$ is easily carried out with high precision, and provides an excellent means for the remote sensing of velocities of any kind of object; including cars, ships, planes, satellites, flows of fluids, or winds.

High-energy physics. Consider a head-on collision of

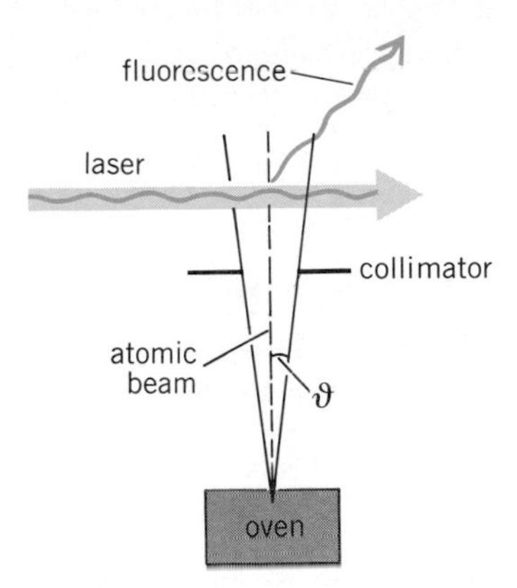

Fig. 3. Doppler-free atomic-beam spectroscopy on thermal atomic beams.

a laser beam, with wavelength λ of approximately 6 $\times$ 10^{-7} m, with a beam of electrons of very high energy, say 5 $\times$ 10^{10} eV, which is therefore extremely relativistic with v/c approximately equal to $1 - 0.5 \times 10^{-10}$. Then the same considerations as in the case of remote sensing, but calculated relativistically, lead to the conclusion that the light, backscattered by the Compton effect, is changed from visible photons into high-energy gamma rays with a wavelength of approximately 4 $\times$ 10^{-17} m and a quantum energy $h\nu$ (where h is Planck's constant) of approximately 3.2 $\times$ 10^{10} eV. The electron has thus transferred most of its kinetic energy to the light quantum. Such experiments have been performed at high-energy accelerators.

Astrophysics. The light from distant stars and galaxies shows a strong Doppler shift to the red, indicating that the universe is rapidly expanding. However, this effect can be mixed up with the gravitational redshift that results from the energy loss which a light quantum suffers when it emerges from a strong gravitational field. *See* Cosmology; Gravitational redshift; Redshift.

The Doppler width and Doppler shift of spectral lines in sunlight (Fraunhofer lines) are important diagnostic tools for the dynamics of the Sun's atmosphere, indicating its temperature and turbulence. *See* Astronomical spectroscopy; Sun.

Doppler-free spectroscopy. Spectral lines from atomic or molecular gases show a Doppler broadening, due to the statistical Maxwell distribution of their velocities. Hence the spectral profile, that is, the intensity I as a function of frequency ν, is a gaussian centered around ν_0, given by Eq. (10), with full half-

$$I(\nu) = I(\nu_0) \exp - \left(\frac{\nu - \nu_0}{\delta\nu}\right)^2 \qquad (10)$$

width given by Eq. (11), where k is the Boltzmann

$$\delta_{1/2}(\nu) = (2 \ln 2)\, \delta\nu = (2 \ln 2) \frac{\nu_0}{c_0} \sqrt{\frac{2kT}{m}} \qquad (11)$$

constant, T is the thermodynamic temperature in kelvins, and m is the mass of a molecule. The Doppler width in Eq. (11) is of order 10^{-6} ν_0 and exceeds the natural linewidth, given by Eq. (12), by orders of

$$\delta\nu_{\text{nat}} = \frac{1}{2\pi\tau} \qquad (12)$$

magnitude. Equation (12) is the ultimate limit of resolution, set by the decay time τ of the quantum states involved in the emission of the line. Thus, precision spectroscopy is seriously hindered by the Doppler width.

Around 1970, tunable, monochromatic, and powerful lasers came into use in spectroscopy and resulted in several different methods for completely overcoming the Doppler-width problem, collectively known as Doppler-free spectroscopy. Lasers also improved the spectroscopic sensitivity by many orders of magnitude, ultimately enabling the spectroscopy of single atoms, and both improvements revolutionized spectroscopy and its application in all fields of science. The most common Doppler-free methods are discussed below.

1. Atomic beams. An atomic beam, emerging from an oven, well collimated and intersecting a laser beam at right angles, absorbs resonant laser light with the Doppler width suppressed by a factor of sin ϑ, where ϑ is the collimation angle (**Fig. 3**). This straightforward method was already used before the advent of lasers with conventional light sources.

2. Collinear laser spectroscopy on fast beams. Another way to reduce the velocity spread in one direction is by acceleration. Consider an ion emerging from a source with a thermal energy spread δE, of the order of the quantity kT, or approximately 0.1 eV, and accelerated by a potential of, say, 50 kV along the x direction (**Fig. 4**). The kinetic-energy spread in this direction remains constant, however, and may be expanded as in Eq. (13), where the overbars represent

$$\delta E \sim kT \sim \delta\left(\frac{m}{2}\,\overline{v_x^2}\right) \approx m\bar{v}_x \cdot \delta v_x$$
$$\approx \frac{m\,c_0^2}{\nu_0^2} \cdot \Delta\nu \cdot \delta\nu = \text{constant} \qquad (13)$$

average values. Since the average velocity $\bar{v}_x$ has been increased by a factor of 700 in the example given, its spread δv_x, and hence the Doppler width with respect to a collinear laser beam, is reduced by the same factor. Consequently, the product of the Doppler shift $\Delta\nu$ of the fast beam and its Doppler width $\delta\nu$ is a constant of the motion. This method is very versatile and extremely sensitive.

3. Laser cooling. The Doppler effect has been used to cool down the thermal motion of atoms or ions by absorption of laser light. For instance, an ion may be trapped in a vacuum by electromagnetic fields (**Fig. 5**). The radial motion is stabilized by a magnetic field B, causing small cyclotron orbits of the ion, whereas axial motion is stabilized by applying a repelling electric potential to the electrodes on the axis. A laser

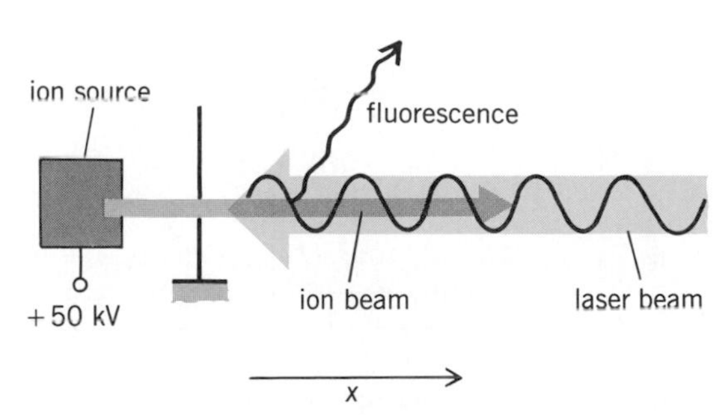

Fig. 4. Collinear laser spectroscopy on accelerated beams.

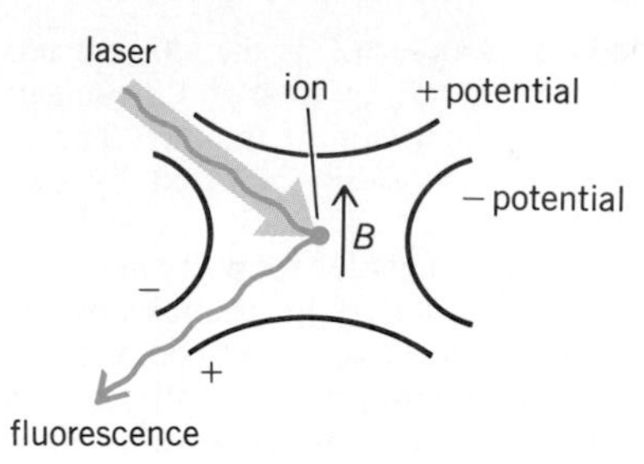

Fig. 5. Laser cooling of a trapped ion.

excites the ion at the low-frequency side of the Doppler profile of a resonance line, thus selecting excitations at which the ion counterpropagates to the light. Therefore the ion is slowed down, since it absorbs the photon momentum $h\nu/c$ (where h is Planck's constant) which is opposite to its own. After the ion emits fluorescent radiation, the process is repeated until the ion comes almost to rest.

The repelling force of the cooling process has also been used to confine a cloud of free atoms in vacuum at the intersection volume of three mutually orthogonal pairs or counterpropagating lasers. Thermodynamic temperatures of the order of 10^{-4} kelvin have been reached in such systems.

4. Saturation spectroscopy. Whereas the velocity distribution of the atoms is manipulated in the above examples, in the following ones this distribution is left untouched, but Doppler broadening is suppressed through the nonlinear optical interaction of the photons with the atoms. The field of nonlinear optics can be roughly characterized by the condition that the atomic system interacts with more than one light quantum during its characteristic decay time τ. In saturation spectroscopy, two almost counterpropagating laser beams of the same frequency ν are used, a strong pumping beam and a weak one (**Fig. 6**). The pumping beam is strong enough to saturate the excitation of those atoms in the vapor which match the resonance condition, that is, which compensate the frequency offset from the center $\nu - \nu_0$ by the proper velocity component, given by Eq. (14). Saturation means that the excitation is fast as compared to the decay time, leading to an equal population of the lower and the upper atomic states. Such a system is completely transparent. The probing beam feeds on atoms with exactly the opposite velocity component given by Eq. (15). It is too weak to saturate, and is

$$v_x = c(\nu - \nu_0) \quad (14)$$

$$-v_x = c(\nu - \nu_0) \quad (15)$$

absorbed in the vapor except at the center frequency $\nu = \nu_0$, where it matches with the saturated part of the velocity spectrum. Resolutions of $\delta\nu/\nu_0 < 10^{-10}$ obtained in this way make possible improved values of fundamental constants and of the standards of length and time.

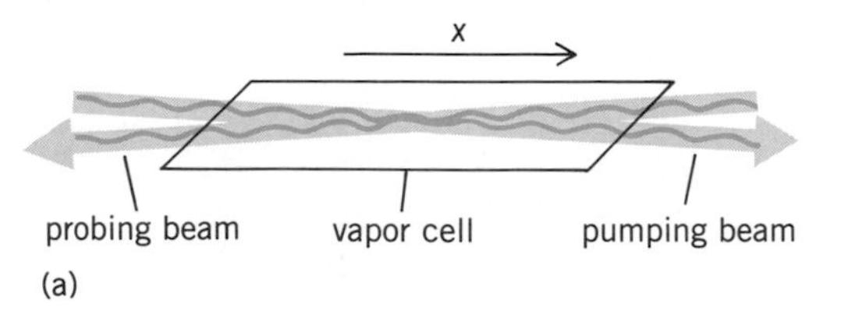

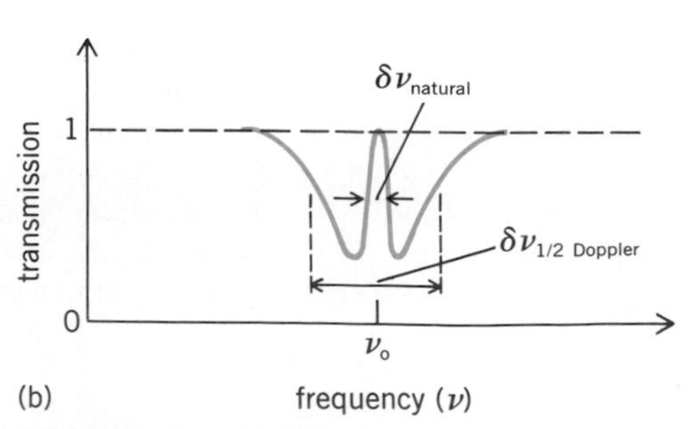

Fig. 6. Saturation spectroscopy. (*a*) Experimental configuration. (*b*) Transmission of probing beam.

5. Two-photon spectroscopy. In this method, two counterpropagating laser beams are operated at half the transition frequency ν_0 of the atomic system. Then Bohr's quantum condition, that the transition energy ΔE equals $h\nu_0$, can be fulfilled by simultaneous absorption of two photons, say, one from each beam

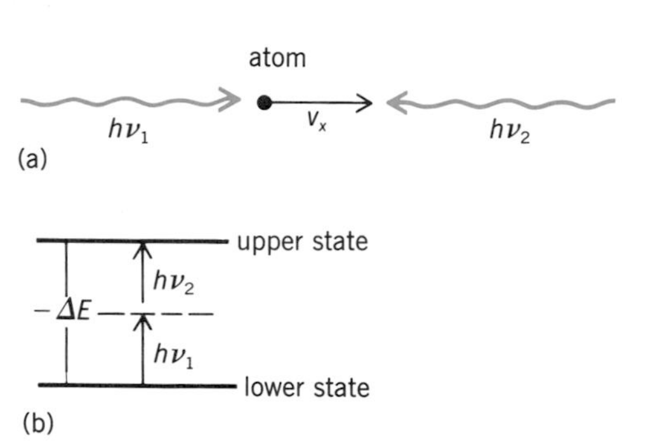

Fig. 7. Two-photon spectroscopy. (*a*) Relation of atom's motion to photons from counterpropagating laser beams. (*b*) Energy-level diagram.

(**Fig. 7**). Including the first-order Doppler effect, the energy balance is then given by Eq. (16). Thus, the

$$h\nu_1 + h\nu_2 = h\nu\left(1 + \frac{v_x}{c}\right) + h\nu\left(1 - \frac{v_x}{c}\right) = 2h\nu = h\nu_0 \quad (16)$$

first-order Doppler effect cancels completely, and the two-photon resonance occurs sharply at half the transition frequency.

Ernst W. Otten

Bibliography. A. Einstein, *The Principle of Relativity*, 1923; D. Halliday and R. Resnick, *Physics*, 3d ed., 1977; S. Haroche et al. (eds.), *Laser Spectroscopy*, Lecture Notes in Physics, vol. 43, 1975; M. Kaivola et al., Measurement of the relativistic Doppler shift in neon, *Phys. Rev. Lett.*, 54:255–258, 1985; W. Persson and S. Svanberg (eds.), *Laser Spectroscopy 8*, Springer Series in Optical Sciences, vol. 55, 1987; K. Shimoda (ed.), *High Resolution Laser Spectroscopy*, Topics in Applied Physics, vol. 13, 1976; H. P. Weber and W. Lüthy (eds.), *Laser Spectroscopy 6*, Springer Series in Optical Sciences, vol. 40, 1983.

Dwarf star

The most common type of star in the Galaxy, also known as a main-sequence star. The Sun is a typical dwarf, with surface temperature of 9900°F (5750 K), radius of 4.32×10^5 mi (6.96×10^5 km), mass of 4.4×10^{30} lbm (2×10^{30} kg), and luminosity of 4×10^{26} W. These numbers are fairly typical as to radius and mass. However, in luminosity, main-sequence stars occur over a range from O and B stars, up to more than 10^4 times brighter than the Sun, with surface temperatures of 63,000–27,000°F (35,000–15,000 K), down to the faintest M dwarfs, 10^{-4} as bright as the Sun, with a temperature of only 4000°F (2500 K). Dwarfs form a single-parameter family in which the significant variable is mass; members range in mass from 20 Suns down to less than 0.1 Sun. Significant variations in composition occur, distinguishing the oldest dwarfs from those more recently formed. Thus, abundance ratios of metals to hydrogen are lower by factors of up to 1000 in some of the fastest-moving stars of the halo population. But all main-sequence stars have the same helium-to-hydrogen ratio, reflecting the primordial helium production in the big bang. The M or red dwarfs are the most common stars in space, and because of their low luminosity have the longest life. Almost all other types of stars have evolved from main-sequence stars. *SEE STAR; SUN.*

Jesse L. Greenstein

Dynamical time

A time scale based on the orbital motions of celestial bodies. It is defined as the independent variable T in the differential equations of motion of bodies subject to the laws of celestial mechanics in a theory of relativity. Dynamical time scales can differ in epoch and rate, but cannot have a relative secular acceleration.

The atomic time scale obtained from atomic clocks is not corrected for periodic relativistic terms, and thus differs from dynamical time by their sum, P_R. Universal Time One (UT1), based on the Earth's rotation, differs from dynamical time because of secular and other variations in the Earth's rotational speed. *SEE ATOMIC TIME; EARTH ROTATION AND ORBITAL MOTION.*

Analytic or numerical integration of the differential equations of motion for a body gives its position as a function of T, tabulated in an ephemeris for equal intervals of time. In 1952 the International Astronomical Union (IAU) defined a dynamical time scale called Ephemeris Time (ET). The orbital ephemeris of a body is entered with an observed position and ET is read out. The primary ET scale was based on the Earth's orbital motion as given by Simon Newcomb's *Tables of the Sun* (NTS) of 1898. (The Earth's position is opposite the Sun's observed position.) In practice, however, ET was obtained from the Moon by using a lunar ephemeris that had been made accordant with NTS by including an empirical quadratic term for tidal friction. *SEE CELESTIAL MECHANICS; EPHEMERIS; MOON; ORBITAL MOTION.*

The frequency of a cesium-beam atomic clock built in 1955 by the National Physical Laboratory in Teddington, England, was found to be 9,192,631,770 cycles per second of ET, as determined by photographic observations of the Moon and stars, made at the U.S. Naval Observatory, Washington, from a joint comparison of the period from mid-1955 to mid-1958. In 1967 the General Conference of Weights and Measures defined the second, the unit of time in the International System of Units (SI), as the duration of this many periods of cesium-133 radiation. This is the second of International Atomic Time (TAI). Standard time differs from TAI by integral numbers of hours and seconds.

In 1976 the IAU defined two time scales to meet needs for increased accuracy of ephemerides and to utilize the high accuracy and availability of atomic time, TAI. Barycentric Dynamical Time (French abbreviation, TDB) is a dynamical scale. It is the independent variable in the differential equations of motion, referred to the barycenter (center of mass) of the solar system (near the center of the Sun). TDB is obtained in the same way as ET, except that modern ephemerides are used. Terrestrial Dynamical Time (TDT) is an auxiliary atomic scale, obtained in practice by letting TDT = TAI + 32.184 s. TDT differs from TDB only by the periodic terms, P_R, included in TDB. P_R is calculable, and has a maximum value of 0.0017 s, but it depends somewhat on the relativity theory that is used to form TDB (although the differences are orders of magnitude less than the maximum).

Analytic or numerical integration of the equations of motion for a planet gives its barycentric position as a function of TDB. This is transformed into an apparent geocentric position as a function of TDT, tabulated in ephemerides for equal intervals of the immediately available TDT. No clock keeps TDB.

Developmental Ephemeris, DE 200, has been used in the *Astronomical Almanac* since January 1, 1984. It was formed at the Jet Propulsion Laboratory in cooperation with the U.S. Naval Observatory by the simultaneous numerical integration of the differential equations of motion of the Sun, Moon, and planets, including five minor planets. Starting coordinates and velocities were based on optical angular observations and on highly accurate distance measurements, using radar, lunar laser, and spacecraft techniques.

The ephemeris of an object may be entered with the TDT of an observation to obtain a calculated ephemeris position or, conversely, with the observed position to obtain a calculated TDT. The differences in the two TDTs would include a T^2 term if dynamical time, TDB, and atomic time, TAI, have a relative secular acceleration. Such an acceleration is predicted by any cosmological theory in which the gravitational constant G varies, such as those proposed by E. A. Milne, by P. A. M. Dirac, and by C. Brans and R. H. Dicke. The comparison has been carried out from observations of the Moon, but because tidal friction also produces a T^2 term of uncertain magnitude, it is not determined, at present, whether TDB and TAI have a relative secular acceleration. If such an acceleration is confirmed, the IAU could redefine TDT to include a T^2 term, which would make the observed and calculated TDTs agree. *SEE GRAVITATION; TIME.*

William Markowitz

Bibliography. *Astronomical Almanac*, annually; *Explanatory Supplement to the Astronomical Ephemeris and the American Ephemeris and Nautical Almanac*, 1961, 1974; T. Fukushima et al., System of astronomical constants in the relativistic framework, *Celest. Mech.*, 36:215–230, 1986; G. Guinot and P. K. Seidelmann, Time scales: Their history, definition and interpretation, *Astron. Astrophys.*, 194:304–308, 1988; G. M. R. Winkler and T. C. Van Flandern, Ephemeris time, relativity, etc., *Astron. J.*, 82:84–92, 1977; E. W. Woolard and G. M. Clemence, *Spherical Astronomy*, 1966.

Earth

The third planet, of nine, from the Sun. The Sun is an average star about two-thirds of the way out from the center of the Milky Way Galaxy, a typical spiral galaxy. The Earth is unique, so far as is known, in having life, although statistics strongly suggest that many similar planets exist, and some probably also have life. The Earth is orbited by the Moon and many artificial satellites.

The Earth's nearest neighbors in space, other than the Moon, are the planets Venus, which is about 67 $\times 10^6$ mi (108 $\times 10^6$ km) from the Sun, and Mars, 141 $\times 10^6$ mi (227 $\times 10^6$ km) from the Sun. Earth is 93 $\times 10^6$ mi (150 $\times 10^6$ km) from the Sun. Certain asteroids move in eccentric orbits around the Sun and occasionally come closer to Earth.

Motions. There are two types of Earth motions, orbital and rotational.

Orbit. The Earth's orbit around the Sun is an ellipse, with the Sun at one focus. Perihelion, or the closest approach to the Sun, occurs about January 3, and aphelion, farthest from the Sun, about July 4. Because of the Earth's orbital precession, these dates will be reversed in about 10,500 years. At perihelion the Earth is about 91.5 $\times 10^6$ mi (147 $\times 10^6$ km) from the Sun, and at aphelion 94.5 $\times 10^6$ mi (152 $\times 10^6$ km) giving an average distance of 93 $\times 10^6$ mi (150 $\times 10^6$ km). The Earth's velocity varies systematically, being greatest at perihelion and least at aphelion. The average velocity is 66,600 mi/h (107,000 km/h) or 18.5 mi/s (29.6 km/s). This changing velocity affects measurements of solar time. SEE CELESTIAL MECHANICS.

The Earth's orbital period is the year, whose length is determined by the average distance between Earth and Sun. The calendar year or tropical year is defined as the time between two successive crossings of the celestial equator by the Sun at the time of the vernal equinox. It is exactly 365 days 5 h 48 min 46 s. The fact that the year is not a whole number of days has affected the development of calendars. SEE CALENDAR.

As the Earth moves in its orbit around the Sun, its north spin axis, or geographic pole, points in the direction of the star Polaris, making it the North Star or polestar. One obvious result is that different parts of the Earth receive differing amounts of sunlight; this is the primary cause of seasons.

The Earth's axis has not always pointed in the direction of Polaris. Because of the gravitational attraction of the Moon and Sun on the Earth's equatorial bulge, the direction in which the Earth's axis points moves very slowly. This motion is called precession, and a complete cycle requires about 25,800 years. Superimposed on the precession is a very much smaller fluctuation, called nutation, that results from the movement of the Sun twice a year and the Moon twice a month across the celestial equator. SEE PRECESSION OF EQUINOXES.

Rotation. The Earth's period of rotation is the day, and the day is used to define the second, which is the basic unit of time. The length of the solar day varies because the Earth's orbital velocity changes predictably with distance from the Sun, and less regularly owing to tidal friction, changes in the Earth's core, and seasonal atmospheric circulation. For this reason, the second was defined as 1/86,400 of a mean solar day averaged over a year. To avoid the obvious problems in determining the length of the mean solar day, the second is now defined atomically. SEE DAY; TIME.

The rotation speed is maximum in late July and early August, and minimum in April; the difference in the length of the day is about 0.0012 s. Since about 1900 the Earth's rotation has been slowing at a rate of about 1.7 s per year. In the geologic past the Earth's rotational period was much faster. Daily, monthly, tidal, and annual growth rings on fossil marine organisms reveal that about 350 $\times 10^6$ years ago (Middle Devonian Period) the year had 400–410 days, and 280 $\times 10^6$ years ago (Pennsylvanian Period) the year had 390 days. These observations are very close to the calculated values of the effect of tidal friction. The slowing of the Earth's rotation is caused by tidal friction between the sea floor and the ocean water. The Moon is the main cause of tides, and the total rotational energy in the Earth-Moon sys-

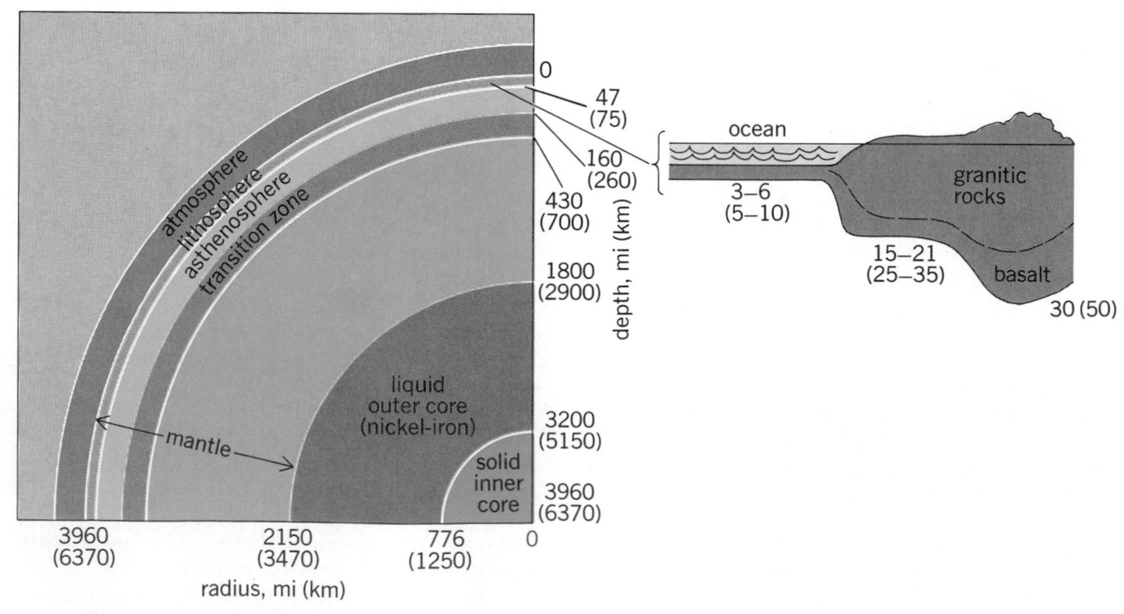

Fig. 1. Principal layers of the Earth. Numbers along small section represent depth in miles (kilometers).

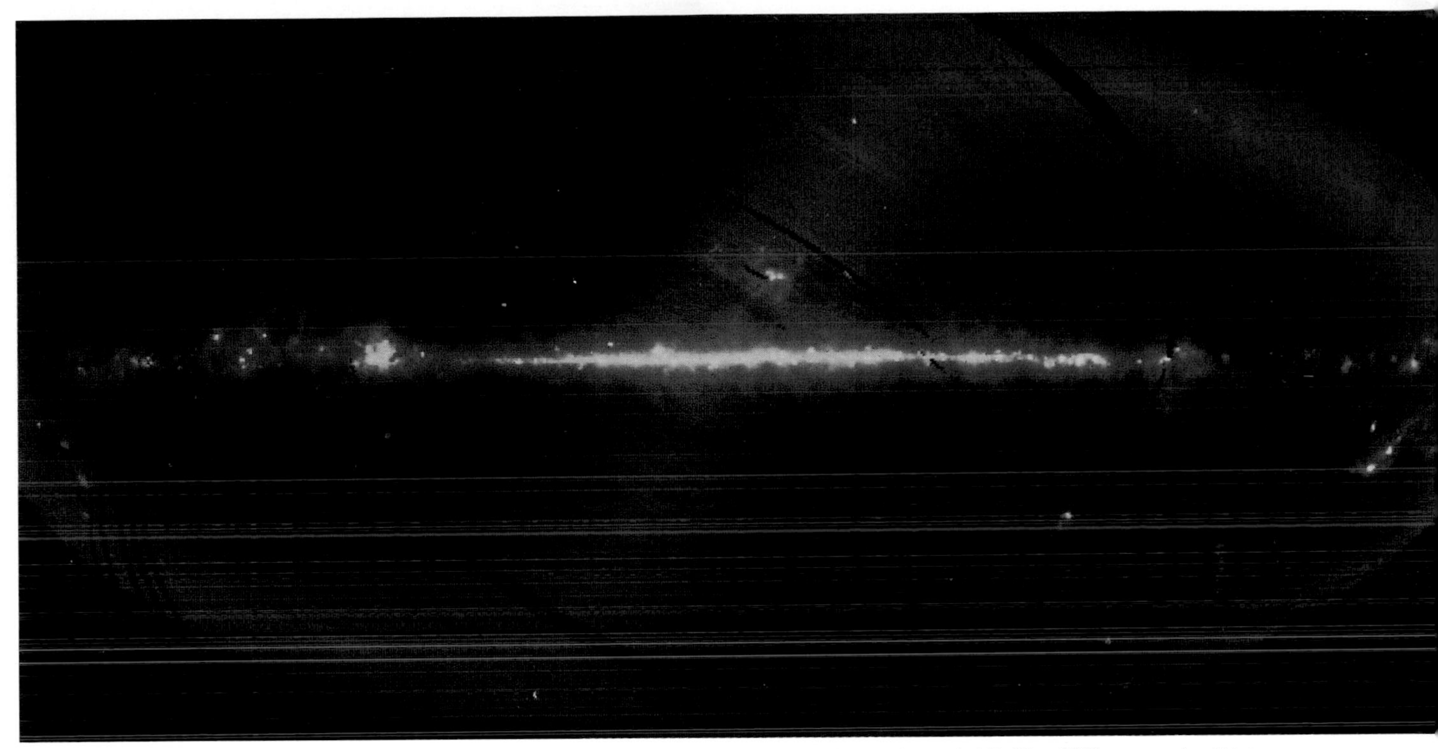

Composite view of the entire sky as viewed by the *Infrared Astronomy Satellite* (*IRAS*). Sources detected at 12, 60, and 100 μm are colored blue, green, and red, respectively. The galactic center is located at the center of the image, and the galactic plane lies along the major axis. The zodiacal dust emission is evident as a blue S-shaped band. Regions of the sky not covered by the survey appear black in this picture.

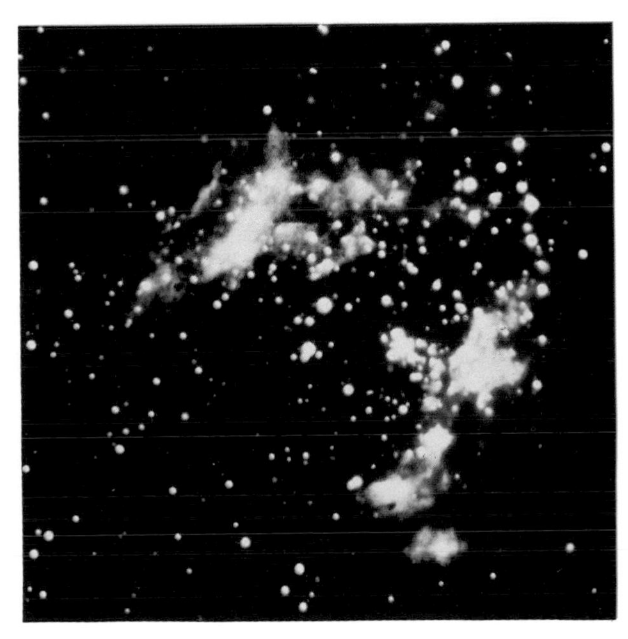

Comparison of (*above*) a picture of the Omega Nebula, Messier 17, seen in optical light with (*right*) a false-color picture obtained by combining images obtained at three near-infrared wavelengths. Sources detected at 1.2, 1.65, and 2.2 μm are colored blue, green, and red, respectively. Most of the stars observed in the infrared do not appear in the optical photograph.

tem is conserved, so that the energy lost by the Earth is gained by the Moon. This causes the Moon to move farther from Earth, and this in turn lengthens the period of the Moon's revolution. *SEE EARTH ROTATION AND ORBITAL MOTION; TIDE.*

Satellites. The Moon is Earth's only natural satellite, although a number of artificial satellites have been put into orbit about the Earth. The Moon's mass is 1/81.3 of Earth's, and its average distance from Earth is 238,247 mi (383,403 km). The center of mass of the Earth-Moon system is within the Earth, about 2886 mi (4645 km) from the Earth's center, and it is about this point that the Earth-Moon system revolves.

The Moon's period of revolution and rotation is 27 days 7 h 43 min 11.5 s. The orbital plane of the Earth-Moon system is inclined to the orbital plane of the Earth-Sun system at an angle of 5°8′33″. At times the Sun, Earth, and Moon are in a line, and an eclipse of the Sun or the Moon occurs. *SEE ECLIPSE; MOON.*

Size, shape, mass, and density. The Earth is not quite a sphere, but has an equatorial bulge caused by its rotation. Its shape therefore is an oblate spheroid. The equatorial radius (semimajor axis) is 3963.205 mi (6378.160 km), and the polar radius (semiminor axis) is 3949.917 mi (6356.775 km).

The Earth's mass is 2.108×10^{26} oz (5.975×10^{27} g), giving it an average density of 3.20 oz/in.^{3} This figure is about twice the density of the common rocks that form the Earth's surface and strongly suggests that the interior is denser than the surface. Seismic studies have confirmed that the Earth's interior has a layered structure (**Fig. 1**).

Internal structure. The deepest layer is the core, which is divided into a solid inner core and a liquid outer core. The core is believed to be composed of nickel-iron and probably lighter elements such as sulfur. Electric currents moving in this liquid conducting core are believed to be the origin of Earth's magnetic field. Above the core is the mantle. The mantle is believed to be composed of silicate minerals, probably the rock peridotite. The composition of the core and mantle are obtained from analogy to meteorites, which are believed to be similar in structure and composition to Earth, and from theoretical calculations.

A number of layers are found in the upper mantle. In the transition zone between about 400 and 160 mi (700 and 260 km) below the surface, the seismic velocity changes in at least two narrow zones. These velocity changes are believed to be caused by changes in mineral phases brought about by variations in pressure and temperature. Above the transition zone is the asthenosphere, which lies between about 160 and 47 mi (260 and 75 km) below the surface. The mantle rocks in the asthenosphere are near their melting point, so this is a weak layer (the low-seismic-velocity zone). The rocks above the asthenosphere are called the lithosphere, and form the tectonic plates discussed below. These plates are believed to move in the weak asthenosphere. The boundary between the mantle and the crust occurs within the lithosphere.

The horizon that separates the mantle from the crust is the Mohorovičić seismic discontinuity, a narrow zone of pronounced change in the velocity of seismic waves believed to be caused by a change in rock type from peridotite in the mantle to basalt in the oceanic crust and the granitic rocks in the continental crust.

The crust is the thinnest of the rock layers, and is composed of 19–37 mi (30–60 km) of basaltic rocks under the oceans. The continents are much older than the oceans, and deformed rocks as old as 3.9×10^9 years indicate an eventful history. No ocean-floor rocks older than about 180×10^6 years have yet been found.

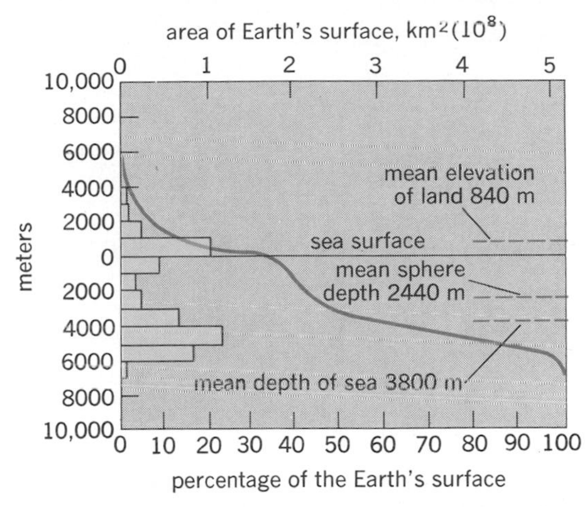

Fig. 2. Hypsographic curve of land surface elevation. 1 m = 3.3 ft; 1 km² = 0.4 mi². (*After E. Kossinna, Die Tiefen des Weltmeeres, Veröff. Inst. Meereskunde Univ. Berlin, Neue Folge A. Geogr. naturwiss., 9:1–70, 1921*)

Crustal structures and surface features. The diverse elevations and shapes of the Earth's surface, as well as the deformed and metamorphosed underlying rocks of various ages that are exposed, record a complex history (**Fig. 2**). On the continents, mountain belts are the most dramatic features. They range in elevation from Mount Everest, 29,030 ft (8848 m), in the Himalaya Mountains to older, rounded, deeply eroded ranges that barely rise above the surrounding plains. Granitic and metamorphic rocks are generally exposed in the cores of mountain ranges. The overlying rocks are generally sedimentary rocks, mainly of shallow marine origin, that have been deformed. The deformation is the result of compression that causes folding and faulting, and may be accompanied by intrusion and metamorphism. Collisions of crustal plates are the probable cause of mountain building. Mountains generally are formed by several phases of deformation spread over several tens of millions of years. The regions deformed in this manner are generally places where shallow-water marine sedimentary rocks accumulated in great thicknesses. Some topographic mountains are formed by volcanoes, and some are the result of erosion of uplifted undeformed rocks, but most mountain ranges are formed of rocks deformed by collision of crustal plates.

Much of the surface of the continents is covered by a thin veneer of sedimentary rocks. Where the underlying rocks of the plains and hills that make up most of the continents are exposed, the rocks and their structures are similar to those found in most mountain ranges. This leads to the theory that continents are formed by the deeply eroded remnants of earlier mountain ranges.

The topographic features underlying the oceans are similarly diverse and reveal more evidence of a dynamic Earth. The continental shelf, an area covered by shallow water, generally less than 500 ft (150 m) deep, surrounds the continents at most places. Such areas are generally underlain by continental, granitic

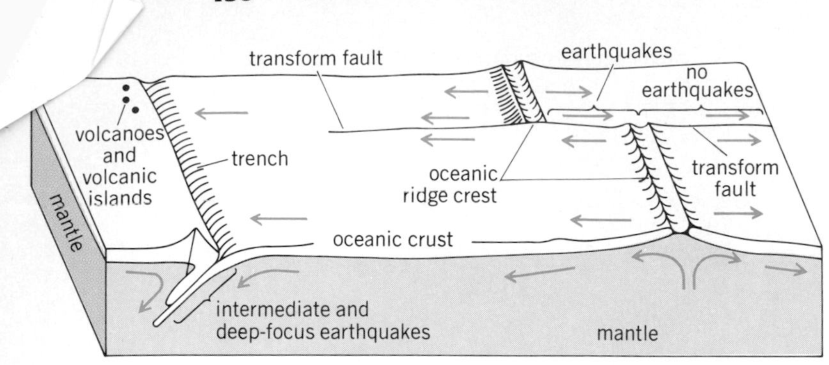

Fig. 3. Sea-floor spreading. (*After G. Gross, Oceanography, 3d ed., Charles E. Merrill, 1976*)

rocks, and are submerged parts of the continents. Continental slopes are the transition between the continental shelf and the ocean floors. Their tops are generally less than 500 ft (150 m) below sea level, and they slope down to about 14,000 ft (4400 m). They are narrow, steep features, with slopes generally between 2 and 6°, but some are up to 45°. They are generally underlain by thick accumulations of sedimentary rocks.

Submarine trenches and their associated volcanic island arcs are another, very different type of border between continent and ocean. The lowest elevations on Earth's surface are found in the submarine trenches. The deepest place on Earth is the Marianas Trench, 36,152 ft (11,022 m). The continental side of such trenches is generally an area of active volcanoes that protrude above the ocean, forming islands. In some cases a shallow sea exists between the volcanic islands and the continent. Much of the western border of the Pacific Ocean is of this type.

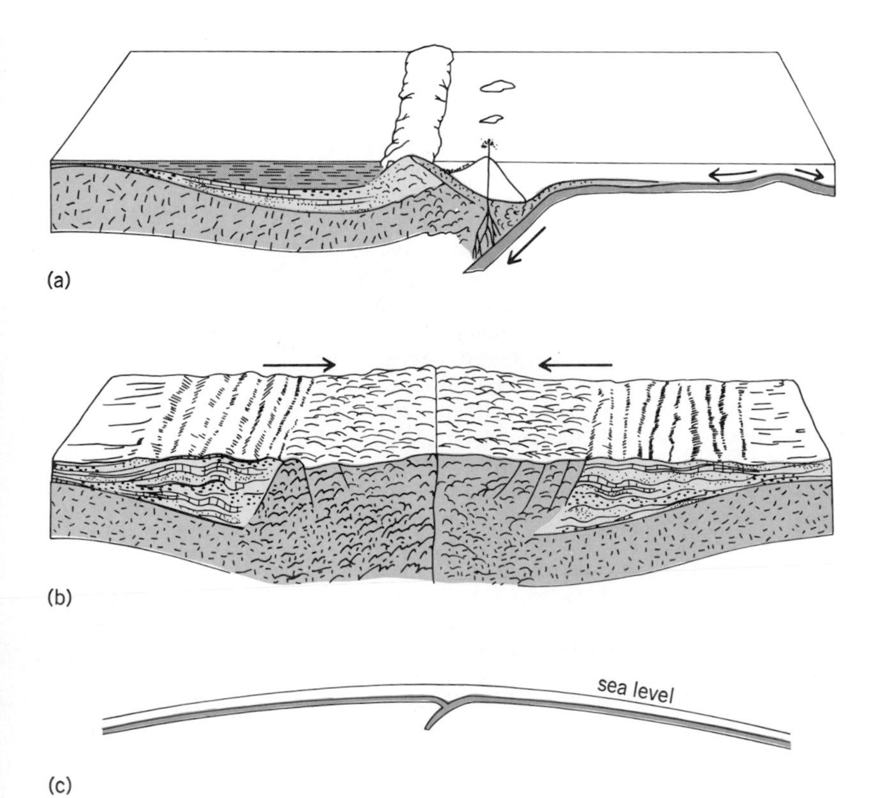

Fig. 4. Plate collisions: (*a*) continent-ocean plate collision; (*b*) continent-continent plate collision; (*c*) ocean-ocean plate collision. (*After R. J. Foster, Physical Geology, 3d ed., Charles M. Merrill, 1979*)

The ocean floors are the most widespread surface feature of Earth. Beneath an average of 2.75 mi (4.4 km) of seawater are about 1.4 mi (2.3 km) of sedimentary rocks with some intercalated basalt, and below that is the oceanic crust, consisting of 3 mi (4.8 km) of basaltic rocks. Interrupting the ocean floor at many places are submarine mountains formed by basalt volcanoes. Some of these volcanoes are very large and form oceanic islands such as the Hawaiian Islands. At other places vertical movements of the ocean floor are revealed by sunken islands whose flat tops were formed by wave erosion (guyots) and coral atolls.

The ocean floors rise gradually to the mid-ocean ridges, a more or less continuous feature through all the oceans with some branches and offsets. The ridges range between 300 to 3000 mi (480 and 4800 km) wide and are much more rugged and irregular than the ocean floors. They rise about 10,000 ft (3000 m) above their base on average. Lines of parallel volcanoes and steep scarps mark the mid-ocean ridges. The central parts of the generally more or less symmetrical ridges are the most active volcanically and seismically.

The source of the energy that deforms the surface rocks is probably crystallization of the liquid core and the decay of radioisotopes within the whole Earth. The energy released at the bottom of the mantle may cause convection in the mantle, which in turn causes plate tectonics and so deforms the crust, but the actual processes are probably much more complex.

Plate tectonics. At places under the oceans new oceanic (basaltic) crust is created at the mid-ocean ridges, and this newly formed thin crustal plate moves away from the ridges (**Fig. 3**). The crustal plates formed in this way may carry continents on them, and are believed to be the mechanism of continental drift. Paleomagnetic data from the continents indicate that the continents have moved relative to each other. The crustal plates are consumed at the trench–volcanic island arc areas. As well as the ridge and the trench, a third type of plate boundary occurs where two plates pass each other at a transform fault. Moving plates may collide in several ways (**Fig. 4**). Such collisions may account for the deformed rocks found in the crust.

The evidence for continental drift in the geologic past includes matching of rock types, ages, fossils, climates, and structures (mountain ranges), as well as the paleomagnetic data. Evidence showing or suggesting present movements consists of shallow earthquakes along mid-ocean ridges and transform faults that offset them; deep earthquakes associated with deep-sea trench–volcanic island arc areas; direct measurement of movement; volcanic activity at mid-ocean ridges; and volcanic activity at trench–island arc areas.

Atmosphere. The Earth's temperature and gravitation are such that an atmosphere is present. The major constituents are nitrogen and oxygen. The atmosphere, especially oxygen, and the presence of water, both at the surface and in the atmosphere, make life possible. Precipitation, mainly rain, results in running water such as streams and rivers on the continents. Running water is the main cause of erosion of the continents, and most of the landscapes are eroded by water, although some are eroded by wind or ice (glaciers).

The atmosphere shields the Earth from most meteorites. Meteors are caused by frictional heat when such high-velocity objects move through the atmosphere. A few of these objects pass through the atmosphere and strike the Earth, causing craters. Other bodies, such as the Moon, are not protected by an atmosphere and so are struck by more meteorites. The craters formed on the Moon are preserved, because without an atmosphere almost no erosion occurs. *See Meteorite.*

A thin ozone layer in the atmosphere also shields the Earth from lethal ultraviolet radiation from the Sun. The structure and composition of the atmosphere is shown in **Fig. 5**.

Age. The Earth with the rest of the solar system is believed to have formed about 4.5×10^9 years ago. This age is determined by dating radioactive isotopes in meteorites. Meteorites are believed to be fragments

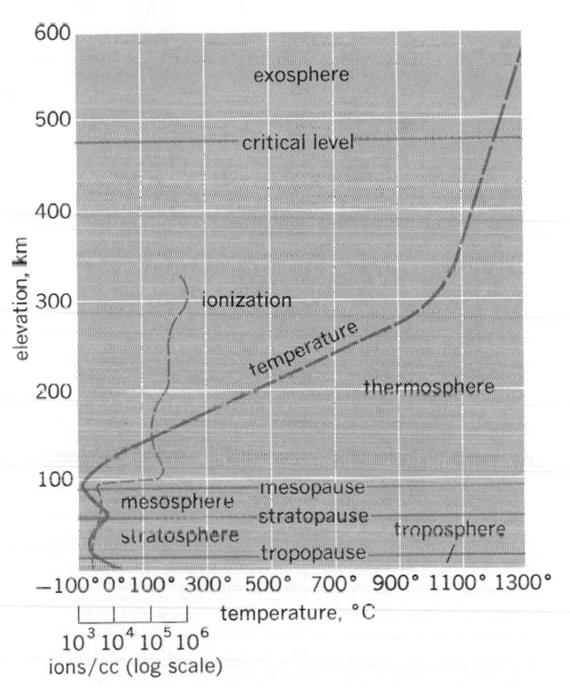

Fig. 5. Structure of the atmosphere. The log scale applies only to the ionization curve. 1 km = 0.6 mi; °F = (°C × 1.8) + 32.

produced by collisions among small bodies formed by the same process that created the solar system. Theoretical studies of the Sun and other studies of radioactive isotopes also suggest a similar age. *See Cosmology; Planet; Solar system.*

Robert J. Foster

Bibliography. M. H. P. Bott, *The Interior of the Earth: Its Structure, Constitution and Evolution*, 1982; B. C. Burchfiel, J. E. Oliver, and L. T. Silver (eds.), *Studies in Geophysics: Continental Tectonics*, National Academy of Sciences, 1980; D. W. Strangway, *The Continental Crust and Its Mineral Deposits*, Geol. Soc. Can. Spec. Pap. 20, 1980.

Earth rotation and orbital motion

The rotation of the Earth about its axis is demonstrated by the classical Foucault pendulum experiment. Its revolution in its orbit around the Sun is shown by the annual parallactic displacement of relatively nearby stars against the background of more distant stars. However, because the Earth is not truly a rigid symmetric body and because it interacts with other members of the solar system gravitationally, these motions vary with time. *See Foucault pendulum; Orbital motion; Parallax.*

Rotation of the Earth

Until recent times the rotation of the Earth has served as the basis for timekeeping. The assumption was made that the rotational speed of the Earth was essentially constant and repeatable, and that the length of the day which resulted from this constant rotational speed was naturally useful as a measure of the passage of time. Astronomical observations, however, have shown that the speed with which the Earth is rotating is not constant with time. It appears that the variations in rotational speed may be classified into three types: secular, irregular, and periodic. The secular variation of the rotational speed refers to the apparently linear increase in the length of the day due chiefly to tidal friction. This effect causes slowing of the Earth's rotational speed and lengthening of the day by about 0.0005 to 0.0035 s per century.

The irregular changes in speed appear to be the result of random accelerations, but may be correlated with physical processes occurring on or within the Earth. These have caused the length of the day to vary by as much as 0.01 s over the past 200 years. Irregular changes consist of so-called decade fluctuations with characteristic periods of 5–10 years as well as variations that occur at shorter time scales. The decade fluctuations are apparently related to processes occurring within the Earth. The higher-frequency variations are now known to be largely related to the changes in the total angular momentum of the atmosphere.

Periodic variations are associated with periodically repeatable physical processes affecting the Earth. Tides raised in the solid Earth by the Moon and the Sun produce periodic variations in the length of the day of the order of 0.0005 s with periods of 1 year, ½ year, 27.55 days, and 13.66 days. Seasonal changes in global weather patterns occurring with approximately annual and semiannual periods also cause variations in the length of the day of this order.

Knowledge of the rotational speed of the Earth is required for observers on the Earth who find it necessary to know the orientation of the Earth in an inertial reference frame. This includes navigators, astronomers, and geodesists. The rotational speed of the Earth remains essentially unpredictable in nature due to the incompletely understood irregular variations. Because of this, astronomical observations continue to be made regularly with increasing accuracy, and the resulting data are the subject of continuing research in the field.

Observations of rotational speed. Astronomical observations of quasars, the Moon, and artificial Earth satellites are used to determine a time scale which is based strictly on the rotation of the Earth within an inertial reference system defined by the positions and motions of the celestial objects. This Universal Time scale (UT1) is compared with time scales known to be more uniform in nature such as that determined from the motion of solar system objects such as the Earth about the Sun (dynamical time) or by atomic clocks (International Atomic Time). Variations in the differences among these types of time scales may be used to determine variations in the rotational speed of the Earth. Astronomical observations of time are made routinely by a number of observatories located around the world for this purpose. Differences between the astronomically determined time scale and a uniform time scale are published routinely by the International Earth Rotation Service, whose Central Bureau is located at the Paris Observatory. One-second adjustments are made at infrequent intervals to the uniform time scale Coordinated Universal Time (UTC) to prevent the difference between the two time scales from becoming greater than 0.9 s. These adjustments are called leap seconds. UTC serves as the basis for civil time in most of the countries of the world. *See Atomic clock; Atomic time; Dynamical time; Time.*

The times and locations of past eclipse observations have also been analyzed to provide information in the length of the day. If the Earth has changed its speed of rotation since ancient times, the path of an eclipse which occurred thousands of years ago would be displaced in longitude with respect to the path that would have occurred if the rotational speed had remained constant. Ancient records of eclipses, while not made with great accuracy, are valuable because they were made long ago. Comparison of very old observations of the longitude of the Sun with current theories of the motion of the Sun based on a uniform time scale has also been used to estimate the increase in the length of the day since ancient times. *See Eclipse.*

Careful analyses of this information reveal the three types of variations in the speed of rotation. These ancient observational data form the basis for estimates of the secular deceleration in the speed of rotation. The more recent information, having been obtained with higher accuracy and more regularity, has shown the changes in the acceleration causing the irregular variations in the length of the day. These data have also been used to detect the periodic variations in the length of the day. **Figure 1** shows a plot of the excess length of day in milliseconds since 1650.

Causes of variations. The conservation of angular momentum of the Earth requires that changes in the Earth's moments of inertia produce changes in the speed with which the Earth is rotating. The moments of inertia are dependent on the distribution of mass on and within the Earth. This includes the mass contained in the atmosphere and in the oceans. As mass is redistributed, the moments of inertia change, producing the subsequent changes in the rotational speed of the Earth. An example of such an effect is that of the spinning skater who is able to change rotational speed by redistributing mass through the changing of the positions of the arms.

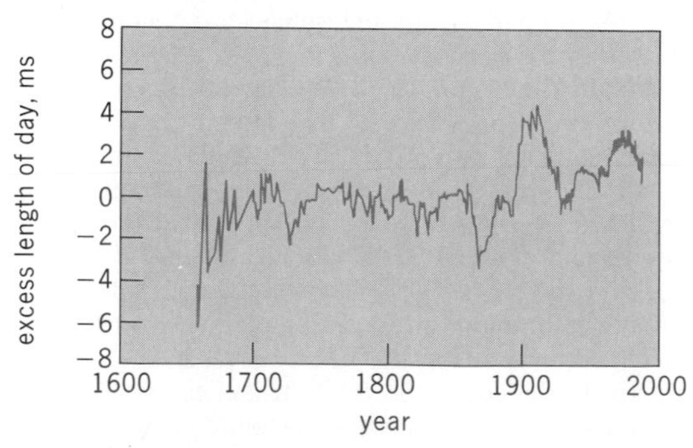

Fig. 1. Excess length of day (length of day − 86,400 s), measured in milliseconds, since 1650.

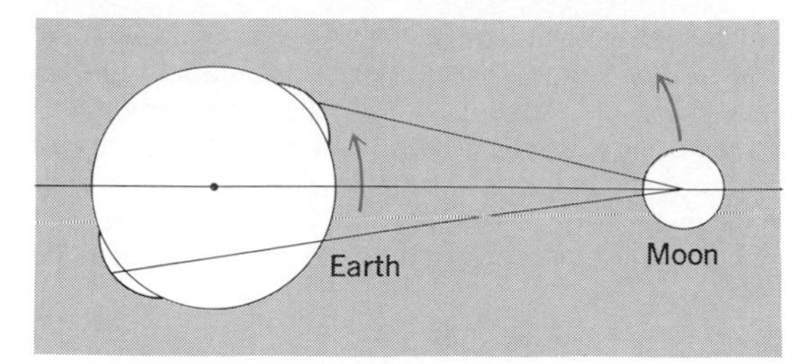

Fig. 2. Couple produced by tidal friction.

The crust and the interior of the Earth are not strictly rigid. Tides generated by the Sun and the Moon change the shape of the Earth. The Earth is thought to have a fluid core with material possibly in motion. Coupling between such a core and the mantle would change the speed of rotation of the mantle from which the observations are made. The Earth may undergo changes in the moments of inertia as landmasses move or rise and fall. All of these motions cause changes in the Earth's moments of inertia and may cause changes in the rotational speed. Seasonal variations are caused by an exchange of momentum between winds and the crust of the Earth.

Changes in the total angular momentum of the atmosphere have also been shown to be correlated with changes in the length of the day. Precise calculations of the total angular momentum using global observations of the wind speed and direction have been used to demonstrate this effect. The variations in the rotational speed are apparently caused by the interaction of the winds with the Earth's topography, and may be the leading cause for all of the irregular variations in the rotation of the Earth.

Another known cause of variations in the rotational speed is tidal friction. The Moon raises tides in the ocean. Friction carries the maximum tide ahead of the line joining the center of the Earth and Moon (**Fig. 2**). The resulting couple diminishes the speed of rotation of the Earth, and this reacts on the Moon to increase its orbital momentum. The sum of the angular momentum of the Earth and the orbital momentum of the Moon remains constant. This produces an increase in the size of the orbit of the Moon and a reduction of its angular speed about the Earth. Tidal friction should be distinguished from actual changes in the moments of inerta of the Earth brought about by the tides. *See Tide.*

The effect of tidal friction is to increase both the distance of the Moon and the length of the lunar month measured with a uniform time scale. Because of the change in the lunar month, the Moon is observed to have an orbital deceleration in terms of a uniform time scale. The proportional change in the distance, however, is greater than in the length of the month. Hence, in terms of Universal Time, the Moon appears to have a secular orbital acceleration, an effect discovered by E. Halley in 1693 from a study of ancient eclipses. In 1853 J. C. E. Adams found that gravitational theory could account for only about half of the acceleration found by Halley. It has become clear since then that the rotational speed is also decreased because of tidal friction.

Revolution about the Sun

The motion of the Earth about the Sun is seen as an apparent annual motion of the Sun along the ecliptic. That the effect is caused by the motion of the

Earth and not that of the Sun is proved by the annual parallactic displacement of nearby stars and by the aberration of light, causing an apparent annual displacement of all stars on the celestial sphere. SEE ABERRATION.

Orbit of the Earth. A large number of astronomical observations of the positions of the Sun and other solar system objects have been made and are being made continuously. This information is required to determine the nature of the motion of the Earth about the Sun. Observations are analyzed using the mathematical methods of celestial mechanics to provide improved estimates of the motions of the solar system objects in the future and to describe the past motions of the objects. The description of the apparent motion of the Sun in the sky provides the determination of the orbit of the Earth. SEE CELESTIAL MECHANICS.

Period of revolution. The true period of the revolution of the Earth around the Sun is determined by the time interval between successive returns of the Sun to the direction of the same star. This interval is the sidereal year of 365 days 6 h 9 min 9.5 s of mean solar time or 365.25636 mean solar days. The period between successive returns to the moving vernal equinox is known as the tropical year of 365 days 5 h 48 min 46.0 s or 365.24220 days. The length of the tropical year is regarded as the length of the year in common usage for calendars. The period of time between successive passages at perihelion (the closest approach of the Earth to the Sun) is called the anomalistic year of 365 days 6 h 13 min 53.0 s or 365.25964 days. The length of each of these years depends on observationally determined astronomical quantities. Improvements in the determination of these quantities will result in slight changes in the numerical values. The lengths of the years listed above are given for the year 1900. These values vary slowly as a consequence of the long-period perturbations of the Earth's orbit by other planets. SEE CALENDAR; PERTURBATION.

Mean radius of orbit. The mean distance from the Earth to the Sun, or the semimajor axis of the Earth's orbit, was the original definition of the astronomical unit (AU) of distance in the solar system. Its absolute value fixes the scale of the solar system and the whole universe in terms of terrestrial standards of length. The distance between the Earth and the Sun can be determined by a variety of methods; the results are usually expressed in terms of solar parallax. The Sun's mean equatorial horizontal parallax p is the angle subtended by the equatorial radius r of the Earth at the mean distance a of the Sun. Mathematically, $a = r/\sin p = 206,265\ r/p$, if p is in seconds of arc. The equatorial radius of the Earth is 3963.182 mi. (6,378,136 m). Geometrical, gravitational, and physical methods have been used at various times to measure the solar parallax. SEE ASTRONOMICAL UNIT; PARALLAX.

Geometrical methods. The geometrical methods involve the direct measurement by optical triangulation of the parallax of a nearby planet (Mars or Venus) or other solar system object such as the minor planet Eros at its closest approach to the Earth. Because the relative distances in the solar system are accurately known in terms of the astronomical unit, the absolute measurement of one very accurate distance can be used to calibrate the scale of the system (**Fig. 3**).

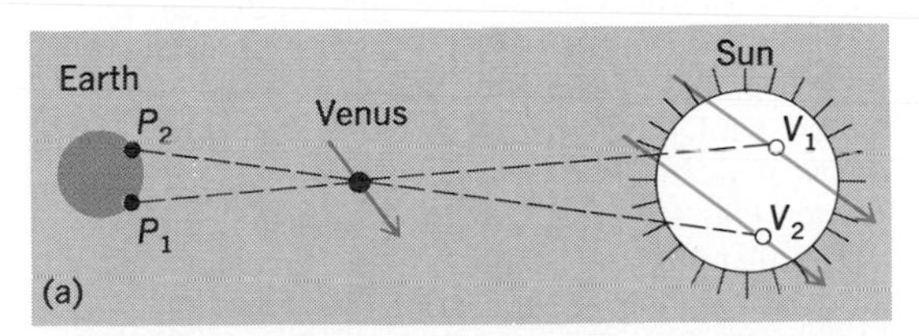

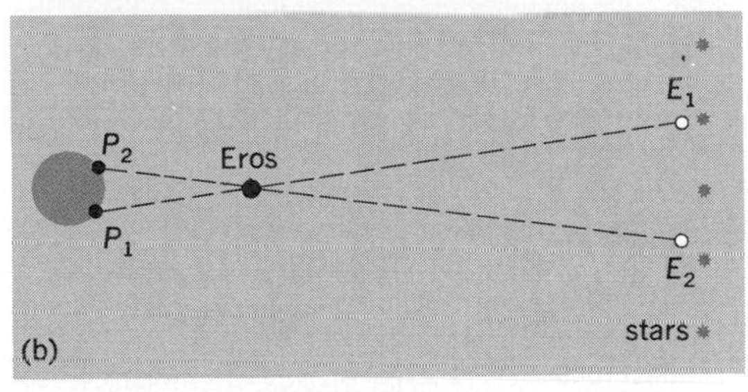

Fig. 3. Determination of solar parallax. (*a*) Observation of Venus in transit. (*b*) Observation of Eros in opposition.

Gravitational methods. The gravitational methods involve the determination of the ratio of the mass m of the Earth to the mass M of the Sun from the perturbations in the motion of a minor planet (such as Eros) caused by the Earth. The method also depends essentially on an accurate determination of the acceleration of gravity g and the length of the sidereal year T in seconds of mean solar time. The distance a is then found from the application of Kepler's third law, $4\pi^2 a^3/T^2 = gr^2[1 + (M/m)]$. Because g and T are known with high accuracy, a determination of M/m gives a directly in terms of terrestrial standards of length. Although it is customary to express this result as a parallax angle, the Earth's radius is not needed to compute a, but it is necessary to make allowances for the mass of the Moon relative to the mass of the Earth. SEE GRAVITY.

Physical methods. One physical method depends on a determination of the ratio of the mean orbital velocity, $V = 2\pi a/T$, to the accurately known velocity of light. This ratio can be derived either from the annual variation of the radial velocities of ecliptic stars (or occasionally planets) determined by observations of the Doppler shift of spectral lines, or with less accuracy, from the constant of aberration. SEE DOPPLER EFFECT.

A far more precise physical method relies on measurement of the travel time of radar signals reflected from objects in the solar system. The distance of the planet is the ratio of half of the round trip travel time of the radar signal to the velocity of the radar waves. Since the distance of the planet is known precisely in astronomical units from the orbital ephemerides, measuring the actual distance in meters at some given time permits the astronomical unit to be calibrated in terms of terrestrial units of length. Also the Doppler shift of radar reflections gives the radial velocity in $m \cdot s^{-1}$, which, on comparison with the known value in $AU \cdot s^{-1}$, gives the astronomical unit in meters. The tracking of artificial space probes to solar system objects has also been used in similar ways to estimate the scale of the solar system. The currently adopted value of the astronomical unit is $1.49597870 \times 10^{11}$ m (92,955,807 mi).

Eccentricity of orbit. The eccentricity of the Earth's orbit can be determined by the variations of the apparent diameter of the Sun's disk. Determinations of higher accuracy are based on the variable speed of the Sun's apparent motion along the ecliptic and the laws of elliptic motion. The nonuniformity of the Sun's motion manifests itself in the equation of time, which is the difference between solar time determined from the actual observation of the Sun, and mean solar time which is based on the motion of a fictitious point having a uniform motion close to the average of the Sun's motion. The difference in the two types of time arises in part from the obliquity of the ecliptic and in part from the eccentricity of the Earth's orbit. The adopted value of the eccentricity is 0.01675. The number varies slowly with time due to perturbations from the planets. The Earth is at perihelion on January 2. It reaches its greatest distance from the Sun on July 2. *SEE EQUATION OF TIME.*

Seasons. The fact that the Equator of the Earth is inclined in space by about 23°5 to the orbital plane of the Earth (the ecliptic) causes the Northern Hemisphere to be exposed to the more direct rays of the Sun during part of the Earth's revolution around the Sun. The Southern Hemisphere receives the more direct rays 6 months, or a half revolution, later. This effect causes the seasons.

OTHER MOTIONS

In addition to the rotation of the Earth and its orbital motion about the Sun, the Earth experiences various small motions about its center of mass. Precession and nutation are examples, and these are caused by the gravitational attraction of the Sun and Moon on the nonspherical Earth. Because the Earth is ellipsoidal in shape, the gravitational attraction of these bodies produces a couple acting on the equatorial bulge, changing the orientation of the Earth about its center of mass. This motion can be predicted with accuracy, based on the knowledge of the shape of the Earth and of the motion of the Earth around the Sun and that of the Moon about the Earth. Precession causes the axis of angular momentum of the Earth to decribe a 23°5 cone in space with a period of about 26,000 years. Nutation causes the axis of angular momentum to "nod" slightly in space as it executes the precessional motion. The main period of the nutational motion is 18.6 years. Other periodic motions due to the gravitational attraction of the Sun and the Moon are also included in nutation. *SEE NUTATION; PRECESSION OF EQUINOXES.*

Because the axis of symmetry of the Earth is not aligned precisely with the axis of rotation, the Earth also executes a motion about its center of mass known as polar motion. This motion, caused by geophysical and meteorological effects on and within the Earth, is not predictable with accuracy, and must be observed continuously to provide the most precise information on the orientation of the Earth. Polar motion is characterized mainly by an approximately 435-day and a 365-day periodic circular motion of the axis of rotation on the surface of the Earth. The radius of the circular motion is of the order of 16 ft (5 m), but this may vary. *SEE EARTH; PLANET; PLANETARY PHYSICS.*

Dennis D. McCarthy

Bibliography. G. O. Abell, *Exploration of the Universe*, 4th ed., 1982; K. Lambeck, *The Earth's Variable Rotation: Geophysical Causes and Consequences*, 1980; H. Moritz and I. Mueller, *Earth Rotation*, 1987; W. H. Munk and G. J. F. MacDonald, *The Rotation of the Earth*, 1975; E. W. Woolard and G. M. Clemence, *Spherical Astronomy*, 1966.

Eclipse

The occultation (obscuring) of one celestial body by another. Solar and lunar eclipses take place at syzygies of the Sun, Earth, and Moon, when the three bodies are in a line. At a solar eclipse, the Moon blocks the view of the Sun as seen from the Earth. At a lunar eclipse, the Earth's shadow falls on the Moon, darkening it, and can be seen from wherever on Earth the Moon is above the horizon.

Eclipses of the Sun could be seen from other planets as their moons are interposed between the planets and the Sun, though their superposition is not as exact as it is for the Earth–Moon system. Eclipses of the moons of Jupiter are well known, occurring whenever the moons pass into Jupiter's shadow. Certain binary stars are known to eclipse each other, and the eclipses can be followed by measuring the total light from the system. *SEE BINARY STAR; ECLIPSING VARIABLE STARS; JUPITER.*

Related phenomena are transits, such as those of Mercury and Venus, which occur when these planets cross the face of the Sun as seen from Earth. They are much too small to hide the solar surface. Transits of the Earth will be seen from spacecraft. Occultations of stars by the Moon are commonly seen from Earth, and are studied to monitor the shape and path of the Moon; a solar eclipse is a special case of such an occultation. Occultations of stars by planets and by asteroids are now increasingly studied; the rings of Uranus were discovered from observations of such an occultation. *SEE OCCULTATION; TRANSIT; URANUS.*

SOLAR ECLIPSES

A solar eclipse can be understood as an occultation of the Sun by the Moon or, equivalently, the Moon's shadow crossing the Earth's surface. The darkest part of the shadow, from which the Sun is entirely hidden, is the umbra (**Fig. 1**). The outer part of the shadow, from which part of the Sun can be seen, is the penumbra.

Solar eclipses can be central, in which the Moon passes entirely onto the solar disk as seen from Earth, or partial, in which one side of the Sun always remains visible. Central eclipses can be total, in which case the Moon entirely covers the solar photosphere, making the corona visible for the period of totality, or annular, in which case the Moon's angular diameter is smaller than that of the Sun because of the positions of the Earth and Moon in their elliptical orbits. At an annular eclipse, a bright annulus of photospheric sunlight remains visible; it is normally thousands of times brighter than the corona, leaving the sky too blue for the corona to be seen.

The plane of the Moon's orbit is inclined by 5° to the plane of the Earth's orbit (the ecliptic), so the Moon's shadow commonly passes above or below the Earth each month at new moon. But three to five times each year, the Moon's shadow reaches the Earth, and a partial, annular, or total eclipse occurs. The Moon is approximately 400 times smaller than the Sun but is also approximately 400 times closer, so its angular diameter in the sky is about the same as the Sun's. Thus the Moon fits approximately exactly over the photosphere, making the phenomenon of a total eclipse especially beautiful.

Phenomena. The partial phases of a total eclipse visible from the path of totality last over an hour. In the minute or two before totality, shadow bands—low-contrast bands of light and dark caused by irregularities in the Earth's upper atmosphere—may be seen to race across the landscape. As the Moon barely covers the Sun, photospheric light shines through valleys on the edge of the Moon, making dots of light—Baily's beads—that are very bright in contrast to the background. The last Baily's bead gleams so brightly that it appears as a jewel on a ring, with the band made of the corona; this appearance is known as the diamond ring effect (**Fig. 2**). It lasts for 5–10 s.

During the diamond ring effect, the solar chromosphere becomes visible around the limb of the Moon, glowing pinkish because most of its radiation is in the form of emission lines of hydrogen, mostly the red hydrogen-alpha line. Its emission-line spectrum apparently flashes into view for a few seconds, and is called the flash spectrum. As the advancing limb of the Moon covers the chromosphere, the corona becomes fully visible (**Fig. 3**). Its shape is governed by the solar magnetic field; common are equatorial streamers and polar tufts. At the maximum of the solar activity cycle, so many streamers exist that the corona appears round when it is seen in projection, as viewed from Earth. At the minimum of the solar activity cycle, only a few streamers exist so that the corona appears more elongated in projection. SEE SUN.

Totality (**Fig. 4**) lasts from an instant up through somewhat over 7 min. At its end, the phenomena repeat, including chromosphere, diamond ring, Baily's beads, shadow bands, and the partial phases.

Positions and timing. The paths of the Sun and Moon in the sky intersect at two points, the ascending node and the descending node. Only when both the Sun and the Moon are near a node can an eclipse occur. Thus eclipse seasons take place each year, whenever the Sun is near enough to the node so that an eclipse is possible. Each eclipse season is 38 days long. Because the Sun's gravity causes the orientation of the Moon's elliptical orbit to change with an 18.6-year cycle, the nodes slide along the ecliptic and a cycle of two eclipse seasons—an eclipse year—has a period of 346.6 days, shorter than a solar year. SEE MOON.

There must be at least one solar eclipse each eclipse season, so there are at least two each year. There may be as many as five solar eclipses in a calendar year (technically, a year), though most of these will be partial. Some of the partials will occur near the ends of the eclipse seasons, and it is also possible for there to be three eclipse seasons during a solar year. Adding lunar eclipses (including penumbral lunar eclipses, which may not be noticeable), there may be seven eclipses in a year.

Saros. An important coincidence relates lunar months and eclipse years. A total of 223 lunar months (technically, synodic months, the period of the phases) takes 6585.32 days. A total of 19 eclipse years—passages of the Sun through the same node of the Moon's orbit—takes 6585.78 days, and 242 nodical months—passages of the Moon through the node—take 6585.36 days. (Nodical months are also called draconic months, after the ancient Chinese dragon once thought to have been devouring the Sun at a solar eclipse.) Thus eclipses appear with this period of 18 years 11⅓ days (plus or minus a day, depending on leap years), a period known as the saros.

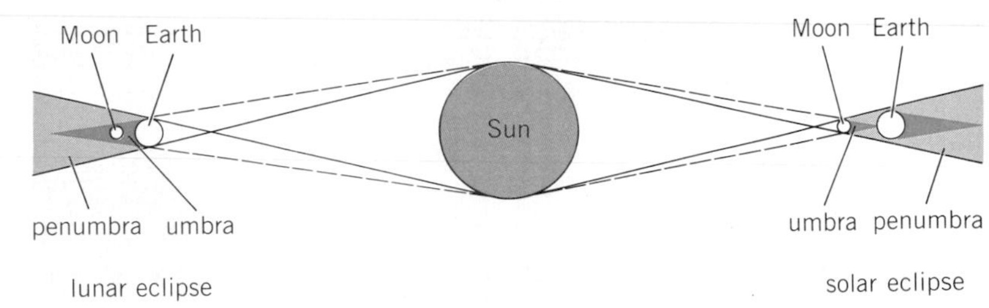

Fig. 1. Circumstances of solar and lunar eclipses.

Further, 239 periods of the variation of distance of the Moon from the Earth, the anomalistic month, is 6585.54 days, so the relative angular sizes of the Sun and Moon are about the same at this interval. (The anomalistic month differs from the nodical and sidereal months because the orientation of the Moon's elliptical orbit drifts around in its orbital plane.)

As a result of the saros, almost identical eclipses recur every 18 years 11⅓ days. The significance of the ⅓ day is that the Earth rotates one-third of the way around, and the eclipse path is shifted on the Earth's surface. Thus the June 30, 1973, 7-min eclipse in Africa was succeeded in a saros series by the July 11, 1991, eclipse in Hawaii, Mexico, and Central and South America, which reached maximum duration of 6 min 54 s in Mexico. After a saros, the Sun is slightly farther west than its original position, and the Moon is slightly north or south, depending on whether it is near an ascending or a descending node, so the eclipses in a saros drift from north to south or from south to north, starting near one pole and departing from the other. A complete series takes 1244 to 1514 years.

Motion over Earth's surface. The Moon's shadow travels at approximately 2100 mi/h (3400 km/h) through space. The Earth rotates in the same direction that the shadow is traveling; at the Equator, the resulting motion of the Earth's surface makes up about 1040 mi/h (1670 km/h), making the speed of the eclipse across the Earth's surface 1060 mi/h (1730 km/h). When eclipses cross higher latitudes, the speed of motion associated with the Earth's rotation is not as high, so the eclipse speed is even higher. The supersonic Concorde took advantage of an equatorial eclipse in 1973 to keep up with totality for 74

Fig. 2. Diamond-ring effect at the beginning of totality during the solar eclipse of November 23, 1984, observed from Papua New Guinea. (*Jay M. Pasachoff, Williams College–Hopkins Observatory*)

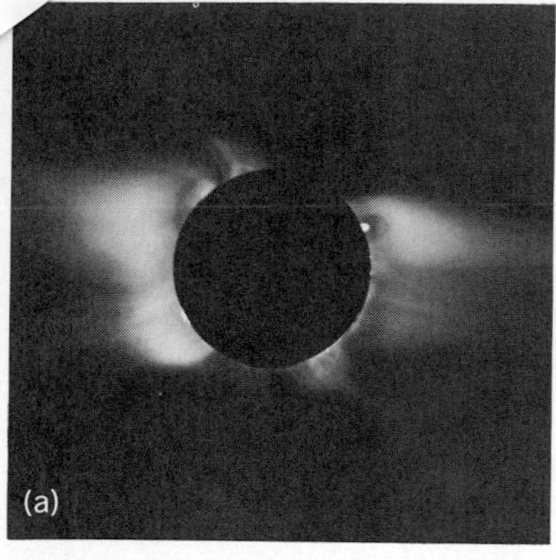

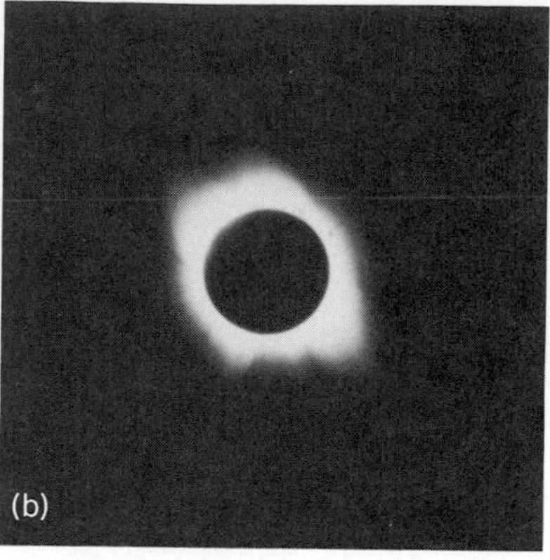

Fig. 3. Corona during a total solar eclipse. (*a*) Eclipse of March 18, 1988, observed from the Philippines, with a filter radially graded in photographic density so as to reduce the intensity of the inner corona so that the entire corona could be recorded on a single piece of film (*High Altitude Observatory, National Center for Atmospheric Research, Boulder, Colorado*). (*b*) Eclipse of November 23, 1984, observed from Papua New Guinea (*Jay M. Pasachoff, Williams College–Hopkins Observatory*).

min. Ordinary jet aircraft cannot keep up with eclipses, so the term eclipse chasing is usually not accurate.

Scientific value. Even with advances in space technology, total solar eclipses are the best way of seeing the lower and middle corona. Coronagraphs, telescopes for which special shielding and internal occulting allow observation of the corona from a few sites in the world on many of the days of the year, are limited to the innermost corona or to use of special filters or polarization. SEE CORONAGRAPH.

Coronagraphs have been sent into orbit, notably aboard *Skylab* and *Solar Maximum Mission*, but limitations in spacecraft control lead to the necessity of overocculting the photosphere, cutting out the inner corona. For example, the coronagraph on *Solar Maximum Mission* occults 1.75 times the solar diameter. Thus, only the outer corona can be studied from spaceborne coronagraphs. Other types of observations from space also apply to the corona, such as imaging x-ray observations that were made from *Skylab* and some rockets. These have been very limited in time coverage. Further, any space observation is extremely expensive, over 100 times more for *Solar Maximum Mission* than for a given extensive eclipse expedition. So it is useful and necessary to carry on ground-based eclipse observations, ground-based coronagraph observations, and space-based observations to get the most complete picture of the Sun. Further, independent derivations of certain basic quantities must be carried out to provide trustworthy results; there have been cases in which data reduced from space observations had to be restudied because the need for different calibrations had shown up in eclipse work.

Related eclipse studies include use of the advancing edge of the Moon to provide high spatial resolution for radio observations of the Sun and, historically, of celestial radio sources. The atmospheric effects of the removal of incident sunlight from the Earth's atmosphere have also been studied at eclipses.

Historically, the test of the deflection of starlight carried out at the eclipse of 1919 and repeated in 1922 was a decisive verification of Einstein's general theory of relativity. These results were restudied for the 1979 Einstein centennial, and their accuracy improved. The experiment is a very difficult one, and has been attempted at some recent eclipses, notably 1970 in Mexico and 1973 in Africa, without improving on the early results. But the effect has been verified to higher accuracy by studies in the radio part of the spectrum, so optical eclipse tests are no longer necessary.

Annular eclipses. Central eclipses in which the Moon is sufficiently far from the Earth that it does not cover the solar photosphere are annular. The Moon's umbra is about 232,000 ± 4000 mi (374,000 ± 6400 km) in length while the Moon's distance is 237,000 ± 16,000 mi (382,000 ± 25,000 km), so the umbra sometimes falls short of reaching the Earth's surface.

Since the corona is 1,000,000 times fainter than the photosphere, if even 1% of photosphere is showing, the corona is overwhelmed by the blue sky and cannot be seen. So most annular eclipses are of limited scientific use.

The annular eclipse of 1984, visible from the southeastern United States, provided 99.8% coverage. The major scientific work carried out involved detailed timing of the Baily's beads in order to assess the size of the Sun. Such work has also been carried out at total eclipses. Some of the results seem to show a possible shrinking of the Sun by a measurable amount in a time of decades or centuries, which would lead to impossibly large effects on geological time scales; this has led to the suggestion that the Sun could be oscillating in size. But the question of whether any real effect is present has not been settled. The 1984 annular eclipse was so close to total that the corona could even be briefly seen and photographed, although no scientific studies of the corona were made.

Fig. 4. Total solar eclipse of November 23, 1984, observed from Papua New Guinea. (*Jay M. Pasachoff, Williams College–Hopkins Observatory*)

Recent and future eclipses. Notable recent total eclipses in terms of duration and favorable weather have been the eclipse of June 30, 1973, which was

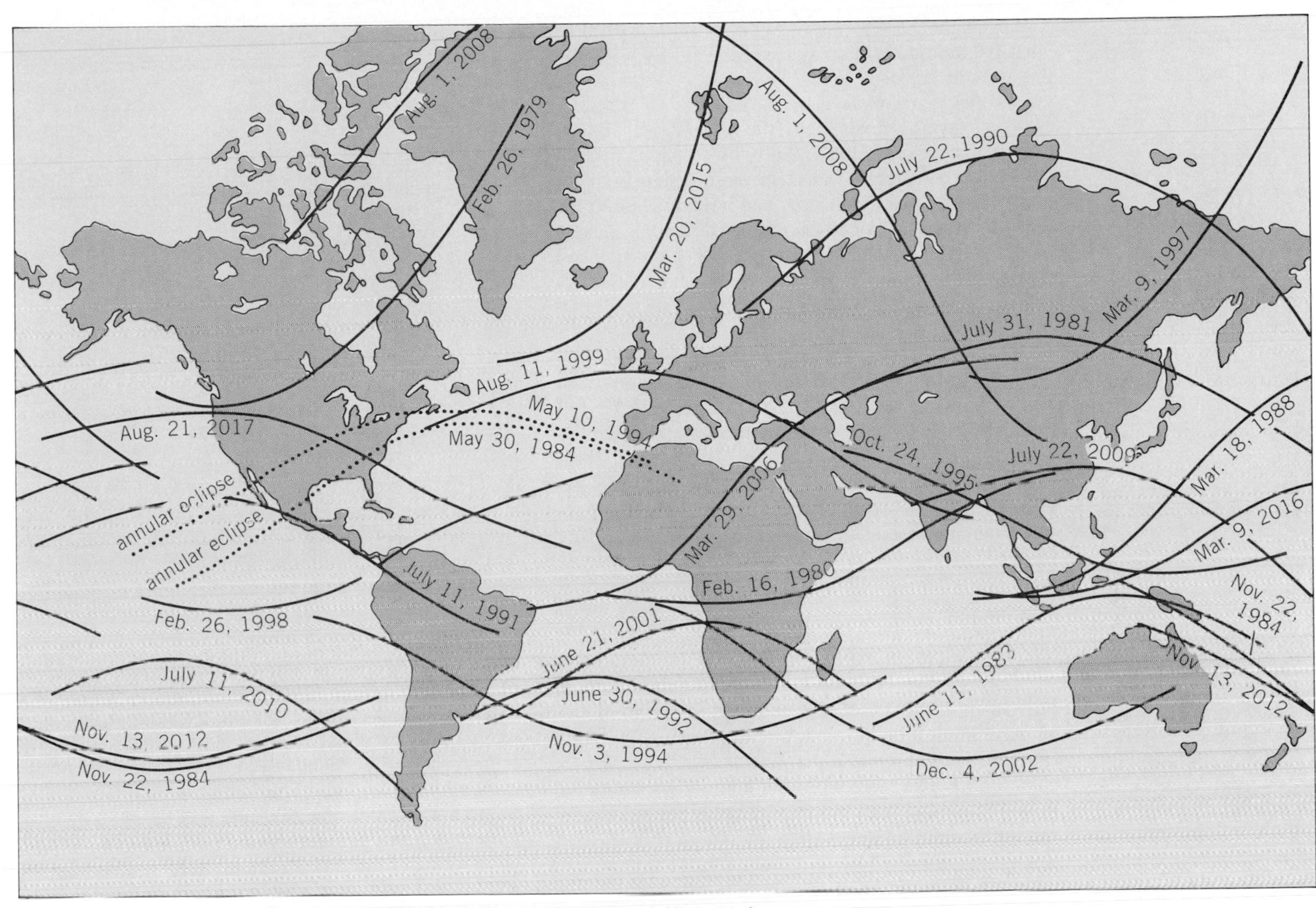

Fig. 5. Paths of total solar eclipses, 1979–2017. (*After Bryan Brewer, Earth View, Inc.*)

longer than 7 min over Africa, the eclipse of February 16, 1980, which crossed Africa and India; the eclipse of May 30, 1983, which crossed Java; and the eclipse of July 11, 1991, in Hawaii, Mexico, and Central and South America, which reached maximum duration of 6 min 54 s in Mexico (see **Table 1**). Major scientific expeditions carried out research on these occasions.

The next total eclipse to cross the continental United States will be in 2017, and the next eclipse to cross Canada will be in 2024. An annular eclipse will cross the United States from Texas through Maine on May 10, 1994 (**Fig. 5**).

Observing a solar eclipse. A solar eclipse is nature's most magnificent phenomenon. The false impression has grown that it is often hazardous to observe a solar eclipse, whereas actually few reports of eye injury exist and the incidence of lasting eye damage is even lower.

The total phase of an eclipse is completely safe to watch with the naked eye. The total brightness of the corona is only that of the full moon, so is equally safe to watch. The darkness at the disappearance of the diamond ring effect comes so abruptly that people have no trouble telling when the safe time begins, and the diamond ring at the end of totality is so relatively bright that it is clear when it is time to look away. A glance at the Sun just before or after totality is not harmful; it is only staring for an extended time (more than a few seconds with the naked eye) or looking at the not totally eclipsed Sun through binoculars or a telescope that can cause harm.

Table 1. Total solar eclipses, 1985–2005*

Date	Maximum duration of totality	Location
Nov. 12, 1985	1m59s	South Pacific Ocean
Oct. 3, 1986	0m00s†	North Atlantic Ocean
Mar. 29, 1987	0m08s†	Atlantic Ocean
Mar. 18, 1988	3m46s	Indonesia, Philippines, Pacific Ocean
July 22, 1990	2m33s	Finland, northernmost Soviet Union, Aleutian Islands
July 11, 1991	6m54s	Hawaii, Mexico, Central America, South America
June 30, 1992	5m20s	Uruguay, Atlantic Ocean
Nov. 3, 1994	4m24s	South America, South Atlantic Ocean
Oct. 24, 1995	2m10s	India, southeastern Asia, Borneo, Pacific Ocean
Mar. 9, 1997	2m50s	Siberia
Feb. 26, 1998	4m08s	Pacific Ocean, north of South America, Atlantic Ocean
Aug. 11, 1999	2m23s	Atlantic Ocean, Europe, southwestern Asia, India
June 21, 2001	4m56s	Atlantic Ocean, South Africa, Madagascar
Dec. 4, 2002	2m04s	South Africa, Indian Ocean, Australia
Nov. 23, 2003	1m57s	Antarctica
April 8, 2005	0m42s†	Pacific Ocean

*Compiled by Jean Meeus.
†These eclipses are annular-total; that is, they are total only near the middle of their central line.

During the partial phases, it is possible to follow what is going on without any special aid by watching the ground under a tree. The spaces between the leaves make a pinhole camera, and project the solar crescent myriad times onto the ground. A pinhole camera can be made individually by punching a small hole (approximately 0.1–0.2 in. or 2–5 mm in diameter) in a piece of cardboard, and holding it up to the Sun. A crescent image is projected onto a second piece of cardboard held 8 to 40 in. (20 cm to 1 m) closer to the ground, or onto the ground itself. An observer looks at the second cardboard, facing away from the Sun.

For direct observation of the partial phases, a special solar filter must be used. Fogged and exposed black-and-white (not color) film developed to full density provides suitable diminution of the solar intensity across the entire spectrum. Inexpensive commercial solar filters made of aluminized Mylar can also be used. Gelatin "neutral-density" filters are actually not neutral in the infrared, and so should not be used, though neutral-density filters made by depositing chromium or other metals on glass are safe if they are ND4 or ND5. *Jay M. Pasachoff*

Lunar Eclipses

A lunar eclipse can occur only when the Moon is full and is near one of the nodes of its orbit. If the Moon enters only the penumbral cone of the Earth (Fig. 1), the eclipse is a penumbral one. If the Moon enters the umbra without being entirely immersed in it, a partial (umbral) eclipse occurs. The eclipse is total if the entire Moon enters the umbra (**Fig. 6**).

The magnitude of an eclipse is the fraction of the diameter of the lunar disk which is eclipsed (in the umbra or in the penumbra) at maximum phase. **Table 2** lists all umbral eclipses taking place during the years 1985 to 2005. If the magnitude is larger than 1, the eclipse is total.

Table 2. Lunar eclipses in the umbra, 1985–2005

Date	Magnitude	Date	Magnitude
May 4, 1985	1.24	Apr. 15, 1995	0.11
Oct. 28, 1985	1.07	Apr. 4, 1996	1.38
Apr. 24, 1986	1.20	Sept. 27, 1996	1.24
Oct. 17, 1986	1.24	Mar. 24, 1997	0.92
Aug. 27, 1988	0.29	Sept. 16, 1997	1.19
Feb. 20, 1989	1.28	July 28, 1999	0.40
Aug. 17, 1989	1.60	Jan. 21, 2000	1.33
Feb. 9, 1990	1.07	July 16, 2000	1.77
Aug. 6, 1990	0.68	Jan. 9, 2001	1.19
Dec. 21, 1991	0.09	July 5, 2001	0.49
June 15, 1992	0.68	May 16, 2003	1.13
Dec. 9, 1992	1.27	Nov. 9, 2003	1.02
June 4, 1993	1.56	May 4, 2004	1.30
Nov. 29, 1993	1.09	Oct. 28, 2004	1.31
May 25, 1994	0.24	Oct. 17, 2005	0.06

Penumbral eclipses are not observable unless their magnitude (in the penumbra) is greater than about 0.7. Small partial penumbral eclipses are undistinguishable, and are calculated and published in the astronomical almanacs because the statistics of eclipses would be incomplete without them.

Frequency. In an average century, 242 lunar eclipses take place, of which 70 are total in the umbra, 84 partial in the umbra, and 88 are penumbral. Hence, the number of lunar eclipses is slightly larger than that of solar eclipses (238 per century). However, if the penumbral eclipses are not taken into consideration, the number of lunar eclipses is much smaller than solar eclipses. Even then, lunar eclipses are seen by many more people than solar eclipses because a lunar eclipse is visible from every point on the Earth's hemisphere where the Moon is above the horizon, while solar eclipses are visible, even in the partial phases, from limited areas of the Earth.

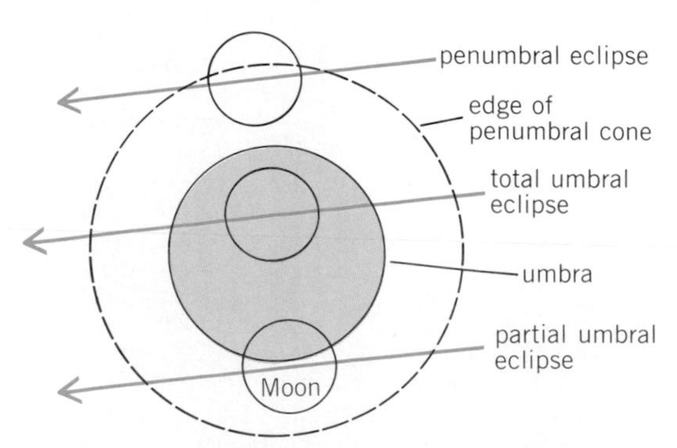

Fig. 6. Cross section through the Earth's shadow cone (umbra) at the Moon's distance, showing the motion of the Moon at three types of lunar eclipses: penumbral eclipse, partial umbral eclipse, and total umbral eclipse.

If penumbral eclipses as well as umbral ones are taken into consideration, the least number of lunar eclipses during a calendar year is two, and the maximum number is five. The last year with five lunar eclipses was 1879, the next one will be 2132. If only umbral eclipses are considered, the least number of lunar eclipses in one calendar year is zero (as in 1987), and the maximum number is three (as in 1982).

Phases of eclipse. When the Moon passes through the center of the Earth's shadow, the entire eclipse takes 5⅓ to 6¼ h, depending on the Moon's position in its orbit at the time of the eclipse. (If the eclipse occurs with the Moon near apogee, the penumbral cone is larger and the Moon's motion is slower, increasing the duration of the eclipse.) The first hour is spent in the penumbra (Fig. 1). No darkening is noticeable until about a quarter hour before the first contact with the umbra, because in the penumbra all of the Moon's side facing Earth is still receiving some direct sunlight. Then as the Moon enters the umbra, the eclipsed part of the Moon appears nearly black by contrast with the bright side of the disk. Approximately the second hour is required for all of the Moon to get into the umbra.

The diameter of the umbra where the Moon crosses it is about 2⅔ times the Moon's diameter. The total phase of a lunar eclipse can last up to 107 min.

While the Moon is going into the umbra or out of it, the edge of the umbra is always a part of a circle. As was realized by some of the ancient Greeks, this is a proof that the Earth is a sphere, since a sphere is the only object that casts a circular shadow, no matter what direction the light comes from.

Appearance of eclipsed Moon. If the Earth had no atmosphere, the Moon would disappear from view while in the umbra. However, the Earth's atmosphere acts like a lens and bends the sunlight into the umbra. The longer waves of red light penetrate the atmosphere better than the short-wave blue light, which is scattered to form the blue of the sky. An observer on the Moon would see the Earth surrounded by a thin ring of bright sunset colors. This explains the usual reddish color of the totally eclipsed Moon. However, if there happened to be many clouds around the Earth's rim, an important fraction of the light would be cut off and the eclipsed Moon would appear dark.

On rare occasions, the totally eclipsed Moon might even drop entirely out of sight to the naked eye, as it almost did on December 30, 1963. These extremely dark eclipses are due to major volcanic eruptions, whose dust temporarily increases the atmosphere's opacity.

A relation has been claimed between the brightness of the eclipsed Moon and the phase of the 11-year period of sunspot activity. However, this effect could not be confirmed by later observations. The claimed correlation, if real, is at best very weak.

Size of umbra. It has been known for nearly three centuries that the umbra is slightly larger than the shadow that geometry predicts should be cast by Earth. This enlargement, which is another effect of the atmosphere, varies slightly in amount from one eclipse to another; its mean value is a little less than 2%.

An excellent way to measure the umbra's size is to time when well-defined lunar craters appear to enter or leave the shadow. Such timings can be made by amateurs. A low-power telescope is best for this purpose, and the observer should estimate to the nearest tenth of a minute, using a clock set to radio time signals, when the most abrupt gradient at the umbra's edge is crossing the center of a lunar feature.

Jean Meeus

Bibliography. G. O. Abell, *Exploration of the Universe*, 4th ed., 1982; J. Meeus, Solar eclipse diary, 1985–1995, *Sky Telesc.*, 68:296–298, 1984; J. Meeus, C. C. Grosjean, and W. Vanderleen, *Canon of Solar Eclipses*, 1966; J. Meeus and H. Mucke, *Canon of Lunar Eclipses*, 1979; D. H. Menzel and J. M. Pasachoff, *A Field Guide to the Stars and Planets*, 2d ed., 1983; S. A. Mitchell, *Eclipses of the Sun*, 5th ed., 1951; J. M. Pasachoff, *Contemporary Astronomy*, 4th ed., 1989.

Eclipsing variable stars

Double star systems in which the two components are too close to be seen separately but which reveal their duplicity by periodic changes in brightness as each star successively passes between the other and the Earth, that is, eclipses the other. Studies of the light changes and the radial velocity changes of each component permit the computation of the radii, masses, and densities of the components—important quantities that cannot be measured directly in single stars. In addition, these close double stars are useful in studies of mass loss and of stellar evolution. Since eclipsing stars are variable in light, they are included in general variable star catalogs under the same system of nomenclature. SEE BINARY STAR; VARIABLE STAR.

Periods. The periods of light variation range from less than 3 h for very close systems to over 27 years for the peculiar system Epsilon Aurigae. However, the majority of the periods lie between 0.5 and 10 days. In many cases the periods are not constant but change with time. In a few cases the variation is caused by a slow change in the orientation of the major axis of an elliptical orbit; in such cases the rate of change combined with other quantities gives information concerning the manner in which the density of the star increases from the outer layers to the center. In most cases, however, the changes are unpredictable and are probably connected with ejections of matter from one of the stars.

Velocity curves. The radial velocity (velocity of approach or recession) of each star can be determined at any time by the displacement of the spectral lines. A plot of velocities against time over one period of orbital revolution is known as a velocity curve. The maximum radial velocity of approach or recession depends on the true orbital velocity of the star and the "inclination," or the amount by which the plane of the orbit is tilted relative to the line of sight to the Earth. (Technically, the inclination is the angle between a perpendicular to the orbit plane and the line of sight; when the inclination is 90°, the eclipses are central.) If the inclination is known, the orbital velocity of each star can be calculated from the radial velocity. The orbital velocity multiplied by the period will give the circumference of the orbit and from this the radius of each star about the center of mass. From the size of the orbit and the period, by using the law of gravitation, the mass of each star can be calculated in terms of the Sun's mass. However, the inclination cannot be determined from the velocity curve alone; the quantities finally determined are $m_1 \sin^3 i$, $m_2 \sin^3 i$, and $a \sin i$, where i is the inclination, m_1 and m_2 the stellar masses, and a the radius of a circular orbit or half the major axis of an elliptical one.

Light curve. The light curve shows the changes in brightness of the system throughout one orbital revolution. The manner in which the light changes during the eclipse of each star by the other depends very strongly on three factors: the size of each star relative to the radius of the orbit (r_1/a and r_2/a) and the inclination. The determination of precisely what relative sizes and inclination give a computed curve which will approximate satisfactorily the observations is one of the more difficult problems of modern astronomy. However, by use of complex sets of tables the problem can be solved in many cases. Numerical values for these three quantities are thus found. Several alternate methods have been developed which make use of the capabilities of electronic computers. These take into account simultaneously the various interaction effects such as distortion of the shape of the components in the earlier methods. These had to be removed from the light curve before a solution could be carried out. SEE ECLIPSE; LIGHT CURVES.

Absolute dimensions. The masses, radii, and densities of the stars, usually expressed in comparison with those of the Sun, are called the absolute dimensions of the system. Combining the results from the light and velocity curves yields this fundamental information which cannot be obtained from either approach alone. The inclination, determined from the shape of the light curve, can be substituted in the quantities $m_1 \sin^3 i$, $m_2 \sin^3 i$, and $a \sin i$ to find m_1, m_2, and a. Thus the masses of the stars in terms of the Sun's mass and the size of the orbit in terms of the Sun's radius can be found. Since the radii of the stars in fractions of the size of the orbit have been determined from the light curve, the sizes of the stars relative to the size of the Sun (either in miles or kilometers) can now be computed. **Figure 1** shows the relative sizes and separation of the stars in a typical eclipsing system.

Complications. Before the light curve can be "solved" to give the above quantities, corrections

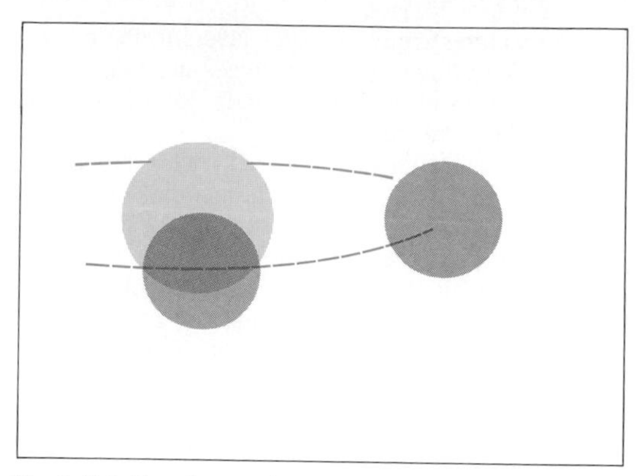

Fig. 1. Relative sizes and separation of components of R Canis Majoris. Shaded star represents cooler component. It is shown in two positions: at the middle of an eclipse and when separation would be greatest as viewed from Earth. Even at the greatest separation, all such systems appear as single stars.

must be made for other factors which influence the light changes. This is done by studying the light between eclipses where, were it not for these complications, the brightness of the system would not change. One of these effects is called ellipticity. The tidal attraction of each star for the other has caused distortions until the stars in extreme cases resemble footballs more than baseballs in shape. The technical term is prolate ellipsoids, although when the stars differ in size and mass they will also differ in shape. Further, the radiation of each star falling on the side of the other nearest it will cause this side to be brighter than the side turned away. The difference will be most marked for the cooler component where the effect of the intense radiation from the nearby hotter star is most strongly evident.

There are other effects, some very poorly understood, which cause light changes between eclipses, and all of these must be carefully studied before the analysis of the eclipse begins.

Evolutionary changes. Studies of single stars indicate that, when the hydrogen in the center of the star has been converted into helium, the star undergoes a relatively rapid expansion in size. The presence of a nearby companion complicates the picture considerably, but it does seem clear that much of the mass of the star must be lost to the system or possibly transferred to the other component.

The mass-losing star eventually becomes a collapsed object—a white dwarf, a neutron star, or a black hole. Each of these types has been identified in at least one binary system. Then when the secondary, originally less massive star begins its expansion, matter from it is transferred to the collapsed object, often with dramatic results. *See Black hole; Neutron star; White dwarf star.*

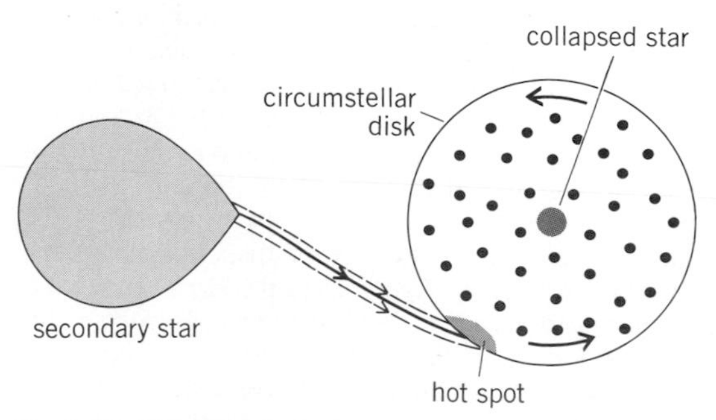

Fig. 2. Model generally accepted for certain types of eclipsing binary stars, in particular those known as dwarf novae.

It is now believed that all explosive variables (novae, recurrent novae, and so forth), with the exception of supernovae, are members of close binary systems. At least some of the x-ray sources are close binaries in this state, although the detection (by instruments carried aboard satellites) of x-radiation from Algol and other systems which are not yet in this state indicates that other physical mechanisms may also be responsible. Some systems show intermittent bursts of radiation at radio frequencies. Evidence indicates the presence of clouds of circumstellar material.

In a few of the eruptive variables, particularly those known as dwarf novae, rapid scintillation is found, presumably from a hot spot where the transferring mass collides violently with a circumstellar disk of relatively low-density material revolving around the collapsed star (**Fig. 2**); the scintillation stops periodically when the spot is eclipsed by the other component. Instruments on satellites have extended observations to the far ultraviolet, as well as the x-ray, regions of the spectrum. Thus, in addition to the classical reasons for studying eclipsing variable stars, observation of them leads into many branches of astrophysics. *See Astrophysics, high-energy; Nova; Star; Stellar evolution; X-ray astronomy.*

Frank Bradshaw Wood

Bibliography. A. H. Batten, *Binary and Multiple Systems of Stars*, 1973; J. Sahade and F. B. Wood, *Interacting Binary Stars*, 1978.

Ecliptic

The path in the sky traced by the Sun in its apparent annual journey as Earth revolves around it. The ecliptic is a great circle on the celestial sphere, inclined about 23°.5 to the celestial equator, the angle of inclination being called the obliquity of the ecliptic. *See Astronomical coordinate systems.*

Gerald M. Clemence

Elements, cosmic abundance of

The abundance of the elements in surface rocks of the Earth, in the Earth as a whole, in meteorites, in the solar system, in the galaxies, or in the total universe corresponds to the average relative amounts of the chemical elements present, or, in other words, to the average chemical composition of the respective object. Element abundances are given in numbers of atoms of one element relative to a certain number of atoms of a reference element. Silicon is commonly taken as the reference element in the study of the composition of the Earth and the meteorites, and the data are given in atoms per 10^6 atoms of silicon (cosmochemical normalization). The results of astronomical determinations of the composition of the Sun and of the stars are often expressed in atoms per 10^{10} atoms of hydrogen (astronomical normalization). Ordinary chemical analyses, including advanced techniques for trace element studies (such as neutron activation or isotope dilution), are used for determination of the composition of rocks and meteorites.

Relative abundance of the elements in the solar system*

Atomic number (Z)	Element	Orgueil, N value	Sun	Most probable value
1	H		2.5×10^{10}	2.5×10^{10}
2	He		2.0×10^{9}	2.0×10^{9}
3	Li	55	0.25	5.5×10
4	Be	0.73	0.35	7.3×10^{-1}
5	B	6.6	5.0	6.6
6	C	7.7×10^{5}	7.9×10^{6}	7.9×10^{6}
7	N		2.1×10^{6}	2.1×10^{6}
8	O	7.3×10^{6}	1.7×10^{7}	1.7×10^{7}
9	F	7.1×10^{2}	1×10^{3}	7.1×10^{2}
10	Ne		2.5×10^{6}	1.4×10^{6}
11	Na	5.7×10^{4}	5.2×10^{4}	5.7×10^{4}
12	Mg	1.01×10^{6}	9.3×10^{5}	1.01×10^{6}
13	Al	8.00×10^{4}	6.3×10^{4}	8.0×10^{4}
14	Si	1.00×10^{6}	1.0×10^{6}	1.00×10^{6}
15	P	8.58×10^{3}	6.8×10^{3}	8.6×10^{3}
16	S	4.8×10^{5}	3.9×10^{5}	4.8×10^{5}
17	Cl	5.0×10^{3}	8×10^{3}	5.0×10^{3}
18	Ar		1×10^{5}	2.2×10^{5}
19	K	3.48×10^{3}	3.2×10^{3}	3.5×10^{3}
20	Ca	5.91×10^{4}	5.7×10^{4}	5.9×10^{4}
21	Sc	35	28	3.5×10
22	Ti	2.42×10^{3}	1.6×10^{3}	2.4×10^{3}
23	V	2.90×10^{2}	2.6×10^{2}	2.9×10^{2}
24	Cr	1.35×10^{4}	1.4×10^{4}	1.35×10^{4}
25	Mn	8.72×10^{3}	6.3×10^{3}	8.7×10^{3}
26	Fe	8.60×10^{5}	8.3×10^{5}	8.6×10^{5}
27	Co	2.24×10^{3}	2.3×10^{3}	2.2×10^{3}
28	Ni	4.83×10^{4}	4.7×10^{4}	4.8×10^{4}
29	Cu	4.60×10^{2}	[illegible]	4.5×10^{2}
30	Zn	1.40×10^{3}	6.6×10^{2}	1.40×10^{3}
31	Ga	34	16	44
32	Ge	1.14	65	113
33	As	6.5		6.5
34	Se	63		63
35	Br	8		8.0
36	Kr			25
37	Rb	6.4	11	6.4
38	Sr	26	17	26
39	Y	4.3	2.5	5.4
40	Zr	11	1.4	11
41	Nb	0.85	2.5	0.85
42	Mo	2.5	4	2.5
43	Tc			
44	Ru	1.8	2	1.8
45	Rh	0.33	0.8	0.33
46	Pd	1.32	0.8	1.32
47	Ag	0.50	0.18	0.5
48	Cd	1.80	1.78	1.30
49	In	0.17	1.12	0.174
50	Sn	3.88	2.51	2.4
51	Sb	0.27	0.4	0.27
52	Te	4.83		4.8
53	I	1.16		1.16
54	Xe			6.1
55	Cs	0.37		0.37
56	Ba	4.22	3.2	4.2
57	La	0.46	0.34	0.46
58	Ce	1.20	0.90	1.20
59	Pr	0.18	0.24	0.18
60	Nd	0.87	0.45	0.87
61	Pm			
62	Sm	0.27	0.14	0.27
63	Eu	0.10	0.13	0.10
64	Gd	0.34	0.32	0.34
65	Tb	0.061		0.061
66	Dy	0.41	0.39	0.41
67	Ho	0.091		0.091
68	Er	0.26	0.16	0.26
69	Tm	0.041	0.035	0.041
70	Yb	0.25	0.16	0.25
71	Lu	0.038	0.16	0.038
72	Hf	0.18	0.20	0.18
73	Ta	2.1×10^{-2}		1.10^{-2}
74	W	0.13	0.16	0.13
75	Re	0.052		0.052
76	Os	0.68	0.16	0.72
77	Ir	0.65	4.0	0.65
78	Pt	1.42	1.6	1.42
79	Au	0.19	0.14	0.19
80	Hg			0.4
81	Tl	0.17	0.2	0.17
82	Pb	3.09	2.0	3.1
83	Bi	0.14		0.14
90	Th	3.2×10^{-2}	0.040	3.2×10^{-2}
92	U	0.91×10^{-2}		9.1×10^{-3}

*Cosmochemical normalization relative to $N(Si) = 10^{6}$.

The composition of the Sun and of stars can be derived by quantitative spectral analysis. On the surface of the Earth, the most abundant elements are oxygen, silicon, magnesium, calcium, aluminum, and iron. In the universe as a whole, hydrogen and helium constitute more than 95% of the total matter. The relative abundance of elements in the solar system is given in the **table**. *SEE ASTRONOMICAL SPECTROSCOPY.*

Abundances in the Sun and stars. The possibility of a quantitative spectral analysis of the Sun, stars, and planetary nebulae is based on the fact that the intensity of the absorption lines in the spectrum, the Fraunhofer lines, depends on the concentration of the atoms causing the absorption. In order to calculate the relation of line intensity with atomic concentration, a number of physical properties of the absorbing atoms, as well as the thermodynamic state of the absorbing stellar matter, have to be known in detail. Furthermore, a knowledge of a depth of the layer in which the absorption occurs, the thermal velocity of the absorbing atoms, their macroscopic turbulent motion, and other characteristics is necessary before the exact functional dependence of line intensity and atomic concentration can be calculated.

The first abundance data by spectral analysis were obtained by C. H. Payne-Gaposchkin in 1925 and by H. N. Russell in 1928. Relatively few stars have been analyzed, and the data from spectral analyses are far from complete. But the data have shown that the chemical composition of the universe is remarkably uniform, although systematic variations in the composition of stars seem to exist, depending on age and position in the galaxies. *SEE SOLAR RADIATION; SUN.*

Abundances in the Earth and meteorites. The abundance of the elements in terrestrial rocks was first investigated by F. W. Clarke and H. S. Washington during the last decade of the nineteenth century. These investigators compared numerous rock analyses, gave average figures for the occurrence of each element in the various types of terrestrial rocks, and hoped that some sort of regularity might become apparent. They expected that the chemical composition of the terrestrial rocks would reflect some fundamental quantity connected with the relative amounts in which the elements occur in nature in general. Since then, it has become obvious that meteorites are better objects for the study of a primeval abundance distribution of the elements. The composition of the Earth

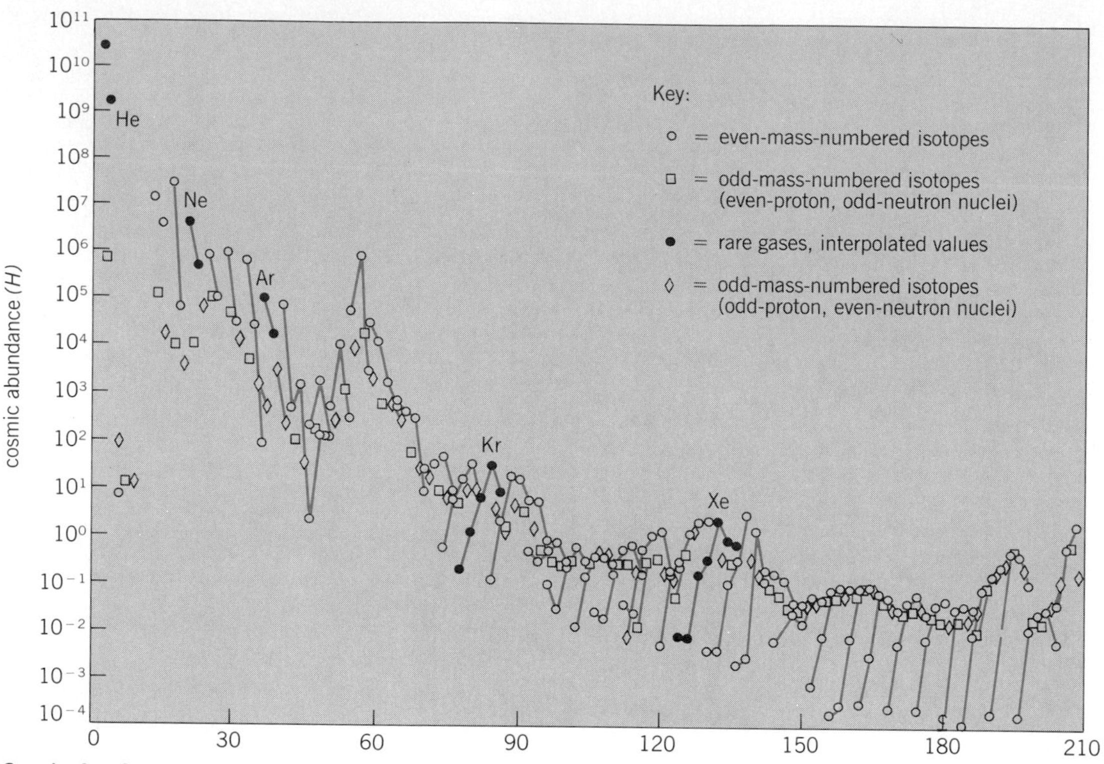

Cosmic abundance *H* of the naturally occurring stable isotopes *H*(Si) ≡ 10⁶, based essentially on chemical analyses of carbonaceous chondrites.

corresponds to the nonvolatile part of a primeval cloud from which the planets originated. The meteorites have formed from the same cloud, but they have undergone less chemical fractionation than any material on Earth. Meteoritic matter shows, in general, separation into three chemical phases—metal, sulfide, and silicate— in a ratio of about 10.6:1:100, respectively. The elements that concentrate in the metal phase are called siderophile, those in the sulfide phase, chalcophile, and those in the silicate phase, lithophile elements. A large fraction of meteorites, the chondrites, contain all three phases in relatively constant proportions. It is generally believed that the chondrites, in particular the type I carbonaceous chondrites, contain the nonvolatile components of the primeval solar matter in essentially unchanged proportions, because it seems improbable that chemically similar elements were separated from each other under conditions that did not lead to an effective separation of the three main phases. The giant planets (Jupiter, Saturn, Uranus, Neptune) have retained much of the volatile substances, including hydrogen and helium, and elements such as carbon, nitrogen, and oxygen, mostly in the form of methane, ammonia, and water, respectively. *See Meteorite*.

Nuclear abundances. Most elements are composed of more than one isotope. The isotopic composition of the elements is practically the same in all terrestrial material and in meteorites. Only small variations can be observed. In light elements variations occur as a consequence of small differences in the chemical properties owing to the difference in mass. Variations also occur if an isotope is produced by radioactive decay or cosmic-ray–produced processes. From the isotopic composition of an element and its cosmic abundance, the nuclear abundances of its isotopes can be calculated. A number of empirical rules exist for the abundances of nuclear species. The most important one is Harkin's rule, which states that elements with an odd mass number are less abundant than their even-mass-numbered neighbors. Another rule postulates that the abundance values of the individual nuclear species, as a function of their mass number, form regular smooth lines for the odd-mass-numbered species, and in a somewhat less regular way, also, for the sum of the abundances of isobars at even mass numbers. Irregularities occur where the number of neutrons or protons reaches a so-called magic number, connected with a nuclear shell closure.

In the **illustration** the abundances of the individual stable isotopes are plotted on a logarithmic scale versus their mass number. The values are derived from chemical analyses of carbonaceous chondrites (specifically, the one named Orgueil). Isotopes, even- and odd-mass-numbered ones separately, belonging to the same elements are connected by a straight line. Values are interpolated for elements not present in solar proportions, such as carbon, nitrogen, oxygen, and the noble gases.

These abundance values show a clear correlation with certain nuclear properties, and can be assumed to represent in good approximation the original yield distribution of the thermonuclear processes that lead to the formation of the elements. The empirical abundance values can therefore serve as the basis for theoretical considerations about the origin of matter and of the universe, and have led to the following conclusion: No simple, single mechanism exists by which the elements in their observed isotopic composition can have formed. The matter of the cosmos appears to be a mixture of material that formed under different conditions by different types of nuclear processes. *See Nucleosynthesis*.

Hans E. Suess

Bibliography. E. Anders and M. Ebihara, *Geochim. Cosmochim. Acta*, 46:2363–2380, 1982; A. G. W.

Cameron, *Space Sci. Rev.*, 15:121–146, 1973; V. M. Goldschmidt, *Geochemistry*, 1954; E. A. Muller, *Transactions*, 16b:118,1977; J. E. Ross and L. M. Aller, *Science*, 191:1223, 1976; H. E. Suess and H. C. Urey, The abundance of the elements, *Rev. Mod. Phys.*, 28:53–74, 1956.

Ephemeris

A tabulation of data pertaining to observable features of bodies, particularly astronomical ones, that depend on time. Such a tabulation can be a printed table or in computer-readable form such as on a magnetic tape. Most ephemerides are for the observed positions of celestial objects; these objects can be as seen from the Earth, or as seen from the Sun, or relative to some other body such as for components of double stars or planetary satellites. There can also be ephemerides of other parameters such as the state of rotation or orientation of a body in space.

An ephemeris is usually derived from what is known as a theory (of motion) of a body, which consists of an extensive series of equations that have to be evaluated at each instant of time of interest. These equations are derived from the laws of celestial mechanics, and they can be quite complicated for bodies in the solar system. An ephemeris can also be derived from tabulations of more fundamental data that may result, for instance, from a computer simulation. Thus, the ephemerides of the planets are now based on tables of rectangular coordinates for these planets. Associated with any ephemeris, whether theoretical or derived, there is a set of physical constants whose values must be derived from observations.

A major source of ephemerides is *The Astronomical Almanac*, an annual volume published by the U.S. Naval Observatory and the Royal Greenwich Observatory. This book gives tables of positions of the major planets and their satellites, some minor planets, and rotational parameters of the planets. It also includes tables of data that are constant with time (at least over the span of a year). With the advent of small computers, there is more interest in generating one's own ephemeris, which has spurred the development of simplified theories that can be used on most available machines. *See* Almanac; Celestial mechanics.

Robert S. Harrington

Bibliography. D. Brouwer and G. M. Clemence, *Methods of Celestial Mechanics*, 1961; R. Green, *Spherical Astronomy*, 1985; L. G. Taff, *Computational Spherical Astronomy*, 1981; E. W. Woolard and G. M. Clemence, *Spherical Astronomy*, 1966.

Equation of time

The annual, cyclic variation between mean solar time shown on uniformly running clocks and apparent solar time displayed on sundials.

In the course of the Sun's daily east-to-west transit of the sky, the Sun crosses the meridian, an imaginary line running from north to south that passes overhead and divides the sky into equal halves. An observer in the middle of a time zone generally thinks of noon as being the moment that the Sun reaches the meridian. This event, however, corresponds to noon recorded by mechanical or electronic clocks on only four dates each year (approximately April 16, June 14, September 1, and December 25). On all other dates the Sun reaches the meridian either early or late, with the extremes being 16.3 min early around November 3 and 14.3 min late around February 12. This difference is the equation of time, and results from the combined effects of Earth's axis of rotation being tipped 23° relative to Earth's orbital plane and the elliptical rather than circular shape of the orbit. *See* Meridian.

The Earth's orbital motion causes the Sun to move eastward against the background stars along a path called the ecliptic. This motion is about 1° each day. Although the Earth turns once on its axis every $23^h56^m4^s$, it must turn slightly more in order for the eastward-moving Sun to return to the meridian. Thus, the length of time from one meridian passage of the Sun to the next averages 24 h over the course of a year.

If the Earth had a perfectly circular orbit and its axis of rotation was perpendicular to the orbital plane, the Sun's motion would be along the celestial equator and uniform throughout the year. It is this idealized case of a mean Sun that shows the mean solar time kept by mechanical and electronic clocks and watches. However, the elliptical orbit brings the Earth closest to the Sun in January. This proximity, as J. Kepler discovered in the sixteenth century, causes the Earth to move more rapidly in its orbit than in July, when the Earth is farthest from the Sun. The changing orbital speed varies the Sun's apparent rate of motion along the ecliptic and is in part responsible for the equation of time. *See* Celestial mechanics; Kepler's laws.

Slightly more influential is Earth's tipped axis, which varies the Sun's position north and south of the celestial equator according to the season. Around the time of the spring and autumn equinoxes the Sun moves at a steep angle relative to the celestial equator. Its daily motion projected onto the equator is less than at the solstices, when the Sun travels parallel to the equator. This situation also creates a departure between the Sun's actual position and that of a mean Sun moving uniformly along the celestial equator. **Figure 1** shows how the effects of the inclination of the Earth's axis and the eccentricity of its orbit are combined.

If the Sun's daily position is considered relative to the meridian at the moment a clock reads noon, from the equation of time it is seen that the Sun will sometimes be west of the meridian (early) and sometimes east of it (late). The coupling of this variation with the north-south movement of the Sun along the eclip-

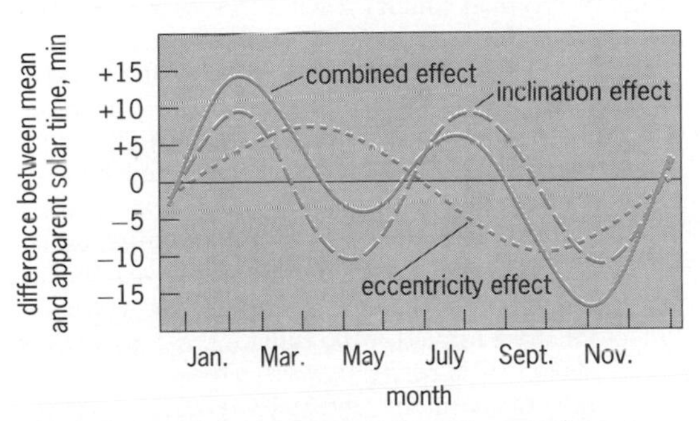

Fig. 1. Graph of the equation of time, showing how the equation results from combining the effects of the inclination of the Earth's axis and the eccentricity of its orbit. (*After B. M. Oliver, The shape of the analemma, Sky Telesc., 44:20–22, July 1972*)

Fig. 2. The analemma recorded in a multiple-exposure, year-long photograph obtained with an east-facing, permanently mounted camera that made an exposure approximately once a week at exactly 8:30 a.m. Eastern Standard Time. On three occasions the shutter was left open from dawn until shortly before 8:30 to show the Sun's path across the sky. All images of the Sun were made through a special dense filter. The foreground was recorded without the filter one afternoon when the Sun was in the western sky and roughly behind the camera (as indicated by the shadow of the chimney). *(Photograph by Dennis di Cicco)*

tic causes the Sun to mark out a large figure-eight known as the analemma (**Fig. 2**). This pattern is sometimes shown on the tropical zones of Earth globes (usually in the Pacific Ocean). The analemma shows the equation of time and the latitude at which the Sun passes directly overhead on any given date. *See* EARTH ROTATION AND ORBITAL MOTION; TIME.

Dennis di Cicco

Bibliography. B. M. Oliver, The shape of the analemma, *Sky Telesc.*, 44:20–22, July 1972.

Equinox

One of the two places in the sky where the ecliptic crosses the celestial equator; or one of the two times of the year when the Sun crosses these points. The ecliptic is the great circle across the sky that marks the path of the Sun; the celestial equator is the great circle that is an extension into the sky of the Earth's Equator. These two great circles meet at two points, one of which is the vernal equinox and the other the autumnal equinox. The Sun passes the vernal equinox each year about March 20, and the autumnal equinox about September 22. The vernal equinox can occur as early as March 19 and as late as March 21; for most of the twenty-first century it will be on March 20. The autumnal equinox can occur as early as September 21 and as late as September 24; for most of the twenty-first century it will be on September 22. The dates and times drift with the difference between the actual solar years and 365 days, and are corrected by leap years. This results in a 4-year variation superimposed on a negative 11-min-per-year slope (see **illus.** *a* and *b*). Since 2000 will be an ordinary leap year in the Gregorian calendar, unlike 1900, the dates will continue to decline through 2100 (illus. *c*). *See* ASTRONOMICAL COORDINATE SYSTEMS; CALENDAR; TIME.

At the vernal equinox, the Sun crosses from southern to northern declinations, marking the beginning of Northern Hemisphere spring and Southern Hemisphere autumn. At the autumnal equinox, the Sun crosses from southern to northern declinations, marking the beginning of Northern Hemisphere autumn and Southern Hemisphere spring. At the equinoxes, the Sun is directly above the Earth's Equator.

The term equinox is derived from the Latin for equal nights, indicating that the day and night are of equal duration. However, the actual duration of daylight is several minutes longer on the days of the equinoxes, and the actual dates of equal days and nights follow the autumnal equinox and precede the vernal equinox by a few days. The equinoctial dates are geometrical constructions in which the Sun is treated as a point; in actuality the top of its disk rises a few minutes ahead of its center. Furthermore, refraction in the Earth's atmosphere makes the Sun appear higher in the sky than it actually is, an effect that also lengthens daylight by several minutes. At dawn, the top of the Sun is actually below the horizon, over which it is elevated by refraction.

The Earth's unsteady orbit creates precession of the equinoxes. Their position moves along the ecliptic about 50 seconds of arc per year, and the equinoxes occur 20 min earlier each year. *See* EARTH ROTATION AND ORBITAL MOTION; PRECESSION OF EQUINOXES.

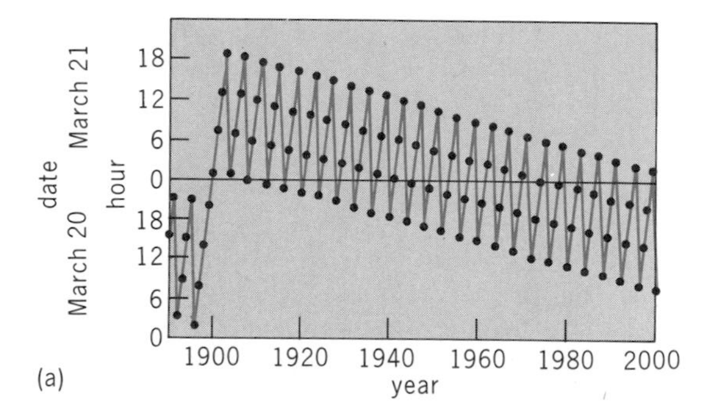

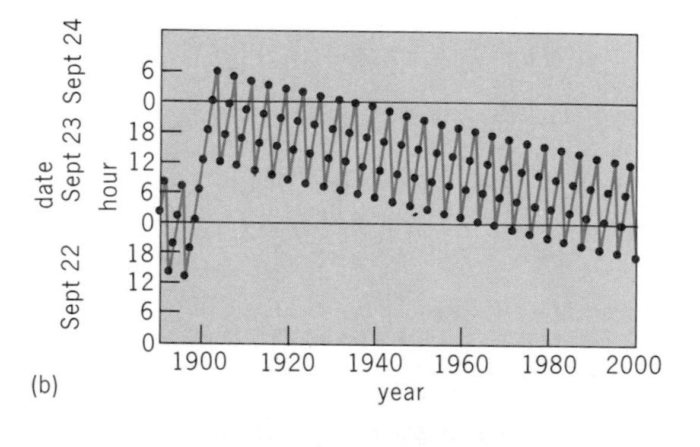

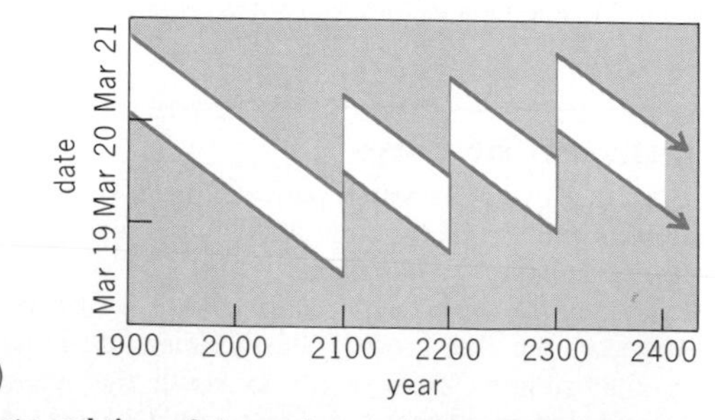

Date and time of the equinoxes (Universal Time or Greenwich mean time). *(a)* Vernal equinox, 1890–2000. *(b)* Autumnal equinox, 1890–2000. *(c)* Approximate 4-year range of times of the vernal equinox, 1900–2400. *(After R. L. Reese and G. V. Chang, The date and time of the vernal equinox: A graphical analysis of the Gregorian calendar, Amer. J. Phys., 55:848-849, 1987)*

Since the vernal equinox was in the constellation Aries when Hipparchus studied it 2000 years ago, it is known as the first point in Aries. It is, however, in Pisces.

Jay M. Pasachoff

Bibliography. R. L. Reese and G. Y. Chang, The date and time of the vernal equinox: A graphical analysis of the Gregorian calendar, *Amer. J. Phys.*, 55:848–849, 1987.

Extraterrestrial intelligence

The potential existence beyond the Earth of other advanced civilizations with a technology at least as developed as that on Earth. The idea that life, especially life with intelligence, might exist in other parts of the universe is a very old one and is found in the writings of Metrodorus of Chios (fifth century B.C.) and Lucretius (first century B.C.). These early ideas were based on an intuitive belief in the enormity of the universe and in what is now called the mediocrity principle, namely that there is nothing special about the Sun, the Earth, and the human race.

Scientific rationale. Present ideas are also based on the mediocrity principle supported by the universality of the laws of physics and chemistry, and by the enormity of the universe, whose exploration has been made possible by large telescopes. The Sun is one of 2×10^{11} stars of the Galaxy (the Milky Way), and there are about 10^{11} galaxies in the visible universe. The Sun, a rather average star in terms of mass, luminosity, and chemical composition, is located about two-thirds of the way from the center to the periphery of the galactic disk. It is a middle-aged star about 4.65×10^9 years old with a life expectancy of about 10^{10} years. The universe is expanding and started with the big bang about 15×10^9 years ago. Every part of it is made of the same 92 chemical elements and obeys the same laws of physics. *See* BIG BANG THEORY; GALAXY, EXTERNAL; MILKY WAY GALAXY; STAR; STELLAR EVOLUTION; SUN; UNIVERSE.

The chemical evolution, that is, the natural formation of complex organic compounds, that led to the origin of life on Earth is quite common in the universe. The presence of such compounds has been observed in certain asteroids (carbonaceous chondrites); comets; satellites such as Titan, the large satellite of Saturn; interstellar clouds; and certain carbonaceous meteorites (chunks of asteroids that fall onto the Earth), where several amino acids and all of the five nitrogen bases that are the key components of deoxyribonucleic acid (DNA) and ribonucleic acid (RNA) have been found. Finally, experiments that simulate the chemistry in the early, oxygen-free, atmosphere of the Earth, as well as in the oceans, have produced consistently a rich yield of amino acids and other organic compounds of great importance to life. *See* ASTEROID; INTERSTELLAR MATTER; METEORITE; SATURN.

It is still not known how chemical evolution led to the origin of life by producing the first replicating systems, but the biological evolution that followed is fairly well understood. Life on Earth started at least 3.5×10^9 years ago, that is, soon after the formation of the oceans, indicating a rather straightforward natural process. Through mutations and Darwinian selection, evolution advanced slowly from primitive unicellular microorganisms to advanced multicellular organisms with intelligence, changing in the process the atmosphere of the Earth to one with free oxygen. Intelligence, which is favored by evolution because it has a high survival value, evolved into a technological society that commands large quantities of energy and can travel into outer space, but is also capable of self-destruction.

Radio searches. Communication with extraterrestrial intelligence (CETI), if indeed such civilizations do exist, has long been a desire of the human race. In the 1820s, Gauss proposed planting a large forest in the shape of an orthogonal triangle to signal to those observing the Earth that the planet is inhabited by intelligent beings familiar with the pythagorean theorem. However, the actual search for extraterrestrial intelligence (SETI) was initiated only after the development of radio astronomy and large radio telescopes, because radio waves seem to be the most efficient means of communication over interstellar distances. *See* RADIO ASTRONOMY; RADIO TELESCOPE.

The basic problem of radio searches is selecting the proper frequency out of an almost infinite range of choices. In 1959 G. Cocconi and P. Morrison proposed the use of the radio line of atomic hydrogen at 21 cm, arguing that if indeed other civilizations want to contact the Earth they will make it as easy as possible by choosing a universal frequency well known to all advanced civilizations practicing radio astronomy. The first radio search was carried out in the United States by Frank Drake in 1960, who, following the suggestion of Cocconi and Morrison, searched at the hydrogen line for signals from two nearby stars that resemble the Sun, Epsilon Eridani, 11 light-years (6.5×10^{13} mi or 1.0×10^{14} km) away, and Tau Ceti, 12 light-years (7×10^{13} mi or 1.1×10^{14} km), away. At least 45 different radio searches have been carried out since then, accumulating more than 10,000 h of observations. Most of these searches have been made at the hydrogen line, although several other characteristic frequencies have also been used. The results continue to be negative, but selecting the proper frequency, bandwidth, polarization, target, and so forth, is an extremely complex problem, and the ranges of these parameters form a multidimensional "cosmic haystack," of which only a small segment has been explored.

Scientific support. SETI was recommended as one of the priorities in astronomy and astrophysics for the 1980s by the U.S. National Academy of Sciences; NASA has been funded to support SETI research. In 1982 the International Astronomical Union established a new commission (section) under the title Search for Extraterrestrial Life, thus formally endorsing this new branch of astronomical work, which has rapidly become known as bioastronomy.

Drake equation. In the early 1960s, Drake developed an equation to estimate the number N of advanced technological civilizations currently active in the Galaxy, based on the scientific knowledge presently available. The Drake equation, $N = R \times P \times L$, gives N in terms of the rate R at which new stars are born in the Galaxy, the probability P (actually a product of probability factors) that any one of these stars will possess the necessary conditions (luminosity, planets at the appropriate distances, and so forth) for life to originate and to slowly evolve to a technological civilization, and the average longevity L of such civilizations.

The values advocated by the proponents of the Drake equation for these parameters are typically: R equal to approximately 20 new stars per year, P equal to approximately 1%, and L equal to approximately

10^6 years—which yield a value of N of approximately 200,000 stellar civilizations, that is, about one advanced civilization per 10^6 stars, placing statistically the nearest extraterrestrial civilization at a distance of about 300 light-years (2×10^{15} mi or 3×10^{15} km). The need to search at least 10^6 stars has been suggested by the supporters of the Drake equation as one of the reasons why radio searches have not yet produced any positive results. However, the uncertainties in P and L are very large. As a result, there is a considerable disagreement among the scientists about the value of N, and some think that the human race is probably the only advanced civilization in the Galaxy; that is, N equals 1. The Drake equation, with the above values of the parameters, predicts that the Galaxy through its history of 10^{10} years must have harbored about 10^9 advanced civilizations, each with an average life of about 10^6 years.

Problem of galactic colonization. The possibility of interstellar travel and galactic colonization introduces a complicating factor, which would invalidate the Drake equation. This equation assumes that all stellar civilizations are the product of indigenous evolution and negates interstellar migration. In the early 1960s, interstellar travel was considered impossible, because round trips to the nearest stars, 5–10 light-years ($3–6 \times 10^{13}$ mi or $5–10 \times 10^{13}$ km) away, in a reasonable fraction of a human life require interstellar speeds close to the speed of light, which are unrealistic even with a matter-antimatter propulsion fuel. In the mid-1970s, however, the possibility of large human colonies in space began to be seriously considered, and the idea that such self-sustained habitats could undertake multigeneration trips of several centuries to other stars began to gain acceptance. Interstellar trips at only a few percent of the speed of light seem entirely feasible with nuclear fusion as a propulsion fuel, and would allow the establishment of space-borne settlements in the vicinity of nearby stars. Such settlements would also eliminate the need for Earth-like planets, which may be rather rare in the Galaxy. The progressive colonization of nearby stars by previously established stellar civilizations can sweep through the entire Galaxy in less than 10^7 years (a very short period compared to the 10^{10}-year history of the Galaxy), establishing a space-borne civilization around every well-behaved star of the Galaxy, including the solar system. Such an event would increase almost instantaneously (in cosmic terms) the value of N predicted by the Drake equation by 10^6 times.

The possibility of galactic colonization, in combination with the natural tendency of life to expand into all available space and with the innate desire of intelligent life to explore all unknown territories, makes many scientists think that such colonization is not only possible but an almost inevitable consequence of the evolution of intelligence and technology. This concept, however, leads to two extreme alternatives: either the Galaxy has already been colonized, in which case N must be very large; or it has not been colonized, because no one was there to do so, in which case N must be very small. Both alternatives are contrary to the results of the Drake equation, which negates galactic colonization and predicts an intermediate value of N.

All three alternatives, however, contain serious contradictions which are not easy to reconcile:

1. If the Galaxy had already been colonized, advanced civilizations would have also colonized the solar system. But, then, where are they? Their apparent absence is often called the Fermi paradox, after Enrico Fermi who supposedly was the first to pose this question, and is interpreted by some scientists as a strong indication that humans must be alone in the Galaxy.

2. If the solar system has not been colonized, the whole Galaxy must not have been colonized, which implies that there were no advanced civilizations to initiate colonization. Consequently, the human race must be one of very few, advanced civilizations if not the only one, in the Galaxy. But, then, what is so special about the Earth and the human race?

3. If advanced stellar civilizations do not engage in interstellar travel and colonization, there must be a reason (economic, social, ethical, and so forth). It is hard to see, however, how any of these reasons could apply to millions of advanced civilizations for billions of years in the past history of the Galaxy, without a single exception which would suffice to trigger the colonization wave.

However, several explanations for these contradictions have been proposed, including:

1. Extraterrestrials may have been present in the solar system and living in space colonies, possibly near sources of raw materials such as the asteroid belt, but may have chosen to avoid making contact with the human race for a variety of reasons, waiting, for example, to see whether humanity will manage to overcome the problems of technological explosion (overpopulation, pollution, exhaustion of resources, nuclear war, and so forth), or will self-destruct.

2. The predictions of the Drake equation may be too high. It is possible, for example, that although advanced civilizations may appear quite frequently in the Galaxy, they survive for only very short intervals (that is, L is grossly overestimated), succumbing to the many problems of technology before they have a chance to initiate interstellar travel.

3. It is possible that although life originates rather easily in many planets, only an extremely small number of them are able to maintain liquid water for billions of years necessary for the slow evolution of life to an advanced technological civilization (that is, P is highly overestimated).

4. There may be a universal barrier, as yet undiscovered, to interstellar travel.

5. Advanced stellar civilizations may choose for some reason not to communicate with emerging civilizations like that on Earth.

Search strategies. There is considerable diversity of opinion on search strategies, primarily because each strategy assumes a certain behavior on the part of the extraterrestrial civilizations for which very little is known. The most prudent approach, therefore, is to maintain the momentum that has been gained for conventional search efforts while also encouraging alternative search strategies, such as observations at different wavelengths and searches inside the solar system.

It is possible that an extensive but unproductive search will lead to the conclusion that the human race is one of very few technological civilizations, and possibly the only one, in the Galaxy. But even this would not represent a failure, because exciting as it might be to find other advanced civilizations, it is equally important to know our place in the universe.

Michael D. Papagiannis

Bibliography. J. Billingham (ed.), *Life in the Universe*, 1981; R. Breuer, *Contact with the Stars*, 1982; D. Goldsmith, *The Quest for Extraterrestrial Life*,

1980; M. H. Hart and B. Zuckerman (eds.), *Extraterrestrials, Where Are They?,* 1982; G. K. O'Neill, *2081: A Hopeful View of the Human Future*, 1982; M. D. Papagiannis (ed.), *Strategies for the Search for Life in the Universe*, 1980; I. Ridpath, *Worlds Beyond*, 1976.

Foucault pendulum

A pendulum or swinging weight, supported by a long wire, by which J. B. L. Foucault demonstrated in 1851 the rotation of Earth on its axis. Foucault used a 62-lb (28-kg) iron ball suspended on about a 200-ft (60-m) wire in the Pantheon in Paris. The upper support of the wire restrains the wire only in the vertical direction. The bob is set swinging along a meridian in pure translation (no lateral or circular motion). In the Northern Hemisphere the plane of swing appears to turn clockwise; in the Southern Hemisphere it appears to turn counterclockwise, the rate being 15 degrees times the sine of the local latitude per sidereal hour. Thus, at the Equator the plane of swing is carried around by Earth and the pendulum shows no apparent rotation; at either pole the plane of swing remains fixed in space while Earth completes one rotation each sidereal day. SEE DAY.

Frank H. Rockett

Bibliography. G. O. Abell, *Exploration of the Universe*, 4th ed., 1982; C. Kittel, W. D. Knight, and M. A. Ruderman, *Mechanics*, Berkeley Physics Course, vol. 1, 2d ed., 1973; O. Struve, B. Lynds, and H. Pillans, *Elementary Astronomy*, 1959.

Fraunhofer lines

Dark absorption features in the solar spectrum. They are named in honor of J. Fraunhofer, who first studied them in 1814. They are found from the ultraviolet at about 180 nanometers to the infrared at 20 micrometers. Each line represents the net absorption of light by a particular atom or molecule. Most form in the Sun's atmosphere, although the Earth's telluric spectrum contributes lines of molecular oxygen (O_2), carbon monoxide (CO), and other molecules. Some lines such as Fraunhofer's C line in the red (hydrogen-alpha) can be seen with a pocket spectroscope. Powerful research instruments reveal millions of lines, most of which are weak and blended together in an almost inextricable tangle. The **table** lists some interesting Fraunhofer lines.

A spectrum line is caused by the absorption of photons of light that excite the atom from a lower to a higher energy level. Spontaneous decay back to the atom's lower level then follows, accompanied by the isotropic emission of light at the wavelength of the line. The result is a loss of light in the Sun-Earth direction. The probability of a given photon being absorbed depends on the quantum state of the atom.

The study of the Fraunhofer spectrum is the principal means of learning about physical conditions in the solar atmosphere. On the resolved solar disk, variations in line strength from point to point convey information about temperature, Doppler shifts of the lines reveal gas motions, and line splitting from the Zeeman effect maps magnetic fields. Because each line represents a chemical element, the composition of the solar atmosphere can be deduced. SEE DOPPLER EFFECT; SOLAR MAGNETIC FIELD; SUPERGRANULATION; ZEEMAN EFFECT.

Profiles. A high-dispersion spectrograph shows that Fraunhofer lines are not sharp; most have a gaussian profile with extended wings. This broadening results mainly from the Doppler motions associated with thermal and unresolved turbulence, together with collisional damping, wherein the atoms interact with their neighbors during the excitation-deexcitation process. Like line intensity and wavelength shifts, line profiles can yield further information about the solar

Fraunhofer lines

Wavelength, nm	Name*	Species†	Cycle variability,‡ %	Comment§
279.54		Mg II	10	Ultraviolet emission, high chromosphere
280.23				
388.36	(CN band-head)	CN	3	High photosphere
393.36	K	Ca II	15	Chromosphere
396.85	H			
486.13	F (Hβ)	H I		Chromosphere
517.27	b_2	Mg I	—	Low chromosphere
518.36	b_1			
525.02		Fe I	0.3	Photosphere, magnetic fields (g = 3)
538.03		C I	0.0	Low photosphere
589.00	D_2	Na I	—	Upper photosphere, low chromosphere
589.59	D_1			
630.25		Fe I	—	Photosphere, magnetic fields (g = 2.5)
656.28	C (Hα)	H I	6	Chromosphere
849.80		Ca II	1	Low chromosphere
854.21				
866.22				
868.86		Fe I	—	Photosphere, magnetic fields (g = 1.7)
1083.03		He I	200	High chromosphere
1281.8	Paschen β	H I	—	Chromosphere
1564.85		Fe I	—	Photosphere, magnetic fields (g = 3)
4665		CO	—	High photosphere
12,320		Mg I	—	High photosphere, magnetic fields (g = 1)

*J. Fraunhofer's original designation, if any.
†Atom or molecule responsible.
‡Any temporal variability observed in integrated light.
§Origin and diagnostic value of the line for solar physics research.

surface, for example, the gas density and pressure. *See Spectrograph*.

Variability. The strength of the weak line of carbon at 538.0 nm is a sensitive indicator of surface temperature. The cores of chromospheric calcium H and K lines at 396.8 nm and 393.3 nm, which originate in active regions, correlate with the Sun's ultraviolet output. A variability of the Sun that might affect the Earth can be looked for by measuring the temporal variation of these lines in integrated sunlight. *See Astronomical spectroscopy; Solar constant; Sun*.

William C. Livingston

Bibliography. P. V. Foukal, *Solar Astrophysics*, 1990; K. H. Schatten and A. Arking (eds.), *Climate Impact of Solar Variability*, 1990; O. R. White, *The Solar Output and Its Variation*, 1977.

Fundamental stars

That relatively small number of the brighter stars distributed as uniformly as possible over the entire sky. For metrical purposes the sky is considered as the inner surface of a sphere; astrometry consists of measuring the spherical coordinates of the stars, analogously to measuring the latitudes and longitudes of points on the surface of the Earth. The fundamental stars are the celestial analog of the first-order triangulation points on the Earth. Thus the coordinates of the fundamental stars are first measured, using elaborate special techniques designed to measure large angular distances; the coordinates of the remaining stars are then inferred by other, less elaborate techniques. *See Astrometry; Astronomical coordinate systems*.

Since the coordinates of the stars are continuously changing as a result of the motion of the Earth and, to a lesser degree, the motions of the stars themselves, it is equally necessary to measure the changes as to measure the coordinates at a definite specified time. Thus the determination of the coordinates is a continuing process, with no definite termination.

From time to time the available knowledge of the coordinates of the fundamental stars, and of their changes, is compiled and published in a fundamental catalog. By international agreement it is used by astronomers throughout the world. The presently accepted one contains the coordinates and their variations of 1535 stars for the years 1950 and 1975; from this information the coordinates at any other time may be derived.

Gerald M. Clemence/Robert S. Harrington

Bibliography. H. K. Eichhorn, *Astronomy of Star Positions*, 1974; E. W. Fricke and A. Kopff (eds.), *Fourth Fundamental Catalogue* (*FK4*), Veroffentlichungen des Astronomisches Rechen-Instituts, Heidelberg, 1963.

Galaxy, external

One of the large self-gravitating aggregates of stars, gas, and dust that contain nearly all of the observed matter in the universe. Typical large galaxies have symmetric and regular forms, are about 50,000 light-years (3×10^{17} mi or 5×10^{17} km) in diameter, and are roughly 5×10^{10} times more luminous than the Sun. The stars and other material within a galaxy move through it, often in regular rotation, with periods of a few hundred million years. The characteristic mass associated with a large galaxy is a few times 10^{12} solar masses. (The solar mass is 4.4×10^{30} lbm or 2×10^{30} kg.) Galaxies often occur in associations containing from two to many thousands of individual galaxies and ranging in size from a few hundred thousand to tens of millions of light-years. Considerable progress has occurred in the study of galactic nuclei, formation, evolution, and interactions. The nearest observed galaxies are about 75,000 light-years (4.5×10^{17} mi or 7×10^{17} km) away; the farthest, almost 10^{10} light-years (6×10^{22} mi or 1×10^{23} km). Galaxies are the landmarks by which cosmologists survey the large-scale structure of the universe.

Composition. The hundreds of billions of stars making up a galaxy are not generally individually observable with current telescope technology because they are too faint and distant. Only the brightest stars in the nearest galaxies can be observed directly with large telescopes. Such stars are of three types: very young, blue, and massive hydrogen-burning (that is, their energy comes from the nuclear fusion of hydrogen into helium) main-sequence stars; very old, red, helium-burning giant stars near the end of their life cycle; and exploding (nova and supernova) or violently variable stars. Although such stars may contribute most of the total visible light from a galaxy, they are few in number; most of the stars and most of a galaxy's mass are in the form of much fainter and lower-individual-mass hydrogen-burning main-sequence stars and in the faint burned-out remnants of dead stars. The faint stars in galaxies can be studied only indirectly through the properties of their combined light. *See Nova; Stellar evolution; Supernova*.

Two general types of stellar populations are distinguished: One type (population I) is characterized by the presence of young stars and by ongoing star formation. It is usually associated with the presence of gas. The second type (population II) shows an absence of gas and young stars as well as other indications that star formation ceased long ago. The Sun is a population I star. *See Star; Stellar population*.

Galaxies contain gas (mostly un-ionized hydrogen) in amounts varying from essentially zero up to a considerable fraction of their total mass. Dust in galaxies, although small in mass (typically 1% of the gas mass), is often dramatic in appearance because it obscures the starlight. *See Interstellar matter*.

Form and size. Galaxies generally display strikingly regular forms. The most common form is a disk with a central bulge. The disk is typically 100,000 light-years (6×10^{17} mi or 1×10^{18} km) in diameter and only about 1000 light-years (6×10^{15} mi or 1×10^{16} km) thick. Its appearance is characterized by radially decreasing brightness with a superposed spiral or bar pattern, or both (**Figs. 1** and **2**). The central bulge may vary in size from hundreds to many thousands of light-years. Such galaxies are classified as spirals (S) and subclassified a, b, or c (for example, Sa) to distinguish increasingly open spiral structure and small bulge size. The disks of these galaxies are dominated by population I stars, while their bulges contain mainly population II stars. The Milky Way Galaxy is an Sb type. *See Milky Way Galaxy*.

Edwin L. Turner; Joyce B. Turner

Many spiral and irregular galaxies have a nearly linear feature in their central regions. Called barred galaxies, these objects can otherwise fit into the general scheme of galaxy types. The letter B is added after the S in the classification of spiral galaxies that contain conspicuous barlike features (for example, M61 is an SBc galaxy). It is likely that almost all galaxies have at least some stars moving in barlike orbits; any source of disturbance, such as a gravitational impulse caused by close passage of another galaxy or even by an uneven internal mass distribution, can set up a temporary barlike distortion of a galaxy's shape.

Paul Hodge

Another common type of galaxy is a featureless ellipsoid with radially decreasing brightness. These galaxies are classified as ellipticals (E) and subclassified according to their axial ratios by a number from 0 (E0 = round; **Fig. 3**) to 7 (E7 = 3-to-1 axial ratio). They may vary in size from thousands to several hundred thousand light-years. They are most commonly found in clusters of galaxies and rarely contain much gas or dust. The brightest galaxies are usually ellipticals. They are dominated by population II stars.

Other, more rare forms of galaxies include a transition class called S0 that has a disk superimposed on an otherwise elliptical type of light distribution, and an irregular (Irr) class composed of galaxies with chaotic forms (Fig. 2). The latter class contains no more than a few percent of all galaxies. Irregulars are generally of low total luminosity and rich in gas, dust, and population I stars.

Edwin L. Turner; Joyce B. Turner

Exotic galaxy types. In addition to the various common types of galaxies, there are some galaxies that lie outside the normal range of morphologies. Most of these galaxies have suffered some disturbing event, such as a gravitational encounter, a merger, or a nuclear explosion. In most cases, the galaxy probably was originally a normal Hubble type, but its (usually recent) history has changed it into an almost unrecognizable form. The exceptions are galaxies that seem to have started out with anomalous characteristics, such as the low surface-brightness galaxies. Those with activity in their nuclei are described below, as are those that have suffered gravitational interactions.

Starbursters. Although their causes are not yet completely known, one of the more spectacular examples of exotic galaxies is the starbursters, galaxies that are presently manufacturing stars at an unusually vigorous rate. An example is the peculiar-looking nearby galaxy NGC 1569, which has furious star formation going on in its inner regions and has huge loops of gas and dust emanating from it, extending far from the main body of the galaxy. The nuclear area of NGC 1569 contains several immense young star clusters. A more distant (and more extreme) example of this phenomenon is the intense radio galaxy NGC 1275, in which the Hubble Space Telescope discovered similar immense young star clusters in formation. These objects may be related to galaxies with nuclear activity, although the connection is not yet clearly established. It is not known whether some gravitational impulse has triggered the unusual star-formation activity or whether the activity has some other cause. It is known that the burst is a temporary condition and that the galaxies now bursting must spend most of their lives in a more quiet condition. SEE SATELLITE ASTRONOMY; STARBURST GALAXY.

Low-surface-brightness galaxies. Another type of exotic galaxy is the low-surface-brightness galaxies, star systems that have such a low spatial density of stars that they are almost invisible. The first to be discovered were the Sculptor dwarfs, which have many characteristics similar to the globular star clusters (very old stars, cluster-type variable stars, and smooth stellar distribution) but which are millions of times less dense. There are about a dozen of these elusive galaxies in the Local Group of galaxies. The most difficult to detect is the Sextans dE galaxy, discovered in 1990. It has never been directly observed; its discovery resulted from the computer detection of a statistical enhancement of the number of very faint stars in a region of the constellation Sextans. Subsequent

Fig. 1. Great spiral galaxy in Andromeda (M31, NGC 224) and its two small elliptical companions (NGC 205 and 221), photographed with the 48-in. (122-cm) Schmidt telescope. *(California Institute of Technology/Palomar Observatory)*

Fig. 2. "Whirlpool" galaxy (NGC 5194), type Sc, and a companion irregular satellite (NGC 5195).

Fig. 3. E0 galaxy M87 (NGC 4486) in the Virgo cluster constellation. This galaxy is a source of radio emission, and it has an active nucleus. (*California Institute of Technology/Palomar Observatory*)

study of the region showed that these stars make up a galaxy of exceedingly low density, lying about 3×10^5 light-years (1.8×10^{18} mi or 2.8×10^{18} km) from the Earth. *See Star clusters.*

Not all low-surface-brightness galaxies are of the Sculptor type. Even among those with an appearance like Sculptor, some are different in their natures. For example, the Carina dwarf has been discovered to have a composite population of stars. At least two star-forming events have occurred there, one about 15×10^9 years ago and one about 7×10^9 years ago. Most of the stars apparently were created in the more recent event. Other well-known examples of composite populations are the Fornax dwarf, with a small 3×10^9 year-old population, and NGC 205 and NGC 185, both of which have a few 1×10^6-year-old stars, although the majority of their stars are about 15×10^9 years old.

Another type of low-surface-brightness galaxy, possibly related to the above, is the extreme irregular and spiral cases, which have some of the structural properties of the normal examples of these types but are so faint that it is difficult to detect them against the sky brightness. The best known is Malin 1, an intrinsically large nearby galaxy that is remarkable because of its extremely low density. It barely shows up as a large, nearly smooth smudge on a photograph. Hundreds of low-surface-brightness galaxies have been found, especially in galaxy clusters, where they are identifiable because of their concentration in the clusters' centers.

Paul Hodge

Internal motions. The motions of the stars and gas within galaxies are of two types, random and rotational. All of the motions are the result of the gravitational interactions of the stars with each other. The galaxies are supported against gravitational collapse by these motions in the same sense in which the planets of the solar system are kept from falling into the Sun by their orbital motions. The random motions are complex and result in highly eccentric and irregular orbits for the individual stars. The rotational motions correspond to ordered and systematic circular orbits. Usually, the inner regions of a galaxy undergo solid-body rotation (velocity proportional to radius), while the outer regions rotate differentially (velocity constant). The total velocity of the material within galaxies varies from one to several hundred miles per second. Typical orbital periods for stars are several hundred million years.

The distribution of kinetic energy into random and rotational motions varies with the galaxy type. The disk of a spiral galaxy may have only about 1% of its total kinetic energy in random motions, while an E galaxy may have essentially all of it there.

Luminosites. The number of galaxies with total luminosities L is roughly proportional to L^{-y}, where y is between 1 and 1.5 for luminosities less than about 3×10^{10} solar luminosities. The number is exponentially cut off for higher luminosities. The brightest observed galaxies are fainter than 2×10^{11} solar luminosities; the faintest, brighter than about 1×10^6. Galaxies with old stellar populations (population II) give off more of their total luminosity in the red than galaxies with young stellar populations (population I). The Milky Way Galaxy's total luminosity is roughly 1×10^{10} times greater than the Sun's.

The distribution of luminosities is such that while there are very many faint galaxies, they do not contribute a large fraction of the total light given off by galaxies. For instance, if $y = 1$, galaxies brighter than 4×10^9 solar luminosities are responsible for almost 90% of all the light from galaxies. This is important because only the brighter galaxies, visible for great distances through space, can be observed easily and in great numbers.

It is generally assumed that the masses of galaxies are roughly proportional to their luminosities. If true, this would mean that the distribution of galaxy masses is described by a relation similar to that for their luminosities (that is, a power law with an exponential cutoff). It would also mean that the bright galaxies contain most of the mass.

Clustering. Although galaxies are scattered through space in all directions for as far as they can be observed, their distribution is not uniform or random. Most galaxies are found in associations (**Fig. 4**) containing from two to hundreds of individual galaxies (with a median number of about five). These figures refer to bright galaxies only; if the fainter dwarf galaxies are included, numbers at least 10 times larger are obtained. The clustering of galaxies may be described by a covariance function that gives the excess probability (above random) of finding two galaxies with a separation r. The observed covariance function is roughly proportional to $r^{-1.8}$. P. J. E. Peebles and collaborators, who pioneered in the measurement of the covariance function, have claimed that this result indicates that galaxy clustering arose from the gravitational growth of random density fluctuations in the early universe.

Like the stars within a galaxy, the galaxies within a cluster move about under the influence of their mutual gravitational attraction. The motions are generally random and show little evidence of rotation. Typical velocities range from about a hundred up to several thousand miles per second.

For reasons that are unknown, there is a tendency for E and SO galaxies to be concentrated in large clusters more strongly than spirals.

On scales larger than individual small groups and rich clusters, the distribution of galaxies through space is still not random. This very large-scale structure in the galaxy distribution is usually referred to as superclustering to indicate that it involves the higher-order clustering of the individual first-order associations of galaxies.

Neither empirical nor theoretical understanding of this very large-scale structure is well established or generally agreed upon. There are two competing views. In the simpler and older theory, superclusters are merely the hierarchical extension of the first-order galaxy clustering, in which simple gravitational attractions first cause galaxies to form, then clusters, and later clusters of clusters, and so on. In this model, the distinction between clusters and superclusters is primarily semantic and not physical. A more complex and recent view of superclustering is that it represents a distinct type of structure in the universe and that it originated through physical processes different from those responsible for the formation of galaxies and small-scale clustering. In such theories, large-scale structures are usually believed to have formed first, before galaxies and small-scale structures. Empirically, the study of superclustering is very difficult, because it necessarily involves mapping the distribution of very large numbers of distant galaxies. The discovery of a preferred flattened or filamentary geometry for superclusters (**Fig. 5**) and of enormous (diameters up to several hundred million light-years) voids, that is, entirely empty regions, in the galaxy distribution supports the modern nonhierarchical view of superclustering, but the reality and ubiquity of these intriguing large-scale structures remain controversial. *See Universe*.

Masses. There are two methods for measuring the masses of galaxies: Either the velocities of stars within a galaxy or of galaxies within a cluster are measured. Then, given the size of the system, it is possible to deduce the amount of mass required to gravitationally generate the observed velocities. When applied to stars within individual galaxies, the method gives typical masses near 2×10^{11} solar masses. However, when applied to galaxies within clusters, typical masses per galaxy come out in the range of 3×10^{12} solar masses. This order-of-magnitude disagreement has led J. P. Ostriker and others to suggest that galaxies may be enclosed in extensive halos containing most of the galaxy's mass but giving off little or no light. Although not directly observable, such halos would explain the mass discrepancy and help to explain certain other properties of galaxy rotation velocities. If correct, the massive halo hypothesis means that the systems normally referred to as galaxies are no more than the bright central regions of much larger and more massive aggregates of dark matter. Another possible explanation of the mass discrepancy is that clusters contain large amounts of unobserved material in some unknown form. Whatever its ultimate solution, the mass problem is one of the most important puzzles in the study of galaxies.

Interactions. As galaxies move about within clusters, they will occasionally pass very near one another or even collide directly. The effect this has on the individual galaxies depends critically on the details of the encounter. Possible outcomes include loss of material from the galaxies' outer regions, transfer of material from one galaxy to another, merger of the two galaxies, modification of the galaxies' forms by tidal perturbations, and loss of gas and dust due to collisional heating. Alar and Juri Toomre have explained the 10 or so most peculiar observed forms of galaxies as tidal perturbations during grazing collisions.

Fig. 4. Clustering characteristics of galaxies. This association of galaxies in the constellation Hercules includes many different types of galaxies. (*California Institute of Technology/Palomar Observatory*)

Active nuclei. In the very central regions (sizes at least as small as a light-year, 6×10^{12} mi or 1×10^{13} km) of galaxies, violent and apparently explosive behavior is often observed. This activity is manifested in many ways, including the high-velocity outflow of gas, strong nonthermal radio emission (implying relativistic particles and magnetic fields), intense and often polarized and highly variable radiation at infrared, optical, ultraviolet, and x-ray wavelengths, and ejection of jets of relativistic material. In the most extreme cases the energy in the nuclear activity surpasses that in the rest of the galaxy combined.

These phenomena are generically referred to as nuclear activity, and the objects that exhibit them are called active galactic nuclei. All of these phenomena are interpreted as indications of the presence of relatively small but extremely powerful energy sources located at the centers of some galaxies. The physical nature of this energy source is generally supposed to

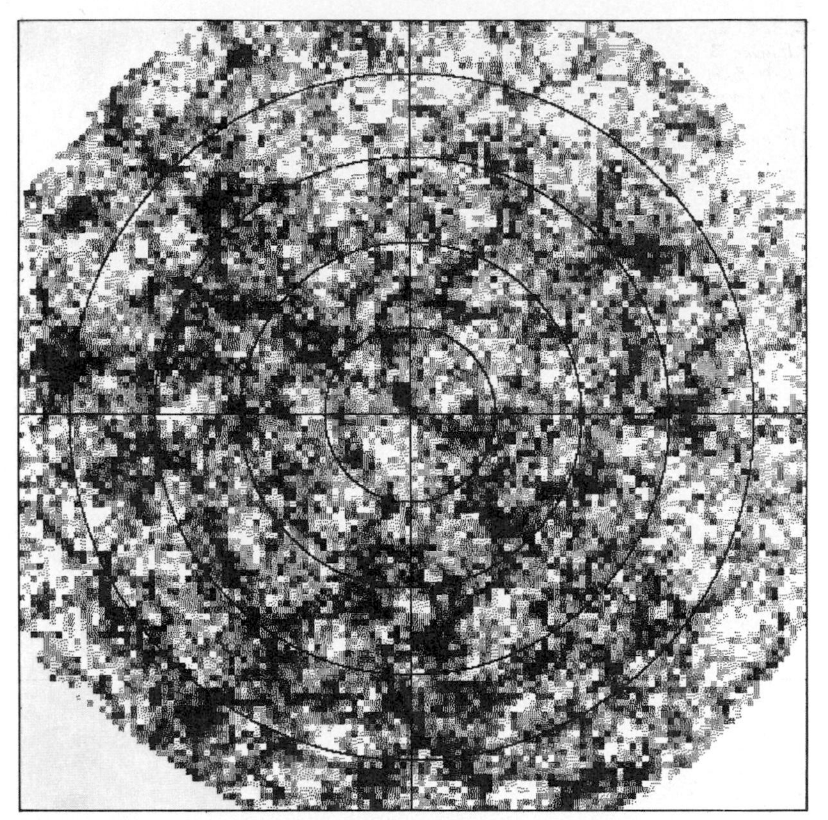

Fig. 5. Computer-generated map of the distribution of nearly a half-million galaxies. Regions of higher galaxy counts, indicated by more darkly shaded pixels, appear to have a filamentary structure. (*From J. E. Moody, E. L. Turner, and J. R. Gott, Filamentary galaxy clustering: A mapping algorithm, Astrophys. J., 273:16–23, October 1, 1983*)

be very different from that of the stars and other objects making up the rest of the galaxy, and the study of active galactic nuclei is, to a considerable extent, pursued separately from the study of galaxies in general.

One of the most prominent characteristic features of active galactic nuclei is the ejection of large masses of high-temperature gas at great velocities. Characteristic temperatures and velocities are in the range of tens to hundreds of thousands of kelvins and thousands to over 10,000 mi/s (16,000 km/s). Total gas masses exceeding 1×10^6 times that of the Sun may be involved. These tremendous and powerful gas flows reveal themselves as bright and broad spectral lines due to the emission of radiation by atoms of common elements at specific wavelengths (that is, colors). The detection of these emission lines is one of the primary empirical indications of nuclear activity; galaxies displaying such lines are referred to as Seyfert galaxies.

A second characteristic feature of active galactic nuclei is the emission of radiation over a wide range of different wavelength bands. An object such as an ordinary star, which emits radiation because it is hot, does so in a characteristic wavelength band. Such radiation is called thermal, and the characteristic wavelength is determined by the object's temperature, with cooler objects radiating predominantly at longer wavelengths. The typical active galactic nucleus emits a quite different sort of radiation, called nonthermal radiation, implying that it is produced by a quite different mechanism. This feature of active galactic nuclei means that astronomers can study them at essentially all wavelength bands, from the radio band out through the infrared, optical, ultraviolet, and x-ray bands and into the gamma rays.

In addition to the wide range of wavelengths covered by radiation from active galactic nuclei, there are other indications that the radiation is of an unusual, nonthermal origin. Chief among these is the observation that the brightness of active galactic nuclei often changes dramatically over relatively (by astronomical standards) short periods of time. Most active galactic nuclei show moderate changes (of a few tens of percent) over time scales of months to years, and some show dramatic variations (by factors of 2 to 10 or more) over times ranging down to a few hours. There is a general tendency for variations to be more rapid for higher-energy (shorter-wavelength) radiation. Since sources of radiation cannot generally change their brightnesses in times much shorter than that required for light to travel across them, these variations imply that the radiation from active galactic nuclei arises in a very small region, in extreme cases no larger than the solar system. This is a fantastically small volume when one considers that the total radiative power output can rival or exceed that of an entire galaxy many tens of thousands of light-years across.

Another important and peculiar phenomenon exhibited by active galactic nuclei is the emission of highly directional jets of relativistic plasmas and magnetic fields. These jets contain elementary particles, particularly electrons, moving at velocities near the speed of light; they are like the high-energy particles produced by particle accelerators on Earth or found in the cosmic rays impinging on the top of the Earth's atmosphere. These jets of material often extend far outside the nucleus and even beyond the whole body of the galaxy. These jets give rise to some of the nonthermal radiation, notably the radio-wavelength radiation, when the fast-moving electrons in the plasma spiral around in the magnetic fields that are embedded in the flows.

There are a variety of classes of active galactic nuclei. The Seyfert galaxies referred to above display the broad emission lines produced by the rapid outflow of hot gas but frequently do not exhibit much radio-wavelength emission. Another complementary class shows strong radio emission but weak or absent emission lines. Yet another class (BL Lac objects, often referred to as blazars) also shows only weak emission lines but is often extremely variable. When active galactic nuclei achieve such great luminosities that they dominate that of the rest of the galaxy, they are sometimes referred to as AGNs. Quasars are widely believed to be the most extreme sort of active galactic nuclei, having emission so intense that the ordinary galaxy in which they exist is entirely lost in the glare of the nuclear emission. It appears that active galactic nuclei in general, and certainly quasars, were much more common during the early history of the universe than they are at present. If so, most or all large galaxies may contain the burned-out remnant of an active nucleus in their centers, although only about 1% still show detectable activity. *See Quasar.*

Perhaps the most intriguing question concerning active galactic nuclei is that of the nature of the energy source that drives all of their diverse phenomena. It is clear that many of the features that define

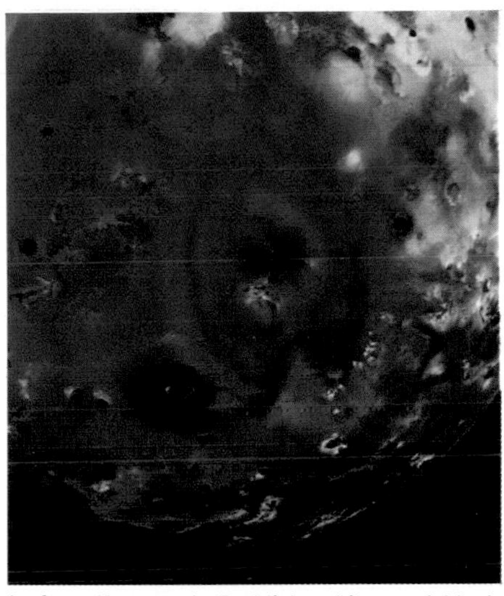

Io from *Voyager 1*. Reddish, white, and black areas are probably surface deposits. Many of the black spots are associated with volcanic craters. (*NASA*)

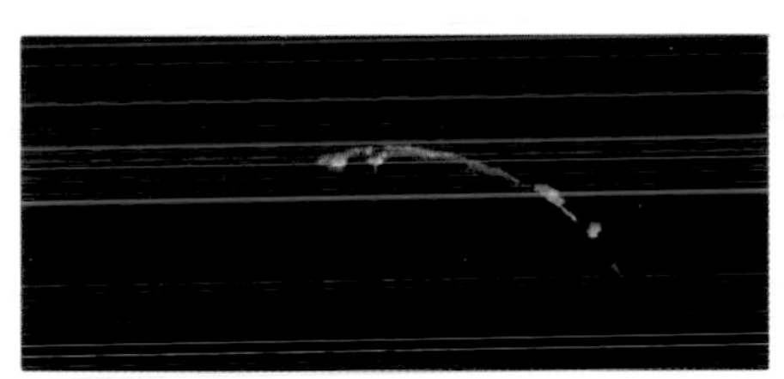

Limb of Io from *Voyager 2*, showing two blue volcanic eruption plumes about 100 km high. (*NASA*)

Jupiter from *Voyager 1*, with the Great Red Spot showing prominently below the center and Ganymede at the lower left. (*NASA*)

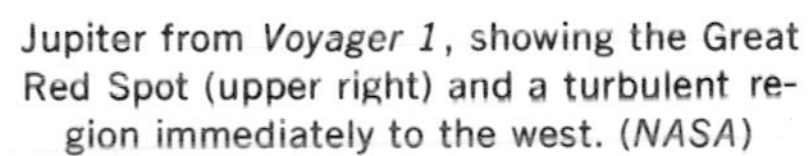

Jupiter from *Voyager 1*, showing the Great Red Spot (upper right) and a turbulent region immediately to the west. (*NASA*)

active galactic nuclei are little more than indirect symptoms of the injection of enormous amounts of energy into the material at the centers of ordinary galaxies and that this energy must have some quite remarkable source. Unfortunately, there is no certain answer to this important question. There is, however, a widely accepted best guess or consensus model that at least appears to be consistent with all that is known about active galactic nuclei.

The basic idea of this best-guess model is that active galactic nuclei are powered by the energy released when matter falls into a massive black hole occupying the center of a galaxy. These black holes are imagined to have masses in the rough range of a million to a billion solar masses and to have formed because of the high density of material expected to accumulate at the center of a galaxy due to its gravitational field. Such a black hole will continue to accrete any gas that finds its way into the vicinity. As such gas falls toward the black hole, its angular momentum will cause it to take up a nearly circular orbit in a disk of material surrounding the black hole. This disk (called an accretion disk) will slowly inject gas into the black hole somewhat like water in a tub flowing out through a drain. As the gas approaches the black hole, the latter's enormous gravitational field will compress and heat the gas to very high temperatures, causing it to radiate. A given mass of gas can release 10 or more times as much energy in this way as it could if it were used as nuclear fuel in a star or a reactor. The mechanisms that convert the thermal radiation generated in this way into the nonthermal radiation and relativistic plasmas observed in active galactic nuclei are not well known. The most plausible scenarios invoke interaction with magnetic fields and confinement of the radiation into collimated beams by the accretion disk itself to account for this conversion. SEE BLACK HOLE.

Whatever the correct explanation for the energy release in active galactic nuclei, it is certain that they represent the most violent and energetic events known in the universe. Understanding their nature, evolution, origin, connection to the galaxies they inhabit, and interaction with other objects in the universe remains one of the outstanding problems in extragalactic astronomy. SEE ASTROPHYSICS, HIGH-ENERGY.

Distances. The nearest galaxies to the Milky Way Galaxy are the Large and Small Magellanic Clouds, two small irregulars lying about 150,000 light-years (9×10^{17} mi or 1.5×10^{18} km) away. They are visible to Southern Hemisphere residents as two faint patches of light in the constellations of Norma and Toucan. The nearest bright galaxy is the Andromeda Nebula, M31 (Fig. 1), visible to the naked eye in the constellation of the same name. It is at a distance of about 2×10^6 light-years (1.2×10^{19} mi or 1.9×10^{19} km). These galaxies and a handful of fainter ones (all within a few million light-years) make up the Local Group, of which the Milky Way Galaxy is a member. SEE ANDROMEDA GALAXY; LOCAL GROUP; MAGELLANIC CLOUDS.

Within a distance of about 6×10^7 light-years (3.5×10^{20} mi or 6×10^{20} km), there are hundreds of galaxies that have large total luminosities but that are too far away to be seen with the unaided eye. Many of these can be seen or photographed with a small telescope. Some of the most spectacular of these are contained in Messier's famous list (M51, M81, M82, M87, and so on). Some are members of the nearest significant cluster of galaxies in the constellation Virgo. SEE MESSIER CATALOG.

The Coma cluster is the best-observed large cluster of galaxies. Its distance of about 3×10^8 light-years (1.8×10^{21} mi or 3×10^{21} km) requires large (usually professional) telescopes. The most distant galaxies detectable with the largest telescopes are nearly 10^{10} light-years (6×10^{22} mi or 1×10^{23} km) away.

Cosmology. Galaxies are closely associated with cosmology because they are the beacons by which the universe is mapped. In 1929, Edwin Hubble discovered that the universe is expanding by showing that galaxies are receding from the Milky Way Galaxy at a speed v proportional to their distance d ($v = H_0 d$, where H_0 is called the Hubble constant). The Hubble relation in turn led to the conclusion that the universe was formed in a big bang about 1×10^{10} years ago. SEE BIG BANG THEORY; HUBBLE CONSTANT.

Studies of very distant galaxies can, in principle, reveal whether or not the mutual gravitational attraction is sufficient to stop and then reverse the Hubble expansion. Despite intense efforts by a number of astronomers, this goal has not been clearly achieved.

The same problem can be attacked by attempting to estimate the density of the universe from a study of relatively nearby galaxies. If the density exceeds a certain critical value (2×10^{-29} g · cm^{-3}), the universe will eventually stop expanding and recollapse. The best available data indicate that the density is a few tenths of the critical density. Thus, barring large amounts of unobserved mass, the universe will continue to expand forever.

Edwin L. Turner; Joyce B. Turner

Evolution. How these diverse objects originated and evolved into their present form is a topic of intense speculation among astrophysicists.

Evidence from structural properties. Some clues can be discerned in certain structural properties of galaxies. One of the most regular forms is the smooth and round light distribution of an elliptical galaxy. Astronomers calculate than an inhomogeneous distribution of stars will take many billions of years to relax into such a uniform pattern. In fact, the most plausible explanation is that the stars formed out of a collapsing gas cloud. The rapidly changing gravitational pull experienced by different stars as the collapse proceeds has been shown by means of sophisticated computer experiments to rearrange the stars into the observed shape of an elliptical galaxy.

The highly flattened disks of spiral galaxies must have formed during a similar collapse, but it is believed that most star formation did not occur until the rotating gas cloud had already flattened into a pancakelike shape. Had the stars formed at an earlier stage of the collapse, their rapid motions would have led to the formation of an elliptical galaxy. However, if the cloud stays gaseous until it flattens, much of the kinetic energy of the collapse is radiated away by gas atoms. Subsequent star formation is found to maintain a highly flattened, disklike shape, characteristic of spiral galaxies.

The flattening occurs in part because of the centrifugal forces in the rotating cloud. A confirmation of this picture has come from the discovery that elliptical galaxies rotate much less rapidly than spiral galaxies. This raises the question of the origin of the rotation itself. A natural explanation seems to lie in the action of the gravitational torques exerted by neighboring galaxies (or rather, protogalaxies: galaxies in the process of formation). Much as the Moon and Sun cause tides in the Earth's ocean, so can a massive neighbor induce a protogalaxy to begin to rotate.

Evidence from composition. Another aspect of galaxies that has evolutionary significance is their composition, and in particular, the actual distribution of heavy elements. The amount of heavy-element enrichment can be inferred by measuring the color of the starlight, blue stars being metal-rich. Galaxies are found to be significantly bluer in their outermost regions and redder toward their central nuclei. Such gradients in color can have developed only during the galaxy formation stage, since most of the starlight from a galaxy displays this trend. The explanation seems to be that galaxies formed out of collapsing gas clouds that formed stars in a piecemeal fashion. As stars formed, they evolved, underwent nuclear reactions, produced heavy elements, and eventually shed enriched material (some stars even exploding as supernovae). Successive generations of stars formed out of the debris of earlier stars, and in this way the stellar content of galaxies systematically became enriched. The greatest enrichment would naturally occur toward the center of a galaxy, where the gaseous stellar debris tended to collect. *See Nucleosynthesis.*

In this manner, theories of galactic evolution have been constructed that explain in outline many of the observed characteristics of galaxies. Detailed models of protogalaxies have even been developed, but because of the very limited knowledge of the fundamental process of star formation, quantitative explanations for most properties of the galaxies have not yet been found.

Origin. An outstanding and unresolved issue concerns the origin of the primordial gas clouds out of which the galaxies evolved. As scientists attempt to look back into the early stages of the universe, the view becomes increasingly obscure. The cosmic background radiation yields a glimpse of the universe prior to the epoch of galaxy formation. This radiation has a blackbody spectrum that corresponds to a temperature of only 3 kelvins (5 Fahrenheit degrees) above absolute zero. It is the remnant of the primeval fireball radiation, created in the early stages of the big bang, when the universe was less than a year old. The universe is now completely transparent to the cosmic background radiation. However, at an epoch corresponding to about 500,000 years after the big bang, the radiation was sufficiently hot that matter was ionized, and the matter was also sufficiently dense to render the universe completely opaque to the radiation. To look out to greater and greater distances is also to look back in time. The cosmic background radiation originates in the most distant regions of the universe that can be "seen" with optical or radio telescopes. To observe the background radiation now (some 10^{10} years later) is to see back to this early epoch, known as the decoupling epoch: at earlier times, matter and radiation were intimately linked, and subsequently the radiation propagates freely to the present time. (However, it is possible that after galaxies have begun to form, there may be sufficient ionizing radiation produced to reionize the intergalactic medium and cause additional scattering of the cosmic background radiation: this could happen at epochs as late as 10^9 years after the big bang.) *See Cosmic background radiation.*

Theory of formation from fluctuations. The cosmic background radiation is found by radio astronomers to be very uniform. This indicates that the matter distribution in the early universe also was uniform, to at least 1 part in 10,000. However, it cannot have been completely without any structure; otherwise galaxies could not have formed. This apparent paradox is resolved by the assumption that small inhomogeneities or fluctuations in the matter distribution were present in the early universe. The mutual action of gravity exerted between these infinitesimal fluctuations results in their gradual enhancement. Eventually, great gas clouds develop that will collapse to form galaxies. Thus the seeds from which galaxies grew by the action of gravitational instability are infinitesimal density fluctuations in the very early universe. The required amplitude for these primordial seed fluctuations must be of the order of 1 part in 10,000.

Numerical simulations of galaxy clustering have enabled the spectrum of fluctuation length-scales and amplitudes to be inferred. It seems likely that these density inhomogeneities were distributed in a random fashion, not unlike that of noise or turbulence. However, the source of the fluctuations remains a mystery. One possibility is that the fluctuations developed from random statistical fluctuations.

The initial stages of the big bang may have been characterized by a period of rapid inflation during the first 10^{-35} s of the expansion. The inflation is triggered by an energy source that can be regarded as similar to a latent heat arising at the transition between a symmetrical state of matter and the asymmetrical universe that now exists, where matter predominates over antimatter. This phase transition is predicted to occur by the grand unification theories of particle physics, which unites the strong, weak, and electromagnetic forces when particle energies attain a level in excess of 10^{14} GeV. One of the consequences of an inflationary epoch is that quantum-statistical fluctuations are amplified up the scales of galaxies and of clusters of galaxies. While current theories of inflationary cosmology appear to predict a prohibitively large level of fluctuations, the fact that fluctuations arise naturally in an initially uniform universe suggests that the most simple of cosmologies may contain the nascent seeds of future galaxies. *See Inflationary universe cosmology.*

Theory of chaotic early universe. According to an alternative viewpoint, the early universe was extremely irregular and chaotic. In this case, cosmologists appeal to physical processes to remove the inhomogeneities in a chaotic universe rather than to create mild fluctuations in a uniform universe, on the grounds that a highly irregular initial state is more likely than the very specific state of the idealized uniform big bang cosmology. This program has not yet met with success, in part because highly inhomogeneous cosmological models are not readily amenable to study. It seems inevitable, however, that primordial black holes form in a chaotic universe, and these could be the nuclei around which galaxies eventually are accreted. There remains the important constraint on this hypothesis of accounting for the isotropy of the cosmic background radiation, for the processes that have yielded structure on galactic scales have evidently conspired to leave the matter and radiation distribution exceedingly uniform on much larger scales. Cosmologists conclude that both approaches, that of the relatively smooth early universe and that of primordial chaos, are confronted with difficulties, and they cannot yet choose between these alternatives.

Isothermal and adiabatic fluctuations. The most notable success of studies of fluctuations in the early universe has been the realization that fluctuations can be categorized into distinct varieties. Of particular importance for galaxy formation are density fluctuations that are found to generally be a combination of two basic types: isothermal and adiabatic. Primordial isothermal fluctuations consist of variations in the matter density, without any corresponding enhancement in the radiation density. Consequently, in the radiation-dominated early phase of the big bang, isothermal fluctuations neither grow nor decay, as the uniform radiation field prevents any motion. Once the universe becomes transparent, the matter fluctuations respond freely to gravity and grow if above a certain critical size. This critical scale (the Jeans length) represents the balance point between attractive gravitational and expansive pressure forces. The smallest isothermal fluctuations that can become enhanced and form gas clouds contain about 10^6 solar masses. On the other hand, primordial adiabatic fluctuations are analogous to a compression of both matter and radiation. The diffusive tendency of the radiation tends to smooth out the smaller adiabatic fluctuations. This process remains effective until the decoupling epoch, and only adiabatic fluctuations that contain upward of 10^{12} solar masses can survive to eventually recollapse into gas clouds and galaxies.

Two alternative schemes are possible, depending on the masses of the first bound structures to form. If small scales of order 10^6 solar masses are the first to develop, then galaxies form hierarchically from clustering of smaller structures. The initial clumps correspond to the masses of globular star clusters, known to contain the oldest stars and speculated to be possible building blocks for the luminous cores of galaxies. If large scales of order 10^{12} to 10^{15} solar masses collapse first, pressure forces are negligible, and random fluctuations in the initial eccentricity of collapsing clouds are amplified. Highly flattened pancakes form, leaving large voids behind, and the compressed sheets of matter fragment to form galaxies. Astronomers have found evidence for highly elongated filaments and sheets of galaxies, as well as giant holes in the galaxy distribution, that lend credence to the pancake theory.

Conceivably, both types of fluctuations may have been present in the early universe. Discovery of small fluctuations in the cosmic microwave background radiation, required at some level by all theories of galaxy formation, should eventually allow a choice between the alternative possibilities. *See* Cosmology.

Joseph Silk

Bibliography. J. Binney and S. Tremaine, *Galactic Dynamics*, 1987; S. M. Fall and D. Lynden-Bell (eds.), *The Structure and Evolution of Normal Galaxies*, 1981; T. Ferris, *Galaxies*, 1980; E. R. Harrison, *Cosmology*, 1981; D. Mihalas and J. Binney, *Galactic Astronomy: Structure and Kinematics*, 1986; A. Sandage, *The Hubble Atlas of Galaxies*, 1961; J. Silk, *The Big Bang*, 1980; J. Silk, A. S. Szalay, and Ya. B. Zel'dovich, The large-scale structure of the universe, *Sci. Amer.*, 249(4):72–80, 1983.

Gamma-ray astronomy

The study of gamma rays of cosmic origin. The field may be divided, according to the energy of the gamma rays, into the study of low-energy (or soft) gamma rays (100 keV up to a few megaelectronvolts); high-energy gamma rays (about 30 MeV up to a few gigaelectronvolts); and very high-energy gamma rays (greater than 3×10^{11} eV). *See* Gamma rays.

Because photons in the gamma-ray regime are completely absorbed by the Earth's atmosphere, it is necessary to place the gamma-ray detectors aboard high-altitude balloons, or better still, artificial satellites. Most of the pioneering work on extraterrestrial gamma rays was done with balloons, starting about 1965; although technical constraints are far more severe, mainly satellites are used now. Many of the results given below were obtained by a first generation of gamma-ray telescopes and spectrometers, including the United States *SAS 2* satellite launched in November 1972, the European *COS-B* satellite launched in August 1975, the scanning experiment on board the United States *HEAO 1* spacecraft launched in June 1977, the high-resolution spectrometer on board the United States *HEAO 3* spacecraft launched in September 1979, and the spectrometer on board the *Solar Maximum Mission (SMM)* satellite launched in February 1980. After a gap of several years, caused in particular by the *Challenger* disaster, an unprecedented series of scientific results has been obtained by new experiments with much greater capabilities, including the French SIGMA telescope aboard the Soviet (now C.I.S.) spacecraft *GRANAT*, launched in December 1989; and the large United States *Compton Gamma-Ray Observatory*, launched from the space shuttle in April 1991. *See* Satellite astronomy.

Telescope limitations. The information obtained by gamma-ray telescopes is limited by various factors, although these limitations should become less restrictive with the development of more advanced detectors. These limitations are both astronomical and instrumental in nature.

Astronomical limitations. The astronomical limitations of gamma-ray telescopes result from the weak intrinsic fluxes of gamma rays and from background problems. The photon fluxes of the brightest gamma-ray sources range from 10^3 m^{-2} s^{-1} at 100 keV to less than 10^{-1} m^{-2} s^{-1} at 100 MeV; as a consequence, a typical gamma-ray observation requires long exposures, up to 1 month in duration, in order to obtain significant data. The main background sources originate outside the instrument: cosmic-ray particles interact with it to give secondary gamma rays. Because the observations are limited by background noise, sensitivity improves only in proportion to the square root of the detector area.

Instrumental limitations. In other regions of the electromagnetic spectrum, sensitivity is increased by the straightforward method of gathering large numbers of photons and concentrating them to form an image, by means of suitable arrangements of reflectors or lenses. This method cannot be used with gamma-ray telescopes, since gamma-ray photons can be neither reflected nor refracted.

Low-Energy Gamma-Ray Astronomy

When the medium is optically thick, the emerging blackbody radiation lies in the soft gamma-ray region if the temperature is sufficiently high. Since the luminosity cannot much exceed the Eddington limit (the level where radiative forces balance that of gravity), only very small celestial bodies, such as neutron stars or stellar-mass black holes, would produce a copious amount of gamma-ray photons. This limitation is no longer effective if the medium is optically thin: nonthermal particles can emit high-energy radiation in a cold, optically thin medium.

Continuum emission. Strongly accelerated fast-moving electrons radiate soft gamma-ray photons. The acceleration is induced by an electric field or a magnetic field. A high-energy electron may also interact with a low-energy photon. In such a case, the photon energy is generally multiplied by the square of the electron Lorentz factor. *SEE BREMSSTRAHLUNG*.

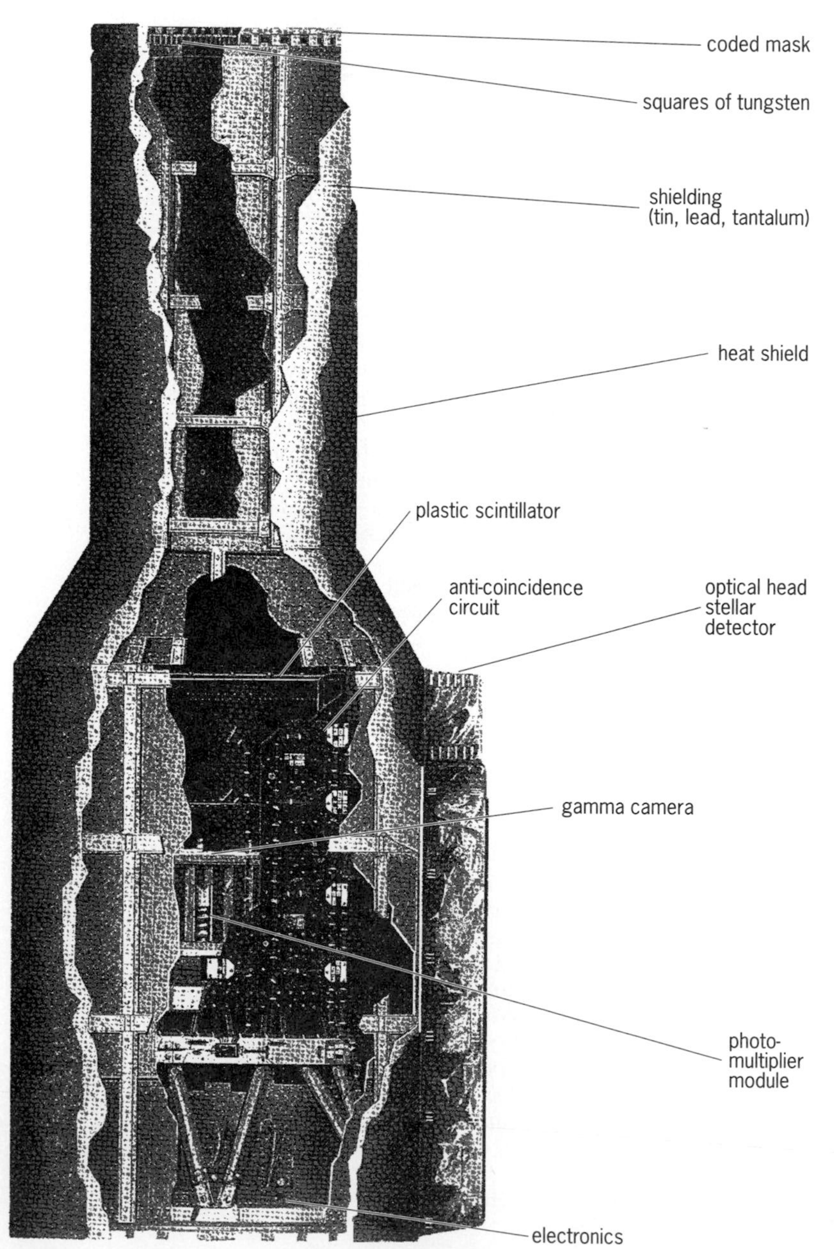

Fig. 1. **Exploded view of the French SIGMA telescope aboard the Soviet (now C.I.S.) *GRANAT* satellite. (*After J. Paul et al., SIGMA: The hard x-ray and soft gamma-ray telescope on board the GRANAT space observatory, Adv. Space Res. 49(8):289–302, 1991*)**

Gamma-ray lines. In the soft gamma-ray regime, the photon energies are of the same order as the binding energies of nuclei, and photons may be emitted by nuclei in a fashion analogous to optical photons emitted by an atom. Processes such as excitation or deexcitation of nuclei exist, giving rise to the emission of gamma-ray lines and therefore to the possibility of nuclear spectroscopy. The 511-keV gamma-ray line is not, strictly speaking, of nuclear origin, but is produced by annihilation between electrons and positrons (positive electrons) that are an essentially unavoidable by-product of all high-energy astrophysical processes.

Low-energy gamma-ray telescopes. Since the early observations of gamma-ray astronomy, telescopes operating in the soft gamma-ray domain, where focusing techniques become totally impracticable, have used the combination of wide-angle (a few degrees) collimation and on-off source chopping to yield source fluxes and location. Because of the poor angular resolution of these techniques, the firm identification of discovered sources with known astronomical objects was often possible only in cases where the emission had a clear time signature.

Coded-aperture telescopes. At the beginning of the 1980s, it was recognized that one possible means of improving existing soft gamma-ray telescopes is the incorporation of the coded-aperture technique to actually image celestial sources. A coded mask is a pattern of elements that absorb gamma-ray photons. In the case of gamma-ray telescopes, it is made of parallellepiped tungsten blocks, arranged so that a given point source at infinity projects on a position-sensitive detector a pattern that is characteristic of the direction of arrival of the photons. The actual position of a source in the sky is determined by comparing the observed pattern with all possible projected patterns. The primary advantage of such a technique is to maintain the angular resolution of a single pinhole camera, while increasing the overall effective area of the instrument. Moreover, the coded-mask principle includes the simultaneous measurement of the sky plus detector background; systematic effects due to temporal variations in the background are minimized. Such a soft gamma-ray coded-aperture telescope (**Fig. 1**) features a coded mask located few meters from a position-sensitive detector and active and passive shielding devices to limit the field of view and to veto the incoming charged particles. The intrinsic angular resolution is about 15 arc-minutes for SIGMA, while the point-source location accuracy is about 1 arc-minute in the case of the brightest sources.

Spectrometers. Gamma-ray spectrometers are generally composed of arrays of sodium iodide (NaI) scintillation detectors plus active and passive collimation devices. The energy resolution of these first-generation experiments is quite poor, making difficult the actual detection of lines. Semiconductor germanium (Ge) detectors offer very high energy resolution; for example, the resolution (full width at half maximum) at 1 MeV is about 2 keV for germanium, compared with about 60 keV for sodium iodide. However, germanium detectors require cooling to temperatures of less than 100 K (−280°F), and whereas many germanium detectors have been flown from balloons, so far the C3 experiment aboard the *HEAO 3* spacecraft has been the only space experiment featuring a germanium detector.

Sun. Low-energy gamma-ray astronomy, as applied to observations of the Sun, has the specific objective of investigating high-energy processes that take place in the Sun's atmosphere and the relationship of these phenomena to the basic problems of solar activity. However, background problems seriously limit the gamma-ray flux that can be observed; with presently available gamma-ray detection systems, solar gamma-ray emission can be detected only during periods of solar flares.

The first results were obtained during two flares in August 1972, when both discrete-line and continuum gamma-ray emissions were observed. All the lines except the 2.223-MeV line from deuterium formation are attributable to interactions with energetic protons and positron–electron interactions. The measurement of the 511-keV emission implies that the density of regions where positron annihilation takes place is greater than 10^{12} electrons per cubic centimeter. The fact that the 2.223-MeV line was observed implies the presence of a significant thermal neutron flux, not absorbed by other processes. Strong lines at 4.438 and 6.129 MeV can be attributed to inelastic scattering of high-energy protons by carbon and oxygen nuclei respectively. Solar gamma-ray emissions have been observed by the *Solar Maximum Mission* satellite and the Japanese *Hinotori* satellite. High-energy emissions are currently monitored by the *Compton Gamma-Ray Observatory*. SEE SUN.

Galactic center. There have been many attempts to observe the galactic center region in the soft gamma-ray band, since this is one of the few windows available for observations through the intervening gas and dust. However, unlike radio, infrared, and x-ray images, early observations performed in the soft gamma-ray band gave a somewhat confused picture of the sky in a 100–200°-square area around the galactic center. This confusion was partly due to the difficulty in achieving high spatial resolution and partly due to the high degree of time variability of the sources. Observations by scanning instruments have led to the discovery of gamma-ray emission from the general direction to the galactic center. The detected emission, with spectrum similar to that of active galactic nuclei, was tentatively attributed to a 10^6-solar-mass massive black hole, which was supposed to be in the dynamic center of the Milky Way Galaxy.

Compact sources. Thanks to the new generation of soft gamma-ray imaging instruments, a clarified picture of the galactic center region is now emerging. Two distinct types of soft gamma-ray sources have been found so far in the galactic center region. The first class includes transient sources, while the second class consists of three persistent, yet strongly variable sources, which alternately dominate the whole galactic center region. In addition to the well-known x-ray pulsar GX 1+4, which manifests sporadic hard emissions at energies beyond 100 keV, two genuine soft gamma-ray sources were discovered by the SIGMA telescope aboard the Soviet spacecraft *GRANAT* (**Fig. 2**). The brightest one, located only 50 arc-minutes from the galactic center, was identified with the faint x-ray source 1E 1740.7-2942. The other source, GRS 1758-258, a newly discovered soft gamma-ray source, is located only about 40 arc-minutes from the well-known x-ray source GX 5-1. Day-to-day variabilities are apparent in the light curves of these two sources, indicating that both are compact with the

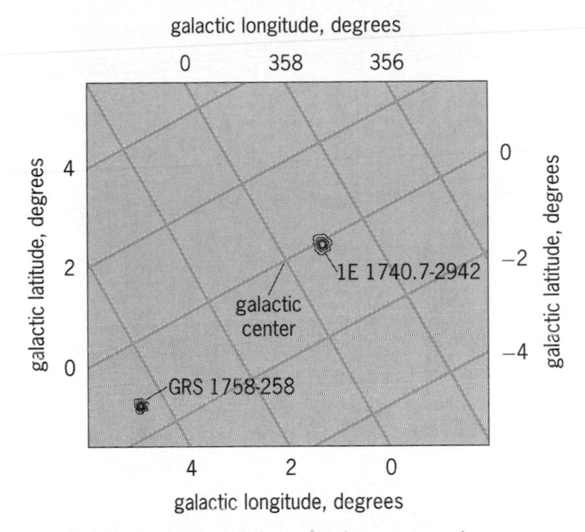

Fig. 2. Contour chart of the galactic center region obtained by SIGMA in September-October 1990, in the 40–150-keV energy band. (*After B. Cordier, Etude du centre galactique dans le domaine des rayons gamma de faible énergie à partir des observations pratiquées par le télescope spatial SIGMA, Ph.D. Thesis, Paris VII University, 1992*)

size of the emission region being less than 1.6×10^{10} mi (2.6×10^{10} km). Their mean spectra are well fitted by a comptonized disk model, with parameters close to those derived in fitting the spectrum of the black-hole candidate Cygnus X-1. These spectra, as well as the long-term temporal behavior of 1E 1740.7-2942 and GRS 1758-258, characterized by at least two distinct states of soft gamma-ray emission, make these two sources galactic black-hole candidates. No soft gamma-ray emission is detected from the dynamic center of the Milky Way Galaxy. SEE ASTROPHYSICS, HIGH-ENERGY; BLACK HOLE.

Observations of the galactic center region by wide-field-of-view high-resolution spectrometers have also suggested the existence of a powerful, compact, and variable source of 511-keV positron-annihilation line radiation within a few degrees of the galactic center. The identification of such a positron-annihilation source with 1E 1740.7-2942 is suggested by the SIGMA observation performed on October 13–14, 1990, during which 1E 1740.7-2942 underwent a dramatic outburst at energies beyond 200 keV, whose spectral shape (**Fig. 3**) probably indicates the presence of a relativistic plasma consisting of electrons and positrons.

Diffuse emission. Spectroscopic observations of the galactic center region have demonstrated the presence of a diffuse component of 511-keV radiation due to the annihilation of positrons in the interstellar medium. The results obtained by the Oriented Scintillation Spectrometer Experiment (OSSE) on the *Compton Gamma-Ray Observatory* suggest that the diffuse 511-keV line emission is concentrated near the galactic center. The source of such a diffuse galactic component is thought to be the beta-decay products from radioactive nuclides produced by novae, red giants, Wolf-Rayet stars, and supernovae.

The discovery and confirmation of the 1.809-MeV line emission from the galactic plane in the region of the galactic center was a milestone for gamma-ray astronomy, being the first direct detection of a cosmic

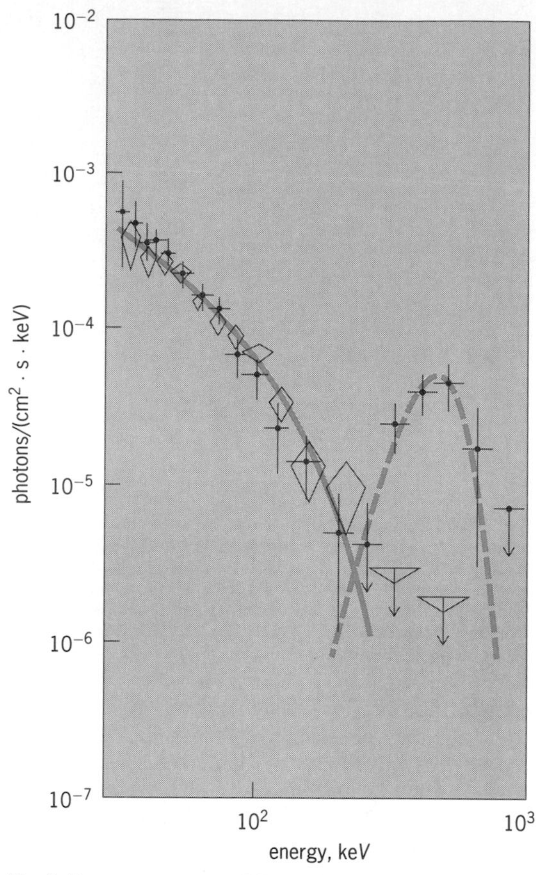

Fig. 3. Energy spectrum of 1E 1740.7-2942 as obtained by SIGMA on October 13–14, 1990 (crosses); the spectrum derived from March-April 1990 (diamonds) is shown for comparison. The most straightforward modelization of the October 13–14, 1990, spectrum consists of superposing a single wide line with a gaussian profile (broken line) onto the comptonized disk emission spectrum (solid line). (*After L. Bouchet el al., SIGMA discovery of variable e^+-e^- annihilation radiation from the near galactic center variable compact source 1E 1740.7-2942, Astrophys. J., 383:L45–L48, 1991*)

radioactive nucleus (aluminum-26) from the characteristic signature of its decay. As the half-life of aluminum-26 (716,000 years) is very short compared to the time scale of galactic chemical evolution, the 1.809-MeV line observations clearly demonstrate that nucleosynthesis is currently taking place in the Milky Way Galaxy. The detected flux (approximately 6×10^{-4} photons cm^{-2} s^{-1} from the central radian of the Milky Way Galaxy) is, however, several times higher than that expected from the decay of the whole production of all the candidate sources of aluminum-26 (novae, red giants, Wolf-Rayet stars, and supernovae). *SEE GIANT STAR; MILKY WAY GALAXY; NOVA; NUCLEOSYNTHESIS; SUPERNOVA; WOLF-RAYET STAR.*

Supernova SN 1987A. Major contributions to the theoretical understanding of stellar evolution and nucleosynthesis have come from the data obtained on supernova SN 1987A, which appeared on February 24, 1987, in the Large Magellanic Cloud, a nearby galaxy, only 135,000 light-years (8×10^{17} mi or 1.3×10^{18} km) from the Milky Way. In particular, in late stages of the evolution of the supernova, several months after the explosion, the envelope became tenuous enough to be transparent to the soft gamma rays resulting from the decay of several radioactive isotopes produced just before or during the explosion. The detection by the *Solar Maximum Mission* and various balloon-borne spectrometers of the 0.847- and 1.238-MeV lines from the decay of cobalt-56, slightly preceded by the soft gamma rays of the corresponding comptonization continuum, confirmed that nickel-56 (decaying into cobalt-56 in about 7 days) was the main power source of the supernova remnant. Their behavior as a function of time gives unprecedented constraints on nucleosynthesis models and on the mixing of freshly synthesized material deep inside the supernova envelope. *SEE MAGELLANIC CLOUDS.*

Jacques Paul

Gamma-ray bursts. Gamma-ray bursts are among the most mysterious and enigmatic phenomena of modern astronomy. It is generally believed that these powerful releases of high-energy photons come from neutron stars, but the mechanism, the distance of the sources, and thus the amount of energy released are very unsettled questions. Gamma-ray bursts were first detected by means of the *Vela* spacecraft in 1967, and their discovery was announced in 1973.

More than a few hundred bursts have been detected with the help of the *Compton Gamma-Ray Observatory*. Its observations considerably strengthen the evidence for an isotropic distribution of bursts; that is, the burst directions appear to be random and show no preference for the galactic plane, in particular. Furthermore, the distribution of intensities implies a falloff in source numbers at the greatest distances. This finding seems to rule out the galactic plane origin preferred by many theorists. Thus, the nearest reasonable source location is probably in a halo surrounding the Milky Way Galaxy; some astrophysicists even favor cosmological distances. The puzzle of the origin of the bursts thus remains unsolved.

Burst of March 5, 1979. The most intense gamma-ray burst ever observed (GB790305) was detected by nine spacecraft on March 5, 1979. It was fortunate that there were so many operational spacecraft able to detect it: three *Vela* spacecraft and *Prognoz 7* near the Earth; *Helios 2*, *ISEE 3*, and *Venera 11* and *12* at considerable distances from the Earth; and *Pioneer Venus Orbiter* in orbit around Venus. The burst rose to its peak strength in less than a millisecond, which allowed the source location to be determined from the detection times for the widely separated spacecraft to an accuracy of less than an arc-minute. By precise timing of the gamma-ray peak as it arrived at the various spacecraft, with differences up to 25 s, the location of the source was determined to be, in galactic coordinates, $l = 276.09°$, $b = -33.24°$. The direction coincided with that of N49, a supernova remnant (**Fig. 4**) in the Large Magellanic Cloud. This was the first time that a gamma-ray burst location error box of small size had included any obvious candidate for the source. If this gamma-ray burst occurred at the distance of the Large Magellanic Cloud, its peak intensity amounted to more than 10^{44} ergs/s (10^{37} W), much larger than had been considered reasonable. Theorists disagree as to whether this event represented a breakthrough in understanding gamma-ray bursts, an untypical gamma-ray burst, or an unfortunate coincidence in direction between an inconspicuous nearby source and a conspicuous galaxy (Large Magellanic Cloud) in the Local Group. *SEE LOCAL GROUP.*

To the astonishment of the various teams of observers, the enormous peak of the outburst, which had a total effective length of 130 milliseconds (**Fig. 5*a***), was followed by an oscillating, softer, and several-hundred-times weaker x-ray yield with a period of 8 s, lasting for at least 200 s (Fig. 5*b*). Thus, this gamma-ray burst source behaved like a temporary x-ray pulsar, with alternating stronger and weaker pulses in the x-ray yield, spaced about 4 s apart. There was also some evidence for a 23-ms period at the end of the initial burst. The 8-s period, if due to rotation of the source, could be produced only by an astronomical body of density at least 2.2×10^6 g/cm^3, which is almost too dense for a white dwarf. Such a period is typical, however, of x-ray pulsars, believed to be relatively slowly rotating neutron stars in binary systems, closely related to the more rapidly rotating radio-emitting pulsars. Presumably, the source in this case had two oppositely placed hot source regions at the magnetic poles, coming alternately into view as it rotated. Surprisingly, GB790305 was soon followed by three considerably weaker recurrences from the same direction, on March 6, April 4, and April 24, 1979. A number of additional recurrences have since been detected. SEE NEUTRON STAR; PULSAR.

This gamma-ray burst was somewhat softer in photon energy than most bursts, which typically emit photons of several hundred kiloelectronvolts of energy. GB790305 had an initial spectrum similar to that of a 30-keV (3.5×10^8 K) blackbody, except for a high-energy tail, with a broad peak at approximately 430 keV. If this corresponds to the 511-keV line produced by the annihilation of positrons and electrons, the lower energy agrees with the gravitational redshift of light produced at the surface of a neutron star of 1 solar mass and a radius of 6 mi (10 km). This spectral line thus furnishes additional evidence that the gamma-ray burst was produced by a neutron star, collapsed to a size only a few times larger than its Schwarzschild radius. Such gravitation-

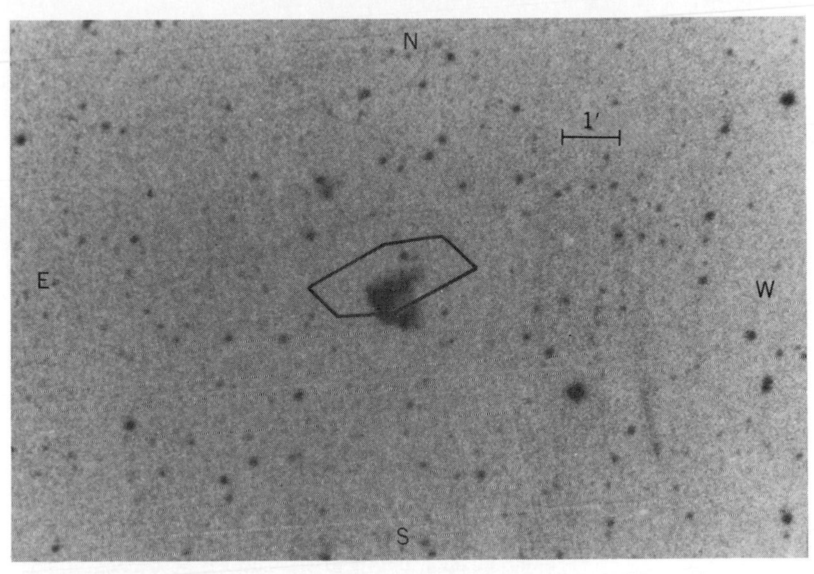

Fig. 4. Original error box (later reduced in size) for GB790305, superimposed on a photo of the supernova remnant N49 in the Large Magellanic Cloud. Photograph is a Cerro Tololo Interamerican Observatory Curtis-Schmidt plate taken by H. C. Epps, University of California, Los Angeles. (*From W. D. Evans et al., Astrophys. J., 237:L7–L9, 1980*)

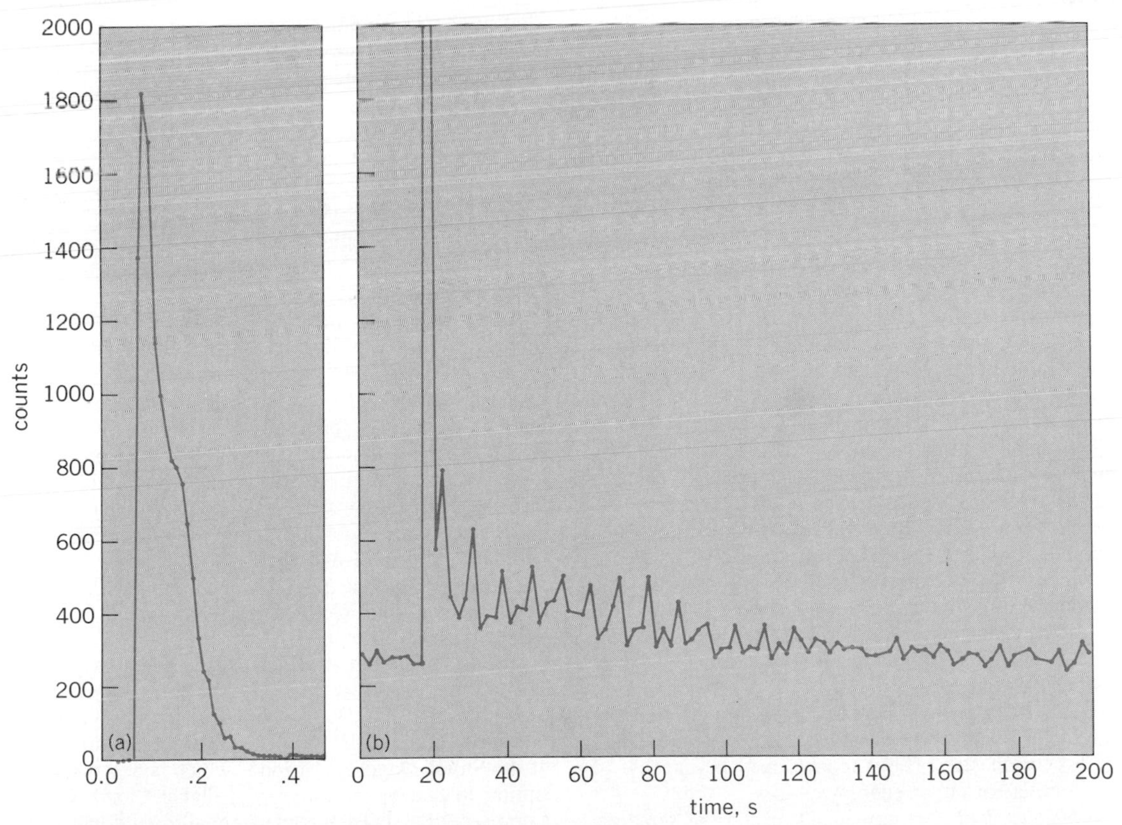

Fig. 5. Time history of the March 5, 1979, gamma-ray burst as seen by *Pioneer Venus Orbiter*. The energy range covered is 60–1200 keV. (*a*) Initial time history, with counts grouped into uniform 11.72-ms time intervals. (*b*) Longer time history, with counts grouped into 2-s bins, showing the 8-s periodicity. For clarity, the actual peak count of 13,000 is not fully shown. (*After J. Terrell et al., Periodicity of the gamma-ray transient event of 5 March 1979, Nature, 285:383–385, 1980*)

ally redshifted lines have also been observed in several other gamma-ray bursts. SEE GRAVITATIONAL REDSHIFT.

Other bursts. A gamma-ray burst on February 5, 1988 (GB880205), detected by the Japanese satellite *Ginga*, showed two sharp absorption lines in the spectrum at 20 and 40 keV. These have been interpreted as cyclotron absorption lines, indicating a magnetic field of approximately 10^8 teslas (10^{12} gauss) on the surface of a neutron star.

Two soft gamma repeaters (SGR) have been discovered, similar to GB790305 in shortness of outburst and in repeated outbursts, but with many repetitions. Their spectra are soft, of the order of 30 keV. These two sources are located in the galactic plane; it is not clear whether they represent a new class of gamma-ray bursters or perhaps something quite different.

Models for origin. A great many models have been proposed for the origin of gamma-ray bursts, from pulsar glitches to thermonuclear explosions or the infall of an asteroid to the surface of a neutron star. Although there is no agreement yet on the nature of these bursts, it seems to be generally accepted that neutron stars are the most likely source. The detonation of accumulated carbon in a thermonuclear explosion on the surface of a neutron star was proposed as a model in 1976, with yield calculated to be about 10^{40} ergs (10^{33} joules). A subsequent proposal involves the collapse of a white dwarf, forming a much denser neutron star while releasing enormous energy.

The infall of a comet (or asteroid) to the surface of a neutron star was suggested as an explanation for gamma-ray bursts in 1973, and has been proposed as the model for GB790305. An energy yield of about 10^{38} ergs (10^{31} J) would be produced by the fall of an asteroid of 10^{18} g mass (2×10^{15} lbm) and a radius of approximately 2 mi (3 km) onto the surface of a neutron star, at the relativistic speed imparted by the enormously strong gravitational field of such a collapsed object. For this and most other models, it seems necessary to assume that the source of GB790305 is not in the Large Magellanic Cloud but is a local object perhaps a thousand times closer, in order to keep the required x-ray and gamma-ray energy from becoming awkwardly large.

Perhaps the only reasonable source for the amount of energy emitted by GB790305, if it did indeed occur in the Large Magellanic Cloud, is gravitational energy released by a phase transition or other internal restructuring within a neutron star. It was proposed in 1974 that gamma-ray bursts could be produced by neutron starquakes or glitches, sudden slight changes in the rotational speed of the star, as could be expected from gradual loss of angular momentum. Such glitches have been observed on several occasions to occur in radio pulsars. It has been proposed that the ensuing readjustments in the immense magnetic field of the neutron star would accelerate charged particles to high energies and thus would produce gamma rays, perhaps as much as 10^{44} ergs (10^{37} J). Another proposal is that up to 10^{39} ergs (10^{32} J) of elastic energy might be released by the crustquake surface heating of a neutron star. These models may be capable of accounting for larger energy releases.

Eddington limit. The principal objection of some astrophysicists to models for gamma-ray bursts at the distance of the Large Magellanic Cloud (or more) is the great amount of gamma-ray emission energy which is required—about 10^{44} ergs/s (10^{37} W). This is a million times larger than the Eddington limit for a neutron star. The Eddington limit is that level of emitted radiation which would repel infalling matter by radiation pressure with a force balancing that of gravity. Thus, an intensity greater than the Eddington limit cannot be powered by accretion of matter, at least not in a steady-state situation. An intensity many orders of magnitude greater than this limit, as was apparently the case for GB790305, would very quickly give relativistic ejection velocities to any infalling matter as well as to the radiating surface.

Two factors which might prevent this are a sufficiently strong magnetic field, such as the 10^8-tesla (10^{12}-gauss) strength believed to be common for neutron stars, or rapid cooling of the surface due to adiabatic expansion. However, if the magnetic field were weaker than about 10^7 teslas (10^{11} gauss), it would not be strong enough to restrain an extremely hot surface from expanding at relativistic speed. Even a modest and brief expansion during which the surface cooled rapidly would allow the emission of x- and gamma-ray flux at a level well above the Eddington limit. It is thus of interest that a number of gamma-ray bursts, in addition to GB790305, consist of a short outburst of the order of 100 ms in length. Some outbursts are as short as a few milliseconds; the total outburst length can be as short as 8 ms. Many others involve a number of short outbursts over a period of seconds, even up to 80 s.

Locations of sources. GB790305 is of special interest in that it was well observed, intense, and well localized in direction. Some of its other interesting properties, such as the 8-s periodicity following the much stronger peak outburst, would not have been observable at the lower intensities of most bursts. A few other gamma-ray bursts have been accurately localized in direction, such as the event of November 19, 1978 (GB781119). Searches of historical photographic records of the optical sky have yielded some evidence of optical flashes in the same direction as recent gamma-ray bursts. In particular, an optical outburst occurred in 1928 within the error box of GB781119. Such optical evidence may lead to identifying gamma-ray burst sources, and will surely be searched for in future events.

An advantageous feature of the *Compton Gamma-Ray Observatory* is that its burst survey instrument (Burst and Transient Source Experiment, or BATSE) gives directional information without the use of timing data from other spacecraft. The angular resolution (mean error in direction) is no better than 6°, however. Precise locations still require accurate times from at least three spacecraft at wide separations. Other spacecraft observing gamma-ray bursts include *GRANAT*, *Ulysses*, and the U.S. Air Force *DMSP* satellites; *Pioneer Venus Orbiter* continued transmitting burst data until October 8, 1992, when it fell into Venus's atmosphere.

Almost no gamma-ray bursts have been identified as coming from known or recurring sources, with the exception of GB790305. Another interesting gamma-ray burst, on July 23, 1974, came from the direction of the Small Magellanic Cloud, which is at a distance similar to that of the Large Magellanic Cloud. Other gamma-ray bursts have come from all directions, with no apparent correlation with the directions of the center of the Milky Way Galaxy or of the galactic plane. This near-isotropy suggests that the sources are either very close by, no more than a few hundred light-years away (probably ruled out by the data from the *Compton Gamma-Ray Observatory*), or considerably farther out, perhaps from a halo distribution around the Milky Way Galaxy at a distance of a few hundred thousand light-years (up to 100 kiloparsecs or 2×10^{18} mi or 3×10^{18} km). The possibility of the larger (halo) distance is now taken seriously by most astrophysicists, in spite of the much larger energy requirements. It is not yet clear whether such gamma-ray bursts as GB790305 are a special class of events, perhaps more powerful than other gamma-ray bursts while softer in photon energy, or are produced by nearby neutron stars which are merely coincidentally in the direction of galaxies in the Local Group.

James Terrell

High-Energy Gamma Rays

The operating principles of high-energy gamma-ray telescopes will be discussed, and the results of observations with these instruments will be described.

High-energy gamma-ray telescopes. Most of the high-energy gamma-ray observations have been performed with devices inspired by those operating in the field of particle physics. Photons in the high-energy gamma-ray regime interact almost exclusively via electron-positron pair production; thus that process is the basis for gamma-ray detection, as in the Energetic Gamma-Ray Experiment Telescope (EGRET) on the *Compton Gamma-Ray Observatory* (**Fig. 6**). In the normal detection mode, a photon enters the spark chamber, passing undetected through the anticoincidence charged-particle scintillator, and converts into an electron-positron pair in one of the high-atomic-number tantalum foils interleaving the closely spaced spark-chamber modules. The trajectory of the electron pair is then recorded by initiating the spark-chamber readout, after the presence of the electrons is detected in the two-scintillation-counter time-of-flight system. The total energy deposited in the underlying large array of sodium iodide scintillator blocks is also recorded, and together with the trajectory data is used to determine the arrival direction and the total energy of the incoming gamma-ray photon. The accuracy with which the genuine direction of a single photon is reconstructed is 2.6° in the 70–150-MeV band, and somewhat better at higher energies (0.5° in the 500–2000-MeV band). An actual observation, however, involves many such photons, so the celestial position of a point source can be determined for optimum conditions with an accuracy of 5–10 arc-minutes.

Galactic gamma radiation. The high-energy gamma-ray sky is dominated by radiation from the galactic plane. The general correlation of high-energy gamma radiation with the galactic structure was clearly shown by the *COS-B* results (**Fig. 7**), which provided the most detailed overall picture of the gamma-ray sky until the completion of the EGRET sky survey.

The gamma-ray image of the Milky Way appears as a bright lane of diffuse emission, much like the optical image, although thinner (2° wide). It spans roughly the same angular extent, from −60° to +60° in galactic longitude (that is, with respect to the galactic center, located in the constellation Sagittarius). Two bright spots, in the constellations Cygnus and Vela, are also visible. There is a fainter emission in other parts of the galactic disk. Since the gamma rays are believed to be due to interactions of cosmic rays with matter, the gamma-ray image is throught to reflect mainly the clumpiness of interstellar matter.

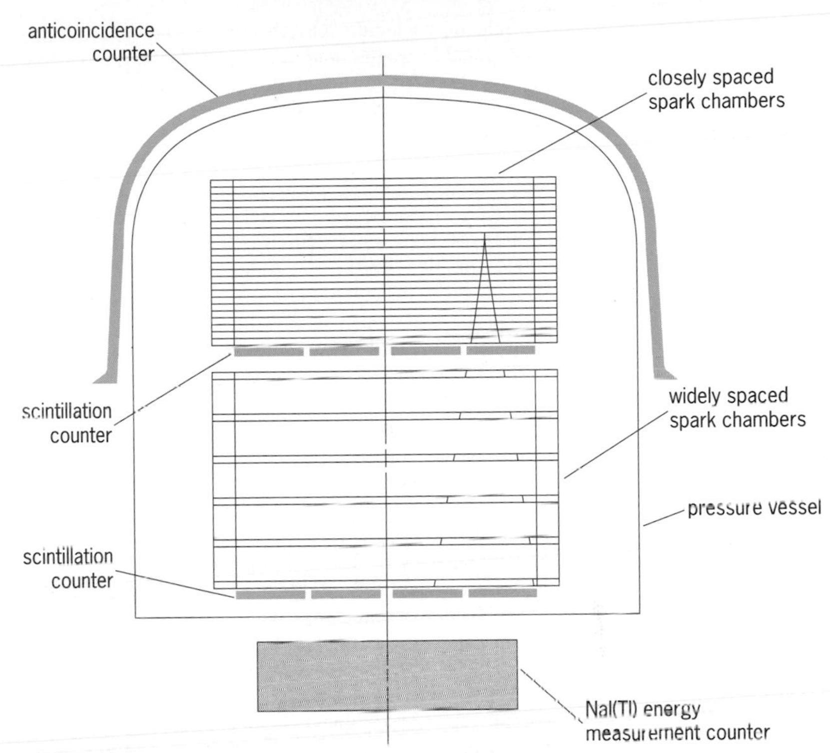

Fig. 6. EGRET telescope aboard the *Compton Gamma-Ray Observatory*. The two scintillation counters form a time-of-flight coincidence system. (*After R. C. Hartman et al., The EGRET high-energy gamma-ray telescope, The Compton Observatory Science Workshop, NASA Conf. Publ. 3137, pp. 116–125, 1991*)

Galactic diffuse emission. The galactic diffuse emission is thought to be mostly the result of the interaction of cosmic rays (electrons and protons) with the interstellar gas throughout the Milky Way Galaxy. Relativistic electrons (of energies above 30 MeV) radiate gamma rays via the bremsstrahlung process (braking radiation) when they cross the electric fields of nuclei; fast protons (of energies above 1 GeV) indirectly generate gamma rays when they collide with protons at rest, because in the course of the reaction elementary particles called π^0 (neutral pions) are created, which decay almost instantaneously in two gamma rays. Other particles, called π^+ and π^- (charged pions), are also created; they decay eventually in neutrinos and electrons. These electrons also generate gamma rays via bremsstrahlung. Bremsstrahlung and π^0 decay contribute about equally to the diffuse gamma-ray emission from the galactic plane.

Given the distribution of the interstelar gas in the Milky Way Galaxy, it is possible in principle to use gamma rays as a tracer of cosmic rays and deduce their density at any point. This application is quite central to the problem of the origin of cosmic rays in general, and to other problems such as the structure of the Milky Way Galaxy. The distribution of the interstellar gas is still controversial, and the little new quantitative information on cosmic rays can be deduced from diffuse gamma rays. However, the data are consistent with the fact that no cosmic-ray density increase beyond a factor of 2 exists toward the center of the Milky Way Galaxy. Also, there is an indication of a decrease of cosmic-ray protons and electrons toward outer regions of the Milky Way Galaxy, giving support to the idea that cosmic rays must be galactic in origin. *See Cosmic rays.*

Galactic molecular clouds contain much of the diffuse galactic matter and are generally believed to be the location for the formation of stars. Although the limited sensitivity and angular resolution of the first generation of high-energy gamma-ray instruments have restricted the gamma-ray information on these objects, two positive identifications were made with the *COS-B* telescope, namely a concentration of clouds near the star Rho Ophiuchi and the Orion cloud complex. In the constellation Aquila, the detection of high-energy gamma-ray emission has in fact led to the discovery of new interstellar clouds, later confirmed by radio observations of the carbon monoxide (CO) molecule. *SEE INTERSTELLAR MATTER; MOLECULAR CLOUD; ORION NEBULA.*

Galactic sources. Also contributing to the galactic emission are localized point sources. Although the point-source contribution cannot be estimated with certainty, several factors suggest that point sources may not be a major contributor. These include the fact that the distribution of the great majority of the high-energy gamma radiation along the plane can be reproduced in great detail with calculations based on cosmic-ray interactions with the interstellar matter. Several localized excesses have been isolated by *COS-B* along the galactic ridge (**Fig. 7**). These sources are remarkable in that the energy emitted in high-energy gamma rays is often several order of magnitude larger than emitted at other wavelengths.

The only certain identifications are those of the Crab and Vela pulsars, because the photon arrival-time analysis gives a light curve in phase with that observed at other wavelengths. The Vela pulsar is the brightest gamma-ray source in the sky. The gamma-ray light curves of the two pulsars are remarkably similar and conspicuously show two peaks, both separated by the same fractions (0.420 ± 0.007) of the pulsar period, which is 33 ms for the Crab and 89 ms for Vela. The Crab light curves at other wavelengths are qualitatively the same, while the Vela light curves are markedly different at different wavelengths (**Fig. 8**). An additional puzzling feature is that the ratio of the intensity of the two peaks of the Crab pulsar is variable over a time scale of years. By contrast, the same ratio for the Vela pulsar has remained constant over the same length of time. *SEE CRAB NEBULA.*

Given that the gamma-ray fluxes for these pulsars are orders of magnitude larger than in the radio range, theoreticians have tried to devise models to explain the observations. The most extensively used models attribute the pulsed gamma-ray emission to a geometrical lighthouse effect. The emission comes from a cone located at the poles of the pulsar (with a different opening angle than for the radio emission) and inclined with respect to the rotation axis. The mechanism is thought to be curvature radiation, somewhat analogous to the synchrotron radiation commonly found in radioastronomy. In the present case, however, the particles are electrons accelerated to very high energies (up to 10^{14} eV) by the huge electric fields induced by the pulsar rotation, and the radiation is emitted as the particles flow along the curved magnetic field lines of the pulsar. These (and other) models account more or less for the basic features of the pulsation, but they do not explain such phenomena as the variation of the intensity ratio of the two peaks.

To find possible identifications to the other sources, extensive searches inside the *COS-B* error boxes have been undertaken at other wavelengths. An interesting finding is the discovery of a very faint star unusually bright in x-rays, proposed to be the counterpart of the source named Geminga, the second brightest high-energy gamma-ray source in the sky, discovered by *SAS 2* in 1973. Thanks to the EGRET telescope and to the

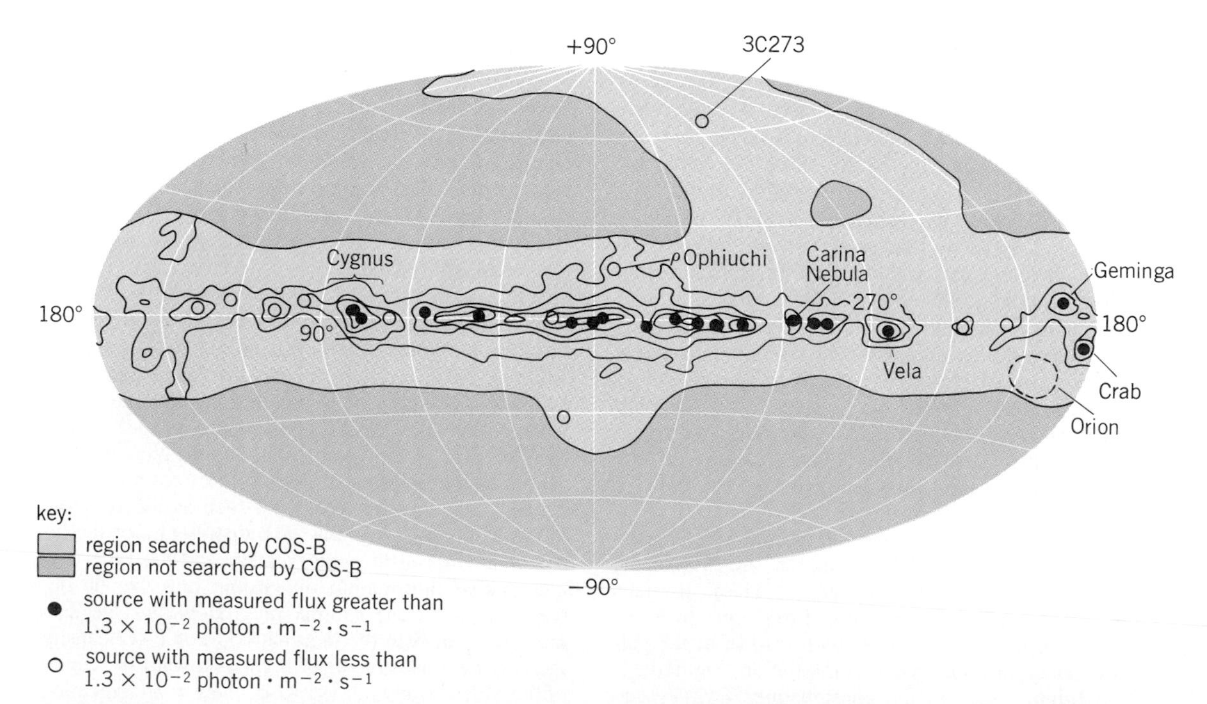

Fig. 7. Regions of the sky searched for gamma-ray sources by *COS-B* and sources detected above 100 MeV as of June 1979. Each source is located within a 1°-radius error box. The diffuse gamma-ray emission of the galactic disk is shown as a schematic contour map. There is a factor of 2 in intensity between each contour, starting from a background of 4×10^{-3} photon s^{-1} sr^{-1} recorded by the telescope. (*After B. N. Swanenburg et al., Second COS-B catalog of high-energy gamma-ray sources, Astrophys. J., 243:L69–L73, 1981*)

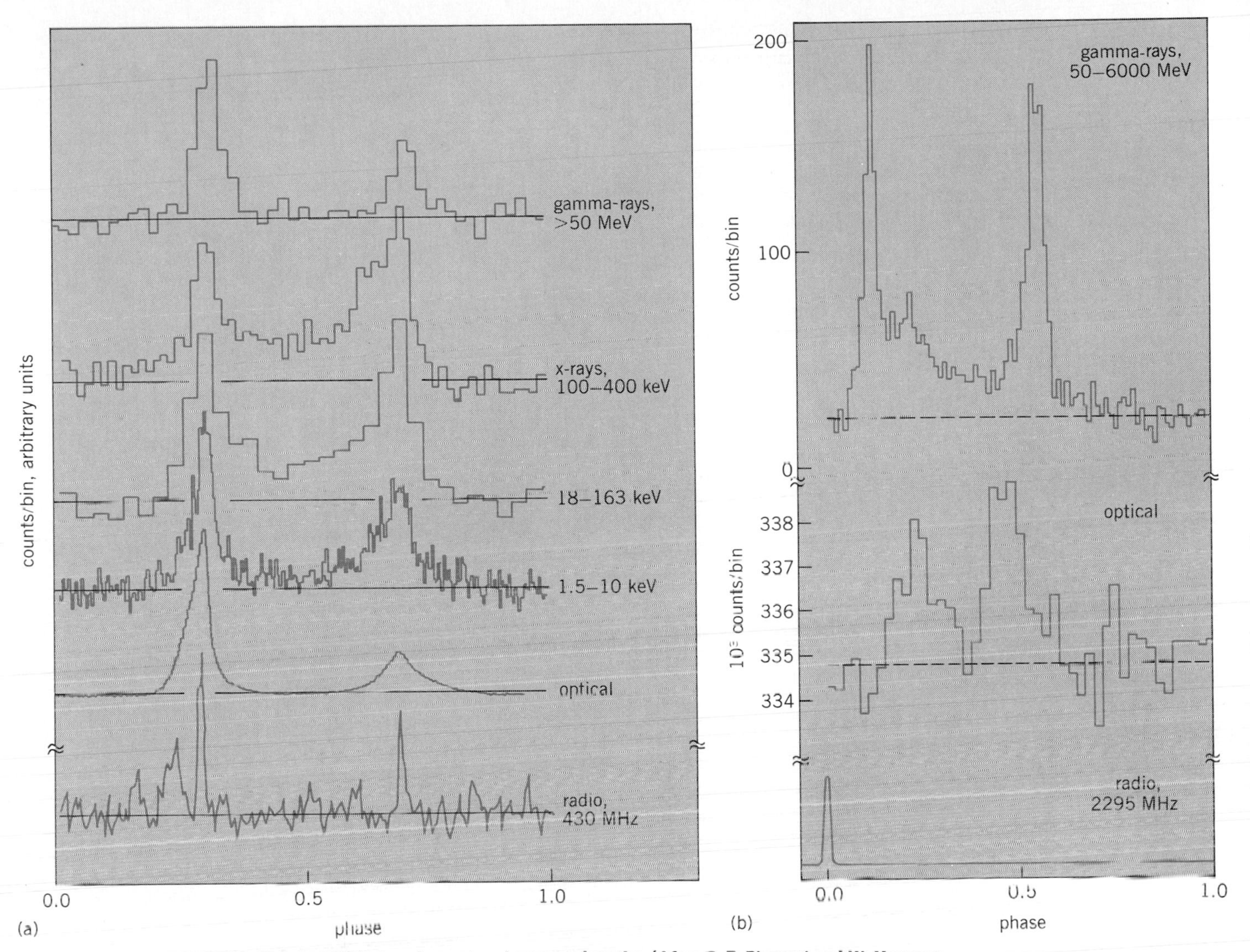

Fig. 8. Light curves of the (a) Crab and (b) Vela pulsars at various wavelengths. (*After G. F. Bignami and W. Hermsen, Galactic gamma-ray sources, Annu. Rev. Astron. Astrophys., 21:67–108, 1983*)

German x-ray observatory *ROSAT*, the identification became irrefutable in 1992, since it has been shown that both the soft x-ray emission of the x-ray source as observed by *ROSAT* and the high-energy gamma-ray emission from Geminga as observed by EGRET feature a similar 237-ms period pulsar behavior. *SEE X-RAY ASTRONOMY*.

Extragalactic sources. Another identification of a *COS-B* source is probable, based on the positional coincidence of 3C 273, one of the nearest quasars, with a point source detected in a low sky-background region, not very far from the north galactic pole (Fig. 7). The earliest EGRET observations confirmed this *COS-B* finding. However, the most surprising result obtained by EGRET during the early portion of the *Compton Gamma-Ray Observatory* mission was the discovery that many other quasars and active galaxies undergo intense high-energy gamma-ray flares, as was the case in the period June 15–28, 1991, for the optically violent variable quasar 3C 279 (**Fig. 9**).

A massive (10^6–10^9-solar-mass) black-hole accretion disk with a collimated perpendicular jet currently represents the basic model for explaining the luminous, broadband radiation emitted from the central engines of active galactic nuclei and quasars. Radiation mechanisms that focus gamma rays in the jet direction, for example, the Compton scattering of accretion-disk photons by relativistic nonthermal electrons in the jet, may account for the observed emission. *SEE GALAXY, EXTERNAL; QUASAR.*

VERY HIGH-ENERGY GAMMA RAYS

Beyond the range currently covered by satellites (a few gigaelectronvolts), no physical reason exists for the absence of electromagnetic radiation, and various groups have tried to detect very high-energy gamma rays. However, above a few gigaelectronvolts, the fluxes are currently undetectable, until new physical processes for the absorption of gamma rays take over. Beyond a threshold of about 3×10^{11} eV, the Earth's atmosphere itself becomes a detector: these energies are so great that the initial collision of a gamma-ray photon with an atmospheric atom results in the successive creation of millions of secondary particles (air showers), mainly muons and electrons (**Fig. 10**).

Detection. Between 3×10^{11} and 1×10^{14} eV, the secondary particles are stopped by the atmosphere before reaching the ground, but, among them, electron-positive pairs are able to generate Cerenkov radiation, that is, ultraviolet photons which can be detected by using large parabolic mirrors. 3–6 ft (1–2 m) in diameter. (Early experiments used modified World War II antiaircraft searchlight mirrors.) The process is so rapid that all photons are emitted almost simultaneously, generating a spherically shaped Cerenkov front, almost 0.6 mi (1 km) in width (Fig. 10), which successively reaches elements of an array of several detectors, spaced up to several miles apart on the ground. Analysis of the signals from the various detectors makes possible the determination of some characteristics of the generating particle (either a gamma-ray or a very high-energy cosmic-ray nucleus): its arrival direction, to within about 1°; and its energy, determined only with respect to some experimental threshold. Because the Cerenkov emission takes place in the ultraviolet and is very weak, the observations must be done on dark, moonless nights.

Beyond about 10^{14} eV, the secondary electrons and muons reach the ground and trigger various detecting devices (such as large scintillators). At these energies, extensive air showers comprise millions of particles, of which only a fraction are detected. The performance of the experiments is comparable to that obtained with the previous technique.

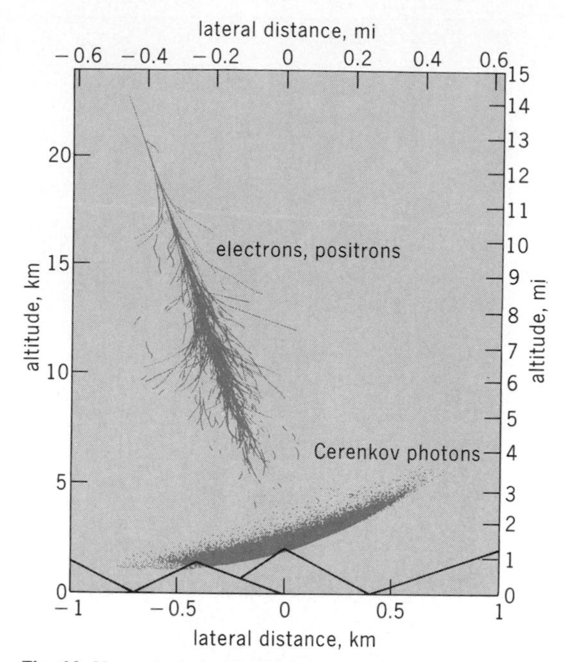

Fig. 10. Numerical simulation of the development of an electromagnetic shower of energy 200 GeV (2 × 10^{11} eV). Cerenkov front is shown schematically. (*After Ph. Goret and S. Basiuk, Centre d'Etudes Nucléaires de Saclay, France*)

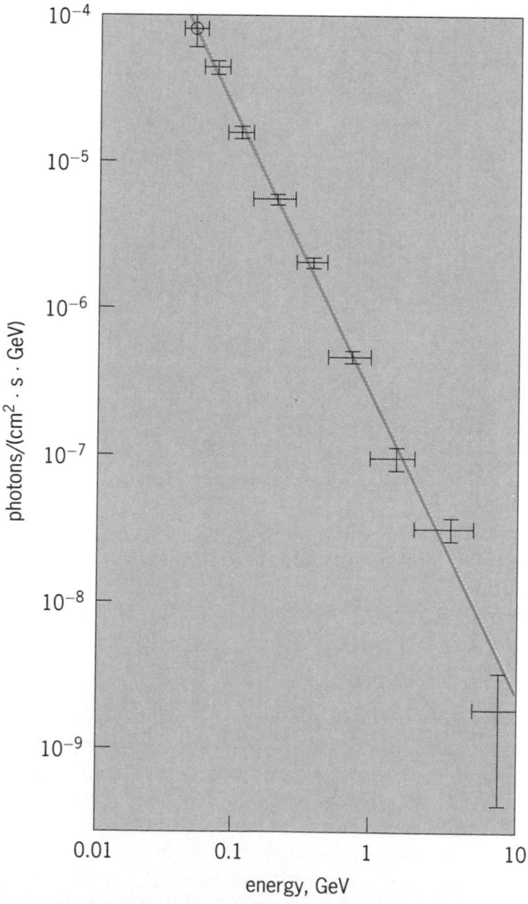

Fig. 9. Differential high-energy gamma-ray spectrum observed by the EGRET telescope aboard the *Compton Gamma-Ray Observatory* for the quasar 3C 279 during the period June 15–28, 1991. The observed spectrum, indicated by crosses, is well represented by a power law. The straight line is the best power law from a least-squares fit. The instrument sensitivity in the 0.05–0.07-GeV region is less well known, so that the top data point, identified by a circle, was not included in the fit. (*After R. C. Hartman et al., Detection of high-energy gamma radiation from quasar 3C 279 by the EGRET telescope on the Compton Gamma-Ray Observatory, Astrophys. J., 385:L1–L4, 1992*)

Very high-energy gamma rays are difficult to detect because of their very weak fluxes (on the order of a few photons per square kilometer per day) and because they must be distinguished from an intense background of very high-energy cosmic-ray nuclei. Major observation sites include Arizona; Utah; Haverah Park, in Great Britain; Kiel, Germany; and the Crimea. Other detector arrays have been built in Australia, France, India, New Zealand, South Africa, and even at the South Pole.

Results. In spite of decades of observations, the results obtained are still ambiguous but may prove highly significant from a theoretical standpoint. The existence of celestial sources of very high-energy gamma rays is controversial, and it is hoped that the results of the improved detectors will help settle the issue. Several positive detections have been announced over the years, including the Crab and Vela pulsars, Cygnus X-3, and Herculis X-1. However, the reported fluxes are always close to the instrument sensitivity, typically 10^{-14} photon cm^{-2} s^{-1}, or slightly less, for energies above 10^{16} eV (about 10 photons km^{-2} day^{-1} or 25 photons mi^{-2} day^{-1}), so that a correct treatment of the celestial (that is, cosmic-ray-induced) and instrumental backgrounds is delicate. The difficulty is that only a few good photons may be detected from a source in an observation lasting several thousands of hours, even in the largest arrays. Although some theoretical suggestions have been made at the time that positive detections were reported, the emission of photons at such high energies is poorly understood, but may take place in the vicinity of a neutron star.

Jacques Paul

Bibliography. D. L. Bertsch, C. E. Fichtel, and J. I. Trombka, Instrumentation for gamma-ray astronomy, *Space Sci. Rev.*, 48:113–168, 1988; G. F. Bignami and W. Hermsen, Galactic gamma-ray sources, *Annu. Rev. Astron. Astrophys.*, 21:67–108, 1983; H. Bloemen, Diffuse galactic gamma-ray emission, *Annu. Rev. Astron. Astrophys.*, 27:469–516, 1989; J.-M. Bonnet-Bidaud and G. Chardin, Cygnus X-3, a critical review, *Phys. Rep.*, 170:325–404, 1988; C. J. Cesarsky and T. Montmerle, Gamma rays from active regions in the Galaxy: The possible contribution of stellar winds, *Space Sci. Rev.*, 36:173–193, 1983; P. Durouchoux and N. Prantzos (eds.), *Gamma-Ray Line Astrophysics*, AIP Conf. Proc. 232, 1991; C. E. Fichtel, Some aspects of the scientific significance of high-energy gamma-ray astrophysics, *Adv. Space Res.*, 49(3):303–312, 1991; G. J. Fishman and W. S. Paciesas (eds.), *Proceedings of the Huntsville Gamma-Ray Burst Workshop, 1991*, AIP Conf. Proc. 265, 1992; T. Murakami et al., Evidence for cyclotron absorption from spectral features in gamma-ray bursts seen with *Ginga, Nature*, 335:234–235, 1988; Problems of Gamma-Ray Astronomy and Planned Experiments, *Space Sci. Rev.*, vol 49, no. 1/2, 1988; D. J. van der Walt and A. W. Wolfendale, Gamma rays and the origin of cosmic rays, *Space Sci. Rev.*, 47:1–45, 1988.

Gemini

The Twins, in astronomy, is a winter zodiacal constellation. Gemini is the third sign of the zodiac. It is conspicuous, containing first-, second-, and third-magnitude stars. These stars, α, β, γ, and μ, form a rough quadrilateral figure (see **illus.**). The constellation is pictured as the figures of the twin heroes Castor and Pollux, with the two brightest stars of the same names representing the heroes' heads. Pollux, slightly brighter than Castor, is a navigational star. The Sun is in this constellation at the time of the summer solstice. The southwest corner of Gemini, being in the Milky Way, contains star fields which are among the finest in the sky. *SEE CONSTELLATION; ZODIAC.*

Ching-Sung Yu

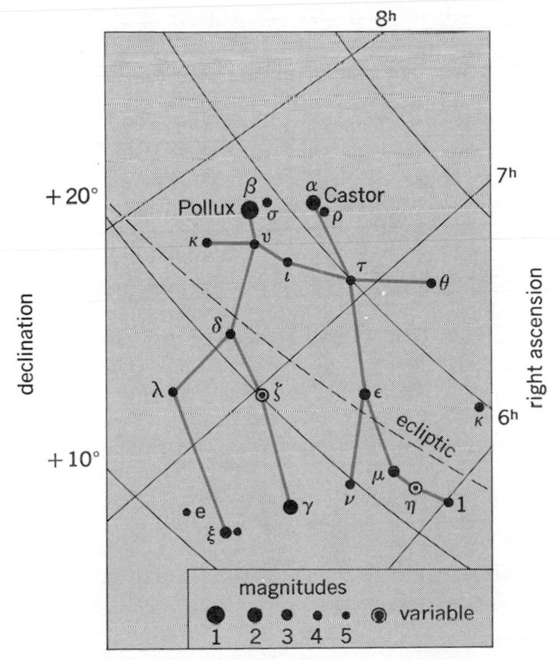

Line pattern of the constellation Gemini. Grid lines represent coordinates of the sky. Apparent brightness, or magnitudes, of stars is shown by sizes of dots graded by appropriate numbers as indicated.

Giant star

An intermediate state in the evolution of a star in which it swells to enormous proportions before its death. During the longest and most stable phase of a star's life, the star, like the Sun, derives its energy from the thermonuclear fusion of hydrogen into helium deep in its dense, hot (10^7 K and up) core. It is then said to be on the main sequence. When the hydrogen fuel is exhausted and the central energy source is exhausted, the core contracts and heats under the action of gravity, fresh hydrogen is ignited in a shell that surrounds the spent core, and the star becomes much more luminous, larger, and cooler at its surface. The lower surface temperature produces a redder color, hence the common term red giant. Stars like the Sun brighten by a factor of 100 and grow in radius by a factor of nearly 50 to about half the size of Mercury's orbit (2×10^7 mi or 3.2×10^7 km).

There are actually two separate giant states. The first, described above, is terminated when the core temperature climbs so high (2×10^8 K) that the helium ignites and fuses into carbon. This event stabilizes the star; though the star is still a giant it then contracts somewhat and dims. When this helium is exhausted, the earlier behavior is repeated. The core contracts and is finally stabilized by electron degeneracy, becoming essentially a white dwarf. Helium then fuses to carbon in a shell around the core, and farther out hydrogen fuses to helium. The star then enters the asymptotic giant branch of the Hertzsprung-Russell diagram and swells to enormous proportions, perhaps two astronomical units (1.8×10^8 mi or 3×10^8 km), becoming even redder than before. It may pulsate and be seen as a long-period variable star, and loses much or most of its mass through a strong wind. *SEE HERTZSPRUNG-RUSSELL DIAGRAM; STELLAR EVOLUTION.*

James B. Kaler

Bibliography. I. Iben, Jr., Stellar evolution within and off the main sequence, *Annu. Rev. Astron. Astrophys.*, 5:571–626, 1967; I. Iben, Jr., and A. Renzini, Asymptotic giant branch evolution and beyond, *Annu. Rev. Astron. Astrophys.*, 21:271–342, 1983; J. B. Kaler, *Stars and Their Spectra: An Introduction to the Spectral Sequence*, 1989; J. M. Pasachoff, *Contemporary Astronomy*, 4th ed., 1989.

Globule

A small, opaque nebula seen in silhouette against a rich star field or a bright nebula. Globules were first cataloged in the 1920s. In 1947, B. J. Bok called attention to their potential significance for star formation, and since then they have been commonly known as Bok globules. A globule is a region of the interstellar medium containing a high density of interstellar grains that obscure the more distant background stars and cause the region to appear as a dark nebula in optical photographs. Only relatively nearby globules can be identified, because if there are many stars in front of the nebula the contrast with the background is too weak. The distances to such nebulae have been estimated by counting the number of stars photographed in the line of sight and comparing this count with an analogous star count in a clearer comparison field. Comparative star counts have also been used to evaluate the degree of extinction produced by the grains in the nebula; if the optical properties of the grains are known, the total number of grains in a cut through the nebula can be estimated. SEE INTERSTELLAR EXTINCTION; NEBULA.

The material contained in interstellar grains represents only a small fraction (about 1%) of the total mass of a globule; most of its mass is in gaseous form. The grains play a key role in shielding the nebular gas from the surrounding starlight, thus creating an environment in which molecules can survive and interact. Radio astronomers have been able to detect carbon monoxide (CO) emission lines in dark nebulae, and in the densest cores of nebulae even heavier molecules have been identified. Dark nebulae are now commonly called molecular clouds: the most abundant molecule is molecular hydrogen, but it is difficult to detect directly. Molecular clouds are of great interest as the regions in which stars are formed. SEE COSMOCHEMISTRY; INFRARED ASTRONOMY.

Nearby globules have been surveyed from the radio emission lines of carbon monoxide, ammonia (NH_3), and other molecules. These data, when combined with optical estimates of distances, have led to fairly accurate determinations of the dimensions of globules. Careful studies of ammonia emissions indicate that a typical (molecular core) globule has a diameter of 0.1 parsec (1 parsec = 1.9×10^{13} mi or 3.1×10^{13} km) and mass four times that of the Sun; larger, more massive globules are also known. The kinetic temperature in a globule is low; it is estimated to be about 10 K (−442°F). SEE RADIO ASTRONOMY.

Calculations predict that in the absence of internal support a typical globule undergoes gravitational collapse in less than a million years to produce one or more protostars. These young objects radiate in the infrared, a region of the spectrum to which the nebular grains are transparent. The *Infrared Astronomical Satellite (IRAS)* has mapped the sky at several infrared wavelengths, and astronomers have identified many protostars in the dense cores of molecular clouds. Some but not all globules have given birth to protostars, and these probably represent the youngest stars of the Milky Way. SEE INFRARED ASTRONOMY; STELLAR EVOLUTION.

Beverly T. Lynds

Bibliography. E. E. Barnard, *Photographic Atlas of Selected Regions of the Milky Way*, 1927; C. A. Beichman, The IRAS view of the Galaxy and the solar system, *Annu. Rev. Astron. Astrophys.*, 25:521–563, 1987; D. Black and M. Mathews (eds.), *Protostars and Planets*, vol. 2, 1985; B. Lynds (ed.), *Dark Nebulae, Globules, and Protostars*, 1971.

Gravitation

The mutual attraction between all masses and particles of matter in the universe. In a sense this is one of the best-known physical phenomena. During the eighteenth and nineteenth centuries gravitational astronomy, based on Newton's laws, attracted many of the leading mathematicians and was brought to such a pitch that it seemed that only extra numerical refinements would be needed in order to account in detail for the motions of all celestial bodies. In the twentieth century, however, A. Einstein shattered this complacency, and the subject is currently in a healthy state of flux.

Until the seventeenth century, the sole recognized evidence of this phenomenon was the gravitational attraction at the surface of the Earth. Only vague speculation existed that some force emanating from the Sun kept the planets in their orbits. Such a view was expressed by J. Kepler, the author of the laws of planetary motion. But a proper formulation for such a force had to wait until I. Newton founded newtonian mechanics, with his three laws of motion, and discovered, in calculus, the necessary mathematical tool. SEE CELESTIAL MECHANICS; PLANET.

Newton's law of gravitation. Newton's law of universal gravitation states that every two particles of matter in the universe attract each other with a force that acts in the line joining them, the intensity of which varies as the product of their masses and inversely as the square of the distance between them. Put into symbols, the gravitational force F exerted between two particles with masses m_1 and m_2 separated by a distance d is given by Eq. (1), where G is called

$$F = \frac{Gm_1m_2}{d^2} \quad (1)$$

the constant of gravitation.

A force varying with the inverse-square power of the distance from the Sun had been already suggested—notably by R. Hooke but also by other contemporaries of Newton, such as E. Halley and C. Wren—but this had been applied only to circular planetary motion. The credit for accounting for, and partially correcting, Kepler's laws and for setting gravitational astronomy on a proper mathematical basis is wholly Newton's.

Newton's theory was first published in the *Principia* in 1686. According to Newton, it was formulated in principle in 1666 when the problem of elliptic motion in the inverse-square force field was solved. But publication was delayed in part because of the difficulty of proceeding from the "particles" of the law to extended bodies such as the Earth. This difficulty was overcome when Newton established that, under his law, bodies having spherically symmetrical distribution of mass attract each other as if all their mass were concentrated at their respective centers.

Newton verified that the gravitational force between the Earth and the Moon, necessary to maintain the Moon in its orbit, and the gravitational attraction at the surface of the Earth were related by an inverse-square law of force. Let E be the mass of the Earth, assumed to be spherically symmetrical with radius R. Then the force exerted by the Earth on a small mass m near the Earth's surface is given by Eq. (2), and

$$F = \frac{GEm}{R^2} \quad (2)$$

the acceleration of gravity on the Earth's surface, g, by Eq. (3).

$$g = \frac{GE}{R^2} \quad (3)$$

Let a be the mean distance of the Moon from the

Earth, M the Moon's mass, and P the Moon's sidereal period of revolution around the Earth. If the motions in the Earth-Moon system are considered to be unaffected by external forces (principally those caused by the Sun's attraction), Kepler's third law applied to this system is given by Eq. (4).

$$\frac{4\pi^2 a^3}{P^2} = G(E + M) \tag{4}$$

Equations (3) and (4), on elimination of G, give Eq. (5).

$$g = 4\pi^2 \frac{E}{E + M} \frac{a^2}{R^2} \frac{a}{P^2} \tag{5}$$

Now the Moon's mean distance from the Earth is $a = 60.27R = 3.84 \times 10^8$ m (2.39×10^5 mi) and the sidereal period of revolution is $P = 27.32$ days $= 2.361 \times 10^6$ s. These data give, with $E/M = 81.35$, $g = 9.77$ m/s^2 (32.1 ft/s^2), which is close to the observed value.

This calculation corresponds in essence to that made by Newton in 1666. At that time the ratio a/R was known to be about 60, but the Moon's distance in miles was not well known because the Earth's radius R was erroneously taken to correspond to 60 mi (97 km) per degree of latitude instead of 69 mi (111 km). As a consequence, the first test was unsatisfactory. But the discordance was removed in 1671 when the measurement of an arc of meridian in France provided a reliable value for the Earth's radius.

Gravitational constant. Equation (3) shows that the measurement of the acceleration due to gravity at the surface of the Earth is equivalent to finding the product G and the mass of the Earth. Determining the gravitational constant by a suitable experiment is therefore equivalent to "weighing the Earth."

In 1774, G was determined by measuring the deflection of the vertical by the attraction of a mountain. This method is much inferior to the laboratory method in which the gravitational force between known masses is measured. In the torsion balance two small spheres, each of mass m, are connected by a light rod, suspended in the middle by a thin wire. The deflection caused by bringing two large spheres each of mass M near the small ones on opposite sides of the rod is measured, and the force is evaluated by observing the period of oscillation of the rod under the influence of the torsion of the wire (see **illus.**). This is known as the Cavendish experiment, in honor of H. Cavendish, who achieved the first reliable results by this method in 1797–1798. More recent determinations using various refinements yield the results: constant of gravitation $G = 6.67 \times 10^{-11}$ SI (mks) units; mass of Earth $= 5.98 \times 10^{24}$ kg. The result of the best available laboratory measurement of G, announced in 1982, is $G = (6.6726 \pm 0.0005) \times 10^{-11}$ in SI (mks) units.

In newtonian gravitation, G is an absolute constant, independent of time, place, and the chemical composition of the masses involved. Partial confirmation of this was provided before Newton's time by the experiment attributed to Galileo in which different weights released simultaneously from the top of the Tower of Pisa reached the ground at the same time. Newton found further confirmation, experimenting with pendulums made out of different materials. Early in this century, R. Eötvös found that different materials fall

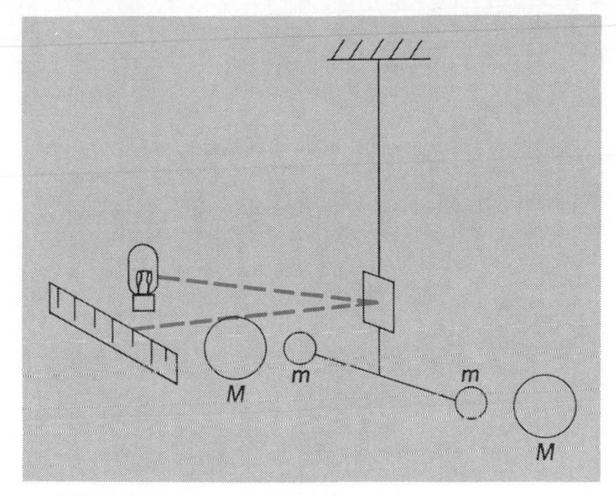

Diagram of the torsion balance.

with the same acceleration to within 1 part in 10^7. The accuracy of this figure has been extended to 1 part in 10^{11}, using aluminum and gold, and to 0.9×10^{-12} with a confidence of 95%, using aluminum and platinum.

With the discovery of antimatter, there was speculation that matter and antimatter would exert a mutual gravitational repulsion. But experimental results indicate that they attract one another according to the same laws as apply to matter of the same kind.

A cosmology with changing physical "constants" was first proposed in 1937 by P. A. M. Dirac. Field theories applying this principle have since been proposed by P. Jordan and D. W. Sciama and, in 1961, by C. Brans and R. H. Dicke. In these theories G is diminishing; for instance Brans and Dicke suggest a change of about 2×10^{-11} per year. This would have profound effects on phenomena ranging from the evolution of the universe to the evolution of the Earth. There is no firm evidence at present to support a time variation of G. For instance, detailed analyses of the motion of the Moon using the lunar laser ranging data, the analysis of solar system data, and especially the ranging data using Viking landers on Mars, and the timing analysis of the binary pulsar PSR 1913+16 have all resulted in only upper limits on a possible temporal variation of G. The present upper limit on the relative rate of time variation of G is a few parts in 10^{11} per year based on data from PSR 1913+16. This result is theory-independent; that is, it is based purely on observations without relying on any particular theory in which G is presumed to vary with time.

Mass and weight. In the equations of motion of newtonian mechanics, the mass of a body appears as inertial mass, a measure of resistance to acceleration, and as gravitational mass in the expression of the gravitational force. The equality of these masses is confirmed by the Eötvös experiment. It justifies the assumption that the motion of a particle in a gravitational field does not depend on its physical composition. In Newton's theory the equality can be said to be a coincidence, but not in Einstein's theory, where this equivalence becomes a cornerstone of relativistic gravitation.

While mass in newtonian mechanics is an intrinsic property of a body, its weight depends on certain forces acting on it. For example, the weight of a body on the Earth depends on the gravitational attraction of the Earth on the body and also on the centrifugal forces due to the Earth's rotation. The body would have lower weight on the Moon, even though its mass would remain the same.

Gravity. This should not be confused with the term gravitation. Gravity is the older term, meaning the quality of having weight, and so came to be applied to the tendency of downward motion on the Earth. Gravity or the force of gravity is today used to describe the intensity of gravitational forces, usually on the surface of the Earth or another celestial body. So gravitation refers to a universal phenomenon, while gravity refers to its local manifestation.

A rotating planet is oblate (or flattened at the poles) to a degree depending on the ratio of the centrifugal to the gravitational forces on its surface and on the distribution of mass in its interior. The variation of gravity on the surface of the Earth depends on these factors and is further complicated by irregular features such as oceans, continents, and mountains. It is investigated by gravity surveys and also through the analysis of the motion of artificial satellites. Because of the irregularities, no mathematical formula has been found that satisfactorily represents the gravitational field of the Earth, even though formulas involving hundreds of terms are used. The problem of representing the gravitational field of the Moon is even harder because the surface irregularities are proportionately much larger.

In describing gravity on the surface of the Earth, a smoothed-out theoretical model is used, to which are added gravity anomalies, produced in the main by the surface irregularities.

Gravity waves are waves in the oceans or atmosphere of the Earth whose motion is dynamically governed by the Earth's gravitational field. They should not be confused with gravitational waves, which are discussed below.

Gravitational potential energy. This describes the energy that a body has by virtue of its position in a gravitational field. If two particles with masses m_1 and m_2 are a distance r apart and if this distance is slightly increased to $r + \Delta r$, then the work done against the gravitational attraction is $Gm_1m_2\Delta r/r^2$. If the distance is increased by a finite amount, say from r_1 to r_2, the work done is given by Eq. (6). If $r_2 \rightarrow \infty$, Eq. (7) holds.

$$W_{r_1,r_2} = Gm_1m_2 \int_{r_1}^{r_2} \frac{dr}{r^2} = Gm_1m_2\left(\frac{1}{r_1} - \frac{1}{r_2}\right) \tag{6}$$

$$W_{r_1,\infty} = \frac{Gm_1m_2}{r_1} \tag{7}$$

If one particle is kept fixed and the other brought to a distance r from a very great distance (infinity), then the work done is given by Eq. (8). This is called

$$-U = \frac{-Gm_1m_2}{r} \tag{8}$$

the gravitational potential energy; it is (arbitrarily) put to zero for infinite separation between the particles. Similarly, for a system of n particles with masses $m_1, m_2, \ldots, m_n$ and mutual distance r_{ij} between m_i and m_j, the gravitational potential energy $-U$ is the work done to assemble the system from infinite separation (or the negative of the work done to bring about an infinite separation), as shown in Eq. (9).

$$-U = -G\sum_{i<j} \frac{m_im_j}{r_{ij}} \tag{9}$$

A closely related quantity is gravitational potential. The gravitational potential of a particle of mass m is given by Eq. (10), where r is distance measured from

$$V = \frac{-Gm}{r} \tag{10}$$

the mass. The gravitational force exerted on another mass M is M times the gradient of V. If the first body is extended or irregular, the formula for V may be extremely complicated, but the latter relation still applies. SEE POTENTIALS.

A good illustration of gravitational potential energy occurs in the motion of an artificial satellite in a nearly circular orbit around the Earth which is affected by atmospheric drag. Because of the frictional drag the total energy of the satellite in its orbit is reduced, but the satellite actually moves faster. The explanation for this is that it moves closer to the Earth and loses more in gravitational potential energy than it gains in kinetic energy.

Similarly, in its early evolution a star contracts, with the gravitational potential energy being transformed partly into radiation, so that it shines, and partly into kinetic energy of the atoms, so that the star heats up until it is hot enough for thermonuclear reactions to start. SEE STELLAR EVOLUTION.

Another related phenomenon is that of speed of escape. A projectile launched from the surface of the Earth with speed less than the speed of escape will return to the surface of the Earth; but it will not return if its initial speed is greater (atmospheric drag is neglected). For a spherical body with mass M and radius R, the speed of escape from its surface is given by Eq. (11). For the Earth V_e is 11.2 km/s (7.0 mi/s); for

$$V_e = \left(\frac{2MG}{R}\right)^{1/2} \tag{11}$$

the Moon it is 2.4 km/s (1.5 mi/s), which explains why the Moon cannot retain an atmosphere such as the Earth's. By analogy, a black hole can be considered a body for which the speed of escape from the surface is greater than the speed of light, so that light cannot escape; however, the analogy is not really exact since newtonian mechanics is not valid. The question as to whether the universe will continue to expand can be considered in the same way. If the density of matter in the universe is great enough, then expansion will eventually cease and the universe will start to contract. At present the density cannot be found with sufficient accuracy to decide the question. SEE BLACK HOLE; COSMOLOGY.

Application of Newton's law. In modern times Eq. (5), in a modified form with appropriate refinements to allow for the Earth's oblateness and for external forces acting on the Earth-Moon system, has been used to compute the distance to the Moon. The results have been superseded in accuracy only by radar measurements and observations of corner reflectors placed on the lunar surface.

Newton's theory passed a much more stringent test than the one described above when he was able to account for the principal departures from Kepler's laws in the motion of the Moon. Such departures are called perturbations. One of the most notable triumphs of the theory occurred when the observed perturbations in the motion of the planet Uranus enabled J. C. Adams in 1845 and U. J. Leverrier in 1846 independently to predict the existence and calculate the position of a hitherto-unobserved planet, later called Neptune. When yet another planet, Pluto, was discovered in 1930, its position and orbit were strikingly similar to predictions based on the method used to discover Neptune. But the discovery of Pluto must be ascribed to the perseverance of the observing astronomers; it is not massive enough to have revealed itself through the perturbations of Uranus and Neptune. SEE PERTURBATION.

F. W. Bessel observed nonuniform proper motions of Sirius and Procyon and inferred that each was gravitationally deflected by an unseen companion. It was only after his death that these bodies were telescopically observed, and they both later proved to be white dwarfs. More recently evidence has been accumulated for the existence of some planetary masses around stars. The discovery of black holes (which will never be directly observed) hinges in part on a visible star showing evidence for having a companion of sufficiently high mass (so that its gravitational collapse can never be arrested). SEE BINARY STAR.

Newton's theory supplies the link between the observed motion of celestial bodies and certain physical properties, such as mass and sometimes shape. Knowledge of stellar masses depends basically on the application of the theory to binary star systems. Analysis of the motions of artificial satellites placed in orbit around the Earth has revealed refined information about the gravitational field of the Earth and of the Earth's atmosphere. Similarly, satellites placed in orbit around the Moon have yielded information about its gravitational field, and other space vehicles have yielded the best information to date on the masses and gravitational fields of other planets.

Newtonian gravitation has been applied without apparent difficulty to the motion in distant star systems. But over very great distance (or over very small distances, when gravitation is swamped by other forces) it has not been confirmed or disproved.

Accuracy of newtonian gravitation. Newton was the first to doubt the accuracy of his law when he was unable to account fully for the motion of the perigee in the motion of the Moon. In this case he eventually found that the discrepancy was largely removed if the solution of the equations were more accurately developed. Further difficulties to do with the motion of the Moon were noted in the nineteenth century, but these were eventually resolved when it was found that there were appreciable fluctuations in the rate of rotation of the Earth, so that it was the system of timekeeping and not the gravitational theory that was at fault. SEE EARTH ROTATION AND ORBITAL MOTION.

A more serious discrepancy was discovered by Leverrier in the orbit of Mercury. Because of the action of the other planets, the perihelion of Mercury's orbit advances. But allowance for all known gravitational effects still left an observed motion of about 43 seconds of arc per century unaccounted for by Newton's theory. Attempts to account for this by adding an unknown planet or by drag with an interplanetary medium were unsatisfactory, and a very small change was suggested in the exponent of the inverse square of force. This particular discordance was accounted for by Einstein's general theory of relativity in 1916, but the final word on the subject has yet to be said.

Gravitational lens. Light is deflected when it passes through a gravitational field, and an analogy can be made to the refraction of light passing through a lens. It has been suggested that a galaxy situated between an observer and a more distant source might have a focusing effect, and that this might account for some of the observed properties of quasi-stellar objects. The multiple images of the quasar (Q0957+561 A,B,) are almost certainly caused by the light from a single body passing through a gravitational lens. While this is the best-studied gravitational lens, about a dozen other examples of this phenomenon have been discovered. SEE GRAVITATIONAL LENS; QUASAR.

Testing of gravitational theories. One of the greatest difficulties in investigating gravitational theories is the weakness of the gravitational coupling of matter. For instance, the gravitational interaction between a proton and an electron is weaker by a factor of about 5×10^{-40} than the electrostatic interaction. (If gravitation alone bound the hydrogen atom, then the radius of the first Bohr orbit would be 10^{13} light-years, or about 1000 times the radius of the Hubble universe!) Of the four basic interactions (that is, strong, electromagnetic, weak, and gravitational) that are known at present, gravitation is by far the weakest force in nature. (A unification of the electromagnetic and weak interactions into the so-called electroweak interaction has been carried out, and further unification of this interaction with the strong and even the gravitational interactions has been suggested.)

Geophysical experiments have raised the possibility of the existence of fifth and sixth forces that would couple to matter with strengths close to that of gravity. These forces are expected to have finite ranges, however, in contrast to the infinite range of universal gravitation. Thus, deviations from Newton's inverse-square law of gravitation could reveal the presence of such interactions. These forces also appear in certain theories that attempt to unify gravity with the other fundamental interactions. Possible ranges include 10^2–10^5 m (300 ft–60 mi), though other ranges cannot be excluded. Geophysical experiments conducted in a mine could be interpreted as indicating the existence of a repulsive fifth force with a range of about 200 m (650 ft) and strength of about 10^{-2} of gravity. On the other hand, a gravimetric experiment performed on a tall television tower has indicated the possibility of existence of a short-range attractive interaction (sixth force). Geophysical experiments conducted in a borehole in the Greenland ice cap have also revealed small deviations from Newton's law that may be ascribed to an attractive force of approximately 1-km (0.6-mi) range with a strength of a few percent of gravity. However, in all these experiments the local density variations of the Earth must be modeled. Incomplete knowledge of these density inhomogeneities might well be the source of the observed anomalies.

It has been suggested that the forces responsible for the deviations from newtonian gravity could be composition-dependent. These forces would, therefore, violate the principle of equivalence of inertial and gravitational masses. Efforts to detect such a violation of the equivalence principle have produced conflicting results. Several experiments (for example, using torsion-balance or free-fall methods) have reported no evidence for a composition-dependent effect, thus setting new limits on the strength and range of a possible new force.

There is no firm evidence at present for any deviation from the newtonian law of gravitation in the nonrelativistic regime. Further experiments to test the inverse-square law are being planned, including space experiments using low-orbiting spacecraft.

Relativistic theories. Before Newton, detailed descriptions were available of the motions of celestial bodies—not just Kepler's laws but also empirical formulas capable of representing with fair accuracy, for their times, the motion of the Moon. Newton replaced description by theory, but in spite of his success and the absence of a reasonable alternative, the theory was heavily criticized, not least with regard to its requirement of "action at a distance" (that is, through a vacuum). Newton himself considered this to be "an absurdity," and he recognized the weaknesses in postulating in his system of mechanics the existence of preferred reference systems (that is, inertial reference systems) and an absolute time. Newton's theory is a superb mathematical one that represents the observed phenomena with remarkable accuracy.

The investigation of electric and magnetic phenomena culminated in the second half of the nineteenth century in the complete formulation of the laws of electromagnetism by J. C. Maxwell. Maxwell based his theory on M. Faraday's field concept. The electromagnetic field propagates with the speed of light; in fact, Maxwell's theory unified the science of optics with electricity and magnetism. Maxwell's theory of the electromagnetic field was extended and strengthened with the subsequent observations of electromagnetic waves by H. Hertz and the successes of the theory of electrons developed by H. A. Lorentz.

The theory of relativity grew from attempts to describe electromagnetic phenomena in moving systems. No physical effect can propagate with a speed exceeding that of light in vacuum; therefore, Newton's theory must be the limiting case of a field theory in which the speed of propagation approaches infinity. Einstein's field theory of gravitation (general relativity) is based on the identification of the gravitational field with the curvature of space-time. The geometry of space-time is affected by the presence of matter and radiation. The relationship between mass-energy and the space-time curvature is therefore a relativistic generalization of the newtonian law of gravitation. The relativistic theory is mathematically far more complicated than Newton's. Instead of the single newtonian potential described above, Einstein worked with 10 quantities that form a tensor.

Principle of equivalence. An important step in Einstein's reasoning is his "principle of equivalence": that a uniformly accelerated reference system imitates completely the behavior of a uniform gravitational field. Imagine, for instance, a scientist in a space capsule infinitely far out in empty space so that the gravitational force on the capsule is negligible. Everything would be weightless; bodies would not fall; and a pendulum clock would not work. But now imagine the capsule to be accelerated by some agency at the uniform rate of 981 cm/s^2 (32.2 ft/s). Everything in the capsule would then behave as if the capsule were stationary on its launching pad on the surface of the Earth and therefore subject to the Earth's gravitational field. But after its original launching, when the capsule is in free flight under the action of gravitational forces exerted by the various bodies in the solar system, its contents will behave as if it were in the complete isolation suggested above. This principle requires that all bodies fall in a gravitational field with precisely the same acceleration, a result that is confirmed by the Eötvös experiment mentioned earlier. Also, if matter and antimatter were to repel one another, it would be a violation of the principle.

Einstein's theory requires that experiments should have the same results irrespective of the location or time. This has been said to amount to the "strong" principle of equivalence.

Classical tests. The ordinary differential equations of motion of newtonian gravitation are replaced in general relativity by a nonlinear system of partial differential equations for which general solutions are not known. Apart from a few special cases, knowledge of solutions comes from methods of approximation. For instance, in the solar system, speeds are low so that the quantity v/c (v is the orbital speed and c is the speed of light) will be small (about 10^{-4} for the Earth). The equations and solutions are expanded in powers of this quantity; for instance, the relativistic correction for the motion of the perihelion of Mercury's orbit is adequately found by considering no terms smaller than $(v/c)^2$. This is called the post-newtonian approximation. (Another approach is the weak-field approximation.)

Einstein's theory has appeared to pass three famous tests. First, it accounted for the full motion of the perihelion of the orbit of Mercury. (Mercury is the most suitable planet, because it is the fastest-moving of the major planets and has a high eccentricity, so that its perihelion is relatively easily studied.) Second, the prediction that light passing a massive body would be deflected has been confirmed with an accuracy of about 5%. Third, Einstein's theory predicted that clocks would run more slowly in strong gravitational fields compared to weak ones; interpreting atoms as clocks, spectral lines would be shifted to the red in a gravitational field. This, again, has been confirmed with moderate accuracy.

Predictions of the theory have been confirmed in an experiment in which radar waves were bounced off Mercury; the theory predicts a delay of about 2×10^{-4} s in the arrival time of a radar echo when Mercury is on the far side of the Sun and close to the solar limb. Tests, similar in principle, have been conducted using observations of the Mariner space vehicles, the accuracy of confirmation being in the region of 4%. A greater level of accuracy has been achieved, by better than an order of magnitude, using data from transponders on Viking orbiters and landers on Mars. Furthermore, the deflection of microwave radiation passing close to the Sun has been observed using radio interferometry with a baseline of 22 mi (35 km). The amount of bending that has been found is 1.015 ± 0.011 times the amount predicted by general relativity. In another test, the precession of a gyroscope in orbit around the Earth is to be studied for evidence of the so-called geodetic precession as well as the precession due to the dragging of the local inertial frames by the rotating Earth. The lunar laser-ranging data have been used to measure the de Sitter precession of Moon's orbit. This is due to the geodetic precession of the Earth-Moon orbital angular momentum in the gravitational field of the Sun. It amounts to an advance in the lunar node and perigee by about 2 seconds of arc per century, a prediction first made by W. de Sitter in 1916 soon after the advent of general relativity. Such small secular effects are suitable for study since they accumulate in time. Other periodic (noncumulative) orbital effects have until recently been too small to observe. But the current revolution in observational techniques and accuracy has changed the situation; post-newtonian terms are now routinely included in many calculations of the orbits of planets and space vehicles, and comparison with observations will furnish tests of the theory.

The observation and analysis of gravitational waves, discussed below, will constitute further tests.

Mach's principle. One of the most penetrating critiques of mechanics is due to E. Mach, toward the end of the nineteenth century. Some of his ideas can be traced back to Bishop G. Berkeley early in the eighteenth century. Out of Mach's work there has arisen Mach's principle; this is philosophical in nature and cannot be stated in precise terms. The idea is that the motion of a particle is meaningful only when referred to the rest of the matter in the universe. Geometrical and inertial properties are meaningless for an empty space, and the motion of a particle in such space is devoid of physical significance. Thus the behavior of a test particle should be determined by the total matter distribution in the universe and should not appear as an intrinsic property of an absolute space. Mach's principle suggests that gravitation and inertia are equivalent. This idea strongly influenced the development of general relativity.

Brans-Dicke theory. This is a classical field theory of gravitation that was developed in 1961 by Brans and Dicke on the basis of an interpretation of Mach's principle. In this theory the gravitational field is described by a tensor and a scalar, the equations of motion being the same as those in general relativity. The addition of a scalar field leads to the appearance of an arbitrary constant, whose value is not known exactly. The Brans-Dicke theory predicts that the relativistic motion of the perihelion of Mercury's orbit is reduced compared with Einstein's value, and also that the light deflection should be less. With regard to the orbit of Mercury, Dicke pointed out that if the Sun were oblate, this might account for some of the motion of the perihelion. In 1967 he announced that measurements showed a solar oblateness of about 5 parts in 100,000 (or a difference in the polar and equatorial radii of about 21 mi or 34 km). His observations and discussion are still subject to some controversy. The difference between the theory and that of general relativity can be parametrized by the number ω, where $\gamma = (1 + \omega)/(2 + \omega)$; for general relativity, $\gamma = 1$. Dicke has proposed $\omega \approx 7.5$; but the results of measurement of deflection of radiation by the Sun indicate a value of ω greater than 23, for which the predictions of the two theories would not be greatly different. Subsequent data from the Viking spacecraft imply that this constant must be greater than about 500, thus rendering the Brans-Dicke theory almost indistinguishable from general relativity.

There are, of course, many other theories not mentioned here.

Supergravity. This is the term applied to a highly mathematical theory of gravitation forming part of a unified field theory in which all types of forces are included.

Gravitational waves. The existence of gravitational waves, or gravitational "radiation," was predicted by Einstein shortly after he formulated his general theory of relativity. They are now a feature of any relativity theory. Gravitational waves are "ripples in the curvature of space-time." In other words, they are propagating gravitational fields, or propagating patterns of strain, traveling at the speed of light. They carry energy and can exert forces on matter in their path, producing, for instance, very small vibrations in elastic bodies. The gravitational wave is produced by change in the distribution of some matter. It is not produced by a rotating sphere, but would result from a rotating body not having symmetry about its axis of rotation: a pulsar, perhaps. In spite of the relatively weak interaction between gravitational radiation and matter, the measurement of this radiation is now technically possible and may already have been achieved. This is due to the pioneering work of Joseph Weber. The present situation contains some uncertainties; but gravitational-wave astronomy has been added to other branches of astronomy, and a new window is opening to the universe.

A classical problem, solved by Einstein, concerns the gravitational radiation from a rod spinning about a perpendicular axis through its center. If the rod has moment of inertia about the axis of spin I ($I = Md^2/3$, where M is the mass of the rod in kilograms and $2d$ its length in meters) and angular velocity ω, the power of the radiation in watts (1 W = 10^7 ergs/s) is given by Eq. (12), where G is the constant

$$P = \frac{32GI^2\omega^6}{5c^5} = 1.73 \times 10^{-52} I^2\omega^6 \qquad (12)$$

of gravitation in mks units and c is the speed of light in meters per second. A calculation using a steel rod of mass 4.9×10^5 kg (1.1×10^6 lb), length 20 m (66 ft), and angular velocity $\omega = 28$ rad/s, limited by the balance between centrifugal force and tensile strength, gives 2.2×10^{-29} W. So the problem of the generation and detection of gravitational waves in the laboratory is at present somewhat academic.

In electromagnetic theory, electric-dipole radiation is dominant. The gravitational analog of the electric dipole is the mass dipole moment whose time rate of change is the total momentum of the system; since this is constant, there is no gravitational dipole radiation; the principal power is in quadrupole radiation. The radiation has fairly elaborate polarization properties.

Binary systems. Consider a binary star system having period P hours and masses m_1, m_2, where the relative orbit is circular. If $M = m_1 + m_2$ and $\mu = m_1 m_2/M$, the power output by gravitational radiation is given by Eq. (13), where $M_\odot$ is the mass of the

$$P_B = \left(\frac{\mu}{M_\odot}\right)^2 \left(\frac{M}{M_\odot}\right)^{4/3} P^{-10/3}\, 3.0 \times 10^{26}\ \text{W} \qquad (13)$$

Sun. For the orbit of the Earth around the Sun, P_B is about 200 W. The gravitational radiation extracts energy from the system. If a binary system has a relative elliptic orbit, then most of the energy is extracted at the closest point of approach and the orbit approaches a circle; then the orbit will gradually shrink, with the bodies colliding after a "spiral time" given by Eq. (14), where a_0 is the initial radius of the rela-

$$\tau_0 = \frac{5c^5}{256G^3} \frac{a_0{}^4}{\mu M^2} \tag{14}$$

tive orbit. Under this mechanism the Earth would have fallen toward the Sun less than a centimeter (0.4 in.) in the lifetime of the solar system!

Clearly it is necessary to look outside the solar system for promising sources. Ordinary binaries are not helpful. Sirius and its companion, with a spiral time of 7×10^{21} years, radiate at 10^8 W, the flux received at the Earth being 10^{-31} W. The closer the members of the system are to each other, the more promising they are; some eclipsing binaries can generate power that would be observed on the Earth at about 10^{-20} W, and have spiral times of the order of 10^{10} years. The shortest periods known are for close pairs consisting of a white dwarf and a main sequence star; here spiral times can be as low as 10^9 years, and the predicted flux at the Earth for the most promising candidate, ι Boo, is 18×10^{-18} W. For these binaries, gravitational radiation appears to play an important part in their physical characteristics. Matter flows toward the white dwarf from the companion star, causing flickering and occasional nova outbursts. The stars are very close, and it seems that the contraction of the orbit, caused by gravitational radiation, plays a crucial part in instigating the flow of matter. *SEE NOVA; WHITE DWARF STAR.*

In 1975 a pulsar (PSR 1913+16) was discovered that appeared to be in orbit about an unseen companion. There is no confirmed optical identification of the pulsar at present. The timing analysis of this binary pulsar has made it possible to test general relativity indirectly beyond the postnewtonian approximation. The first binary pulsar has a pulse period of about 59 ms and an orbital period of about 8 h. The parameters describing the binary pulsar have been obtained from fitting the pulse arrival times (accumulated since 1974) to models based on a general relativistic two-body system. Five keplerian and five so-called post-keplerian orbital parameters can be determined from the observations. The data are consistent with a pair of point masses that slowly spiral toward each other as a consequence of emission of gravitational radiation. Each member of the binary has a mass approximately equal to 1.4 solar masses, the orbit has a high eccentricity of 0.617, and the orbital period decays at a rate of about 2.4 parts in 10^{12}. This rate of decay is consistent with the prediction of general relativity at the level of 1%. As further arrival-time data accumulate, the consistency of the scheme can be checked and the amount of gravitational radiation damping can be obtained with greater accuracy. The binary pulsar PSR 1913+16 has provided the only compelling, though indirect, evidence for the existence of gravitational waves. *SEE PULSAR.*

Pulsars. The most rapidly rotating single objects that have been observed are pulsars. These are neutron stars rotating with periods mostly less than 1 s. From their irregular light curves it is reasonable to suppose that they do not possess symmetry about the axis of rotation. Suppose that they are assumed homogeneous and the equatorial section is an ellipse with axes a and b, and that the ellipticity of the equator is $\epsilon = (a - b)/a$. If the star rotates with angular velocity ω, then the power radiated is given by Eq. (15), where I is the moment of inertia about the axis

$$P_R = \frac{32G\omega^6 I^2 \epsilon^2}{5c^5} \tag{15}$$

of rotation. A promising candidate here is the pulsar in the Crab Nebula, remnant of a supernova; the period of rotation is 0.033 s; the moment of inertia is likely to be of the order of 4×10^{37} kg m^2, and the power output can be estimated by writing Eq. (15) as Eq. (16), where P is the period. Clearly it is impor-

$$P_R = \left(\frac{I}{4 \times 10^{37}\ \mathrm{kg\ m^2}}\right)^2 \cdot \left(\frac{P}{0.033\ \mathrm{s}}\right)^{-6} \left(\frac{\epsilon}{10^{-3}}\right)^2 10^{31}\ \mathrm{W} \tag{16}$$

tant to estimate ϵ. The periods of rotation are known to be increasing, and this puts an upper limit on ϵ. It is estimated that the flux received from the Crab pulsar would be less than 3×10^{-20} W. Some pulsars occasionally show sudden changes or glitches in their rotational period. These could be due to starquakes (neuron stars have solid surfaces) and might lead to strong bursts of gravitational radiation. *SEE CRAB NEBULA.*

Explosive events. The gravitational collapse involved in a supernova explosion might produce the strongest radiation that can be observed. The processes involved can only be tentatively estimated, and unfortunately supernova occur only about once every 100 years in the Milky Way Galaxy. But their radiation, probably in short bursts, could be sufficiently powerful for them to be observed from other galaxies. It is possible that stellar collapse takes place without the display of a supernova, and so estimates of frequency may be much too low. *SEE SUPERNOVA.*

Many galaxies show evidence of explosive activity. For quasars gravitational radiation at 10^{38} W has been suggested, and for explosions in galactic centers, 10^{30} W; but these estimates are not at all definitive. *SEE GALAXY, EXTERNAL.*

As matter falls into a black hole, it will release a burst of gravitational radiation; the energy released is proportional to the square of the mass captured and inversely proportional to the mass of the black hole; the time of the outburst is proportional to the mass of the black hole. It has been suggested that there might be a large black hole at the center of the galaxy; if its mass were 10^8 solar masses and it captured a star of 1 solar mass, a burst of energy 10^{37} W might be produced. If the black hole were rotating or the infalling star somehow had a speed greater than that acquired from falling from infinity, then the energy could be greater. *SEE MILKY WAY GALAXY.*

Nature of radiation. The radiation discussed could be continuous or in bursts. The radiation would have a spectrum that might be discrete, as in the case of rotation or orbital revolution, where the fundamental frequency is 2ω (there will also be harmonics), or broadband in the case of explosive events. The longest wavelengths suggested are from the primordial history of the universe, when they could be greater than the size of a galaxy; the shortest, in supernovae and stellar collapse.

Dirac worked out a quantum theory for this radiation; the graviton is a theoretically deduced particle postulated as the quantum of the gravitational field.

Detection of gravitational waves. When a gravitational wave interacts with a system of particles, the particles wiggle slightly; in the case of a solid body, strains are set up in the body; what is actually measured is a sort of tidal effect. On this basis, Weber developed bar antennas for the detection of gravitational waves in the 1960s. Weber's detectors involved strains that are produced in a bar in response to an incident gravitational wave with a frequency compatible with the natural frequency of vibration of the bar. Weber worked with aluminum cylinders suspended in vacuum. They are directional, being most strongly sensitive to radiation traveling perpendicular to the axis of the cylinder. The strains in the cylinder are converted into measurable voltages by piezoelectric crystals bonded around the girth of the cylinder. Weber's room-temperature antennas were limited by the background thermal noise, a random effect due to the thermal motion of the individual molecules.

A second generation of bar antennas is in operation in several laboratories. Greater sensitivity has been achieved by making the detectors more massive, by operating at liquid-helium temperatures, or by using large crystals such as sapphire that naturally have less noise. The detection of gravitational radiation by Weber-type antennas is based on the principle of resonance; the frequency of radiation that could be detected by such antennas is typically of order 10^3Hz. A strain sensitivity of a few parts in 10^{18} has been achieved in a low-temperature bar detector. This constitutes an improvement in sensitivity of several orders of magnitude on the original Weber antennas. The strain sensitivity corresponds to the detectable amplitude of an incident gravitational wave, that is, the amplitude of deviation of the space-time geometry in the neighborhood of the antenna from flat Minkowski space-time.

Another approach to the detection of gravitational waves is based on laser interferometry. A laser beam is split into two parts along suspended perpendicular arms, each beam being reflected many times by mirrors at the ends. If the length of one arm changes relative to the other as a consequence of the passage of a gravitational wave pulse, this will be shown through interference when the beams are recombined. Greater sensitivity can be achieved by constructing large-scale interferometers. Large-scale devices of high sensitivity with arm lengths equal to or exceeding 10 m (33 ft) have been built in a few laboratories, and space-based laser interferometers are planned. Compared to Weber-type detectors, the laser interferometers have the advantage of being broadband detectors with a frequency range from about 10 Hz to about 10^4 Hz. In a ground-based detector with a powerful laser and interferometer arms that are several kilometers long, a sensitivity of 1 part in 10^{21} may be achieved for detecting an incident gravitational wave burst. However, many technical difficulties need to be overcome.

There is no conclusive evidence at present that gravitational waves have ever been detected in the laboratory. The various detectors can, therefore, place upper limits on possible fluxes of gravitational radiation within certain frequency ranges.

Many other methods have been suggested to search for gravitational waves, including the use of the Earth as a resonant gravitational-wave antenna, the analysis of timing data of pulsars, and the Doppler tracking of interplanetary spacecraft. Experiments have provided only upper limits on the energy density of an isotropic cosmic background of gravitational radiation. That is, gravitational waves generated by all the sources during the Hubble expansion as well as primordial waves left over from the very beginning of the expansion are expected to form a stochastic background of gravitational radiation. This could be part of the cosmological missing mass that is required in order to have a spatially closed universe. According to studies of the background Earth noise, the energy density of random waves that could excite the normal modes of the Earth must be much less than the critical density for spatial closure. Stronger upper limits have been obtained at lower frequencies by using the timing noise of pulsars. Furthermore, the Doppler tracking of *Pioneer 10* spacecraft has resulted in useful upper limits on the energy density of a cosmic background in the microhertz region of the frequency spectrum. *SEE DOPPLER EFFECT; GRAVITATIONAL RADIATION DETECTOR.*

J. M. A. Danby; Bahram Mashhoon

Bibliography. P. C. W. Davies, *The Search for Gravity Waves*, 1980; R. H. Dicke, *Gravitation and the Universe*, 1970; G. Gamow, *Gravity*, 1962; C. W. Misner, K. S. Thorne, and J. A. Wheeler, *Gravitation*, 1973; J. V. Narlikar, *The Lighter Side of Gravity*, 1982; B. Schwarzschild, From mine shafts to cliffs: The "fifth force" remains elusive, *Phys. Today*, 41(7):21–24, July 1988; K. S. Thorne, Gravitational-wave research: Current status and future prospects, *Rev. Mod. Phys.*, 52:285–297, 1980; S. Weinberg, The forces of nature, *Amer. Sci.*, 65:171–176, 1977; C. M. Will, The confrontation between general relativity and experiment: An update, *Phys. Rep.*, 113:345–422, 1984.

Gravitational collapse

The stage in the evolution of a star in which the pressure in the star is insufficient to maintain the star at a stable size. The material in the star or in the core of the star then falls inward under its own gravitational attraction. Depending on the mass, composition, and spin of the star, the collapse may proceed to the formation of a neutron star or black hole, possibly accompanied by a supernova explosion. *SEE STELLAR EVOLUTION.*

Stability of stars. Stars similar to the Sun maintain a stable size by continually burning nuclear fuel. In most stars, as for the Sun, this burning involves the conversion of hydrogen to helium with a release of energy as the nuclear reactions take place. This energy supplies heat to the interior of the star, which in turn keeps the core of the star hot enough so that nuclear reactions can continue. Heat is also continually transferred from the interior to the exterior of the star, where it is lost, primarily in the form of radiation.

Just as the temperature decreases from the interior to the exterior of the star, so does the pressure. The change in the pressure over an interval of distance is called a pressure gradient. A pressure gradient is required to keep the star in equilibrium, as can be seen by the following argument. First, any chunk of gas in a stable star is in equilibrium, on the average, if the star is stable. Second, there is a net gravitational force acting on any chunk of gas as a result of the attraction to the rest of the matter in the star, and this force, which is normally called the weight of the gas, must be directed toward the center of the star. Finally, the pressure gradient supplies a force to balance the gravitational force, since the pressure on the bottom of the chunk of gas (bottom meaning closer to the center of the star) is larger than the pressure on the top of the chunk. In equilibrium, the variation of pressure in the star is just that for which all chunks of gas are in equilibrium.

Contraction and the onset of collapse. Suppose now that the nuclear burning is turned off, as is the case if the hydrogen in the core is used up and only helium remains. Then, no more heat is supplied to the core of the star, and its temperature drops as remaining heat is transported to the surface of the star and is lost. Likewise, since the pressure in the gas depends on the temperature, the pressure in the core also drops, and more importantly, the pressure gradient decreases throughout the star. The various chunks of gas are no longer in equilibrium, and they move inward. This compression of the gas then causes a rise in temperature and the temporary reestablishment of the required temperature gradient. However, as heat is transported outward, the temperature again drops, the gas is further compressed, and so on. In this stage, the star is undergoing contraction, the heating coming from the gravitational potential energy of the star rather than nuclear reactions.

Eventually, the star condenses, and the temperature rises to a sufficiently high value that helium burning can take place. Once again, the star is in stable equilibrium until the reservoir of fuel in the core of the star is used up, after which contraction takes place until conditions are right for nuclear burning of higher elements. The contraction discussed here refers to what is going on in the core of the star; the atmosphere may actually be expanding in the process.

When the star has a core which is composed of iron, no further nuclear burning can take place, since iron is the most stable element. Continued contraction must then take place. In fact, theoretical calculations indicate that under normal conditions the temperature does not rise high enough for nuclear burning to proceed beyond carbon, at which point the star effectively runs out of fuel, and continued contraction also takes place. If, after ejection of material by the star, the mass of the star is less than about 1.2–1.4 solar masses, the limiting value known as the Chandrasekhar limit, the contraction takes place to a white dwarf. Theoretical studies indicate that normal stars smaller than about 4 solar masses will eject enough matter in the process of evolution to end up as white dwarfs. A white dwarf is stable without any nuclear burning taking place. The pressure gradient is produced by the same kind of quantum interactions among the electrons as those which make atoms stable. The white dwarf then can only cool off with time, since it has no source of heat. The cooling period is very long, however, up to 1 billion years. As a result, many white dwarfs are observed in the sky, since the radiation continues long after the star, in some sense, has died. *See White dwarf star.*

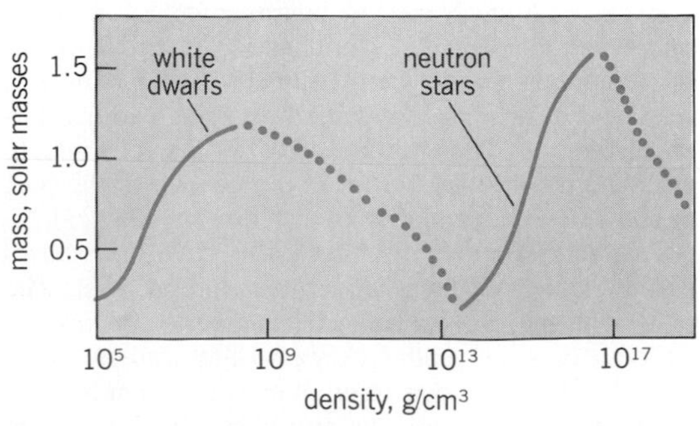

Masses of white dwarfs and neutron stars as a function of central density. The solid lines indicate possible stable configurations.

The fate of stars in the range of around 4 to 8 solar masses is not well understood. However, theoretical studies indicate that stars of more than about 8 solar masses will not lose enough mass in their evolution to become white dwarfs. What happens then is that the contracting star at some point becomes unstable; that is, the heating as a result of contraction is insufficient to produce the required pressure gradient to support the gas against gravity. This instability occurs first in the core of the star, and the core of the star starts to fall inward on itself. This is the stage of gravitational collapse. During collapse the released gravitational potential energy goes into kinetic energy of motion of the core rather than into heating the core, since a near-equilibrium situation is required for efficient transfer of gravitational energy into heat energy. In essence, the core is then freely falling inward under its own gravitational attraction.

Collapse to a neutron star. There are several possible end points to the gravitational collapse of a star. White dwarfs have already been mentioned as a possible end point of stellar evolution, but a collapsing core will be too condensed to form a white dwarf, even if its mass is less than the maximum allowed mass. The only other known possibilities are collapse to a neutron star and collapse to a black hole. A neutron star is a highly condensed star in which the predominant constituent of matter is in the form of neutrons. The stability of such stars is a result of the quantum-mechanical interaction of neutrons, the same kind of interaction which, for electrons, leads to stability of white dwarfs. *See Neutron star.*

A neutron star is so condensed that the density of matter in the neutron star is comparable with the density of matter in the nucleus of an atom. For example, a neutron star of 1 solar mass has a radius of about 6 mi (10 km). As with white dwarfs, there is a maximum mass for stable neutron stars, in the range of 1.4 to 3.0 solar masses, the actual value being somewhat uncertain because nuclear forces are important at such a high density. The upper limit of 3 solar masses is fairly certain, however, since beyond that limit gravity will make the neutron star unstable, independent of the nature of the nuclear forces. The **illustration** shows how the mass of white dwarfs and neutron stars depends on the central density of the star. The actual curve for white dwarfs depends on the composition of the star, and the actual curve for neutron stars is uncertain because of lack of knowledge of nuclear forces at high density.

Theoretical studies of collapsing stars indicate that the star develops a collapsing iron core as result of nuclear reactions induced in the compressed matter. When the core is sufficiently compressed, protons in the nuclei of the ions in the core are converted to neutrons, with the emission of neutrinos in the process. This takes place because electrons, due to quantum-mechanical interactions, are being forced into higher and higher energy states as the core of the star gets smaller and smaller. Protons can then absorb electrons to become neutrons. This is just the inverse of the process by which a free neutron decays to a proton plus an electron (and an antineutrino), a process which is prohibited in the interior of a neutron star because the electron does not have sufficient energy to be emitted.

Once the core reaches nuclear densities, repulsive nuclear forces become important and, if the mass of the core is small enough, the collapse is stopped and the core "bounces." The current thinking is that the bounce produces a shock wave which blows off the outer core and envelope of the star while a neutron star may be formed at the center. The large kinetic energy of the collapsing core is then converted to heat which ends up in radiation and in the kinetic energy of the material blown off the star. This process is thought to be the origin of some supernovae, although current models of collapsing stars do not follow this scenario in detail. The problems with the models are thought to arise from simplifying assumptions, for example, no rotation or magnetic fields, which are needed to make the calculations tractable. *SEE SUPERNOVA*.

Collapse to a black hole. Some collapsing stars are expected to end up with a core which has a mass larger than the maximum mass of a neutron star. In addition, neutron stars (and white dwarfs as well) which had been formed earlier could accrete enough matter to become more massive than the maximum mass of a neutron star. In these cases, the only known alternative is for such stars to collapse to a black hole. A black hole is a region in space in which gravity is so strong that even light cannot escape from its surface. Although black holes had been conjectured earlier, Karl Schwarzschild found the first black-hole solution of general relativity in 1916, although the significance of the solution as a black hole was not realized at the time. After suggestions that black holes might be an end point of stellar evolution, J. R. Oppenheimer and H. Snyder showed in 1939 that a black hole must result from spherically symmetric gravitational collapse if the mass of the collapsing body is large enough. At present, a black hole is believed to be the only result of a collapsing star that cannot lose enough matter to become a white dwarf or neutron star. Black holes are even more condensed than neutron stars. For example, a 1-solar-mass black hole would have a radius of about 2 mi (3 km). *SEE BLACK HOLE*.

The collapse of a star to a black hole could take place without a supernova occurring, in which case the disappearance of the star might not be seen as a visible event. However, one potentially observable characteristic of such a collapse would be the emission of gravitational waves as the black hole settled down to its equilibrium configuration. Gravitational waves interact with matter even more weakly than neutrinos, but detectors being planned and built would be sensitive enough to measure the gravitational radiation coming from collapse of a nearby star to a black hole. *SEE GRAVITATION*.

Because black holes emit no light, they are intrinsically dark and directly unobservable. However, they can still have an effect on nearby matter because their gravitational field is still present. In fact, there is good evidence that one condensed body in a binary star system, Cygnus X-1, is a black hole. Its gravitational field results in matter heating up sufficiently to emit x-rays, which can only be done by a highly condensed body, and its mass is determined from the orbital motion to be larger than 5 solar masses. From theory, the only known body which has these properties is a black hole. Black holes have also been identified by similar evidence in two other x-ray binaries, LMC X-3 and A0620-00. There is no proof that such black holes came from the gravitational collapse of a star, but that is the most reasonable hypothesis. *SEE ASTROPHYSICS, HIGH-ENERGY; BINARY STAR; X-RAY ASTRONOMY; X-RAY STAR*.

One might think that the range of masses of black holes formed from collapse is limited, since a normal star can burn nuclear fuel under stable conditions only if its mass is less than about 60 solar masses. Of course, black holes of larger than 60 solar masses could exist if sufficient additional matter fell into existing black holes or if many black holes coalesced to form a larger black hole. However, it has been conjectured that supermassive stars could exist, with masses between 10^3 and 10^7 solar masses, in which the pressure gradient arises from radiation pressure, and the release of gravitational energy during contraction provides the major or, for the more massive stars, the only energy source. Theoretical studies have shown that the more massive stars would become, at some stage in their evolution, unstable to gravitational collapse, ending up as supermassive black holes. There is some evidence for supermassive, condensed bodies at the centers of some galaxies, including the Milky Way Galaxy, and supermassive black holes have been proposed to be at the center of quasars. Even if the existence of such large black holes is confirmed, it would not prove that the black holes necessarily came from the collapse of a supermassive star. *SEE GALAXY, EXTERNAL; MILKY WAY GALAXY; QUASAR; SUPERMASSIVE STARS*.

General relativistic collapse. All scenarios of gravitational collapse involve Einstein's theory of gravitation, general relativity, in an intimate way. In some cases, general relativity is responsible for the instability in the star which starts the collapse. In other cases, general relativity prevents the formation of any final state of matter other than a black hole. While these effects are important, they do not illustrate the dramatic way in which general relativity changes perception of the geometry of space and time. This is best illustrated by imagining what gravitational collapse to a black hole would look like, both to an observer looking at the event from outside the system and to an observer riding down on the surface of the collapsing star. In order to lessen the effect of the gravitational tidal forces, the same kind of forces responsible for the tides on the Earth, it is assumed that the collapse is of a supermassive star.

The outside observer sees the star initially falling inward, not unlike what one would expect from newtonian gravity. However, as the radius of the star decreases to a value close to the Schwarzschild radius (the radius of the black hole which will be formed), the rate of falling inward slows. In fact, the star never quite reaches the Schwarzschild radius, even after an infinite time, as far as the outside observer is concerned. Another important effect comes about because of the gravitational redshift, the shifting of light toward the red part of the spectrum when the emitting atoms are in a region of large gravitational potential.

The light from the star is shifted more and more toward longer wavelengths as the star approaches the Schwarzschild radius. To the outside observer, any light emitted by the star becomes shifted quickly outside the visible range of the spectrum and the star becomes dark. Finally, since the redshifting of the light just reflects the apparent slowing down of the atomic clocks in the atoms in the star, other clocks, including the life processes of the observer riding down on the surface of the star, will be similarly slowed down. As the images of the star and the observer on the surface fade out of view, there will be left a picture of the star and the observer at some definite time, as when a motion picture slows to a halt. *See Gravitational redshift.*

The observer riding down on the surface has a quite different view of the process. Nothing particularly unusual is observed as the star collapses past its Schwarzschild radius, which occurs at the same time that the outside observer sees the star frozen in place as it fades out of view. However, crossing that radius does limit the options of the observer on the star; no matter what the observer does, escape past the Schwarzschild radius into the outside universe is impossible. Moreover, no observer, riding on the surface of a collapsing star, can escape the fate that awaits him or her. Initially, as the star gets smaller, unusual optical effects are noticed—the surface of the star appears to rise around the observer and the sky becomes smaller. More ominous is the increase in the tidal force on the observer. At the same time, this force stretches the observer from head to toe and squeezes the observer from the sides. After a short time this force rises to an infinite value, crushing both the observer and the star to infinite density. At this point the star is said to have reached a singularity. Once inside the Schwarzschild radius, the observer has no way of avoiding the singularity. Even if the observer were able to move outward at the speed of light, he or she would be crushed by the singularity in a short time.

Observation of collapse. Although gravitational collapse does not necessarily imply a violent astronomical event, the existence of supernovae in the Milky Way Galaxy and in other galaxies is taken to be evidence for the collapse of the cores of stars. Moreover, in the Milky Way Galaxy the Crab and Vela supernovae now show the remnants of a dramatic explosion. Both remnants also have pulsars approximately at their center, pulsars being identified as rapidly rotating neutron stars. The Crab supernova was noted when it occurred, so there is no question that the remnant seen today came from an explosive event. Prior to 1987, these examples provided the best evidence that at least some stars undergo gravitational collapse to a neutron star, in the process generating a supernova explosion. However, many questions could not be answered, in particular the nature of the star before collapse and the details of the collapse itself. *See Pulsar.*

On February 24, 1987, a supernova was observed in the Large Magellanic Cloud, a small galaxy associated with the Milky Way Galaxy, and therefore relatively close to Earth as far as supernovae are concerned. This supernova, named 1987A, must be considered one of the most important astronomical events of the twentieth century. For the first time a supernova occurred in a star that could be identified and studied on existing plates, and the record of the evolution of the supernova has been obtained with remarkable detail, using all parts of the spectrum from radio waves to x-rays and gamma rays. Moreover, prior to the appearance of the supernova optically, neutrinos from the supernova were observed in two widely separated detectors, confirming for the first time the predicted production of neutrinos in supernova explosions. Information from the light curve yields information on the radioactive elements produced in the explosion. The gravitational collapse of the core of the star was expected to end up as a neutron star, and a submillisecond pulsar was found in 1989 at the location of the supernova. However, not all the details corresponded to existing models. For example, the progenitor star was a blue supergiant, not a red supergiant, which is the kind of star, based on models, thought most likely to explode. Perhaps connected with this is the fact that the supernova was initially much dimmer than a normal supernova would have been at that distance. As these puzzles are unraveled, an increasingly accurate picture of the process of gravitational collapse should emerge. *See Supernova.*

Philip C. Peters

Bibliography. H. A. Bethe and G. Brown, How a supernova explodes, *Sci. Amer.*, 252(5):60–68, 1985; D. Helfand, Bang: The supernova of 1987, *Phys. Today*, 40(8):24–32, 1987; P. C. Peters, Black holes: New horizons in gravitational theory, *Amer. Sci.*, 62(5):575–583, 1974; S. L. Shapiro and S. A. Teukolsky, *Black Holes, White Dwarfs, and Neutron Stars*, 1983; S. Shklovskii, *Stars: Their Birth, Life, and Death*, 1978; J. C. Wheeler and R. P. Harkness, Helium-rich supernovas, *Sci. Amer.*, 257(5):50–58, 1987.

Gravitational lens

A massive body producing distorted, magnified, or multiple images of more distant objects when its gravitational fields bend the paths of light rays. Lenses have been observed when the light from very distant quasars is affected by intervening galaxies and clusters of galaxies, producing several different images of the same quasar. A. Einstein predicted the occurrence of this phenomenon in 1936, but the discovery of real gravitational lenses did not occur until 1979. Gravitational lenses, in addition to being intrinsically interesting, can reveal the intrinsic properties of galaxies, active galaxies, and quasars, and provide information on the overall geometry and scale of the universe.

Action of gravity. The lens phenomenon exists because gravity bends the paths of light rays, which is predicted by Einstein's general theory of relativity. Since photons, the carriers of light energy, have no mass, Newton's theory of gravity indicates that light would always travel in a straight line even if there were heavy, massive objects between the source and the observer. (Even if photons are given mass in Newton's theory, the predicted bending of light is different from the result in general relativity.) But in general relativity, gravity acts by producing curvature in space-time, and the paths of all objects, whether or not they have mass, are also curved if they pass near a massive body. *See Gravitation.*

Numerous eclipse observations have confirmed Einstein's prediction with modest accuracies of 20–

(Right) Large Magellanic Cloud; the large red patch is the 30 Doradus nebula, one of the largest H II regions known. *(Below)* Small Magellanic Cloud. *(Photographed at Anglo-Australian Observatory, © Royal Observatory, Edinburgh, 1984)*

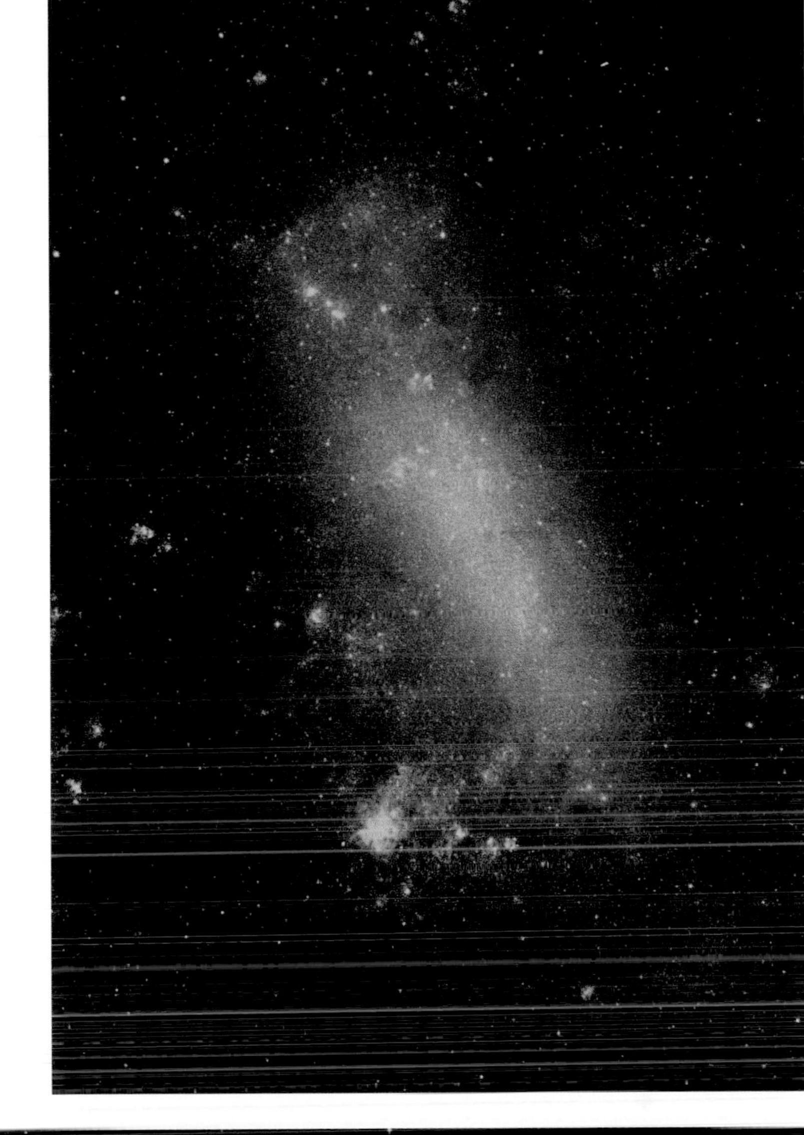

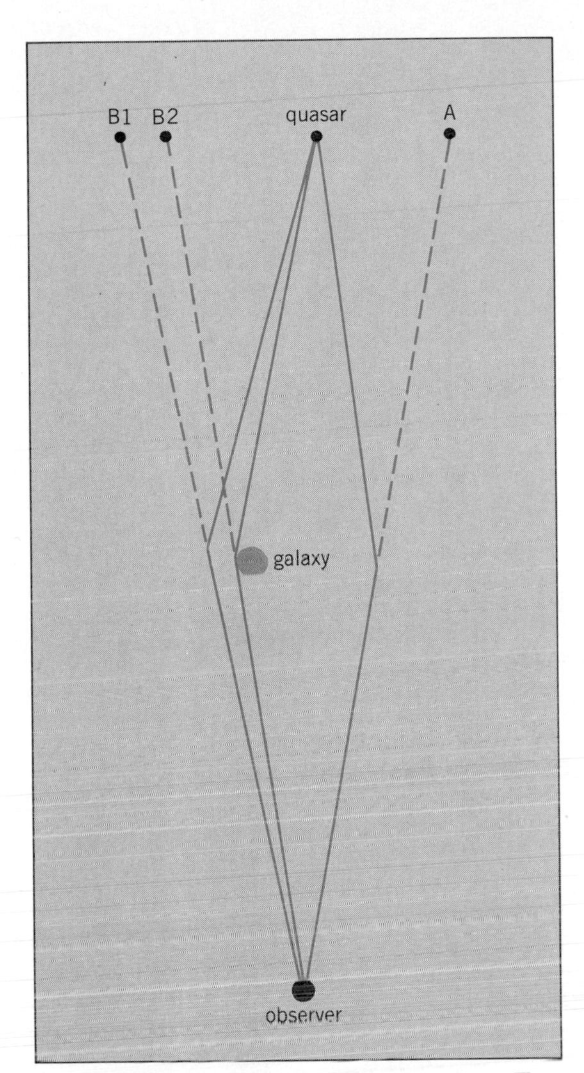

Fig. 1. Schematic illustration of a gravitational lens. The angles are exaggerated for clarity. Here, the lens action produces three images of a quasar (A, B1, and B2).

30%. Radio astronomers have measured changes in the positions of quasars that occur when the Sun passes near them in the sky. The precision of these experiments, which fit Einstein's predictions, is now at the level of tenths of a percent.

A massive object acts as a gravitational lens when light rays from a distant quasar are bent around or through it and are focused to form an image, which can be seen or photographed by an astronomer on Earth, as shown in **Fig. 1**. Here, three different images of the quasar at A, B1, and B2 are seen, since light from the quasar can travel along three different curved paths and still reach the observer.

Sometimes a lens can amplify the total intensity of light in a quasar image, making it considerably brighter than the quasar would appear to be in the absence of a lens. If the galaxy and quasar are sufficiently well aligned, several images of the same quasar will appear, since light can travel on many different paths and still arrive at the detecting telescope. It has been shown that if there are multiple images of the same quasar, there must be an odd number of them, as long as the galaxy is big enough so that it does not act as a point mass.

Discovery of lenses. Astronomers treated gravitational lenses as curiosities for a long period of time. Space is so empty that the probability of two stars being aligned accurately enough is extremely small. But the discovery of quasars, hyperactive galaxies which are bright enough to be visible even though they are nearly 10^{10} light-years (1 ly is equal to 5.88×10^{12} mi or 9.46×10^{12} km) away, made it possible to probe a much larger volume of space. Now, there was a reasonable chance that a galaxy might lie in the path of light traveling from a quasar to the Earth. *See Quasar.*

In March 1979, spectra were obtained of a pair of quasars located very close to each other, only 6 seconds of arc apart (the size of a dime at 2000 ft or 600 m away; **Fig. 2***a*). The similarity of their spectra indicated that the twin quasars might be two images of the same object. More measurements of the radio, infrared, and ultraviolet radiation from the twin quasars eventually confirmed the gravitational lens interpretation. This object is known as Q0957+561 A,B, where the numbers serve to indicate the position of the object in the sky, and the letter Q indicates that the object is a quasar. The letters A and B show that there are two objects at that location.

In the next 3 years two more gravitational lenses were discovered. A group of astronomers obtained measurements of a triple quasar, PG1115+080 A,B,C (PG = Palomar-Green; its unusual nature was first recognized in a survey of faint blue objects made at Palomar Observatory by R. Green; **Fig. 3**), and another double, designated 2345+007, was discovered in August 1981. The survey techniques which were responsible for the discovery of the second and

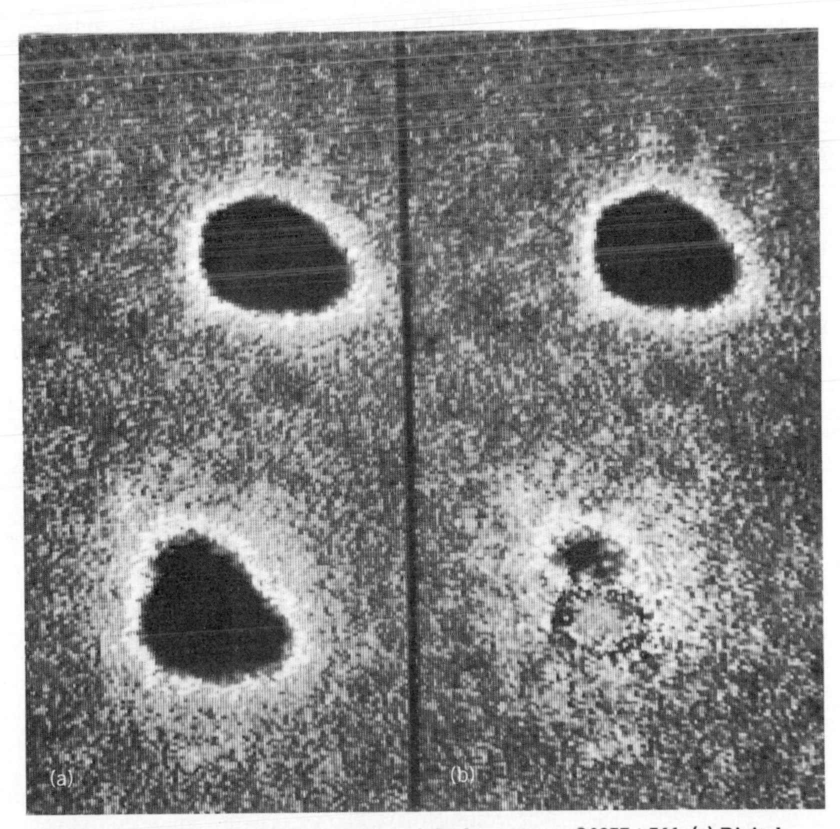

Fig. 2. Computer-processed photographs of the lens quasar Q0957+561. (*a*) Digital superposition of five 1-min exposures. (*b*) Same superposition with 0.7 of the northern (A) quasar image subtracted from the southern image, showing the galaxy just above the southern image. (*Institute of Astronomy, University of Hawaii*)

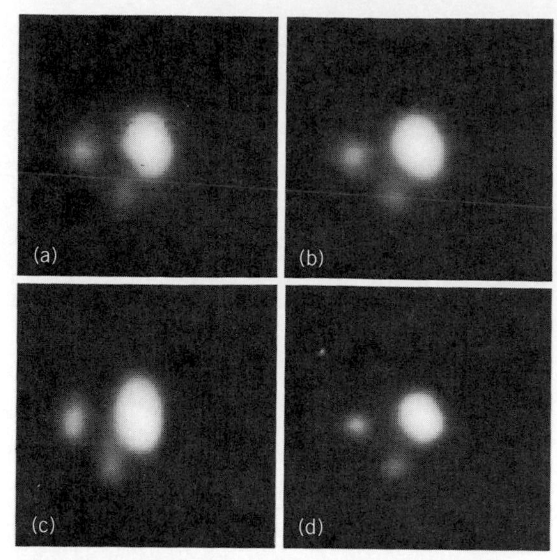

Fig. 3. Images of the triple quasar PG1115+08. (*a*) Blue filter. (*b*) *V* band (yellow) filter. (*c*) Unfiltered image. (*d*) Red filter. (*From E. K. Hege et al., Morphology of the triple QSO PG1115+08, Nature, 287:416–417, 1980*)

third lenses have been responsible for the discovery of more than a thousand quasars.

Double quasar. The best-studied gravitational lens consists of the first double quasar Q0957+561 and an unnamed galaxy which lies between it and the Earth. The first evidence that this was indeed a lens came from optical spectroscopy. The redshifts, the strength of the emission lines, and the position of the absorption lines in the two quasars were identical. The southern, or B, quasar was about 0.7 times as bright as the northern, or A, quasar. Since the redshifts, spectra, and line strengths of different quasars vary considerably, it would be a remarkable coincidence if two quasars with identical properties were very close to each other. *See Redshift.*

But coincidences do happen, and more observations were needed in order to confirm the lens interpretation. It was important to determine whether the two images really were identical twins and whether a reasonable type of galaxy could produce the bending needed to produce the observed positions of the images. Since lens galaxies are supposed to produce an odd number of images, the location of the third one was a mystery, and the possibility arose that the intervening object was a huge black hole instead of a galaxy. It was also important to detect the galaxy which was responsible for bending the light from the quasar and producing the multiple images. *See Black hole; Galaxy, external.*

Further evidence in support of the lens model comes from observations in the infrared and ultraviolet parts of the spectrum. If this object is a lens, then the intensity of the B image should always be a certain fixed percentage of the intensity of the A image, as long as the measurements are taken at the same time. Allowing for possible long-term variations in the brightness of the quasar, this constraint indicates that the B image should be 70–80% as bright as the A image in all wavelengths. Infrared and ultraviolet observations have shown that this prediction is correct.

Detailed radio maps of the region surrounding the double quasar, once they were correctly interpreted, confirmed that a gravitational lens was indeed at work in the double quasar. Astrophysicists used radio maps, shown in **Fig. 4**, to predict where the lensing galaxy should be. The third optical image of the quasar, designated B2, lies right on top of the lensing galaxy. In Fig. 2*b*, 0.7 of the A image is substracted from the B1 image so that the image of the lensing galaxy can be seen more easily. The third quasar image, B2, is probably submerged by the light from the lensing galaxy.

About a dozen other examples of gravitational lenses were discovered in the 1980s. Many cases resemble the double quasar in that light from a distant quasar is being lensed by an intervening galaxy. In some cases, the distant object is a galaxy, with a fuzzy image built up from the individual light from billions of stars. Quasars have pointlike, stellar images, which usually retain their pointlike nature when being imaged by an intervening galaxy.

The most spectacular example of a gravitational lens vindicates Einstein's original picture of the result of gravitational lensing, a ringlike image produced by a near-perfect alignment of the Earth, the lensing galaxy, and the lensed object. Some clusters of galaxies are surrounded by faint, circular arcs of light, centered on the cluster. Detailed analysis of one of these arcs, associated with the cluster of galaxies named Abell 370, proved fairly conclusively that it was the image of a distant galaxy, gravitationally lensed by the cluster.

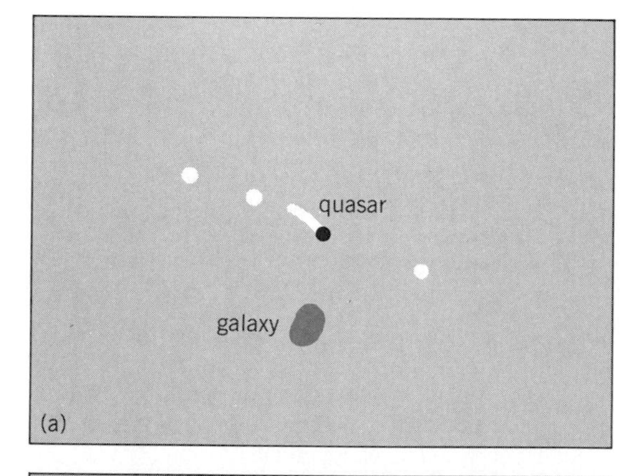

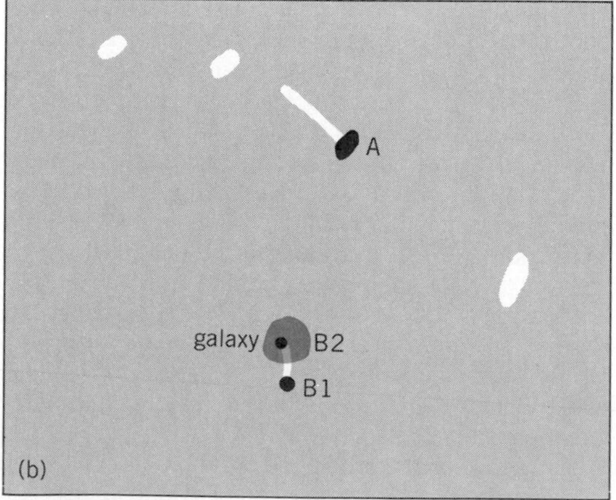

Fig. 4. Configuration of the double quasar Q0957+561. White areas represent areas of radio emission; dark gray and black areas represent optical images. (*a*) Configuration which would appear with the lens action of the galaxy turned off. (*b*) Configuration which is actually observed, with the lens action on.

Implications. The discovery of gravitational lenses affects astronomers' understanding of the universe as a whole. The very existence of this phenomenon demonstrates that a dozen quasars are more distant than the galaxies which are focusing their light. Thus lenses provide one more way of showing that quasars are extremely distant objects, and that their redshifts are produced by the expansion of the universe, rather than by some kind of exotic new physics as some astrophysicists have argued.

Gravitational lenses can make distant quasars appear brighter than they would otherwise be, since their focusing action beams more light in the direction of the Earth. This effect could cloud the interpretation of the quasar statistics, which currently argues for a "quasar era" early in cosmic evolution. Further, lenses acting as quasar amplifiers might explain the high luminosity of some extreme quasars. But they cannot account for the high power of quasars considered as a class, since it is very unlikely that all of the 1500 known quasars are single-image lenses.

When the brightness of the multiple images in gravitational lenses has been monitored for a long time, lenses may reveal the cosmos in some more fundamental ways. If the energy-emitting regions of quasars are small (as is believed), then stars in the halos of galaxies could also act as lenses, changing the intensity of the light from quasar images when the stars pass between the Earth and the quasar. The lens phenomenon could be used to detect the existence of these stars, even if they were too faint to see optically. It is even possible, in principle, to detect Jupiter-sized objects in the halo of a galaxy acting as a gravitational lens, though other variations in the intensity of radiation from quasars might make it very difficult to unscramble the amplification produced by the low-mass objects from other fluctuations in the quasar brightness.

Variations in the brightness of multiple images of the same quasar could also determine the distance to the lensing galaxy and thus the scale of the universe. The distance to the quasar along the path that corresponds to one image differs from the distance along the path corresponding to the other image. Thus, if the quasar suddenly increased its power, one image would brighten first, before the other changed. The time lag could be measured, and the difference in path lengths thus determined, since light follows both paths at the same speed. The difference in path lengths is related to the distance to the quasar and to the lens galaxy. By measuring the distance to these very distant objects, a much better measurement of the overall cosmic distance scale would be obtained. *See* Cosmology.

Harry L. Shipman

Bibliography. N. Cohen, *Gravity's Lens: Views of the New Cosmology*, 1988; J. K. Lawrence, Gravitational lenses and the double quasar, *Mercury*, 9(3):66–72, May-June 1980; P. Young et al., Q0957 + 561: Detailed models of the gravitational lens effect, *Astrophys. J.*, 244:736–755, March 15, 1981.

Gravitational radiation detector

A research instrument for receiving gravitational radiation predicted to come from astronomical sources such as supernova explosions or rotating binary stars. The two classes of detectors are resonant-mass and free-mass antennas. Both types comprise a mechanical element and a transducer that converts mechanical motion to an electronic signal. A gravitational wave from a supernova explosion in the Milky Way Galaxy would exert a weak tidal force on the mechanical elements of the antenna, which causes a deformation smaller than 10^{-18}. The deformation is converted to an electronic signal by an electromechanical transducer on the resonant-mass antennas and a laser interferometer in the case of the free-mass detectors. *See* Binary star; Supernova.

Resonant-mass antennas. The resonant-mass antennas are commonly cylindrical bars of a few tons mass. The material most commonly used is aluminum, followed by niobium. To reduce thermal noise, the bar is brought to cryogenic temperatures in a vacuum vessel cooled by liquid helium and is isolated against vibrations from the terrestrial environment by a series of mechanical filters. The lowest-frequency longitudinal mode of the antenna, in which the two end faces of the cylinder are displaced in opposition, is most strongly excited by a passing gravitational wave. An electromechanical transducer is attached to one end face of the bar. Typically it comprises an electrical readout circuit and a less massive mechanical resonator tuned to the frequency of the bar's lowest longitudinal mode. The energy of the bar vibration is transferred to the transducer mechanical resonator, the motion of which is converted to an electronic signal by the readout circuit. The ratio of the vibrational amplitude of the transducer resonator to the antenna end face is proportional to the square root of the ratio of the masses, $\sqrt{M/m}$, where M is the effective antenna mass and m the transducer resonator mass. The ratio of masses is usually close to 10,000, giving a factor of 100 for amplitude amplification.

Transducer electrical readout circuits typically incorporate the transducer resonator into an energy-storage component, either an inductor which stores magnetic energy or a capacitor which stores energy in the form of an electric field. The vibration of the transducer resonator modulates the value of the inductance or capacitance, transferring some of the stored electromagnetic energy to another part of the readout circuit, where it is detected by a low-noise amplifier. Readout circuits are also classified according to the nature of the stored electromagnetic energy. In one type the energy is stored in the form of a static magnetic or electric field; the modulation of the energy storage element at the antenna frequency produces an electrical signal at that frequency. In the other type of readout the stored energy is in a radio-frequency or higher-frequency oscillating field. Such transducers are called parametric or active devices.

The minimum detectable gravitational wave strength is determined by the noise in the detector. Sources of noise include the brownian motion of the antennas, which is the thermally induced fluctuations of the bar amplitude similar to the brownian motion of small particles observed under a microscope. The thermally induced electrical fluctuations (the electrical analog of brownian motion) in the readout circuit and the noise of the amplifier which senses the readout signal are the other major sources of noise. Ultimately the quantum-mechanical uncertainty principle limits the precision of the measurement of the antennas' end-face position. The back-action-evasion or quantum-nondemolition technique has been proposed to circumvent the quantum limit. In this technique the antenna bar is forced into a so-called squeezed state by interaction with the transducer. In a squeezed state the fluctuations of the antenna are reduced at certain instants at the expense of increased fluctuations at other times. If a weak gravitational-wave impulse arrives at the time of minimum noise, the probability of detection is enhanced by the squeezing technique.

Resonant-bar gravitational-wave detectors are being developed in the United States, Italy, Japan, and China.

Free-mass antennas. The free-mass antennas are composed of almost inertial masses which are actually very low-frequency pendulums. A common design is the arrangement of three such masses at the vertices of a right triangle with equal legs lying on the surface of the Earth. The passage of a gravitational wave in a direction perpendicular to the plane of the free-mass antenna lengthens one leg of the triangle relative to the perpendicular leg. This change in length can be detected by a laser interferometer which is composed of mirrors mounted on the masses and a high-power visible laser light source. Each mass is rigorously isolated from vibrations in the environment, and the entire system is placed in a high-vacuum chamber to eliminate light scattering by gas and dust particles. There are two types of interferometers being developed: the multiple-pass Michelson type and the Fabry-Perot type. In both the laser beam makes multiple passes before emerging from either arm of the interferometer. The interference pattern of the two beams is monitored close to an intensity minimum with a photodetector that creates an electrical signal proportional to the light intensity. Aside from terrestrial vibration, the major source of noise is the shot-noise fluctuation of the laser beam intensity arising from the discrete arrival times of the photons composing the laser light. Ultimately, when highly intense lasers are used to illuminate the interferometer, the fluctuating pressure of the radiation in the interferometer on the free masses will limit the performance.

A United States laser interferometer gravitational observatory and very large European free-mass antennas are in the planning stage. The interferometers in these detectors will have baselines of 2–3 mi (3–5 km).

Gravitational radiation detectors do not yet have sufficient sensitivity to detect gravitational waves from outer space. The development of more sensitive detectors for gravitational radiation from astrophysical sources is an active area of research.

Mark F. Bocko; David H. Douglass

Bibliography. *General Relativity: An Einstein Centenary Survey*, 1979; D. Jeffries et al., Gravitational wave observatories, *Sci. Amer.*, 256:(b)50–56, June 1987; P. F. Michelson, J. C. Price, and R. C. Taber, Resonant mass detectors of gravitational radiation, *Science*, 237:150–156, 1987; *Noise in Physical Systems and 1/f Noise 1985*, 1986.

Gravitational redshift

A shift toward longer wavelengths of spectral lines emitted by atoms in strong gravitational fields. It is also known as the Einstein shift. One of three famous predictions of the general theory of relativity, this shift results from the slowing down of all periodic processes in a gravitational field. The amount of the shift is proportional to the difference in gravitational potential between the source and the receiver. For starlight received at the Earth the shift is proportional to the mass of the star divided by its radius. In the solar spectrum the shift amounts to about 0.001 nanometer at a wavelength of 500 nanometers. In the spectra of white dwarfs, whose ratio of mass to radius is about 30 times that of the Sun, the shift is about 0.03 nm, which can easily be measured if it can be separated from the Doppler effect. This was first done by W. S. Adams for the companion of Sirius, a white dwarf whose true velocity relative to the Earth can be deduced from the observed Doppler effect in the spectrum of Sirius. The measured shift agreed with the prediction based on Einstein's theory and on independent determinations of the mass and radius of Sirius B. A more accurate measurement was carried out in 1954 by D. M. Popper, who measured the gravitational redshift in the spectrum of the white dwarf 40 Eridani B. Similar measurements, all confirming Einstein's theory, have since been carried out for other white dwarfs. Attempts to demonstrate the gravitational redshift in the solar spectrum have thus far proved inconclusive, because it is difficult to distinguish the gravitational redshift from so-called pressure shifts resulting from perturbations of the emitting atoms by neighboring atoms. SEE WHITE DWARF STAR.

In 1960 R. V. Pound and G. A. Rebka, Jr., succeeded in measuring the gravitational redshift in a laboratory experiment. ^{57}Fe nuclei bound in a solid emit and absorb gamma rays in an exceedingly narrow frequency range ($\Delta\nu/\nu = 3 \times 10^{-13}$) centered on a sharply defined frequency near 14.4 keV. A source and an absorber of this accurately monochromatic radiation were separated by a vertical distance h of 74 ft (22.5 m), so that the radiation incident on the absorber was shifted in frequency by an amount given by the equation below, according to Einstein's

$$\frac{\Delta\nu}{\nu} = \frac{gh}{c^2} = 2.5 \times 10^{-15}$$

theory, where g is the acceleration of gravity and c is the speed of light. The rate of absorption was measurably diminished relative to the rate that would have been observed in the absence of the gravitational redshift. A suitable Doppler shift introduced by a small, accurately measurable relative motion of the source and the absorber restored the measured absorption rate to its normal value. The Doppler shift needed to accomplish this restoration agreed with the value predicted by Einstein's theory to within 10%. Subsequently, the error has been reduced to 1%.

In 1976 R. F. C. Vessot and collaborators used a rocket-borne hydrogen maser with a frequency stability of 1 part in 10^{14} to measure a gravitational redshift $\Delta\nu/\nu = 4 \times 10^{-10}$, corresponding to the difference in gravitational potential between the ground and the apogee of the rocket. This experiment confirmed the predicted redshift to better than 1 part in 10^4. SEE MASER.

Attempts have also been made to deduce stellar masses from measurements of the gravitational redshift, but the difficulty of allowing properly for the Doppler effect and for pressure shifts renders these determinations very uncertain.

David Layzer

Greenhouse effect

The effect created by a planet's or satellite's atmosphere acting as the glass walls of a greenhouse in trapping heat from the Sun. Plants are raised in greenhouses because the glass walls help to keep heat inside, so that temperatures are higher inside than outside. In a crudely similar way, the Earth's atmosphere acts like the walls of a greenhouse and partially blocks the loss of surface heat to space, thereby raising the Earth's surface temperature above the freezing point of water. The greenhouse effect has greatly varying strengths for other objects in the solar system. It is extremely strong for Venus and very weak for Mars, the Earth's neighbors in the solar system; it has a modest strength for Titan, Saturn's largest satellite; and it is nonexistent for airless bodies, such as the Moon and Mercury. SEE MERCURY; MOON.

Mechanism. On an airless body, the globally averaged surface temperature is determined by a balance between the amount of sunlight that its surface absorbs and the amount of heat or thermal radiation that its surface emits to space. Since thermal radiation depends very sensitively on temperature (it varies as the fourth power of the temperature), the balance between sunlight absorbed and thermal radiation emitted acts as a powerful thermostat on the surface temperature. If the Earth absorbed the same fraction of sunlight that it does now (70%) but lacked an atmosphere, its surface temperature would be about 255 K (0°F). In that case, all the oceans would freeze and life would not exist.

However, the Earth does have a substantial atmosphere made mostly of nitrogen and oxygen, together with a number of less abundant gases, including water vapor and carbon dioxide. Although nitrogen and oxygen do not absorb or emit thermal radiation to any appreciable degree, water vapor and carbon dioxide do. Therefore, the Earth's surface is warmed not only by sunlight but also by thermal radiation emitted by the atmosphere. It loses energy by thermal radiation and to a lesser extent by turbulent (convective) heat exchange with the lowest parts of the atmosphere. This new energy balance (Fig. 1) results in a mean surface temperature of 288 K (59°F), which is 33 K (59°F) warmer than its airless counterpart. This difference allows water to be present in its liquid phase, one of the essential conditions for life. SEE EARTH.

Effect in the solar system. The magnitude of the greenhouse effect depends on how large the atmosphere is and what gases it contains.

Mars. Only a very small greenhouse warming occurs for a planet with a very thin atmosphere containing only one type of gas that absorbs and emits at a limited set of thermal infrared wavelengths. The Martian atmosphere is an example. The atmospheric pressure at the surface of Mars is only about 0.7% that at the Earth's surface. Carbon dioxide, the major constituent of the Martian atmosphere, is the only gas species that absorbs and emits significant amounts of thermal radiation. At the low temperature and pressure conditions of the Martian atmosphere, it does so only at wavelengths near 15 micrometers. As a result, the mean surface temperature of Mars, 218 K (−67°F), is augmented by only about 7 K (13°F) by the greenhouse effect. SEE MARS.

Venus. A massive atmosphere with several gases that absorb and emit thermal radiation over a wide range of infrared wavelengths can produce a very large greenhouse effect. Venus's atmosphere is an example. The atmospheric pressure at Venus's surface is about 90 times that at the Earth's surface. Carbon dioxide, the major atmospheric gas, absorbs over a broad range of infrared wavelengths because of the high temperatures in Venus's atmosphere and because

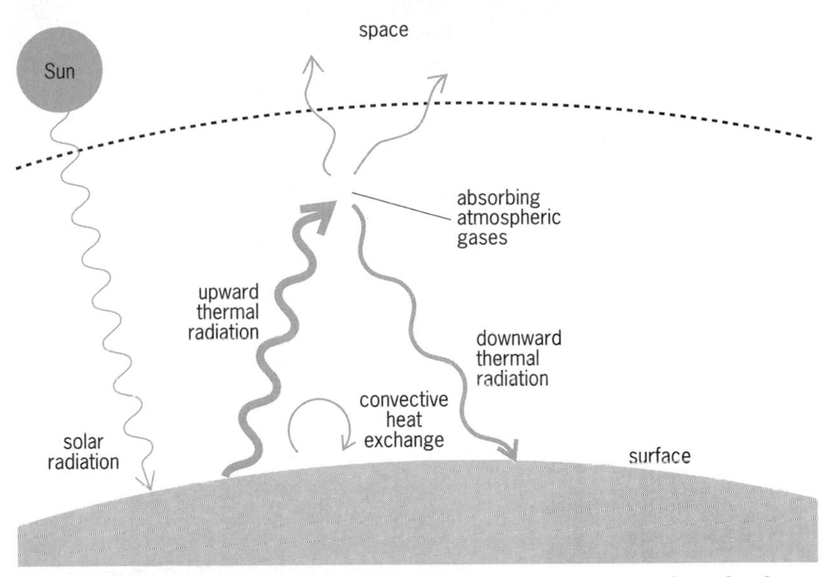

Fig. 1. Schematic diagram showing the greenhouse warming of the surface of a planet that has an atmosphere. The surface is heated by a combination of solar radiation and thermal radiation from the atmosphere. It is cooled chiefly by the thermal radiation it emits, as well as by heat lost through convective exchange with the somewhat cooler overlying atmosphere.

of the very large amounts of carbon dioxide that permit weak absorption bands to absorb radiation effectively. In addition, water vapor and sulfur dioxide, each constituting about 0.01% of Venus's air, absorb and emit effectively at wavelengths where carbon dioxide does so only weakly at best. Thus, collectively, the lower atmosphere of Venus absorbs and emits at essentially all infrared wavelengths carrying significant thermal radiation. Consequently, Venus's surface receives almost as much thermal radiation from the atmosphere as the amount it radiates upward, leading to a surface temperature of about 730 K (855°F). This greenhouse warming of 500 K (900°F) occurs even though Venus absorbs less sunlight than does the Earth: although Venus is closer to the Sun than the Earth is, it reflects a much larger fraction of sunlight (75%) to space than the Earth (30%). SEE VENUS.

Titan. Titan provides an example of a modest greenhouse effect, whose magnitude is limited by the occurrence of a major window region, a range of infrared wavelengths where the atmosphere absorbs and emits ineffectively. The pressure at Titan's surface is about 1.5 times that at the Earth's surface. As on the Earth, nitrogen is the most abundant gas species, with minor but significant amounts of methane (a few percent) and molecular hydrogen (a few tenths of a percent). At Titan's very cold temperatures, isolated nitrogen, methane, and hydrogen gas molecules do not absorb or emit radiation at infrared wavelengths carrying appreciable thermal energy. However, they do absorb well at longer infrared wavelengths when pairs of them are close together, mutually inducing dipole moments that allow them to absorb radiation. This pressure-induced absorption is ineffective at wavelengths shorter than about 25 μm. Because of this short-wavelength window, thermal radiation from the surface is only partially compensated by downward thermal radiation from the atmosphere. Titan's surface temperature of 94 K (−290°F) is about 12 K (22°F) warmer because of its modest greenhouse effect. SEE SATURN.

Earth. The Earth's atmosphere contains several gases that absorb in portions of the thermal infrared. Of these, water vapor and carbon dioxide are the most important. Since carbon dioxide absorbs well at wavelengths near 15 μm while water vapor absorbs well at wavelengths longer than 20 μm, these two gases collectively encompass a larger fraction of the infrared spectral domain than either one does separately. However, the magnitude of the greenhouse warming of the Earth's surface is limited by the occurrence of a window region between 8 and 12 μm. While no gas absorbs well throughout this entire region, ozone does absorb strongly near 9.6 μm, while trace constituents, such as nitrogen dioxide, methane, and chlorofluorocarbons (CFCs), absorb in limited portions of the region.

Figure 2 summarizes the surface temperatures and amount of greenhouse warming present on Venus, Earth, Mars, and Titan.

Past and future climates. The Earth's climate has varied on time scales that range from decades to many millions of years. Changes in the amounts of greenhouse gases in the atmosphere have contributed to many such variations. For example, the Earth experienced a succession of ice ages and ice-free periods over the last several million years. These climatic variations appear to have been driven by quasiperiodic changes in the degree of circularity of the Earth's orbit and the tilt and orientation of its axis of rotation. However, examination of ice cores shows that the abundances of such greenhouse gases as carbon dioxide and methane varied by several tens of percent between glacial and interglacial periods. These gases were less abundant during the glacial periods, thereby helping to lower temperatures then. *SEE EARTH ROTATION AND ORBITAL MOTION.*

Human activities are injecting into the atmosphere steadily increasing amounts of greenhouse gases, such as carbon dioxide, methane, nitrogen dioxide, and chlorofluorocarbons. Continuous measurements of the abundances of these gases over several decades provide unambiguous evidence of their increase at unprecedented rates. For example, there is about 25% more carbon dioxide in the atmosphere now than there was at the start of the industrial revolution, and its abundance is increasing at about 0.5% per year.

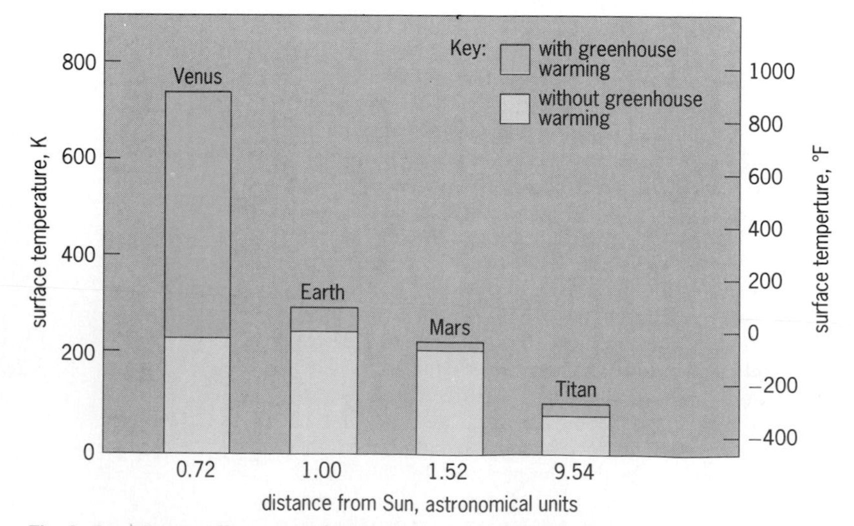

Fig. 2. Greenhouse effect on surface temperatures of Venus, Earth, Mars, and Titan, showing the actual surface temperatures at the tops of the bars.

This increase is widely thought to be due almost entirely to such activities as the burning of fossil fuels and the clearing and burning of forests. The increases in greenhouse gases will enhance the greenhouse effect by narrowing the atmospheric greenhouse windows and hence increasing the downwardly directed thermal radiation emitted by the atmosphere.

If greenhouse gases keep increasing at current rates, they would produce, in the absence of other factors, an increase in the globally averaged surface temperature of 1 to 4 K (2 to 7°F) by the latter part of the twenty-first century. This change would be accompanied by larger changes in local temperatures, sizable shifts in precipitation patterns, and, eventually, a partial melting of the ice caps at high latitudes, which would elevate the sea level. An important challenge is how to deal with or avoid such climate changes. *SEE PLANETARY PHYSICS.*

James B. Pollack

Bibliography. J. F. Kasting, O. B. Toon, and J. B. Pollack, How climate evolved on the terrestrial planets, *Sci. Amer.*, 256(2):90–97, February 1988; C. P. McKay, J. B. Pollack, and R. Courtin, The greenhouse and antigreenhouse effects on Titan, *Science*, 253:1118–1121, 1991; J. B. Pollack, Kuiper Prize Lecture: Present and past climates of the terrestrial planets, *Icarus*, 91(2):173–198, June 1991; S. Schneider, The changing climate, *Sci. Amer.*, 261(2):70–79, September 1989.

Halley's Comet

The most famous of comets, associated with many important events in history. Records of Halley's Comet appear at least as far back as 240 B.C., and they are found in the Bayeux Tapestry (the apparition of 1066) and in the *Nuremberg Chronicle* (the apparition of 1456 and probably those of 684 and 1301). The comet's size, activity, and favorably placed orbit, with the perihelion roughly halfway between the Sun and the Earth's orbit, ensure its visibility to the naked eye at each apparition.

History. This comet was the first to have its return predicted, a feat accomplished by Edmond Halley in 1705. He computed the orbits of several comets with Isaac Newton's new gravitational theory. The orbits of comets observed in 1531, 1607, and 1682 were remarkably similar. Halley assumed that the sightings were of a single comet and predicted its return in 1758–1759. The prediction was verified, and the comet was named in his honor. Halley's Comet was observed in 1835 by F. W. Bessel and by numerous astronomers in 1910 and 1986. Halley's Comet displays the gamut of known cometary phenomena, including a long tail when sufficiently close to the Sun (**Fig. 1**). Because of its predictable orbit, brightness, and extensive activity, it was the prime target of the six spacecraft making up the Halley Armada in March 1986. *SEE COMET.*

Orbital properties. The comet's orbit is a very elongated ellipse, with an eccentricity of 0.967, which has a perihelion of 0.59 astronomical unit (1 AU = the average Earth–Sun distance = 9.30×10^7 mi = 1.496×10^8 km) and an aphelion of 35 AU, between the orbits of Neptune and Pluto. Halley's Comet was farthest from the Sun in 1948, and has been moving away from the Sun since its perihelion on February 9, 1986. Halley's Comet will return to perihelion in 2061. The average period of revolution is 76 years, and the comet's

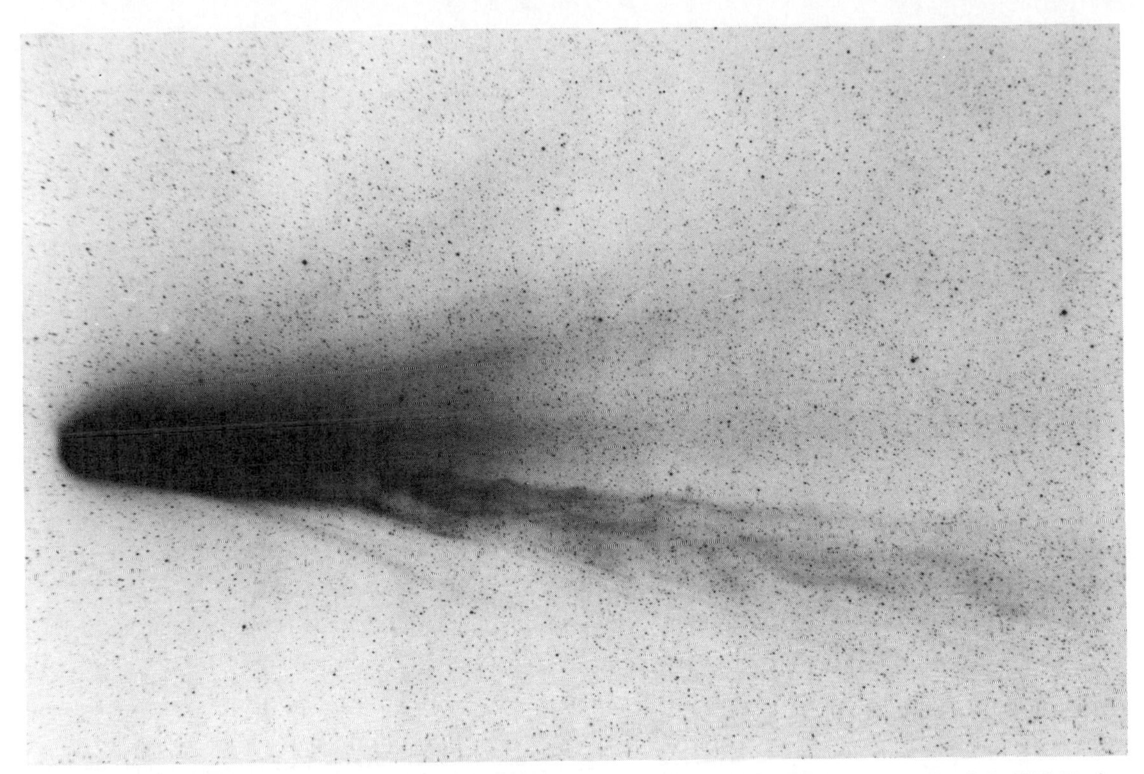

Fig. 1. Halley's Comet as photographed by the United Kingdom Schmidt telescope in Australia on March 9, 1986. Dust-tail structures are visible (above), and the plasma tail (below) also shows a completely detached portion called a disconnection event. *(Copyright © by Photolabs, Royal Observatory, Edinburgh)*

motion is retrograde, that is, opposite the planets' motion. The present relative positions of the comet's orbit and the Earth's orbit mean that the comet can approach Earth as close as 0.15 AU at the descending node. *See* CELESTIAL MECHANICS.

1986 apparition. Halley's Comet was detected for the first time on its most recent approach to the Sun by D. C. Jewitt and G. E. Danielson of the California Institute of Technology. The comet was recovered on October 16, 1982, by using the 200-in. (5-m) telescope on Palomar Mountain and an advanced electronic detector originally designed for the Hubble Space Telescope.

Casual viewers, if they were far enough south and well away from city lights, saw the comet in March and April 1986. However, Halley's Comet was not the spectacular object for public viewing that it was in 1910. The excitement for the 1985–1986 apparition lay primarily in the massive cooperative effort that scientists launched to observe the comet.

In March 1986, six uncrewed spacecraft successfully encountered Halley's Comet on the sunward side and made measurements in its vicinity (see **table**). These missions produced data that have greatly enhanced the understanding of comets. Observations of Halley's Comet were also made by spacecraft in orbit around the Earth and the planet Venus, and by ground-based instruments. The ground-based observations were coordinated by the International Halley Watch, composed of networks of astronomers and institutions worldwide formed to coordinate the total observing effort and to archive results.

Results from both space and ground-based observations focus on three general areas. The first concerns the interaction of the comet with the solar wind. H. Alfvén's basic picture has been confirmed: the comet's plasma tail is indeed formed by the comet–solar wind interaction. Molecular ions from the comet are trapped onto solar-wind magnetic field lines, causing the magnetic field lines to drape around the comet to form the plasma tail. This process has been confirmed in detail by the spacecraft observations. Somewhat surprising is the immense distance over which the interaction takes place, up to approximately 1×10^7 km (6×10^6 mi) or more from the comet, as measured by the spacecraft. Plasma processes in

Space missions to Halley's Comet

Spacecraft (sponsor)	Closest approach, approx. distance (date)
*Vega 1** (Soviet Union)	8890 km or 5520 mi (March 6, 1986)
Suisei (Japan)	151,000 km or 94,000 mi (March 8, 1986)
*Vega 2** (Soviet Union)	8030 km or 4990 mi (March 9, 1986)
Sakigake (Japan)	7,000,000 km or 4,350,000 mi (March 11, 1986)
*Giotto** (European Space Agency)	605 km or 375 mi (March 14, 1986)
International Cometary Explorer (NASA, United States)	28,000,000 km or 17,400,000 mi (March 25, 1986)

*Produced imaging of comet nucleus.

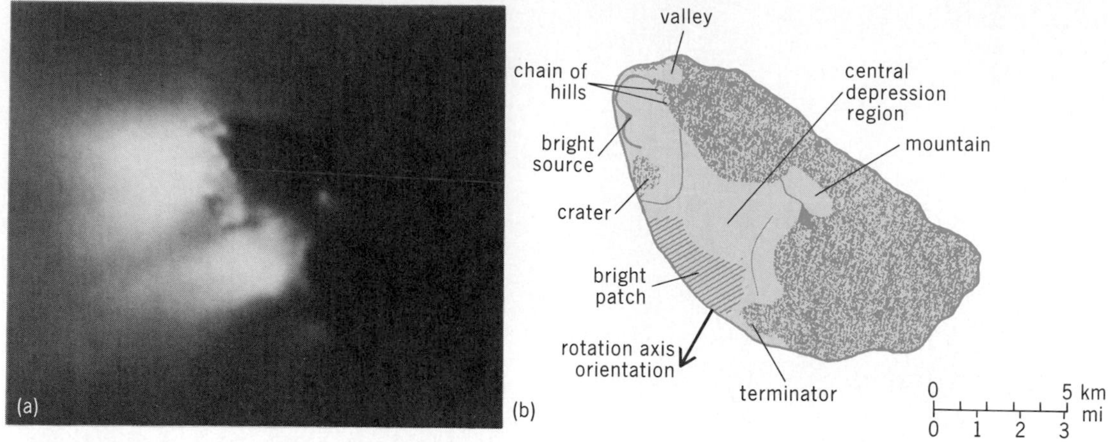

Fig. 2. Nucleus of Halley's Comet from *Giotto* spacecraft. (*a*) Composite image formed from 60 individual images. The resolution varies from about 800 m (2600 ft) at the lower right to about 80 m (260 ft) at the upper left. The material in the bright jets streams toward the Sun, leftward. (*b*) Matching drawing labeling the features on the nucleus. (*Harold Reitsema, Ball Aerospace; copyright © 1986 by Max Planck Institut für Aeronomie*)

comets can produce spectacular results, as illustrated by the disconnection event shown in Fig. 1.

The second area is chemical composition. The composition of the gas as measured in the inner coma is approximately 80% water (H_2O), roughly 10% carbon monoxide (CO) as determined from a rocket observation, approximately 3.5% carbon dioxide (CO_2), a few percent in complex organic compounds such as polymerized formaldehyde (H_2CO), and the remainder in trace elements. The deuterium-to-hydrogen ratio (D/H) was found to be close to the value for terrestrial ocean water. These results support the view that comets may have supplied an important fraction of the volatile elements to the terrestrial planets, possibly including prebiotic molecules to Earth. Note that the composition values refer to a comet that has passed through the inner solar system many times. The values for the deep interior of Halley's Comet or for other comets may be different and would probably have a higher value of carbon dioxide.

The dust composition was found to be as follows. Some particles are composed primarily of the light atoms hydrogen (H), carbon (C), nitrogen (N), and oxygen (O); these are "CHON" particles. Another kind has a silicate composition similar to the rocks that make up the crusts of Earth, Moon, and Mars, and of most meteorites. Most dust particles resemble a mixture of these two types, that is, they resemble carbonaceous chrondrites enriched in the light elements (hydrogen, carbon, nitrogen, and oxygen). These should resemble the Brownlee particles collected in the Earth's upper atmosphere.

The third general area on which the results focus is the cometary nucleus. Researchers have long based their understanding of the nucleus on F. Whipple's dirty, water–ice snowball model. According to this model, the Sun's heating of the nuclear surface produces sublimation of the ices, causing gas and dust to be released to form the cometary atmosphere and tails. All available evidence from the Halley observations, including images of the nucleus obtained by *Giotto* and the *Vega*s, confirms Whipple's model. The spacecraft observations also show that the nuclear body, shaped like a potato or peanut, is roughly 15 km (9 mi) long and 8 km (5 mi) across, somewhat larger than expected (**Fig. 2**). The surface of the comet, found to be darker than expected, is comparable to black velvet or coal. The nuclear surface is likely a crust of dust covering the sublimating ices. Solar energy heats the surface, and energy is conducted downward to the ices, where sublimation occurs. Some of the gases produced by the sublimation escape through thin areas or openings in the crust, forming the jets that were observed originating from the surface of Halley's nucleus.

A major area of controversy concerning Halley's Comet persists. Solid evidence for the rotation period of the nucleus has been presented to support values of 2.2 days and 7.4 days. In retrospect, the rotation of the highly asymmetrical nucleus is likely to be complex. Observations of the bare nucleus as it recedes from the Sun may resolve this problem.

Some insight into the active comet's dusty environment can be inferred from the dust's effect on the spacecraft. The three spacecraft that passed closest to Halley—*Giotto* and the two *Vegas*—survived but were seriously damaged because of the density of the dust and the high speed of the spacecraft. Several of the spacecraft, including the damaged ones, may be retargeted to carry out further observations in the solar system.

Associated meteors. Halley's Comet is associated with two meteor showers, the η-Aquarids and the Orionids. The origin of these showers undoubtedly involves meteoroids from the comet's nucleus that were gravitationally perturbed into their current orbits, which intersect the Earth's orbit. SEE METEOR.

John C. Brandt

Bibliography. J. C. Brandt, *Comets: A Scientific American Reader*, 1981; R. D. Chapman and J. C. Brandt, *The Comet Book*, 1984; M. Grewing, F. Praderie, and R. Reinhard (eds.), *Exploration of Halley's Comet*, 1988; F. L. Whipple, *The Mystery of Comets*, 1985.

Helioseismology

A technique for probing the interior of the Sun, using methods akin to terrestrial seismology. Because the Sun, although the nearest star by far, is a typical star, what can be learned of its interior through helioseismology is of broad importance to the stars in general.

Solar waves. Like terrestrial seismology, helioseismology entails the analysis of many "seismic" wave modes to determine the structure of the interior.

However, although terrestrial seismic waves are initiated by a singular event such as an earthquake, waves within the Sun are continuously excited, probably by the turbulent convective motions in its outer layers. Thus the solar waves are always present at all points within the Sun and on its surface. The Sun is "ringing" like a bell, but not like one struck by a clapper; it vibrates more like a bell suspended in a sandstorm, continuously struck by tiny grains of sand.

The solar waves are seen at the surface as up-and-down motions of the gases with a speed of about 0.3 mi/s (0.5 km/s) and a vertical displacement of about 30 mi (50 km). These waves are detected through the Doppler shift of the wavelength of absorption lines in the solar spectrum. They have periods clustering near 5 min (that is, with a frequency of one cycle in 5 min or about 0.003 cycle per second). As a result, the solar surface undulates up and down in a so-called five-minute oscillation (**Fig. 1**). The oscillation is actually the superposition of as many as 10^7 individual modes of oscillation of the Sun as a whole, where each mode has its own characteristic frequency (near, but not exactly at, 0.003 cycle per second) and spatial pattern on the solar surface.

Precise observations of the solar oscillations are difficult. The individual oscillation modes have velocities at the solar surface of only 12 in./s (30 cm/s) or less, and only extremely sensitive and stable spectrographs can measure such small motions. In addition, a nearly continuous stream of data extending over days is needed to separate the many individual modes with nearly identical oscillation frequencies. The day–night cycle is a formidable obstacle to continuous observations, although pairs of stations at different longitudes can reduce the nightly data gaps. Also, observations from the South Pole during the austral summer (**Fig. 2**) have yielded stretches of uninterrupted data extending over several days.

Each solar oscillation mode is produced within a specified region of the interior. The top of this resonant cavity is near the solar surface, which reflects the waves downward when they strike the surface from below. The bottom occurs at the depth where upward refraction reverses the propagation direction of descending waves. Waves with long horizontal wavelength, which travel nearly straight downward into the Sun, penetrate deeply, whereas waves with short horizontal wavelength remain near the surface. Within each cavity there can be both a lowest mode of oscillation and many higher frequency overtones, or harmonics. Thus, a two-dimensional spectrum, displaying the oscillation power as a function of frequency and horizontal wavelength, shows a number of "ridges" of power, each corresponding to a separate overtone number (Fig. 2).

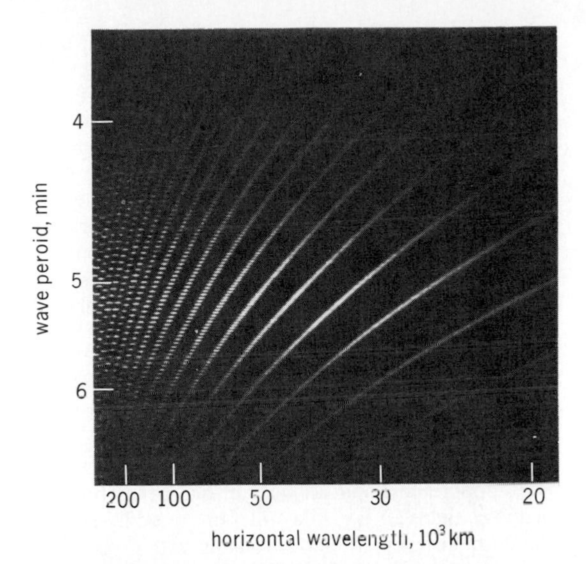

Fig. 2. Space-time spectrum of solar oscillations from observations obtained at the South Pole during the Antarctic summer, when the Sun was observed continuously for several days. Ridges of power show how oscillation periods vary with horizontal wavelength; from lower right to upper left, each ridge is a higher overtone. 1 mi = 1.6 km. (*J. Harvey, T. Duvall, M. Pomerantz, and the National Optical Astronomy Observatories*)

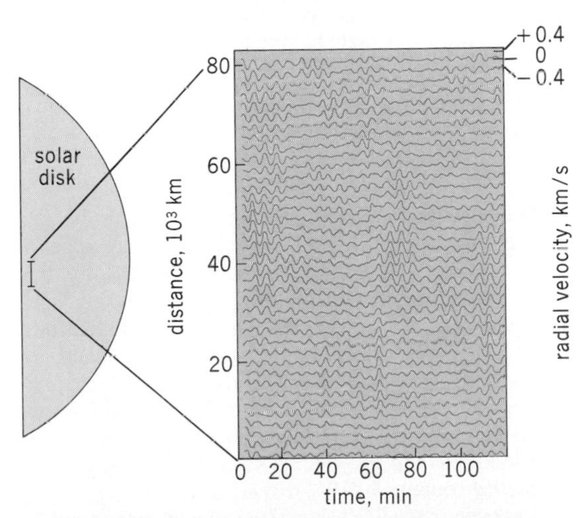

Fig. 1. Variation of radial velocity with time along an 80,000-km (50,000-mi) line on the solar surface. The 5-min oscillations, visible as wave packets localized in space and time, are actually the superposition of millions of discrete oscillation modes. 1 km = 0.6 mi. (*Adapted from S. Musman and D. Rust, Vertical velocities and horizontal wave propagation in the solar photosphere, Solar Phys., 13:261–286, 1970*)

Structure of solar interior. It is possible to predict the frequencies of oscillation modes in the Sun, given a model of the interior, that is, a numerical tabulation of temperature, density, pressure, and composition as a function of distance from the surface. When the oscillation frequencies predicted by the standard model in general use before the advent of helioseismology were compared with the observed frequencies, small but significant discrepancies were found.

Solar models, however, depend on two little-known parameters: the primordial helium abundance and the depth of the Sun's convection zone (or equivalently, the efficiency of convective energy transport). Adjustments to these parameters result in a model with excellent agreement with the oscillation data. It appears that the primordial helium abundance is about 25% by mass and that the outer convection zone of the Sun extends somewhat deeper than previously thought, to a depth of about 125,000 mi (200,000 km) or 30% of the solar radius. The helium abundance is close to the value predicted by standard big bang models of the early universe. This high abundance, however, means that the core is less rich in the hydrogen that fuels nuclear reactions. As a result, in order to create the Sun's energy the nuclear-burning core must have a higher temperature, sufficient to create high-energy neutrinos in the core at a rate some three times that observed. Thus, the first results of helioseismology are in serious conflict with solar neutrino measurements. *See Solar neutrinos.*

Rotation in solar interior. Helioseismology offers insight into the structure of the solar interior and also into its rotation. Waves propagating with or against the direction of rotation are carried by it, and their effective propagation speed and frequency are increased or decreased. The frequency shift for any mode depends on the average rotation rate within the resonant cavity for that mode, and comparison of the shift for many modes with different cavities makes it possible to determine how the rotation varies with depth.

Without helioseismology, it had been speculated that the deep interior rotates much faster than the surface, as a consequence of the much more rapid rotation of the Sun when originally formed. This would have important implications for the generation of magnetic fields in the solar interior. In addition, the gravity field from an interior rotating with a period as short as a few days would affect the precession of the perihelion of the planet Mercury, and thus bring into question a key observation confirming Einstein's general theory of relativity. However, the measured splittings of solar oscillation modes now indicate that throughout much of the interior the rotation rate is close to the surface rate, although rapid rotation near the very center cannot be ruled out. *See Satellite astronomy; Stellar rotation; Sun.*

Robert W. Noyes

Bibliography. J. Christensen-Dalsgaard, D. O. Gough, and J. Toomre, Seismology of the sun, *Science*, 229:923–931, 1985; D. O. Gough (ed.), *Seismology of the Sun and the Distant Stars*, 1986; J. W. Leibacher et al., Helioseismology, *Sci. Amer.*, 253(3):48–57, September 1985.

Herbig-Haro objects

A small, bright, semistellar knot of nebular emission in one of the dark interstellar clouds of gas and dust from which stars form. Herbig-Haro objects are named for G. Herbig and G. Haro, who independently discovered them in the early 1950s. They range in size from 300 to 1000 astronomical units and can vary in intensity over periods of only a few years. Because of their variability and their locations in star-forming clouds, these objects were originally thought to be new stars. However, their spectra show the presence of the emission lines characteristically formed behind a radiative shock wave, and suggest masses only about a factor of 10 greater than the mass of the Earth. It is now believed that Herbig-Haro objects are manifestations of the mass-loss phenomenon associated with very young stars. *See Interstellar matter; Molecular cloud.*

Since the late 1970s, evidence of bipolar outflows has been detected in star-forming regions of the interstellar medium. In this dramatic phase of early stellar evolution, oppositely directed jets of high-speed gas are observed emanating from visible, pre-main-sequence T Tauri stars and, more frequently, from objects so young that their presence within their obscuring, parent clouds can be inferred only from infrared measurements. When these visible jets, which have speeds up to 400 km/s (250 mi/s), collide with the ambient interstellar gas, the violent heating and compression known as a shock wave results. Herbig-Haro objects are fequently the hot spots where the jets hit the surrounding material. Their luminous appearance is the result of excitation of the gas by the shock. The general morphology of this kind of Herbig-Haro object region is shown in the **illustration**. Herbig-Haro objects 34N and 34S are bow shock regions at the sites where two well-collimated and oppositely directed jets collide with the surrounding interstellar clouds. Where bow shocks like these are detected, the convex side always points away from the stellar source. *See Protostar; T Tauri star.*

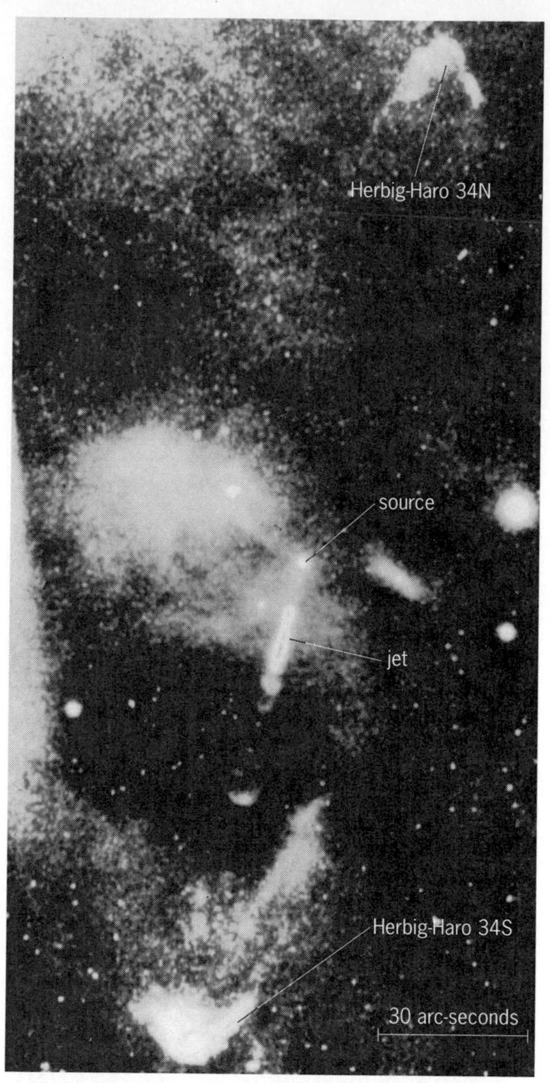

Image of Herbig-Haro 34 taken through an [SII] filter with the 3.5-m telescope on Calar Alta, Spain. A southerly directed jet can be seen emerging from the central source. Although dense, foreground, interstellar clouds partially obscure this jet, Herbig-Haro 34S is clearly seen as a bow shock at the point of impact with the surrounding medium. The expected, oppositely directed jet is completely hidden by dark clouds, but again Herbig-Haro 34N indicates where this jet collides with the ambient clouds. (*From T. Bührke, R. Mundt, and T. P. Ray, A detailed study of HH 34 and its associated jet, Astron. Astrophys., 200:99–119, 1988*)

However, not all Herbig-Haro objects are found at the terminal points of bipolar jets. Particularly when the source of the outflow is a much more luminous star than the Sun, Herbig-Haro objects can be scattered over a wide angular region rather than restricted to a well-defined outflow axis. Alternatively, several well-known Herbig-Haro objects with characteristic spectral emission lines are the brightest knots within individual jets, and probably reflect the presence of internal shocks. While it is generally agreed that Herbig-Haro objects reflect the presence of cooling regions behind shocks, there is no consensus as to how the shocks are produced and no unique formation mechanism has been postulated. *SEE STELLAR EVOLUTION.*

Anneila I. Sargent

Bibliography. I. Appenzeller and C. Jordan (eds.), *Circumstellar Matter*, IAU Symp. 122, 1987; A. K. Dupree and M. T. Lago (eds.), *Formation and Evolution of Low Mass Stars*, 1988; E. H. Levy, J. I. Lunine, and M. S. Matthews (eds.), *Protostars and Planets III*, 1992; R. D. Schwartz, Herbig-Haro objects, *Annu. Rev. Astron. Astrophys.*, 21:209–237, 1983.

Hertzsprung-Russell diagram

A graphical representation of the absolute magnitudes and the surface temperatures of stars. It is named after E. Hertzsprung and H. N. Russell, who independently discovered the importance and usefulness of such a representation. In 1911 Hertzsprung made plots of magnitude versus color for stars in several star clusters. In 1913 Russell examined the stars in the solar neighborhood by plotting their absolute magnitudes versus their spectral classes. These kinds of plots were prototypes of contemporary Hertzsprung-Russell (H-R) diagrams. The ordinate is commonly the absolute visual magnitude; in the case of a star cluster, all stars may be assumed to lie at a common distance, and apparent magnitude may be used instead. For comparison with theory, the logarithm of the luminosity (in units of the Sun's luminosity) is used. The abscissa is the spectral type or color index, both of which are related to the star's surface temperature. In order to make a comparison with theory, the logarithm of the effective temperature (which is closely related to surface temperature) is used. When a Hertzsprung-Russell diagram has apparent or absolute magnitude plotted versus color index, the diagram is also called a color–magnitude diagram. *SEE MAGNITUDE.*

General features. A schematic Hertzsprung-Russell diagram is shown in **Fig. 1**. It is obvious that stars do not fall at random in this diagram; rather, they are found in particular regions. This indicates that the intrinsic brightness of a star and its surface temperature are intimately related, a conclusion supported by stellar structure and evolution theory. By far the majority of stars fall within a band running from upper left to lower right; these are called main-sequence stars. The region above and to the right of the main sequence is also populated, but more sparsely; the stars found here are giants and supergiants. The lower left region of the Hertzsprung-Russell diagram contains even fewer stars; these are the white dwarfs. *SEE DWARF STAR; GIANT STAR; RED DWARF STAR; SUPERGIANT STAR; WHITE DWARF STAR.*

The distribution of stars in Fig. 1 does not represent the true distribution of stars in a given volume of space because the intrinsically faint stars, like the white dwarfs and fainter red main-sequence stars, are difficult to observe at great distances. **Figure 2*a*** is the Hertzsprung-Russell diagram of the nearest stars, those within 5 parsecs (1.55×10^{14} km or 0.96×10^{14} mi) of the Sun. Most of the stars in this sample are fainter than the Sun (whose position is shown), and, despite their proximity to the Sun, only a few of them are visible to the unaided eye. In contrast, Fig. 2*b* shows the Hertzsprung-Russell diagram of the 20 brightest-appearing stars in the sky. These stars are intrinsically very luminous; the Sun is the faintest star on the diagram. Except for Sirius, Procyon, Alpha Centauri, and Altair, all the other stars are located at distances much greater than the 5-parsec limit in Fig. 2*a*, yet they are so intrinsically luminous that they still appear bright in the sky.

The total range in absolute magnitude is about 27 magnitudes, from +18 to −9, corresponding to a factor of 10^{11} in brightness. The range in surface temperature is from 2300 to about 50,000 K (4100 to 90,000°F). The Sun, for example, is an average star with a G2 main-sequence spectral type, an absolute visual magnitude of about +5, and a surface temperature of about 6000 K (10,300°F). Main-sequence stars represent roughly 90% of all stars in the solar neighborhood; 10% are white dwarfs, and about 1% are giants or supergiants. The number of stars on the main sequence increases greatly going from upper left to lower right, indicating that faint red main-sequence stars are the most common. Stellar structure theory indicates that the more massive a main-sequence star is, the hotter its surface temperature and the higher its luminosity will be. Stellar masses range from about one-tenth of the Sun's mass (typical faint red stars on the lower right part of the main sequence) to almost 100 times the Sun's mass (typical hot blue stars on the upper left part of the main sequence).

Stellar evolution. The Hertzsprung-Russell diagram is an indispensable tool for demonstrating the effects of stellar evolution. Stars shine because of nu-

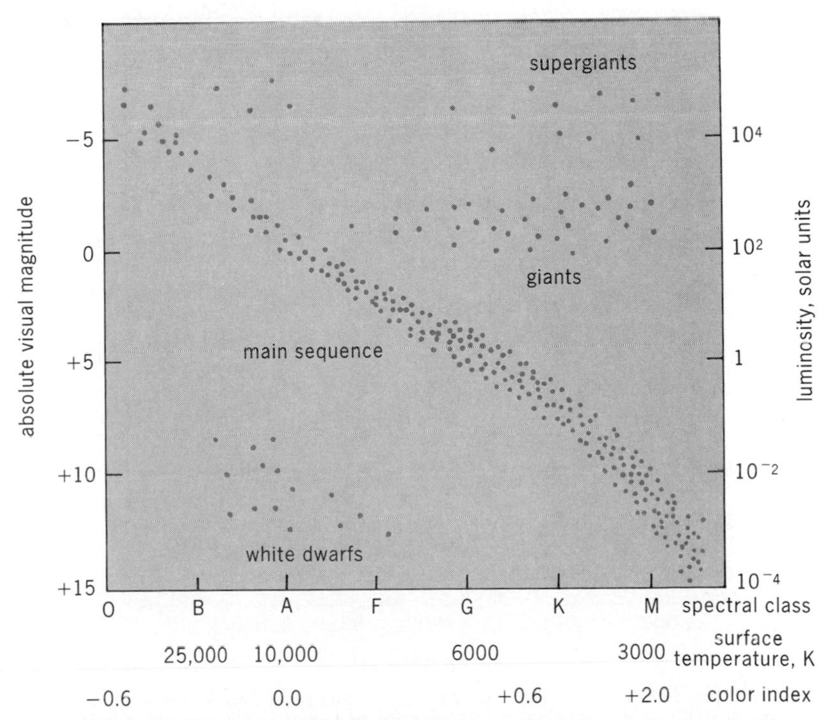

Fig. 1. Schematic Hertzsprung-Russell diagram. The main-sequence, giant, supergiant, and white dwarf regions are indicated, and values are shown for luminosities, spectral classes, and color indices. °F = (K × 1.8) − 460.

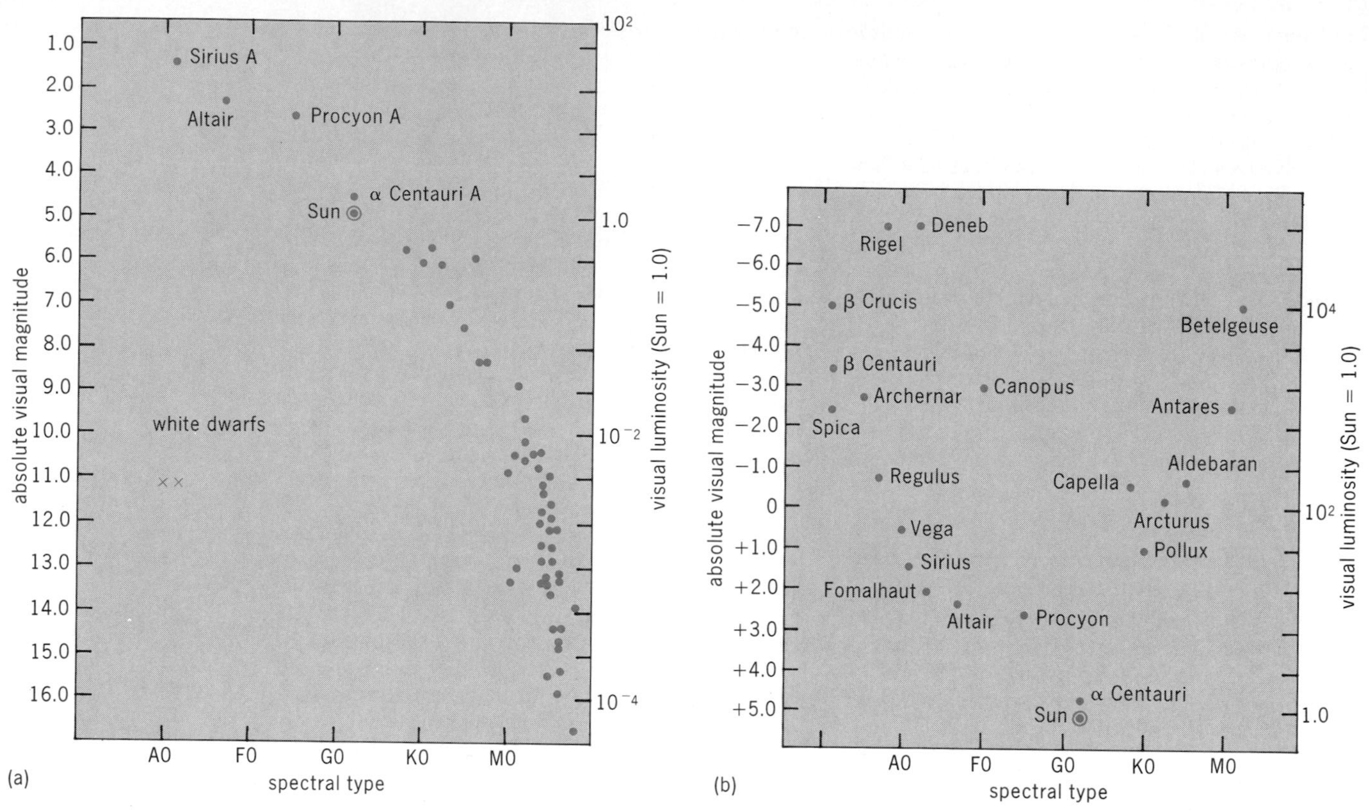

Fig. 2. Hertzsprung-Russell diagrams. (*a*) Stars within 5 parsecs of the Sun. (*b*) The 20 brightest-appearing stars. These stars are bright enough to have been given individual names, as indicated. (*After M. Zeilik and E. v. P. Smith, Introductory Astronomy and Astrophysics, Saunders, 1987*)

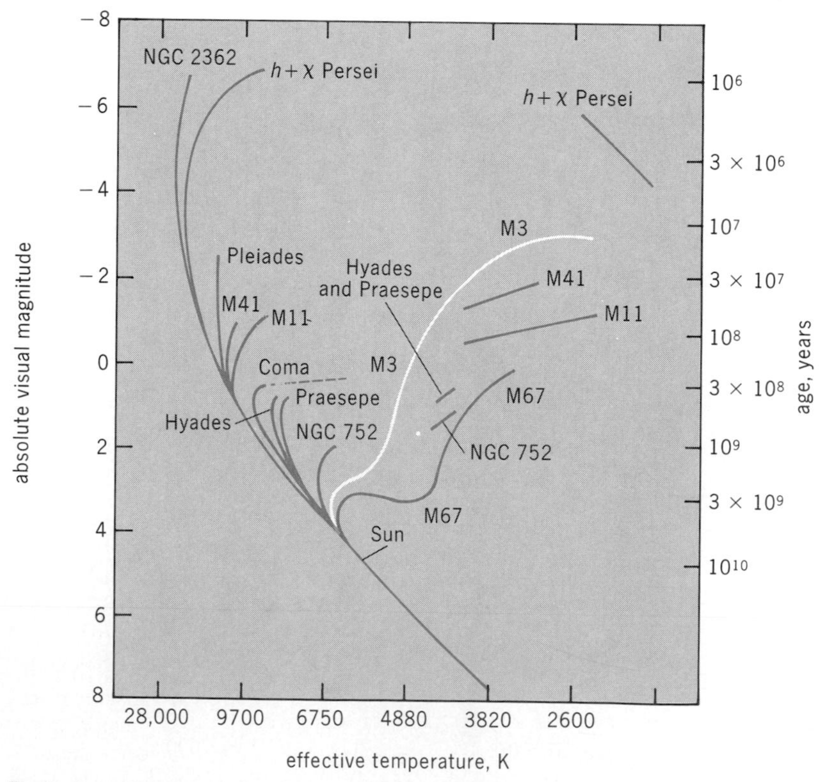

Fig. 3. Composite Hertzsprung-Russell diagram for 10 open clusters and one globular cluster. °F = (K × 1.8) − 460. (*After A. Sandage, Observational approach to evolution, I. Luminosity functions, Astrophys. J., 125:435–444, 1957*)

clear reactions in their central regions. When a star's supply of nuclear fuel is depleted, its structure is altered, producing changes in its surface temperature and luminosity, which thus affects its location on the Hertzsprung-Russell diagram. Therefore, a star's changing position on the Hertzsprung-Russell diagram tracks its evolution. For example, a main-sequence star like the Sun produces energy by nuclear fusion of hydrogen into helium; eventually, the supply of hydrogen in the central region is depleted, and the star must readjust its structure. These structural adjustments make the star swell and make its outer layers cool: it becomes a red giant, and the point representing it on the Hertzsprung-Russell diagram moves up and to the right of the main sequence. Other such changes occur at successive stages in the star's evolution, which generally takes place much more slowly than human time scales: for example, the main-sequence lifetime of the Sun is approximately 10^{10} years, half of which is yet to come. SEE STELLAR EVOLUTION.

Star clusters. While the evolution of individual stars cannot be observed, nevertheless, it is possible to see the effects of stellar evolution by studying star clusters. In addition to their all being essentially the same distance from the Sun, all stars in a given cluster are the same age. Theory predicts that the rate of stellar evolution is a strong function of a star's mass. Therefore, the Hertzsprung-Russell diagrams of star clusters, which contain stars with many different masses, show the more massive stars in a cluster to be in more advanced evolutionary stages than the low-mass stars. **Figure 3** is a composite Hertzsprung-Russell diagram for 11 star clusters: 10 open star clusters whose names are indicated and a single globular cluster named M3. Several of the open clusters have discontinuous Hertzsprung-Russell diagrams, with some stars falling in the region up and to the right of the main sequence; these are the giant and supergiant stars, those that have finished the main-sequence part of their evolution. The empty, wedge-shaped region between the main sequence and the giant and supergiant regions is called the Hertzsprung gap, and its existence is understood in terms of the dependence of the rate of evolution on a star's mass. The transition time from main sequence to giant or supergiant for massive, hot stars is very short; therefore the chances of observing these stars during the transition are very small, and the Hertzsprung-Russell diagram appears discontinuous. For less massive stars, which go through all of their evolution at a slower pace than the stars of higher mass, the transition time from main sequence to giant is slower, and it is possible to observe these stars as they evolve; thus the Hertzsprung-Russell diagram for clusters with less massive stars evolving off the main sequence is not discontinuous.

The age of a star cluster can be inferred from the appearance of its Hertzsprung-Russell diagram. The relevant characteristic is the turn-off point, the upper termination of the main sequence. The principal determinant of the turn-off point is the age of the cluster: as the cluster ages, the turn-off point moves down the main sequence to stars of progressively lower mass. The cluster age is thus presumed equal to the age of the most massive (shortest-lived) star still observed on the main sequence, that is, the main-sequence lifetime of a star just at the turn-off point. Thus the youngest clusters are those with massive, hot blue stars still on the main sequence. In Fig. 3, NGC 2362 and h + χ Persei are two such young clusters, with ages of about 2×10^6 and 2×10^7 years, respectively. In contrast, the oldest open cluster in Fig. 3 is M67, at about 5×10^9 years. The Hertzsprung gap is evident in the Hertzsprung-Russell diagrams of all but the oldest open cluster. Although the globular cluster M3 appears younger than M67, it is considerably older. The reason for this discrepancy lies in the effects of chemical composition on stellar evolution. *See* Star clusters.

Populations. Open clusters belong to population I in the Milky Way Galaxy. Population I consists of stars and clusters that formed after the initial collapse of the Milky Way Galaxy to its current state, all the way up to the present. Population I objects formed from material that had already been enriched in metals, with typical abundances of 1% by number compared with hydrogen. (In astronomy, metals include all chemical elements other than hydrogen and helium.)

In contrast, population II objects are the oldest in the Milky Way Galaxy, having formed early in its history from primordial material with very low metal abundance (from one-tenth down to less than one-hundredth of population I abundance). Globular clusters are typical population II objects. The position on the Hertzsprung-Russell diagram of a population II star is different from that of a population I star, even if the two stars have the same mass; therefore the Hertzsprung-Russell diagram of M3 cannot be compared directly with those of open clusters to infer its age. *See* Milky Way Galaxy; Star.

Karen B. Kwitter

Bibliography. L. H. Aller, *Atoms, Stars, and Nebulae*, rev. ed., 1971; J. M. Pasachoff, *Contemporary Astronomy*, 4th ed., 1989; A. G. D. Philip and D. S. Hayes, (eds.), *The H-R Diagram*, International Astronomical Union Symposium No. 80, 1978; M. Zeilik and E. v. P. Smith, *Introductory Astronomy and Astrophysics*, 1987.

Horizon

Traditionally, the apparent boundary between the sky and the Earth or the sea. Without complicating factors, it would be a line located 90° from the zenith or overhead point; because of trees, hills, or buildings, the visible horizon differs from the ideal one. *See* Astronomical coordinate systems; Celestial sphere.

Black holes. The term horizon also refers to the imaginary boundary of a black hole. The boundary, often referred to as the event horizon, is a spherical surface, with a radius equal to 1.8 mi (3 km) times the number of solar masses in the mass of the hole. A black hole forms when an object becomes so small that its gravity prevents nearby objects from escaping from its gravitational attraction. Objects close enough to a black hole, within its horizon, would have to move faster than light in order to escape its gravitational field. Because this is impossible according to the theory of relativity, escape cannot occur. *See* Black hole.

Cosmology. In cosmology the term horizon takes on still another meaning. According to the widely accepted big bang theory, cosmic evolution began from an explosion 10 to 20 billion years ago. Because nothing can travel faster than light, there is a limit to how far an observer can see or influence the surroundings; this limit is referred to as the horizon or horizon length. For example, when the universe is a million years old, the horizon length is a million light-years or about 3×10^5 parsecs. In classical cosmology the horizon length is the distance that light can travel since the beginning of time.

The uniformity of the universe is difficult to understand, considering the small size of the horizon when the universe was young. How is it possible that two parts of the universe which were separated by several million light-years or several horizon lengths when the universe was a million years old were at the same temperature at that time? Neither light nor heat could be transmitted from one region to the other in the short time that had passed since the big bang, and yet temperature uniformity is shown by observation. The so-called inflationary scenario presents a slightly modified version of the early stages of the big bang model and provides an explanation. *See* Big bang theory; Cosmology; Inflationary universe cosmology; Universe.

Harry L. Shipman

Horology

Measurement of the time dimension. In practice, horology is the search for a steady or repetitive action, and the design of an instrument to perform that action and to indicate (read out) a measure of the action. Until early in the twentieth century, horology dealt with mechanical instruments, with effort distributed between improving accuracy and decreasing size of timepieces. Increasingly, however, electronic instruments provided means for meeting these objectives.

Early instruments. The rotation of the Earth on its axis provides a naturally repeating action that produces a directly accessible indication of the flow of time. A sundial increases the precision of readout in subdivisions of the diurnal unit. In modern times the photographic zenith tube is used to identify the orientation of the Earth against the background of the fixed stars relative to the Sun. *See* EARTH ROTATION AND ORBITAL MOTION; OPTICAL TELESCOPE; SUNDIAL.

Among the earliest instruments for indicating the flow of time was a burning candle. Although highly sensitive to its own structural variations and to such environmental influences as air currents, the candle transformed the measurement of time into the measurement of height, which could be done accurately. A somewhat related currently used action is radioactive decay of an isotope.

Mechanical clocks. Galileo first observed the isochronism of the pendulum, its period depending primarily on its length. By choice of configuration and of material, the length of a pendulum from its hinge point to its center of gravity was made relatively independent of temperature. Many mechanisms were devised to transfer energy from a falling weight to the pendulum in synchronism with its swing, to make up for frictional losses and drive the readout mechanism. Modern pendulum clocks can be accurate to 1 part in 10^8 during a day.

The swinging of the pendulum about its pivotal support required that the clock be fixed vertically. To overcome this restriction, an alternative timing mechanism was developed in the form of the balance wheel and hairspring. The moment of inertia of the balance wheel and the compliance of the spring jointly determine the frequency of oscillation. Again, by temperature compensation and by release of energy from a main spring through an escapement, the clock is isolated from its environment. In the shipboard chronometer, the gimbals isolate the movement from pitch and roll and a massive balance wheel decouples it from yaw. A fusee between main spring and great wheel, and a helical hairspring further improve accuracy.

Electronic clocks. The piezoelectric crystal replaced the assembly of balance wheel and hairspring with a naturally oscillatory structure. The piezoelectric property of the crystal couples its mechanical vibration into an electrical circuit. Its mechanical frequency is the consequence of its physical constants: density and elastic compliance; the crystal is precisely ground to produce the desired frequency, in the region of the kilohertz to megahertz. The electronic circuit around the crystal replaces the escapement to drive the crystal in forced oscillation with an alternating voltage that the circuit develops from a direct-current source such as a battery to sustain vibration and to power the readout display. Thus, time measurement becomes essentially frequency measurement. A quartz clock may drift as little as about 1 part in 10^{11} during a day. *See* QUARTZ CLOCK.

Atomic clocks. The next advance in accurate measurement of time came by replacement of dynamic mechanical oscillators with quantum energy transitions. Two standards in common use are the cesium atomic beam clock and the rubidium gas cell. The cesium clock is based on quantum emissions that accompany the transition in the spin axis of the outer unpaired electron of vaporized cesium-133. A beam of cesium-133 atoms is passed through a magnetic field. The field deflects each atom in which the valence electron has the desired one of two possible spins into a cavity. There, microwave energy derived from an auxiliary crystal oscillator and frequency-multiplier chain aligns the electrons to the opposite spin. The cesium beam exits the cavity through a second magnetic field that directs the realigned electrons to a detector whose output slaves the auxiliary oscillator so that the energy fed to the cavity is at the precise frequency required to stimulate the quantum transition in spin direction. The auxiliary oscillator also feeds a counter and readout. The transition frequency is inherently invariant and highly independent of environmental influence. As a consequence, the best cesium-beam oscillators are accurate to 2 parts in 10^{14} over long time periods, and the transition frequency is used to define the second. *See* ATOMIC CLOCK.

In any sensitive system, mechanical or electronic, thermal fluctuations set a noise threshold to short-term stability. To combat this limitation, various electronic amplifiers and oscillators operate in cryogenic enclosures. Also, at low temperature, hydrogen atoms exist predominantly at two energy states in which the spins of the electron and nucleus (proton) are, respectively, in opposition to and parallel to each other. Consequently operation of a hydrogen maser at low temperature acquires both the advantages of lower-noise electronics and more discrete energy states. A hydrogen maser clock with the cavity cooled by carbon tetrafluoride to $-414°F$ ($-248°C$) provides a stability of 1 part in 10^{15} on time scales of 10^4–10^5 s. *See* TIME.

Atomic clocks based on ion traps, which use electric and magnetic fields to trap ions in a vacuum, are under development, and might attain an accuracy of 1 part in 10^{18}.

Frank H. Rockett

Astronomical precision timekeeping. Radio astronomers require precise time for two areas of experimentation: very long-baseline interferometry (VLBI) and pulsars. Both areas have the capability of providing precise time information.

Very long-baseline interferometry involves multiplying samples of the electric field that were recorded independently at telescopes situated around the globe while being trained on the same object in the sky. Precisely synchronized atomic frequency standards such as hydrogen masers are required at each site so that the same observing frequency is sampled at the same time with respect to the incoming wavefront. A time transfer system such as the Global Position Satellites that is coordinated by various national time standards laboratories such as the Naval Observatory and the National Institute of Standards and Technology in the United States, is used to provide synchronization. The coherence between a pair of electric field samples measures a Fourier component of the object's brightness distribution. The combination of many pairs creates an image that is equivalent to that which would be produced by a telescope with the angular resolution given in radians by the ratio of the observing wavelength to the maximum telescope spacing. Solar time, or Coordinated Universal Time

(UTC), is offset from International Atomic Time (TAI) to allow for the variable rotation of the Earth. The most precise measurement of Earth rotation comes from VLBI measurements.

Pulsars are highly magnetized and rapidly rotating neutron stars that emit intense beams of radio emission. Astronomers keep track of the rotations of these stars by referencing pulse arrival times to TAI, but not even the TAI time scale is sufficiently accurate because of the relativistic effects, gravitational redshift, and transverse Doppler effect that result from the motion of the Earth. A new time scale, Barycentric Dynamical Time (TDB), is derived from TAI without the relativistic effects. Timing measurements of the fastest pulsars, which rotate more than 600 times per second, are now as precise as the best Earth clocks over durations of a year or more. Time, which in prehistory was reckoned solely by astronomical events, has again become the province of astronomical observations. SEE PULSAR; RADIO ASTRONOMY; RADIO TELESCOPE.

Donald C. Backer

Bibliography. J. T. Fraser et al. (eds.), *The Study of Time,* vols. 1–5, 1972–1986; J. L. Jesperson et al., Special issue on time and frequency, *Proc. IEEE,* (60)5:476–638, 1972; J. L. Jesperson and J. Fitz-Randolph, *From Sundials to Atomic Clocks: Understanding Time and Frequency*, 1977, reprint 1982; P. Kartaschoff, *Frequency and Time*, 1978; J. D. Kraus, *Radio Astronomy*, 1986; A. R. Thompson, J. M. Moran, and G. W. Swenson, *Interferometry and Synthesis in Radio Astronomy*, 1986; M. E. Whitney, *The Ship's Chronometer*, 1985.

Hubble constant

The rate at which the velocity of recession of the galaxies increases with distance. Edwin Hubble established the existence of this motion observationally in 1929. SEE REDSHIFT.

The value of the expansion rate is still uncertain. Most values lie in the range 16 to 32 km per second per 100 light-years (50 to 100 km per second per megaparsec, or $1.7–3.4 \times 10^{-18}$ s^{-1}). The determination of this important number is difficult because of (1) the corrections for the absorption of light by interstellar matter both in the Galaxy and in the others observed; (2) the largely unknown influence of varying chemical composition on the intrinsic luminosity of the objects studied; (3) the lack of a dependable photometric scale standard at the faint end of the scale; and (4) the perturbations in the local velocity field that may result from the gravitational attraction of the Ursa Major and Virgo clusters of galaxies, that is, from the inhomogeneous distribution of galaxies in the Local Supercluster.

A. Sandage and G. A. Tammann used as distance indicators the period-luminosity law for Cepheids for the nearest objects, H II regions for those at intermediate distances, and the established properties of the brightest class of late-type spirals for those farthest away. They obtained a value of 16 km per second per 10^6 light-years (50 km per second per megaparsec, or 1.7×10^{-18} s^{-1}). They found no significant difference in the Hubble constant (a) for objects with recessional velocities below 2000 km per second and for those above 6000 km per second; (b) for galaxies in the hemisphere toward the Virgo cluster and those in the opposite hemisphere; and (c) for various regions along the equator of the supercluster. SEE CEPHEIDS; INTERSTELLAR MATTER.

G. de Vaucouleurs and colleagues, using a number of overlapping distance indicators, find 30 km per second per 10^6 light-years (95 km per second per megaparsec, or 3.2×10^{-18} s^{-1}).

M. Aaronson, J. Huchra, and J. Mould first established the existence of a close correlation between the infrared magnitudes of late-type galaxies and line-width profile of the 21-cm line of hydrogen. This correlation is to be expected since the width increases with the rate of rotation, which itself increases with total mass, and the total luminosity in the infrared also increases with mass. Using this correlation and known recessional velocities for relatively nearby galaxies, they obtained a value of the Hubble constant in agreement with that of Sandage and Tammann. Next Aaronson and colleagues extended their work to galaxies beyond the Virgo cluster, obtaining 26 km per second per 10^6 light-years (80 km per second per megaparsec, or 2.7×10^{-18} s^{-1}). Aaronson attributed the difference in the two results to the Milky Way Galaxy falling toward Virgo. SEE COSMIC BACKGROUND RADIATION.

If the expansion rate has been constant, then the age of the universe is the reciprocal of the Hubble constant. If, as seems more probable, the expansion has decelerated due to the gravitational attraction of the galaxies, the reciprocal of the Hubble constant gives an upper limit to the age of $10–20 \times 10^9$ years. Attempts to determine the deceleration from observations of galaxies so distant that they are observed at a substantially earlier time, when they may have been receding more rapidly, are vitiated by the possible evolutionary changes in luminosity which may have occurred. The age of the universe as determined from the oldest globular clusters in the Galaxy is approximately 18×10^9 years, and from the abundances of the various radioactive elements it is 15×10^9 years. Therefore, the lower end of the present range of values of the Hubble constant is consistent with the other estimates of the age of the universe. SEE COSMOLOGY; GALAXY, EXTERNAL; STAR CLUSTERS.

Louis C. Green

Bibliography. M. Aaronson and J. Mould, The distance scale: Present status and future prospects, *Astrophys. J.*, 303:1–9, 1986; A. Dressler, The D_n-σ relation for bulges and disk galaxies: A new independent measure of the Hubble constant, *Astrophys. J.*, 317:1–10, 1987; P. W. Hodge, The cosmic distance scale, *Amer. Sci.*, 72:474–482, 1984; P. W. Hodge, The extragalactic distance scale, *Annu. Rev. Astron. Astrophys.*, 19:357–372, 1981; *13th Texas Symposium on Relativistic Astrophysics, 1986*, 1987.

Hyades

A group of stars known from early times, scattered through the V-shaped asterism in the constellation of Taurus. An estimated 350 stars form the Hyades cluster, with about half of its members in a sphere of 40 light-years (2.4×10^{14} mi or 3.8×10^{14} km) diameter. Its brightest stars are yellow to red. The cluster age is estimated at 6.5×10^8 years. The Hyades is the nearest star cluster and much studied. Its distance is of great importance as a basis for the galactic and extragalactic distance scales. Its hydrogen-burning stars are used as luminosity standards for the zero-age main sequence, from which the ages and distances of other clusters may be deterimed. SEE HERTZSPRUNG-RUSSELL DIAGRAM; STELLAR EVOLUTION.

Different methods of distance determination give a spread in results. The geometric methods (trigonometric parallax and moving cluster) may be affected by observational errors; photometric methods require assumptions about chemical composition. Observations of binary stars in the Hyades have proved useful, especially the star HD 27130, which is both a spectroscopic and an eclipsing binary. A reliable estimate from a combination of results yields a distance of 147 light-years (8.64×10^{14} mi or 1.39×10^{15} km) with an uncertainty of 10 light-years (5.8×10^{12} mi or 9.4×10^{13} km). SEE BINARY STAR; ECLIPSING VARIABLE STARS.

Sophisticated techniques have produced much new information on the Hyades stars. In 1981, nine M-type dwarfs were detected as x-ray sources by using the *Einstein Observatory;* subsequently 66 Hyades members have been detected with it. No microwave emission has yet been found with the Very Large Array in New Mexico. The Hyades stars have moderate to small rotational velocities and show very slight enhancement of metals with respect to the Sun. SEE RADIO ASTRONOMY; STELLAR ROTATION; X-RAY ASTRONOMY.

The Hyades is also the most prominent moving cluster. Known as the Taurus moving cluster, the group has a velocity of 28 mi/s (45 km/s) toward a convergent point in the constellation of Orion. SEE CONSTELLATION; STAR; STAR CLUSTERS; TAURUS.

Helen S. Hogg

Bibliography. P. Hodge, How far are the Hyades?, *Sky Telesc.*, 75(2):138–140, 1988.

Image tube

A photoelectric device for intensifying faint astronomical images. In these devices a photoemissive surface, called the photocathode, emits electrons through the photoelectric effect. In most tubes the photocathode is semitransparent and is deposited on the inside of a transparent window that is mounted on the end of an evacuated glass or ceramic cylinder. When light from a telescope or spectrograph is imaged on the photocathode, electrons are ejected into the vacuum inside the tube. Electric fields or electric and magnetic fields in combination then accelerate and direct the photoelectrons through the device. In most image tubes the photoelectrons are ultimately reimaged on a phosphor-coated output window that converts them back into a visible image. Optical gain is provided by the phosphor, which may release as many as a thousand photons for each photoelectron which crashes into it. A separate television camera or charge-coupled device (CCD) is used to record the intensified output image.

The quantum efficiency of an image tube is simply the probability that an incoming photon of light will release a detectable photoelectron inside the tube. Although a single photon of light is all that is necessary to create a photoelectron, some photons pass through the photocathode without being absorbed. Others are absorbed but do not produce a free electron. In spite of these losses, many low-light imaging systems are capable of detecting about one-fifth of the photons incident upon them. Since individual photons are recorded by most of these systems, measurement uncertainties are primarily caused by statistical fluctuations in the photon arrival rate.

Two types of photocathode are commonly used in astronomical image tubes. Bialkali photocathodes are composed of antimony and two alkali metals. Photocathodes of this type are most sensitive to ultraviolet, blue, and green light at wavelengths below 600 nanometers. Multialkali photocathodes have three or more alkali metals in combination with antimony and have red responses that extend beyond 800 nm. Both types eventually lose sensitivity at long wavelengths because light photons at long wavelengths have very little energy to eject a photoelectron. Their responses at short wavelengths are usually limited only by the transparency of the photocathode window or by the Earth's atmosphere. The Earth's atmosphere becomes opaque at about 320 nm. Satellite-borne image tubes with magnesium fluoride windows, however, can be used down to 120 nm in the far ultraviolet. SEE SATELLITE ASTRONOMY; ULTRAVIOLET ASTRONOMY.

All photocathodes tend to emit photoelectrons at a modest rate even in total darkness. This spontaneous emission in the absence of incident light, or dark emission, makes observations of faint astronomical sources more difficult. Since bialkali photocathodes have somewhat lower dark emission rates than multialkali photocathodes, they are used whenever the extended red response of the multialkali photocathode is not required.

Charge-coupled devices are now used without image tubes in many astronomical applications. In long integrations they have a high quantum efficiency and relatively low noise. Image tubes continue to be useful, however, in applications that require the image to be read out at a high frame rate. They are thus particularly useful for telescope guiding and speckle interferometry, a technique by which the adverse effects of atmospheric turbulence are removed by recording both the positions and arrival times of individual photons. SEE CHARGE-COUPLED DEVICE; SPECKLE.

Proximity-focused tubes. In a proximity-focused image intensifier the photocathode and phosphor screen are both deposited on plane-parallel optical windows. The tube is very short, and the phosphor screen is only 0.06–0.14 in. (1.5–3.5 mm) from the photocathode. Electrons emitted from the photocathode encounter a very strong electric field and are rapidly accelerated across the gap between the photocathode and the phosphor screen. Although photoelectrons leave the photocathode at all angles to its surface, very little defocusing takes place. Unfortunately, most proximity-focused intensifiers have unacceptably high dark emission rates.

Electrostatically focused tubes. Electrostatically focused image intensifiers are constructed with fiber-optic input and output windows to optimize and simplify the electron imaging. The fiber-optic windows at both ends of such a tube are flat on the outside and spherical on the inside. Electrons emitted from the spherical surface of the photocathode are accelerated through a small hole at the tip of a conical-shaped electrode. They then come back to focus on the spherical surface of the output window. A phosphor screen on the output window converts the electron image back into light. Because electrostatically focused image tubes have fiber-optic photocathode windows made of glass, they cannot be used in the ultraviolet. SEE OPTICAL FIBERS.

A two-stage image intensifier containing a pair of electrostatically focused intensifiers is shown in

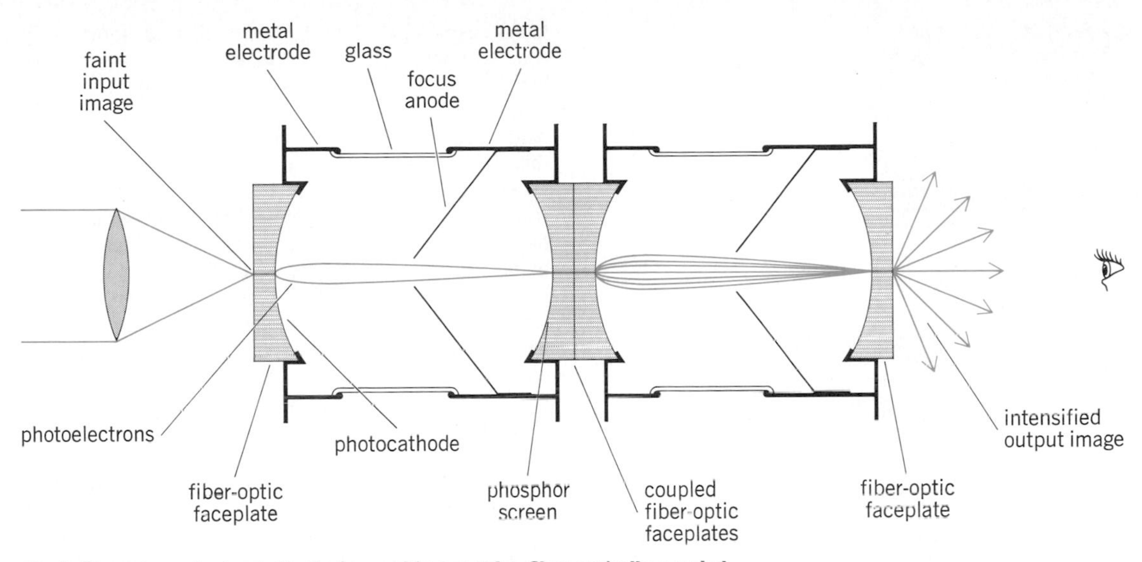

Fig. 1. Two-stage electrostatically focused image tube, fiber-optically coupled.

Fig. 1. The output light of the first tube travels through the contacted fiber-optic windows and falls on the photocathode of the second. Light from the first tube is thus amplified and reimaged by the tube behind it. The overall light gain of this intensifier would be approximately 2500. A similar four-stage intensifier would have a light gain of over a million. A single photon falling on such an intensifier would produce a pulse of light that is easily visible to the naked eye.

Magnetically focused tubes. A magnetically focused image intensifier uses both electric and magnetic fields for electron imaging. Electrons from the photocathode are accelerated by a series of ring-shaped electrodes that are spaced out at regular intervals down the length of the tube. A uniform magnetic field down the center of the intensifier forces the photoelectrons into a tight spiral as they are accelerated. By balancing the electric and magnetic fields properly, the photoelectrons are brought to focus on a phosphor-coated output window at the opposite end of the tube. The input and output windows of a magnetically focused intensifier are flat on both sides. Since magnetically focused tubes can be made without fiber-optic input windows, they are useful in the ultraviolet.

A two-stage magnetically focused cascade intensifier is shown in **Fig. 2**. Additional gain is achieved in this tube by focusing the primary photoelectrons on a thin transparent membrane which has a phosphor screen on one side and a photocathode on the other.

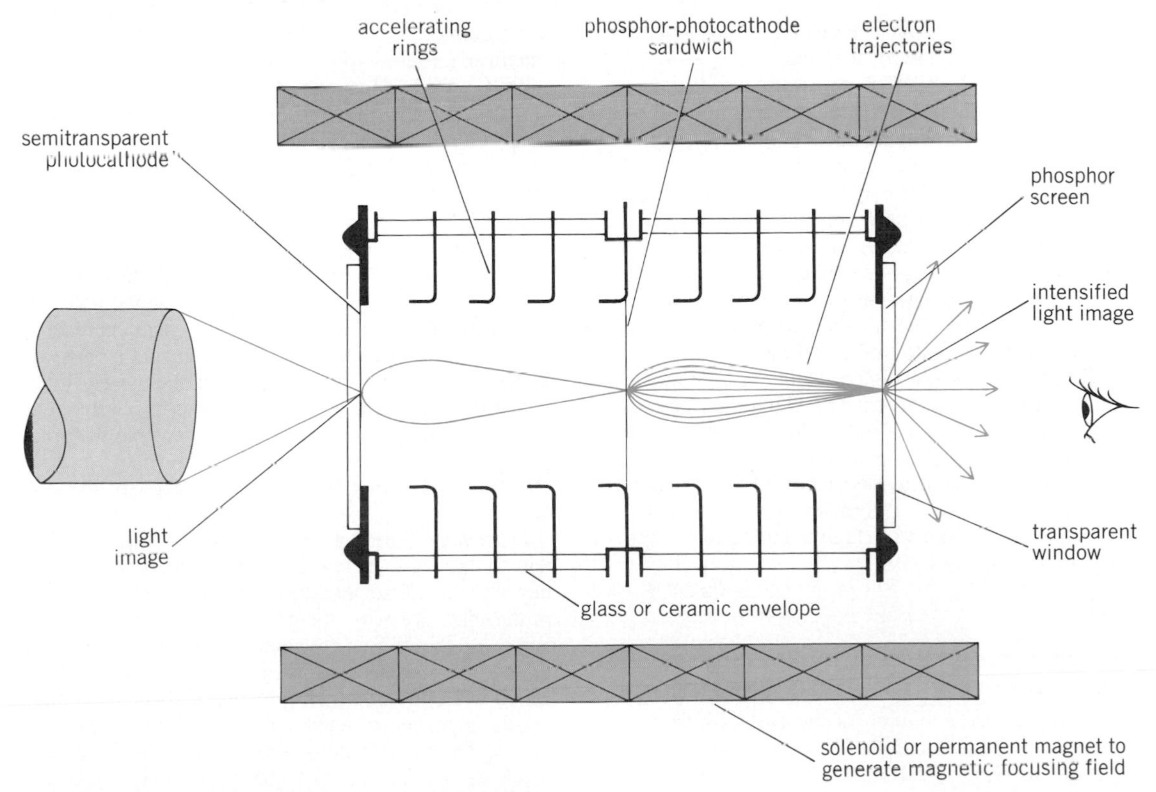

Fig. 2. Two-stage magnetically focused cascade image tube.

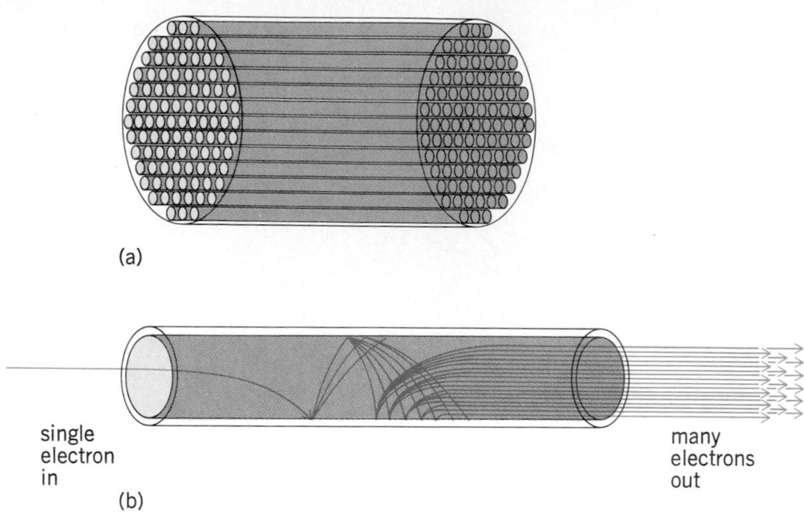

Fig. 3. Microchannel plate showing (*a*) array of tubes and (*b*) details of a single channel.

The electron gain with such a sandwich is usually somewhat higher than that obtained by coupling single-stage tubes. High gain is also achieved without serious loss of resolution.

Microchannel-plate tubes. An effective method of obtaining high electron gain within an image tube is to focus the photoelectrons on a microchannel plate. A microchannel plate (**Fig. 3**) consists of an array of closely spaced tubes, each of which is coated on the inside with a secondary-emitting material. When a primary electron enters a channel, it collides with the side wall of the tube and generates several secondary electrons. The secondary electrons travel farther down the tube and eventually produce secondary electrons of their own. Since this process is repeated several times down the channel, thousands of electrons exit the rear face of the microchannel plate. These electrons are then brought to focus on a phosphor screen on the output window of the tube.

Microchannel plates are most often added to proximity-focused intensifiers. Such tubes are barely over 0.4 in. (1 cm) in length, but they can have approximately the same gain as conventional two-stage intensifiers. No permanent magnet or solenoid is required to operate one, and they are almost entirely free of distortion. Unfortunately, individual optical pulses from a microchannel-plate tube vary enormously in amplitude, and integration times often have to be increased to compensate for the added noise. Lifetimes of microchannel plates are also limited.

Digicon tubes. A digicon tube is a single-stage magnetically focused device in which the electron image is focused directly on a silicon diode array. Incoming photoelectrons thus produce a pulse of electrons in the diode array instead of a burst of photons. Since each diode is a completely separate detector, digicons can handle very high counting rates. *SEE ASTRONOMICAL PHOTOGRAPHY.*

Richard G. Allen

Inflationary universe cosmology

A theory of the evolution of the early universe, motivated by considerations from elementary particle physics as well as certain paradoxes of standard big bang cosmology, which asserts that at some early time the observable universe underwent a period of exponential, or otherwise superluminal, expansion.

In order to resolve the various paradoxes of the big bang theory, inflationary theories predict that during this inflationary epoch the scale of the universe increased by at least 28 orders of magnitude. The period of exponential expansion (inflation) is caused by the appearance of a nonzero constant energy density in the universe associated with the postulated existence of a phase transition between different ground-state configurations of matter that occurs as the universe expands and cools. After the transition is completed, the constant energy density is converted into the energy density of a gas of relativistic particles. At this point, inflationary scenarios match the standard big bang cosmological model.

Origin of inflationary models. The suggestion of an inflationary period during the early universe, in connection with a specific model, was first made in 1980 by A. Guth. (Somewhat earlier a less concrete but not entirely unrelated possibility was discussed by A. A. Starobinskii.) Guth proposed, based on a consideration of recently proposed grand unification theories in particle physics, that phase transitions—associated with the breaking of certain symmetries of the dynamics governing the interactions of matter—could occur at sufficiently high temperatures and have important consequences in the early universe.

Symmetry breaking and phase transitions. The concept of symmetry is of fundamental importance in physics. In particular, a symmetry can be associated with each of the known forces in nature, and each symmetry is, in turn, associated with the existence of certain conserved quantities, such as electric charge in the case of electromagnetism. Surprisingly, a physical system in its ground state may not possess a symmetry that could be fundamental to the basic physics governing its dynamics. The most familiar example is a ferromagnet made up of many individual elementary magnetic dipoles. The equations of electromagnetism are manifestly rotationally invariant: there is no intrinsic north or south. However, the ground state of such a spin system will involve all the spins aligned in one particular direction, yielding the familiar case of a permanent magnet. In this case, it is said that the rotational symmetry of the equations of electromagnetism has been spontaneously broken. The signal for this is that the average value of the total spin of the system points in a certain direction. If the system is heated, so that all the spins become excited and randomly oriented, the net spin of the system approaches zero, its magnetic field vanishes, and rotational symmetry is restored. The total spin of the system is referred to as an order parameter, because its value tells which state the system is in, and a change in the order parameter is an indication of the presence of a phase transition between different ground states of the system.

Phase transitions in particle physics. A situation similar to the case described above occurs in particle physics. At the temperatures and energies that occur in the universe now, there is a vast difference in the nature of the known fundamental forces. For two of these forces—the electromagnetic force, and the so-called weak force, which governs the beta decay processes important to energy production in the Sun—it is now known that the perceived differences are the result of spontaneous symmetry breaking. Above a certain energy or temperature, the weak and electromagnetic interactions appear exactly the same. However, at a critical temperature, the ground state of matter, which in particle physics is called the vacuum state, breaks the symmetry relating the two interactions, and below this temperature they appear quite different. The signal for this symmetry breaking is again the appearance of a nonzero value in a certain order parameter of the system; in this case it is referred to as a vacuum expectation value. In the case of the ferromagnet, the order parameter was a spin, and the ground state after symmetry breaking had a net magnetization. In the case of the particle physics system, the order parameter describes the ground state expectation value of a certain elementary particle field, and the vacuum state after symmetry breaking has a net charge that would have been zero had the symmetry not been broken.

In particle physics, if a symmetry is broken at a certain scale of energy, particles that transmit the force of nature associated with that symmetry can have masses characteristic of this energy scale. This is indeed what happens in the case of the weak interactions, which are weak precisely because the particles that transmit the weak force are so massive. However, once the temperature of a system is so great that the energy it takes to produce such a massive particle is readily available in thermal energy, the distinction between this force and electromagnetism, which is transmitted by massless particles (photons), disappears and the two forces become unified into one.

Shortly after it was recognized that these two forces could be unified into one, it was proposed that all three observed forces in nature outside of gravity might be combined into one grand unified theory. There are reasons to believe that the scale of energy at which all the forces would begin to appear the same is truly astronomical—some 16 orders of magnitude greater than the mass of the proton. Such energies are not achieved in any terrestrial environment, even in the laboratory. The only time when such a scale of energy was common was in the earliest moments of the big bang fireball explosion.

"Old" inflationary model. Based on these facts, Guth reasoned that if grand unification symmetries are indeed broken at some large energy scale, then a phase transition could occur in the early universe as the temperature cooled below the critical temperature where symmetry breaking occurs. According to the standard big bang model of expansion, the time at which this would occur would be about 10^{-35} s after the initial expansion had begun (**Fig. 1**). This time is far earlier than the times previously considered in cosmological analyses using the big bang model. For instance, the time at which it is predicted that many of the light elements were first formed in the process of primordial nucleosynthesis is of the order of 1 s, some 35 orders of magnitude later in time.

First-order phase transitions. As Guth demonstrated, the effects of such a phase transition in the early universe could be profound. In order to calculate the dynamics of a phase transition, it is necessary to follow the behavior of the relevant order parameter for the transition. This is done by determining the energy (actually the thermodynamic free energy in the case of systems at finite temperatures) of a system as a function of the order parameter, and following the changes in this energy as the temperature changes.

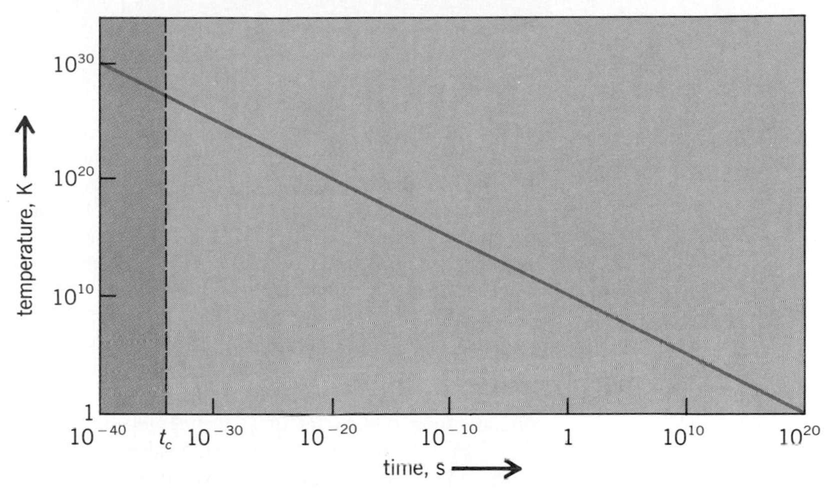

Fig. 1. Temperature versus time after beginning of expansion for the standard big bang model. Before T_c, inflationary scenarios differ from the standard model.

Figure 2 shows a typical example of what might be expected for the case of symmetry breaking in grand unification theories. At some high temperature T, the minimum of the relevant energy function is at zero value of the order parameter, the vacuum expectation value of a certain field. Thus the ground state of the system, which occurs when this energy is a minimum, will be the symmetric ground state. This case is analogous to the spin system, which at high temperatures is disordered, so that no preferred direction is picked out. As the temperature is decreased, however, at a certain critical temperature T_c a new minimum of the energy appears at a nonzero value of the order parameter. This is the symmetry-breaking ground state of the system.

How the system makes a transition between the original, symmetric ground state and the new ground state depends on the shape of the energy curve as a function of the order parameter. In Fig. 2 it is seen that there is a barrier between the two minima. This means that classically the system cannot make a transition between the two states. However, it is a well-

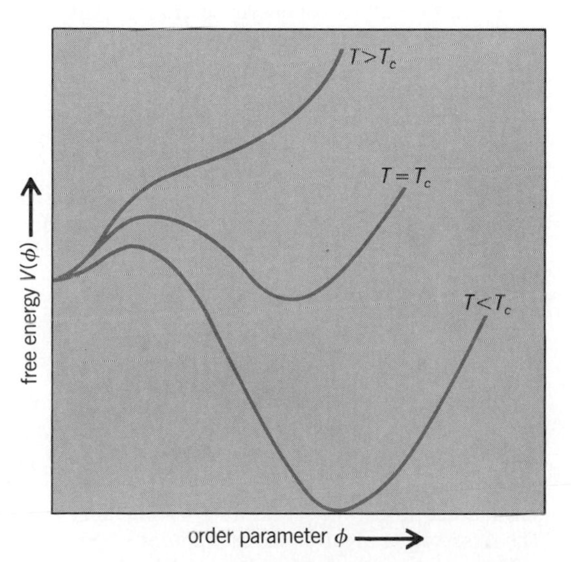

Fig. 2. Free energy as a function of the order parameter for the case of a first-order transition.

known property of quantum mechanics that the system can, with a certain very small probability, tunnel through the barrier and arrive in the new phase. Such a transition is called a first-order phase transition. Because the probability of such a tunneling process is small, the system can remain for a long time in the symmetric phase before the transition occurs. This phenomenon is called supercooling. When the transition finally begins, "bubbles" of new phase locally appear throughout the original phase. As more and more bubbles form, and the original bubbles grow (in the case of such a phase transition in particle physics the bubbles grow at the speed of light), eventually they combine and coalesce until all of the system is in the new phase. Another example of this type of phenomenon is the case of a supercooled liquid such as water that has been continually stirred as it has been cooled. At a certain point, ice crystals will spontaneously form throughout the volume of water, and grow until all of the water has turned to ice.

Inflation. Such a phase transition, when treated in the context of an expanding universe, can result in remarkable behavior. The dynamics of an expanding universe are governed by Einstein's equations from the theory of general relativity. When these equations are solved by using the properties of ordinary matter, it is seen that the rate of expansion of the universe, characterized by the Hubble parameter H—which gives the velocity of separation of objects—slows down over time in a way that is dependent on the temperature of the universe, T. (Specifically, in a radiation-dominated universe, H is dependent on T^2.) This slowdown occurs because the energy density of matter, which is driving the expansion, is decreased as the matter becomes more dilute because of the effects of expansion.

An examination of Fig. 2, however, shows that once the temperature is below the critical temperature for the transition, the metastable symmetric phase has a higher energy than the new lower-energy symmetry-breaking phase. Until the transition occurs, this means that the symmetric phase has associated with it a large constant energy density, independent of temperature. When this constant energy density is placed on the right-hand side of Einstein's equations, where the energy density of matter appears, it is found that the resultant Hubble parameter describing expansion is a constant. Mathematically, this implies that the scale size of the universe increases exponentially during this supercooling phase. This rapid expansion is what is referred to as inflation. Once the phase transition is completed, the constant energy density of the original phase is converted into the energy density of normal matter in the new phase. This energy density decreases with expansion, so that the universe reverts to the slower rate of expansion characteristic of the standard big bang model. SEE HUBBLE CONSTANT.

Successes of inflation. Guth pointed out that a period of exponential expansion at some very early time could solve a number of outstanding paradoxes associated with standard big bang cosmology.

Flatness problem. In 1979, R. H. Dicke and P. J. E. Peebles pointed out that present observations seem to imply that either the present era is a unique time in the big bang expansion or the initial conditions of expansion had to be fine-tuned to an incredible degree. The observation involves the question of whether the universe is now open, closed, or flat. A closed universe is one where there is sufficient mass in the universe to eventually halt and reverse the observed expansion because of net gravitational attraction; an open universe will go on expanding at a finite rate forever; and a flat universe forms the boundary between these two cases, where the expansion rate will continually slow down and approach zero asymptotically. Measurements of the observed expansion rate, combined with measurements of the observed mass density of the universe, yield a value for the density parameter Ω that rapidly approaches 0 for an open universe, infinity for a closed universe, and is exactly equal to 1 for a flat universe. All measurements of Ω yield values between about 0.1 and 2. What is so strange about this measurement is that theory suggests that once the value of Ω deviates even slightly from 1, it very quickly approaches its asymptotic value far away from 1 for open or closed universes. Thus, it is difficult to understand why, after 10^{10} years of expansion, the value of Ω is now so close to 1. At a quantitative level, the problem is even more mysterious. In order to have its present value in the range given above, at the time of nucleosynthesis the value of Ω would have had to have been equal to 1 within 1 part in 10^{15}.

Inflation naturally explains why Ω should exactly equal 1 in the observable universe today. Einstein's equations for an expanding universe can in the cases of interest be written in the form of the equation below, where T is the time after beginning of expansion,

$$1 + \frac{K}{R(t)^2 \rho(t)} = \Omega(t)$$

R is the cosmic scale factor, related to the size of the universe in closed universe models, ρ is the energy density of matter, and K is a constant that is equal to 0 for flat universe models. Normally, for an expanding universe, the energy density of matter decreases with scale at least as fast as $1/R^3$, so that the left-hand side deviates from 1 as time goes on. This is the quantitative origin of the flatness problem. However, in an inflationary phase, $\rho(t)$ remains constant, while R increases exponentially. Thus, during inflation $\Omega(t)$ is driven arbitrarily close to 1 within the inflated region. If there are some 28 orders of magnitude of exponential expansion, which is possible if supercooling lasts for some time, then $\Omega(t)$ need not have been finely tuned to be close to 1, even if inflation occurred as early as the Planck time (10^{-45} s).

Horizon problem. An equally puzzling problem that inflationary cosmology circumvents has to do with the observed large-scale uniformity of the universe. On the largest observable scales, the universe appears to be largely isotropic and homogeneous. In particular, the 3-K microwave radiation background, which in different directions has propagated from distances separated by 10^{10} light-years, is known to be uniform in temperature to about 1 part in 10,000. SEE COSMIC BACKGROUND RADIATION.

This observed uniformity may not seem so puzzling, until an attempt is made to derive it in the context of any set of reasonable initial conditions for the big bang expansion. Physical effects can propagate at best at the speed of light. The distance a light ray could have traveled since the big bang explosion, which is also thus the farthest distance at which one object can affect another, is called the horizon. The horizon size increases linearly with time, since light travels with a constant velocity even in an expanding universe. Thus, the size of the horizon today, which is about 10^{10} light-years, is much larger than the horizon size at the time the radiation in the microwave background was emitted. In particular, in the standard big bang model the sources of the radiation observed coming from opposite directions in the sky were separated by more than 90 times the horizon distance at the time of emission. Since these regions could not possibly have been in physical contact, it is difficult to see why the temperature at the time of emission was so uniform in all directions.

Even if some very isotropic initial conditions are postulated for the big bang expansion to account for this uniformity, a quantitative problem similar to the flatness problem is encountered in attempting to account for the degree of uniformity now observed. Small fluctuations in energy density tend to grow as the universe evolves because of gravitational clumping. Indeed, observed clumping on the scale of galaxies and smaller attests both to this fact and to the existence of some initial irregularities in the distribution of matter. However, since irregularities grow but the universe is relatively smooth on large scales, any initial conditions for the big bang expansion imposed at early times, say 10^{-45} s, would have to involve an absurdly uniform distribution of matter on the largest scales.

Inflation solves the horizon problem very simply. If the observed universe expanded by 28 orders of magnitude in a short period at a very early time, then in inflationary cosmology it originated from a region 10^{28} times smaller than the comparable region in the standard big bang model extrapolated back beyond that time. As shown in **Fig. 3**, this makes it quite possible that at early times the entire observed universe was contained in a single horizon volume.

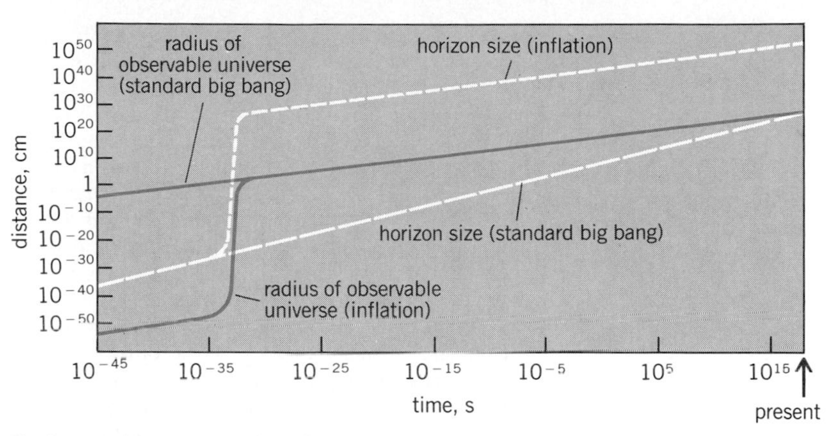

Fig. 3. Horizon size and radius of the observable universe as functions of time after beginning of expansion, in the standard big bang and inflationary models. ***(After A. H. Guth and P. J. Steinhardt, The inflationary universe, Sci. Amer., 250(5):116–128, May 1984)***

Problems with "old" inflation. The list of cosmological problems resolved by the original inflationary universe model is quite impressive. Moreover, its origin in current elementary-particle-physics ideas made it more than an artificial device designed specifically to avoid cosmological paradoxes. Unfortunately, the original inflationary scenario was fundamentally flawed. The central problem for this scenario was how to complete the phase transition in a uniform way. The phase transition began by the formation of bubbles of one phase nucleating amidst the initial metastable phase of matter. While these bubbles grow at the speed of light once they form, the space between the bubbles is expanding exponentially. Thus, it is extremely difficult for bubbles to eventually occupy all of space, as is required for the completion of the transition. Moreover, even if bubbles are forming at a constant rate, each region of space quickly becomes dominated by the largest bubble, which formed earliest. Collisions with much smaller bubbles in this region will not adequately or uniformly dissipate the energy of this largest bubble, so that even if the transition does manage to percolate through space, the final state will be very nonuniform on all scales. Finally, topologically stable objects such as magnetic monopoles and domain walls can form at the intersection of bubbles, and end up with densities after inflation well above those allowed by observation.

While these problems motivated the development of a "new" inflationary scenario, described below, in 1989 P. J. Steinhardt and collaborators pointed out that in certain previously examined special models related to so-called Brans-Dicke models of gravitation, the gravitational constant can itself vary during an inflationary phase transition. In this case, the inflationary expansion rate, while very fast, need not be exponentially fast, so that bubbles growing at the speed of light can eventually percolate throughout all of space, thereby completing the phase transition and ending the inflationary expansion. However, it has not been determined whether these rather special scenarios, which may still leave large remnant inhomogeneities on the scale where bubbles coalesce, are viable. *See Gravitation.*

"New" inflationary cosmology. For almost 2 years the problems of the original inflationary scenario remained unresolved. Near the end of 1981, however, a new approach was developed by A. D. Linde in the Soviet Union, and independently by A. Albrecht and Steinhardt in the United States, which has since become known as new inflation. They suggested that if the energy function (potential) of Fig. 2 were slightly changed, then it might be possible to maintain the successful phenomenology of the old inflationary model while avoiding its problems. In particular, they considered a special form of the potential that is extremely flat at the origin (**Fig. 4**; such functions were considered in connection with particle theory by S. Coleman and E. J. Weinberg). Most important, such functions have essentially no barrier separating the

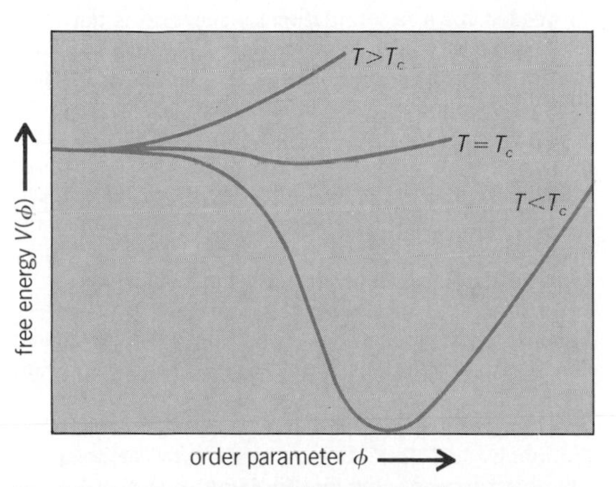

Fig. 4. Free energy as a function of the order parameter for the case of a so-called slow-rollover transition that appears in the new inflationary cosmology.

metastable from the stable phase at low temperature. Now, when the universe cooled down below the critical temperature, the order parameter could continuously increase from zero instead of tunneling discretely to a large nonzero value. As long as the potential is sufficiently flat near the origin, however, it can take a long time before the order parameter approaches its value at the true minimum of the potential (a so-called slow-rollover transition), at which point the region will be in the new phase. During this time the region of interest can again be expanding exponentially because of the large constant energy density that is maintained while the order parameter remains near the origin. Thus, in some sense a single bubble can undergo inflation in this scenario. If the amount of inflation in this region is sufficient, the whole observable universe could have originated inside a single inflating bubble.

This key difference between old and new inflationary cosmology accounts for the ability of the latter to bypass the problems of the former. Because in new inflation the observable universe grew from a single inflating bubble, problems of inhomogeneity, and topological defects resulting from bubble percolation and collisions are avoided, at least as far as observational cosmology is concerned. Moreover, the problem of completing the phase transition is naturally avoided. Once the order parameter in the inflating region approaches the steep part of the potential, the large constant energy density of the initial phase is reduced, and inflation slows. This energy density is converted into a form of kinetic energy of the changing order parameter. This motion of the order parameter is reflected in a changing charge of the new ground state of matter. Just as a time-varying electric charge produces electromagnetic radiation, so the variation of the order parameter produces a thermal background of real particles. By the time the order parameter has settled at the new minimum of the potential, all of the original constant energy density of the symmetric phase of matter has been converted into energy of real particles at a finite temperature. (Any particle density that existed before inflation has, of course, been diluted away by the vast expansion of the region during inflation.) The evolution from this point on is that of the standard big bang model, with the initial conditions being a uniform density of matter and radiation at some finite temperature.

Problems with new inflation. New inflation, too, is not without its problems. The type of potential (Fig. 4) needed for a new inflationary scenario is not at all as generic as that shown in Fig. 2. In order to have such a slow-rollover transition, the parameters of particle physics models must be finely tuned to some degree. Moreover, the temperature to which the universe is reheated after inflation must be large enough so that, at the very least, big bang nucleosynthesis can proceed afterward. This again rules out some possible models. No clear candidate model for new inflation has emerged from particle physics.

Another potential problem for any inflationary scenario concerns initial conditions. As discussed above, if an inflationary phase precedes the standard big bang expansion, then it is possible to resolve problems of the standard big bang model related to the unphysical fine tunings that seem necessary at time zero in order for the big bang to evolve into its presently observed form. However, there is also the question of how generic such an inflationary phase is—namely, how special the initial preinflationary conditions must be so that space-time will undergo an inflationary transition in the first place. If these too are unphysical, then inflation may not have really solved any fundamental problems. While Guth suggested in his initial work that the preconditions for inflation were not severe, this view has been questioned by some. R. Penrose has attempted to develop an existence proof that not all initial configurations of the universe can be made isotropic by inflation. This, of course, does not address the fundamental issue of whether such configurations are at all generic.

In 1983, Linde proposed a version of inflation, which he called chaotic inflation, that may in principle address this issue. He suggested that the early universe may have been arbitrarily inhomogeneous. In some regions, inflation may have successfully taken place, even without the spontaneous symmetry breaking associated with a grand unified theory, and in other regions it may not have. He then suggested that the regions in which inflation did take place are in fact most probable. In particular he argued that those values of parameters that allow large amounts of inflation to take place are most probable, if an initial random distribution of regions in the preinflation universe governed by different metric parameters is allowed. In addition, he pointed out that life would form only in those regions that became sufficiently isotropic so that, if many different regions of the universe now exist, it is not surprising that humans live in one that has undergone inflation. This argument, a reformulation of the so-called anthropic principle, is, however, difficult to quantify. In any case, the issue of initial conditions for inflation is the subject of much research, but may require an understanding of quantum gravity for its eventual resolution. SEE CHAOS; QUANTUM GRAVITATION.

Predictions. Whether or not there exists a specific model for new inflationary cosmology, and whether or not understanding inflation may require further refinements in the understanding of quantum gravity, the implications of an inflationary phase in the early universe are profound. As discussed above, many of the fundamental paradoxes of standard big bang cosmology can be resolved. Moreover, it has been demonstrated that new inflationary cosmology naturally allows a derivation from first principles of the spectrum of primordial energy density fluctuations responsible for galaxy formation. Before new inflation, there existed no method, even in principle, for deriving this spectrum for the standard big bang model; it always had to be input as an artificial addition. Remarkably, the spectrum that emerges from inflationary models is a so-called scale-invariant spectrum of perturbations. It is exactly this type of spectrum that had previously been postulated by Y. B. Zel'dovich and E. R. Harrison 10 years before the formulation of the new inflationary model to account for the observed properties of galaxies while maintaining agreement with the observed anisotropy of the microwave background.

In a remarkable discovery, the first observation of anisotropies in the microwave background was announced in April 1992, based on the analysis of 2 years of data on the microwave background structure using the differential microwave radiometer experiment aboard the *Cosmic Background Explorer (COBE)* satellite launched by the National Aeronautics and Space Administration (NASA) in 1989. A quadrupole anisotropy in the background was observed at a level of about 5×10^{-6}, just in the range that might be expected for primordial fluctuations from inflation that might also result in the observed distribution of galaxies. Also, a quadrupole anisotropy of this magnitude could come from gravitational waves generated during inflation if it occurred at a scale comparable to that at which it is now expected that grand unification might occur. Moreover, the correlation of temperature deviations observed across the sky on scales greater than about 10° is remarkably consistent with the scale-invariant spectrum predicted by inflationary models. While neither of these observations conclusively proves the existence of an inflationary phase in the early universe, the fact that they are clearly consistent with such a possibility, and at present with no other scenario, gives great confidence that inflationary models may provide the correct description of the universe in the era preceding the present observed isotropic expansion.

Thus, inflationary models are attractive for many reasons. In particular, they make definite predictions about the universe today. Specifically, if inflation is correct, (1) the observable universe should have an exactly critical density now; and (2) the anisotropy of the microwave background should have a specific magnitude, consistent with that observed by *COBE*, and likely to be directly related to the observed clustering of galaxies. With the prediction of microwave anisotropy now confirmed by *COBE*, possible confirmation of the prediction of the density of the universe is awaited. Such confirmation would require the existence of dark matter, invisible to telescopes, that would have to make up about 99% of the matter in the universe. *SEE BIG BANG THEORY; COSMOLOGY; UNIVERSE.*

Lawrence M. Krauss

Bibliography. A. H. Guth, Inflationary universe: A possible solution to the horizon and flatness problems, *Phys. Rev.*, D23:347–356, 1981; A. H. Guth and P. J. Steinhardt, The inflationary universe, *Sci. Amer.*, 250(5):116–128, May 1984; L. Krauss, *The Fifth Essence: The Search for Dark Matter in the Universe*, 1989; A. D. Linde, The inflationary universe, *Rep. Prog. Phys.*, 47:925–986, 1984; G. Smoot et al., *Astrophys. J.*, 396:L1–L5, 1992.

Infrared astronomy

The branch of astronomy that employs data in the heat radiation region of the electromagnetic spectrum to investigate the nature and physical characteristics of astronomical sources. The infrared region borders the visible region at a wavelength of 1 micrometer (3×10^{14} Hz), and the radio region at 1000 μm (3×10^{11} Hz).

Heat radiation was discovered in 1800 by W. Herschel as he performed the first infrared solar astronomy experiment. To facilitate his study of the Sun and sunspots, Herschel had used various combinations of colored darkening glasses; he found that with some combinations he felt heat but saw little light, and vice versa. Using the clue that the glass combinations had different colors, he set up an experiment. He placed thermometers behind a slitted board so that one would be visible through the slit and all the others shaded by the board. Using a prism to disperse sunlight, he moved the board to select which part of the solar spectrum illuminated the exposed thermometer, and he noted the temperature difference between the illuminated and shaded thermometers after 8 min of illumination at each position. In this way, invisible heat radiation was discovered beyond the red end of the spectrum.

A century later, M. Planck introduced the quantum hypothesis to derive the formula that encompassed all that had been learned piecemeal about heat radiation from a cavity; this radiation is also called planckian, thermal, or blackbody radiation. Planck's formula expresses the spectral power of cavity radiation—the thermal energy emitted each second from a unit area of a blackbody into a unit wavelength interval. It provides the theoretical basis for understanding Herschel's experiment, and for evaluating the fundamental and technical limitations of infrared astronomy, as well as for interpreting many of the current observational results. For blackbodies, the total power emitted from a unit area is proportional to the fourth power of the absolute temperature in kelvins, and the spectral power that is radiated at any given wavelength is higher for the hotter blackbody. There is a peak in the spectral power at the point where the product of the wavelength and the absolute temperature is 2900 μm · K. At wavelengths that are somewhat longer than this peak, the spectral power is directly proportional to the blackbody temperature and inversely proportional to the fourth power of the wavelength. Shortward of this peak, the spectral power is inversely proportional to the fifth power of the wavelength, but dominated by an exponential cut off.

Observational technology. The Earth's atmosphere is opaque throughout most of the infrared, because of absorption by water (H_2O), carbon dioxide (CO_2), and occasionally a trace species like ozone (O_3). The **illustration** shows wavelength regions in which the atmosphere is relatively transparent; observations can be made in these windows: near-infrared (1–3 μm), mid-infrared (3–30 μm), and (on those rare occasions that the atmosphere above is particularly dry) submillimeter (300–1000 μm). The atmosphere is opaque

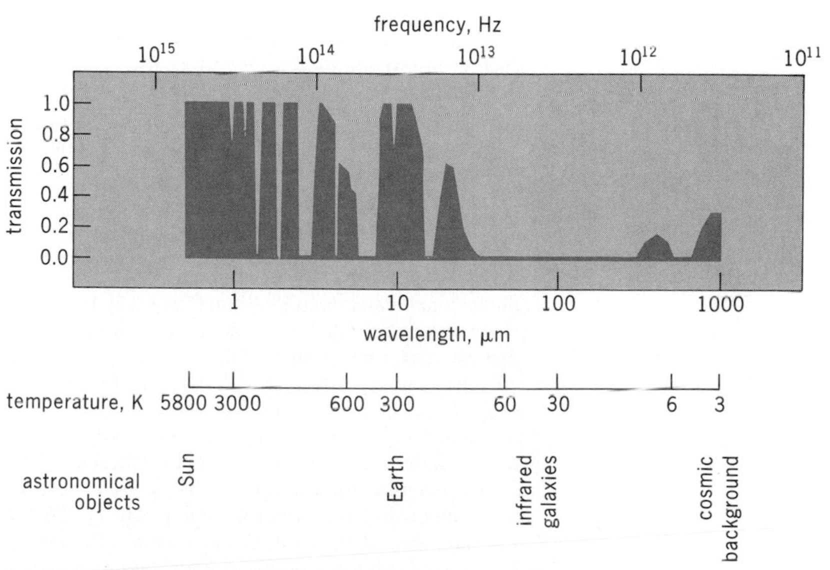

Location of infrared atmospheric windows in terms of frequency and wavelength. The temperatures of various blackbodies are shown where the peak in their spectral power distribution occurs. Types of astronomical objects with characteristic temperatures in each range are indicated.

Power radiated by a blackbody at 300 K (80°F) in infrared spectral regions

Region	Wavelength limits, μm	Ratio of power radiated in region to total radiated power, %
Near-infrared	1–3	0.1
Mid-infrared	3–30	83
Far-infrared	30–300	17
Submillimeter	300–1000	0.02

from the ground in the far-infrared (30–300 μm). However, at 45,000 ft (14 km), the height of the tropopause, atmospheric transmission is about 0.80, and at 90,000 ft (27 km) the transmission is about 0.95. Thus, far-infrared observations can be made from telescopes on aircraft, balloons, or spacecraft.

Background interference. To the extent that the telescope and the atmosphere do not perfectly transmit astronomical infrared radiation, they themselves act like 300 K (80°F) sources of heat energy, generating a bright background. The fractions of the total power radiated by a 300 K (80°F) blackbody in the various infrared spectral regions are given in the **table**. Clearly, the observation site must be high and dry, especially for observations in the mid-infrared, and particular care must be exercised to reduce extraneous emission from the telescope. Even so, the contrast for observing Venus at noon with the unaided eye is about 600 times greater than for measuring Vega at 10 μm with even an optimized 98-in.-diameter (2.5-m) telescope. *SEE TELESCOPE.*

Detectors and filters. For both detectors and spectral filters, variety and capability are greater at shorter infrared wavelengths. Many photon and thermal detectors are available, most of which can be operated at temperatures of 77 K (−321°F) or higher. (77 K is particularly convenient because it is the boiling point of liquid nitrogen at 1 atm pressure.) Out to about 30 μm, high-performance interference filters based on multilayer dielectric coatings are available. In the far-infrared, the most sensitive detector is a bolometer operating at less than 2 K (−456°F). Filters based on interference between wire grids, scattering, or true absorption provide only relatively coarse spectral resolution. In the submillimeter region, some radio techniques have begun to be introduced—for example, heterodyne radiometry, which provides good sensitivity and narrow spectral bandpass. *SEE SUBMILLIMETER ASTRONOMY.*

Large two-dimensional detectors. The principal technology change required to enable future astronomical advances will be the development of large two-dimensional detectors. Until the late 1980s, all infrared astronomy had been done with either single-element detectors, a small linear array, or a mosaic collection of several tens of single detectors. However, large-area detectors subdivided into several thousand picture elements (pixels) are rapidly becoming available. In mercury-cadmium-tellurium (MgCdTe) material, for example, a 64 × 64 format has been demonstrated at 10 μm, and a 256 × 256 format has been demonstrated for wavelengths shorter than 5 μm. Mosaics of these large two-dimensional detectors have been assembled to make even larger focal planes. Simply because of the relative number of pixels, these large two-dimensional detectors can in a given amount of observing time provide larger area coverage with more sensitivity and greater angular resolution. A focal plane with the same number of discrete detectors would have problems of high data rates and a large number of wires. To circumvent these problems, the two-dimensional detectors are generally mated to integrating, parallel-to-serial multiplexers. In some cases, the detector and the multiplexer are made from the same material, as a monolithic device; usually, however, the two are dissimilar materials joined together by a matrix of indium bumps, one at each pixel. Frequently the multiplexer is a charge-coupled device (CCD), whose wells can collect charge from a given pixel, and then transfer that charge when the detector is being read out. Increasingly, however, multiplexers utilize special-purpose circuitry whose features have been miniaturized to about 1 μm so that the circuit can fit into the area of a small (25 μm × 25 μm) pixel. This direct-readout (DRO) circuitry can provide optimized low-noise readout, and the potential for more extensive signal processing than can be supported by a charge-coupled device. *SEE CHARGE-COUPLED DEVICES.*

Sky surveys. Knowledge in infrared astronomy has been gained rapidly as a result of the success of several large-area sky surveys. The most sensitive was carried out by the *Infrared Astronomy Satellite (IRAS),* which carried a 24-in. (0.6-m) telescope cooled by liquid helium. An array of 62 infrared detectors at its focal plane covered wavebands centered at 12, 25, 60, and 100 μm. The *IRAS* field of view was scanned across the sky as the satellite traversed its orbit, in such a way that 95% of the sky was covered during the 11-month operational lifetime of the mission. More than 250,000 infrared sources were discovered.

Celestial infrared sources. These include stars at wavelengths in the near-infrared region, and dust at longer wavelengths.

Cool stars. At near-infrared wavelengths, stars are the primary celestial sources of infrared radiation. Stars with surface temperatures less than 4000 K (6700°F), which is approximately the temperature of sunspots, emit most of their energy in the infrared. The majority of stars in the Milky Way Galaxy are this cool.

There are two large classes of stars cooler than the Sun. The first, and by far the most numerous, are simply low-mass stars that burn their nuclear fuels more slowly than the Sun and never reach the Sun's temperature. An object with a mass less than 0.08 times that of the Sun could not sustain hydrogen burning and therefore would never become a star. Such an object could, however, burn deuterium for a short time, after which it would cool forever. Such low-mass objects are expected to spend most of their lives at temperatures of a few hundred degrees, and to have radii not much larger than that of Jupiter; they are therefore called brown dwarfs. Intensive searches for these low-luminosity objects are motivated by the possibility that they could constitute a large fraction of the mass of the Milky Way Galaxy. Whether any true brown dwarf has yet been found is still a matter of controversy. *SEE BROWN DWARF.*

The second large class of stars cooler than the Sun comprises red giants, whose radii are about equal to the orbital radius of the Earth about the Sun. These stars, with masses equal to or greater than that of the Sun, are burning helium in their cores, having con-

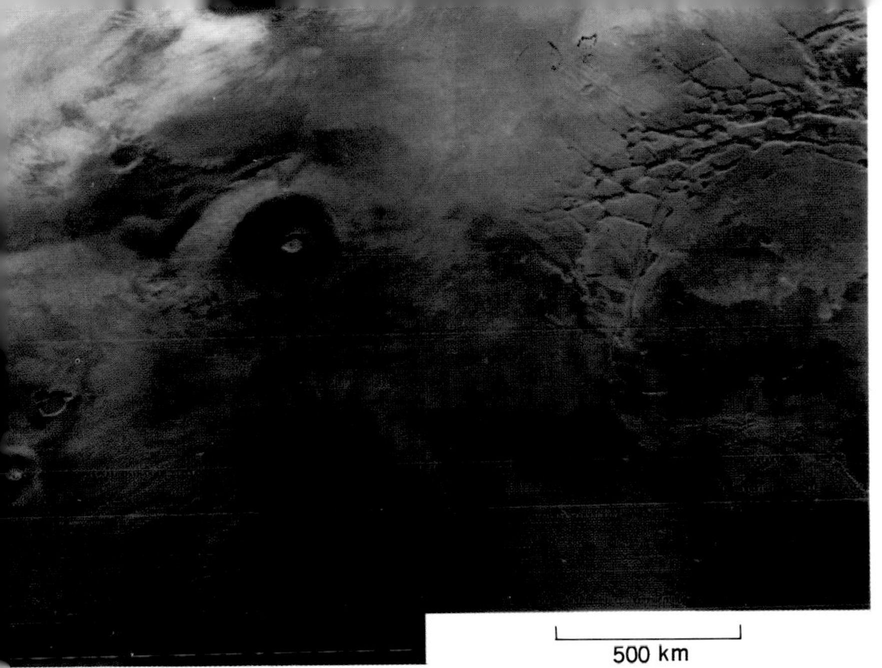

Oblique view of Mars along Tharsis ridge from *Viking Orbiter1*, showing the three Tharsis volcanoes. (*NASA*)

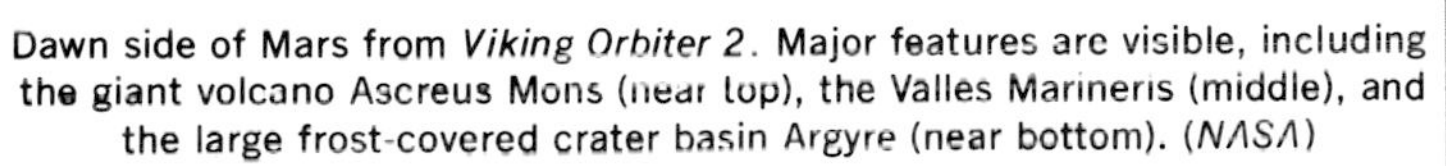

Dawn side of Mars from *Viking Orbiter 2*. Major features are visible, including the giant volcano Ascreus Mons (near top), the Valles Marineris (middle), and the large frost-covered crater basin Argyre (near bottom). (*NASA*)

Artist's conception of Olympus Mons wreathed in clouds. A well-defined cloud train extends several hundred kilometers beyond the mountain (upper left). (*NASA*)

sumed most of their hydrogen. This is the last stage of nuclear-burning evolution for stars like the Sun, and is characterized by such high luminosity (equal to 10^3 times that of the Sun, or more) that the outer layers of the star must swell in order to release it. Studies of these stars, which can be detected at very large distances from the Sun, provide a powerful probe of the spatial distribution of solar-type stars throughout the Milky Way Galaxy. *See Star; Stellar evolution.*

Reddened stars. Stars hotter than the Sun also appear to be relatively brighter at infrared wavelengths than at optical wavelengths. This reddening happens because interstellar space contains particles of dust. Even though the average separation between these particles is nearly 1000 mi (1600 km), the distances between stars are so great that even such a small concentration is enough to dim the visible light from distant stars. In some regions of the sky the effects of dust obscuration are dramatic, with the light from stars being completely obliterated. However, the existence of dust eluded astronomers until the 1930s, when R. Trumpler measured the angular sizes and brightnesses of a large number of star clusters, and analyzed his measurements on the assumption that all clusters had about the same linear dimension and the same number and types of stars. He found that clusters with a small angular size were systematically fainter than expected by comparison to clusters with large angular sizes, just the effect that could be produced by interstellar soot. Further observations soon showed that the effects of dust obscuration were much reduced if the clusters could be viewed in red light. At infrared wavelengths, the effects of dust are smaller still. The *IRAS* image of the sky shows that the Milky Way Galaxy is more brilliant, flattened, and centrally concentrated than indicated from optical wavelength images. *See Interstellar extinction; Interstellar matter; Milky Way Galaxy.*

Circumstellar dust envelopes. At wavelengths longer than about 3 μm, nonstellar sources of infrared radiation become relatively more prominent. The most important mechanism operating in the mid- and far-infrared wavelength range is emission by solid material. The amount of energy emitted in the infrared depends on the amount of light that is absorbed, and the wavelength dependence of the emitted energy depends on the equilibrium temperature reached by the absorber, and on its physical characteristics (size, chemical makeup).

Dust in the solar system, called zodiacal dust, is concentrated in the ecliptic, the plane near which the planets orbit the Sun. This dust, heated by the Sun, reradiates the energy it absorbs at mid-infrared wavelengths between 5 and 30 μm. Thus, it is so pervasive that, within this spectral range, the total energy radiated by zodiacal dust exceeds that of all of the planets combined. The wavelength dependence of the emission from zodiacal dust indicates that it is mostly silicate; this kind of dust is thought to originate from the fragmentation of asteroids. Therefore, as the asteroid population becomes depleted, the formation of zodiacal dust will diminish. If all stars form with planetary systems, the youngest ones should have more prominent zodiacal dust emission. An important result of the *IRAS* flight was the discovery of excess infrared emission from a nearby young star, Vega. If this excess emission is due to dust analogous to the zodiacal dust, then this discovery represents the first direct detection of a component of a planetary system associated with another star. Some of the goals of work in infrared astronomy are to test this interpretation and to expand evidence for planetary systems. *See Asteroid; Interplanetary matter.*

Excess infrared radiation, attributable to thermal emission from dust, had been observed in the spectral energy distributions of red giants long before *IRAS* detected Vega. However, the dust associated with these stars is not related to the existence of a planetary system. Rather, it is condensed from gas that has drifted away from the outer, distended atmospheres of these stars. The most intense dust emission comes from the inner regions of the circumstellar envelopes of these stars, where the dust density and temperature are highest. Thus, the infrared emission from circumstellar envelopes of cool stars is hotter than the excess infrared emission from hot stars like Vega. Since red giant stars have extremely high mid-infrared luminosities (where obscuration due to interstellar dust is nearly negligible), they can be viewed at very large distances, and therefore provide a means of tracing the large-scale structure of the Milky Way Galaxy. These stars also play a key role in the chemical evolution of the Galaxy, since the matter they eject into the interstellar medium, enriched by nuclear burning in the stellar interior, is eventually incorporated into new stars.

Outer planets. At far-infrared wavelengths, the brightest celestial sources are the outer planets—Jupiter, Saturn, Uranus, and Neptune. Their many satellites are also far-infrared sources. The major planets Jupiter, Saturn, and Neptune emit significantly more heat energy than they absorb from the Sun. This excess energy is remnant heat of formation (stored gravitational energy), an indication that the structures of these planets have not yet stabilized. *See Jupiter; Neptune; Planetary physics; Saturn; Uranus.*

Interstellar clouds. Beyond the solar system, the brightest far-infrared sources are interstellar clouds. The largest of these, even with an average density less than 10^{-12} that of the Earth's atmosphere, can reach masses of 10^6 times the mass of the Sun, and are the most massive objects in the Milky Way Galaxy. In the interior of such a giant cloud, most external starlight is blocked out. This provides the conditions required (high density, low temperature) for density perturbations to undergo runaway collapse, resulting eventually in the formation of new stars. The stars heat the clouds from within to temperatures of about 30 K (−406°F) or more. One of the most prodigious star-forming regions in the Galaxy is the nebula M17, whose large star population was discovered when a sensitive infrared camera was used to peer through the overlying layers of dust.

Most interstellar clouds, and the more evenly dispersed dust in the interstellar medium, are not sufficiently dense and cool that star formation is occurring within them. These inactive clouds are heated mainly by diffused starlight from the Milky Way Galaxy, and the dust in them therefore rarely exceeds temperatures of 20 K (−424°F). Despite their low density and low temperature, the ubiquity of these clouds makes them powerful sources of far-infrared emission.

Normal external galaxies. Of the more than 20,000 galaxies detected in the *IRAS* survey, most are spirals like the Milky Way. Their far-infrared emission is thought to be produced by a combination of warm, discrete clouds and a cooler, diffuse interstellar medium. Half or more of the starlight of normal spiral galaxies is reprocessed by dust and emerges in the far-infrared.

For some galaxies, almost all of the luminosity appears in the far-infrared. One explanation for this anomalous behavior seems to be that collisions and near-collisions between galaxies distort their shapes, particularly the distributions of interstellar matter, stimulating the formation of large-scale density enhancements and clouds and, within them, new clusters of stars. These galaxies must be in a transient stage since the star formation rates deduced from their infrared luminosities are so high that they would deplete their entire supply of interstellar gas and dust in a fraction of their lifetimes.

Though infrared-bright galaxies are only a fraction of all galaxies, their luminosities are so great that they can be used to trace out the large-scale structure of the universe. Initial analyses of *IRAS* data suggest that there is an anisotropy in the distribution of galaxies, which has been attributed to a very large-scale anisotropy in the distribution of matter. If confirmed, this discovery should lead to a new understanding of the fundamental laws that shaped the universe. *See Cosmology; Galaxy, external; Universe.*

Peculiar galaxies. The high ratio of infrared to optical luminosities of some galaxies suggests that there may exist a significant class of galaxies that, until now, have gone totally unseen. Soon after *IRAS* was launched, several notable exceptions were found to the rule that most far-infrared sources (other than those associated with clouds in the plane of the Milky Way) could be readily associated with external galaxies that could be observed at optical wavelengths. Subsequent studies of these sources showed that they were distant galaxies with especially faint optical counterparts. Such large distances, coupled with their observed infrared brightness, means that these galaxies have luminosities as large as those of the most luminous quasars. Studies of these objects are motivated by the possibility that there may be an evolutionary link between these two classes of peculiar galaxies.

IRAS has opened a new window on the understanding of quasars themselves. These sources are known to be the highly luminous nuclei of external galaxies and are thought to be powered by some supermassive object, presumably a black hole. The intensity of the nuclei had previously completely overwhelmed the ability to detect the surrounding "host" galaxies. Whether quasars constitute a distinct class of galaxies or whether all galaxies contain similar but lower-luminosity objects (such as the peculiar nonthermal source at the center of the Milky Way Galaxy) has long been a matter of controversy. *IRAS* observations of quasars showed that many of them are brighter than expected in the infrared, especially at mid-infrared wavelengths. Though the infrared emission from most quasars is due to thermal emission from dust, as in other galaxies, this dust is heated to higher temperatures and is probably located near the nucleus. The discovery of infrared-bright quasars suggests that they may be both more numerous and more luminous than previously thought, deepening their mystery. *See Quasar.*

Prospects. Further studies of the known types of celestial infrared sources have the potential to provide much information about the evolution of the planetary system, planetary systems associated with other stars, the early life history and later disintegration of stars, the structural and chemical evolution of the Milky Way Galaxy and other galaxies, and the evolution of quasars and of the universe as a whole. Technical advances in infrared detector development enable a thousandfold improvement in the capabilities of ground-based telescopes to obtain critical details for the understanding of these sources. Far more dramatic is the expected outcome of planned satellite-borne infrared observing facilities. These include the United States' *Cosmic Background Explorer (COBE)*, a sensitive satellite-borne probe of diffuse far-infrared and submillimeter radiation presumably associated with the primeval fireball; Near-Infrared Camera and Multi-Object Spectrometer (NICMOS), a highly sensitive camera and spectrometer to fly on the Hubble Space Telescope about five years after its launch; Space Infrared Telescope Facility (SIRTF), a cooled, satellite-borne, infrared space observatory; and the European Space Agency's *Infrared Space Observatory (ISO)*. *See Cosmic background radiation; Satellite astronomy.*

Susan Kleinmann; Douglas E. Kleinmann

Bibliography. C. A. Beichman, The IRAS view of the Galaxy and the solar system, *Annu. Rev. Astron. Astrophys.*, 25:521–563, 1987; R. M. Catchpole, A window on our Galaxy's core, *Sky Telesc.*, 75:154–155, 1988; I. Gatley, D. L. DePoy, and A. M. Fowler, Astronomical imaging with infrared array detectors, *Science*, 242:1264–1270, 1988; M. Hoskin, William Herschel and the making of modern astronomy, *Sci. Amer.*, 254:106–112, 1986; S. D. Price, The infrared sky: A survey of surveys, *Publ. Astron. Soc. Pacific*, 100:171–186, 1988; B. T. Soifer, J. R. Houck, and G. Neugebauer, The IRAS view of the extragalactic sky, *Annu. Rev. Astron. Astrophys.*, 25:187–230, 1987.

Interferometry

The design and use of optical interferometers. An interferometer is an instrument in which two or more beams of light from the same source are united after traveling over different paths, and display interference. Optical interferometers based on both two-beam and multiple-beam interference of light are extremely powerful tools for metrology, spectroscopy, and imaging and are used to perform a wide variety of measurements. This article focuses on long-baseline stellar interferometers, which combine starlight from two or more telescopes to achieve very high angular resolution.

The angular resolution of the human eye is approximately 3 arc-minutes; if two objects are closer together than 3 arc-minutes, the eye cannot perceive them as distinct objects. With the invention of the telescope around 1600, angular resolution took a dramatic jump. Even telescopes of modest size (2 in. or 5 cm in diameter) can resolve objects as small as 3 arc-seconds, 60 times better than the unaided eye. The Hubble Space Telescope will have a resolution of 0.1 arc-second, 2000 times the resolution of the eye, after it is repaired in 1993. Interferometers represent the next major step in obtaining high angular resolution. Existing interferometers such as the Mark III on Mount Wilson in California have resolved objects as small as 0.001 arc-second, 100 times the resolution of the Hubble Telescope. Interferometers under construction, such as the Big Optical Array at Lowell Observatory in Arizona and the Sydney University Stellar Interferometer in Australia, will have an angular resolution of 0.0001 arc-second, or 100 microarc-seconds. *SEE OPTICAL TELESCOPE; SATELLITE ASTRONOMY.*

Limitations to angular resolution. Ground-based telescopes have a limitation to angular resolution imposed by turbulence in the atmosphere. At the best astronomical sites, in Chile and Hawaii, the turbulence-limited resolution is almost never better than 0.25 arc-second. The angular resolution of a telescope is also limited by the wave nature of light. The effect of diffraction limits the resolution of a telescope 4 in. (10 cm) in diameter to approximately 1 arc-second, the diameter of a dime seen from a distance of 2.2 mi (3.5 km). The larger the telescope, the higher the angular resolution.

The new generation of large telescopes has diameters between 320 and 400 in. (8 and 10 m), with a diffraction-limited resolution of about 0.01 arc-second. There are significant technological and economic barriers to building filled-aperture telescopes that are much larger than 400 in. (10 m). In order to obtain higher angular resolution, astronomers coherently combine the light from two or more telescopes separated by tens to hundreds of meters (1 m = 3.3 ft). The use of interferometry to obtain high angular resolution started in the 1910s, when A. A. Michelson built an interferometer to measure the diameter of Betelgeuse, a red supergiant star. The technique of interferometry, however, was perfected by radio astronomers in the 1960s, 1970s, and 1980s.

Principles of stellar interferometry. Coherent combination of light from several telescopes requires that the path length of light from the star to the beam combiner via the separate telescopes be equal within a fraction of a wavelength. At radio wavelengths, this requirement means that the lengths of cable connecting two radio antennas must be measured to an accuracy of about 1 in. (25 mm). At optical wavelengths the required accuracy in the placement of mirrors is approximately 1 micrometer, or 1/25,400 of an inch. The factor of difference of approximately 10,000 in the wavelengths of radio waves and light waves is the reason that radio interferometry was successfully developed several decades before optical interferometry. *SEE RADIO TELESCOPE.*

Figure 1 is a diagram of a two-element long-baseline interferometer. Light from two telescopes approximately 100 ft (30 m) apart is brought together with mirrors. The light is directed through a set of movable mirrors. By using laser interferometry and servo control, the optical path of the light can be controlled with an accuracy of a few nanometers (approximately 10^{-7} in.). When the mirrors are properly positioned, the light of one star imaged by the two telescopes will interfere. The interference pattern gives information on the size of the star, the location of the star on the sky, and the presence of star spots (similar to sunspots but on other stars).

Stellar diameter measurements. To the unaided eye, stars appear as points of light. Atmospheric tur-

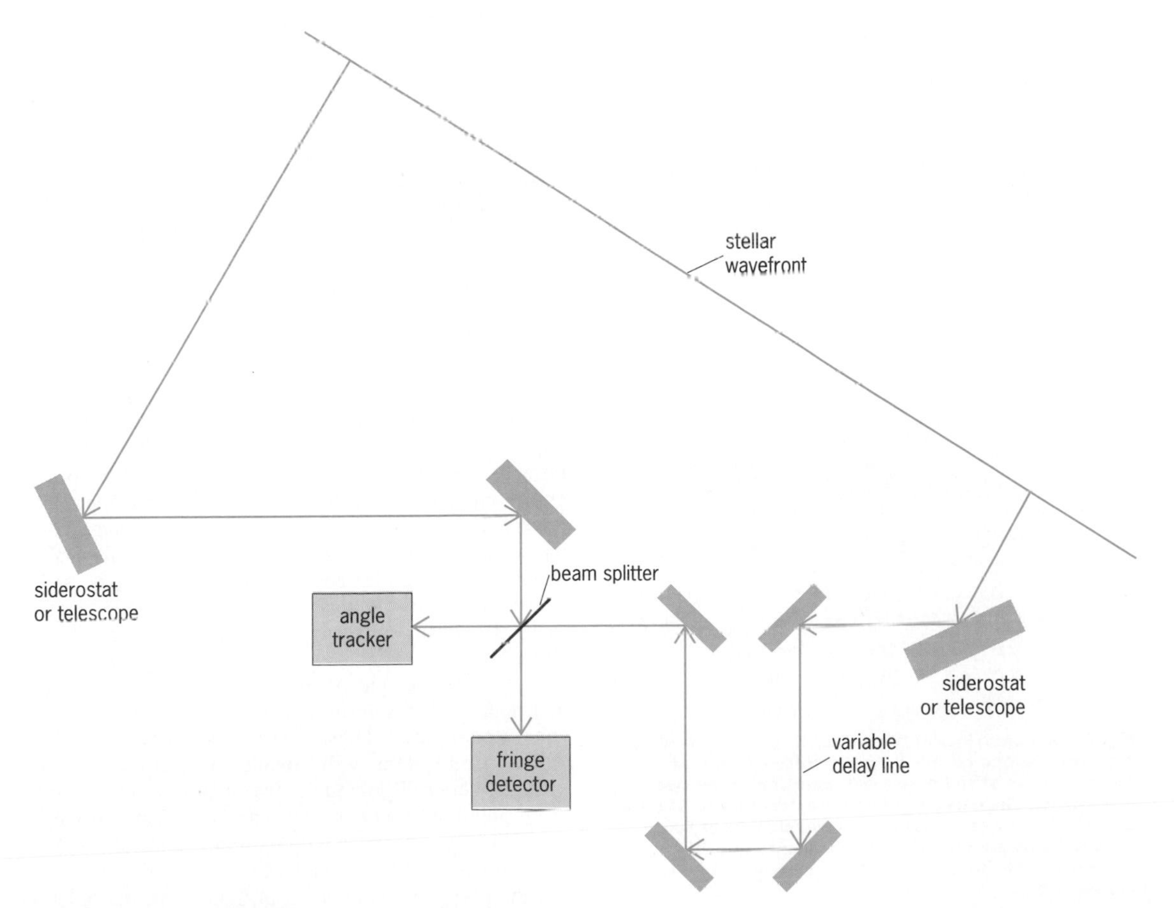

Fig. 1. Two-element long-baseline stellar interferometer.

bulence prevents an observer looking through the largest ground-based telescope from seeing a star as it really is, a round disk with perhaps a few star spots. In space, the Hubble Telescope is not quite large enough to resolve the largest (angular diameter) star into a disk. Interferometers with long baselines can measure the diameter of a star through the depth of the interference pattern formed when the light from two separate telescopes is combined. The depth of the interference pattern, called the fringe visibility, decreases as the two telescopes are moved apart. **Figure 2** shows some measurements of fringe visibility at several values of the baseline (separation of the two telescopes). The curve is a fit to the data points. Analysis of these data showed the diameter of the star to be 4.351 milliarc-seconds. The probable error of this measurement is approximately 20 microarc-seconds.

Some of the most interesting stars have diameters that change with time, pulsating with a period of a few days to a few weeks. Some Cepheid variables change their diameters by 20–30% over a few weeks. The next generation of interferometers, now under construction, will be able to measure this pulsation with very high accuracy. *See* Cepheids; Star; Variable star.

Star position measurements. To the average person, the stars in the sky are fixed and never move. In reality, stars move for a number of reasons. This motion, however, is extremely small. As the Earth moves around the Sun, the apparent positions of nearby stars change with respect to distant stars because of the parallax effect. By measuring this motion, it is possible to determine the distances to nearby stars. Another reason that stars move is galactic rotation. All the individual stars seen in the sky are part of the Milky Way Galaxy. The Sun is about 25,000 light-years from the center of the Galaxy, and orbits the Galaxy in about 10^8 years. Precise measurements of the positions of the stars will make it possible to determine the size and mass distribution of the Galaxy. *See* Milky Way Galaxy; Parallax.

Astrometry, the precise measurement of the positions of stars, is one of the oldest branches of astronomy. In the early 1900s, measurements of the positions of stars near the Sun in the sky during an eclipse verified one of the predictions of A. Einstein's general theory of relativity, the gravitational deflection of light. *See* Astrometry.

Interferometers can be used to measure the position of a star with extreme accuracy. The phase of the interference pattern is directly related to the position of the star. Interferometers are currently the most accurate instruments on the ground for wide-angle or fundamental astrometry, the measurement of the position of a star in an inertial coordinate system. For the measurement of very small angles and the relative positions of double stars, long-baseline interferometers again have made the most accurate measurements, with errors in the 100-microarc-second range. At this level of accuracy, the position of an astronaut on the Moon could be measured with approximately 6-in. (15-cm) accuracy. *See* Binary star.

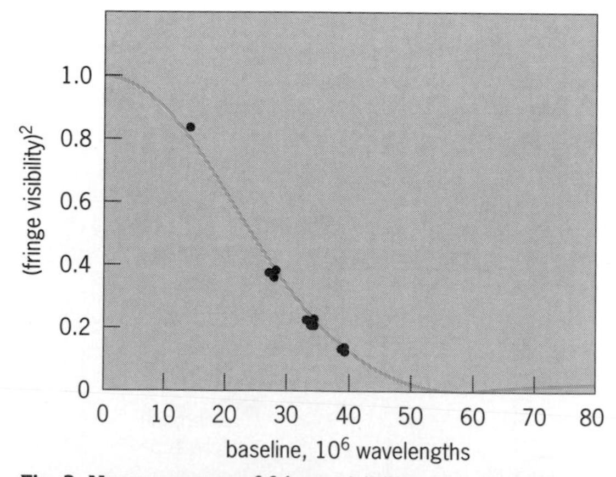

Fig. 2. Measurements of fringe visibility of interference pattern of a long-baseline stellar interferometer at several values of the baseline (separation of the two telescopes). The curve is a fit to the data points. The star Epsilon Cygni was observed at a wavelength of 800 nm. Analysis of these data showed the diameter of the star to be 4.351 milliarc-seconds, with an error of 17 microarc-seconds.

Prospects. The main reason for building interferometers is economic. It is far cheaper to build two telescopes and place them 1000 ft (300-m) apart than it is to build one telescope with a 1000-ft-diameter (300-m) primary mirror. Because the cost of an interferometer is only weakly dependent on the separation of the telescopes, very high resolution is possible at moderate cost. There are two interferometers in routine operation in the United States, and seven under construction in the United States, Australia, France, and England. There are also two very large interferometer projects in the planning stage.

The Very Large Telescope (VLT) interferometer, in an advanced stage of planning, it to be built by the European Southern Observatory. The VLT interferometer will consist of an array of four 320-in. (8-m) telescopes and initially three mobile 72-in. (1.8-m) telescopes with a maximum baseline of 1150 ft (350 m). The light would be combined coherently in a 460-ft-long (140-m) underground building. Initial operation of the interferometer with two elements is expected in 1997 on Cerro Paranal in Chile. The Keck interferometer is in an early stage of planning. The inteferometer is an extension of the two 400-in. (10-m) Keck telescopes. The interferometric array would add four mobile 60–80 in. (1.5–2.0-m) telescopes to the two fixed 400-in. (10-m) telescopes to form a six-element array on top of Mauna Kea in Hawaii.

These two instruments, the Keck and the VLT, represent the two largest telescopes as well as the two largest interferometric arrays planned for the 1990s at infrared and optic wavelengths. Key scientific programs include a search for planets around approximately 100 nearby stars, high-resolution images of the areas where stars and planets are being formed, images of the central energy sources of quasars, and measurements of the mass of compact massive objects such as neutron stars and black holes in binary systems.

In the first decade of the twenty-first century, the first space-based interferometers are expected to become operational. These instruments above the turbulent atmosphere will enable interferometers to achieve their full potential. Astrometric accuracies of a few microarc-seconds will make it possible to map the Milky Way Galaxy with extreme precision, make direct distance measurements to nearby galaxies, and test the gravitational bending of light to second order,

1000 times more accurately than current tests at radio wavelengths.

Michael Shao

Bibliography. D. Mozurkewich et al., Angular diameter measurements of stars, *Astron. J.*, 101:2207–2219, 1991; National Research Council, *The Decade of Discovery in Astronomy and Astrophysics*, 1991; M. Shao and M. Colavita, Long baseline optical and IR stellar interferometry, *Annu. Rev. Astron. Astrophys.*, 30:457–498, 1992.

Interplanetary matter

Gas and small solid particles in the space between the planets in the solar system. Most of the interplanetary gas originates in the Sun, flowing outward in the solar wind. Prior to its direct detection by spacecraft, the existence of such a wind had been hypothesized as an explanation of why gaseous tails of comets were directed radially outward from the Sun. Multiwavelength diagnostics of stellar rotation, magnetically controlled surface activity, and stellar winds indicate that the Sun is not unusual in these characteristics, though better studied because of its proximity.

The widespread presence of microscopic particulate matter has been inferred from observations of the scattered zodiacal light. This interplanetary dust is also warm; its thermal emission at midinfrared wavelengths makes the zodiacal cloud the most prominent component of diffuse emission in the sky at 12 and 25 micrometers, as surveyed by the *Infrared Astronomical Satellite* (*IRAS*). Somewhat larger meteoroids, in their chance encounters with the Earth, are manifested as meteors and fireballs, and, for the most massive, as meteorites. The solid material originates in the asteroids and comets. SEE INFRARED ASTRONOMY.

IRAS spectral measurements of stars like the Sun and somewhat hotter often reveal excess thermal emission attributable to orbiting solid material. Flattening suggestive of a disk geometry has been found where measurements of the shape have been possible. Confirmation of this has been obtained for β Pictoris, where scattered light observations reveal an edge-on disk. Whether this is potential protoplanetary material or subsequent evolutionary debris is under discussion. SEE STELLAR EVOLUTION.

Interplanetary gas. The interplanetary gas is essentially the solar wind, which sweeps out and replenishes interplanetary space at an average rate of 10^9 kg · s^{-1} (5×10^8 lbm · s^{-1}). Enhancements in the flow are associated with increased solar activity, particularly solar flares. On average, the contribution by the wasting of comets is much smaller, probably by several orders of magnitude, although locally the cometary gas can dominate for a time. Even in the gaseous tail of the comet, however, the sweeping action of the solar wind can already be seen, and ultimately all the cometary gas is cleared out of interplanetary space.

The solar wind, which originates in the solar corona, consists primarily of protons, reflecting the composition of the Sun. It is a supersonic flow, with a typical speed of 250 mi · s^{-1} (400 km · s^{-1}). At 1 astronomical unit (1 AU is equal to 9.3×10^7 mi or 1.496×10^8 km, the mean distance from Earth to Sun), the particle density in this spherically symmetric wind has fallen (as r^{-2}, where r is the distance from the Sun) to 80 in.$^{-3}$ (5 cm^{-3}). Being very hot (temperature on the order of 10^5 K), this gas is highly ionized and carries out the solar magnetic field, which also falls as r^{-2} to about 6×10^{-5} gauss (6 nanoteslas) at 1 AU. The wind is divided into unipolar magnetic sectors probably rooted in coronal holes. Beyond about 5 AU (the size of the orbit of Jupiter), both the particle density and field strength have fallen below the values typical of the interstellar medium. The question therefore arises of the radial extent of the solar wind or, alternatively, of the penetration of interstellar gas into the solar system. An estimate of the size of the spherical interstellar bubble carved out by the solar wind can be obtained by balancing the ram pressure of the wind with the thermal pressure of the interstellar gas. The result is about 50 AU, which is comparable to the size of the orbit of Pluto (40 AU), but is much smaller than the solar system cometary cloud and the distances to the nearest stars (10^5 AU). The transition from wind to interstellar medium takes place through a shock. As *Pioneer 10* and the *Voyager* spacecraft continue outward in the solar system, it might become possible to study this region directly. SEE INTERSTELLAR MATTER; SOLAR WIND.

The solar wind, or interplanetary gas, does not impact directly on the Earth's atmosphere, but is forced to flow around by the magnetic pressure of the terrestrial field. The size of the resulting magnetosphere in the solar direction, obtained by balance with the local ram pressure, is about 10 earth radii. (For comparison, the size of the exosphere is only 300 mi or 500 km or 0.1 earth radius.) The interaction zone just beyond this distance is a collisionless shock. In the antisolar direction the wake has a much greater extent. SEE MAGNETOSPHERE.

The magnetic field carried in the solar wind is responsible for excluding low-rigidity galactic cosmic rays from the solar system. Significant solar modulation occurs for energies less than 20 GeV per nucleon. There is evidence that particles can be accelerated in interplanetary shocks up to energies of 10 MeV per nucleon. SEE COSMIC RAYS.

Interplanetary dust. The largest nonplanetary bodies in the solar system are the asteroids (or minor planets) and the comets. Collisions between asteroids and the breakup of short-period (captured) comets result in the smaller solid particles, or meteroids, to be discussed below. Because the orbits of asteroids and periodic comets lie near the mean plane of the solar system, the meteoric complex is a flattened system, unlike the spherically symmetric solar wind. Its limited radial extent, only a few astronomical units, also reflects the sources of dust. SEE ASTEROID; COMET.

Composition. The largest pieces (masses greater than 1 kg or 2 lbm) of rubble are potential survivors of a fiery passage through the Earth's atmosphere, to become meteorites. However, the observed meteorites are not necessarily representative of the interplanetary population. The most common belt asteroids, those of C type, are thought to create the largest number of interplanetary fragments, a few of which are ultimately seen on Earth as the rare carbonaceous chondrites. On the other hand, the most numerous meteorites, ordinary chondrites, seem to be fragments from a rarer type of asteroid, among which are some with favorable Earth-approaching orbits. Cometary debris, seen in one manifestation as fireballs, is apparently so friable that even massive bodies break up in the atmosphere. *SEE METEORITE.*

There are much larger numbers of smaller meteoroids. Particles of mass on the order of 1 g produce visual meteors in the Earth's atmosphere, whereas meteoroids of about 10^{-5} g produce ionization trails which are detectable by radar. Micrometeorites (10^{-9} g) are detected by impact experiments on spacecraft and have been collected in the upper atmosphere. Scattering of solar radiation by particles of this size (10–30 μm) is responsible for the zodiacal light. Modeling of observations of the brightness and polarization distributions of this diffuse light as a function of elongation angle and heliocentric distance, and of the thermal emission, indicates an oblate zodiacal cloud, with an axial ratio of about 7, lying in the plane of the solar system. The particle density falls off as $r^{-1.5}$ to r^{-1} out to a cutoff near 3.3 AU. *SEE METEOR; MICROMETEORITE; ZODIACAL LIGHT.*

Evolution of meteoric complex. There are a number of different clues for finding the sources of the meteoroids. For example, on the basis of their orbits and composition, meteorites appear to originate in asteroids, whereas shower meteors are clearly associated with the dissipation of short-period comets, and with their dust tails. Collected micrometeorites are aggregates of submicrometer-sized particles similar to the matrix material of the most primitive meteorites, type C1 carbonaceous chondrites. The relative contributions of primary particles, which ultimately produce the smaller-sized population through mutual collisions of fragments, are difficult to assess. Furthermore, some classes of asteroid are probably old cometary nuclei. Micrometeoroids are subject to nongravitational forces which soon smooth out all traces of their origins. Therefore, the extent of the zodiacal cloud, which lies interior to the asteroid belts, is not necessarily conclusive support for an asteroidal origin, just as the lack of an enhancement in the density of micrometeoroids within the asteroid belts is not a firm rebuttal. However, discrete bands were found in the diffuse zodiacal thermal emission mapped by *IRAS*, the most obvious being about 5% enhancements encircling the ecliptic 9° above and below the ecliptic plane, probably from relatively cool particles at the distance of the asteroid belt. These enhancements might be maintained by continuous collisions among a family of asteroids, or might be the signature of a recent major collision which has not decayed away. The overall mass requirements are not extreme, so that such collisions might account for the bulk, if not all, of the zodiacal cloud.

What does appear reasonable is that the steep power-law mass distribution observed down to 10^{-5} g is a steady state resulting from successive mutual collisions of primary particles. The flattening of the mass distribution function below 10^{-5} g indicates that particles with masses around 10^{-5} g dominate the meteoric complex, whose integrated mass is 10^{16}–10^{17} kg (2×10^{16}–2×10^{17} lbm). Although there is a further increase down to 10^{-14} g, it is clear that additional destructive processes are depleting the distribution at smaller masses.

The indicated mechanism, which first becomes important in just this mass range, involves the Poynting-Robertson effect. These meteoroids lose a substantial amount of their orbital momentum to the solar photons and the solar wind which they intercept, and thus continuously spiral in toward the Sun. So dramatic is this effect that the zodiacal cloud would become devoid of small particles (masses less than 10^{-6} g) on a time scale of only 10^5 years, were it not for continuous replenishment by collisions between more massive meteoroids. To maintain the observed density, injection of primary particles at a rate near 10^4 kg · s^{-1} (2×10^4 lbm · s^{-1}) is required. This rate is close to some estimates of the contribution available from the steady destruction of comets, whereas others suggest an episodic history in which the present density is unusually high and must be due to a single, relatively recent massive comet. Over the age of the solar system, the accumulated mass requirement would be 10^{21} kg (2×10^{21} lbm), the mass of a large asteroid, so that an asteroidal origin is not ruled out either.

The evolution of the particles as they spiral into the inner solar system can be modeled. In addition to the destructive effects of mutual collisions, which become increasingly important, further degradation by sputtering by the solar wind and by sublimation must be considered, especially at heliocentric distances less than 1 AU. As the characteristic particle size decreases, the relative importance of radiation pressure force to gravity (the ratio β) grows, possibly leading to a timely increase in orbit size. No particle would survive an orbital decay to less than 0.1 AU. It is encouraging that steady-state size distributions and radial distributions predicted by models including these processes are at least qualitatively like those observed for micrometeoroids in the zodiacal cloud.

Submicrometer-sized meteoroids have been discovered streaming out of the inner solar system into the interstellar medium at speeds of 60 mi · s^{-1} (100 km · s^{-1}). It is hypothesized that these are fragments which have become so small that β exceeds unity and they experience a net outward force. They have been named the β-meteoroids. However, the flux appears to be too large for this source alone, and it has been suggested that these are interstellar grains whose trajectories have passed close to the Sun.

P. G. Martin

Bibliography. I. Halliday and B. A. McIntosh (eds.), *Solid Particles in the Solar System*, 1980; P. W. Hodge, *Interplanetary Dust*, 1981; J. A. M. McDonnell et al. (eds.), *Cosmic Dust and Space Debris*, 1989; J. A. Wood, *The Solar System*, 1979.

Interstellar extinction

Dimming of light from the stars due to absorption and scattering by grains of dust in the interstellar medium. In the absorption process the radiation disappears and is converted into heat energy in the interstellar dust grains. In the scattering process the direction of the radiation is altered. Interstellar extinction produces a dimming of the light from stars situated beyond interstellar clouds of dust according to the equation below,

$$F_\lambda = F_\lambda(0)\, e_{-\tau_\lambda}$$

where F_λ is the observed flux of radiation from the star at a wavelength λ, $F_\lambda(0)$ the flux that would be observed in the absence of interstellar extinction, and τ_λ the dimensionless optical depth for interstellar extinction at λ. Measures of the radiation from pairs of stars of similar intrinsic properties but with differing amounts of interstellar extinction can be used to obtain information about τ_λ, which can then be used to

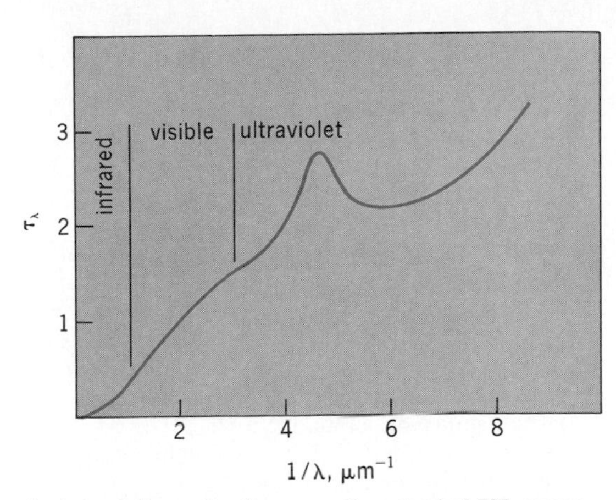

An interstellar extinction curve for a typical 500-parsec path through the interstellar medium of the Milky Way disk. Extinction optical depth τ_λ at wavelength λ is plotted versus $1/\lambda$ in μm^{-1}. The curve spans the infrared, visible, and ultraviolet regions of the spectrum.

provide clues about the nature of the interstellar dust grains.

An interstellar extinction curve is a plot of the extinction optical depth τ_λ versus the wavelength or frequency of the radiation. A typical interstellar extinction curve for a 500-parsec path (1630 light-years, 9.58×10^{15} mi, or 1.542×10^{16} km) through the interstellar medium of the Milky Way disk is shown in the **illustration.** The spectral regions spanned by the curve are indicated. The extinction curve exhibits a nearly linear increase from infrared to visual wavelengths. In the ultraviolet there is a pronounced extinction peak near $1/\lambda = 4.6\ \mu m^{-1}$ or $\lambda = 217.5$ nm, followed by an extinction minimum and rapidly rising extinction to the shortest ultraviolet wavelengths for which data exist. Details not shown in the illustration include weak enhancements in extinction in the infrared near 20 and 9.7 μm and a very large number of such enhancements at visible wavelengths which are referred to as the diffuse interstellar features. A feature near 3.1 μm appears in the extinction curves for sight lines passing through exceptionally dense interstellar clouds.

The absorbing and scattering properties of solid particles depend on their size and composition. A detailed interpretation of the interstellar extinction curve of the illustration and other data relating to interstellar dust suggests that the interstellar grains of dust range in size from about 0.01 to 1 μm and are composed of silicate grains (to explain the 20- and 9.7-μm features) and probably some type of carbon grain (to explain the 217.5-nm extinction peak). The 3.1-μm feature in the extinction curve implies that the interstellar dust acquires coatings of water ice and ammonia ice in the densest regions of interstellar space. Other substances are also likely to contribute to the extinction, but their identification is made difficult by the lack of spectroscopic structure in the extinction produced by many common materials. A comparison of interstellar extinction with the absorption by interstellar atomic hydrogen reveals that the dust contains about 1% of the mass of the interstellar medium. *SEE INTERSTELLAR MATTER.*

Blair D. Savage

Bibliography. J. M. Greenberg, The structure and evolution of interstellar grains, *Sci. Amer.*, 250(6):124–135, June 1984; P. G. Martin, *Cosmic Dust*, 1978; B. D. Savage and J. S. Mathis, The observed properties of interstellar dust, 1978, *Annu. Rev. Astron. Astrophys.*, 17:73–112, 1979.

Interstellar matter

The material between the stars, constituting several percent of the mass of stars in the Galaxy. Being the reservoir from which new stars are born in the Galaxy, interstellar matter is of fundamental importance in understanding both the processes leading to the formation of stars, including the solar system, and ultimately the origin of life in the universe. Among the many ways in which interstellar matter is detected, perhaps the most familiar are attractive photographs of bright patches of emission-line or reflection nebulosity. However, these nebulae furnish an incomplete view of the large-scale distribution of material, because they depend on the proximity of one or more bright stars for their illumination. Radio observations of hydrogen, the dominant form of interstellar matter, reveal a widespread distribution throughout the thin disk of the Galaxy, with concentrations in the spiral arms. The disk is very thin (scale height 135 parsecs for the cold material, where 1 pc is equal to 3.26 light-years, 1.92×10^{13} mi, or 3.09×10^{13} km) compared to its radial extent (the distance from the Sun to the galactic center is about 8000 pc, for example). Mixed in with the gas are small solid particles, called dust grains, of characteristic radius 0.1 micrometer. Although by mass the grains constitute less than 1% of the material, they have a pronounced effect through the extinction of starlight. Striking examples of this obscuration are the dark rifts seen in the Milky Way. On average, the density of matter is only 15 hydrogen atoms per cubic inch (1 hydrogen atom per cubic centimeter; in total, 2×10^{-24} g · cm^{-3}), but because of the long path lengths over which the material is sampled, this tenuous medium is detectable. Radio and optical observations of other spiral galaxies show a similar distribution of interstellar matter in the galactic plane.

A hierarchy of interstellar clouds, concentrations of gas and dust, exists within the spiral arms. Many such clouds or cloud complexes are recorded photographically. However, the most dense, which contain interstellar molecules, are often totally obscured by the dust grains and so are detectable only through their infrared and radio emission. These molecular clouds, which account for about half of the interstellar mass, contain the birthplaces of stars. *SEE GALAXY, EXTERNAL; MILKY WAY GALAXY.*

Gas. Except in the vicinity of hot stars, the interstellar gas is cold, neutral, and virtually invisible. However, collisions between atoms lead to the production of the 21-cm radio emission line of atomic hydrogen. Because the Milky Way Galaxy is quite transparent at 21 cm, surveys with large radio telescopes have produced a hydrogen map of the entire Galaxy. Different emission regions in the Galaxy along the same line of sight are moving with systematically different velocities relative to Earth and so are distinguishable by their different Doppler shifts. Supplemental information is obtained from 21-cm absorption-line measurements when hydrogen is located in front of a strong source of radio emission. These radio studies show that the gas is concentrated in clouds within the spiral arms, with densities typically 20 times the average of 15 atoms per cubic inch (1 atom per cubic centimeter). A typical size is 10 pc, encompassing the equivalent of 500 solar masses. These cold clouds (−316°F or 80 K) appear to be in near pressure equilibrium with a more tenuous warmer (14,000°F or 8000 K) phase of neutral hydrogen (whose mass totals that in clouds) and with even hotter ionized coronal-type gas (discussed below) in which they are embedded. SEE RADIO ASTRONOMY.

Other species in the gas are detected by the absorption lines they produce in the spectra of stars, and so are observable only in more local regions of the Milky Way Galaxy. Interstellar lines are distinguished from stellar atmospheric lines by their extreme narrowness and different Doppler shifts. High-dispersion spectra show the lines are composed of several components possessing unique velocities which correspond to the individual clouds detected with the 21-cm line. Elements such as calcium, sodium, and iron are detected in optical spectra, but the more abundant species in the cold gas, such as hydrogen, carbon, nitrogen, and oxygen, which produce lines only in the ultraviolet require observations from satellites outside the Earth's atmosphere.

A broad ultraviolet line of O VI (oxygen atoms from which five electrons have been removed) has also been discovered; collisional ionization to such an extent requires a very hot gas ($\sim 3 \times 10^5$ K). Such hot gas emits soft x-rays; a Local Bubble of hot gas, extending to about 100 pc from the Sun, is a dominant contributor to the diffuse x-ray background radiation. A similar hot ionized component, called the coronal gas, occupies about half of the volume of interstellar space, but because of its low density (0.045 $in.^{-3}$ or 0.003 cm^{-3}) its contribution to the total mass density is small. SEE X-RAY ASTRONOMY.

When the Galaxy formed, only the elements hydrogen and helium were present. In the course of the evolution of a star, heavier elements are built up by nuclear burning processes in the hot interior. Obviously, the interstellar gas will be gradually enriched in these heavy elements if some of this processed material can be expelled by the star. The most significant event identified is the supernova explosion, which can occur late in the lifetime of a star whose original mass exceeds about 8 solar masses. A large fraction of the stellar mass is ejected, and in the course of the violent detonation further nucleosynthesis can occur. Also important are supernova explosions of white dwarfs and perhaps nova explosions on white dwarf surfaces. Computer simulations of the burning in supernova shocks predict remarkably large relative abundances of the isotopes. Lower-mass stars might be needed to explain the abundance of some elements like carbon and nitrogen, the relevant mass loss being by stellar winds from red giants, culminating in the planetary nebula phase. SEE NOVA; NUCLEOSYNTHESIS; PLANETARY NEBULA; SUPERNOVA.

Through many cycles of star formation and mass loss from dying stars, the present interstellar element abundances have been built up to the levels seen in the atmospheres of relatively recently formed stars, including the Sun. The elements other than hydrogen and helium (principally carbon, nitrogen, and oxygen) constitute about 2% by mass. Direct examination of the cool interstellar gas using interstellar absorption lines reveals a considerable depletion of heavy elements relative to hydrogen. The atoms missing in the interstellar gas can be accounted for by the interstellar matter seen in solid particle form.

Molecules. In interstellar space, molecules form either on the surfaces of dust particles or by gas-phase reactions. In dense (greater than 5000 $in.^{-3}$ or 300 cm^{-3}) clouds where molecules are effectively shielded from the ultraviolet radiation that would dissociate them, the abundances become appreciable. These are called molecular clouds. Hydrogen is converted almost completely to its molecular form, H_2. The next most abundant molecule is carbon monoxide (CO), whose high abundance may be attributed to the high cosmic abundances of carbon and oxygen and the great stability of carbon monoxide.

The presence of molecules in interstellar space was first revealed by optical absorption lines. Unfortunately, most species produce no lines at optical wavelengths, and so the discovery of large numbers of molecules had to await advances in radio astronomy, for it is in this spectral band that molecular rotational transitions take place. Most lines are seen in emission. Dozens of molecules, with isotopic variants, are now known, the most widespread being carbon monoxide (CO), hydroxyl (OH), and formaldehyde (H_2CO). The most abundant molecule, H_2, has no transitions at radio wavelengths, but its presence in molecular clouds is inferred by the high density required for the excitation of the emission lines of other species and the lack of 21-cm emission from atomic hydrogen. In localized regions which have been heated by a shock or stellar wind to several thousand kelvins, the quadrupole infrared rotational vibrational transitions of H_2 have been detected. Molecular hydrogen has also been observed in less opaque clouds by ultraviolet electronic transitions seen in absorption (carbon monoxide is also seen this way).

The existence of molecular clouds throughout the galactic disk has been traced by using the 2.6-mm emission line of carbon monoxide, in a manner analogous to mapping with the 21-cm line. About half of the interstellar hydrogen is in molecular rather than atomic form, with enhancements in the inner 500 pc of the Galaxy and in a 3–7 kiloparsec ring.

The molecular clouds are quite cold (−441°F or 10 K), unless they have an internal energy source such as a newly forming protostar. They exist in complexes with the following typical characteristics: size, 20 pc; density, 1.5×10^4 $in.^{-3}$ (1×10^3 cm^{-3}); mass, 10^5 solar masses. Within the complexes are more dense regions, whose counterparts are locally identified with the optically defined dark nebulae (discussed below). Only in the largest and most dense condensations can the rarer polyatomic (largely organic) molecules be seen. One of the chief such regions is the Orion Molecular Cloud OMC1, of which the Orion Nebula H II region is a fragment ionized by newly formed massive stars. The other region is the 5–10-pc core of the Sagittarius B2 cloud near the galactic center. Altogether, about 70 molecules and molecular ions have been detected, many of which could be classed as prebiotic. It is interesting that the first stages of organic evolution can occur in interstellar space. SEE ORION NEBULA.

An intriguing phenomenon seen in some molecular emission lines is maser amplification to very high intensities. The relative populations of the energy levels of a particular molecule are determined by a combination of collisional and radiative excitation. The population distribution is often not in equilibrium with either the thermal gas or the radiation field, and if a higher energy sublevel comes to have a higher population than a lower one, the electromagnetic transition between the states is amplified through the process of stimulated emission. Such nonequilibrium occurs in interstellar space in two distinct environments.

The best-understood masers are the hydroxyl (OH), water (H_2O), and silicon monoxide (SiO) masers in the circumstellar envelopes of cool mass-losing red supergiant stars, because the relevant physical conditions relating to density, temperature, and radiation and velocity fields can be estimated independently from other observations. Positional measurements made possible by very long-baseline interferometry show that the silicon monoxide and water masing regions are within a few stellar radii of the star, whereas the hydroxyl masers arise in less dense regions more than 100 stellar radii out. Even here the identification of the detailed pump (inversion-producing) mechanism is difficult, but it seems that the 1612-MHz hydroxyl maser is pumped by far-infrared emission by warm circumstellar dust grains and that the silicon monoxide masers are pumped by collisions.

The other masers are found in dense molecular clouds, in particular near compact sources of infrared and radio continuum emission identified as massive stars just being formed. Interstellar hydroxyl and water masers are widespread throughout the Galaxy, methyl alcohol (CH_3OH) masing is seen in OMC1, and silicon monoxide masers are seen in OMC1 and two other star-forming regions. The strongest hydroxyl masers are identified with underlying compact H II regions, probably occurring near the boundary. Although water masers are usually close to such compact H II regions, they are not coincident with the regions of radio or infrared continuum emission. Interstellar masers appear to require collisional pumping.

Particles. The light of stars near the galactic disk is dimmed by dust grains which both absorb the radiation and scatter it away from the line of sight. The amount of extinction at optical wavelengths varies approximately as the reciprocal of the wavelength, resulting in a reddening of the color of a star, much as molecular scattering in the Earth's atmosphere reddens the Sun, especially near sunrise and sunset. The dependence of extinction on wavelength is much less steep than it is for molecular scattering, indicating the solid particles have radii about 0.1 micrometer. Satellite observations show a continued rise in the extinction at ultraviolet wavelengths which seems to require a size distribution extending to smaller interstellar grains. SEE INTERSTELLAR EXTINCTION.

By comparison of the observed color of a star with that predicted from its spectral features, the degree of reddening or selective extinction can be fairly accurately determined, but the total extinction at any given wavelength is more difficult to measure. On average, extinction by dust over a pathlength of 1000 pc in the galactic plane reduces a star's visual brightness by 80%. This requires the mass density of grains to be about 1% of the gas density. Since pure hydrogen or helium grains cannot exist in the interstellar environment, a major fraction of the heavier elements must be in the solid particles. The number density of 0.1-μm grains would be about 8000 mi^{-3} (2000 km^{-3}). Studies of reddening in conjunction with measurements of the 21-cm and ultraviolet Lyman-α lines of hydrogen and the 2.6-mm line of carbon dioxide show that dust and gas concentrations are well correlated. This correlation is borne out in infrared maps of diffuse 100-μm thermal emission from dust, made in the all-sky survey by the *Infrared Astronomical Satellite (IRAS)*.

Because of the high concentration of interstellar material toward the galactic plane, it is extremely difficult to detect radiation with a wavelength less than 1 μm coming a large distance through the plane. Conversely, a line of sight to a distant object viewed out of the plane is much less obscured because the disk is so thin. The zone of avoidance, corresponding roughly to the area occupied by the Milky Way, is that region of the sky in which essentially no extragalactic object can be seen because of intervening dust. The dark rifts in the Milky Way result from the same obscuration. The component of starlight scattered rather than absorbed by the grains can be detected as a diffuse glow in the night sky near the Milky Way. However, it must be carefully separated from other contributions to the night sky brightness: the integrated effect of faint stars, zodiacal light from dust scattering within the solar system, and airglow (permanent aurora). SEE INTERPLANETARY MATTER; ZODIACAL LIGHT.

Interstellar dust particles are heated by the optical and ultraviolet interstellar radiation field and, being relatively cool, reradiate this absorbed energy in the infrared. Such thermal radiation accounts for a third of the bolometric luminosity of the Milky Way Galaxy. Measurements by *IRAS* at 100 and 60 μm are in accord with emission expected on the basis of grain properties deduced from extinction. On the other hand, the unexpectedly large portion (25%) at shorter wavelengths (25 and 12 μm) has been taken to indicate a significant population of very small (~0.001 μm) grains or large molecules. Being of small heat capacity, these can be heated transiently to a relatively high temperature by the absorption of a single photon.

The light of reddened stars is partially linearly polarized, typically by 1% but reaching 10% for the most obscured stars. The broad peak in polarization at yellow light, together with the correlation of the degree of polarization with reddening, suggests that the polarization and extinction are caused by the same dust grains. The grains must be both nonspherical and spinning about a preferred direction in space to produce polarization. The agent for the large-scale ordering required to explain the strong tendency of the planes of polarization of stars in some directions in the Milky Way to lie parallel to the galactic plane is believed to be the galactic magnetic field. The field strength is very small; measurements based on the Zeeman effect in the 21-cm line of atomic hydrogen and the 18-cm line of hydroxyl indicate a few microgauss (1 μG = 10^{-10} tesla), with some compression in regions of higher density. Minute amounts (0.01%) of circular polarization have also been used to study the topology of the magnetic field.

The possible types of grain material can be restricted through considerations of relative cosmic abundances, but detailed identification is difficult because of the paucity of spectral features in the extinction curve. Silicates are suggested by 10-μm absorption in front of strong infrared sources, and in the ultraviolet an absorption peak at 220 nanometers could be explained by a component of small graphite particles. Spatially extended red emission in excess of

Fig. 1. Quadrant of the shell-shaped Rosette Nebula, showing dense globules of obscuring dust and gas silhouetted on the bright emission-line background of an H II region. Central hole may have been swept clear of gas by radiation pressure from central star (lower left) acting on the dust grains. Photographed in red light with the 48-in. (122-cm) Schmidt telescope of the Palomar Observatory. (*California Institute of Technology/Palomar Observatory*)

that expected from scattering in reflection nebulae has been interpreted as evidence for fluorescence of hydrogenated amorphous carbon solids. A popular theory of grain formation begins with the production of small silicate and carbon particles in the extended atmospheres of red supergiant stars. While in the circumstellar region, these grains are warmed by the starlight, so that they are detectable by their thermal emission on the near infrared (10 μm). Radiation pressure ejects these particles into the interstellar gas, where they become much colder (−433°F or 15 K). A dielectric mantle is then built up by accretion of the most abundant elements, hydrogen, carbon, nitrogen, and oxygen. Ices of water and carbon monoxide are detected by 3.1- and 4.67-μm absorption bands, respectively, but only deep within molecular clouds; these are probably in the form of volatile coatings accreted on more refractory grains.

Dark nebulae. A cloud of interstellar gas and dust can be photographed in silhouette if it appears against a rich star field. The largest and most dense clouds are most easily detected because of the large contrast produced in the apparent star density. A distant cloud is difficult to find because of many foregound stars. Many large dark nebulae or groups of nebulae can be seen in the Milky Way where the material is concentrated. They are coincident with molecular clouds. The distance to a dark nebula can be estimated by using the assumption that statistically all stars are of the same intrinsic brightness. When counts of stars within a small brightness range are made in the nebula and an adjacent clear region, the dimming effect of the cloud will appear as a sudden relative decrease in the density of stars fainter than a certain apparent brightness, which corresponds statistically to a certain distance. Alternatively, a lower limit to the distance is provided by the distance to the most distant unreddened stars in the same direction. One of the best-known and nearest dark nebulae is the Coal Sack, situated at a distance of 175 pc. Another example is the "Gulf of Mexico" area in the North America Nebula. *SEE COAL SACK*.

Obscuring clouds of all sizes can be seen against the bright H II regions described below. In many cases the H II regions and dark nebulae are part of the same cloud. The bay in the Orion Nebula is one such region, but perhaps even more familiar is the spectacular Horsehead Nebula. Even smaller condensations, called globules, are seen in the Rosette Nebula (**Fig. 1**) and NGC 6611 (M16; **Fig. 2**). The globules, which are almost completely opaque, have masses and sizes which suggest they might be the last fragments accompanying the birth of stars. *SEE GLOBULE; NEBULA*.

Bright nebulae. An interstellar cloud can also become visible as a bright nebula if illuminated by a nearby bright star. Whether or not an H II region or a reflection nebula results depends on the quantity of ionizing radiation available from the star. To be distinguished from H II regions, but often also called bright gaseous nebulae, are shells of gas that have been ejected from stars. Included in this latter category are planetary nebulae and nova shells which have a bright emission-line spectrum similar to that of an H II region, and supernova remnants. *SEE CRAB NEBULA*.

H II regions. A star whose temperature exceeds about 45,000°F (25,000 K) emits sufficient ultraviolet radiation to completely ionize a large volume of the surrounding hydrogen. The ionized regions (Figs. 1 and 2), called H II regions, have a characteristic red hue resulting from fluorescence in which hydrogen, ionized by the ultraviolet radiation, recombines and emits the Hα line at 656.3 nm. Optical emission lines from many other elements have been detected, including the "nebulium" line of oxygen at 500.7 nm.

An H II region can be extended with a relatively low surface brightness if the local density is low, as in the North America Nebula. However, the best known regions, such as the Orion Nebula, are in clouds that are quite dense (1.5×10^4 to 1.5×10^5 in.$^{-3}$ or 1×10^3 to 1×10^4 cm^{-3}) compared to the average; dense clouds use up the ionizing radiation in a region closer to the star and consequently are smaller with a higher surface brightness. Since the brightest stars are also the youngest, it is not surprising to find them still embedded in the dense regions from which they formed. Later in its evolution an H

Fig. 2. NGC 6611 (M16), a complex H II region in which the exciting stars are members of a cluster. Note dark globules and elephant-trunk structures, and the bright rims where ionizing radiation is advancing into more dense neutral gas. Photographed in Hα + [N II] with the 200-in. (508-cm) telescope of the Palomar Observatory. (*California Institute of Technology/Palomar Observatory*)

II region can develop a central hole if radiation pressure on the dust grains is sufficient to blow the dust and gas away from the star, as in the Rosette Nebula (Fig. 1). H II regions are also conspicuous sources of free–free-radio emission characteristic of close electron-proton encounters in the 14,000°F (8000 K) gas. Some H II regions are seen only as radio sources because their optical emission is obscured by dust grains. Radio recombination lines of hydrogen, helium, and carbon, which result when the respective ions recombine to highly excited atoms, are also important.

Reflection nebulae. In the absence of sufficient ionizing flux, the cloud may still be seen by the light reflected from the dust particles in the cloud. The scattering is more efficient at short wavelengths, so that if the illuminating star is white the nebula appears blue. The absorption or emission lines of the illuminating star appear in the nebular spectrum as well. Reflection nebulae are strongly polarized, by as much as 40%. Both the color and the polarization can be explained by dust grains similar to those which cause interstellar reddening. Extended near-infrared (2-μm) emission in some bright reflection nebulae provides evidence for thermal emission from transiently heated small grains. There are also numerous near-infrared spectral emission features consistent with C–H and C–C bending and stretching vibrations and suggestive of polycylic aromatic hydrocarbons (PAHs). Whether such compounds exist as free-flying molecules (or small grains), as coatings on larger dust grains, or both, is undecided.

Some reflection nebulae, such as those in the Pleiades, result from a chance close passage of a cloud and an unrelated bright star, providing a unique look at interstellar cloud structure. Other reflection nebulae appear to be intimately related to stars in early or late stages of stellar evolution. The Orion Nebula has an underlying reflection nebula arising from the dust in the gas cloud which produced the H II region. However, in this and other H II regions the emission-line radiation rather than the reflected light dominates the nebulosity.

Supernova remnants. Long after an exploding star has faded away, the expelled gas, called the supernova remnant (SNR), can still be seen. Supernova remnants provide material which continuously enriches the interstellar gas in heavy elements. Initially the gas shell expands at up to 6000 mi · s^{-1} (10,000 km · s^{-1}); during this phase the supernova remnant is a strong emitter of radio waves through the synchrotron emission process in which internally accelerated relativistic electrons spiral in a magnetic field which has been amplified by turbulent and convective motions. Later the supernova remnant is decelerated as it plows up more and more of the surrounding interstellar gas. Compression of the ambient magnetic field and cosmic-ray electron components of this gas leads again to a radio synchrotron source, with a characteristic shell structure. The supernova remnant generates a shock where it impinges on the interstellar medium, and because of the high relative velocity, temperatures of 10^5–10^6 K are reached in the postshock gas. This hot gas has been detected by its x-radiation, both in emission lines and the free-free continuum. As the gas cools behind the shock, it produces optical emission lines too, giving rise to the bright shells seen in wide-field optical photographs (for example, the Veil Nebula which is part of the Cygnus Loop supernova remnant). Supernova remnants provide an important source of energy for heating and driving turbulent motions in the cooler interstellar gas, and perhaps for accelerating cosmic rays. In low-density regions their influence can propagate through large regions of space, and they are probably responsible for maintaining the high-temperature coronal gas seen throughout the galactic disk. *See Cosmic rays.*

Star formation. Superluminous stars such as those exciting H II regions cannot be very old because of the tremendous rate at which they are exhausting their supply of hydrogen for nuclear burning; the most luminous are under 100,000 years old. With this clear evidence for recent star formation, observations have been directed toward discovering stars even closer to their time of formation. Compact H II regions, such as the Orion Nebula, appear to be the first fragments of much larger molecular clouds to have formed stars. Ultracompact H II regions, seen at radio wavelengths in molecular clouds but totally obscured optically, appear to be an earlier stage in which the protostellar core has just become highly luminous. These are often called cocoon stars. Even earlier still are the compact infrared sources in which hot dust grains in a protostellar cloud are being detected at wavelengths of 5–100 μm. These earliest phases are often associated with intense water and hydroxyl molecular maser emission. Examples of all of these stages are often found in the same region of space. In addition to the Orion molecular cloud, many regions such as the W3 radio and infrared sources associated with the visible H II region IC 1795 have been studied extensively.

Both the broad spectral energy distribution of thermal infrared emission and the spatial distribution of molecular emission from dense gas indicate that the final gravitational collapse of an interstellar cloud fragment commonly proceeds through an accretion disk. A flattened geometry is indeed expected from conservation of angular momentum during the collapse. In the protostar stage the energy radiated is derived from gravitational potential energy released as material is added to the disk and to the forming star at the center. An accompanying common property is that young stellar objects eject substantial amounts of gas in high-velocity bipolar outflows along the axial direction, perpendicular to the disk. These are observed as optical emission lines from collimated jets and from shocked knots called Herbig-Haro objects, and as radio emission from carbon monoxide in extended lobes.

Although the details are far from understood, there is evidence that star formation is a bimodal process, with more massive stars forming only in some giant molecular cloud complexes which aggregate in the spiral arms of the Milky Way Galaxy, and less massive stars occurring independently over the whole range of molecular cloud sizes. Low-mass stars are less luminous and cooler and so do not produce the diagnostic compact H II regions. In the nearby Taurus molecular cloud, numerous dense (greater than 1.5 × 10^5 in.$^{-3}$ or 10^4 cm^{-3}) stellar mass cores detected by the ammonia (NH_3) molecule appear to be the immediate precursors of low-mass stars, as evidenced by their coincidence with other signposts of star formation: *IRAS* infrared sources, bipolar outflows, and T Tauri stars, which are low-mass pre-main-sequence stars.

Molecular clouds are supported against wholesale

gravitational collapse by turbulent motions, probably associated with waves in the interstellar magnetic field, and are eventually dispersed by mechanical energy which results from the star formation process itself (bipolar outflows, expanding H II regions, stellar winds). On average, the star formation rate corresponds to a 2% conversion efficiency of molecular cloud material, but occasionally a higher efficiency, in excess of 30%, can lead to a gravitationally bound cluster of stars such as the Pleiades or NGC 6811 (M16; Fig. 2). Overall, the Milky Way Galaxy processes about five solar masses of interstellar matter per year into new stars, somewhat larger than the rate at which dying stars replenish the gas, so that the present interstellar medium is decaying on a time scale of about 1×10^9 years. *See Infrared astronomy; Stellar evolution.*

P. G. Martin

Bibliography. L. J. Allamandola and A. G. G. M. Tielens (eds.), *Interstellar Dust*, 1989; D. C. Black and M. S. Mathews (eds.), *Protostars and Planets II*, 1985; D. J. Hollenbach and H. A. Thronson (eds.), *Interstellar Processes*, 1987; D. E. Osterbrock, *Astrophysics of Gaseous Nebulae and Active Galactic Nuclei*, 1989; F. H. Shu, F. C. Adams, and S. Lizano, Star formation in molecular clouds: Observations and theory, *Annu. Rev. Astron. Astrophys.*, 25:23–81, 1987; G. L. Verschuur, *Interstellar Matter*, 1988.

Jupiter

The largest planet in the solar system, and the fifth in the order of distance from the Sun. It is visible to the naked eye, except for short periods when in near conjunction with the Sun. Usually it is the second brightest planet in the sky; only Mars at its maximum luminosity and Venus appear brighter. Jupiter is brighter than Sirius, the brightest star.

Planet and its orbit. The main orbital elements are a semimajor axis, or mean distance to the Sun, of 484×10^6 mi (778×10^6 km); an eccentricity of 0.048, causing the distance to the Sun to vary about 47×10^6 mi (75×10^6 km) between perihelion and aphelion; sidereal revolution period of 11.86 years; mean orbital velocity of 8.1 mi/s (13.1 km/s); and inclination of orbital plane to ecliptic of 1°.3. *See Planet.*

The apparent equatorial diameter of its disk varies from about 47″ at mean opposition (50″ at perihelic opposition, 44″ at aphelic opposition) to 32″ at conjunction. The polar flattening due to its rapid rotation is considerable and is easily detected by visual inspection; the ellipticity is $(r_e - r_p)/r_e = 0.065$, where r_e is the equatorial radius and r_p is the polar radius. The equatorial diameter is about 88,700 mi (142,700 km), and the polar diameter is 82,800 mi (133,300 km). The volume is about 1317 (Earth = 1) with an uncertainty of several percent. The mass is about 318.4 (Earth = 1), and is accurately determined from the motion of the four major satellites. The mean density is 1.34 g/cm^3, a low value characteristic of the four giant planets; the corresponding value of the mean acceleration of gravity at the visible surface is about 85 ft/s^2 (26 m/s^2); however, because of the large radius and rapid rotation, the centrifugal force at the equator amounts to 7.37 ft/s^2 (2.25 m/s^2), reducing the effective acceleration of gravity to about 78 ft/s^2 (24 m/s^2).

Phases. As an exterior planet, Jupiter shows only gibbous phases from Earth. Because of the large size of Jupiter's orbit compared with that of the Earth, the maximum phase angle is only 12° at quadratures and the phase effect shows up only as a slightly increased darkening of the edge at the terminator. The apparent visual magnitude at mean opposition is −2.4, and the corresponding value of the reflectivity (geometrical albedo) is about 0.3; the physical albedo is 0.34, with some uncertainty due to the small range of phase angle observable. The high value of the albedo, characteristic of the four giant planets, indicates the presence of a dense, cloud-laden atmosphere. *See Albedo.*

Telescopic appearance. Through an optical telescope Jupiter appears as an elliptical disk, strongly

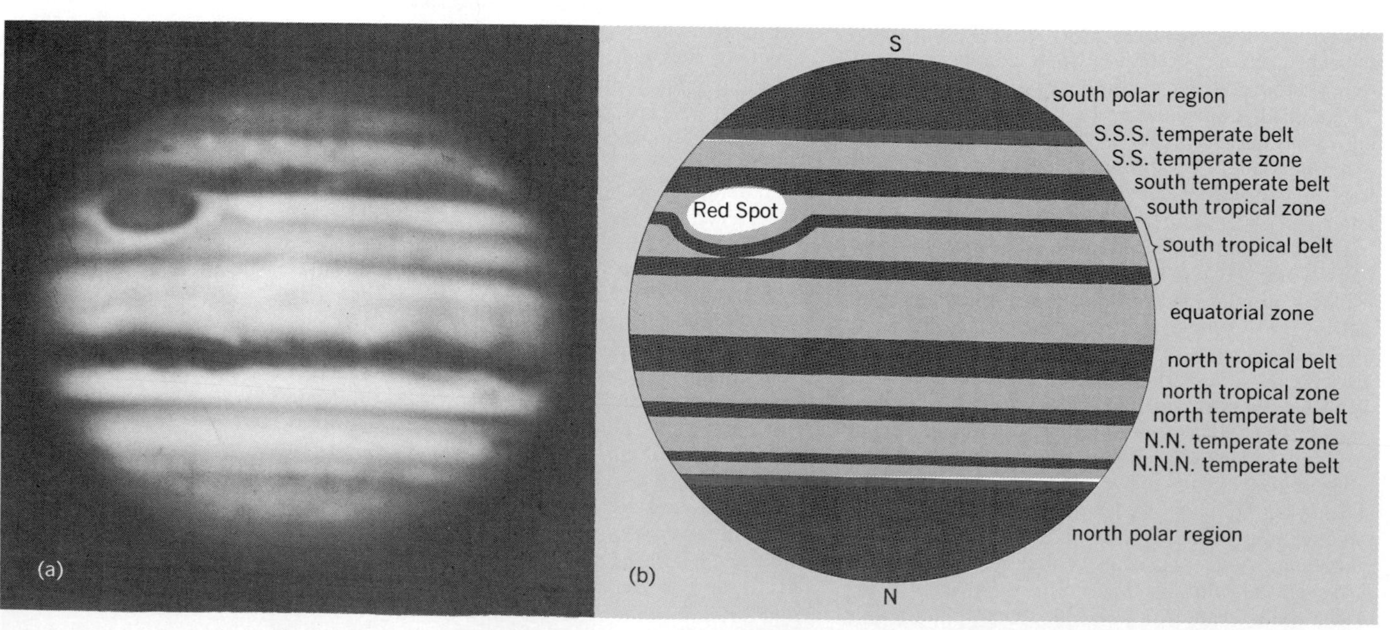

Fig. 1. Jupiter. (*a*) Telescopic appearance (*California Institute of Technology/Palomar Observatory*). (*b*) Principal bands.

Table 1. Mean latitudes and rotation periods of Jupiter's bands

Band	Region	Latitude	Period, 9h+
North polar region		>+48°	55m42s
N.N.N. temperate belt	Center	+43°	55m20s
N.N. temperate belt	Center	+38°	55m42s
	South edge	+35°	53m55s
North temperate belt	North edge	+31°	56m05s
	Center	+27°	53m17s
	South edge	+23°	49m07s
North tropical zone	Center	+18°	55m29s
North tropical belt	Center	+13°	54m09s
	South edge	+6°	50m24s
Equatorial zone	Middle	0°	50m24s
South tropical belt	North edge	−6°	50m26s
	Center	−10°	51m21s
	South edge	−19°	55m39s
South tropical zone	Center	−23°	55m36s
Great Red Spot	Center	−22°	55m38s
South temperate belt	North edge	−27°	55m02s
	Center	−29°	55m20s
	South edge	−31°	55m07s
S.S. temperate zone	Center	−38°	55m07s
S.S.S. temperate zone	Center	−45°	55m30s
South polar region		<−45°	55m30s

darkened near the limb, and crossed by a series of bands parallel to the equator (**Fig. 1**). Even fairly small telescopes show a great deal of complex structure in the bands and disclose the rapid rotation of the planet. The period of rotation, determined from long series of observations of the transits of spots at the central meridian, is very short, about 9h55m, the shortest of the giant planets. The details observed, however, do not correspond to the solid body of a planet but to clouds in its atmosphere, and the rotation period varies markedly with latitude. The nomenclature and mean rotation periods of the main belts of the clouds are given in **Table 1**. The rotation period of any given zone is not exactly constant but suffers continual fluctuations about a mean value. Occasionally, short-lived atmospheric phenomena may depart more strongly from the mean rotation period of the zone in which they appear and thus drift rapidly with respect to other details in the zone. The rotation axis is inclined only 3° to the perpendicular to the orbital plane, so that seasonal effects are practically negligible.

Red Spot. Apart from the constantly changing details of the belts, some permanent or semipermanent markings have been observed to last for decades or even centuries, with some fluctuations in visibility. The most conspicuous and permanent marking is the great Red Spot, intermittently recorded since the middle of the seventeenth century and observed continually since 1878, when its striking reddish color attracted general attention. It was conspicuous and strongly colored again in 1879–1882, 1893–1894, 1903–1907, 1911, 1914, 1919–1920, 1926–1927, and especially in 1936–1937, 1961–1968, and 1973–1974; at other times it has been faint and only slightly colored, and occasionally only its outline or that of the bright "hollow" of the south temperate zone which surrounds it has remained visible.

The mean rotation period of the Red Spot between 1831 and 1982 was 9h55m38.2s, with a range of variation of about ±6 s (**Fig. 2**). The mean dimensions of the Red Spot are about 15,000 mi (25,000 km) in longitude and 10,000 mi (16,000 km) in latitude. The center of the Red Spot has an appearance that suggests little or no circulation, while its outer rim is estimated to possess 220-mi/s (360-km/s) winds.

The *Pioneer 10* and *11* and *Voyager 1* and *2* spacecraft flybys have shown that the clouds making up the Red Spot rotate counterclockwise approximately every 6 days, and that the spot itself consists of an area of uplifted clouds that are about 10 mi (16 km) higher than the surrounding atmospheric levels and about 18°F (10°C) colder (**Fig. 3**). Its distinctive coloration is probably due to a phosphorus compound which has been transported from deeper in the atmosphere.

Many theories have been suggested to explain the origin and longevity of the Red Spot. The most likely concludes that it is an eddy of atmospheric gases which is driven by the strong Coriolis force of planetary rotation. Perturbations and whirlpools in the turbulent atmosphere tend to coalesce, forming a vortex of increasing size, which eventually reaches a state of equilibrium. In this

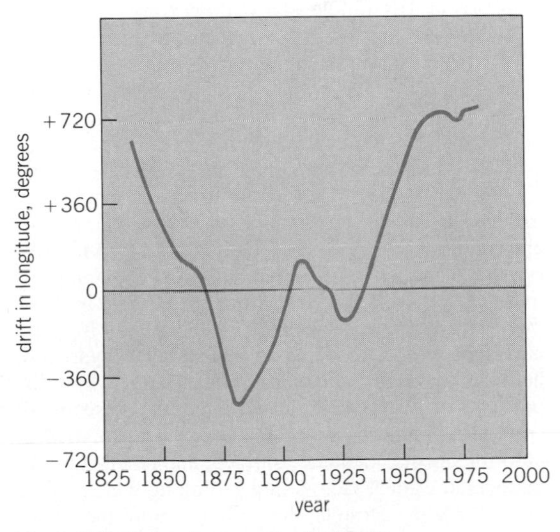

Fig. 2. Drift in longitude of the Red Spot.

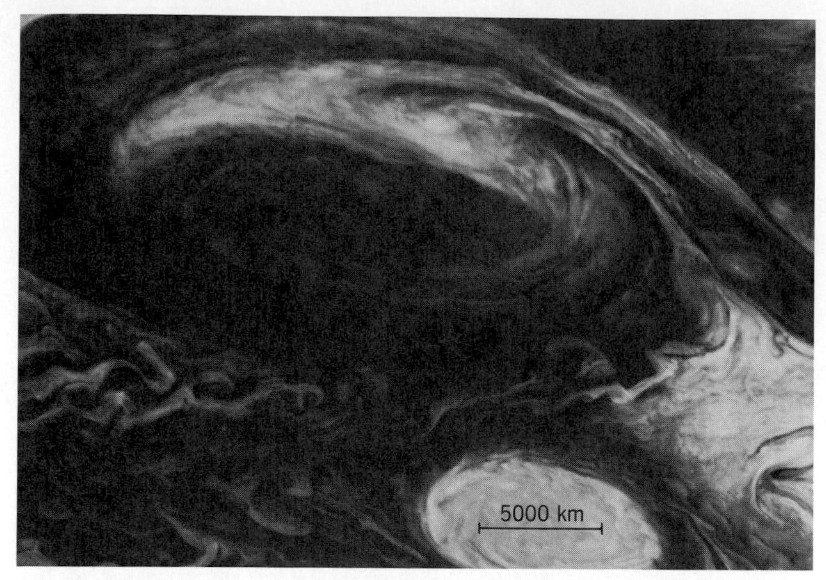

Fig. 3. Details of the Red Spot as seen by the *Voyager 1* flyby. Visible features include a white oval with a wake of counterrotating vortices, puffy features inside the Red Spot, and reverse-S spirals inside both the Red Spot and the oval. The large white feature extending over the northern part of the Red Spot was observed to revolve about the Red Spot center with a period of 6 days. 5000 km = 3000 mi. (*NASA*)

model, the lifetime of the Red Spot vortex is indefinite and its formation inevitable.

Other markings. Another remarkable, semipermanent marking, the South Tropical Disturbance, was intermittently observed on Jupiter between 1901 and 1935 and possibly also in 1940. This marking circulated in the same zone as the Red Spot, but with a shorter mean rotation period, about 9h55m16s, and periodically came into conjunction with it. A large number of temporary features of shorter duration have been observed. Their mechanisms of formation and laws of motions are not understood, but are likely similar to those which cause the Red Spot. It has been observed that the major white ovals are hot-spot regions of strong infrared emission.

Atmosphere. Limits on atmospheric pressure at the top of the clouds can be set by observing pressure broadening of spectral lines; the pressure is probably a bit less than 3 atm (300 kilopascals). The atmosphere must be largely composed of hydrogen and helium, as is shown from the detection of weak spectral lines arising from molecular hydrogen.

The temperature of the visible disk of Jupiter determined from measurement of infrared radiation, about −236°F (124 K), is in fairly good agreement with the value theoretically estimated from the assumption that the visible cloud layer is mainly composed of ice crystals of solidified ammonia (about −172°F or 160 K). Jupiter actually emits 1.668 times as much thermal energy as it absorbs from the Sun. *Pioneer 10* and *11* flybys showed that Jupiter's zones and belts consist of gases at different altitudes and temperatures. The coloring agents of the bands are unknown; the only substances positively identified in Jupiter's atmosphere spectroscopically have been hydrogen, helium, ammonia, methane, and water, while the presence of hydrogen sulfide is inferred. These are all colorless, and other molecules have been proposed as coloring agents. Latitude, altitude, and dwell-time are undoubted critical factors in determining which colors appear where. Despite the turbulence of the atmosphere, the correlations between cloud color and certain latitudes have been maintained for decades, suggesting the importance of an internal energy source and deep circulation in generating some of the chromophores. Infrared images show that the cloud colors are related to altitude. The blue features (made up of hydrogen, which is visible in this case due to Rayleigh scattering) are at the deepest levels and appear through holes in the overlying clouds. Browns are next highest, followed by white, and finally the highest features of all are red. The most likely coloring agent is elemental sulfur and its compounds which produce the necessary variety of colorings. The chromophores are undoubtedly activated when the chemical equilibrium is disturbed by ions, energetic photons, lightning, or rapid vertical motion throughout different temperature regions.

The *Voyager* probes showed that the Jovian clouds (**Fig. 4**) are generally made up of counterflowing easterly and westerly winds (**Fig. 5**). The strongest winds (up to 250 mi/h or 400 km/h) are found at the boundaries between the belts and the zones. The clouds are also in vertical motion with the vertical velocities of the plumes in the equatorial regions between 8 and 16 in./s (20 and 40 cm/s). The zonal jet streams have not changed position in latitude for at least the 80 years of telescopic observations. A typical midzonal wind velocity is nearly 110 mi/h (50 m/s), and it is probably maintained by the continual pulling apart of the many atmospheric eddies which form and whose kinetic energy then goes into maintaining the zonal jets. A considerable portion of the energy necessary also undoubtedly comes from the interior heat source.

There are probably three distinct cloud layers, the lowest of which is made up of water ice or possibly liquid water droplets. The next highest is a layer of crystals of ammonium hydrosulfide (NH_4SH), a combination of ammonia (NH_3) and hydrogen sulfide (H_2S). The highest of all is an ammonia ice cloud layer.

Interior composition and structure. Jupiter is primarily made up of liquid and metallic hydrogen. Its composition is similar to that of the Sun, namely 10 times as much hydrogen as helium (by numbers of atoms), and undoubtedly reflects the primordial composition of the solar system. This is not surprising

Fig. 4. Fine details of Jovian atmospheric features as photographed by *Voyager 1*. The dark halo surrounding the smaller bright spot to the right of large oval is a hot spot, a warmer region of Jupiter's atmosphere. (*NASA*)

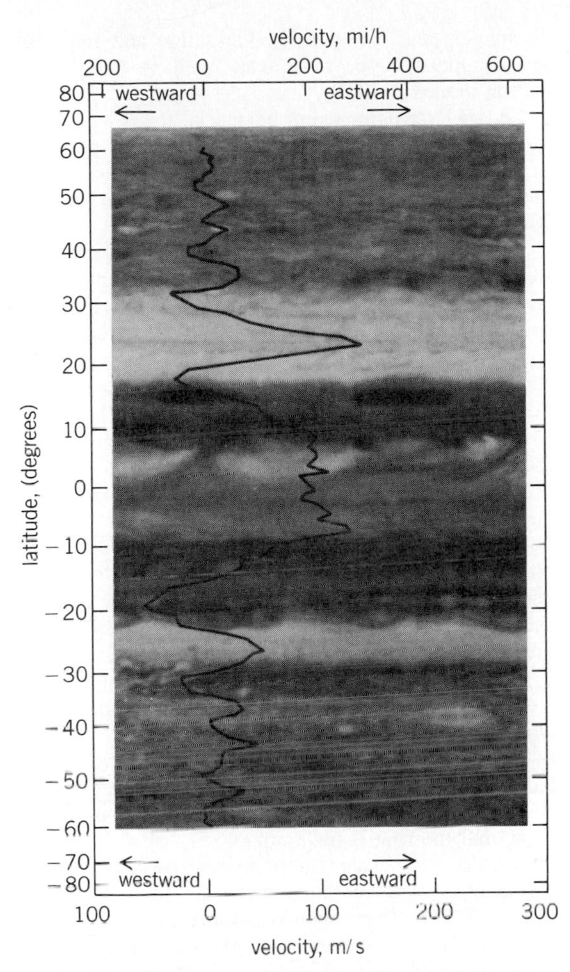

Fig. 5. Graph of variation of wind speed on Jupiter with latitude as measured by *Voyager* probes, superimposed in *Voyager* image of clouds. (*NASA*)

since its strong gravitational forces would have captured and retained most of the primitive material.

The cloud zone thickness actually extends only 0.1–0.3% of the Jovian radius. Beneath that, the atmosphere is clear and gradually metamorphoses into the liquid hydrogen molecular fluid envelope which makes up approximately 60% of the radius. The transition at lower depths to metallic hydrogen is abrupt, as is the change to the small ice-silicate core which has a radius of perhaps a few thousand miles. Pressure and temperature properties of the interior are listed in **Table 2**.

Jovian magnetosphere. Jupiter possesses the strongest magnetic field intrinsic to a planet in the solar system. This field rotates with the rotational period of the planet and contains an embedded plasma trapped in the field. At the distance of the satellite Io, the field revolves faster than the satellite, and so numerous collisions occur with the atmospheric gas of that body, resulting in the stripping away of 10^{28}–10^{29} ions (mainly atomic sulfur and atomic oxygen) per second. The energy involved slows the magnetic field, and so, beyond Io, the magnetic field no longer rotates synchronously with the planet. The ions removed from Io spiral around the magnetic lines of force, oscillating above and below the plane of Io's orbit. This ring of sulfur and oxygen ions is known as the Io plasma torus and emits strongly in the ultraviolet. *Voyager* data have revealed that the Io flux tube carries nearly 3×10^6 amperes of electrical current but with very little voltage.

The torus probably powers a series of belts of auroras (discovered by *Voyager 1*) which encircle both the Jovian poles. Eventually some of the particles move out into the extended magnetospheric disk which is approximately 2×10^6 mi (3.2×10^6 km) in radius, and eventually escape into interplanetary space through the magnetotail (**Fig. 6**). Evidence has been found that the Jovian magnetotail extends to at least 8700 Jovian radii (about 4.5 astronomical units or 4×10^8 mi or 6.5×10^8 km) from the planet and as a consequence affects the magnetosphere of the next-distant planet, Saturn. *Pioneer* and *Voyager* observations have indicated the existence of an equatorial anomaly in the ionosphere, with the peak electron concentration showing a minimum value near the equator. SEE MAGNETOSPHERE.

Under the influence of the solar wind, the Jovian magnetic field is severely disturbed. When solar wind particles encounter the magnetic field, they form a bow shock wave, some 2×10^6 mi (3.2×10^6 km) from the cloud surfaces. The solar wind particles are then deflected around the planet. SEE SOLAR WIND.

Except near the planet, the major component of the Jovian magnetic field is dipolar and opposite in direction from that of the Earth. At a distance of three Jovian radii, the field strength is 0.16 gauss (16 microteslas). Closer to the planet, the *Pioneer 11* probe has shown the field to be quadrupolar and octopolar. There the field strength varies from 3 to 14 gauss (0.3 to 1.4 milliteslas).

Radio astronomy. Jupiter is known to produce three distinct types of radio emission. Thermal radiation from the high stratosphere is detectable at wavelengths below about 10 cm and indicates temperatures in the upper emitting layers of −280 to −225°F (100 to 130 K). Microwave nonthermal emission is observed in the band from about 3 to 70 cm, and is known to arise from synchrotron radiation from relativistic electrons in extended Jovian Van Allen belts

Table 2. Internal properties of Jupiter*

Property	Values			
Fractional radius, %	100 (surface)	76	15	0 (center)
Pressure, 10^6 bars or 10^{11} Pa	10^{-6}	3.0	42	80
Density, g/cm³	2×10^{-4}	1.1/1.2	4.4/15	20
Temperature, K (°F)	165 (−163)	10,000 (18,000)	20,000 (36,000)	25,000 (45,000)
Fractional mass, %	100	77	8	0

*After J. K. Beatty, B. O'Leary, and A. Chaikin (eds.), *The New Solar System*, 1981.

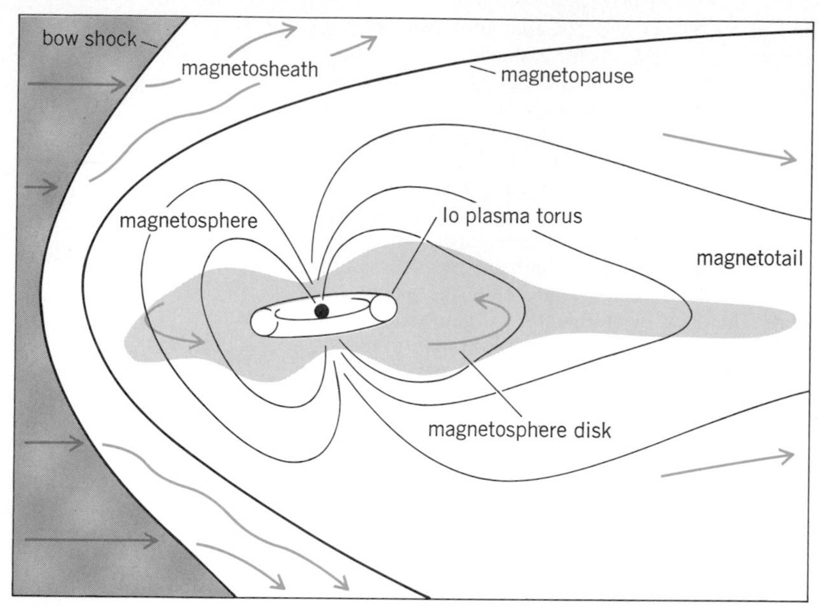

Fig. 6. Diagram of the Jovian magnetic environment. (*From Return to Jupiter, Astronomy, 7(9):6–24, 1979*)

(**Fig. 7**). Sporadic decametric radio noise storms are head on wavelengths longer than 8 m; these seem consistent with gyrofrequency radiation produced by spiraling electrons, perhaps near the magnetic poles of Jupiter. The existence of both kinds of these nonthermal emissions suggests a Jovian polar surface magnetic field.

Some radio noise is thought to originate in the lightning storms which were observed on the dark side of the planet by the cameras of *Voyager 2*. The *Voyager* data suggest a stroke rate that is significantly less than that of the earth. No indication has been found for high-altitude lightning near the ammonia cloud level. Therefore, the rate of production of complex organic molecules by lightning and thunder shock waves is negligible compared to the rates of known photochemical processes for forming the cloud chromophores. The principal effect of the lightning discharges near the water condensation and freezing levels appears to be to convert methane (CH_4) into carbon monoxide (CO).

Certain transient radio bursts in the 8–30-MHz range are strongly correlated with the orbital position of the satellite Io and are most easily explained if the electrons generating these bursts are confined to the Io plasma torus.

Periodicities in the radio-noise storms and rocking of the polarization plane of the microwave nonthermal emission led to the well-determined radio rotation period 9h55m29.7s. The difference between the radio and the various other observed Jovian rotation periods suggests that the core of Jupiter is rotating about 13 s faster than the mantle, that the atmosphere has vast wind currents with relative velocities up to 300 mi/h (500 km/h), and that angular momentum may be significantly exchanged among these regions over periods of years. *See Radio astronomy.*

Jovian ring. *Voyager 1* and *2* detected a faint ring encircling Jupiter (**Fig. 8**). It appears to be made up of three separate components: a bright ring, a faint sheet, and an out-of-plane halo. The outer radius of the bright ring is 1.81 Jovian radii, the inner radius 1.72. The faint sheet (visible in Fig. 8) extends from the inner edge of the bright ring to the top of the cloud surfaces. The outer edge of the ring is defined by the satellite XIV Thebe. There probably exists a halo which extends 6×10^3 mi (10^4 km) above the ring plane. The reflection spectra of the ring particles show that the ring is made up entirely of rock bodies with little or no water, ammonia, or methane frosts. The average size of the particles is small, approximately 0.1–0.5 μm, with the upper limit being approximately a kilometer. The particles of small dimension are likely micrometeorites which come from dust particles ejected from Io's volcanoes or debris excavated from the surfaces of the parent boulders in the ring.

Satellites. Jupiter has 16 known satellites of which the four largest, I Io, II Europa, III Ganymede, and IV Callisto, discovered by Galileo in 1610, are by far the most important (**Table 3**).

The four galilean satellites are of fifth and sixth

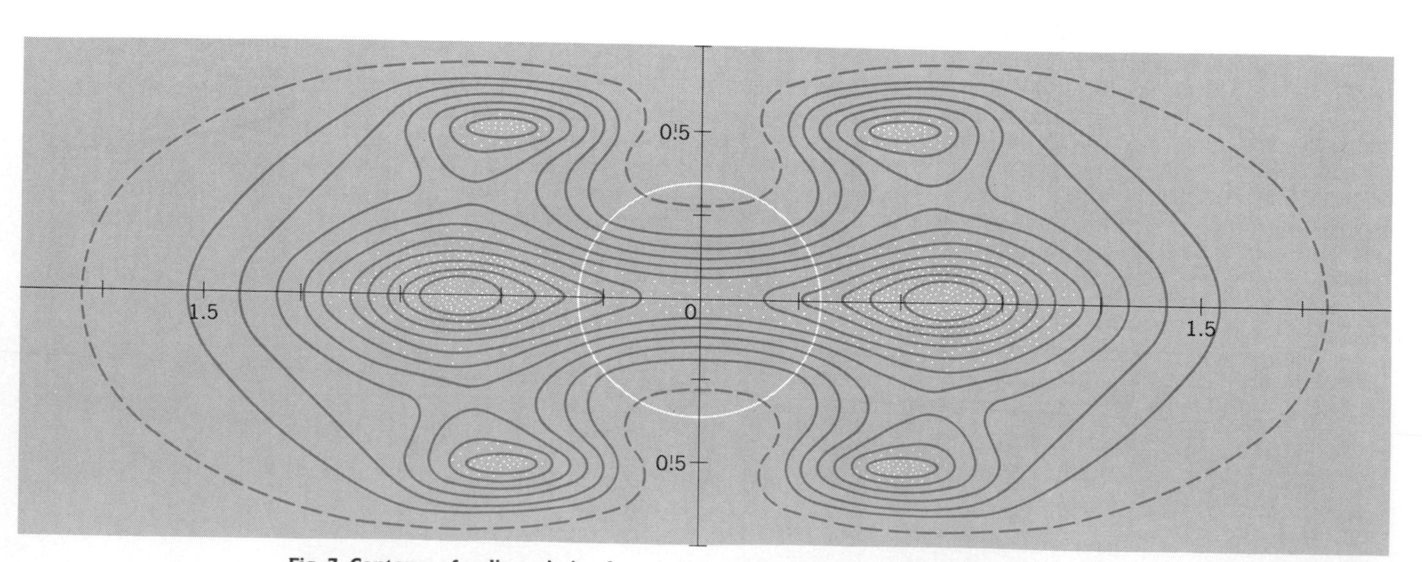

Fig. 7. Contours of radio emission from Jupiter originating from the Van Allen belts, and extending far beyond the disk of the planets.

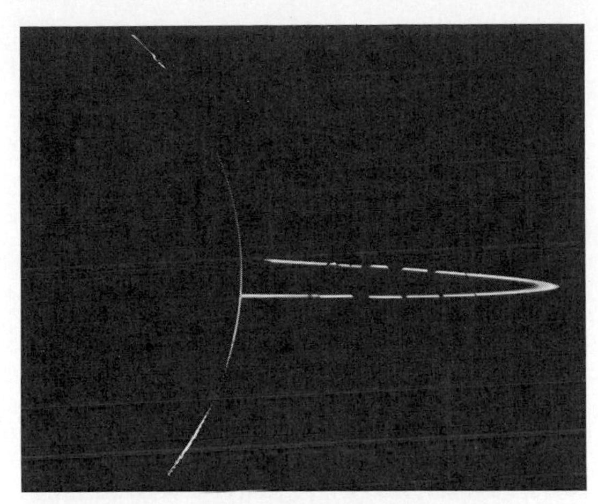

Fig. 8. Portion of Jupiter's ring seen in mosaic of photographs from *Voyager 2*, taken as the spacecraft looked back from within the planet's shadow at a distance of 9.5 × 10⁵ mi (1.55 × 10⁶ km). Apparent gaps in the ring are spaces between photographs. Highlighting of the ring and Jupiter's limb is due to strong forward scattering by small particles in the ring and upper Jovian atmosphere, respectively. Jupiter's shadow is cast on some of upper ring segment, closest to the spacecraft. (*NASA*)

stellar magnitudes and would be visible to the naked eye if they were not so close to the much brighter parent planet. All the others are faint telescopic objects.

The masses of the major satellites can be roughly estimated from their mutual perturbations; in terms of the mass of the Moon as a unit, the mass of III is about 2, of I about 1, of IV about ⅔, and of II about ⅔. The apparent diameters are on the order of 1–2″, and the linear radii measured from the Voyager probes are listed in Table 3.

The planes of the orbits of the major satellites are inclined less than 0.°5 to the equatorial plane of Jupiter, so that with the occasional exception of IV, they are eclipsed in Jupiter's shadows on Jupiter and transit in front of its disk near conjunction. The eclipses, transits, and occultations of Jupiter's satellites led to the discovery of the finite velocity of propagation of light by O. Roemer in 1675. Satellites VIII, IX, XI, and XII have retrograde motion. The very small outer satellites of Jupiter are probably captured asteroids; their orbits are subject to large perturbations by the Sun. Thus at least the more distant satellites probably form part of a fluctuating population gained and lost over very long time spans. *See Eclipse; Occultation; Transit*.

The close approaches of the *Voyager* spacecraft have shown that the four galilean moons are quite different from each other (**Fig. 9**). I Io was discovered to possess the solar system's most active volcanoes. There have been observed 11 active volcanoes on Io's surface, the largest of which, Pele, is temporarily inactive. As *Voyager 1* arrived, 9 were in eruption. The volcanoes appear to be more or less uniformly distributed in longitude, but are concentrated in the equatorial regions. Over 200 calderas and several volcanic lakes (notably Loki Lake) with rafts of floating elemental sulfur have been observed. The volcanic activity can be attributed to tidal energy transformation into heat. Forced into an eccentric orbit by the gravitational influence of the other galilean satellites, Io is subject to varying tidal stresses under the influence of Jupiter's gravity, and its interior is heated by the physical bending of several hundred feet of the satellite's surface. The molten material then reaches the surface through the volcanoes and their plumes. There are two classes of volcanoes, one possessing smaller and longer-lived plumes, concentrated at 240° longitude, and the other centered in one small area that produces giant eruptions. The smaller plumes appear to be driven by ejections of sulfur dioxide heated to about 260°F (400 K), while the larger feed on molten and highly fluid black sulfur in excess of 710°F (650 K). The lifetimes of the longer eruptions can be up to several years. The structure of Io is thought to consist mainly of a silicate mantle and crust over which is lain molten and solid sulfur and sulfur dioxide. The surface is geologically young and has no craters of impact origin. The average accumulation of volcanic material is about 0.04 in./year (1 mm/year), and as a consequence, in or-

Table 3. Satellites of Jupiter

	Satellite	Orbital radius, 10^3 km (10^3 mi)	Orbital period, days	Radius,* km (mi)	Magnitude at mean opposition
XVI	Metis	127.96 (79.51)	0.295	20 (12)	17.4
XV	Adrastea	128.98 (80.14)	0.298	12, 10, 8 (7, 6, 5)	18.9
V	Amalthea	181.3 (112.7)	0.498	75, 85, 135 (47, 53, 85)	14.1
XIV	Thebe	221.9 (137.9)	0.675	45–55 (28–34)	15.5
I	Io	421.6 (262.0)	1.769	1815 (1128)	5.0
II	Europa	670.9 (416.9)	3.551	1569 (975)	5.3
III	Ganymede	1070 (665)	7.155	2631 (1635)	4.6
IV	Callisto	1880 (1168)	16.689	2400 (1491)	5.6
XIII	Leda	11,094 (6893)	238.7	5 (3)	20.2
VI	Himalia	11,480 (7133)	250.6	90 (55)	14.8
X	Lysithea	11,720 (7282)	259.2	10 (6)	18.4
VII	Elara	11,737 (7293)	259.7	40 (25)	16.7
XII	Ananke	21,200 (13,200)	631	10 (6)	18.9
XI	Carme	22,600 (14,000)	692	15 (9)	18.0
VIII	Pasiphae	23,500 (14,600)	735	20 (12)	17.7
IX	Sinope	23,700 (14,700)	758	15 (9)	18.3

*Multiple entries refer to the dimensions of a nonspherical object.

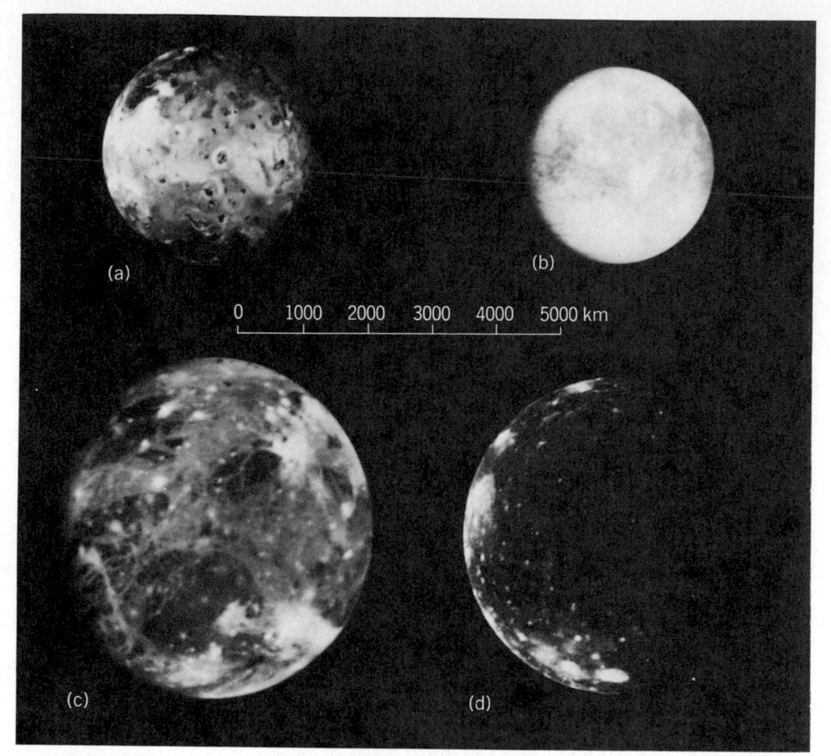

Fig. 9. ***Voyager 1*** **photographs of the Galilean satellites, shown to scale. (*a*) Io. (*b*) Europa. (*c*) Ganymede. (*d*) Callisto. 1 km = 0.6 mi. (*NASA*)**

der to maintain this very high rate, the upper portion of Io's mantle and crust has been recycled many times throughout Io's history. Because of the abundant volcanic plumes, there exists a cloud of sodium and potassium around the satellite.

II Europa and Io differ from the remaining two galilean moons by being nearly twice as dense. Europa is primarily a rocky body which may be rich in silicates and lightweight water ices. The *Voyager* probes showed a satellitewide system of cracks on the surface, running for thousands of miles, about 10–25 mi (16–40 km) in width. There are few elevations on the satellite, and the surface is remarkably free from craters. It is thought that the fractures are caused by the cyclical tidal deformation resulting from Europa's orbital eccentricity. There are also numerous dark and mottled regions which seem to be areas of upwelling silicate-laden water and ice. Some of the trailing side of Europa's surface shows evidence of sulfur ejected from Io deposited on the surface.

III Ganymede and IV Callisto probably share the same structure: a liquid core and an icy crust. The surface is composed of two distinctly different terrain types which show abundant evidence of global-scale tectonics. The younger and brighter terrain is probably formed by the emplacement of relatively clean ice into rift zones developed in more silicate-rich dark material. There are grooves on the surface 2–6 mi (3–10 km) apart, 1000–1300 ft (300–400 m) tall, and concave upward, but no large craters.

IV Callisto is perhaps the most heavily cratered body in the solar system, thereby showing that its surface is very old. Large body impacts are shown by the presence of immense concentric crater basins. Callisto's low density requires a composition of liquid water or slush with a thin, rigid ice crust. There also exists a class of blue-haloed craters which may reveal material from deeper within.

V Amalthea is an unusual object in that it is very nonspherical. It is in synchronous rotation, and the surface is very red (probably from deposition of sulfur from Io) with several bright greenish spots 6–30 mi (10–50 km) in diameter.

VI Himalia has been shown from photometry to consist of carbonaceous chondritic material, and probably the remainder of the satellites of Jupiter are constructed of similar material. *See* PLANETARY PHYSICS.

Elaine M. Halbedel

Bibliography. J. F. Baugher, *The Space-Age Solar System*, 1988; J. K. Beatty, B. O'Leary, and A. Chaikin (eds.), *The New Solar System*, 3d ed., 1989; T. V. Johnson and L. A. Soderblom, Io, *Sci. Amer.*, 249(6):56–68, 1983; G. P. Kuiper and B. M. Middlehurst (eds.), *Planet and Satellites*, 1961; B. M. Peek, *The Planet Jupiter*, 1958; Reports of *Voyager 1* encounter with Jupiter, *Science,* 204:945–1008, 1979; Reports of *Voyager 2* encounter with Jupiter, *Science*, 206:925–996, 1979; Return to Jupiter, *Astronomy,* 7(9):6–24, 1979; J. H. Wolfe, Jupiter, *Sci. Amer.*, 233(3):118–126, 1975.

Kepler's equation

The mathematical relationship between two different systems of angular measurement of the position of a body in an ellipse; specifically the relation between the mean anomaly M and eccentric anomaly θ', $M = \theta' - e \sin \theta'$, where e is the eccentricity of the ellipse.

The true position of a planet P in an elliptical orbit can be represented by the angle θ (true anomaly) measured at the focus S between the line directed to the planet and the line directed to the perihelion A (see **illus.**). The radius vector r from the focus to the planet can be expressed by $r = a(1 - e^2)/(1 + e \cos \theta)$, where a is the semimajor axis. The radius vector may also be expressed in terms of the eccentric anomaly θ' by $r = a(1 - e \cos \theta')$, where θ' is the angle at the center of the ellipse measured from perihelion along the circumscribed circle of radius a to

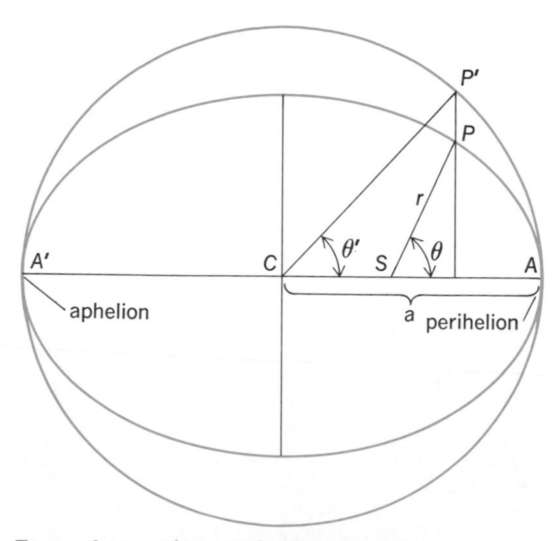

True and eccentric anomaly for a planet *P*.

the point whose projection perpendicular to the major axis passes through the planet. The true anomaly θ may be expressed in terms of the eccentric anomaly θ' by $\tan(\theta/2) = [(1 + e)/(1 - e)]^{1/2} \tan(\theta'/2)$. In actual practice, however, it is more convenient to describe the angular position of a planet in an elliptical orbit at any time t by means of its average angular velocity n (called mean motion) and the time T of last perihelion passage. This angle M (mean anomaly) is expressed by $M = n(t - T)$. Therefore, given the orbital elements a, e, n, and T, it is possible by means of Kepler's equation and the intermediary angle θ' to evaluate the true anomaly θ and the actual position of the planet in the orbit for any instant t. SEE PLANETARY PHYSICS.

Several practical methods for the iterative solution of this transcendental equation exist: (1) Starting with the value of M and denoting approximate values of θ' and $\sin\theta'$ by θ'_0 and $\sin M$, solve $\theta'_0 = M + e \sin M$. Denoting a second approximation to θ' by θ'_1, solve $\theta'_1 = M + e \sin \theta'_0$. Indicating the third approximation to θ' by θ'_2, solve $\theta'_2 = M + e \sin \theta'_1$, continuing the iteration until the required convergence of θ'_i is obtained. (2) Starting with M and an approximate value of θ', solve $M_0 = \theta'_0 - e \sin \theta'_0$, where M_0 corresponds to θ'_0. Set $M - M_0 = \Delta M_0$ and let $\Delta\theta'_0$ be a first correction to θ'_0, then $\Delta\theta'_0 = \Delta M_0/1 - e \cos \theta'_0$. Set $\theta'_1 = \Delta\theta'_0 + \Delta\theta'_0$, evaluate M_1 corresponding to θ'_1, form $M - M_1 = \Delta M_1$, and solve $\Delta\theta'_1 = \Delta M_1/1 - e \cos \theta'_1$. Then $\theta'_2 = \theta'_1 + \Delta\theta'_1$, and the process is repeated until the required convergence of $\Delta\theta'_i$ is reached. SEE CELESTIAL MECHANICS.

Raynor L. Duncombe

Bibliography. J. M. A. Danby, *Fundamentals of Celestial Mechanics*, 1962; G. R. Smith, A simple efficient starting value for the iterative solution of Kepler's equation, *Celestial Mech.*, 19(2):163–166, 1979.

Kepler's laws

The three laws of planetary motion discovered by Johannes Kepler during the early years of the seventeenth century.

First law. The first law of Kepler states that a planet moves in an elliptical orbit around the Sun that is located at one of the two foci of the ellipse. An ellipse is one of the conic curves originally studied by Greek geometers. It is formed when a cone is cut by a plane that is neither parallel nor perpendicular to the axis and not parallel to the side of the cone. As the plane approaches perpendicularity with the axis, the ellipse approaches a circle; Kepler's first law accepts a circular orbit as the limiting case of an ellipse. As the cutting plane approaches a position parallel to the side of the cone, the ellipse approaches a parabola; although a body can move in a parabolic path around another placed at its focus, it would not be a planet in the accepted sense because it would never return. An ellipse is also defined as the locus of points the sum of whose distances from two fixed points (the foci) is a constant.

Kepler discovered the first law by analyzing the observations that Tycho Brahe had made of the planet Mars in the late sixteenth century. An earlier astronomer of that century, Nicolaus Copernicus, had broken with the commonsense notion, apparently supported by daily experience, that the Earth is the center of the universe around which all other bodies move. However, he had accepted another tradition in astronomy, that all heavenly appearances are to be explained by combinations of circles, which were held to be perfect figures. Kepler's ellipses shattered that tradition. They also challenged another premise of earlier astronomy, that the heavens are fundamentally different from the Earth. Although Kepler discovered the ellipse by studying Mars, he generalized the shape to the orbits of all the planets.

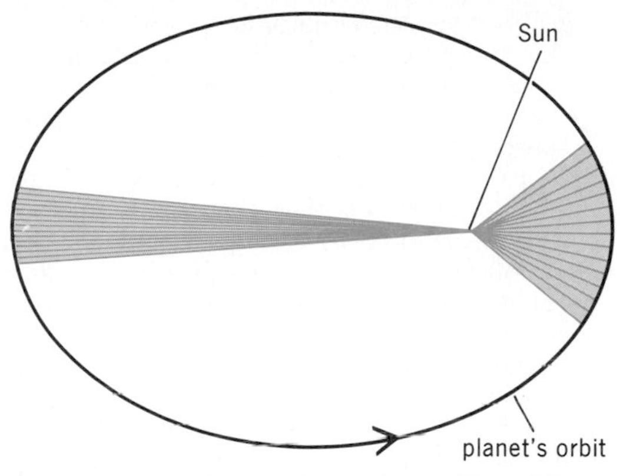

Demonstration of Kepler's first and second laws. The planet moves along an elliptical orbit at a nonuniform rate, so that the radius vector drawn to the Sun, which is located at one focus of the ellipse, sweeps out areas that are proportional to time. Thus, the planet would take equal times (corresponding to the equal areas) to traverse the unequal distances along the ellipse that correspond to the two shaded areas. The diagram greatly exaggerates the eccentricity of any orbital ellipse in the solar system.

In 1610, only one year after Kepler published his first law, Galileo discovered the satellites of Jupiter, and later in the seventeenth century satellites were observed around Saturn. All of these bodies obey Kepler's first law. In 1687, Isaac Newton demonstrated that any body, moving in an orbit around another body that attracts it with a force that varies inversely as the square of the distance between them, must move in a conic section. This path will be an ellipse when the velocity is below a certain limit in relation to the attracting force. Thus the first law is not limited to primary planets around the Sun or to heavenly bodies. It is a general law that applies to all satellites held in orbit by an inverse-square force.

Second law. Astronomers needed to predict the location of planets at given times. The system of combining circles, which were assumed to turn uniformly, had made it possible to calculate a position at any time by the vectorial addition of the radii. In rejecting the system of circles, Kepler needed some other method by which to predict locations, and this he provided with his second law. It states that the radius vector of the ellipse (the imaginary line between the planet and the Sun) sweeps out areas that are proportional to time (see **illus.**). A planet does not move along its elliptical path at a uniform velocity; it moves more swiftly when it is closer to the Sun and more slowly when it is farther removed. Behind the nonuniform velocities the second law finds a uniformity in the areas described.

Again Newton demonstrated the dynamic cause behind Kepler's second law. In this case, it is not restricted to forces that vary inversely as the square of the distance; rather it is valid for all forces of attraction between the two bodies, regardless of the law the forces obey. The second law expresses the principle of the conservation of angular momentum. If body B moves in relation to body A in a straight line with a uniform velocity, it has a constant angular momentum in relation to A, and the line joining B to A sweeps out equal areas in equal increments of time. No force of attraction (or repulsion) between the two bodies can alter their angular momentum about each other. The second law then expresses a relation that holds for all pairs of bodies with radial forces between them.

Third law. Kepler's first two laws govern the orbits of individual planets around the Sun. His third law defines the relations that hold within the system of planets. It states that the ratio between the square of a planet's period (the time required to complete one orbit) to the cube of the mean radius (the average distance from the Sun during one orbit) is a constant. The four satellites that Galileo had discovered around Jupiter were found to obey the third law, as did the satellites later found around Saturn. Newton demonstrated once again that the third law is valid for every system of satellites around a central body that attracts them, as the Sun attracts the planets, with a force that varies inversely as the square of the distance.

In fact, none of the three laws precisely describes the motions of the planets around the Sun, for the laws assume the only forces that act are those between the Sun and individual planets. But Newton's law of universal gravitation asserts that all planets also attract each other. In the case of the Moon orbiting the Earth, the attraction that the Sun exerts on the Moon is large enough to introduce perturbations, which astronomers had identified empirically before Kepler stated his three laws. Because the Sun is immensely more massive than the planets, the perturbations caused by the planets' mutual attractions are quite small, and Kepler's laws offer an accurate first approximation of their motions. Kepler's laws of planetary motion were the first of the mathematical laws of modern science; they did much to determine the pattern that science has continued to pursue. SEE CELESTIAL MECHANICS; GRAVITATION; ORBITAL MOTION; PERTURBATION.

Richard S. Westfall

Bibliography. M. Caspar, *Kepler*, trans. by C. D. Hellman, 1959; G. Holton, Johannes Kepler's universe: Its physics and metaphysics, *Amer. J. Phys.*, vol. 24, 1956; A. Koyré, *The Astronomical Revolution*, trans. by R. E. W. Maddison, 1973; T. Kuhn, *The Copernican Revolution*, 1957; W. Pauli, *The Influence of Archetypal Ideas on the Scientific Theories of Kepler*, 1955.

Latitude and longitude

The latitude of a location specifies the angle between an imaginary line directed generally toward the center of the Earth and the Equator. The longitude measures the angle between the meridian (the plane defined by the Earth's axis and this local reference direction) and the plane of the Greenwich meridian (**illus.** *a*).

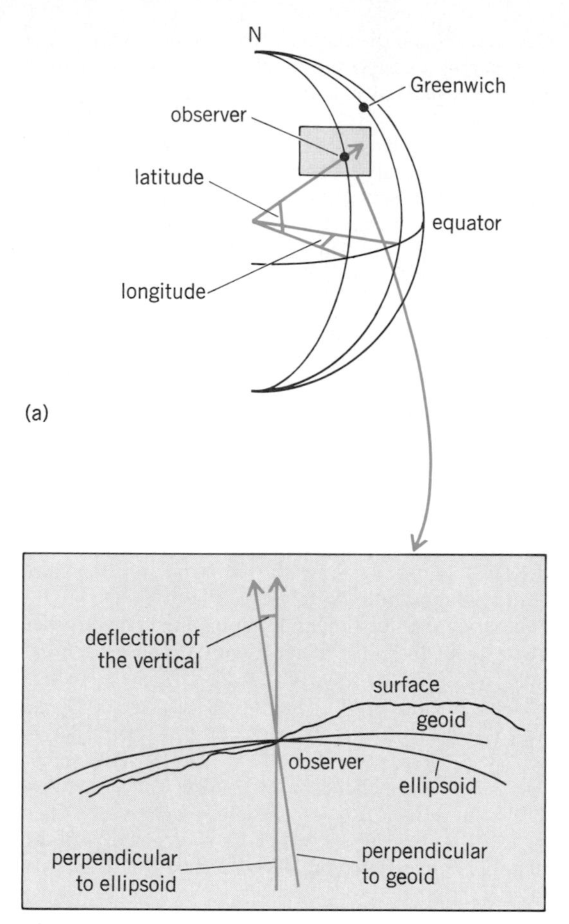

Latitude and longitude. (*a*) Angles formed by reference direction, defining latitude and longitude. (*b*) Detail, showing angle between reference directions in astronomical and geodetic coordinates, the deflection of the vertical.

Astronomical coordinates. Astronomical (or astronomic) latitude and longitude use the direction of gravity for the reference direction. This direction, known as the astronomical vertical, is perpendicular to the equipotential surface of the Earth's gravitational field at the location of the observer. SEE ASTRONOMICAL COORDINATE SYSTEMS.

Geodetic coordinates. A particular geopotential surface approximating mean sea level in the open ocean is called the geoid. A mathematical surface in the form of an oblate ellipsoid may be constructed to approximate the geoid. The direction perpendicular to this reference ellipsoid at the observer's location is used as the reference direction in defining geodetic latitude and longitude. A geodetic datum is defined by its reference ellipsoid, the adopted coordinates of a reference station, and the azimuth of a reference line. Many geodetic datums are in use.

Geocentric coordinates. Geocentric latitude and longitude are defined by a reference direction which passes precisely through the center of mass of the Earth. These coordinates are determined mathematically from the geodetic latitude and longitude, assuming a fixed relationship between the center of the geodetic datum and the center of mass and knowing the mathematical shape of the ellipsoid.

Determination. Astronomical latitude is determined by observing the altitude on the meridian of a celestial object whose declination is known. Astronomical longitude is identical with the difference between Universal Time and local mean time. Local mean time is determined by observing the time of meridian transits of celestial objects with known positions. The direction of gravity is established with the aid of a spirit level, a liquid surface, or observation of the horizon corrected for the elevation of the observer. It varies locally, and is generally not directed toward the center of the Earth. *See Time.*

Geodetic latitude and longitude are determined by referring precise measurements of distance and direction, made with geodetic instruments on the surface of the Earth, to the datum. Geodetic coordinates are used for location of sites, while astronomical latitude and longitude are used to determine the angular orientation of celestial objects with respect to observers on the Earth. The difference between astronomical and geodetic coordinates is called the deflection of the vertical (illus. *b*), and commonly amounts to some seconds of arc, occasionally reaching a minute of arc.

Dennis D. McCarthy

Bibliography. I. I. Mueller, *Spherical and Practical Astronomy as Applied to Geodesy*, 1969.

Leo

The Lion, in astronomy, is a magnificent zodiacal constellation appearing during spring and early summer. It is the fifth sign of the zodiac. Leo is well defined and bears a close resemblance to the creature it represents. The head is outlined by a group of six stars, α, η, γ, ζ, μ, and ε, called the Sickle. The first-magnitude star, Regulus (Little Ruler), forms the handle of the sickle. Three stars to the east, β, δ, and θ, forming a small triangle, constitute the Lion's haunches, with the bright star, Denebola (tail of the lion), a navigational star (see **illus.**). Associated with this constellation are the famous Leonids shower of meteors, which can be seen radiating from Leo in November of each year and appearing especially brilliant at intervals of about 33 years. *See Constellation; Meteor; Zodiac.*

Ching-Sung Yu

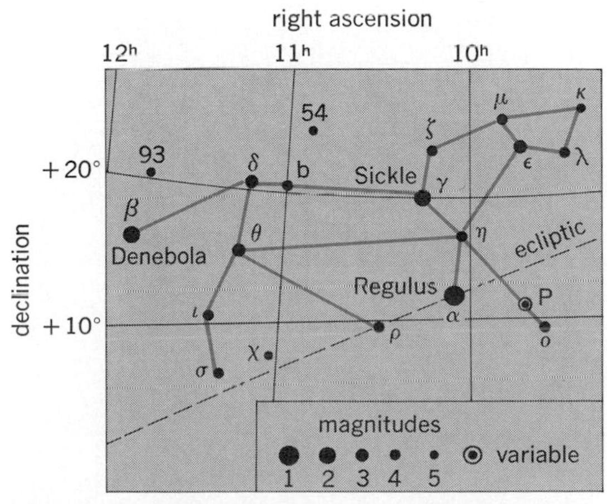

Line pattern of the constellation Leo. The grid lines represent the coordinates of the sky. Right ascension (E-W) in hours, and declination (N-S) in degrees, corresponding to the longitude and latitude of the Earth. The apparent brightness, or magnitudes, of the stars is shown by the sizes of the dots, which are graded by appropriate numbers as indicated.

Libra

The Balance, in astronomy, appearing in the evening sky during the spring. It is the seventh sign of the zodiac. The constellation consists of faint stars and is

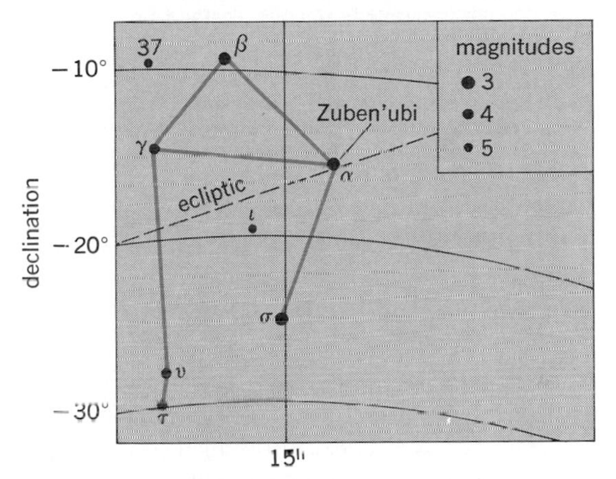

Line pattern of the constellation Libra. The grid lines represent the coordinates of the sky. The apparent brightness, or magnitude, of the stars is shown by the sizes of the dots, graded by appropriate numbers.

not conspicuous. It lies just west of the claws of Scorpius. The principal stars, α, β, γ, and σ, outline a four-sided figure resembling a balance with beam and pans (see **illus.**). The balance might have been held originally in the hand of Virgo, a zodiacal constellation nearby, who was identified with the Goddess of Justice. Another possible reason for identifying the constellation with the balance is that 2000 years ago the Sun was in Libra at the then autumnal equinox, at which time days and nights are of equal length. *See Constellation; Virgo; Zodiac.*

Ching-Sung Yu

Light curves

Graphs of the intensity of radiation from astronomical objects as they change with time. Variations may be caused by the changing perspective from the Earth of two stars in orbit around each other, by pulsations that change an individual star's size and surface temperature, by mass ejection or accretion, by explosions, by beams of radiation sweeping across the line of sight from the Earth, or by clouds of very high-energy electrons in powerful magnetic fields. The information contained in the light curve includes the timing of events, such as eclipses or pulses, and the amplitude of changes in the radiation received at Earth.

Each data point in a light curve is a photometric measurement, recorded at a particular time. These points represent measurements of the amount of radiation from the source received at Earth per second per area in a particular bandpass, for example, through a blue filter. In optical light, photometry is

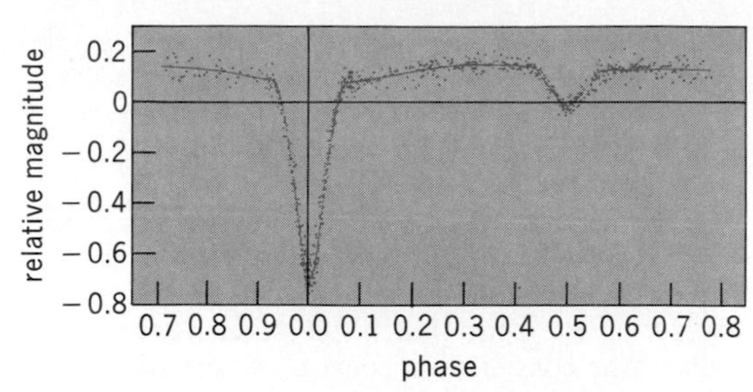

Fig. 1. Light curve of the eclipsing binary Algol. The measured infrared intensity on the vertical axis is given in magnitudes relative to the nearby star κ Persei. The phase is the fraction of the 2.9-day period elapsed relative to the deep primary minimum. Also apparent are the secondary minimum, when the K giant is eclipsed by the B star, and the brightening around secondary minimum caused primarily by heating of one face of the cooler star. (*After K.-Y. Chen and G. Reuning, Astron. J., 71:283–296, 1966*)

affected by the varying transparency of the atmosphere, so that light curves are often obtained as ratios to the intensities of nearby comparison stars. Simultaneous measurements can be made visually, with photometers that have two channels or with imagers, such as charge-coupled devices (CCDs) or photographic emulsions. These relative light curves are put on an absolute scale by means of calibrating measures of the comparison stars taken on clear nights. *SEE ASTRONOMICAL PHOTOGRAPHY; CHARGE-COUPLED DEVICES; MAGNITUDE.*

Binary stars. If the orbital plane of a binary star system lies nearly along the line of sight from the Earth, the stars appear to eclipse each other periodically, either partially or totally, depending on the tilt of the orbit (**Fig. 1**). The larger drop in intensity represents the primary eclipse, when the dimmer star passes in front of the brighter; the secondary eclipse is shallower, produced when the light from the dimmer star is blocked. If one face of the cooler star is strongly irradiated by a hot primary or gravity distorts the star into a nonspherical shape, the light curve also shows a hump that is centered on the secondary eclipse. The timing of the eclipses yields the ratio of the radii of the two stars and the period and tilt of the orbit. Temperature differences between the two stars make the drop in intensity during an eclipse differ among color bands. *SEE BINARY STAR; ECLIPSING VARIABLE STARS.*

Variable stars. Intrinsically variable single stars have distinct signatures in their light curves. Some giant stars have an instability in their outer layers that acts as a heat engine driving radial pulsations of the star's surface. A light curve such as **Fig. 2** reveals the amount of extra flux received at Earth during each pulsation. Spectroscopy gives the temperature and change in radius of the star during the same cycle. When combined, this information yields the distance to the variable star and its intrinsic power output. For example, Cepheid variables can be identified by their light curve signatures and used as distance indicators for nearby galaxies. *SEE CEPHEIDS.*

Other types of stars also show surface pulsations that appear as variations in their light curves. These types include early B stars on the main sequence, and stellar remnants such as planetary nebula nuclei and white dwarfs. The amplitudes of these pulsations can be very low (<1% of the white light intensity) and can consist of a complicated blend of many frequencies. These variations are interpreted as nonradial pulsations that act like waves near the star's surface. The central stars of planetary nebulae are evolving so fast that the frequencies of the pulsations show tiny but measurable shifts over the course of 5 to 10 years that are produced by the collapsing cores as they become white dwarf stars. *SEE PLANETARY NEBULA; WHITE DWARF STAR.*

Many single stars show irregular variations brought about by mass loss. Illuminated ejecta add to the brightness until the material disperses. Newly formed stars may be accreting new gas from disks at their equators and ejecting material in jets at the poles. These T Tauri stars show strong variability. Hot stars of type B with emission lines (Be stars) on the main sequence also have disks or shells that vary in light with irregular mass-loss episodes. Red supergiant stars eject tremendous amounts of matter from their outer envelopes; long-period variables, Miras, and nebular variables all fall in this class. Light curves for these objects in both the optical and infrared provide the time history of mass ejection. *SEE STAR; STELLAR EVOLUTION; VARIABLE STAR.*

Interacting binaries. The stars in some binary systems are so close together that gas flows between the components. If the primary is a high-density stellar remnant, the gas may spiral in, forming an accretion disk. The accretion flow itself may radiate much of the power from the system, especially in the ultraviolet and x-rays. These high-energy light curves show evidence for flickering and for a hot spot where the stream of material impacts the disk. If the compact star is a white dwarf, the buildup of gas on its surface can cause a nuclear explosion, which is a nova out-

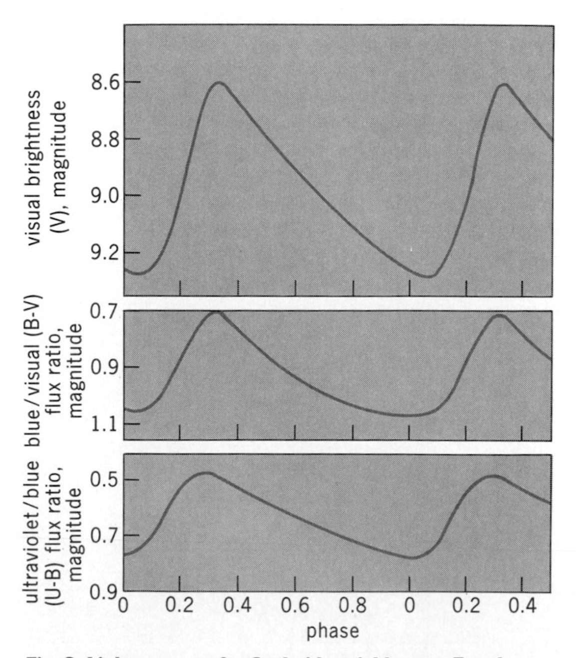

Fig. 2. Light curves of a Cepheid variable star. For the lower two curves, bluer colors are plotted toward the top. The phase is the fraction of the 4.9-day pulsation period, relative to the time of minimum brightness in the cycle. (*After R. I., Mitchell et al., Boletin de los Observatorios Tonantzintla y Tacubaya, 3:153–304, 1964*)

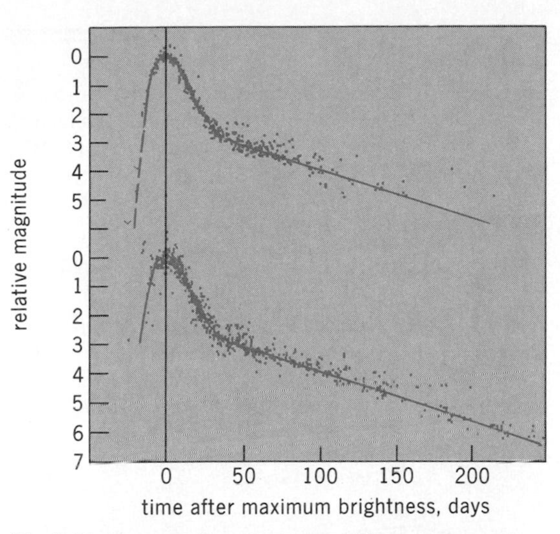

Fig. 3. Light curves for supernovae. The upper curve is a composite of blue photoelectric measurements for 22 supernovae, while the lower curve is a composite of 16 objects measured in the blue photographic magnitude system. (*After N. Bartel, ed., Workshop on Supernovae as Distance Indicators, Lectures Notes in Physics Series, Springer-Verlag, 1985*)

burst. The light curve shows the object brightening by 10 to 100,000 times its quiescent level. X-ray light curves have shown that some unseen compact companions must be neutron stars or black holes. The ultimate explosion is when an entire star (or pair of stars) detonates; this explosion is a supernova, which can be brighter than a whole galaxy of stars combined. Supernovae fade by a factor of about 15 in the first 30 days and then dim more gradually; the shape of their light curves (**Fig. 3**) demonstrates that after the initial explosion the energy input is dominated by the decay of radioactive elements produced by the thermonuclear runaway. *See* ASTROPHYSICS, HIGH-ENERGY; BLACK HOLE; NEUTRON STAR; NOVA; SUPERNOVA; X-RAY ASTRONOMY; X-RAY STAR.

Exotic objects. Variable radiation can be emitted by clouds of very high-energy electrons moving in powerful magnetic fields. Beams and jets of radiated energy may result. Pulsars are an example, rapidly rotating neutron stars that sweep the Earth with such beams up to hundreds of times per second. Their light curves are commonly observed at radio frequencies, but young pulsars also show pulses in optical light. Blazars are actively varying powerful nuclei of galaxies. They may have giant beams and jets produced by supermassive black holes, and present strongly variable light curves in radio, optical, ultraviolet, and x-rays. Quasars are modeled as having both accretion disks and jets from supermassive black holes. Their light curves show more modest variability than those of blazars. The time delays between variations in the continuous radiation from the central engine and the spectral lines emitted by the surrounding clouds of gas are interpreted to yield the physical size of these active galactic cores. *See* GALAXY, EXTERNAL; PULSAR; QUASAR.

Richard F. Green

Bibliography. J. S. Glasby, *The Dwarf Novae*, 1970; C. Hoffmeister, G. Richter, and W. Wenzel, *Variable Stars*, 1985; R. N. Manchester and J. H. Taylor, *Pulsars*, 1977; J. R. Percy (ed.), *The Study of Variable Stars Using Small Telescopes*, 1986; G. Swarup and V. K. Kapahi (eds.), *Quasars*, Int. Astron. Union Symp. 119, 1986.

Light pollution

A general glow in the night sky that results from the light emitted from artificial fixtures and significantly obscures the view of the stars.

Astronomy is suffering from many adverse environmental impacts, one of the worst of which is light pollution. The increasing urban sky glow has already compromised astronomical research at many observatories worldwide, severely impacted most amateur astronomers, and removed, perhaps forever, the view of the prime nighttime skies that previous generations enjoyed. A great deal of the frontier research in astronomy consists of observations of very faint objects, such as galaxies and quasars, at such distances that the light has been traveling for more than the age of the solar system before reaching the Earth, only to be lost in the artificial sky glow. The large telescope at Mount Wilson Observatory, near Los Angeles, California, has already been closed down because of the increased sky glow at the site. At Palomar Observatory, near San Diego, California, the sky glow is about double the natural background, while at Lick Observatory, near San Jose, California, it is triple the natural background. So far, conditions are much better at Kitt Peak National Observatory in Arizona, McDonald Observatory in Texas, and the Mauna Kea Observatories in Hawaii. *See* ASTRONOMICAL OBSERVATORY.

Wasted light. This sky glow is caused, to a considerable extent, by totally wasted light going upward, never having been an aid to visibility at night (see **illus.**). It is this wasted light from inefficient fixtures (many with less than half the light output directed toward the ground) that is the major offender in light pollution. Fortunately, solutions to the problem exist and can help minimize the adverse effects greatly. Quality nighttime lighting is the key. Everything done to help solve the problem maximizes the

Satellite photograph of the United States at local midnight, taken in 1979, showing how poor-quality lighting wastes energy by casting light skyward, where it is not needed. (*U. S. Air Force*)

quality of nighttime lighting, enabling people to see better, to be safer and more secure, and to live in a more desirable nighttime environment. Light is used, not wasted, and large amounts of energy and expense can be saved. Wasted lighting also produces a great deal of glare, which never helps visibility, and light trespass (bothersome localized lighting directed where it is not wanted or needed). For astronomy interests, quality lighting has little or no direct upgoing light, and sky glow is minimized.

Solutions. Solutions to the problem of light pollution can be summarized as follows.

1. Light should be directed downward, where it is needed. The use and effective installation of fixtures with well-designed reflectors or refractor elements will achieve excellent lighting control. Poor housings should be retrofitted with quality ones whenever possible. In all cases, the goal is to minimize the up light, reduce glare (to zero in many cases), and eliminate light trespass and energy waste.

2. Night lighting should be used only when needed. Otherwise lights should be turned off when they are no longer needed; and only the correct amount of light should be used.

3. Low-pressure sodium light sources should be used whenever possible. As the light output is nearly monochromatic (the two sodium resonance lines at wavelength 589 nanometers), it greatly minimizes any adverse effects on professional observatories. Fortunately, low-pressure sodium is also the most energy-efficient light source in existence, giving up to 180 lumens per watt efficiency. It also produces the best light for visual acuity, because the eye is most sensitive to its yellow emission lines. It should be used in efficient fixtures, of course. Areas where low-pressure sodium lighting is especially effective include street lighting, parking lots, security lighting, and any application where color rendering is not critical. Even in the latter cases, creative lighting designs exist that can offer excellent color rendering.

4. Growth nearest the major observatories should be avoided, and rigid controls should be applied on the outdoor lighting in their vicinity. Such controls do not compromise in any way safety or utility; they enforce the use of quality lighting and thus improve visibility at night and maximize energy savings. The enactment and enforcement of lighting codes have probably rescued Palomar Observatory, and have enabled observatories on Kitt Peak and Mount Hopkins, in Arizona, to operate relatively efficiently.

Many organizations have formed committees to address the issues, including most of the major astronomical societies. The International Dark-Sky Association, in Tucson, Arizona, is devoted to building awareness of these problems so as to preserve the nighttime sky and maximize the quality and effectiveness of nighttime outdoor lighting.

David L. Crawford

Bibliography. D. L. Crawford, *The American Astronomical Society's Position on Light Pollution*, IDA Inform. Sheet 19, 1990; D. L. Crawford, *Astronomy's Problem with Light Pollution*, IDA Inform. Sheet 1, 1988; D. L. Crawford, *Light Pollution, Radio Interference, and Space Debris*, Astronomical Society of the Pacific Conference Series, vol. 17, 1991; D. L. Crawford and T. Hunter, The battle against light pollution, *Sky Telesc.*, 80(1):23–29, July 1990.

Light-year

A unit of measurement of astronomical distance. A light-year is the distance light travels in 1 sidereal year. One light-year is equivalent to 9.461×10^{12} km, or 5.879×10^{12} mi. Distances to some of the nearer celestial objects, measured in units of light time, are shown in the **table**.

Distances from the Earth to some celestial objects

Object	Distance from Earth (in light time)
Moon (mean)	1.3 s
Sun (mean)	8.3 min
Mars (closest)	3.1 min
Jupiter (closest)	33 min
Pluto (closest)	5.3 h
Nearest star (Proxima Centauri)	4.3 years
Andromeda Galaxy (M31)	2.3×10^6 years

This unit, while useful for its graphic presentation of the enormous scale of stellar distances, is seldom used technically except in cosmology. SEE ASTRONOMICAL UNIT; COSMOLOGY; PARALLAX; PARSEC.

Jesse L. Greenstein

Local Group

The small cluster of galaxies that contains the Milky Way Galaxy. Galaxies exhibit a pronounced tendency to clump together on a variety of scales. It is assumed that gravitational attraction draws galaxies together. On a very large scale, this attractive process may still be at an early stage, resulting in filamentary structures which are only mildly perturbed in their motions from the general expansion of the universe. On a smaller scale, the process has led to collapse. Galaxies have fallen together, though with enough angular momentum that they usually orbit each other rather than collide.

Collapsed structures that contain a few dozen to a few thousand substantial galaxies are called clusters; collapsed structures that contain a few but less than 10 or 20 big galaxies are called groups. The Milky Way Galaxy is a large but not exceptional spiral galaxy and one of two dominant members of a small assemblage referred to as the Local Group.

Environment. On successively larger scales, the Local Group is a member of the Coma-Sculptor Cloud and the Local Supercluster. The Coma-Sculptor Cloud contains several hundred galaxies in an extremely flattened distribution that extends across 4×10^7 light-years. (1 light-year equals 9.46×10^{12} km or 5.48×10^{12} mi), in the general direction toward the Virgo cluster at the center of the Local Supercluster. Its immediate vicinity is shown in the **illustration.** The dominant members of the nearest-neighboring groups are New General Catalogue (NGC) 253, Maffei 1, Index Catalogue (IC) 342, and Messier (M) 81. The closest galaxies beyond the Local Group are at a distance of 7×10^6 light-years.

Members. The **table** identifies 35 known or suspected members of the Local Group, ordered from

(*Below; top*) Chainpur, India, chondrite. Photomicrograph of a typical chondritic texture with a small intact chondrule composed of olivine and glass adjacent to two larger broken olivine-plus-pyroxene chondrules. The chondrules are set in a matrix of fine-grained silicate fragments.

(*Center*) Bununu, Nigeria, achondrite. Typical brecciated texture of the howardite achondrites. The dark angular glass fragments are mixed with mineral and rock fragments derived from basaltic crystalline rocks by impact brecciation.

(*Bottom*) Isna, Egypt, chondrite. The texture of a fine-grained carbonaceous chondrite displays chondrules, angular to rounded rock fragments, and black metal and sulfide particles set in a dark carbon-rich matrix.

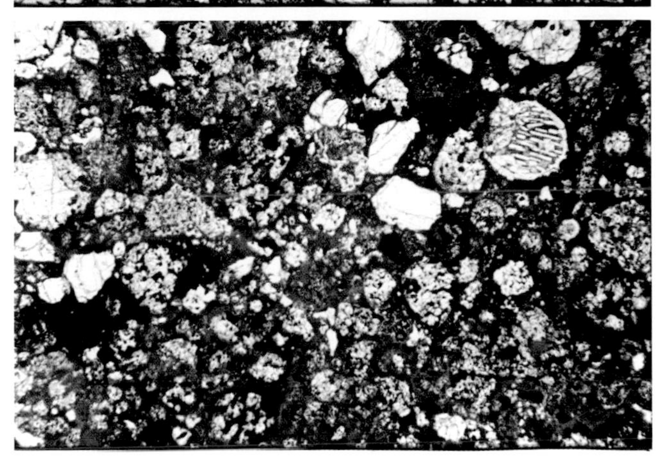

(*Above; top*) Luotolax, Finland, achondrite. The striations in the fusion crust covering this specimen show that it fell through the air in a fixed position.

(*Center*) Inclusion AL3S4 of the Allende carbonaceous chondrite. This inclusion is approximately 1.5 cm in diameter. It is composed of an outer rim of large melilite crystals and an interior of pyroxene, spinel, and an orthite. The pyroxene and spinel are highly enriched in nucleogenetic ^{16}O, but oxygen in the melilite is normal.

(*Bottom*) Moore County, North Carolina, achondrite. This coarse-grained eucrite, which contains interlocking grains of twinned plagioclase and pyroxene, is similar to many terrestrial basalts.

(*Photographs courtesy of Edward P. Henderson, except Allende chondrite courtesy of Dr. I. D. Hutcheon*)

brightest to faintest. The only system larger than the Milky Way Galaxy is M31, the Andromeda Nebula. These two giant galaxies generate 80% of the light of the group. There are also two intermediate-scale galaxies: M 33, the Triangulum Nebula near Andromeda, and the Large Magellanic Cloud, a close companion of the Milky Way and a conspicuous feature of the night sky in the southern hemisphere. Fainter than these are 31 small systems in or peripheral to the group. The smallest among them are identified only because they are so close. Many would not be detected in even the nearest adjacent groups. The census of Local Group members may be very incomplete at the faint end and in the zones obscured by the plane of the Milky Way. *See Andromeda Galaxy; Magellanic Clouds; Milky Way Galaxy.*

Subclustering. The two giant galaxies have entourages of small systems. The Large and Small Magellanic clouds are so close that they may be in the process of merging with the Milky Way Galaxy. The Milky Way also has the seven dwarf satellites: Fornax, Sculptor, Leo I, Leo II, Draco, Carina, and Ursa Minor. These are gas-depleted spheroidal systems with essentially no ongoing star formation.

In any photograph of M31 the high surface-brightness companions M32 and NGC 205 are clearly visible. In addition, M33 is within 6.5 × 10^5 light-years of M31, and in proximity there are NGC 147, NGC 185, IC 10, and the very small systems LGS 3 and Andromeda I, II, and III.

In another part of the sky, NGC 6822, David Dunlap Observatory (DDO) 210, and SAGDIG (Sagittarius Dwarf Irregular Galaxy) are close together. Furthermore, six other dwarf irregular systems are reasonably close to each other but sufficiently removed from the rest of the Local Group that they could also be thought of as a distinct entity: NGC 3109, Sextans A, Sextans B, Leo A, DDO 155, and DDO 187. A rigorous criterion for group membership might require total or continuing collapse toward the common potential of the group. However, the distances and three-dimensional motions of each of the systems or of the underlying mass distribution are not well enough known to identify with certainty which systems are bound and which are separating. The distance to NGC 404 is known only approximately; this galaxy might be outside the group.

Types of galaxies. Galaxies within the Local Group, as with most others, can be assigned to one of two families. Some have substantial reservoirs of hydrogen and are actively forming new stars; others are gas-depleted and contain only old stars. The three spirals and the Magellanic or dwarf irregulars all have young stars and gas as well as old constituents. The elliptical, lenticular, and dwarf spheroidal systems are all deficient in interstellar gas and have the smooth textures of old, dynamically relaxed galaxies.

Except for NGC 404, all the known gas-poor systems lie close to one of the giant galaxies and to each other. In contrast, the gas-rich systems tend to be farther apart even though still identified with the group. Tidal or direct collisional stripping has probably affected the systems in crowded regions and deprived them of the gas reserves necessary for continued growth.

Distances. The galaxies in the Local Group are so close that individual stars can be resolved with modern detectors on large telescopes. More massive

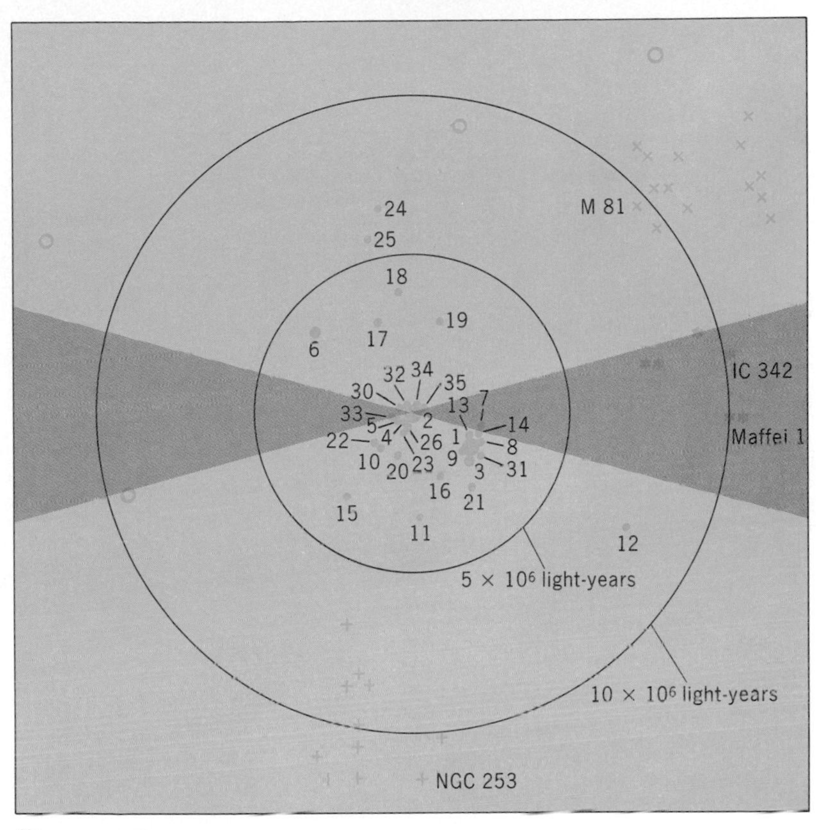

Key:
- ● potential members of the Local Group
- + members of the Sculptor Group
- ✱ members of the Maffei 1/IC 342 Group
- × members of the M 81 Group
- ○ other galaxies

Neighborhood of the Local Group. Shaded wedges indicate the region obscured by the plane of the Milky Way Galaxy. The concentric circles are centered on the Milky Way and indicate distances of 5 and 10 × 10^4 light-years.

young stars are easily observed in the spiral and irregular galaxies. Among the populations of old stars, only the brightest of those in a certain phase of their evolution called red giants are visible in the more distant parts of the Local Group. In each case, all well-studied systems contain stars with familiar properties that give an indication of their distances. The most useful are the Cepheid and RR Lyrae stars because they pulsate with frequencies that are related to their intrinsic luminosities. It has been established that the distances measured in the first half of the century were badly underestimated. The current estimates (see table) are fundamental for the calibration of the extragalactic distance scale. *See Cepheids.*

Mass. Assuming that the galaxies within 3 × 10^6 light-years of either the Milky Way or M31 are part of a dynamically collapsed system, the mass of the unit can be estimated with the virial theorem. This theorem specifies the relationship between the mass, dimensions, and motions in an isolated, relaxed system of gravitating particles. In practice, there are problems because not all the velocity and position information is available, and the assumption that the system has collapsed and relaxed is probably poorly satisfied. Nevertheless, the results of such an analysis are consistent with similar analyses of other groups and can be taken to be valid in a statistical sense. It is estimated that the mass of the Local Group is

Local Group members

Name	Type	Distance, 10^6 light-years	Solar luminosities
1. M31 = Andromeda	Spiral	2.4	3×10^{10}
2. Milky Way Galaxy	Spiral	—	2×10^{10}
3. M33 = Triangulum	Spiral	2.6	6×10^{9}
4. Large Magellanic Cloud	Irregular	0.16	3×10^{9}
5. Small Magellanic Cloud	Irregular	0.19	7×10^{8}
6. NGC 3109*	Irregular	6.	6×10^{8}
7. IC 10	Irregular	2.	6×10^{8}
8. NGC 205	Elliptical	2.4	3×10^{8}
9. M32	Elliptical	2.4	2×10^{8}
10. NGC 6822	Irregular	1.6	2×10^{8}
11. WLM	Irregular	3.1	2×10^{8}
12. NGC 404*	Elliptical	8.	2×10^{8}
13. NGC 185	Elliptical	2.0	2×10^{8}
14. NGC 147	Elliptical	2.1	1×10^{8}
15. IC 5152	Irregular	3.	1×10^{8}
16. IC 1613	Irregular	2.3	1×10^{8}
17. Sextans B*	Dwarf irregular	5.	8×10^{7}
18. Sextans A*	Dwarf irregular	4.	8×10^{7}
19. Leo A*	Dwarf irregular	3.	4×10^{7}
20. DDO 210	Dwarf irregular	2.	3×10^{7}
21. Pegasus	Dwarf irregular	3.	3×10^{7}
22. SAGDIG	Dwarf irregular	2.	2×10^{7}
23. Fornax	Dwarf spheroidal	0.8	2×10^{7}
24. DDO 187*	Dwarf irregular	7.	7×10^{6}
25. DDO 155 = GR 8*	Dwarf irregular	6.	6×10^{6}
26. Sculptor	Dwarf spheroidal	0.27	4×10^{6}
27. Andromeda I	Dwarf spheroidal	2.	2×10^{6}
28. Andromeda II	Dwarf spheroidal	2.	2×10^{6}
29. Andromeda III	Dwarf spheroidal	2.	2×10^{6}
30. Leo I	Dwarf spheroidal	0.9	1×10^{6}
31. LGS 3	Dwarf irregular	2.	9×10^{5}
32. Leo II	Dwarf spheroidal	0.75	8×10^{5}
33. Carina	Dwarf spheroidal	0.55	8×10^{5}
34. Draco	Dwarf spheroidal	0.33	8×10^{5}
35. Ursa Minor	Dwarf spheroidal	0.22	3×10^{5}

*Doubtful members.

roughly 10^{12} times the mass of the Sun (2×10^{42} kg or 4×10^{42} lb). Hence, the ratio of the mass to light of the Local Group is 20 in solar units. This value is lower than a more typical value of 100 for galaxy groups, but much larger than the value of about 5 for the luminous parts of individual galaxies. It is supposed that in the Local Group, as in many other environments, the potential well is dominated by a component of invisible matter, so-called dark matter. The mass contributed by the stars that produce the light and by the interstellar hydrogen are relatively insignificant. *See Cosmology; Galaxy, external.*

R. Brent Tully

Bibliography. A. R. Sandage, *The Hubble Atlas of Galaxies*, 1961; R. B. Tully and J. R. Fisher, *Nearby Galaxies Atlas*, 1987.

Lyra

The Lyre, in astronomy, a summer constellation, small but important. Lyra has a first-magnitude star, Vega, a navigational star and the most brilliant star in this part of the sky. Vega forms, with two faint stars ε and ζ to the east, an almost perfect equilateral triangle (see **illus.**). The southern one in turn forms, with three brighter stars, β, γ, and δ, to the south, an approximate parallelogram. The resulting overall figure resembles a tortoise more than a stringed musical instrument. However, according to legend, Mercury made the first lyre from a turtleshell by placing strings across it. Hence the two different representations are not incompatible. In this constellation lies the famous Ring Nebula, M57. *See Constellation; Nebula.*

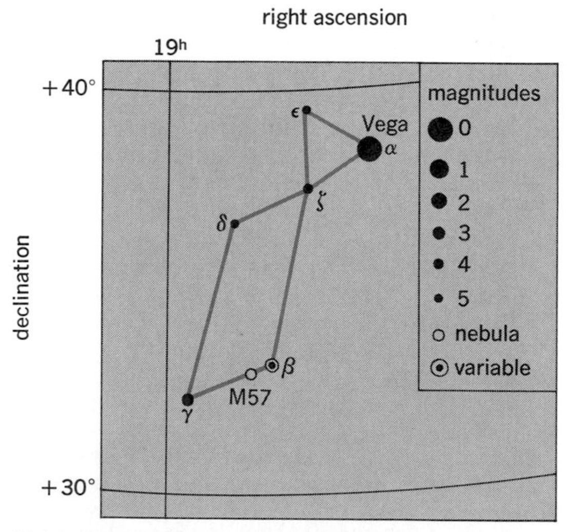

Line pattern of the constellation Lyra. The grid lines represent the coordinates of the sky. The apparent brightness, or magnitudes, of the stars is shown by the sizes of the dots, graded by appropriate numbers.

Ching-Sung Yu

Magellanic Clouds

The easily observable external galaxies nearest to the Milky Way galactic system. The distance of the Large Magellanic Cloud (LMC) is 40,000 parsecs (130,000 light-years, 8.3×10^9 astronomical units, 7.7×10^{17} mi, or 1.23×10^{18} km); that of the Small Magellanic Cloud (SMC) is 59,000 pc (192,000 light-years, 1.22×10^{10} astronomical units, 1.13×10^{18} mi, or 1.82×10^{18} km); these distances are uncertain by about 10%. Because of their proximity, many types, such as bright main-sequence stars, supergiants, classical cepheid variables, novae, and planetary nebulae, can be studied. Since all these objects are at nearly the same distance from the observer, differences in apparent brightness correspond to differences in true brightness. Thus, the period-luminosity relation for classical cepheids was first established in the Small Magellanic Cloud by H. Leavitt. Once the zero point of that relation (that is, the true brightness of at least one cepheid) was found, the distance of any stellar system containing classical cepheids could be established. Distances are now found by using different types of objects such as bright main-sequence stars similar to those in the Milky Way Galaxy. Thus, the Clouds served as a "Rosetta Stone" to astronomy. *See Cepheids; Galaxy, external; Variable star.*

Both Clouds are visible to the naked eye; the Large Magellanic Cloud has an angular diameter of about 7°, the Small Magellanic Cloud an approximate diameter of 3°. E. P. Hubble classified them as irregular galaxies, although it has been suggested that the Large Magellanic Cloud may be a barred spiral. The Large Magellanic Cloud is much more massive than the Small Magellanic Cloud; estimates of 4×10^9 and 4×10^8 solar masses have been given.

Both Clouds contain populations of old stars and also numerous young stars and associations. Many of the remarkable young and intermediate-age clusters are similar in appearance to conventional globular clusters. Some have ages less than 1×10^8 years. Massive, true globular clusters may range in age up to 8×10^9 years. They are probably as old as the Clouds themselves. *See Star clusters.*

Nebulosities are recorded strikingly on narrow-wavelength-range photographs centered on the Hα line. The Large Magellanic Cloud shows many wispy nebulosities many times larger than the Orion Nebula. Most outstanding is the complex centered on 30 Doradus (see **illus.**), an intricate, extremely filamentary gaseous nebula with graceful loops suggestive of supernova remnants or stellar winds. Its angular diameter of 7′ corresponds to a linear diameter of 80 pc, 265 light-years, 1.7×10^7 astronomical units, 1.6×10^{15} mi, or 2.5×10^{15} km). It is larger than any optically observed H II (ionized hydrogen) region in the Milky Way Galaxy or local group. The entire complex is probably excited by a compact group of very massive stars. The Small Magellanic Cloud nebulosities are less impressive, but here, as in the Large Magellanic Cloud, the presence of luminous bright blue stars in H II regions and of supergiants shows that a star formation is occurring. *See Nebula; Supergiant star; Supermassive stars.*

Nebula centered on 30 Doradus in the Large Magellanic Cloud. (*Photograph by S. J. Czyzak and L. H. Aller at Cerno Tololo International Observatory*)

Other evidence, such as chemical composition determinations of stars and nebulosities, shows that the rate of element building, and therefore of star formation, is slower in the Large Magellanic Cloud than in the Galaxy and slower yet in the Small Magellanic Cloud (see **table**). Element building in the Clouds and in the Galaxy differs, and although the same kinds of stars are found in each, chemical composition differences promise to throw significant light on the problem of stellar evolution. *See Nucleosynthesis; Star; Stellar evolution.*

The most exciting event observed in the clouds has been supernova 1987A, a stellar explosion in a normal-appearing blue supergiant, Sanduleak −60°202. It attained a luminosity of 2×10^8 suns about 80 days after the outburst in the late February 1987. Intensive observational studies have confirmed elaborate theories of the supernova process. The light curve shows that in the late stages energy is derived from the decay of cobalt-56 to iron-56. A neutrino flux

Chemical composition of Magellanic Clouds and Milky Way Galaxy, in numbers of atoms per 10^6 hydrogen atoms

Galaxy	He	C	N	O	Ne	S	Ar	Fe
Large Magellanic Cloud	85,000	80	9.5	256	54	9	1.8	15
Small Magellanic Cloud	83,000	14.5	3.3	118	30	3	0.6	4
Milky Way Galaxy*	100,000	460	95	810	112	17	3.7	35

*Milky Way Galaxy values are taken mostly from solar data.

consistent with theoretical predictions was also found. The supernova radiated x-rays produced by the degradation of gamma rays as cobalt-56 decayed to iron-56. SEE SUPERNOVA.

Lawrence H. Aller

Bibliography. P. Hodge and F. Wright, *Large Magellanic Cloud*, Smiths. Publ. 4699, 1967; P. Hodge and F. Wright, *Small Magellanic Cloud*, 1977; F. Kerr and A. Rodgers (eds.), *Galaxy and the Magellanic Clouds*, 20th International Astronomical Union Symposium, 1964; A. B. Muller (ed.), *Magellanic Clouds*, 1971; S. Van den Bergh and K. DeBoer (eds.), *Structure and Evolution of the Magellanic Clouds*, International Astronomical Union Symposium 108, 1983.

Magnetohydrodynamics

The study of the slow, large-scale motions that occur in electrically conducting fluids, usually plasmas.

Definition. A plasma is a large collection of charged particles of different species such as electrons and different kinds of ions. In general, it is necessary to describe a plasma by completely specifying how each species of particle is distributed with respect to position and velocity. This rather complicated description can be greatly reduced if the plasma is sufficiently collisional. Then for each type of particle the distribution at each point is a maxwellian distribution characterized by a local pressure, mass velocity, and density. The magnetohydrodynamic description of the plasma characterizes it as a fluid with a density, mass velocity, and temperature. The main difference between a magnetohydrodynamic fluid and an ordinary fluid is the presence of a magnetic force. Thus, in addition to normal fluid properties, magnetohydrodynamics involves the magnetic field. However, this magnetic field is controlled by plasma motions in a simple way.

Magnetic fields are describable by magnetic lines of force that thread the magnetohydrodynamic fluid, and whose direction gives the direction of the field, and whose density gives the field strength. The great simplicity introduced by magnetohydrodynamics, at least in its ideal limit, is that these lines can be thought of as frozen into the fluid and bodily carried around by fluid motions (**Fig. 1**). In addition, the magnetic force on the fluid can be pictured in terms of the field lines. The field lines tend to move the fluid in such a way as to straighten themselves out when they are curved and to smooth out their density when they are crowded together (**Fig. 2**). Thus, magnetohydrodynamics provides a very qualitative and useful method for interpreting complicated situations.

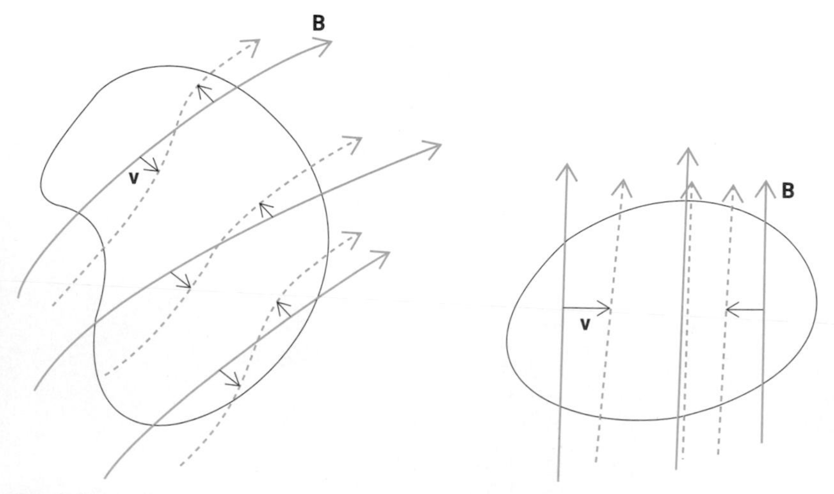

Fig. 1. Two examples of motion of magnetic field lines in a plasma. Initial magnetic field lines B (solid lines) are moved by fluid motions v (arrows) to new magnetic field lines (broken lines).

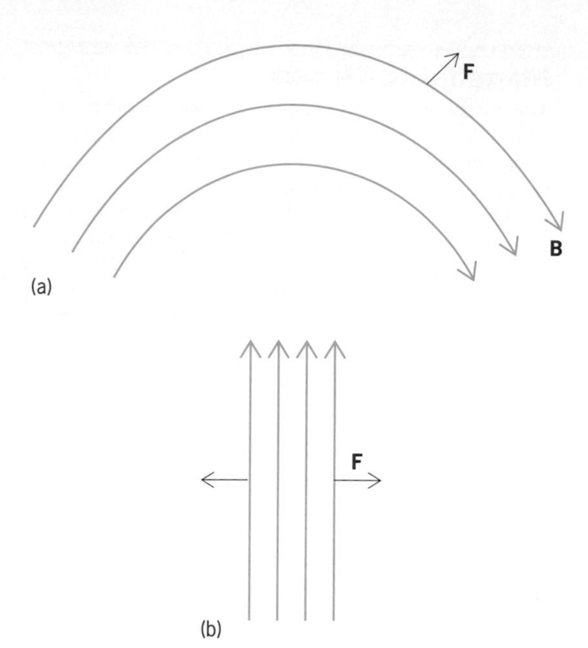

Fig. 2. Examples of magnetic field lines B exerting a force F (arrows) on a plasma. (*a*) Sharply curved lines of force exert a force in the convex direction. (*b*) Compressed field lines exert an outward force.

Examples. Magnetohydrodynamics plays an important role in many astrophysical processes.

Star formation. An example of magnetohydrodynamics is the process of magnetic braking that occurs during formation of a star by collapse of an interstellar cloud of gas. The cloud begins to spin rapidly as it collapses because of conservation of angular momentum. This spinning is fast enough to stop the collapse and to prevent the formation of a star. However, the gas is initially threaded by lines of force that are straight and connect the cloud to the surrounding interstellar medium. As the cloud collapses and rotates relative to the surrounding media, the lines of force become twisted like a spring. The magnetic force on the cloud resulting from the twisted lines then reduces the cloud's rate of rotation and allows the cloud to collapse. In this manner, magnetic fields aid the formation of a star (**Fig. 3**).

On the other hand, in later stages of collapse the field lines are drawn closer and closer together in the cloud relative to their spacing in the surrounding region. The magnetic field then exerts an outward force that resists this intensification described by the compression of its lines. This outward force would halt the cloud collapse at too early a phase for star formation. However, the cloud itself is only partly ionized. The true magnetohydrodynamic fluid in which the field is embedded is actually only the ionized part of the cloud medium. Under the normal conditions for stellar formation, the neutral material can still collapse by slipping through the ionized material at a rate controlled by friction between the neutral and ionized components, leaving the magnetic field behind. This process, called ambipolar diffusion, is considered to be essential to star formation. *SEE PROTOSTAR; SOLAR SYSTEM; STELLAR EVOLUTION*.

This example illustrates very clearly the manner in which the behavior of magnetohydrodynamic fluids can be understood by purely qualitative means. The behavior of the field is traced through the freezing of the magnetic field lines in the fluid. The magnetic forces are understood through their curvature or winding up as a spring and through their compression.

Coronal heating and solar flares. A very important physical concept in magnetohydrodynamics is that of magnetic energy. When magnetic fields are present, large amounts of energy can be transferred from kinetic, pressure, and gravitational energy into magnetic energy and stored in this form, possibly for later explosive release.

This process is illustrated nicely by the phenomena of coronal heating and the solar flare. The coronal region of the Sun is threaded by magnetic field lines that are anchored in the photosphere. In general, the magnetic energy density of the magnetic field in the corona is much larger than the thermal and gravitational energy densities, so that the total magnetic forces must nearly balance among themselves with only a small net force exerted on the coronal fluid. This state is called force-free equilibrium. That is to say, the forces arising from the curvature of the lines must everywhere nearly balance forces arising from compression of lines. Now, as motions occur in the photosphere, the magnetic lines, which are anchored in the photospheric material, are forced to move with it by flux freezing. By contrast, in the photosphere motions dominate the magnetic forces. As a result, the equilibrium balance of magnetic forces in the corona is disturbed and new force-free equilibria are set up. These equilibria generally have more twisted lines of force, with larger magnetic energy density. In this manner, energy is transferred from the photosphere to the corona and stored in magnetic form.

If there were nonmagnetohydrodynamic processes by which this magnetic energy could be released through violation of the flux-freezing constraint, then considerable heating of the corona would occur. The search for such processes is an active field. Of even greater interest is the question whether energy could be released on a very short time scale. If this release were possible, it would lead to a viable explanation of the solar flare phenomena. Such processes, if they exist, are termed magnetic reconnection because, during any such transformation of the magnetic field, there will be many places where a magnetic line will break and its two ends will be reconnected to other neighboring magnetic lines of force. The actual mechanism that leads to a breakdown of normal magnetohydrodynamics is believed to be associated with instabilities that lead to rapid evolution of the field on small length scales. Magnetohydrodynamics can break down on such small scales and allow rapid reconnection of the lines of force. *SEE SOLAR CORONA; SUN*.

Validity. The role that magnetic fields play in astrophysics can be understood through the simple and elegant picture presented by normal magnetohydrodynamics. However, the really interesting phenomena are associated with conditions under which the flux-freezing constraint of magnetohydrodynamics is violated. The conditions under which magnetohydrodynamics is valid are that (1) there are sufficient collisions of the charged particles so that the plasma behaves as a fluid, and (2) the electrical conductivity of the plasma is large enough so that the field lines are controlled by flux freezing. Astrophysical plasmas have approximately the same conductivity as metals. The electrical conductivity is large for flux freezing if, during any relevant time t, the corresponding skin depth of the plasma is small, since this is the distance that the lines of force can slip relative to the plasma. For a plasma of 10,000 K, the skin depth δ is approximately $3 \times 10^{-2}\sqrt{t}$ kilometers ($2 \times 10^{-2}\sqrt{t}$ miles), where t is expressed in seconds. If this dis-

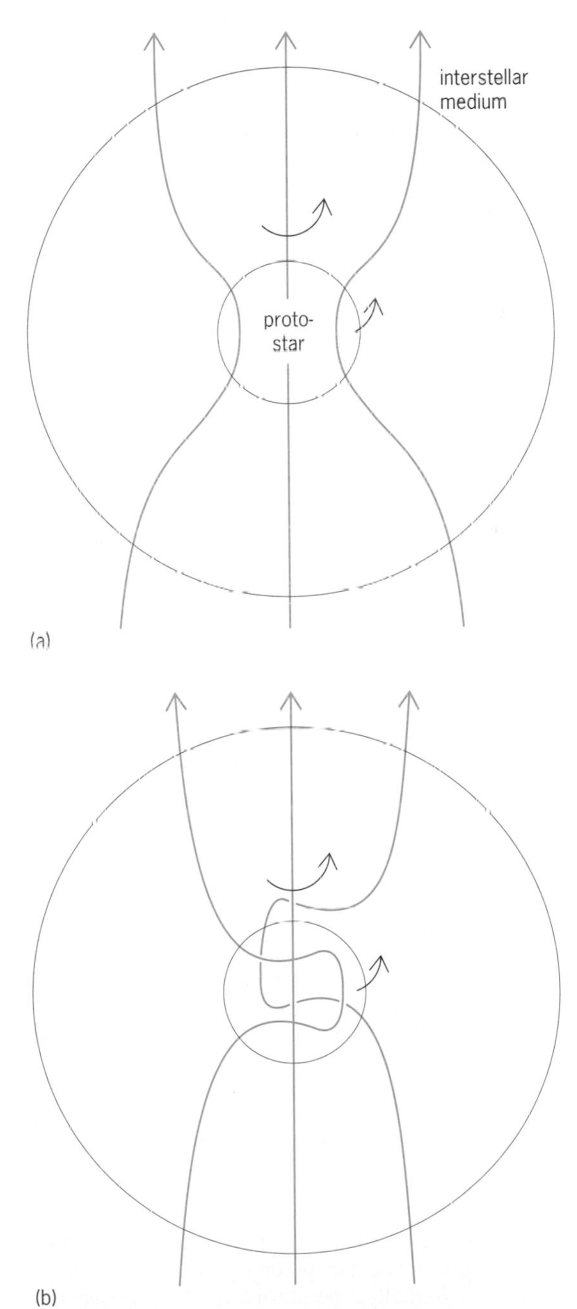

Fig. 3. Magnetic braking during star formation. (*a*) Initially, magnetic field lines B are connected between the rotating protostar and the interstellar medium. (*b*) After a short time these lines become twisted, transferring angular momentum from the rotating protostar to the interstellar medium, so that further collapse occurs.

tance is unimportantly small, the lines may be considered as frozen in the plasma. As an example, for an interstellar cloud at 10,000 K, and t equal to the Hubble time (10^{10} years or 3×10^{17} s), the skin depth δ is of the order of 10^7 km (10^7 mi). Since this is an extremely small distance on interstellar scales, flux freezing is generally valid for the interstellar medium. For the solar corona, the temperature is 10^6 K, and if t is taken to be a solar rotation period, δ is about 5 km (3 mi). (The skin depth is proportional to the temperature to the $-3/4$ power.) Only the presence of instabilities that can cause variations on such a small scale could lead to the breakdown of magnetohydrodynamics. Theories of magnetic reconnection are concerned with finding such instabilities. SEE INTERSTELLAR MATTER.

Instabilities. The subject of magnetohydrodynamic instabilities is very important. The most significant ones are the interchange instability, the kink instability, and the Parker instability.

The interchange instability, which is very similar to the Raleigh-Taylor instability in fluids, occurs when a plasma is supported against gravitational forces by magnetic fields. Since it is legitimate for this purpose to consider the magnetic field as a light fluid, in effect a heavy fluid, the plasma, is supported by a light fluid, the magnetic field. A motion that interchanges two magnetic tubes of force, and the fluids tied to them, will lower the plasma in the gravitational potential while leaving the magnetic energy unchanged, leading to a transformation of energy into kinetic energy. Even a small initial motion will be amplified exponentially on a free-fall time scale. Small-scale motions tend to be amplified fastest, and a very chaotic situation quickly results, with variations of the magnetic field on very short scales. This situation can rapidly lead to breakdown of magnetohydrodynamics. The interchange instability also occurs when plasma pressure is supported by curved field lines, the role of gravity being taken by the centrifugal acceleration of thermal particles moving along the curved field lines. Such instabilities occur in the corona when prominences are supported by coronal magnetic fields, or around magnetized neutron stars when plasma is being rained down on them too rapidly by the surrounding accretion disk.

The kink instability is associated with lines of force too tightly twisted about each other. It characteristically happens in the corona when lines become twisted by motions in the photosphere. The instability is very similar to a buckling instability that attempts to relieve the too-tight twisting of flexible wires. Very likely this instability is closely associated with the solar-flare phenomena.

The Parker instability is large-scale and occurs in the galactic disk if the interstellar material is unusually compressible. It is very similar to the instability that occurs when one places clothes on a line with hangers. The hangers tend to slip toward the center of the clothes line, depressing it further so that the hangers slip more. The magnetic field plays the role of the clothes line while the plasma plays the role of the hangers. Normally, the plasma is able to resist slipping down the depressed magnetic field by the increase in pressure produced by the compression. However, if cooling of the plasma is rapid enough, the plasma becomes very compressible and this resistance is removed. The Parker instability can then occur. This instability is believed to be one of the chief mechanisms for the formation of new giant molecular clouds. SEE MOLECULAR CLOUD.

The above examples indicate how various magnetohydrodynamic processes can be reasonably understood in a qualitative manner by employing the concepts of flux freezing and magnetic lines of force. In terms of them, many astrophysical phenomena become comprehensible. SEE ALFVÉN WAVES.

Russel M. Kulsrud

Bibliography. H. Alfvén and C.-G. Fälthámmar, *Cosmical Electrodynamics,* 2d ed., 1963; T. G. Cowling, *Magnetohydrodynamics,* 2d ed., 1976; E. N. Parker, *Cosmical Magnetic Fields,* 1979; E. R. Priest, *Solar Magneto-hydrodynamics,* 1982.

Magnetosphere

A comet-shaped cavity or bubble around the Earth, carved in the solar wind. This cavity is formed because the Earth's magnetic field represents an obstacle to the solar wind, which is a supersonic flow of plasma blowing away from the Sun. As a result, the solar wind flows around the Earth, confining the Earth and its magnetic field into a long cylindrical cavity with a blunt nose (**Fig. 1**). Since the solar wind is a supersonic flow, it also forms a bow shock a few earth radii away from the front of the cavity. The boundary of the cavity is called the magnetopause. The region between the bow shock and the magnetopause is called the magnetosheath. The Earth is located about 10 earth radii from the blunt-nosed front of the magnetopause. The long cylindrical section of the cavity is called the magnetotail, which is on the order of a few thousand earth radii in length, extending approximately radially away from the Sun. SEE SOLAR WIND.

The concept of the magnetosphere was first formulated by S. Chapman and V. C. A. Ferraro in 1931, but the term magnetosphere was introduced in 1959 by T. Gold, who defined it as "the region above the ionosphere in which the magnetic field of the Earth has a dominant control over the motions of gas and fast charged particles." It was, however, only during the 1970s that the magnetosphere was extensively explored by a number of satellites carrying sophisticated instruments. This exploration was a part of the effort to understand that region of the Earth's environment above the level where most meteorological phenomena take place, namely, the space between the Sun and the stratosphere/mesosphere. The space which surrounds the magnetosphere is often referred to as geospace.

Plasma. The satellite observations have indicated that the cavity is not an empty one, but is filled with plasmas of different characteristics (Fig. 1). Just inside the magnetopause some solar wind plasma blows; this particular plasma region is called the plasma mantle. There are also two funnel-shaped regions (one in each hemisphere), extending from the front magnetopause to the polar ionosphere, which are also filled with solar wind plasma; these regions are called the cusp. The extensive tail region of the magnetosphere is divided into the northern and southern halves by a thin sheet of plasma, called the plasma sheet. The distant tail region (beyond the lunar distance) has been infrequently explored when it was traversed by space probes on their way to the planets. The inner magnetosphere is occupied by the Van Allen belts, the ring current belt, and the plasmasphere. All of them are doughnut shaped and surround the Earth. The ring current belt is colocated with the Van Allen belts and is so named because it carries a large amount of a westward-directed electric current around the Earth.

It is generally believed that the main part of the plasma sheet is of solar origin; the mantle plasma may be an important source of it. The plasmasphere is formed mainly by ionospheric plasma which diffuses outward along magnetic field lines. At times the ring current belt contains heavy ions of ionospheric origin, such as helium and oxgen ions, as well as protons of solar origin. The heavy ions are found also in the plasma sheet.

The Earth's dipolar magnetic field is considerably deformed by these plasmas and the electric currents generated by them. The dipolar field is compressed in the dayside part of the magnetosphere, mainly by the action of the impacting solar wind, whereas it is considerably stretched along the equatorial plane in the magnetotail (Fig. 1).

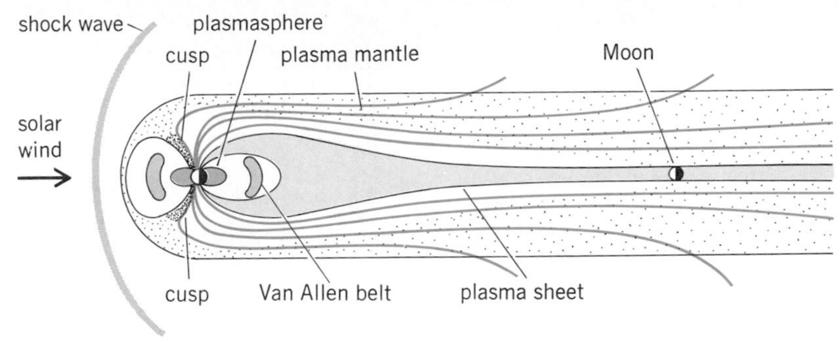

Fig. 1. Noon-midnight cross section of the magnetosphere and the distribution of various plasmas inside it.

Solar wind–magnetosphere generator. If there were no magnetic field in the solar wind, the Earth's magnetic field would simply be confined by the solar wind into a teardrop-shaped cavity (rather than a comet-shaped cavity), and the cavity would be almost empty. It is the magnetic field of the solar wind which introduces considerable complexity in the interaction between the solar wind and the magnetosphere. Some of the magnetic field lines originating from the polar region are almost always connected to the solar wind magnetic field lines across the magnetopause (**Fig. 2**). In fact, because the solar wind magnetic field lines originate from the Sun, it may be said that the Sun and the Earth are connected by magnetic field lines. More specifically, the connected field lines originate from an oval area in the polar region. This area is called the polar cap and is bounded by a belt of the aurora called the auroral oval. The magnetic field lines originating from the northern polar cap are stretched along the northern half of the magnetotail as a bundle of magnetic field lines and are connected to the solar wind magnetic field lines across the magnetopause. Similarly, the magnetic flux from the southern polar cap occupies the southern half of the magnetotail.

This connection between the solar wind magnetic field lines and the Earth's magnetic field lines provides the most important aspect of the solar wind–magnetosphere interaction. Since the solar wind (a fully ionized plasma and thus a conductor) must blow across the connected magnetic field lines as it flows along the magnetopause (Fig. 2), an electromotive force is generated. This force powers the solar wind–magnetosphere generator and thus provides practically all the electric power for various magnetospheric processes, such as auroral phenomena and magnetospheric disturbances, including geomagnetic storms. It is precisely in this way that the solar wind (a flow of magnetized plasma) couples its energy to the magnetosphere. Therefore, if there were no connection of the field lines, there would be no energy transfer from the solar wind to the magnetosphere, and thus there would be no auroral phenomena, which are one of the most spectacular wonders of nature.

Because the solar wind–magnetosphere generator is operated on the magnetopause, where the solar wind blows across the connected field lines, the whole cylindrical surface of the magnetotail can be considered to be the generator in which the current flows from the duskside magnetopause to the dawnside magnetopause along the cylindrical surface. Much of the current thus generated flows in the plasma sheet across the magnetotail; this part of the circuit may be called the cross-tail circuit (**Fig. 3**). Thus, the magnetotail may be viewed as two long solenoids, producing two bundles of magnetic field lines separated by the plasma sheet. As a result, the Earth's dipolar field is stretched along the magnetotail in the antisolar direction. The generator also powers a large-scale motion of plasmas in the magnetosphere. In the main part of the plasma sheet, the plasma flows toward the Earth.

Auroral phenomena. Another important circuit that is connected to the generator consists of a current flow along magnetic field lines from the morning side of the magnetotail to the morning half of the auroral oval, a flow along the auroral ionosphere (the auroral oval), and a flow along magnetic field lines from the evening half of the auroral oval to the evening side of the magnetotail. This discharge circuit may be called the auroral circuit because it is responsible for causing acceleration of current-carrying charged particles

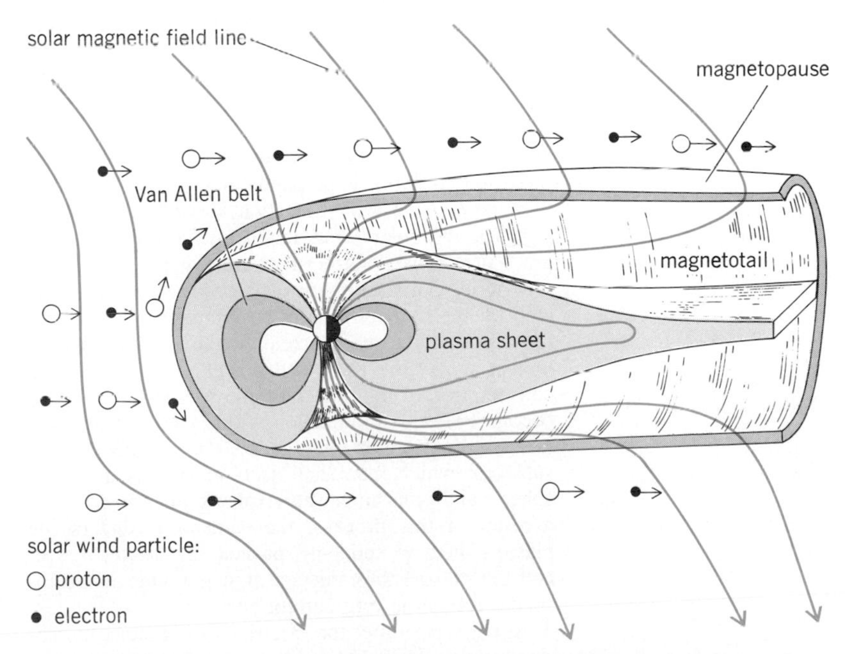

Fig. 2. Internal structure of the magnetosphere (noon-midnight cut). Motions of solar wind protons and electrons around the magnetopause and the linkage of the solar wind magnetic field lines and Earth's magnetic field lines are also shown.

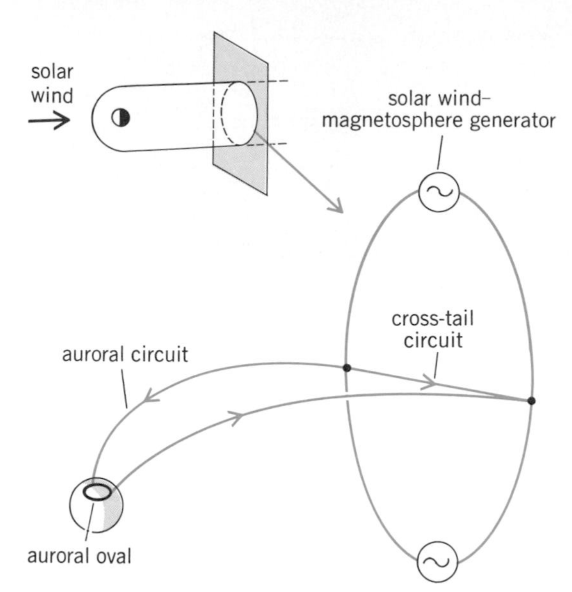

Fig. 3. The solar wind–magnetosphere generator located on the magnetopause (one cross section is shown) and its two main circuits: the cross-tail circuit and the auroral circuit.

(Fig. 3). When the accelerated particles collide with upper atmospheric particles, they excite or ionize them. The auroral lights are emitted by those excited or ionized particles when they return to a lower or the ground state.

The total power generated by the solar wind–magnetosphere generator depends on the magnitude and direction of the solar wind magnetic field, as well as on the speed of the solar wind. In particular, when the solar wind magnetic field is directed southward, the generated power becomes maximum. On the other hand, when the solar wind magnetic field is directed northward, the power becomes minimum. The power is also greater if the magnitude of the solar wind magnetic field B and the speed of the solar wind V are higher. The power P is given by the equation below,

$$P = VB^2 \sin^4 \left(\frac{\theta}{2}\right) l_0^2$$

where θ is the polar angle of the solar wind vector projected to a plane perpendicular to the Sun-Earth line and l_0 is a constant of about 7 earth radii.

When the solar wind magnetic field vector turns gradually from a northern to a southern direction in the vicinity of the magnetosphere, the power generated by the dynamo begins to increase. As a result, the cross-tail electric current in the plasma sheet is increased. As the generated power is very high, the consumption of the power by the magnetosphere also becomes high. An indication of this high consumption rate is the phenomenon called the magnetospheric substorm which manifests itself in various magnetospheric and polar upper-atmospheric disturbances. As a result of this process, the earthward edge of the plasma sheet is suddenly pushed toward the Earth, and the plasma thus injected into the inner magnetosphere forms the ring current belt.

At the same time, the electric current along the auroral circuit is also intensified. As a result, the aurora becomes bright and active. The enhanced ionospheric portion of the auroral circuit is called the auroral electrojet. Intense magnetic field variations are experienced under and near the auroral electrojet. In this way, effects of the enhanced power of the solar wind–magnetosphere generator are manifested most conspicuously in the polar region. An intense auroral activity associated with this phenomenon is called the auroral substorm. The associated intense magnetic disturbances are called polar magnetic substorms. The ionosphere is also greatly disturbed in this period, and this phenomenon is called the ionospheric substorm. The ionospheric disturbances greatly affect high-frequency and very high-frequency radio communication in high latitudes. Even satellite microwave communication is at times disturbed by auroral activity.

Because the Earth is a good conductor, the auroral electrojets induce a potential difference on the Earth's surface. Therefore, if two points, along which the potential difference is present, are connected by a conductor, an electric current flows along it. It is for this reason that electric currents are induced in oil and gas pipelines and in power transmission lines in high latitudes during auroral activity. In the power lines there are severe power fluctuations when the electrojets are particularly intense.

Various solar activities are responsible for the time variations of the three solar wind quantities which determine the power of the generator. In particular, an intense explosion on the solar surface, a solar flare, causes a gusty solar wind and a large-scale deformation of the solar wind magnetic field. If this deformation happens to produce a large southward-directed magnetic field around the magnetosphere for an extended period (approximately 6 to 24 h), the generated power is considerably enhanced, and intense magnetospheric substorms occur. As a result, an unusually intense ring current grows, carrying a westward current surrounding the Earth and producing a southward-directed magnetic field around the Earth. Since the Earth's magnetic field is directed northward, this addition of the southward field is observed as a decrease of the Earth's magnetic field, one typical aspect of geomagnetic storms.

The Sun has also a few regions from which a high-speed stream is generated, particularly during the declining epoch of the 11-year sunspot cycle. Such a stream lasts for an extended period, from several months to more than 1 year. The solar wind magnetic field is markedly variable in the high-speed stream. As the Sun and the stream rotate with a period of about 27 days, the Earth is immersed in the stream once each period. As a result, the power of the solar wind–magnetosphere dynamo is increased and fluctuates considerably, causing magnetospheric substorms and auroral activity. *SEE SUN; SUNSPOT.*

Planets. All other magnetic planets, such as Mercury, Jupiter, and Saturn, have magnetospheres which are similar in many respects to the magnetosphere of the Earth. The magnetosphere of Mercury is much smaller than that of the Earth. Since the solar wind pressure is higher at Mercury's than at the Earth's distance and since the magnetic field of Mercury is weaker than that of the Earth, the blunt-nosed front of the magnetosphere of Mercury is located at a distance of only 0.5 Mercury radius from its surface. On the other hand, because of a weak solar wind, a very strong magnetic field, and a fast rotation, the magnetosphere of Jupiter is gigantic and has an extensive current disk which lies along its equator. The distance to the blunt-nosed front is about 100 Jupiter radii. The

Voyager 2 space probe found that the magnetosphere of Jupiter is filled with sulfuric ions, presumably injected from one of its satellites, Io. The space probe also traversed its gigantic magnetotail. The magnetosphere of Saturn is a little smaller than that of Jupiter, but extends well beyond its rings. *SEE JUPITER; MERCURY; PLANETARY PHYSICS; SATURN.*

A great variety of plasma processes, plasma waves, and plasma instabilities are involved in all magnetospheric phenomena. Some of them occur commonly in a thermonuclear machine. The magnetosphere provides a natural laboratory in which these plasma behaviors can be studied without the so-called wall effects. Among a variety of astrophysical conditions, the magnetospheres of the Earth and the planets are the only regions in which plasma behaviors can be studied in natural conditions. Therefore, a study of the magnetosphere provides an important foundation upon which astrophysical plasma theories can be tested and constructed. It has also been speculated that many magnetic stars, pulsars, and even some galaxies have structures similar to the Earth's magnetosphere. Since a magnetosphere results from an interaction of magnetized plasma flow and a magnetized celestial body, it is likely that it is of common occurrence in galaxies. *SEE MILKY WAY GALAXY; PLASMA PHYSICS; PULSAR.*

S.-I. Akasofu

Bibliography. S.-I. Akasofu, *Physics of Magnetospheric Substorms*, 1977; S.-I. Akasofu and S. Chapman, *Solar-Terrestrial Physics*, 1972.

Magnitude

The numerical scale on which the brightnesses of astronomical objects are expressed. The stellar magnitude scale is a logarithmic scale and is inverted in that fainter objects have larger magnitudes. The term magnitude in astronomy refers to brightness only and has nothing to do with size or other quantities. A distinction is made between the apparent magnitude of an object viewed from the Earth and its absolute magnitude, which measures the object's intrinsic luminosity by indicating its apparent magnitude when viewed from a standard distance. Since the brightness of any object varies with wavelength, many different magnitude scales have been defined corresponding to different spectral regions, bandwidths, and methods of observation; visual magnitudes, corresponding to the sensitivity of the human eye centered in the yellow part of the spectrum, are usually implied if the type is unspecified. Brightness measurements of the Sun, planets, asteroids, nebulae, galaxies, and even background radiation are often expressed in terms of stellar magnitudes.

Development. The star catalog of Hipparchus (about 150 B.C.) is thought to have contained approximately 850 naked-eye stars classified according to brightness. The 15 or so brightest stars were referred to as stars of the first magnitude, while second-magnitude stars were on the average two or three times fainter, and so on. The scale is logarithmic because intervals that are perceived by the human eye as equal intervals are, in fact, equal brightness ratios.

Measurements of brightness ratios in the nineteenth century showed that, on average, stars of the sixth magnitude (near the limit of naked-eye vision) were about 100 times fainter than those of the first. On the scale introduced by N. R. Pogson in 1856 and universally adopted, an interval of 5 magnitudes corresponds to a factor of exactly 100, so that each magnitude corresponds to a factor of $\sqrt[5]{100} \approx 2.512 \ldots$. The zero point of the Pogson scale was set so that most stars retained their customary magnitudes.

Mathematical properties. The stellar magnitude scale has many convenient properties. Its logarithmic character means that brightness ratios can be calculated by simple subtraction, and that the entire range of astronomical brightnesses can be represented in terms of small numbers, ranging from -27 (the apparent magnitude of the Sun) to approximately $+27$ [the faintest objects appearing on deep images taken with charged-coupled devices (CCDs) on large telescopes]. Its inverted nature seems natural to observers who are apt to think in terms of a star's faintness, and it is particularly convenient in that the most familiar stars have magnitudes near zero. *SEE CHARGE-COUPLED DEVICES.*

The correspondence between magnitude interval and brightness ratio is shown in **Table 1**. In general, an interval of n magnitudes corresponds to a factor of 2.512^n in brightness. A factor of 10 corresponds to exactly 2.5 magnitudes, a factor of 100 to exactly 5 magnitudes. Any linear measure of brightness X can be expressed as a magnitude m through Eq. (1),

$$m = -2.5 \log_{10} X + \text{constant} \qquad (1)$$

where the negative sign is needed to produce an inverted scale and the constant can be evaluated by measuring X for any star of known magnitude.

An attractive feature of the magnitude scale is the ease with which fractional magnitudes can be interpreted. Each change of 1% in the brightness of an object corresponds to a change of 0.01 in the magnitude, and this numerical correspondence holds to good accuracy for changes up to about 30%.

Absolute magnitude. The apparent magnitude m of a star observed from the Earth depends upon both its absolute magnitude M—an intrinsic property of the star—and its distance d from Earth. The difference $m - M$ between the apparent and absolute magnitude is thus a function of the distance, given by Eq. (2),

$$m - M = 5 \log_{10} d - 5 \qquad (2)$$

Table 1. Magnitude scale

Magnitude difference	Brightness ratio	
0.00	1	
0.01	1.01	
0.05	1.05	
0.30	1.32	
0.753	2	
1.000	2.512 . . .	$(= \sqrt[5]{100})$
1.5	~4	
2	6.31 . . .	$(= 2.512^2)$
2.5	10	exactly
3	~16	$(= 2.512^3)$
3.5	25.12 . . .	
4	~40	
5	100	exactly
10	10^4	
15	10^6	
50	10^{20}	

and is sometimes called the distance modulus. Here the distance d is expressed in parsecs (1 pc = 3.26 light-years = 1.92×10^{13} mi = 3.09×10^{13} km). The constants in the formula come from the inverse-square law of light, the definition of a magnitude, and the choice of $d = 10$ pc as the standard distance at which the apparent and absolute magnitudes are defined to be equal. Thus the absolute magnitude may be defined as the apparent magnitude an object would have if viewed from a distance of 10 pc. *SEE PARSEC.*

The absolute magnitude can thus be found for any astronomical object whose distance has been measured (**Table 2**). The Sun has an absolute visual magnitude of +4.79; that is, it would appear as about a fifth-magnitude star to an observer 10 pc away. The brightest stable stars in the Milky Way Galaxy and in external galaxies have absolute magnitudes about −9, while the faintest are near +20 or even fainter. Supernova 1987A in the Large Magellanic Cloud ($m - M \approx 18.5$) reached an absolute magnitude of −15.5 at its maximum of May 1987; more typical supernovae attain even higher luminosities. *SEE STAR; SUPERNOVA.*

It is often possible to assign an absolute magnitude to a star on the basis of its spectral classification, in which case its distance can be calculated from Eq. (2). The problem of determining absolute magnitudes is thus equivalent to that of determining distances.

Absorption by interstellar dust affects the apparent magnitudes of distant stars, particularly those lying in the plane of the Milky Way. If the absorption A is expressed in magnitudes, Eq. (2) may be rewritten as Eq. (3).

$$m - M = 5 \log_{10} d - 5 + A \tag{3}$$

SEE INTERSTELLAR EXTINCTION.

Visual magnitudes. The scale of visual magnitudes was originally established with the dark-adapted eye as the detector. Visual magnitudes can be measured photographically (in which case they are called photovisual) by using an emulsion sensitive to yellow light and a filter to block the shorter wavelengths. However, most modern measurements of visual magnitudes are done photoelectrically with a combination of photocell and filter that mimics the response of the human eye; the symbol V is reserved for photoelectrically determined visual magnitudes. Eye estimates are often employed to monitor the visual magnitudes of variable stars. *SEE VARIABLE STAR.*

Photographic magnitudes. Magnitudes of stars can be determined from image sizes on astronomical photographs. Magnitudes can be found by interpolation if they are known for a sequence of stars appearing on the same photograph. The advantage of photographic photometry is that many thousands of stellar images can often be recorded on a single exposure.

The basic photographic emulsion is sensitive to blue light, and the term photographic magnitude (designated m_{pg}) specifically means a wideband blue magnitude that includes the near-ultraviolet but excludes the yellow region and all longer wavelengths. Star catalogs usually list either m_{pg} or the visual magnitude m_v. When both are known, the difference can be used as a quantitative measure of the color of the star, given by Eq. (4). The zero point of the m_{pg} scale has

$$\text{Color index} = m_{pg} - m_v \tag{4}$$

been set so that $m_{pg} - m_v = 0.0$ for stars similar to Vega, of spectral type A. The color index is positive for stars redder than Vega, negative for bluer stars. Although colors can be affected by interstellar absorption, they serve primarily as indicators of surface temperature; cooler stars have redder (numerically larger) colors. *SEE ASTRONOMICAL PHOTOGRAPHY.*

Multicolor photoelectric photometry. Magnitudes can be measured accurately and conveniently with a photomultiplier tube and a filter to limit the bandpass. Stars must normally be measured one at a time, but it is straightforward to measure each star through several (say n) filters, obtaining a magnitude and $n - 1$ color indices. The various color indices of a star are normally well correlated, but they often provide additional information because of the presence of absorption features in the stellar spectrum. An accuracy of 0.01 magnitude or better is routinely achieved in this work.

Many multicolor photometric systems have been defined with wide, intermediate, and narrow bandpass filters. One of the first and still the most widely used is the wideband UBV (ultraviolet, blue, visual) system introduced by H. L. Johnson about 1950. The B and V filters are the photoelectric equivalents of m_{pg} and m_v, respectively (the most significant difference being that the B filter excludes the ultraviolet light that contributes to the photographic magnitude), and B − V has replaced $m_{pg} - m_v$ as the general-purpose color index most commonly employed. The addition of the U filter, placed in a spectral region strongly affected by absorption by neutral hydrogen in hot stars, makes it possible to determine the intrinsic properties of stars from UBV photometry alone, even in the presence of interstellar absorption.

Measurements in red and near-infrared (R,I) filters are often added to UBV photometry, especially when

Table 2. Apparent and absolute visual magnitudes and distances of familiar stars

Name	m	M	d, pc	Comment
Sun	−26.78	4.79	$\frac{1}{206,265}$	
Barnard's star	9.54	13.25	1.81	Nearby red dwarf
Sirius	−1.46	1.42	2.65	The brightest star
Arcturus	−0.06	−0.3	11	Red giant
Betelgeuse	0.5 var.	−6.0	200	Cool supergiant
Deneb	1.25	−7.2	500	Very bright supergiant
SN 1987A (maximum)	3.0	−15.5	50,000	Supernova in Large Magellanic Cloud

Table 3. Filter characteristics for standard wideband multicolor photometric system

Filter	Central wavelength, μm	Bandwidth, μm	Flux density* for magnitude = 0.00	Thermal sources peaking in filter	
				Temperature, K†	Example
U	0.36	0.07	4350	8500	Fairly hot star
B	0.44	0.10	7200	7000	
V	0.55	0.08	3920	5500	Solar-type star
R	0.70	0.21	1760	4300	
I	0.90	0.22	830	3300	Typical red giant
J	1.25	0.4	340	2400	
H	1.6	0.5	120	1900	Very cool star
K	2.2	0.6	39	1400	
L	3.4	0.7	8.1	900	Circumstellar dust
M	5.0	1.2	2.2	600	
N	10	5	0.12	300	Room temperature

*Unit = 10^{-15} W·cm^{-2}·μm^{-1}.
†°F = (K × 1.8) − 460.

the object is to determine stellar temperatures without relying on the U and B filters, which are strongly affected by absorption lines in the stellar spectrum.

Infrared magnitudes. With the improvement of infrared detectors and the development of telescopes optimized for infrared work, Johnson was able to extend the wideband UBVRI photometric system to much longer wavelengths. Filters were chosen to match the windows between bands of atmospheric absorption. The 11 filters of the standard multicolor system are listed in **Table 3** with their central wavelengths and bandwidths (these are full widths at one-half the peak transmission) as originally defined by Johnson. Subsequent observers have often used somewhat narrower bands. The magnitude scale for each filter has an arbitrary zero point, and these have been set by requiring a typical star of type A0 (such as Vega) to have the same magnitude in each filter.

Applications of multicolor infrared photometry have included refinement of the stellar temperature scale, especially for cool stars, and the determination of bolometric magnitudes. Unlike other magnitudes, which refer to particular wavelength regions, the bolometric magnitude is an integrated quantity referring to the total radiative energy output of the star. That is, it represents, on the magnitude scale, the area under the flux curve. The absolute bolometric magnitude (referred to a distance of 10 pc) represents the intrinsic luminosity of the star. SEE INFRARED ASTRONOMY.

Robert F. Wing

Bibliography. A. A. Henden and R. H. Kaitchuck, *Astronomical Photometry*, 1982; H. L. Johnson, Astronomical measurements in the infrared, *Annu. Rev. Astron. Astrophys.*, 4:193–206, 1966; H. L. Johnson and W. W. Morgan, Fundamental stellar photometry, *Astrophys. J.*, 117:313–352, 1953; G. Neugebauer and R. B. Leighton, *Two-Micron Sky Survey*, NASA SP-3047, 1969; J. R. Percy (ed.), *The Study of Variable Stars Using Small Telescopes*, 1986.

Mars

The planet fourth in distance from the Sun. It is visible to the naked eye as a bright red star, except for short periods near its conjunctions with the Sun. Having a mean (synodic) period between oppositions of 780 days, Mars is in opposition with the Sun every other year. The main orbital elements are the semimajor axis (mean distance to the Sun) = 1.524 astronomical units = 142 × 10^6 mi = 228 × 10^6 km; eccentricity = 0.093, one of the largest of the main planets (this eccentricity causes the distance to the Sun to vary from 128 × 10^6 mi or 207 × 10^6 km at perihelion to 155 × 10^6 mi or 249 × 10^6 km at aphelion); sidereal period of revolution = 1.881 years = 686.98 days; mean orbital velocity = 15.0 mi/s = 24.1 km/s; and inclination of the orbital plane to the ecliptic = 1°51′. SEE PLANET.

Elements of the globe. The mean diameter of the globe of Mars is 4202 mi (6762 km), or about 53% that of the Earth. The planet's volume is therefore about 15% of Earth's. The mass of Mars, about 11% of Earth's, has been well determined, first by the motions of its natural satellites, and more recently with greater precision by the motion of spacecraft orbiting about the planet. The mean density is about 3.94 g/cm^3 (3.94 times that of water), less than that of any other terrestrial planet. The surface gravity is about 38% of the Earth's, or more precisely 12.3 ft/s^2 (375 cm/s^2) at the equator and 12.5 ft/s^2 (380 cm/s^2) at the poles. The escape velocity is about 3.1 mi/s (5.0 km/s). The rotation period is very accurately determined from several centuries of observation of the times of transits of surface markings at the central meridian of the disk; the sidereal period is 24h37m22.66s; the corresponding length of the mean solar day is 24h39m25.0s. It follows that there are about 646 Martian sidereal days and 645 mean Martian solar days in the planet's year. The inclination of the polar axis to a line perpendicular to the orbital plane is 25°.0, only slightly greater than that of the Earth, so the seasons are similar but longer and also more unequal because of the greater eccentricity of the orbit. In the northern hemisphere of the planet the spring lasts 199 (terrestrial) days; summer, 182 days; autumn, 146 days; and winter, 160 days. Further, because the passage at perihelion (which takes place at heliocentric longitude 357°), the cold season is longer and colder, and the warm season is shorter and hotter in this hemisphere. This asymmetry has a pronounced effect on the Martian climates and seasonal variations, especially in the polar regions.

Telescopic appearance. In a telescope, Mars usually appears as a bright reddish disk marked by complex, semipermanent dark regions and variable white polar caps. In general, surface albedo features are not visible on photographs taken in violet and ultraviolet light because of decreasing contrast between the light and dark regions at these shorter wavelengths. Scientists have long known of atmospheric phenomena, variable polar caps of ice and clouds, and large bright and dark regions on the surface with dimensions of several hundred kilometers or more. But a multitude of other small surface formations with characteristic scales less than 125 mi (200 km) manifest their existence by the optical illusion of ''canals'' and ''oases.'' These ephemeral features, some of which vary with the Martian seasons, were long thought by some astronomers to be due to vegetation responding to the Martian seasons. As revealed by Mariner and Viking spacecraft, however, these features are now known to be a complex of smaller, dark splotchy areas whose contrast changes with the shifting Martian sands driven by seasonal winds.

The apparent diameter of the disk varies from a minimum of 3″.5 at conjunction to a maximum of 25″.1 at the most favorable perihelic opposition, when the distance to the Earth is only 35×10^6 mi (56×10^6 km). At aphelic oppositions the apparent diameter may not exceed 13″.8. The most favorable oppositions with Mars (near its perihelion) occur every 15 years; those of the past century, during which astronomers closely scrutinized its surface, took place in 1877, 1892, 1909, 1924, 1939, 1956, 1971, and 1986.

As a ''superior'' planet (orbiting beyond the Earth), Mars presents only gibbous phases when it is not in opposition to or conjunction with the Sun; the maximum phase angle (Sun-Mars-Earth angle) is 48°, when Mars is in quadrature. The apparent visual magnitude of Mars in conjunction averages −1.5, but it brightens to −1.2 at perihelic opposition (Mars is then brighter than any other planet except Venus). SEE MAGNITUDE.

The visual albedo of Mars is about 0.15, half again that of the Moon or Mercury, but less than half the values for the planets surrounded by dense atmospheres. The reflectivity of Mars increases rapidly from a low value of only 5% in the near ultraviolet to more than 30% in the near infrared. This variation accounts for the reddish color of Mars. At ultraviolet wavelengths shorter than 0.3 micrometer, observable only from high-altitude rockets or satellites, the albedo increases toward shorter wavelengths because of an increasing contribution of scattering in the atmosphere from molecules and aerosol particles. In the infrared, at wavelengths between 1 and 2 μm, the albedo curve of Mars shows only slight variations. These resemble the spectral reflectivity curve of iron oxides, especially geothite and hematite. SEE ALBEDO.

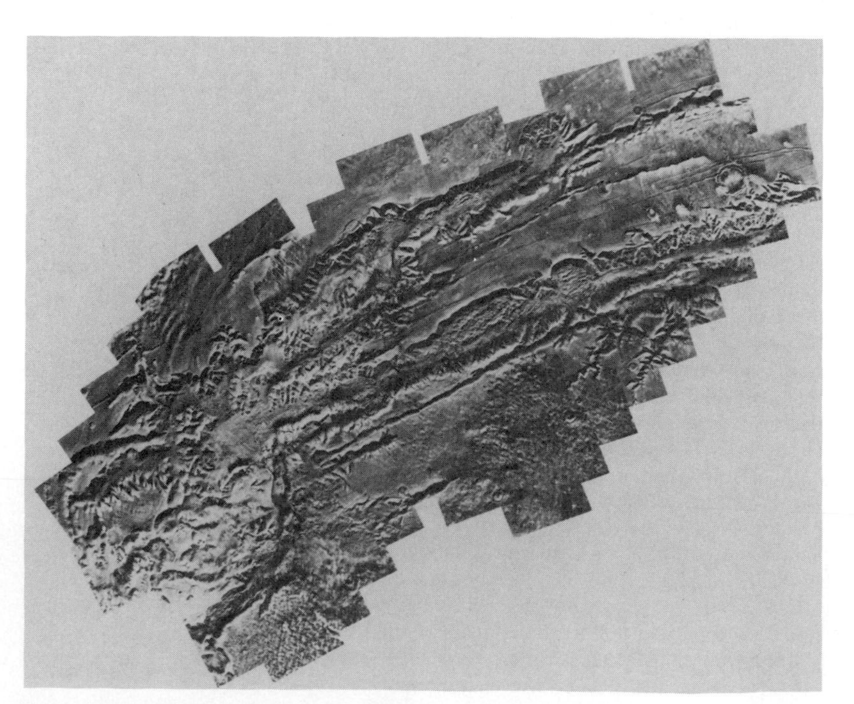

Fig. 1. **Valles Marineris, a huge canyon system on Mars, over 3100 mi (5000 km) long; photographed from *Viking Orbiter 1*. (NASA)**

Surface. The Martian surface, which initially resembled the heavily cratered lunar uplands, has been modified extensively by the processes of volcanism, faulting, and the effects of wind and water. The northern hemisphere tends to be characterized by large volcanic lava plains, whereas the southern hemisphere appears to be older and more heavily cratered. Several large impact basins associated with the final accretion of the planet are easily seen even in Earth-based telescopes. The largest of these, Hellas, is approximately 1250 mi (2000 km) across and at least 2.5 mi (4 km) deep. SEE MOON.

Volcanoes. Among the more impressive features on the Martian surface are several enormous shield volcanoes, larger than any mountains on Earth. The largest and perhaps the youngest is Olympus Mons, which is nearly 370 mi (600 km) across at its base and stands approximately 16 mi (26 km) above the surrounding terrain. Three other large shield volcanoes, Ascraeus Mons, Pavonis Mons, and Arsia Mons, lie along the nearby Tharsis Ridge. The Tharsis volcanoes are approximately 250 mi (400 km) across and 12 mi (20 km) high. These features may be compared with the largest volcanic constructs on Earth, Mauna Loa and Mauna Kea in Hawaii, which together are approximately 125 mi (200 km) across and stand 5.5 mi (9 km) above the ocean floor.

In shape and structure the Martian shield volcanoes bear a strong resemblance to their Hawaiian counterparts. At the summit of each shield is a complex of calderas, collapsed craterlike features that were once vents for lava. The central smooth-floored caldera of Arsia Mons is the largest, about 80 mi (130 km) in diameter. Studies of Viking data suggest this caldera may be the youngest surface feature on the planet, perhaps less than 300×10^6 years old. About 3000 mi (5000 km) to the west of the Tharsis area is another volcanic field, dominated by the symmetrical shield volcano, Elysium Mons, approximately 140 mi (225 km) across. Many other volcanoes with dimensions of up to 60 mi (100 km) and in various states of preservation are found scattered over the Martian surface, mostly in the northern hemisphere. However, while no large, young volcanoes appear in the southern hemisphere, volcanic activity has, nonetheless, played a significant role there. Close examination of spacecraft images shows that nearly one-third of this hemisphere is covered by relatively smooth plains of volcanic origin.

A volcanic formation known as a patera (Latin for ''saucer''), a shallow, complex crater with scalloped edges, occurs on Mars but has not been observed on any other planet. Alba Patera, northeast of Olympus Mons, is between 900 and 1200 mi (1500 and 2000 km) across but rises only 2 to 3 mi (3 to 5 km) above the surrounding plains. Enormous quantities of highly fluid lava must have emerged from its central caldera, flowing to great distances along channels and tubes.

The Tharsis ridge, or bulge, is an uplifted portion of the surface that stands more than several kilometers above the mean elevation of the planet. Most of Mars's tectonic features are associated with Tharsis, and it affects approximately one-quarter of the entire surface. Its formation appears to have begun some 3 to 4 $\times$ 10^9 years ago, long before the volcanoes that now dot its crest and surroundings. A fact puzzling to scientists is that Tharsis is, in effect, top-heavy—its bulk should not be sitting so high on the crust, but rather sink in (much as an iceberg submerges most of its volume beneath the ocean's surface).

Canyon lands. Perhaps the most spectacular features on the Martian surface are the huge canyons located primarily in the equatorial regions. Valles Marineris, actually a system of canyons, extends for over 3000 mi (5000 km) along the equatorial belt (**Fig. 1**). In places, the canyon complex is as much as 300 mi (500 km) wide and drops to more than 4 mi (6 km) below the surrounding surface. Dwarfing the Grand Canyon, Valles Marineris is comparable in size to the great East African Rift Valley. In general, the walls of the canyon are precipitous, have well-defined edges, and show evidence of slumping and landslide activity (**Fig. 2**). An intricate system of smaller canyons extending back from the main rim may have developed during melting and evaporation of subsurface ice, or it may have been caused by wind or water erosion. The principal canyons of Valles Marineris appear to be linked with the formation of the Tharsis uplift. They may have formed by faulting along lines of tension within the crust. In essence, such canyons (called graben) are pairs of parallel faults with a sunken area between them. While erosion continues to eat away at the canyon walls, that is an unlikely origin for Valles Marineris and similar complexes. These enormous chasms have no outlets, and the only obvious way to remove debris is by wind; yet the material to be transported out of the canyons is so great as to cast serious doubt on the effectiveness of this mechanism operating by itself.

Channels. An astonishing class of features on the Martian surface is a widespread network of channels that bear a very strong resemblance to dry river beds (**Fig. 3**). Ranging in size from broad, sinuous features nearly 40 mi (60 km) wide, to small, narrow networks less than 300 ft (100 m) wide, the channels present the strong impression that they were created by water erosion. Early analysis assumed the apparently dendritic nature of some channels to be evidence of watershed rain collection and episodic flow. On closer inspection, however, many systems previously considered dendritic now appear to result from more gradual formation through groundwater sapping, not rainfall. The largest channels must have been formed by enormous torrents of water, presumably released in some catastrophic manner. Yet, liquid water on Mars, under conditions that prevail today, cannot exist, because the atmospheric pressure and surface temperature keep most available water trapped as subsurface ice or locked up in the permanent polar caps. As a way out of this enigma, it has been suggested that the long-term climate of Mars is variable, and that in the past the planet might have had conditions suitable for free running water over its surface. Long-term changes in the inclination of Mars's axis to its orbital plane and in the eccentricity of its orbit tend to support this supposition.

These channels do not coincide with the "channels" or "canals" which, in the past, were frequently reported by visual observers of Mars. Only Valles Marineris can be associated with a feature previously seen from Earth.

Polar caps. The seasonal cycle of growth and decay of the bright polar caps has long been taken as evidence of the presence of water on Mars. However,

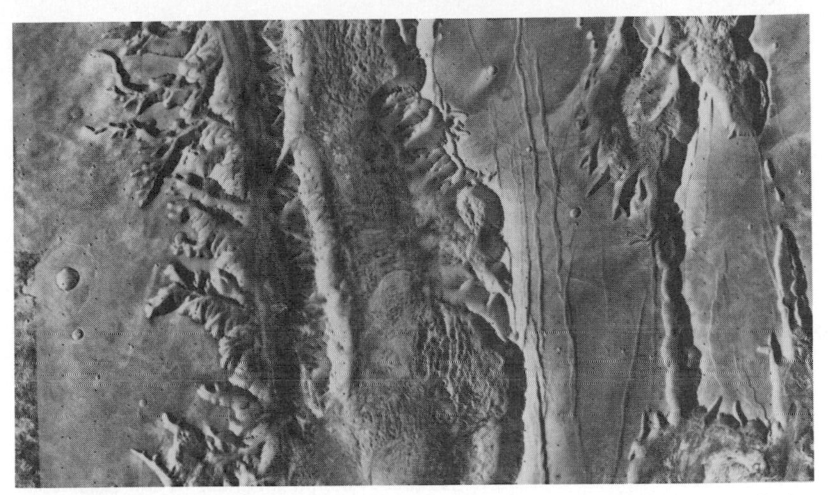

Fig. 2. Part of western end of Valles Marineris, showing canyon walls; photographed by *Viking Orbiter 1*. (*NASA*)

Fig. 3. Channel system in the Mangala Vallis region of Mars; photographed by *Viking Orbiter 1*. (*NASA*)

evidence from the Mariner spacecraft identified carbon dioxide as the principal constituent of the polar snow, with lesser amounts of water ice also present. Formed during autumn and winter by condensation and deposition of the icy mist covering the polar regions, the polar caps at the end of winter cover a vast area extending down to latitude 60° in the southern hemisphere and 70° in the northern hemisphere. The surfaces covered are, respectively, 0.4 and 1.7×10^6 mi^2 (1 and 4.5×10^6 km^2). The thickness of the cap is unknown but presumably decreases radially outward from the pole. During spring and summer the edges of the caps retreat, reaching minimum diameters of about 200 and 600 mi (a few hundred and 1000 km), respectively. The residual caps (**Fig. 4**), which never completely disappear, are believed to be composed almost entirely of water ice. It is estimated that, if the total water component of the polar caps were distributed uniformly over the surface of Mars, it would produce a layer 30 ft (10 m) deep. A far greater amount of water may exist as subsurface ice in nonpolar regions.

The polar regions are characterized by layered terrain, consisting of narrow, evenly spaced strata from 60 to 150 ft (20 to 50 m) in thickness. A sequence composed of more than 100 such layers has been measured in the south polar regions. The absence of craters in these layered terrains suggests that they represent either the youngest areas on Mars or, more likely, the most heavily eroded.

Atmosphere. The Martian atmosphere is very thin, with surface pressure (measured at the Viking landing sites) varying between 7 and 10 millibars (700 and 1000 Pa) and averaging about 8 mb (800 Pa), or less than 1% of the pressure at the Earth's surface. It is composed principally of carbon dioxide, but contains nitrogen, argon, oxygen, and a trace of water vapor totaling about 5%. If all the water remaining in the atmosphere of Mars were to condense onto the surface, it would form a film only 10 μm thick.

Clouds. Both carbon dioxide and ice form clouds in the Martian atmosphere. Certain water-ice condensations are associated with the largest volcanic peaks, while carbon dioxide clouds occur frequently in the polar regions. Often, clouds will appear in the hours just after sunrise, when frosts formed during the night sublime back into the atmosphere.

Dust storms. Localized dust storms appear quite frequently on Mars. Because of the extreme thinness of the Martian atmosphere, wind velocities greater than 90–110 mi/h (40–50 m/s) are needed to set surface dust grains in motion, although they can then be carried to high altitudes by winds of lesser velocity. The most abundant particles in a typical Martian dust storm are estimated to be about 10 μm in diameter, similar to finely powdered talc. The very smallest particles may be carried to heights as great as 30 mi (50 km) above the surface.

Some dust storms develop with such intensity that their total extent may be hemispheric or even global. Such was the case of the great dust storm of 1971, which began in September and completely enshrouded the planet by the time *Mariner 9* arrived in November. Major dust storms tend to begin at the end of spring in the southern hemisphere, when Mars is near perihelion. The initial dust-raising disturbance is probably related to temperature instabilities in the Martian atmosphere at a time when solar heating is at a maximum. Once the dust is in the air, increased absorption of solar energy causes the dust-laden air to

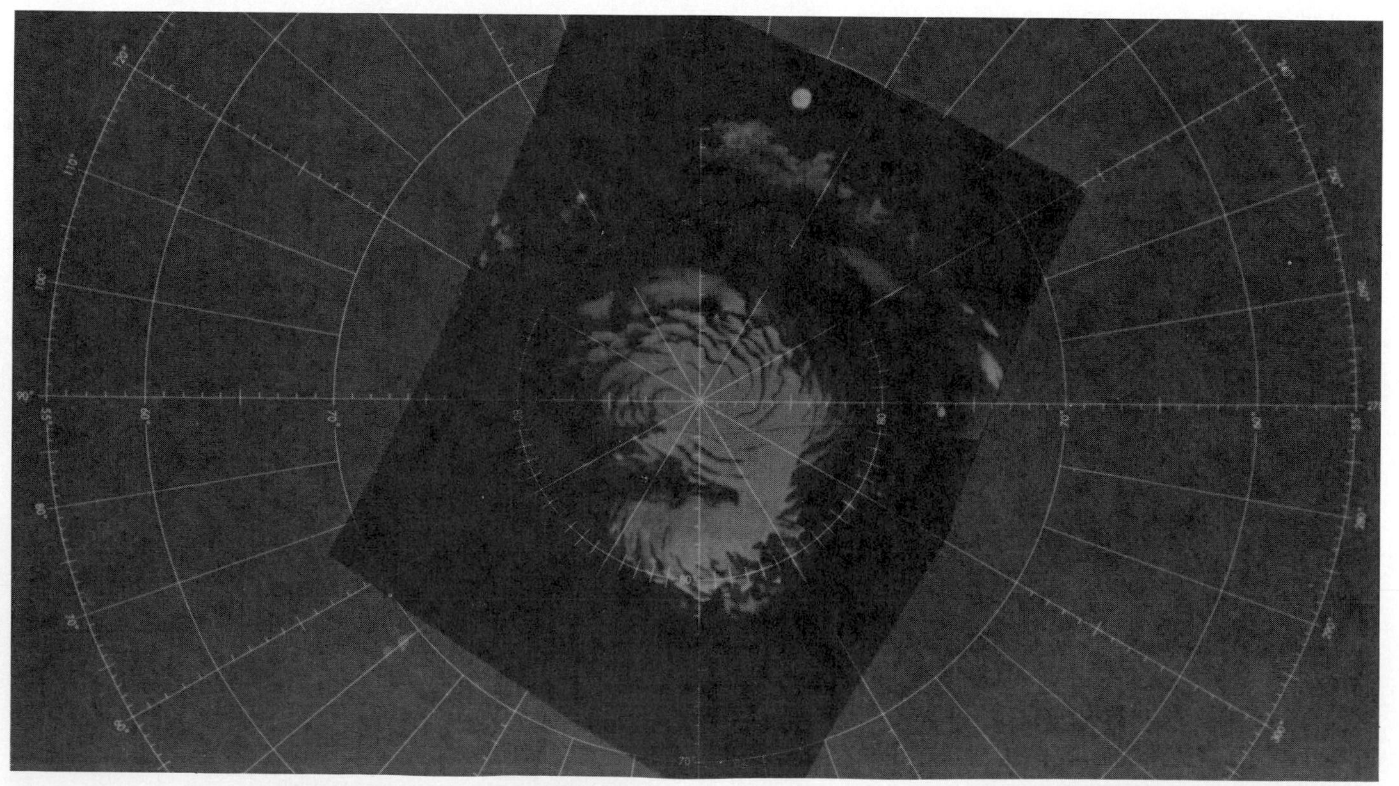

Fig. 4. North polar region of Mars, photographed by *Mariner 9*, showing polar cap nearing its minimum extent approximately 2 weeks after the summer solstice. Crater at upper right trapped and shielded frost from sun, leaving a large patch on its floor. (*NASA*)

Fig. 5. Martian landscape, photographed by *Viking Lander 1*. Picture covers 100°, looking northeast at left, southeast at right. (*NASA*)

rise, producing strong surface winds which rush in to replace the rising air mass. Thus the storm propagates, growing in intensity and extent. When the distribution of dust finally reaches global dimensions, the temperature differences over the planet are diminished, winds correspondingly subside, and dust begins to settle out of the atmosphere. The settling process may take weeks, or in the case of a severe storm, several months. From beginning to end, the 1971 storm lasted for approximately 5 months.

Viking landing sites. In 1976 two American spacecraft—the Viking landers—successfully dropped onto the surface of Mars. *Lander 1* set down in a region known as Chryse Planitia (22°27N, 47°97W) on July 20, and *Lander 2* arrived at Utopia Planitia (47°67N, 225°74W) on September 3. From these locations, the spacecraft collected more than 1400 photographs and large quantities of other data concerning the character, composition, and organic content of the surface, and the composition and climatic patterns of the lower atmosphere. Both landers operated long after the completion of their basic 90-day missions, and one continued to collect and relay data for over six years.

Lander 1 sits in a moderately cratered, low-lying volcanic plain near the mouth of a large outflow channel. In **Fig. 5**, craters up to 2000 ft (600 m) across near the horizon are probably the source of rocks littering the scene. These rocks range in size and color, and some bedrock may also be exposed; fine-grained material occasionally collects into large drifts. *Lander 2*, situated 900 mi (1500 km) farther north, is in a region of fractured plains about 125 mi (200 km) south of the large crater Mie; it probably sits atop a lobe of the crater's ejecta blanket. This site (**Fig. 6**) is very flat and sparsely cratered, subdivided by polygonal fractures and troughs crisscrossing the area. While no drifts are visible and rocks cover a higher proportion of the surface than at the other site, rocks surrounding *Lander 2* look remarkably similar to those near *Lander 1*, with numerous pits giving them a spongy appearance. During the course of a full Martian year, the landers recorded only slight variations in the brightness and color of the surrounding surface, which can be explained by the deposition and removal of a thin veneer of dust and, for *Lander 2*, frost.

Five experiments sampled the soil within reach of the 10.5-ft (3.2-m) mechanical arm aboard each spacecraft. The arms themselves were used to test the properties of rocks and soils, and these ranged in character from loose piles of drift material near *Lander 1* to rocks at both sites that were too hard to be chipped or scratched. Three biological investigations and a gas chromatograph/mass spectrometer attempted to settle the question of whether life existed on the planet. The latter instrument detected no organic compounds at all (within the limits of its sensitivity). One of the trio of biological experiments gave a positive result—the release of oxygen from a soil sample when humidified—but there is no consensus as to the source of this reaction. Some scientists have suggested inorganic superoxides or other catalysts as the substance involved. Most of the data concerning the composition of the surface came from x-ray fluorescence spectrometers, which showed that small particles were remarkably similar at both sites, with silicon and iron accounting for about two-thirds of the samples' content. The sulfur content was unexpectedly high, 100 times greater than in the Earth's crust, while potassium was found to be present at one-fifth of Earth's proportion. Small particles also clung to magnets on each spacecraft, and these are probably composed of the mineral maghemite. In general, the surface material at both sites can be characterized as ion-rich clays.

The weather (atmospheric pressure, temperature, wind velocity, and direction) was monitored by both landers for more than a full Martian year. Temperatures ranged from a low of −190°F (150 K) at *Lander 2* to a high above −10°F (250 K) at *Lander 1*; on any given day, the diurnal variation at each site ranged between 55 and 90°F (35 and 50 K), and dust storms tended to moderate temperatures. In the Martian summer, as atmospheric pressure dropped toward its minimum (because of frost condensing onto the south polar cap), the weather became rather monotonous: diurnal variations were very consistent and winds averaged 2–4 mi/h (1–2 m/s). But toward au-

Fig. 6. Martian landscape, photographed by *Viking Lander 2*, looking northeast. (*NASA*)

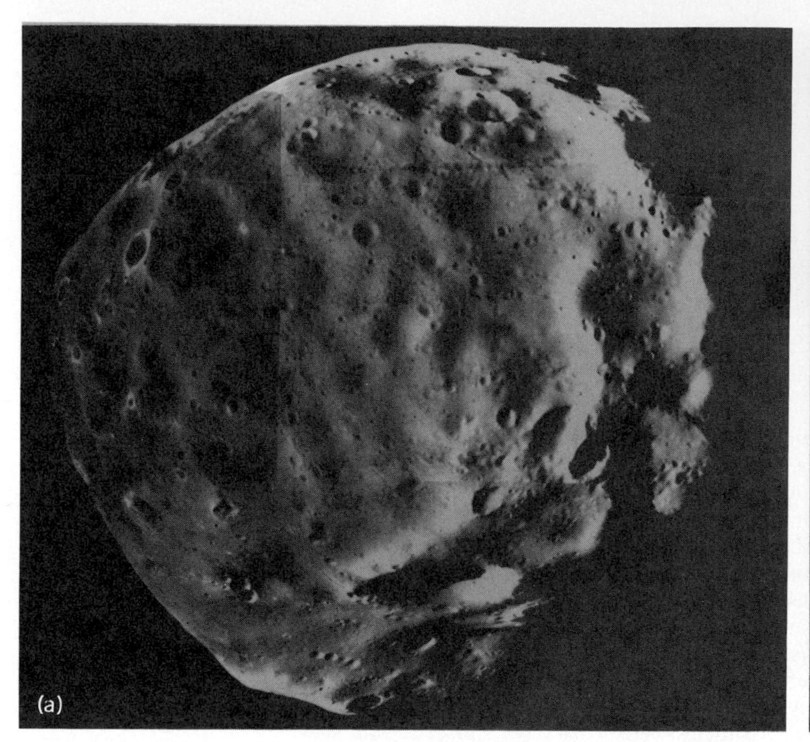

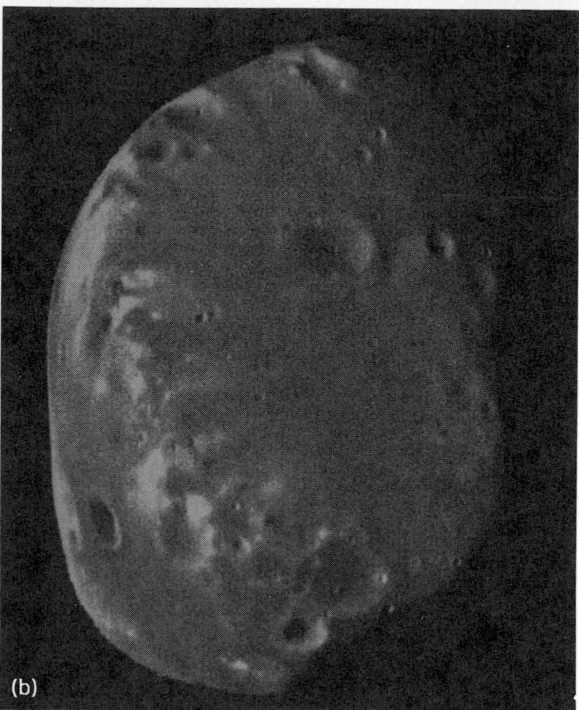

Fig. 7. Martian satellites. (*a*) Phobos, photographed by *Viking Orbiter 1*. (*b*) Deimos, photographed by *Viking Orbiter 2* (*NASA*)

tumn, day-to-day variations increased (especially at *Lander 2*), and a regular sequence of cyclones and anticyclones were observed to pass eastward over the more northerly lander with a frequency of about once per week. Some of these moving systems were apparently associated with frontal systems, such as the advance of the so-called north polar hood.

Internal constitution. Accurate measurements of the planet's gravitational field made by three orbiting spacecraft have shown that Mars has a dense core and thus is differentiated. The crust, composed of lighter rocks, enriched in silicon and aluminum and deficient in magnesium, is believed to be about 30 mi (50 km) thick. Melting of the crust or upper mantle, over geologic time, has caused lighter materials to float to the surface, creating the observed lava plains and volcanic structures. This volcanism is believed to have continued episodically up to the present.

Satellites. Mars has two small satellites, Phobos and Deimos, discovered by A. Hall in 1877. The apparent visual magnitudes at mean opposition are 11.5 and 12.5, respectively. Although they are not intrinsically very faint, a large telescope is required to see them clearly against the bright glare surrounding the planet. The outer satellite, Deimos, moves at a mean distance from the planet's center of 14,600 mi (23,500 km) in a sidereal period of 30h18m, only slightly longer than the period of rotation of Mars. The orbital plane is inclined about 1° to the planet.

The inner satellite, Phobos, moves at a mean distance from the planet's center of 5800 mi (9350 km) in a sidereal period of 7h39m, which is less than one-third of the mean Martian solar day. It is the only known satellite whose period of revolution is less than the period of revolution of the planet; it follows that, as seen from the surface of Mars, Phobos rises in the west and sets in the east twice daily, moving apparently in the opposite direction to all other celestial bodies, including Deimos. The inclination of the orbital plane to the equatorial plane of the planet is 1°.7. The mean distance of Phobos to the surface of Mars is only 3700 mi (6000 km), and the apparent diameter of Mars seen from Phobos is 42°; conversely, Phobos cannot be seen from regions of Mars within 21° of the poles. Seen overhead near the Martian equator, Phobos would appear less than half the size of the Moon viewed from the Earth, and Deimos would be a barely discernible disk.

The attraction of the equatorial bulge of Mars causes the lines of the apsides of the satellites' orbits to advance and the lines of the nodes to retrograde in a period of about 55 years for Deimos and 2.2 years for Phobos. Neither satellite is massive enough to be gravitationally contracted to a spherical shape (**Fig. 7**). Phobos measures only 12 × 14× 17 mi (20 × 23 × 28 km), and Deimos just 6 × 7 × 10 mi (10 × 12 × 16 km). The very low surface gravity of these tiny satellites leads to an escape velocity of only 30 ft/s (10 m/s). A person standing on Deimos or Phobos could quite easily throw rocks which would leave the satellite forever.

Both satellites are saturated with impact craters; the largest is Stickney, a 5-mi-diameter (8-km) crater located on Phobos. Had the impacting object which excavated Stickney been much larger, it probably would have shattered Phobos. In fact, it has been suggested that Deimos and Phobos may be the shattered and, perhaps, partially reconstituted remains, of an earlier, single, large satellite of Mars. Both satellites have the same low albedo, approximately 0.05, surfaces that are as dark as the very darkest regions of the Moon. This low reflectivity suggests that the Martian satellites may have had their origin in the asteroid belt. SEE ASTEROID.

J. Kelly Beatty

Bibliography. J. K. Beatty and B. O'Leary, *The New Solar System*, 1981; M. H. Carr, *The Surface of Mars*, 1981; M. H. Carr, The volcanoes of Mars, *Sci.*

Amer., 234:33–43, 1976; C. W. Snyder, The planet Mars as seen at the end of the Viking mission, *J. Geophys. Res.*, 84:8487–8519, 1979; Viking Lander Imaging Team, *The Martian Landscape*, 1978; J. A. Wood, *The Solar System*, 1979.

Mass-luminosity relation

The relation, observed or predicted by theory, between the quantity of matter a star contains (its mass) and the amount of energy generated in its interior (its luminosity). Because of the great sensitivity of the rate of energy production in a stellar interior to the mass of the star, the mass-luminosity relation provides an important test of theories of stellar interiors. This predicted relation is a corollary of the more general principle that if a star is in equilibrium, that is, if it is neither expanding nor contracting, all its other properties depend only on its mass and the distribution of chemical elements within it. For a family of stars with different masses but with the same mixture of chemical elements uniformly distributed throughout the stellar volumes, there should be a unique mass-luminosity relation. The observed relation, obtained from binary stars for which masses and luminosities can be observationally evaluated, is shown in the **illustration**. It conforms reasonably well with theory. It was this agreement that showed the pioneering studies of stellar structure by A. S. Eddington in the 1920s to be on a sound basis. SEE BINARY STAR; STAR.

There are classes of stars not on the main sequence which depart markedly from the mass-luminosity relation for main-sequence stars. The theory of stellar evolution predicts that changes in interior chemical composition will result from conversion of hydrogen into helium and heavier elements in the process of nuclear energy generation, which is the source of stellar luminosity. Accompanying these changes in composition there should be changes in luminosity. Examples are the three stars in the illustration lying below the smooth relation. These are white dwarf stars, which are known to be extremely underluminous for their masses, and are thought to be stars in which all sources of nuclear energy have been exhausted. Giants and supergiants are thought to be in an intermediate state of evolution, between main sequence stars and white dwarfs.

If a star is a member of a close binary system, it may be prevented by the presence of its companion from evolving by expansion into a giant star. The evolutionary histories of stars in such systems are predicted to involve mass exchange between the components. The relation between masses and luminosities will depend upon the state of evolution of the individual system. In some cases, gross departures from the mass-luminosity relation for main-sequence stars are predicted and observed. A star which has lost the bulk of its mass to its companion may radiate as much energy as a normal star, even though its mass is much less. X-ray binaries are thought to be extreme examples of advanced evolution of close binary systems. Application of the mass-luminosity relation to the visible components in x-ray binaries leads to the conclusion that, in a couple of cases, the invisible component may well be a black hole, the best candidates for this extreme among stellar configurations. A number of the predictions of the evolution of close binaries appear to be verified observationally. SEE BLACK HOLE; GIANT STAR; STELLAR EVOLUTION; SUPERGIANT STAR; WHITE DWARF STAR; X-RAY STAR.

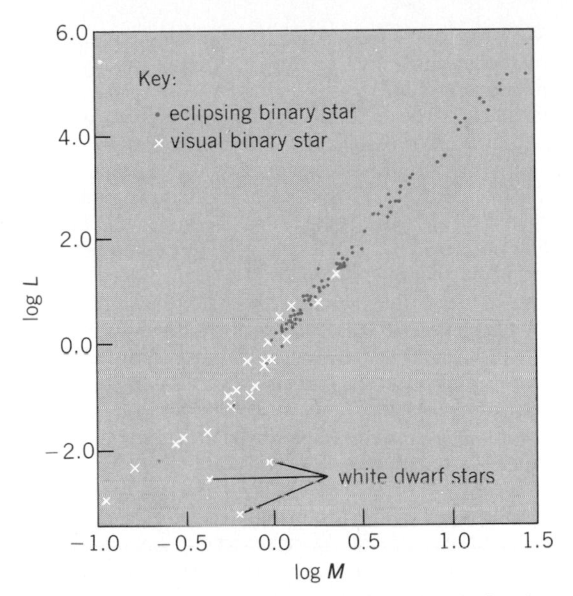

Diagram of the mass-luminosity relation. The abscissa is the logarithm of mass *M* and the ordinate is the logarithm of luminosity *L*, both relative to the Sun as unit.

Daniel M. Popper

Bibliography. W. D. Heintz, *Double Stars*, 1978; E. Novotny, *Introduction to Stellar Atmospheres and Interiors*, 1973; D. M. Popper, Stellar masses, *Annu. Rev. Astron. Astrophys.*, 18:115–164, 1980; J. Sahade and F. B. Wood, *Interacting Binary Stars*, 1978.

Mercury

The planet closest to the Sun. It is visible to the unaided eye only shortly after sunset or shortly before sunrise, when it is near its greatest angular distance from the Sun (28°). Mercury is smaller than any other planet except Pluto. Its diameter is 3031 mi (4878 km), and its mass is 7.28×10^{23} lb (3.301×10^{23} kg), or 0.055 times the mass of the Earth. Most current knowledge of Mercury is derived from data returned by the *Mariner 10* spacecraft, which flew by the planet three times in 1974 and 1975. *Mariner 10* imaged only about 45% of the surface at an average resolution of about 0.6 mi (1 km), and less than 1% at resolutions between about 300 and 1500 ft (100 and 500 m). This coverage and resolution is somewhat comparable to Earth-based telescopic coverage and resolution of the Moon before the advent of spaceflight. As a consequence, there are still many uncertainties and questions concerning this unusual planet. Mercury represents an end member in solar system origin and evolution because it formed closer to the Sun than any other planet and, therefore, in the hottest part of the solar nebula from which the entire solar system formed. SEE SOLAR SYSTEM.

Motions. Mercury has the most eccentric (0.205) and inclined (7°) orbit of any planet in the solar system except Pluto. Its average distance from the Sun is 3.60×10^7 mi (5.79×10^7 km), but distance varies from 2.86×10^7 mi (4.60×10^7 km) at perihelion to 4.34×10^7 mi (6.98×10^7 km) at aphelion because of the large eccentricity. The rotation period is 58.646 Earth days, and the orbital period, 87.969. Therefore,

Mercury has a 3:2 resonant relationship between its rotational and orbital periods; it makes exactly three rotations around its axis for every two orbits around the Sun. Thus, a solar day (sunrise to sunrise) lasts two mercurian years (176 Earth days).

The motion of Mercury played an important role in the development of the general theory of relativity. Mercury's elliptical orbit is pulled around the Sun so that the perihelion point changes its position in space by about 5600 arc-seconds per century. When all known forces that act on the planet are taken into account, Mercury's perihelion-point motion exceeds that calculated by newtonian gravitational theory by 43 arc-seconds per century. This discrepancy remained unexplained until 1915, when Einstein's general theory of relativity predicted a perihelion motion in exact agreement with observation.

Temperature. Although Mercury is closest to the Sun, it is not the hottest planet. The surface of Venus is hotter because of the atmospheric greenhouse effect. However, Mercury experiences the greatest range in surface temperature (1130°F or 610°C) of any planet or satellite in the solar system because of its proximity to the Sun, its long solar day, and its lack of an insulating atmosphere. Its maximum surface temperature is 800°F (427°C) at perihelion on the equator, hot enough to melt zinc. At night, however, the unshielded surface plunges to below −300°F (−183°C). Three feet (1 m) below the surface, the temperature remains at an average value of about 170°F (350 K). *See* Venus.

Magnetic field. *Mariner 10* discovered an intrinsic dipole magnetic field with a dipole moment equal to about 0.004 that of the Earth. Mercury is the only terrestrial planet besides the Earth with a magnetic field. Although weak compared to that of the Earth, the field has sufficient strength to hold off the solar wind, creating a bow shock and accelerating charged particles from the solar wind. As with the Earth, the magnetic axis is inclined about 11° from the rotation axis. Mercury occupies a much larger fraction of the volume of its magnetosphere than do other planets, and the solar wind actually reaches the surface at times of highest solar activity. Because of the small size of Mercury's magnetosphere, magnetic events happen more quickly and repeat more often than in Earth's magnetosphere. The maintenance of planetary magnetic fields apparently requires an electrically conducting fluid outer core. Therefore, Mercury's dipolar magnetic field is taken as strong evidence that Mercury currently has a fluid outer core of unknown thickness. *See* Magnetosphere.

Interior. Mercury's internal structure is unique in the solar system. The planet's mean density is 5.44 g/cm^3 (5.44 times that of water), which is larger than that of any other planet or satellite except Earth (5.52 g/cm^3). Because of Earth's large internal pressures, however, its uncompressed density is only 4.4 g/cm^3 compared to Mercury's uncompressed density of 5.3 g/cm^3. This means that Mercury contains a much larger fraction of iron than any other planet or satellite in the solar system. The iron core must be about 75% of the planet diameter, or 42% of Mercury's volume. It is surrounded by a silicate mantle and crust only about 370 mi (600 km) thick. Earth's core is only 54% of the planet diameter, or just 16% of the total volume. Because the dipole magnetic field indicates that the core is at least partly molten at present, a light alloying element in the core must have lowered the melting point and retained a partially molten core over geologic history (4.5×10^9 years); otherwise the core would have solidified long ago. Although oxygen is such an element, it is not sufficiently soluble in iron at Mercury's low internal pressures. Therefore, sulfur is the most reasonable candidate for this alloying element. Mercury probably has between about 0.2 and 7% sulfur in its core, since a sulfur abundance of less than 0.2 would result in an entirely solid core at the present time, whereas an abundance of 7% would result in an entirely fluid core at present. *See* Earth.

Origin. The origin of Mercury and how it acquired such a large percentage of iron is a major unsolved problem. Although chemical equilibrium condensation models successfully explain the densities of most planets in the solar system, they cannot explain Mercury. For Mercury's position in the solar nebula, these models predict an uncompressed density of only about 4 g/cm^3 rather than the observed 5.3 g/cm^3; they also predict the complete absence of sulfur.

Three different hypotheses have been proposed to account for the large discrepancy between the iron

Fig. 1. Photomosaic of Mercury as seen by the outgoing *Mariner 10* spacecraft in March 1974. The terminator, the boundary between the lighted and unlighted halves of the planet, runs vertically along the left side of the photograph. The Caloris Basin is on the terminator, slightly above the middle. It is surrounded and filled by younger smooth plains. (*Jet Propulsion Laboratory*)

Fig. 2. Photomosaic showing the 800-mi-diameter (1300-km) Caloris Basin, the largest impact structure observed by *Mariner 10*, on the left, midway between top and bottom. (*Jet Propulsion Laboratory*)

abundance indicated by Mercury's high density and that predicted by equilibrium condensation models. In the selective accretion model, a differential response of iron and silicates to impact fragmentation and aerodynamic sorting leads to iron enrichment owing to higher gas densities and shorter dynamical time scales in the innermost part of the solar nebula. The postaccretion vaporization model proposes that intense bombardment by solar electromagnetic and corpuscular radiation in the earliest phases of the Sun's evolution (the T-Tauri phase) vaporized and drove off much of the planet's silicate mantle. In the giant impact hypothesis, a collision of a planet-sized object with Mercury ejected much of the mantle. Each of these hypotheses has major consequences for the formation of the other terrestrial planets, and each predicts a significantly different chemical composition for the silicate portion of Mercury, which could be tested by a future space mission to Mercury.

Atmosphere. Mercury's atmosphere is very tenuous and is essentially exospheric in that its atoms rarely collide with each other. The atmospheric surface pressure is 10^{12} times less than Earth's. *Mariner 10*'s ultraviolet spectrometer identified hydrogen, helium, oxygen, and argon in the atmosphere, all of which are probably derived largely from the solar wind. Earth-based telescopic observations in 1985 discovered that Mercury is surrounded by a tenuous atmosphere of sodium and potassium that is probably derived from its surface. At the subsolar point the atmospheric density of sodium is 280,000 to 620,000 atoms per cubic inch (17,000 to 38,000 atoms per cubic centimeter), and for potassium it is 8000 atoms per cubic inch (500 atoms per cubic centimeter). The mechanism for removing these elements from the surface and injecting them into the atmosphere is uncertain, but it may involve micrometeorite vaporization or solar-wind sputtering of surface material.

Geology. The surface of Mercury superficially resembles that of the Moon (**Fig. 1**). It is heavily cratered, with large expanses of younger smooth plains (similar to the lunar maria) that fill and surround major impact basins. Unlike the Moon surface, Mercury's heavily cratered terrain is interspersed with large regions of gently rolling intercrater plains, the major terrain type on the planet. Also unlike the Moon surface, a system of thrust faults, unique in the solar system, transects the surface viewed by *Mariner 10*. The largest structure viewed by *Mariner 10* is the 800-mi-diameter (1300-km) Caloris impact basin (**Fig. 2**). It is filled and surrounded by smooth plains remembling the Moon's maria, which also fill impact basins. Directly opposite (antipodal to) the Caloris Basin, on the other side of Mercury, is a peculiar terrain consisting of hills and valleys, called the Hilly and Lineated Terrain (**Fig. 3**). This terrain is thought to be the result of surface disruption by seismic waves focused at the antipodal region and caused by the Caloris Basin impact. Infrared temperature measurements from *Mariner 10* indicate that the surface is a good thermal insulator and, therefore, must be covered with porous soil or rock powder like the lunar regolith. This is expected on a planet whose surface is shattered and stirred by meteorite impacts.

The heavily cratered terrain records the period of heavy bombardment that occurred throughout the entire inner solar system and ended about 3.8×10^9

Fig. 3. View of the Hilly and Lineated Terrain, taken by *Mariner 10* in March 1974. (*Jet Propulsion Laboratory*)

years ago on the Moon, and presumably at about the same time on Mercury. The origin of the objects responsible for it is very uncertain, but they may have been remnants left over from the accretion of the terrestrial planets.

The oldest plains (intercrater plains) occur in the heavily cratered terrain and are thought to be primarily volcanic deposits erupted at various times during the period of heavy bombardment. The younger smooth plains are also thought to be volcanic deposits that were erupted within or around large impact basins such as Caloris. Based on the size–frequency distribution and density of impact craters superimposed on the smooth plains, they appear to have been formed near the end of heavy bombardment and may be on average about 3.8×10^9 years old. If so, they are, in general, older than the lava deposits that constitute the lunar maria. The unique tectonic framework of Mercury, expressed as a system of thrust faults, is probably the result of global compressive stresses in Mercury's crust caused by a decrease in the planet's radius due to cooling of the core and mantle (**Fig. 4**). *SEE MOON.*

Geologic history. The general picture of Mercury's geologic history which has emerged from analyses of *Mariner 10* data is that soon after Mercury formed, it became almost completely melted by heating from the decay of radioactive elements and the inward migration of the large amount of iron to form its enormous core. This led to expansion of the planet and tensional fracturing of a thin solid lithosphere that provided egress for lavas to reach the surface and form the intercrater plains during the period of heavy bombardment. As the core and mantle began to cool, Mercury's radius decreased by about 1.2 mi (2 km) or more, and the crust was subjected to compressive stresses that resulted in the system of thrust faults. At about this time, the Caloris Basin was formed by a gigantic impact, which caused the Hilly and Lineated Terrain from focused seismic waves in the antipodal regions. Further eruptions of lava within and surrounding the Caloris and other large impact basins formed the smooth plains. Volcanism finally ceased when compressive stresses in the lithosphere became strong enough to close off magma sources. All of these events probably happened very early, perhaps during the first $7–8 \times 10^8$ years, in Mercury's history. Since that time, only occasional impacts of comets and asteroids have occurred. *SEE PLANET; PLANETARY PHYSICS.*

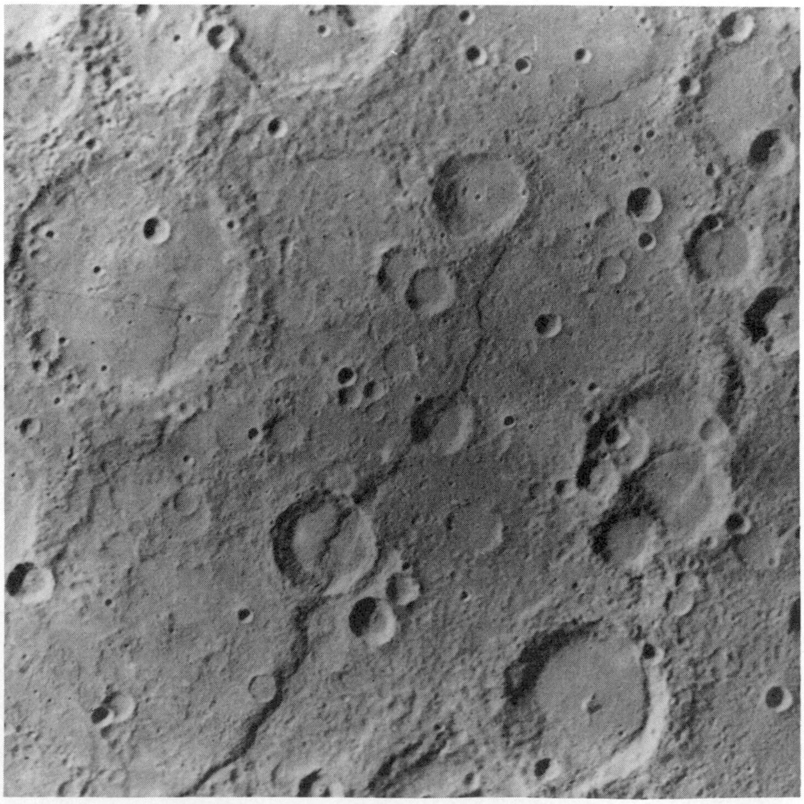

Robert G. Strom

Bibliography. R. G. Strom, *Mercury: The Elusive Planet*, 1987; Various papers reporting *Mariner 10* results, *Science*, 185:141–180, July 12, 1974, and *J. Geophys. Res.*, 80:2341–2514, June 10, 1975; F. Vilas, C. R. Chapman, and M. S. Matthews (eds.), *Mercury*, 1988.

Fig. 4. Close view of Mercury from *Mariner 10* showing a heavily cratered region interspersed with intercrater plains. Running down the center is Discovery Scarp, a major compressive thrust fault about 300 mi (500 km) long and up to 1.8 mi (3 km) high. (*Jet Propulsion Laboratory*)

Meridian

That half of a great circle on Earth that passes through points having the same longitude and terminates at the North and South poles. The meridian from which longitudes are measured is the one passing through Greenwich, near London, England; it is called the prime meridian. *SEE LATITUDE AND LONGITUDE.*

The celestial meridian is a great circle on the celestial sphere passing through the two celestial poles and the observer's zenith. Two branches of it are distinguished, each extending from pole to pole, the upper branch containing the zenith and the lower the nadir. The meridian passage, or culmination, of a celestial object is its crossing over the celestial meridian, upper and lower culminations referring to the upper and lower branches. Stars nearer to the celestial pole than a distance equal to the latitude of the observer are above the horizon at both upper and lower culminations, and are called circumpolar stars. *SEE ASTRONOMICAL COORDINATE SYSTEMS.*

Gerald M. Clemence

Messier catalog

An early listing of nebulae and star clusters. Charles Messier (1730–1817) was primarily interested in discovering comets, so he compiled a list of objects that might be confused with comets in small telescopes. His first catalog was published in the Royal Academy of Sciences of France in 1771; later lists appeared in the *Connaissance des Temps*. Some of the later objects were discovered by Pierre Mechain. The compilation (see **table**) contains both open and globular star clusters, diffuse galactic nebulae, so-called planetary nebulae, and external galaxies. A star cluster in Scorpio, denoted as M7, and the Praesepe cluster, M44, have been known since classical antiquity, while the Pleiades, M45, have been known since prehistoric times. Messier's catalog included the supernova remnant, the Crab Nebula, M1, the Andromeda spiral galaxy, M31, the Triangulum spiral galaxy,

The Messier catalog of nebulae and star clusters*

Messier number	NGC or (IC)	Right ascension (2000)	Declination (2000)	Apparent visual magnitude	Description
1	1952	5h34.5m	+22°01′	8.4	Crab Nebula in Taurus; remains of SN 1054
2	7089	21h33.5m	−0°49′	6.4	Globular cluster in Aquarius
3	5272	13h42.2m	+28°23′	6.3	Globular cluster in Canes Venatici
4	6121	16h23.6m	−26°31′	6.5	Globular cluster in Scorpio
5	5904	15h18.5m	+2°05′	6.1	Globular cluster in Serpens
6	6405	17h40.0m	−32°12′	5.5	Open cluster in Scorpio
7	6475	17h54.0m	−34°49′	3.3	Open cluster in Scorpio
8	6523	18h03.7m	−24°23′	5.1	Lagoon Nebula in Sagittarius
9	6333	17h19.2m	−18°31′	8.0	Globular cluster in Ophiuchus
10	6254	16h57.2m	−4°06′	6.7	Globular cluster in Ophiuchus
11	6705	18h51.1m	−6°16′	6.8	Open cluster in Scutum Sobieskii
12	6218	16h47.2m	−1°57′	6.6	Globular cluster in Ophiuchus
13	6205	16h41.7m	+36°28′	5.9	Globular cluster in Hercules
14	6402	17h37.6m	−3°15′	8.0	Globular cluster in Ophiuchus
15	7078	21h30.0m	+12°10′	6.4	Globular cluster in Pegasus
16	6611	18h18.9m	−13°47′	6.6	Open cluster with nebulosity in Serpens
17	6618	18h20.8m	−16°10′	7.5	Swan or Omega Nebula in Sagittarius
18	6613	18h19.9m	−17°08′	7.2	Open cluster in Sagittarius
19	6273	17h02.6m	−26°16′	6.9	Globular cluster in Ophiuchus
20	6514	18h02.4m	−23°02′	8.5	Trifid Nebula in Sagittarius
21	6531	18h04.7m	−22°30′	6.5	Open cluster in Sagittarius
22	6656	18h36.4m	−23°54′	5.6	Globular cluster in Sagittarius
23	6494	17h56.9m	−19°01′	5.9	Open cluster in Sagittarius
24	6603	18h18.4m	−18°25′	4.6	Open cluster in Sagittarius
25	(4725)	18h31.7m	−19°14′	6.2	Open cluster in Sagittarius
26	6694	18h45.2m	−9°24′	9.3	Open cluster in Scutum Sobieskii
27	6853	19h59.6m	+22°43′	8.2	"Dumbbell" planetary nebula in Vulpecula
28	6626	18h24.6m	−24°52′	7.6	Globular cluster in Sagittarius
29	6913	20h24.0m	+38°31′	8.0	Open cluster in Cygnus
30	7099	21h40.4m	−23°11′	7.7	Globular cluster in Capricornus
31	224	0h42.7m	+41°16′	3.5	Andromeda Galaxy
32	221	0h42.7m	+40°52′	8.2	Elliptical galaxy; companion to M31
33	598	1h33.8m	+30°39′	5.8	Spiral galaxy in Triangulum
34	1039	2h42.0m	+42°47′	5.8	Open cluster in Perseus
35	2168	6h08.8m	+24°20′	5.6	Open cluster in Gemini
36	1960	5h36.3m	+34°08′	6.5	Open cluster in Auriga
37	2099	5h53.0m	+32°33′	6.2	Open cluster in Auriga
38	1912	5h28.7m	+35°50′	7.0	Open cluster in Auriga
39	7092	21h32.3m	+48°26′	5.3	Open cluster in Cygnus
40		12h22.2m	+58°05′		Close double star in Ursa Major
41	2287	6h47.0m	−20°44′	5.0	Loose open cluster in Canis Major
42	1976	5h35.3m	−5°23′	4	Orion Nebula
43	1982	5h35.5m	−5°16′	9	Northeast portion of Orion Nebula
44	2632	8h40.0m	+20°00′	3.9	Praesepe; open cluster in Cancer
45		3h47.5m	+24°07′	1.6	The Pleiades; open cluster in Taurus
46	2437	7h41.8m	−14°49′	6.6	Open cluster in Puppis
47	2422	7h36.6m	−14°29′	5	Loose group of stars in Puppis
48	2548	8h13.8m	−5°48′	6	"Cluster of very small stars"; identifiable
49	4472	12h29.8m	+8°00′	8.5	Elliptical galaxy in Virgo
50	2323	7h03.0m	−8°21′	6.3	Loose open cluster in Monoceros
51	5194	13h29.9m	+47°12′	8.4	Whirlpool spiral galaxy in Canes Venatici
52	7654	23h24.2m	+61°36′	8.2	Loose open cluster in Cassiopeia
53	5024	13h12.9m	+18°10′	7.8	Globular cluster in Coma Berenices
54	6715	18h55.1m	−30°28′	7.8	Globular cluster in Sagittarius
55	6809	19h40.0m	−30°57′	6.2	Globular cluster in Sagittarius
56	6779	19h16.6m	+30°11′	8.7	Globular cluster in Lyra
57	6720	18h53.6m	+33°02′	9.0	Ring Nebula; planetary nebula in Lyra
58	4579	12h37.7m	+11°49′	9.9	Spiral galaxy in Virgo
59	4621	12h42.0m	+11°39′	10.0	Spiral galaxy in Virgo
60	4649	12h43.7m	+11°33′	9.0	Elliptical galaxy in Virgo
61	4303	12h21.9m	+4°28′	9.6	Spiral galaxy in Virgo
62	6266	17h01.2m	−30°07′	6.6	Globular cluster in Scorpio
63	5055	13h15.8m	+42°02′	8.9	Spiral galaxy in Canes Venatici
64	4826	12h56.7m	+21°41′	8.5	Spiral galaxy in Coma Berenices
65	3623	11h18.9m	+13°06′	9.4	Spiral galaxy in Leo

*See footnotes on next page.

The Messier catalog of nebulae and star clusters* ***(cont.)***

Messier number	NGC or (IC)	Right ascension (2000)	Declination (2000)	Apparent visual magnitude	Description
66	3627	11h20.3m	+13°00′	9.0	Spiral galaxy in Leo; companion to M65
67	2682	8h51.3m	+11°48′	6.1	Open cluster in Cancer
68	4590	12h39.5m	−26°45′	8.2	Globular cluster in Hydra
69	6637	18h31.4m	−32°21′	8.0	Globular cluster in Sagittarius
70	6681	18h43.2m	−32°17′	8.1	Globular cluster in Sagittarius
71	6838	19h53.7m	+18°47′	7.6	Globular cluster in Sagitta
72	6981	20h53.5m	−12°32′	9.3	Globular cluster in Aquarius
73	6994	20h59.0m	−12°38′	9.1	Open cluster in Aquarius
74	628	1h36.7m	−15°47′	9.3	Spiral galaxy in Pisces
75	6864	20h06.1m	−21°55′	8.6	Globular cluster in Sagittarius
76	650	1h42.2m	+51°34′	11.4	Planetary nebula in Perseus
77	1068	2h42.7m	−0°01′	8.9	Spiral galaxy in Cetus
78	2068	5h46.7m	0°04′	8.3	Small emission nebula in Orion
79	1904	5h24.2m	−24°31′	7.5	Globular cluster in Lepus
80	6093	16h17.0m	−22°59′	7.5	Globular cluster in Scorpio
81	3031	9h55.8m	+69°04′	7.0	Spiral galaxy in Ursa Major
82	3034	9h56.2m	+69°42′	8.4	Irregular galaxy in Ursa Major
83	5236	13h37.7m	−29°52′	7.6	Spiral galaxy in Hydra
84	4374	12h25.1m	+12°53′	9.4	Elliptical galaxy in Virgo
85	4382	12h25.4m	+18°11′	9.3	Elliptical galaxy in Coma Berenices
86	4406	12h26.2m	+12°57′	9.2	Elliptical galaxy in Virgo
87	4486	12h30.8m	+12°23′	8.7	Elliptical galaxy in Virgo
88	4501	12h32.0m	+14°25′	9.5	Spiral galaxy in Coma Berenices
89	4552	12h35.7m	+12°33′	10.3	Elliptical galaxy in Virgo
90	4569	12h36.8m	+13°10′	9.6	Spiral galaxy in Virgo
91	Omitted				
92	6341	17h17.1m	+43°08′	6.4	Globular cluster in Hercules
93	2447	7h44.6m	−23°53′	6.5	Open cluster in Puppis
94	4736	12h50.9m	+41°07′	8.3	Spiral galaxy in Canes Venatici
95	3351	10h44.0m	+11°42′	9.8	Barred spiral galaxy in Leo
96	3368	10h46.8m	+11°49′	9.3	Spiral galaxy in Leo
97	3587	11h14.9m	+55°01′	11.1	Owl Nebula; planetary nebula in Ursa Major
98	4192	12h13.8m	+14°54′	10.2	Spiral galaxy in Coma Berenices
99	4254	12h18.8m	+14°25′	9.9	Spiral galaxy in Coma Berenices
100	4321	12h22.9m	+15°49′	9.4	Spiral galaxy in Coma Berenices
101	5457	14h03.5m	+54°21′	7.9	Spiral galaxy in Ursa Major
102	5866(?)	15h06.9m	+55°44′	10.5	Spiral galaxy (identification as M102 in doubt)
103	581	1h33.1m	+60°42′	6.9	Open cluster in Cassiopeia
104†	4594	12h40.0m	−11°42′	8.3	Spiral galaxy in Virgo
105†	3379	10h47.9m	+12°43′	9.7	Elliptical galaxy in Leo
106†	4258	12h19.0m	+47°18′	8.4	Spiral galaxy in Canes Venatici
107†	6171	16h32.5m	−13°03′	9.2	Globular cluster in Ophiuchus
108†	3556	11h11.6m	+55°40′	10.5	Spiral galaxy in Ursa Major
109†	3992	11h57.7m	+53°22′	10.0	Spiral galaxy in Ursa Major
110†	205	0h40.3m	+41°41′	9.4	Elliptical galaxy (companion to M31)

*From G. O. Abell, *Drama of the Universe,* Holt, Rinehart and Winston, 1978; and J. M. Pasachoff and D. H. Menzel, *A Field Guide to the Stars and Planets*, Houghton Mifflin 1983.
†Not in Messier's original (1781) list; added later by others.

M33, the giant elliptical galaxy in Virgo, M87, the globular clusters, M13 and M15, the Great Orion Nebula, M42, and the Ring Nebula, M57 or NGC 6720.

The New General Catalog (NGC) and the supplementary Index Catalogs (IC) published by J. L. E. Dreyer at the end of the nineteenth century were far more extensive than that of Messier. Consequently the NGC numbers are generally used for all except a few objects whose Messier numbers had become firmly established. Some Messier objects have not been identified. For example, M102 is probably a repeat observation of the great spiral galaxy M101, although Dreyer suggested it could be identified with a nearby, much fainter galaxy, NGC 5928. SEE GALAXY, EXTERNAL; NEBULA; STAR CLUSTERS.

Lawrence H. Aller

Bibliography. K. G. Jones, *Messier Nebulae and Star Clusters*, 1968; J. H. Mallas and E. Kreimer, *Messier Album*, 1978; C. Messier, *The Messier Catalogue*, ed. by P. H. Niles, 1981.

Meteor

The luminous streak lasting seconds or fractions of a second and seen at night when a solid, natural body plunges into the Earth's (or another planet's) atmosphere. The entering object is called a meteoroid and, if any of it survives atmospheric passage, the remainder is called a meteorite. If the remainder is very small (less than a fraction of a millimeter), it is sometimes called a micrometeorite. If the apparent brightness of a meteor exceeds that of the planet Venus as seen from Earth, it is called a fireball, and when a bright meteor is seen to explode, it is called a bolide. SEE METEORITE; MICROMETEORITE.

Visual observation. Under normal, clear atmospheric conditions and dark skies (no moonlight or artificial lights), an observer will see an average of five meteors per hour. The spatial distribution of meteoroid orbits relative to the Sun and the circumstances of their intersections with the moving Earth are responsible for pronounced variations in meteor rates.

As the Earth moves in its orbit, its velocity points toward that part of the sky which is visible from local midnight through morning to local noon and points away from that part of the sky which is visible from local noon through evening to local midnight. On average, an observer sees more meteors during the early

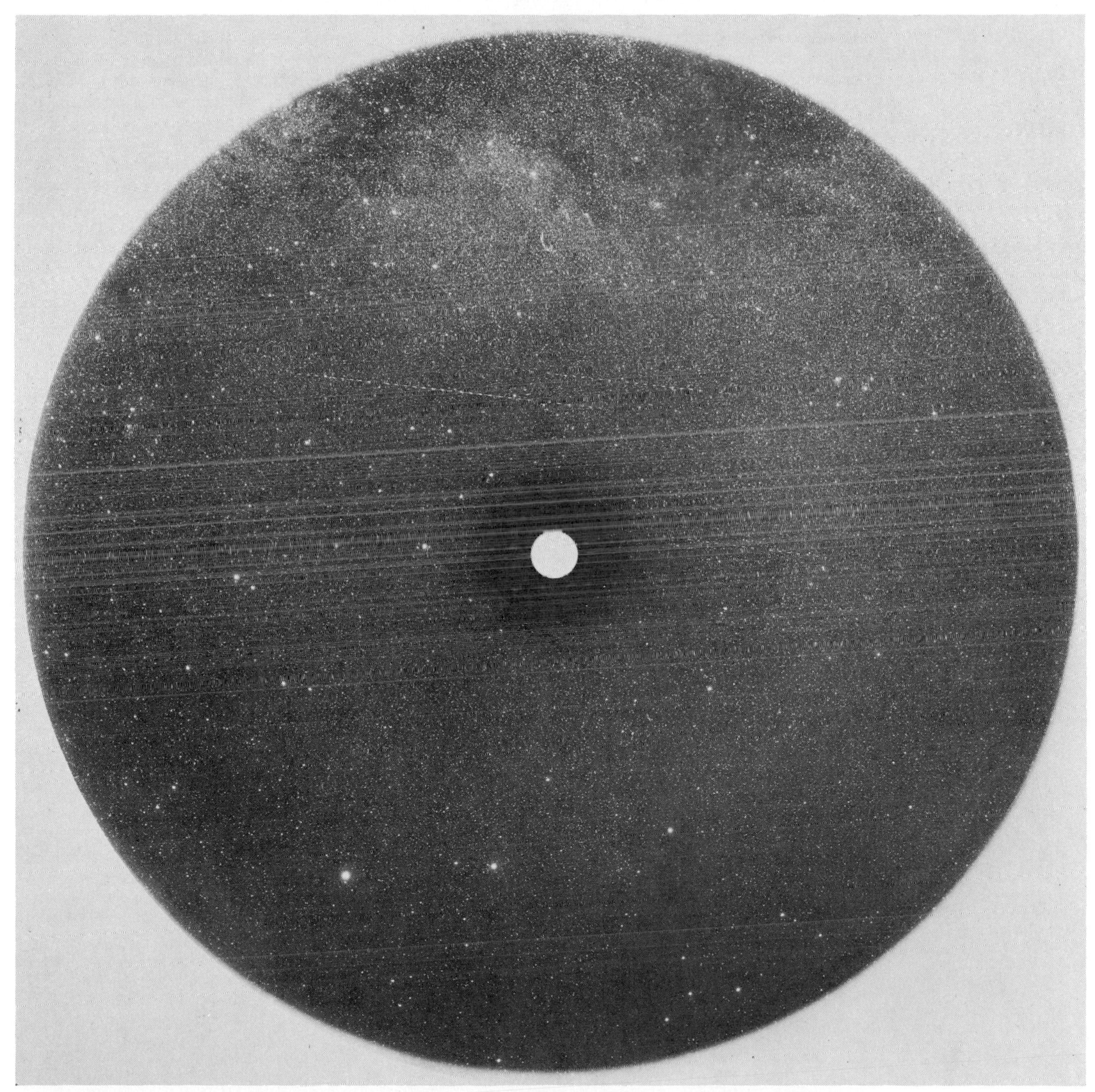

Photographic trails of three meteors of the Perseid shower, extended back to the left, converge toward a common radiant. A rotating focal-plane shutter in the Baker Super-Schmidt meteor camera interrupted the trails each 1/60 s. Hub of shutter caused unexposed center. (*Harvard University Meteor Project*)

morning hours as the Earth sweeps up objects in its path than in the early evening hours when meteor-producing objects must catch up with the Earth.

Physical characteristics. The heights of appearance and disappearance of a meteor depend on meteoroid initial velocity, angle of entry with respect to the vertical, initial meteoroid mass, and meteoroid material strength. The average meteor seen by the unaided eye starts with a meteoroid velocity of 18 mi/s (30 km/s) and leaves a luminous trail from 67 to 50 mi (110 to 80 km) high. The fainter the meteor (the smaller the meteoroid mass), the shorter is the meteor length. The faster the meteoroid travels just before hitting the atmosphere, the higher in the atmosphere the meteor trail occurs.

In the end, most, if not all, of the meteoroid material is vaporized, leaving a deposit of metallic atoms (predominantly sodium, calcium, silicon, and iron) in the upper atmosphere. This deposition is an important mechanism in the wind-shear formation of certain types of highly ionized, radio-reflecting, upper-atmospheric phenomena called the sporadic E layers.

The meteor trails themselves are rapidly expanding columns of atoms, ions, and electrons dislodged from the meteoroid by collisions with air molecules, and can be excited to temperatures of several thousand degrees Celsius. For a time after trail formation, the free electrons are dense enough to reflect radio waves in the very high-frequency range, and therefore can be used to transmit radio messages that are brief (0.1–15 s) but high in information content (on account of their large bandwidth) for up to 1300 mi (2200 km). Since meteor-reflected signals are not as subject to ionospheric and other disturbing influences as are other means of radio communications, there is increasing interest in using meteors for certain military and commercial purposes.

Some meteors show relatively long-lasting glows along their paths. These are called meteor trains. Novice observers often confuse this glow with the brief meteor wake that is sometimes seen just behind the luminous gases surrounding the meteoroid itself. These trains are subject to winds in the upper atmosphere, and for many years their apparent motions (twisting and drifting) were the only probes of wind velocities in the mesosphere, or middle atmosphere (42–72 mi or 70–120 km high). The motions of meteor trains and trails detected through the Doppler effect on reflected radar signals have been extensively studied, and global mesospheric wind patterns have been derived. *SEE DOPPLER EFFECT.*

Meteoroid orbits and velocities. The Earth moves around the Sun with an average speed of 18 mi/s (30 km/s). According to the laws of celestial mechanics, if a meteoroid comes from beyond the solar system, its velocity at the Earth's distance from the Sun must be greater than 26 mi/s (42 km/s). If such a meteoroid hits the Earth head-on, indications of preatmospheric speeds in excess of 45 mi/s (72 km/s) would be observed. The fact that so far there are no verifiable cases of meteoroids traveling in solar orbits much in excess of 45 mi/s (72 km/s) indicates that most meteors are caused by fragments that are long-term members of the solar system. The slowest velocity of Earth–meteoroid encounter occurs when the meteoroid has to catch up with the Earth. The gravitational attraction of the Earth keeps such an encounter from producing a zero atmospheric velocity. The velocity that an object will achieve by falling toward the Earth from an infinite distance and in the absence of the Sun's gravitational attraction is 7 mi/s (11 km/s). Thus the maximum and minimum preatmospheric velocities attained are 44 and 7 mi/s (72 and 11 km/s). *SEE CELESTIAL MECHANICS; ORBITAL MOTION.*

Radiants and showers. A combination of the meteoroid's and Earth's velocities of travel around the Sun make the meteor itself seem to originate from a specific direction in the sky called the radiant. If there are numerous meteoroids in nearly the same orbit (sometimes incorrectly called meteor streams), the Earth sweeps them up at specific times of the year and a so-called meteor shower is observed. Meteor showers are named after the constellation or single star in the sky from which they appear to radiate. While shower meteoroids are really moving nearly parallel through space and result in nearly parallel meteor

Table 1. Major meteor showers

Shower	Duration	Approx. date of maximum	Approx. radiant coordinates, degrees		Meteoroid orbital speed		Strength*	Suggested parent body	Notes
			Right ascension	Declination	mi/s	km/s			
Quadrantids	Jan. 1–6	Jan. 3	230	+49	27	42	M	—	Sharp maximum
Lyrids	Apr. 20–23	Apr. 22	271	+34	30	48	M–W	1861 I	Good in 1982
π Puppids	Apr. 16–25	Apr. 23	110	−45	—	—	W–M	Grigg-Skjellerup	Highly variable†
η Aquarids	Apr. 21 – May 12	May. 4	336	−2	40	64	M	Halley	Second peak May 6
Arietids	May 29 – June 19	June 7	44	+23	24	39	S	—	Daytime shower
ζ Perseids	June 1–17	June 7	62	+23	18	29	S	—	Daytime shower
β Taurids	June 24 – July 6	June 29	86	+19	20	32	S	Encke	Daytime shower
S. δ Aquarids	July 21 – Aug. 25	July 30	333	−16	27	43	M	—	Primary radiant
S. ι Aquarids	July 15 – Aug. 25	Aug. 6	333	−15	19	31	W	—	Primary radiant
Perseids	July 23 – Aug. 23	Aug. 12	46	+57	37	60	S	1862 III	Best-known shower
Orionids	Oct. 2 – Nov. 7	Oct. 21	94	+16	41	66	M	Halley	Trains common
S. Taurids	Sept. 15 – Nov. 26	Nov. 3	50	+14	17	27	M	Encke	Known fireball producer
Leonids	Nov. 14–20	Nov. 17	152	+22	45	72	W–S	1866 I	Next peak due 1999?
Puppids-Velids	Nov. 27 – Jan.	Dec. 9	135	−48	—	—	M	2102 Tantalus?	Many radiants in region†
Geminids	Dec. 4–16	Dec. 13	112	+32	23	36	S	3200 Phaethon	Many bright meteors

*Estimate of relative meteor hourly rate for visual observers: S = strong (sometimes above 30 per hour at peak); M = moderate (10 to 30 per hour at peak); W = weak (5 to 10 per hour at peak).

†Radiant coordinates are those of the apparent radiant rather than the geocentric radiant.

Section of lunar rock in crossed Nicols showing zoned pyroxene, plagioclase, and ilmenite. (*NASA*)

Rock from surface of the Moon. (*NASA*)

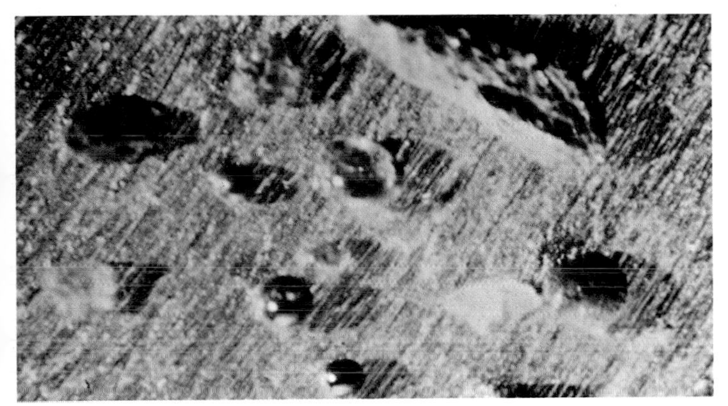

Microscopic view of spherules from lunar soil which was collected by the *Apollo 11* astronauts. (*NASA*)

Photomicrographs of thin sections of lunar rock in polarized white light. Colors are caused by the interaction of the polarized light with the crystalline structure of the various minerals. Each color usually represents a different mineral. (*NASA*)

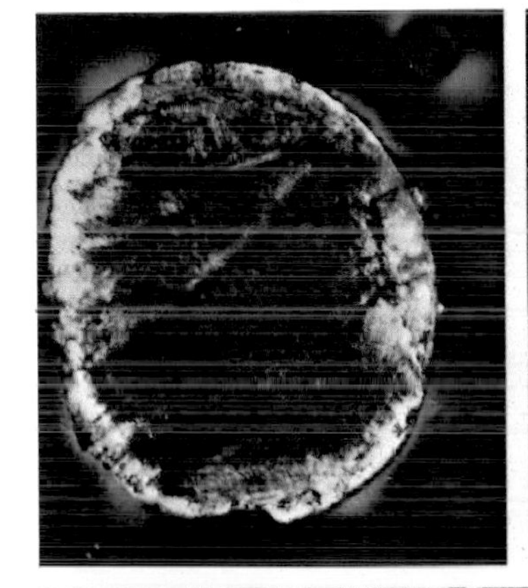

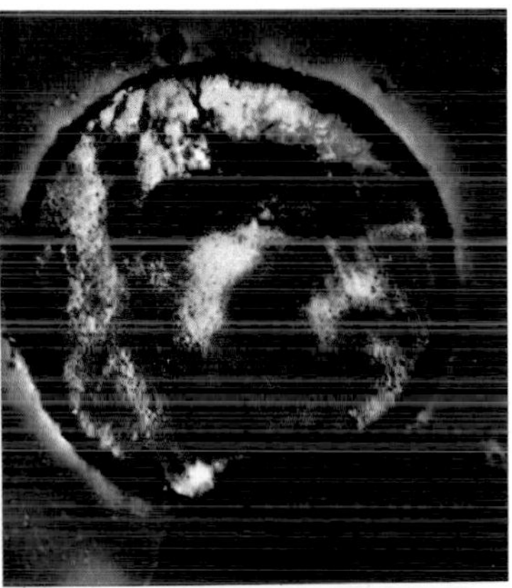

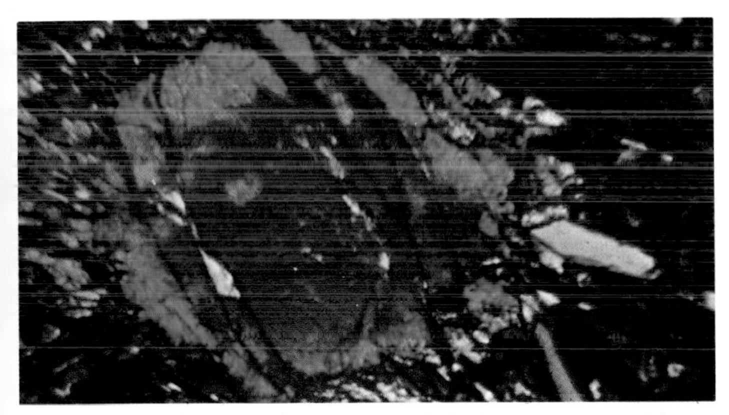

Thin section of lunar rock showing relatively large equant crystals. Porphyritic texture suggests differing rates of growth of the large crystals versus the groundmass minerals. Large crystals are olivine and the groundmass minerals are pyroxenes, feldspars, and metal compounds. Large crystals are approximately 1 mm across. The texture and mineralogy are both common in Earth volcanic rocks. (*NASA*)

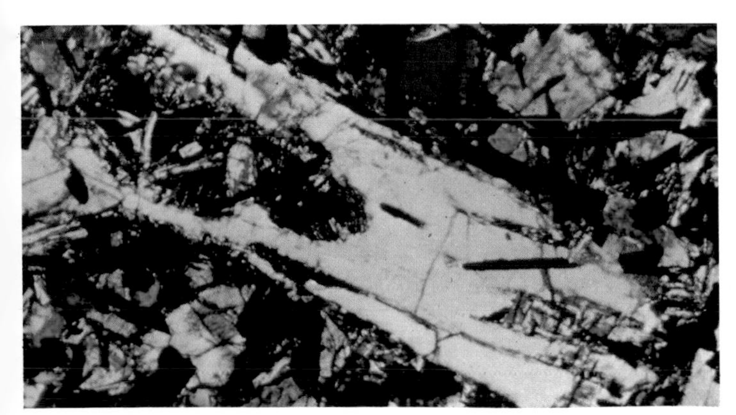

Photomicrograph of lunar sample. Light blue and white mineral is plagioclase; black mineral is ilmenite; blue and/or green and/or orange and/or yellow and/or red mineral is pyroxene. The large pyroxene is a phenocryst that had been partially resorbed. (*NASA*)

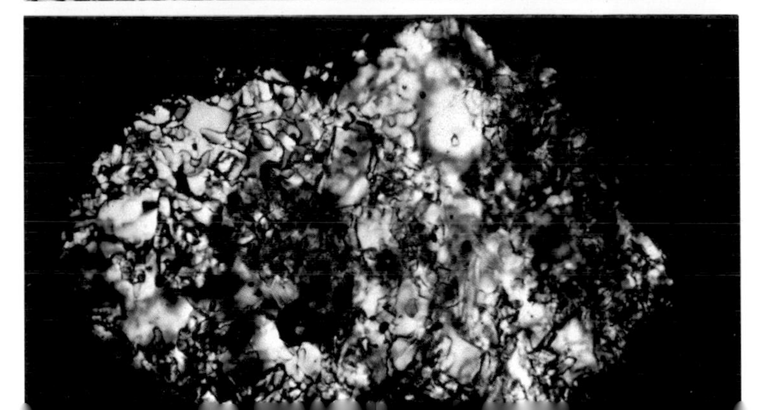

Table 2. Minor meteor showers*

Shower	Duration	Approx. date of maximum	Approx. radiant coordinates, degrees		Meteoroid orbital speed		Suggested parent body	Notes
			Right ascension	Declination	mi/s	km/s		
Coma Berenicids	Dec. 12 – Jan. 23	Jan. 17	186	+20	40	65	1913 I	Uncertain radiant position
α Centaurids	Jan. 28 – Feb. 23	Feb. 8	209	−59	—	—	—	Colors in bright meteors[†]
δ Leonids	Feb. 5 – Mar. 19	Feb. 26	159	+19	14	23	—	Slow, bright meteors
Virginids	Feb. 3 – Apr. 15	Mar. 13?	186	+00	21	35	—	Other radiants in region
δ Normids	Feb. 25 – Mar. 22	Mar. 14	245	−49	—	—	—	Sharp maximum[†]
δ Pavonids	Mar. 11 – Apr. 16	Apr. 6	305	−63	—	—	Grigg-Mellish	Rich in bright meteors[†]
σ Leonids	Mar. 21 – May 13	Apr. 17	195	−05	12	20	—	Slow, bright meteors
α Scorpids	Apr. 11 – May 12	May 3	240	−22	21	35	—	Other radiants in region
τ Herculids	May 19 – June 14	June 3	228	+39	9	15	—	Very slow meteors
Ophiuchids	May 19 – July	June 10	270	−23	—	—	—	One of many in region[†]
Corvids	June 25–30	June 26	192	−19	6	11	—	Very low speed
June Draconids	June 5 – July 19?	June 28	219	+49	8	14	Pons-Winnecke	Maximum only 1916
Capricornids	July – Aug.	July 8	311	−15	—	—	—	May be multiple[†]
Picis-Australids	July 15 – Aug. 20	July 31	340	−30	—	—	—	Poorly known[†]
α Capricornids	July 15 – Aug. 25	Aug. 2	307	−10	14	23	1948 n	Bright meteors
N. δ Aquarids	July 14 – Aug. 25	Aug. 12	327	−06	26	42	—	Secondary radiant
κ Cygnids	Aug. 9 – Oct. 6	Aug. 18	286	+59	15	25	—	Bursts of activity
N. ι Aquarids	July 15 – Sept. 20	Aug. 20	327	−06	19	31	—	Secondary radiant
S. Piscids	Aug. 31 – Nov. 2	Sept. 20	6	+00	16	26	—	Primary radiant
Andromedids	Sept. 25 – Nov. 12	Oct. 3	20	+34	11	18	—	"Annual" version
October Draconids	Oct. 10	Oct. 10	262	+54	14	23	Giacobini-Zinner	Can be spectacular
N. Piscids	Sept. 25 – Oct. 19	Oct. 12	26	+14	18	29	—	Secondary radiant
Leo Minorids	Oct. 22–24	Oct. 24	162	+37	38	62	1739	Probable comet association
μ Pegasids	Oct. 29 – Nov. 12	Nov. 12	335	+21	7	11	1819 IV	Probable comet association
Andromedids	Nov. 25	Nov. 25	25	+44	10	17	Biela	Once only, 1885
Phoenicids	Dec. 5	Dec. 5	15	−50	—	—	—	Once only, 1956[†]
Ursids	Dec. 17 – 24	Dec. 22	217	+76	20	33	—	Good in 1986

*Peak strength for visual observers usually less than 5 per hour.
†Radiant coordinates are those of the apparent radiant rather than the geocentric radiant.

trails, the effects of perspective make the meteors appear to diverge from the radiant (see **illus.**). If the meteor shower is particularly long-lasting, the radiant will appear to drift slowly in position from night to night as the relative directions of the Earth–meteoroid velocities change. Radiants are not geometric points, but areas in the sky that can be several degrees in diameter. Hence, quoted radiant positions are averages over such areas and often differ somewhat from one compilation to another. Meteors that cannot be shown to be associated with a known shower are termed sporadic meteors.

While all meteor showers show both day-to-day and year-to-year variations, some are more reliable than others, and these are called annual or major showers. A list of major showers is given in **Table 1** along with the approximate defining parameters of each, including a crude estimate of visual strength at maximum for a single observer. A list of minor showers of special interest is given in **Table 2**. In addition to those listed, there are about 100 other minor showers. Many minor showers have hourly rates so low that they are often not noticed except by experienced observers or when there is an unusual burst of activity. Some showers have been discovered by radio or radar methods to occur only during the hours of daylight, with no visible counterpart seen at night. Most of the radiants listed refer to geocentric radiants where the effects of the Earth's gravitational attraction have been removed. Radiant positions that have not been corrected for the Earth's attraction (usually because the meteoroid velocities are unknown) are called apparent radiants.

Stationary meteors. Occasionally, a meteor is seen coming directly toward an observer and shows a bright point of light rather than a trail. Such meteors are called, somewhat inaccurately, stationary meteors. The positions of these head-on meteors define the radiant precisely and are quite important to observe, particularly during showers. However, short glints of light from Earth-orbiting spacecraft and space debris are sometimes mistaken for stationary meteors. It has been suggested that a certain number of visible light pulses from cosmic x-ray and gamma-ray sources are also mistaken for stationary meteors. A thorough search through historical meteor records, however, fails to show clear evidence for this possibility. *SEE GAMMA-RAY ASTRONOMY; X-RAY ASTRONOMY.*

Bright meteors. Extremely bright meteors rivaling even the full moon (often bolides with associated sonic phenomena) are usually not associated with the major showers. Instead, they have meteoroid orbits that are more characteristic of those minor planets and short-period comets that are in highly eccentric orbits in the inner solar system. If the observed luminous-trail end points of these events are less than 18 mi (30 km) high in the atmosphere, there is a good chance that recognizable fragments (meteorites) will fall to Earth and perhaps be recovered. Thus most types of meteorites are believed to be samples of minor planets and possibly certain short-period comets. This connection with minor planets is verified by the fact that the reflecting properties of many minor planets resemble reflections from known meteoritic materials. *SEE ASTEROID; COMET.*

Origins of shower meteors. A number of meteor showers have been observed to be in orbits that are similar to those traveled by known comets. Thus an association between shower meteors and comets has gradually become a firmly entrenched concept, and a number of the best-proven cases are listed in Tables 1 and 2. There are numerous theoretical scenarios where vaporization of the more volatile cometary ices ejects small solid particles from the surface of the nucleus. A fair proportion of these fragments, particularly the smaller dust-sized ones, escape and take up their own orbits as meteoroids. Cometary nuclei have been known to split into two or more pieces and, when this occurs, it is likely that particles larger than dust size are released as well.

Gravitational attractions of the major planets and the disturbing effect of solar radiation pressure on individual particles tend to spread meteoroids out from the parent object position. Thus ''young'' showers are those that last only briefly (some as short as an hour or less), while ''old'' showers may show a few meteors per night but last a month or more. Many showers have a nonuniform structure along their orbits with highest meteoroid densities near the parent body. Lacking orbital synchronism with the Earth's position, these showers do not have an annual appearance at a reliable level. Instead, they show a tendency for strong showings to be separated by intervals roughly equal to the meteoroid orbital periods. The concentration of particles in these orbiting clumps can be relatively high, giving rise to brief deluges called meteor storms, where equivalent rates of thousands of meteors per hour have been noted for times that are at most an hour or so long. These numbers, however, give a false impression of the actual number density of meteoroids in space. With relative velocities on the order of tens of miles per second and a collecting area for each observer of a few hundreds of miles in diameter, the average separation between individual meteoroids is still a few thousand miles. Away from these maxima, however, the meteoroid number densities and hence the observed meteor hourly rates are quite low.

There are some instances where the Earth crosses the meteoroid stream twice per year, giving rise to two separate meteor showers. For example, Comet Halley gives rise to the May Aquarids and the October Orionids, while Comet Encke gives rise to the June Taurids and the November Taurids.

While the parent comet idea nicely explains many features of meteor showers, there are problems with this simple picture. First, certain minor planets resemble what might be termed extinct comet nuclei. Some of these have been observed at times with faint atmospheres, a main characteristic of comets. These identifications have been strengthened by the spacecraft observation of the properties of the nucleus of Comet Halley. Second, a perfectly respectable minor planet, 3200 Phaethon (which was discovered by the *Infrared Astronomical Satellite*), has an orbit nearly the same as the prominent Geminid annual shower visible in December. Evidence has also been found for several other minor planet connections with minor meteor showers. The meteoroid parent body question is apparently more complicated than previously thought.

Studies and research. Ground-based scientific interest in meteor studies has declined in North America since around 1970, when spacecraft methods of investigating the interplanetary dust became available. Outside North America, however, particularly in the Soviet Union, there still is considerable interest in meteor physics. In 1982, the Soviet Geophysical Committee of the Academy of Sciences of the U.S.S.R. proposed an international program of meteor observation called GLOBMET (Global Meteor Observation System), and this is being carried out as a joint project of several professional geophysical and astronomical groups. SEE INTERPLANETARY MATTER.

Photographic and electronic observations. The strategy of photographic measurements is to place at least two cameras 10–52 mi (15–85 km) apart over a known baseline, but arranged to examine the same volume of space at a height of about 56 mi (90 km). Each camera has a rotating shutter so that the meteor trail consists of a line of bright dashes (see illus.). Since meteor durations are fairly short, the shutters must spin at speeds of up to 10 times per second to accurately determine the meteor speed. While the early techniques utilized photographic film, ultrasensitive television and other electronic imaging devices have been increasingly used. It has also been customary to place a prism or grating over the front aperture of at least one of the shutter-equipped cameras, so that spectroscopic information can be obtained as well. Television observation of the evolution of meteor trains and wakes has led to a better understanding of upper-atmosphere photochemistry.

Meteor photography is one of the most difficult areas of astrophotography, even with ultrafast cameras and film. Chances for getting an image are better during major showers, because bright meteors are generally more numerous then. Meteor spectroscopy is even more difficult since the light is spread out over areas hundreds of times larger than the meteor trail itself. In spite of these difficulties, several amateur and professional astronomy groups worldwide operate successful bright meteor patrols using photography. SEE ASTRONOMICAL PHOTOGRAPHY; ASTRONOMICAL SPECTROSCOPY.

Radio and radar observations. Radio and radar observations depend on the fact that the initial ion-electron densities in a meteor trail are considerably higher than the average for the ionosphere at an altitude of 56 mi (90 km). For a radar system, the maximum reflected signal occurs when the meteor trail is at right angles to the outgoing wave, and for a system where the transmitter and receiver are separated, the maximum signal occurs when the meteor trail makes equal angles with the transmitter and receiver lines of sight. As the meteor trail forms, the ions and electrons pick up the speed of the original meteoroid and therefore show a Doppler effect that can be measured at the receiver. Thus, original meteoroid speeds can be estimated and, in certain cases, meteoroid radiants and orbits can be determined without reference to optical images. Radio and radar observations generally require a considerable amount of expertise to obtain scientifically valuable results. SEE RADIO ASTRONOMY.

David D. Meisel

Bibliography. V. A. Bronshten, *Physics of Meteoric Phenomena*, 1983; C. L. Hemenway, P. M. Millman, and A. F. Cook, *Evolutionary and Physical Properties of Meteoroids*, NASA SP-319, 1973; A. B. C. Lovell, *Meteor Astronomy*, 1954; D. W. R. McKinley, *Meteor Science and Engineering*, 1961; P. Roggemans (ed.), *Handbook for Visual Meteor Observations*, 1989; R. G. Roper (ed.), *The First GLOBMET Symposium*, vol. 25 of *Handbook for Middle Atmosphere Program*, International Council of Scientific Unions Scientific Committee on Solar

Terrestrial Physics and NASA, 1987; G. R. Sugar, Radio propagation by reflection from meteor trails, *Proc. IEEE*, 52(2):117–136, 1964.

Meteorite

A naturally occurring solid object from interplanetary space that survives impact on a planetary surface. While in space, the object is called a meteoroid, and a meteor if it produces light or other visual effects as it passes through a planetary atmosphere. Various sounds, including hissing and thunderous detonations, have also been reported for large meteors arriving at Earth. Explosive surface impacts by large meteorites are believed to have created the plethora of craters on the solid planets and moons of the solar system. Meteor Crater, Arizona, is Earth's most famous example of an impact crater. *See Meteor; Micrometeorite.*

A meteorite seen to strike a surface is known as a fall, whereas a meteorite discovered by chance is known as a find. In both cases, meteorites are named after their geographic places of recovery. Although meteorites were observed and even collected by people for thousands of years, their extraterrestrial origin was not accepted until after the insightful writings of E. F. F. Chladni (1794) and the fall of the L'Aigle (France) meteorite shower in 1803, which was witnessed by numerous, well-educated observers.

Meteorites have been broadly classified into stony, stony-iron, and iron varieties in recognition of their compositions that are dominated by silicate minerals and iron-nickel alloys either alone or as admixtures. Within each of the three categories, detailed classifications are based on distinctive mineralogical and chemical compositions and physical structures (see **Table 1**). Group names for unusual types have been chosen based on the established names of first-recognized specimens. For example, shergottites are so named because they belong to the same variety as the meteorite that fell at Shergotty, India, in 1865.

Meteorites represent the most ancient rocks known. Their ages, as determined by radiometric dating, extend to more than 4.5×10^9 years, which is thought to be near the time of solar system formation. As samples of primordial material, stony meteorites known as chondrites are studied for clues about how the solar system formed. In contrast, achondrites, stony-irons, and irons are samples of melt products formed during processing of solid material in planetary or preplanetary bodies. Chemical analyses of meteorites by geochemists in the 1930s through 1950s supplied first knowledge about abundances of chemical elements (other than hydrogen and helium) in the solar system. Modern research includes dissection of meteorites into their many complex components and subsequent analyses of their mineral, microchemical, and isotopic compositions, with the aim of learning when, where, and how the meteorites formed. *See Solar system.*

Since 1969, most newly discovered meteorite specimens have been recovered from glacial "blue ice" localities in Antarctica that seem to favor surface concentration of meteorites that have fallen on that continent over the past million years. Although other finds and, of course, the rare falls are still highly valued, the large numbers and new types found among the Antarctic meteorites have stimulated further me-

Table 1. Classification of meteorites

STONY	Chondrite*	C (petrologic types 1 through 4) or CI, CM, CO, CV, and CR E (1 and 2) H (3 through 6) L (3 through 6) LL (3 through 6)	
	Achondrite†	Aubrites Diogenites Eucrites Howardites "Lunar"	Angrites Chassignites Nakhlites Shergottites Ureilites
STONY-IRON		Lodranites Mesosiderites Pallasites Siderophyres	
IRON‡	Structural groups	Hexahedrites Octahedrites Nickel-rich ataxites	
	Chemical groups	IA through IC II through IIE IIIA through IIIF IVA and IVB	

*For C (carbonaceous) chondrites, two different but generally equivalent systems are in common use. Numbers 1–4 refer to a grid defined by characteristics observable with a petrographic microscope; I, M, O V, and R designate subgroups that share common features with particular specimens. For example, CO designates a meteorite that shows kinship with the Ornans (France) meteorite. CI is generally equivalent to C1, CM to C2, CO/CV to C3; CR ("Renazzo" subgroup) is distinct from other groups.

†Achondrite names in far-right column denote groups named after specific meteorite specimens; other names are based on petrological properties. To date, no special group name has been suggested for "lunar" meteorites.

‡For iron meteorites, the two different systems are alternatives rather than equivalents. Structural groups denote differences observed in sizes and compositions of metal crystals, whereas chemical groups (IA through IVB) are based on different concentrations of elements such as gallium and germanium.

teorite research. Asteroids are believed to be the sources of most meteorites. In 1982, however, it was conclusively demonstrated that a small achondrite found in Antarctica in 1981 was from the Moon—apparently propelled to Earth at some undetermined time by a large lunar impact event. Since then, a total of seven specimens of lunar rocks have been recovered as meteorites from Antarctica. Even more exciting is the prospect that eight closely related achondrites (four shergottites, three nakhlites, and the Chassigny meteorite), from various recovery locations around the world, are from Mars; one of the eight contains trapped gases that are nearly identical to those measured for the Martian atmosphere by the Viking lander in 1976. *See Asteroid; Mars; Moon.*

Meteoritics, the study of meteors and meteorites, is a premier example of interdisciplinary science, involving chemists, physicists, geologists, and astronomers allied through international research projects. Many advanced laboratory methods, especially for isotope-ratio measurements, have been motivated in large part by meteorite research problems.

James L. Gooding

Stony Meteorites

Stony meteorites include a large class known as chondrites and a smaller class known as achondrites.

Chondrites. The stony meteorites called chondrites, which are the most abundant class of known meteorites, constitute approximately 92% of all meteorite falls. Chrondrites are divided into three major categories: ordinary, carbonaceous, and enstatite. These categories are based on relative mineral abundances and compositions, aggregate mineral shapes and sizes (known as textures), and bulk chemical compositions. Most important, all chondrites contain various amounts of small (generally 0.5–2 mm) breadlike objects known as chondrules.

Ordinary chondrites. These are the most abundant chondrites, constituting 93% of all chondrite falls. They are composed mainly of the minerals olivine, low-calcium pyroxene, plagioclase, iron-nickel (Fe-Ni) metal, and troilite. They may contain silicate glass.

Chondrules are contained in a matrix of the same minerals that make up the chondrites. The difference is in texture. Chondrule minerals crystallized within

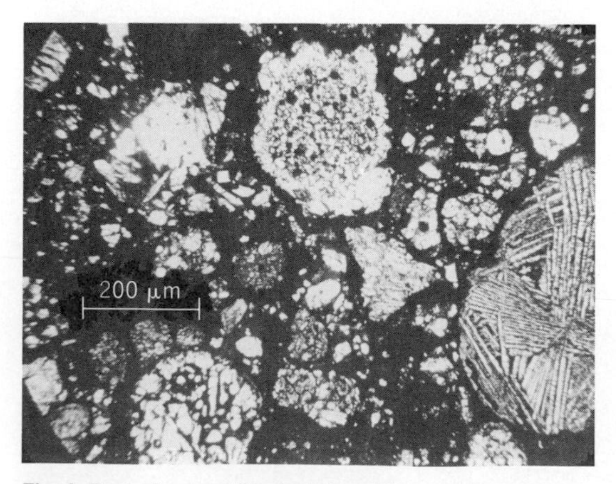

Fig. 1. Typical unequilibrated chondrite: many chondrules are present in a fine-grained mineral matrix.

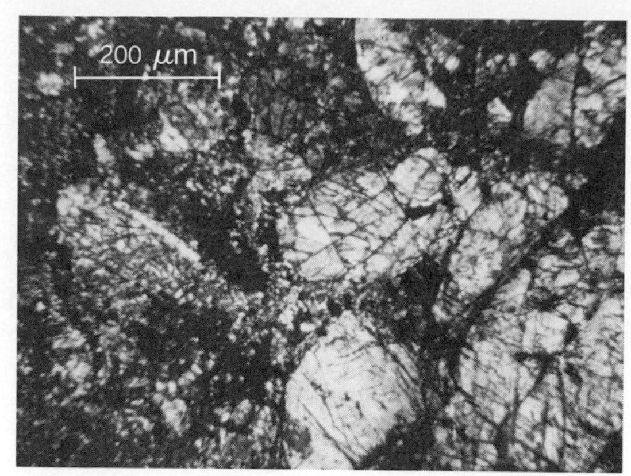

Fig. 2. Typical equilibrated chondrite: no chondrules are evident, and the matrix has relatively coarse mineral grain sizes.

molten droplets, and they show a variety of shapes consistent with a molten origin. Matrix minerals are granular and of small grain sizes (**Fig. 1**).

Olivine and low-calcium pyroxene are solid-solution minerals with respect to the iron (Fe^{2+}) and magnesium (Mg^{2+}) ions. Their compositions vary among the ordinary chondrites from fayalite 16 (16 mole %; Fa16) to Fa33 and ferrosilite 15 (Fs15) to Fs25. The variation is not continuous, and two distinct 1–2% Fa gaps divide ordinary chondrites into three groups: H group (Fa16 to Fa20; high iron), L group (Fa22 to Fa25; low iron), and LL group (Fa26 to Fa33; low low iron). Correlated with these Fa composition groups is the ratio of iron combined in metallic minerals (Fe-Ni metal and troilite) relative to the total iron content (which includes iron in silicate and oxide minerals). This ratio is 0.6 in the H group, 0.3 in the L group, and 0.1 in the LL group.

Within each iron group there is a continuous series of textural and mineralogical changes designated by numbers: H3 to H6, L3 to L6, LL3 to LL6. Each subgroup value shows similar characteristics. For example, type 3 ordinary chondrites (H, L, or LL, Fig. 1) contain abundant chondrules consisting of fine-grained minerals (olivine, low-calcium pyroxene, with or without metal and troilite) held together in silicate glass. The glass varies from chondrule to chondrule; however, it has a plagioclaselike composition. Within chondrules, and in the granular matrix surrounding chondrules, olivine and low-calcium pyroxene grains exhibit great differences in composition from grain to grain. Adjacent grains have different Fa and Fs contents, and so these meteorites are known as unequilibrated; this feature is a characteristic of type-3 ordinary chondrites. Variations also occur in nickel content of metal grains. Crystalline feldspar is not present; there is only chondrule glass, which has a plagioclaselike composition.

Each numerical iron subgroup shows a transition in these characteristics into the next subgroup. Ultimately, type 6 has no glass but has crystalline plagioclase; olivine and low-calcium pyroxene have uniform compositions from grain to grain, and so these meteorites are referred to as equilibrated. The average grain size is larger than in lower numerical groups, and chondrules are almost entirely absent (**Fig. 2**). These changes are attributed to a process of metamorphism.

Carbonaceous chondrites. These are the most primitive of all meteorites. In addition to the major chemical elements that occur in all of the chondrite meteorites, carbonaceous chondrites contain significant amounts of carbon, hydrogen, and nitrogen, which are present in only trace amounts in ordinary chondrites. In addition to chondrules, they have a large number of other types of inclusions. Most important is the fact that the minerals that make up chondrules and inclusions are different in composition and kind from minerals that make up the surrounding matrix. Carbonaceous chondrites fall into types 1, 2, and 3.

In type 1 carbonaceous chondrites, chondrules are extremely rare. Matrix constitutes 99% of these meteorites and consists of clay minerals and similar layer-lattice silicates. Water is a structural component in these minerals; the type 1 carbonaceous chondrites contain an average of 20% water by weight.

Within the matrix are about 1% small crystals (≤100 micrometers) of olivine and pyroxene, which could have formed only at high temperatures (750°–1500°C or 1380–2730°F). These are known as refractory minerals. Metal is entirely absent. The clay matrix could have formed only at low temperatures (<120°C or 250°F). Thus, these meteorites are extremely unequilibrated.

Running through the matrix are veins of epsomite and gypsum. Invisible within the matrix is a complex of organic molecules, most of them poorly characterized polymers, designated kerogen, that were formed by nonbiological processes. These compounds, along with graphite, account for much of the high carbon content (3.6%) of these meteorites.

Type 2 carbonaceous chondrites consist of approximately one-half clay and layer-lattice matrix minerals, which contain water and are associated with organic compounds. The remaining half consists of refractory minerals in several forms, including single crystals, crystal fragments, chondrules, and irregularly shaped inclusions. Olivine and low-calcium pyroxene are the most common refractory minerals, and chondrules make up only about 2% by volume. Metal is very rare. The textures of some inclusions suggest that they are aggregations of loose grains assembled together before incorporation into the matrix. A small number of inclusions consist of extremely refractory minerals, such as oxides and silicates of titanium (Ti), aluminum (Al), and calcium (Ca); these minerals formed at temperatures above 1500°C (2730°F). The combination of clay-rich matrix and refractory inclusions means these meteorites are highly unequilibrated (**Fig. 3**).

Type 3 carbonaceous chondrites have very much less water (~1%) and carbon (~0.5%) than other carbonaceous chondrites. This is because clays and layer-lattice silicates are rare. The matrix consists, instead, of submicrometer blades of high fayalite olivine (~Fa50). The matrix is only 35–40% by volume, with 60–65% consisting of larger single crystals, fragments, chondrules, and inclusions of refractory minerals (olivine and low-calcium pyroxene). Very refractory inclusions are relatively abundant. Thus, these meteorites are also unequilibrated.

Enstatite chondrites. These consist of 60–80 vol % enstatite (FsO) with 10–30 vol % metal and 5–15 vol % troilite. Enstatite chondrites range from those with many chondrules to those that are free of chondrules. Like the ordinary chondrites, they are divided into subgroups E3 to E6, with decreasing chondrule abundance. The lower numbers contain glass and no plagioclase; type 6 has plagioclase. Common minor components are cristobalite, tridymite, and quartz.

Most extraordinary are minor and trace minerals in the matrix: for example, oldhamite (CaS), sinoite (Si_2ON_2), and osbornite (TiN). The elements calcium, silicon, and titanium are always combined with oxygen in the silicate and oxide minerals of other chondrites. These unusual minerals can exist only under conditions of low oxygen activity, that is, under extremely reducing conditions. This is consistent with the absence of ferrous iron (Fe^{2+}) in low-calcium pyroxene. The matrix contains abundant microscopic graphite (0.4–0.8 wt %); organic compounds are unknown.

Origin. Chondrites contain components from two environments of origin: as results of processes that occurred under dispersed conditions in space, known as nebular; and as results of processes that occurred within parent bodies, known as planetary.

The carbonaceous chondrites have the most evident nebular components in the form of refractory mineral inclusions. These minerals formed by direct condensation from a gas cloud surrounding the primitive Sun. The minerals were accreted into small asteroidal bodies and never subjected to subsequent heat or pressure, that is, metamorphism that would have erased their primitive characteristics. During late stages of formation, as temperature in the nebular gas fell, lower-temperature minerals formed and were accreted together with high temperature, refractory minerals. Thus, carbonaceous chondrites contain the highest content of low-temperature volatile chemical elements among all meteorites.

Carbonaceous chondrites, however, are not entirely pristine. They have undergone some low-temperature alteration. It is not certain how much of this was nebular and how much was planetary.

The early formation and accretion histories of the ordinary chondrites and enstatite chondrites into asteroidal bodies are unknown. Postformation (planetary) processes are broadly understood. The progression from type 3 through type 6 is interpreted as reaction to metamorphism in parent bodies, destroying chondrules and homogenizing the compositions of minerals so that compositional disequilibrium disappears. Most ordinary chondrites and some enstatite chon-

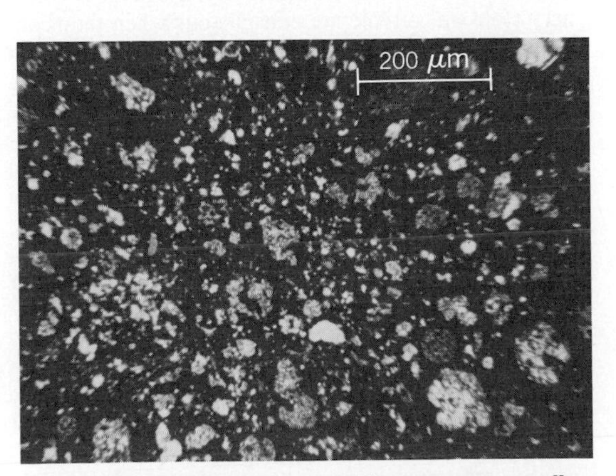

Fig. 3. Type 2 carbonaceous chondrite: numerous small grains of refractory minerals occur in a dense matrix of clay and other layer-lattice silicate minerals.

drites exhibit brecciation and shock effects caused by impacts on their parent bodies. The impacting bodies were, in most cases, other chondrites.

It is not known how the LL, L, H, and E groups are specifically related to each other. The LL-chondrite minerals show the highest oxygen activity, and the E-chondrite minerals show the least. This has been interpreted as formation at different distances from the primitive Sun: E-chondrites form nearest the Sun, and the H, L, LL, and carbonaceous chondrites form progressively farther away. This is a hypothetical model. One conclusion is certain: all chondrites could not have come from a single parent body.

Edward Olsen

Chondrules. These are the most abundant particles in chondrites, generally being ~1 mm (0.04 in.) in diameter. They contain iron-magnesium silicates that crystallized from a melt. Chondrules are distinguished from other components in chondrites—fine-grained matrix, unmelted aggregates, and refractory (Ca-Al-rich) melted particles—by textural and compositional criteria. The properties of chondrules, established from unequilibrated chondrites that have escaped modification in asteroidal parent bodies, indicate that extensive melting took place in the solar accretion disk (solar nebula).

Mineralogy and textures. The most abundant minerals of chondrules are olivine ($Fo_{100} - Fo_{40}$), pyroxene (mainly inverted protopyroxene), glass rich in a feldspar component, iron metal, and iron sulfide (FeS). Olivine crystals in many unequilibrated chondrules show enrichment in iron, calcium, and other elements toward their rims. The main textural categories of chondrules are granular, porphyritic (more strictly microporphyritic), barred, and radial/excentroradial, based on abundance, size, and shape of crystals. There are also cryptocrystalline and glassy chondrules. A shorthand notation for chondrules combines the texture and dominant mineral(s): thus there are barred olivine (BO), radial pyroxene (RP), and porphyritic olivine pyroxene (POP) chondrules.

Chondrules contain a number of anomalous inclusions that did not crystallize from the chondrule melt. These include olivine (forsterite), which is almost free of iron but contains high concentrations of calcium and aluminum. The refractory composition and complex zonation patterns seen in cathodoluminescence images suggest an origin by several cycles of evaporation–condensation. So-called relict grains of ordinary (ferroan) olivine are conspicuous when they contain inclusions of iron metal due to reduction, and clear rims richer in magnesium caused by interaction with the chondrule melt. Additional inclusions inside chondrules are other chondrules and patches of fine-grained matrix material. The anomalous inclusions in chondrules show that chondrules were formed by melting solid precursors, in part nebular condensate material.

Chondrules experienced additional events between crystallization and final incorporation into chondrites, for example, they acquired several kinds of rims. Very fine-grained rims containing ferroan olivine, pyroxene, metal, and sulfide resemble opaque matrix material. Apart from some metal and sulfide that may have been lost from molten chondrules, the rims represent grains that were sintered onto the chondrules. The compositional variation in rims and in matrix samples suggests that they formed by nebular processes rather than in a regolith. Rim grains could be condensates or possibly annealed interstellar dust grains. Coarse-grained rims suggest more extensive heating or slower cooling, either before or after sintering onto chondrules. In some cases, fine-grained rims are superimposed on coarse-grained rims. Ferroan olivine also forms rims on chondrules in the Allende (Mexico) meteorite as overgrowths on individual olivine crystals, with compositions explicable as a result of gas–chondrule reaction (condensation).

Chondrules were abraded and fractured, in some cases before rim deposition. Compound chondrules, and pits on chondrule surfaces, also suggest collisions between chondrules. Fractured chondrules are richer in iron than round chondrules, suggesting that collisions took place in a warm environment where magnesium in the chondrule could be replaced by iron from outside.

Chemical and isotopic data. Despite their igneous textures, the major element concentrations in chondrules do not vary with the simple, systematic trends associated with igneous rocks in response to processes such as fractional crystallization. The rather random compositions cannot be explained by simple vapor condensation processes, either. The precursors were apparently assembled from random mixtures of refractory and volatile materials. The bulk compositions can be rationalized by the random deposition of segments of a fractional condensation sequence on relatively refractory grains migrating because of gas turbulence and by the mixing of condensates with little-processed interstellar dust and debris from early chondrules. Volatile components may have condensed to varying extents in porous precursor or aggregate particles. Statistical analysis of major- and trace-element data has led to the identification of mineral precursors, which differ in composition for different major chondrite groups. This suggests different nebular processing in different regions of the accretion disk.

Radiometric data indicate that chondrules formed at the same time as the solar system, but the techniques are not sensitive enough to distinguish nebular events from early parent body events. Oxygen isotopes show that, although the chondrules of carbonaceous and ordinary chondrites have similar textures, minerals, and bulk compositions, they come from two distinct reservoirs. Ordinary chondrite chondrules contain a component relatively poor in ^{16}O, and carbonaceous chondrite chondrules contain a component relatively rich in ^{16}O. Both of these precursor phases were modified isotopically during processing in nebular gas with a terrestrial oxygen isotopic composition. Either the precursors partially exchanged oxygen with nebular gas during melting or the precursors were mixtures of presolar relics and nebular condensates.

Crystallization experiments. An effort has been made to define the melting and crystallization conditions of chondrules by duplicating their textures and mineral compositions experimentally (**Fig. 4**). Synthetic chondrules corresponding to a range of chondrule compositions have been produced as droplets at 1500–1600°C (2700–2900°F) suspended on platinum wire loops and crystallized in an atmosphere giving appropriate redox conditions. The influence of melting time coupled to grain size and initial temperature have not been investigated systematically, but the most satisfactory textural matches have been produced by melting for 30 min or greater. The combined influences of initial heating temperature, cooling rate, and bulk composition have been studied.

Fig. 4. Porphyritic olivine. (*a*) Natural specimen, consisting of clear olivine crystals in dark devitrified glass surrounded by fine-grained dark matrix in unequilibrated ordinary chondrite. (*b*) Synthetic spherule produced experimentally by melting at 30°C (54°F) below the liquidus temperature, showing gravitational settling; synthetic chondrules match natural chondrules in composition zoning as well as textures for cooling rates between 100 and 1000°C (180 and 1800°F) per hour.

The dynamic crystallization experiments have shown that chondrule textures are controlled primarily by extent of initial melting, that is, abundance of heterogeneous nuclei when cooling is initiated. If a charge is modestly superheated, a glassy spherule may result. If the initial temperature is close to the liquidus, barred olivine and radial pyroxene textures may result. With initial temperatures increasingly below liquidus temperature, more and more nuclei lead to more and smaller crystals, yielding microporphyritic to granular textures. Cooling rates much below 100°C (180°F) per hour produce olivine crystals larger than, and lacking the composition zoning of those in natural chondrules. Cooling rates much above 1000°C (1800°F) per hour tend to produce more skeletal or dendritic crystals than those observed in chondrules.

Texture–composition–temperature relationships. The extent of melting, and thus the texture, can be controlled by changing the initial temperature or the bulk composition. The whole range of chondrule textures has been produced experimentally in a suite of charges varying in iron/magnesium ratio, with fixed initial temperature and fixed cooling rate. Since a single hot droplet would cool very rapidly by radiation, it appears that chondrules were produced in large batches. Therefore, natural chondrules may have experienced much the same initial temperatures, and their wide range of compositions may have been a specific factor controlling their texture.

Although chondrules with average compositions can show a variety of textures, there is a correlation between textures and compositions. Magnesium-rich chondrules, which have high temperatures of total melting, are dominantly melted incompletely (porphyritic texture; **Fig. 5**). Compositions poor in magnesium, which have low calculated liquidus temperatures, show a high percentage of totally melted chondrule types (barred, radial, and cryptocrystalline textures). Fewer than 10% of chondrules formed above 1700°C (3090°F) are totally melted, and fewer than 10% below 1400°C (2550°F) are incompletely melted (Fig. 5); and so 1700–1400°C (3090–2550°F) is the temperature range for chondrule formation. The temperature 1700°C (3090°F) is considered to be the strict upper heating limit, but precursors could have been heated to temperatures below 1400°C (2550°F), which would cause so little melting that the objects would be classified as aggregates rather than chondrules.

Origin. Although the exact cause of chondrules is unknown, many of the various processes that might be suggested can be ruled out. Impact on planetary surfaces produces melt droplets, but they are mixed with abundant angular fragments. Volcanism produces melt droplets, but cannot explain bulk chemical and oxygen isotope heterogeneities or rims. Condensation of liquids from nebular gas is unlikely in view of solid inclusions in chondrules, as well as compositional data.

It has become widely agreed that chondrules were formed by nebular melting of solid precursors, consisting of randomly assembled condensates, presolar relics, and chondrule debris. They were hot for only a few hours in an environment with a low ambient temperature. This requires a local heating event, rather than a nebula-wide process. The high iron/magnesium ratios of chondrules are not consistent with formation in normal hydrogen-rich nebula, in which volatile elements like sodium would also be expected to boil away. This suggests that chondrules formed where oxygen pressures were high, perhaps as a result of concentration and evaporation of solids. The resulting high gas pressure in heated dusty re-

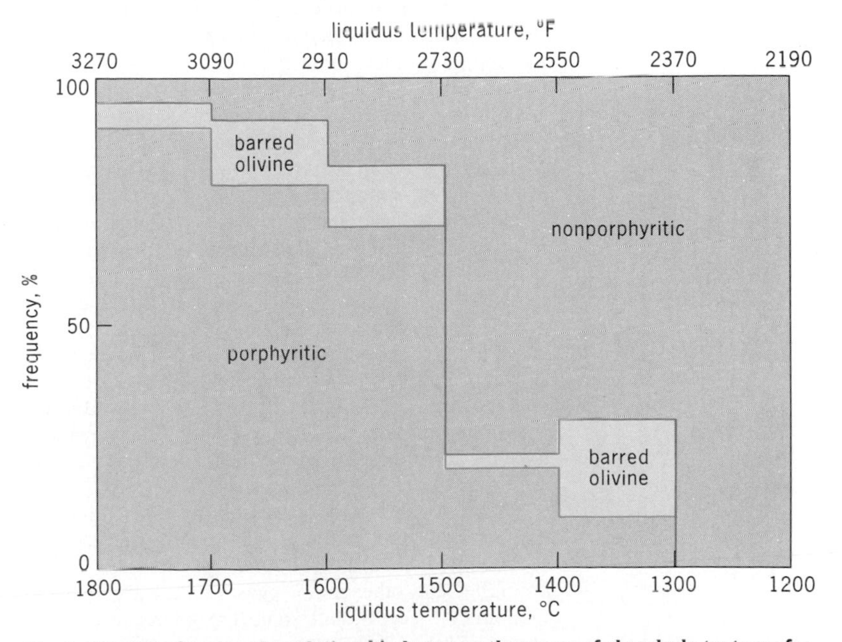

Fig. 5. Diagram showing the relationship between the range of chondrule textures for a given composition and calculated liquidus temperature.

gions would tend to prevent melts from evaporating. Solids could be expected to be concentrated beneath the surface of the disk after infall and subsequently in the midplane prior to accretion. Specific suggestions for melting mechanisms have ranged from lightning to frictional heating of infalling interstellar gains.

Roger H. Hewins

Achondrites. These are stony meteorites that have few, if any, chondrules and differ chemically from chondrites. They constitute about 8% of all meteorite falls and 1% of all finds. Although achondrites can be divided into several distinct groups based on chemical and isotopic composition, they are generally believed, based on aspects of their textures and composition, to have formed as the result of igneous processes on asteroidal or planetary bodies. Much of the interest in these meteorites derives from the fact that they provide clues into the nature of igneous processes and planetary differentiation early in the history of the solar system on planetary bodies outside the Earth–Moon system and on bodies presumed to be much smaller than the Earth and Moon.

Basaltic achondrites. The eucrites, howardites, and diogenites—often collectively referred to as the basaltic achondrites—are the most abundant achondritic meteorites. They appear to be samples of a series of related igneous rocks and of regolith breccias composed of fragments of these igneous rocks. They (along with the stony-iron meteorites, that is, mesosiderites and certain pallasites, and the iron meteorites) define a coherent group in terms of their oxygen isotope compositions, suggesting they are closely related. With ages near 4.5 billion years, they are products of igneous activity from the earliest history of planetary bodies in the solar system. They have been studied vigorously since the early days of the Apollo program, for they contain the most detailed record of early planetary differentiation, of igneous activity outside the Earth–Moon system, and of igneous activity on asteroid-sized bodies, from which they are presumed to have been derived.

Eucrites are dominantly composed of low-calcium pyroxene (usually pigeonite) and anorthitic plagioclase, with subordinate amounts of chromite, iron-rich metal, high-calcium pyroxene, silica polymorphs, and other minor phases. They are often, but not invariably, brecciated; most of the brecciated samples are monomict, but many of the more recently discovered samples from Antarctica are polymict. Many of the eucrites have pyroxene and plagioclase in ophitic and subophitic textures similar to those observed in terrestrial and lunar basalts. The textures of most of these eucrites and aspects of their chemical compositions, which cover a small range compared to terrestrial and lunar basalts, suggest that they are samples of crystal-poor magmas. Some of these eucrites have vesicles, suggesting an extrusive or hypabyssal origin. A small number have gabbroic textures and appear to be cumulates from liquids related to the more common eucrites. Many eucrites show features attributed to shock metamorphism and to thermal metamorphism.

The diogenites are composed nearly entirely of magnesian orthopyroxene, with subordinate amounts of olivine, silica polymorphs, plagioclase, metal, chromite, and other minor phases. They are typically monomict breccias, but some are unbrecciated, and others contain minor amounts of eucritic material. They range texturally from coarse- to fine-grained and show evidence of thermal metamorphism. They are generally believed to represent cumulates from magmas related to, but more magnesian than, known eucrites.

Based on comparison with lunar regolith breccias, with which they have some similarities, the howardites are generally considered to be samples of the regoliths of the parent planets of the eucrites and diogenites. They are polymict breccias that consist dominantly of angular to subrounded basaltic and pyroxenitic clasts and mineral fragments set in a finer matrix. They also contain chondritic fragments and rare glass beads (presumed to result from impact melting). Most of the fragments found in the howardites resemble eucritic and diogenitic material, and howardite bulk chemical compositions can be approximated as mixtures of these two types of meteorites. However, howardites also contain igneous fragments with textures and chemical compositions unknown among eucrites and diogenites. Many lithic and mineral clasts in howardites experienced a complex range of shock and metamorphic processes prior to incorporation into howardites. Ages of clasts in howardites extend from close to those of the oldest eucrites to 2×10^9 years younger; the young ages are usually interpreted as due to secondary, nonigneous processes.

The eucrites, diogenites, and individual fragments in howardites contain the record of extensive early igneous activity on what are generally assumed to be asteroidal parent bodies. The magmas from which this series of igneous rocks formed are thought to have been produced by partial melting of volatile-depleted, metal-depleted, olivine-rich planetary interiors. Although the relative roles of differing degrees of partial melting and fractional crystallization in the formation of the suite of igneous rocks sampled by these meteorite groups are controversial, it is clear that igneous processes as they are known from study of terrestrial and lunar rocks were active on small bodies very soon after their formation. The heat source for such igneous activity is still under investigation, but it could be the decay of ^{26}Al or perhaps heating by electric currents induced in small planets by the passage of an intense solar wind associated with a very active early Sun (T-Tauri phase). The reflectance spectrum of the surface of the asteroid 4 Vesta closely resembles those of eucritic meteorites, and it has been suggested that this could be the source of the basaltic achondrites, although there are dynamical difficulties with such a source. *See Solar wind.*

Shergottites, nakhlites, and chassignites. These are rare meteorites, but there is considerable interest in them because it has been suggested that they may have come from Mars. The four shergottites are composed primarily of pigeonite and augite pyroxenes plus glass with the composition of an intermediate plagioclase feldspar. The plagioclase glass is known as maskelynite, formed from crystalline plagioclase by shock metamorphism. The shergottites also contain titanomagnetite and rare hydrous amphibole; some contain olivine. The textures are similar to terrestrial diabases, and they have been interpreted as partial cumulates. Determination of their ages is controversial and may be complicated by the effects of shock metamorphism, but the ages are generally between 1.8×10^8 years and 1.3×10^9 years. Aspects of the chemical and isotopic composition of the shergottites suggest that they are derived from a parent body that experienced a complex, multistage history extending over much of the history of the solar system. Although they are basaltic like the eucrites, the shergottites are distinguished by their much higher volatile contents, the presence in them of iron(III) ion (Fe^{3+}; as opposed to metallic iron in the eucrites), and their much younger ages. In these respects the shergottites are more similar to terrestrial basaltic rocks.

The three nakhlites are augite-rich rocks whose textures strongly suggest they are cumulates. Augite appears to be the only cumulus phase; other phases include iron-rich olivine, pigeonite, sodic plagioclase feldspar, alkali feldspar, and titanomagnetite. The nakhlites contain hydrous alteration phases that may be preterrestrial in origin. Like the shergottites, the nakhlites have relatively young crystallization ages ($\sim 1.3 \times 10^9$ years) and appear to have formed in oxidizing, volatile-rich, Earth-like environments compared to the basaltic achondrites. Unlike the shergottites, the nakhlites show only minor effects of shock metamorphism.

The one chassignite is a lightly to moderately shocked dunite. Like the shergottites and nakhlites, it is volatile-rich and oxidized and also contains hydrous amphibole. Its age is not well constrained, but a single potassium–argon age is consistent with the young ages of the shergottites and nakhlites.

The young ages of the shergottite, nakhlite, and chassignite meteorites (often referred to as the SNC group), plus the similarity of the bulk compositions of the shergottites to that of the Martian soil as determined by the Viking landers, first led to the suggestion that these meteorites could be derived from Mars. It is difficult to conceive of a heat source for endogenous igneous activity (these meteorites have no features resembling known impact melts) on an asteroidal parent body at about 1.3×10^9 years ago; and given the limited choice of available larger planets, Mars seemed the most likely choice. The similarity of relative noble gas and nitrogen abundances and isotopic ratios in the Martian atmosphere and shock-produced glass in one shergottite provide strong support for this hypothesis. The very low paleomagnetic intensities of the shergottites are also consistent with a Martian origin. It is still a subject of controversy whether fragments of sufficient size to explain measured cosmic-ray exposure ages could be ejected more or less intact from Mars by impact and subsequently delivered to Earth. It is generally accepted that the question of whether or not these meteorites are from Mars will be resolved only after a sample-return mission to Mars.

Angrites. Until the discovery of two additional examples of this class of meteorite in Antarctica, Angra dos Reis was the only meteorite of this type. Nevertheless, it has been of great interest because of its unusual mineralogy and very primitive isotopic characteristics. Angra dos Reis is composed nearly entirely of an iron-bearing, aluminous clinopyroxene (fassaite), with minor amounts of plagioclase, spinel, calcic olivine, and a variety of other phases. The more recently discovered members of this group are less enriched in fassaite, but their mineralogies are similar. Angra dos Reis is generally interpreted as having formed by accumulation of fassaite from a relatively silica-poor magma, while the Antarctic examples have been interpreted as more closely resembling magmatic compositions. With an age of 4.55×10^9 years and evidence for extinct short-lived radionuclides, Angra dos Reis provides another case of magmatic activity from the very earliest history of the solar system. However, based on the inferred characteristics of the parent liquids of the angrites, this igneous activity was distinct from that recorded by the basaltic achondrites. Due to their rarity, it has been difficult to constrain the details of their petrogenesis, but efforts to do so have generally suggested complex igneous histories.

Ureilites. The dominant silicates of this strange meteorite type are olivine and low-calcium pyroxene (typically pigeonite), usually present as millimeter-sized mosaic aggregates. The silicates are enclosed in a dark matrix that consists mainly of metallic iron, graphite, diamond, lonsdaleite (another high-pressure form of carbon), plus other minor phases. Where the matrix and silicates are in contact, they have reacted at low pressures to form iron-rich metal and more magnesian silicates. Some of the ureilites also exhibit long, oriented cracks or voids. In terms of their oxygen isotopic compositions, the ureilites are distinct from other achondrites, but are similar to certain carbonaceous chondrites.

The silicates often are preferentially oriented. This, along with their relatively coarse grain size, suggests that they are plutonic rocks. However, it is generally difficult to distinguish between accumulations of crystals settled from a magma chamber and residual crystals left behind after the extraction of partial melts from planetary interiors, and both origins have been suggested for the ureilites. In either case, aspects of their trace-element and isotopic compositions suggest a complex igneous history for the ureilites if these features are all igneous in origin. The carbon in the ureilites could be a primary igneous feature, or it could have been introduced subsequent to their igneous history, perhaps during the shock events that led to the mosaicism of the silicates and the presence of the high-pressure phases diamond and lonsdaleite in the matrix. Aspects of the textures of these meteorites tend to favor the former possibility. If so, the coexistence of graphite plus iron-bearing silicates during igneous activity suggests somewhat elevated pressures (that is, more than a few hundred bars). Although no direct connection has been made, many researchers in this field believe that the parent bodies of the ureilites were related to those of carbonaceous chondrites.

Aubrites. These meteorites, also known as enstatite achondrites, are composed almost entirely of polymorphs of enstatite ($MgSiO_3$), with grain sizes sometimes exceeding several centimeters. Most are monomict breccias, but unbrecciated and polymict examples are known. At least some of these meteorites appear to be regolith breccias, based on their textures and the presence of foreign meteorite clasts and solar rare gases. Like the enstatite chondrites, they contain silica polymorphs, metallic iron with dissolved silicon, and an assortment of rare, highly reduced minerals, some of which are known only from these meteorites. The assemblages found in these meteorites indicate that they formed under conditions more reducing than expected in a gas of solar composition.

It is generally accepted that the aubrites are related to the enstatite chondrites, but the relationship is not clear. They are chemically similar to enstatite chondrites, but they are distinctive; in most cases they extend to the more extreme fractionations the same trends observed among the enstatite chondrites. The very coarse grain size of the aubrites suggest that they formed as cumulates, perhaps during differentiation of an enstatite chondrite parent body. Alternatively, it has been suggested that they originated by fractionation processes occurring in the solar nebula and that they represent a continuation of the same processes by which the various nonigneous enstatite chondrites are related. According to this view, the coarse grain size of the aubrites would have resulted from annealing in high-temperature nebular environments. SEE NEBULA.

Edward Stolper

Iron Meteorites

Iron meteorites are pieces of once molten metallic cores and pools in asteroids that were subsequently eroded and fragmented by impacts after slow cooling. About 650 different iron meteorites have been identified; 30 were seen to fall, and the rest fell during the last million years. The smallest iron meteorites, which weigh only 5–30 g (0.18–1.1 oz), were found in Antarctica and are aerodynamically shaped to resemble buttonlike tektites. The largest single iron meteorite weighs about 60 metric tons (66 tons) and still lies in Namibia. The second largest is from Greenland and weighs 31 metric tons (34 tons). It is on display in the American Museum of Natural History in New York. SEE TEKTITE.

Nine much larger iron masses also hit the Earth during the last million years, forming craters 100 to 1200 m (330 to 3900 ft) in diameter. However, each of the fragments surviving from these meteoroids weighs less than a ton. The largest and most famous crater, which is in Arizona, was formed about 50,000 years ago by the impact of a meteoroid weighing around 300,000 metric tons (330,000 tons) and measuring about 40 m (130 ft) across. The impact released energy equivalent to about 15 megatons of TNT.

Mineralogy. When iron meteorites are sawed, ground, polished, and etched, they typically show a striking geometrical array of oriented crystals known as a Widmanstätten pattern. This pattern results from the crystallographically controlled formation of kamacite plates parallel to the four sets of octahedral planes in the precursor taenite crystals (**Fig. 6**). Kamacite and taenite are the mineral names for the body-centered cubic and face-centered cubic crystal structures of iron alloys. Iron meteorites that show such an oriented array of kamacite plates in taenite are called octahedrites and contain 6–15% nickel. A few iron meteorites known as hexahedrites contain 5–6% nickel and are almost entirely composed of kamacite. Even rarer are the ataxites which have more than 15% nickel and contain only microscopic kamacite plates.

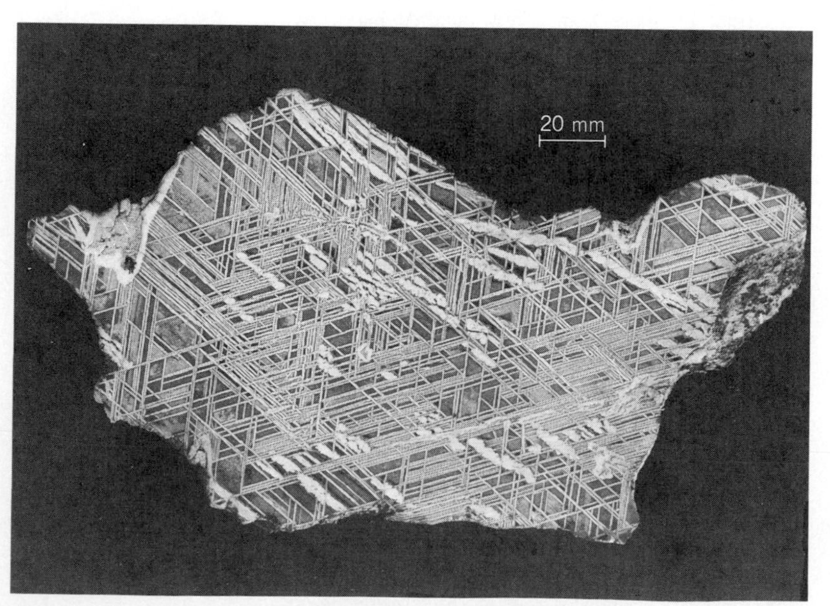

Fig. 6. A piece of the Edmonton (Kentucky) iron meteorite that has been sliced, polished, and etched to show the Widmanstätten pattern. The plates of kamacite 0.3 mm (0.01 in.) wide are oriented parallel to the faces of an octahedron. (*Smithsonian Institution*)

Table 2. Common minerals in iron meteorites

Mineral	Composition
Kamacite	Fe-Ni; Ni $<7.5\%$
Taenite	Fe-Ni; Ni $>30\%$
Tetrataenite	FeNi (ordered)
Troilite	FeS
Daubreelite	$FeCr_2S_4$
Schreibersite	$(Fe,Ni)_3P$
Cohenite	Fe_3C
Haxonite	$Fe_{23}C_6$
Graphite	C
Carlsbergite	CrN
Chromite	$FeCr_2O_4$
Whitlockite	$Ca_3(PO_4)_2$
Chlorapatite	$Ca_5(PO_4)_3Cl$

The large sizes of the precursor taenite crystals and the oriented kamacite plates indicate that iron meteorites were hot when they formed and that they cooled at an extremely slow rate. Thus at one time iron meteorites must have been buried deep inside a large volume of poorly conducting rocky material. The cooling rate in the temperature range 700–400°C (1290–750°F) can be estimated from the thickness of the kamacite plates, the bulk nickel concentration, knowledge of the stability fields of kamacite and taenite, and the rate at which iron and nickel atoms diffuse in these minerals. From these data, computer models have been developed for the diffusion-controlled growth of kamacite plates; these models yield typical cooling rates of 10–100°C (18–180°F) per million years. Therefore, most iron meteorites must have been buried under tens of kilometers of rocky material when they cooled from 700 to 400°C (1290 to 750°F).

Most iron meteorites contain small amounts of a wide variety of other minerals; sulfides, phosphides, carbides, oxides, phosphates, and silicates together make up no more than a few percent by volume of most iron meteorites (see **Table 2**).

Although the mineralogy of iron meteorites largely reflects slow cooling at depth, some features result from shock waves of high intensity caused by impacts at velocities of several kilometers (1 km = 0.6 mi) per second. Kamacite in many iron meteorites has a distorted structure caused by transient formation of a hexagonal close-packed structure at shock pressures above 13 gigapascals (130 kilobars). Two iron meteorites contain diamonds formed from graphite by shock. Those in the Canyon Diablo iron formed during the impact responsible for the 1200-m (3900–ft) crater in Arizona, whereas diamonds in a closely related octahedrite from Antarctica probably formed during an impact on the parent asteroid. SEE SHOCK WAVE.

Chemical composition. The chemical compositions of iron meteorites provide important clues to their classification and the processes that formed them in molten asteroids. The two trace elements, gallium and germanium, which are present at concentrations by weight of between 0.01 and 1000 ppm (that is, between 10^{-6} and 10^{-1} wt %), are especially useful in classifying iron meteorites. Most iron meteorites (84%) belong to one of 13 chemical groups, all but two of which have concentrations of gallium and germanium that vary by less than a factor of two (Table 1). Each of the 13 groups, which are identified as IAB, IC, IIAB, and so forth, have between 5 and 190 members. This chemical classification of iron meteorites has not made the older, structural classification obsolete, because most iron meteorites (perhaps 70%) can be classified into chemical groups on the basis of structure alone. For example, nearly all hexahedrites belong to group IIA, the medium octahedrites (those having kamacite bandwidths of 0.5 to 1.3 mm or 0.02 to 0.068 in.) largely belong to group IIIAB, and ataxites to group IVB.

Analyses for other trace elements, such as iridium, gold, and tungsten and the minor elements phosphorus and nickel can also be used to classify iron meteorites, but they show larger variations within groups. For example, iridium concentrations vary by a factor of 6000 in group IIAB alone. The iridium variations within groups, which are very much larger than the variation in chondritic metal, were produced when molten metal solidified. Thus iridium, tungsten, and other elements that tend to reside in solid metal were enriched in the first metal to solidify, whereas elements that tend to reside in molten metal, such as nickel, phosphorus, and gold, became concentrated in the last metal that solidified.

Laboratory measurements of the concentrations of these elements in coexisting solid and liquid metallic iron-nickel allow the chemical trends found within the groups to be modeled accurately. Thus, nearly all of the 13 groups of iron meteorites formed from separate pools of molten metal; the 96 ungrouped iron meteorites probably come from another 80-odd pools. Cooling rates for samples from a single metallic pool should be uniform, and this is observed for most groups.

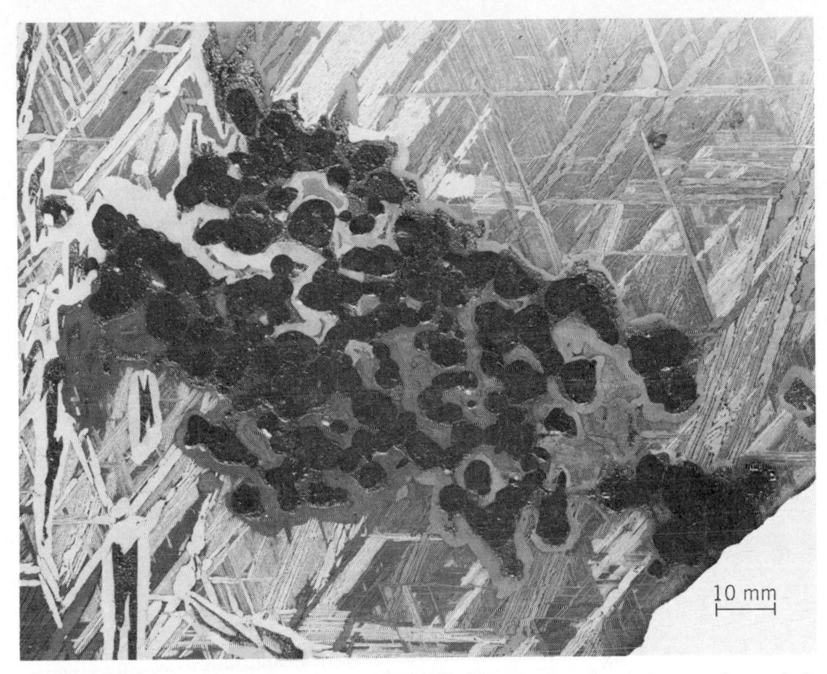

Fig. 8. A part of a slice of the Brenham (Texas) pallasite, showing a cluster of rounded, dark crystals of olivine enclosed in iron-nickel metal that shows a Widmanstätten pattern. Pallasites formed at the core–mantle boundary of asteroids that melted. (*Smithsonian Institution*)

Fig. 7. A slice of the Four Corners (southwestern United States) iron meteorite showing angular black silicate inclusions separating regions with independently oriented Widmanstätten patterns. Like other group IAB members, this meteorite did not form in the core of an asteroid. Impacts probably mixed silicate with metallic iron-nickel before the Widmanstätten pattern formed. (*Smithsonian Institution*)

Origins. The chemical and mineralogical evidence discussed above shows that iron meteorites formed from molten pools of metal that solidified and then cooled over many millions of years. This evidence is consistent with an origin for iron meteorites in the cores of asteroids that melted and differentiated. When an asteroid is partly melted, iron-nickel and iron sulfide, being denser than the associated silicates, will begin to sink to the center. With sufficient heating, a core of molten sulfur-rich metal will form. Since most iron meteorites contain no silicates and most achondrites have only trivial amounts of metal, it is likely that metallic cores are the source of many iron meteorites.

Two groups of iron meteorites (IAB and IIE) contain several volume percent of silicate inclusions and probably did not form in cores of asteroids. Instead these meteorites are probably derived from many metallic pools distributed within asteroids. Group IAB iron meteorites have angular silicate inclusions (**Fig. 7**), and it is likely that impacts mixed their metal and silicate components while they were hot. In addition, if the planetesimals from which the asteroids formed by accretion had already melted and solidified, it is possible that several metallic cores could have been combined into one asteroid.

Some stony iron meteorites known as pallasites are closely related to the iron meteorites. Pallasites consist of 5–80 vol % of olivine [$(Fe,Mg)_2SiO_4$] embedded in metallic iron-nickel (**Fig. 8**). They probably formed at the core–mantle boundary of differentiated asteroids when molten metal was intruded into the surrounding olivine mantle. Most pallasites have metal compositions that match those of the nickel-rich members of the IIIAB group of iron meteorites, implying that they formed in the same asteroid.

The metallic cores and pools that formed after the asteroids melted and solidified were subsequently stripped of surrounding silicate layers by impacts between asteroids and broken up. Astronomers believe they have identified many metallic asteroids 2 to 200 km (1.2 to 120 mi) in diameter that were produced in this way. Evidence for the timing and nature of the breakup of the meteorite parent bodies is given by the cosmic-ray exposure ages of the meteorites. Since galactic cosmic rays can penetrate only to depths of about a meter, meteoroids or asteroidal fragments are not significantly exposed to cosmic rays until they are a meter or less in size. Exposure ages of most iron meteorites are $0.1–1 \times 10^9$ years, but two groups show a significant clustering of ages: group IIIAB and IVA iron meteorites have ages of 6.5×10^8 and 4.5×10^8 years, respectively. Thus the 230 iron meteorites that make up these groups are fragments of just two giant collisions between asteroids that occurred at these times. *See* Cosmic rays.

Edward R. D. Scott

Isotopic Anomalies

In contrast to materials from differentiated planetary bodies such as the Earth and the Moon, primitive meteorites exhibit isotopic anomalies, that is, deviations from the average solar system composition (= "normal" composition) that are not the result of processes taking place in the solar system but are of presolar origin. These anomalies provide information about the nucleosynthetic sources of the material that formed the solar system. *See* Nucleosynthesis.

Carbon and all heavier elements are produced in stars, and their isotopic compositions reflect different nucleosynthetic reactions taking place in different stellar sources. Many different stars must have contributed to the mixture of gas and dust from which the solar system formed, and it is assumed that this mixture was originally chemically and isotopically heterogeneous. Before 1970 it was generally believed that presolar material had been completely vaporized and isotopically homogenized before the condensation of minerals and the accretion of planets. This dogma of a homogeneous solar nebula was shattered by the discovery of isotopic anomalies in an increasing number of different elements. Ample evidence has been found not only for the incomplete mixing of distinct isotopic components but also for the survival of presolar matter in primitive meteorites.

Isotopically anomalous material constitutes only a small fraction of primitive meteorites. The largest isotopic variations are found through the analysis of small samples where the effects are not diluted by isotopically normal material.

Isotopic effects in meteorites can be divided into four classes: (1) mass-dependent fractionation due to physicochemical processes (diffusion, evaporation, condensation), although certain chemical processes can also lead to non-mass-dependent fractionation that mimics isotopic effects of nuclear origin; (2) effects due to the decay of radioactive isotopes—while effects from the decay of long-lived isotopes are also seen in terrestrial samples, meteorites in addition exhibit effects from the decay of short-lived, now extinct isotopes; (3) nuclear effects reflecting different nucleosynthetic processes in stellar sources; (4) effects due to the irradiation of meteoritic samples by galactic and solar cosmic rays, which provide information on the exposure history of samples on their parent bodies and in interplanetary space.

Isotopic fractionation effects. Fractionation (F) effects believed to be due to evaporation and condensation processes are observed in refractory inclusions found in primitive meteorites, notably carbonaceous and unequilibrated ordinary chondrites. Elements exhibiting fractionation effects include oxygen, magnesium, silicon, calcium, and titanium. In so-called FUN inclusions, relatively large fractionation effects of oxygen, magnesium, and silicon are associated with unusually large unknown nuclear (UN) effects, but there exists no unique correlation between fractionation and unknown nuclear effects; that is, there are inclusions with large fractionation effects only with large unknown nuclear effects only. Fractionations in magnesium and silicon are correlated, with coarse-grained inclusions showing heavy-isotope enrichments characteristic of evaporation residues and fine-grained inclusions with light-isotope enrichments indicating a condensation origin. The lack of any correlation in fractionation effects between the more volatile magnesium and silicon and the more refractory elements calcium and titanium is evidence for multistage processing of many refractory inclusions.

Short-lived isotopes in the early solar system. The existence of a short-lived isotope at the time of the formation of a given sample of solid matter is indicated by an excess of the daughter isotope that is proportional to the concentration of the parent element. From meteoritic studies, there is evidence for the presence of certain radioisotopes in the early solar system (**Table 3**).

In contrast to stable isotopes, radioisotopes provide information on the time interval between nucleosynthetic production and solar system formation. The presence of the radioisotope ^{26}Al with a half-life of only 720,000 years in primitive meteorites sets stringent time constraints of at most a few million years between its production and the formation of solids in the early solar system.

Figure 9 shows $^{26}Mg/^{24}Mg$ versus $^{27}Al/^{24}Mg$ diagrams for two refractory inclusions. The good correlation between the two ratios indicates that the variation in the $^{26}Mg/^{24}Mg$ ratios is due to the decay of ^{26}Al. The slopes of the correlation lines give the initial $^{26}Al/^{27}Al$ ratios. They are different for a refractory inclusion from the Allende carbonaceous chondrite and a hibonite inclusion from the Dhajala ordinary chondrite. Generally, while many inclusions have initial ratios of 5×10^{-5}, in many other samples the ratio of $^{26}Al/^{27}Al$ is smaller or there is no indication of ^{26}Mg excesses from the decay of ^{26}Al. It is unclear

Table 3. Short-lived isotopes in the early solar system

Radio-isotope	Half-life years	Daughter product	Abundance
^{26}Al	7.2×10^5	^{26}Mg	$^{26}Al/^{27}Al \approx 5 \times 10^{-5}$
^{53}Mn	5.3×10^6	^{53}Cr	$^{53}Mn/^{55}Mn \approx 4 \times 10^{-5}$
^{107}Pd	6.5×10^6	^{107}Ag	$^{207}Pd/^{108}Pd \approx 2 \times 10^{-5}$
^{129}I	1.7×10^7	^{129}Xe	$^{129}I/^{127}I \approx 10^{-4}$
^{244}Pu	8.2×10^7	Fission Xe tracks	$^{244}Pu/^{238}U \approx 5 \times 10^{-3}$
^{146}Sm	1.03×10^8	^{142}Nd	$^{146}Sm/^{144}Sm \approx 5 \times 10^{-3}$

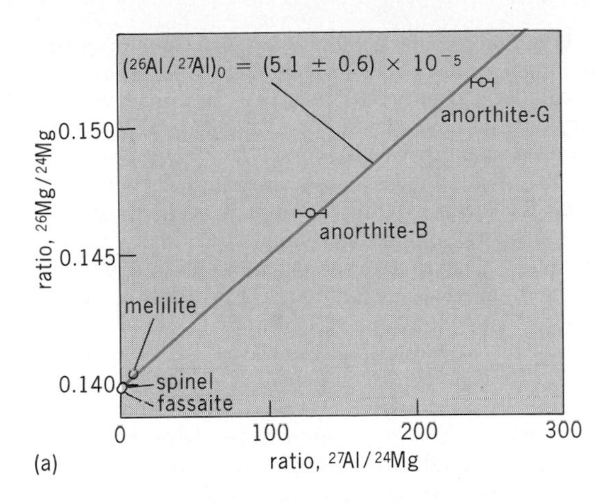

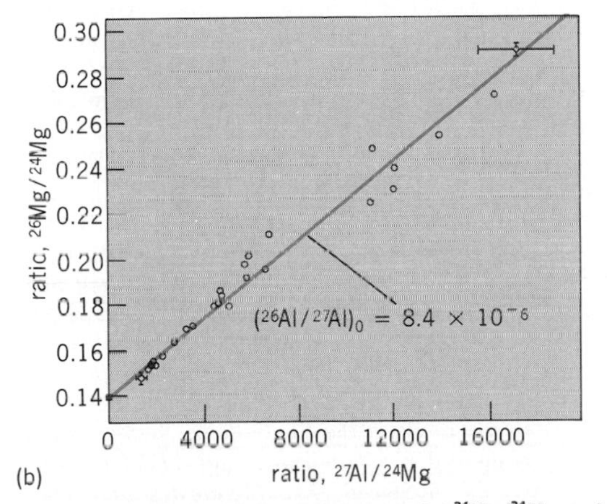

Fig. 9. Correlation diagrams between the $^{26}Mg/^{24}Mg$ and the $^{27}Al/^{24}Mg$ ratios measured in two different meteoritic samples. (*a*) Allende inclusion (*after T. Lee, D. A. Papanastassiou, and G. J. Wasserburg, Aluminum-26 in the early solar system: Fossil or Fuel?, Astrophys. J., 211:L107–L110, 1977*). (*b*) Sample of Dhajala hibonite inclusion (*after R. W. Hinton and A. Bischoff, Ion microprobe magnesium isotope analysis of plagioclase and hibonite from ordinary chondrites, Nature, 308:169–172, 1984*).

whether this reflects time intervals of several million years in the formation of refractory inclusions or a spatially heterogeneous distribution of ^{26}Al. Aluminum-26 has been considered as a possible heat source for melting small planetesimals, but it has not yet been established if it was distributed widely enough to play such a role.

Oxygen. The discovery of large ^{16}O excesses in refractory minerals from the Allende carbonaceous chondrite in 1973 triggered a reevaluation of the hot solar nebular model. Large variations in ^{16}O dominate the oxygen isotopic effects in meteorites with only small variations in the $^{17}O/^{18}O$ ratio. The observed patterns indicate mixing between a component rich in ^{16}O and one that is isotopically normal. Refractory inclusions were formed from a reservoir enriched in ^{16}O by about 5%. The ^{16}O enrichment may have a nucleosynthetic origin (hydrostatic burning of helium or explosive burning of carbon in supernovae), or it may result from non-mass-dependent physicochemical processes in the early solar system.

Nuclear anomalies in refractory rock-forming elements. As for ^{26}Al and ^{16}O effects, the largest effects in this category are found in refractory inclusions containing high-temperature phases. The most prominent nuclear anomalies are observed in the neutron-rich isotopes of the elements in the vicinity of iron in the periodic table, ^{48}Ca, ^{50}Ti, ^{54}Cr, and ^{64}Ni. Effects are pronounced in FUN inclusions, but are largest in hibonite ($CaO[Al_2O_3]_6$)–bearing inclusions and are qualitatively correlated in the sense that excesses and depletions in several of these isotopes accompany one another (**Fig. 10**). This correlation and the dominance of variations in the isotopes richest in neutrons is evidence for a nucleosynthetic origin of the anomalies, because these isotopes can be produced by nuclear processes taking place in the cores of massive stars undergoing supernova explosions. Smaller effects in other isotopes of calcium, titanium, and chromium due to other nucleosynthetic sources are also present, and at least four different nucleosynthetic components are required to explain the isotopic variation of these elements in primitive meteorites.

In refractory inclusions, there are also small effects in magnesium (not associated with ^{26}Al) and silicon, but they cannot be assigned unambiguously to a specific isotope. FUN inclusions also exhibit anomalies (magnitude of a few per thousand) in the elements strontium, barium, neodymium, and samarium. These anomalies indicate contributions from so-called r-, s-, and p-processes. The r-process involves synthesis of certain isotopes of elements heavier than iron by the rapid addition of neutrons in the high-neutron-density environments encountered during supernova explosions, while the s-process involves synthesis of isotopes by slow neutron addition under low neutron densities found in certain types of stable stars. The p-process synthesizes the proton-rich isotopes of the heavy elements. The meteoritic data show that the products of these processes are decoupled from one another, originating from distinct sources.

Circumstellar dust grains. Meteoritic materials are the carriers of isotopically anomalous noble gas components. There are many potential components; the following have the clearest nucleosynthetic signatures. Neon-E is essentially pure ^{22}Ne, believed to be the decay product of ^{22}Na (half-life 2.6 years), which in turn had its origin in nova explosions. Xenon-HL is characterized by enrichments in the light (^{124}Xe, ^{126}Xe, ^{128}Xe) and heavy (^{132}Xe, ^{134}Xe, ^{136}Xe) isotopes of xenon and probably results from explosive nucleosynthesis taking place during supernova explosions. Xenon-S exhibits characteristic excesses in the iso-

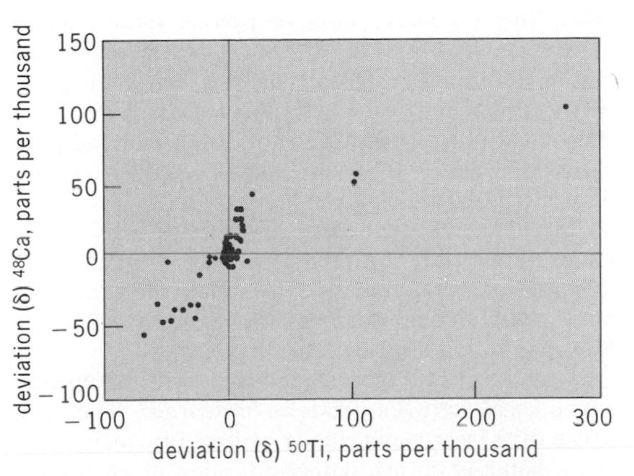

Fig. 10. Diagram showing deviations from terrestrial compositions of calcium and titanium in refractory inclusions; correlated anomalies in calcium-48 and titanium-50 indicate a nucleosynthetic origin.

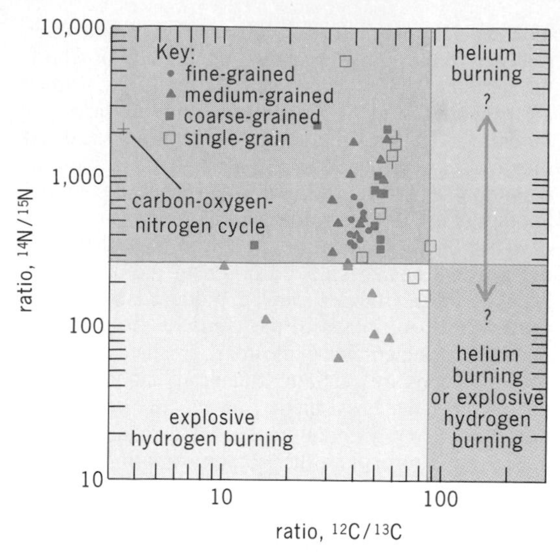

Fig. 11. Carbon and nitrogen isotopic variations measured in different silicon carbide samples of increasing grain size (0.2 to 10 μm) from the Murchison carbonaceous chondrite. The solid data points denote measurements on agglomerates averaging over many grains; for the largest sizes, measurements were made on single grains. Thin lines designate terrestrial compositions. Nucleosynthetic processes resulting in different isotopic ratios are indicated in the four quadrants. (*After M. Tang et al., Meteoritic silicon carbide and its stellar sources: Implications for galactic chemical evolution, Nature, 339:351–354, 1989*)

topes ^{128}Xe, ^{130}Xe, and ^{132}Xe that are produced by the s-process.

It has been known for some time that the carriers of these exotic components are carbonaceous, and chemical and physical separation in the laboratory has led to their purification and identification. The carrier of Xenon-HL was shown to be diamond with an average grain size of only 2.0 nanometers. In addition to xenon, it contains anomalous nitrogen (depleted in ^{15}N by about 34%), but its carbon isotopic composition is surprisingly normal. Neon-E and Xenon-S are carried by silicon carbide (SiC) 0.1 to 10 μm in size, the former apparently by the coarser grains, the latter by the finer grains. The silicon carbide itself is highly anomalous in its silicon and carbon isotopic compositions and carries anomalous nitrogen. Silicon isotope ratios vary by about 30%, showing ^{28}Si excesses and deficits, but also effects in at least a second isotope. The carbon and nitrogen isotopic ratios vary by large factors (**Fig. 11**). Silicon carbide grains could not have formed in the solar nebula with its high oxygen/carbon ratio, and the silicon carbide probably originated in the atmospheres of carbon-rich red giant stars or of novae. The first source would account for low $^{12}C/^{13}C$ and high $^{14}N/^{15}N$ ratios (left upper region in Fig. 11), the second for low $^{12}C/^{12}C$ and $^{14}N/^{15}N$ ratios (lower left region in Fig. 11). Infrared astronomical observations have identified SiC grains in dust shells around carbon-rich stars. *SEE CARBON-NITROGEN-OXYGEN CYCLES; STELLAR EVOLUTION.*

A number of other components with anomalous carbon and nitrogen exists in meteorites. The existence of these components is indicated by thermal release patterns during stepwise heating of chemically processed material, but identification of carrier phases has not yet been achieved.

Interstellar cloud material. Deuterium excesses (up to 10×) have been measured in carbonaceous and unequilibrated ordinary chondrites and in interplanetary dust particles. These anomalies appear to be associated mainly with organic matter. Since deuterium is destroyed in stars, these anomalies cannot be of nucleosynthetic origin. They are most likely due to the incorporation of interstellar cloud material into primitive meteorites. This material aquired high deuterium/hydrogen ratios through fractionation during ion-molecule exchange reactions that can take place at the very low temperatures (10 to 30 K or −442 to −406°F) characteristic of dense molecular clouds. Extremely large (up to 100,000×) enrichments of deuterium relative to the interstellar medium are observed astronomically in simple molecules from interstellar clouds. *SEE INTERSTELLAR MATTER.*

Ernst Zinner

Bibliography. C. A. Barnes, D. D. Clayton, and D. N. Schramm (eds.), *Essays in Nuclear Astrophysics*, 1982; F. Begemann, Isotopic anomalies in meteorites, *Rep. Prog. Phys.*, 43:1309–1356, 1980; V. F. Buchwald, *Handbook of Iron Meteorites*, 1975; R. N. Clayton, R. W. Hinton, and A. M. Davis, Isotopic variations in the rock-forming elements in meteorites, *Phil. Trans. Roy. Soc. Lond.*, A325:483–501, 1988; R. T. Dodd, *Meteorites: A Petrologic-Chemical Synthesis*, 1981; R. T. Dodd, *Thunderstones and Shooting Stars*, 1986; T. Gehrels and M. S. Matthews, *Asteroids*, 1979; R. Hutchison, *The Search for Our Beginnings*, 1983; J. F. Kerridge and M. S. Matthews (eds.), *Meteorites and the Early Solar System*, 1988; E. A. King (ed.), *Chondrules and Their Origins*, 1983; H. Y. McSween, Jr., *Meteorites and Their Parent Planets*, 1987; E. R. D. Scott and G. J. Taylor, Chondrules and other components in C, O and E chondrites: Similarities in their properties and origins, Proceedings of the 14th Lunar and Planetary Science Conference, *J. Geophys. Res.*, Suppl. 88, pp. B275–B286, 1983; J. T. Wasson, *Meteorites: Their Record of Early Solar System History*, 1985; K. F. Weaver, Meteorites, invaders from space, *Nat. Geog.*, 170(3):390–418, 1986.

Micrometeorite

A submillimeter extraterrestrial particle that has survived entry into the atmosphere without melting. Meteoroids are natural interplanetary objects that orbit the Sun, and they range in size from small dust grains to objects that are miles (kilometers) in diameter. Particles below 0.04 in. (1 mm) in diameter are considered micrometeoroids, and the micrometeoroids that enter the atmosphere without melting are called micrometeorites. Meteoroids of all sizes enter the atmosphere with velocities in excess of the Earth's escape velocity of 7.0 mi/s (11.2 km/s), and all but the smallest ones are heated sufficiently by air friction to produce at least partial melting. Micrometeorites survive entry without severe heating because they are small and they totally decelerate from cosmic velocity at high altitudes near 55 mi (90 km). In the thin air at such altitudes the power generated by frictional heating is low enough to be radiated away without a particle reaching its melting point, typically about 2400°F (1300°C) for common meteoritic samples. Larger objects penetrate deeper into the atmosphere before slowing down, and are melted and partially vaporized by friction with the comparatively dense air. Most of the mass of extraterrestrial matter that annually collides with the Earth is in the micrometeoroid size range, a total of about 10^4 tons (10^7 kg), but only a small fraction survives as micrometeorites. Usually only the particles smaller than 0.1 mm survive as true unmelted micrometeorites, although the survival of an individual micrometeorite depends on entry velocity, angle of entry, melting point, and density as well as size. The flux of micrometeorites falling onto the Earth's surface is approximately 1 per square meter per day (0.1 per square foot per day) for particles with diameters of at least 10 micrometers and approximately 1 per square meter per year (0.1 per square foot per year) for particles with diameters of at least 100 μm. *See Meteorite.*

Sources. Micrometeorites are of particular interest because they are samples of comets and asteroids, small primitive bodies that have survived without major change since the earliest history of the solar system. Some of these particles are generated by collisions in the asteroid belt, while others are released from comets when these bodies approach the Sun and ice volatilization releases dust grains and propels them into space. Once released from a parent comet or asteroid, particles survive only for a few thousand to a hundred thousand years, depending on size, before they are either destroyed or collide with a planet. Particles are destroyed either when they collide with other particles or when they spiral into the Sun because of the Poynting Robertson drag, an effect of sunlight that causes the orbits of small particles to decay. During exposure in space, the small particles accumulate large amounts of helium implanted by the solar wind, and they also are riddled with radiation damage tracks produced by solar cosmic rays, high-energy particles accelerated from solar flares. *See Asteroid; Comet; Cosmic rays; Solar wind; Sun*

Although it is certain that micrometeorites have both cometary and asteroidal sources, the relative importance of the two sources is not well understood. The dust released from comets can be seen directly as sunlight reflected from the comet tail, and in some cases infrared observations show paths of relatively large particles in the wake of the comet. Most meteors or shooting stars in the night sky are millimeter-sized pieces of comet dust entering the atmosphere at high velocity. Debris from asteroid collisions is observable in the infrared as bands that stretch completely across the sky. Dust from both comets and asteroids can also be seen with the naked eye as the zodiacal light, a glow preceding sunrise and following sunset that is caused by sunlight reflecting off micrometeoroids in interplanetary space. *See Interplanetary matter; Meteor; Zodiacal light.*

Collection. The collection and laboratory analysis of micrometeorites provide an important source of information on the nature of materials in comets and asteroids. This work complements research on larger conventional meteorites because the dust particles are probably a more representative sampling of early solar-system materials than are the conventional meteorites. Most and possibly all conventional meteorites are believed to be samples of a small number of asteroids from the inner regions of the asteroid belt. These samples are strongly biased because they have to be strong to survive atmospheric entry without

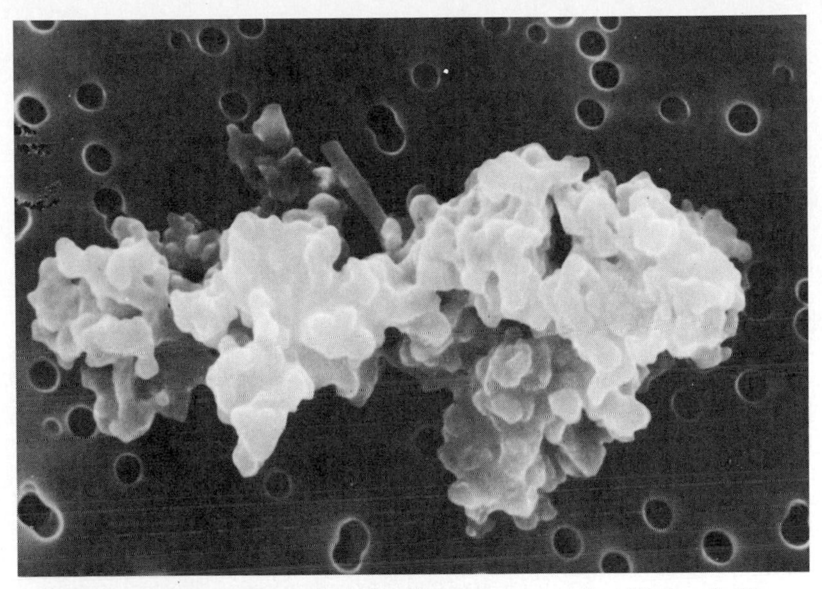

A 10-mm-long micrometeorite collected in the stratosphere with a U2 aircraft. The image was taken with a scanning electron microscope. The holes on the mounting substrate are unrelated to the particle.

fragmenting into dust and they are perturbed to orbits that intersect that of the Earth's only after rare gravitational perturbations, whereas dust particles do not have to be strong to survive atmospheric entry and light-pressure effects allow dust to diffuse through the solar system. Thus, collected micrometeorites include fragile cometary samples and samples of a diverse set of asteroids.

Most micrometeorites are collected in the stratosphere with aircraft such as the U2, which is capable of flying at an altitude of 12 mi (20 km) where terrestrial particles as large as 10 μm are rare. The spatial density of micrometeorites in the stratosphere is exceedingly low, but it is a million times larger than in space because of the low atmospheric fall speeds of micrometeorites relative to their original velocity in space. Micrometeorites are collected from the stratosphere by direct impact onto sticky plates that are extended from aircraft wings into the ambient airstream. With a collection surface area of 5 in.^{2} (30 cm^2), the collection rate is one 10-μm particle per hour of flight time. After a cumulative exposure of many hours, the plates are returned to a clean room where the microscopic particles are picked off with needles and placed onto mounts where they can be studied by electron microscopes (see **illus.**), mass spectrometers, and other instruments. Because of limitations imposed by flux and interference by terrestrial particles, the collection of micrometeorites in the stratosphere is usually limited to the size range from 2 to 100 μm in diameter. Most particles larger than this limit melt to form cosmic spherules during atmospheric entry and are not true micrometeorites. A small fraction of particles up to 1 mm in diameter do manage to survive as giant micrometeorites. The flux of these larger particles is too low to collect in the air, but they can be collected from a few locations on the Earth's surface where they accumulate without being severely diluted with terrestrial particles. The best location on Earth for collecting particles of this type is ultrapure polar ice that forms in regions that are distant from rock outcrops or other sources of submillimeter particles. SEE COSMIC SPHERULES.

Properties. Typical micrometeorites are small black particulates. Some are composed of a few relatively large mineral grains, but most are aggregates of large numbers of submicrometer mineral grains, plus glass and carbonaceous matter. This latter group is often called chondritic micrometeorites because their elemental composition matches that of chondritic meteorites. The composition matches that of the Sun for condensable elements such as magnesium, iron, silicon, aluminum, sulfur, and sodium. Particles dominated by a small number of mineral grains have elemental compositions similar to that of the largest constituent grain. Most of the chondritic particles have similar elemental compositions, but they vary significantly in mineralogical composition. The two most common mineralogical groups are dominated respectively by hydrous minerals such as serpentine and smectite and anhydrous minerals such as olivine, pyroxene, and iron sulfide. Some of the hydrous particles are similar to carbonaceous chondrite meteorites and are thought to have asteroidal origins. They show evidence of moderate aqueous alteration, which is most likely to have occurred inside a moderately warm asteroidal parent body. Comets are smaller, cold bodies that are close to the Sun only for small fractions of their lifetimes. In comets, the damp and moderately warm internal environments required for aqueous alteration must be very rare. The anhydrous particles are often very porous, show no evidence of aqueous alteration, and are likely to have cometary origins. These particles are unlike any material found in conventional meteorites, and apparently they are samples of meteoroid types that are too fragile to survive atmospheric entry as bodies larger than the size of dust. These particles are being investigated with a broad range of laboratory instruments to determine elemental, chemical, mineralogical, and isotopic compositions and provide clues on the nature and origin of the materials that formed comets and asteroids in the early solar system. Detection of regions inside the particles approximately 1 μm in diameter that are highly enriched in deuterium provide evidence that micrometeorites may also contain records of interstellar materials that predate the origin of the Sun and planets. SEE COSMOCHEMISTRY; ELEMENTS, COSMIC ABUNDANCE OF; SOLAR SYSTEM.

Donald E. Brownlee

Bibliography. D. E. Brownlee, Cosmic dust: Collection and research, *Annu. Rev. Earth Planet. Sci.*, 13:147–173, 1985; J. F. Kerridge and M. S. Matthews (eds.), *Meteorites and the Early Solar System*, 1988; I. D. R. Mackinnon and F. J. M. Rietmeijer, Mineralogy of chondritic interplanetary dust particles, *Rev. Geophys.*, 25:1527–1553, 1987; S. A. Sandford, The collection of extraterrestrial dust particles, *Fundamentals of Cosmic Physics*, 12:1–73, 1987.

Midnight sun

A phenomenon observed in the polar zones of the Earth near the time of the summer solstice, when the Sun remains visible above the horizon at midnight and reaches its minimum altitude without setting. The midnight sun is a consequence of the inclination of the rotation axis of the Earth, by which the Earth presents in turn each pole to the Sun for 6 months. The length of the period of uninterrupted daylight decreases as one goes away from the poles and, in principle, would vanish at the Arctic and Antarctic circles (latitude ±66°33′). However, because atmospheric refraction at the horizon raises the Sun's image by 34′, the midnight sun can be seen for a few days around the summer solstice from all points north of +66° or south of −66° latitude. SEE EARTH ROTATION AND ORBITAL MOTION.

Gerard de Vaucouleurs

Milky Way Galaxy

The large disk-shaped aggregation of stars, gas, and dust in which the solar system is located. The term Milky Way is used to refer to the diffuse band of light visible in the night sky emanating from the Milky Way Galaxy. Although the two terms are frequently used interchangeably, Milky Way Galaxy, or simply the Galaxy, refers to the physical object rather than its appearance in the night sky.

Appearance. The Milky Way is visible in the night sky to the unaided eye as a broad diffuse band of light stretching from horizon to horizon when viewed from locations away from bright city lights (**Fig. 1**). Nearly all of the visible light is due to individual stars, which in many directions are too numerous to be resolved without a telescope. The patchy appearance of the Milky Way is due to collections of microscopic dust particles, which block the light of more distant stars. All of the stars seen by the unaided eye are part of the Milky Way Galaxy and lie relatively close to the Sun. The overall appearance is due to the Sun's location near the midplane of the galactic disk; the diffuse band of light is seen toward directions close to the midplane where there are many more stars along the line of sight than in directions away from the plane. The Milky Way Galaxy also has two small companion galaxies in orbit around its center: the Large and Small Magellanic Clouds. These are seen as diffuse patches of light roughly 7° and 5° in diameter and separated from the plane of the Milky Way by about 30° and 40° degrees, respectively. They are visible mainly from the Earth's southernmost latitudes. SEE MAGELLANIC CLOUDS.

Structure and contents. The Milky Way Galaxy contains about 2×10^{11} solar masses of visible matter. (The Sun's mass is 4.4×10^{30} lb or 2×10^{30} kg.) Roughly 97% is in the form of stars, and about 3% is in the form of interstellar gas. The gas both inside the stars and in the interstellar medium is primarily hydrogen (roughly 87–90% by number of atoms) and helium (about 10%) with a small admixture of all of the heavier atoms (0–3% depending strongly on the location within the Galaxy). The mass of dust is about 1% of the interstellar gas mass and is therefore an insignificant fraction of the total mass of the Galaxy. The presence of this dust, however, limits the view from the Earth in the plane of the Milky Way Galaxy to a small fraction of the Galaxy's diameter in most directions. SEE INTERSTELLAR MATTER.

The Milky Way Galaxy contains four major structural subdivisions: the nucleus, the bulge, the disk, and the halo. The Sun is located in the disk about half way between the center and the distance at which the cdisk of stars becomes undetectable. The currently accepted value of the distance of the Sun from the galactic center is 8.5 kiloparsecs, although some mea-

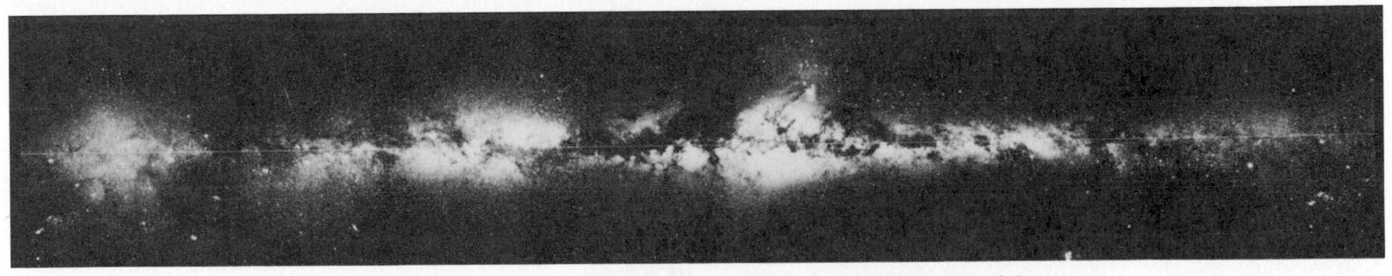

Fig. 1. Panoramic photograph of the entire Milky Way. The direction of the galactic center is in the center of the photograph, but it is hidden by foreground dust. The galactic bulge can be seen as the thickening of the disk in the center of the photograph. (*European Southern Observatory*)

surements suggest that the distance may be as small as 7 kpc. (A parsec is equal to 2.06×10^5 times the average distance between the Sun and the Earth, and is approximately equal to the average distance between stars in the solar neighborhood; it is also equal to 3.26 light-years, 1.91×10^{13} mi, or 3.09×10^{13} km.) **Figures 2** and **3** show edge-on and inclined views of two galaxies that are structurally similar to the Milky Way. *See Parsec*.

The nucleus of the Mily Way Galaxy is an ill-defined region within a few tens of parsecs of the geometric center. The nucleus is teh source of very energetic activity detected by means of radio waves, infrared radiation, and gamma radiation, some of which is generally believed to be powered by matter interacting with a black hole at the very center of the Milky Way. No unambiguous evidence has yet been found, however, to confirm the presence of a black hole. The nucleus also contains a very dense cluster of stars observed by means of their infrared radiation. At a distance of about 10 pc from the center is a ring of gas consisting primarily of molecular hydrogen orbiting the nucleus. A representation of a radio emission from one of the molecular constituents of the ring is shown in **Fig. 4**. Molecular hydrogen gas, which is characterized by low temperatures and relatively high densities, is found in great profusion in the inner few hundred parsecs of the Milky Way. Great arcs of gas resulting from the interaction of cosmic rays and magnetic fields have been mapped in

Fig. 2. NGC 4565, a galaxy of similar morphological type to the Milky Way, seen edge-on. The bulge at the center and dark dust lanes across the midplane of the disk are apparent. The dust lanes are probably coincident with spiral arms. (*The Observatories of the Carnegie Institution of Washington*)

Fig. 3. M58, another galaxy with a gross morphology similar to the Milky Way, seen in an inclined view. The spiral arms and the bulge in the center can be seen. The bulge appears to be elongated because the galaxy also contains a bar, an extended feature containing primarily old stars that rotate collectively about the center as if they were a solid body. (*The Observatories of the Carnegie Institution of Washington*)

the central tens of parsecs of the Milky Way; some of these are from the nucleus itself. SEE BLACK HOLE; COSMIC RAYS; GAMMA-RAY ASTRONOMY; INFRARED ASTRONOMY; RADIO ASTRONOMY.

The bulge is a spheroidal distribution of stars centered on the nucleus that extends to a distance of about 3 kpc from the center. It contains a relatively old population of stars, very nearly as old as the Milky Way itself. The bulge can be seen by the unaided eye as a thickening of the diffuse band of light that constitutes the Milky Way in the direction of the center of the Galaxy, toward the direction of the constellation Sagittarius. There is little gas and dust in the bulge. SEE STAR.

The disk consists of a disk of stars and a disk of gas. The disk of stars can be identified to about 17 kpc from the center of the Galaxy, and the disk of gas can be identified to distances of at least 22 kpc from the center. The thicknesses of the disks are characterized by a scale height: the distance from the midplane at which the density of gas and stars falls by a factor of the transcendental number e (2.718). In the solar vicinity, the scale height varies for the various components of the disk from about 75 pc for the molecular gas to about 350 pc for the lowest-mass stars, the component containing most of the mass of the disk. The disk is thus very thin and has the relative proportions of a long-playing phonograph record. At this scale, the bulge would have the size of a tennis ball superimposed on the nucleus.

This disk is warped and deviates in its outermost parts by about 3 kpc from a true plane. The reason for the warping is not yet understood but is a common feature of spiral galaxies. There is also a bridge of gas between the disk and the Magellanic Clouds that has been drawn out by the tidal forces of the clouds on the Milky Way.

The disk is the location of the spiral arms that are characteristic of most disk-shaped galaxies, and are the sites of present-day star formation. Attempts to map the location of the spiral arms are hampered by the Sun's location in the disk of the Galaxy, but in the parts of the Galaxy beyond the Sun's distance from the center, several long coherent spiral arms have been identified. The inner regions appear to be more chaotic, and no agreed-upon spiral-arm structure has been identified. The spiral arms are where giant clouds of molecules are primarily found. These clouds are the most massive entities in the Milky Way, up to about 5×10^6 solar masses, and they are the sites of all present-day star formation. The Sun apparently once formed in a giant molecular cloud. SEE STELLAR EVOLUTION.

The halo is a rarefied spheroidal distribution of stars that is very nearly devoid of interstellar gas and dust and that surrounds the disk. The stars found in the halo are the oldest stars in the Galaxy. The stars are found individually as so-called field stars as well as in globular clusters, that is, spherical clusters of up to about a million stars with very low abundances of elements heavier than helium. The extent of the halo is not well determined, but globular clusters with distances of about 40 kpc from the center have been definitely identified. Dynamical evidence suggests that the halo contains nonluminous matter in some unknown form, commonly referred to as dark matter. The dark matter, if it exists, contains most of the mass of the Galaxy, dominating that even in the form of stars. SEE STAR CLUSTERS.

Dynamics and kinematics. The stars and gas in the disk of the Milky Way Galaxy rotate around the galactic center in nearly circular orbits in the plane of the disk. The deviations from circularity are generally small, and give rise to oscillations of stellar motions perpendicular to the disk, epicyclic motions in the plane of the disk, and other, more complex motions.

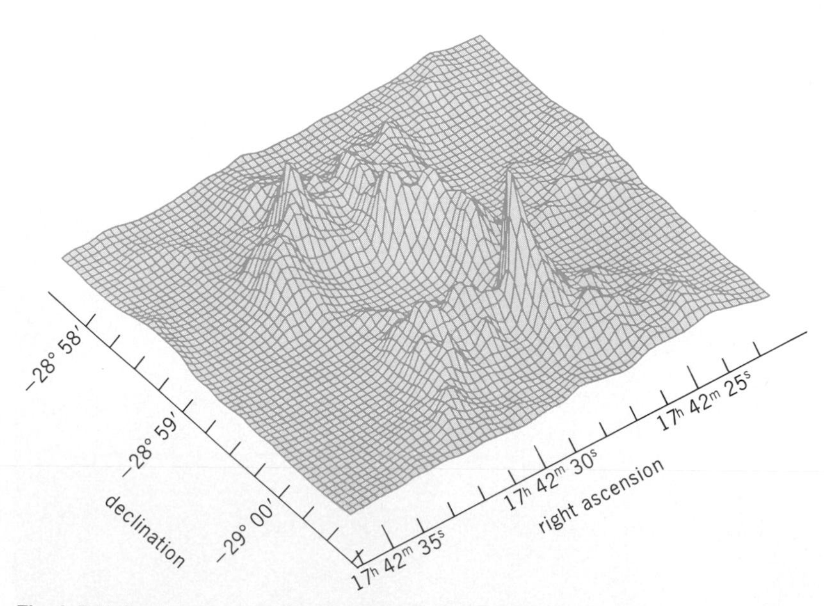

Fig. 4. Relief-map representation of radio emission from the hydrogen cyanide (HCN) molecule within 3 parsecs of the galactic center. Hydrogen cyanide traces the location of the much more abundant hydrogen molecule, and displays the ringlike morphology of the gas that is rotating about the galactic center. (*After R. Güsten et al., Aperture synthesis observations of the circumnuclear ring in the galactic center, Astrophys. J., 318:124–138, 1987*)

The overall rotation of the disk is differential rather than like a solid body in that the rotation period of the stars and the gas decreases with increasing distance from the center. Large-scale deviations from circularity are associated with the spiral arms of the disk.

Large noncircular motions in the inner part of the disk, as well as smaller very large-scale noncircular motions in the outer parts of the disk, have suggested to many astronomers that the Milky Way Galaxy contains a bar in its inner parts. A bar is a large collection of stars in the inner part of a galaxy in which the stars collectively rotate about the center as a solid bar would, even though the individual stellar orbits are quite complex. Such bars are common in other galaxies.

The orbits of the stars and clusters in the bulge and halo are not confined to the galactic plane and are, in general, fully three-dimensional. Although the motions of the stars are too small for the orbits to be determined from their changes in position alone, orbits can be inferred from the measured radial velocities of the stars (the velocities measured along the line of sight), their positions in the Galaxy, and, for stars close to the Sun, their apparent motions in the sky measured over many years.

In the outermost parts of the Milky Way Galaxy, the orbital velocities of stars and gas around the galactic center would be expected to decrease in a well-determined way, as they do, for example, in the solar system. Measurements show, however, that as far out in the Galaxy as measurements are possible, the velocities of the gas and stars do not change with increasing distance from the galactic center (**Fig. 5**). These measurements of what is referred to as the flat rotation curve of the Milky Way (contrasted with a falling rotation curve) are generally cited as the most compelling evidence for large quantities of material in the Milky Way in a form not yet detected by means of radiation from the material itself. This dark matter is common to most galaxies for which rotation measurements are possible.

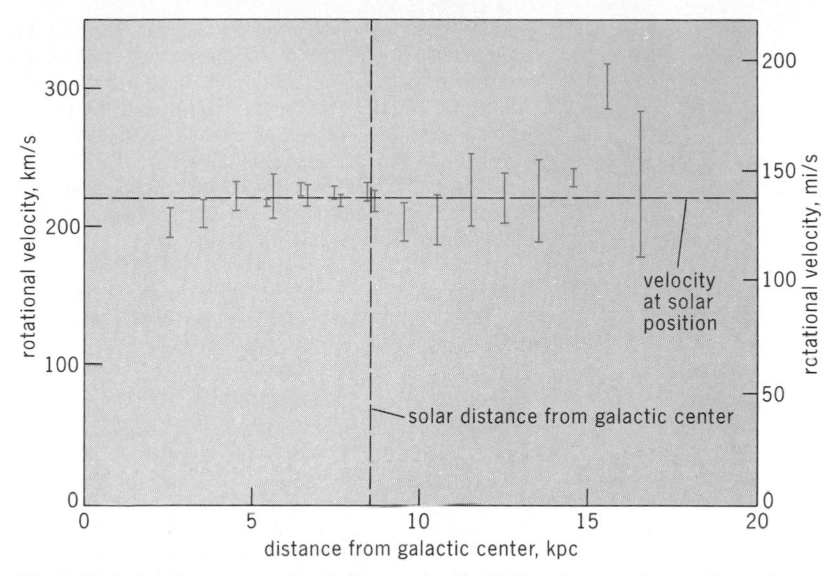

Fig. 5. Plot of the measurement of the rotational velocity of gas and stars about the center of the Milky Way Galaxy as a function of distance from the galactic center, showing that the rotational velocities are essentially constant. (*Courtesy of M. Fich, L. Blitz, and A. A. Stark*)

Dark matter. The dark matter in the Milky Way Galaxy constitutes by various estimates 2½ to 10 times the total amount of known matter in the Milky Way, and is consequently the dominant component of the mass. Because the kinematics of most galaxies indicate that they, too, contain dark matter as their largest mass constituent, the dark matter appears to be the dominant form of matter in the universe. The identification of the composition of the dark matter is one of the preeminent problems of astronomy.

In the Milky Way, various forms tht the dark matter may take have been ruled out. It cannot be in the form of ordinary stars or in remnants such as white dwarfs, neutron stars, or black holes that are the end products of ordinary stellar evolution. It cannot be in the form of gas or small, solid dust particles. Although a small fraction of the dark matter may reside in the disk, most of it appears to reside in the halo. *SEE NEUTRON STAR; WHITE DWARF STAR.*

Two possibilities that have not yet been ruled out are that the darkl matter consists of small planet-sized bodies that are insufficiently luminous to be detected with present instruments (commonly called brown dwarfs), and that the dark matter consists of primordial black holes in the halo. Still other possibilities include weakly interacting massive particles (WIMPs), that are predicted by various elementary particle theories. Observations of the supernova 1987A have apparently ruled out that neutrinos, massless or very low-mass neutral particles, constitute most of the dark matter in the universe. An alternative theory to newtonian gravitation has also been proposed to explain the rotation curve of the Milky Way and other galaxies, but this idea is generally considered to be a particularly radical explanation. *SEE BROWN DWARF; COSMOLOGY; GRAVITATION; SUPERNOVA.*

Location. The Milky Way Galaxy is part of a small grouping of galaxies known as the Local Group. The Local Group contains two large spiral galaxies, the great nebula in Andromeda (M31) and M33, and at least 18 small irregular and elliptical galaxies. The Local Group is itself part of a large supercluster of galaxies known as the Virgo supercluster containing about 1000 known galaxies. The Local Group is an outlying collection of galaxies in the Virgo supercluster and is about 10 megaparsecs from its center. The Virgo supercluster is one of many such groupings in the universe and occupies no special location within it. *SEE ANDROMEDA GALAXY; GALAXY, EXTERNAL; LOCAL GROUP; UNIVERSE.*

Evolution. Inferences about the evolution of the Milky Way Galaxy can be drawn from a large variety of sources, including the theory and observation of the kinematics and dynamics of the stars in the Milky Way, the chemical abundances of the stars, their locations in the Galaxy, and stellar evolution theory. Evidence suggests that the ages of the oldest stars in the Milky Way are within about 10% of the age of the universe as a whole; thus the Milky Way must have formed early in the history of the universe, about $12–16 \times 10^9$ years ago.

The Milky Way Galaxy presumably formed from a very large (about 10^{15} times the mass of the Sun) cloud of gas of hydrogen and helium that was essentially devoid of all heavier elements. The material in this gas cloud formed the Virgo supercluster, and one portion of it formed the Milky Way and the Local Group. The first generation of stars are those that make up the population in the halo. Presumably, as the gas cloud collapsed under the influence of its self-gravity, viscous dissipation in the gas allowed the protogalaxy to collapse along the rotation axis, forming the disk at a later evolutionary stage than the halo. About 4.6 × 10^9 years ago, the Sun and the planets formed from the interstellar medium in the disk, the Sun being a rather unremarkable star in an unremarkable location. That part of the disk that has not yet been used up in the process of star formation continues to form stars. *SEE SOLAR SYSTEM; SUN.*

Leo Blitz

Bibliography. L. Blitz, M. Fich, and S. Kulkarni, The new Milky Way, *Science*, 220:1233–1240, 1983; B. J. Bok, The Milky Way Galaxy, *Sci. Amer.*, 244(3):92–94, March 1981; B. J. Bok and P. F. Bok, *The Milky Way*, 5th ed., 1981; G. Gilmore and B. Carswell, *The Galaxy*, 1987; K. Y. Lo, The galactic center, *Science*, 233:1394–1403, 1986; W. L. Shuter, *Kinematics, Dynamics, Structure of the Milky Way*, 1983.

Molecular cloud

A large and relatively dense cloud of cold gas and dust in interstellar space from which new stars are born. Molecular clouds consist primarily of molecular hydrogen (H_2) gas, have temperatures in the range 10–100 K, and contain 10^{31}–10^{36} kg of mass (for comparison the mass of the Sun is 2 × 10^{30} kg). Molecular clouds are among the most massive gravitationally bound objects in the Milky Way Galaxy.

Molecules. Molecular hydrogen is not directly observable under most conditions in molecular clouds. Therefore, almost all current knowledge about the properties of molecular clouds has been deduced from observations of trace constituents, mostly simple molecules such as carbon monoxide (CO), which have strong emission lines in the centimeter- and millimeter-wavelength portions of the electromagnetic spectrum. The first molecules were detected as optical absorption lines in the spectra of bright stars lying behind molecular clouds. During the 1960s, centimeter-wavelength radio emission lines of hydroxyl (OH), water (H_2O), and ammonia (NH_3) were detected from the vicinity of bright optical nebulae such as the Orion Nebula (Messier 42). In 1970, the most important tracer of molecular gas in space, carbon monoxide, was discovered at a wavelength of 2.6 mm. *SEE RADIO ASTRONOMY.*

Distribution. During the 1970s, surveys of carbon monoxide emission from the sky demonstrated that molecular clouds are widespread in the plane of the Milky Way Galaxy, with the greatest concentration of clouds lying in a broad ring encircling the galactic center. This molecular ring has an inner radius of about 3 kiloparsecs (1 parsec = 3 × 10^{13} km or 2 × 10^{13} mi) and an ill-defined outer radius of about 6 kpc. Within the ring, molecular clouds are concentrated in the major spiral arms of the Milky Way Galaxy. About 10% of all molecular gas in the Milky Way Galaxy is found within a concentration of high-density molecular clouds lying within 500 parsecs of the galactic center. Nearly half of all the mass of gas found in the interstellar medium that fills the space between the stars is in molecular clouds.

Properties. Most molecular gas in the Milky Way Galaxy is found in giant molecular clouds which have masses of about 100,000 times the mass of the Sun. About 10,000 giant molecular clouds have been found in the Milky Way Galaxy. The closest and best-studied giant molecular clouds lie at a distance of about 450 parsecs in the constellation of Orion (see **illus.**) where, over the last 10^7 years, they have given birth to several groups of hot and massive stars known collectively as the Orion OB Association. In addition to these hot massive stars, several thousand lower-mass stars (with masses closer to the mass of the Sun) were born during the same period. *SEE ORION NEBULA.*

Many smaller molecular clouds, such as those in the constellation of Taurus, have masses ranging from one to several thousand times the mass of the Sun. These clouds tend to form only low-mass Sun-type stars. In the mid-1980s, a category of even smaller molecular clouds was found to lie far from the plane of the Milky Way Galaxy. These high-latitude clouds were found to be the nearest molecular clouds, with the nearest one (MBM-12) located only about 50 parsecs from the Sun. High-latitude clouds are, as a class, too small to form stars and, in some cases, have masses only several times that of the Sun. *SEE PROTOSTAR.*

Structure. Molecular clouds typically have average gas densities ranging from about 10 molecules per cubic centimeter on large scales to more than 10^6 molecules per cubic centimeter in cloud cores. The clouds have a very complex internal structure consisting of clumps and filaments of dense gas surrounded by in-

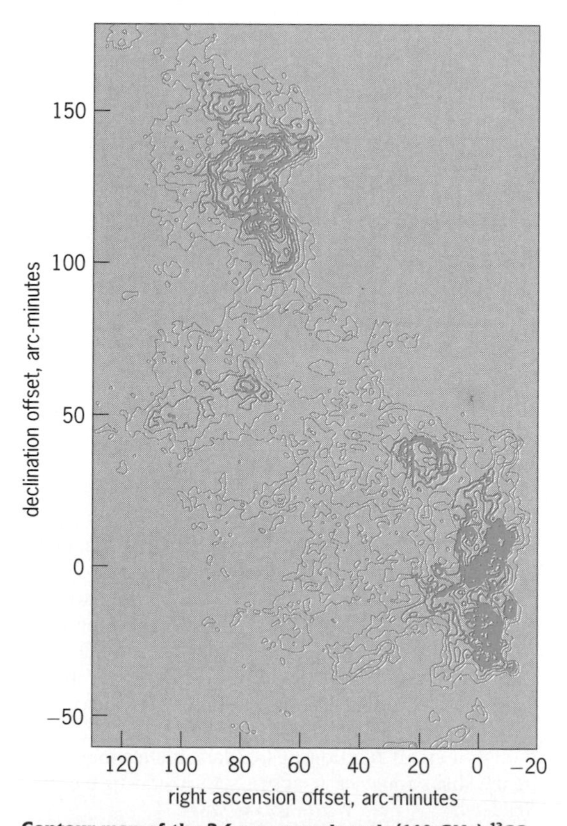

Contour map of the 2.6-mm-wavelength (110-GHz) ^{13}CO emission from the giant molecular cloud in the northern portion of the constellation Orion. The bright Belt Stars of Orion are located just off the right edge of the map, while the star Betelgeuse lies to the upper left. The axes are labeled in arc-minute offsets from the H II region NGC 2024. Contours show the intensity of the emission in brightness temperature units, a measure of surface brightness used in radio astronomy. Contours are separated by 3 K·km·s^{-1}.

terclump gas of much lower density. Individual clumps usually have supersonic internal motions with a velocity of several kilometers per second. Clumps within a cloud frequently exhibit even larger relative motions produced by the gravitational potential of the entire cloud. Clouds near the galactic center have very large internal motions, frequently with internal velocities of more than 30 km/s (20 mi/s).

Composition. Carbon monoxide is the second most abundant molecule, after hydrogen. There is 1 carbon monoxide molecule for about every 10,000 hydrogen molecules. About 100 different chemical species have been so far identified within molecular clouds, indicating that there is a rich chemistry taking place. About 1% of the mass of molecular clouds is in the form of interstellar dust grains that absorb starlight, making molecular clouds opaque at visible and ultraviolet wavelengths. Therefore, most nearby clouds can be seen in silhouette against the background of stars. For example, the protrusion near the lower-right corner of the illustration is the Horse Head Nebula, which in visible-wavelength photographs is seen in absorption against a background screen of ionized gas. *SEE INTERSTELLAR EXTINCTION.*

Molecular hydrogen is formed primarily from atomic hydrogen by chemical reactions taking place on the surfaces of dust grains. However, most other chemical species are produced by gas-phase chemical reactions involving the interactions of neutral species with ions. Energetic cosmic-ray particles moving through the molecular gas ionize a small fraction (about 1 part in 10^7) of the molecules, resulting in the formation of the highly reactive H_3^+ ion, which drives the ion-neutral reactions leading to the formation of other observed chemical species. The small abundance of ions (and electrons) in molecular clouds is sufficient to couple cosmic magnetic fields to the gas. The magnetic field, together with the gravitational force, probably regulates the rate at which gas can undergo gravitational collapse to form stars. *SEE COSMOCHEMISTRY; MAGNETOHYDRODYNAMICS.*

Evolution. Molecular clouds are believed to survive for several times 10^7 years. The ultraviolet radiation produced by massive stars born from the clouds dissociates molecules, fully ionizes atoms and molecules, and heats the remaining gas. The resulting pressure gradients accelerate and disperse the molecular cloud, leading to its destruction.

The formation of molecular clouds is poorly understood at present. Compression of low-density atomic gas by the passage of a large-scale shock wave or a spiral arm of the Milky Way Galaxy, the tendency of atomic gas to cool by radiation, and the self-gravity of the gas may all play a role in the formation of molecular clouds.

Clouds in galaxies. Most spiral and irregular galaxies that form stars contain molecular clouds. On the other hand, elliptical galaxies contain very little molecular gas. Certain peculiar galaxies that are extremely luminous at infrared wavelengths contain very large amounts of molecular gas and are believed to be forming stars at a rate of many hundreds of solar masses per year. There is evidence that these starburst galaxies are produced by the merging of two gas-rich spiral galaxies. *SEE STARBURST GALAXY.*

In isolated spiral galaxies such as the Milky Way Galaxy, the total mass of molecular gas in the interstellar medium (at present about 2×10^9 times the mass of the Sun) probably decreases with time as gas is consumed by star formation. In the absence of substantial infall of fresh gas into the galaxy from the intergalactic medium, the mass of gas in the interstellar medium is halved approximately every few billion years. *SEE GALAXY, EXTERNAL; INTERSTELLAR MATTER.*

John Bally

Bibliography. E. Falgarone, F. Boulanger, and G. Duvert (eds.), *Fragmentation of Molecular Clouds and Star Formation,* 1991; D. J. Hollenbach and H. A. Thronson, Jr. (eds.), *Interstellar Processes,* 1987; C. J. Lada and N. D. Kylafis (eds.), *The Physics of Star Formation and Early Stellar Evolution,* 1991.

Month

Any of several units of time based on the revolution of the Moon around Earth.

The calendar month is one of the 12 arbitrary periods into which the calendar year is divided. *SEE CALENDAR.*

The synodic month is the average period of revolution of the Moon with respect to the Sun, the same as the average interval between successive full moons. Its duration is 29.531 days.

The tropical month is the period required for the mean longitude of the Moon to increase 360°, or 27.322 days.

The sidereal month, 7 s longer than the tropical month, is the average period of revolution of the Moon with respect to a fixed direction in space.

The anomalistic month, 27.555 days in duration, is the average interval between closest approaches of the Moon to Earth.

The nodical month, 27.212 days in duration, is the average interval between successive northward passages of the Moon across the ecliptic. *SEE TIME.*

Gerald M. Clemence

Moon

The Earth's natural satellite. United States and Soviet spacecraft have obtained lunar data and samples, and Americans have orbited, landed, and roved upon the Moon (**Fig. 1**). Though the first wave of exploration has passed, it left a store of information whose meanings are still being deciphered. Many of the Moon's properties are now well understood, but its origin and relations to other planets remain obscure. Theories of its origin include: independent condensation and then capture by the Earth; formation in the same cloud of preplanetary matter with the Earth; fission from the Earth; and formation after the impact of a Mars-sized body on the proto-Earth. Because many of the Moon's geologic processes stopped long ago, its surface preserves a record of very ancient events. However, because the Moon's rocks and soils were reworked by geochemical and impact processes, their origins are partly obscured, so that working out the Moon's early history remains a fascinating puzzle. Major characteristics of the Moon are listed in **Table 1**.

The apparent motions of the Moon, its waxing and waning, and the visible markings on its face (Fig. 1), are reflected in stories and legends from every early civilization. At the beginning of recorded history on the Earth, it was already known that time could be

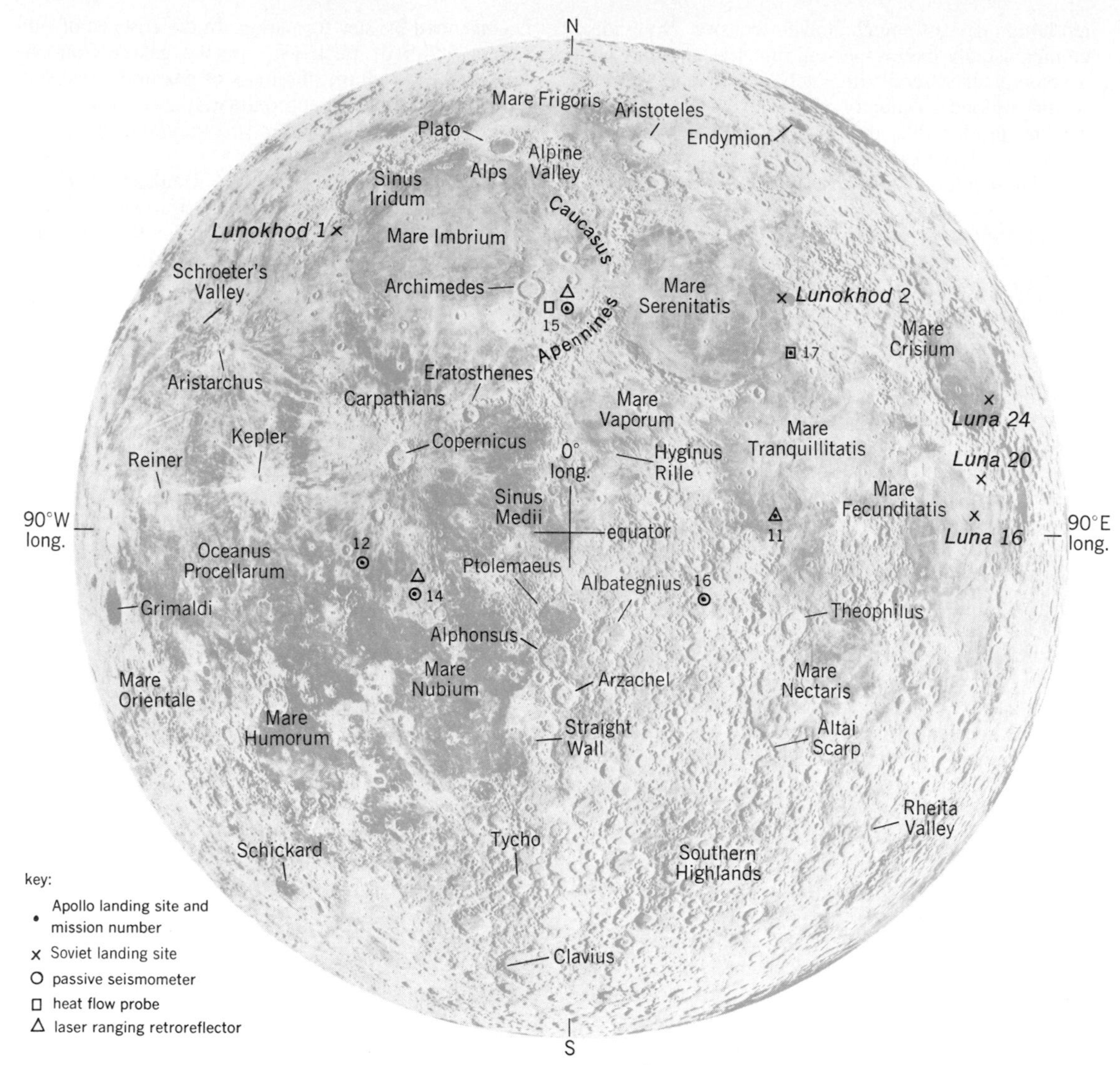

Fig. 1. Map of near side of Moon, showing principal features and American and Soviet landing sites.

Table 1. Characteristics of the Moon

Characteristics	Values and remarks
Diameter (approximate)	2160 mi (3476 km)
Mass	1/81.301 Earth's mass, or 1.62×10^{23} lb (7348×10^{22} kg)
Mean density	0.604 Earth's, or 209 lb/ft^3 (3.34 g/cm^3)
Mean surface gravity	0.165 Earth's, or 5.3 ft/s^2 (162 cm/s^2)
Surface escape velocity	0.213 Earth's, or 1.48 mi/s (2.38 km/s)
Atmosphere	Surface pressure 10^{-12} torr (1.3×10^{-10} Pa); hints of some charged dust particles and occasional venting of volatiles
Magnetic field	Dipole field less than $\sim 0.5 \times 10^{-5}$ Earth's; remanent magnetism in rocks shows past field was much stronger
Dielectric properties	Surface material has apparent dielectric constant of 2.8 or less; bulk apparent conductivity is 10^{-5} mho/m or less
Natural radioactivity	Mainly due to solar- and cosmic-ray-induced background (about 1 milliroentgen per hour for quiet Sun)
Seismic activity	Much lower than Earth's; deep moonquakes occur more frequently when the Moon is near perigee; subsurface layer evident
Heat flow	3×10^{-2} W/m^2 (*Apollo 15* site)
Surface composition and properties	Basic silicates, three sites (Table 4); some magnetic material present; soil grain size is 2–60 μm and 50% is less than 10 μm; soil-bearing strength 15 lb/in.^{2} (1 kg/cm^2) at depth of 1–2 in. (a few centimeters)
Rocks	All sizes up to tens of meters present, concentrated in strewn fields; rock samples from Mare Tranquillitatis include fine- and medium-grained igneous and breccia
Surface temperature range	At equator 260°F (400 K) at noon; −315 to −280°F (80–100 K) night minimum; 3 ft (1 m) below surface, −45 °F (230 K); at poles −280°F (∼100 K)

reckoned by observing the position and phases of the Moon. Attempts to reconcile the repetitive but incommensurate motions of the Moon and Sun led to the construction of calendars in ancient Chinese and Mesopotamian societies and also, a thousand years later, by the Maya. By about 300 B.C., the Babylonian astronomer-priests had accumulated long spans of observational data and so were able to predict eclipses. Major events in the subsequent development of human knowledge of the Moon are summarized in **Table 2**.

Space flight experiments have now confirmed and vastly extended understanding of the Moon; however, they have also opened many new questions for future lunar explorers.

Motions. The Earth and Moon now make one revolution about their barycenter, or common center of mass (a point about 2900 mi or 4670 km from the Earth's center), in $27^d7^h43^m11.6^s$. This sidereal period is slowly lengthening, and the distance (now about 60.27 earth radii) between centers of mass is increasing, because of tidal friction in the oceans of the

Table 2. Growth of human understanding of the Moon

Date	Event
Prehistory	Markings and phases observed, legends created connecting Moon with silver, dark markings with rabbit (shape of maria) or with mud.
~300 B.C.	Apparent lunar motions recorded and forecast by Babylonians and Chaldeans.
~150 B.C.	Phases and eclipses correctly explained, distance to Moon and Sun measured by Hipparchus.
~A.D. 150	Ancient observations compiled and extended by C. Ptolemy.
~700	Ephemeris refined by Arabs.
~1600	Empirical laws of planetary motion derived by J. Kepler.
1609	Lunar craters observed with telescopes by T. Harriot and Galileo.
1650	Moon mapped by J. Hevelius and G. Riccioli; features named by them in system still in use.
1667	Experiments by R. Hooke simulating cratering through impact and vulcanism.
1687	Moon's motion ascribed to gravity by I. Newton.
1692	Empirical laws of lunar motion stated by J. D. Cassini.
1700–1800	Lunar librations measured, lunar ephemeris computed using perturbation theory by T. Mayer. Secular changes computed by J. L. Lagrange and P. S. de Laplace. Theory of planetary evolution propounded by I. Kant and Laplace. Many lunar surface features described by J. H. Schroeter and other observers.
1800–1920	Lunar motion theory and observations further refined, leading to understanding of tidal interaction and irregularities in Earth's rotation rate. Photography, photometry, and bolometry applied to description of lunar surface and environment. Lunar atmosphere proved absent. New disciplines of geology and evolution applied to Moon, providing impetus to theories of its origin.
1924	Polarization measured by B. F. Lyot, showing surface to be composed of small particles.
1927–1930	Lunar day, night, and eclipse temperatures measured by E. Pettit and S. B. Nicholson.
1946	First radar return from Moon.
1950–1957	New photographic lunar atlases and geologic reasoning; renewed interest in theories of lunar origin by G. P. Kuiper, H. C. Urey, and E. M. Shoemaker. New methods (for example, isotope dating) applied to meteorites, concepts extended to planetology of Moon. Low subsurface temperatures confirmed by Earth-based microwave radiometry.
1959	Absence of lunar magnetic field (on sunlit side) shown by *Luna 2*.
1960	Eastern far side photographed by *Luna 3*. Slower cooling of Tycho detected during lunar eclipse.
1961	United States commitment to crewed lunar flight.
1962	Earth-Moon mass ratio measured by *Mariner 2*.
1964	High-resolution pictures sent by *Ranger 7*. Surface temperatures during eclipse measured by Earth-based infrared scan.
1965	Western far side photographed by *Zond 3*.
1966	Surface pictures produced by *Luna 9* and *Surveyor 1*. Radiation dose at surface measured by *Luna 9*. Gamma radioactivity measured by *Luna 10*. High-resolution, broad-area photographs taken by *Lunar Orbiter 1*. Surface strength and density measurements made by *Luna 13*.
1967	Mare soil properties and chemistry measured by *Surveyor 3*, *5*, and *6*. Whole front face mapped by *Lunar Orbiter 4*, sites of special scientific interest examined by *Lunar Orbiter 5*. Particle-and-field environment in lunar orbit measured by *Explorer 35*.
1968	Highland soil and rock properties and chemistry measured by *Surveyor 7*. Mass concentrations at circular maria discovered.
1968	Astronauts orbit Moon, return with photographs.
1969	Astronauts land and emplace instruments on Moon, return with lunar samples and photographs.
1969–1972	Lunar seismic and laser retroreflector networks established. Heat flow measured at two sites. Remanent magnetism discovered in lunar rocks. Geologic traverses accomplished. Orbital surveys of natural gamma radioactivity, x-ray fluorescence, gravity, magnetic field, surface elevation, and subsurface electromagnetic properties made at low latitudes. Metric mapping photos obtained. Samples returned by both piloted (United States) and automated (Soviet) missions; sample analyses confirmed early heating and chemical differentiation of Moon, with surface rocks enriched in refractory elements and depleted in volatiles. Age dating of lunar rocks and soils showed that most of the Moon's activity (meteoritic, tectonic, volcanic) occurred more than 3×10^9 years ago.
1975	Giant-impact hypothesis for lunar origin advanced by W. K. Hartmann and D. R. Davis.
1982	Earth-based spectrometry reveals mineral variations over Moon's near side; central peaks of crater Copernicus found to be rich in olivine.
1983	Antarctic meteorite, ALHA 81005, proved to have come from the Moon.

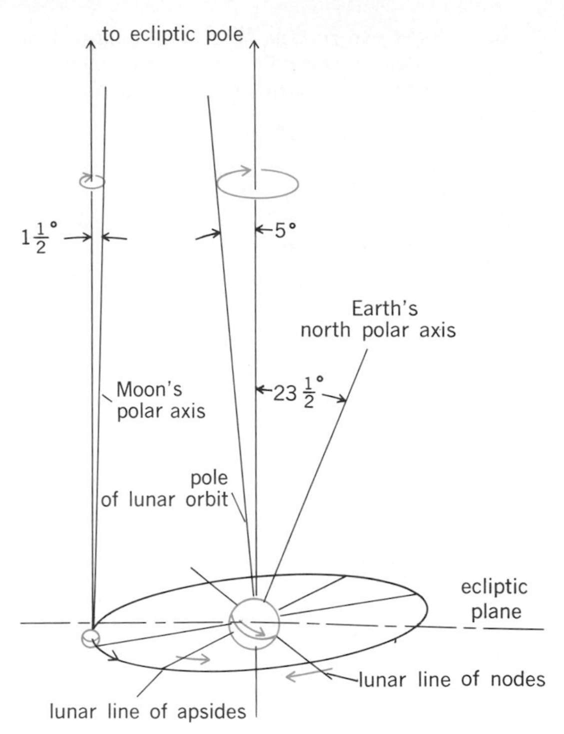

Fig. 2. Sketch of Moon's orbit.

Earth. The tidal bulges raised by the Moon are dragged eastward by the Earth's daily rotation. The displaced water masses exert a gravitational force on the Moon, with a component along its direction of motion, causing the Moon to spiral slowly outward. The Moon, through this same tidal friction, acts to slow the Earth's rotation, lengthening the day. Tidal effects on the Moon itself have caused its rotation to become synchronous with its orbital period, so that it always turns the same face toward the Earth.

Tracing lunar motions backward in time is very difficult, because small errors in the recent data propagate through the lengthy calculations, and because the Earth's own moment of inertia may not have been constant over geologic time. Nevertheless, the attempt is being made by using diverse data sources, such as the old Babylonian eclipse records and the growth rings of fossil shellfish. At its present rate of departure, the Moon would have been quite close to the Earth about 4.6×10^9 years ago, a time which other evidence suggests as the approximate epoch of formation of the Earth.

The Moon's present orbit (**Fig. 2**) is inclined about 5° to the plane of the ecliptic. **Table 3** gives the dimensions of the orbit (in conventional coordinates with origin at the center of the Earth, rather than the Earth-Moon barycenter). As a result of differential attraction by the Sun on the Earth-Moon system, the Moon's orbital plane rotates slowly relative to the ecliptic (the line of nodes regresses in an average period of 18.60 years) and the Moon's apogee and perigee rotate slowly in the plane of the orbit (the line of apsides advances in a period of 8.850 years). Looking down on the system from the north, the Moon moves counterclockwise. It travels along its orbit at an average speed of nearly 0.6 mi/s (1 km/s) or about 1 lunar diameter per hour; as seen from Earth, its mean motion eastward among the stars is 13°11′ per day.

As a result of the Earth's annual motion around the Sun, the direction of solar illumination changes about 1° per day, so the lunar phases do not repeat in the sidereal period given above but in the synodic period, which averages $29^d\ 12^h\ 44^m$ and varies some 13 h because of the eccentricity of the Moon's orbit. *See Earth rotation and orbital motion; Orbital motion.*

When the lunar line of nodes (Fig. 2) coincides with the direction to the Sun, and the Moon happens to be near a node, eclipses can occur. Because of the 18.6-year regression of the nodes, groups of eclipses recur with this period. When it passes through the Earth's shadow in a lunar eclipse, the Moon remains dimly visible because of the reddish light scattered through the atmosphere around the limbs of the Earth. When the Moon passes between the Earth and Sun, the solar eclipse may be total or annular. As seen from Earth, the angular diameter of the Moon (31′) is almost the same as that of the Sun, but both apparent diameters vary because of the eccentricities of the orbits of Moon and Earth. Eclipses are annular when the Moon is near apogee and the Earth is near perihelion at the time of eclipse. A partial solar eclipse is seen from places on Earth that are not directly along the track of the Moon's shadow. *See Eclipse.*

The Moon's polar axis is inclined slightly to the pole of the lunar orbit (Fig. 2) and rotates with the same 18.6-year period about the ecliptic pole. The

Table 3. Dimensions of Moon's orbit

Characteristics	Values
Sidereal period (true period of rotation and revolution)	(27.32166140 + 0.000000167) ephemeris days, where T is in centuries from 1900
Synodic period (new Moon to new Moon)	(29.5305882 + 0.000000167) ephemeris days
Apogee	252,700 mi or 406,700 km (largest); 251,971 mi or 405,508 km (mean)
Perigee	221,500 mi or 356,400 km (smallest); 225,744 mi or 363,300 km (mean)
Period of rotation of perigee	8.8503 years direct ("direct" meaning that the motion of perigee is in the direction of Moon's motion about the Earth)
Period of regression of nodes	18.5995 years
Eccentricity of orbit	0.054900489 (mean)
Inclination of orbit to ecliptic	5°8′43″ (oscillating ±9′ with period of 173 days)
Inclination of orbit to Earth's equator	Maximum 28°35′, minimum 18°21′
Inclination of lunar equator	
to ecliptic	1°32′40″
to orbit	6°41′

Most gaseous nebulae derive their energy from stars near or within them. The Trifid Nebula (*above*) contains a number of young, hot stars that cause the gas, mostly hydrogen, to emit its characteristic red light. One side of the nebula contains numerous dust grains which preferentially reflect blue star light. Some regions contain so many dust grains that they hide the glowing gas, producing dark lanes which give the object its name. The Lagoon Nebula (*below*), excited by a star cluster, also displays dark lanes and globules resulting from solid grains. Both photographs were made with the 150-in. (3.9-m) Anglo-Australian Telescope. (*Copyright by Royal Observatory of Edinburgh*)

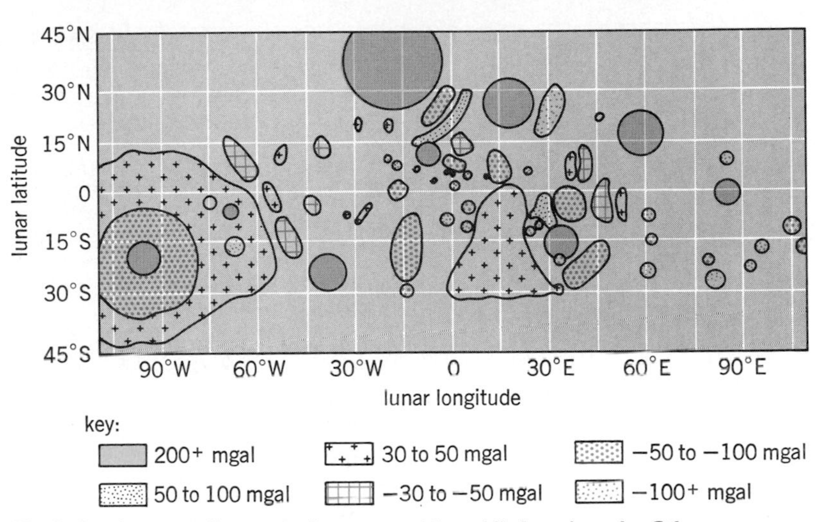

Fig. 3. Gravity anomalies on the lunar near side and limb regions. 1 mGal = gravitational acceleration of 10^{-5} m/s^2 = 3.3 × 10^{-5} ft/s. Circular areas correspond to mass concentrations in circular maria. (*W. L. Sjogren, Jet Propulsion Laboratory, NASA*)

rotation of the Moon about its polar axis is nearly uniform, but its orbital motion is not, owing to the finite eccentricity and Kepler's law of equal areas, so that the face of the Moon appears to swing east and west about 8° from its central position every month. This is the apparent libration in longitude. The Moon does rock to and fro in a very small oscillation about its mean rotation rate; this is called the physical libration. There is also a libration in latitude because of the inclination of the Moon's polar axis. The librations make it possible to see about 59% of the Moon's surface from the Earth.

The lunar ephemeris, derived from precise astronomical observations and refined through lengthy computations of the effects perturbing the movements of the Moon, has now reached a high degree of accuracy in forecasting lunar motions and events such as eclipses. Laser ranging to retroreflectors landed on the Moon, aided by radio ranging to spacecraft, provides measurements of Earth-Moon distances to a precision of the order of meters. *See Perturbation.*

Selenodesy. The problem of determining the Moon's true size and shape and its gravitational and inertial properties has been under attack by various methods for centuries (Tables 1 and 2). However, results from space flights have invalidated some of the premises on which the earlier methods were based, and have revealed discrepancies in the older data. The relation between the Moon's shape and its mass distribution is very important to theories of lunar origin and the history of the Earth-Moon system. Radio-tracking data from Lunar Orbiters indicate that the Moon's gravitational field is ellipsoidal, with the short axis being the polar one (as expected for any rotating body), and with the equatorial section being an ellipse possibly slightly elongated in the Earth-Moon direction. But the Earth-based radar measurements and tracking data from Rangers and Surveyors showed that the Moon's actual surface at the points of landing is about 1.2 mi (2 km) farther from the Earth than expected. Further evidence of an anomalous relationship between mass and shape for the Moon is provided by the mass concentrations in circular maria, discovered through analysis of short-term variations in the Lunar Orbiter tracking data and then mapped in detail by Apollo tracking (see **Fig. 3**). By radio altimetry, Apollo confirmed that the Moon's surface on the far side is higher on the average than the near side; that is, the center of mass is offset from the center of figure. The offset is about 1.2 mi (2 km) toward the Earth. These observations suggest that the Moon's crust is thicker on the far side than on the near side, as shown (not to scale) in **Fig. 4**.

Body properties. The Moon's small size and low mean density (Table 1) result in surface gravity too low to hold a permanent atmosphere, and therefore it was to be expected that lunar surface characteristics would be very different from those of Earth. However, the bulk properties of the Moon are also quite different—the density alone is evidence of that—and the unraveling of the Moon's internal history and constitution is a great challenge to planetologists. *See Earth.*

The Earth, with its dense metallic fluid core, convective mantle, strong and variable magnetic field with trapped radiation belts, widespread seismic tremors, volcanoes and folded mountain ranges, moving lithospheric plates, and highly differentiated radioactive rocks, is plainly a planet seething with inner activity. Is the Moon also an active, evolving world or is it something very different? The answer lies in a group of related experiments: seismic investigations, heat-flow measurements, surface magnetic and gravity profiles, determination of abundances and ages of the radioactive isotopes in lunar material, and comparison of the latter with those found in the Earth and meteorites. Present theories and experimental data yield the following clues to the problem.

1. The Moon is too small to have compressed its silicates into a metallic phase by gravity; therefore, if it has a dense core at all, the core should be of nickel-iron. But the low mass of the whole Moon does not permit a large core unless the outer layers are of very light material; available data suggest that the Moon's iron core may have a diameter of at most a few hundred kilometers.

2. The Moon has no radiation belts, and behaves as a nonconductor in the presence of the interplanetary field. Moon rocks are magnetized, but the source of the magnetism remains a mystery as there is now little or no general lunar magnetic field.

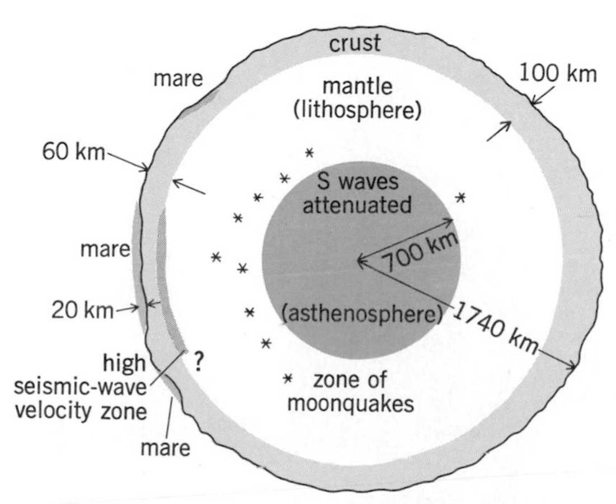

Fig. 4. Schematic diagram of lunar structure. The near side of the Moon is to the left of the figure. 1 km = 0.6 mi. (*After S. R. Taylor, Lunar Science: A Post-Apollo View, Pergamon Press, 1975*)

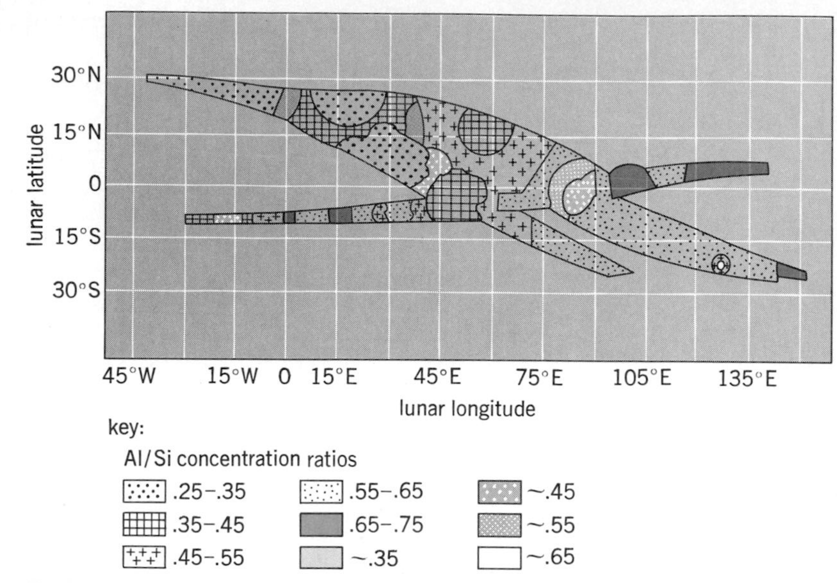

Fig. 5. Aluminum-silicon concentration ratios as detected by x-ray experiments on *Apollo 15* and *16*. (*I. Adler, University of Maryland*)

3. The Moon's natural radioactivity from long-lived isotopes of potassium, thorium, and uranium, expected to provide internal heat sufficient for partial melting, was roughly measured from orbit by *Luna 10*, and the component above the cosmic-ray-induced background radiation was found to be at most that of basic or ultrabasic earthly rock, rather than that of more highly radioactive, differentiated rocks such as granites. *Apollo 11* and *12* rock samples confirmed this result; *Apollo 15, 16*, and *17* mapped lunar composition and radioactivity from orbit (Fig. 3). X-ray experiments showed higher aluminum-silicon concentration ratios over highland areas and lower values over maria (**Fig. 5**), while magnesium-silicon ratios showed a converse relationship—higher values over maria and lower values over highlands.

4. The Moon is seismically much quieter than the Earth. Moonquakes are small, many of them originate deep in the interior (Fig. 4), and activity is correlated with tidal stress: more quakes occur when the Moon is near perigee.

When all of the Apollo observations are taken together, it is evident that the Moon was melted to an unknown depth and chemically differentiated about 4.5×10^9 years ago, leaving the highlands relatively rich in aluminum and an underlying mantle relatively rich in iron and magnesium, with all known lunar materials depleted in volatiles. The subsequent history of impacts and lava flooding includes further episodes of partial melting until about 3.9×10^9 years ago, with the final result being a thick, rigid crust with only minor evidence of recent basaltic extrusions. The temperature profile and physical properties of the Moon's deep interior are, despite the Apollo seismic and heat-flow data, under active debate. Figure 4 shows a rough sketch of the Moon as revealed by the data.

Fig. 7. Mare Tsiolkovski on far side of Moon. Crater, partly flooded by dark mare material, is about 120 mi (200 km) across. (*Langley Research Center, NASA*)

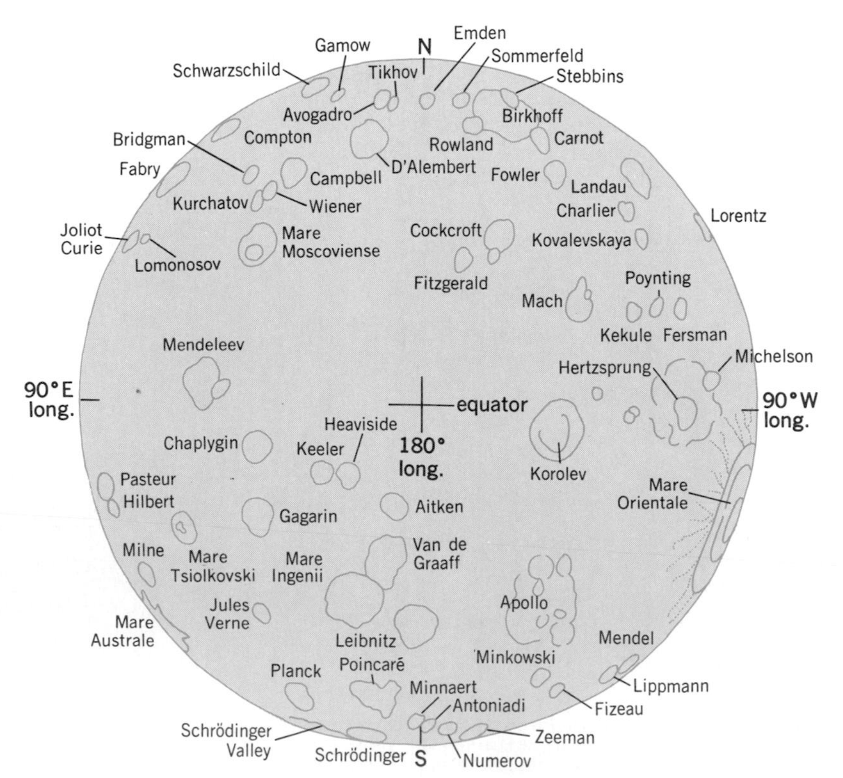

Fig. 6. Map of far side of Moon.

Large-scale surface features. As can be seen from the Earth with the unaided eye, the Moon has two major types of surface: the dark, smooth maria and the lighter, rougher highlands (Fig. 1 and **Fig. 6**). Photography by spacecraft shows that, for some unknown reason, the Moon's far side consists mainly of highlands (**Fig. 7**). Both maria and highlands are covered with craters of all sizes. Craters are more numerous in the highlands than in the maria, except on the steeper slopes, where downhill movement of material apparently tends to obliterate them. Numerous different types of craters can be recognized. Some of them appear very similar to the craters made by explosions on the Earth; they have raised rims, sometimes have central peaks, and are surrounded by fields of hummocky, blocky ejecta. Others are rimless and tend to occur in lines along cracks in the lunar surface. Some of the rimless craters, particularly those with dark halos, may be gas vents; others may be just the result of surface material funneling down into subsurface voids. Most prominent at full moon are the bright ray craters (Fig. 1) whose grayish ejecta appear to have traveled for hundreds of kilometers across the lunar surface. Observers have long recognized that some erosive process has been and may still be active on the Moon. For example, when craters overlap so that their relative ages are evident, the younger ones are seen to have sharper outlines than the older ones. Bombardment of the airless Moon by meteoritic matter and solar particles, and extreme temperature cycling, are now considered the most likely erosive agents, but local internal activity is also a possibility. Rocks returned by the Apollo astronauts are covered with tiny glass-lined pits, confirming erosion by small high-speed particles.

The lunar mountains, though very high (26,000 ft or 8000 m or more), are not extremely steep, and lunar explorers see rolling rather than jagged scenery (**Fig. 8**). There are steep slopes (30–40°) on the inside walls and central peaks of recent craters, where the lunar material appears to be resting at its maximum angle of repose, and rocks can be seen to have rolled down to the crater bottoms.

Fig. 8. The crater Copernicus, showing the central peaks, slump terraces, patterned crater walls, and (background) slopes of the Carpathian Mountains. (*Langley Research Center, NASA*)

Though widespread networks of cracks are visible, there is no evidence on the Moon of the great mountain-building processes seen on the Earth. There are some low domes suggestive of volcanic activity, but the higher mountains are all part of the gently rolling highlands or the vast circular structures surrounding major basins. **Figure 9** shows one of these, the Mare Orientale, as revealed by *Lunar Orbiter 4*. This large concentric structure is almost invisible from Earth because it lies just past the Moon's western limb; at favorable librations parts of its basin and mountain ramparts can be seen. The great region of radial sculpture surrounding the Orientale basin strongly suggests a catastrophic origin, with huge masses of matter thrown outward from the center. Note, however, the gentle appearance of the flooding by the dark mare material, which seems to lie only in the lowest parts of the concentric rings. Other basins, namely, Imbrium, Serenitatis, and Crisium, appear more fully flooded (Fig. 1). These maria were created by giant impacts, followed by subsidence of the ejecta and (probably much later) upwelling of lava from inside the Moon. Examination of small variations in Lunar Orbiter motions has revealed that each of the

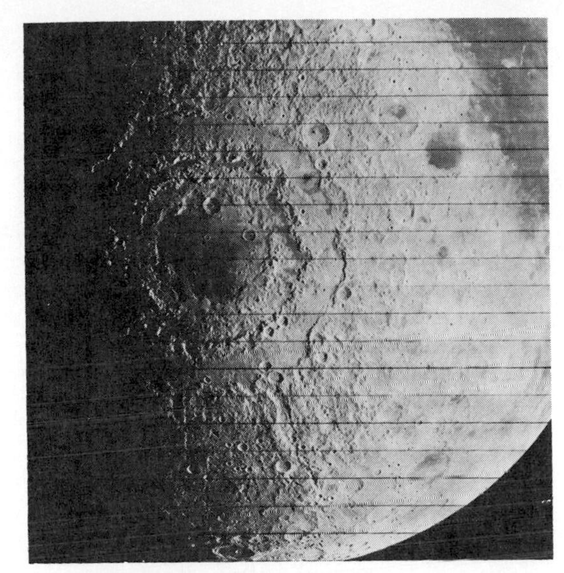

Fig. 9. Mare Orientale. (*Langley Research Center, NASA*)

great circular maria is the site of a positive gravity anomaly (excess mass), shown in Fig. 3. The old argument about impact versus volcanism as the primary agent in forming the lunar relief, reflected in lunar literature over the past 100 years, appears to be entering a new, more complicated phase with the confirmation of extensive flooding of impact craters by lava on the Moon's near side, while on the far side, where

Fig. 10. Aristarchus-Harbinger region of the Moon, photographed from the *Apollo 15* spacecraft in lunar orbit, with the craters Aristarchus and Herodotus and Schroeter's Valley, the largest sinuous rille on the Moon. The impact crater Aristarchus, about 25 mi (40 km) in diameter and more than 2.5 mi (4 km) deep, lies at the edge of a mountainous region that shows evidence of volcanic activity. (*NASA*)

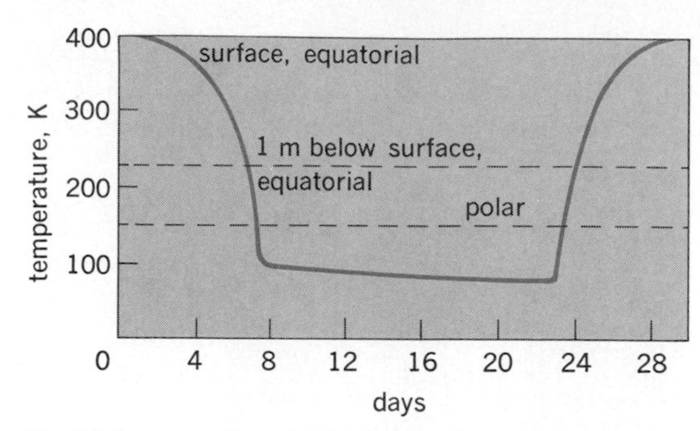

Fig. 11. Lunar surface and subsurface temperatures.

the crust is thicker, the great basins remain mostly empty.

In some of the Moon's mountainous regions bordering on the maria are found sinuous rilles (**Fig. 10**). These winding valleys, some of them known since the eighteenth century, were shown in Lunar Orbiter pictures to have an exquisite fineness of detail. Some of them originate in small circular pits and then wriggle delicately across the Moon's gentle slopes for hundreds of kilometers, detouring around even slight obstacles, before vanishing on the plains. Though their resemblance to meandering rivers is strong, the sinuous rilles have no tributaries or deltas. No explanation for them yet offered (for example, dust flows, lava channels, or subsurface ducts made by water eroding ice) has proved entirely convincing.

Other strange large-scale features, observed by telescope and then revealed in more detail by spacecraft cameras, are the ghost craters, circular structures protruding slightly from the maria, and the low, ropy wrinkle ridges that stretch for hundreds of kilometers around some mare borders.

Small-scale surface features. Careful observations, some of them made decades before the beginning of space flight, revealed much about the fine-scale nature of the lunar surface. Since the smallest lunar feature telescopically observable from the Earth is some hundreds of meters in extent, methods other than direct visual observation had to be used. Photometry, polarimetry, and later radiometry and radar probing gave the early fine-scale data. Some results of these investigations suggested bizarre characteristics for the Moon. Nevertheless, many of their findings have now been confirmed by spacecraft. The Moon seems to be totally covered, to a depth of at least tens of meters, by a layer of rubble and soil with very peculiar optical and thermal properties. This layer is called the regolith. The observed optical and radio properties are as follows.

1. The Moon reflects only a small portion of the light incident on it (the average albedo of the maria is only 7%, darker than any familiar object except things like soot or black velvet).

Fig. 13. Lunar soil and rocks, and the trenches made by *Surveyor 7*. (*Jet Propulsion Laboratory, NASA*)

Fig. 12. Lunar patterned ground, a common feature on moderate slopes. (*Langley Research Center, NASA*)

Fig. 14. Crater showing appearance of upwelling on its floor. (*Langley Research Center, NASA*)

2. The full moon is more than 10 times as bright as the half moon.

3. At full moon, the disk is almost equally bright all the way to the edge; that is, there is no "limb darkening" such as is observed for ordinary spheres, whether they be specular or diffuse reflectors.

4. Color variations are slight; the Moon is a uniform dark gray with a small yellowish cast. Some of the maria are a little redder, some a little bluer, and these differences do correlate with large-scale surface morphology, but the visible color differences are so slight that they are detectable only with special filters. (Infrared spectral differences are more pronounced and have provided a method for mapping variations in the Moon's surface composition.)

5. The Moon's polarization properties are those of a surface covered completely by small, opaque grains in the size range of a few micrometers.

6. The material at the lunar surface is an extremely good thermal insulator, better than the most porous terrestrial rocks. The cooling rate as measured by infrared observations during a lunar eclipse is strongly variable; the bright ray craters cool more slowly than their surroundings.

7. The Moon emits thermal radiation in the radio wavelength range; interpretations of this and the infrared data yield estimates of surface and shallow subsurface temperatures as shown in **Fig. 11**.

Fig. 15. Region near the crater Tycho, including *Surveyor 7* landing site. 20 km = 12 mi. (*Langley Research Center, NASA*)

8. At wavelengths in the meter range, the Moon appears smooth to radar, with a dielectric constant lower than that of most dry terrestrial rocks. To centimeter waves, the Moon appears rather rough, and at visible light wavelengths it is extremely rough (a conclusion from observations 1–5 above).

These observations all point to a highly porous or underdense structure for at least the top few millimeters of the lunar surface material. The so-called backscatter peak in the photometric function, which describes the sudden brightening near full Moon, is characteristic of surfaces with deep holes or with other roughness elements that are shadowed when the lighting is oblique.

The Ranger, Luna, Surveyor, and Lunar Orbiter missions made it clear that these strange electromagnetic properties are generic characteristics of the dark-gray, fine soil that appears to mantle the entire Moon, softening most surface contours and covering everything except occasional fields of rocks (**Figs. 12** and **13**). This soil, with a slightly cohesive character like that of damp sand and a chemical composition similar to that of some basic silicates on the Earth, is a product of the radiation, meteoroid, and thermal environment at the lunar surface. Figure 12 shows a surface texture, called patterned ground, that is common on the moderate slopes of the Moon. This widespread phenomenon is unexplained, though there are some similar surfaces developed on the Earth when unconsolidated rock, lava, or glacial ice moves downhill beneath an overburden. At many places on the Moon, there is unmistakable evidence of downward sliding or slumping of material and rolling rocks. There are also a few instances of apparent upwelling (**Fig. 14**), as well as numerous "lakes" where material has collected in depressions. **Figures 15, 16,** and **17** show, at different scales, the landing site of *Surveyor 7* near the great ray crater Tycho (Fig. 1). *Surveyor 7* found the soil of the highlands near Tycho to be rockier than, and slightly different in chemical composition from, the mare materials sampled by earlier Surveyors. Figure 13 shows trenches made in the lunar soil during soil mechanics experiments on *Surveyor 7*. Magnets on the Surveyors collected magnetic particles from the soil, demonstrating the presence of either meteoritic or native iron minerals at the sites ex-

Fig. 16. *Orbiter* photograph of *Surveyor 7* site. (*Langley Research Center, NASA*)

Fig. 17. Surface view at *Surveyor 7* site. (*Jet Propulsion Laboratory*)

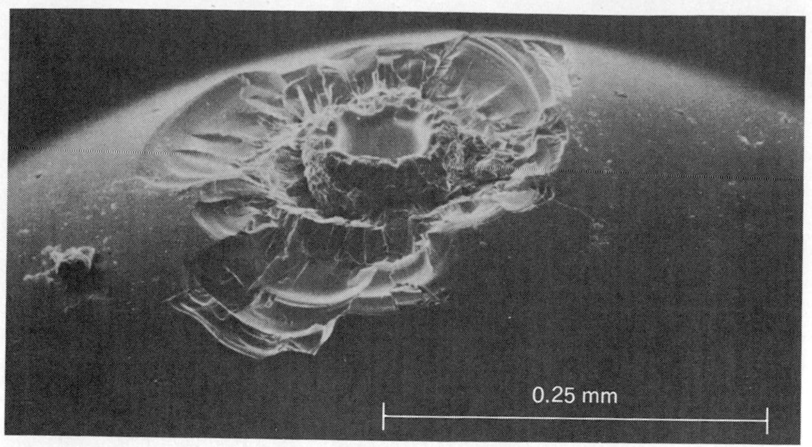

Fig. 18. Example of microcratering, caused by hypervelocity impact of tiny particles, on a dark-brown glass sphere. The diameter of the sphere is approximately 0.75 mm, and the diameter of the inner crater, inside the raised rim, is about 50 μm. This photograph was taken through a scanning electron microscope.

amined. Meteoroid experiments on the Lunar Orbiters showed about the same flux of small particles as is observed at the Earth, so that the lunar soil would be expected to contain a representative sample of meteoritic and possibly also cometary matter. Apollo results confirmed and extended the Surveyor data and also indicated that glassy particles are abundant in and on the soil. Evidence of micrometeoroid bombardment is seen in the many glass-lined microcraters found on lunar rocks (**Fig. 18**). *See Comet; Meteor*.

Chemical, mineral, and isotopic analyses of minerals from the *Apollo 11* site showed that mare rocks there are indeed of the basic igneous class and are very ancient (3–4 × 10^9 years). The *Apollo 12* samples are significantly younger, suggesting that Mare Tranquillitatis and Oceanus Procellarum were formed during a long and complex lunar history. The *Apollo 12* astronauts visited *Surveyor 3* and brought back parts of that spacecraft to permit analysis of the effects of its 2½-year exposure on the surface of the Moon. The lunar rock and soil samples returned by the Apollo and Luna missions have yielded much new information on the composition and history of the Moon. Among the dominant characteristics of these rocks are enrichment in refractories, depletion in volatiles, much evidence of repeated breaking up and rewelding into breccias, and ages since solidification extending back from the mare flows of 3–4 × 10^9 years ago into the period of highland formation more than 4 × 10^9 years ago, but not as yet including the time of the Moon's original accretion. Some characteristics of the lunar samples are summarized in **Table 4**.

In 1983 an Antarctic meteorite was found to resemble some of the lunar samples, and it was determined by analysis to have come from the Moon (**Fig. 19**). *See Meteorite*.

As the Apollo missions progressed, each new landing site was selected with the aim of elucidating more of the Moon's history. A main objective was to sample each of the geologic units mapped by remote observation, either by landing on it or by collecting materials naturally transported from it to the landing site. Although this process did result in collection of both mare and highland materials with a wide range of ages and chemical compositions, it did not result in a complete unraveling of the history of the Moon. Apparently, the great impacts of 3–4 × 10^9 years ago erased much of the previous record, resetting radioactive clocks and scrambling minerals of diverse origins into the complicated soils and breccias found today.

Atmosphere. Though the Moon may at one time have contained appreciable quantities of the volatile elements and compounds (for example, hydrogen, helium, argon, water, sulfur, and carbon compounds) found in meteorites and on the Earth, its high daytime surface temperature and low gravity would cause rapid escape of the lighter elements. Solar ultraviolet and x-ray irradiation would tend to break down volatile compounds at the surface, and solar charged-par-

Table 4. Some selected data from Apollo and Luna missions

Mission	Main sample properties	Other data
Apollo 11	Mare basalts, differentiated from melt at depth 3.7 × 10^9 years ago. Some crystalline highland fragments in soils. Unexpected abundance of glass. Much evidence of impact shock and microcratering. No water or organic materials.	Study of seismic properties showed low background, much scattering, and low attenuation.
Apollo 12	Basalts 3.2 × 10^9 years old. One sample 4.0 × 10^9 years old includes granitic component. Some samples with high potassium, rare-earth elements, and phosphorus (KREEP) may be Copernicus crater ejecta.	Surveyor parts returned showed effects of solar and cosmic bombardment.
Apollo 13	Spacecraft failure—no samples.	Despite emergency, some lunar photos returned.
Luna 16	Basalt 3.4 × 10^9 years old, relatively high Al content.	
Apollo 14	Shocked highland basalts, probably Imbrium ejecta, 3.95 × 10^9 years old, higher Al and lower Fe than mare materials.	Deep moonquakes.
Apollo 15	Highland anorthosites including one sample 4.1 × 10^9 years old, mare basalts similar to *Apollo 11* samples.	Orbital remote sensing began mapping of surface compositions.
Luna 20	Possibly Crisium ejecta, 3.9 × 10^9 years old.	
Apollo 16	Highland anorthosite breccias 3.9–4 × 10^9 years old, also possibly Imbrium ejecta.	Seismic network began recording locations of impacts and deep moonquakes; orbital compositional mapping extended.
Apollo 17	Variety of basalts and anorthosites 3.7–4 × 10^9 years old, possibly volcanic glass, few dunite fragments 4.48 × 10^9 years old, possibly surviving from before the great highland bombardment.	Orbital mapping, and study of seismic, particle-and-field, and subsurface electrical properties yielded comprehensive (but still unexplained) picture of Moon.
Luna 24	Very low titanium basalt from Mare Crisium. Sample includes rock 3.3 × 10^9 years old.	

Fig. 19. Meteorite collected in Antarctica in 1982, designated ALHA 81005, proved by analysis to be a lunar rock. (*NASA*)

ticle bombardment would ionize and sweep away even the heavier gas species. Observations from the Earth, looking for a twilight glow of the lunar atmosphere just past the terminator on the Moon, and watching radio-star occultations have all been negative, setting an upper limit of 10^{-12} times the Earth's sea-level atmospheric density for any lunar gas envelope. Therefore, either the lunar volatile compounds have vanished into space or they are trapped beneath the surface. The samples returned by Apollo are enriched in refractory elements, depleted in volatiles, and impregnated with rare gases from the solar wind. No water appears to have ever been present at any Apollo site, and carbonaceous materials were present, if at all, only in very small amounts.

Occasional luminescent events reported by reliable observers suggest that some volcanic gases are vented from time to time on the Moon, particularly in the regions of the craters Aristarchus and Alphonsus. A slight, transient atmosphere does exist on the night side of the Moon as a result of the trapping and release of gas molecules at the very low temperatures prevailing there; also frozen liquids or gases could exist in permanently shadowed crater bottoms near the lunar poles. No experiment to detect such accumulations of volatiles has been made. *Lunokhod 2*, a Soviet roving spacecraft, measured a slight glow attributed to a very thin cloud of small particles near the surface, which could explain the Surveyor observations of a slight horizon glow after sunset. Also, the ALSEP (Apollo lunar surface experiments packages) experiments landed by Apollo have occasionally detected small gas emanations, including water, from unknown sources.

Lunar resources. Enough is known about the Moon to show that it is a huge storehouse of metals, oxygen (bound into silicates), and other materials potentially available for future human use in space. Because of the Moon's weak gravity, lunar materials could be placed into orbit at less than one-twentieth of the energy cost for delivering them from Earth. At the sites so far explored, no water exists, and the only available hydrogen is the small amount implanted in soil by the solar wind. It will be a task for future explorations to find the polar ices if they exist, to discover concentrations of meteoritic or cometary materials, and to investigate atypical geologic phenomena such as the seemingly volcanic regions. Any of these sites might yield additional treasures, but it is already known that the Moon could be an important resource for humanity in space.

James D. Burke

Bibliography. J. R. Arnold, Ice at the lunar poles, *J. Geophys. Res.*, 84(B10):5659–5668, 1979; J. R. Arnold (ed.), *Workshop on Near-Earth Resources*, La Jolla, NASA Conf. Publ. 2039, 1978; H. S. F. Cooper, Jr., *Moon Rocks*, 1970; A. de Visscher (ed.), *Atlas of the Moon*, trans. by R. G. Lascelles, 1964; F. El-Baz, The Moon after Apollo, *Icarus*, 25:495–537, 1975; W. K. Hartmann, R. J. Phillips, and G. J. Taylor (eds.), *Origin of the Moon*, 1986; Z. Kopal, *The Moon in the Post-Apollo Era*, 1974; L. J. Kosofsky and F. El-Baz, *The Moon as Viewed by Lunar Orbiter*, NASA SP-200, 1970; A. A. Levinson and S. R. Taylor, *Moon Rocks and Minerals*, 1972; Lunar and Planetary Institute, *Proceedings of the Lunar and Planetary Science Conferences*, 1970–1989; *The Moon: A New Appraisal*, Royal Society of London, 1979; S. R. Taylor, *Lunar Science: A Post-Apollo View*, 1975; S. R. Taylor, *Planetary Science: A Lunar Perspective*, 1982; S. R. Taylor, Structure and evolution of the Moon, *Nature*, 281:105, 1979; F. L. Whipple, *Earth, Moon and Planets*, 3d ed., 1968; D. Wilhelms et al., *Geologic History of the Moon*, U.S. Geol. Sur. Prof. Pap. 1348, 1988.

Nebula

Originally, any fixed, extended, and usually fuzzy luminous object seen in a telescope. Nebulae are now distinguished from star clouds that can be resolved into individual stars, but earlier workers were unable to differentiate between white nebulae, which are stellar systems so remote as to show no individual stars, and gaseous or diffuse nebulae in the Milky Way Galaxy. *See Star clouds.*

Extragalactic nebulae are stellar systems comparable with the Milky Way Galaxy or the Magellanic Clouds in size and number of stars, and are more properly termed external galaxies. They are grouped as spirals, ellipticals, or irregulars, and various classification systems have been devised. *See Galaxy, external.*

Types of nebulae. This article deals with gaseous nebulae. This class of objects includes diffuse nebulae which contain dust and gas of the interstellar medium, excited and caused to fluoresce by embedded stars, for example, the Great Orion Nebula; so-called planetary nebulae; and supernova remnants such as the Network or Veil Nebula in Cygnus. Gaseous nebulae are members of the Milky Way galactic system, and small compared with its overall dimensions. Various types of gaseous nebulae have been identified. *See Interstellar matter.*

Diffuse nebulae. These range in density from a few atoms per cubic centimeter to 10,000 or more atoms per cubic centimeter (as in the Orion Nebula). Some are compact objects less than a parsec in diameter. Both dust and gas are excited by ultraviolet radiation of stars. Some diffuse nebulae such as Orion occur at the edges of large clouds of cool dust and gas, mostly in molecular form. Those of lower density are found from the faint glow in the red hydrogen line produced as hydrogen ions recapture electrons. For this reason they are also called H II regions, indicating regions of ionized hydrogen. *See Orion Nebula.*

Reflection nebulae. These show no bright line spectra. Dust grains simply reflect the light of nearby embedded stars. Hydrogen gas is present but mostly neutral. The Pleiades nebulosity is an example of this type. *SEE PLEIADES.*

Nebulae associated with star formation. These include the so-called fan-shaped nebulae associated with T Tauri stars, certain bipolar nebulae, and Herbig-Haro Objects. Some, such as Hubble's variable nebula, associated with the variable star R Monocerotis, show brightness fluctuations. In many instances, a newly formed star excites and ionizes the gas in its immediate neighborhood, although the star itself is quite concealed by its dusty surroundings. *SEE STELLAR EVOLUTION.*

Planetary nebulae. Planetary nebulae are so denoted because they often show small greenish disks in the telescope, not unlike the images of the planets Uranus and Neptune. The best-known of this class is the Ring Nebula in Lyra, M57 or NGC 6720 (**Fig. 1**). The energy emitted by planetary nebulae is derived from the rich ultraviolet radiation of stars embedded within them. *SEE PLANETARY NEBULA.*

Supernova remnants. The detonation of a star in a supernova event causes the ejection of the outer layers into the surrounding interstellar medium. In early stages as in the Crab Nebula, the radiating material consists of ejecta from the star. In the later stages this rapidly moving material is slowed down as it mixes with the surrounding dust and gas of the interstellar medium. Such a phase is illustrated, for example, by the Network Nebula in Cygnus or IC 433. Heating by shock waves causes the material to radiate optically. Sometimes, the temperature behind the shock front can rise to more than 10^6 K, but the gas is so rarefied that the intensity of the emitted radiation is extremely low. Supernovae remnants characteristically emit nonthermal radio-frequency emission, whereby they are often detected in nearby galaxies as well as in the Milky Way system. *SEE CRAB NEBULA; SUPERNOVA.*

Cocoon nebulae. These are associated with very massive stars. At a late stage in its evolution a massive star may eject a dense shell of material that effectively hides it from view temporarily. Although the extended Carina Nebula appears to be a normal H II region, η Carinae itself is a remarkable dense, compact object (with a diameter of about 2 arc-seconds) which hides the central star and emits a remarkable spectrum dominated by forbidden lines of ionized iron.

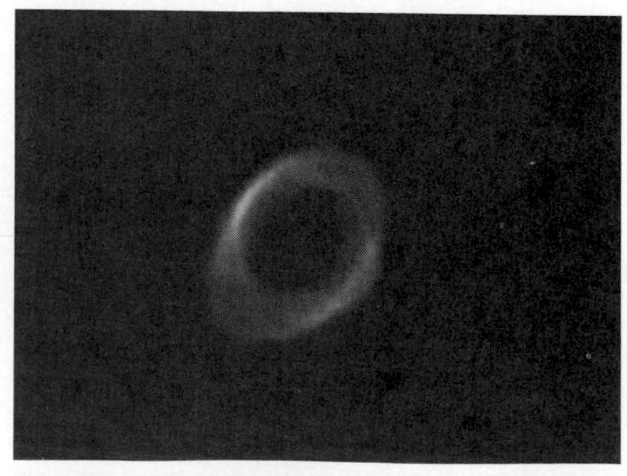

Fig. 1. Ring Nebula NGC 6720, in a U.S. Navy electrograph made with 61-in. (2-m) astrometric reflector at Flagstaff, Arizona.

Catalogs. Nebulae are cataloged according to various systems. The first list was made by C. Messier. A much more complete list was made by J. L. E. Dreyer in the *New General Catalogue* (abbreviated NGC) and the two Index catalogs. These lists include also galaxies and star clusters. A revised catalog has been compiled, and special lists have been published for planetary nebulae, diffuse nebulae and H II regions, supernova remnants, and faint H II regions in the Milky Way. *SEE ASTRONOMICAL CATALOGS; MESSIER CATALOG.*

Methods and types of observations. Measurements of positions and sizes are straightforward, except for irregular structures that are hard to describe quantitatively. Early observations were visual, but by the unaided eye only the brightest nebulae or their most conspicuous features could be detected. Photography has contributed greatly to nebular observation. Most emission nebulae radiate strongly in the red hydrogen line Hα. Hence, by using red-sensitive plates and narrow-bandpass filters, it is possible to suppress the sky background and register nebulosities of low surface brightness. Even better results are obtained when photographic plates are replaced by charge-coupled devices (CCDs). Then the sky subtraction can be done quantitatively. Gaseous nebulae of both thermal and nonthermal types are observable with radio telescopes. Energy fluxes from nonthermal sources persist and even rise as one goes to the very lowest frequencies, while radiation from thermal sources eventually declines with decreasing frequency in accordance with Rayleigh-Jeans law. Infrared measurements have revealed a number of important emission lines of ions whose presence cannot be detected in ordinary spectral regions; they often indicate the presence of great quantities of thermally emitting dust in diffuse nebulae as well as in planetaries. Ultraviolet and x-ray observations provide invaluable supplements to data obtained from ordinary spectral regions, sometimes revealing attenuated gases at temperatures of hundreds of thousands of degrees, particularly in shocked gas in supernova remnants and large regions of the Milky Way. *SEE CHARGE-COUPLED DEVICES; INFRARED ASTRONOMY; RADIO ASTRONOMY; ULTRAVIOLET ASTRONOMY; X-RAY ASTRONOMY.*

Brightness. Because surface brightness is independent of the distance as long as the eye perceives the object as an extended area, no advantage is gained on objects such as Orion or the Trifid nebula by using large telescopes, unless one wishes to examine small details. For small diffuse nebulae and planetaries, a large telescope has considerable advantage. For monochromatic radiation, the surface brightness may be expressed in terms of ergs/(s)(cm^2)(unit solid angle), although other units such as SI units [W/m^2(sr)] or even magnitudes per square minute of arc have also been used.

The brightness of a nebula can also be measured in the radio-frequency region, although it is necessary to take into consideration the limited resolving power of radio telescopes. Surface brightness may be measured by photographic photometry, but more accurate work is done by photoelectric methods, using a spectrum scanner or narrow-band-pass filters to select monochromatic radiations, or with charge-coupled device detectors. Particularly effective opportunities are offered by infrared areal detectors since with them it is often possible to observe directly through intervening clouds of interstellar dust. *SEE ASTRONOMICAL PHOTOGRAPHY.*

The measurement of the brightness of a nebula is more complicated than that of a star. The nebula is an extended surface of nonuniform brightness; hence, the complete description of a nebula in monochromatic radiation would consist of a set of isophotic contours calibrated in terms of intensity units. Gaseous and diffuse nebulae show a huge range in surface brightness from objects like Orion to faint wisps barely visible on long exposures with narrow-band-pass filters.

Distances. If the nebula is associated with a star or a star cluster, its distance may be found by measuring the stellar distances. For example, the distance of the Orion Nebula is found by establishing the absolute luminosities of the illuminating stars and the amount of space absorption. Then the distance is found from a comparision of the apparent and intrinsic brightnesses of the stars. Similar methods may be applied to the Lagoon and Trifid nebulae or any other nebula for which the intrinsic luminosities of embedded stars can be determined. *SEE MAGNITUDE.*

In some instances, such as the Network Nebula in Cygnus, it is possible to measure the angular rate of expansion and also the velocity in the line of sight, which gives the radial rate of expansion in kilometers per second. The method cannot be applied indiscriminately to the planetaries because the rates of expansion may represent not just a lateral motion of material but a change in size of the ionized volume. Often, statistical methods are used in which radial velocities and proper motions are compared, together with correlation between angular diameter and distance. Direct determinations are possible for a few objects—the nucleus of NGC 246 which has a dwarf companion of absolute magnitude $M = 7.0$, the planetary nebula in the globular cluster M15, the planetaries near the central bulge of the Galaxy, and those in external galaxies such as the Magellanic Clouds, in the Andromeda Spiral and its companions, and in other members of the local group. Astrophysical methods of limited accuracy are often used for planetaries; often these procedures involve some assumption such as constancy of the mass within the emitting volume or constancy of luminosity. A powerful technique has been developed for obtaining reliable distances of planetary nebulae with absorption-line spectra. *SEE ANDROMEDA GALAXY; LOCAL GROUP; MAGELLANIC CLOUDS.*

Spectra. When small gaseous nebulae (which have diameters of a few seconds of arc) are observed with a slitless spectrograph, an image of the nebula is formed in each of its monochromatic radiations. It is found that the radiations of ions of higher excitation, such as neon, Ne^{4+}, are always concentrated closer to the central star than are the radiations of ions of lower excitation, such as oxygen, O^{+}. The reason is that the higher-energy quanta capable of producing highly ionized atoms are exhausted before they reach the outer layers. *SEE ASTRONOMICAL SPECTROSCOPY.*

For the spectroscopic studies of large nebulae, weak lines, or lines that fall close together, it is necessary to use a slit spectrograph, equipped with appropriate detectors.

The spectra of the gaseous nebulae show the recombination lines of hydrogen and helium which are observed in optical ultraviolet and radio-frequency

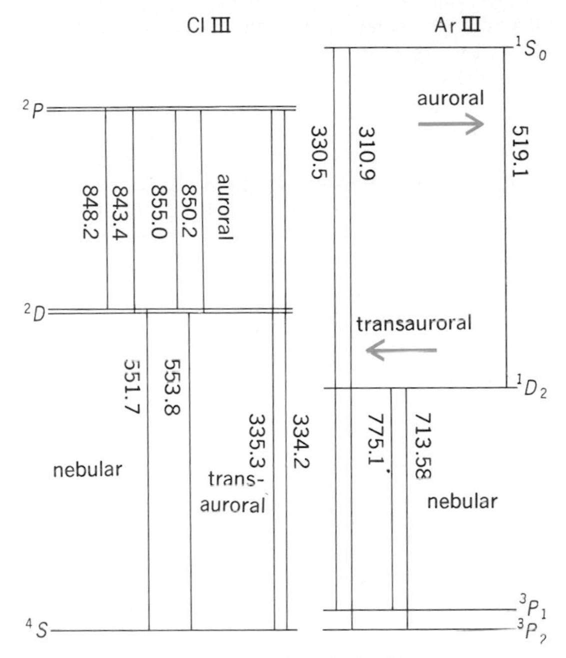

Fig. 2. Forbidden-line transitions in doubly ionized chlorine and argon with the wavelength of each transition given in nanometers.

ranges. The strongest lines in the optical region are often some that have never been produced in any terrestrial laboratory. These are the so-called forbidden lines of ions of various abundant elements. They represent transitions between the metastable levels of the ground configuration. **Figure 2** illustrates these transitions for the ions argon [Ar III] and chlorine [Cl III], where numeral III designates a doubly ionized atom, the ionized atoms having ground configurations $3p^4$ and $3p^3$, respectively. Transitions of the type ^2P–^2D in p^3 ions or ^1S–^1D in p^2 or p^4 ions are called auroral transitions, because this type of forbidden transition is the most important in the Earth's aurora. Transitions between the middle metastable term and the ground term give the nebular lines; jumps from the highest metastable terms to the ground term give transauroral lines. The nebular transitions of both Ar III and Cl III are observed in many planetary nebulae, but the auroral transitions such as 519.1 [Ar III] and 848.2 [Cl III] are very weak. *SEE ATOMIC STRUCTURE AND SPECTRA.*

Weak recombination lines of oxygen, nitrogen, and carbon are observed in gaseous nebulae, but (aside from H and He) the strongest permitted lines are certain O III transitions observed in high-excitation planetaries. The O III lines are produced by a remarkable fluorescent mechanism discovered by I. S. Bowen. Ions of O^{2+} in the $2p^2\ ^3P_2$ level of the ground configuration absorb the 30.378-nanometer resonance line of ionized helium and are excited to the $3d\ ^3P_2$ level, from which they cascade downward with the emission of observable lines.

The visible and radio-frequency lines of hydrogen and helium are produced by a process of photoionization from the ground level, followed by recombination in one of the highly excited levels with subsequent cascade and the emission of observable lines. For example, the red hydrogen line Hα may be produced by the recapture of an electron on the third level with a subsequent jump from the third to the second level. Finally, the atom goes from the second level to the ground level.

On the other hand, the forbidden lines are excited by electron impacts which cause the atoms to rise from the ground term to one of the nearby metastable terms. They return to the ground level with the emission of a forbidden line. The transition probabilities, such as the probability of an atom making the jump in a unit time, are very low. An atom may remain in a metastable level for seconds or even minutes, whereas it will leave an ordinary excited level in 10^{-7} s or even less time. The forbidden lines attain such great strength in gaseous nebulae because of their vast extent and because the processes operating to produce the ordinary permitted lines are enormously reduced in efficiency. In a typical planetary nebula the green forbidden lines of oxygen [O III] may be 10 times as strong as the hydrogen Hβ line, yet the concentration of hydrogen ions per unit volume may be 10,000 times as great as the concentration of O^{2+} ions.

Great advances have been made by extending observations to the ultraviolet with the *International Ultraviolet Explorer* satellite and to the infrared with detectors flown in satellites such as *IRAS* and with the Kuiper Airborne Observatory. Virtually all data on the abundance of the important element carbon comes from ultraviolet observations, while the infrared measurements provide accurate information on elements such as nitrogen and argon. *See* Satellite astronomy.

The diffuse galactic nebulae always show relatively low excitation, the Balmer lines are strong, the [O II] lines are often more intense than the [O III] lines, and lines such as [Ar IV] and [Ne V] are absent. In nebulae of low surface brightness, only a few lines can be observed: Hα, and sometimes the 372.7-nm [O II], and occasionally [O III]. In external galaxies, there sometimes exist high-excitation, extended gaseous nebulae. In the so-called Seyfert galaxies, broad high-excitation lines are observed.

Quasistellar sources show emission lines that resemble those of ordinary planetary nebulae, but strongly shifted toward the red. Lines that fall normally in the far ultraviolet are often observed in these objects, for example, Lyα of hydrogen. *See* Quasar.

Masses, densities, and temperatures. The radius R of a nebula of angular radius R^{seconds} is obtained as soon as its distance r is known. Consider a uniform spherical nebula, where N_ϵ is the electronic density, $N(H^+)$ is the density of hydrogen ions, and T_ϵ is the electron temperature. The emission per unit volume in the Hβ line can be written as the equations below.

$$E(H\beta) = N(H^+)N_\epsilon A^0\ 10^{-25} \quad \text{cgs units}$$
$$E(H\beta) = N(H^+)N_\epsilon A^0\ 10^{-36} \quad \text{SI units}$$

Here A^0 is a factor that depends solely on temperature, with the value 2.22 at 5000 K (8500°F), 1.24 at 10,000 K (17,500°F), and 0.66 at 20,000 K (35,500°F).

The entire energy emitted by the nebula in the Hβ line is $\frac{4}{3}\pi R^3 E(H\beta)$, so that the flux passing through each unit area of surface is $RE/3$, and therefore the flux received by an observer at distance r will be $\frac{1}{3}RE(H\beta)(R/r)^2$.

To establish the electron density, not only measurements of the surface brightness are required but also a knowledge of the temperature T_ϵ. The electron temperature could be estimated from the width of the spectral lines, but it would be necessary to separate the effects of the gas kinetic motion from those of the large-scale mass motion.

The best optical method involves the use of the relative intensities of the auroral and nebular transitions of a given ion, such as the ^1S-^1D (436.3-nm) and ^1D-^3P (500.7-nm, 495.9-nm) transitions of [O III]. The relative number of collisional excitations to the $^1D_2$ level and to the ^1S level depends on the velocity distribution of the electrons and, hence, on the temperature. If the target areas for collisional excitation and the transition probabilities are known, a relation involving temperature, density, and intensity ratio can be found. If the nebular and auroral lines of two ions, both occurring in the same region in the nebula, can be measured, both N_ϵ and T_ϵ can be found independently of the surface brightness measurements. If the nebula has a filamentary structure, the electron density found in this way will be greater than that found from the surface brightness, which represents an average over the space occupied by the nebula.

From radio-frequency observations carried out at different wavelengths, both the optical thickness and the gas kinetic temperature of the radiating gas can be determined. Additional information is provided by combining radio and optical observations. The radio-frequency observations are not affected by space absorption, but they give the temperatures in the cooler portions of the radiating gas. In the past, radio-frequency observations of small nebulae suffered from lack of spatial resolution, but with instruments such as the Very Large Array (VLA) in New Mexico, this difficulty has been overcome and now radio observations often give better resolution than optical ones.

The masses of the planetary nebulae are typically of the order of 0.1 to 0.5 solar mass. H II regions often have masses many times that of the Sun, while the neutral hydrogen clouds in Orion (which are not ionized by the hot stars) have a total mass about 100,000 times that of the Sun.

Internal motions. The motions of the gases perpendicular to the line of sight have been found in only a few nebulae, notably the Network Nebula and the Crab Nebula, but motions along the line of sight can be detected by radial velocity measurements with a slit spectrograph and by a Fabry-Pérot etalon. Use of a multislit consisting of a series of closely placed slits parallel to one another is the most efficient way to observe radial velocity shifts in small nebular regions. With this device, many planetaries have been observed. They appear to be expanding, the rate of expansion depending on the degree of ionization of the ion. Thus [Ne V] lines show the smallest expansion rate and [O II] lines show the largest, suggesting that because the degree of ionization depends on the distance from the central star, the material is accelerated on the outer side. Subsequent measurements made with charge-coupled device detectors show that the expansion velocities in planetary nebulae such as NGC 2392 (the Eskimo Nebula) are often far from simple. Nonsymmetrical ejection of shells or blobs of gas at different epochs from the central star appears to be common. *See* Interferometry.

Studies of motions in diffuse nebulae, particularly Orion, show a variety of phenomena. Mass streaming motions and perhaps shock waves appear to occur. Also, the conventional theory of incompressible turbulence, which gives a definite relation between eddy size and velocity, cannot be applied. In some nebulae, internal motions are almost certainly complicated by the influence of magnetic fields.

Relation to illuminating stars. Except for the nonthermal radio-frequency sources, gaseous nebulae derive their energy from stars near or within them. If the star is relatively faint (like T Tauri), the nebula is small; large luminous nebulae are necessarily excited by high-temperature bright stars. The hot star ionizes the surrounding gas up to a boundary that is more or less sharp, depending on the density inhomogeneities in the gas. Beyond that boundary the gas is neutral and nonluminous, although it may be detectable from its 21-cm radio-frequency emission. An example in point is the Lagoon Nebula NGC 6523, which is excited by the star cluster NGC 6530 (**Fig. 3**). The patchy luminous nebula appears to be surrounded by a much larger region of cold, largely molecular or neutral hydrogen and other gases. *SEE STAR; STAR CLUSTERS.*

Dust is an omnipresent nuisance in work on gaseous nebulae. In H II regions, it comes from the cool clouds of the interstellar medium. It persists even as these clouds are ionized and heated by hot stars. The dust in planetary nebulae has been manufactured in the cool atmospheres of the parent giant stars. Often it is mixed thoroughly into the ionized gas, resisting destruction by the radiation of a hot star.

Brisk winds from rapidly evolving (for example, Wolf-Rayet) stars may supply energy and certainly affect the kinematics and structures of many nebulae. NGC 604 in M33 and 30 Doradus in the Large Magellanic Cloud are examples of objects where steller windblown filaments dominate the structure.

The bright O- and B-type stars that excite typical diffuse gaseous nebulae have effective temperatures of 25,000–40,000 K (45,000–72,000°F) and luminosities 1000–10,000 times that of the Sun. The central stars of planetary nebulae are by comparison dwarf stars, although their temperatures range from 25,000 K (45,000°F) to perhaps 200,000 K (360,000°F). Their luminosities range from less than that of the Sun to more than a thousand times that of the Sun, in terms of total energy output.

Supernova remnants are excited by the dissipation of the kinetic energy of the ejected material. Spectra produced in shock phenomena differ from those caused by absorption of direct ultraviolet radiation from stars. Forbidden lines of ionized sulfur [S II] are usually prominent; sometimes in the wake of the shock as in the Cygnus loop, even lines of coronal excitation [Fe X] can be produced.

Fig. 3. Lagoon Nebula NGC 6523 and star cluster NGC 6530. The photograph was made with the Curtis Schmidt Telescope, University of Michigan; a red-sensitive filter and emulsion were used to isolate the spectral region about the red hydrogen line Hα. Note the irregular structure. The dark lanes and the globules are caused by solid grains in the neighborhood.

Gaseous nebulae in external galaxies. H II regions are observed in external galaxies, for example, in the Andromeda Spiral M31, NGC 6822, and Messier 101; the largest number has been cataloged in the Triangulum Spiral M33. H II regions have been used as distance indicators for external galaxies. Planetary nebulae have also been found in nearby galaxies and appear to provide good distance estimates for these galaxies.

Lawrence H. Aller

Bibliography. L. H. Aller, *Physics of Thermal Gaseous Nebulae*, 1984, paper, 1987; S. A. Kaplan and S. B. Pikelner, *The Interstellar Medium*, 1970; Y. Kondo (ed.), *Exploring the Universe with the IUE Satellite*, 1987; K. R. Lang, *Astrophysical Formulae*, 1978, paper, 1986; S. Maran (ed.), *The Gum Nebula*, NASA Spec. Publ. 332, 1973; B. Middlehurst and L. H. Aller (eds.), *Interstellar Medium and Gaseous Nebulae*, vol. 7 of *Stars and Stellar Systems*, 1968; D. E. Osterbrock, *Astrophysics of Gaseous Nebulae and Active Galactic Nuclei*, 1989; R. A. R. Parker, T. R. Gull, and R. P. Kirshner, NASA Spec. Publ. 434, 1979; L. Perek and L. Kohoutek, *Catalogue of Galactic Planetary Nebulae*, 1967; R. Sinnott (ed.), *NGC 2000.0: The Complete New General Catalogue (NGC) and Index Catalogue (IC) of Nebulae and Star Clusters*, 1989; L. Spitzer, *Physical Processes in the Interstellar Medium*, 1978.

Neptune

The outermost of the four giant planets, a near twin of Uranus in size, mass, and composition. Its discovery in 1846 within a degree from the theoretically predicted position was one of the great achievements of celestial mechanics. Difficulties in accounting for the observed motion of Uranus by means of perturbations by the other known planets led early in the nineteenth century to the suspicion that a new planet, beyond the orbit of Uranus, might be causing the deviation from the predicted path. The difficult problem of deriving the mass and orbital elements of the unknown planet was solved independently in 1845–1846, first by J. C. Adams in Cambridge, England, and then by U. J. Leverrier in Paris. Adams's result did not receive immediate attention, and so it was Leverrier's solution that led to the discovery of Neptune by J. G. Galle, in Berlin, who found the planet on September 23, 1846, only 55′ from its calculated position. *SEE CELESTIAL MECHANICS; PERTURBATION.*

The planet and its orbit. The actual mass and orbit of Neptune differ considerably from the values predicted by Adams and by Leverrier, since both assumed that the mean distance of the planet to the Sun would be that predicted by the Titius-Bode relation, namely, 38.8 astronomical units, whereas it is only 30.1 AU or 2.8 × 10^9 mi (4.5 × 10^9 km). The eccentricity of the orbit is only 0.009, the second smallest (after that of Venus) among the planets; the inclination is 1.8°; the period of revolution is 164.8 years; and the mean orbital velocity of Neptune is 3.4 mi/s (5.45 km/s). *SEE PLANET.*

Through a small telescope, Neptune appears as a

tiny greenish disk, with a mean apparent diameter of about 2.1″ (the Moon has an apparent diameter of 31′). This corresponds to a linear equatorial diameter of 30,770 mi (49,520 km)—very similar to that of Uranus. The mass of Neptune is 17.15 times the mass of Earth, corresponding to a mean density of 1.64, somewhat above that of its sister planet. This suggests that the enrichment of heavy elements is somewhat greater in Neptune than in Uranus. *SEE URANUS.*

The apparent visual magnitude of Neptune at mean opposition, that is, when closest to Earth, is +7.8, too faint to be seen by the naked eye. The corresponding albedo is 0.4, a relatively high value characteristic of a planet with a dense atmosphere. Photographs taken through powerful telescopes under excellent visual conditions with special filters and cameras revealed the presence of discrete cloud systems in Neptune's atmosphere, again in contrast to Uranus, which is essentially featureless.

The atmosphere. Most of what is known about Neptune is the result of the flyby of the planet by the *Voyager 2* spacecraft in August 1989. The cloud features that were dimly glimpsed from Earth were recorded in great detail (**Fig. 1**). They include a large dark oval (about the size of Earth), reminiscent of Jupiter's Great Red Spot, as well as the white clouds of condensed methane whose brilliant contrast with the blue-green atmosphere made them visible from Earth. By following the clouds from frame to frame, scientists were able to deduce the presence of currents at different latitudes, with a tendency for the high-latitude winds to be faster than those near the equator. The reference frame is established by the rotation period of the deep interior, 16^h7^m, as determined from periodic radio emissions.

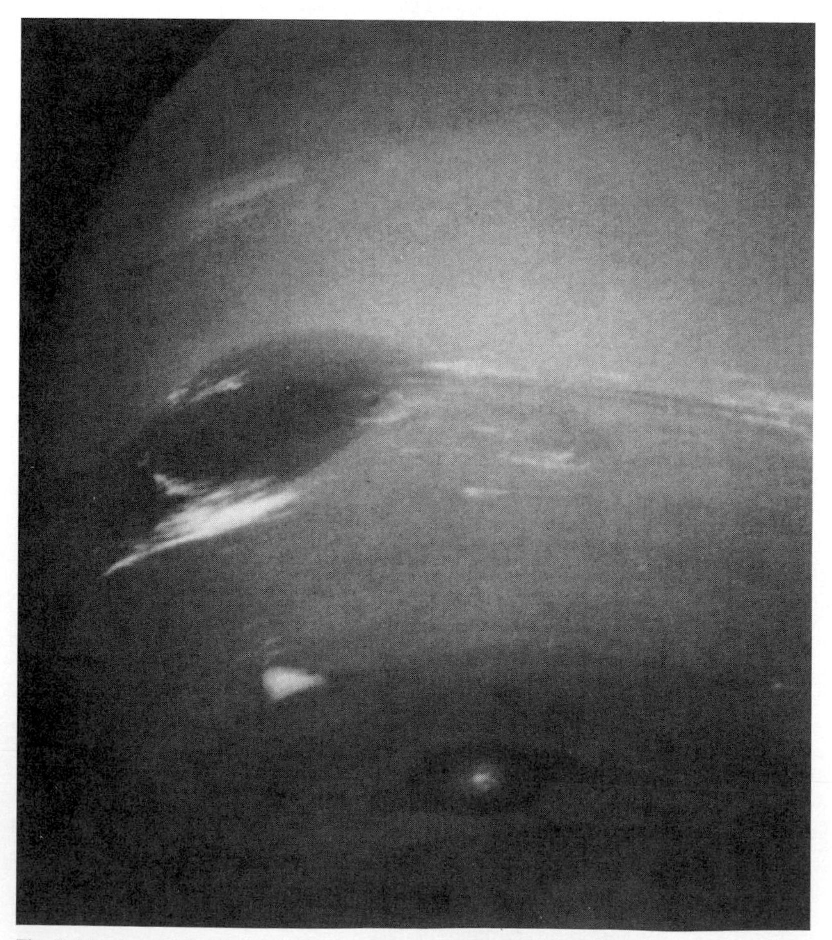

Fig. 1. *Voyager 2* **picture of Neptune, showing the huge dark oval at 22° south latitude, with bright, white methane clouds above it. Farther to the south is a triangular cloud system called the scooter because of its high relative velocity, and still farther south is another dark oval with a bright core. (*NASA*)**

This circulation pattern resembles that of Uranus, despite the fact that the inclinations of the rotational axes of the two planets are very different (that of Neptune is 29.6°, while that of Uranus is 97.9°, and that of Earth is 23.5°). This active meteorology on Neptune may well be driven by the escaping internal heat, some 2.7 times the magnitude of the heat absorbed from the Sun. Of the four giant planets, only Uranus exhibits no sign of an internal heat source, which may partially explain its lack of clouds.

The atmosphere of Neptune, like those of the other giant planets, is composed predominantly of hydrogen and helium. The relative abundance of methane is the highest of any outer planet, approaching 3%. This gas contributes to Neptune's color, by absorbing red and orange sunlight that otherwise would be reflected to the observer. Ammonia and other reduced gases must be present at great depths where the atmosphere is warm enough for them to be in the gaseous state. There are indications from Earth-based radio observations that detect thermal radiation emerging from the planet at these deep levels that ammonia may be deficient on Neptune relative to methane, but this conclusion is uncertain. *SEE PLANETARY PHYSICS.*

Magnetic field. The orientation of Neptune's magnetic field is surprisingly similar to that of Uranus. It can be represented by a bar magnet inclined at an angle of 46.8° with respect to the axis of rotation and offset by 0.55 planetary radius. (For comparison, the Earth's field is inclined by only 11° and offset by 0.07 radius.) Because of the offset, the field strength varies from a minimum of less than 0.1 gauss (10^{-5} tesla) in the northern hemisphere of Neptune to a maximum of greater than 1.0 gauss (10^{-4} T) in the southern hemisphere.

This field has trapped a plasma of ionized and neutral gases in the planet's magnetosphere. The maximum plasma density is only 1.4 particles per cubic centimeter (0.09 particle per cubic inch), the lowest of any of the giant planets. There are ions of molecular hydrogen and helium as well as atomic hydrogen and heavier ions that have probably escaped from Triton. These should include nitrogen from the satellite's atmosphere.

A faint aurora was observed on Neptune, caused by charged particles from the radiation belts (the trapped plasma) spiraling around planetary magnetic field lines and bombarding the planet's atmosphere. Because of the unusual orientation of the field, the maximum auroral activity occurs at midlatitudes rather than near the poles as it does on Earth. *SEE MAGNETOSPHERE.*

Satellites. Before the *Voyager* encounter, only two satellites of Neptune were known, both of them in highly irregular orbits. Triton was discovered visually by W. Lassell in 1846. It is moving in a retrograde direction around Neptune with a period of 5.9 days in a nearly circular orbit. Nereid was found almost 100 years later as a result of a photographic search by G. P. Kuiper. It has the most eccentric orbit of any known satellite. The eccentricity e measures the flattening of the ellipse; a circle has $e = 0$, while a par-

Table 1. Neptune satellites

Satellite name	Distance to Neptune, 10^3 mi (10^3 km)	Sidereal period, h	Diameter, mi (km)	Albedo
Naiad	29.9 (48.0)	7.1	34 ± 10 (54 ± 16)	0.06 ?
Thalassa	31.1 (50.0)	7.5	50 ± 10 (80 ± 16)	0.06 ?
Despoina	32.6 (52.5)	8.0	112 ± 12 (180 ± 20)	0.06 ?
Galatea	38.5 (62.0)	10.3	93 ± 19 (150 ± 30)	0.054
Larissa	45.7 (73.6)	13.3	118 ± 12 (190 ± 20)	0.056
Proteus	73.1 (117.6)	26.9	249 ± 12 (400 ± 20)	0.060
Triton	220.5 (354.8)	141.0	1681 ± 4 (2705 ± 6)	0.6–0.9
Nereid	3425.9 (5513.4)	8643.1	211 ± 31 (340 ± 50)	0.14

abola, representing an unclosed orbit, has $e = 1$. While the orbits of most satellites and planets have eccentricities close to zero, Nereid has $e = 0.7$. In sharp contrast to these two bodies, the six satellites discovered by the *Voyager* cameras all have very regular orbits: in the plane of the planet's equator and nearly circular. They are all close to the planet (**Table 1**). *See* ORBITAL MOTION.

Observations from Earth during the 1980s revealed that Triton has a tenuous atmosphere containing both methane and nitrogen, with evidence for condensed nitrogen on the satellite's surface, but even the size of Triton was unknown. *Voyager* revealed that this remarkable object has a diameter of only 1681 ± 4 mi (2705 ± 6 km), making it considerably smaller than the Earth's Moon (2086 mi or 3476 km). The surface temperature of Triton is −391 ± 7°F (38 ± 4 K), and so it is the coldest sunlit surface in the solar system.

This surface has an appearance quite different from that of any other satellite. The scarcity of impact craters means that the surface is geologically young. There are long intersecting valleys and ridges crossing various types of terrain. The illuminated part of the satellite at the time of the *Voyager* encounter was the southern hemisphere. Most of it appeared to be covered with layers of highly reflective ice, on which darker splotches appeared, many clearly organized by near-surface winds. All of this surface has a slight pinkish color, probably the result of organic materials produced by reactions in the lower atmosphere of Triton and subsequently deposited with the ice. The ice itself is most likely predominantly frozen nitrogen, with some admixture of frozen methane and carbon monoxide. Near the equator, there is a bluish deposit that is probably fresh nitrogen ice, while at higher latitudes the surface resembles the skin of a cantaloupe.

Triton's atmosphere has a surface pressure of only 1.6 ± 0.3 pascals or 16 ± 3 microbars (Earth's atmospheric pressure is approximately 10^5 Pa or 1 bar), dominated by molecular nitrogen. Only 1 part in 10,000 is methane, while carbon monoxide must be less than 4%. Yet reactions in this tenuous atmosphere (and in the surface ices) apparently produce the organic compounds that give the surface its characteristic pinkish color.

Perhaps the most unusual aspect of Triton is the presence of eruptive plumes at several places on its surface. These plumes consist of columns of dark material ejected some 5 mi (8 km) upward into the atmosphere, where they form small clouds that are then blown into narrow wind trails extending hundreds of miles from their sources. These plumes may be powered by a so-called solid greenhouse effect within Triton's icy surface, in which solar radiation that penetrates the ice encounters dark material that it warms. This causes the surrounding nitrogen ice to sublime until sufficient gas pressure is produced to cause gas to break through to the surface, forming the plume. The dark material carried upward by these jets contributes to the hazes seen in Triton's atmosphere.

Rings. Earth-based observations of Neptune when it passed in front of distant stars indicated the presence of material in orbit about the planet. Similar observations of stellar occultations by Uranus led to the discovery of that planet's rings. In some cases, the light from the star would briefly disappear before Neptune reached it, but this behavior would not be repeated on the other side of the planet as would be expected for a planetary ring. In other cases, no dimming of the star's light on either side of the planet was observed. These observations led to the idea that Neptune might be surrounded by incomplete rings or arcs. *See* OCCULTATION.

The *Voyager* cameras showed that in fact there are three well-defined, complete rings around Neptune, accompanied by a sheet of material that itself constitutes a broad ring. In order of increasing distance from the planet, the three discrete rings have been

Table 2. Neptune ring data

Feature	Distance from Neptune's center, 10^3 mi (10^3 km)	Comments
Pressure level of 10^5 Pa (1 bar) in Neptune's atmosphere	15.4 (24.8)	Equatorial radius of Neptune
1989N3R	26.1 (42.0)	Outer edge of this ring
1989N2R	32.9 (53.0)	High dust content
1989N4R	36.0 (58.0)	Outer edge of broad ring
1989N1R	39.1 (63.0)	9 mi (15 km) wide; contains three arcs

Fig. 2. The two outer discrete rings of Neptune (1989N1R and 1989N2R), with the three arcs of concentrated material in the outer ring. (*NASA*)

designated 1989N3R, -N2R, and -N1R, while the sheet of material is known as 1989N4R (**Table 2**). The outermost of these rings, 1989N1R, contains three concentrated clumps of material (**Fig. 2**), rather like sausages strung on a wire, and these plus a chance occultation by one of the inner satellites were apparently responsible for the confusing ground-based observations. The outer two discrete rings are very narrow, reminiscent of the rings of Uranus, with an average width of 9 mi (15 km).

The confinement of these narrow rings is commonly assumed to require the presence of small shepherding satellites. Galatea and Despoina orbit, respectively, just inside the outer two narrow rings, but the corresponding outer shepherds have not been found. Similarly, the persistence of the three arcs within the outer ring remains an enigma. In the absence of some gravitational control, the material in these arcs would be expected to spread out around the planet, simply contributing to the ring itself. *SEE PLANETARY RINGS.*

Origin and evolution. Like the other giant planets, Neptune is thought to have formed in a two-stage process. First a large core of solid material accumulated, growing as a result of collisions with smaller so-called planetesimals. This core was dominated by ices, although it contained rocky material as well. As the core grew to the size of several Earth masses, it developed an atmosphere, the result of impact vaporization of the icy bodies that were crashing into it. This atmosphere consisted of gases such as molecular nitrogen, methane, carbon monoxide, and ammonia. As the core grew, it began to attract gas from the surrounding solar nebula, which added hydrogen and helium to the atmosphere. The process stopped when the supply of materials in the vicinity of the growing planet was exhausted.

In the case of Jupiter and Saturn, there was enough nebular gas available to produce huge planets with deep atmospheres dominated by hydrogen and helium. Even in these atmospheres, however, the heavy elements were enriched. At the great distances where Uranus and Neptune formed, the solar nebula evidently had less material to offer the growing planets, with the result that these two objects do not have such deep atmospheres. In other words, they exhibit a much higher proportion of core mass to total mass than Jupiter and Saturn do. This difference in the proportion of core to atmosphere is reflected in the higher proportion of carbon to hydrogen in the atmospheres of Uranus and Neptune, exemplified by the large relative abundance of methane discussed above. *SEE JUPITER; SATURN.*

The six inner regular satellites of Neptune resemble the ten that *Voyager* found around Uranus, in size, albedo, and orbital characteristics. In both cases, it is thought that these bodies formed out of the planetary subnebula, in much the same way that the planets themselves formed from material in the solar nebula in orbit about the Sun. Triton and Nereid must have had a different beginning, however. Both are probably captured objects that accreted independently. In the case of Triton, the process that led to capture evidently dissipated enough energy to melt the satellite completely, obliterating the record of impact craters which should have been left on the surface. Further study of this satellite with Earth-based telescopes should be able to demonstrate whether or not carbon monoxide is present on the surface or in the atmosphere, as would be expected for an icy body with this history. *SEE SOLAR SYSTEM.*

Tobias C. Owen

Bibliography. J. K. Beatty (ed.), *The New Solar System*, 3d ed., 1990; M. Grosser, *The Discovery of Neptune*, 1962; D. Morrison and T. Owen, *The Planetary System*, 1988; Reports of *Voyager 2* encounter with Neptune, *Science*, 246:1417–1501, 1989.

Neutron star

A star containing about 1½ solar masses of material compressed into a volume approximately 6 mi (10 km) in radius. (1 solar mass equals 4.4×10^{33} lbm or 2.0×10^{33} kg.) Neutron stars are one of the end points of stellar evolution and are the final states of stars that begin their lives with considerably more mass than the Sun. The density of neutron star material is 10^{14} to 10^{15} times the density of water and exceeds the density of matter in the nuclei of atoms. Neutron stars are pulsars (pulsating radio sources) if they rotate sufficiently rapidly and have strong enough magnetic fields.

Neutron stars play a role in astrophysics which extends beyond their status as strange, unusual types of stellar bodies. The interior of a neutron star is a cosmic laboratory in which matter is compressed to densities which are found nowhere else in the universe. Precise measurements of the rotation of neutron stars can probe the behavior of matter at such densities. Neutron stars in double-star systems can emit x-rays when matter flows toward the neutron star, swirls around it, and heats up. Neutron stars are probably formed in supernova explosions, events in which a dying star becomes more luminous than an entire galaxy, up to 10^{12} times as powerful as the Sun. A few pulsars are found in double-star systems, and careful timing of the pulses they emit can test Einstein's general theory of relativity. *SEE GRAVITATION.*

Dimensions. The radius of a neutron star, while not precisely known, is about 6 mi (10 km). A few neutron stars exist in double-star systems, wherein it is possible to measure the strength of the gravitational pull of the neutron star on the other star in the system and hence determine the mass of each. The masses of these neutron stars are slightly more than the mass of the Sun, with the measured values ranging from 1.4 to 1.8 solar masses. *SEE BINARY STAR.*

The highest mass that a neutron star can have is about 2 solar masses. More massive neutron stars might exist if current ideas about the behavior of neutron star matter turn out to be wrong. If Einstein's theory of gravitation is the correct one, though, a neutron star with a mass larger than some limiting value will collapse catastrophically, because its internal pressure will be insufficient, and become a black hole. The exact value of this limiting mass is not known precisely, but lies between 3 and 5 solar masses. *SEE BLACK HOLE.*

Internal structure. The state of matter beneath the surface of a neutron star is completely different from matter on the surface of the Earth and cannot be reproduced in terrestrial laboratories. The outer crust of neutron stars is solid. The gaseous surface, only several feet thick, is different from ordinary gases, for the structure of atoms is dominated by the strong magnetic field that certainly exists in all pulsars and probably is present in all neutron stars. The magnetic field distorts the shape of the electron orbits which form the outer part of the atoms. Several feet below the surface the material solidifies. The solid crust is about 0.6 mi (1 km) thick (see **illus.**) and is approximately 10^{17} times stiffer than steel. At greater depths, the densities increase, and a progressively greater fraction of the electrons are forced into atomic nuclei, where they combine with protons to form neutrons. *SEE ATOMIC STRUCTURE AND SPECTRA.*

Most of the interior of a neutron star consists of matter which is almost entirely composed of neutrons. In the bulk of the star, this matter is in a superfluid state, where circulation currents can flow without resistance. In rotating superfluids maintained in physics laboratories, small vortices form; such vortices may also exist in neutron stars. This material is under pressure, since it must be able to support the tremendous weight of the overlying layers at each point in the neutron star. This pressure, called degeneracy pressure, is caused by the close packing of the neutrons rather than by the motion of the particles. As a result, neutron stars can be stable no matter what the internal temperature is, because the pressure that supports the star is independent of temperature.

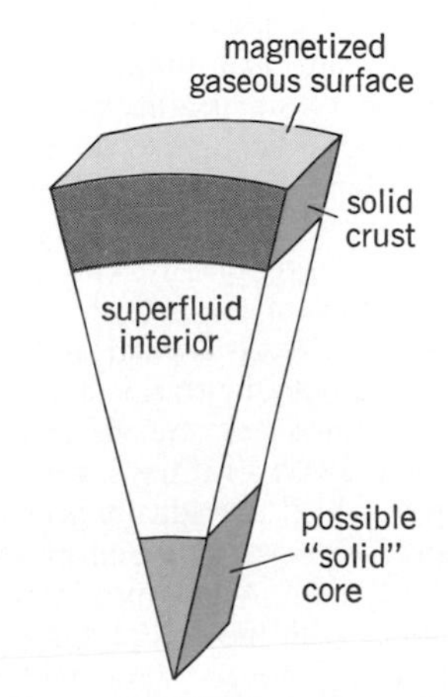

Cross section of a neutron star, with various layers.

The central regions of neutron stars are poorly understood. For example, the role played by subnuclear particles such as the Δ-hyperon is unknown. Some calculations indicate that pi mesons, particles which carry the strong nuclear force, are created by the strong forces between the neutrons and form some kind of condensation. It is remotely possible that quark matter, material composed of the postulated fundamental particles of matter, exists in the cores of neutron stars.

Observations. Theoretical calculations form the basis for the above description of a neutron star. A variety of astronomical observations may eventually confirm or disprove various aspects of this model. The pulses from a pulsar (a rotating neutron star) can be timed very precisely and can suggest the way that neutron stars change their shape as they spin more slowly. A single, isolated neutron star loses energy in the form of high-speed particles emitted from its surface, and its rotation rate slows down. A newly formed pulsar spins very fast—the pulsar in the Crab Nebula, formed a little less than 1000 years ago as observed from Earth, rotates 30 times every second. Eventually it will rotate more slowly, and its structure will readjust, occasionally producing abrupt changes in the rotation period which are called glitches. *SEE CRAB NEBULA.*

Some neutron stars exist in double-star systems, where matter flows from one star to the neutron star, forms a whirling disk around it, and causes it to rotate faster as it gains more mass. These neutron stars emit pulses of x-rays, which can be detected and timed by satellite observatories orbiting above the Earth's atmosphere. Sudden and irregular changes in the arrival times of these x-ray pulses show that solid matter exists somewhere in the interior of neutron stars, but the data are sufficiently poor that it is not yet possible to test detailed models of neutron stars. *SEE SATELLITE ASTRONOMY; X-RAY ASTRONOMY.*

In the late 1980s, the *EXOSAT* satellite, launched by the European Space Agency and containing an x-ray telescope, detected what were described as quasiperiodic oscillations from a number of x-ray-emitting double stars. The x-ray intensity increased and decreased with a reasonably well-defined period, though the frequency of these oscillations changed from one observation time to another. It is possible that these oscillations can indicate something about the neutron star itself, though it is more likely that they originate from the disk surrounding the neutron star and that analysis of them will provide information on accretion disk physics rather than neutron star physics.

In the early 1980s, neutron stars were identified as the source of sudden outbursts of gamma radiation. Material falling onto the surface of a neutron star is heated to the point where nuclear fusion reactions occur, suddenly releasing large quantities of energy. For a minute or so, the neutron star becomes a source of high-energy radiation hundreds of thousands of times as powerful as the Sun, and it also emits light. *SEE GAMMA-RAY ASTRONOMY.*

Another puzzling observation was the discovery of two pulsars with pulse periods of a few milliseconds. These objects pulsate so fast that distinct pulses would not be heard if they were listened to; rather, a musical tone would be heard (as would be the case for the fastest conventional pulsar). It is not certain whether these objects are newly born pulsars, or very old pulsars spun up to high speeds because they accrete matter in a binary system. However, the second alternative seems more likely. One ultrafast pulsar has a companion intermediate in mass between a star and a planet.

Origin. The explosion of a supernova in the Large Magellanic Cloud, seen on Earth in February 1987, provided considerable new information about the connection between supernovae and neutron stars and the way that neutron stars are formed. The detection of neutrinos from this event dramatically confirmed one picture of neutron star origin, in which the collapse of a massive star's core releases energy that ejects the star's outer envelope. These models predict that much of the energy should be ejected in the form of neutrinos, as was observed by underground detectors located in the United States and Japan. SEE GRAVITATIONAL COLLAPSE; PULSAR; STELLAR EVOLUTION; SUPERNOVA.

Harry L. Shipman

Bibliography. G. Greenstein, *Frozen Star: Of Pulsars, Black Holes, and the Fate of Stars*, 1983; L. Marschall, *The Supernova Story*, 1988; S. Shapiro and S. Teukolsky, *Black Holes, White Dwarfs, and Neutron Stars*, 1983; H. Shipman, Quasi periodic oscillations in galactic x-ray sources, *Phys. Today*, 40(1):S-9–S-11, January 1987.

Nova

The brightening of a seemingly undistinguished star by a factor of 10,000 or more in a few days, followed by fading in weeks to years. A typical nova explosion releases more than 10^{38} joules of electromagnetic radiation (as much as the Sun radiates in 10,000 years) and still more in the kinetic energy of a shell blown off during the explosion at speeds of 200–2000 mi/s (300–3000 km/s). The basic energy source is rapid nuclear burning of hydrogen, whose products are seen in the ejecta.

Novae are the second most powerful kind of stellar event known, exceeded only by supernovae. In the Milky Way Galaxy, 25–75 novae go off each year, of which a few are detected; about one per decade becomes bright enough to be visible to the unaided eye. The most recent, Nova Cygni 1975, was the brightest star in its constellation for a couple of days. SEE SUPERNOVA.

Novae have been observed (and more or less understood) as sources of radio, infrared, ultraviolet, and soft x-ray radiation as well as of visible light. Some of the infrared radiation comes from dust formed and heated in the ejecta, beginning within a few weeks after the explosion. The outflow of gas with constantly changing ranges of temperature, density, and chemical composition results in spectra with rapidly changing and very complex emission and absorption lines. Analyses of these reveal the amount, composition, and velocity of the gas expelled. SEE ASTRONOMICAL SPECTROSCOPY; INFRARED ASTRONOMY.

Between 1954 and 1964, work by M. F. Walker and R. P. Kraft established the single most important fact about novae: the explosions occur only in binary stars. The specific progenitors invariably consist of a white or degenerate dwarf in orbit with a cool, red normal star. Most orbit periods are shorter than a day and many shorter than a few hours (with a curious deficiency, not fully understood, of periods between 2 and 3 h). A number of other kinds of excitement and variability also occur in similar binaries, and the systems are collectively called cataclysmic variables or cataclysmic binaries. SEE BINARY STAR; STAR.

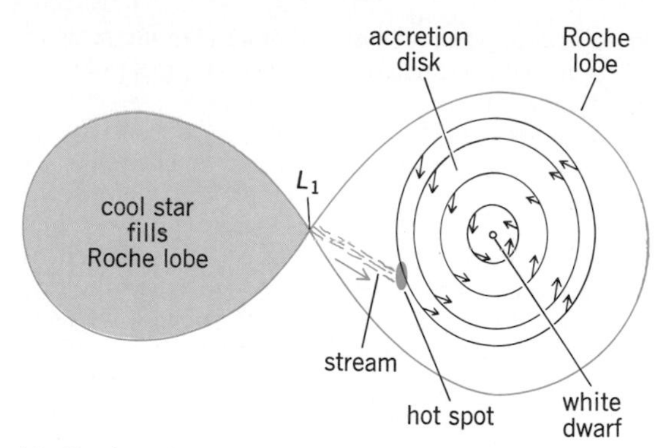

Idealization of a typical cataclysmic binary. Light curves, especially those of dwarf novae, provide evidence for all the components shown. (*After V. Trimble, How to survive the cataclysmic binaries, Mercury, 9(1):8, Astronomical Society of the Pacific, 1980*)

Origin and evolution of cataclysmic variables. Something like half of all stars form in pairs close enough together for the stars eventually to influence each other's evolution. Because the lifetime of a star is roughly 10^{10} years divided by the square of its mass in units of the Sun's mass, the larger member of the pair will normally complete its evolution while the smaller is still a core-hydrogen-burning, main-sequence star. If that larger star (called the primary) is smaller than about eight solar masses to begin with, it ends its life as a white dwarf, about 10,000 mi (16,000 km) across, and consisting of helium, or carbon and oxygen (most often), or oxygen, neon, and magnesium at very high density. More massive primaries end as neutron stars (or black holes), and their binary systems become x-ray sources rather than cataclysmic variables. SEE BLACK HOLE; NEUTRON STAR; STELLAR EVOLUTION; WHITE DWARF STAR.

During the primary's evolution it expands, and its outer layers can envelop the secondary star. This drives off both matter and angular momentum from the system, so that the stars spiral together, ending up with orbit periods of hours to days rather than the weeks to years they started with. The binary stars V471 in the constellation Taurus represent the prototype of systems that have evolved this far.

Eventually (it can take millions to billions of years), the secondary star begins to evolve and expand, until it is large enough to fill its Roche lobe (see **illus.**), a surface on which the gravitational and Coriolis forces are equal and gas can move freely without gaining or losing energy. Gas at L_1 therefore falls in a stream toward the hot white dwarf. However, the gas has too much angular momentum (rotation) to fall straight onto the star, and thus spirals into a disk (or complicated gas streams at the magnetic poles of the white dwarf, if it has a very strong magnetic field), from which it gradually accretes onto the star along paths like those indicated by the small arrows in the illustration. A hot spot is formed where the stream collides with the disk. Light curves, especially those of dwarf novae, show evidence for the presence of all these components, and they can all change their brightnesses and temperatures, giving rise to the various kinds of cataclysmic variables. *SEE LIGHT CURVES; ROCHE LIMIT; STELLAR MAGNETIC FIELD.*

The main types are (1) dwarf novae, which flare up in brightness by a factor of 10 to 100 every few weeks, months, or years, because the rate of gas flow through the disk onto the white dwarf increases, liberating more gravitational potential energy; (2) the novalike variables, with a continuously high rate of flow onto the white dwarf, but no explosion in recent years; (3) the AM Her stars and intermediate polars, like the previous classes, but with white dwarf magnetic fields of a few thousand teslas and a few hundred teslas, respectively; (4) symbiotic stars, where the secondary is a giant rather than a main-sequence star, the orbit period long, and no accretion disk forms; and (5) the classical novae and recurrent novae, in which transferred gas has accumulated on the white dwarf until hydrogen burning ignites explosively and blows off the accumulated layer. In recurrent novae, the layer has built up and burned more than once in historical times and is not always completely expelled by the explosion. Classical novae also recur, but only after a thousand to a million years. The classes are not fully disjoint. Nova Herculis 1934 is also an intermediate polar; and Nova Cygni 1975 has the strong field of an AM Her star. In addition, one sort of system will evolve into another if the rate of mass transfer changes, including return to the V471 Tauri condition if the transfer essentially stops. *SEE SYMBIOTIC STAR; VARIABLE STAR.*

Explosion mechanisms. A nuclear reaction that ignites when the fuel is degenerate (meaning that, at sufficiently high density, pressure depends only on density and not on temperature) necessarily explodes. In a normal gas, when a reaction releases some extra energy, the increased temperature raises the pressure and the gas expands and cools, reducing the reaction rate. In a degenerate gas, the increased temperature does not affect the pressure, but only increases the rate of the nuclear reaction until the gas is no longer held stably in place by gravitation and the burning layer overturns or bursts out, releasing all the energy at once.

In the case of classical novae, the exploding fuel is hydrogen, which begins to burn when the layer transferred from the secondary star to the white dwarf becomes deep enough to be degenerate and hot at its bottom. The details of what happens depend on the mass and surface temperature of the white dwarf, the rate of mass transfer, and the amount of carbon and oxygen (or oxygen, neon, and magnesium) from the white dwarf that gets mixed into the hydrogen layer. These rates and amounts determine the time that elapses before an explosion occurs, and, once it happens, its duration and brightness, and the amount of ejected material. In general, a massive, hot white dwarf with rapid transfer will explode first (recurrent novae being the extreme case), while the largest amount of mixed-in elements from the white dwarf yields the fastest, brightest explosions with the most efficient ejection of material at highest speed.

Thus, the main phases of a nova cycle are (1) accretion of a hydrogen-rich layer (with diffusion and rotation-driven mixing in of material from the white dwarf) up to 10^{-6} to 10^{-4} of the mass of the white dwarf; (2) onset of proton–proton chain hydrogen burning, quickly followed by the turn-on of hydrogen burning catalyzed by nuclei of carbon, nitrogen, and oxygen, leading to an explosion; (3) convective overturn, which carries the energy to the surface, releasing it at first mostly in the form of ultraviolet and soft x-ray radiation, at a rate of about 10^4 solar luminosities; (4) onset of expansion of the outer layers, which thus cool so that the luminosity comes out increasingly as visible light, peaking within a day or two, and fading again in weeks as the layers continue to expand and deeper, hotter regions become visible; (5) a constant luminosity phase, lasting years to decades (but with most of the energy coming out as infrared and ultraviolet radiation) as the rest of the hydrogen reacts with the carbon, nitrogen, and oxygen catalysts; (6) fading as the nuclear energy is exhausted; and (7) resumption of mass transfer from the secondary star. *SEE CARBON-NITROGEN-OXYGEN CYCLES; PROTON-PROTON CHAIN.*

While the white dwarf remains hot and bright, this transfer is fairly rapid and the system is seen as a novalike variable. As the white dwarf fades, the mass-transfer rate gradually declines into the range where instabilities cause dwarf-nova outbursts. The rate may decline almost to zero; modern observations show that the novae of 1670 and 1783, called CK Vul and WY Sge, are now no brighter than V471 Tauri stars. The secondary star continues to evolve, expand, and lose angular momentum in its wind until it once again fills the critical surface (see illus.). Gas then streams down again onto the white dwarf. The system gradually brightens again, first to a dwarf nova and then to a novalike variable. The brighter transfer phases may last a few hundred years, the fainter ones 10^4–10^5 years.

Eventually a sufficient hydrogen layer builds up, and another explosion occurs. Computer simulations show that successive explosions in the same system give rise to similar-looking novae. The most powerful explosions remove the entire hydrogen layer and the elements that were mixed in from the white dwarf (thus leaving the primary less massive after the explosion than before), and can even strip off material from the secondary star. Less powerful explosions (especially the recurrent novae, which show no excess of carbon, nitrogen, and oxygen in their ejecta) can leave a layer of burned material (mostly helium) on the white dwarf and thus gradually increase its mass.

Relationship to other astronomical objects. The idea that a white dwarf in a cataclysmic variable can grow in mass immediately suggests growth up to the maximum stable mass for a star supported by degeneracy pressure (the Chandrasekhar mass). At this limiting mass, the white dwarf must either ignite carbon and burn explosively or collapse to a neutron star. The energy released, 10^{44}–10^{46} J, far exceeds even that of a nova explosion, and the star is expected to resemble some kind of supernova.

The cataclysmic variables are closely related to several other kinds of astronomical problems and objects as well. Because the two stars are close together, they often eclipse each other, permitting detailed study of the temperature and density structures of the accretion disks. These accretion disks are important miniatures of the similar disks around black holes that are thought to be the energy sources for quasars and active galactic nuclei. *SEE ASTROPHYSICS, HIGH-ENERGY; GALAXY; QUASAR.*

The chemical composition of nova ejecta is important in two ways. First, some material has been dug out of the white dwarf. Thus the excesses of carbon, nitrogen, and oxygen seen in most novae and of neon in a few provide confirmation of theoretical models of stellar evolution leading up to white dwarfs made of carbon and oxygen and of oxygen, neon, and magnesium. Furthermore, it is confirmed that carbon-oxygen white dwarfs are much more numerous but oxygen-neon-magnesium white dwarfs are more massive and so able to explode as novae more often. The explosive reactions produce significant amounts of a number of isotopes that are not otherwise produced in astrophysical processes. These isotopes are then ejected into the interstellar medium from which later generations of stars form. Novae thus contribute to nucleosynthesis and the chemical evolution of the galaxy. SEE COSMOCHEMISTRY; NUCLEOSYNTHESIS.

Finally, the strong (and comprehensible) correlation between the maximum luminosity of a nova, the rate at which its brightness rises and falls, and the speed of its ejecta means that the distance to a well-studied nova can be determined from its apparent brightness. Since novae are bright enough to be seen in galaxies at least as far away as the Virgo cluster, they are potential yardsticks on a cosmic scale. The first detection of novae in the Virgo cluster in 1986 suggested a cosmic distance scale consistent with a Hubble constant of about 43.5 mi/s (70 km/s) per megaparsec. Use of this yardstick will undoubtedly continue, especially with the observations of the Hubble Space Telescope. SEE COSMOLOGY; HUBBLE CONSTANT; SATELLITE ASTRONOMY.

Virginia Trimble

Bibliography. M. F. Bode and A. Evans (eds.), *Classical Novae*, 1989; F. Cordova (ed.), *Multifrequency Astrophysics*, 1988; H. Drechsel, Y. Kondo, and J. Rahe (eds.), *Cataclysmic Variables: Recent Multifrequency Observations and Theoretical Developments*, 1987; J. Patterson and D. Q. Lamb (eds.), *Cataclysmic Variables and Low Mass X-Ray Binaries*, 1984; C. H. Payne-Gaposchkin, *The Galactic Novae*, 1957; M. Shara, Recent progress in understanding the eruptions of classical novae, *Pub. Astron. Soc. Pacif.*, 101:5, 1989.

Nucleosynthesis

Theories of the origin of the elements involve synthesis with nucleons (neutrons and protons), the elementary building blocks of the nucleus. The nuclear theory comprises nine distinct processes. Any acceptable theory of nucleosynthesis must lead to an understanding of the cosmic abundance of the elements observed in the solar system, stars, and the interstellar medium. The curve of these abundances is shown in **Fig. 1**. Hydrogen and helium comprise about 98% of the total element content, and there is an almost exponential decrease with increasing nuclear mass number *A*. The processes of nucleosynthesis described in this article attempt to understand these facts and the peaks that occur at certain mass regions. SEE ELEMENTS, COSMIC ABUNDANCE OF.

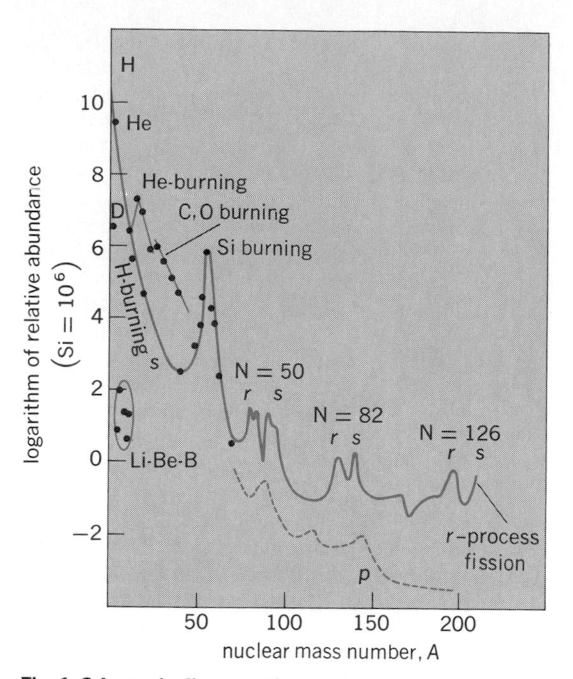

Fig. 1. Schematic diagram of cosmic abundance as a function of nuclear mass number *A*, based on data of H. E. Suess and H. C. Urey. Predominant nucleosynthetic processes are indicated. (*After E. M. Burbidge et al., Synthesis of elements in stars, Rev. Mod. Phys., 29:547–650, 1957*)

Observations of the expanding universe and of the 3-K background radiation lend credence to the belief that the universe originated in a primordial event known as the big bang some 1–2 × 10^{10} years ago. Absence of stable mass 5 and mass 8 nuclei preclude the possibility of synthesizing the major portion of the nuclei of masses greater than 4 in those first few minutes of the universe, when density and temperature were sufficiently high to support the necessary nuclear reactions to synthesize elements. SEE COSMOLOGY.

The principal source of energy in stars is certainly nuclear reactions which release energy by fusion of lighter nuclei to form more massive nuclei. Since the masses of the products are less than those of the constituent parts, the energy released is $E = mc^2$, where m is the mass difference and c is the velocity of light. In 1957 E. M. Burbidge, G. R. Burbidge, W. A. Fowler, and F. Hoyle laid the foundations for astrophysical research into nucleosynthesis, stellar energy generation, stellar structure, and stellar evolution.

There is considerable evidence that nucleosynthesis is going on in stars, and has been doing so for billions of years. Observations make it possible to determine the relative ages of some stars, and it is found that the oldest stars show a ratio of iron and heavier elements to hydrogen that is 100–1000 times smaller than in the Sun. This is understood on the basis of nuclear reactions occurring in previous-generation stars that evolved to the point of exploding as supernovae thus enriching the interstellar medium with the massive nuclei which the stars had synthesized during their lifetimes. Later-generation stars were then formed from the interstellar gas and dust that is still primarily hydrogen and helium, but with small amounts of the more massive nuclei added. In certain of the carbon stars, the ratio of ^{12}C to ^{13}C is as low as 4, much less than the nearly universal ratio of 90 for those isotopes of carbon. This is believed to reflect the synthesis of ^{13}C in the central regions of those stars. Some heavy-metal stars show lines of technetium in their spectra. That technetium must have been produced in those stars themselves or have accreted there from relatively recent supernovae in their neighborhood. No technetium is found naturally on Earth since there is no stable nucleus of technetium, and the stars showing technetium spectra are obviously older than its less than a million years half-life.

Hydrogen burning. The first of the nine processes of nucleosynthesis converts hydrogen nuclei into helium. In stars of 1.2 or less solar masses, hydrogen burning proceeds by the proton-proton chain. Such reactions occurring during the big bang are believed to be responsible for the helium and deuterium, and possibly some of the ^{7}Li observed today. In more massive stars where temperatures are equal to or greater than 2×10^{7} K, hydrogen burning is accomplished through proton captures by carbon, nitrogen, and oxygen nuclei, in the carbon-nitrogen-oxygen (CNO) cycles, to form ^{4}He. Hydrogen burning is responsible for the synthesis of helium, but much of the helium produced in stars may be consumed in later stages of nucleosynthesis in those stars. The observed abundances of some carbon, nitrogen, oxygen, and fluorine nuclei are attributed to hydrogen burning in the carbon-nitrogen-oxygen cycles. The cycles are believed to be responsible for the anomalous abundances of ^{13}C observed in some carbon stars. SEE CARBON-NITROGEN-OXYGEN CYCLES; PROTON-PROTON CHAIN.

Helium burning. When the hydrogen fuel is exhausted in the cental region of the star, the core contracts and its temperature and density increase. When the core density is some 10^{4} g · cm^{-3} and its temperature is of the order 10^{8} K, helium becomes the fuel for further energy generation and nucleosynthesis. The basic reaction in this thermonuclear phase is the three-alpha process in which three ^{4}He nuclei (three alpha particles) fuse to form a carbon nucleus of mass 12 (^{12}C). Capture of an alpha particle by ^{12}C forms oxygen-16. This reaction can be represented by the expression $^{12}C + ^{4}He \rightarrow ^{16}O + \gamma$, where γ represents energy in the form of electromagnetic radiation. It is convenient to express the reaction in shorthand notation as $^{12}C(\alpha,\gamma)^{16}O$. Other reactions that are usually included in helium burning are $^{16}O(\alpha,\gamma)^{20}Ne$, $^{20}Ne(\alpha,\gamma)^{24}Mg$, $^{14}N(\alpha,\gamma)^{18}F$, and $^{18}O(\alpha,\gamma)^{22}Ne$. The fluorine-18 that is produced when ^{14}N captures an alpha particle is unstable and decays by emitting a positron (e^{+}) and a neutrino (ν) to form oxygen-18 [that is, $^{18}F(e^{+}\nu)^{18}O$]. Because there is likely to be ^{13}C in the core of the star if hydrogen burning proceeded by the carbon-nitrogen-oxygen cycles, one should also include the neutron-producing reaction $^{13}C(\alpha,n)^{16}O$ with the helium-burning reactions. Helium burning is probably responsible for much of the ^{12}C observed in the cosmic abundances, although in more massive stars the later stages of nucleosynthesis will consume the ^{12}C produced in their interiors.

Carbon burning. Upon exhaustion of the helium supply, if the star has an initial mass of at least 7.5 solar masses, gravitational contraction of the core can lead to a temperature of about 5×10^{8} K, where it becomes possible for two ^{12}C nuclei to overcome their high mutual Coulomb-potential barrier and fuse to form ^{20}Ne, ^{23}Na, and ^{24}Mg through the reactions $^{12}C(^{12}C,\alpha)^{20}Ne$, $^{12}C(^{12}C,p)^{23}Na$, and $^{12}C(^{12}C,\gamma)^{24}Mg$. Carbon burning can produce a number of nuclei with masses less than or equal to 28 through proton and alpha-particle captures by nuclei that are present.

Oxygen burning. Carbon burning is immediately followed by a short-lived stage, sometimes referred to as neon burning, in which ^{20}Ne is photodisintegrated via the reaction $^{20}Ne(\gamma,\alpha)^{16}O$. The eventual result is that most of the carbon from helium burning becomes oxygen, which supplements the original oxygen formed in helium burning. This is all followed by the fusion of oxygen nuclei at significantly higher temperatures. Temperatures of about 10^{9} K are probably required to overcome the high Coulomb barrier. Fusion reactions of oxygen are $^{16}O(^{16}O,\alpha)^{28}Si$, $^{16}O(^{16}O,p)^{31}P$, and $^{16}O(^{16}O,\gamma)^{32}S$. Nuclei of masses up to $A = 40$ may be produced in this phase through proton, neutron, and alpha-particle captures by nuclei from oxygen fusions.

Silicon burning. This process commences when the temperature reaches 3×10^{9} K. In this phase, photodisintegration of ^{28}Si and other intermediate-mass nuclei in the neighborhood of $A = 28$ produces copious supplies of protons, alpha particles, and neutrons. Capture of these by other intermediate-mass nuclei synthesizes nuclei up to a mass A of about 60, and results in the buildup of the iron-group peak near $A = 56$ (Fig. 1).

Synthesis of nuclei by charged-particle reactions beyond mass $A \sim 60$ is difficult because of increasing Coulomb barriers and the fact that the maximum binding energy per nucleon occurs in that mass range. Additional nucleosynthetic processes must be invoked to understand the cosmic abundances of the more massive nuclei.

The s-process. Because neutrons are neutral particles, their capture is not affected by the Coulomb barrier that inhibits charged-particle capture by the massive nuclei. If the number of neutrons per seed nucleus is small, so that time intervals between neutron captures are long compared to the beta-decay lifetimes of unstable nuclei that are formed, the *s*-process (slow process) takes place. The seed nuclei are probably primarily those in the iron peak, but neutron processes probably alter some of the lighter-mass nuclei as well. In this slow process, after neutron capture a nucleus that is unstable because of excess neutrons decays by electron and antineutrino emission to a stable isobar which may capture another neutron, leading to a chain of captures and decays to synthesize nuclei of masses up to about $A = 209$, when alpha decay becomes a deterrent to further synthesis by the *s*-process. The *s*-process is believed to occur in red giant stars. The reactions $^{13}C(\alpha,n)^{16}O$, $^{21}Ne(\alpha,n)^{24}Mg$, and $^{22}Ne(\alpha,n)^{25}Mg$ are probable sources of neutrons during phases when both hydrogen burning and helium burning are occurring in concentric shells around the hydrogen- and helium-exhausted cores of the red giant stars. The technetium observed in some heavy-metal stars is probably synthesized by the *s*-process. Since the cross sections for neutron capture by nuclei with magic numbers of neutrons ($N = 50$, 82, and 126) are small, the *s*-process builds high abundances of these nuclei and produces the peaks at $A = 90$, 138, and 208, labeled *s* in the atomic abundance curve in Fig. 1. The *s*-process path among nuclei is represented by the single line in **Fig. 2**, and lies near the beta stability line through the nuclei. Capture of a neutron increases the number of neutrons (N) in the nucleus by one, beta decay increases the number of protons (Z) by one, and the path follows a stepwise curve near the line of beta stability in the N-Z plane, since time between neutron captures allows the unstable nuclei to decay.

The r-process. This process occurs when a large neutron flux allows rapid captures of neutrons. Up to 100 captures can occur in 1–100 s. Only in the massive explosions of supernovae are such large neutron fluxes conceivable. With such rapid successive cap-

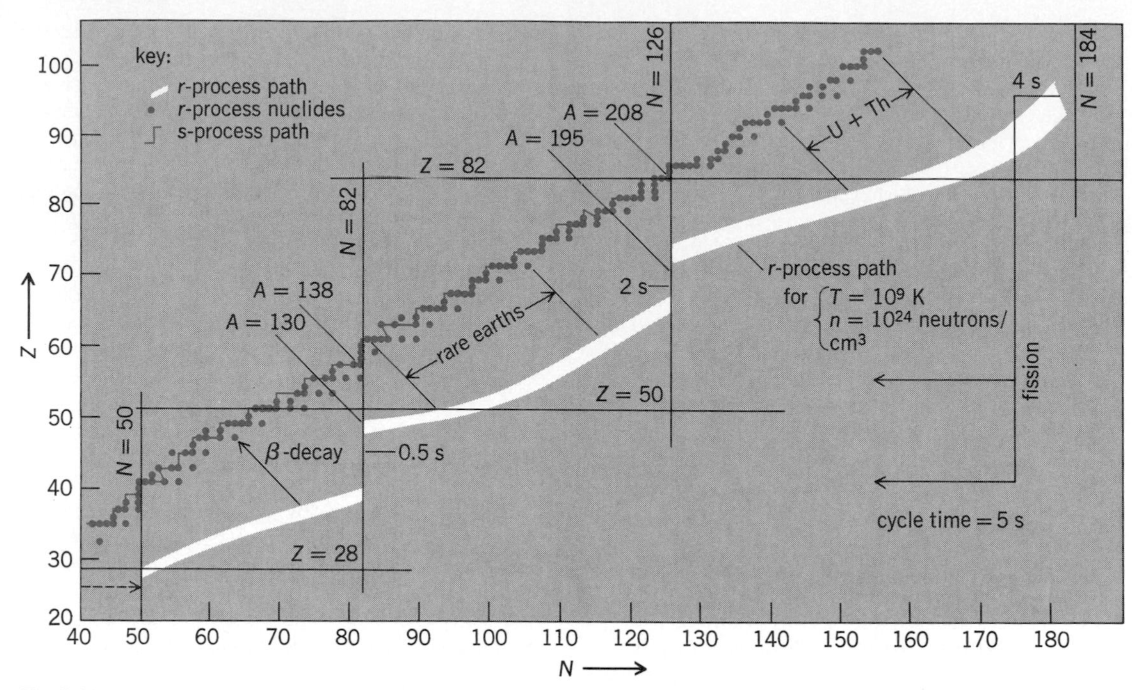

Fig. 2. Neutron-capture paths among the nuclei plotted in the *N-Z* plane, showing nuclei identified by the number of neutrons (*N*) and number of protons (*Z*) in the nucleus. (*After P. A. Seeger, W. A. Fowler, and D. D. Clayton, Nucleosynthesis of heavy elements by neutron capture, Astrophys. J., Suppl., 11(97):121–166, 1965*)

tures, it is possible to proceed through unstable nuclei before they can either beta or alpha decay. In this manner it is possible to synthesize nuclei all the way into the transuranic elements. The *r*-process path is shown by a band in Fig. 2. The path depends on the temperature and the neutron density at the site of the explosion. The path in Fig. 2 is based on a reasonable assumption of a temperature $T = 10^9$ K and neutron density $n = 10^{24}$ neutrons cm^{-3}. The path lies to the right of the line of beta stability and represents the progenitors of the nuclei that result from the *r*-process. The circles in Fig. 2 show the *r*-process nuclei that result after the unstable progenitor nuclei beta decay under these conditions. Diagonal lines indicate the beta-decay lines along isobars to the region of stability. The times shown are intervals to reach locations along the *r*-process path for these conditions. SEE SUPERNOVA.

At $Z = 94$, the transuranic elements are unstable to neutron-induced fission, and *r*-process synthesis of more massive nuclei ceases. The resultant fission leads to production of intermediate-mass nuclei which may then act as seed nuclei to introduce a cycling effect. The peaks at $A = 80$, 130, and 195 labeled *r* in Fig. 1 are due to the *r*-process effects at magic numbers of neutrons ($N = 50$, 82, and 126) and the subsequent decays of unstable nuclei.

The p-process. This process produces proton-rich heavier elements probably by modifying a small fraction of the *s*-process and *r*-process nuclei through (p,γ), (p,n), and (γ,n) reactions. This would be expected to occur in the red giant stars and in phases of explosive nucleosynthesis. Nuclei produced in this process are indicated by the curve marked *p* in Fig. 1.

The l-process. The lithium, beryllium, and boron found in the cosmic abundances cannot have survived processing in stellar interiors because they are readily destroyed by proton bombardment. Although some ^{7}Li may have originated in the big bang, that primordial event is probably not responsible for all the ^{7}Li in existence. Spallation of more abundant nuclei such as carbon, nitrogen, and oxygen by protons can account for the low-abundance nuclides ^{6}Li, ^{9}Be, ^{10}B, and ^{11}B and for some ^{7}Li. This breaking up of the more massive nuclei to form the light nuclei is called the *l*-process. The most probable site for the *l*-process is in the interstellar medium, where high-energy galactic cosmic rays bombard the interstellar gas. The surfaces of magnetic stars have also been suggested as possible sites for the *l*-process, but it is difficult to show that sufficient energy is available there to produce the observed abundances. The nuclei produced in the *l*-process are indicated by Li-Be-B in Fig. 1.

Conclusions. Although stars spend the majority of their lifetimes in hydrogen-burning phases and most of the remaining time in burning helium, the vast majority of the elements are synthesized in the relatively brief time spans of later stages of evolution and the processes described. Figure 1 serves as a very brief summary of those processes and the mass ranges of elements for which they are believed to be responsible.

Nucleosynthesis remains a developing field of research. There are many unanswered questions regarding which processes are mainly responsible for certain elements. In Fig. 1 the schematic diagram is a modification of one that was developed by Burbidge and colleagues in 1957. The C-, O-, and Si-burning processes replace the phases originally called alpha process (because of the number of alpha particles and $A = 2Z$ nuclei involved) and the *e* process (because of the quasiequilibrium extant during silicon burning). Research has shown that the terms C-, O-, and Si-burning processes are more realistic. Although many of the nuclear reaction rates for intermediate-mass nuclei ($A = 20$–60) are still theoretical estimates, nuclear physicists are continuing to improve knowledge of the rates at which nuclear reactions occur under conditions believed to be present at the sites of stellar nucleosynthesis. Astrophysicists are increasing the knowledge of the probable thermodynamic conditions prevailing in stars. As a result, further modifications in the theory of the processes of nucleosynthesis may occur in the future. Clarification of some of the issues that are cloudy today can be expected, but new puzzles will undoubtedly arise. *See* STELLAR EVOLUTION.

Georgeanne R. Caughlan

Bibliography. W. D. Arnett and J. W. Truran (eds.), *Nucleosynthesis: Challenges and New Developments*, 1985; W. D. Arnett et al. (eds.), *Nucleosynthesis*, 1968; J. Audouze and S. Vanclair, *An Introduction to Nuclear Astrophysics,* 1980; E. M. Burbidge et al., Synthesis of elements in stars, *Rev. Mod. Phys.,* 29:547–650, 1957; D. D. Clayton, *Principles of Stellar Evolution and Nucleosynthesis*, 1984; W. A. Fowler, *Nuclear Astrophysics,* 1967; C. E. Rolfs and W. S. Rodney, *Cauldrons in the Cosmos: Nuclear Astrophysics*, 1988; V. Trimble, The origin and abundance of the chemical elements, *Rev. Mod. Phys.*, 47:877–976, 1975.

Nutation

In mechanics, a bobbing motion that accompanies the precession of a spinning rigid body, such as a top. Astronomical nutation refers to irregularities in the precessional motion of the equinoxes caused by the varying torque applied to the Earth by the Sun and Moon. Astronomical nutation, which is sometimes called nutational wandering of the terrestrial poles, should not be confused with nutation as defined in mechanics; the latter is present even if the source of the torques is unvarying.

Nutation of tops In simple precession, the axis of a top with a fixed point of contact sweeps out a cone, whose axis is the vertical direction. In the general motion, the angle between the axis of the top and the vertical varies with time (**Fig. 1**). This motion of the top's axis, bobbing up and down as it precesses, is known as nutation.

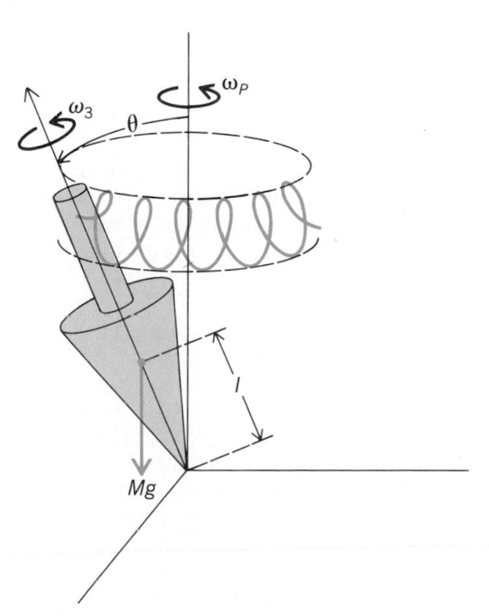

Fig. 1. Nutation and precession of a spinning top with a fixed point of contact.

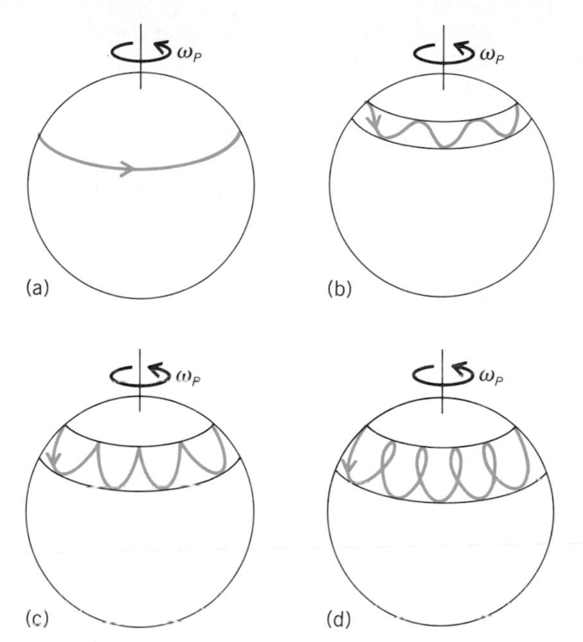

Fig. 2. Path traced out by the symmetry axis of a top with a fixed point of contact for different initial angular velocities ω_0 of the ϕ motion. (*a*) $\omega_0 = \omega_P$. (*b*) $\omega_0 = \frac{1}{2}\omega_P$. (*c*) $\omega_0 = 0$. (*d*) $\omega_0 = -\omega_P$.

Slow precession and fast nutation are commonly observed in the motion of a rapidly spinning top. For this case, the equations of motion have a simple approximate solution, which will be presented below. The angular frequency of the precessional motion is given by Eq. (1), where ω_3 is the angular velocity

$$\omega_P \simeq \frac{Mgl}{I_3\omega_3} \tag{1}$$

about the spin axis of the top, I_3 is the moment of inertia about this axis, M is the mass of the top, l is the distance of the center of mass from the point of contact, and g is the gravitational acceleration. The nutation is governed by the angular frequency given in Eq. (2), where I is the moment of inertia about

$$\omega_N = \frac{I_3\omega_3}{I} \tag{2}$$

an axis that is perpendicular to the spin axis of the top.

The location of the spin axis can be specified by an azimuthal angle ϕ and a polar angle θ. Suppose that the top is started at a polar angle $\theta(t = 0) = \theta_0$ with an initial azimuthal angular velocity $\phi(t = 0) = \omega_0$. Then the approximate solution to the equations of motion are given by Eqs. (3), where C is given by

$$\theta(t) = \theta_0 + C \sin \theta_0(1 - \cos \omega_N t) \tag{3a}$$

$$\phi(t) = \omega_P t - C \sin \omega_N t \tag{3b}$$

Eq. (4). This solution for $\theta(t)$ oscillates between lim-

$$C = \frac{\omega_P - \omega_0}{\omega_N} \tag{4}$$

its of θ_0 and $\theta_0 + 2C \sin \theta_0$; the top's spin axis bobs up and down between cones of these angles. The maximal excursion in θ depends on C, and hence can be changed by the initial conditions on ω_0 and ω_3. The precession angle ϕ has a sinusoidal time dependence determined by ω_N in addition to the steady precession term. When the top is set in motion with $\omega_0 = \omega_P$, it will undergo precession with no nutation. Typical curves traced out by the end of the top's symmetry axis are shown in **Fig. 2**.

The nutation frequency ω_N in Eq. (2) is proportional to the spin ω_3 of the top while the precession frequency ω_P in Eq. (1) is inversely proportional to ω_3. Also the nutation amplitude C is inversely proportional to ω_N. Thus it is difficult to observe the nutation of a very fast top. However, a buzzing tone with the nutation frequency can be heard when a fast top is spun on a resonant surface.

Vernon D. Barger

Astronomical nutation. The rotating Earth can be regarded as a spinning symmetrical top with small angular speed but large angular momentum, the latter due to its large mass. The gravitational attractions of the Sun and Moon cause the Earth's axis to describe a cone about the normal to the plane of its orbit. However, the magnitude of these gravitational attractions is continually varying, due to the changing positions in space of Sun, Moon, and Earth. The Moon's orbit is continually changing its position in such a way that the celestial pole undergoes a nodding (nutation) as well as a periodic variation in the rate of advance. The largest nutation is about 9.″2, and occurs in a period of a little less than 19 years; that is, the celestial pole completes a small ellipse of semimajor axis 9.″2 in about 19 years. *See* PRECESSION OF EQUINOXES.

There are lesser nutation effects which are due to the motion of the Moon's nodes, the changing declination of the Sun, and the changing declination of the Moon. *See* CELESTIAL MECHANICS.

Ray E. Bolz

Bibliography. A. P. Arya, *Introduction to Classical Mechanics*, 1990; V. Barger and M. Olsson, *Classical Mechanics: A Modern Perspective*, 1973; E. P. Federov and M. L. Smith (eds.), *Nutation and the Earth's Rotation: Proceedings of the I.A.U. Symposium*, no. 78, Kiev, May 23–28, 1980; H. Goldstein, *Classical Mechanics,* 2d ed., 1980; K. Symon, *Mechanics,* 3d ed., 1971.

Occultation

The temporary apparent disappearance from view of a celestial body as another body passes across the line of sight. Usually, it refers to stars occulted by the Moon, but it also applies to stars occulted by major or minor planets, minor planets occulting each other, Jupiter occulting its Galilean satellites, and so forth. Some occultations are called eclipses for historical reasons, for example, solar eclipses. An occultation refers strictly to a blockage in the line of sight, and not to disappearance in a shadow. *See* ECLIPSE.

Observational techniques. Since both ends of the occultation line of sight are usually firmly established (for example, the position of an occulted star and the coordinates of each observer), and the speed and distance of the occulting body are known, it follows that when the limb (apparent edge) of the occulting body crosses the line of sight, the event (disappearance or reappearance) observed provides a precise instantaneous directional mapping of a point on the cross-section limb of the occulting object. For example, two observers who see the limb of the Moon occult a star obtain two points on the profile of a mountain at the edge of the Moon. The statistical combination of many such observations from one or more occultation passages effectively form a fine grid that can yield the position, size, and shape of the occulting body (and companions, if any) with resolution not attainable by direct-image methods. Depending on the brightness of the occulted object, events can be observed by eye, video camera, or high-speed photoelectric detectors.

To understand how shape is observed, it is useful to think of a star as casting a "same-size" shadow of a body on the Earth's surface. (Since stars are so far away, their light rays are parallel by the time they reach the solar system.) As the shadow progresses from west to east, it traces a track. Observers in the track who time the disappearance and reappearance of the star are measuring the lengths of chords of the shadow parallel to the track. The lengths depend on the distance of the observer from the center of the track, going (nonlinearly) to zero toward the edge of the track. The set of all measurements, analyzed together, gives an outline of the shadow, and hence of the casting body, of accuracy proportional to the number and separation of chords. If the shape is very irregular, the chords for some observers may be interrupted, the observer seeing multiple events.

Applications. The study of lunar occultations of stars over several centuries was used to improve the theory of the Moon's motion, and also contributed to detecting irregularities of the Earth's rotation. With that knowledge as a basis, lunar occultations are of more importance now in studying the figure of the Moon. Observations of stars whose occultation paths appear to lie just on the north or south limb of the Moon (grazing occultation) yield precise information on the rugged terrain seen in profile. *See* EARTH ROTATION AND ORBITAL MOTION; MOON.

Occultations are used to study the size and shape of bodies whose disk cannot be seen. In such cases, if a planet is predicted to pass within its provisional radius of a star, and no occultation occurs, an upper limit is set. It was in this way that Pluto's small size was established. *See* PLUTO.

Application of observational techniques has given very accurate profiles of minor planets. Successful observations are still quite rare, however, because the orbits of most minor planets are known only to low precision and because the possibility of an occultation occurs only a few times per year. The method is also being used to seek suspected binary minor planets, though none has yet been proven conclusively. *See* ASTEROID.

The occultation of a star by a planet with an atmosphere is not an instantaneous event. As the star is seen through the different layers of the planet's atmosphere, high-speed photoelectric observations of the light in several distinct wavelengths will yield information on the temperature of the atmosphere.

Occultations may be used in some instances to gather information about objects or structures that are not otherwise visible. The very thin, faint rings of Uranus were discovered accidentally in 1977 when they occulted a star during calibration observations made in preparation for observing an occultation of the star by Uranus. Two separate observers detected the events. This discovery led to the discovery of rings around Jupiter, and perhaps Neptune. Now such observations are routinely made, from ground and space, both to search for more rings and to determine the fine structure of known ones. The existence and structure of such ring systems has important implications for the origin of the solar system. SEE JUPITER; NEPTUNE; SOLAR SYSTEM; URANUS.

Alan D. Fiala

Bibliography. A. Brahic and W. B. Hubbard, The baffling rings of Neptune, *Sky Telesc.*, 77:606–609, 1989; D. DiCicco, Occultations and the amateur, *Sky Telesc.*, 76:480–481, 1988; J. Elliot and R. Kerr, *Rings: Discoveries from Galileo to Voyager*, 1984; J. L. Elliott, E. Dunham, and R. L. Millis, Discovering the rings of Uranus, *Sky Telesc.*, 53:412, 1977; T. Gehrel (ed.), *Asteroids*, 1979; Joint discussion: Photo-electric observations of stellar occultations, in C. De Jager (ed.), *Highlights of Astronomy*, vol. 2, pp. 587–723, 1971.

Olbers' paradox

The riddle of cosmic darkness. The obvious explanation for the darkness of the night sky, that the Sun is on the other side of the Earth, does not account for the fact that, in space far from any star, the universe is full of darkness and not of light.

The hypothesis of a boundless universe full of stars is reasonable, but astronomers since Johannes Kepler have realized that this hypothesis leads to the startling conclusion that the sky at every point should blaze with starlight and that the universe should therefore be full of light.

In a boundless universe of stars, with no interstellar absorption, every line of sight from the eye must eventually intercept the surface of a star. If most stars are similar to the Sun, the sky at every point should shine as bright as the Sun's disk. The sky (or celestial sphere) is 180,000 times larger than the Sun's disk, and the starlight incident on the Earth should therefore be 180,000 times more intense than sunlight, which obviously is not the case. Hermann Bondi resurrected the riddle of cosmic darkness in 1952 and attributed it to the nineteenth-century astronomer Wilhelm Olbers, although, as is now known, it had previously been discussed by Edmund Halley and other astronomers.

Missing starlight. The riddle has two versions. In the first, the sky is actually covered with stars, most of which are invisible, and the question is: where is the missing starlight? Olbers and other astronomers replied that the starlight is absorbed by an interstellar medium. This solution fails because the absorbing medium quickly heats up and then reemits all the absorbed radiation. Bondi argued that in an expanding universe the redshift renders the light from distant stars invisible. This solution became popular, and it was widely believed that darkness at night was evidence for the expanding universe. The redshift solution, however, applied to the steady-state universe that was disproved in 1965 by the discovery of the low-temperature background radiation. SEE COSMIC BACKGROUND RADIATION.

Missing stars. In the second version the sky is not covered with stars and the question is: where are the missing ones? Why are there gaps between the ones observed? The nineteenth-century astronomer John Herschel suggested a hierarchical universe in which stars form clusters that are members of larger clusters that are members of still larger clusters, and so on, so arranged that a line of sight has little chance of intercepting the surface of a star. Edgar Allan Poe suggested in 1848 that the universe is not old enough for the light from very distant stars to have reached the Earth; this was investigated by Lord Kelvin in 1901. Modern calculations confirm Kelvin's results: light travels at approximately 186,000 mi/s (300,000 km/s) and, in a static universe 10–20 $\times$ 10^9 years old, stars cannot shine long enough for their light to reach the Earth from regions sufficiently distant for the visible stars to cover the entire sky. This means that stars cannot shine long enough to fill the universe with radiation in equilibrium with their surfaces. Clearly, if the sky at night is dark in a static universe of finite age, then in an expanding universe of similar age the night sky is even darker because of the redshift.

Big bang universe. In the universe, roughly 15 $\times$ 10^9 years old, it is impossible to see enough stars to cover the sky. Instead, between the stars is seen the big bang of the early universe, and the expansion of the universe has shifted the incandescent light of the big bang into the invisible infrared. SEE BIG BANG THEORY; COSMOLOGY; UNIVERSE.

Edward Harrison

Bibliography. H. Bondi, *Cosmology*, 2d ed., 1960; E. Harrison, *Cosmology: The Science of the Universe*, 1981; E. Harrison, *Darkness at Night: A Riddle of the Universe*, 1987.

Optical telescope

An instrument that collects light energy from a distant source and focuses it into an image that can then be studied by a number of different techniques. This definition of an optical telescope must be narrowed somewhat. Satellite-borne telescopes operated to study celestial x-rays or ultraviolet radiation and ground-based telescopes used to study radio radiation can be called optical telescopes in the sense that they operate according to the principles of geometrical optics. The following discussion stresses ground-based telescopes which are used to study radiation from celestial objects in the wavelength range from the Earth's atmospheric cutoff in the near ultraviolet to the infrared, that is, from 300 to 1000 nanometers. Such instruments may be classified as (1) large astronomical telescopes, used to study the nature of astronomical objects themselves, and (2) astronomical transit instruments, used to study positions and motions of astronomical objects and in the accurate determination of time. SEE RADIO TELESCOPE; TELESCOPE.

Large Telescopes

A large astronomical telescope is used by astronomers to study the fundamental problems in the field. The size required for a telescope to be considered large depends on its type, as discussed below.

Types of telescopes. There are basically three types of optical systems in use in astronomical telescopes: refracting systems whose main optical elements are lenses which focus light by refraction; reflecting systems, whose main imaging elements are mirrors which focus light by reflection; and catadioptric systems, whose main elements are a combination

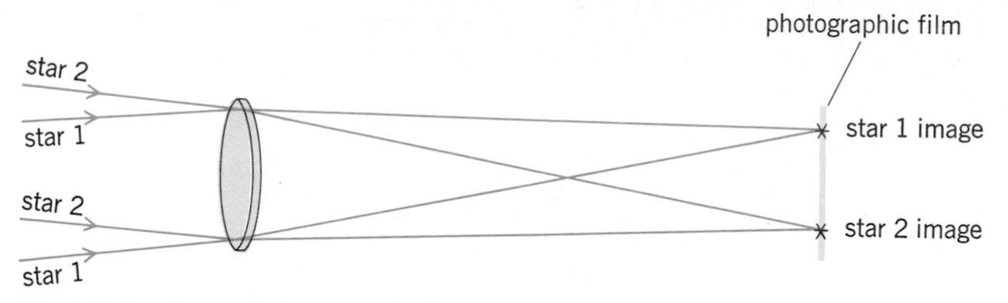

Fig. 1. Refracting optical system used to photograph a star field.

of a lens and a mirror. The most notable example of the last type is the Schmidt camera.

In each case, the main optical element, or objective, collects the light from a distant object and focuses it into an image that can then be examined by some means. Specific types of tools that are frequently employed to study astronomical objects are discussed below.

Refracting telescopes. The main optical element, or objective, of a refracting telescope is usually a long-focal-length lens. The objective lens is typically compound; that is, it is made up of two or more pieces of glass, of different types, designed to correct for aberrations such as chromatic aberration. **Figure 1** shows a refractor lens imaging the light of two stars onto a photographic plate. To construct a visual refractor, a lens is placed beyond the images and viewed with the eye. To construct a photographic refractor or simply a camera, a photographic plate is placed at the position of the image. The characteristics of the components of a photographic lens may differ from those of a visual lens.

Generally, refracting telescopes are used in applications where great magnification is required, namely, in planetary studies and in astrometry, the measurement of star positions and motions. For example, most stellar parallaxes have been measured with refractors such as the Sproul Observatory 24-in. (61-cm) *f*/18 refractor. However, this practice is changing, and the traditional roles of refractors are being carried out effectively by a few new reflecting telescopes. This changing role has come about in part because of effective limitations on the size of refracting telescopes. The largest refractor, located at Yerkes Observatory at Williams Bay, Wisconsin, has a 40-in. (1.02-m) objective lens.

A refractor lens must be relatively thin to avoid excessive absorption of light in the glass. On the other hand, the lens can be supported only around its edge and thus is subject to sagging distortions which change as the telescope is pointed from the horizon to the zenith; thus its thickness must be great enough to give it mechanical rigidity. An effective compromise between these two demands is extremely difficult, if not impossible, for a lens over 40 in. (1 m) in diameter, making larger refractors unfeasible.

Reflecting telescopes. The principal optical element, or objective, of a reflecting telescope is a mirror. The mirror forms an image of a celestial object (**Fig. 2**) which is then examined with an eyepiece, photographed, or studied in some other manner.

Reflecting telescopes generally do not suffer from the size limitations of refracting telescopes. The mirrors in these telescopes can be as thick as necessary and can be supported by mechanisms which prevent sagging and thus inhibit excessive distortion. In addition, mirror materials having vanishingly small expansion coefficients (Cer-Vit, ultralow-expansion-fused silica, and others), together with ribbing techniques which allow rapid equalization of thermal gradients in a mirror, have eliminated the major thermal problems plaguing telescope mirrors. Telescopes with monolithic mirrors up to 236 in. (6 m) in diameter have been built. A concept being used for large reflecting telescopes under development for the 1990s and beyond is the use of segmented mirrors, composed of many separate pieces.

The reflecting telescope has other advantages which make it an attractive system. By using a second mirror (and even a third one, in some telescopes), the optical path in a reflector can be folded back on itself (**Fig. 3*a***), permitting a long focal length to be attained with an instrument housed in a short tube. A short tube can be held by a smaller mounting system and can be housed in a smaller dome than a long-tube refractor, thus decreasing costs.

Finally, a large mirror has only one surface to be figured—that is, to be ground and polished to the desired optical shape—whereas a refractor lens usually consists of two pieces of glass that must be laboriously figured. The secondary mirror of a reflector is a second surface that needs to be figured, but it is much smaller than the primary and thus is much easier and cheaper to fabricate. The multiple segmented mirrors to be used in many modern telescopes present new problems, such as the proper alignment of all the mirror components.

A variety of optical arrangements are possible in large reflecting telescopes, including the prime focus, the newtonian focus, the Cassegrain focus, and the coudé focus.

The newtonian focus is probably most widely used by amateur astronomers in reflectors having apertures on the order of 6 in. (15 cm; Fig. 3*b*). A flat mirror placed at 45° to the optical axis of the primary mirror diverts the focused beam to the side of the telescope, the image being formed by the paraboloidal primary mirror alone. An eyepiece, camera, or other accessories can be attached to the side of the telescope tube to study the image. In the largest telescopes, with apertures over 100 in. (2.5 m), provision is not usually made for a newtonian focus. Instead, an observing cage is placed inside the tube, where the observer can take accessories to observe the image formed by the primary mirror. This prime focus is identical to the newtonian focus optically, since the newtonian flat does nothing more than divert the light beam. The modern reflectors have fast primary mirrors, so that the focal ratio at the prime focus is *f*/2.5 to *f*/6. Lower focal ratios permit shorter exposures on extended objects such as comets and nebulae.

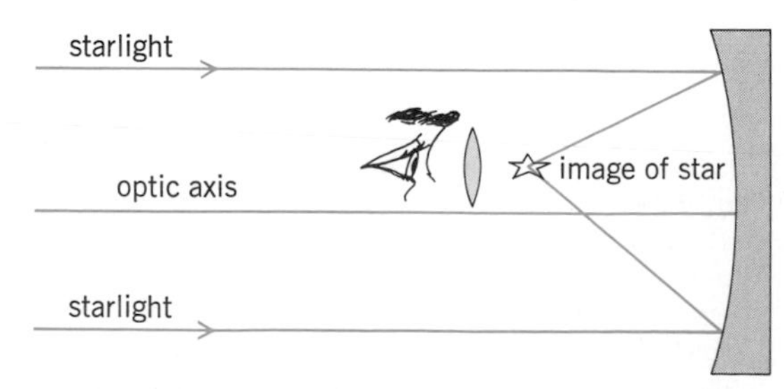

Fig. 2. Viewing a star with a reflecting telescope. In this configuration the observer may block the mirror, unless it is a very large telescope.

Images of the Neptune system from *Voyager 2*. *(Right)* The planet, with cloud features including a large dark oval near the western limb (the left edge) and a second dark spot with a bright core near the terminator (the lower right edge). (*Below*) Composite picture of Triton showing the illuminated southern hemisphere, with patches of dark, windblown material directed away from the subsolar point. The "cantaloupe" terrain northward of the equator (which runs left to right in this picture) stretches away into the darkness. It is crossed by intersecting ridges and valleys. The absence of impact craters shows that this surface is relatively young. (*NASA*)

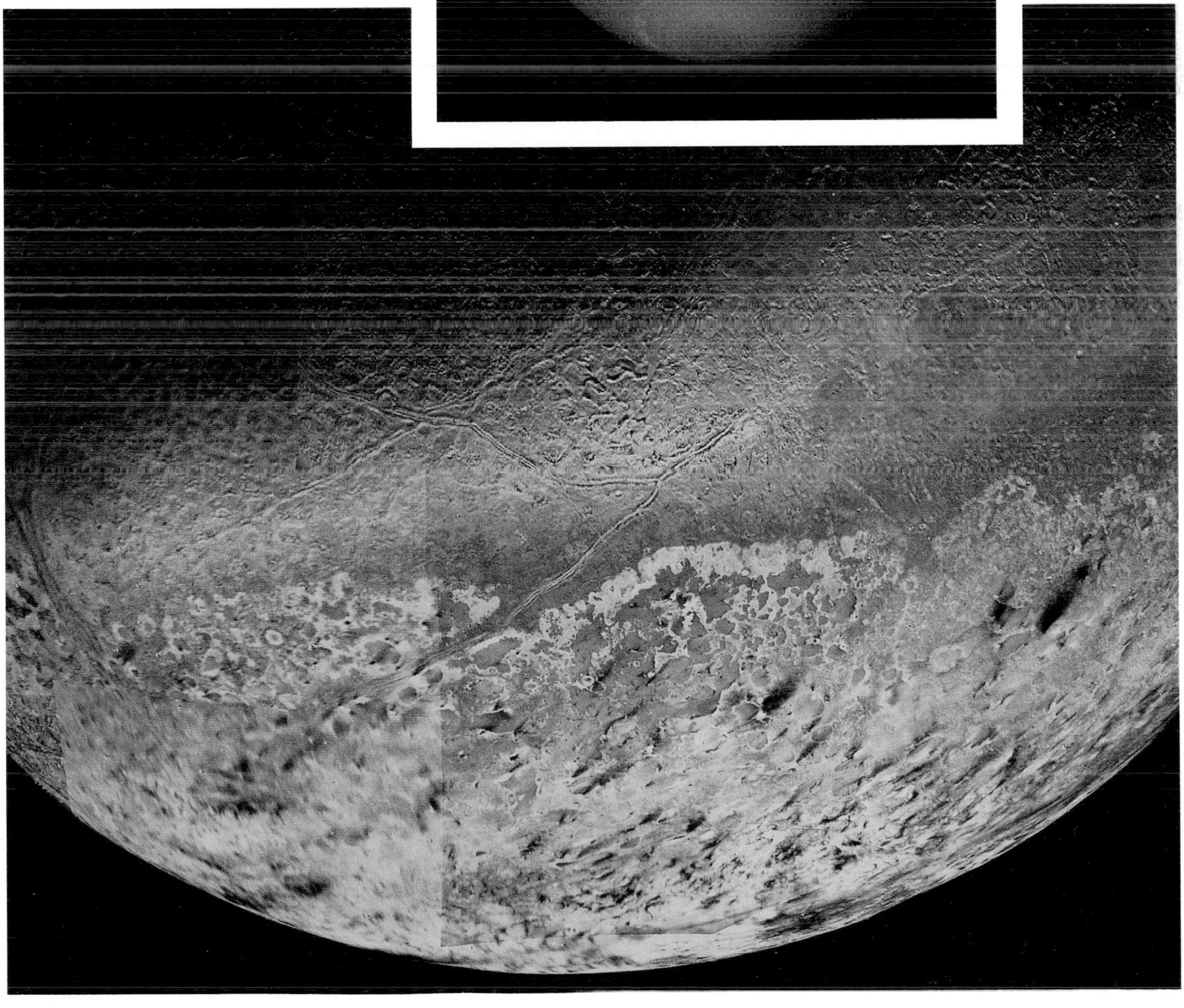

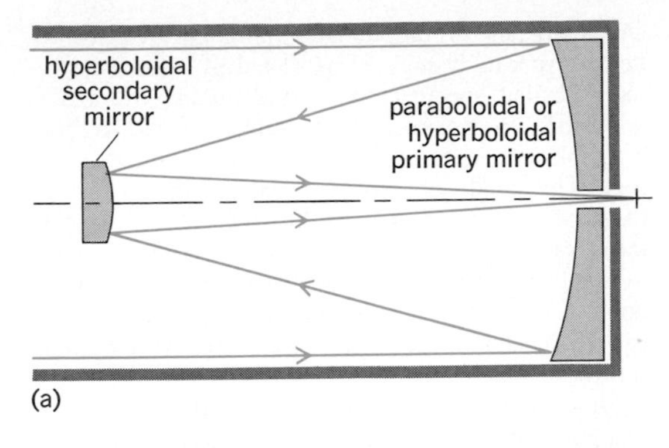

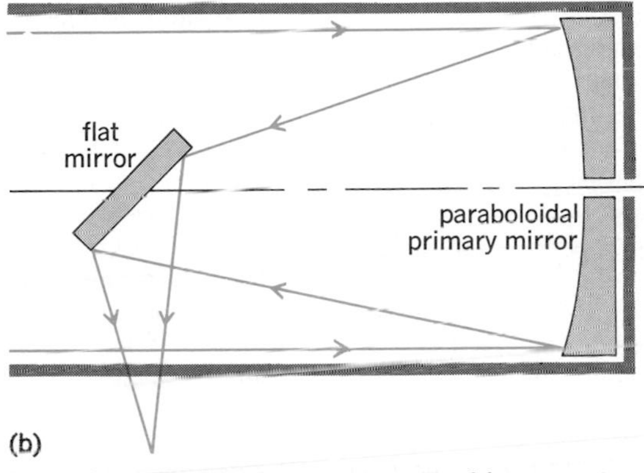

Fig. 3. Diagrams of reflecting telescopes. (*a*) Cassegrain telescope, with either classical or Ritchey-Chrétien optics. (*b*) Newtonian telescope.

A Cassegrain system consists of a primary mirror with a hole bored through its center, and a convex secondary mirror which reflects the light beam back through the central hole to be observed behind the primary mirror (Fig. 3*a*). Since the secondary mirror is convex, it decreases the convergence of the light beam and increases the focal length of the system as a whole. The higher focal ratios (*f*/8 to *f*/13) of Cassegrain systems permit the astronomer to observe extended objects, like planets, at higher spatial resolution and to isolate individual stars from their neighbors for detailed studies.

The classical Cassegrain system consists of a paraboloidal primary mirror and a hyperboloidal secondary mirror. The newtonian focus, prime focus, and Cassegrain focus are not affected by spherical aberration. However, all of the systems are plagued by coma, an optical aberration of the paraboloidal primary. Coma causes a point source off the center of the field of view to be spread out into a comet-shaped image. To correct for the effect of coma, a corrector lens is often used in front of the photographic plate. The design of prime-focus corrector lenses is a major consideration in large telescope design, since the corrector lens itself can introduce additional aberrations. *See Aberration; Astronomical photography.*

To avoid complicated corrector lenses at the Cassegrain focus, the Ritchey-Chrétien system, an alternate design with both a hyperboloidal primary and a hyperboloidal secondary, is used in modern telescopes. This arrangement is not affected by coma or spherical aberration, so it has a wider field of view than the classical Cassegrain.

Schmidt camera. This is an optical system used almost exclusively for photographic applications such as sky surveys, monitoring of galaxies for supernova explosions, and studies of comet tails. The primary mirror of a Schmidt camera has a spherical shape, and therefore suffers from spherical aberration. To correct for this problem, the light passes through a thin corrector plate as it enters the tube, as illustrated in **Fig. 4**. Schmidt camera correcting plates are among the largest lenses made for astronomical applications. The special features of this system combine to produce good images over a far larger angular field than can be obtained in a Cassegrain. The Schmidt camera at the California Institute of Technology/Palomar Observatory, which carried out the National Geographic Society–Palomar Observatory Sky Survey, can photograph a 6° by 6° field. *See Schmidt camera.*

Solar telescopes. Solar instrumentation differs from that designed to study other celestial objects, since the Sun emits great amounts of light energy. One solar instrument, the Robert R. McMath Solar Telescope located at Kitt Peak National Observatory near Tucson, Arizona (**Fig. 5**), consists of an 80-in. (2.03-m) heliostat that reflects sunlight down the fixed telescope tube to a spectrograph. This spectrograph is evacuated to avoid problems that would be created by hot air currents. An alternate design has been used at the Sacramento Peak Observatory, in which the heliostat feeds light into a vertical telescope that is evacuated.

Site selection. Efficient use of a telescope requires a site with a clear, steady atmosphere. Site selection for a large telescope is given considerable attention, every effort being made to locate the instrument in an area that is climatologically and geologically favorable—preferably an area at high elevation, on solid footing, and having a proved record for number of clear nights per year and maximum mean atmospheric stability. The site should be far from the lights and polluted air of large population centers. Today, it is difficult to find excellent sites in the continental United States. The great observatories in California, for instance, all feel the impact of civilization. The most sought-after sites are in the high mountains of Chile and Hawaii (see **table**).

Tools. Astronomers seldom use large telescopes for visual observations. Instead, they record their data for future study. Modern developments in photoelectric imaging devices are supplanting photographic techniques for many applications. The great advantages of detectors such as charge-coupled devices is their high sensitivity, and the images can be read out in a computer-compatible format for immediate analysis. *See Charge-coupled devices.*

Light received from most astronomical objects is made up of radiation of all wavelengths. The spectral

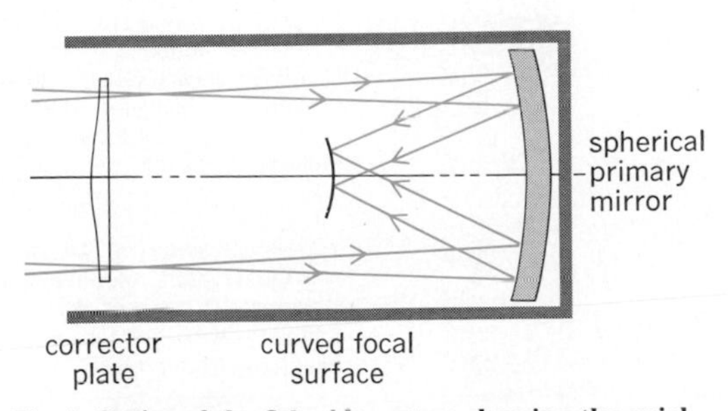

Fig. 4. Optics of the Schmidt system showing the axial and the extra-axial light paths.

(a)

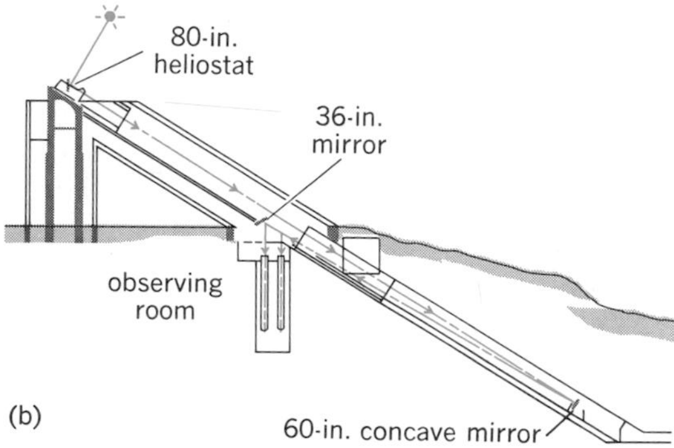

(b)

Fig. 5. Unconventional configuration. (*a*) McMath 60-in. solar telescope, Kitt Peak, Arizona. (*b*) Diagram of optical elements in the telescope. 1 in. = 2.5 cm.

characteristics of the radiation emitted from or reflected by a body may be extracted by special instruments called spectrographs. Wide field coverage is not critical in spectroscopy of stellar objects, so spectrographs are mounted at the Cassegrain and coudé foci. The coudé position has the advantage that its focal point is fixed in position, regardless of where the telescope points in the sky (**Fig. 6**). Thus a spectrograph or other instrument that is too heavy or too delicate to be mounted on the moving telescope tube can be placed at the coudé focus. *SEE ASTRONOMICAL SPECTROSCOPY*.

Photoelectric imaging devices may be used in conjunction with spectrographs to record spectral information. Photoelectric detectors, usually photomultiplier tubes, are useful tools for classifying stars, monitoring variable stars, and quantitatively measuring the light flux from any source at which the telescope is pointing. The photocathode of the detector is placed just behind the focal plane of the telescope, preceded by a small diaphragm which permits a view of only a very small area of the sky. The phototube converts the incident light energy into an electrical signal, which is subsequently amplified and recorded. Photometry carried out with different filters yields basic information about the source with shorter observing time than that required for a complete spectroscopic analysis.

Limitations. The largest telescope in operation is the 236-in. (6-m) reflector in the Caucasus Mountains in the Soviet Union. For many applications the Earth's atmosphere limits the effectiveness of larger telescopes.

Large telescopes

Mirror diameter			
Meters	Inches	Observatory	Year completed
		Some of the world's largest reflecting telescopes	
6.0	236	Special Astrophysical Observatory, Zelenchukskaya, Crimea	1976
5.1	200	California Institute of Technology/Palomar Observatory, Palomar Mountain, California	1950
4.2	165	William Herschel Telescope, La Palma, Canary Islands	1987
4.0	158	Kitt Peak National Observatory, Arizona	1973
4.0	158	Cerro Tololo Inter-American Observatory, Chile	1976
3.9	153	Anglo-Australian Telescope, Siding Spring Observatory, Australia	1975
3.8	150	United Kingdom Infrared Telescope, Mauna Kea Observatory, Hawaii	1978
3.6	144	Canada-France-Hawaii Telescope, Mauna Kea Observatory, Hawaii	1979
3.6	142	Cerro La Silla European Southern Observatory, Chile	1976
3.6	141	New Technology Telescope, European Southern Observatory, Chile	1989
3.2	126	NASA Infrared Telescope, Mauna Kea Observatory, Hawaii	1979
3.0	120	Lick Observatory, Mount Hamilton, California	1959
2.7	107	McDonald Observatory, Fort Davis, Texas	1968
2.6	102	Crimean Astrophysical Observatory, Crimea	1960
2.6	102	Byurakan Observatory, Yerevan, Armenia	1976
2.5	100	Mount Wilson and Las Campanas, Mount Wilson, California	1917
2.5	100	Cerro Las Campanas, Carnegie Southern Observatory, Chile	1976
2.5	98	Royal Greenwich Observatory, United Kingdom	1967
2.4	94	University of Michigan–Dartmouth College–Massachusetts Institute of Technology, Kitt Peak, Arizona	1986
6 × 1.8	6 × 71	Multi-Mirror Telescope, Mount Hopkins, Arizona	1979
		World's largest refracting telescopes	
1.02	40	Yerkes Observatory, Williams Bay, Wisconsin	1897
0.91	36	Lick Observatory, Mount Hamilton, California	1888
0.83	33	Observatoire de Paris, Meudon, France	1893
0.80	32	Astrophysikalisches Observatory, Potsdam, Germany	1899
0.76	30	Allegheny Observatory, Pittsburgh, Pennsylvania	1914

The most obvious deleterious effect of the Earth's atmosphere is image scintillation and motion, collectively known as poor seeing. Atmospheric turbulence produces an extremely rapid motion of the image resulting in a smearing of the image. On the very best nights at ideal observing sites, the image of a star will be spread out over a 0.25-arc-second seeing disk; on an average night, the seeing disk may be between 0.5 and 2.0 arc-seconds. It has been demonstrated that most of the air currents that cause poor seeing occur within the observatory buildings themselves. To counteract the problem, new techniques of insulation and cooling have been adopted, and substantial improvements in seeing have been achieved.

One of the chief uses of large telescopes is to study phenomena in quasars and galaxies in distant regions of the observable universe. Once again, however, the atmosphere limits what can be observed. The upper atmosphere glows faintly because of the constant influx of charged particles from the Sun. This airglow is a phenomenon similar to the aurora borealis, although it is typically fainter than a visible aurora display. Airglow adds a background exposure or fog to photographic plates that depends on the length of the exposure and the speed (*f*-ratio) of the telescope. The combination of the finite size of the seeing disk of stars and the presence of airglow means that a 33-ft (10-m) telescope could not see an object $(10/6)^2 = 2.8$ times fainter than the faintest object seen by a 20-ft (6-m) telescope. In fact, the gain is much less than that figure. On the other hand, a 33-ft telescope might cost as much as 10 times more than a 20-ft telescope. One solution is placing a large telescope in orbit above the atmosphere.

In practice, the effects of air pollution and light pollution from large cities outweigh the effect of airglow at most observatories in the United States. There are few unspoiled observatory sites left in the continental United States.

Notable telescopes. The definition of a large telescope depends on its type. For a refracting telescope to be considered a large telescope, its objective lens must be larger than about 24 in. (0.6 m), whereas a reflecting telescope will have to exceed 79 in. (2 m) to be considered large. The table lists a few of the largest telescopes. It is of interest to compare the dates of the reflecting telescopes and refracting telescopes. The limitations of refractors were recognized very early. A few telescopes that are notable for their historical importance, large size, or innovative design will be discussed. *See* ASTRONOMICAL OBSERVATORY.

Although Galileo's telescopes were small in size, they were notable because of the tremendously important discoveries made with them.

In 1790, William Herschel built a telescope whose mirror was 48 in. (1.2 m) in diameter. Also, William Parsons, the third Earl of Rosse, built a telescope with a 72-in. (1.8-m) mirror about 1840. However, these telescopes were difficult to operate and led to relatively few discoveries. Both were built with alt-azimuth mountings, which made it very difficult to follow celestial objects as the Earth rotated. In an equatorial mounting, found in almost all modern telescopes, rotation about one axis, the polar axis, compensates for the Earth's rotation. However, in an alt-azimuth mounting, motion in both altitude and azimuth is required to follow an object.

The 40-in. (1.02-m) refractor at the Yerkes Observatory was completed in 1897. The 40-in. objective lens has a focal length of 62 ft (19 m) and is housed in a 90-ft (23-m) dome. With its delicately balanced equatorial mounting and clock drives, the Yerkes refractor has been a major contributor to astronomy.

The 200-in. (5-m) Hale telescope at Palomar Mountain, California, was completed in 1950. The primary mirror is 200 in. in diameter with a 40-in. (1.02-m) hole in the center. Its focal length is 660 in. (14 m) and it has a paraboloidal figure. The focal ratio of the prime focus is *f*/3.3, and of the Cassegrain focus *f*/16. The Hale telescope was the first that was large enough to have a prime-focus cage where the observer could sit inside the main tube.

Since the completion in 1950 of the Hale reflector, the number of telescopes over 100 in. (2.5 m) in aperture has steadily grown, and 14 such instruments have been built since 1960. It is not feasible to describe each of these instruments in detail, although it is worth mentioning the modern instruments that represent the most advanced systems in operation.

The 158-in. (4-m) Mayall reflector at the Kitt Peak National Observatory was dedicated in 1973 (**Fig. 7**). The 158-in. mirror is made from a 24-in.-thick (61-cm) fused quartz disk which is supported in an advanced design mirror cell. It took three years to grind and polish the mirror to its *f*/2.7 hyperboloidal shape. The prime focus has a field of view six times greater than that of the Hale reflector. The first photographs with the 158-in. mirror showed its outstanding optical characteristics. An identical telescope was subsequently installed at Cerro Tololo Inter-American Observatory, in Chile.

Another notable instrument is the 236-in. (6-m) Soviet reflector. It is supported in an alt-azimuth mounting and like the great telescopes of Herschel and the Earl of Rosse, must be moved in both altitude and azimuth to compensate for the Earth's rotation.

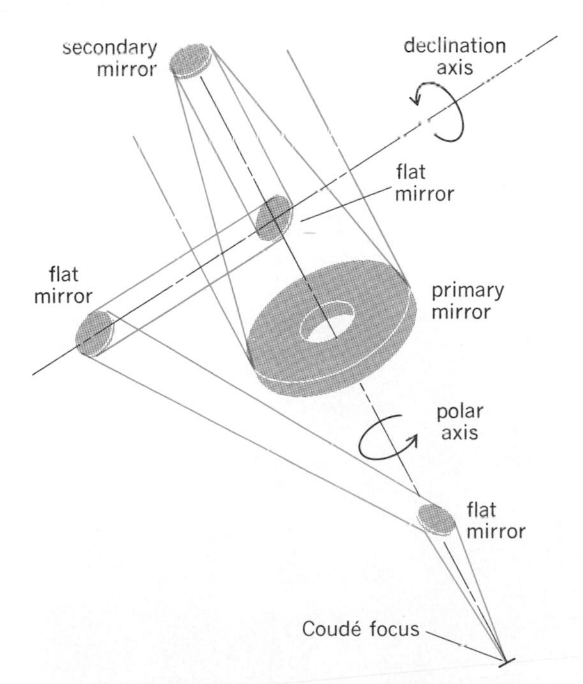

Fig. 6. Optical diagram for a common coudé configuration, showing axes of rotation of the telescope, viewed in isometric projection for clarity.

Fig. 7. The 158-in. (4-m) Mayall reflector of the Kitt Peak National Observatory. The dark tube at the upper end of the telescope is the prime focus cage. The large horseshoe bearing floats on oil pads, seen near each end of the mounting walkway. (*Kitt Peak National Observatory Photograph*)

Fig. 8. Multi-Mirror Telescope. (*Multi-Mirror Telescope Observatory, Smithsonian Institution/University of Arizona*)

This delicately balanced instrument is driven by a computer which calculates the relative rates of motion in azimuth and in altitude required to follow an object, and controls the motors which provide the motion. However, there is yet a third motion required. In an alt-azimuth system, the sky rotates relative to the telescope tube as the telescope tracks a celestial object. This rotation would smear a photograph of the object. To avoid this smearing, the photographic plate (or other observing device) must be rotated by the computer at the proper rate. The Soviet astronomers estimate the alt-azimuth mounting cost half that of an equatorial mounting.

The mirrors for these traditional large telescopes were all produced using the same general methodology. A large, thick glass mirror blank was first cast; then the top surface of the mirror was laborously ground and polished to the requisite shape. In the process, great quantities of glass were removed from the blank. The remaining glass helps to give the mirror the required structural rigidity as it is tilted at different angles as the telescope is pointed to different parts of the sky.

In 1989 the European Southern Observatory put into operation their New Technology Telescope, with a 141-in. (3.58-m) mirror. This mirror was produced by a technique known as spincasting, where molten glass is poured into a rotating mold. As the glass cools and solidifies, the surface of the relatively thin mirror takes on a shape that is close to the desired one, reducing substantially the need for grinding away excess glass. In operation, computers monitor the mirror's shape, and direct a support sustem that will maintain the proper shape. The first images made with the telescope were outstanding.

The Smithsonian Astrophysical Observatory/ University of Arizona Multi-Mirror Telescope (MMT; **Fig. 8**) is one of the most innovative of the new telescopes. It uses six 72-in. (1.8-m) mirrors working together to provide light-gathering power equivalent to a 177-in. (4.5-m) mirror. The six images from the telescopes are brought to a common focus by a mirror system. The coalignment of the six optical systems is maintained by means of an active laser and computer system that continually moves the secondary mirrors to maintain the alignment.

Observations with the Multi-Mirror Telescope were begun in early 1979, and it quickly demonstrated its capabilities by producing a number of interesting discoveries about stars and quasars. Plans are under way for a conversion of the Multi-Mirror Telescope, in which the six existing 72-in. (1.8-m) mirrors will be replaced by a single 260-in. (6.5-m) mirror.

Two space-based telescopes have been highly successful. The *International Ultraviolet Explorer* (*IUE*) was launched into geosynchronous orbit in early 1978, and has operated as one of NASA's most successful space observatories. It is not a giant telescope; its mirror is only 18 in. (45 cm) in diameter. The satellite hangs over the Atlantic Ocean, where it can be operated in real time from both the United States and Europe to obtain high-dispersion ultraviolet spectra of solar system objects, stars, nebulae, and extragalactic objects. The *Einstein Observatory* (*High Energy Astronomy Observatory 2*), launched in November 1978, carries a telescope that images soft x-rays in the 0.3–5-nm wavelength range. The data that scientists have obtained for stars from the *IUE* and the *Einstein* are revolutionizing understanding of stellar physics. *See* SATELLITE ASTRONOMY; ULTRAVIOLET ASTRONOMY; X-RAY TELESCOPE.

The Hubble Space Telescope (HST), which was launched into Earth orbit by the space shuttle on April 24, 1990, is a 2.4-m (94.5-in.) Cassegrain reflector, which feeds light into a number of scientific instruments. The instruments on the Hubble Space Telescope can see objects many times fainter than achieved so far. If the results from the small telescope on the *IUE* combined with the *Einstein* are any clue, the Hubble Space Telescope should begin a major revolution in understanding the universe. The scientific instruments flown on the Hubble Space Telescope include a wide-field and planetary camera, a faint-object camera, a faint-object spectrograph, a high-resolution spectrograph, and a high-speed photometer. After several years of operation the instruments will be changed out in orbit and replaced by innovative new instruments. The second round of instruments will stress infrared science. NASA plans to operate the Hubble Space Telescope through the remainder of the twentieth century.

New generation. Worldwide efforts are under way on a new generation of large ground-based telescopes, using both the spincasting method and the segmented method to produce large mirrors. In the design of the 394-in. (10 m) Keck telescope, sited at 13,788 ft (4203 m) on Hawaii's Mauna Kea volcano, the mirror is made up of 36 hexagonal segments, each about 72 in. (1.8 m) across and only 3 in. (76 mm) thick. As with the mirror of the new Technology Telescope, a sophisticated computer system is designed to control the positioning of the mirror segments.

The Very Large Telescope being developed by the European Southern Observatory for a site in Chile is planned to consist of four 315-in. (8-m) telescopes with spincast mirrors. The light of the four telescopes will be combined to give the equivalent light-gathering power of a 630-in. (16 m) telescope.

Robert D. Chapman

Astronomical Transit Instruments

Astronomical transit instruments are telescopic instruments adapted to the observation of the passage, or transit, of an astronomical object across the meridian of the observer. The astronomical transit instrument is the classic instrument of positional astronomy, which is the study of the positions and motions of astronomical objects and the related determination of positions by observation of these astronomical bodies from the Earth (the specific categories of astronomy concerned with these investigations are astrometry and celestial mechanics). The chief variants of the classic design include the meridian circle, the vertical circle, the horizontal transit circle, the broken or prism transit, and the photographic zenith tube.

The astronomical transit instrument was first developed by the Danish astronomer Ole Roemer in 1689. The modern transit instrument has a telescopic objective with a diameter of 6–10 in. (15–25 cm) and a focal length of 72–90 in. (180–230 cm). The instrument consists of a telescope mounted on a single fixed horizontal axis of rotation. The horizontal axis has a central hollow cube (or sometimes a sphere) and two conical semiaxes ending in cylindrical pivots. The objective and eyepiece halves of the telescope are also fastened to the cube of the instrument, perpendicular to the horizontal axis. Rotation of the instrument in its bearings, or wyes, permits the optical axis to sweep only in the plane of the meridian. An accurate clock is the essential ancillary scale by which the transits of the astronomical objects are observed. *See* ASTRONOMICAL COORDINATE SYSTEMS.

The large transit instrument of Pulkovo Observatory, Russia, in use since 1838, is seen in **Fig. 9**.

Applications. The astronomical transit instrument has three interrelated uses. (1) From a known position on the Earth, observations of transits of stars lead to the determination of their right ascension with respect to the astronomical coordinate system. (2) The determination of corrections to the clock may be made by the observation of stars of known position with an instrument situated at a known longitude. (3) Finally, with a knowledge of the positions of the stars observed and of Greenwich time, the longitude of the observer can be computed from observations of the time of transit of a star.

The astronomical transit instrument takes advantage of a special case of the astronomical triangle which is composed of arcs of great circles on the celestial sphere. Its vertices are, respectively, the north celestial pole, the zenith point of the observer, and the celestial object under observation. The angle at the north celestial pole represents the hour angle of the object; hence when the object is on the meridian, the hour angle is zero, and the triangle degenerates to a single arc, a segment of the meridian. At that instant the local sidereal time equals the right ascension of the celestial object.

Fig. 9. Large transit instrument, Pulkovo Observatory, Russia. (*Courtesy of B. L. Klock*)

A divided circle, graduated into fractions of a degree, is seated on the horizontal axis of the transit instrument and is used to set the instrument at the required zenith distance. The instrument has a fastening clamp also situated on the horizontal axis. The clamp may have a fine-motion adjustment mechanism to improve alignment for an observation.

The micrometer, or eyepiece part of the instrument, contains a movable wire, or pair of wires, and a stationary grid of vertical wires for use in registering the transit. At the Bordeaux, Perth, Tokyo, La Palma, and Pulkovo observatories, photoelectric observations have been made, thereby replacing the observer and yielding higher-quality observations.

An accurate quartz crystal or atomic clock, with an error rate of less than 0.001 s per day, is used as the scale for recording the transit data in conjunction with some form of data storage, such as punched tape, a printing chronograph, or a computer. *See Atomic clock; Quartz clock.*

Corrections. It is extremely difficult to adjust the instrument to the point of perfection, where the mean wire will trace the true meridian as the instrument is rotated on its pivots; therefore corrections must be determined and applied to the observational data. The three principal instrument errors that require correction are azimuth, collimation, and level. The azimuth correction is the horizontal angle between the axis of rotation and the true east-west direction. The collimation correction is the angle between the line from the optical center of the telescope objective to the mean wire in the micrometer and the plane perpendicular to the horizontal axis of rotation. The level correction is the angle that the axis of rotation makes with the plane of the horizon.

These errors do not remain constant even though the instrument may be well constructed and mounted. The principal cause for their change is attributed to variations of temperature in the environment of the pavilion housing the instrument. This temperature is ambient since the roof of the pavilion must be open to make observations. Modern design of transit instruments attempts to utilize advances in optics and metallurgy to minimize the thermal influence on the instrument.

Another error, the clock correction, represents the error between the true sidereal time that the star should transit the local meridian and the time of transit recorded by the local sidereal clock. At one time the clocks had error rates that were significant enough to warrant a correction; however, today they have sufficient accuracy so that this correction may be generally regarded as a constant for each night's work. This correction is evaluated through observations on bright stars whose positions are already well known.

Meridian circle. The major astrometrical observatories of the world have astronomical transit instruments called meridian or transit circles. These instruments are similar to the transit instrument previously described, except they have a micrometer eyepiece that has an extra pair of moving wires perpendicular to the vertical set. These wires are used to measure the zenith distance or declination of the celestial object in conjunction with readings taken from a large, accurately calibrated circle attached to the horizontal axis. The circle may be read photographically, photoelectrically, or electronically.

The 6-in. (15-cm) transit circle of the U.S. Naval Observatory (**Fig. 10**) was designed and constructed at the end of the nineteenth century, but it has been improved continuously with the latest technological developments. It is the first transit circle in the world to have an electronic circle and the first to have the micrometer data read directly into an electronic computer. In addition to the electronic circle data, the divided glass circle is scanned photoelectrically with six scanning micrometers. The data from the scanners are entered into the computer for real-time processing. The probable error of a single set of circle readings is 0″.07. However, the declination of a star may contain uncertainties of 0″.25–0″.50 due to errors from other sources, such as atmospheric refraction, mechanical flexure of the instrument, and residual errors in the divided circle.

The 6-in. (15-cm) transit circle is used not only to observe the brighter stars (as faint as ninth magnitude) but also the Sun, Moon, planets, and several of the brighter asteroids. These observations are made visually, with the observer seated on a couch with an adjustable back to lend head support during an observation. The probable error of a single observation of the right ascension data of an equatorial star is about 0.012 s with the use of the motor-driven micrometer. A computer presets the basic speed of the right-ascen-

Fig. 10. Six-in. (150-mm) transit circle, U.S. Naval Observatory. (*Official U.S. Naval Observatory photograph*)

sion wires and then the observer adds or subtracts to this speed through pushbuttons on a hand keyboard as the celestial object transits the field of view.

The U.S. Naval Observatory has refurbished its 7-in. (18-cm) transit circle with new instrumentation featuring an automatic micrometer, automatic setting, and a circle scanning system similar to the 6-in. (15-cm) transit circle system. The 7-in. transit circle was sent to New Zealand in the southern hemisphere in 1984 in order to improve knowledge of the positions and motions of celestial objects in that part of the sky.

The Tokyo Astronomical Observatory at Mitaka has constructed a transit circle with automatic tracking and automatic reading of the circle. The circle reading system features a unique charge-coupled-device linear array. SEE CHARGE-COUPLED DEVICES.

Benny L. Klock

Bibliography. J. Cornell, Six new eyes peer from Mount Hopkins, *Sky Telesc.*, 58:23–24, 1979; D. L. Crawford (ed.), *The Construction of Large Telescopes*, 1966; H. Eichorn, *Astronomy of Star Positions*, 1974; H. K. Eichorn and R. J. Leacock (eds.), *Astrometric Techniques*, 1986; W. C. Gliese et al. (eds.), *New Problems in Astrometry*, 1974; R. Green, *Spherical Astronomy*, 1985; H. C. King, *The History of the Telescope*, 1955, reprint 1979; G. P. Kuiper and B. M. Middlehurst (eds.), *Telescopes*, 1960, reprint 1977; S. P. Maran, Beyond Galileo's "vast crowd of stars," *Smithsonian*, 18:40–53, 1987; J. B. Oke, Palomar's Hale telescope: The first 50 years, *Sky Telesc.*, 58:505–509, 1979; T. Page and L. W. Page, *Telescopes*, 1966; T. J. Rafferty and B. L. Klock, Circle scanning systems of the U.S. Naval Observatory, *Astron. Astrophys.*, 164:428–432, 1986; P. van de Kamp, *Principles of Astrometry*, 1967; R. M. West, Europe's astronomy machine, *Sky Telesc.*, 75:471–481, 1988.

Orbital motion

In astronomy the motion of a material body through space under the influence of its own inertia, a central force, and other forces. Johannes Kepler found empirically that the orbital motions of the planets about the Sun are ellipses. Isaac Newton, starting from his laws of motion, proved that an inverse-square gravitationalfield of force requires a body to move in an orbit that is a circle, ellipse, parabola, or hyperbola.

Elliptical orbit. Two bodies revolving under their mutual gravitational attraction, but otherwise undisturbed, describe orbits of the same shape about a common center of mass. The less massive body has the larger orbit. In the solar system, the Sun and Jupiter have a center of mass just outside the visible disk of the Sun. For each of the other planets, the center of mass of Sun and planet lies within the Sun.

For this reason, it is convenient to consider only the relative motion of a planet of mass m about the Sun of mass M as though the planet had no mass and moved about a center of mass $M + m$. The orbit so determined is exactly the same shape as the true orbits of planet and Sun about their common center of mass, but it is enlarged in the ratio $(M + m)/M$. SEE PLANET.

Parameters of elliptical orbit. The **illustration** shows the elements or parameters of an elliptic orbit. Major axis AP intersects the ellipse AOP at the apsides; the extension of the major axis is the line of apsides. The body is nearest the center of mass at one apside, called perihelion P, and is farthest away at the other, called aphelion A.

Shape and size of an orbit are defined by two elements: length of semimajor axis and departure of the orbit from a circle. Semimajor axis a equals CP; this length is expressed in units of the mean distance from the Earth to the Sun. Eccentricity e equals CS/CP where C is the center of the ellipse and S is a focus. For elliptical orbits e is always less than unity.

Position of a body in its orbit at time t can be computed if a, e, and time of perihelion passage p and period of revolution T are known. Let O be the position of a planet at time t and OSP be the area swept out in time $t - p$. From Kepler's area law, area OSP equals $(t - p)/T$ multiplied by the area of the full ellipse. SEE AREAL VELOCITY.

To describe the orientation of an orbit in space, several other parameters are required. All orbits in the solar system are referred to the plane of the ecliptic, this being the plane of the orbit of Earth about the Sun. The reference point for measurement of celestial longitude in the plane of the ecliptic is the vernal equinox ♈, the first point of Aries. This is the point where the apparent path of the Sun crosses the Earth's equator from south to north. The two points of intersection of the orbit plane with the plane of the ecliptic (N and N') are called the nodes, and the line joining them is the line of nodes. Ascending node N is the one where the planet crosses the plane of the ecliptic in a northward direction; N' is the descending node. The angle as seen from the Sun S measured in the plane of the ecliptic from the vernal equinox to the ascending node is ♈SN; it is termed the longitude of the ascending node ☊ and fixes the orbit plane with respect to the zero point of longitude. The angle at the ascending node between the plane of the ecliptic

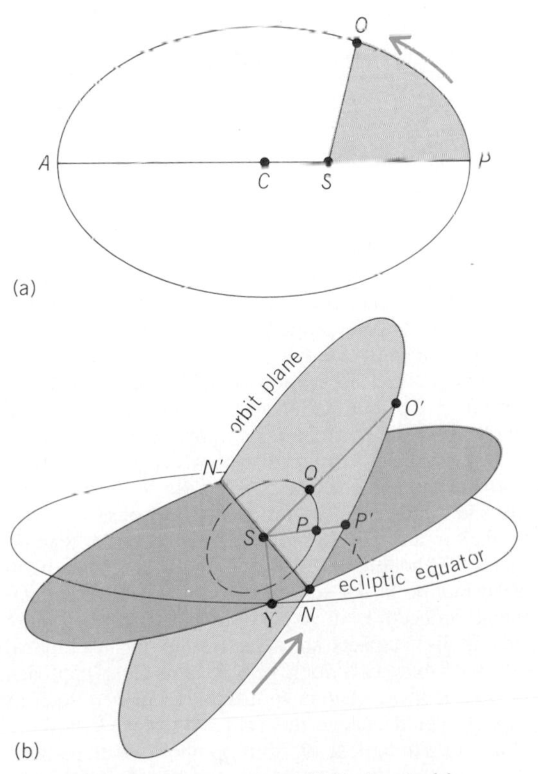

Parameters, or elements, of an elliptical orbit. (*a*) Relative orbit. (*b*) Orbit in space.

and the orbit plane is called the inclination i and defines the orientation of the orbit plane with respect to the fundamental plane. The angle as seen from the Sun, measured in the orbit plane from the ascending node to perihelion, is NSP' and is referred to as the argument of perihelion; it defines the orientation of the ellipse within the orbit plane. The angle $NSP' + \Omega$, measured in two different planes, is called the longitude of perihelion $\tilde{\omega}$. Because dynamically the semimajor axis a and period T of a planet of mass m revolving under the influence of gravitation G about the Sun of mass M are related by Eq. (1), only six ele-

$$\frac{4\pi^2}{T} = \frac{G(M + m)}{a^3} \tag{1}$$

ments, a, e, i, Ω, $\tilde{\omega}$, and p, are required to fix the position of a planet in space. Instead of these elements, however, a position vector x, y, z and the associated velocity vector $\dot{x}$, $\dot{y}$, $\dot{z}$ at a given instant of time would serve equally well to define the path of a planet in a rectangular coordinate system with origin at the Sun.

Orbital velocity. Orbital velocity v of a planet moving in a relative orbit about the Sun may be expressed by Eq. (2) where a is the semimajor axis and

$$v^2 = G(M + m)\left(\frac{2}{r} - \frac{1}{a}\right) \tag{2}$$

r is the distance from the planet to the Sun. In the special case of a circular orbit, $r = a$, and the expression becomes Eq. (3). When the eccentricity of an

$$v^2 = \frac{G(M + m)}{a} \tag{3}$$

orbit is exactly unity, the length of the major axis becomes infinite and the ellipse degenerates into a parabola. The expression for the velocity then becomes Eq. (4). This parabolic velocity is referred to

$$v^2 = G(M + m)\left(\frac{2}{r}\right) \tag{4}$$

as the velocity of escape, since it is the minimum velocity required for a particle to escape from the gravitational attraction of its parent body.

Eccentricities greater than unity occur with hyperbolic orbits. Because in a hyperbola the semimajor axis a is negative, hyperbolic velocities are greater than the escape velocity.

Parabolic and hyperbolic velocities seem to be observed in the motions of some comets and meteors. Aside from the periodic ones, most comets appear to be visitors from cosmic distances, as do about two-thirds of the fainter meteors. For ease of computation, the short arcs of these orbits that are observed near perihelion are represented by parabolas rather than ellipses. Although the observed deviation from parabolic motion is not sufficient to vitiate this computational procedure, it is possible that many of these "parabolic" comets are actually moving in elliptical orbits of extremely long period. The close approach of one of these visitors to a massive planet, such as Jupiter, could change the velocity from parabolic to elliptical if retarded, or from parabolic to hyperbolic if accelerated. It is possible that many of the periodic comets, especially those with periods under 9 years, have been captured in this way. SEE CELESTIAL MECHANICS; COMET; GRAVITATION; PERTURBATION; STELLAR ROTATION.

Raynor L. Duncombe

Bibliography. G. O. Abell, D. Morrison, and S. C. Wolff, *Exploration of the Universe*, 5th ed., 1987; A. E. Roy, *Orbital Motion*, 3d ed., 1988; V. Szebehely, *Adventures in Celestral Mechanics: A First Course in the Theory of Orbits*, 1989.

Orion

The Warrior, in astronomy, and undoubtedly the finest of all constellations in the sky. Orion is a winter group near the celestial equator. Four of the most prominent stars, α, γ, β, and κ, form a huge crude rectangle (see **illus.**). The group is pictured as the figure of a warrior, holding a shield with his left hand and swinging a club with his raised right arm ready

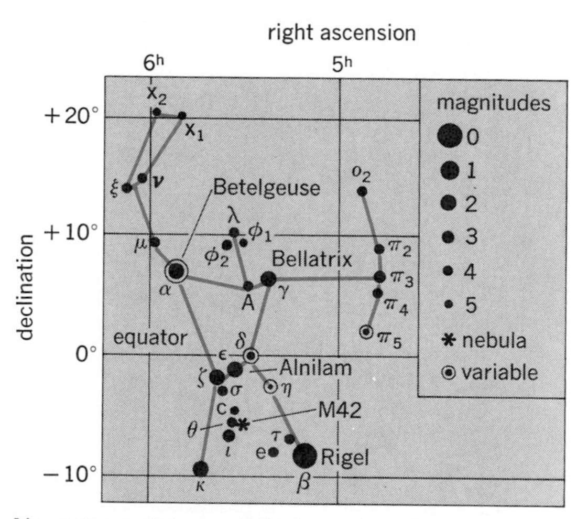

Line pattern of the constellation Orion. The grid lines represent the coordinates of the sky. The apparent brightness, or magnitude, of the stars is shown by the size of the dots, graded by appropriate numbers.

to strike the charging Bull. Betelgeuse (meaning armpit), one of the largest stars known, is the red star at the right shoulder, Bellatrix is at the left shoulder, and Rigel, the blue-white star, is at the left leg. Three bright stars, δ, ε, and κ, in a straight line in the middle of the rectangle represent the warrior's belt. The center star is Alnilam. These four stars are all navigational stars. Three faint stars below the belt form the Sword of Orion, the middle one of which is actually four very hot stars, the Trapezium, embedded in the Great Nebula of Orion (M42). SEE CONSTELLATION; ORION NEBULA; TAURUS.

Ching-Sung Yu

Orion Nebula

The brightest emission nebula in the sky, designated M42 in Messier's catalog. The Great Nebula in Orion consists of ionized hydrogen and other trace elements (see **illus.**). The nebula belongs to a category of objects known as H II regions (the Roman numeral II indicates that hydrogen is in the ionized state), which mark sites of recent massive star formation. Located in Orion's Sword at a distance of 500 parsecs or 1600

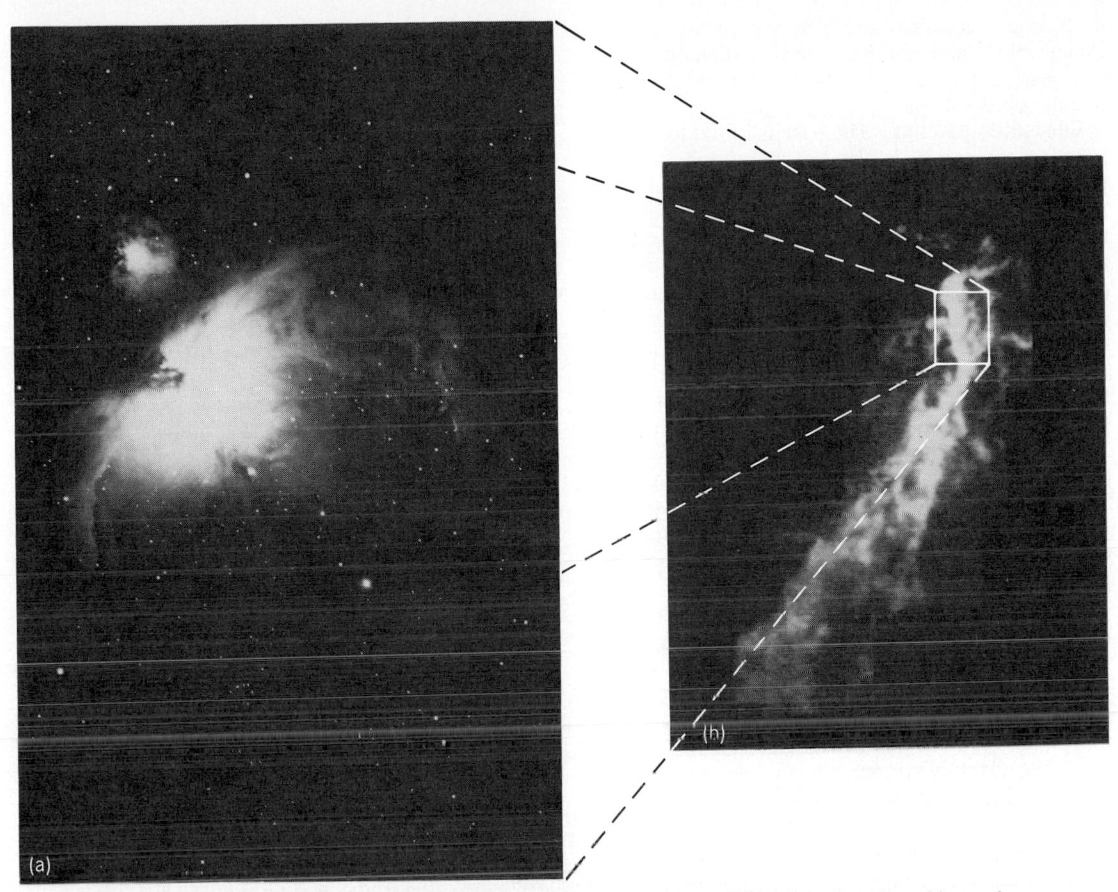

Orion Nebula and its neighborhood. (*a*) Visual-wavelength view of the Orion Nebula obtained by taking a time exposure on a photographic plate with the 4-m-diameter (158-in.) telescope on Kitt Peak, Arizona (*National Optical Astronomy Observatories*). (*b*) Orion A giant molecular cloud as seen in the 2.6-mm-wavelength emission line of ^{13}CO. Data used to construct this image were obtained with a 7-m-diameter (23-ft) millimeter-wave telescope at Crawford Hill, New Jersey. The image shows a region which extends 5° north-south and 2° east-west. The Orion Nebula, located in front of the northern part of the molecular cloud, is not visible in this image. The approximate dimensions of the optical view are shown by the inset box. (*AT&T Bell Laboratories*)

light-years (9.5×10^{15} mi or 1.5×10^{16} km), the Orion Nebula consists of dense plasma, ionized by the ultraviolet radiation of a group of hot stars less than 100,000 years old known as the Trapezium cluster. The nebula covers an area slightly smaller than the full moon and is visible with the aid of binoculars or a small telescope. *See* Orion.

Stars form by the gravitational collapse and fragmentation of dense interstellar molecular clouds. A star-forming molecular cloud core (known as OMC1 for Orion Molecular Cloud 1) is hidden behind the Orion Nebula by a shroud of dust. A group of infrared sources, believed to be stars too young to have emerged from their dusty birth sites, is buried within OMC1. Although invisible at optical wavelengths, OMC1 has been investigated in the infrared and radio portions of the spectrum. Observations of the massive young stars in OMC1 have shown that stellar birth is associated with powerful outflows of gas, naturally occurring maser emission, and shock waves that disrupt the remaining cloud core.

The cloud core behind the Orion Nebula is only a small part of a giant molecular cloud (the Orion A cloud), 100,000 times more massive than the Sun (see illus. *b*). Although mostly made of molecular hydrogen (H_2), the cloud is best traced in the 2.6-mm-wavelength emission line ^{13}CO, the rarer isotopic variant of carbon monoxide (CO). Over the past 1×10^7 years, the Orion A cloud, together with a second giant molecular cloud in the northern portion of Orion (the Orion B cloud), have given birth to the Orion OB association, a loose, gravitationally unbound grouping of hot massive stars of spectral types A, B, and O. The collective effect of the ionizing radiation and winds produced by these stars, and the supernova explosions occurring at their death, has generated an expanding bubble 70 by 200 pc (1.3 by 3.8×10^{15} mi or 2.2 by 6.2×10^{15} km) in extent in the interstellar medium. The eastern portion of this bubble is running into dense gas near the plane of the Milky Way Galaxy, where faint optical emission is seen as the 8° radius arc of nebulosity called Barnard's Loop. *See* INTERSTELLAR MATTER; NEBULA; RADIO ASTRONOMY; STAR; STELLAR EVOLUTION; SUPERNOVA.

John Bally

Bibliography. A. E. Glassgold, P. J. Huggins, and E. L. Schucking (eds.), *Symposium on the Orion Nebula to Honor Henry Draper*, *Ann. N. Y. Acad. Sci.*, 395: 1–338, 1982; J. M. Pasachoff, *Contemporary Astronomy*, 4th ed., 1989.

Parallax

The difference in direction to an astronomical object as seen from two different locations in space. With a known distance between observation sites, the distance to the object can be directly determined; and so, in practice, parallax and distance are used interchangeably. There are three kinds of direct parallaxes (see **illus.**), depending on the baseline used to separate the observations.

Geocentric parallax. The baseline used to measure geocentric parallax is the equatorial radius of the Earth (3963 mi or 6378 km). Directions observed from other points on the Earth will be displaced by lesser amounts, but the derived parallax is scaled to that which would be observed from the Equator. This parallax is of importance only for objects within the solar system, since it is completely negligible for even the nearest star. The largest geocentric parallax is that of the Moon, which, for its mean distance, is 57′02″.61. The most important geocentric parallax is that of the Sun, since the average distance from the Sun to the Earth (known as the astronomical unit, or AU) is the unit of distance for astronomical objects. This distance can now be determined by tracking interplanetary space probes, and it is equal to 92,955,807 mi (149,597,870 km), equivalent to a parallax of 8″.794148. *See Astronomical unit.*

Stellar parallax. The baseline used to measure stellar parallax is the astronomical unit, although in practice the direction to a star is observed from opposite sides of the Earth's orbit about the Sun. This angle is very small for even the nearest star, and the first stellar parallax was not measured until 1838 (indeed, the inability to detect stellar parallaxes before then was used as an argument for the geocentric universe). Parallaxes are expressed in seconds of arc, and the largest one known—that of Alpha Centauri—is only 0″.751. The inverse of the parallax gives the distance, and if the parallax is in seconds of arc, the distance is in parsecs: 1 parsec (pc) is the distance at which 1 AU subtends an angle of 1″, and it thus is equal to 206,265 AU, or 3.2616 light-years, or 1.9174×10^{13} mi, or 3.0857×10^{13} km.

Reliable parallaxes became available only in the twentieth century, with the advent of the large, long-focus telescope and the astronomical application of the photographic process. Until about 1960, almost all parallaxes were determined with refractors of the sub-meter-aperture class. Although in principle only three observations of direction are required to determine a parallax (three are needed to permit simultaneous determination of the star's intrinsic, or proper, motion along with the parallactic motion), some two dozen or more plates are taken over several years, because of a variety of unavoidable sources of error, to permit a good parallax determination. These plates are measured on high-precision, often semiautomatic measuring machines that permit the determination of the relative position of the star of interest against a background of presumably very much more distant objects (statistically, only 1 star in 5000 is likely to have a significant parallax). Unfortunately, even the most sophisticated analysis cannot eliminate some errors that are intrinsic to refractors, and hence it is impossible by traditional means to reduce the error of a parallax below about 0″.02. *See Optical telescope.*

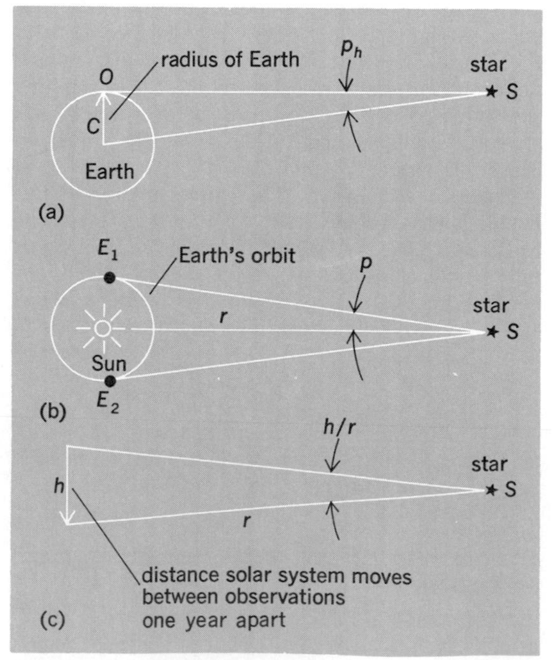

Three types of parallax. *(a)* **Geocentric or horizontal.** *(b)* **Annual.** *(c)* **Secular.**

However, several reflectors have been built for the sole purpose of determining parallaxes and related parameters, and they have been designed to eliminate the usual sources of errors of normal reflectors (chromatic aberration, lens tilt, lens rotation, lens decentering, lens inhomogeneity, and so on). By their very nature, they do not have the problems that plague refractors, since they do not have lenses. The most active such instrument today is the 61-in. (1.55-m) astrometric reflector of the U.S. Naval Observatory. Because the major sources of error have been effectively eliminated, parallaxes are routinely determined with real precisions around 0″.004, and for special stars of greater interest the errors can be brought down to 0″.001 or less with enough material. Further, because of the larger aperture, fainter stars, such as the red dwarfs that dominate the immediate solar neighborhood, and the intriguing white dwarfs, can have their distances determined. Thus, the distance out to which stars can have distances reliably determined has increased from something like 20 pc to perhaps 200 pc, and meaningful parallaxes might be obtainable with enough effort for objects out to 1000 pc, a distance that takes in some of the exotic stars whose distances heretofore could only be estimated indirectly. *See Star; White dwarf star.*

Modern technology provides new means of astronomical parallax determination in terms of both instrumentation and detectors. The use of charge-coupled devices and interferometric detectors makes possible the determination of individual star positions with far greater precision than that possible with the photographic plate. The possibility of placing instruments in Earth orbit and thus above atmospheric influence will mean yet more precise positional measurements. Notable is the European satellite *HIPPARCOS* (*High Precision Parallax Collecting Satellite*), which was originally designed to determine the parallaxes of approximately 115,000 stars brighter than 12th magnitude to a precision of 0″.002 during a 2½-year mission. The satellite was launched in 1989, but because it did not achieve geosynchronous orbit its mission was modified and its goals were somewhat reduced. *See Charge-coupled devices; Satellite astronomy.*

A benefit of the parallax determination process is that several other parameters of interest are obtained at the same time. Proper motions have already been mentioned. It is also possible in some cases to measure the change of proper motion with time (the acceleration). That change, combined with the motion itself and the distance, can give the speed along the line of sight (the radial velocity). This has been applied to one or two nearby white dwarfs to separate the true radial velocity from the relativistic Doppler shift in the observed spectra. It is also possible to deduce the presence of an unseen companion, since its orbital motion will produce a motion of the visible star about the center of mass of the system, and this will show up as a wobble on top of the parallactic and proper motion. The size of the wobble indicates something of the mass of the companion, and it may be possible to detect bodies as small as planets this way. *See* Gravitational redshift.

Secular parallax. The baseline for measuring secular parallax is determined by the linear motion of the Sun with respect to the stars, and hence it gets longer with time. Unfortunately, because all of the stars are themselves moving (the proper motion), it is impossible to determine from a single star how much of the observed change in position is due to the star's motion and how much is due to that of the Sun. However, with enough stars of apparently generally similar characteristics, it is possible to obtain statistical data on the distance to the stars as a group, and hence these are often referred to as statistical parallaxes.

Indirect parallaxes. Besides these directly determined parallaxes, there are other kinds of parallaxes that are derived from the distance that has been determined by another approach. Photometric, or spectroscopic, parallaxes are derived from a well-known relationship between the color (temperature) of a star and its intrinsic luminosity. If a color or spectral type and an apparent luminosity can be obtained, the distance follows at once. The star has to be close enough that its apparent brightness has not been decreased by absorption in the interstellar matter.

Cluster parallaxes are determined by measuring the radial velocity of the cluster members, and by determining where the proper motions of these members converge to or diverge from. This information allows determination of the actual velocity across the line of sight, and therefore the distance, by comparison with the average proper motion. *See* Star clusters.

Dynamical parallaxes are determined for binary star systems in which enough relative orbit motion has been observed that the period of revolution and angular average separation are well known. Since there is also a relationship between color and mass of a star (unfortunately not well determined), and one between the mass of a binary, its revolution period, and its actual mean separation, the comparison of observed and computed mean separation yields the distance. *See* Astrometry; Binary star; Mass-luminosity relation.

Robert S. Harrington

Bibliography. H. K. Eichhorn and R. J. Leacock (eds.), *Astrometric Techniques*, 1986; L. G. Taff, *Computational Spherical Astronomy*, 1981; W. van Altena, *General Catalog of Trigonometric Stellar Parallaxes*, 3d ed., 1990; P. van de Kamp, *Stellar Paths*, 1981.

Parsec

A unit of measure of astronomical distances. One parsec is equivalent to 3.084×10^{13} km or 1.916×10^{13} mi. There are 3.26 light-years in 1 parsec. The parsec is defined as the distance at which the semimajor axis of Earth's orbit around the Sun (1 astronomical unit) subtends 1 second of arc. Thus, because the angle is small, the equation below holds. A parsec

$$\frac{1 \text{ astronomical unit}}{1 \text{ parsec}} = 1 \text{ second} = \frac{1}{206{,}265}$$

is then 206,265 astronomical units; its accuracy depends on the precision with which the distance from Earth to Sun is measured. At a distance of 1 parsec, the parallax is 1 second of arc. The nearest star is approximately 1.3 parsecs distant; the farthest known galaxy is several billion parsecs. *See* Parallax.

Jesse L. Greenstein

Pegasus

The Winged Horse, in astronomy, an autumnal constellation. Pegasus is usually identified by the four bright stars α, β, γ, and α situated on the corners of a large square known as the Great Square in Pegasus (see **illus.**) The constellation is represented by a winged horse. Markab (the Saddle), a navigational star, occupies the southwestern corner of the square. The star Alpheratz at the opposite corner is really in the constellation Andromeda. The star at the northwestern corner is a red star, known as Scheat, a giant irregular variable. Diagonally opposite on the southeastern corner of the square is Algenib. Enif, another navigational star, lies in the nose of the horse. *See* Constellation.

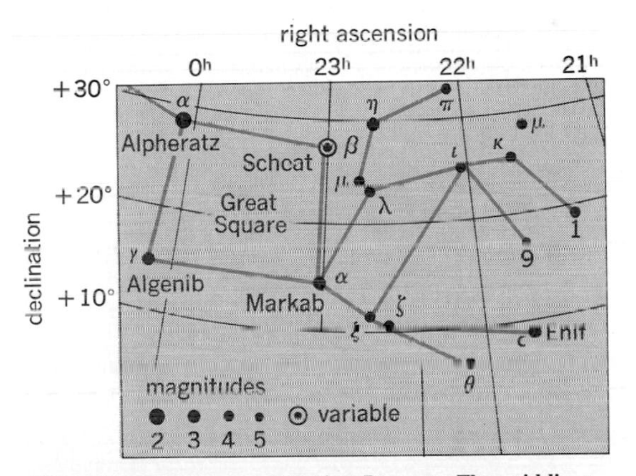

Line pattern of the constellation Pegasus. The grid lines represent the coordinates of the sky. The apparent brightness, or magnitude, of the stars is shown by the sizes of the dots, graded by appropriate numbers.

Ching-Sung Yu

Perigee

The point nearest the Earth in the orbit of the Moon or of an artificial satellite. At perigee the Moon is 5% closer to Earth than at its mean distance, the orbital eccentricity being 0.055. Because, on the average, the Moon and Sun subtend nearly equal angles, a solar eclipse near perigee lasts about 5 min; an eclipse near apogee is annular. The line perigee-Earth-apogee is the major axis of the orbital ellipse on the line of apsides. The differential attraction of the Sun on Earth and Moon causes the line of apsides of the

Moon to move forward in the orbital plane with a period of 8.85 years. *SEE MOON; PERIHELION.*

Gerald P. Kuiper

Perihelion

In astronomy, that point at one extremity of the major axis of the elliptical, parabolic, or hyperbolic orbit of a planet or comet about the Sun where the planet or comet is closest to the Sun. The instant when a planet or comet is at perihelion is referred to as the time of perihelion passage. For Earth this occurs about January 3, at which time Earth is some 1.55×10^6 mi (2.5×10^6 km) closer to the Sun than its mean distance of 93.0×10^6 mi (149.6×10^6 km). *SEE CELESTIAL MECHANICS; ORBITAL MOTION.*

Raynor L. Duncombe

Perseus

A compact circumpolar constellation of the northern sky, like its neighbor, Cassiopeia, on the east. Both constellations lie in a brilliant part of the Milky Way. The prominent stars in Perseus form the capital letter *A (see* **illus**.). This group is represented by the figure of the hero Perseus. The conspicuous curved arc of

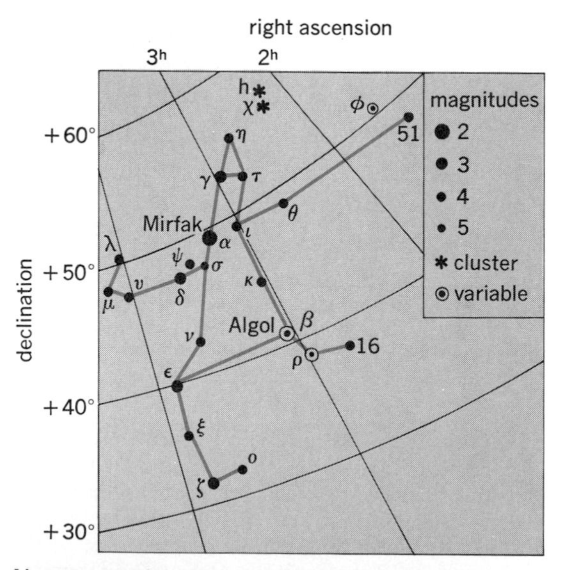

Line pattern of the constellation Perseus. The grid lines in the chart represent the coordinates of the sky. The apparent brightness, or magnitude, of the stars is shown by the size of the dots, which are graded by appropriate numbers as indicated.

stars, bright and easy to identify, is commonly known as the Segment of Perseus. Mirfak, a navigational star, lies in the right shoulder. The constellation is noted for its clusters of stars. Just above the head are the famous double clusters *h* and χ in Perseus. Algol, the Demon Star, which is an eclipsing variable, is located in this constellation. *SEE CASSIOPEIA; CONSTELLATION.*

Ching-Sung Yu

Perturbation

Departure of a celestial body from the trajectory it would follow if moving only under the action of a single central force. Perturbations may be caused by either gravitational or nongravitational forces.

Corrections to elliptic orbits. In the solar system, orbits of planets may be adequately represented by mean elliptical elements to which are added small corrections due to the mutual planetary attractions. Although such motion is referred to as disturbed, it is as much a consequence of the law of gravitation as is undisturbed elliptic motion.

Another method of representing perturbed motion is to augment the position derived from the mean ellipse by the actual displacements in the coordinates due to the disturbing forces. These perturbations of the elements, and perturbations of the coordinates, are represented by infinite series; usually many terms are required to represent the disturbed motion accurately. These analytical expressions are referred to as general perturbations and, with their associated mean elements, form a general theory of the motion. In some instances, such as the orbit of an outer satellite of Jupiter or the motion of a comet moving with nearly parabolic velocity, the analytical expressions representing the perturbing function become so involved (mainly because of lack of convergence of the Fourier series) that general perturbations are not attempted. Instead, the perturbed positions are computed from a step-by-step numerical integration of the equations of motion; this is known as the method of special perturbations.

Long- and short-term disturbances. Planetary orbits are subject to two classes of disturbances: secular, or long-term, perturbations; and periodic, or relatively short-term perturbations. Secular perturbations, so called because they are either progressive or have excessively long periods, arise because of the relative orientation of the orbits in space. They cause slow oscillatory changes of eccentricities and inclinations about their mean values with accompanying changes in the motions of the nodes and perihelia. The periods of time involved in these oscillations may extend from 50,000 to 2,000,000 years. Periods and major axes of orbits are not affected by secular change. For the orbit of the Earth, the present inclination to the invariable plane is 1°35′. This will diminish to a minimum of 47′ in approximately 20,000 years. The eccentricity, presently 0.017, is diminishing also and will reach a minimum of 0.003 in about 24,000 years.

Periodic perturbations arise from the relative positions of the planets in their orbits. When the disturbed and disturbing planets are aligned on the same side of the Sun, the perturbation reaches a maximum, and reduces to minimum when alignment is reached on opposite sides of the Sun. The size of a periodic perturbation is a function of the mass of the disturbing body and of the length of time the two planets remain near the point of closest approach. Periodic perturbations continually shift a planet away from the position it would occupy in undisturbed motion, moving it above or below the orbital plane, nearer to or farther from the Sun, and forward or backward in the orbit.

Commensurable motions. If the mean motion of the disturbed planet were exactly a submultiple, say ½, of the mean motion of the disturbing planet, the maximum perturbation produced by their close approach would always occur in the same part of the disturbed orbit. The displacement in position of the disturbed planet would increase with each coincidence until the character of the orbit became modified to the point where exact commensurability of the mean motions would cease to exist.

Because the solar system is middle-aged, cosmically speaking, few examples of commensurability of mean motions exist today. None is found in the motions of the major planets. Cases of near commensurability exist which give rise to long-period periodic terms of large amplitude. As an example, the periods of Jupiter and Saturn are nearly in the ratio of 2:5. Thus, after nearly five revolutions of Jupiter, the two planets return to approximately the same juxtaposition. Their line of coincidence, however, sweeps slowly around Jupiter's orbit, completing a circuit in about 850 years and thus producing a perturbation of this period.

Among the four inner planets, the periodic perturbations are small, amounting in orbital longitude at most of 0'.25 for Mercury, 0'.5 for Venus, 1' for Earth, and 2' for Mars. Periodic perturbations of the outer planets are larger, reaching in the case of the long-period terms to 30' for Jupiter, 70' for Saturn, 60' for Uranus, and 35' for Neptune.

Because the amplitude of a periodic perturbation depends on the mass of the disturbing planet, observational measurement of this amplitude affords a method of determining the disturbing mass. For the planets Mercury, Venus, and Pluto, which do not have satellites, this is the only method of determining the mass. As a consequence of the mutual perturbations of the planets, the distance of a planet from the Sun is, on the average, decreased by the action of planets closer to the Sun, and increased by planets farther from the Sun; this mean effect represents a perturbation of the radius vector with a constant value.

The orbits of the minor planets are affected in varying degree by the attractions of the major planets. Those orbits passing close to Jupiter suffer large perturbations which, if the mean motions were commensurable with that of Jupiter, would be augmented at each close approach until the trajectories were sufficiently altered to reduce the commensurability. In the overall distribution of mean motions of the minor planets there are noticeable gaps near the points where the period would be an exact submultiple (½, ⅓, ⅖, . . .) of the period of Jupiter. In cases of near commensurability, observational determination of the amplitude of the long-period perturbation affords a method for measuring Jupiter's mass. A small group of minor planets, called the Trojan asteroids, has been so completely captured by Jupiter that they oscillate about the 60° points which form equilateral triangles with Jupiter and the Sun. *See* TROJAN ASTEROIDS.

Effect on comets. Planetary perturbations also affect the orbits of comets. Studies of the motion of Halley's comet indicate that the time from one perihelion passage to the next has varied by almost 5 years because of perturbations. Most comets approach the Sun at nearly parabolic speeds in randomly oriented orbits, but if a comet approaches close to one of the more massive major planets, the planet may so alter the trajectory that the comet pursues an elliptical orbit thereafter. A number of short-period comets whose orbits agree only in that they all pass close to Jupiter illustrate the perturbing effect of this planet on cometary orbits.

Nongravitational causes. Material forming the tails of comets is subject to a nongravitational type of perturbation. This rarefied matter which is given off by the head of the comet is forced into a trajectory away from the Sun by the pressure of solar radiation.

Associated with many of the periodic comets are swarms of smaller particles which appear as meteors upon collision with the upper atmosphere of the Earth. The density of these swarms is so tenuous that they cannot hold themselves together by their own gravitation, and planetary perturbations of speed and direction soon spread the components completely around the orbit. The annual meteor showers, such as the Perseids, reflect this dispersal of particles along the orbit. The effect of the Earth's attraction on a meteor trajectory depends on the relative velocity, that is, whether the Earth is overtaking the meteor or meeting it head on. Once the meteor enters the upper reaches of the Earth's atmosphere its motion is subject to a nongravitational perturbation caused by atmospheric drag. This resistance to the passage of the particle is evidenced by the trail of incandescent gas and vapor which forms until the particle is consumed or continues in its trajectory greatly decelerated. *See* METEOR.

Perturbations of satellite orbits. The motions of planetary satellites, natural and artificial, reflect both gravitational and nongravitational perturbations. The centrifugal force arising from the rotation of a planet causes a deformation or oblateness of figure. In such a case the central mass does not attract as if it were concentrated at its center. For a close satellite the principal perturbation arises from the attraction of this equatorial bulge. The effect of this attraction on an otherwise undisturbed satellite orbit is a gradual regression of the line of nodes on the equatorial plane and a rotation of the line of apsides. Both rotations vary with the inclination of the satellite orbit. Nearer to the primary, the tidal forces may become so great that a satellite would be literally torn to pieces. For a fluid satellite of the same density as the planet, the limit within which this disruptive perturbation occurs is about 2½ times the radius of the planet. *See* SATURN.

Satellite motions are also disturbed by the direct attraction of other satellites, the Sun, and, to a lesser amount, by other planets. Observation of the orbital displacements caused by the mutual perturbations of satellites in the systems of Jupiter and Saturn makes possible the determination of the masses of these satellites. The solar attraction is significant in the orbits of the outer satellites of Jupiter and Saturn, reaching to one-ninth the planet's attraction for the eighth satellite of Jupiter. So greatly disturbed is this satellite that it is not possible to derive a general theory for its motion.

The orbit of the Moon is disturbed mainly by the Sun, with some changes in motion due to the oblateness of the Earth, the figure of the moon, and smaller perturbations caused by the planets. The attraction of the Sun on the Moon is more than twice the Earth's attraction, but because both the Earth and Moon are free to move it is only their relative acceleration with respect to the Sun which determines the motion. This relative acceleration toward the Sun is always less than 1/80 of the acceleration of the Moon toward the Earth. The eccentricity and inclination of the Moon's orbit oscillate slowly about their mean values, while the line of apsides advances with an average period of almost 9 years and the nodes regress through one revolution in 18.6 years.

The observed motion of the lunar node and perigee affords one means of measuring the oblateness of the Earth. The present lunar theory incorporates the value 1/294. The *International Astronomical Union System of Astronomical Constants* (1976) contains a reference ellipsoid of revolution for the Earth having a flattening of 1/298.257. This value has been derived mainly from measures of the motions of the nodes and apsides of artificial Earth satellites. Lunar and solar perturbations of artificial Earth satellite orbits are minor for orbits 500 mi (800 km) above the surface but grow with increasing distance from the Earth. Atmospheric drag perturbations are significant at this altitude, but decrease with increasing altitude. SEE CELESTIAL MECHANICS.

Raynor L. Duncombe

Bibliography. G. Brouwer and G. M. Clemence, *Methods of Celestial Mechanics*, 1961; J. M. A. Danby, *Fundamentals of Celestial Mechanics*, 2d ed., 1988; A. E. Roy, *Orbital Motion*, 3d ed., 1988; V. Szebehely, *Adventures in Celestial Mechanics: A First Course in the Theory of Orbits*, 1989.

Phase

The changing fraction of the disk of an astronomical object that is illuminated, as seen from some particular location. The monthly phases of the Moon are a familiar example (see **illus.**). When the Sun is approximately on the far side of the Moon as seen from Earth (conjunction), the dark side of the Moon faces the Earth and there is a new moon. The phase waxes, beginning with crescent phases, as an increasing fraction of the illuminated face of the Moon is seen. At quadrature, when half the visible face of the Moon is illuminated, the phase is called the first-quarter moon, since the Moon is now one-quarter of the way through its cycle of phases. The waxing moon continues through its gibbous phases until it is in opposition; the entire visible face of the Moon is illuminated, the full moon. During the full moon, the Moon and the Sun are on opposite sides of the Earth, a configuration known as a syzygy. Then the Moon wanes, going through waning gibbous, third-quarter, and waning crescent phases until it is new again. The cycle of moon phases takes approximately 29.53 days and explains the origin of the word month.

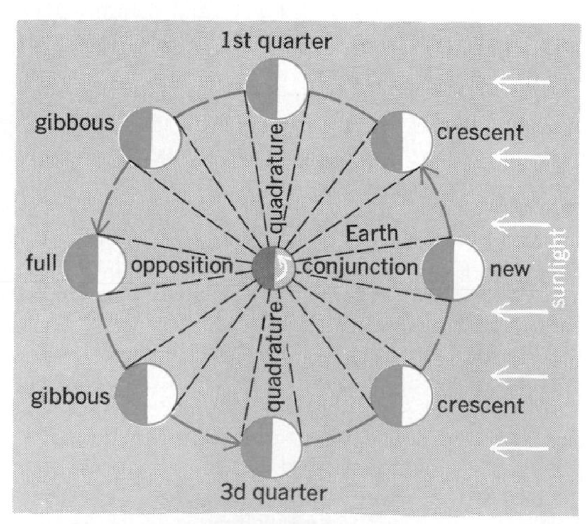

From Earth, different fractions of the illuminated half of the Moon are seen at different times as the Moon goes through a 29.53-day cycle of phases.

The Earth, Moon, and Sun are not directly in line at the times of new moon and full moon, because the Moon's orbit around the Earth is inclined 5° to the plane of the Earth's orbit around the Sun. When they are directly in line, a lunar eclipse occurs when the Earth's shadow falls on the Moon, and a solar eclipse when the Moon's shadow falls on the Earth. SEE ECLIPSE; MOON.

Galileo discovered the phases of the planet Venus when he observed the sky with his telescope in 1610. Giovanni Zupus discovered the phases of the planet Mercury in 1639. Because of the angle at which the outer planets are seen from Earth, and because of their great distance, they do not appear to go through phases as seen from Earth. Mars, Jupiter, Saturn, Uranus, and Neptune have now all been seen by spacecraft as crescents.

Jay M. Pasachoff

Photosphere

The apparent, visible surface of the Sun. The photosphere is a gaseous atmospheric layer a few hundred miles deep with a diameter of 864,000 mi (1,391,000 km; usually considered the diameter of the Sun) and an average temperature of approximately 5800 K (10,500°F). Radiation emitted from the photosphere accounts for most of the solar energy flux at the Earth. The solar photosphere provides a natural laboratory for the study of dynamical and magnetic processes in a hot, highly ionized gas.

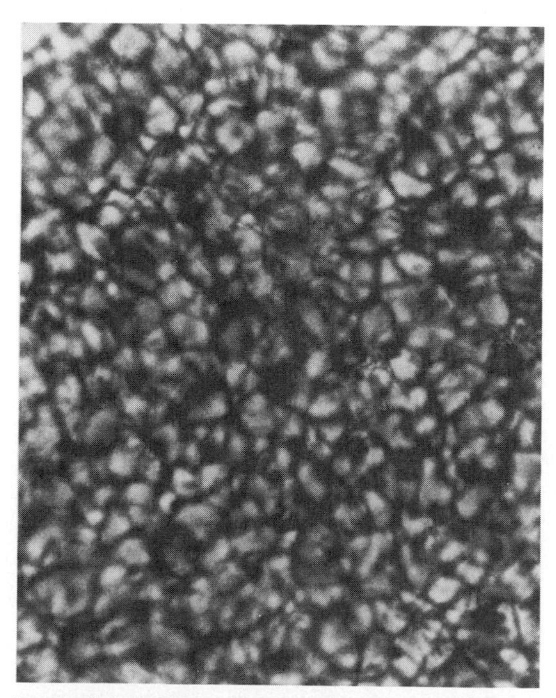

Image of the solar photosphere made in white light near 500 nm, using the 75-cm (30-in.) Vacuum Tower Telescope of the National Solar Observatory at Sacramento Peak, Sunspot, New Mexico. (*National Solar Observatory*.)

When studied at high resolution, the photosphere displays a rich structure. Just below it, convective motions in the Sun's gas transport most of the solar energy flux. Convective cells penetrate into the stable photosphere, giving it a granular appearance (see **illus.**) with bright cells (hot rising gas) surrounded by dark intergranular lanes (cool descending gas). A typical granule is approximately 600 mi (1000 km) in diameter. Measurements of horizontal velocity reveal a larger convective pattern, the supergranulation, with a scale of approximately 20,000 mi (30,000 km); the horizontal motion of individual granules reveals intermediate-scale (3000-mi or 5000-km) convective flows.

Magnetic fields also play an important role in shaping photospheric structure. Magnetic flux tubes, smaller than the intergranular lanes, penetrate the photosphere vertically. When many of these tubes are forced together, probably by convective motions deep within the Sun, they suppress the convective motions near the surface, cool the atmosphere, and form dark magnetic pores and sunspots.

Gravitational, acoustic, and magnetic waves are propagated through, and at some frequencies trapped, in the photosphere. The trapped waves, most of which have periods of about 5 min, represent global oscillations of the Sun and provide information about its interior when analyzed by seismological techniques. SEE HELIOSEISMOLOGY; SUN.

Stephen L. Keil

Bibliography. R. J. Bray, R. E. Loughhead, and C. J. Durrant, *The Solar Granulation*, 2d ed., 1984; R. Giovanelli, *Secrets of the Sun*, 1984; R. W. Noyes, *The Sun, Our Star*, 1982; H. Zirin, *Astrophysics of the Sun*, 1988.

Pisces

The Fishes, in astronomy, a zodiacal constellation appearing in the autumn evening sky. Pisces is the twelfth and last sign of the zodiac. It is inconspicuous, having no star brighter than the fourth magnitude. But it is an important constellation because the vernal equinox, which marks the beginning of the astronomical year, is now located in it. Its most distinctive feature is a V-shaped figure, with the fishes' tails toward the point of the V tied together by a ribbon (see **illus.**). The northern fish is poorly defined, but the western one is marked with a group of stars forming an irregular pentagon, known as the Circlet in Pisces. SEE CONSTELLATION; ZODIAC.

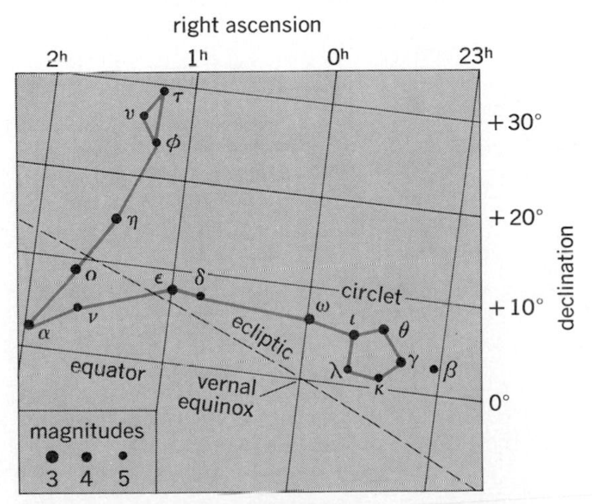

Line pattern in the constellation Pisces. The grid lines represent the coordinates of the sky. The apparent brightness, or magnitude, of the various stars is shown by the sizes of the dots, which are graded by appropriate numbers as indicated.

Ching-Sung Yu

Planet

A relatively small, solid celestial body moving in orbit around a star, in particular the Sun. Besides Earth, the eight known planets of the solar system are Mercury, Venus, Mars, Jupiter, Saturn, Uranus, Neptune, and Pluto; in addition, over 4000 minor planets, or asteroids, mostly located between the orbits of Mars and Jupiter, are known (**Fig. 1**).

Classification. There are two basic groups of planets: the small, dense, terrestrial planets—Mercury, Venus, Earth, Mars, and Pluto—and the giant or Jovian planets—Jupiter, Saturn, Uranus, and Neptune. With the exception of Pluto, the terrestrial planets are all located within the inner solar system. The low-density Jovian planets extend outward from Jupiter to the remote outer reaches of the solar system. This distribution is not accidental, but is related to the fractionation of rocky, icy, and gaseous materials during the early stages of formation of the solar system.

Each of the main planets from the Earth to Pluto is accompanied by one or more secondary bodies called satellites. Many of the smallest satellites are not observable from Earth, but were discovered during spacecraft flybys. SEE SATELLITE.

The planets may also be divided into inferior planets, Mercury and Venus, located inside the Earth's orbit, and superior planets, from Mars to Pluto, circulating outside the Earth's orbit.

Kepler's laws. The motions of the planets in their orbits around the Sun are governed by three laws discovered by J. Kepler at the beginning of the seventeenth century.

First law: The orbits of the planets are ellipses of which the Sun occupies a focus.

Second law (law of areas): Equal areas of the ellipse are described by the radius vector from the Sun to the planet in equal intervals of time.

Third law (harmonic law): The squares of the periods of revolution P are proportional to the cubes of the semimajor axes of the orbits a; that is, for all planets the ratio P^2/a^3 is equal to a constant. The constant is equal to unity if a is given in astronomical units and P in sidereal years. One astronomical unit (AU) is the mean distance from Earth to the Sun and is approximately equal to 1.496×10^8 km (9.30×10^7 mi). SEE ASTRONOMICAL UNIT.

Otherwise, the constant of the harmonic law is given by Newton's law of gravitation as $G(M + m)/4\pi^2$, where M and m are the masses of the Sun and the planet, and G is the constant of gravitation. SEE GRAVITATION.

Kepler's laws are true only when the mutual perturbations of the motions of the planets by the others are neglected.

Planetary configurations. In the course of their motions around the Sun, the Earth and other planets occupy a variety of relative positions or configurations (**Fig. 2*a***), the principal of which are designated as follows: The inferior planets are in conjunction

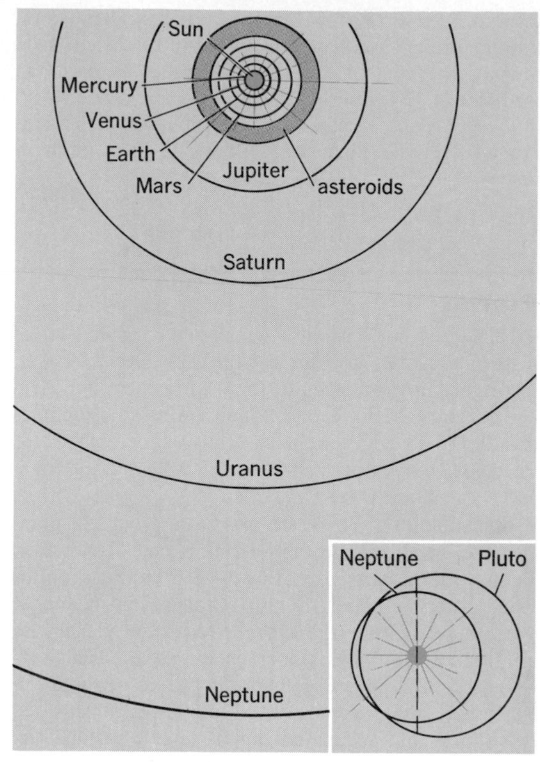

Fig. 1. Plan of the solar system. (*After L. Rudaux and G. de Vaucouleurs, Astronomie, Larousse, 1948*)

with the Sun when closest to the Earth-Sun line, either between the Earth and the Sun (inferior conjunction) or beyond the Sun (superior conjunction). On rare occasions when the planet is very close to the plane of the Earth's orbit at the time of an inferior conjunction, a transit in front of the Sun is observed. *SEE TRANSIT.*

Between conjunctions, the geocentric angular distance from the planet to the Sun, or the elongation, varies up to a maximum value; the greatest or maximum elongations of Mercury and Venus are 28° and 47°, respectively. The superior planets are not so limited, and their elongations can reach up to 180° when they are in opposition with the Sun; when the elongation is ±90°, they are in quadrature (eastern or western) with the Sun.

The telescopic aspect of the disks of the planets varies according to their configurations, which determine the angle between the directions of illumination and observation, or the phase angle. Between inferior conjunction and greatest elongations, the interior planets show crescent phases, like the Moon between new moon and first or last quarters (Fig. 2*b*); between greatest elongations and superior conjunction, they show a gibbous phase, like the Moon between quarters and full moon. At superior conjunction, they show a circular disk, fully illuminated and seen face on, while during transits, the dark side is profiled against the Sun. The superior planets show their full phase at both conjunction and opposition and a gibbous phase near quadrature, at which time the unilluminated portion of the disk is at a maximum.

Apparent motions. The combinations of the orbital motions of Earth and of any other planet give rise to complicated apparent motions of the planets as seen from the Earth. Because the orbits of the main planets are, except for Pluto, only slightly inclined to the plane of the orbit of Earth, the apparent paths of the planets (except Pluto) are restricted to the zodiac, a belt 16° wide centered on the ecliptic. The ecliptic is the path in the sky traced out by the Sun in its apparent annual journey as the Earth revolves around it. Along this path, the apparent motions of the inferior planets with respect to the Sun are alternatively westward, from greatest elongation through inferior conjunction to greatest elongation, then eastward, from greatest elongation through superior conjunction to greatest elongation (Fig. 2). The mean motion of the superior planets is always westward. *SEE ASTRONOMICAL COORDINATE SYSTEMS.*

The apparent motions with respect to the celestial sphere, that is, to the fixed stars, appear for the inferior planets as oscillations back and forth about the position of the Sun steadily moving eastward among the stars. For the superior planets, the apparent motion is generally eastward or direct, but for short periods near the time of opposition it is westward or retrograde (**Fig. 3**). At times when the direction of the apparent motion on the sphere reverses, the planet appears to be stationary.

The mean interval of time between successive returns to the same place with respect to the stars is the sidereal period, which is established by the true motion of revolution of the planet in its orbit around the

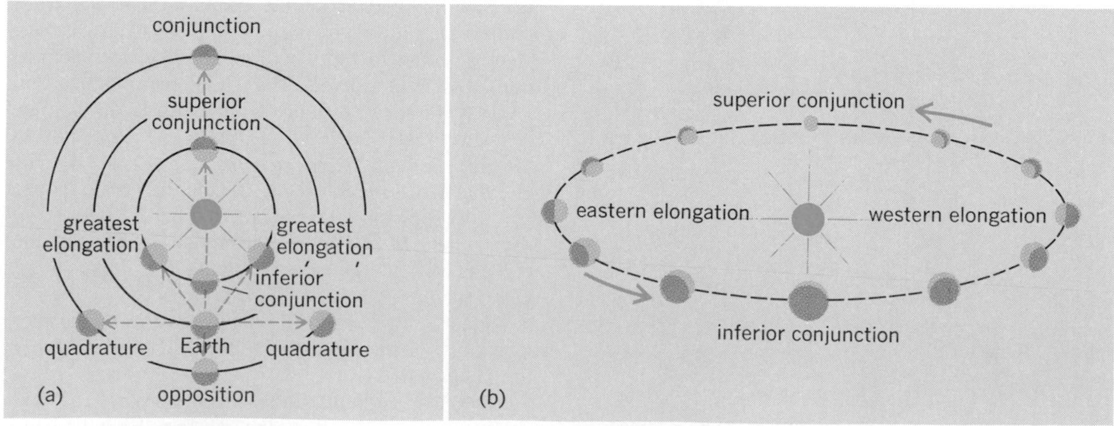

Fig. 2. Planetary configurations and phases. (*a*) Positions of Earth and other planets relative to Sun. (*b*) Phases of inferior planets (Mercury and Venus). (*After L. Rudaux and G. de Vaucouleurs, Larousse Encyclopedia of Astronomy, Prometheus Press, 1959*)

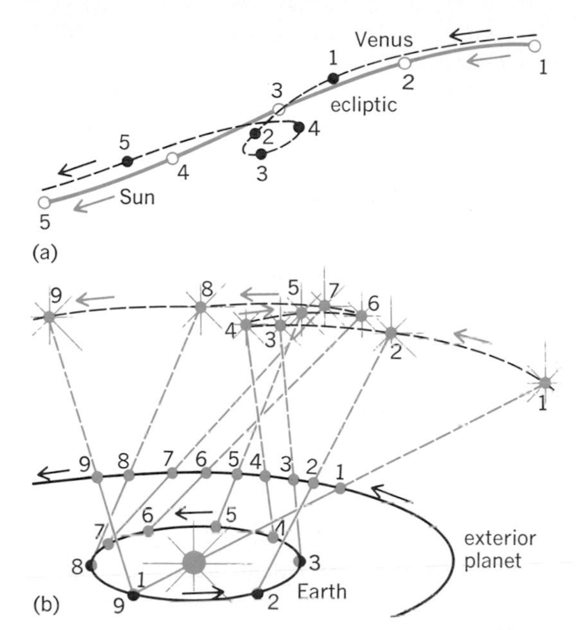

Fig. 3. Apparent motions as observed from Earth (*a*) of an interior planet with respect to the Sun and (*b*) of an exterior planet with respect to the fixed stars. (*After L. Rudaux and G. de Vaucouleurs, Larousse Encyclopedia of Astronomy, Prometheus Press, 1959*)

Sun. The mean interval of time between successive returns of the same configuration with respect to the Sun (for example, conjunctions or oppositions) is the synodic period, which governs the apparent motion of the planet as seen from Earth (**Table 1**).

Elliptic motion. The motion of a planet having an elliptical orbit of semimajor axis a, with the Sun at the focus S, brings it each revolution to the perihelion A and to the aphelion A', the points of the orbit respectively nearest to and farthest from S. If C is the center of the ellipse, the semimajor axis is $a = CA = CA'$; the eccentricity of the ellipse is $e = CS/CA = CS/a$, whence $SA = a(1 - e)$, $SA' = a(1 + e)$. The distance $SP = r$ of the planet to the Sun at any other point is $r = a(1 - e^2)/(1 + e \cos \theta)$, where the angle $\theta = \angle ASP$ is the true anomaly. If P' is the point on the principal circle of radius $CA = a$ whose projection in the ellipse is P (**Fig. 4**), the eccentric anomaly is the angle $\theta' = \angle ACP'$, so that $r = a(1 - e \cos \theta')$. If the planet is at perihelion at time T and returns to it at time $T + P$, the mean angular velocity (or mean motion) is $n = 2\pi/P$, and the mean anomaly at any time t is $M = n(t - T)$.

The relation between the mean and eccentric anomalies, $\theta' - e \sin \theta' = M$, is known as Kepler's equation; its solution gives θ' and, consequently, r at any time t when the orbital elements a, e, n, T are known. SEE KEPLER'S EQUATION.

Orbital elements. The position of a planet in its orbit and the orientation of the orbit in space are completely defined by seven orbital elements (**Fig. 5**): (1) the semimajor axis a, (2) the eccentricity e, (3) the inclination i of the plane of the orbit to the plane of the ecliptic, (4) the longitude ☊ of the ascending node N, (5) the angle ω from the ascending node N to the perihelion A, (6) the sidereal period of revolution P, or the mean (daily) motion $n = 2\pi/P$, and (7) the date of perihelion passage T, or epoch E.

If the plane of a planet's orbit is inclined to the plane of the ecliptic, their intersection NN' is the line of nodes; in its motion, the planet crosses the plane of the ecliptic from south to north at the ascending node N and from north to south at the descending node N'. The longitude of the ascending node is the angle ☊ = ∠♈ SN, measured in the plane of the ecliptic from the vernal equinox ♈. The longitude of perihelion is $\tilde{\omega}$ = ☊ + ω = ∠♈ SN + ∠NSA, the second angle being measured in the plane of the planet's orbit (Fig. 5). The location of the plane of the orbit in space is defined by i and ☊, the orientation of the ellipse in this plane by ω, its form by e, and its size by a, and the position of the planet on the ellipse by P and T (and by the time t). SEE ORBITAL MOTION.

Determination of orbital elements. Accurate observations of the positions of the planets with respect to the stars (for example, as measured on photographs) or with respect to the celestial coordinates (for example, by means of the meridian circle instrument) are used to determine the elements of their orbits. In principle, three observations of two coordinates (right ascension and declination) and the laws of elliptic motion are sufficient to determine the six independent elements of a planetary orbit, since by Kepler's third law $a^3 \propto P^2$. In practice, as many observations as possible are combined, and the equations solved by the method of least squares; the elements for a given epoch so obtained are subject to variations and corrections allowing for planetary perturbations. Tables of the motions of the planets for several centuries past and future have been established, from which the yearly ephemerides are extracted in a form convenient

Table 1. Elements of planetary orbits

Planet	Symbol	Mean distance (semimajor axis)			Sidereal period of revolution		Synodic period, days	Mean velocity		Eccentricity	Inclination
		AU	10^6 mi	10^6 km	Years	Days		mi/s	km/s		
Mercury	☿	0.387	36.0	57.9	0.241	87.97	115.88	29.76	47.89	0.206	7°00′
Venus	♀	0.723	67.2	108.2	0.615	224.70	583.92	21.77	35.03	0.007	3°24′
Earth	⊕	1.000	93.0	149.6	1.000	365.26		18.51	29.79	0.017	0°00′
Mars	♂	1.524	141.7	227.9	1.881	686.98	779.94	14.99	24.13	0.093	1°51′
Jupiter	♃	5.203	483.6	778.3	11.862	4332.71	398.88	8.12	13.06	0.048	1°18′
Saturn	♄	9.539	886.7	1427.	29.458	10759.	378.09	5.99	9.64	0.056	2°30′
Uranus	⛢	19.19	1784.	2871.	84.014	30685.	369.66	4.23	6.81	0.046	0°46′
Neptune	♆	30.06	2794.	4497.	164.79	60189.	367.49	3.37	5.43	0.010	1°46′
Pluto	♇	39.53	3674.	5913.	248.5	90800.	366.73	2.95	4.74	0.248	17°09′

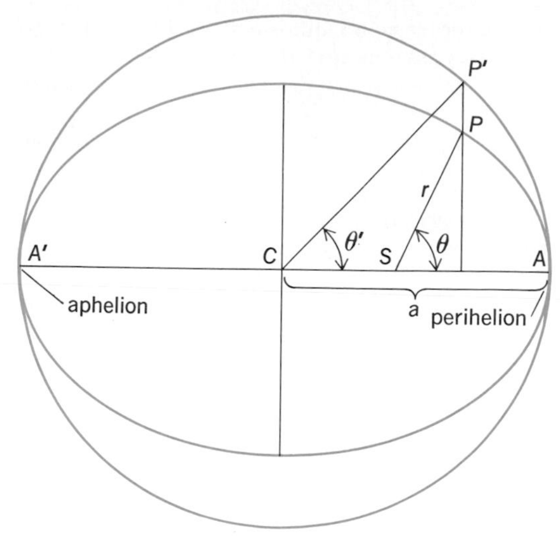

Fig. 4. Elliptic motion of a planet. Symbols explained in text.

for immediate use. *See* Ephemeris; Least-squares method.

The elements of the planetary orbits are given in Table 1.

Planetary sizes. The apparent diameter of a planet may be determined visually by means of a filar micrometer or, preferably, a birefringent or double-image micrometer attached to a telescope, or it may be measured on large-scale photographs taken through telescopes or by planetary spacecraft. If the apparent diameter of a planet is d'' when its distance to the Earth is Δ, the linear diameter is $D = \Delta \sin d'' = \Delta d''/206{,}265$, where d'' is measured in seconds of arc. The linear diameter is expressed in the same units as Δ, which is given by the ephemerides in astronomical units; conversion to kilometers or miles is given by the adopted value of the astronomical unit: 1 AU = 149.6×10^6 km = 93.0×10^6 mi.

When polar flattening is perceptible, both the polar and equatorial radii r_p, r_e can be determined or, as in **Table 2**, the mean radius $r = (r_p + r_e)/2$ and the ellipticity $\epsilon = 1 - (r_p/r_e)$. The mean radius may also be expressed in terms of the mean radius of the Earth

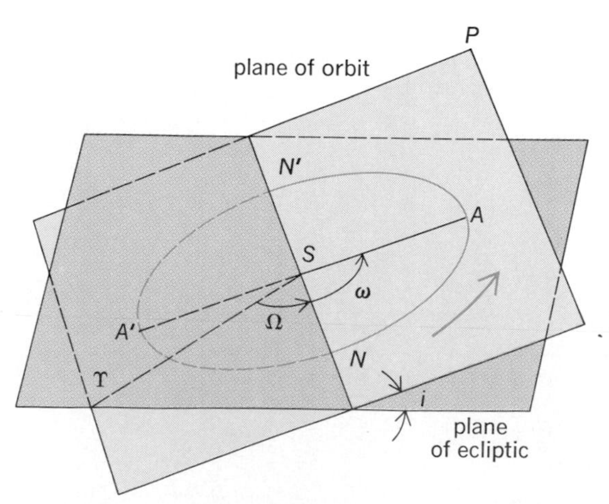

Fig. 5. Orbital elements, which determine the position of a planet in its orbit and the orientation of the orbit in space. Symbols explained in text.

(6378 km) as a unit. The relative area is then very nearly equal to r^2 and the relative volume to r^3. For the nearer planets with a solid surface (Mercury, Venus, and Mars), the linear diameter can be determined more precisely by radar, using the time-delay technique.

Masses, gravity, and density. The mass of a planet is found easily if it has one or more satellites. If a is the mean distance (semimajor axis) of the satellite's orbit and P its period of revolution expressed respectively in astronomical units and sidereal years, the mass m of a planet is given through Newton's law of gravitation by $m = A^3/P^2$, in terms of the mass of the Sun as a unit. This assumes that the mass of the satellite relative to that of the planet, and of the Earth relative to that of the Sun, may be neglected, which is nearly always the case within the accuracy of the data. Since the ratio m_e/M_s of the masses of the Earth and the Sun can be determined similarly, planetary masses are also known in terms of m_e.

If the planet has no satellite (Mercury and Venus), its mass can be derived only from the perturbations that it causes in the motions of the other planets, occasional comets, or passing spacecraft. Since the perturbations are small, the masses so obtained are generally of low accuracy. The most precise values are those derived from the perturbations of the orbits of space probes. *See* Perturbation.

Once the mass m and the radius r of a planet are known in terms of the Earth's mass and radius, its surface gravity and mean density relative to the Earth are given by $g = m/r^2$ and $\rho = m/r^3$, respectively. Multiplication by 981 and 5.552 gives the corresponding values in cgs units.

From r and m follows also the escape velocity V_1 that permits a projectile (or a molecule) to leave the planet on a parabolic orbit: $V_1 = (2Gm/r)^{1/2}$; this is $\sqrt{2}$ times the velocity of an hypothetical satellite moving in a circular orbit close to the surface of the planet. These elements are listed in Table 2.

Rotation periods. The period of rotation of a planet can be determined by several methods: (1) Direct telescopic observation of the permanent markings on its surface (Mars, Mercury) or of the semipermanent cloud formations in its atmosphere (Jupiter, Saturn) is the best method. (2) The spectroscopic determination of the velocity difference between the opposite equatorial limbs can give, in combination with the linear diameter, an approximate value of the rotation period (Uranus, Neptune). (3) Since the giant planets lack solid surfaces, their rotation rates are found by timing the cyclic pattern of radio energy emanating from the planets' magnetospheres, which rotate in synchrony with their deep interiors. (4) When the apparent diameter of the disk is too small for any of these methods (Pluto, asteroids), a determination of the periodicity of the light variations, if any, due to the changing presentation of bright and dark regions of the surface may give a fairly accurate value of the rotation period. The rotation periods of all the main planets are now well determined, as indicated in Table 2.

Planetary radiations. The electromagnetic radiation received from a planet is made up of three main components: the visible reflected sunlight, including some ultraviolet and near-infrared radiation; the thermal radiation due to the planet's heat, including both infrared radiation and ultrashort radio waves; and the

Table 2. Physical elements of planets

Planet	Equatorial radius, r_e (Earth = 1)	mi	km	Ellipticity	Volume (Earth = 1)	Mass (Earth = 1)	Density, g/cm³	Escape velocity mi/s	km/s	Rotation period	Inclination of axis*
Mercury	0.38	1515	2439	0.000	0.055	0.053	5.6	2.6	4.2	58.65d	2
Venus	0.95	3760	6051	0.000	0.87	0.815	5.2	6.4	10.3	243d	177.3
Earth	1.00	3963	6378	0.0034	1.00	1.000	5.52	7.0	11.2	23h56m22.7s	23.45
Mars	0.53	2110	3395	0.006	0.150	0.107	3.95	3.1	5.0	24h37m22.6s	23.98
Jupiter	11.21	44423	71492	0.064	1408.	317.93	1.33	37.0	59.5	9h50m30s†	3.1
Saturn	9.45	37449	60268	0.098	844.	95.18	0.69	22.1	35.6	10h14m‡	26.73
Uranus	4.01	15882	25559	0.023	64.	14.53	1.20	13.2	21.2	17h54m§	97.9
Neptune	3.89	15410	24800	0.020	59.	17.07	1.64	14.7	23.3	17h48m¶	29.6
Pluto	0.18	715	1150	?	0.006	0.002	2.0	0.7	1.1	6.39d	122.5

*To perpendicular to orbit, in arc-degrees.
†Latitude < 12° (system I); 9h55m40.6s, latitude > 12° (system II).
‡Near Equator; 10h38m at intermediate latitudes.
§16h00m near the pole.
¶At midlatitudes.

nonthermal radio emission due to electrical phenomena, if any, in the planet's atmosphere or in its radiation belts.

Planetary brightness. The apparent brightness of a planet can be measured by visual, photographic, and photoelectric photometry and is usually expressed in the stellar magnitude scale; it varies in inverse proportion to the squares of the distances r from the Sun and Δ from the Earth. The fraction of the incident light reflected at full phase compared with the fraction that would be reflected under the same conditions by an equivalent perfect diffuse reflecting disk is called the geometrical albedo. It is a measure of the backscattering reflectivity of the planet's visible surface. The visual albedos of the planets vary between 5 and 70%. SEE ALBEDO.

Thermal radiation. The thermal radiation from a planet can be measured either with a radiometer at wavelengths of 8–14 micrometers, 17–25 μm, and 30–40 μm (which are partially transmitted by the Earth's atmosphere) or with a radio telescope at wavelengths between 1 mm and 30 cm. In either case, the amount of energy corresponds to that which would be received under the same conditions from a perfect radiator of the same size at a certain temperature T, called the blackbody temperature of the planet. Its relation to the actual temperature depends on the thermal and radiative properties of the atmosphere and surface of the planet. Jupiter has been found to radiate twice as much energy as it receives from the Sun. From this it is inferred that the planet has an internal source, perhaps primordial heat continuing to escape long after Jupiter's formation 4.6×10^9 years ago. Saturn has also been observed to emit more energy than it receives from the Sun, as verified both from Earth and by the *Pioneer 11* spacecraft in 1979. The output is some two to four times higher than incoming sunlight. Unlike Jupiter, however, Saturn may be too old to still possess a primordial heat source. While gravitational separation of the planet's hydrogen and helium has been suggested as one possible mechanism for this excess energy, no single solution appears to account for the entire energy outflow. SEE JUPITER; SATURN.

Nonthermal radiation. Nonthermal radio emission at decimeter and dekameter wavelengths has been received from Jupiter by means of large radio telescopes. The dekameter emission takes the form of irregular bursts of noise originating in the planets atmosphere from subjacent sources. Voyager spacecraft revealed that powerful electric currents exist inside the Jovian magnetosphere, particularly one called a flux tube linking higher latitudes on the planet with satellite Io. Since the observed dekametric radiation is modulated by the orbital position of Io, this current loop may be responsible for the outbursts, as well as for auroral discharges observed by the spacecraft in the planet's polar regions.

Planetary atmospheres. The principal constituents of the atmospheres of the terrestrial planets (Pluto, a special case, is discussed below) are carbon dioxide, argon, nitrogen, water, and (on Earth only) oxygen; Mercury has a very tenuous envelope dominated by atoms of sodium and potassium. The atmospheres of the giant planets are composed primarily of hydrogen and helium, with lesser amounts of methane, ammonia, and water. Atmospheric motions are driven by temperature gradients—in general, those existing between the warm equatorial regions and the cooler polar areas. An atmosphere thus tends to redistribute heat over the planetary surface, lessening the temperature extremes found on airless bodies.

On planets having relatively dense atmospheres, heat from the Sun is trapped by the greenhouse effect, wherein visible radiation from the Sun passes readily through the atmosphere to heat the planetary surface, but infrared radiation reemitted from the surface is constrained from escaping back to space by the lower transparency of the atmosphere to longer wavelengths. Although Venus absorbs approximately the same amount of energy from the Sun as does Earth, the greenhouse effect is responsible for heating the surface of Venus to a much higher temperature, approximately 750 K (900° F). SEE VENUS.

Directly through wind erosion or indirectly by carrying water vapor which can precipitate as rain, the atmospheres of the terrestrial planets are a major factor in modifying surface structure and rearranging the distribution of surface materials. Mercury, being virtually airless, exhibits a relatively unmodified surface, very similar in appearance to that of the Moon. Pluto, however, is covered with a methane ice frost and thus maintains a tenuous atmosphere of methane vapor when in the part of its orbit nearest the Sun. *See* MERCURY; PLUTO.

Possible unknown planets. During the nineteenth century, an unexplained irregularity in the motion of Mercury was thought by some investigators to be caused by an unknown planet circulating between the Sun and Mercury, called Vulcan, which was looked for in vain. This irregularity was satisfactorily explained in 1915 by Einstein's general theory of relativity. It is now certain that no intra-Mercurial planet of size comparable to the terrestrial planets exists. The possibility of one or more planets circulating beyond the orbits of Neptune and Pluto has also been discussed, but there is no compelling evidence for the existence of such planets.

Planets outside solar system. Minute perturbations in the motion of some nearby stars have indicated the existence of minor components of small mass. The masses of these satellite bodies, although larger than planetary masses in the solar system, are considered to be too small to be self-luminous; they are consequently more like planets than dwarf stars. From this evidence it is inferred that planetary systems are not as uncommon in the universe as was previously believed, and studies of multiple-star systems suggest that one star in every three may be attended by planetary companions. Whether some of these planets are inhabited by advanced beings or sustain lower life forms is unknown, but it would perhaps be more surprising to learn that life on Earth is unique. Presumably, whenever temperature, composition, and other conditions are favorable, the chemical evolution of complex organic molecules will begin, and the crucial transition from chemical to biochemical evolution may be just a matter of time. *See* ASTEROID; CELESTIAL MECHANICS; CERES; EARTH ROTATION AND ORBITAL MOTION; EXTRATERRESTRIAL INTELLIGENCE; MARS; NEPTUNE; OPTICAL TELESCOPE; RADIO ASTRONOMY; SOLAR SYSTEM; TROJAN ASTEROIDS; URANUS.

J. Kelly Beatty

Bibliography. J. K. Beatty and A. Chaikin (eds.), *The New Solar System*, 3d ed., 1990; C. R. Chapman, *Planets of Rock and Ice*, 1982; R. M. Goody and J. C. G. Walker, *Atmospheres*, 1972; W. K. Hartmann, *Moons and Planets*, 2d ed., 1983; D. Morrison and T. Owen, *The Planetary System*, 1988; J. A. Wood, *The Solar System*, 1979.

Planetarium

An instrument that projects the stars, Sun, Moon, planets, and other celestial objects upon a large hemispherical dome, showing their motions as viewed from the Earth or space near the Earth. Days and years may be compressed into minutes. There are over 100 major planetariums around the world with domes 50 ft (15 m) or more in diameter; and there are also over 1000 smaller planetariums in communities, schools, and colleges.

The term planetarium originally applied to a mechanical model (also known as an orrery) that depicted the motions of the planets. Today the term refers to an optical projector. Most planetariums now have mechanical movements, but planetariums projecting computer-generated displays have also been developed. Additional optical devices and computer controls are common. The term planetarium also refers to the theater or building that houses the projector.

Development. Models of the sky date from early times. A celestial globe was made by Claudius Ptolemy of Alexandria around A.D. 150. The Gottorp Globe, a 10-ft (3-m) globe built in the mid-sixteenth century, is hollow, and up to 10 people may sit inside to see the sky which is painted on the interior. This globe is on display at the Lomonosov Museum, St. Petersburg, Russia. A larger globe, constructed in 1912, and which is on display at the Chicago Academy of Sciences, is 15 ft (4.5 m) in diameter and seats 17 spectators. The metal sphere has 692 holes to depict the stars.

Motions of the Earth, Moon, and planets have also been represented from early times with various mechanical models. About 1682, C. Huygens designed an elaborate model that showed the planets out to Saturn. Around 1712, J. Rowley built the original "orrery," in which, by turning a crank, the Moon's motion around the Earth could be seen as well as the Earth's motion around the Sun, thus explaining the lunar phases and the Earth's seasonal relations. *See* EARTH ROTATION AND ORBITAL MOTION; MOON; PLANET.

The invention of the projection planetarium by W. Bauersfeld in 1919 solved the problem of presenting stars and planetary motions in a realistic fashion in a domed theater. The first such planetarium, built by the Carl Zeiss Company of Germany, opened in Munich in 1923.

Types of projectors. Many projectors are patterned after the basic design of the Carl Zeiss Company. Star spheres at each end of the projector show 8900 stars down to magnitude 6.5; 32 (16 located in each globe) lenses are used to project the stars. Cages between the two star spheres contain projectors for the Sun, Moon, and planets. The center part of the machine houses the driving motors. Additional projectors show such effects as variable stars, solar and lunar eclipses, the Milky Way, comets, and various circles and coordinates. Depending on the manufacturer, there may be constellation outlines, clouds, and built-in zoom effects for the planets.

Of the 25 large-model projectors built by Zeiss prior to 1940, five are still in service and have been joined by many new models. These include the ZKP-2, the Spacemaster, and the GP-85 (**Fig. 1***a–c*), for dome diameters of 20–33 ft (6–10 m), 33–57 ft (10–17.5 m), and 59–75 ft (18–23 m), respectively. The Mark VI projector (Fig. 1*d*) is designed for domes 60–80 ft (18–25 m) in diameter. It has a device that makes the stars twinkle, and can be mounted on a hydraulic hoist, permitting it to descend below the floor for parts of the program. Two smaller projectors based upon the same design principles are the M 1015 (Fig. 1*e*), for domes 33–50 ft (10–15 m), and the M 1518 (Fig. 1*f*), for domes 49–59 ft (15–18 m). Still other projectors of the same basic design cover dome diameters of 9–66 ft (3–20 m) and 16–66 ft (5–20 m); an example of the latter group is the MS-15 (Fig. 1*g*), for 33–49-ft (10–15-m) domes.

Another design philosophy, begun in the late 1940s when A. Spitz designed a small planetarium for school classrooms and museums, was to manufacture a small and relatively inexpensive projector that would do for schools and small communities what the larger machines had done for the cities. Early models

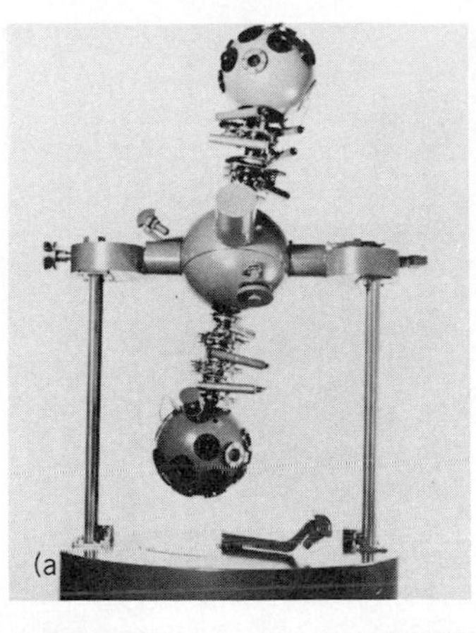

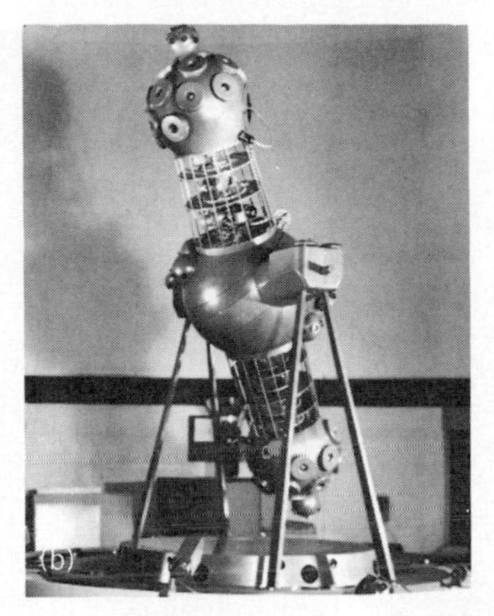

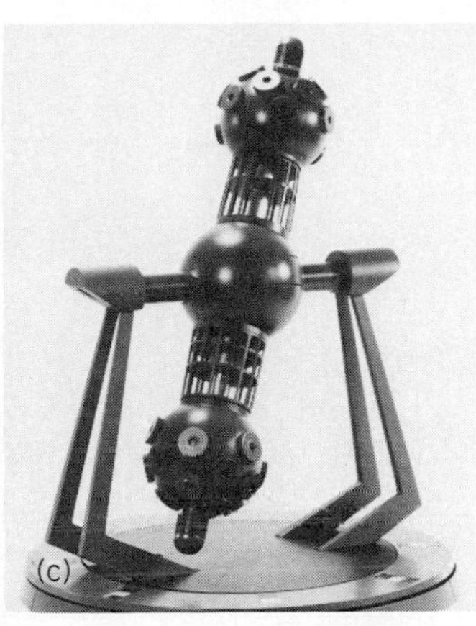

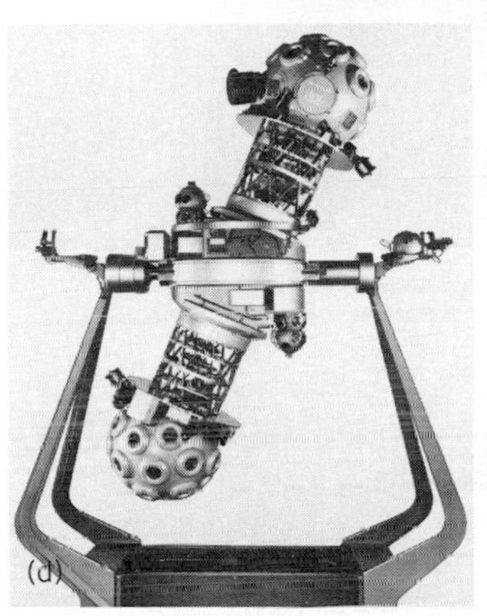

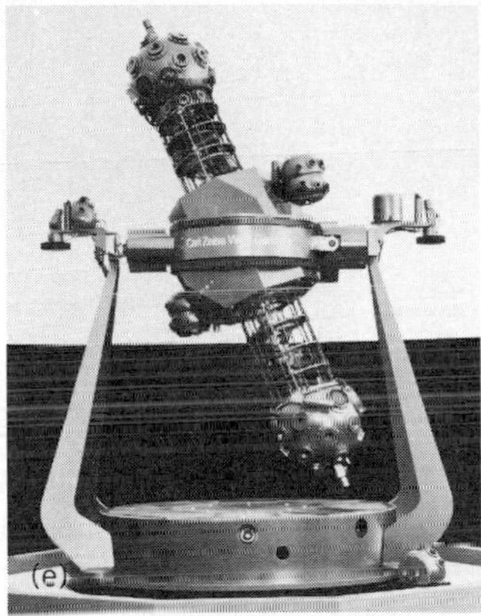

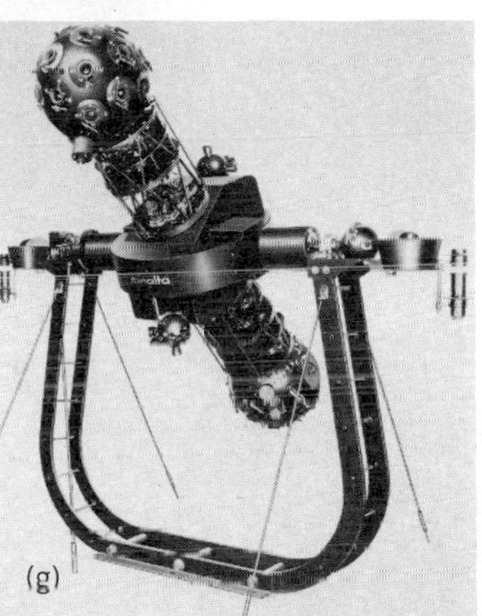

Fig. 1. Zeiss-type planetarium projectors. (*a*) ZKP-2. (*b*) Spacemaster. (*c*) GP-85. (*d*) Mark VI. (*e*) M 1015. (*f*) M 1518. (*g*) MS-15. (*Parts a–c from Jenoptik Jena GmbH; d–f from Carl Zeiss; g from Minolta Corp.*)

of Spitz Space Systems planetariums featured planetary motions through innovative planet analog mechanisms. One of these models became the most common projector for 24–30-ft (7–9-m) domes, while another was manufactured for dome diameters of 50 ft (15 m). Later models include System 512 (**Fig. 2***a*), a medium-sized projector designed for the 24–40-ft (7–12-m) dome size. Spitz's STS (Space Transit Simulator), designed for major theaters (Fig. 2*b*), has a star-ball which is 4 ft (1.2 m) in diameter. These STS projectors are generally used in combination with an all-sky 70-mm motion picture projection system in a theater whose hemispherical dome is oriented at an angle.

Another innovation in planetarium design based on television projection is the Digistar. The projector, which stands only 42 in. (about 1 m) high, utilizes a high-resolution cathode-ray tube with a special wide-angle lens for projection onto a 76-ft (23-m) dome. The theater has a tipped dome to facilitate a 70-mm motion picture projector. Software and documentation include stellar, planetary, and constellation data files. The star and planet positions are fed into the high-intensity cathode-ray tube projector by a computer and thence onto the dome via the wide-angle lens. The system allows the stars to be seen not only from any place on the Earth but from any place out to about 500 light-years from the Sun.

Mechanical aspects. The Digistar has, essentially, no moving parts, whereas the other projectors described above all feature similar mechanical operation derived from the basic Zeiss design. These mechanical-optical projectors all show the apparent motions of the sky which are, in reality, a reflection of the Earth's motions, or the motion of the observer. These mechanical projectors are geocentric, that is, they show the sky from an Earth-centered viewpoint. For example, daily motion depicts the turning of the sky (rising and setting of the Sun, Moon, planets, and stars) caused by the Earth's rotation on its axis. A variable-speed motor turns the entire projector. Latitude motion shows the changing aspect of the sky as the observer travels north or south over the Earth. Thus as one moves toward the North Pole, the North Star rises higher. This is accomplished by rotation of the entire projector around a horizontal (east-west) axis.

Annual motion uses auxiliary projectors to the star globe to depict the changing monthly phases of the Moon, the eastward motion of the Sun along the ecliptic, as well as the annual (yearly) motion of the planets in the zodiac. More complicated mechanical linkages are required to represent these motions, because of the more complex mathematical ratios of planetary periods. Numerous gears are required to produce the desired accuracy. Some instruments also approximate the variable planetary speeds. To project a planet, for example, Mars, in its correct position in the zodiac requires not only a gear train to represent the period of Mars, but also a gear train to represent the Earth's position (because the planet is observed from the moving Earth). A sliding rod links the Earthdrive to the planetdrive, which in turn aligns the

Fig. 2. Spitz planetarium. (*a*) System 512 (*Spitz Space Systems, Inc.*). (*b*) STS, or Space Transit Simulator, in a theater that has a 76-ft (23-m) dome tipped at a 25° angle and seats 344 (*R. H. Fleet Space Theater, Balboa Park, San Diego*).

planet projector in the correct direction in space. The Moon-phase projector not only shows where the Moon is among the stars each hour but also the correct phase. The projector also depicts the regression of the lunar nodes. Some planetariums also have built-in solar and lunar eclipse projectors as well as planet zoom effects.

In addition to the rotation of the Earth upon its axis and its annual journey around the Sun, the Earth has a wobbling motion much like that of a top. The wobbling motion of the Earth's axis, called precession, is caused by the Moon and the Sun pulling on the equatorial bulge of the Earth. The Earth takes 25,800 years for one precession cycle. Planetariums reproduce this effect by rotating the projector around an axis which is inclined 23.5° to the daily-motion axis, to show how the position of the pole star changes as well as the movement of zodiacal constellations in relation to the equinox. SEE PRECESSION OF EQUINOXES.

Spitz planetariums also utilize mechanical analogs to reproduce the aforementioned motions of the planetarium in their small and medium-sized projectors. They do not follow the basic Zeiss design, however, because of the necessity for miniaturization. The daily, latitude, and precession motions are similar to those of the Zeiss, but the annual motion for the planets is based upon the Tychonic system rather than the Copernican or Keplerian. For the STS, mechanical linkages are replaced by computer control. Digistar is also fully computer-controlled with no moving parts, and is not limited to geocentric projection but may depict space views out to 500 light-years from the Sun (a limitation governed by knowledge of accurate star positions and motions to that distance).

Charles F. Hagar

Bibliography. C. F. Hagar, I.P.S. Survey of the world's planetariums, pts. I–IV, *Planetarian*, International Planetarium Society, vol. 11, no. 4, 1982, and vol. 12, nos. 1–3, 1983; C. F. Hagar, *Planetarium: Window to the Universe*, 1980; H. C. King, *Geared to the Stars*, 1978.

Planetary nebula

A gaseous shell thrown off by a dying star just before it settles down to become a degenerate white dwarf (see **illus.**). Planetary nebulae are so called because many display small greenish disks, akin to those of the planets Uranus and Neptune as seen in small telescopes. They have nothing to do with planets but are associated with late stages of stellar evolution.

These nebulae do not congregate in spiral arms, but show a spatial distribution similar to that of most stars: a concentration toward the galactic center and some preference for the galactic disk. Well over a thousand have been cataloged, and the total number in the Galaxy must be about 20,000. They are also found in neighboring galaxies, such as the Magellanic Clouds, the Fornax galaxy, M31, and M32. SEE GALAXY, EXTERNAL; MAGELLANIC CLOUDS; MILKY WAY GALAXY.

Determinations of distances of individual galactic planetaries are extremely difficult. All are too remote for trigonometric parallaxes; proper motions give only statistical results. A few nebular central stars have companions of known luminosity. Astrophysical methods based, for example, on the assumption that ejected shells have constant masses are likely to fail for individuals. The best distance determinations are by a method in which the shapes of spectral lines are correlated with the theory of stellar atmospheres to obtain the star's temperature and surface gravity, and these can be used to determine the luminosity and distance.

Although a planetary nebula is often symmetrical about the central star, which is usually seen (see illus.), a great variety of forms appear. These include double rings, bilaterally symmetrical envelopes, amorphous structures, and even apparent central holes. Delicate filamentary wisps and knotlike concentrations are the rule, indicating that the ejected shells must be usually very inhomogeneous.

Spectra of planetary nebulae are dominated by emission lines of abundant elements, in various ioni-

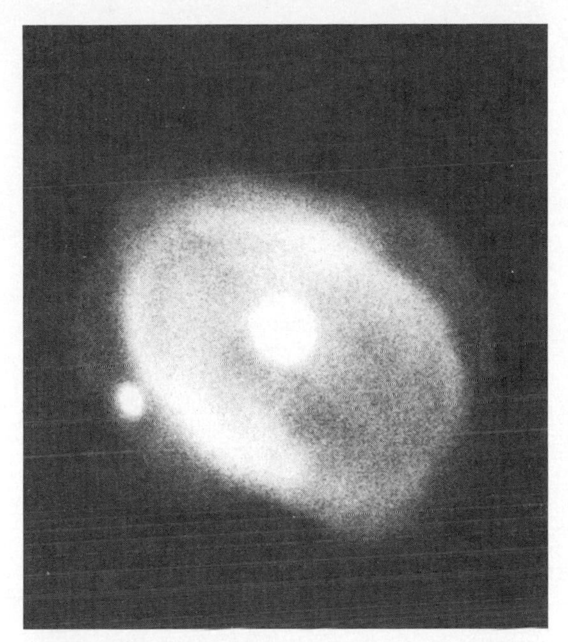

Planetary nebula NGC 3132. The central star is clearly visible. (*Photographed by R. Minkowski, Mount Wilson and Palomar Observatories*)

zation stages, all excited because of the rich ultraviolet radiation emitted by the central star. In the usual optical range, the principal lines are those of hydrogen and helium (produced by recombination and subsequent cascade) and forbidden lines of ions of nitrogen, oxygen, neon, sulfur, and argon. Prominent lines of some elements such as carbon appear only in the ultraviolet. There is also a continuous spectrum due mainly to recombination of protons and electrons, double photon emission, and bremsstrahlung. Many objects show a strong infrared continuum caused by thermal radiation of solid grains, presumably often carbon. *SEE ASTRONOMICAL SPECTROSCOPY; BREMSSTRAHLUNG.*

The chemical composition of the nebular gas (see **table**) is predestined by the material from which the parent star was formed, the effects of element-building processes in those stars, and the removal of refractory elements such as iron and calcium by grains. Some planetary nebulae show compositions not greatly different from that of the Sun; some which originate from massive stars are nitrogen-rich, while yet others are carbon-rich. *SEE NUCLEOSYNTHESIS.*

Possibly, most stars with masses between one and five times that of the Sun produce planetary nebulae. When hydrogen becomes exhausted in the core of a solar-type main-sequence star, it evolves into a red giant and possibly into a Mira variable or OH-IR (infrared) star. The outer envelope is eventually ejected. As the shell density declines, the material is exposed to energetic photons of the hot residual core. Hence, the characteristic bright-line nebular spectrum is produced. As the nebula expands further, this spectrum gradually fades away. The hot, blue-white core, deprived of all nuclear fuel, slowly evolves to a white dwarf. *SEE STELLAR EVOLUTION; WHITE DWARF STAR.*

The chemical composition of shells ejected by low-mass stars reflects that of the interstellar material from which they were formed. In more massive stars, however, products of nuclear reactions in the core get blended in with surface material, and the resultant planetary nebula is enriched in helium, carbon, and nitrogen. Some of the ejecta remains as a gas, and some must condense into solid grains in the inhomogeneous escaping shell, which eventually returns to the interstellar medium. The cool blobs of material contain molecules as well as solid particles and neutral gases. Planetary nebulae appear to be important suppliers of dust and gas to the interstellar medium. They may be useful distance indicators for galaxies, as their maximum luminosity is fixed at about 3000 times that of the Sun. *SEE INTERSTELLAR MATTER; NEBULA.*

Lawrence H. Aller

Bibliography. L. H. Aller, *Planetary Nebulae*, 1971; D. R. Flower (ed.), *Planetary Nebulae*, Internat. Astron. Union Symp. 103, 1983; J. B. Kaler, Planetary nebulae, *Annu. Rev. Astron. Astrophys.*, 23:89–117, 1985; M. Peimbert and S. Torres-Peimbert (eds.), *Planetary Nebulae*, Internat. Astron. Union Symp. 131, 1988; S. R. Pottasch, *Planetary Nebulae*, 1983.

Chemical composition of a typical planetary nebula compared with the Sun*

Element	Nebula	Sun
He	110,000	100,000
C	690	460
N	243	95
O	440	810
F	0.04	0.04
Ne	103	112
Na	1.5	2.04
S	11	17
Cl	0.21	0.3
Ar	2.9	3.7
K	0.08	0.14
Ca	0.11	2.2

*Numbers of atoms on scale H = 1,000,000.

Planetary physics

The study of the structure, composition, and physical and chemical properties of the planets of the solar system, including their atmospheres and their immediate cosmic environment.

Extent of knowledge. While the atmosphere and the oceans have been explored quite thoroughly, probes in the form of deep holes have barely sampled the outermost layer of the Earth's crust. A great deal has been deduced about the interior of the Earth as the result of the application of principles of physics and chemistry and knowledge of the properties of materials. Knowledge of the structure of the deep interior of the Earth principally depends upon studies of the properties of the seismic (earthquake) waves which penetrate through the innermost regions and provide knowledge of conditions there. The Earth's magnetic field also has very deep roots in the interior, and the knowledge obtained from studies of the variations in the magnetic field are supplementary to that obtained from seismic waves.

Knowledge of the other planets in the solar system is much less than that of the Earth. Quite a bit is known about the Earth's Moon as the result of the lunar samples which have been brought back for study in terrestrial laboratories, and of the instruments which astronauts left operating upon the surface of the Moon, which continued to gather a variety of physical information for a period of years following the lunar landings in the Apollo program. A great deal is known about the surface features of Mars, owing to the long series of images taken of the surface by Mars-orbiting spacecraft such as *Mariner 9* and the two *Viking* orbiters. Both the former Soviet Union and the United States have landed spacecraft upon the surface of Mars, but so far these have given relatively little information about the interior of the planet. In the case of Venus, the very thick hazy atmosphere obscures the surface of the planet, and detailed knowledge of the planetary surface depends on increasingly better-resolution radar measurements. Extensive low-resolution radar maps were obtained with terrestrial radar, particularly at Arecibo. Improved resolution was obtained with the radar altimeter on board the *Pioneer* Venus orbiting spacecraft, and the highest-resolution images (about 0.6 mi or 1 km) have been produced for part of the planet by radar on board a Soviet *Venera* spacecraft. Both the Soviet Union and the United States have landed spacecraft on the planet, but these have been primarily concerned with atmospheric investigations. For Mercury, reasonably good images of one of the hemispheres were taken by a fly-by spacecraft. For the outer solar system, images of Jupiter and its four large Galilean satellites, and of Saturn and its larger satellites and ring system, have been returned by the *Voyager* spacecraft.

Construction of models. Planetary scientists attempt to synthesize their information about the structure and properties of each of the planets by constructing models of them. These models make use of the laws of physics and chemistry and are considered to be successful when they reproduce all of the known measured information about a planet. Sometimes it is possible to fit all of this information using more than one model, indicating a considerable uncertainty about the interior properties of the planet, but nevertheless such model building is a useful exercise, because it tends to limit many of the properties of the planet to certain ranges of values.

The most obvious gross properties of the planet are its mass and its radius. In constructing a model, it is required that the model be in hydrostatic equilibrium. This means that at any interior point in the model, the pressure must be great enough to sustain the weight of the overlying mass of material. Thus, given the mass and the radius, estimating the interior pressures through the principles of hydrostatic equilibrium, and knowing something about the compressibilities of materials, it is generally possible to place constraints upon the interior composition of the planets.

All of the planets rotate to some degree, and some of them spin quite rapidly. The effect of the spin is equivalent to a force pressing outward in the equatorial plane of the planet, which helps to sustain the weight of the material in that plane, producing an equatorial bulge. Knowledge of the equatorial bulge and the rate of spin provides additional information about the distribution of mass in the interior of the planet.

Another property which is difficult to measure, but which is known in some detail for the Earth and very crudely for the Moon, is the rate of heat flow from the interior of the planet. For the Earth and the Moon this heat flow appears to result from radioactive heating of the interior of the body, and the knowledge of this heat flow gives information about the distribution of temperature in the interior. In the case of the Earth and the Moon, this interior heat flow is very small compared to the heat which is received from the Sun. In the case of Jupiter and Saturn, the heat flow from the interior is comparable to the heat received from the Sun, so that it has been possible fairly easily to measure this heat flow for those planets because their temperatures are significantly greater than would be expected on the basis of the heat received from the Sun alone.

Classes of chemical composition. The planets in the solar system have an extremely wide range of properties. This distribution of characteristics can be understood in part from a knowledge of the more abundant elements in nature and their volatility properties.

Approximately 98% of matter in the Sun, and therefore also presumably in the matter from which the Sun and the solar system were formed, consists of the gases hydrogen and helium. Most of the remaining material consists of carbon, nitrogen, and oxygen, which in the presence of very large amounts of hydrogen tends to form methane, ammonia, and water. These substances are collectively called ices, and they evaporate at relatively low temperatures. Both the light gases hydrogen and helium and the ices are of quite low abundance on the Earth and the other inner planets in the solar system. What comprises the bulk of the material in these planets is the rocky material, constituting only about 3 parts in 1000 of the solar mix of elements, and among the rocks the most abundant elements are magnesium, silicon, iron, aluminum, calcium, and chromium, all of which are present in rocks predominantly in the form of oxides. In the relatively undifferentiated rocks which fall to Earth from outer space, called meteorites, iron appears both in the form of oxide and as the metal. In the surface rocks of the Earth, iron is almost entirely in the form of the oxide, since the metallic iron has predominantly collected near the center of the Earth to form the core. *SEE ELEMENTS, COSMIC ABUNDANCE OF; METEORITE.*

The differences in the volatilities of these materials, which correlate with the properties of the planetary bodies in the solar system, give information about the properties of the environment in which the planets formed in the solar system. The inner planets, composed predominantly of rocks, evidently formed in a rather hot environment, so that the volatile gases and ices were not condensed and did not collect along with the rocky material, which presumably was condensed. The comets, residing at very large distances from the Sun in the solar system, appear to be mixtures of rocky materials and of the ices. The outer giant planets, Uranus and Neptune, appear to be primarily composed of materials heavier than hydrogen and helium, probably mixtures of rocky and icy materials. The two largest planets in the solar system, Jupiter and Saturn, are much closer in composition to that of the Sun itself, although studies of the interior structures tend to indicate that there is some degree of enrichment in the heavier elements. These differences in composition thus indicate that the tendency to collect hydrogen and helium depends not upon the ability to condense hydrogen and helium into solid form,

Images of Saturn system from *Voyager 1*. (*a*) The planet, color-enhanced. (*b*) Rings and their shadows. (*c*) The satellite Dione. (*d*) Layers of haze covering the satellite Titan; colors are false and are used to show details of the haze. (*e*) The satellite Rhea; colors have been exaggerated to bring out differences. (*NASA*)

(a)

(b)

(c)

(d)

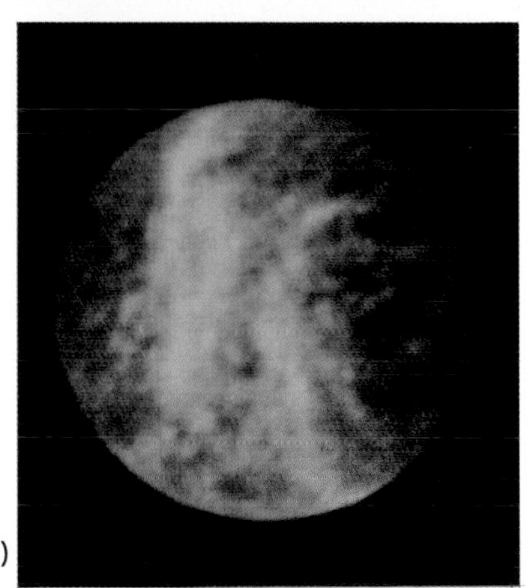

(e)

which would require extremely low temperatures, but rather upon the size of the body which was formed, the larger bodies being more successful in gravitationally capturing the elusive hydrogen and helium. Comets are quite small and are devoid of these materials; Uranus and Neptune are intermediate in mass between the Earth on the one hand and Jupiter and Saturn on the other, and they have been only moderately successful in obtaining hydrogen and helium. However, Jupiter and Saturn were very successful in obtaining these gases. *SEE COMET.*

Since these compositional classes provide a natural means for dividing the planetary objects within the solar system into separate groups, the structure of the various planetary groups will be discussed in turn.

Giant planets. The giant planets are Jupiter, Saturn, Uranus, and Neptune.

Jupiter. Jupiter, the most massive planet in the solar system, is only about 1/1000 of the mass of the Sun. It comes closest in composition to that of the Sun itself. If the composition of Jupiter truly matches that of the Sun, then it would contain in its total mass the equivalent of about one earth mass of rocky material. However, the best attempts to construct models of the interior of Jupiter indicate the amount of material heavier than hydrogen and helium is significantly in excess of that which would be expected for the solar composition. There are probably something like 10 to 20 earth masses of rock and ice in the interior of Jupiter, which is an enrichment of a factor of three to six over the solar composition if the ice-to-rock ratio in the interior of Jupiter is the solar ratio, which is not known. Even this enhanced amount of material amounts to only a few percent of the total mass of Jupiter. The considerable uncertainty in the amount of heavy-element enrichment in the Jovian interior results from the uncertainties in the extrapolation of the properties of hydrogen and helium to very high pressures and temperatures such as those in the interior of Jupiter. It is not even clear whether these heavier materials have settled to the center of Jupiter or are suspended in the atmosphere which is being continually mixed throughout the different interior levels of Jupiter due to convective motions.

One of the interesting properties of hydrogen at higher pressures is its tendency to form a conducting metal, metallic hydrogen. Because hydrogen is a simple substance, the physical calculations that lead to the expected transformation from molecular to metallic hydrogen are reasonably certain, but the precise pressure at which this transformation takes place is still quite uncertain. It appears to be somewhat in excess of 10^6 atm (10^{11} pascals). Most of the mass of Jupiter exists at a pressure considerably in excess of this amount, so that metallic hydrogen is anticipated to form a substantial portion of the interior mass of the planet. *SEE JUPITER.*

Saturn. Saturn has about only one-third of the mass of Jupiter, but nevertheless it also is predominantly composed of hydrogen and helium, and in this case it is definitely clearer that there are heavier elements in excess of solar composition within the interior of Saturn. Again, it is not known whether these heavy elements maintain the solar composition ratio between the ices and rocky materials, and the precise amount of enrichment is therefore uncertain, depending upon this ratio. However, the total amount of heavy materials in the interior of Saturn is comparable to the excess amount in Jupiter. *SEE SATURN.*

Heat flow and helium segregation. Attempts have been made to construct evolutionary sequences of models of Jupiter and Saturn which would follow the changes in structure that take place as the planets cool off after their formation. The research has suggested that Jupiter should still be radiating away its interior heat of formation at about the rate which is actually observed as an excess heat flow from the interior, wheras the amount of primordial heat still emerging from Saturn is expected to be much less than is observed. The explanation of this discrepancy may lie in another interesting property expected for a mixture of helium and hydrogen at higher pressures. Below some temperature which is still quite uncertain, it is expected that helium will collect to form small bubbles within the hydrogen; these bubbles, being heavier, will then sink through the hydrogen toward the center of the planet. Not only does this lead to a greater mass concentration toward the center of the planet, but it also releases additional gravitational potential energy, thereby enhancing the heat flow from the interior. It has been suggested that the interior of Jupiter is still sufficiently hot to have prevented this segregation of helium from hydrogen, whereas the interior of Saturn is sufficiently cooler so that a significant amount of such segregation has and is continuing to occur, thus leading to the observed heat outflow from Saturn.

Uranus and Neptune. Uranus and Neptune are quite similar planets, being 14.5 and 17.2 times the mass of the Earth, respectively. Approximately three-quarters of this mass is expected, on the basis of model building, to consist of materials heavier than hydrogen and helium. The precise numbers will depend upon whether these materials are in the solar ratio of ices to rock, which is not known. If one assumes this ratio to be valid, then each of the planets contains approximately four earth masses of rock and approximately twice that much in the form of ices. The remaining hydrogen and helium form a very deep atmosphere. *SEE NEPTUNE; URANUS.*

Physical composition. Nowhere in the interiors of the giant planets can anything resembling a solid surface be expected. The temperatures in the interiors are very uncertain and can be estimated only as the result of model construction, but they tend to be thousands to tens of thousands of degrees Celsius. The pressures range up to the order of 10^7 atm (10^{12} Pa) and higher. Under these circumstances all materials behave like fluids. There may be a certain amount of compositional stratification, with denser fluids underlying lighter ones.

This issue of stratification is significant in connection with one of the interesting properties of the interior, the transport of gravitational potential energy released in the deep interior to the surface. The thermal conductivity within the interiors of these planets appears to be much too small to do this job efficiently, even in regions of metallic hydrogen. Conduction may be required to transport heat from a layer of one composition to a neighboring layer of different composition. But within a layer of any given composition, the transport of heat appears to require convection. Convection consists of an irregular pattern of overturning motions within a fluid, similar to that which occurs when one boils water within a pot. It has been argued on this basis that the interiors of the giant planets are primarily engaged in convective motions which transport heat outward.

Terrestrial planets. The terrestrial planets include Mercury, Venus, Earth, and Mars. The Earth's Moon may also be considered a terrestrial planet.

Earth. The prototype for the terrestrial planets, and the one about which the most is known, is the Earth. The Earth consists of a thin upper crust composed of rocks of relatively low density and low melting points, overlying a much thicker mantle composed predominantly of metallic silicates and oxides, which in turn overlies a substantial core, which is composed of much denser materials, believed predominantly to be iron with other elements, either alloyed or in solution. Most of the core is liquid, but there is a smaller inner core which appears once again to be solid, and which probably has some compositional differences relative to the outer core.

On the scale of volatility, the Earth is a very refractory place. Most of the materials in its composition condense at quite high temperatures in a gas of solar composition, usually considerably in excess of 2200°F (1200°C). Under such circumstances, most of the iron is expected to be metallic, and since metallic iron is so much heavier than other typical rocky material, such as magnesium silicates, it is natural for the metallic iron to collect at the center of the planet. The detailed seismic evidence indicates that the core of the Earth is not pure iron, but also has some admixture of lighter elements, probably some combination of oxygen, silicon, and sulfur. Several percent of the core must also be nickel, which has properties very similar to that of iron.

The overlying mantle is composed of the oxides and silicates of the metals which are more abundant in nature. Many phase changes take place as such material is subjected to increasing pressure, and some of the increasing density with depth in the Earth's mantle is due to such phase changes.

Among the many different mineral phases which are present within the Earth, there is a natural sorting process for those minerals which combine a relatively low melting point with low density. Such minerals melt easily and tend to find their way to the surface of the Earth through such cracks or pores as become available. In this way the crust of the Earth is formed predominantly of such materials through tectonic activity.

One of the major revolutions in thinking in the earth sciences has come with the realization that the Earth is a very dynamic place. The position of the Earth's pole has changed dramatically in location with respect to the surface throughout the history of the Earth, and the land masses themselves have drifted about from one part of the surface to another. This continental drift is rendered somewhat easier by the relatively large mass of the Earth and hence the fairly rapid rate with which the temperature increases into the Earth's interior, thereby weakening the materials and allowing them to deform and flow more easily. *See Earth.*

Venus. The next most massive planet within the inner solar system is Venus, which has slightly more than four-fifths of the mass of the Earth. Venus has a very thick atmosphere, and the temperature at its surface is very much higher than is typical of the Earth's surface. The conditions make it very difficult to land spacecraft which can operate for appreciable lengths of time such as would be required to obtain seismic signals from the interior of the planet. On these grounds it can only be conjectured that the interior of the planet is probably much like that of the Earth, with a core, a mantle, and a crust. The *Pioneer* Venus orbiter radar altimeter has found some major structural features on the surface of the planet, suggestive of extensive tectonic activity, but also, to the extent that some of the features are correctly determined to be large craters, indicating that surface weathering processes take place very slowly. The extent to which the crust of Venus is subject to extensive continental drift motions is quite unknown. *See Venus.*

Mars. The mass of Mars is approximately one-tenth that of the Earth, and hence significant differences in the internal structure are to be expected. There appears to be less of a density contrast between the core of Mars and that of its mantle, suggesting that the amount of lighter material allied with the core—say, possibly sulfur—is increased relative to the Earth. Because the planet is smaller, the temperature increases less rapidly with depth than in the case of the Earth, and hence Mars should have a somewhat more rigid outer mantle and crust than the Earth. There is no indication that large amounts of continental drift have taken place on Mars. On the other hand, tectonic activity has clearly played a large role in the history of Mars, since the planet can be roughly divided into a hemisphere which is of predominantly ancient and heavily cratered terrain, and another hemisphere which is of much younger and less heavily cratered material. The density of craters on the surface of a planet such as Mars, with so little atmosphere that incoming massive bodies are not significantly impeded in striking the planet, is a measure of the relative age of the surface which has been exposed to space. Since the cratering rate apparently fell off rapidly throughout the first few hundred million years of the history of the inner solar system, differences in crater density frequently represent age differences of some few hundred million years back in the heavy cratering epoch. *See Mars.*

Mercury. Mercury has only about half the mass of Mars, but has several distinct planetary characteristics. The mean density of Mercury is very high, indicating that Mercury probably has an abnormally large core predominantly composed of metallic iron. There is much evidence of extensive tectonic activity, although, like Mars, the increase of temperature below the surface of Mercury probably occurs sufficiently slowly so that the crust and upper mantle are relatively rigid, and nothing resembling continental drift has probably taken place. Mercury is a very heavily cratered planet, with the craters of a given size apparently having been produced by smaller projectiles than in the case of Mars. The reason is that at the distance of Mercury from the Sun such infalling projectiles tend to have higher velocities than they do near the orbit of Mars, so that the resulting impacts are more energetic. *See Mercury.*

Moon. Although the Earth's Moon is technically a satellite, it makes sense to describe it as a planetary body, and planetary scientists consider the twin bodies of the Earth and the Moon as interesting examples of the extremes of planetary physics ranging from relatively large bodies to relatively small but still chemically differentiated objects. The Moon has a history which includes extensive episodes of melting and differentiation, much of which can be reconstructed on the basis of the returned lunar samples. The upper layers of the Moon, which is only just over 1% of the mass of the Earth, are quite rigid, and there is no evidence for extensive horizontal motions of the structural units.

The Moon is unique in the solar system in having a relatively low density among the inner planets, and at best a very small core, indicating that the planet is practically devoid of metallic iron. Relative to the Earth, it is also highly depleted in the more volatile elements. This unusual compositional pattern presumably requires an explanation in the mechanisms which resulted in the formation of the Moon, about which there has been much controversy. *SEE MOON.*

Major satellites. The four Galilean satellites of Jupiter—Io, Europa, Ganymede, and Callisto—have masses which are all roughly comparable to the mass of the Earth's Moon. It is therefore quite clear that they should be considered as planetary bodies in their own right by planetary scientists. The detailed images of these satellites returned by the *Voyager* spacecraft which passed through the Jupiter system revealed them to be very interesting places with many rich, complex, and exotic properties.

The most spectacular of these planetary bodies is undoubtedly Io. This satellite has a surface characterized by large deposits of sulfur and sulfur dioxide, which is in a state of continual change. It appears to have at any time several active volcanos, each of which is likely to be spewing a stream of gas and entrained rocky particles about 60 mi (100 km) or so above the surface (see **illus.**). Such volcanic plumes spread the gases and rocky material from the volcano over a considerable portion of the surrounding terrain. This vigorous tectonic activity is understood to arise from a combination of orbital perturbations of Io by the other Galilean satellites and tidal damping by Jupiter, which results in the dissipation in the interior of Io of very large amounts of heat.

The Galilean satellites appear to represent a composition class which is slightly more volatile-rich than the pure rocky materials characteristic of the inner solar system. In particular, the sulfur content is likely to be considerably higher. The water and carbonaceous contents may also be much higher. If Io ever had much water, it appears to have been lost from the body quite early in its history. It has been suggested that Io has a sufficiently large reservoir of sulfur that this may form an effective fluid layer, or ocean, underlying the solid surface crust.

Europa, Ganymede, and Callisto all appear to have outer crusts composed of water ice. There are a variety of surface markings which indicate a history of cracking, cratering, and in some instances renewal of the icy surfaces. The mean density of Europa is sufficiently high that the planetary body is probably primarily composed of rock. On the other hand, both Ganymede and Callisto have a significantly reduced density, suggesting that quite thick layers of an icy mantle are likely to be part of those planetary bodies. It would not be surprising if the amount of tidal and radioactive heating in the interiors of these bodies was sufficient to maintain a substantial portion of this icy mantle in the form of a liquid brine. Nothing is known about the character of the underlying rocky core.

The Saturnian satellite system contains only one satellite comparable in mass to the Galilean satellites, Titan (the many other satellites are all much smaller in mass). Titan has a significantly higher volatile content than the Galilean satellites. It has an extensive atmosphere (virtually unique in the solar system) largely composed of methane. It is not known what lies at the bottom of this atmosphere, but it has been reasonably speculated that there is a transition layer of heavier hydrocarbons. The satellite has a relatively low density, characteristic of an extensive content of ices, quite likely more than just water ice as in the Galilean satellites. The atmosphere is completely opaque, and hence it is not known whether Titan has surface relief.

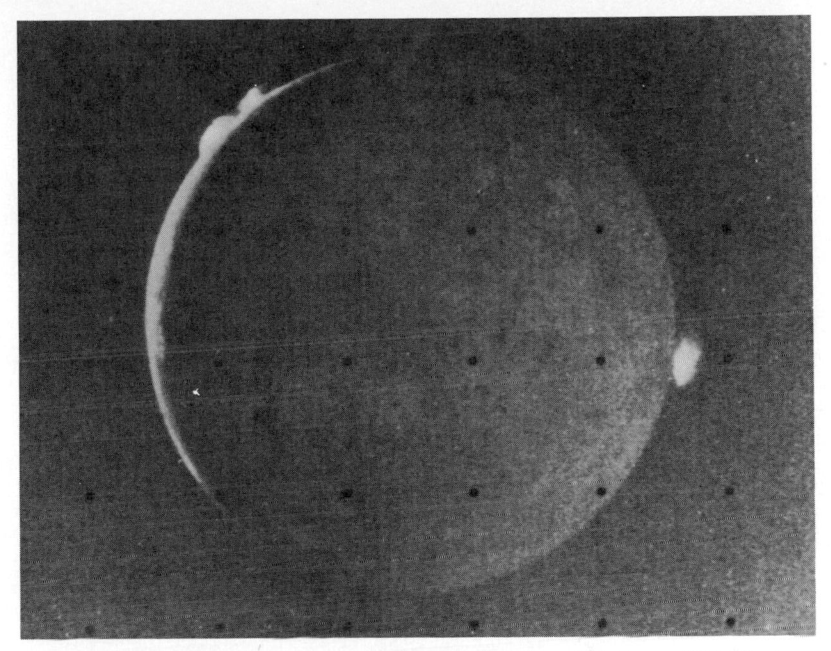

Photograph of Io taken by *Voyager 2* on July 10, 1979, from a distance of 74,000 mi (1,200,000 km). Three volcanic eruption plumes are visible on the limb, all previously seen by *Voyager 1*, 4 months earlier. (*NASA*)

Oceans. With the possible exception of the Galilean satellites, the Earth is the only planet in the solar system having oceans. Mars is too cold for substantial bodies of liquid water to exist upon its surface, although there is evidence in its surface features that water once ran through a number of channels for at least short periods of time. Mars also has a substantial amount of ice in its polar caps. The atmosphere of Venus is sufficiently hot that if any liquid water were to be placed upon its surface, it would quickly be evaporated into steam. Venus contains very little water in its atmosphere, giving rise to the question of whether Venus has ever had substantial amounts of water at any time in its past, or whether it has found some mechanisms for getting rid of the bulk of it. Both the Moon and Mercury are very dry.

Within the oceans of the Earth, a complex set of currents and motions takes place. Many of these currents are driven by slight differences of density within the oceans, which in turn arise because of variations of the amount of dissolved salt, or salinity. Some differences in salinity arise from the evaporation of water into the atmosphere from the oceans, followed by rain upon the land and the runoff of salt-depleted water from the rivers. Other differences in salinity arise from the formation of ice from the oceans, which results in a concentration of salt within a liquid phase.

Large-scale currents are also set up in the oceans as the result of temperature differences within the ocean waters, resulting from the preferential heating of the water in the tropical regions of the Earth. The oceans play an important role in the transport of heat from the region of the Equator toward the poles of the Earth.

Atmospheres. There are certain general principles which govern the structure and dynamics of planetary atmospheres. In most cases these atmospheres receive their primary heat input from above, resulting from heating due to the Sun. Most atmospheres contain some form of haze or condensed layers in the form of clouds, which results in a reflection of a portion of the incident sunlight back into space where it has not contributed to the deposition of heat within the atmosphere. The remainder of sunlight is either absorbed within the atmosphere or transmitted or scattered downward to the ground where absorption takes place. The heat thus received by the ground must be reradiated into the atmosphere, which will transmit some of it and absorb some of it. The absorbed radiation from the ground will in part be reradiated by the atmosphere toward the ground, adding to the heating effect that has taken place as the result of the original receipt of the corresponding energy. This enhancement of the heating effect is commonly called the greenhouse effect, despite the fact that the mechanism by which a greenhouse keeps warm is somewhat different.

The temperature at the surface of the planetary body therefore depends in a complex manner on the properties of the overlying atmosphere, as well as upon the distance of the planet from the Sun. The atmosphere of Venus is very much hotter relative to the Earth than would be expected purely on the basis of the relative distances from the Sun. The difference appears to arise from the extensive operation of the greenhouse effect within the very thick atmosphere of Venus. Soviet spacecraft landed upon the surface of Venus have found quite large amounts of illumination by sunlight there, indicating that significant amounts of solar energy do manage to penetrate to the ground of that planet.

The only terrestrial planets with atmospheres are Venus, Earth, and Mars. Both Mars and Venus have atmospheres composed predominantly of carbon dioxide. If all of the carbonate rocks of the Earth had the carbon dioxide extracted from them, the Earth also would have a thick predominantly carbon dioxide atmosphere very similar to that of Venus. Thus the difference between these two planets arises to a large extent from the ability to form carbonate rocks, which is a function of temperature. At the high temperature of the ground surface of Venus, carbonate rocks are broken down, and the carbon dioxide is released to the atmosphere; the thick atmosphere of Venus is therefore the stable state under the circumstances. In the case of the Earth, it appears that water has played an important role in the formation of carbonate rocks from carbon dioxide, and such water has not been present significantly in liquid form on Mars. There is no evidence that Mars has a substantial reservoir of carbon dioxide in the form of carbonate rocks.

The element of next greatest abundance in the atmospheres of Mars and Venus is nitrogen. This happens to be the predominant element in the atmosphere of the Earth. The next most abundant element in the terrestrial atmosphere is oxygen, which is maintained there predominantly as the result of the operation of life upon the surface of the Earth. Any planet with a large content of oxygen in the atmosphere is very likely to be extensively populated by living organisms.

The atmospheres of the terrestrial planets appear to be substantially mixed as a result, in part, of convective processes which transport heat, and in part from winds which are produced by pressure differences and which cause mixing by stirring up the atmosphere. At a sufficiently great height in the atmosphere, mixing is no longer effective, and a gravitational stratification of the components of the atmosphere takes place, with the lighter components of the atmosphere extending to greater heights. At these great altitudes, solar ultraviolet radiation produces an extensive amount of ionization of these atmospheric constituents, producing a plasma layer at the top of the atmosphere which is called the ionosphere. At a sufficiently great height the molecules of the atmosphere are in free ballistic trajectories; this region is called the exosphere.

The same principles of physics and photochemistry also apply to the atmospheres of giant planets, but the details are considerably different, because the predominant constituents are hydrogen, helium, and methane. At significantly lower levels in the atmosphere, a substantial amount of ammonia appears, and this forms a layer of ammonia clouds, probably contaminated by some amount of hydrogen sulfide which forms a compound with ammonia. At a still lower level in the atmosphere it is expected that water clouds will be present.

In the case of the giant planets, the rapid rotation gives rise to a distinctive banded structure along parallels of latitude within the atmosphere, because Coriolis forces make it difficult for the convective motions within the atmosphere to transport material significantly across parallels of latitude.

Magnetospheres. Some of the planets contain substantial magnetic fields; others do not. Within the inner solar system, the Earth possesses a relatively strong field, Mercury a relatively weak one, and if Venus and Mars contain significant intrinsic fields, they are sufficiently weak that they have not been confirmed. On the other hand, it is known by direct measurement that Jupiter and Saturn have very strong magnetic fields.

The generation of planetary magnetic fields appears to depend upon a combination of planetary rotation with an inner convecting layer having significant electrical conductivity. These conditions appear to be met in the core of the Earth and in the metallic hydrogen mantles of Jupiter and Saturn.

One of the most striking features of planetary magnetospheres is the trapping of energetic particles within them. This gives rise to a great variety of phenomena within the Earth's magnetic field, but in the case of Jupiter the effects are so strong that radiation damage significantly affects the operational lifetime of any space probe inserted into the magnetosphere. SEE MAGNETOSPHERE; PLANET.

A. G. W. Cameron

Bibliography. J. K. Beatty and A. Chaikin (eds.), *The New Solar System*, 3d ed., 1990; T. Gehrels (ed.), *Jupiter*, 1976; T. Gehrels and M. S. Matthews (eds.), *Saturn*, 1984; H. Hunter, T. M. Donahue, and V. I. Moroz (eds.), *Venus*, 1983; H. Jeffreys, *The Earth*, 6th ed., 1976; D Morrison and T. Owen, *The Planetary System*, 1988; F. Vilas, C. R. Chapman, and M. S. Mathews (eds.), *Mercury*, 1988.

Planetary rings

Bands of particles orbiting a planet. Planetary rings have fascinated astronomers and physicists since Galileo first observed Saturn's rings in 1610. The interpretation of Galileo's observations raised many questions about the nature of planetary rings, as did the much more recent observations of rings around Uranus, Jupiter, and Neptune. Ring systems exhibit a diverse array of phenomena, and physicists from Galileo onward have met the challenge of understanding how fundamental physical processes can sculpt the often bizarre forms. The study of rings also provides information about processes that have formed the planets and satellites of the solar system.

Four unique ring systems. Although Saturn's rings were initially observed by their reflected sunlight with small telescopes, the present knowledge of ring systems has been gleaned through a much broader use of the electromagnetic spectrum, from ultraviolet to radio wavelengths. Also, spacecraft fly-bys, notably the *Voyager* missions, have provided information on the structure of these systems at over 10,000 times the spatial resolution possible with ground-based telescopes. The occultation technique, practiced both from spacecraft and from the Earth, provides extremely high spatial resolution. An occultation observation involves recording the intensity of the light from a star or the level of a radio signal from a spacecraft as a ring system passes between the source and the observer. The occultation technique is so potent, even when practiced from Earth, that two of the four ring systems were discovered with it. SEE OCCULTATION.

Although the particles in all planetary ring systems obey the same physical laws, the four known ring systems appear strikingly different from one another.

Saturn. Visible from Earth with high-power binoculars, Saturn's rings are the most visually spectacular in the solar system. Broad and brilliant, the system is composed of numerous ringlets of various breadths and brightnesses, and is the most complex of the four systems (**Fig. 1**). The main components are denoted

Fig. 1. *Voyager 2* photograph of the lit face of Saturn's B ring, obtained on August 25, 1981, from a distance of 743,000 km (461,000 mi). It covers a range of about 6000 km (3700 mi) and shows the ring structure broken up into about 10 times more ringlets than was previously suspected. The narrowest features are due to a combination of differences in ring-particle number density and light-scattering properties. (*NASA*)

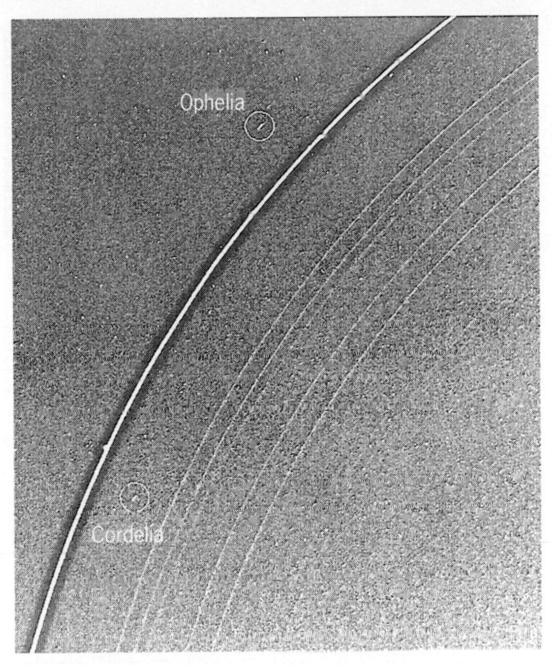

Fig. 2. *Voyager 2* image of the Uranian rings, taken January 21, 1986, at a distance of 4.1×10^6 km (2.5×10^6 mi) and resolution of about 36 km (22 mi). Two shepherd satellites associated with the rings, designated Cordelia and Ophelia, are visible on either side of the bright epsilon ring; all nine of the known Uranian rings are visible. (*NASA*)

the A, B, and C rings, with A and B separated by a band visible from Earth and known as the Cassini division. These rings are extremely flat, with a vertical thickness about 10^{-6} of their diameter. The particles composing these rings are mainly centimeter-sized ice balls. SEE SATURN.

Uranus. Discovered by accident during a 1977 occultation of a star, these rings are narrow with well-defined boundaries and exhibit small deviations from perfect circles (**Fig. 2**). The average Uranian ring particle is larger and much darker than that of Saturnian rings. Interspersed among the narrow rings is a tenuous sheet of small particles. SEE URANUS.

Jupiter. The rings of Jupiter, discovered during the *Voyager 2* fly-by in 1979, are faint and tenuous, in contrast to Saturn's bountiful rings. Broad ringlets with sharply defined boundaries characterize Jupiter's system. These rings are composed mainly of micrometer-sized particles, and *Voyager* images show three distinct bands. SEE JUPITER.

Neptune. Neptune's rings were discovered with occultation observations in 1984. With Earth-based occultations, however, only sections of Neptune's rings are dense enough to be detectable, and are called ring-arcs. The *Voyager* spacecraft found that the rings completely encircle the planet but some sections are much denser than others (**Fig. 3**). These dense sections are the only parts of the rings that were detected from Earth. The rings appear bright in Fig. 3 because microscopic ring particles scatter sunlight toward the camera. The particle-size distribution in Neptune's rings is thus quite different from that of Uranus's rings, which contain few dust-size grains. SEE NEPTUNE.

Early work to understand Saturn's rings. When Galileo observed Saturn's rings through his small and

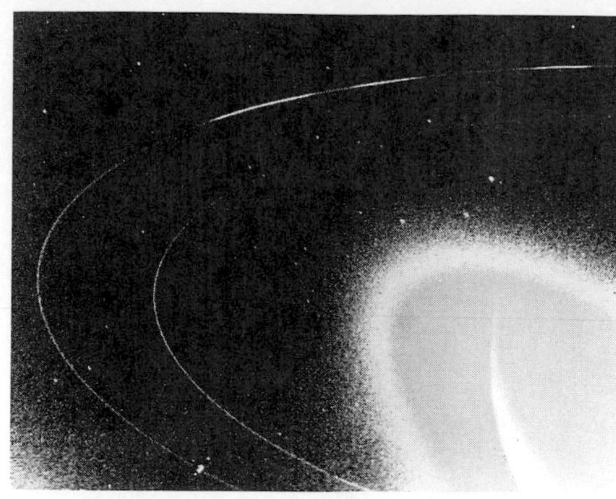

Fig. 3. ***Voyager 2*** **image of Neptune's two main rings, about 53,000 km (33,000 mi) and 63,000 km (39,000 mi) from the center of the planet, seen backlit by the Sun. The image of the planet was greatly overexposed to capture detail in the rings. At the right in the outer ring is the main clumpy arc, composed of three features, each about 6–8° long, that were first detected from Earth-based occultations. (*NASA*)**

imperfect telescope in 1610, he interpreted them to be two large satellites of Saturn. In 1612, when the rings were edge-on to the Earth and therefore not visible, Galileo had no explanation for their apparent disappearance. The idea that the Saturnian rings were a set of satellites persisted until 1655, when C. Huygens applied a working theory of how satellites behave to his own observations of Saturn and determined that an inclined, rotating, solid, and relatively thick ring was orbiting Saturn.

Although there had been speculation that Saturn's ring was composed of particles, it was not until 1857 that J. C. Maxwell demonstrated that under newtonian physics even minutely subdivided rings could not exist: "The only system of rings which can exist is one composed of an indefinite number of unconnected particles revolving around the planet with different velocities according to their respective distances." Subsequently, spectroscopic data were obtained and indicated that the orbital velocities of the rings decreased with increasing distance from the planet, as expected for particulate rings. The opposite trend of velocity would have been observed for solid rings.

The extreme flatness of Saturn's rings can be understood in terms of what would be expected if a large collection of particles were orbiting a planet. If initially there were a spherical cloud of particles, collisions would occur, since the individual particle orbits would precess at different rates. These collisions would cause the ring to become flat. Next, spreading of rings would occur because particles closer to the planet would move faster than particles farther from the planet. The difference in particle speed, called keplerian shear, would cause collisions between the particles. These collisions would cause the faster particles to lose angular momentum and therefore to drop closer to the planet. The slower particles would tend to be bumped farther out from the planet. Thus, the rings would spread out in a thin sheet.

Another concept in the historical development of the understanding of Saturn's rings is the Roche limit. Within a radius $2.46R\ (\rho_p/\rho_o)$, where R is the radius of the planet, ρ_p the density of the planet, and ρ_o the density of the orbiting body, a fluid body would break up because of the tidal force from the planet. This tidal force increases with decreasing distance to the planet, so that even a solid body, if it was large enough, would break into pieces. Collisions between these pieces would carry the disintegration further, so that a ring of orbiting particles would be formed. SEE ROCHE LIMIT.

The physical effects just discussed are sufficient to understand what can be seen of Saturn's rings through an Earth-based telescope. However, in order to understand the results of the high-resolution observations of all the ring systems made from spacecraft and with Earth-based occultations, more subtle physical effects must be considered.

Gravitational forces from satellites. In analyzing the dynamics of satellites and ring particles around a planet, a useful approach is to think in terms of gains and losses of angular momentum by the satellites and particles. A force in the direction of motion will cause a particle or satellite to gain angular momentum: the semimajor axis of orbit will be increased, but in this larger orbit the velocity of the body will be less. The opposite happens when a force is applied that opposes a body's motion: the orbit becomes smaller and the body speeds up. The total angular momentum of a planet's rotation and the orbital motion of its rings and satellites is conserved, but angular momentum can be transferred between bodies through their gravitational interaction. An example of angular momentum transfer is seen in the tides raised on the Earth by the Moon: the Moon's orbit is expanding through its gain of angular momentum, while the Earth's rotation is gradually slowing down. SEE TIDE.

The synchronous orbit is that in which a satellite's angular motion exactly matches that of the surface of the planet. For orbits lying within the synchronous orbit, the satellite is moving faster than the tidal bulge it raises on the surface of the planet; hence its orbit evolves inward, until it gets far enough inside the Roche limit to break up or to impact the planet. The distance for this occurrence depends on the strength of the material composing the satellite. For orbits outside the synchronous orbit, the satellite's orbit will evolve outward until the planetary rotation is slowed enough to match the satellite's orbital period.

Gravitational forces from small satellites can affect the configuration of ring particles because, although satellite mass is small, the distances between the satellites and particles are small also. The most dramatic effect of small satellites on rings is the formation of narrow rings by so-called shepherding satellites. The first known narrow rings, those of Uranus, initially defied explanation. After all, ring particles should spread out through collisions, not clump into narrow rings. However, two nearby small satellites, one orbiting inside the ring and the other orbiting outside, can inhibit the spreading and produce narrow rings. The outer satellite gains angular momentum from the ring particles, while the inner satellite loses angular momentum to the ring particles; the net result is an effective repulsion of the ring particles by both satellites, and prevention of the spreading of the ring. (Two shepherd satellites are shown in Fig. 2.)

Ring particles whose orbital period is in an even-integer ratio with a satellite are said to be in resonance with the satellite. The persistent, periodic tugs from such a satellite can introduce eccentricities into the ring particle orbits, which lead to collisions with other particles until all particle orbits near the resonance are depleted.

Waves in a sheet of ring particles. A large disk of particles, such as Saturn's rings, exhibits collective behavior, analogous to the wave motions that can be induced in a thin sheet of elastic material. Two kinds of waves occur. First, there are wave motions within the plane of the ring particles. These are known as spiral density-waves, since they have a spiral structure when viewed from above, analogous to the spiral structure of some galaxies. A second type of particle wave produces motions perpendicular to the plane of the particles, known as bending waves, since they resemble waves induced in a flapping bed sheet. Both types of waves are excited at locations of resonances with satellites outside the rings. The spiral density-wave propagates outward from the resonance location, and the bending wave propagates toward the planet.

Behavior of small particles. Because of their much larger surface area in proportion to their mass, micrometer-sized particles are significantly affected by weak forces that depend on the particle area. Poynting-Robertson drag describes the loss of particle angular momentum in response to collisions with photons from the planet and the Sun. Plasma drag describes the loss of angular momentum from collisions with the charged particles of the plasma trapped by the planetary magnetic field. If a small ring particle acquires a charge, the influence of the planetary magnetic field can significantly affect its motion. The transient, dark "spokes" discovered by *Voyager* in Saturn's rings are believed to be primarily a charged-particle phenomenon.

Prospects. Since the mid 1970s there have been great strides in the understanding of the physics of planetary rings, but several important questions remain. For example, there is no generally accepted explanation for the sharp inner edges of Saturn's A, B, and C rings and the clumpiness of Neptune's ring-arcs. Also, the age of the rings remains unknown. It is not known whether the rings were formed concurrently with the planets themselves from particles inside the Roche limit that could not gravitationally bind into a satellite, or whether the ring systems have formed and dissipated several times during the 4.5×10^9 history of the solar system through the breakup of bodies that orbited within the Roche limit.

At least one ring system can be expected to form around Mars within the next 5×10^7 years or so, when the orbit of its moon Phobos decays sufficiently for this satellite to break up from tidal forces. Neptune's ring system may become much more extensive when the orbit of Triton decays and this massive satellite meets the same fate. The time scale for this, however, will be much longer than for Phobos and is uncertain. *See Mars.*

Two spacecraft missions have been designed to study the ring systems of Jupiter and Saturn. The *Galileo* spacecraft will tour the Jovian system; and the *Cassini* mission, a joint venture for the National Aeronautics and Space Administration (NASA) and the European Space Agency (ESA), is scheduled to be launched for a tour of the Saturn system after the year 2000. Meanwhile, Earth-based occultation observations will continue to provide high-resolution data for the ring systems of Uranus, Saturn, and Neptune. These data alone may lead to answers to several of the dynamical questions about the rings.

James L. Elliot; Lyn E. Elliot

Bibliography. J. K. Beatty and A. Chaiken (eds.), *The New Solar System*, 3d ed., 1990; J. Elliot and R. Kerr, *Rings: Discoveries from Galileo to Voyager*, 1984; R. Greenberg and A. Brahic (eds.), *Planetary Rings* , 1984.

Pleiades

A beautiful group of stars resembling a little dipper, in the constellation of Taurus, known since earliest records. The Pleiades is a typical open cluster (see **illus.**); it contains several hundred stars within a radius of 1° from Alcyone. Its distance is 410 light-years (2.4×10^{15} mi or 3.9×10^{15} km), its linear diameter about 15 light-years (9×10^{13} mi or 1.4×10^{14} km), its age 8×10^7 years. The brightest stars are blue, of B type. The cluster is permeated with diffuse nebulosity. Though early accounts refer to the Pleiades in terms of seven stars, only six are now conspicuous to the unaided eye, which raises a theory that one, the lost Pleiad, has faded. The observation of flare stars and x-ray emission has increased interest in this cluster. *See Constellation; Star clusters.*

The Pleiades. (*Lick Observatory photograph*)

Helen S. Hogg

Pluto

The most distant known planet in the solar system. Pluto was discovered on February 18, 1930, by C. W. Tombaugh at the Lowell Observatory, Flagstaff, Arizona, on photographic plates taken as part of a systematic search. The presence of a planet beyond Neptune's orbit had been predicted independently by P. Lowell and W. H. Pickering (among others) on the basis of analyses similar to those which had led U. J. Leverrier and J. C. Adams to the prediction of Neptune. It was thought that there were perturbations in the motion of Uranus that Neptune alone could not explain. Pluto was found surprisingly near its predicted position, since modern investigations have shown that the mass of Pluto is far smaller than was assumed in the calculations.

Pluto's orbit (**Fig. 1**) has a semimajor axis (mean distance to the Sun) of 3.7×10^9 mi (5.9×10^9 km), an eccentricity of 0.25, and an inclination of 17.2°. The inclination and eccentricity are the largest of any of the planets. At a mean orbital velocity of 2.96 mi/s (4.7 km/s), it takes Pluto 247.7 years to make one revolution around the Sun. The large orbital eccentricity means that at perihelion Pluto is closer to the Sun than Neptune, but the orbits of the two planets do not intersect. *SEE NEPTUNE; PLANET.*

Pluto is visible only through fairly large telescopes, since its visual magnitude at mean opposition is 14.7. Periodic variations in its brightness demonstrate that the surface of Pluto is covered with bright and dark markings and indicate a period of rotation of 6.3 days. This is the longest rotational period of any of the outer planets and is similar to the orbital period of Triton about Neptune. This fact and the high inclination and eccentricity of Pluto's orbit around the Sun have led several scientists to propose that Pluto was originally a satellite of Neptune which escaped from the control of the planet, but it may also represent an icy planetesimal that was never accreted by a giant planet.

Charon. Pluto has a small satellite named Charon, discovered by J. W. Christy and R. Harrington in 1978 (**Fig. 2**). The orbit of Charon is unique in the solar system in that it is synchronous; that is, the satellite's period is identical to the rotational period of the planet. Thus an inhabitant of Pluto who lived on the appropriate hemisphere would see Charon hanging motionless in the sky. From 1985 to 1991, the geometry of Charon's orbit was such that occultations and eclipses could be observed from the Earth as Charon moved around Pluto (Fig. 2*b*). Careful observations of these events have greatly improved knowledge about this distant system. *SEE OCCULTATION; SATELLITE.*

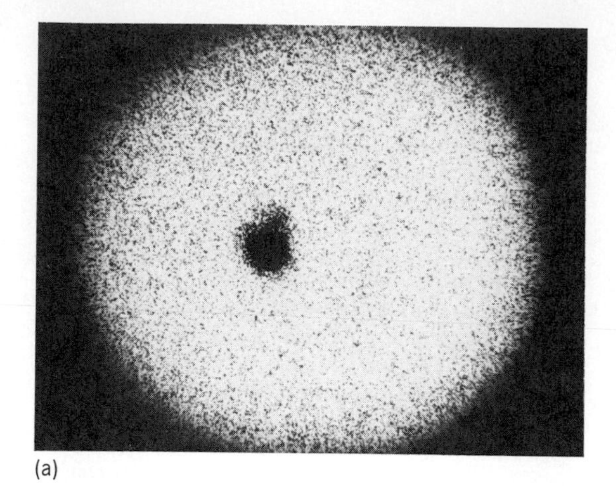

(a)

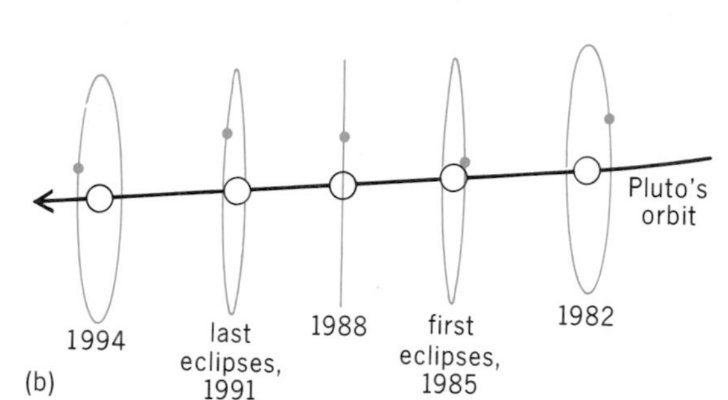

(b)

Fig. 2. Pluto's satellite Charon. (*a*) Discovery photograph, obtained on July 2, 1978, by J. W. Christy and R. Harrington with the 60-in. (1.54-m) reflector of the U.S. Naval Observatory. Charon appears as a small bulge at the top of Pluto's image. (*b*) Changing aspect of Charon's orbit as viewed from Earth. (*After D. Morrison and T. Owen, The Planetary System, Addison-Wesley, 1988; Official U.S. Navy Photograph*)

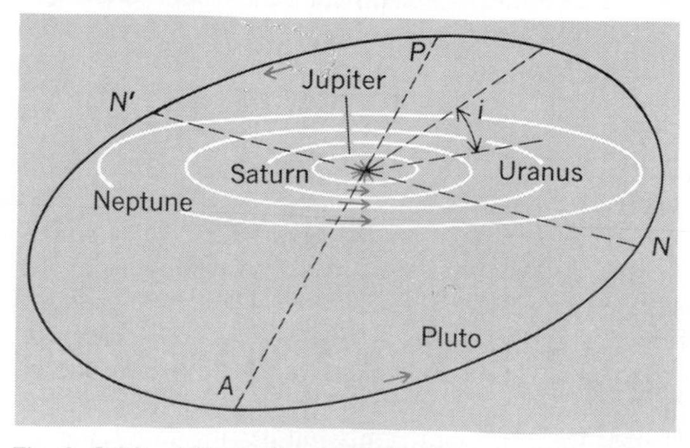

Fig. 1. Orbit of Pluto. A perspective view to show the inclination *i* and eccentricity of the orbit. *A*, aphelion; *P*, perihelion; NN′, line of nodes. (*After L. Rudaux and G. de Vaucouleurs, Astronomie, Larousse, 1948*)

Size and mass. The radius of Pluto was found to be 715 ± 12 mi (1150 ± 20 km), and so it is significantly smaller than the Earth's Moon (whose radius is 1080 mi or 1738 km), while Charon's radius is 368 ± 9 mi (593 ± 14 km). Charon is thus about half the size of Pluto itself, making this the most closely matched pair in the solar system. The combined mass of Pluto and Charon is 0.0025 times the mass of the Earth. Their densities are close to 2.0 g/cm^3, indicating a composition of ice and rock, similar to that of the large icy satellites of Jupiter and Saturn, but surprisingly dense for bodies this small.

Surface and atmosphere. Solid methane has been detected on the surface of Pluto by means of spectroscopy, which has also indicated the presence of methane gas in the planet's atmosphere. No methane has been detected on Charon. The average surface temperature of Pluto is close to 50 K (−370°F), which is consistent with the formation of methane frost on the surface. The atmosphere has also been studied by means of a stellar occultation, which indicated a total surface pressure of 10^{-5} times the sea-level pressure on Earth. Possibly some carbon monoxide, nitrogen, and argon are present in addition to the methane. *SEE PLANETARY PHYSICS.*

Seasons on Pluto are modulated by the large eccentricity of the orbit. Pluto reached perihelion in 1989, which brought it closer to the sun than Neptune. This means that Pluto is now observed at the warmest time in its seasonal cycle, when the atmospheric pressure should be at its maximum value. No spacecraft are scheduled to fly by this distant world, but some results on atmospheric composition and surface properties may be expected from observations by the Hubble Space Telescope.

Search for further planets. Tombaugh continued his photographic survey at Lowell after discovery of Pluto, with negative results. Planets at still larger distances must be fainter than sixteenth magnitude. Observations with the *Infrared Astronomical Satellite* in 1983 provided a more sensitive survey. No new planets have been reported. *SEE INFRARED ASTRONOMY.*

Tobias C. Owen

Bibliography. W. K. Hartmann, *Moons and Planets,* 2d ed., 1983; W. G. Hoyt, *Planets X and Pluto,* 1980; D. Morrison and T. Owen, *The Planetary System,* 1988.

Polarimetry

The science of determining the polarization state of electromagnetic radiation (x-rays, light or radio waves). Radiation is said to be linearly polarized when the electric vector oscillates in only one plane. It is circularly polarized when the x-plane component of the electric vector oscillates 90° out of phase with the y-plane component. In experimental work, the polarization state is usually expressed in terms of the Stokes parameters I (the total intensity of the radiation), Q (the preference for light to be linearly polarized in the x plane or at 0° in the instrument reference system), U (the preference for linear polarization at 45°), and V (the preference for left-circular polarization). When $Q/I = 1$, for example, the radiation is 100% linearly polarized along an axis at 0°. Similarly, $V/I = -1$ means that the light is 100% right-circularly polarized.

To completely specify the polarization state, it is necessary to make six intensity measurements of the light passed by a quarter-wave retarder and a rotatable linear polarizer, such as a Polaroid or a Nicol prism. In the **table**, the parameter pass angle refers to the angle, in the instrument reference system, of the plane of oscillation of the radiation exiting the linear polarizer. This angle, of course, rotates as the polarizer rotates, and the intensity must be measured with a polarization-insensitive detector.

The retarder converts circular light into linear light.

Determination of polarization state by intensity measurements

Polarizer pass angle	Stokes interpretation of measured intensity: Retarder positioned out	Retarder positioned in
0°	$I + Q$	
45°	$I + U$	$I + V$
90°	$I - Q$	
135°	$I - U$	$I - V$

It is placed upstream of the polarizer and is moved into the beam to determine Stokes V. Then I, Q, U, and V can be calculated from the six intensity measurements of the table. For fully polarized light, $I^2 = Q^2 + U^2 + V^2$. If the light is only partially polarized, the degree of polarization P is given by the equation below.

$$P = \sqrt{\frac{Q^2 + U^2 + V^2}{I^2}}$$

Polarization of radiation. An ensemble of atoms in a hot gas, such as a stellar atmosphere, usually radiates in no preferred state of polarization. Most starlight is unpolarized. However, atoms in the presence of a magnetic field align themselves at fixed, quantized angles to the field direction. Then the spectral lines they emit are circularly polarized when the magnetic field is parallel to the line of sight and linearly polarized when the field is perpendicular. In 1908, G. E. Hale used this property in solar spectral lines to prove that sunspots have magnetic fields of 1000–3000 gauss (0.1–0.3 tesla). For comparison, the Earth's magnetic field is 0.3 gauss (30 microteslas). Fields as high as 100 megagauss (10,000 T) have been measured in white dwarf stars. *SEE SOLAR MAGNETIC FIELD; SUN; WHITE DWARF STAR; ZEEMAN EFFECT.*

The light from sunspots is polarized because the magnetic fields impose some direction in the emitting gas. Other phenomena also remove isotropy and produce polarization. Sunlight scattered by electrons in the solar corona is partially linearly polarized because the radiation from the solar surface is much greater than the radiation into the surface (anisotropic radiation). Similarly, beams of electrons or protons will excite partially polarized light when they strike a gaseous cloud. Polarimetry can reveal the presence of the atomic beams.

Synchrotron emission and scattering are important polarizers of light. Lord Rayleigh showed that scattering by the barbell-shaped molecules of the Earth's atmosphere produces partial linear polarization of the blue sky. Similarly, needle-shaped grains in interstellar dust clouds polarize the starlight that passes through them whenever magnetic fields have aligned the grains. By measuring the angle and degree of polarization in many stars, astronomers have been able to map the magnetic fields of the Milky Way Galaxy and many other galaxies. Galactic magnetic fields are typically only a few microgauss (a fraction of a nanotesla). *SEE INTERSTELLAR MATTER; MILKY WAY GALAXY.*

Synchrotron radiation is produced when electrons moving at nearly the speed of light spiral rapidly in a magnetic field. Because of the centripetal acceleration, the electrons emit linearly polarized light or radio waves. The Crab Nebula is a well-known source of highly polarized synchrotron radiation from the ultraviolet to the radio regime. The inferred field strength in the nebula is approximately 100 microgauss (10 nT). Radio galaxies, quasars, and active galactic nuclei also produce polarized synchrotron radiation. Sometimes the degree of polarization P is as high as 50%. *SEE CRAB NEBULA; GALAXY, EXTERNAL; QUASAR; RADIO ASTRONOMY.*

Magnetic fields penetrate every corner of the universe. They are generated by dynamos deep inside stars and planets and possibly in galaxies, but the dynamo mechanism is not well understood. And magnetic fields must have played some role in the evolution of the universe. These are good reasons to continue to develop more efficient and accurate polarimetry. *See* MAGNETOHYDRODYNAMICS.

Measurement of polarization. Electrooptical devices are rapidly replacing rotating polarizers and fixed retarders. To make daily maps of the weak magnetic fields on the Sun, H. W. Babcock invented the magnetograph, which consists of a spectrograph to isolate the atomic spectral line for study; a Pockels cell, an electrooptic crystal whose retardance depends on an applied voltage; a polarizing prism to isolate the polarization state passed by the retarder; a pair of photocells to detect the transmitted light; and a scanning mechanism to sweep the solar image across the spectrograph entrance slit. *See* SPECTROGRAPH.

Since Babcock's magnetograph measured only circular polarization, it recorded only the line-of-sight, or longitudinal, component of the field. The sensitivity was about 5 gauss (500 μT). A magnetograph can be made sensitive to linear polarization, but the signal levels are about 100 times weaker for the inferred transverse fields than for longitudinal fields of comparable strength. In this case, longer measurement times are required to obtain useful signal-to-noise levels.

To improve signal-to-noise levels while measuring weak linear or circular polarization over a wide field, the spectrograph can be replaced with an optical filter having a narrow passband, and the photocells can be replaced with an array of photosensitive picture elements (pixels). In this manner, the narrow portion of the spectrum most sensitive to the effect (Zeeman effect) of magnetic fields can be isolated and recorded at many points simultaneously.

Most classical polarizing optical devices are effective over only narrow fields of view, sometimes as little as 1°. This characteristic severely limits the amount of light that can be analyzed at once. The introduction of polarizing dichroic films (Polaroids) was a great improvement over classical polarizing prisms (for example, the Nicol prism) because of their wide acceptance angle. It is more difficult to make a wide-angle retarder. The widest-angle device is the photoelastic modulator, invented by J. Kemp. The photoelastic modulator can be made to alternately pass right- and left-circular light over a field with a half-angle of 20°. The photoelastic modulator aperture is generally limited to about 5 cm (2 in.), but this dimension is about twice that of the aperture of classical prisms.

Some devices for detecting linear and circular polarization depend on scattering. To detect circular polarization in sunlight at the wavelength of the sodium line (589 nanometers), a cell filled with sodium vapor is sometimes used. Such a resonance cell is placed inside a laboratory magnet. Then the atoms preferentially absorb and reemit the circularly polarized component of the incident light. The resonance cell is very stable and accepts light from a very large angle.

In the infrared, glass containing tiny aligned silver needles can discriminate efficiently between the two senses of linear polarization because the needles scatter electromagnetic waves aligned with their long axis far more efficiently than they scatter nonaligned waves.

David M. Rust

Bibliography. R. J. Bray and R. E. Loughhead, *Sunspots*, 1965; D. Clarke and J. F. Grainger, *Polarized Light and Optical Measurement*, 1971; W. G. Driscoll and W. Vaughan (eds.), *Handbook of Optics*, 1978; T. Gehrels (ed.), *Planets, Stars and Nebulae Studied with Photopolarimetry*, 1974; D. J. Saikia and C. J. Salter, Polarization properties of extragalactic radio sources, *Annu. Rev. Astron. Astrophys.*, 26:93–143, 1988; P. Zeeman, *Researches in Magneto-optics*, 1913.

Precession of equinoxes

A slow change in the direction of the axis of rotation of the Earth that results in a slow movement on the sky of the vernal equinox, which is the intersection of the celestial equator (the projection of the terrestrial equator) with the ecliptic (the apparent path of the Sun across the sky).

Lunisolar precession. The major component of this precession is caused primarily by the Moon and secondarily by the Sun. The Earth is an oblate spheroid, bulging outward around the Equator by about 1 part in 300. Because the Earth's axis of rotation is tilted by about 23.4° to the plane of its orbit about the Sun, this bulge is also tilted. The extra gravitational pull of the Moon and Sun on this bulge produces a revolution, or precession, of the axis of rotation about the perpendicular to the orbit (see **illus.**); one complete revolution requires just under 26,000 years. This is called lunisolar precession and has been recognized for over 2000 years.

In addition to uniform revolution, there are very small-amplitude periodic oscillations; the primary one has a period of 18.6 years. These oscillations are collectively called nutation. They are caused by the same forces as precession, but are treated observationally and computationally as separate phenomena. *See* NUTATION.

Planetary precession. A separate and much less significant precession is caused by the other planets in the solar system. The motion of Earth and Moon about the Sun would be on a simple fixed ellipse if it were not for the perturbing effects of other bodies. The principal effect is to produce an extremely small rotation of the axis of revolution of the Earth around an axis parallel to the total-angular-momentum axis of the solar system. This therefore produces a very small motion of the ecliptic itself and thus an additional small change in the apparent position of the equinox. This precession is called planetary precession.

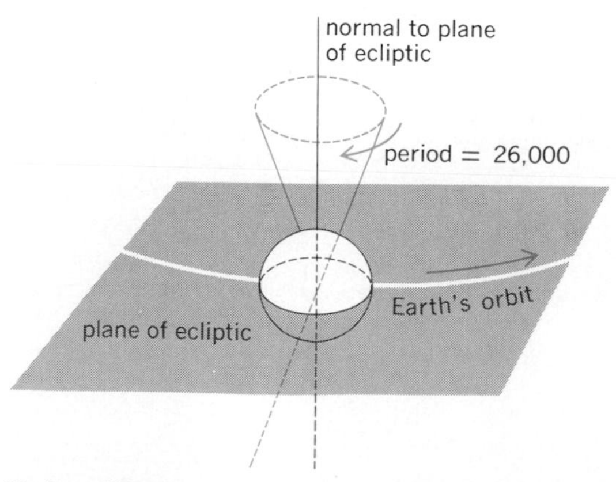

Motion of Earth's axis of rotation in lunisolar precession.

Effects. The effect of precession is to produce apparent changes in the positions of celestial bodies as determined from the surface of the Earth. The direction of motion of general precession (the combination of the two above precessions) is westward, producing an apparent eastward motion of all objects apart from those in certain regions near the celestial poles. As a result, Polaris is not always close to the North Pole, and the existence of the North Star is fortuitous. In time, Vega (the brightest star in the present northern hemisphere) will be closer to the pole than Polaris and will again, as in the past, serve as a pole star of true significance. SEE ASTRONOMICAL COORDINATE SYSTEMS; EARTH ROTATION AND ORBITAL MOTION.

Robert S. Harrington

Bibliography. R. M. Green, *Spherical Astronomy*, 1985; H. Moritz and I. I. Mueller, *Earth Rotation*, 1987; K. Lambeck, *The Earth's Variable Rotation*, 1980; I. I. Mueller, *Spherical and Practical Astronomy, as Applied to Geodesy*, 1969; W. H. Munk and G. F. MacDonald, *Rotation of the Earth: A Geophysical Discussion*, 2d ed., 1975; F. D. Stacey, *Physics of the Earth*, 2d ed., 1977.

Proton-proton chain

A group of nuclear reactions involving fusion of light nuclei that converts hydrogen into helium. It is believed to be the principal source of energy in main sequence stars of a little more than a solar mass and of less massive stars. Completion of a chain results in the consumption of four protons (hydrogen-1 nuclei, designated ^{1}H), and the production of a helium (^{4}He) nucleus plus two positrons (e^+) and two neutrinos (ν). The two positrons are annihilated along with two electrons (e^-), and the total energy release is 26.73 MeV. Approximately 0.58 MeV is released as neutrino energy and is not available as thermal energy in a star. The chain can be thought of as the conversion of four hydrogen atoms into a helium atom plus energy in the form of photons or neutrinos, or the kinetic energy of particles. The energy E = 26.73 MeV arises from the mass difference between four hydrogen atoms and the helium atom, and is calculated from the Einstein mass-energy equation $E = \Delta mc^2$, where Δm is the mass difference and c^2 is the square of the velocity of light. Because hydrogen is the fuel consumed in the process, it is referred to as hydrogen burning by means of the proton-proton chain.

Reactions. The first reaction involves the fusion of two protons to form a nucleus of heavy hydrogen (the deuteron, ^{2}H) with the release of a positron and a neutrino. In a relatively rare alternate reaction, one proton may capture an electron and fuse with a second proton to form the deuteron with release of a monoenergetic neutrino. This is followed by the fusion of a proton and a deuteron to form a helium nucleus of mass 3 (^{3}He) with the release of a 5.494-MeV photon (γ). The electromagnetic energy of the photon is promptly converted into thermal energy. Most often, the chain is completed by fusion of two ^{3}He nuclei to form a ^{4}He nucleus with the release of two protons. These reactions are represented by reactions (1)–(4) in **Tables 1** and **2**. In Table 1, the energy release, Q, of each reaction is indicated following the reaction. The energy convertible into thermal heat, Q (thermal), is also given, as is the maximum neutrino energy E_ν^{max}. The average neutrino energy loss is given by $Q - Q$(thermal).

If there is sufficient ^{4}He already present at the site of hydrogen burning, it is possible to complete the proton-proton chain by reactions (5)–(7) in Tables 1 and 2, or at higher hydrogen-burning temperatures by reactions (8)–(10). Reaction (6) in Tables 1 and 2 represents the capture of an electron by the beryllium-7 nucleus to produce a lithium-7 nucleus and a neutrino. Boron-8 is unstable to β^+ decay and is shown releasing a positron (e^+) and a neutrino in its decay to an excited state in ^{8}Be (the asterisk on ^{8}Be* shows that it is an excited state rather than the ground state of that nucleus). The ^{8}Be* then decays [reaction (10)] to two alpha particles (^{4}He nuclei) releasing 3.03 MeV, which represents the 2.94-MeV excitation energy plus the 0.092 MeV by which the mass of two alpha particles differs from that of the ground state of ^{8}Be.

Reaction rates and mean lifetimes τ in the Sun for the reactions in the proton-proton chain, calculated from the reaction rate equations of M. J. Harris and colleagues using the conditions of temperature, density, and mass fractions of hydrogen and helium from a standard model of the Sun developed by J. N. Bahcall and colleagues, are shown in Table 2. In this model, the principal site of hydrogen burning is just outside the central core. At a distance of 0.0511 solar

Table 1. Energies involved in the proton–proton chain reaction

Reaction	Q, MeV	Q (thermal) MeV	E_ν^{max}, MeV	
$^1H + {}^1H \rightarrow {}^2H + e^+ + \nu$	1.442	1.192	0.420	(1)
or				
$^1H + e^- + {}^1H \rightarrow {}^2H + \nu$	1.442	0.001	1.441	(2)
$^2H + {}^1H \rightarrow {}^3He + \gamma$	5.494	5.494		(3)
$^3He + {}^3He \rightarrow {}^4He + 2{}^1H$	12.859	12.859		(4)
or				
$^3He + {}^4He \rightarrow {}^7Be + \gamma$	1.586	1.586		(5)
$^7Be + e^- \rightarrow {}^7Li + \nu$	0.862	0.050	0.862 (89.5%) 0.384 (10.5%)	(6)
$^7Li + {}^1H \rightarrow 2{}^4He$	17.347	17.347		(7)
or				
$^7Be + {}^1H \rightarrow {}^8B + \gamma$	0.137	0.137		(8)
$^8B \rightarrow {}^8Be^* + e^+ + \nu$	15.04	7.41	14.06	(9)
$^8Be^* \rightarrow 2{}^4He$	3.03	3.03		(10)

Table 2. Reaction rates and mean lifetimes in the Sun

Reaction	R^*	$\lambda^\dagger$, s^{-1}	$\tau^\ddagger$, y	%[§] termination	
$^1H + {}^1H \rightarrow {}^2H + e^+ + \nu$	5.90×10^{-20}	3.17×10^{-18}	1.00×10^{10}	99.75	(1)
or					
$^1H + e^- + {}^1H \rightarrow {}^2H + \nu$	2.10×10^{-24}	9.33×10^{-21}	3.40×10^{12}	0.25	(2)
$^2H + {}^1H \rightarrow {}^3He + \gamma$	9.63×10^{-3}	5.18×10^{-1}	6.12×10^{-8}	100.	(3)
$^3He + {}^3He \rightarrow {}^4He + 2{}^1H$	7.74×10^{-11}	7.76×10^{-14}	4.08×10^{5}	86.	(4)
or					
$^3He + {}^4He \rightarrow {}^7Be + \gamma$	7.61×10^{-16}	1.06×10^{-14}	2.98×10^{6}		(5)
$^7Be + e^- \rightarrow {}^7Li + \nu$	1.52×10^{-9}	1.26×10^{-7}	2.52×10^{-1}	14.	(6)
$^7Li + {}^1H \rightarrow 2{}^4He$	7.73×10^{-6}	4.16×10^{-4}	7.63×10^{-5}		(7)
or					
$^7Be + {}^1H \rightarrow {}^8B + \gamma$	1.71×10^{-12}	9.19×10^{-11}	3.45×10^{2}		(8)
$^8B \rightarrow {}^8Be^* + e^+ + \nu$		9.01×10^{-1}	3.52×10^{-8}	0.015	(9)
$^8Be^* \rightarrow 2{}^4He$		2.37×10^{21}	1.34×10^{-29}		(10)

*The units of the thermonuclear reaction rate R are reactions per second per mole/cm^3 for reactions involving two interacting nuclei and in reactions per second per $(mole/cm^3)^2$ for three interacting particles. (After M. J. Harris et al., Thermonuclear reaction rates, III, *Annu. Rev. Astron. Astrophys.*, 1983.)

†The quantity λ is the destruction rate per second for the first nucleus in column one.

‡The mean lifetime of the first nucleus in the first column is $\tau = 1/(3.1558 \times 10^7 \lambda)$ in years.

§After J. N. Bahcall et al., Standard solar models and uncertainties in predicted capture rates of solar neutrinos, *Rev. Mod. Phys.*, 54:767–799, 1982.

radius from the center, chosen to lie in the most active part of that site, the mass fraction of ^1H is calculated to be $X(^1H) = 0.483$, that of ^3He is $X(^3He) = 2.70 \times 10^{-5}$, and that of ^4He is $X(^4He) = 0.500$. The remainder $X(A, \geq 12) = 0.017$ is in the form of heavy elements. The moles per gram of electrons is $Y(e^-) = 0.738$; this value is approximately the number of electrons per nucleon (in this context, nucleons include neutrons bound in nuclei plus free and bound protons). The density at this distance is $\rho = 112$ g/cm^3 (112 times the density of water), and the temperature is 14×10^6 K (25×10^6 °F). The thermonuclear reaction rates R, calculated for this temperature, are shown in Table 2. Small corrections for screening of the Coulomb field between nuclei by electrons and for partial electron degeneracy have been neglected.

The quantity λ in Table 2 is calculated from $\rho RX(^1H)/1.0078$ for two-body reactions induced by ^1H, from $\rho RX(^3He)/3.0160$ for reactions induced by ^3He, and from $\rho RX(^4He)/4.0026$ for reactions induced by ^4He. In the three-body reaction $^1H + e^- + {}^1H \rightarrow {}^2H + \nu$, $\lambda = \rho RY(e^-)X(^1H)/1.0078$. Proper allowance has been made for interactions involving identical particles.

Using the rates of the nuclear reactions in the chain under conditions believed to be prevalent in the hottest part of the core of the Sun, Bahcall and colleagues calculated the probable percentages of the different modes of completion of the *p-p* chain. They determined that about 86% of the reaction chains in the Sun are completed through the $^3He + {}^3He$ reaction, and 14% through the $^3He + {}^4He$ reaction. This 14% divides up with only 0.015% through $^7Be + {}^1H$, and the remainder, approximately 14%, through $^7Be + e^-$.

Neutrino emission. The electromagnetic photon energies and the particle kinetic energies produced in these nuclear reactions are promptly converted into thermal energy in the central region of a star. This energy is transported rather slowly to the surface of the star by radiative transfer or by convection. In the Sun, for example, several millions of years are required for the energy to reach the surface. In contrast, the neutrinos that are produced interact only weakly with matter and thus travel directly out of the Sun at the velocity of light in about 2 s. Although the neutrinos carry only a small percentage of the energy produced in the nuclear reactions, their penetrability provides a means of determining what is happening at the center of the Sun now, rather than millions of years ago. R. Davis and his collaborators at the Brookhaven National Laboratory devised an ingenious experiment to measure the influx of these neutrinos reaching the Earth from the Sun. Their detector is ^{37}Cl, the heavy isotope of chlorine. This detector is not sensitive to the abundant low-energy neutrinos from reaction (1) in the tables, but is particularly sensitive to the relatively high-energy neutrinos of the ^8B decay, so the analysis of the importance of the infrequent branch of the chain which produces ^8B is of particular interest. Scientists are puzzled by the fact that the flux of neutrinos expected on the basis of laboratory reaction-rate measurements and theories of the present structure and evolutionary state of the Sun is several times larger than that found in Davis's very careful measurements. Astrophysicists are reexamining all aspects of their theories of stellar structure and conditions of the core of the Sun in their attempts to solve the solar neutrino problem. Nuclear and elementary particle physicists are studying the properties of the electron-type neutrinos emitted by the Sun, particularly the remote possibility that they may transform into undetectable muon and tauon neutrinos on their journey from the Sun to the Earth. An experiment has been designed using ^{71}Ga, the heavy isotope of gallium. This isotope is sensitive to the abundant low-energy neutrinos from the Sun and should provide a definitive solution to the solar neutrino problem. SEE CARBON-NITROGEN-OXYGEN CYCLES; SOLAR NEUTRINOS; STELLAR EVOLUTION; SUN.

Georgeanne R. Caughlan

Bibliography. J. N. Bahcall et al., Standard solar models and the uncertainties in predicted capture rates of solar neutrinos, *Rev. Mod. Phys.*, 54:767–799, 1982; C. A. Barnes, Nucleosynthesis by charged particle reactions, *Advan. Nucl. Phys.*, 4:133–203, 1971; C. A. Barnes, D. D. Clayton, and D. N.

Schramm (eds.), *Essays in Nuclear Astrophysics*, 1982; D. D. Clayton, *Principles of Stellar Evolution and Nucleosynthesis*, 1984; W. A. Fowler, *Nuclear Astrophysics*, 1967; M. J. Harris et al., Thermonuclear reaction rates, III, *Annu. Rev. Astron. Astrophys.*, 1983; P. D. Parker, J. N. Bahcall, and W. A. Fowler, Termination of the proton-proton chain in stellar interiors, *Astrophys, J.*, 139:602–621, 1964; F. Reines (ed.), *Cosmology, Fusion, and Other Matters*, George Gamow Memorial Volume, 1972.

Protostar

A dense condensation of material that is still in the process of accreting matter to form a star. Protostars are expected in dense interstellar complexes of gas and dust. Since the gas is mostly in the form of molecules, especially molecular hydrogen (H_2), these complexes are called molecular clouds. The cloud dust is actually tiny solid particles, probably a mixture of silicates and carbon-containing compounds such as graphite. Because these dust particles absorb starlight, some of the nearest molecular clouds can be observed in the night sky as extended dark patches against the bright band of stars that form the Milky Way. *SEE INTERSTELLAR MATTER; MOLECULAR CLOUD.*

Observation. Most of the radiation emitted by a protostar is absorbed by the surrounding dust. This radiation heats the dust, causing it to reradiate at infrared wavelengths. Strong infrared emission from a compact region in a molecular cloud is therefore one sign of the presence of a protostar. The gas around the protostar will also be heated and excited. Molecular hydrogen itself does not emit spectral lines at wavelengths convenient for detection by astronomers, but other molecules such as carbon monoxide (CO) are also present in significant amounts. When these molecules are excited, they emit lines that can be detected by radio telescopes sensitive to millimeter- and submillimeter-wavelength radiation. The Doppler shifts of these spectral lines can be measured and the motions of the gas thereby determined. Thus, while the protostars themselves are completely hidden from the view of optical telescopes, their properties can, in principle, be inferred from observations of the obscuring dust and gas. *SEE DOPPLER EFFECT; RADIO ASTRONOMY.*

Evolution. The millimeter-wavelength radio observations of molecular clouds show that they are far from homogeneous. In particular, they contain clumps of material with relatively low temperatures, typically 10 to 50 K (−442 to −370°F), and densities significantly higher than in the surrounding medium, between 10,000 and 100,000 atoms per cubic centimeter (**illus.** *a*). In the early part of the twentieth century, J. Jeans demonstrated that such condensations would collapse under the effect of gravity if they were more massive than some value, now called the Jeans mass, $M_J = 30{,}000(\sqrt{T^3/n})M_\odot$. Here, $M_\odot$ is the mass of the Sun, T is the temperature of the gas in kelvins, and n is the density in atoms per cubic meter. A combination of low temperature and high density clearly makes M_J smaller and so encourages cloud collapse and ultimately the formation of a star.

The protostar is the central core that forms as a cloud condensation contracts. According to Newton's theory of gravitation, the denser material at the center of the condensation collapses faster. Because the collapsing core is also rotating, its outer parts are flattened into a disk. At this stage, the protostar's mass is very small, perhaps only 1/1000 of the mass of the present Sun. Over a period of 10^4–10^6 years, it continues to accrete material from the infalling envelope and eventually attains a stellarlike mass. Core collapse is checked, leaving a central protostar, when the gravitational heating produced by the contraction can no longer be radiated away as quickly as it is generated. Eventually, the core becomes sufficiently hot that the nuclear reactions that sustain stars can begin. The next step in the evolutionary sequence from a molecular cloud core to a protostar surrounded by a disk and embedded in an infalling envelope is shown schematically in illus *b*. *SEE GRAVITATION.*

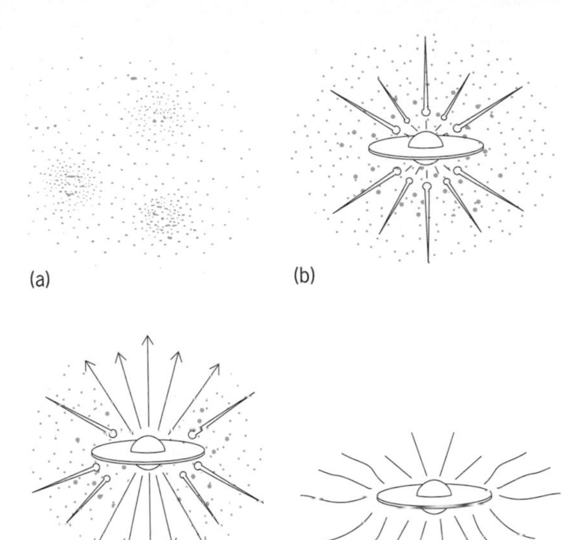

Schematic diagram of various stages of protostellar evolution. (a) Denser cores occur in a larger molecular cloud. (*b*) A protostar surrounded by a disk has formed in a collapsing core. (*c*) A well-collimated wind has broken through at the opposite poles of the system. (*d*) A visible star with a circumstellar disk is revealed. (*After F. H. Shu, F. C. Adams, and S. Lizano, Star formation in molecular clouds: Observations and theory, Annu. Rev. Astron. Astrophys., 25:23–84, 1987*)

Evidence. Observations of molecular spectral lines at millimeter wavelengths have established the properties of condensations in molecular clouds and shown that the conditions for collapse are readily met. The situation depicted in illus. *a* is not in doubt. Nevertheless, the scenario shown in illus. *b* has not yet been observed. Much of the description of protostellar evolution presented here is based on theoretical considerations. Although the velocity patterns of the gas around many objects have been measured by using the presence of small, intense, infrared emission regions in molecular clouds as signatures of protostars, infalling material has not been detected. By contrast, there is frequently evidence that material is being shed by the central object. Since the early 1980s, many examples of the situation in illus. *c*, where gas is seen emanating from an embedded source in two well-collimated and oppositely directed outflows, have been observed. Because the true protostar stage of illus. *b* has not yet been unambiguously detected, the objects at the cores of bipolar outflows (illus. *c*) are loosely referred to as protostars.

High-mass and low-mass stars. The initial phases of formation are the same for both low- and high-mass stars. Here, the term massive implies a stellar mass more than 10 times that of the Sun. The lack of evidence for infalling material around massive stars is not surprising. For these stars, nuclear fusion begins, and therefore ultraviolet radiation is emitted, before the accretion phase ends. The ultraviolet radiation ionizes the surrounding gas and produces H II regions that expand and help clear away the remnant envelope. These processes and the rapidity with which massive stars evolve—a visible object emerges in about 10,000 years—combine to make the protostellar stage very difficult to observe directly. Bipolar outflows centered on compact H II regions and, by implication, massive protostars are not, however, uncommon.

Low-mass stars develop more slowly, and there is insufficient ultraviolet radiation to cause an H II region. Under the relatively quiescent circumstances, the collapsing envelopes should be detectable. Indeed, when these objects become visible as T Tauri stars, they still lie above and to the right of the main sequence in the Hertzsprung-Russell diagram, and their associated disks remain undissipated (illus. *d*). The earlier evolutionary phase in illus. *c*—a bipolar gas outflow centered on a pre-main-sequence star with a surrounding disk—is also a well-observed phase of low-mass protostellar evolution. Nevertheless, the phase in illus. *b* remains undiscovered, and the search for the elusive protostar in a collapsing envelope is one of the greatest challenges facing observational astronomers. *SEE HERTZSPRUNG-RUSSELL DIAGRAM; STELLAR EVOLUTION; T TAURI STAR.*

Anneila I. Sargent

Bibliography. D. J. Hollenbach (ed.), *Interstellar Processes*, 1987; C. J. Lada, Cold outflows, energetic winds, and enigmatic jets around young stellar objects, *Annu. Rev. Astron. Astrophys.*, 23:267–317, 1985; C. J. Lada and F. H. Shu, The formation of sunlike stars, *Science*, 248:564–572, 1990; M. Peimbert and J. Jugaku (eds.), *Star Forming Regions*, I.A.U. Symp. 115, 1987; F. H. Shu, F. C. Adams, and S. Lizano, Star formation in molecular clouds: Observations and theory, *Annu. Rev. Astron. Astrophys.*, 25:23–84, 1987.

Pulsar

A celestial radio source producing intense short bursts of radio emission. The unforeseen discovery of pulsars was made by A. Hewish, J. C. Bell, and colleagues at Cambridge University in early 1968. Since that time, 458 pulsars have been found, and it has become clear that 200,000 pulsars must exist in the Milky Way Galaxy—most of them too distant to be detected with existing radio telescopes. *SEE RADIO ASTRONOMY.*

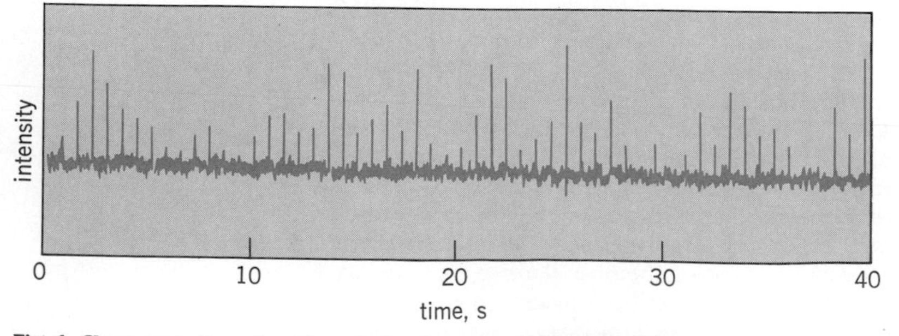

Fig. 1. Chart recording of radio emission from the pulsar PSR 0329+54, observed at a frequency of 400 MHz. This pulsar lies at a distance of about 7500 light-years (4.5 × 10^{16} mi or 7 × 10^{16} km) from Earth.

Pulsars are distinguished from most other types of celestial radio sources in that their emission, instead of being constant over time scales of years or longer, consists of periodic sequences of brief pulses. The interval between pulses, or pulse period, is nearly constant for a given pulsar, but for different sources ranges from 0.002 to 4 s. The bursts of emission are generally confined to a window whose width is a few percent of the interpulse period. A tracing of the signal received from a pulsar by a large radio telescope is shown in **Fig. 1**. Individual pulses can vary widely in intensity; however, their periodic spacing is accurately maintained.

Neutron stars. The association of pulsars with neutron stars, the collapsed cores left behind when moderate- to high-mass stars become unstable and collapse, is supported by many arguments. Prior to the discovery of pulsars, neutron stars were thought to be unobservable owing to their extremely small size (radius less than 10 mi or 15 km). The realization that a rapidly spinning, highly magnetized neutron star might be responsible for the luminosity of the Crab Nebula predated the discovery of pulsars by months. The standard model for pulsars is a spinning neutron star with an intense dipole magnetic field (surface field of 10^{12} gauss or 10^{8} teslas) misaligned with the rotation axis. This combination of properties evidently gives rise to highly directive radio emission at meter and centimeter wavelengths that is seen as pulses owing to the rotation, just like a rotating searchlight.

The most important observation that supports this model is the remarkable stability of the basic pulsation periods, which typically remain constant to a few tens of nanoseconds over a year. This stability is natural to the free rotation of a compact, rigid object like a neutron star, but is extremely difficult to produce by any other known physical process. A small emitting body is also required by the observed rapid variations within pulses. A spinning, magnetized body will gradually slow down because it emits low-frequency electromagnetic radiation at harmonics of its rotation period. This slowing down is observed in all objects where observations of sufficient precision are available.

Further support of the neutron star model comes from the discovery of a pulsar in the center Crab Nebula, the remnant of the supernova observed in A.D 1054. The lengthening of this pulsar's short period on a time scale of 1000 years, matches the age of the remnant. The star's properties match the earlier neutron star prediction. The half dozen associations that have been found between pulsars and supernova remnants lend further support to these ideas. *SEE CRAB NEBULA; NEUTRON STAR; SUPERNOVA.*

Pulsed emission. The waveform of a pulsar's periodic emission, averaged over several hundred pulses or more, has a distinctive shape characteristic of each individual source. Several examples of average pulse profiles are shown in **Fig. 2**. Variations in pulse shape from one pulsar to another are probably the result of differences in detailed structure of the neutron star's magnetic field, or of a different orientation of the rotating beam relative to the Earth, or both. Pulse

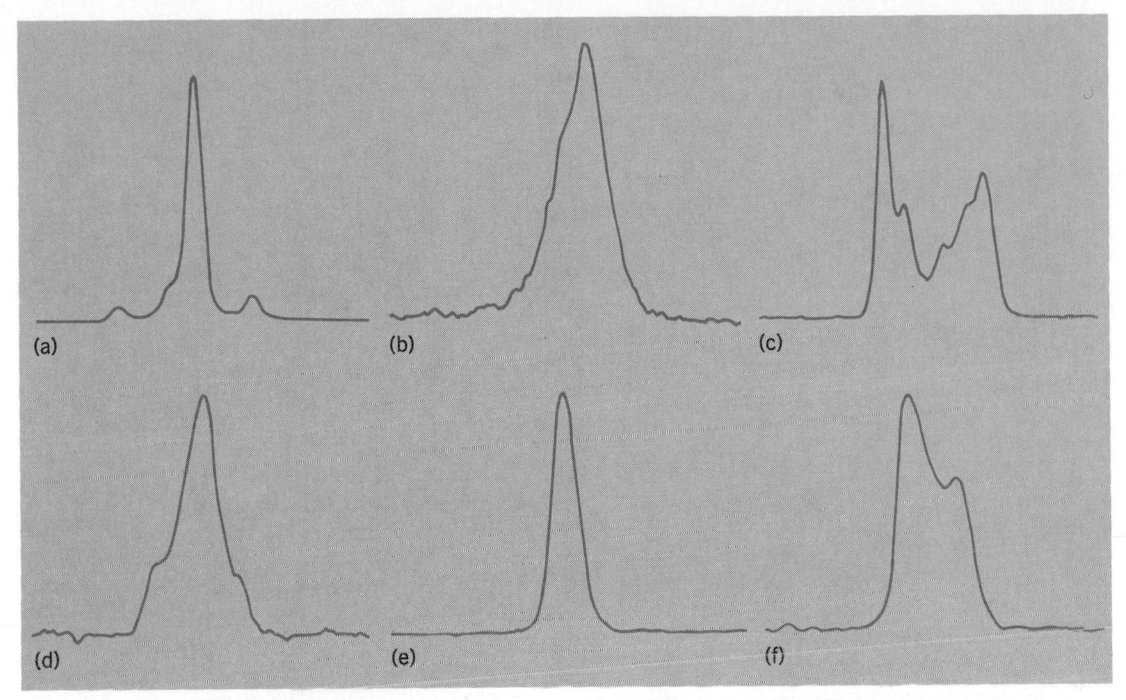

Fig. 2. Average pulse shapes of six pulsars. (*a*) PSR 0329+54. (*b*) PSR 0950+08. (*c*) PSR 1237+25. (*d*) PSR 1508+55. (*e*) PSR 1642−03. (*f*) PSR 1919+21.

shapes generally exhibit approximate mirror symmetry about a central point. The emission is usually highly polarized, and observations with high time resolution show that the angle of the polarization vector (projected onto the plane of the sky) rotates smoothly across the pulse profile. The total change in angle is always less than or about 180°, and this behavior is taken to be the straightforward result of the changing projection of the pulsar's magnetic field lines as the star spins relative to the line of sight.

Individual pulses from many of the strong, relatively nearby pulsars have been studied in great detail, and they are found to vary rapidly in intensity, shape, and polarization. Intensity fluctuations on time scales down to 300 microseconds are shown for one pulsar, PSR 1133+16, in **Fig. 3**. Individual pulses normally consist of one or more subpulses, and the subpulses may in turn contain "microstructure" of even shorter duration. In some pulsars, subpulses vary in intensity in an obviously nonrandom way, and sometimes exhibit secondary periodicities incommensurate with the basic pulsar period. Individual pulses are more highly polarized than the average pulse profiles, and frequently undergo abrupt changes of polarization angle of almost exactly 90°. Many attempts have been made to incorporate such complications as these into models of the pulse emission mechanism, but no one model has won general acceptance, and many uncertainties remain.

Spectra. With a few notable exceptions, pulsars are observable only at radio frequencies, typically between 50 and 5000 MHz. Over this range the average flux density decreases rapidly, usually in approximate proportion to f^{-1} or f^{-2}, where f is the observing frequency. At frequencies below a few hundred megahertz, a combination of galactic background radiation and terrestrial interference makes observations very difficult; at frequencies above a few thousand megahertz, most pulsars become too weak to be observed with existing instruments.

When extrapolated to optical or higher frequencies, the steep radio-frequency spectra of pulsars predict flux densities far below detectable limits. However, a few of the youngest pulsars, including the pulsar in the Crab Nebula, have been detected at optical, x-ray, and gamma-ray frequencies. Evidently, another emission mechanism (or mechanisms) is responsible for the high-frequency radiation from these pulsars. Several dozen pulsarlike objects have also been observed in x-rays, but not at radio frequencies. Invariably these sources are found to be in gravitationally bound orbits around large, evolving stars which are transferring some of their atmospheric gas onto their companions, presumably neutron stars. The transferred gas becomes so hot that it emits thermal x-rays, but at the same time interferes with the radio emission process and prevents the neutron star from becoming an observable radio-frequency pulsar. *SEE ASTROPHYSICS, HIGH-ENERGY; BINARY STAR; X-RAY ASTRONOMY.*

Interstellar medium. Pulsars have provided astronomers with a unique set of probes for the investigation of the diffuse gas and magnetic fields in interstellar space. Measurement of absorption at 1420 MHz, the frequency of the hyperfine transition in ground-state neutral hydrogen atoms, gives information on the structure of gas clouds, and in many cases provides an estimate of the pulsar distance. The index of refraction for radio waves in the ionized interstellar gas is strongly frequency-dependent, and low-frequency signals propagate more slowly than those at high frequencies. The broadband, pulsed nature of pulsar signals makes them ideal for measurements of this dispersion. When independent estimates of individual pulsar distances are available, the dispersion may be used to determine the mean interstellar electron density. Conversely, when a value for the mean

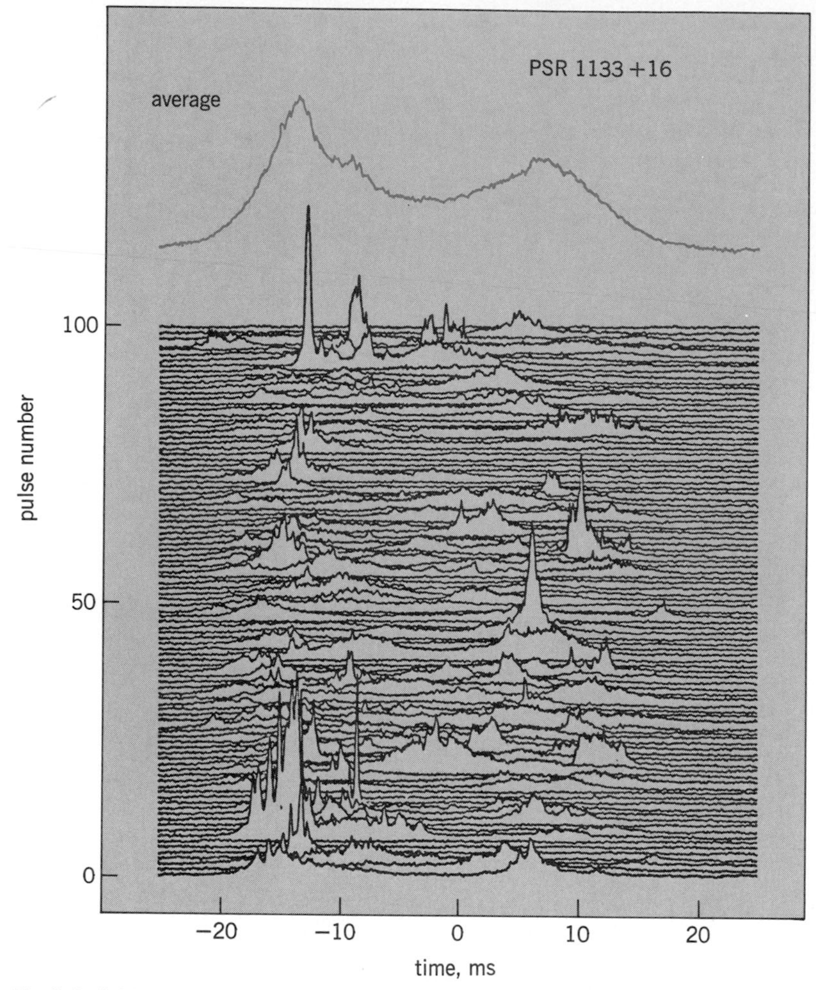

Fig. 3. Individual pulses from pulsar PSR 1133+16, illustrating the variations which take place within pulses and from one pulse to the next.

density is assumed, individual pulsar distances may be derived from observed dispersion measures. It has been found in this way that the average interstellar electron density is approximately 0.03 cm^{-3} in the galactic disk, and that most of the known pulsars lie within 10,000 light-years of the Earth (1 light-year equals 5.88×10^{12} mi or 9.46×10^{12} km).

Because of the large-scale galactic magnetic field, the index of refraction in the interstellar medium is different for the two circular components of a linearly polarized wave. For this reason, the plane of polarization rotates along the propagation path, an effect known as Faraday rotation. Measurements of this effect for some 60 pulsars have shown that the magnetic field strength in the solar neighborhood of the Milky Way Galaxy is approximately 2×10^{-6} G (2×10^{-10} T) in a direction along the local spiral arm. *See* Interstellar matter; Milky Way Galaxy.

Galactic distribution. Over 90% of the sky has been searched for pulsars down to a uniform minimum flux density of approximately 10^{-28} W m^{-2} Hz^{-1}. Some 326 pulsars have been detected above this limit, and their distribution in the sky is shown in the galactic coordinates map in **Fig. 4**. The figure also shows 40 pulsars with flux densities as much as 10 times weaker, detected in a limited deep survey of the galactic plane at longitudes 40–60°. The obvious concentration of pulsars within ±10° of the galactic equator and in the hemisphere toward the galactic center (longitudes 0–90° and 270–360°) demonstrates that pulsars are members of the Milky Way Galaxy and that their distances are large compared to the thickness of the galactic disk. When distance estimates are combined with the two-dimensional information in Fig. 4, a more elaborate analysis shows that pulsars occupy a disk-shaped region of space approximately 2000 light-years thick and 40,000 light-years in radius. (The Sun, at a distance of about 30,000 light-years from the galactic center, lies near the outer edge of the pulsar distribution.)

The massive, luminous stars which are believed to be the progenitors of pulsars have a galactic distribution similar to that of the pulsars, except that the thickness of their disk is only about 500 light-years. It appears that pulsars are found at much larger distances from the galactic plane because most pulsars acquire a substantial velocity (60 mi s^{-1} or 100 km s^{-1} or more) at birth, as a result of the supernova explosion. During a typical pulsar's active lifetime, roughly 5×10^6 years, it has time to move as much as 1500 light-years away from its birthplace. The equivalent thickness of the pulsar disk will thus be several times greater than that of the parent stars.

If pulsars typically remain active for 5×10^6 years, and if the present inferred galactic population of 200,000 pulsars is to be maintained, then a new pulsar must be formed somewhere in the Galaxy approximately every 25 years. This birth rate is reasonably close to the estimated rate of occurrence of galactic supernovas, if allowance is made for those that are optically obscured from view by interstellar dust.

Millisecond pulsars. No pulsars spinning faster than the Crab Nebula pulsar were found in the all-sky surveys conducted during the 1970s. The resulting view that few if any pulsars spin faster than 0.030 s was shattered with the discovery in 1982 of PSR 1937+21 with the remarkably short period of 1.6 ms, the first millisecond pulsar. This period is within a factor of 2 or so of the stability limit of neutron stars of 1 solar mass. The period is lengthening, but on time scales of 10^8 years, not 10^6 years or less. Evidently, this object is old and has a low magnetic field. A dozen millisecond-period pulsars are now known.

Long-term timing observations of millisecond pulsars have shown that they are extremely accurate clocks which exhibit few or none of the rotation irregularities seen in most pulsars. The observed stabilities approach and may exceed the best terrestrial atomic time scale, which is derived from cesium clocks maintained around the world. Timing of an array of millisecond pulsars will provide the best time scale for durations exceeding 1 year. The data may also improve knowledge of the orbit of the Earth in the solar system. After local effects are removed, pulsar timing array data can be used to set limits on, or even detect, a background of gravitational radiation left over from chaotic events in the early universe. *See* Atomic clock; Atomic time; Gravitation; Horology.

Binary pulsars. Most stars in the Milky Way Galaxy are members of binary systems. Many compact, x-ray-emitting objects in the Galaxy, mentioned above, are also binary systems. The first pulsar in a binary, PSR 1913+16, was found in 1975. A dozen binary pulsars are now known. Orbital periods range from hours to months. Timing observations of the binary pulsars, which provide indications of the companion masses, suggest that there are two categories

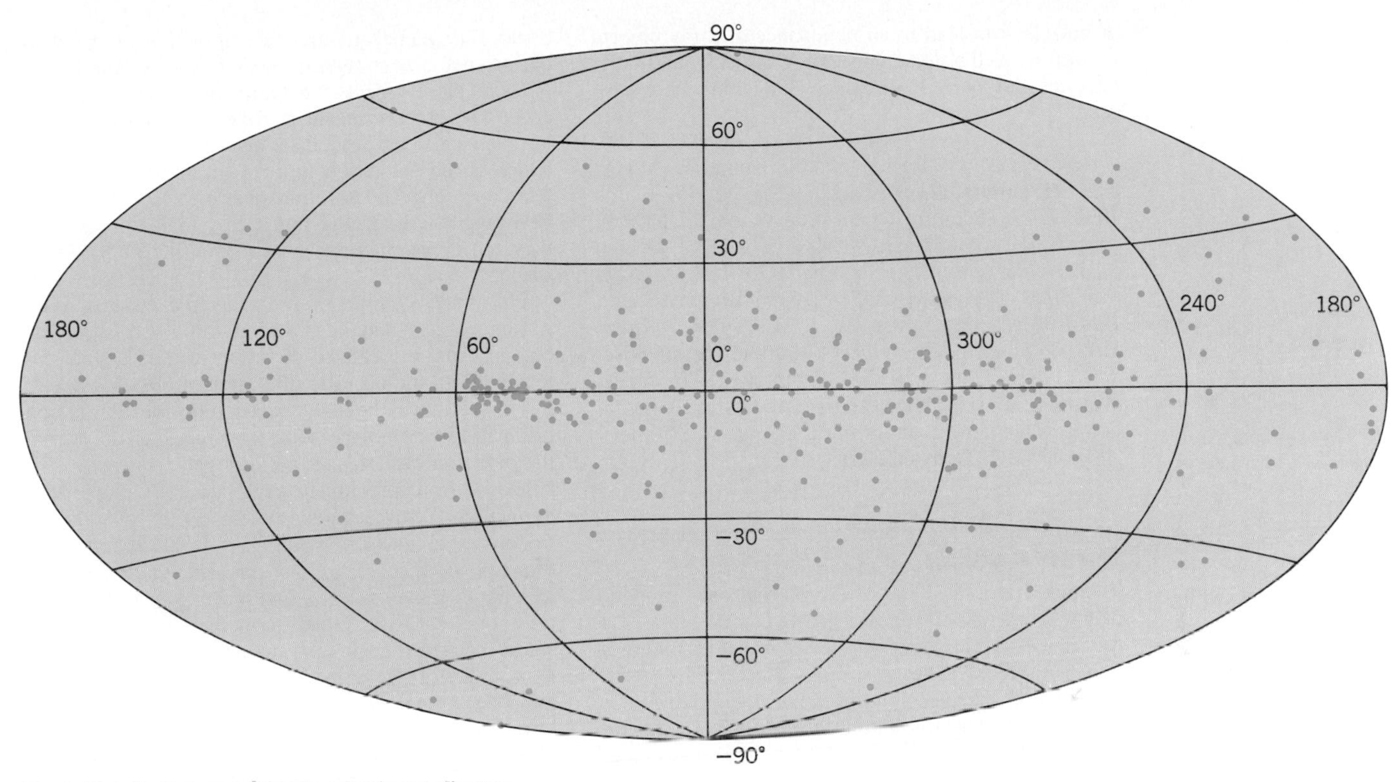

Fig. 4. Distribution of pulsars in galactic coordinates.

of pulsar binaries: low-mass and high-mass. High-mass systems are probably a pair of neutron stars, only one of which has its emission beam passing over the Earth. These have managed to survive the supernova explosions of both stars, which is unlikely since the expulsion of matter in a supernova can disrupt a binary system.

Mass and angular momentum transfer from a pre-white dwarf star onto the neutron star can spin-up the neutron star to millisecond periods. The companions of the pulsars in two low-mass binaries have been found to be white dwarf stars. The old age of the white dwarfs confirms the ages obtained from the pulsar's period decay, as well as the hypothesis that the dipole magnetic field decays with age. While the spin-up scenario explains the origin of most millisecond pulsars, PSR 1937+21 and a few others are still a puzzle since they have no companion. One binary millisecond pulsar shows eclipses by matter surrounding the companion, which suggests that a pulsar can ablate its companion out of existence.

The binary pulsar PSR 1913+16 has an orbital speed near one-thousandth the speed of light. This large speed and the intense gravitational field of the nearby companion neutron star, when combined with the accurate clock mechanism provided by the pulsations, make this system an ideal testing ground for relativistic gravitation theories. One especially important prediction of the general theory of relativity is that a close binary star system should gradually lose energy by the radiation of gravitational waves, and consequently the two stars should slowly spiral closer together. Observations of Doppler shifts of the periodic pulsar signals over more than 10 years have established that the two masses in the PSR 1913+16 system are each approximately 1.4 times the mass of the Sun. The quantitative prediction of general relativity is, then, that the orbital period should diminish by about 10^{-7} s per orbit, amounting to a cumulative orbital phase shift of 8 s after 14 years. Just such an effect has been found (**Fig. 5**), and the observations provide the first (and only) experimental evidence in support of the existence of gravitational waves. SEE GRAVITATION.

Globular-cluster pulsars. Globular clusters are ancient groups of stars, as many as 10^6, that formed in the early stages of the Milky Way Galaxy. In their centers, stars are only a few light-weeks apart so that frequent stellar encounters occur. Neutron stars in clusters have a 10% chance of capturing a companion during the life of the cluster. This constant production

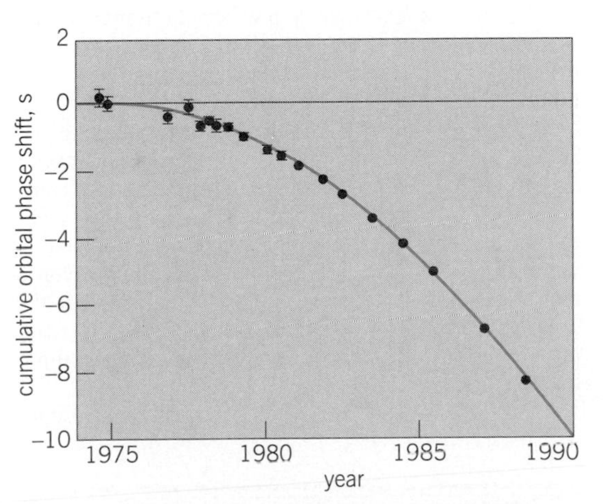

Fig. 5. Accumulating phase shift in the orbit of binary pulsar PSR 1913+16. The parabolic curve gives the prediction of general relativity for energy loss by the radiation of gravitational waves.

of binaries can lead to an abundance of x-ray objects as well as millisecond pulsars. A rush of pulsar discoveries has been made since 1987. SEE STAR CLUSTERS.

Donald C. Backer

Bibliography. D. Backer and S. Kulkarni, A new class of pulsars, *Phys. Today*, 43(3):26–35, March 1990; D. J. Helfand, Recent observations of pulsars, *Amer. Sci.*, 66:332–339, 1978; D. J. Helfand and J.-H. Huang (eds.), *The Origin and Evolution of Neutron Stars: Proceedings of IAU Symposium No. 125*, 1986; R. N. Manchester and J. H. Taylor, *Pulsars*, 1977; J. Shaham, The oldest pulsars in the universe, *Sci. Amer.*, 256(2):50–56, February 1987; J. M. Weisberg, J. H. Taylor, and L. A. Fowler, Gravitational waves from an orbiting pulsar, *Sci. Amer.*, 245(4):74–82, October 1981.

Quartz clock

A clock that makes use of the piezoelectric property of a quartz crystal. When a quartz crystal vibrates, a difference of electric potential is produced between two of its faces. The crystal has a natural frequency of vibration that depends on its size and shape. If it is placed in an oscillating electric circuit having nearly the same frequency as the crystal, it is caused to vibrate at its natural frequency, and the frequency of the entire circuit becomes the same as the natural frequency of the crystal.

In the quartz oscillator, this natural frequency may be used to produce such other frequencies as 1 or 5 MHz. A clock displaying the time of day can also be driven by using one of these frequencies.

The natural frequency of a quartz crystal is nearly constant if precautions are taken when it is cut and polished and it is maintained at nearly constant temperature and pressure. After a crystal has been placed in operation, its frequency usually varies slowly as a result of physical changes. If allowance is made for changes, laboratory quartz-crystal clocks may run for a year with accumulated errors of less than a few thousandths of a second. However, quartz crystals typically used in watches may accumulate errors of several tens of seconds in one year.

For comparison, clocks using rubidium as a frequency standard might be expected to have accumulated errors of less than a few ten-thousandths of a second, while those using cesium might be expected to have accumulated errors better than a few millionths of a second. SEE ATOMIC CLOCK.

The advantage of quartz clocks is that they are relatively inexpensive and easy to use in various applications such as computers and microprocessors. Thus, despite their inaccuracy relative to some other types of clocks, they enjoy wide popularity, particularly in applications requiring accurate timekeeping over a relatively short time span. In these applications, the rates and epochs of the quartz clocks may be readjusted periodically to account for possible accumulated errors. SEE HOROLOGY; TIME.

Dennis D. McCarthy

Quasar

An astronomical object that appears starlike on a photographic plate but possesses many other characteristics, such as a large redshift, that prove that it is not a star. The name quasar is a contraction of the term quasistellar object (QSO), which was originally applied to these objects for their photographic appearance. The objects appear starlike because their angular diameters are less than about 1 second of arc, which is the resolution limit of ground-based optical telescopes imposed by atmospheric effects. Stars also have angular diameters much less than this, and so they too appear unresolved or pointlike on a photograph.

Discovery. Quasars were discovered in 1961 when it was noticed that very strong radio emission was coming from a localized direction in the sky that coincided with the position of a starlike object. Prior to this time, it was believed that strong radio emission originating beyond the solar system came only from the direction of certain exploding galaxies (radio galaxies) or from previously exploded stars (supernova remnants). Two techniques were used by radio astronomers to measure within a few arc-seconds the direction of the radio emission. The more common technique, called interferometry, makes use of two (or more) antennas pointing in the direction of the source. After the signals are amplified at each antenna, they are combined via cables or a microwave radio link. Another method, very long-baseline interferometry (VLBI), records the signals on magnetic tapes which are then synchronously played back so that the signals can be combined. The combined signals yield sinusoidally varying output voltages called fringes. Fringes represent the beating in and out of phase of the incoming radio waves arriving at different times at the two antennas. The phase of the fringes yields an accurate determination of the direction of arrival of the radio waves. SEE INTERFEROMETRY; SUPERNOVA.

The second technique makes use of the motion of the Moon through the sky in its orbit around the Earth. Occasionally the Moon will pass in front of a radio source and block its emission from reaching the Earth. The exact time of such an occultation is observed by using a radio telescope and is compared with the known position of the edge of the Moon at that instant. When the source reappears from behind the Moon several minutes later, the time of reappearance is also measured and compared with the location of the Moon's edge. These two fixes on the radio source are usually sufficient to permit its position to be measured accurately. SEE OCCULTATION.

When these techniques were used to make accurate position measurements of small-angular-diameter radio sources, the coincidence with starlike objects on optical photographs led to the discovery of a new, hitherto-unsuspected class of objects in the universe, the quasars. The full significance of the discovery was not appreciated until 1963, when it was noted that the hydrogen emission lines seen in the optical spectrum of the quasar 3C273 were shifted by about 16% to the red from their normal laboratory wavelength. This redshift of the spectral lines is characteristic of galaxies whose spectra are redshifted because of the expansion of the universe and is not characteristic of stars in the Milky Way Galaxy, whose spectra are only slightly shifted to the red or blue from the Doppler effect of their small velocities relative to the Earth and Sun.

Optical characteristics. The color of quasars is generally much bluer than that of most stars with the exception of white dwarf stars. The optical continuum spectrum of stars has a blackbody distribution, while the energy distribution of quasars is generally a power law of the form f^{α}, where f is the frequency and α is the spectral index and is usually negative. The blueness of quasars as an identifying characteristic led to the discovery that many blue starlike objects have a large redshift and are therefore quasars. The quasistellar objects discovered this way turned out to emit little or no radio radiation and to be about 20 times more numerous than the radio-emitting quasistellar radio sources (QSSs). Why some should be strong radio emitters and most others not is unknown at the present time. The orbiting Einstein x-ray observatory has found that most quasars also emit strongly at x-ray frequencies. *SEE X-RAY ASTRONOMY.*

Most of the luminous energy emitted by quasars is in the far-infrared at about 100-micrometer wavelengths. Unfortunately, this emission is difficult to observe since it is absorbed by the Earth's atmosphere. A few quasars have been observed in the far-infrared from the Kuiper Airborne Observatory and by the *Infrared Astronomical Satellite (IRAS). SEE INFRARED ASTRONOMY.*

The emission from quasars also varies with time. The shortest time scale of variability ranges from years to months at short radio wavelengths, to days at optical wavelengths, to hours at x-ray wavelengths. These different time scales suggest that the emission from the different bands originate from different regions in the quasar. Very highly active quasars are sometimes referred to as optically violent variables (OVVs) or BL Lac's, after the prototype BL Lacertae, a well-known variable "star" that turned out to be a quasar. The optically violent variables have no or very weak emission lines in their optical spectrum.

The rapid fluctuations indicate that there are some components in quasars that have diameters less than a light-hour or of the order of 10^9 km (10^9 mi), the size of the solar system. The radiation mechanism responsible for the optical and x-ray continuum emission is not known for certain, but the most likely possibilities are synchrotron emission or inverse Compton radiation.

Emission lines characteristic of a hot ionized gas are superimposed on the continuum. The emission lines are highly Doppler-broadened, indicating internal velocities of hundreds to thousands of kilometers per second. The most common permitted lines seen are those of H, He, C IV, Si IV, and Mg II, while forbidden lines originating from less dense gaseous regions in the quasar are emitted by the ions [C III], [O II], [O III], [Ne V], and others. *SEE DOPPLER EFFECT.*

In addition to the emission lines, many quasars, especially those with large redshifts, show very narrow absorption lines. In contrast to the emission-line regions, the absorbing regions have internal velocities of about 10 km/s (6 mi/s). Frequently, many absorption-line systems are found with differing redshifts. The absorption-line redshifts are typically less than the emission-line redshifts by amounts corresponding to relative velocities of several thousand kilometers per second. In some quasars, there are absorption-line redshifts substantially less than the emission-line redshift. The most likely interpretation of the latter is that the light from the quasar is passing through and being absorbed by a distant galaxy that just happens to lie along the line of sight. The more numerous other lines could be absorptions by clouds in the vicinity of the quasar. Multiple images due to gravitational lensing by the mass of an intervening galaxy have been observed for a few quasars. *SEE GRAVITATIONAL LENS.*

For the more than 1000 quasars measured, the emission-line redshift z ranges from near 0 to 4.2. For quasars having a z greater than about 0.5, many normally ultraviolet spectral lines are seen that are redshifted into the visible region of the spectrum. These lines, such as the Lyman α line at 121.6 nanometers—seen when $z > 1.6$—cannot otherwise be seen in astronomical objects observed from the surface of the Earth because of the atmospheric absorption of all radiation less than about 310 nm.

Not all quasarimages are entirely point sources of light. Quasar 3C273 (**Fig. 1**) has an elongated jet of optical emission 20 arc-seconds from the quasar but pointing radially away from it as if the gas in the jet had been ejected from the quasar. The image of the quasar has been intentionally overexposed to reveal the ejected jet of material. Other quasars show faint nebulous emission surrounding the starlike image. The spectrum of this emission is difficult to observe, but has been found in some cases to be an emission-line spectrum indicating the presence of heated gas extending thousands of light-years from the quasar. In other quasars the nebulous emission has a blackbody spectrum with superimposed absorption lines indicating the existence of a galaxylike distribution of stars surrounding the quasar. Furthermore, objects having quasar characteristics but with small redshifts are sometimes found surrounded by a faint distribution of stars. These are called N galaxies and may be similar to the Seyfert galaxies, which are spiral galaxies having a very bright quasarlike nucleus. The existence of these galaxy types and the starlike spectrum of the faint emission seen around some quasars strongly suggest that quasars are exceedingly luminous nuclei of galaxies. Thus, they are often referred to as active galactic nuclei (AGNs). *SEE GALAXY, EXTERNAL.*

Distances of quasars. When the quasar redshift is plotted against the quasar's brightness or magnitude, as is done for galaxies, there is not a strong correlation between large z and faint magnitudes. Such a correlation is observed for galaxies and indicates that the fainter, more distant galaxies have progressively larger redshifts. This correlation, called Hubble's law, is considered to be proof that the universe is ex-

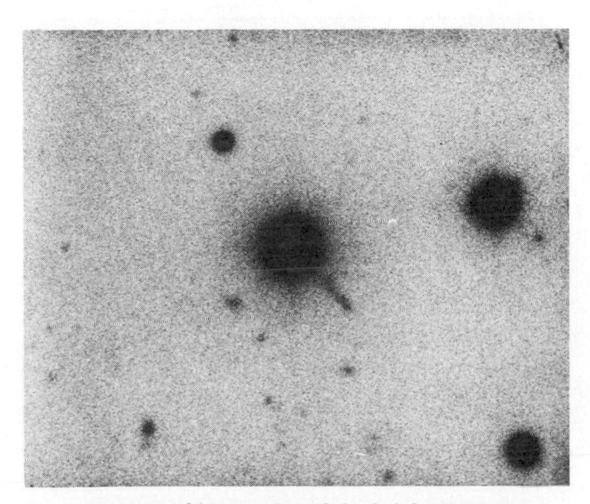

Fig. 1. Photographic negative of the brightest quasar, 3C273. The cross-shape pattern is due to diffraction within the telescope. The jet extends toward the lower right. (*Palomar Observatory, California Institute of Technology*)

panding. In the case of the quasars the scatter in the plot is so great that it does not prove that the redshift can be used as a distance indicator. This does not prove that the large-z quasars are not the most distant objects known, since the scatter in the diagram could be due to large differences in the intrinsic luminosities of the quasars. *SEE HUBBLE CONSTANT; REDSHIFT.*

If the quasars exhibited the Hubble law relationship between their magnitude and redshift, then their redshifts would unambiguously show them to be the most distant known objects in the universe, ranging up to about 10^{10} light-years away. The statistics of the redshift number distribution could then be interpreted as showing a greater abundance of quasars at earlier epochs in the universe. However, the paucity of known quasars with z greater than 3 implies that they may have been absent at even earlier epochs when the density of the universe was 64 times greater than its present density.

A few astronomers have argued that the redshifts are not cosmological in origin and hence do not indicate the distances of quasars. One of these arguments rests on the observations that there are examples where two quasars of very different redshift are seen closer together than would be expected by chance if the two quasars were not physically related. On the other hand, the redshifts of a few quasistellar objects seen in nearby clusters of galaxies have been found to agree with the measured redshifts of the associated galaxies, implying that the redshifts are cosmological. Models postulating noncosmological redshifts, such as gravitational or Doppler redshifts, have met with difficulties. If the redshifts are not cosmological, then they are probably a result of some new unknown physical effect. *SEE COSMOLOGY.*

Radio characteristics. Some quasars emit a significant fraction of their radiated energy at radio frequencies ranging from about 30 MHz (wavelength, 10 m) to 300 GHz (wavelength, 1 mm). There appear to be two general categories of radio emitters. Those that emit predominantly at the lower radio frequencies (less than 1 GHz) have spectral indices α of about -0.8, showing that the flux of radio emission is rapidly decreasing with increasing frequency. Interferometers have shown that the radio emission from these types of quasistellar radio sources does not originate from the optical quasar but comes from two considerably distant regions symmetrically displaced on either side of the optical object. Such radio-frequency characteristics are nearly identical to those of radio galaxies, which suggests that the two may be generically related.

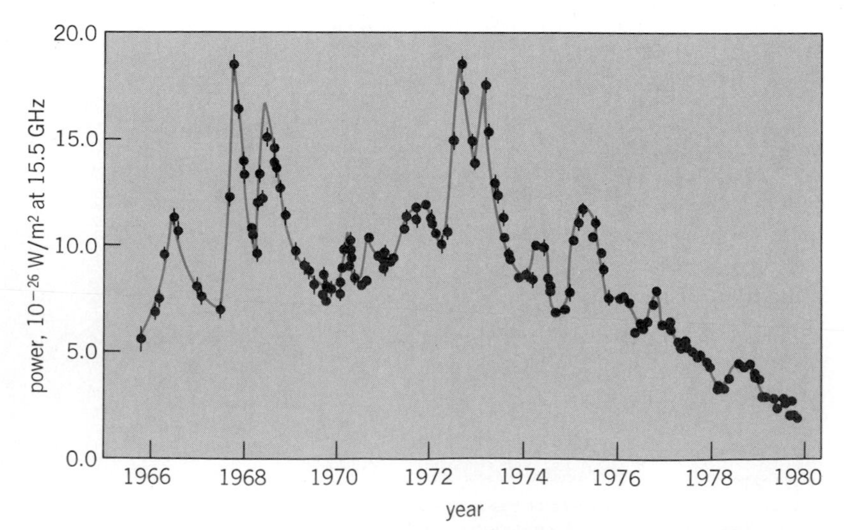

Fig. 2. Outbursts in radio emission which were observed in the quasarlike source 3C120.

The other, more interesting type of quasistellar radio source emits primarily at centimeter and millimeter wavelengths (1 GHz to 300 GHz) and originates in the same region as the optical quasar. The flux of emission of these sources is nearly independent of frequency ($\alpha = 0$) except that there is an abrupt drop-off in the emission spectrum at the low-frequency end. There are a few examples of quasistellar radio sources that exhibit both sets of characteristics, suggesting that one type may evolve into another. In 1965 large variations in the flux of radio emission from the second type of quasistellar radio source were discovered, indicating that, since these types are active, they are young objects while those of the first type are probably decaying remnants. **Figure 2** shows an example of measurements of the active galaxy source 3C120 made at a frequency of 15.5 GHz (1.9 cm) with the 120-ft (37-m) radio telescope of the Haystack Observatory in Massachusetts. Sudden outbursts of emission are evident, particularly the pair in late 1972 and early 1973. Outbursts such as these have enormous luminosities sometimes approaching 10^{45} ergs/s (10^{38} joules/s), nearly 10^{12} times the luminosity of the Sun.

The nature of the origin of this energy is not known and represents one of the most important unanswered questions concerning quasars. The radio emission is probably synchrotron radiation produced by electrons with velocities very near the speed of light that are spiraling in magnetic fields. The electrons can absorb their own photons at lower frequencies, producing the drop-off in their emission spectrum. From the observed frequency of this self-absorption, the magnetic field strengths can be estimated as 0.01 to 1 gauss.

There have been VLBI observations of some of the active quasars. These observations show that very compact radio components with angular diameters of about 0.001 arc-second exist in quasars of the second type. Furthermore, successive observations seem to show that these compact components are expanding rapidly and separating at speeds greater than the speed of light, if cosmological distances are assumed. A number of theories have been advanced to explain the phenomena without requiring that the quasars be relatively nearby. The most likely explanation is that one of the separating radio components is moving almost directly toward the Earth with a velocity close to the speed of light. At high velocities of approach, effects predicted by the special theory of relativity become important. Intervals of time in a rapidly approaching object would appear to be much shorter, giving the appearance that the object is moving much faster than it really is.

Theories of quasars. The many similarities of the observed characteristics of quasars with radio galaxies, Seyfert galaxies, and BL Lacertae objects strongly suggest that quasars are active nuclei of galaxies. There is good statistical evidence which shows that quasars with large redshifts are spatially much more numerous than those with small redshifts. Because high-redshift objects are very distant and emitted their radiation at an earlier epoch, quasars must have been much more common in the universe about

10^{10} years ago. Since this is about the same epoch when galaxies are thought to have formed, it is possible that quasars may be associated with the birth of some galaxies.

More than 10^{60} ergs (10^{53} J) of energy are released in quasars over their approximately 10^6-year lifetime. Of the known energy sources, only gravitational potential energy associated with a mass about 10^9 times the mass of the Sun can provide this energy, but it is unknown how this gravitational energy produces jets of particles that are accelerated to very near the speed of light.

Several theories have been proposed for quasars. One theory proposes supernovalike explosions of supermassive objects following their gravitational collapse; another theory postulates that a massive black hole is somehow driving the quasar; still others assume the particles are accelerated either by magnetic fields associated with a giant rotating object or by many smaller rotating neutron stars. Yet another theory postulates that the energy is released in collisions between rapidly moving stars in a very dense cluster at a galactic nucleus. Currently, the most favored interpretation of quasars is a massive black hole surrounded by a rapidly spinning disk of gas in the nucleus of some galaxies. *See Astronomical spectroscopy; Astrophysics, high-energy; Black hole; Infrared astronomy; Neutron star; Radio astronomy; Supermassive stars.*

William Dent

Bibliography. G. Burbidge and M. Burbidge, *Quasi-Stellar Objects*, 1967; C. Hazard and S. Mitton (eds.), *Active Galactic Nuclei*, 1979; H. L. Shipman, *Black Holes, Quasars, and the Universe*, 1980; D. Weedman, *Quasar Astronomy*, 1986, paperback 1988.

Radar astronomy

A powerful astronomical technique that furnishes otherwise-unavailable information about bodies in the solar system. By comparing a radar echo to the transmitted signal, information can be obtained about the target's size, shape, topography, surface bulk density, spin vector, and orbital elements. While other astronomical techniques rely on passive measurement of reflected sunlight or naturally emitted radiation, the illumination used in radar astronomy is a coherent signal whose polarization and time modulation or frequency modulation are tailored to meet specific scientific objectives. Through measurements of the distribution of echo power in time delay or Doppler frequency, radar achieves spatial resolution of a planetary target despite the fact that the radar beam is typically much larger than the angular extent of the target. This capability is particularly valuable for asteroids and planetary satellites, which appear as unresolved point sources through optical telescopes. Moreover, the centimeter-to-meter wavelengths used in radar astronomy readily penetrate cometary comas and the optically opaque clouds that conceal Venus and Titan, and also permit determination of near-surface roughness (abundance of wavelength-scale rocks), bulk density, and metal concentration in planetary regoliths. *See Asteroid; Satellite; Saturn; Venus.*

Telescopes. A radar telescope is essentially a radio telescope equipped with a high-power transmitter (a klystron vacuum-tube amplifier) and specialized instrumentation that links the transmitter, low-noise maser receiver, high-speed data-acquisition computer, and antenna together in an integrated radar system. Planetary radars, which must detect echoes from targets at distances from about 10^6 km (10^6 mi) for closely approaching asteroids and comets to more than 10^9 km (nearly 10^9 mi) for Saturn's rings and satellites, are the largest and most sensitive radars on Earth. *See Radio telescope.*

The two active planetary radar facilities are the Arecibo (radar wavelengths of 13 and 70 cm) and Goldstone (3.5 and 13 cm) instruments; for each, the shorter wavelength provides the greater sensitivity. The Arecibo radio-radar telescope in Puerto Rico consists of a 305-m-diameter (1000-ft) fixed reflector, the surface of which is a section of a 265-m-radius (870-ft) sphere. Movable line feeds suspended from a triangular platform 130 m (427 ft) above the reflector can be aimed toward various positions on the reflector, enabling the telescope, located at 18°N latitude, to point up to 20° from the zenith. The Goldstone radar in California is part of the National Aeronautics and Space Administration's Deep Space Network. The Goldstone main antenna is a steerable, 70-m (230 ft), parabolic reflector with horn feeds. Arecibo is roughly twice as sensitive at a wavelength of 13 cm (corresponding to a frequency of 2380 MHz) as Goldstone is at 3.5 cm (8510 MHz), but Goldstone can track targets continuously for much longer periods and has access to the whole sky north of −40° declination. Experiments in aperture synthesis radar astronomy, which employ transmission from Goldstone and reception of echoes at the Very Large Array in New Mexico, achieve 50% more sensitivity than the Arecibo telescope in its 1992 configuration, and also synthesize a beam whose width (as small as 0.25 arcsecond, versus 2 arc-minutes for single-dish observations) provides useful angular resolution of planets and natural satellites. By 1994, improvements in the Arecibo radar telescope are expected to have made that instrument over an order of magnitude more sensitive than in 1992, dramatically extending its reach to smaller and more distant objects.

Techniques. A typical transmit-receive cycle consists of signal transmission for a duration close to the round-trip light time between the radar and the target, that is, until the first echoes are about to return, followed by reception of echoes for a similar duration. The target's apparent radial motion introduces a continuously changing Doppler shift into the echoes. This shift is removed by continuously tuning the receiver according to an ephemeris based on an orbit calculated from optical (and perhaps radar) astrometric observations. In continuous-wave observations, a nearly monochromatic signal is transmitted and the distribution of echo power is measured as a function of frequency. The resultant echo spectrum can be thought of as a one-dimensional image, or a brightness scan across the target through a slit parallel to the target's apparent spin vector. *See Doppler effect.*

Range resolution can be obtained by using a coherent pulsed continuous waveform, but in practice, because of engineering considerations associated with klystrons, a pulsed waveform is simulated by encoding a continuous-wave signal with a preset sequence of 180° phase reversals. Fourier transformation of time samples taken at the same position within each of many successive range profiles yields the echo

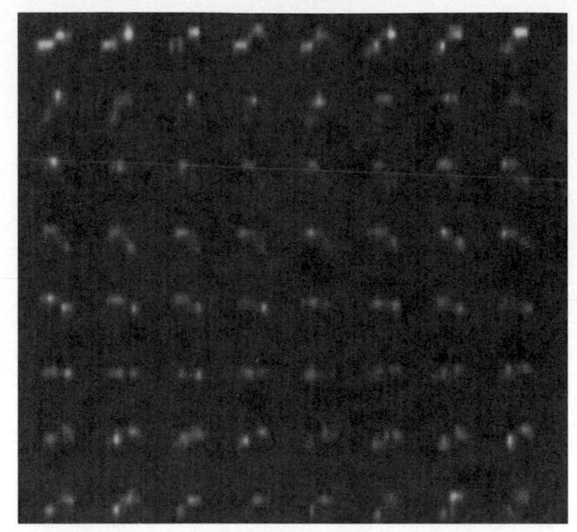

Radar images of the near-Earth asteroid Castalia (originally designated 1989 PB). Two distinct, 1-km-diameter (0.6-mi) lobes are revealed that appear to be in contact. It seems likely that the lobes once were separate and then collided gently to produce the current contact-binary shape. This radar movie provides a nearly pole-on view of the asteroid turning counterclockwise through 60% of a rotation in 2.5 h. The radar illumination comes from the top of the page, so that parts of the asteroid facing toward the bottom are not seen in these images. (*After S. J. Ostro et al., Radar images of asteroid 1989 PB, Science, 248:1523–1528, 1990*)

power spectrum for the corresponding range cell on the target. Spectral analysis of echoes from multiple range cells yields a delay-Doppler image. If the target's shape is known, it is a straightforward matter to convert that image into a radar brightness map. Heterogeneities in brightness can be introduced by variations in large-scale surface tilts, small-scale roughness, or surface bulk density. By measuring the echo's polarization state, the nature of the scattering process and the severity of wavelength-scale roughness near the surface can be deduced, thereby refining the interpretation of reflectivity measurements. *See Polarimetry.*

Observational results. Radar-detected planetary targets include the Moon, Mercury, Mars, Venus, Phobos, the Galilean satellites, Saturn's rings, Titan, five comets, three dozen main-belt asteroids, and 30 near-Earth asteroids. The radar signatures of these bodies are extraordinarily diverse. For example, for each of the two most basic radar properties (albedo and circular polarization ratio), estimated values span two orders of magnitude. *See Albedo; Comet; Jupiter; Mars; Mercury.*

Radar observations have revealed that one Earth-approaching asteroid is a contact binary (see **illus.**), that another is metallic, and that at least two comets (including Halley) are accompanied by swarms of large particles. Other experiments have ruled out global coverage of Titan by a deep hydrocarbon ocean, probed far below the visible surfaces of the icy Galilean satellites, disclosed the extraordinary diversity of Mars's surface at decimeter-to-meter scales, and produced startling evidence for radar-bright polar caps on Mercury. Thirty years after the radar discovery of Venus's slow retrograde rotation, a radar instrument on the Venus-orbiting *Magellan* spacecraft has mapped almost all of that planet's surface with striking geologic clarity. Another space-borne radar mapper is targeted for Titan as part of the *Cassini* Saturn Orbiter, scheduled for launch in 1997. *See Halley's Comet; Planetary physics.*

Steven J. Ostro

Bibliography. D. B. Campbell et al., Styles of volcanism on Venus: New Arecibo high resolution radar data, *Science,* 246:373–377, 1989; J. K. Harmon et al., Radar observations of comet IRAS-Araki-Alcock 1983d, *Astrophys. J.*, 338:1071–1093, 1989; D. O. Muhleman et al., Radar reflectivity of Titan, *Science,* 248:975–980, 1990; G. H. Pettengill et al., *Magellan*: Radar performance and data products, *Science,* 252:260–265, 1991; J. H. Taylor and M. M. Davis (eds.), *Proceedings of the Arecibo Upgrading Workshop*, National Astronomy and Ionosphere Center, Arecibo, Puerto Rico, 1987; D. K. Yeomans et al., Asteroid and comet orbits using radar data, *Astron. J.*, 103:303–317, 1992.

Radio astronomy

The study of celestial objects by measurement and analysis of the electromagnetic radiation they emit in the wavelength range from 1 mm to 30 m. Karl Jansky was the first to determine that radio emission from the heavens could be detected. At the Bell Telephone Laboratories in 1932, Jansky built a rotating antenna array operable at 14.6-m wavelength and attempted to investigate the source of short-wave interference. He found that, in addition to the intermittent interference resulting from thunderstorms, a steady hiss-type static was most intense when his antenna was directed toward the center of the Galaxy. With these and subsequent observations he concluded that the Galaxy itself was an intense source of cosmic radio radiation at frequencies of a few tens of megahertz.

Jansky's work was not continued seriously until 1940 when an electronics engineer, Grote Reber, built a 31-ft (9.4-m) parabolic reflector in his backyard in Wheaton, Illinois, and studied the galactic radio emission at 162 MHz. He confirmed Jansky's impression that the radio emission was strongest toward the galactic center; he further succeeded in detecting radio emission from the Sun. Solar radio emission was also discovered during World War II by radar researchers in Britain, who found that the severe interference they had noted each day at dawn was in fact emanating from the rising Sun. After World War II the radar scientists made use of their expertise in astronomical pursuits, and a concerted study of astronomical objects at radio wavelengths began. Today radio astronomers study an entire range of celestial objects, including the normal stars, planets, galaxies, and the exotic quasars, pulsars, and x-ray sources.

Radio universe. Since the late 1940s, radio telescopes have been used to map the skies and determine the positions and intensities, or fluxes, of individual sources of radio emission. Such maps have been made with increasing sensitivity and angular resolution; the latter property enables astronomers to determine the position of the radio sources accurately. Knowing the position, astronomers can refer to optical photographs of the sky and establish precisely which object is emitting radio waves. This procedure has led to the identification of radio sources with many bright galaxies and even with the most distant objects in the universe, the quasistellar objects (or quasars). However, nearly one-fifth of all radio sources are unidentified; that is, excellent photographs taken at the radio source positions show no object at all from which the radiation could arise. One concludes from this that these unidentified sources are "normal" galaxies and quasars at such great distances that they cannot be seen optically. Hence, radio astronomy seems to provide a more complete sampling of objects in the universe than does optical astronomy; as such it may be a tool for investigating cosmological questions, in particular questions about the origin of the universe. *See Radio telescope*.

Cosmological questions in radio astronomy can be addressed by a simple procedure that makes use only of the fact that the farther away a radio source is, the weaker it appears. Specifically, the measured radio flux of a source decreases inversely with distance R squared, $1/R^2$. If the universe is filled uniformly with radio sources, then as one looks to greater and greater distances one can sample more and more sources; that is, the volume of space surveyed by the observation increases as the distance cubed, R^3. Therefore, if one simply counts the number of sources at a given flux level S, the number should increase as the inverse 3/2 power of the received flux; that is, the number of sources is proportional to $S^{-3/2}$. This proportionality is not found. Rather it appears that there are more sources at great distances per unit volume of space than there are nearby, a result that means the universe is expanding, and that those distant sources are, in general, stronger than ones nearby. These two conclusions are fundamentally linked. Since the radio waves received on Earth from distant sources were emitted billions of years prior to the radio waves that are received from nearby sources, the finding that the more distant (and hence younger) objects are also more luminous means that the radio sources change with time. Thus, if cosmological questions are to be answered, the evolution of individual radio sources must first be understood. *See Cosmology*.

Galaxies and radio galaxies. The morphological structure of most galaxies resembles either a spiral (or whirlpool) or an ellipse. Spiral galaxies, such as the Milky Way Galaxy, are seen to be laced with gas, dust, and newly formed stars; in addition, they are often radio sources, although most are quite weak. Elliptical galaxies, on the other hand, consist of older stars and show no evidence of gas and dust; they are usually not radio sources. However, the few elliptical galaxies that are radio sources are very spectacular ones, being among the most energetic radio objects in the sky. These are known as the radio galaxies.

The radio emission from spiral galaxies is typically confined to a small nuclear region supplemented by much weaker extended emission from the disk of the galaxy. In elliptical radio galaxies, on the other hand, the radio emission emanates from a small region at the center of the galaxy. Radio-emitting material is expelled from the nuclear region in two oppositely directed collimated streams, or jets, that extend out to distances many times the size of the visible galaxy. The enormous radio energy involved, 10^{10}–10^{12} times the solar luminosity, together with the very small size of the nuclear region from which this energy arises, less than 10^{14} km, leads to the conclusion that the source of energy is a black hole at the center of the radio galaxy. Such a black hole would be 10^7–10^9 times as massive as the Sun. *See Black hole*.

A computer-processed radio image of the giant radio galaxy Cygnus A is shown in **Fig. 1**. The nucleus of the radio galaxy is the bright spot midway between the two lobes of radio emission. The thin jet of material can be seen flowing from the nucleus to power the bright lobes.

The radio emission from galaxies and radio galaxies is generated by the electron synchrotron process, in which relativistic electrons spiral around magnetic field lines and emit a continuous radiation spectrum throughout the band accessible to radio astronomers. *See Galaxy, external*.

Quasars. Quasars are perhaps the most enigmatic objects in the universe. To the radio astronomer they are essentially indistinguishable from the radio galaxies, with regard to both their spectrum and apparent radio flux. However, optical measurement of the redshift of the quasars reveals that they are receding from Earth at velocities approaching that of light. This

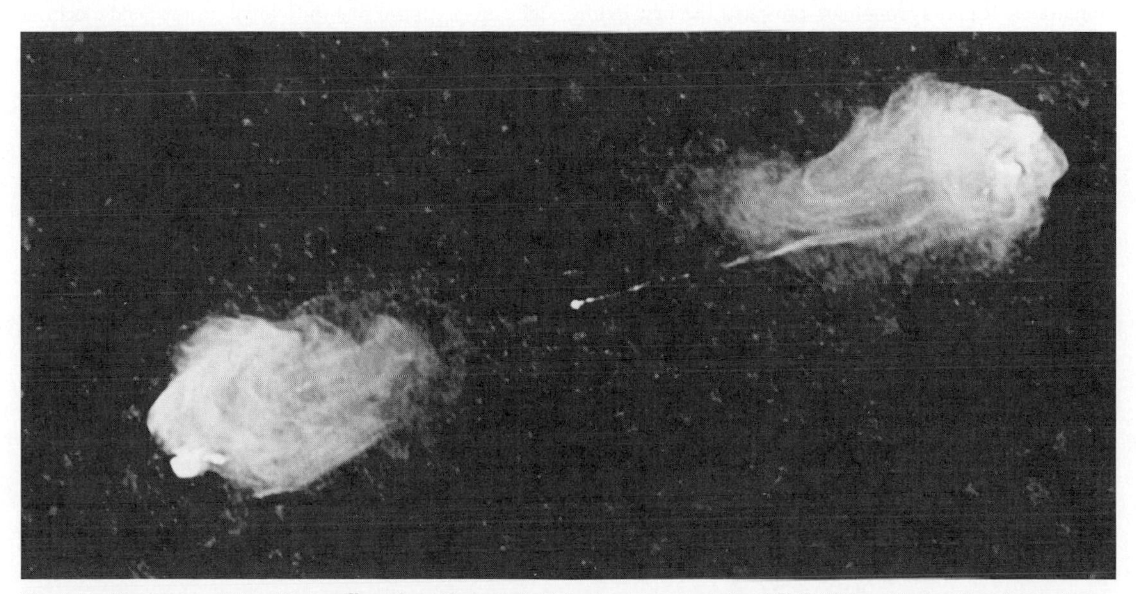

Fig. 1. Radio picture of the giant radio galaxy Cygnus A. This picture was made at 5000 MHz with the Very Large Array Synthesis Radio Telescope. (*National Radio Astronomy Observatory*)

means, according to the work of Edwin Hubble, who showed that the distance to extragalactic objects is proportional to their velocity, that quasars must be the most distant objects in the universe. If quasars are so much more distant than radio galaxies, yet produce radio emission that is observed at a flux level comparable to nearby galaxies, their intrinsic radio power must be much greater than that of the already very energetic radio galaxies. *See Redshift*.

The inference that a massive black hole supplies the energy is even stronger for the quasars than for the radio galaxies. In both types of sources the radio emission mechanism is synchrotron radiation from relativistic electrons. *See Quasar*.

Big bang. The universe itself, as distinguished from all of the discrete objects individually or collectively within it, is also a source of radio radiation. In 1965 Arno Penzias and Robert Wilson of the Bell Telephone Laboratories measured the radio temperature of the sky to be 3 K (5°F above absolute zero) and showed that the universe is uniformly permeated by an electromagnetic flux of radiation at this temperature. This radiation is a relic of the initial evolutionary phase of the universe. When the universe was a small fraction of its present size, the "heat" of its formation was contained in subnuclear particles and electromagnetic waves. As time passed, the particles coalesced into atoms and ultimately into galaxies and stars, whereas the radiation simply cooled by the expansion of the universe to its present temperature near 3 K. The detection of this radiation along with its interpretation as direct evidence for the big bang origin of the universe is the most significant achievement of radio astronomy. *See Big bang theory; Cosmic background radiation*.

Solar system astronomy. The Sun is an intense radio source, but only because it is so close to Earth. If it were at the distance of the nearest stars, its radio emission could not be detected. Solar radio emission tends to be intermittent: solar flares that produce cosmic rays and plasma streams that interact with Earth are visible as radio bursts; these are most frequent during the peak of the 11-year solar cycle. *See Cosmic rays; Sun*.

Radio observations of the planets have revealed a great deal. Radio astronomers provided the first indication that Mercury does not keep one face constantly toward the Sun, as had been believed for nearly a century by astronomers, by showing that the dark side of the planet had a high temperature. Radar astronomers subsequently proved this point conclusively. Venus was first discovered to have an extremely hot (roughly 850°F or 450°C) surface by radio measurements of its temperature; measurements at different wavelengths have yielded information on the atmospheric composition of Venus. High-resolution measurements of Venus and the Moon by large radio telescopes and interferometers have provided information on the surface composition and roughness, as well as on the distribution of temperature on the surface. *See Mercury; Moon; Venus*.

Jupiter is a most interesting planet, being a much stronger radio source than had been expected from estimates of its surface temperature by optical astronomers. Most of its radio emission is caused by electron synchrotron emission in its very strong magnetic field. The very long-wavelength emission of Jupiter is impulsive, and its strength depends upon the position of Io, one of Jupiter's moons. These processes are not completely understood. Similar impulsive long-wavelength bursts have also been discovered from Saturn. *See Jupiter; Saturn*.

Radio stars. Several nearby, apparently normal stars are detectable at radio wavelengths. Such stars as Algol, β Persei, and AR Lacerta are multiple star systems, long studied by optical astronomers, that have been found to be radio sources. The radio emission from these stars is dominated by radio bursts in which the radio fluxes may increase by a factor of 100 or more. It is clear that the radio bursts are initiated by or are a product of mass exchange processes going on between (at least) two closely bound stars.

An extreme example of radio emission from stars comes from stars that are also x-ray sources. Again, these objects are usually binary systems in which mass exchange plays a deciding role in their continuing evolution, but with x-ray sources one of the component stars appears to be a star that has exhausted its reservoir of nuclear fuel and is collapsing to its final state. *See Binary star; X-ray astronomy*.

Pulsars. In 1968 a completely new class of continuum radio sources was discovered. Instead of emitting energy continuously, these sources emit bursts of energy at extremely regular intervals. Precise measurements have shown that the pulse period is highly constant, better than 1 part in 10^{12}, but the pulse intensity varies unpredictably with time. In all pulsars the pulses are stronger at the longer wavelengths; often the pulse strength varies rapidly and unpredictably with wavelength. The length of individual pulses ranges from 0.1 to about 20 milliseconds, depending on the particular pulsar; time structure of less than 1 ms also appears and exhibits both linear and circular polarization. The arrival time of the pulses at long wavelengths is later than at short wavelengths, because of dispersion effects caused by ionized gas between the pulsar and Earth; analysis of such data yields otherwise unattainable information concerning the electron density, magnetic field, and cosmic-ray intensity in interstellar space.

Pulsars appear to be very highly evolved stars that have exhausted their inner energy sources and are slowly collapsing. The size inferred for these stars (6 mi or 10 km) and their mass (10^{30} kg) means that they are so dense that individual protons and electrons cannot exist as separate entities. Rather, the extreme pressure inside the pulsars will cause the protons and electrons to "coalesce" into neutrons. Pulsars are therefore true neutron stars with enormous magnetic fields, $B = 10^{12}$ gauss (or 10^8 teslas), which emit radio radiation most likely by cyclotron radiation from electrons moving in these intense magnetic fields. *See Neutron star; Pulsar*.

Supernova remnants. The final neutron star phase of stellar evolution is thought to be preceded by a phase of catastrophic stellar instability in which the core of the star suddenly collapses, causing the outer envelope of the star to be explosively ejected. During this process the luminosity of the star increases briefly by a factor of 10^{12} or more, and the star is said to be a supernova. The atmospheric envelope of the star that is ejected in this process becomes a rapidly expanding cloud of relativistic particles, magnetic field, and filaments of ionized gas. These conditions are precisely those necessary for the generation of radio emission through electron synchrotron radiation, and very intense radio sources indeed exist at the positions of old supernovae chronicled by ancient astronomers.

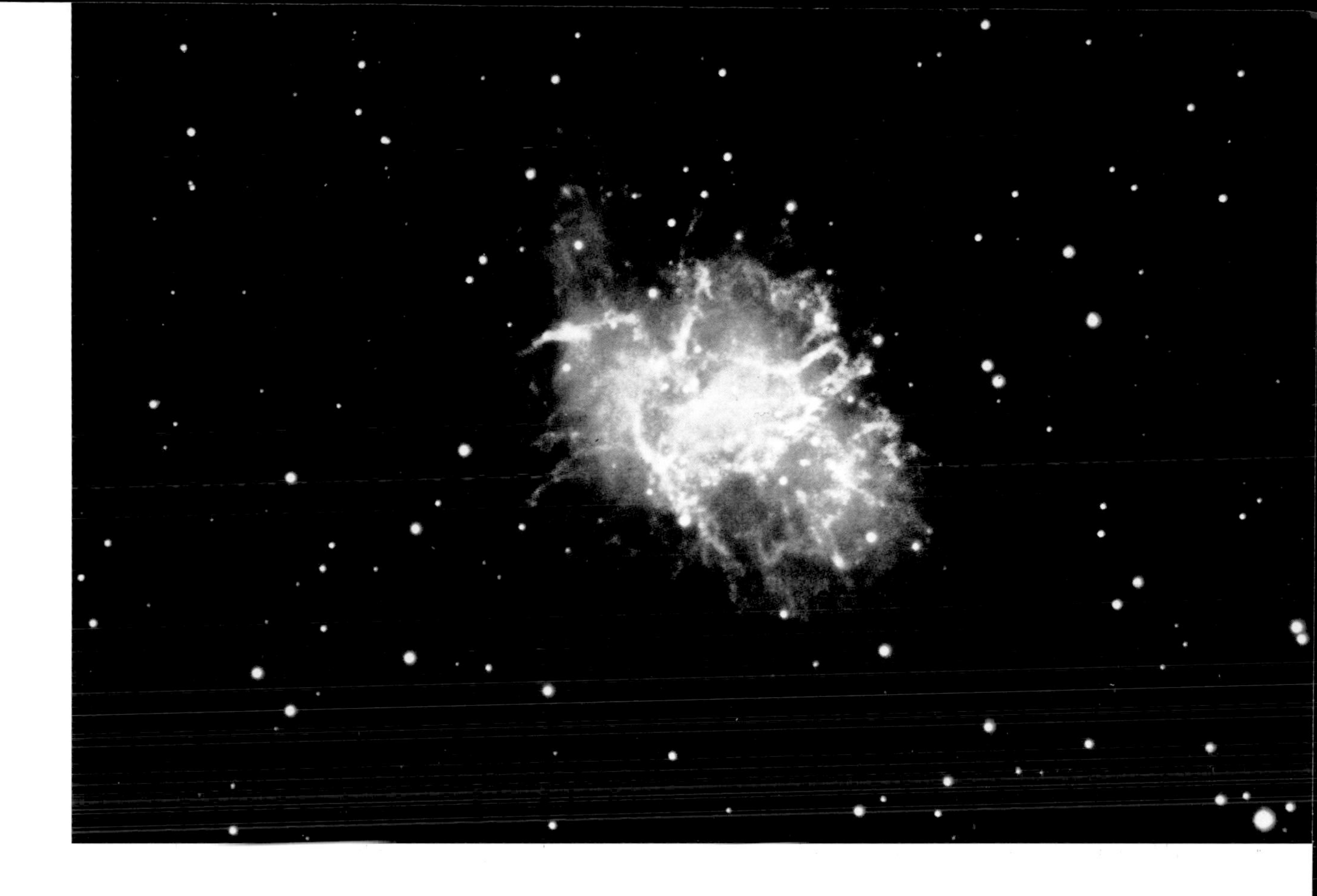

Crab Nebula in Taurus (*above*) is debris from a stellar explosion seen as a nova by Oriental astronomers in 1054. High-energy electrons cause the gas to glow. Photograph was made with 36-in. (0.9-m) reflector at Lick Observatory (*copyright by U. C. Regents*). Great Nebula in Orion (*below*), visible as the middle "star" in the sword of Orion, is a gas cloud excited to incandescence by hot stars in its center. Stars may form in such a region. Photograph was made with 150-in. (3.9-m) Anglo-Australian Telescope (*copyright by Anglo-Australian Telescope Board, 1981*).

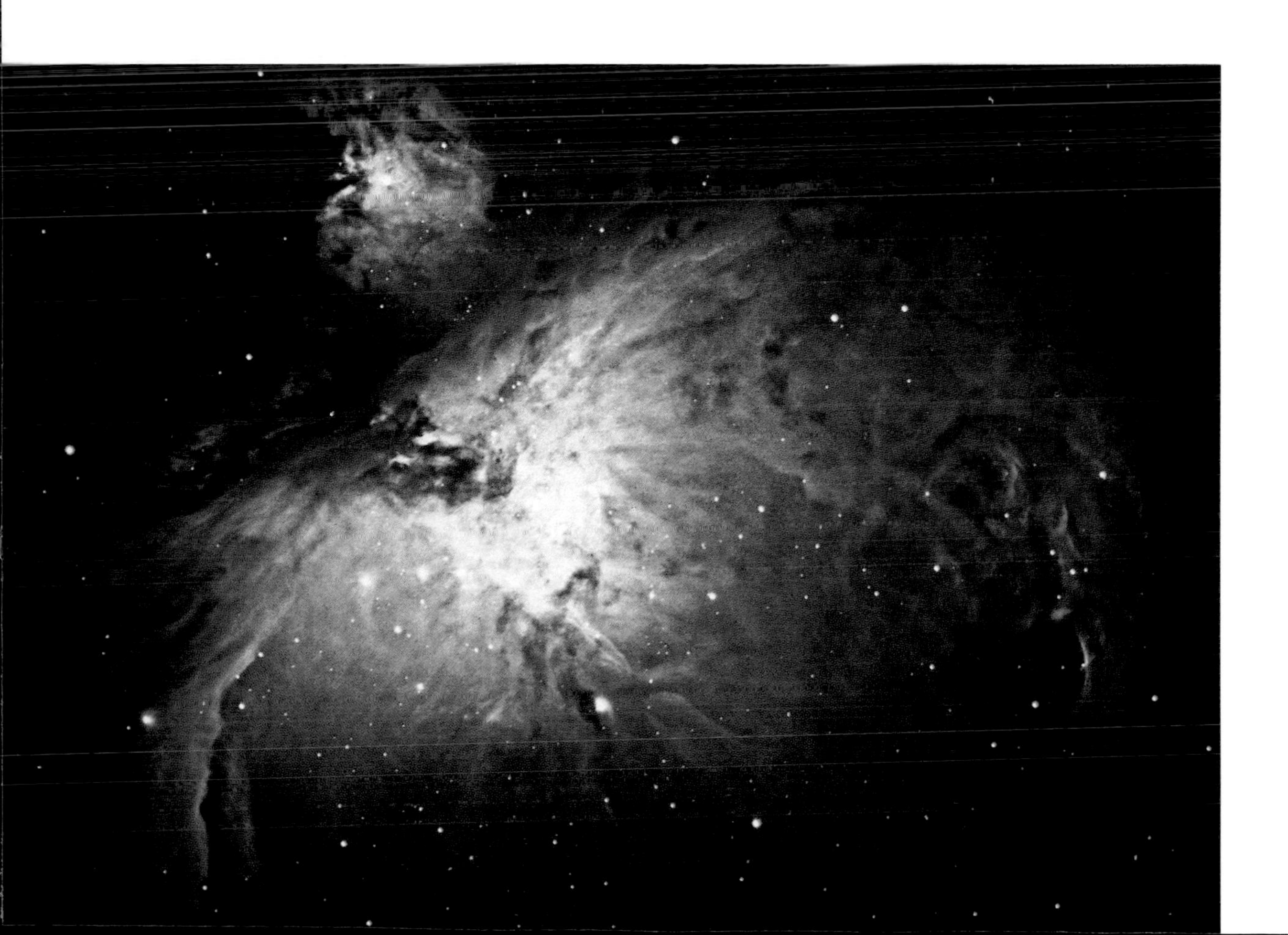

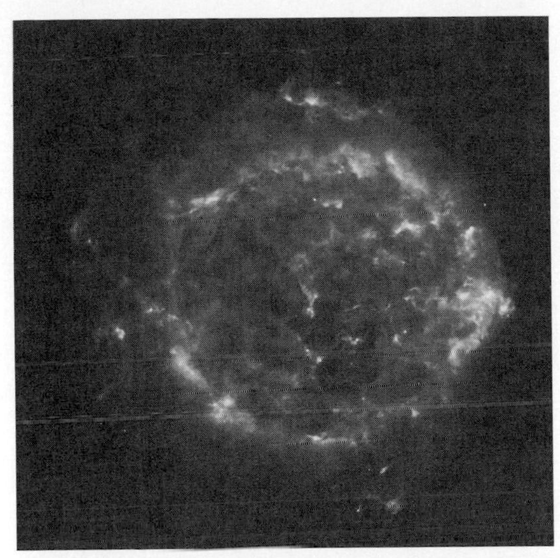

Fig. 2. Radio image of the supernova remnant Cassiopeia A. The image was made with the Very Large Array. (*National Radio Astronomy Observatory*)

The strongest radio source in the sky at microwave frequencies is the supernova remnant Cassiopeia A (**Fig. 2**). With an age of approximately 500 years, this expanding cloud of relativistic gas is seen to be a structured web of shells and filaments, the interpretation of which provides clues to the origin of cosmic rays. *SEE COSMIC RAYS*.

One of the most interesting of the radio sources associated with a supernova remnant is the Crab Nebula. It first appeared as a visible supernova in A.D. 1054 when its position and brightness were recorded by Chinese and Japanese astronomers; today the Crab Nebula is one of the strongest radio sources, although it is slowly dimming as a result of its expansion. The polarization of the radio and optical radiation indicates that both are produced by synchrotron radiation. The polarization and intensity of the optical radiation vary with time; such variations are most pronounced at the position of the pulsar (presumably the remains of the star that exploded and produced the supernova remnant) in this nebula. Evidently the pulsar supplies much of the energy for the nebular radio, optical and x-ray emission in the form of relativistic particles. *SEE CRAB NEBULA; SUPERNOVA*.

H II regions. An H II region (a region of ionized hydrogen) is a large cloud of interstellar gas that has been ionized and heated by one or more bright, hot stars located within. These nebulae are sources of both continuum and line energy at radio and optical wavelengths. Since cosmic matter consists mostly of hydrogen, the ionized gas consists mainly of protons and electrons that emit continuum energy by bremsstrahlung. *SEE BREMSSTRAHLUNG; NEBULA*.

The gas also emits recombination line radiation. In the radio region these lines arise from transitions between two high quantum levels in the hydrogen atom. Since many such levels exist, many lines at different wavelengths appear. Similar transitions exist for elements other than hydrogen; comparison of the intensities of lines from two elements yields a direct measure of the abundance of the respective elements.

Measurement of the line wavelength usually reveals that it is displaced slightly from the value that would be arrived at in a terrestrial laboratory. This difference arises from the Doppler effect, caused by relative motion of the H II region and Earth. In a similar way the wavelength width of the line yields information on the relative motion among the hydrogen atoms within the H II region; if there were no relative motion, the width of the line would be very small. This motion arises from a combination of thermal motions of individual atoms, turbulence, and large-scale velocity fields within the nebula. *SEE DOPPLER EFFECT*.

Comparison of the strengths of lines arising from different levels, along with comparison of these strengths with the continuum intensity, yields information on the number of atoms per unit volume, the temperature, and the degree of fluctuation of both of these quantities within the nebula. It is of course of great value to be able to measure the variation of these quantities with position within an H II region; but this is a difficult task requiring many observations with the largest telescopes to extract a reasonably complete physical picture. *SEE INTERSTELLAR MATTER*.

Hydrogen line. In 1944 Henk van de Hulst suggested that radiation from individual hydrogen atoms could be detected at 21.1-cm wavelength. The radiation he had in mind was radiation in a hyperfine spectral line, which arises when a hydrogen atom with the proton and electron spinning in opposite directions suddenly rearranges itself so that the electron and proton spin in the same direction; a discrete radio-frequency quantum or photon is emitted at 21.1 cm in this rearrangement process. In 1951 H. I. Ewen and E. M. Purcell at Harvard detected this spectral line and introduced a new field of research into the nature of the gaseous content of the Milky Way Galaxy and other galaxies.

Study of the 21-cm line of neutral atomic hydrogen has been exceptionally rewarding in its contribution to the knowledge of galactic structure and of the physical characteristics of interstellar gas. Line intensity normally reflects the amount of gas in the line of sight; line wavelength and width indicate the line-of-sight velocity of the gas and the state of internal motion, just as with recombination lines in H II regions. If the gas overlies a strong radio source, the gas temperature can be inferred by observing the 21-cm line in absorption.

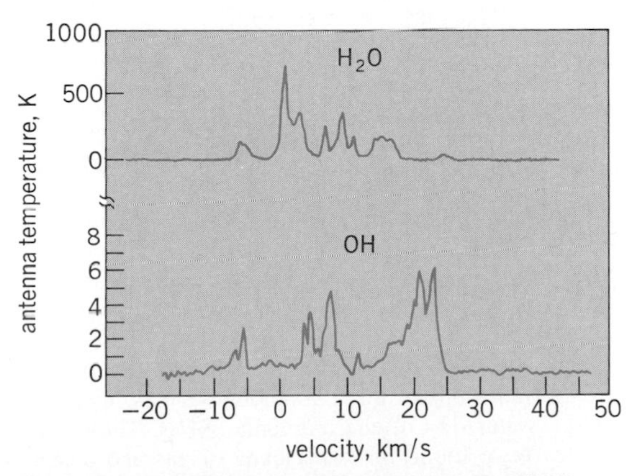

Fig. 3. Maser-amplified spectral lines of hydroxyl (OH) and water (H_2O) in the great nebula in Orion. Antenna temperature, a measure of the intensity of radiation received by the radio telescope, is plotted against the radial velocity inferred from the Doppler shift of the source. °F = (K × 1.8) −459.67. 1 km/s = 0.6 mi/s. (*National Radio Astronomy Observatory*)

Number of atoms									
2	3	4	5	6	7	8	9	11	12
H_2	H_2O	NH_3	SiH_4	CH_3OH	CH_3CHO	CHOOCH	CH_3CH_2OH	$H(C\equiv C)_4CN$	$H(C\equiv C)_5CN$
OH	H_2S	H_3O^+	CH_4	NH_2CHO	CH_3NH_2		$(CH_3)_2O$		
SO	SO_2	H_2CO	CHOOH	CH_3CN	CH_3CCH		CH_3CH_2CN		
SO^+	HN_2^+	H_2CS	$HC\equiv CCN$	CH_3NC	CH_2CHCN		$H(C\equiv C)_3CN$		
SiO	HNO	HNCO	CH_2NH	CH_3SH	$H(C\equiv C)_2CN$		$H(C\equiv C)_2CH_3$		
SiS	H_2D^+	HNCS	NH_2CN	C_5H	CH_3CCN				
NO	HCN	CCCN	H_2CCO	HC_2CHO	C_6H				
NS	HNC	HCO_2^+	C_4H	$H_2C{=}CH$					
HCl	HCO	Si_2CC	C_3H_2						
NaCl	HCO^+	CCCH	CH_2CN						
KCl	HOC^+	c-CCCH	C_5						
AlCl	OCS	CCCO							
AlF	CCH	CCCS							
PN	HCS^+	HCCH							
CH	CCO	$HCNH^+$							
CH^+	CCS								
CN	C_3								
CO									
CS									
C_2									
SiC									

Fig. 4. Molecules discovered in interstellar space.

The structure of the Galaxy has been elucidated by the study of the amount and velocity of the hydrogen within it. A prime advantage of this method is that the very distant gas is just as visible as nearby gas, whereas optical studies of the whole Galaxy are impossible because the very distant stars are made invisible by intervening clouds of dust. The results of these radio studies indicate that the Milky Way Galaxy is a spiral, as are many of the galaxies in the sky. About 5% of the total mass of the Galaxy exists in the form of interstellar gas; it is spread throughout most of the Galaxy, resulting in an average density of about one hydrogen atom per cubic centimeter. The hydrogen in other galaxies can also be studied by its 21-cm line radiation. The structure of hydrogen in this Galaxy is similar to that in other spiral galaxies; an important result is that, since spiral galaxies have a deficiency of gas in their central portions, the gas distribution is doughnutlike. Elliptical galaxies contain very little hydrogen gas. *See Milky Way Galaxy.*

Study of the small-scale structure of interstellar gas in the Galaxy has revealed that the gas is clumped into structures of various sizes and shapes. These include clouds that are tens of light-years in diameter, smaller cloudlets, and larger flat sheets of gas. The intensity of the magnetic field has been measured in some of these clouds by observations of the Zeeman effect in the 21-cm line. *See Zeeman effect.*

Molecular lines. It has long been believed that simple molecules could not exist in the tenuous gas between stars, because the starlight radiation field would be sufficiently intense to break apart even the simplest molecular species. In spite of these arguments, by 1968 radio astronomers had found rotational transitions of three simple molecules, hydroxyl (OH), water (H_2O), and ammonia (NH_3). These molecules were found in dark clouds of gas and dust in the interstellar medium. Such clouds are believed to be the sites of recent and continuing star formation. *See Interstellar matter.*

The hydroxyl and water emission lines were also found to be peculiar in two respects: first, they were unexpectedly strong; and second, they were composed of many components separated slightly in frequency, presumably as a result of the Doppler effect (**Fig. 3**). Radio interferometric observations of these individual line components show that each component arises in a discrete region spatially separated from the others, and moreover that the diameter of each of these discrete regions is quite small, usually only a few astronomical units. (An astronomical unit is the distance between the Sun and Earth, 9.3×10^7 mi or 1.5×10^8 km.) The great intensity of these sources combined with their small sizes implies that the equivalent temperatures of the emitting gas is unreasonably high; it is on the order of 10^{13} K. Since this cannot represent a real temperature, the line emission must be amplified in the source; that is, microwave amplification by stimulated emission of the hydroxyl and water lines is being directly observed. These lines are interstellar masers. The source which energizes or "pumps" the masers may be collisional processes with molecular hydrogen (H_2) or free electrons; or it may be intense infrared, optical, or ultraviolet radiation. This energy ultimately comes from the rapid gravitational collapse of part of the molecular cloud. The collapse will culminate 10^4–10^6 years later in the "birth" of a new star. Interstellar hydroxyl and water masers are the earliest observable signs of star formation; masers arise from true protostars.

Since 1968, more than 87 molecular species have been detected in interstellar space principally by observations of millimeter rotational lines or lines arising from the interaction of the rotation of the molecule's nuclei with the spin of its electrons (lambda-doubling transitions; **Fig. 4**). More complicated organic molecules containing as many as 12 atoms have been discovered. Observations of these molecular species may make it possible to establish the chemistry and thermodynamics in the interstellar clouds from which, ultimately, stars, planets, and life itself must form. *See Molecular structure and spectra.*

Robert L. Brown

Bibliography. J. S. Hey, *The Radio Universe*, 3d ed., 1983; J. D. Kraus, *Radio Astronomy*, 1986; G. L. Verschuur, *The Invisible Universe Revealed*, 1987; G. L. Verschuur and K. I. Kellermann, *Galactic and Extragalactic Radio Astronomy*, 1988.

Radio telescope

An instrument used in astronomical research to detect and measure the radio-frequency power coming from various directions in the sky. It consists of three complementary parts: the large reflecting surface that collects and focuses the incident radiation; the electronic receiver that amplifies and detects cosmic radio signals; and a data display device. From the ground, observations with radio telescopes must be made at wavelengths shorter than 30 m, because of ionospheric attenuation, and longer than 1 mm, because the very short-wavelength radio radiation is absorbed by atmospheric water (H_2O), oxygen (O_2), and ozone (O_3). Large radio telescopes and arrays are listed in the **table**.

Principle of operation. The fundamental principle of a radio telescope is identical to that of a reflecting telescope used at visual wavelengths. The incoming waves (radio or optical) are intercepted by a precise mirror and reflected to a common focal point. The shape of the reflecting surface or "dish" is important: the radio waves must arrive "in phase" at the focal point following their reflection from the dish; that is, the path length from the point of reflection to the focus must be exactly the same for all points on the dish. This restriction can be most simply satisfied if the shape of the reflecting surface is made paraboloidal; consequently, most modern radio telescopes have a paraboloidal shape (**Fig. 1**). *SEE OPTICAL TELESCOPE; TELESCOPE.*

Fig. 1. The 330-ft-diameter (100-m) radio telescope operated by the Max Planck Institut für Radioastronomie at Effelsberg, Germany. (*Max Planck Institut für Radioastronomie*)

Large radio telescopes and synthesis arrays

Institution	Location	Size of reflector, ft (m)
Radio telescopes for meter and centimeter wavelengths		
Fully steerable paraboloids		
Max Planck Institut für Radioastronomie	Effelsberg, Germany	330 (100)
Nuffield Radio Astronomy Laboratory	Jodrell Bank, England	250 (76)
CSIRO	Parkes, N.S.W., Australia	211 (64)
Jet Propulsion Laboratory	Goldstone, California	211 (64)
Algonquin Radio Observatory	Lake Traverse, Ontario	152 (46)
National Radio Astronomy Observatory	Green Bank, West Virginia	142 (43)
California Institute of Technology	Big Pine, California	132 (40)
Haystack Observatory	Westford, Massachusetts	122 (37)
Crimean Astrophysical Observatory	Crimea	73 (22)
Limited-tracking transit telescopes		
Special Astrophysical Observatory	Crimea	33 × 6221 (10 × 1885)
Tata Institute	Ootacamund, India	99 × 1746 (30 × 529)
National Astronomy and Ionosphere Center	Arecibo, Puerto Rico	1007 (305)
Observatory of Paris	Nancy, France	132 × 660 (40 × 200)
Radio telescopes for millimeter wavelengths		
Nobeyama Radio Observatory†	Nobeyama, Japan	148 (45)
Institut de Radio Astronomie Millimetrique	Pico de Veleta, Spain	99 (30)
Onsala Observatory	Gothenburg, Sweden	66 (20)
University of Massachusetts	Amherst, Massachusetts	46 (14)
National Radio Astronomy Observatory	Kitt Peak, Arizona	40 (12)
California Institute of Technology*	Big Pine, California	33 (10)
University of Texas	Fort Davis, Texas	16 (5)
University of California‡	Hat Creek, California	13 (4)
Synthesis arrays		
National Radio Astronomy Observatory (Very Large Array, VLA)	Socorro, New Mexico	Resolution 0".1
Nuffield Radio Astronomy Laboratory (MERLIN)	Jodrell Bank, England	Resolution 0".1
Mullard Radio Astronomy Observatory (5-km array)	Cambridge, England	Resolution 0".5
Westerbork Radio Observatory (WSRT)	Westerbork, Netherlands	Resolution 1"

*Also three-element millimeter-wave interferometer.
†Also five-element millimeter wave interferometer.
‡Four-element millimeter-wave interferometer.

Fig. 2. Very Large Array (VLA) radio telescope near Socorro, New Mexico. (*a*) Some of the antennas. (*b*) Aerial view looking down the southwestern arm. Prominent structure in the center of the photograph is the antenna assembly building; the northern arm branches to the lower right, and the southeastern arm extends to the left. (*National Radio Astronomy Observatory*)

The requirement that the radio waves arrive at the focus in phase means that the entire surface of the dish must be very accurate. As a general rule, observations at a particular radio wavelength λ can be made efficiently with a particular radio telescope if the mean deviation of the surface of that telescope from a perfect parabola is no greater than $\lambda/10$ at each point on the surface. Hence, for example, the 330-ft-diameter (100-m) telescope shown in Fig. 1, which is designed for observations at 1-cm wavelength, must have a surface that is paraboloidal to better than 0.04 in. (1 mm) across the whole dish. These restrictions can place severe demands on the design, construction, and financing of large telescopes. Of course, the smaller the telescope, the more feasible it is to make the surface highly precise; thus, one radio astronomy observatory may have several telescopes of different sizes for use at different wavelengths. SEE ASTRONOMICAL OBSERVATORY.

Once the radio waves are collected and brought together at the focal point of the telescope, they are in general still extremely weak; received power levels near 10^{-18} W m^{-2} are common for astronomical objects. The incoming radio-frequency (rf) signals are first amplified at the focus 10 to 1000 times and then converted to a lower frequency, the intermediate frequency (i-f), that can be easily transmitted by cables from the focal point to the telescope-control building. There the i-f is further amplified, and the signal is detected and displayed in the manner the astronomer finds most suited to the particular investigation.

Astronomical considerations. The types of astronomical objects that emit radio-frequency radiation and hence can be studied by radio astronomers are of such a diverse nature that a variety of radio telescopes and receiving equipment are necessary for a modern radio observatory. Two general astronomical considerations dictate what instruments are needed: first, radio telescopes should have the highest possible angular resolution so that the small-scale details of radio sources can be studied; second, the radio receivers should be extremely sensitive to the very weak signals emitted by cosmic radio sources.

The angular resolution obtainable by a particular radio telescope is limited by diffraction, a blurring of the image that will, for example, cause two radio sources that are close together to appear as if they were a single source. In the presence of this diffraction limit, the minimum angular separation between two sources that a radio telescope would resolve as two sources and not as a single blurred image depends only on the observing wavelength and the diameter D of the telescope as λ/D (radians). To obtain better angular resolution, observations may be made at a shorter wavelength (reduce λ) or a larger telescope (increase D) may be employed. There are, of course, practical limitations to both of these proposals: there is a finite limit to how large a single telescope could be built (as well as concomitant financial limitations); whereas shorter wavelengths require more precise telescope construction, and such instruments must necessarily be smaller. The highest angular resolution obtainable by single-dish radio telescopes is about 30 seconds of arc.

Since nearly all sources of radio emission have structure on a scale much smaller than this, it is necessary to realize better angular resolution; this can be done by employing the principles of interferometry. One can imagine replacing a single large radio telescope with many small dishes connected electrically by cables and separated by distances ranging up to the diameter of the original large dish. This concatenation of small dishes set up to simulate the single large dish is called an array. An array provides an angular resolution defined by λ/d, where d is the maximum separation between two of the array members. Since two such small dishes can be separated by reasonable distances (1–10 mi, or 1.6–16 km, is common) and connected electrically by cables or waveguides, very high angular resolution on the order of 1 second of arc is obtainable.

To adequately reproduce a single large dish with an array would require a very large number of small dishes separated by roughly the diameter of each small dish, which is not feasible. However, the simulation of such an array using a small number of elements is feasible with the techniques of Earth-rotation aperture synthesis, a concept pioneered by Martin Ryle at Cambridge, who was awarded the Nobel Prize in Physics in 1974 for this work. Imagine an array with a small number of radio telescopes seen from a fixed radio source. As the Earth rotates, the relative positions and displacements of the individual dishes seem (to the radio source) to be changing. If the original array had a reasonable number of elements with various initial separations, the net effect after 12 h of observing would be to nearly simulate an observation by a single very large telescope. These aperture synthesis techniques provide extremely detailed pictures of radio sources. The latest generation of arrays, such as the Very Large Array (VLA) in New Mexico with 27 separate radio telescopes along the arms of a Y (**Fig. 2**), are designed to provide radio pictures which have an angular resolution comparable with that of the largest optical telescopes.

The ultimate in angular resolution is achieved by the technique of very long-baseline interferometry (VLBI), in which radio telescopes separated by thousands of miles are utilized simultaneously. It is clearly impossible to connect such telescopes by cables to operate them as a single interferometer, and so data are acquired independently at each telescope and recorded on video tape. Precise time markings are also made on the tape by using hydrogen maser clocks that are so accurate that they neither gain nor lose more than 1 s in 10^6 years. After the data are recorded, the video tapes from the separate telescopes are brought together: the time markings on the individual tapes are aligned, and the data taken at precisely the same times can be compared and analyzed. Such VLBI techniques have achieved angular resolutions of about 0.0003 second of arc.

Electronics. The radio receiver that is mounted on the telescope and used to detect the incoming radio waves is generally composed of two parts, both of which are optimized so as to permit detection of extremely weak signals. The "front end" of the receiver is a sensitive preamplifier mounted at the telescope focus that amplifies the incoming rf signal 10 to 1000 times. The rf is then mixed with the lower i-f, and the i-f is further amplified in the "backend" of the receiver.

The sensitivity of a radio receiver depends critically on the quality of the front-end preamplifier. The front end generates its own radio noise, which is added to the radio power from the sky; since the receiver noise far exceeds the power coming from the sky, the sensitivity of the receiver is increased if the front-end noise is reduced as much as possible. A variety of low-noise amplifiers have been developed that are suitable as radio astronomy front ends. These amplifiers utilize negative-resistance circuits: the most common ones are parametric amplifiers, gallium-arsenide field-effect transistor (GaAs FET) amplifiers, and maser amplifiers. Parametric amplifiers, high-electron-mobility transistor (HEMT) amplifiers, and GaAs FETs rely on variable-capacitance diodes to achieve amplification, whereas maser amplifiers make use of the quantum of the energy states of atomic particles to provide amplification. In either case, the amount of radio noise generated by the amplifiers is further reduced by cooling them cryogenically with liquid helium.

The back-end configuration of the receiver may be varied to suit the purposes of the type of measurement undertaken. For continuum observations in which the goal is merely to identify the total radio power emitted by a source in the sky, the back end consists of a calibrated laboratory source of radio power and a switch enabling comparison of the difference between the source in the sky and the laboratory source. However, when a spectral line is being observed, it is necessary to measure the radio power at many frequency points at and around the frequency of the line. Thus a spectral-line receiver must perform all the functions of a continuum receiver, but it must do so simultaneously at many closely spaced frequency points (typically 100 to 1000). The output of the radio receiver is recorded on magnetic tape suitable for later processing with an electronic computer. *See Radio astronomy.*

Robert L. Brown

Bibliography. N. G. Basov (ed.), *Radio, Submillimeter, and X-ray Telescopes*, 1976; N. Henbest, *The New Astronomy*, 1983; J. D. Kraus, *Radio Astronomy*, 1986; G. L. Verschuur, *The Invisible Universe Revealed*, 1987.

Red dwarf star

A red star of low luminosity, so designated by E. Hertzsprung. Dwarf stars are commonly those main-sequence stars fainter than an absolute magnitude of about +1. Red dwarfs are the faintest and coldest of the dwarfs. They are present among both old and young stars. Red dwarfs are the most numerous class of stars in space, although they are so faint that their presence in remote parts of the Galaxy must be inferred from their frequency near the Sun and their contribution to the total mass of the Galaxy. They have a low-energy output per unit mass and a long nuclear lifetime.

Theoretical work has shown it is improbable that red dwarfs will have appreciable energy generation or be on the main sequence if their masses are less than 0.06 times that of the Sun. Furthermore, infrared observations have defined the total energy emitted, which in cool red dwarfs near 3000°F (2000 K) is mostly in the infrared. Thus, the total luminosity of the faintest and lowest-mass stars is about 3×10^{-4} times that of the Sun. *See Dwarf star; Star.*

Jesse L. Greenstein

Redshift

A systematic displacement toward longer wavelengths of lines in the spectra of distant galaxies, and also of the continuous part of the spectrum. First studied systematically by E. Hubble, redshift is central to observational cosmology, in which it provides the basis for the modern picture of an expanding universe. There are two fundamental properties of redshifts. *See Cosmology.*

First, the fractional redshift $\Delta\lambda/\lambda$ is independent of wavelength. ($\Delta\lambda$ is the shift in wavelength of radiation of wavelength λ.) This rule has been verified from 21 cm (radio radiation from neutral hydrogen atoms) to about 6×10^{-5} cm (the visible region of the electromagnetic spectrum) and leads to the interpretation of redshift as resulting from a recession of distant galaxies. Though this interpretation has been questioned, no other mechanism is known that would explain the observed effect.

Second, redshift is correlated with apparent magnitude in such a way that when redshift is translated into recession speed and apparent magnitude into distance, the recession speed is found to be nearly proportional to the distance. This rule was formulated by Hubble in 1929, and the constant of proportionality bears his name. Hubble's constant is currently estimated to lie between 20 and 35 km/s (10^6 light-years) or 1.9 and 3.3 × 10^{-18} s^{-1}. SEE HUBBLE CONSTANT; MAGNITUDE.

Until 1975 the largest redshifts that could be routinely measured for optical galaxies were around $\Delta\lambda/\lambda = 0.2$. Since that time the use of image tubes and techniques for automatically subtracting the sky background have made it possible to measure redshifts of order $\Delta\lambda/\lambda = 1.5$ for bright spiral and elliptical galaxies. Quasars, being brighter than ordinary galaxies, can be observed at considerably greater distances and correspondingly greater redshifts. Quasar redshifts as large as 4.5 have been measured. In quasars the far-ultraviolet spectrum is shifted into the infrared. SEE IMAGE TUBE; QUASAR.

The recession speed indicated by the redshift in the spectrum of a given galaxy is not the current value for that galaxy but the value appropriate to the epoch when the light now reaching the Earth was emitted. Consequently, the observed relation between redshift and apparent magnitude contains information about past values of Hubble's constant, as well as about the present value. If this information could be extracted from the record it would enable astronomers to choose among various model universes that have been proposed by cosmologists. At present, however, this cannot be done.

The spectrum of the cosmic microwave background approximates that of blackbody radiation at a temperature of about 3 K (5°F above absolute zero). This radiation must have been thermalized (acquired the spectral distribution appropriate to blackbody radiation) when the universe was opaque in the appropriate wavelength range, hence much denser than it is now. The prevailing view among astronomers is that the radiation now observed has been redshifted by a factor of 10^4, but there is some evidence that at least a substantial part of the radiation, and perhaps all of it, was produced much later and has been redshifted by factors of only 50 to 100. The most direct evidence pointing in this direction comes from a 1978 experiment by D. P. Woody and P. L. Richards, who measured the intensity of the background radiation at millimeter wavelengths and found significant departures from a blackbody spectrum. Subsequent experimental attempts to confirm or refute this finding have been inconclusive. SEE COSMIC BACKGROUND RADIATION.

David Layzer

Retrograde motion

In astronomy, either an apparent east-to-west motion of a planet or comet with respect to the background stars or a real east-to-west orbital motion of a comet about the Sun or of a satellite about its primary. The majority of the objects in the solar system revolve from west to east about their primaries. However, near the time of closest approach of Earth and a superior planet, such as Jupiter, because of their relative motion, the superior planet appears to move from east to west with respect to the background stars. The same apparent motion occurs for an inferior planet, such as Venus, near the time of closest approach to Earth.

Actual, rather than apparent, retrograde motion occurs among the satellites and comets; the eighth and ninth satellites of Jupiter and the ninth satellite of Saturn are examples. SEE ORBITAL MOTION.

Raynor L. Duncombe

Roche limit

The closest distance which a satellite, revolving around a parent body, can approach the parent without being pulled apart tidally. The simplest formal definition is that the Roche limit is the minimum distance at which a satellite can be in equilibrium under the influence of its own gravitation and that of the central mass about which it is describing a circular orbit. If the satellite is in a circular orbit and has negligible mass, the same density as the primary, and zero tensile strength, the Roche limit is 2.46 times the radius of the primary.

The concept of an equilibrium state was first given by Édouard Roche in 1849. The radius r of the Roche limit appropriate to satellite formation is given by the equation below, where B is the radius and ρ the mean

$$r = 2.46\left(\frac{\rho}{\rho_s}\right)^{1/3} B$$

density of the parent planet, and ρ_s is the mean density of the satellite. This is the radius at which the surface of a homogeneous, fluid satellite can remain in stable hydrostatic equilibrium.

The popular definition of Roche's limit, that is, the distance from the planet at which a satellite would suffer tidal loss of particles, depends on properties of the material and the shape and density of the satellite. If, for instance, the material can flow to adopt the hydrostatic equilibrium shape, then the body will be elongated in the direction of the radius vector. For a binary system, the Roche limit is defined by the value of the Jacobi constant such that the zero velocity surfaces around the two bodies intersect in the lagrangian libration point L_2 between the finite masses. This implies an upper limit to the size of the components. While magnetohydrodynamic forces act on the outer layers of stars and the theory holds only for the circular restricted problem with the masses acting as point masses, the Roche limit seems to correspond to the observational data. When a star has exhausted the supply of hydrogen in its core, its radius will increase by a factor of 10 to 100. A star in a binary system may then exceed its Roche limit, material will thus escape from that star, and its companion will receive the excess material. SEE BINARY STAR; CELESTIAL MECHANICS.

P. K. Seidelmann

Rocket astronomy

The discipline that makes use of sounding rockets that fly near-vertical paths carrying scientific instruments to altitudes ranging from 25 to more than 900 mi (40 to 1500 km). Altitudes up to 30 mi (48 km) can be reached by balloons, so sounding rockets are typically used for higher altitudes in order to measure emissions from the Sun or other celestial sources that do not penetrate the Earth's atmosphere. Sounding rockets do not achieve orbital velocity; after completion of the launch phase, the payload follows a ballistic trajectory that permits 5–15 min of data taking before reentry.

Mission profile. Rocket astronomy began in the United States in 1946, using captured German V-2 rockets with no pointing capability and very simple instrumentation. A typical modern launch is comparatively sophisticated, involving engine ignition, vehicle spinup, payload separation and despin, target acquisition, attitude control system (ACS) operation, telemetry and data command via the ground station, door closure or boom retraction, reentry and chute deployment, impact and recovery, data retrieval, and reuse of the scientific payload.

In comparison with experiments launched on satellites, rockets offer the advantages of simplicity, in that the payload is usually composed of a single instrument package; relatively frequent access to launch opportunities, typically once per year in solar physics; a shorter time scale from conception to reality, 3–5 years for a rocket versus 5–20 years for satellites; lower cost; and recoverability of the payload and the possibility of postflight instrument calibration, refurbishment, and reflight. Major disadvantages are short observing time, minutes versus months or years for satellites; localized coverage, restricted to an area adjacent to the launch site; and size and weight restrictions on payloads. *SEE SATELLITE ASTRONOMY*.

Over a dozen different launch vehicles are in common use (**Fig. 1**), ranging in length from 10 to 65 ft (3 to 20 m) and in diameter from 4.5 to 44 in. (11 to 112 cm). Payloads up to 2200 lb (1000 kg) can be carried to altitudes that are typical of those at which the space shuttle operates, 150–200 mi (250–320 km). For each launch vehicle the maximum altitude is strongly dependent upon payload weight. For instance, the Black Brant X can carry a 200-lb (91-kg) payload to an altitude of 620 mi (1000 km) and can boost an 800-lb (364-kg) payload to 250 mi (400 km). A performance summary of apogee versus payload weight for National Aeronautics and Space Administration (NASA) sounding rockets is shown in **Fig. 2**.

Sounding rockets are routinely launched in the United States from Wallops Island, Virginia; Poker Flats Research Range, Alaska; and White Sands Missile Range, New Mexico. Sites in Norway, Sweden, and Canada are also used. Temporary launch ranges are arranged when needed, and scientific requirements may dictate heavy usage of a particular facility. This was the case in 1988 and 1989 when numerous observations of Supernova 1987*a* were carried out from Woomera, in Australia. The United States sounding rocket program makes extensive use of surplus military rockets as an economical source of engines. NASA's Orion was formerly a Hawk ground-to-air missile, while Nike, Taurus, and Terrier engines are used as boosters in two- and three-stage flights. The largest sounding rocket in current use is the 44-in.-diameter (112-cm) Aries, formerly the upper stage of a Minuteman I missile. These components are usually supplied without guidance control systems.

The United States, Canada, Japan, and many European countries, including Germany, France, Sweden, Norway, Switzerland, and the United Kingdom, maintain vigorous scientific rocket programs. These programs focus on the disciplines of aeronomy, magnetospheric physics, meteorology, and material sciences, as well as astronomy and astrophysics. The

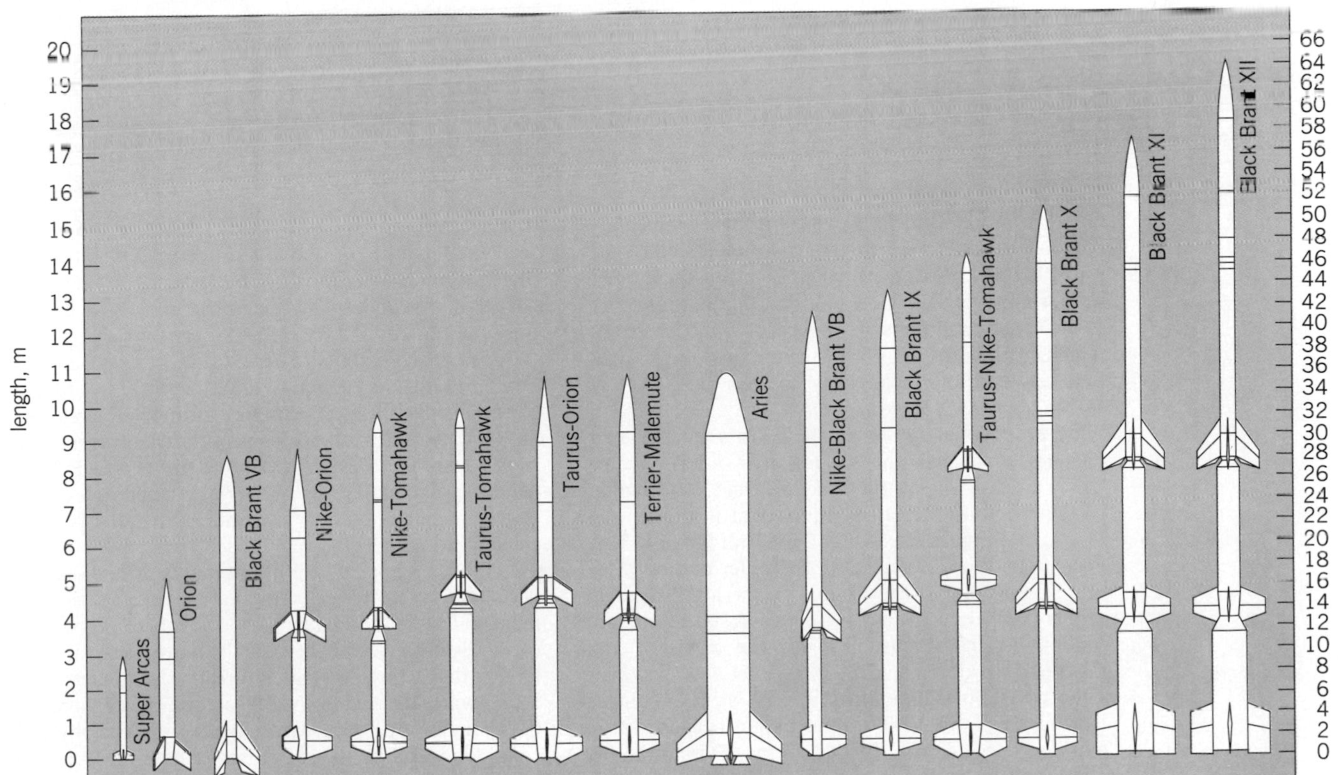

Fig. 1. NASA sounding rockets.

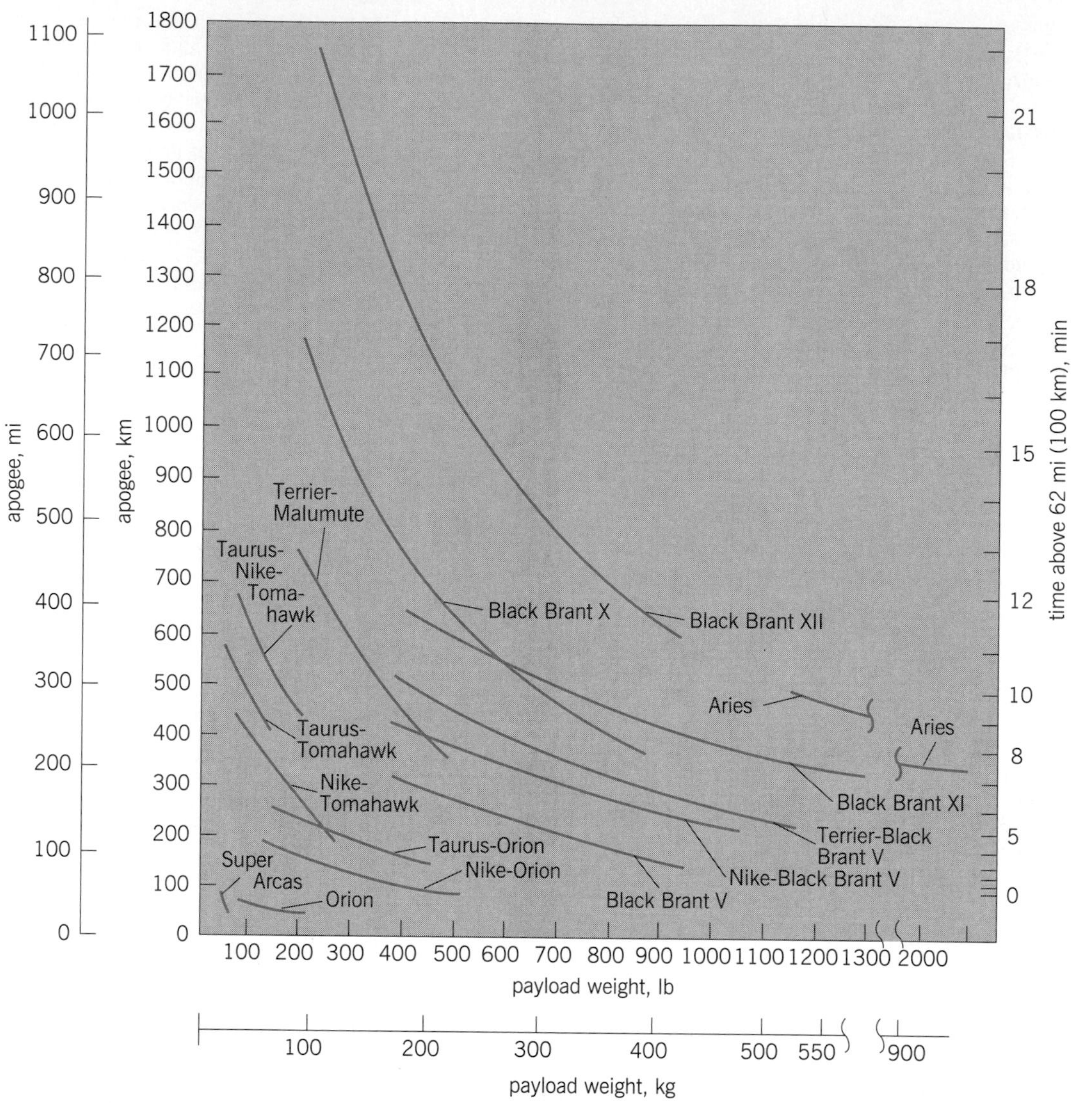

Fig. 2. Lift capability of NASA sounding rockets. Apogee is shown as function of payload weight.

ability to carry out vertical profile measurements of relevant atmospheric parameters at heights of 25–125 mi (40–200 km) is essential in many of these scientific disciplines. To the extent that the Sun influences or controls conditions in the upper atmosphere and magnetosphere of the Earth, there is a strong connection between solar astronomy and the more local research areas. *SEE MAGNETOSPHERE.*

Solar rocket astronomy. The Earth's atmosphere is opaque to wavelengths shorter than ~300 nanometers. Observations in the ultraviolet and x-ray regions of the spectrum therefore require that instruments be placed at altitudes exceeding 25–87 mi (40–140 km), depending upon wavelength. The outer portion of the Sun's atmosphere, the corona, is at a substantially higher temperature than is the photosphere, several million kelvins versus 6000 K. The emission from this region is dominated by ultraviolet and x-ray photons, whose total luminosity is very weak in comparison to the Sun's visible light output because of the extremely low density of the corona. During the late 1950s and early 1960s, techniques were developed for focusing x-rays and thereby providing direct imaging of the corona. These early studies revealed the highly structured nature of the atmosphere, with approximately semicircular loops of hot plasma outlining the shape of the underlying magnetic field, which confines the hot gas. *SEE X-RAY TELESCOPE.*

Areas of the solar surface at which strong magnetic fields emerge are known as active regions; in addition to producing strong x-ray and ultraviolet emission, they are the sites of explosive solar flare events. In such events the plasma is temporarily heated to temperatures exceeding 10^7 K by the rapid release of stored magnetic energy; in addition, material is often ejected into interplanetary space, and these events may have associated with them bursts of high-energy particles, microwave emission, gamma rays, and magnetic storms. Flares have long been known to cause disturbances of the Earth's atmosphere, and the largest flares pose a serious threat to astronauts situated above the atmosphere; forecasting of flares is thus a major area of research. Analysis of the physical conditions leading to and during a flare involves observation of the entire three-dimensional atmospheric structure before a flare and the location and evolution

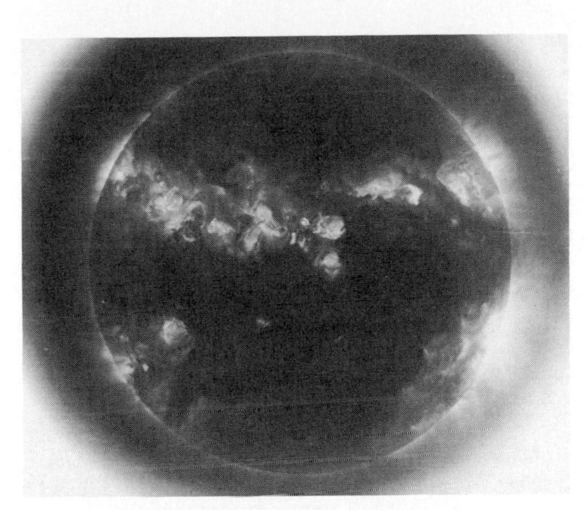

Fig. 3. Soft x-ray image of the solar corona obtained by a sounding rocket on September 11, 1989. (*IBM Research and Smithsonian Astrophysical Observatory*)

of the hot plasma during the flare. Such studies require recording of all temperature regimes from 3000 to 10^7 K, which implies the ability to observe wavelengths from the infrared, through the ultraviolet and soft x ray, into the hard x-ray (10-keV) regime, plus microwave and other radio emission. Such data cannot be obtained by a single technique, and solar research satellites are therefore composed of multiple co-observing instruments combining both imaging and spectroscopy, and including ground-based observations as a necessary part of the overall scientific program.

Among the notable advances in coronal studies from rockets has been the development of a new technique for x-ray imaging: the use of multilayer coatings for enhanced x-ray reflectivity. The technique may be viewed as depositing an artificial Bragg crystal onto a figured substrate and offers the advantage of substantially higher image sharpness (**Fig. 3**), and provides as well simultaneous spectroscopy because of the narrow x-ray bandwidth of the coatings. The major technological limitation of this technique is the inability to produce adequate normal-incidence reflectivity at short wavelengths (less than 2 nm); in this spectral region, grazing-incidence methods continue to be the most appropriate. *SEE SUN; X-RAY DIFFRACTION.*

Other celestial sources. Observations of nonsolar sources from sounding rockets encounter two major difficulties: the low intensity of the emission and the technological problem of pointing the payload at the source during the flight. The two problems are related in that the low flux levels generally lead to a requirement for long exposure or integration times, and the faintness of the source leads to difficulty in constructing a sensor that can provide the appropriate detection signal to an attitude control system. The development of a gas-jet attitude control system for Aerobee rockets in 1964 and the addition of gyroscopic stabilization in 1965 permitted ultraviolet spectroscopy of stars with high enough dispersion to record interstellar absorption features in the 130–160-nm range. Such data permit measurement of interstellar hydrogen, carbon, nitrogen, oxygen, silicon, and sulfur. Spectra in the 100–250-nm region permit studies of stars with surface temperatures of 10,000–30,000 K; in addition to total-luminosity determination, a major discovery from line-profile measurements of hot-star absorption features was the finding that such stars have large mass loss due to high-velocity (greater than 600 mi/s or 10^3 km/s) stellar winds. In 1970, a sounding rocket detected the absorption line of interstellar molecular hydrogen; this detection was the forerunner of later extensive studies from the *Copernicus* satellite, which have had a major impact on the field of interstellar chemistry. *SEE ASTRONOMICAL SPECTROSCOPY; COSMOCHEMISTRY; INTERSTELLAR MATTER; ULTRAVIOLET ASTRONOMY.*

In 1962, a rocket experiment detected nonsolar x-rays for the first time, at a level many orders of magnitude higher than would be produced by an equivalently placed solar-strength source. Subsequent flights confirmed the initial result and discovered other x-ray sources, as well as a diffuse x-ray background and an emission source in the Crab Nebula. The field of x-ray astronomy is now dominated by observations from satellites, which have the ability to carry out all-sky surveys and to integrate for many hours in order to study low-intensity sources. *SEE X-RAY ASTRONOMY.*

Leon Golub

Bibliography. American Institute of Aeronautics and Astronautics, *Proceedings of the 7th Conference on Sounding Rockets, Balloons, and Related Space Systems*, 1986; G. V. Groves (ed.), *Dynamics of Rockets and Satellites*, 1965; P. A. Hanle (ed.), *Space Science Comes of Age*, 1981; H. E. Newell, Jr., *Beyond the Atmosphere*, NASA Publ. SP-4211, 1980; H. E. Newell, Jr., *Sounding Rockets*, 1959.

Sagittarius

The Archer, in astronomy, a zodiacal and summer constellation, the major portion of which lies directly in the Milky Way. Sagittarius is the ninth sign and

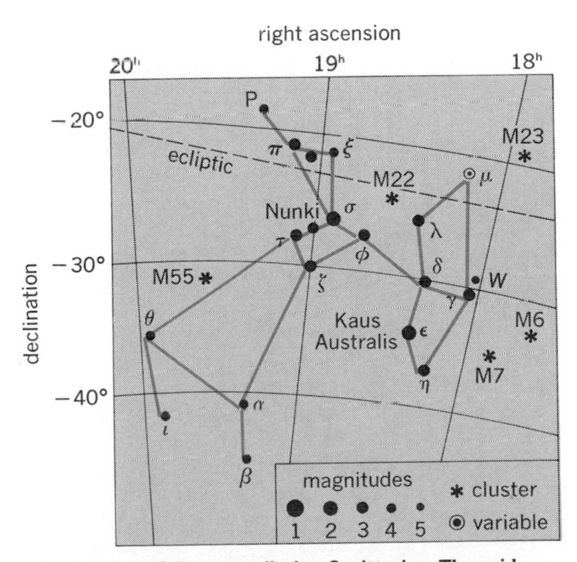

Line pattern of the constellation Sagittarius. The grid lines in the chart represent the coordinates of the sky. The apparent brightness, or magnitude, of the stars is shown by the sizes of the dots, which are graded by appropriate numbers as indicated.

the southernmost constellation of the zodiac. In mythology, it is represented by a centaur, Chiron, drawing his bow to release an arrow. Its most prominent feature is a star group commonly called the Little Milk Dipper. It is an inverted dipper with four stars to form the bowl and one to form the handle (see **illus.**). The Milky Way in Sagittarius is very bright, containing rich star fields and clusters, because its direction lies in the center of the Milky Way stellar system. *See Constellation; Zodiac.*

Ching-Sung Yu

Satellite

A relatively small body orbiting a planet. In the solar system, all of the planets except Mercury and Venus have satellites. A total of 61 have been observed and their orbits identified, distributed as follows: Earth 1, Mars 2, Jupiter 16, Saturn 18, Uranus 15, Neptune 8, and Pluto 1. Additional satellites have been glimpsed in both the Jupiter and Saturn systems; the more distant planets may also have yet undetected companions.

It is customary to distinguish between regular satellites that have nearly circular orbits lying essentially in the plane of a planet's equator and irregular satellites whose orbits are highly inclined, very elliptical, or both. The former almost certainly originated with the parent planet, while the latter are likely to be captured objects. The Earth's Moon is a special case. The most widely favored hypothesis for its origin invokes an impact with Earth by a Mars-sized planetesimal, and ejection of material that first formed a ring around the Earth and then coalesced to form the Moon. *See Moon.*

The two satellites of Mars are both tiny objects, less than 9 mi (15 km) in their longest dimension. They are irregularly shaped because the gravitational fields produced by their small masses are too weak to deform the material into spheres. The same is true of the small satellites of the outer planets. *See Mars.*

Jupiter has two groups of small outer satellites, both of which are irregular. Four are in direct orbits with high eccentricities; the other four are in retrograde (clockwise) orbits. Both groups were probably captured. The four largest satellites were discovered by Galileo and Simon Marius in 1610. One of these, Io, exhibits an astonishing degree of active volcanic activity, driven by tidal forces. Another, Europa, may have an ocean of liquid water trapped beneath an icy crust. The remaining two, Ganymede and Callisto, are both larger than the planet Mercury. But unlike the planet, they consist of 50% ice by mass, which surrounds a rocky core. *See Jupiter.*

Saturn has one highly irregular satellite, Phoebe, which revolves in a retrograde direction. The forward hemisphere of Iapetus is 10 times darker than the trailing side. Titan has a dense atmosphere with a surface pressure greater than that on Earth. Composed mainly of nitrogen, Titan's atmosphere also contains methane and at least 10 other gases that are forming as a result of photochemical reactions. Enceladus may be the source of Saturn's E ring, while Tethys and Dione each have their own accompanying Trojan satellites. *See Saturn; Trojan asteroids.*

Voyager 2 discovered 10 small satellites of Uranus during its 1985–1986 encounter with the planet. The five previously known satellites of Uranus are unusually regular, except for Miranda, whose surface reveals evidence of a surprisingly vigorous geological history. Neptune has two highly irregular satellites, Triton and Nereid, plus six small regular moons discovered by *Voyager 2*, in 1989. Triton has a tenuous atmosphere of nitrogen (a few millionths of the Earth's surface pressure) with a trace of methane, but its surface reveals streaks of windblown material and active, eruptive plumes. Pluto's Charon appears to occupy an orbit whose period of rotation is the same as the planet's. Thus, Charon is like one of the Earth's artificial geosynchronous satellites: it would appear to hang motionless in the dark skies of Pluto. *See Asteroid; Neptune; Planet; Pluto; Uranus.*

Tobias C. Owen

Bibliography. J. K. Beatty (ed.), *The New Solar System*, 3d ed., 1990; J. Burns and M. Matthews, *Planetary Satellites*, 1987; D. Morrison and T. Owen, *The Planetary System*, 1988.

Satellite astronomy

The study of astronomical objects by using detectors mounted on Earth-orbiting satellites or deep-space probes so that observations unobstructed by the Earth's atmosphere can be made. Many astronomical satellites carrying detectors to record electromagnetic radiation at wavelengths shorter than visible light (ultraviolet, x-, and gamma rays) were launched in the 1960s and 1970s. Noteworthy were the *OSO*, *Uhuru* (*SAS 1*), *OAO*, *SAS 2*, *ANS*, *SAS 3*, *COS-B*, *IUE*, and *HEAO* satellites. In general, one or two wavelength regions were observed from each satellite. Successive satellites incorporated significant developments in operating modes and improvements in detector sensitivity and resolution such that new discoveries were possible. Although most satellite observatories launched before 1980 were from the United States, this has not been the case since then, with major satellite experiments for astronomy launched by the Japanese and Europeans. These include the x-ray astronomy satellites from Europe (*EXOSAT*) and Japan (*Hakucho*, *Tenma*, and *Ginga*). Considerable x-ray and low-energy gamma-ray astronomy has been conducted with telescopes on the Soviet (now C.I.S.) *Mir* space station and the Soviet-French satellite *GRANAT* (launched in 1989). In 1990 and 1991 a group of even more powerful astronomical laboratories were launched: the Hubble Space Telescope, the German–United States x-ray telescope *ROSAT*, and the *Compton Gamma-Ray Observatory*.

Ultraviolet astronomy. Since the Earth's atmosphere is opaque to ultraviolet (uv) light with wavelengths shorter than about 310 nanometers, ultraviolet astronomy had to await the space age. In the 1940s Lyman Spitzer of Princeton University pointed out some of the astrophysical problems that could be addressed with a large telescope above the Earth's atmosphere. Many of these problems have now been explored with the several ultraviolet detector systems flown on the *OSO* (*Orbiting Solar Observatory*) and *OAO* (*Orbiting Astronomical Observatory*) satellite series in the 1960s and early 1970s, as well as with the subsequent *IUE* (*International Ultraviolet Explorer*) satellite. A much more powerful tool became available in 1990 with the launch of the Hubble Space Telescope. In June 1990 the *ROSAT* satellite was launched by the National Aeronautics and Space Administration (NASA) for the Germans, and carried out the first far-ultraviolet sky survey with the Wide Field Camera supplied by United Kingdom investigators. The first dedicated ultraviolet astronomy satellite, the *Extreme Ultraviolet Explorer* (*EUVE*), was scheduled for launch in 1992.

Solar ultraviolet astronomy. The *OSO* satellites were the first to be devoted primarily to astronomical ultraviolet and x-ray observations. *OSO 1* through *8* were launched over a 12-year period beginning in 1963. Each was of a similar design and contained a pointed section mounted above and across a wheel section which spun and maintained stable pointing of the satellite toward the Sun. The experiments mounted in the wheel section scanned a great circle in the sky including the Sun.

The solar ultraviolet experiments included both low- and high-spectral-resolution spectrometers. Small entrance slits were used so that high-spatial-resolution (several arc-seconds) raster scans could be made to map the entire Sun in the light of a given ultraviolet emission line. Different spectral lines, or narrow wavelength intervals, could then be chosen by changing the tilt of the spectrograph grating. The resulting detailed maps of the Sun greatly increased understanding of the temperature-versus-height profile of the solar atmosphere. This, in turn, enabled much more detailed theoretical models of the transfer of radiation out through the solar atmosphere. SEE ASTRONOMICAL SPECTROSCOPY.

Time-resolved spectra and images were also recorded so that the first ultraviolet observations of solar flares were conducted. Although superseded in quality by the higher-resolution observations made from *Skylab* in 1972 and 1973, the first solid evidence for the existence of coronal holes came from experiments on *OSO 4*. These are apparent holes in the solar corona, or hot (1×10^6 K or 2×10^6 °F) outer low-density region of the solar atmosphere, through which the solar wind is escaping. Another fundamental discovery made from *OSO 8* was that the solar corona could not be heated by acoustic waves driving up from below (in the chromosphere) as had been commonly assumed. SEE SUN.

Galactic ultraviolet astronomy. Extensive ultraviolet observations of stars and nebulae in the Galaxy but outside the solar system were first conducted with ultraviolet telescopes on the *Orbiting Astronomical Observatory* satellites. Two of these spacecraft were successfully launched. *OAO 2*, launched in December 1968 (*OAO 1* did not achieve orbit), carried two principal experiments—a spectrophotometer with 2-nm resolution to study ultraviolet emission lines from bright stars (as well as the first ultraviolet absorption lines from interstellar gas) and a broadband photometer with a 12-in. (31-cm) telescope to measure the ultraviolet fluxes in broader bands of stars down to fainter magnitudes. More than 5000 stars were observed at 140 and 200 nm by the latter instrument, which provided the first large sample of ultraviolet observations of stars for comparison with predictions from model stellar atmospheres derived from visible-light observations.

Whereas *OAO 2* operated for about 16 months (and the individual *OSO* satellites typically for 1–2 years), the *OAO 3*, called *Copernicus*, was operated as an ultraviolet and x-ray observatory until early 1981 after a 1972 launch. This satellite contained a 32-in.-aperture (80-cm) ultraviolet telescope and high-resolution

Fig. 1. ***International Ultraviolet Explorer*** **spacecraft in geosynchronous orbit. (*IUE Observatory, NASA Goddard Space Flight Center*)**

spectrograph with sensitivity and resolution such that fractional angstrom (1 Å = 0.1 nanometer) resolution could be obtained on stars brighter than about 7th magnitude. The spectrograph was actually a scanning monochromator (as on the *OSO* satellites) so that detailed line profiles produced by absorption from the fractional percent composition of heavy elements (primarily carbon, nitrogen, oxygen, and silicon) in the interstellar medium could be studied. Surprisingly, these heavy elements and "metals" were found to be much lower in their total apparent abundance in the interstellar medium than the average cosmic composition which they might otherwise be expected to show. The explanation seems to be that the heavier elements are "locked up" in the interstellar grains. These grains, which are typically 10 micrometer particles, could not themselves produce the ultraviolet absorption spectra seen (as would single atoms); instead they can totally absorb the ultraviolet and reradiate it as heat or infrared radiation. SEE INTERSTELLAR MATTER.

The Wide Field Camera ultraviolet telescope on *ROSAT* has detected some 400 far-ultraviolet objects in several bands in the wavelength range 300–900 nm. Most of the objects were very hot stars and white dwarfs. Severe interstellar absorption at these extreme ultraviolet wavelengths means that only relatively nearby (out to 100–200 parsecs) or unobscured objects are detectable. SEE STAR; WHITE DWARF STAR.

Extragalactic ultraviolet astronomy. The *International Ultraviolet Explorer* was launched in January 1978 into a geosynchronous orbit such that it can be continuously viewed from both the NASA Goddard Space Flight Center and the European Space Agency (ESA) Operations Control Center near Madrid, Spain. United States astronomers use the facility 16 h per day whereas ESA controls the satellite for 8 h. The *IUE* spacecraft (**Fig. 1**), like *OAO 3*, is three-axis-stabilized and can be pointed at any object outside a 45° cone centered on the Sun.

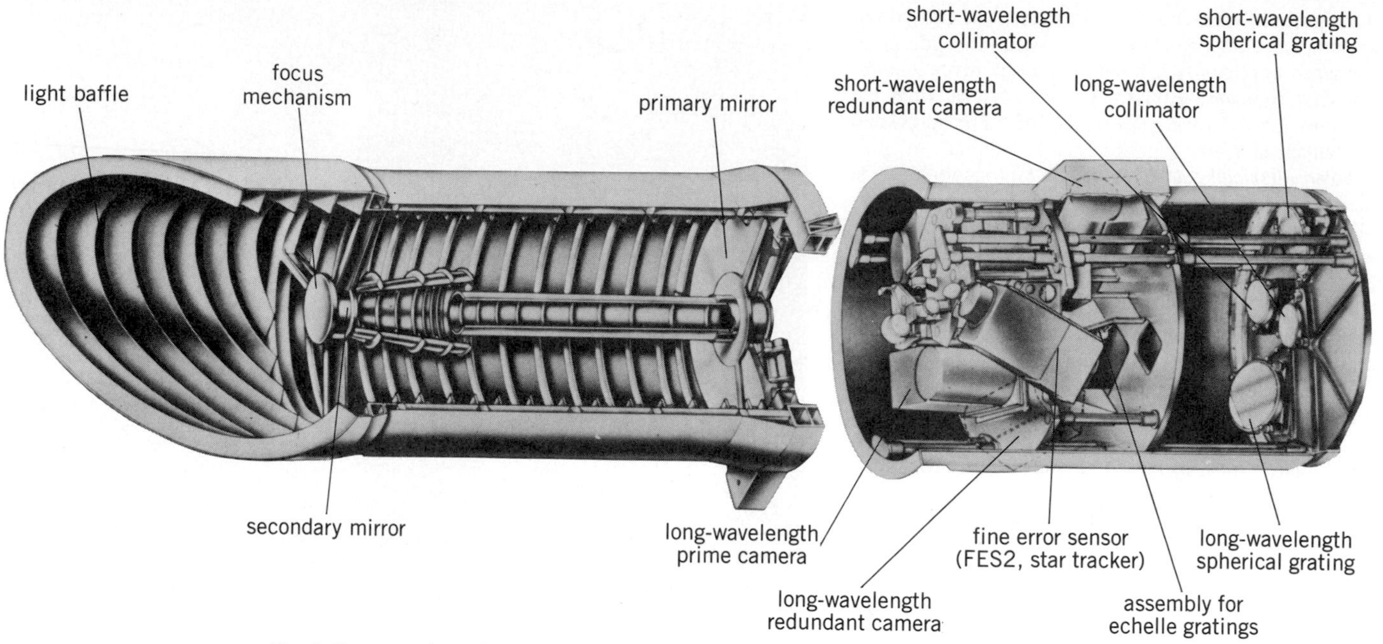

Fig. 2. Cutaway view of the *International Ultraviolet Explorer* telescope. (*IUE Observatory, NASA Goddard Space Flight Center*)

The *IUE* telescope (**Fig. 2**) is an 18-in.-aperture (45-cm) Ritchey-Chrétien with an effective focal ratio of *f*/15 and image quality of about 3 arc-seconds within a 16-arc-minute field. At the telescope focus is an echelle spectrograph, allowing very high spectral resolution ($\lambda/\Delta\lambda$ = 10,000) over a large spectral range (120 nm) by stacking multiple orders of the dispersed spectrum on a two-dimensional detector.

IUE has, like *OAO 3*, been used for a great many observations of objects within the Galaxy, such as galactic x-ray binaries, hot stars, and globular clusters, to name but a few. However, its greater sensitivity has also permitted for the first time ultraviolet observations of extragalactic objects. Active galaxies and nearby bright quasars such as 3C 273 have been some of the primary targets. These observations have enabled the first studies of the ultraviolet emission line spectrum in nearby quasars for comparison with the many ground-based optical observations of these same lines from higher redshift (more distant) quasars. This in turn allows the evolution of quasars, which are the most distant objects known, to be studied. *IUE* also provided critical ultraviolet spectra of the bright supernova SN 1987A, beginning shortly after its discovery in February 1987 and continuing for 2 years thereafter. *See* Galaxy, external; Quasar; Supernova; Ultraviolet astronomy.

X-ray astronomy. X-ray astronomy can be done only from above the Earth's atmosphere. The first astronomical x-ray detectors, launched with captured V2 rockets in the late 1940s, discovered x-ray emission from the Sun. Cosmic (that is, nonsolar) x-ray sources were discovered in 1962 by detectors carried above the Earth's atmosphere on sounding rocket flights of typically 5 min duration. *See* Rocket astronomy.

Nonimaging x-ray astronomy. X-ray astronomy really came of age with the launch of the first *Small Astronomy Satellite (SAS 1)* on December 12, 1970, designated *Uhuru*. It carried two proportional-counter x-ray detectors in which a pulse of electric charge proportional to the energy of the incident x-ray photon is detected. The satellite was spinning so that the two detectors, with differing fields of view, alternately scanned over a given source, and the first survey of the entire sky could be conducted. *Uhuru* enabled the key discovery of x-ray binary systems in which a collapsed object (neutron star or black hole) accretes gas from the atmosphere of a "normal" companion star. These systems are often 10,000 times as luminous in x-rays (alone) as the entire output of the Sun, and they are probably involved with most (if not all) of the bright galactic x-ray sources. *Uhuru* also made the initial discoveries that active galaxies (Seyferts) and galaxy clusters are also prodigious sources of x-rays. *See* Binary star; Black hole; Neutron star.

The *OSO* satellites (described above) also carried cosmic x-ray detectors (proportional counters similar to *Uhuru*) and contributed much to the detailed understanding of individual sources. Qualitatively different cosmic x-ray satellite experiments, however, were launched in 1974 and 1975 with the *Astronomical Netherlands Satellite* (*ANS*) and *SAS 3* satellite, respectively.

ANS was the first x-ray observatory: it was a pointed instrument carrying a variety of x-ray detectors that could be operated by an on-board computer in a variety of modes. A schematic view of the United States experiments on *ANS* is shown in **Fig. 3**. (*ANS*, a joint Netherlands–United States venture, also carried two Dutch experiments.) This experiment consisted of two proportional counters with fields of view slightly inclined to each other (so that accurate source locations in one dimension could be derived from the ratio of fluxes recorded in the two detectors) and two Bragg crystal spectrometers. The Bragg spectrometer was tuned to reflect (from the planes of atoms in a crystal) onto proportional counter detectors only x-rays at the wavelengths (about 0.67 nm) produced by K-shell transitions in the element silicon stripped of all but one or two electrons—that is, Si(XIV) or Si(XIII), respectively. The upper limits found for silicon-line emission from a number of bright x-ray sources were able to restrict the physical conditions in the source. The most significant discovery made with *ANS*, however, was that of the x-ray burst sources. Intense bursts of x-rays were found to be occasionally emitted by certain strong x-ray sources, which are often located in globular clusters. Although it was originally thought that these might be produced by gas falling into a massive black hole, subsequent studies indicated that the sources are (again) binary systems containing a low-mass "normal" star in orbit around a neutron star. The discovery of the comparatively rare x-ray bursts was due in large part to the ability of *ANS* to point continuously and thus observe an object for much longer times than the earlier scanning experiments. SEE STAR CLUSTERS.

The *SAS 3* satellite, launched in May 1975, contained a variety of x-ray detectors, including a low-energy x-ray flux concentrator (a similar device was included on *ANS* as one of the two Dutch experiments), which detected x-rays from "normal" stars, and a rotating modulation collimator (RMC). The RMC was the first to be flown on a satellite. It was able to determine relatively precise x-ray source positions (to about 30 arc-seconds) by measuring, on a proportional counter detector, the modulation of the detected x-ray flux due to the shadow cast by a rotating grid of closely spaced wires. These precise source positions in turn allowed many x-ray sources to be identified optically for the first time. *SAS 3* was able to both scan and point, and detailed timing studies of x-ray pulsars and bursters were emphasized. SEE PULSAR.

A major increase in sensitivity for x-ray astronomy was achieved with the *High Energy Astronomical Observatory* (*HEAO*) satellite. *HEAO 1*, launched in June 1977, carried four major experiments: a large-area (nearly 1 m^2 or 10 ft^2) proportional counter; a broad-energy-range and wide-field-of-view set of proportional counters; a modulation collimator; and a crystal scintillator-detector system to extend the energy range to much higher values (that is, to low-energy gamma rays). The total number of x-ray sources known was quadrupled to some 1500 as a result of *HEAO 1*, and high-quality continuum spectra were obtained for many sources. The increased sensitivity meant that fainter objects (primarily extragalactic) such as active galaxies and quasars could be well observed for the first time. *HEAO 1* was a scanning instrument (though some pointed observations were also done) designed to survey the entire sky.

In the 1980s the Japanese launched and successfully operated three x-ray astronomy satellites: *Hakucho, Tenma,* and *Ginga*. Whereas *Hakucho* continued studies of x-ray burst sources with sensitivity comparable to *SAS 3*, *Tenma* provided significantly improved spectral resolution for the first detections of line features in a variety of x-rays sources. *Ginga*, with its very large collecting area, carried out high-time-resolution studies, and also measured spectra at energies up to 20–30 keV for a variety of quasars.

The European satellite *EXOSAT*, which was launched into an 80-h orbit, was able to conduct the first studies of x-ray sources on time scales of 1–80 h. Many discoveries were possible.

Imaging x-ray astronomy. The *HEAO 2* satellite, or *Einstein Observatory*, marked the start of a new era

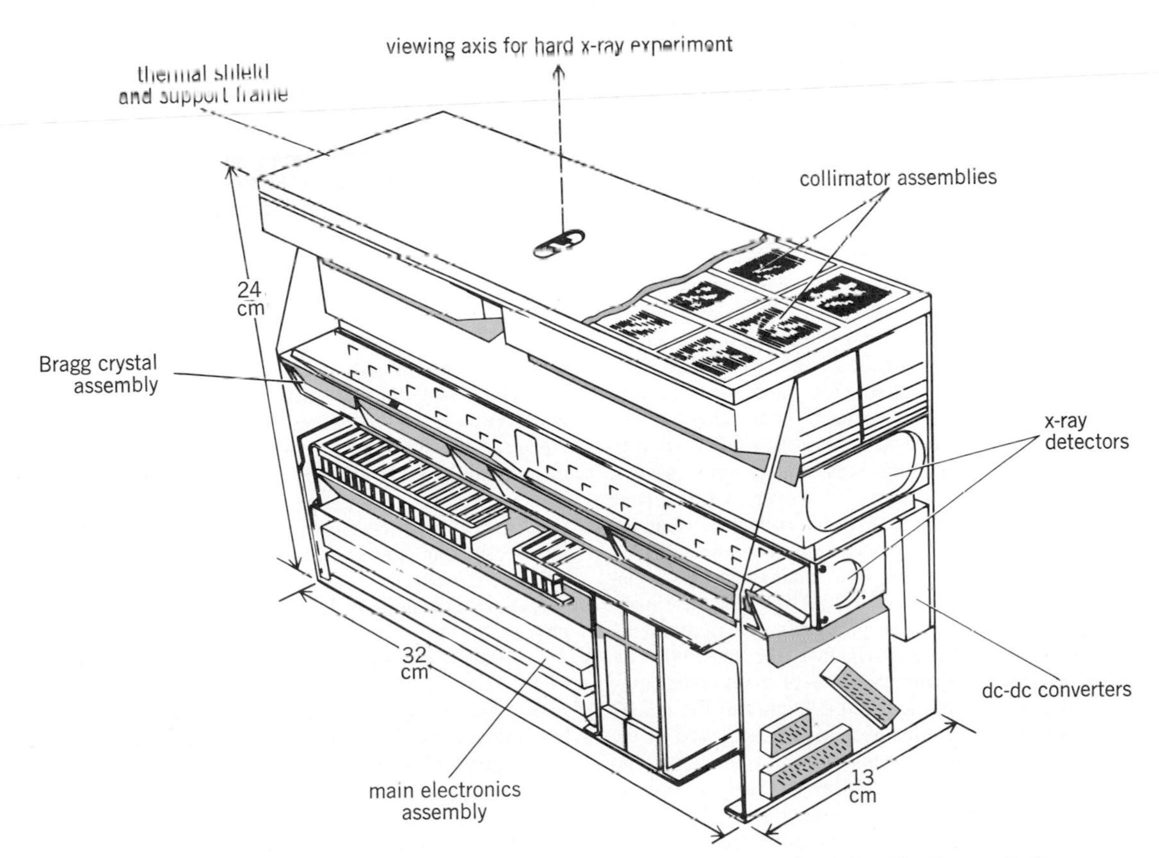

Fig. 3. United States hard x-ray (1–20 keV) experiment on the *Astronomical Netherlands Satellite*. 1 cm = 0.4 in.

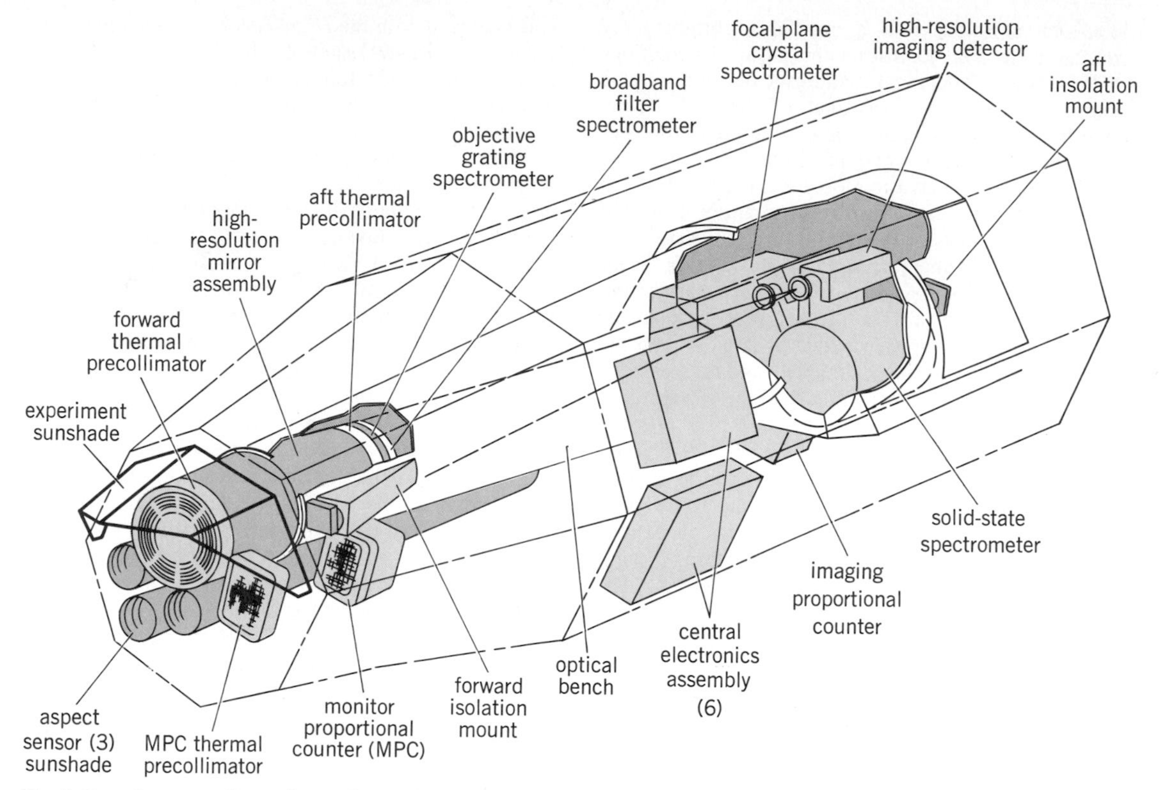

Fig. 4. Experiment configuration and overall layout of the *HEAO 2* (*Einstein Observatory*).

in x-ray astronomy and in satellite astronomy in general. Launched in November 1978 and operated through April 1981, *Einstein* provided the first x-ray images of celestial objects and detected thousands of new sources. The satellite carried the largest x-ray telescope yet constructed (with 24-in. or 60-cm aperture), and the first to be used exclusively for studies of cosmic x-ray sources (a smaller x-ray telescope was included on the *Skylab* satellite and used to take x-ray pictures of the Sun). The overall length of the satellite was about 23 ft (7 m). *Einstein*, like *IUE*, was used extensively by guest observers as if it were a ground-based astronomical observatory.

In the grazing-incidence x-ray telescope on *HEAO 2*, x-rays were reflected from highly polished, nickel-coated surfaces of first a paraboloid and then a hyperboloid. The incidence angles had to be small (less than 2°) for x-rays to reflect from, and not be instead absorbed by, the mirror surface. Thus, although the *HEAO 2* telescope had a large surface area, its projected area for imaging was small, only several hundred square centimeters. However, a true image of the object (extended or point source or sources) being observed was formed, and thus the detected background was very low in a small-image pixel element. Thus, the sensitivity of *HEAO 2* or any imaging detector is very much greater than a nonimaging detector of the same size. In fact, the sensitivity increase achieved in the 16 years of x-ray astronomy from the first rocket-launched detectors to the *HEAO 2* instrument was comparable to that achieved in the 300 years of optical astronomy from the first telescope of Galileo to the 200-in. (5-m) telescope on Palomar Mountain. *SEE X-RAY TELESCOPE.*

The *Einstein Observatory* included four different detectors which could be individually positioned at the focus of the telescope (**Fig. 4**). These were a low-resolution (arc-minute) and high-resolution (arc-second) imaging detector (that is, ''camera'') for recording x-ray images with and without spectral information and a moderate-resolution and high-resolution spectrometer for measuring x-ray spectral lines in a broad versus narrow spectral range. Major discoveries were made by each of these instruments. Perhaps foremost of these discoveries is the finding that even the most distant quasars known (and many previously unknown) are detected as strong x-ray sources and that collectively these may contribute most of the mysterious cosmic x-ray background. Another important result is that ''normal'' stars can be much more powerful x-ray emitters than was expected from comparisons with the Sun.

The *EXOSAT* observatory provided the first follow-up x-ray imaging and spectroscopy (with gratings) to *Einstein*. *EXOSAT* had two small grazing-incidence telescopes. It was launched by the European Space Agency in 1982, and provided very long exposures on individual x-ray sources through mid-1986, when it reentered the atmosphere. It was particularly sensitive to very low energy x-ray sources, although its total collecting area was much less than that of the *Einstein Observatory*.

ROSAT, a relatively large x-ray telescope with approximately twice the sensitivity of the *Einstein Observatory*, was launched in 1990, as noted above. *ROSAT* has carried out the first all-sky imaging x-ray survey and thus is as revealing as the Palomar survey for optical astronomy. It is used by a very large number of guest observers as a very sensitive x-ray telescope, with a heavily oversubscribed program of pointed observations.

A permanent x-ray observatory, the Advanced X-ray Astrophysics Facility (AXAF), employing a much larger (48-in. or 1.2-m aperture) imaging telescope, is planned for launch in 1999 to carry on these highly productive studies. *See* X-RAY ASTRONOMY.

Gamma-ray astronomy. Gamma rays, which are more energetic than x-rays, still do not penetrate the Earth's atmosphere and, except at the very highest energies, can be detected only from high-altitude balloons, rockets, or satellites. The first cosmic gamma-ray detectors were flown in the early 1960s on the *Explorer 11* satellite and a Ranger spacecraft; the first to detect the clear signature of a cosmic gamma-ray source (in this case, gamma rays from the disk of the Galaxy) was an experiment carried on the *OSO 3*. The *SAS 2*, launched in 1972, first established the existence of gamma-ray point sources (such as the Crab and Vela pulsars) at energies of about 100 MeV. The detector used was a digitized spark chamber in which gamma rays are detected by the secondary electron pair they produce upon interacting in the detector. *SAS 2* conducted pointed observations but, unfortunately, operated for only about 6 months.

A more sensitive follow-up mission, *COS-B*, was launched by the European Space Agency in 1975. *COS-B* mapped the diffuse gamma-ray emission from the Milky Way Galaxy in much more detail than *SAS 2*, and also detected a variety of new point sources in the Milky Way Galaxy, as well as the first quasar detected at 100 MeV (3C 273). The mission ceased operation when fill gas for the spark chamber was exhausted in 1981.

In April 1991 was launched the *Compton Gamma-Ray Observatory*, the second of NASA's so-called great observatories (after the Hubble Space Telescope, and to be followed by AXAF and the Space Infrared Telescope Facility, or SIRTF). The *Compton Observatory* carries four high-sensitivity gamma-ray detectors: EGRET (Energetic Gamma-Ray Experiment Telescope, a much larger and more sensitive spark chamber than on *COS-B*), for the energy range 30 MeV–10 GeV; COMPTEL (Compton Telescope), a Compton-scattering telescope sensitive in the poorly explored 1–30-MeV band; OSSE (Oriented Scintillation Spectrometer Experiment), a scintillator detector sensitive in the 100-keV–10-MeV band, and thus able to detect hard x-ray sources; and BATSE (Burst and Transient Source Experiment), a detector system specifically designed to study gamma-ray bursts and determine their arrival directions. While striking results were obtained with all four instruments during the first year of operation, the most dramatic came from EGRET, which discovered 13 new gamma-ray quasars (virtually all bright and beamed radio quasars seem to be powerful gamma-ray sources); and BATSE, which discovered that the mysterious gamma-ray bursts (discovered originally in 1967 but still not identified) appear to be isotropic on the sky. The latter discovery supports models where the bursts are produced by colliding neutron stars in very distant galaxies. *See* GAMMA-RAY ASTRONOMY.

Jonathan E. Grindlay

Hubble Space Telescope. The Hubble Space Telescope is a large telescope operated in orbit above the disturbing effects of the Earth's atmosphere. It was

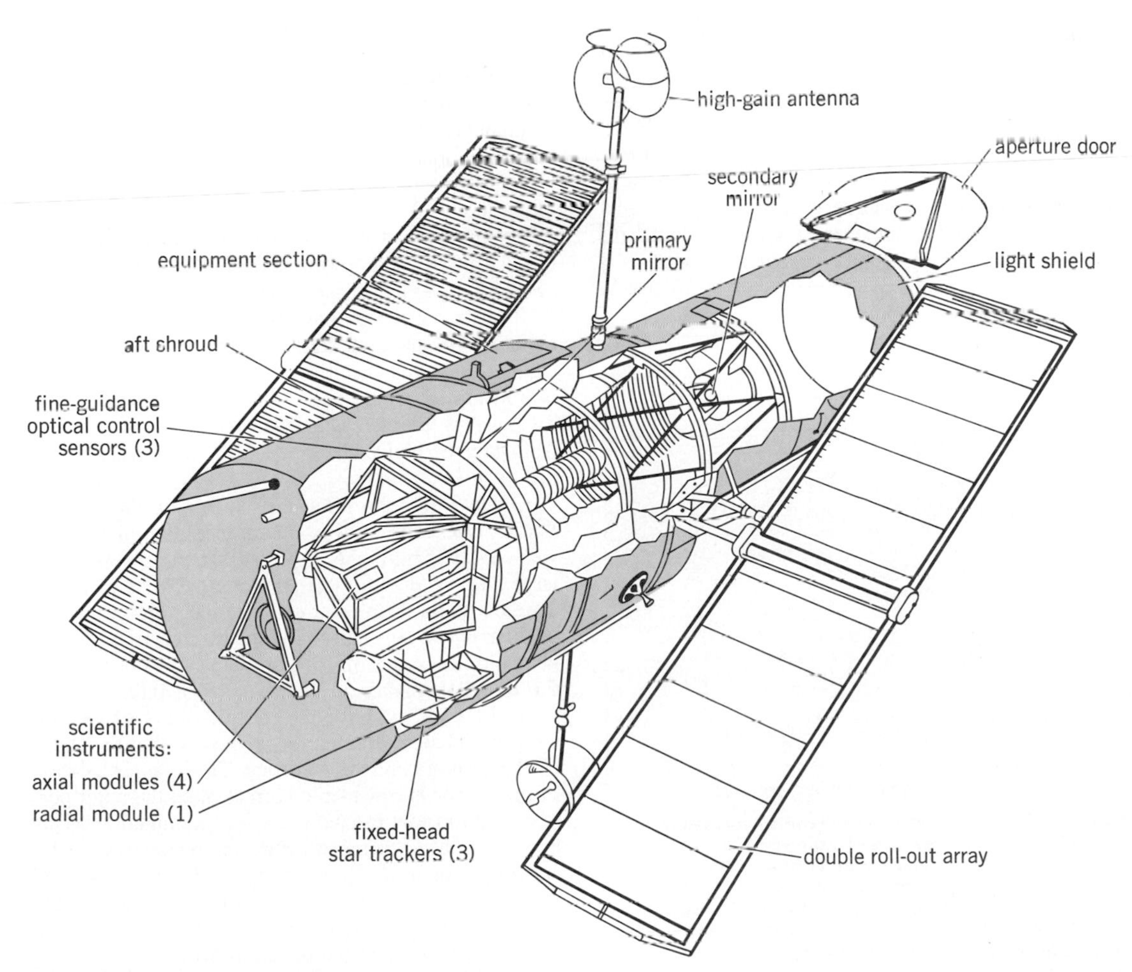

Fig. 5. Configuration of the Hubble Space Telescope.

placed in orbit by the space shuttle *Discovery* on April 24, 1990. It was developed by NASA with European Space Agency participation, and builds on experience gained in previous astronomical satellite programs such as the *SAS* and *OAO* series. The Hubble Space Telescope program involves several major departures and advances relative to previous satellites. The telescope provides broad wavelength coverage and can be used to collect light with wavelengths from 115 nm in the ultraviolet, through the optical (or visible) portion of the spectrum, and on to wavelengths of 1 mm in the far-infrared. With its 94-in. (2.4-m) primary mirror diameter, it is the largest astronomical telescope ever placed above the atmosphere. The telescope is a long-lived mission and has been designed so that shuttle astronauts can replace failed or obsolete components and upgrade the instruments. The telescope is managed as a guest observer facility by the Space Telescope Science Institute, located on the campus of the Johns Hopkins University in Baltimore, Maryland. The observing schedule is arranged at the institute from proposals submitted by astronomers, with selections being based on scientific merit.

Optics. The configuration of the Hubble Space Telescope is shown in **Fig. 5**. Its overall length is about 40 ft (12 m). The telescope optics are of the Ritchey-Chrétien type with an *f*/24 Cassegrain focus. Light enters the telescope when the aperture door is open, strikes the 94-in.-diameter (2.4-m) primary mirror, and reflects to the secondary mirror, where it is reflected again, back through a hole in the primary mirror, to a focus behind the primary mirror.

Unfortunately, the primary mirror of the Hubble Space Telescope was manufactured with a gross error of about 1 mm in its curvature. Images therefore have only 15% of their light in the intended core, which is about 0.07 arc-second in diameter, with the remainder in a broad halo, whose diameter is approximately 1–2 arc-seconds. This spherical aberration will be possible to correct with a shuttle visit to the Space Telescope scheduled for late 1993, at which time the COSTAR instrument will be inserted (as well as a new Wide Field Camera) to correct the beam sent to the three other instruments.

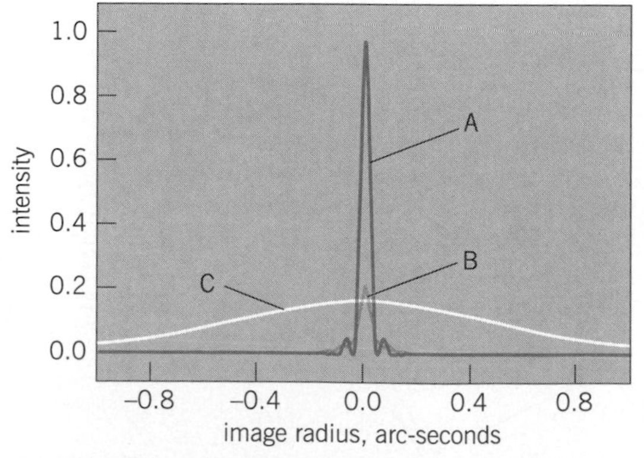

Fig. 6. Comparison of the stellar profile expected from the Hubble Space Telescope (A), the actual Space Telescope profile (B), and a ground-based telescope profile (C). The ground-based profile is so spread out that it had to be multiplied by 100 times to be seen on this diagram. The maximum intensity expected from the Hubble Space Telescope is normalized to 1.0.

Instruments. At the focal plane, the light is shared among at most eight instruments, and only four may be operating simultaneously. There are four radial instrument modules and four axial modules. All of the modules are replaceable in orbit so that instruments may be changed in the event of component failure or as instrument technology improves. Three of the radial modules are occupied by fine-guidance sensors. During an observation, two of the sensors lock onto guide stars and use these stars to generate error signals. The signals are processed by the pointing control system, which generates small corrections to the telescope pointing in order to keep a target object accurately centered in the field of view of one of the other instruments. The pointing stability achieved by this system can be as good as 0.007 arc-second, much smaller than has been achieved with any previous spacecraft or telescope.

The remaining radial module and the four axial modules are available for scientific instruments. The initial set of scientific instruments includes two cameras, two spectrographs, and a photometer, forming a powerful complement of instruments capable of carrying out most types of observations performed by ground-based observatories. In addition, the third fine-guidance sensor (the one not being used to supply pointing information) can be used to measure accurate relative positions of stars, providing an astrometric capability. All of the instruments take advantage of the ultraviolet capability of the telescope, but because of detector limitations, the maximum wavelength that can be reached by any instrument is only 1 μm. To extend the wavelength coverage, NASA is developing an infrared instrument. Also under development are a second-generation spectrograph and a spare Wide Field and Planetary Camera. SEE INFRARED ASTRONOMY.

Other equipment. The rest of the spacecraft includes computers, communications equipment, star trackers, batteries, solar arrays, and all the other equipment needed to operate a spacecraft by remote control from the ground.

Problems. Shortly after launch it was discovered that the telescope pointing can oscillate with a period of about 10 s and an amplitude of 0.1 arc-second or more. The oscillation has been traced to the solar arrays. It appears that thermal gradients in the arrays can set them into oscillation and that this oscillation is transmitted to the spacecraft. The problem occurs for several minutes whenever the telescope crosses from day to night or night to day and occasionally occurs at other times.

About 2 months after the launch of the Hubble Space Telescope it was determined that the optics were not performing as expected. The telescope has a large amount of spherical aberration. This means that different parts of the mirror bring the light to a focus in different focal planes. Light reflected from the inner portions of the mirror is brought to a focus about 1.6 in. (40 mm) ahead of the focus of light reflected from the outer parts of the mirror. Study of the images produced by the telescope showed that the shape of the primary mirror is wrong. If the inner portion of the mirror is considered correct, then the outer portions are too low by about 2 μm. Examination of the equipment used to manufacture the mirror revealed a spacing error in the components of the null lens used to measure the shape of the mirror as it was being polished. The spacing error is consistent with the inferred error in the shape of the mirror.

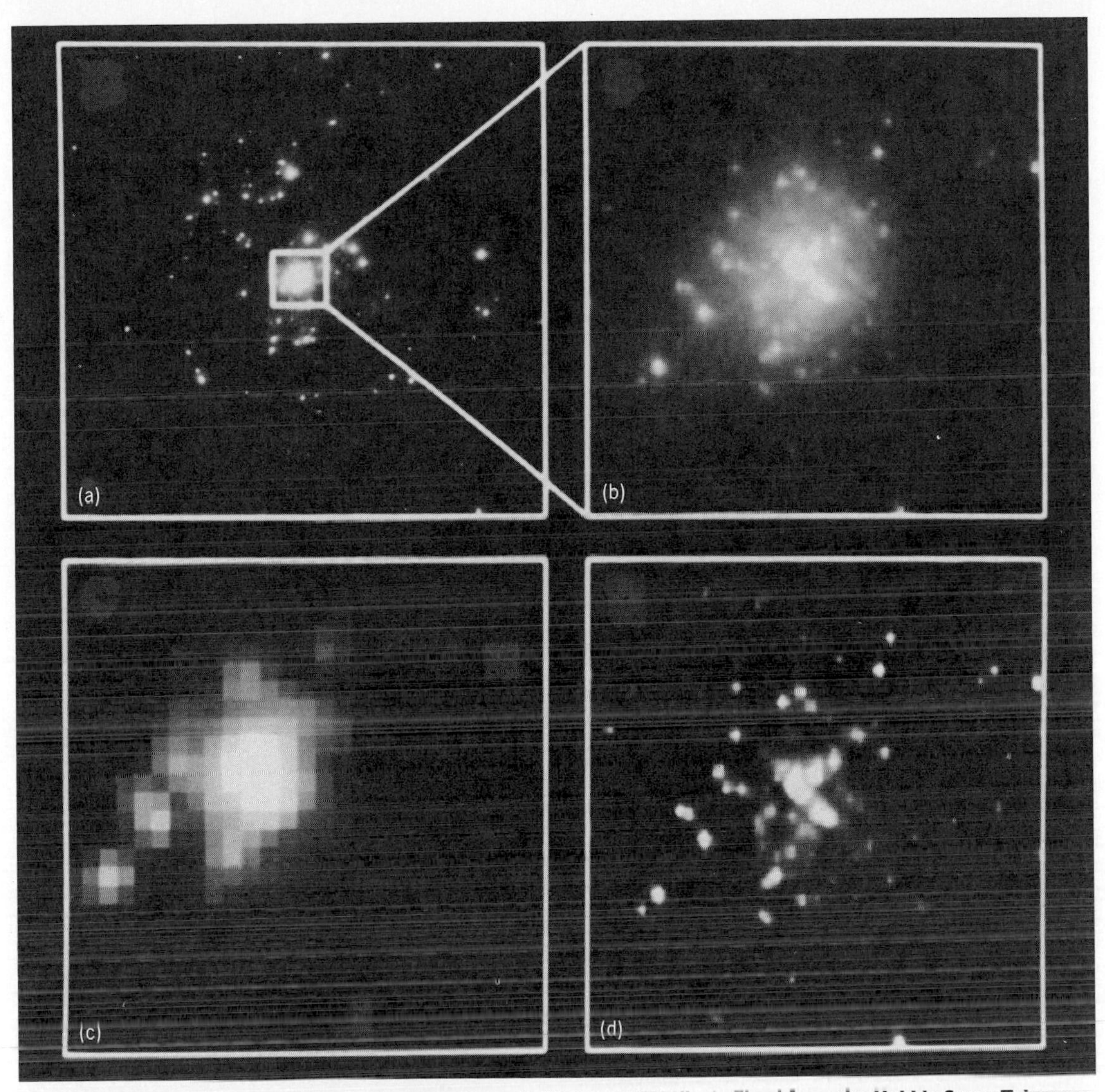

Fig. 7. Comparison of images of the 30 Doradus cluster in the Large Magellanic Cloud from the Hubble Space Telescope and a ground-based telescope. (*a*) Space Telescope mosaic of the cluster. (*b*) Expanded view of the center of the cluster. (*c*) Ground-based picture of the same region obtained by G. Meylan, with the Max Planck 87-in. (2.2-m) telescope at the European Southern Observatory in Chile. (*d*) Result of applying computer processing to the image of *b*. (*NASA*)

Solutions. In the near term, it has been possible to largely correct the solar array oscillation problem by a software filter in the on-board pointing control system. The arrays still oscillate, but when the oscillations are detected, compensating oscillations are introduced in the body of the spacecraft so that the pointing remains steady. Since the arrays degrade over time, it had been planned that the arrays would be replaced about 5 years after the beginning of the mission. It is expected that the replacement solar arrays can be modified to eliminate the oscillations and provide a long-term solution to the problem. Replacement arrays are to be provided during the major shuttle repair mission in 1993.

A solution for the spherical aberration problem has been devised. The COSTAR instrument will be able to compensate for the spherical aberration and thus "correct" the image shape before it is directed to the three remaining instruments. This correction is possible because, although the primary mirror is the wrong shape, the shape error appears to be very simple and symmetric so that it introduces a well-defined error in the convergence of the light to a focus. Such an error can be corrected with modifications in the internal optics of the instruments. The spare Wide Field and Planetary Camera, which was already under construction at the time the spherical aberration was discovered, has internal optics with mirrors at the right place to make the correction. By slightly changing the shape of the mirrors in the camera, the aberration should be completely corrected as far as the camera is concerned. It is expected that the camera can be launched as early as 1993. The internal optics of the infrared instrument and the second-generation spectrograph (which were in the design stage at the time the aberration was discovered) can also be modified to correct the aberration. These instruments may be launched as early as 1996. Until the new instruments are in place, the science program of the Hubble Space Telescope will be directed toward projects that are not severely compromised by the spherical aberrations.

Capabilities. A large telescope above the atmosphere is a very powerful tool for astronomy. Although the Hubble Space Telescope is not as large as the largest ground-based telescopes, the absence of the atmosphere means that for many problems the telescope can substantially outperform the largest United States ground-based telescope, the 200-in. (5-m) Hale Telescope on Palomar Mountain, even though the Hale Telescope has about four times the light-gathering power of the Space Telescope.

Above the atmosphere, the Hubble Space Telescope can observe at wavelengths where the atmosphere is opaque or only partially transparent, as in the ultraviolet and the infrared. Very hot objects and very cool objects radiate most of their energy in the ultraviolet and infrared, respectively. Ground-based studies of these objects, made with visible light, can see only the "tip of the iceberg." The atmosphere itself is a source of background light, especially in the infrared. Absence of this background makes it easier to study faint objects. Furthermore, the atmospheric absorption and background light fluctuate with time, making it difficult to determine whether observed changes in the brightness of an object are due to the atmosphere or to intrinsic changes in the object.

All of the foregoing provide a strong case for operating a telescope in space. But by far the most compelling reason to place a telescope above the atmosphere is the greatly improved angular resolution that can be achieved. As a telescope is made larger, it is capable of forming smaller images of stars (in inverse proportion to the diameter of the primary mirror). Yet this capability is never used on ground-based telescopes because irregularities in the atmosphere distort the light and smear the images—a phenomenon called astronomical seeing. Typically, images with a diameter of about 1 arc-second may be formed with ground-based telescopes. At visible wavelengths, the Space Telescope should form images with a diameter of about 0.07 arc-second (**Fig. 6**). When the Space Telescope observes a faint star, the area of the image occupied by the star should be 200 times smaller and include 400 times less background light (because the sky is twice as dark above the atmosphere) than if the star is observed by the same-sized ground-based telescope. This means that the Space Telescope should be able to detect stars 20 times fainter than the faintest stars that can be detected by a similar ground-based telescope and 10 times fainter than can be detected with the Hale Telescope. These expectations will not be met until the installation of replacement instruments that correct the spherical aberration. Star images that are formed by the telescope in its present condition have a small core as expected, but this core contains only about 15% of the light. The rest of the light is spread out in large wings of several arc-seconds diameter. With such images, the Hubble Space Telescope is able to detect faint stars only about as well as the Hale Telescope can.

When the Space Telescope is used to observe objects that are not points but that are extended and have observable structure such as planets, star clusters, nebulae, or galaxies, the improved angular resolution will provide a much clearer and more detailed image. This is true even with the spherical aberration. **Figure 7*a*** shows a 90 × 90 arc-second portion of a Wide Field Camera mosaic of the 30 Doradus star cluster. (The Wide Field Camera images four adjoining regions of the sky simultaneously; the images are joined into a single image by ground processing.) This image was taken by the Hubble Space Telescope on August 3, 1990, through a near-ultraviolet filter (368-nm wavelength) in order to make a so-called finding chart for the alignment of one of the spectrographs. The 30 Doradus cluster is a star-forming region in the nearby galaxy (about 160,000 light-years distant) known as the Large Magellanic Cloud. The cluster contains young, hot, massive stars. Figure 7*b* shows a 9 × 9 arc-second picture of the center of the cluster. The bright points of light are individual stars. The diffuse glow between the stars results from the spherical aberration and is formed from the overlapping wings of the star images. For comparison, Fig. 7*c* shows a ground-based picture of the same region and at the same scale. This picture, with 0.6 arc-second diameter star images, was obtained with an 87-in. (2.2-m) telescope at the European Southern Observatory in Chile. Figure 7*d* shows the results of computer processing applied to the image in Fig. 7*b*. The computer, using a technique known as deconvolution, can remove most of the undesirable wings of the star images. Prior to the Hubble Space Telescope image, laborious ground-based work on this object, including speckle interferometry, had resolved about two dozen stars within the central region of the cluster. About 100 stars can be found in the Space Telescope image. Work with the Hubble Space Telescope will make possible a detailed study of the star formation process in this cluster. *See Magellanic Clouds; Speckle.*

The spherical aberration is a serious setback for the Hubble Space Telescope program. There is no reason to doubt that in the long term it can be fully corrected as new instruments are installed in orbit. In the meantime, observations with the telescope will concentrate on ultraviolet spectroscopy and imaging, which cannot be done at all from the ground, and on high-resolution imaging, which reveals details that are only hinted at by ground-based observations. As new instruments are installed, the Hubble Space Telescope will fullfil its promise of becoming one of the most powerful tools for astronomy. *See Optical telescope; Telescope.*

Edward J. Groth; Jonathan E. Grindlay

Bibliography. N. A. Bahcall, The science program of the Hubble Space Telescope, *Highlights of Astronomy,* 8:435–439, 1989; J. Cornell and P. Gorenstein (eds.), *Astronomy from Space,* 1983; J. K. Davies, *Satellite Astronomy,* 1988; R. Giacconi, The Einstein x-ray observatory, *Sci. Amer.*, 242(2):80–102, February 1980; D. N. B. Hall (ed.), *The Space Telescope Observatory*, NASA CP-2244, 1982; D. S. Leckrone, The Space Telescope scientific instruments, *Publ. Astron. Soc. Pacific*, 92:5–21, 1980; M. S. Longair and J. W. Warner (eds.), *Scientific Research with the Space Telescope*, 1979; F. Macchetto, F. Pacini, and M. Tarenghi (eds.), *Astronomical Uses of the Space Telescope*, 1979; S. Maran and A. Boggess III, Ultraviolet astronomy enters the eighties, *Phys. Today*, 33(9):40–46, September 1980; C. A. Norman, The Hubble Space Telescope and other initiatives, *14th Texas Symposium on Relativistic Astrophysics,* 1989.

Saturn

The second-largest planet in the solar system and the sixth in order of distance to the Sun. The outermost planet known prior to 1781, Saturn is surrounded by a beautiful system of rings. Despite the planet's huge size, its mean density is so low it could float in water. Saturn is also the only planet that has a satellite (Titan) with a dense atmosphere. This distant planetary system has been visited by three NASA spacecraft: a preliminary survey by *Pioneer 11* in September 1979, and a more sophisticated reconnaissance by *Voyager 1* in November 1980 and *Voyager 2* in August 1981.

Orbit and physical elements. The orbit of Saturn has a semimajor axis or mean distance to the Sun of 8.95×10^8 mi (1.43×10^9 km), an eccentricity of 0.056, and its plane is inclined to the plane of the ecliptic at an angle of 2.5°. With a mean orbital velocity of 6 mi/s (9.65 km/s), Saturn makes one revolution about the Sun in 29.46 years. SEE PLANET.

The equatorial diameter of Saturn is about 75,000 mi (120,660 km), and the polar diameter about 67,300 mi (108,350 km). The volume is 769 (Earth = 1) with a few percent uncertainty. The polar flattening caused by the rapid rotation is the largest of all the planets, and the ellipticity, $(r_e - r_p)/r_e = 0.102$, is 30 times the value for Earth (r_e is the equatorial radius and r_p is the polar radius).

The mass, about 95.1 (Earth = 1) or 1/3500 (Sun = 1), is accurately determined from the motions of the planet's brighter satellites. The mean density is 0.69 g/cm^3, the lowest mean density of all the planets. The corresponding value of the mean gravity at the visible surface (superimposed cloud layers) is 0.93 (Earth = 1) or 9.1 m/s^2.

Photometric properties. The apparent visual magnitude of Saturn at mean opposition is +0.7 when the ring is seen edgewise, and the corresponding value of the reflectivity (geometric albedo) is about 0.44. SEE ALBEDO.

Appearance. Observed through a telescope, Saturn appears as an elliptical disk, darkened near the limb and crossed by a series of barely discernible bands parallel to the equator. Frequently only the bright equatorial zone and the two darker tropical bands on either side of it are visible (**Fig. 1**). The rings may be seen even with a relatively small telescope, but their visibility changes with the position of the planet in its orbit, because the rotation axis (of both the planet and the rings) is inclined 29° to the perpendicular to the orbital plane. Thus Saturn has seasons similar to those on Earth, but each season lasts 7.36 years (**Fig. 2**).

The rotation period of Saturn's interior is 10^h40^m, as determined by radio emissions controlled by the planet's magnetic field. The disk of Saturn is much more homogeneous in appearance than that of Jupiter, even when seen at close range with the cameras of the *Pioneer* and *Voyager* spacecraft. There is no feature comparable to the Great Red Spot, and the contrast of the features that are visible is very low (Fig. 1). By studying the movement of these features, it has been possible to determine that the circulation patterns on Saturn are also very different from those on Jupiter. At a latitude of ±40°, the atmosphere rotates with the same velocity as the interior, but wind velocities increase smoothly toward the equator, where they reach a value of 1100 mi/h (500 m/s), about four times faster than Jupiter's equatorial jet. There is no alternation of easterly and westerly currents corresponding to transitions between belts and zones, as there is on Jupiter, except at latitudes above ±40°. This difference between Saturn and Jupiter represents a fundamental difference in global circulation, perhaps related to the relative sizes of the cores of the two planets. SEE JUPITER.

Atmosphere. The optical spectrum of Saturn is characterized by strong absorption bands of methane

Fig. 1. Saturn, viewed from *Voyager 1*. The soft, velvety appearance of the low-contrast banded structure is due to scattering by a haze layer above the planet's cloud deck. (*NASA*)

(CH_4) and by much weaker bands of ammonia (NH_3). Absorption lines of molecular hydrogen (H_2) have also been detected. The estimated quantities of these gases that are present above the clouds are equivalent to STP path lengths of about 30 mi or 50 km (H_2), 200 ft or 60 m (CH_4), and 6 ft or 2 m (NH_3). Here STP refers to standard temperature (273 K) and pressure (1 atm or 1.01325×10^5 pascals). In these units, the entire atmosphere of the Earth would be equivalent to an 5-mi (8-km) path length. The presence of about 2.5 mi (4 km) of helium (He) has been deduced indirectly from infrared observations of pressure-broadened hydrogen emission lines.

The temperature the planet should assume in response to solar heating is calculated to be about 76 K, somewhat lower than the measured value of 92 K. This suggests that Saturn has an internal heat source of roughly the same magnitude as that on Jupiter. Relatively intense thermal emission near a wavelength of 12 micrometers has been identified as resulting from ethane (C_2H_6) formed in the upper atmosphere from the dissociation of methane. As in the case of Jupiter, a thermal inversion exists in this region of Saturn's atmosphere. Emission bands of methane and phosphine have also been detected here. This region is well above the main cloud layer, which is thought

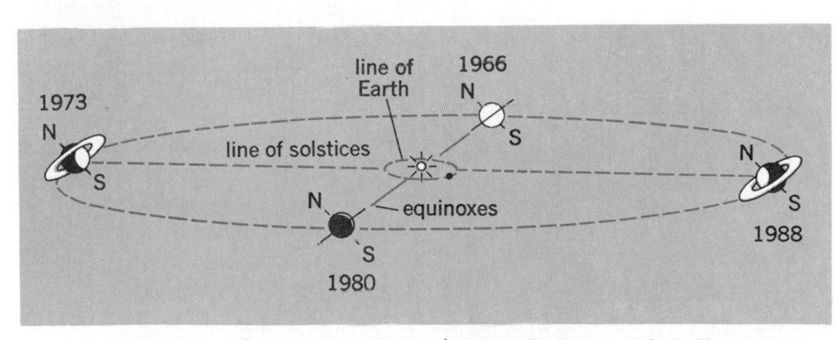

Fig. 2. Presentations of Saturn's ring system. (*After L. Rudaux and G. de Vaucouleurs, Larousse Encyclopedia of Astronomy, Prometheus Press, 1959*)

to consist primarily of frozen ammonia crystals, with an admixture of some other substances to provide the yellowish color sometimes observed in the equatorial zone. The ammonia cirrus on Saturn is apparently denser and more ubiquitous than on Jupiter, since one does not see through it to lower cloud layers (Fig. 1). This difference probably results from the lower atmospheric temperature and smaller gravity of Saturn, which will act together to spread out and increase the density of the cloud layer.

Internal structure and radiation belts. Observations of Saturn at radio frequencies indicate that the temperature steadily increases with depth into the atmosphere. Theoretical models for the internal structure of Saturn are similar to those for Jupiter, that is, a dense core surrounded by hydrogen compressed to a metallic state which gradually merges into an extremely deep atmosphere. The fact that the two planets radiate comparable amounts of energy despite their difference in size means that smaller Saturn must have some additional energy source besides gravitational contraction. The gradual solution of helium in the liquid hydrogen surrounding the core would suffice, and would also explain the smaller abundance of helium (relative to hydrogen) found in Saturn's atmosphere. The existence of a magnetic field and belts of trapped electrons was initially deduced from observations of nonthermal radiation at dekameter wavelengths and was mapped out in detail by the *Pioneer* and *Voyager* spacecraft. Saturn's magnetic field has the same polarity as Jupiter's; a terrestrial compass taken to either planet would point south instead of north. The magnetic moment of Saturn is 4.3×10^{28} gauss cm^3 (4.3×10^{18} teslas · m^3), 500 times Earth's and 34 times smaller than Jupiter's. But the magnetic field at the equator is only 0.2 gauss (2×10^{-5} tesla), two-thirds the value of Earth's field, because of Saturn's greater size. The magnetosphere contains a plasma of charged particles, but no distinct torus like the one associated with Io in the Jupiter system. The charged particle belts are absent in the region of the rings. Radio signals resembling radiation from lightning discharges in the atmosphere were detected by the *Voyager* spacecraft.

Jupiter and Saturn are relatively similar bodies. Both seem to have bulk compositions close to that of the Sun and the other stars, and both are rich in hydrogen and helium. In that sense, they may represent the primitive material from which the entire solar system was formed, whereas the other planets have undergone fractionation processes resulting in the loss of most of the light gases. However, both Jupiter and Saturn show an enhancement of C/H (as determined from methane and hydrogen) compared with the Sun. This suggests that both of these planets formed in a two-stage process that led initially to formation of a large core and secondary atmosphere, with subsequent collapse of an envelope of gases from the surrounding nebula. SEE PLANETARY PHYSICS.

Ring system. The most remarkable feature associated with Saturn is the complex ring system that surrounds the planet (**Fig. 3**). The system is divided into six main regions, designated A through F. Observations by several instruments on the *Voyager* spacecraft demonstrated that each of these six regions is subdivided into many individual "ringlets," so that Saturn is actually surrounded by thousands of rings.

Both theory and observations prove that the ring system is made up of myriad separate particles that move independently in flat, mostly circular orbits in Saturn's equatorial plane. The discontinuous, meteoric nature of the rings is demonstrated directly by spectroscopic observations, which show that the inner edge of the ring revolves about the planet faster than the outer edge, and the material in each ringlet moves precisely with the velocity that independent satellites would have at the same distance from the planet.

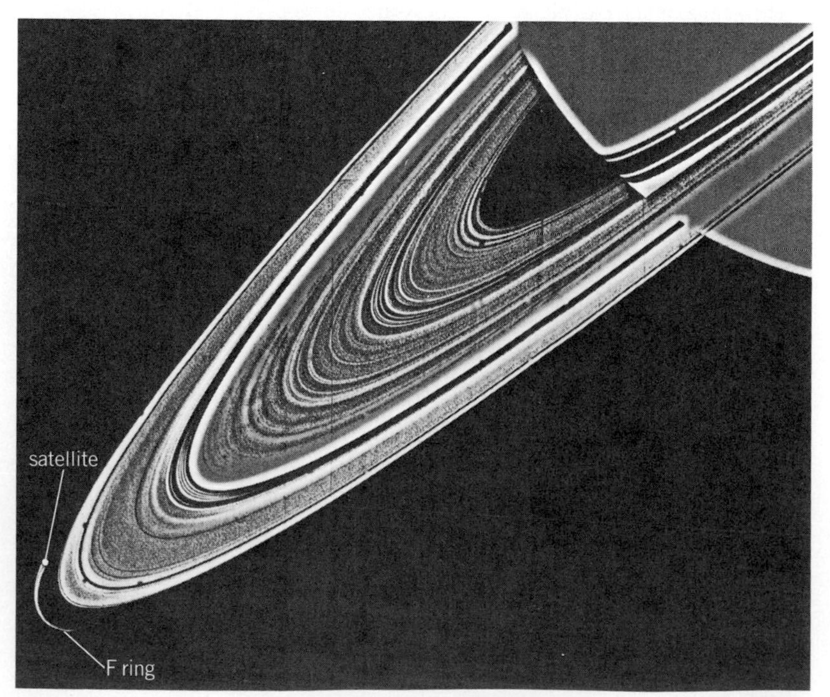

Fig. 3. Saturn's rings, viewed from *Voyager 1*. Approximately 95 individual concentric features are visible. One of the satellites discovered by *Voyager 1* is visible just inside the narrow F ring. (*NASA*)

Appearance from Earth. As viewed with telescopes from Earth, the nearly circular ring system looks like an ellipse whose appearance changes with the relative positions of Earth and Saturn (Fig. 2). The maximum opening of the rings occurs when the system is tilted 27° to the line of sight. The rings are seen as a thin line when Earth crosses Saturn's equatorial plane. The slightly variable point of view, depending on the position of Earth in its orbit, causes a slight annual oscillation of the tilt angle. Thus when the tilt is near 0°, Earth may cross the plane of the ring system either once or three times. During these edge-on configurations, it is possible to search for faint inner satellites and distant rings that ordinarily are hidden by the bright light reflected by the main ring system.

Nomenclature and structure. The bright outer ring, A, has an outside diameter of 169,000 mi (272,000 km) and an inner diameter of 150,000 mi (242,000 km). In the high-resolution pictures obtained by the *Voyager* cameras, this ring is seen to contain many wavelike patterns, indicating gravitational disturbances by the satellites. A dark gap in the A ring, often called the Encke division, was found to contain a narrow, elliptical, discontinuous and kinky ring. The broad (2200 mi or 4030 km wide) division separating ring A from B is called the Cassini division after its discoverer. It too was found by the *Voyager* instruments to contain material comprising at least five discrete rings, each of which shows some internal structure.

Ring B, which is the brightest segment of the system, has an outer diameter of 146,000 mi (235,000 km) and an inner diameter of 114,000 mi (183,000 km). This ring contains the largest density of material in the system. It is slightly less bright in its inner regions, where it is also more transparent.

Ring C, sometimes called the crepe ring, is much fainter and more transparent. It appears as a dusky band in projection against the disk of the planet and only faintly against the sky. Its outer diameter is 114,000 mi (183,000 km), and its inner diameter 91,000 mi (146,000 km). This means the inner edge of ring C is just 7500 mi (11,500 km) above the visible cloud deck on the planet.

A fourth zone, D, of the ring of Saturn between C and the globe was discovered by P. Guerin at Pic-du-Midi Observatory in October 1969. It is fainter than the crepe ring, but its reality has been confirmed by *Voyager* pictures.

The E ring has been observed well only when the Earth passed through the ring plane in 1966 and especially again in 1980 (Fig. 2). It begins inside the orbit of Enceladus at a distance of about 112,000 mi (181,000 km) from Saturn, and extends outward past the orbit of Dione, fading from view at about 300,000 mi (480,000 km) from the planet. The ring is very much brighter just at the orbit of Enceladus, suggesting that this satellite is somehow responsible for the production of the material seen in the ring.

The F ring was discovered by the *Pioneer* spacecraft and seen in detail by *Voyager* (**Fig. 4**). It lies just outside the A ring with an average diameter of 174,600 mi (281,000 km), and actually consists of more than five separate strands, which may not all be in the same plane. The two brightest strands deviate markedly from ellipses, showing the effects of nongravitational forces and perhaps the influence of embedded moonlets. The F ring is held in place by two small satellites, one inside and one outside. A third guards the outer edge of the A ring, and a fourth is responsible for the Encke division.

Finally, *Voyager* established the existence of the tenuous G ring with an average diameter of 211,000 mi (340,000 km), 21,500 mi (34,500 km) outside the A ring. Although this feature resembles the E ring, no known satellites are associated with it, so both its source and its ability to persist are not understood.

To place this complex system in perspective, the mean distance from the Earth to its Moon is 240,000 mi (384,000 km), so Saturn and its rings (except for the outer edge of the E ring) would just fit in this space.

Pictures obtained by the *Voyager* spacecraft showed that in addition to these nearly circular rings, transient radial features exist in the B ring. These features, called spokes, appear to be clouds of micrometer-size charged particles that are initially controlled by the planet's magnetic field but soon begin moving in keplerian orbits like the larger particles.

Origin and nature. E. Roche proved in 1849 that a liquid satellite of a planet with the same density cannot form if it is closer to its planet than 2.44 times the planetary radius. Within this distance, disruptive tidal forces (the gradient of the planet's gravitational field) will be greater than the self-gravity of the satellite. But a rocky or icy satellite is held together by stronger forces, and will, therefore, not be disrupted by the planet's field outside a distance of 1.35 radii. In fact, the F ring and its guardian satellites are at 2.33 radii, just inside Roche's limit.

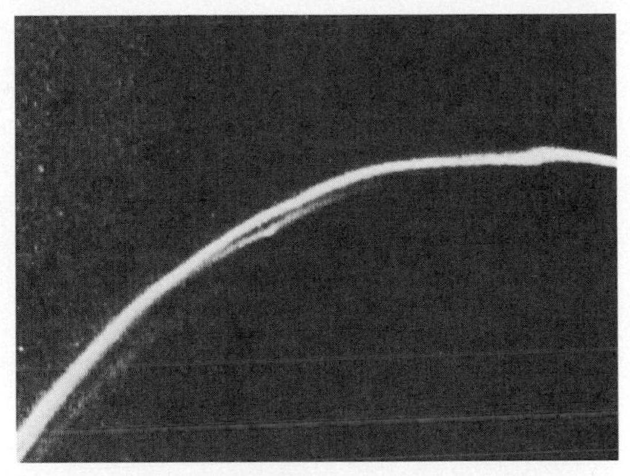

Fig. 4. Saturn's F ring viewed from *Voyager 1*. Two narrow, braided bright rings that trace distinct orbits, and a broadened, very diffuse component about 22 mi (35 km) in width, are visible. Also visible are knots, which probably are local clumps of ring material, but which may be minisatellites. (*NASA*)

In 1859 J. C. Maxwell was able to show that a ring system of small mass, formed of a large number of particles, is quite stable against external perturbations such as those caused by the larger satellites. The aggregate mass of the ring system is quite small, probably less than one-half the mass of the Earth's Moon. Periodic perturbations by the major satellites are responsible, in part, for the main divisions of Saturn's rings in the same way that perturbations by Jupiter cause the Kirkwood gaps in the asteroid belt. The cause of the many small divisions discovered by *Voyager* (Fig. 3) is still obscure. Following a theory of P. Goldreich and S. Tremaine, it has been proposed that tiny moonlets may be present in these gaps, stabilizing individual rings even as the F ring is stabilized by its two satellites. But a careful search has failed to reveal any small satellites (with diameters of 5–10 km) in any of the gaps. *See* Asteroid; Perturbation.

G. P. Kuiper demonstrated in 1952 that the reflection spectrum of the ring was identical with that of water ice, a result that has been confirmed by subsequent measurements. In 1974 the ring system was found by R. M. Goldstein to be a surprisingly good reflector of radar waves, suggesting that some relatively large (diameter approximately 1 km) particles may be present. The scattering of radio waves from the *Voyager* spacecraft by the C ring indicated an average effective particle size of 1 m. The mean thickness of the ring system is not well determined, but *Voyager* observations of the occultation of a star by the rings demonstrated that the outer edge of the A ring is only 500 ft (150 m) thick.

Satellites. Saturn has 22 known satellites, more than any other planet (see **table**). The largest and brightest, Titan, was discovered by C. Huygens in 1655 and is visible with small telescopes; the other satellites are much fainter. The outermost satellite was discovered photographically by W. H. Pickering in 1898; two inner satellites (X Janus and XI Epimetheus) were discovered in 1966, when Saturn's rings were seen nearly edgewise, but it was not until the next such configuration, in 1980, that the true nature of these objects was determined: they are very nearly in the same orbit. Three additional satellites were discovered from the ground in 1980 and 1981; one of them is in the leading lagrangian point of Dione's or-

Saturn's satellites[a]

Satellite	Mean distance from Saturn, 10^3 km	Revolution period			Mean density, g/cm^3	Diameter, km	Visual magnitude at mean opposition
		d	h	m			
1981 S 13[b]	134	0	13	49	?	20(?)	19
XV Atlas	137	0	14	26	?	30[c]	(18)
Prometheus	139	0	14	43	?	100[c]	(13.5)
Pandora	142	0	15	05	?	90[c]	(14)
X Janus	151	0	16	41	?	190[c]	(14)
XI Epimetheus	151	0	16	41	?	120[c]	(14.5)
I Mimas	187	0	22	37	(1.4)	390	12.9
II Enceladus	238	1	08	53	(1.2)	500	11.8
III Tethys	295	1	21	18	1.21	1,060	10.3
XIII Telesto	295	1	21	18[d]	?	25[c]	(18)
XIV Calypso	295	1	21	18[e]	?	25[c]	(18)
IV Dione	378	2	17	41	1.43	1,120	10.4
XII Electra	378	2	17	41[f]	?	30[c]	(17.5)
V Rhea	526	4	12	25	1.33	1,530	9.7
VI Titan	1,221	15	22	41	1.88	5,800[g]	8.4
VII Hyperion	1,481	21	06	38	?	300[c]	14.2
VIII Iapetus	3,561	79	07	56	1.16	1,460	10.2–11.9
IX Phoebe	12,960	550	11		?	220	16.5

[a]Values in parentheses still uncertain. Satellites with no roman numerals require additional observations to improve the orbits. The orbits of four satellites have not been identified, and therefore they are not listed in the table. 1 km = 0.62 mi.
[b]Temporary designation, pending assignment of permanent name.
[c]Irregular in shape.
[d]Librates about trailing (L_5) lagrangian point of Tethys's orbit.
[e]Librates about leading (L_4) lagrangian point of Tethys's orbit.
[f]Librates about leading (L_4) point of Dione's orbit.
[g]This diameter refers to top of haze layer. Diameter of solid body is 5150 km.

bit, and two are in the lagrangian points of Tethys's orbit. The four satellites associated with the rings that were discovered by *Voyager* have already been mentioned. The one responsible for the Encke division was identified from *Voyager* photographs in 1990 with the help of computer analysis. The four other satellites discovered on *Voyager* pictures consist of two in the lagrangian points of Mimas's orbit, an additional member of the Dione family, and a satellite in orbit between Tethys and Dione. The orbits of these four satellites have not been identified. All of these objects are small (less than 60 mi or 100 km long) and irregular in shape. *See Celestial mechanics.*

The small outermost satellite, Phoebe, moves in a retrograde direction (opposite to that of the inner 20 satellites and to the direction of the planets about the Sun) in an orbit of relatively high eccentricity (0.16). These two characteristics place Phoebe in the class of irregular satellites that have probably been captured by the planet they orbit, instead of being formed with it from the original solar nebula. Phoebe's rotation period is only 9 h, completely out of synchrony with its 550-day period of revolution. *See Retrograde motion.*

Titan shows a measurable disk in large telescopes; the mean apparent diameter corresponds to a linear diameter of approximately 3600 mi (5800 km). But this diameter refers to the satellite's atmosphere, which is filled with a dense aerosol produced photochemically by incident sunlight. The solid surface of Titan has a diameter of 3200 mi (5150 km), making this satellite larger than Mercury but smaller than Jupiter's giant Ganymede. This large satellite has a mass about two times the mass of the Moon, with a corresponding mean density of 1.9 g/cm^3. The low density (Moon = 3.3) means that this object contains a large fraction of icy material and is thus quite different from the Moon or the inner planets in composition. Furthermore, it is large and cold enough to retain an atmosphere of gases with relatively high molecular weights. In fact, the existence of a methane atmosphere about Titan was established through spectroscopic observations by Kuiper in 1944. Subsequent studies have indicated that a thermal inversion exists in the satellite's upper atmosphere, where temperatures as high as 175 K have been measured, produced by absorption of solar ultraviolet radiation in the aerosol. The main constituent of this atmosphere is molecular nitrogen (N_2), which produces a surface pressure of 1.5 bars (1.5×10^5 Pa), or 1.5 times the sea-level pressure on Earth. The surface of Titan is so cold (94 ± 2 K) that methane and ethane can liquify. Since ethane is likely to be a major product of atmospheric chemistry, oceans of this hydrocarbon may be present on Titan's surface. The presence in the atmosphere of a variety of organic compounds such as cyanoacetylene (HC_3N), hydrogen cyanide (HCN), and propane (C_3H_8) in addition to the aerosol should allow future investigators to use this satellite as a natural laboratory for testing ideas about chemical evolution on the primitive Earth.

The other satellites encompass a variety of characteristics. All have densities near unity, indicating a predominantly icy composition. Yet the small variations in density that do occur appear random, rather than showing a radial trend as is the case for the Jupiter system. It has frequently been suggested that these small, icy satellites are similar in composition to the nuclei of comets. The slight differences in density would reflect differences in the amount of rocky material embedded in the ices. The surfaces of these objects are covered with impact craters, with the exception of Enceladus and Iapetus. Large regions of the surface of Euceladus are free of craters, indicating reworking of the surface in recent times. Coupled with the unusually high reflectivity of this satellite (close to 100%), the modified surface suggests internal activity leading to partial melting and the production and expulsion of the tiny ice grains that populate the E ring. Iapetus is unique in the solar system in that its trailing hemisphere is six times brighter than the leading one. The cause for this anomaly is not known. All of these satellites except Hyperion and Phoebe appear to keep the same hemisphere facing Saturn as they revolve around it, meaning that their rotational periods are equal to the periods of revolution as a result of tidal frictions, just as is the case for the Moon. Hyperion exhibits an aperiodic, ''chaotic'' rotation, which, with its irregular shape, may be an indication of a recent collision. *See Comet; Moon.*

Tobias C. Owen

Bibliography. J. K. Beatty, B. O'Leary, and A. Chaikan (eds.), *The New Solar System*, 2d ed., 1982; W. K. Hartmann, *Moons and Planets*, 1982; Special issue on *Voyager 1* encounter with the Saturnian system, *Science*, 212(4491):159–243, 1981; Special issue on *Voyage 2* encounter, *Science*, 215(4532):499–594, 1981.

Fig. 2. European Southern Observatory Schmidt in La Silla, Chile. (*European Southern Observatory*)

Schmidt camera

A wide-field telescope that uses a thin aspheric front lens and a larger concave spherical mirror to focus the image **(Fig. 1)**; it is also known as a Schmidt telescope. The German optician Bernhard Schmidt devised the scheme in 1931. The field of best focus is located midway between the lens and the mirror and is curved convexly toward the mirror, with a radius of curvature equal to the focal length. Usually film or photographic plates are bent to match this curved focus. With shorter focal lengths, a field-flattening lens may be used. These telescopes are known as Schmidt cameras because they are always used photographically, and no focus accessible to the eye is provided in basic Schmidts.

Schmidt telescopes are very fast, some having focal ratios in the vicinity of *f*/1. Thus, they are sensitive to objects of low surface brightness. Schmidt telescopes have no coma; because the only lens element is so thin, they suffer only slightly from chromatic aberration and astigmatism. The front element is sometimes known as a corrector plate. Often the outer surface of the corrector plate is plane, with the inner surface bearing the figure. This corrector plate reduces the spherical aberration severely, giving extremely sharp images. The Schmidt design became widely used after 1936, when the secret of the fabrication of the corrector plate was released. Schmidt is credited as much with his skill in figuring the fourth-degree curves on the corrector plate as with the design itself. The largest Schmidt cameras **(Fig. 2)** are listed in the **table**. *See Aberration.*

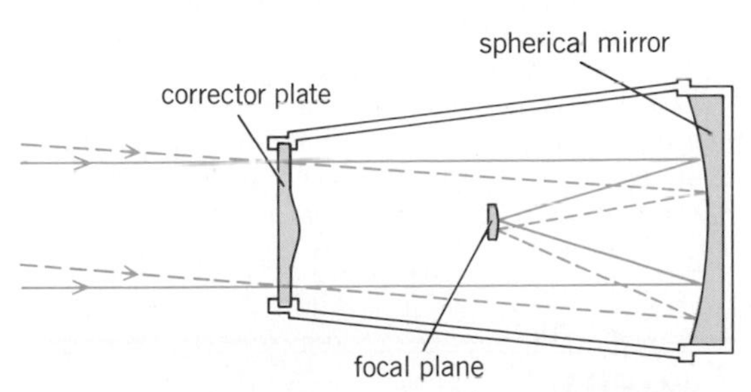

Fig. 1. Cross section of Schmidt camera with aspherical corrector plate. (*After J. M. Pasachoff, Astronomy: From the Earth to the Universe, 2d ed., Saunders College Publishing, 1983*)

Sky surveys. The Schmidt camera at the Palomar Observatory in California was used in the 1950s to survey the northern two-thirds of the sky in the Palomar Observatory–National Geographic Society Sky Survey (POSS). The plate pairs—one in the red and one in the blue—provide a first-epoch coverage on which many nebulae, galaxies, clusters of galaxies, and other objects were discovered. Each plate, about 14 in. (35 cm) square, covers an area about 6° square. The POSS is a basic reference in most observatories. It can be obtained as glass, film, or paper copies, often as negatives to preserve detail. A smaller Schmidt telescope (18 in. or 46 cm) at the Palomar Observatory has discovered many asteroids with orbits that cross that of the Earth.

The European Southern Observatory (ESO) Schmidt in Chile and the United Kingdom Schmidt in Australia have been used to continue the survey to the southern hemisphere, with improved emulsions to survey from declination −20° to the south pole. The project is jointly conducted by the European Southern Observatory and the Science Research Council of the United Kingdom. The European Southern Observatory telescope carried out the blue survey, and the United Kingdom telescope carried out the red survey. The survey includes 606 blue and 606 red plates to cover the one-quarter of the sky that cannot be reached by the Palomar Schmidt. The United Kingdom Schmidt has also completed an infrared survey of the Milky Way and Magellanic Clouds, and has undertaken a blue-red study of the region just south of the celestial equator to provide images showing stars about 1.5 magnitudes fainter than those on the Palomar survey. In addition to mapping, the plates are used to help make optical identifications of southern radio and x-ray sources. Study of the plates has been speeded by the COSMOS automatic plate-measuring machine.

In 1987, a new achromatic doublet corrector plate was installed in the Palomar Schmidt, which was renamed the Oschin Telescope. With it, a second-epoch sky survey was undertaken. An infrared color (800–900 nanometers, limiting magnitude 19) has been

The largest Schmidt cameras

Telescope or institution	Location	Diameter: Corrector plate, in. (m)	Diameter: Spherical mirror, in. (m)	Focal ratio	Date completed
Karl Schwarzschild Observatory	Tautenberg (near Jena), Germany	53 (1.3) (removable)	79 (2.0)	*f*/2	1960
Oschin Telescope, Palomar Observatory	Palomar Mountain, California	48 (1.2)	72 (1.8)	*f*/2.5	1948, 1987
United Kingdom Schmidt, Siding Spring Observatory	Warrumbungle National Park, New South Wales, Australia	48 (1.2)	72 (1.8)	*f*/2.5	1973
Tokyo Astronomical Observatory	Kiso Mountains, Japan	41 (1.1)	60 (1.5)	*f*/3.1	1975
European Southern Observatory Schmidt	La Silla, Chile	39 (1.0)	64 (1.6)	*f*/3	1972

added to the blue (385–550 nm) and red (610–690 nm) colors, which are slightly different from those of the first survey but are similar to those used by the United Kingdom and European Southern Observatory Schmidts for their southern survey. Advances in film technology have allowed a limiting magnitude fainter by a factor of 2 in the blue (to magnitude 22) and 6 in the red (to magnitude 23) to be reached and images to be recorded with smaller grain. The infrared work has been made possible by developments in hypersensitization of infrared emulsions, which has increased their speed by a factor of 200 for the long exposures that astronomers need. The survey includes 894 fields between the celestial equator and the north celestial pole, each covering a region 5° across. SEE ASTRONOMICAL ATLASES; ASTRONOMICAL PHOTOGRAPHY.

Schmidt-Cassegrain design. Many amateur astronomers use telescopes of the Schmidt-Cassegrain design, in which a small mirror attached to the rear of the corrector plate reflects and refocuses the image through a hole in the center of the primary mirror. These telescopes, often in sizes of 5 in. (12.5 cm), 8 in. (20 cm), and 14 in. (35 cm), are portable and relatively inexpensive for the quality of image. Their fields are much narrower than a standard Schmidt design, however. Schmidt cameras are also available for amateurs.

Related systems. The success of the Schmidt design has led to many other types of catadioptric systems, with a combination of lenses and mirrors. The desire for a wide field has led to most modern large telescopes being built to the Ritchey-Chrétien design instead of the traditional paraboloid, though these fields are perhaps one-third the diameter of those of Schmidts.

Schmidt optics are often used in microscopes and in projection televisions. SEE OPTICAL TELESCOPE.

Jay M. Pasachoff

Bibliography. British astronomers look south, *Sky Telesc.*, 64:543, December 1982; G. B. Kuiper and B. Middlehurst (eds.), *Stars and Stellar Systems*, vol. 1: *Telescopes*, 1960; J. J. Labrecque, Testing a Schmidt corrector at a finite distance, *Sky Telesc.*, 37:250–251, April 1969; The largest Schmidt's first 20 years, *Sky Telesc.*, 62:554–557, December 1981; S. Laustsen, C. Madsen, and R. M. West, *Exploring the Southern Sky*, 1987; D. Malin and P. Murdin, *The Colours of the Stars*, 1984; J. M. Pasachoff, *Contemporary Astronomy*, 4th ed., 1989.

Scorpius

The Scorpion, in astronomy, one of the most beautiful and vivid constellations in the sky. Scorpius is the eighth sign of the zodiac. The constellation resembles a scorpion even to the sting (see **illus.**). The bright red star Antares is situated at the heart. Its name (Ant-Ares) means the Rival of Mars, since both the plant and the star are bright and red in color, and the two are often found near each other. Antares is one of the largest stars known, having a diameter over 450 times that of the Sun. As in Sagittarius, the Milky Way in Scorpius is bright and rich in star clouds and clusters. SEE ANTARES; CONSTELLATION; ZODIAC.

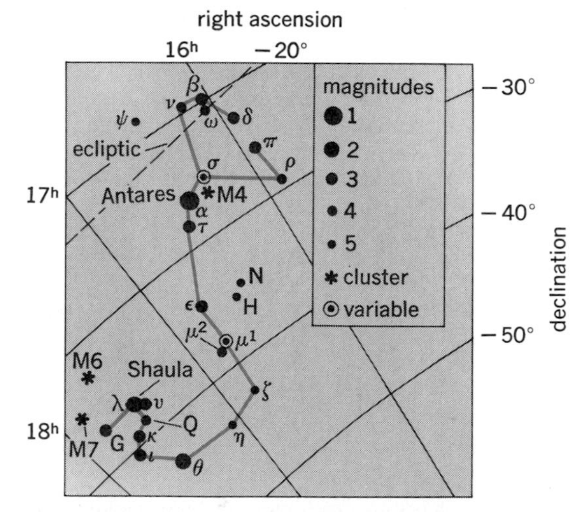

Line pattern of the constellation Scorpius. The grid lines represent the coordinates of the sky. The apparent brightness, or magnitudes, of the stars is shown by the sizes of the dots, which are graded by appropriate numbers as indicated.

Ching-Sung Yu

Seasons

The four divisions of the year based upon variations of sunlight intensity (solar energy per unit area at the Earth's surface) at local solar noon (noontime) and

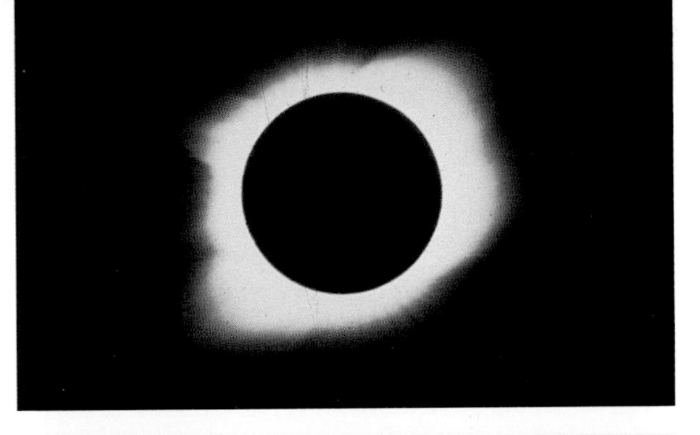

Solar corona, during eclipse of the Sun on November 23, 1984, observed in Papua New Guinea. Long equatorial streamers and short polar brushes are typical of solar activity during the minimum portion of the sunspot cycle. (*Jay M. Pasachoff*)

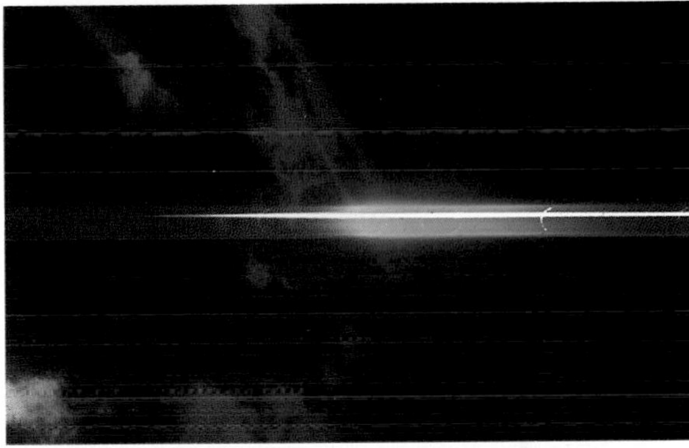

A few seconds before the Moon completely covers the Sun, the solar chromosphere flashes into view. This spectrogram taken at that time reveals the chemical elements in this region of the solar atmosphere. The phenomenon is called the flash spectrum. The bright horizontal line is photospheric continuum from a Baily's bead. (*Dennis DiCicco, Eclipse at sea, Sky Telesc., 54(6):470–474, December 1977*)

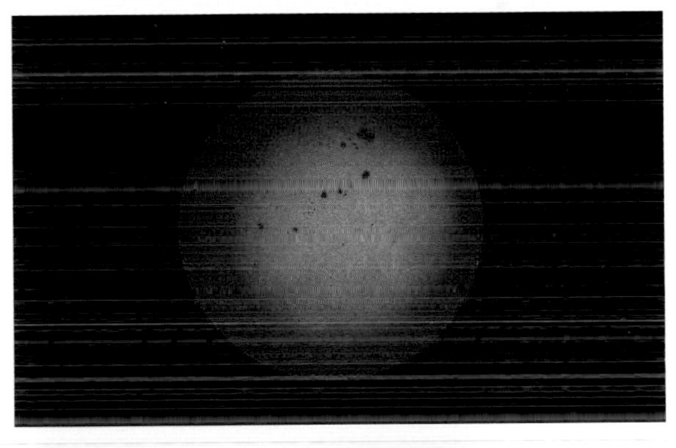

Photograph of the Sun taken in visible light shows the majority of the solar activity on the northern hemisphere. Note the limb darkening and complexity of the spot groups. Eight spot groups are visible, containing a total of 210 individual spots. (*Tersch Enterprises, Terry Schmidt*)

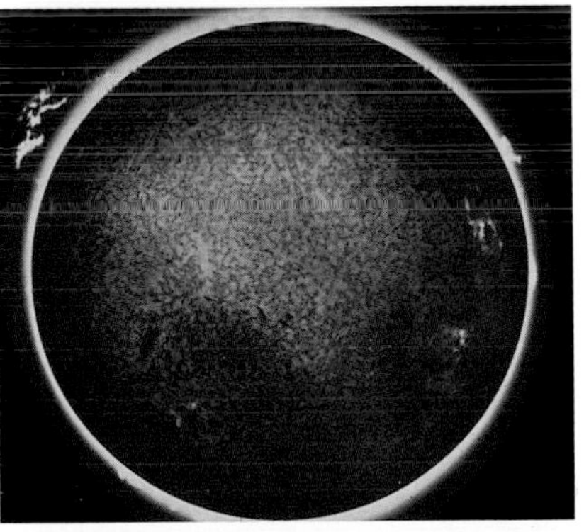

Spectroheliogram of the Sun in H-alpha light. Sunspots appear as dark centers, and hydrogen prominences projected against the solar disk as dark filaments. The bright patches are hydrogen flocculi. Prominences are visible at the solar limb. (*R. G. Poole*)

Chromosphere of the Sun showing a very active surge prominence developing at the solar limb. (*Sacramento Peak Observatory, Air Force Cambridge Research Laboratories*)

These giant loop prominences on the surface of the Sun were photographed on June 28, 1957. (*Sacramento Peak Observatory, Air Force Cambridge Research Laboratories*)

daylight period. The variations in noontime intensity and daylight period are the result of the Earth's rotational axis being tilted 23°.5 from the perpendicular to the plane of the Earth's orbit around the Sun. The direction of the Earth's axis with respect to the stars remains fixed as the Earth orbits the Sun. If the Earth's axis were not tilted from the perpendicular, there would be no variation in noontime sunlight intensity or daylight period and no seasons.

A common misconception is that the seasons are caused by variation of the Earth-Sun distance: many people think that the Earth is nearest to the Sun in summer. However, the Earth is farthest from the Sun during the first week of July (early summer in the Northern Hemisphere) and nearest to the Sun during the first week of January; also, winter and summer occur simultaneously on opposite sides of the Equator. The Earth-Sun distance does not influence the seasons because it varies only slightly, and it is overwhelmed by the effects of variations in sunlight intensity and daylight period due to the alignment of the Earth's axis.

The more hours the Sun is above the horizon, the greater the heating effect produced by the Sun. Daylight period varies for all locations except the Equator, which has 12 h of daylight every day. The poles have the widest variation in daylight period, experiencing 6 months of continuous daylight and 6 months of continuous night.

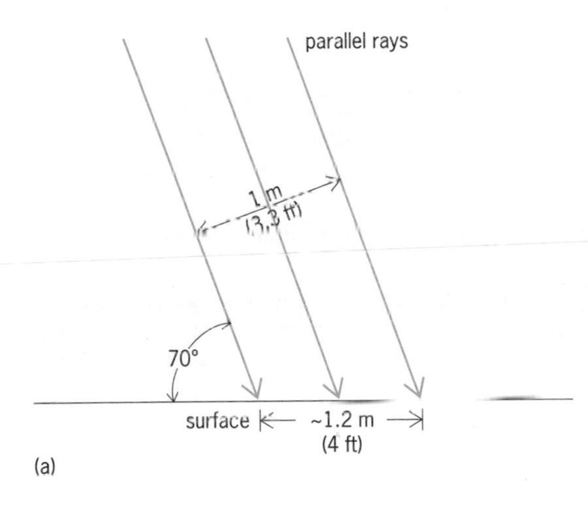

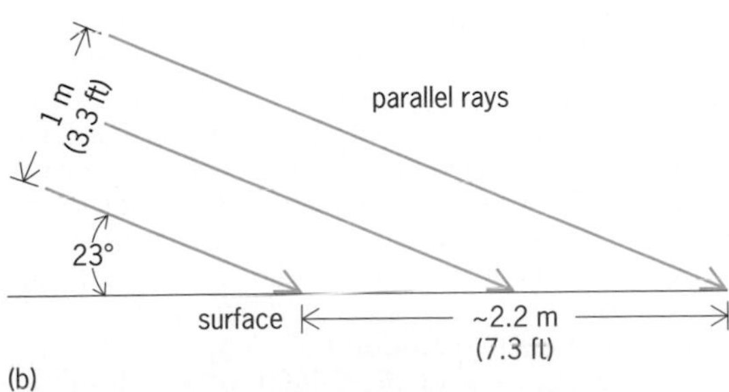

Relation of the angle of the Sun above the horizon to the area covered by a beam of sunlight 1 m (3.3 ft) on a side at 43° latitude. (*a*) The beam at solar noon on the summer solstice; the beam covers an area 1.2 m (4 ft) on a side, or about 1.4 m² (16 ft²). (*b*) The beam at solar noon on the winter solstice; the beam covers an area 2.2 m (7.3 ft) on a side, or about 4.8 m² (53 ft²). Since sunlight at noon on the winter solstice is spread over an area more than three times larger than that covered on the summer solstice, less solar heating occurs in winter than in summer.

Sunlight intensity at a location depends upon the angle from the horizon to the Sun at local solar noon; this angle in turn depends upon the location's latitude and the position of the Earth in its orbit. At increased angles, a given amount of sunlight is spread over smaller surface areas, resulting in a greater concentration of solar energy, which produces increased surface heating (see **illus.**). Intensity is a more important factor than the number of daylight hours in determining the heating effect at the Earth's surface. For example, the poles receive 6 months of continuous daylight, a situation that would seem likely to produce significant heating. However, the maximum sunlight intensity at the poles is equal only to the noontime intensity on the first day of winter at 43° latitude (for example, the northern United States). The many hours of daylight at the poles provide relatively little heating effect because of the low sunlight intensity. By contrast, the maximum noontime intensity at 43° latitude equals the solar energy per unit area received at the Equator's surface in late November.

For locations on or north of the Tropic of Cancer (23°.5 north) and on or south of the Tropic of Capricorn (23°.5 south), maximum sunlight intensity and daylight period occur on the summer solstice; minimum intensity and period occur on the winter solstice. The solstices occur on about June 21 and December 21, the June solstice marking the start of summer in the Northern Hemisphere and the beginning of winter in the Southern Hemisphere. *SEE EARTH ROTATION AND ORBITAL MOTION.*

Harold P. Coyle

Bibliography. G. O. Abell, D. Morrison, and S. C. Wolff, *Exploration of the Universe*, 5th ed., 1987; J. Kaufmann, *Universe*, 3d ed., 1991; J. S. Pickering, *1001 Questions Answered about Astronomy*, 1976; T. P. Snow, *The Dynamic Universe*, 3d ed., 1988.

Sidereal time

One of several kinds of time scales used in astronomy, whose primary application is as part of the coordinate system to locate objects in the sky. It is also the basis for determining the solar time commonly used in everyday living.

Time measurement. Time is measured with respect to some repetitious event. The measure of time is a parameter whose scale and units are adopted for specific purposes at hand. The most common measurements of time are based on the motions of the Earth that most affect everyday life: rotation on its axis, and revolution in orbit around the Sun. Objects in the sky reflect these motions and appear to move westward, crossing the meridian each day. A particular object or point is chosen as a marker, and the interval between its successive crossings of the local meridian is defined to be a day, divided into 24 equal parts called hours. The actual length of the day for comparison between systems depends on the reference object chosen. The time of day is reckoned by the angular distance around the sky that the reference object has moved westward since it last crossed the meridian. In fact, the angular distance west of the meridian is called the hour angle. *SEE MERIDIAN.*

Sidereal day and year. The reference point for marking sidereal time is the vernal equinox, one of the two points where the planes of the Earth's Equator and orbit appear to intersect on the celestial sphere.

The sidereal day is the interval of time required for the hour angle of the equinox to increase by 360°. One rotation of the Earth with respect to the Sun is a little longer, because the Earth has moved in its orbit as it rotates and hence must turn approximately 361° to complete a solar day. A sidereal year is the time required for the mean longitude of the Sun to increase 360°, or for the Sun to make one circuit around the sky with respect to a fixed reference point. SEE EQUINOX.

Astronomical coordinate system. Sidereal time is essential in astronomy and navigation because it is part of the most common coordinate system for describing the position of objects in the sky. These coordinates are called right ascension and declination, analogous to longitude and latitude on the surface of the Earth, with the vernal equinox playing the same role as the Greenwich meridian for the zero point. The local sidereal time is the hour angle of the vernal equinox past (westward of) the local meridian, and also the right ascension of any object on the meridian. By a sidereal clock, a star rises and crosses the local meridian at the same times every day. Greenwich sidereal time is local sidereal time plus the longitude of the observer west of Greenwich. SEE ASTRONOMICAL COORDINATE SYSTEMS.

These concepts are further described in the **illustration**, in which the reader is looking down on the North Pole as the Earth moves counterclockwise in orbit around the Sun. An observer at point *O* is considered. When the Earth is at position *A*, the Sun is taken to be in line with the vernal equinox and overhead for the observer. At position *B*, the Earth has rotated counterclockwise through the shaded angle, which represents the hour angle of the vernal equinox, or the local sidereal time for the observer and also the right ascension of a star overhead for the observer. At position *C*, the Earth has made a complete rotation with respect to the vernal equinox but not with respect to the Sun. The right ascension of the Sun has increased by approximately 1°.

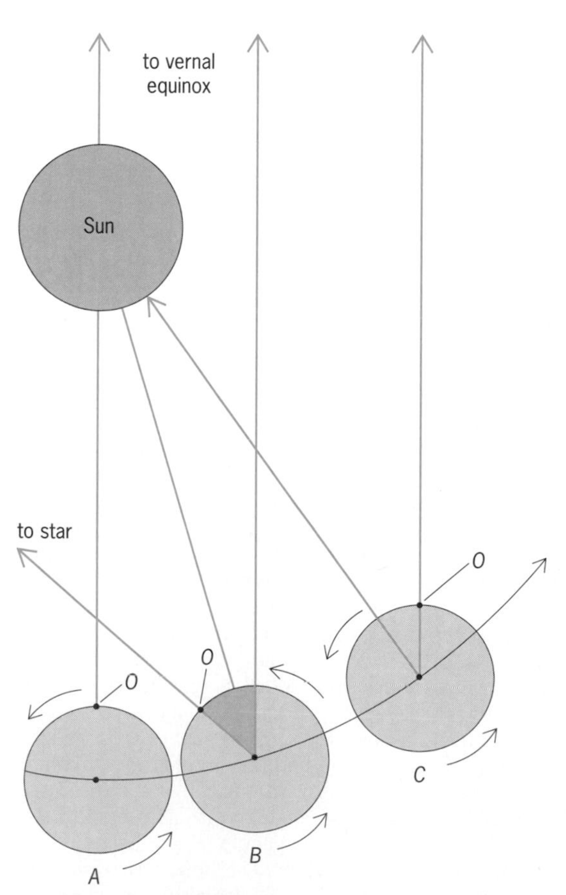

Diagram of the Earth's motions, illustrating the concepts of sidereal time, hour angle, and right ascension. Symbols are explained in text.

Determination. For a measure of time to be of practical use, determination of it by observation must be possible. Sidereal time, defined by the daily motion of the vernal equinox, is determined in practice from observations of the daily motions of stars. Solar time, which is used in the conduct of everyday life, is not directly observable. It is determined from the sidereal time by a known relationship. SEE OPTICAL TELESCOPE.

Effect of motion of Earth's axis. Upon more sophisticated examination, the sidereal day is not exactly the period of rotation of the Earth, being shorter by about 0.0084 s. The Earth's axis of rotation moves very slowly but regularly in space because of the gravitational effects of other bodies in the solar system, and the Earth's crust also shifts very slowly and irregularly about the axis of rotation. As a result, neither the equinox nor the local meridian plane is absolutely fixed. Consequently, the relation of sidereal time to the measure of time defined by the rotational motion of the Earth alone is highly complex. SEE EARTH ROTATION AND ORBITAL MOTION; PRECESSION OF EQUINOXES; TIME.

Alan D. Fiala

Bibliography. G. O. Abell, *Exploration of the Universe*, 4th ed., 1982; R. M. Green, *Spherical Astronomy*, 1985; P. K. Seidelmann (ed.), *Explanatory Supplement to the Astronomical Almanac*, 1992.

Solar constant

The total solar radiant energy flux incident upon the top of the Earth's atmosphere at a standard distance (1 astronomical unit, 1.496×10^8 km or 9.3×10^7 mi) from the Sun. In 1980 it was discovered that the so-called solar constant actually varies with time, though only by small amounts, around a value of about 1367 $W \cdot m^{-2}$ (1.96 $cal \cdot cm^{-2} \cdot min^{-1}$). The solar radiant energy flux received at the Earth is more than 10^{10} times that of the next brightest star, and almost 10^6 times that of the full moon. This solar energy, averaged over its apparent variation (almost 7%) due to the Earth's orbital motions, maintains the climate on Earth. The solar constant represents a good measure of the solar luminosity (its total radiant energy), although anisotropic components (for example, a dark sunspot), which produce directional solar-constant variations, must be taken into account. SEE EARTH ROTATION AND ORBITAL MOTION; STAR.

An important test of the validity of any theory of stellar structure and evolution is the accuracy with which it predicts the steady luminosity of the Sun. However, the main astrophysical interest in the solar constant centers on the ideas that the variations of luminosity have generated about dynamical phenomena in the solar interior (rotation, convection, and dynamo action to create magnetism). These ideas are welcome in view of the apparent failure of classical stellar theory and particle physics theory to predict the solar

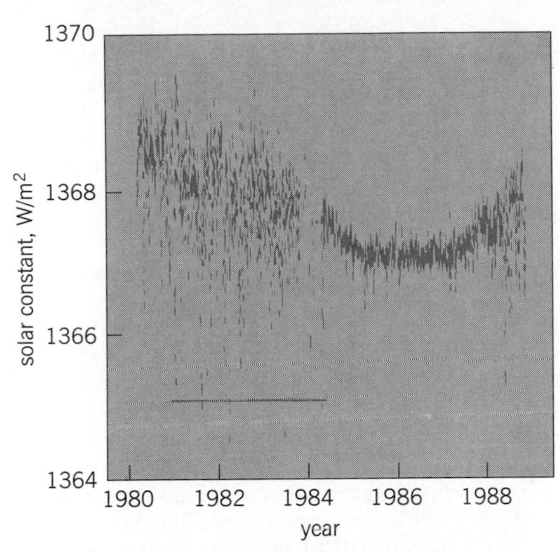

Time series of daily values of the so-called solar constant as observed by the *Solar Maximum Mission* satellite, 1980–1989. A long-term variation (approximately 0.1% amplitude) is clearly visible, its minimum coinciding with the sunspot minimum of 1985–1988. The horizontal line near the bottom indicates the period of decreased sampling before the *Challenger* astronauts repaired the satellite in 1984. (*After R. C. Willson, NASA, Jet Propulsion Laboratory*)

neutrino flux successfully. *See* Solar neutrinos; Stellar evolution.

Causes of variability. Both expected and unexpected items contribute to the variability of the so-called solar constant. Sunspots can produce deficits of up to a few tenths of 1% of the solar constant, on typical time scales of 1 week. Other surface manifestations of solar magnetic activity, faculae, contribute excesses rather than deficits. These features show modulation near a 27-day period due to solar rotation. Similar types of variability (due, for example, to so-called starspots) occur on other stars. These have larger relative amplitudes. Global oscillations of the solar interior, analogous to seismic waves on the Earth, produce variations of a few parts per million on time scales of a few hundred seconds; these waves are used to study solar interior structure and dynamics. Finally, and unexpectedly, there is an apparent 11-year sunspot cycle variation amounting to an approximately 0.1% increase of the solar constant during the sunspot maxima (see **illus.**). This long-term effect has the opposite dependence from that found for individual sunspots, which block the solar radiant energy and cause decreases rather than increases. If such small changes in solar energy flux could affect the Earth's climate directly, this solar-cycle variation would be consistent with an association noted some time ago: the so-called Maunder minimum period of greatly reduced sunspot activity in the seventeenth century corresponded with decades of cool climate in Europe, known as the Little Ice Age. *See* Helioseismology.

Measurement. Data on solar-constant variations come from the *Solar Maximum Mission* spacecraft (1980–1989). This satellite was repaired in orbit by the *Challenger* astronauts in 1984, making it possible to obtain the unique decade of data displayed in the illustration. The measurements of the solar constant were made by an active cavity radiometer, consisting of a small black cone operated as a thermostat; fluctuations in the solar energy absorbed by the cone were compensated by a servo-controlled heater circuit. This instrument provided an absolute accuracy approaching 0.1% and a relative precision, even on long time scales of 0.003%. The smallest variations detected had amplitudes less than 1 part per million of the steady flux. *See* Sun.

Hugh S. Hudson

Bibliography. J. A. Eddy, R. L. Gilliland, and D. V. Hoyt, Changes in the solar constant and climatic effects, *Nature*, 300:689–693, 1982; R. C. Willson, Measurements of solar total irradiance and its variability, *Space Sci. Rev.*, 38:203–242, 1984; R. C. Willson et al., Observations of solar irradiance variability, *Science*, 211:700–702, 1981; M. Woodard and H. S. Hudson, Frequencies, amplitudes, and linewidths of solar oscillations from total irradiance observations, *Nature*, 305:589–593, 1983.

Solar corona

The outer atmosphere of the Sun. The corona is dominated by intense magnetic forces, which penetrate it from denser regions of the Sun. Coronal gas accumulates around these magnetized regions to produce the shapes seen during a solar eclipse, with a coronagraph, or in x-rays (see **illus.**). These shapes include long streamers that penetrate interplanetary space, looplike tubes over the strongest fields, and vast regions of very low density called coronal holes. The general magnetic field of the Sun, about 1 gauss (0.1 millitesla), is revealed near the north and south poles by polar rays that resemble the pattern formed by iron filings near a bar magnet. The corona is hot enough to emit x-rays, and x-ray telescopes in space can form images of the corona. Such images display the magnetic loops connecting bright regions in the lower corona (dark in the inner portion of the illustration, which is an x-ray negative).

Outer part of the figure is a contrast-enhanced photograph of the total solar eclipse of June 30, 1973; coronal structures such as streamers are visible. Inner portion of the figure is a negative x-ray image of the lower corona made from a rocket at the same time and oriented so that the features in the lower corona correspond to the outer features seen at eclipse. (*After R. C. Altrock, ed., Solar and Stellar Coronal Structure and Dynamics, Proceedings of the 9th Sacramento Peak Summer Symposium, National Solar Observatory, Sunspot, New Mexico, Association of Universities for Research in Astronomy, 1988*)

Coronal holes are the source of solar matter streaming into interplanetary space. These streams, and other coronal matter flung into space by twisting and snapping magnetic fields, often strike the Earth to cause aurorae and other geophysical phenomena. The gyrating magnetic fields in the corona are sometimes forced together and "short out" (in a so-called reconnection) with a huge explosion, or flare, which can have geophysical effects similar to those of the streams and ejecta. *See Magnetosphere; Solar wind.*

From a minimum in the lower atmosphere (approximately 4000 K or 7000°F), the temperature rises to values in the corona of 1–5 × 10^6 K (2–9 × 10^6 °F) caused by dissipation of mechanical or magnetic energy produced by turbulent flows in the lower atmosphere. The mechanisms by which the corona is heated are uncertain. Densities range from ~10^{-17}g/cm^3 at the bottom of the corona to ~10^{-23}g/cm^3 in the highest visible parts. *See Sun.*

Richard C. Altrock

Bibliography. R. C. Altrock (ed.), *Solar and Stellar Coronal Structure and Dynamics*, Proceedings of the 9th Sacramento Peak Summer Symposium, National Solar Observatory, Sunspot, New Mexico, 1988; D. E. Billings, *A Guide to the Solar Corona*, 1966; A. I. Poland (ed.), *Coronal and Prominence Plasmas*, NASA Conf. Publ. 2442, 1986.

Solar magnetic field

The magnetic field embedded in the Sun. The magnetic field is detected at the surface through its distortion of the electronic structure of the light-emitting atoms (the Zeeman effect). The strong magnetic fields (3000 gauss or 0.3 tesla) of sunspots were first detected and measured by G. E. Hale in 1908. It was not until 1952 that an instrument with sufficient sensitivity to detect the weaker fields that occur everywhere else over the surface of the Sun was developed. *See Zeeman effect.*

Polar fields and bipolar regions. The magnetic field is made up of many different structures. On a large scale, there are polar fields, within about 35° of each pole, of some 10 G (1 millitesla), inward at one pole and outward at the other, representing a magnetic dipole. At lower latitudes, the surface is speckled with localized irregular, usually bipolar, regions of magnetic field of 20–200 G (2–20 mT). These bipolar regions occur in all sizes from the 100,000–200,000 km (60,000–120,000 mi) of the so-called normal magnetic active regions down to the smallest transient bipolar regions that can be resolved in the telescope, at 2000 km (1200 mi). The smallest magnetic regions are detected over the entire surface outside the polar regions. The normal active regions define a latitudinal belt around the Sun in each hemisphere, where strong bands of east-west subsurface fields circling the Sun erupt through the surface. The east-west fields have opposite directions in the northern and southern hemispheres. Sunspots form in the normal bipolar magnetic regions by a puzzling spontaneous gathering together of magnetic field, in opposition to the magnetic pressure, during the periods of erupting field. *See Sunspot.*

The individual normal active region is sustained by intermittent emergence of fresh field from the subsurface east-west field, at intervals from a few days to a few weeks, sustaining the normal active region anywhere from a month to a year. The old magnetic flux in each region continually fades away so that the region gradually breaks up and disappears when fresh eruptions of magnetic field cease to maintain it, while new eruptions elsewhere create new regions. On the other hand, the small bipolar magnetic regions have much shorter lives, on the order of hours or a few days, and appear to be the result of a single small eruption.

Twenty-two-year cycle. The normal active regions appear first at latitudes of about 40° north and south. The bands of active regions gradually migrate (at about 2 m/s or 4 mi/h) toward the equator over a period of 10–12 years, where they disappear at about the same time that new magnetic bands appear at latitude ±40°. The new bands are oppositely oriented, as compared to the old disappearing magnetic bands, and follow the same pattern of equatorward migration. Thus in a period of about 22 years the Sun completes one whole magnetic cycle. Each 11-year sunspot cycle is just half of the total 22-year magnetic cycle. The polar fields are part of the cycle, reversing in an irregular manner within a year or two of the peak of each sunspot cycle.

Magnetic flux bundles. The patches of field on the surface of the Sun, just described, are composed of many concentrated magnetic flux bundles of 1000–2000 G (0.1–0.2 T) with diameters of 100–500 km (60–300 mi), which is too small to be individually resolved in most magnetic observations. The spacing of the individual bundles determines the mean field strength (stated above) in any given area. Sunspots are composed of many such bundles firmly packed together.

Generation and effects. It is believed that the magnetic field of the Sun is generated by a combination of cyclonic convection and nonuniform rotation (the surface of the Sun rotates about 40% faster at the equator than at the poles) at depths on the order of 200,000 km (120,000 mi) below the surface (to be compared with the solar radius of 700,000 km or 400,000 mi). The generation process involves the opposite fields in the northern and southern hemispheres as well as the polar fields, and the net result is the 22-year magnetic cycle with equatorward migration of the strong east-west fields, estimated at 3000 G (0.3 T) or more, far below the surface.

The magnetic fields that erupt through the surface of the Sun are the primary cause of solar activity, producing such suprathermal effects as the x-ray corona (2–3 × 10^6 K), flares, coronal mass ejections, and prominences. *See Solar corona.*

Other stars. It is to be assumed that the magnetic fields and the associated activity of the Sun are typical of the magnetic fields and associated activity of most other stars (all too far away to be resolved in the telescope). The Sun, then, is the main laboratory for studying and eventually understanding the general phenomenon of stellar activity, x-ray emission, and magnetic fields. *See Sun.*

Eugene N. Parker

Bibliography. H. W. Babcock, The Sun's magnetic field, *Annu. Rev. Astron. Astrophys.*, 1:41–58, 1963; R. Howard, Magnetic field of the sun (observational), *Annu. Rev. Astron. Astrophys.*, 5:1–24, 1967; E. N. Parker, The Sun, *Sci. Amer.*, 233 (3):42–50, September 1975; R. W. Noyes, *The Sun, Our Star*, 1982; E. R. Priest, *Solar Magnetohydrodynamics*, 1982; D. G. Wentzel, *The Restless Sun*, 1989; C. Zwaan, The emergence of magnetic flux, *Solar Phys.*, 100:397–414, 1985.

Solar neutrinos

Neutrinos produced in nuclear reactions inside the Sun. The first direct test of how the Sun produces its luminosity (observed most conspicuously on Earth as sunlight) has been carried out by observing these particles. The results of this experiment are in disagreement with the joint theoretical predictions based upon the well-established astronomical theory of stellar evolution or the well-established physical theory of neutrinos. This discrepancy between theory and observation suggests that either the process by which the Sun shines is not understood as well as previously believed, or some modification in the physical theory of neutrinos is required. *See Sun.*

Nuclear fusion in the Sun. The Sun shines because of fusion reactions similar to those envisioned for terrestrial fusion reactors. The basic solar process is the fusion of four protons to form an alpha particle, two positrons (e^+), and two neutrinos (ν); that is, $4p \rightarrow \alpha + 2e^+ + 2\nu_e$. The principal reactions are shown in the **table**, with a column indicating in what percentage of the solar terminations of the proton-proton chain each reaction occurs. The rate for the initiating proton-proton (PP) reaction, number 1 in the table, is largely determined by the total luminosity of the Sun. Unfortunately, these neutrinos are below the energy thresholds, for the first two experiments to detect solar neutrinos with chlorine-37 (^{37}Cl) and with ultrapure water (H_2O). Two experiments in progress, one in Russia and one in Italy, use gallium-71 (^{71}Ga) to detect neutrinos from the PP reaction and from other reactions listed in the table.

The proton-electron-proton (PEP) reaction (number 2), which is the same as the familiar PP reaction except for having the electron in the initial state, is detectable in the ^{37}Cl experiment. The ratio of PEP to PP neutrinos is approximately independent of which solar model is used for the solar predictions. Two other reactions in the table are of special interest. The capture of electrons by ^{7}Be (number 6) produces detectable neutrinos in the ^{37}Cl experiment. The ^{8}B decay (number 9) was expected to be the main source of neutrinos for the ^{37}Cl experiment because of their relatively high energy (14 MeV), although it is a rare reaction in the Sun. The ^{8}B neutrinos are the only substantial contributions to the ultrapure water experiment. There are also some less important neutrino-producing reactions from the carbon-nitrogen-oxygen (CNO) cycle, which will not be discussed here in detail since this cycle is believed to play a rather small role in the energy-production budget of the Sun. *See Carbon-nitrogen-oxygen cycles; Proton-proton chain.*

Experimental test. The first solar neutrino detector was based on the reaction $\nu_{solar} + {}^{37}\text{Cl} \rightarrow {}^{37}\text{Ar} + e^-$, which is the inverse of the electron-capture decay of argon-37 (^{37}Ar). This reaction was chosen for the first experiment because of its unique combination of physical and chemical characteristics, which were favorable for building a large-scale solar neutrino detector. Neutrino capture to form ^{37}Ar in the ground state also has a relatively low energy threshold (0.81 MeV) and a high sensitivity, nuclear properties that are important for observing neutrinos from ^{7}Be, ^{13}N, and ^{15}O decay and the PEP reaction.

The ^{37}Cl detector was built deep underground to avoid the production of ^{37}Ar in the detector by cosmic rays. The final detector system consists of an approximately 100,000-gallon (400,000-liter) tank of perchloroethylene, a pair of pumps to circulate helium through the liquid, and a small building to house the extraction equipment, all in the deep Homestake gold mine in South Dakota, 5000 ft (1500 m) below the surface.

A set of experimental runs were carried out in the ^{37}Cl experiment during the 1970s and 1980s, and have continued into the 1990s. They have shown that the ^{37}Ar production rate in the tank is 0.5 ± 0.05 ^{37}Ar atoms per day. Even though the tank is nearly a mile (1.6 km) underground, a small amount of ^{37}Ar is produced by cosmic rays. This background rate has been estimated to be of order 0.1 ^{37}Ar atoms per day, but could be larger.

If the above-mentioned background rate is assumed, then a positive signal of (2.2 ± 0.3) solar neutrino units (SNU) is inferred; 1 SNU = 10^{-36} capture per target particle per second.

The predicted capture rates for the ^{37}Cl experiment (in SNU) for one solar model are as follows: PP reaction: 0; ^{8}B beta decay: 6; PEP reaction: 0.2; ^{7}Be electron capture: 1; ^{13}N decay: 0.1; and ^{15}O decay: 0.3. The total theoretical prediction is 8 SNU. Many investigations have been undertaken of the best values to use for various parameters, and the total predicted rate may well differ from 8 SNU by 2 or 3 SNU.

Solar neutrinos have also been observed with the aid of electrons that the neutrinos scatter in a 660-short-ton (600-metric-ton) detector of pure water lo-

Proton-proton chain in the Sun

Number	Reaction		Percentage of solar terminations	Maximum neutrino energy, MeV
1	$p + p \rightarrow {}^2\text{H} + e^+ + \nu$	or	99.75	0.420
2	$p + e^- + p \rightarrow {}^2\text{H} + \nu$		0.3	1.44 (monoenergetic)
3	${}^2\text{H} + p \rightarrow {}^3\text{He} + \nu$		100	
4	${}^3\text{He} + {}^3\text{He} \rightarrow {}^4\text{He} + 2p$	or	85	
5	${}^3\text{He} + {}^4\text{He} \rightarrow {}^7\text{Be} + \nu$		15	
6	${}^7\text{Be} + e^- \rightarrow {}^7\text{Li} + \nu$		—	0.861 (90%), 0.383 (10%) (both monoenergetic)
7	${}^7\text{Li} + p \rightarrow 2\,{}^4\text{He}$	or	—	
8	${}^7\text{Be} + p \rightarrow {}^8\text{B} + \nu$		0.02	
9	${}^8\text{B} \rightarrow {}^8\text{Be}^* + e^+ + \nu$		—	14
10	${}^8\text{Be}^* \rightarrow 2\,{}^4\text{He}$		—	

cated in the Kamioka metal mine about 200 mi (300 km) west of Tokyo in the Japanese Alps. These results are of great importance since they provide direct evidence that the observed neutrinos originate in the Sun; the electrons are scattered by neutrinos in the forward direction between the Earth and the Sun. Moreover, the neutrino events are registered at the exact time they are detected, making possible a sensitive search for possible time dependences. The first measurement of the Kamioka collaboration yielded the result given by the equation below, where the first

$$\frac{\text{Observed}}{\text{Predicted}} = 0.47 \pm 0.06 \text{ (stat.)} \pm 0.06 \text{ (syst.)}$$

error is a 1-standard-deviation statistical uncertainty and the second error is a systematic uncertainty. The predicted event rate in the equation stands for the best estimate of the number of events calculated to occur according to the standard model of the Sun and the standard model of how neutrinos behave. The Kamioka result refers to high-energy neutrinos from ^{8}B decay (number 9 in the table).

Possible explanations. Many explanations have been advanced for the discrepancy between the observed and the predicted event rates in the ^{37}Cl and Kamioka solar neutrino experiments. These explanations can be divided into two general classes: (1) the standard solar model must be significantly modified; (2) the standard model of how neutrinos behave must be significantly modified. There is considerable evidence that the theory of neutrino behavior must be changed, but this conclusion will be tested by future experiments.

Further experiments. Further experiments are required to settle the issue of whether present concepts in astronomy or physics are at fault. Fortunately, a testable distinction can be made. The flux of low-energy neutrinos from the PP and PEP reactions (numbers 1 and 2 in the table) is almost entirely independent of astronomical uncertainties and can be calculated from the observed solar luminosity, provided only that the basic physical ideas of nuclear fusion as the energy source for the Sun and of stable neutrinos are correct. If these low-energy solar neutrinos are detected at the expected rate in a future experiment, then it will be known that the present crisis is caused by a lack of astronomical understanding. If the low-energy neutrinos are absent, it will be established that the present discrepancy between theory and observation is due, at least in part, to faulty physics, and not just to poorly understood astrophysics.

Since gallium is especially sensitive to the fundamental, low-energy PP neutrinos, the results of experiments using gallium as a detector will establish whether the fundamental low-energy PP neutrinos reach the Earth in the numbers calculated from the standard solar model. If the observed rate with gallium detectors is much less than that predicted by the standard solar model—assuming nothing happens to the neutrinos on the way to Earth from the Sun—then this would be strong evidence for the standard model of neutrinos being modified. The first results from the gallium experiment in Russia and an independent experiment are not conclusive; the statistical uncertainties, which will improve with time, are large in the initial studies.

A different type of experiment would measure the relative number of solar neutrinos of different energies that reach the Earth. A mistake in the solar model can change the total number of neutrinos that are observed but not the relative number of neutrinos of different energies that are produced. On the other hand, modifications of the standard model of neutrinos can cause the observed number of neutrinos of different energies to change from the expected proportions to some different ratios. An experiment being developed in Canada will measure the energies of individual neutrinos and will test directly whether a new model of the Sun or of neutrinos is required. SEE STELLAR EVOLUTION.

John N. Bahcall

Bibliography. A. I. Abazov et al., Search for neutrinos from the Sun using the reaction ^{71}Ga(ν_e, e^-)^{71}Ge, *Phys. Rev. Lett.*, 67:3332–3335, 1991; P. Anselmann et al., Solar neutrinos observed by GALLEX at Gran Sasso, *Phys. Lett. B*, 285:376, 1992. J. N. Bahcall, *Neutrino Astrophysics*, 1989; J. N. Bahcall and H. A. Bethe, A solution of the solar neutrino problem, *Phys. Rev. Lett.*, 65:2233–2235, 1990; J. N. Bahcall and R. Davis, Jr., Solar neutrinos: A scientific puzzle, *Science*, 191:264–267, 1976; R. Davis, Jr., Results of the ^{37}Cl experiment, *Proceedings of the Brookhaven Solar Neutrino Conference*, BNL 50879, 1:1–54, 1978; K. S. Hirata et al., Real-time directional measurement of ^{8}B solar neutrinos in the Kamiokande-II Detector, *Phys. Rev. D*, 44:2241–2260, 1991.

Solar radiation

The electromagnetic radiation and particles (electrons, protons, and rarer heavy atomic nuclei) emitted by the Sun. Electromagnetic energy has been observed over the whole spectrum with wavelengths varying from 0.01 nanometer to 30 km. The bulk of the energy is in the spectrum of visible light (400–800 nm). The solar spectrum doubtless extends far beyond the observed limits in both directions. The total power is 3.86×10^{26} W.

The Sun also emits a continuous stream of electrons with shorter bursts of electron and proton showers sufficiently intense to affect the ionization of the upper terrestrial atmosphere. These sporadic particles have energies from a few thousand to a few billion electronvolts. The lower-energy particles are much more abundant, but those of high energy are sufficient to occasionally damage the solid-state circuitry of spacecraft. The physical mechanism of high-energy particle emission is not understood, but is closely associated with the more energetic forms of solar activity. SEE SOLAR WIND; SUN.

John W. Evans

Solar system

The Sun and the bodies moving in orbit around it. The most massive body in the solar system is the Sun, a typical single star that is itself in orbit about the center of the Milky Way Galaxy. Nearly all of the other bodies in the solar system—the terrestrial planets, outer planets, asteroids, and comets—revolve on orbits about the Sun. Various types of satellites revolve around the planets; in addition, the giant planets all have orbiting rings. The orbits for the planets appear to be fairly stable over long time periods and hence have undergone little change since the formation of the solar system. It is thought that some 4.56 $\times 10^9$ years ago a rotating cloud of gas and dust collapsed to form a flattened disk (the solar nebula) in which the Sun and other bodies formed. The bulk of the gas in the solar nebula moved inward to form the Sun, while the remaining gas and dust is thought to have formed all the other solar system bodies by accumulation proceeding through collisions of intermediate-sized bodies called planetesimals. Planetary systems are believed to exist around many other stars in the Milky Way Galaxy, and evidence for the existence of other solar systems is likely to increase in the near future.

Composition. The Sun is a gaseous sphere with a radius of about 7×10^5 km (4×10^5 mi), composed primarily of hydrogen and helium and small amounts of the other elements. The Sun is just one of about 2×10^{11} stars in the Milky Way Galaxy. The solar system lies in the disk of the Galaxy and moves around the galactic center about once every 2×10^8 years on a circular orbit with a radius close to 3×10^4 light-years (about 3×10^{17} km or 2×10^{17} mi). The Sun is somewhat peculiar in that it is a single star, unaccompanied by another star; most stars in the disk of the Galaxy are in double or multiple systems where two or more stars orbit about their common center of mass. The Sun's mass of 2×10^{30} kg (4.4×10^{30} lb), however, is quite typical; stellar masses range from about 0.08 to about 60 times the Sun's mass, with the great majority of stars being similar in mass (and hence in other characteristics) to the Sun. *SEE BINARY STAR; MILKY WAY GALAXY; STAR; SUN.*

The terrestrial planets (Mercury, Venus, Earth, and Mars) are the closest to the Sun, with orbital radii ranging from 0.39 AU for Mercury to 1.5 AU for Mars. One astronomical unit (AU), the distance from Earth to Sun, equals 1.5×10^8 km or 9.3×10^7 mi. The terrestrial planets are composed primarily of silicate rock (mantles) and iron (cores). The Earth is the largest terrestrial planet, with an equatorial radius of 6378 km (3963 mi) and a mass of 5.974×10^{24} kg (1.317×10^{25} lb); Mercury is the smallest, with a mass of 0.053 times that of Earth. *SEE EARTH; MARS; MERCURY; PLANET; PLANETARY PHYSICS; VENUS.*

The outer planet region begins with Jupiter at a distance of 5.2 AU from the Sun. The outer planets are subdivided into the giant or Jovian planets (Jupiter, Saturn, Uranus, and Neptune) and Pluto. By far the largest planet is Jupiter, with a mass 318 times that of the Earth, while the other giant planets are more massive by a factor of 15 or more than Earth. Jupiter and Saturn are composed primarily of hydrogen and helium gas, like the Sun, but with rock and ices, such as frozen water, methane, and ammonia, concentrated in their cores. Uranus and Neptune also have rock and ice cores surrounded by envelopes with smaller amounts of hydrogen and helium. Pluto, slightly smaller than the Earth's Moon, is probably composed primarily of rock and ice. *SEE JUPITER; NEPTUNE; PLUTO; SATURN; URANUS.*

The region between Mars and Jupiter is populated by a large number of rocky bodies called asteroids. The asteroids are smaller than the terrestrial planets, with most known asteroids being about 1 km (0.6 mi) in radius, though a few have radii of hundreds of kilometers. Some asteroids have orbits that take them within the orbits of Earth and the other terrestrial planets, leading to the possibility of impacts with these planets; asteroids in the main belt beyond Mars are occasionally perturbed into these Earth-crossing orbits. Small fragments of asteroids (or comets) that impact the Earth first appear as meteors in the sky; any meteoric material that survives the passage through the Earth's atmosphere and reaches the surface is called a meteorite. The Allende meteorite, which fell in Mexico in 1969, contains inclusions that have been radiometrically dated at 4.56×10^9 years; this age, the oldest found so far, is presumably the age of the solar system. Along with the lunar rocks returned by the Apollo missions, Allende and other meteorites are samples of other bodies in the solar system. In addition, primitive components of meteorites such as Allende give important information about conditions in the very early solar system. A few meteorites are thought to have come from Mars or the Moon. *SEE ASTEROID; METEOR; METEORITE.*

Comets are icy bodies (so-called dirty snowballs) with diameters on the order of 10 km (6 mi). In contrast to the orbits of most planets, cometary orbits often are highly elliptical and have large inclinations that take them far above and below the plane where the planets orbit. The region well beyond Pluto's orbit is populated with a very large number (perhaps 10^{12}) of comets, out to a limiting distance of about 10^5 AU, at which point external forces (due to nearby stars, passing molecular clouds, and the Milky Way Galaxy itself) disturb any body trying to stay in a stable orbit around the Sun. The distribution of comets within this huge volume, the Oort Cloud, is uncertain; there may be an inner cloud extending outward to about 2×10^4 AU, and an outer cloud beyond that. The outer cloud is the source of the new comets that appear in the inner solar system; the inner cloud may replenish the outer cloud, and may also supply some comets directly to the inner solar system. Comets that enter the inner solar system are eventually evaporated by solar radiation, leaving behind interplanetary dust particles and perhaps rocky cores similar to certain asteroids. *SEE COMET.*

Satellites orbit around all of the planets except Mercury and Venus. The Earth and its rocky Moon are unusual in that the Moon is about one-fourth the size of the Earth; nearly all other satellites are much smaller than their planet, with the exception of Pluto and its satellite Charon, which is half the size of Pluto. While Mars has only two tiny, rocky satellites, the Voyager missions have shown that the giant planets all have large systems of satellites (Saturn has at least 20), ranging in size from a few kilometers to 2700 km (1700 mi) in radius and composed of varying amounts of rock and ice. Each giant planet also has a number of rings of particles whose orbits are interspersed with and controlled by the planet's innermost satellites. *SEE MOON; SATELLITE.*

Origin. A number of imaginative theories of the origin of the solar system have been presented in the last several hundred years, but nearly everyone presently working on solar system cosmogony agrees on one fundamental concept, advanced in 1796 by P. S. de Laplace. This concept, the nebular hypothesis, holds that the Sun and the rest of the bodies in the solar system formed from the same rotating, flattened cloud of gas and dust, now called the solar nebula. The nebular hypothesis at once explains the gross orbital properties of the solar system: all planets orbit (and most rotate) in the same sense as the Sun rotates, with their nearly circular orbits being confined largely to a single plane almost perpendicular to the Sun's rotation axis.

While the grand experiment that led to the formation of the solar system can never be repeated, present-day regions of star formation in the Milky Way Galaxy can be observed to gain insight into the processes involved in the formation of stars similar in mass to the Sun. Assuming that present-day star formation proceeds much as it did 4.56×10^9 years ago, this approach should yield many constraints on the general process. These observations confirm the stellar implications of Laplace's nebular hypothesis: very young stars (protostars) are indeed found embedded in dense clouds of gas and dust that often show evidence for flattening and rotation. While understanding of solar system formation is almost certain to undergo further evolution, the following summarizes the basic framework in which cosmogonists are working.

Collapse phase. The solar nebula was produced by the collapse of a dense interstellar cloud (**illus**. *a*) composed primarily of molecular hydrogen (about 77% by mass) and helium gas (about 21%), with about 2% of the cloud mass being in the form of the other elements, mostly frozen into dust grains. Elements heavier than hydrogen and helium were formed by nucleosynthesis in previous generations of stars and then ejected into interstellar clouds through events such as supernovae. *SEE NUCLEOSYNTHESIS; SUPERNOVA.*

Radio telescopes have shown that dense interstellar clouds exist with masses comparable to that of the Sun, containing about 10^4–10^6 molecules/cm^3, and with quite low temperatures, about 10 K (−263°C or −441°F). Because of their low temperatures, gas pressure generally is not sufficient to support these clouds against their own self-gravity, which tries to pull the cloud into a smaller configuration. Many clouds appear to be supported primarily by magnetic fields; however, magnetic field support requires a continual contraction of the largely neutral cloud past the field lines, and eventually magnetic forces can no longer support the increasingly dense cloud. At this point, dense clouds enter the collapse phase, where supersonic inward motions develop that lead to the formation of a stellar-sized core at the center of the cloud in about 10^5–10^6 years. Some of the gas in the cloud continues to fall onto this protostellar core, where it is suddenly stopped and its kinetic energy is converted into thermal energy. This heat leaves the cloud as infrared radiation; infrared telescopes have shown that many dense cloud cores contain deeply embedded young stars. *SEE INFRARED ASTRONOMY; INTERSTELLAR MATTER; RADIO ASTRONOMY; STELLAR EVOLUTION.*

Solar nebula. In a rotating cloud, not all of the infalling gas and dust falls directly onto the central protostar, because of the conservation of angular momentum. Instead, a disklike solar nebula forms, where the inward pull of gravity is expended in maintaining the rotational motion of the nebula. If too much angular momentum is present in the initial cloud, the collapse process is likely to result in the formation of a double or multiple star system instead of a single star and a planetary system; if extremely little angular momentum is present, only a single star forms.

Because the Sun contains 99.9% of the mass of the solar system but only about 2% of the angular momentum, the planetary system has considerably more angular momentum per unit mass than the Sun. This discrepancy is more than can be accounted for by forming the Sun out of the lowest-angular-momentum portions of a dense cloud, or by allowing for the fact that the Sun has lost substantial angular momentum during the age of the solar system through magnetic braking associated with the solar wind. *SEE SOLAR WIND.*

The solution to this angular momentum problem appears to lie in substantial evolution of the solar nebula after its formation. In order to have enough angular momentum to lead to planetary system formation, much of the mass of the newly formed solar nebula must reside in the disk rather than in the early Sun. The disk must then evolve in such a way as to transfer mass inward to feed the growing Sun, while transporting outward the excess angular momentum undesired by the Sun but required for the planets. While this sort of evolution may appear to be contrived if not miraculous, it is actually to be expected on very general grounds for any viscous disk that is undergoing a loss of energy, as the solar nebula will, through radiation to space. Several different physical mechanisms have been identified that potentially are capable of driving solar nebula evolution: turbulence caused by convective instability or velocity shear, gravitational torques between a bar-shaped protosun and spiral arms in the nebula, and magnetic torques associated with remanent magnetic fields or with a nebula magnetic field generated by the dynamo mechanism. These mechanisms are likely to lead to evolution of the bulk of the nebula gas inward onto the protosun (illus. *b*), on a time scale on the order of 10^5–10^7 years.

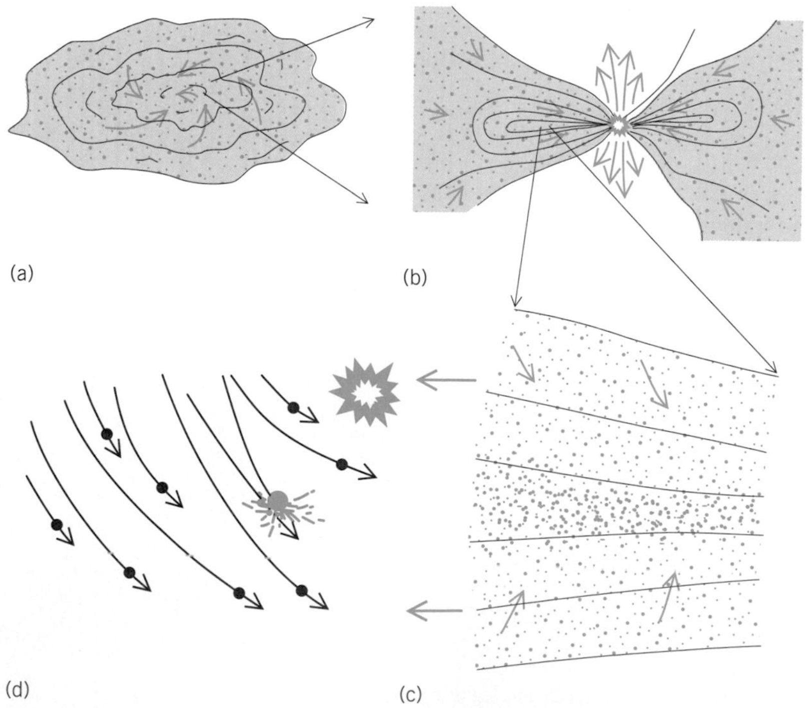

(a) (b) (d) (c)

Schematic diagram of four phases of the formation of the solar system. (*a*) A dense, rotating interstellar cloud collapsed to form the solar nebula, a flattened cloud of gas and dust. (*b*) Most of the gas in the solar nebula flowed onto the growing protosun, whose energetic wind first flowed outward in a bipolar pattern and later removed the last vestiges of the nebula. (*c*) Dust grains sedimented to the nebula midplane and coagulated into kilometer-sized planetesimals. (*d*) Gravitational forces between the planetesimals resulted in collisions and their accumulation into planetary-sized bodies.

Planetesimal formation. The portion of the nebula that is to form the planets must decouple from the gaseous nebula to avoid being swallowed by the Sun. This occurs by the process of coagulation of dust grains through mutual collisions; when solid bodies become large enough (roughly kilometer-sized), they will no longer be tied to the nebula through brownian motion (as is the case with dust grains) or gas drag (as happens with smaller bodies).

The interstellar dust grains that eventually form the solid bodies in the solar system start out with sizes on the order of 0.1 micrometer. These miniscule particles must stick together in order to begin the planet formation process; sticking can be caused by van der Waals forces. Coagulation is enhanced by the increase in the spatial density of dust grains, which occurs as the dust grains sediment down through the gaseous nebula to form a dense subdisk composed largely of solid particles (illus. *c*). Both dust grain sedimentation and coagulation are inhibited by the vigorous stirring expected in a highly turbulent nebula (the latter because grains may hit each other at relative velocities that are high enough to produce fragmentation rather than coagulation), but once the turbulence dies down, sedimentation and coagulation can produce centimeter- to meter-sized bodies in about 10^3–10^4 years.

The next phase of growth has long been thought to involve a collective gravitational instability of the dust subdisk that would rapidly produce bodies of kilometer size or larger, termed planetesimals. However, this instability may be prevented by turbulence induced by the difference in orbital speed between the gaseous nebula and the dust subdisk. (The gas orbits the Sun somewhat more slowly than the solid bodies because the gaseous disk is partially supported by gas pressure.) If the gravitational instability is prevented, then growth to planetesimal size may occur through further collisional coagulation; gas drag preferentially slows down smaller bodies, producing relative motions between larger and smaller bodies that can lead to collisions. In this case, growth to kilometer-sized planetesimals is expected to occur in about 10^4 years.

Planetary accumulation. About 10^{12} kilometer–sized planetesimals are needed to form just the terrestrial planets; significantly greater numbers of similarly sized bodies would be needed to form the giant planets. These plantesimals are already roughly the size of many asteroids and comets, suggesting that many of these bodies are simply leftovers from intermediate phases of the planet formation process.

Planetesimals are massive enough for gravitational forces to determine their collision probabilities; self-gravity is also needed to prevent debris from escaping following collisions. The subsequent growth of the planetesimals through gravitational accumulation (illus. *d*) is in two distinct phases. In the first phase, planetesimals grew by accumulation of other planetesimals at essentially the same distance from the Sun. Once the nearby planetesimals were all swept up, this phase ended. The first phase lasted about 10^4 years in the inner solar system and produced planetesimals about 500 km (300 mi) in size. This phase is likely to have been characterized by the runaway growth of relatively few bodies, because once one body becomes larger than its neighbors (as must happen after any collision occurs), its increased gravitational pull increases the chances that another body will impact it, and so on.

In the second phase, accumulation requires bodies at significantly different distances from the Sun to collide. This can happen only if the planetesimal orbits become highly elliptical and hence intersecting. Gravitational forces between the orbiting planetesimals can produce elliptical orbits, but only over relatively long time periods; about 10^7–10^8 years is required for accumulation to proceed by this means in the inner solar system. The second phase is currently thought to have proceeded differently from the first phase, with collisions occurring primarily between more or less equal-sized bodies, rather than the highly unequal-sized collisions that occur in a runaway accumulation process. The second phase may then have involved violent collisions between planetary-sized bodies, a spectacular finale to the entire planet formation process. A glancing collision between a Mars-sized and an Earth-sized body appears to be the best explanation for the formation of the Earth-Moon system; debris from the giant impact would end up in orbit around the Earth and later form the Moon.

Astronomical observations of young stars similar to the Sun imply that the gaseous portion of the solar nebula was removed some time between 10^5 and 10^7 years after the Sun began to form. While most of the gas presumably was added to the Sun through nebula evolution, the residual nebula gas and dust was probably removed from the solar system by the early solar wind, which had a mass loss rate roughly 10^6 times larger than at present (illus. *b*). The final formation of the terrestrial planets, lasting 10^7–10^8 years, then occurred in the absence of appreciable nebula gas, which explains the absence of significant hydrogen and helium in the inner solar system. However, formation of the giant planets must have occurred prior to the loss of the nebula gas; otherwise their rock and ice cores would not have been able to capture the gas needed to account for their present compositions. Formation of the giant planets within about 10^6 years appears to require that the second phase of planetesimal accumulation proceeded through runaway accretion all the way to bodies about 10 Earth masses in size in the outer solar system. The satellite systems of the giant planets were largely formed in minisolar nebulae orbiting around each protoplanet, through processes that are similar to those described for planet formation.

The alternative means of forming the giant planets rapidly, gravitational instability of the gaseous nebula, suffers from several severe difficulties. First, a massive nebula is required for the instability to occur, but subsequent evolution of this massive nebula is likely to remove any giant protoplanets. Second, a giant planet formed in this fashion could not have the rock and ice cores inferred to exist in the giant planets. Formation of the giant planets by accumulation of a rock and ice core followed by capture of a gaseous envelope provides a superior means of explaining the structure of the giant planets.

Extrasolar planetary systems. Confirmation of the implications of the nebular hypothesis for planetary formation requires the detection of planetary systems orbiting around other stars; if the basic theory of solar system formation is correct, planetary systems should not be rare in the Milky Way Galaxy.

Several hints of the existence of extrasolar planetary systems already exist. Images of optical radiation reflected by dust particles aligned in a thin disk around the nearby star Beta Pictoris are the most graphic evidence in support of the existence of flattened systems of solid bodies orbiting other stars. A few observations of radial velocity shifts induced by the orbital motions of stars around the center of mass of their systems suggest the existence of planets with masses of 1–10 times that of Jupiter. Detection of extrasolar planets is an exceedingly difficult task, but a number of inventive schemes for planetary detection have been developed and some may well be implemented in the near future.

Alan P. Boss

Bibliography. D. C. Black and M. S. Matthews (eds.), *Protostars and Planets II*, 1985; A. P. Boss, Low mass star and planet formation, *Publ. Astron. Soc. Pacific*, 101:767–786, 1989; A. G. W. Cameron, Origin of the solar system, *Annu. Rev. Astron. Astrophys.*, 26:441–472, 1988; V. S. Safronov, *Evolution of the Protoplanetary Nebula and Formation of the Earth and the Planets*, 1969, transl., NASA TTF-677, 1972; H. A. Weaver and L. Danly (eds.), *The Formation and Evolution of Planetary Systems*, 1989; G. W. Wetherill, Formation of the Earth, *Annu. Rev. Earth Planet. Sci.*, 18:205–256, 1990.

Solar wind

The continuous outward flow of ionized solar gas and a "frozen-in" remnant of the solar magnetic field through the solar system. This flow arises from strong outward pressure in the solar corona, becomes supersonic at a few solar radii (1 solar radius = 4.32×10^5 mi or 6.96×10^5 km) above the visible surface of the Sun (the photosphere), and attains speeds in the range 150–450 mi s^{-1} (250–750 km s^{-1}) in interplanetary space. The solar wind is believed to remain supersonic out to a distance from the Sun of 50–100 astronomical units (AU), where it is slowed by interaction with the interstellar gas and magnetic field.

Studies of geomagnetic activity led to the first suggestions (as early as the 1850s) that the Sun emits particles in connection with solar activity. In the late 1950s E. N. Parker demonstrated that the extended, high temperature of the solar corona (above a million degrees) precluded the nearly static equilibrium state that prevails in the atmospheres of planets like the Earth. Rather, the corona attains equilibrium by expanding continuously into interplanetary space. Simplified theoretical models of this expansion predicted a transition from very slow motions in the lower corona to a supersonic outward flow at 120–600 mi s^{-1} (200–1000 km s^{-1}) in the interplanetary region.

Continuous emission. This concept of a continuous emission of particles (as an ionized but electrically neutral plasma), or a "solar wind," unified many of the ideas about solar particle emission that had been deduced from indirect sources. By 1970 direct observations made with spacecraft-borne instruments had led to a fairly complete and detailed description of the basic characteristics of the solar wind near the orbit of Earth. The range of observed values and the long-term averages of most important solar wind properties are given in the **table**.

In the early 1970s, much attention was given to the study of inhomogeneities in the flow, in particular of streams of abnormally fast plasma. In the mid-1970s, studies of coronal holes, regions of exceptionally low density in the corona, led to the identification of the solar sources of these streams. Further understanding of the relationship between holes and streams and of both to the solar magnetic field then led to a remarkably unified view of the three-dimensional structure of the corona and interplanetary medium. The study of time-dependent solar wind flows, in particular those associated with solar activity, remains an active (and sometimes controversial) aspect of solar wind research. SEE INTERSTELLAR MATTER; SOLAR MAGNETIC FIELD.

Solar wind streams and magnetic sectors. In 1962, sampling instruments on the *Mariner 2* spacecraft revealed an important pattern in the variations of solar wind speed with time. The observed speed rose systematically from low values, 180–240 mi s^{-1} (300–400 km s^{-1}), to high values, 360–420 mi s^{-1} (600–700 km s^{-1}), in 1 or 2 days and then returned to low values during the next 3 to 5 days (**Fig. 1***a*) Each of these high-speed streams tended to be seen at approximately 27-day intervals or to recur with the rotation period of the Sun. A similar recurrent tendency had been noted in geomagnetic activity, and was widely interpreted as the effect of localized, long-lived streams of particles emitted from the Sun and swept past the Earth once during each solar rotation. The high-speed solar wind streams have been linked to recurrent geomagnetic activity and found to be prominent features of the solar wind much of the time since 1962. SEE GEOMAGNETISM; MAGNETOSPHERE.

Interplanetary magnetic sectors. Variations in numerous other solar wind properties have been found to be organized within these streams. For example, the interplanetary magnetic field points predominantly toward or away from the Sun within a stream, normally changing this polarity in the low-speed wind between streams. Indirect, geomagnetic evidence strongly suggests the presence of these interplanetary magnetic sectors for the past five 11-year sunspot cycles. Another important, stream-organized variation involves the density. The solar wind particle density tends to increase near the front of a stream, attain a high, maximum value where the speed is increasing rap-

Ranges and average values of observed solar wind properties

Property	Range*	Average value
Flow speed	180–420 mi s^{-1} (300–700 km s^{-1})	270 mi s^{-1} (450 km s^{-1})
Flow direction	±5° of direction from Sun	Within 2° of direction from Sun
Proton density	3–20 cm^{-3}	8 cm^{-3}
Proton temperature	$0.1–3 \times 10^5$ K	1×10^5 K
Electron temperature	$0.9–2 \times 10^5$ K	1.5×10^5 K
Magnetic field intensity	$2–10 \times 10^{-5}$ gauss ($2–10 \times 10^{-9}$ tesla)	6×10^{-5} gauss (6×10^{-9} tesla)

*Determined by excluding the lowest and highest 5% of observed values.

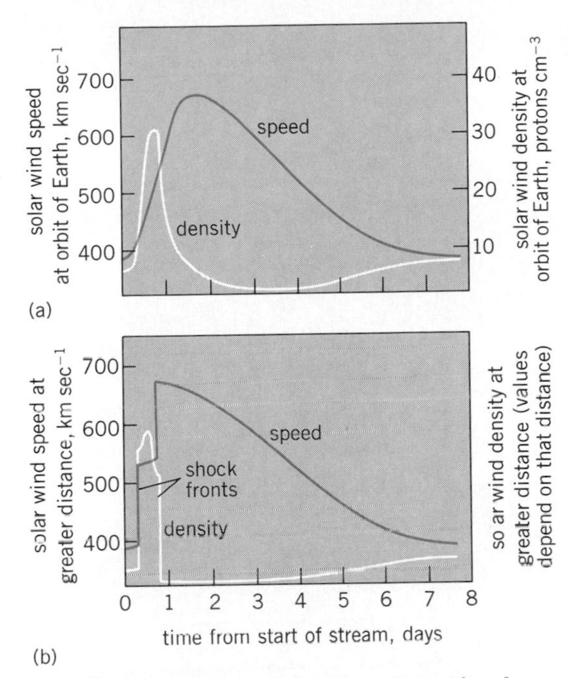

Fig. 1. Pattern of speed and density variations in solar wind streams (*a*) observed near orbit of Earth and (*b*) predicted and later observed farther from the Sun. 1 km = 0.6 mi; 1 cm^3 = 0.06 in.^{3}

idly, and then decrease to abnormally low values as the speed peaks and falls (Fig. 1).

Stream steepening. Straightforward physical arguments suggest that high-speed streams will evolve as they move outward from the Sun through a process known as stream steepening. The fast-moving material at the crest of the stream should overtake the slower-moving material in front of it and outrun the slower-moving material behind it. This will result in a fast-rise, slow-decay profile for the speed variation, and produce a compression of the plasma ahead of the stream crest and a rarefaction in the plasma behind the crest. The typical variations in solar wind speed and density observed near the orbit of Earth (Fig. 1) follow these patterns and are widely interpreted as evidence that the streams have evolved in this fashion between the Sun and 1 AU.

Smoothing process. The rearrangement of material (compression and rarefaction) produced by stream steepening implies an additional effect that resists further steepening and that must ultimately smooth out the speed variations at the root of the process. The compressed plasma between the stream front and crest will be at an elevated pressure, and thus pressure forces will act away from this region on both sides. These forces will accelerate the slow-moving material near the front of the stream and decelerate the fast-moving material near the crest, tending to diminish the speed difference that produced the compression. An important complication arises from the large amplitudes, 60–240 mi s^{-1} (100–400 km s^{-1}), of the speed variations in solar wind streams. As these differences are much larger than the speed of sound in the interplanetary plasma (typically about 24 mi s^{-1} or 40 km s^{-1} near 1 AU), the steepening will continue until very large spatial gradients are produced; dissipative processes at these gradients lead to the formation of shock fronts on both sides of the compressed region (Fig. 1*b*). Strong heating and the acceleration and deceleration of the plasma will then occur mainly at these fronts.

Evolution of high-speed streams. These rather complex steepening and smoothing processes combine to determine the evolution of high-speed streams as they move outward through the solar system. Detailed theoretical models of this evolution, applied to the streams actually observed in the solar wind during the mid-1970s, predicted that the steepening would dominate until shock fronts formed between 1 and 2 AU and that the smoothing out of the streams would then occur very slowly, over a distance scale of about 10 AU. Thus shells of highly compressed plasma, bounded by shock fronts (Fig. 1*b*), should arise in each recurrent solar wind stream, and these shells should be the dominant structure in the interplanetary plasma between about 2 and 10 AU from the Sun. SEE PLASMA PHYSICS.

The flights of *Pioneer* and *Voyager* spacecraft to the outer solar system during the mid and late 1970s provided opportunities to test the general concept and detailed models of stream evolution. The solar wind structure observed as these spacecraft moved slowly outward from the orbit of Earth showed an increasing concentration of plasma and frozen-in magnetic field in the stream associated compressions. By 2–3 AU, many streams showed the predicted pair of shock fronts, and these shocks persisted as the spacecraft moved to the orbit of Jupiter (at 5 AU) and beyond. The solar wind observed near Jupiter was far more inhomogeneous than that near 1 AU, as a result of this expected evolution of high-speed streams. Observations made beyond 10 AU have shown a solar wind in which most of the high-speed streams are "smoothed out." The shock fronts that were formed in the stream steepening process remain, with shocks from different streams having passed through one another to produce a very complex structure characterized by the remaining pressure variations rather than the streams seen near the orbit of Earth. In addition to confirming these predictions and determining the details of the resulting solar wind structure that fall beyond the capabilities of theoretical prediction, these spacecraft have also found that the stream-associated compressions and shock waves are unexpectedly important sources of energetic particles. The very complicated structures produced by the interactions of different streams or shocks seem to play a crucial role in the modulation of cosmic rays entering the solar system. SEE COSMIC RAYS.

Source of streams. A number of solar features were proposed as sources of the particle streams thought to be responsible for recurrent geomagnetic activity. Early solar wind observations established the basic characteristic of the streams, that is, abnormally fast flow speeds. By the early 1970s, it could be reasonably argued that high-speed streams emanated from the central portions of open magnetic structures in the corona, with the associated magnetic sector pattern resulting from the basically unipolar nature of the magnetic field underlying and permeating such a region.

Coronal holes. In the early 1970s x-ray and ultraviolet images of the Sun brought attention to the features known as coronal holes. These are regions of abnormally low coronal density that appear dim in any radiation emitted (x-rays and the ultraviolet) or scattered (the white-light corona observed at eclipse)

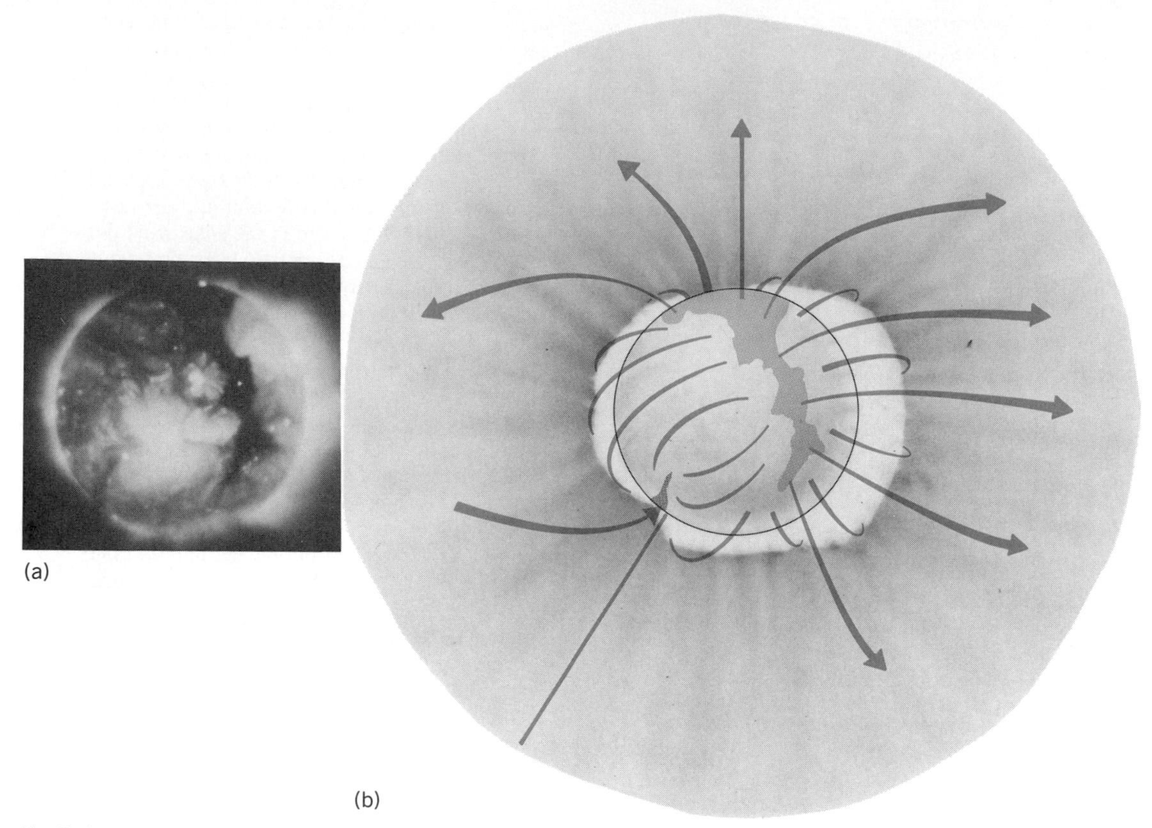

Fig. 2. Coronal holes and solar magnetic geometry. (*a*) X-ray image of the Sun obtained by an instrument on the *Skylab* mission. The dark regions near the north pole of the Sun and extending in a long north-south lane to the right of center are coronal holes (*American Science and Engineering Co.*). (*b*) Magnetic geometry associated with coronal holes. Open magnetic field lines extend from the holes (with a polarity as indicated by the arrows), while the neighboring regions are threaded by closed magnetic field lines.

from the corona. For example, in the x-ray image (**Fig. 2***a*) made at wavelengths of nanometers (the characteristic radiation of the million-degree corona), holes appear as dark patches outlined by emission from neighboring, dense, hot regions. Detailed examination of these images suggests that the holes are regions where the solar magnetic field is open (Fig. 2*b*), with field lines reaching from their roots in the photosphere out into interplanetary space. In contrast, the structure seen in the bright or dense corona suggests closed magnetic fields, with field lines connecting separated locations in the photosphere. Numerous studies, using coronal holes observed by a variety of techniques, have established that the appearance of a hole near the equator of the Sun leads to the appearance in the solar wind observed in the ecliptic plane (near the solar equator) of a high-speed stream, embedded in a magnetic sector with the polarity matching that observed in the photospheric region underlying the hole. The low density of the holes can be attributed directly to their open magnetic character and the resulting escape of coronal material to interplanetary space. In contrast, the closed magnetic fields in other parts of the corona inhibit the escape of plasma and lead to a confined, dense corona.

Polar holes. The largest and longest-lived coronal holes have been observed to occur in the polar regions of the Sun (as defined by its rotation axis); conspicuous polar holes existed for at least 8 years during the past 11-year sunspot cycle. These polar holes are related to the weak, dipolelike, general magnetic field of the Sun and are thus of opposite polarity in the two hemispheres. Near-equatorial holes of a given magnetic polarity tend to occur in the same hemisphere as the polar hole of the same polarity; at some time during their lifetime most near-equatorial holes ''connect'' to the polar hole of the same polarity and thus appear as an equatorward extension of a polar hole. This suggests a rather simple large-scale geometry for magnetically open regions of the corona and their extension into interplanetary space—two regions of opposite magnetic polarity, each arising from a polar hole and its equatorward extensions, separated by a neutral sheet or surface at which the outwardly extended magnetic field changes its sign or polarity. This neutral sheet must separate all holes of opposite polarity.

Sector boundaries. Were the solar magnetic field a simple dipole aligned with the axis of rotation, the northern and southern hemispheres would be expected to contain the field lines of opposite polarity, each arising from a polar hole, and these hemispheres to be separated by a neutral sheet in the equatorial plane. The complexities of the actual solar magnetic field distort this idealized geometry, producing a neutral sheet that is warped out of the equatorial plane (**Fig. 3**). The major warps in the sheet would correspond to the positions of equatorward extensions of the polar holes (and perhaps other near-equatorial holes). The rotation of this spatial structure with the Sun then produces the changing pattern of magnetic polarity seen by a stationary interplanetary observer. Sector boundaries occur when the neutral sheet sweeps past the observer, with the unipolar magnetic sectors between such boundaries corresponding to residence on one side of the neutral sheet. The solar wind near sector boundaries has been found to have numerous special properties including an unusual chemical composition.

Three-dimensional model. The observed pattern of recurrent variations in solar wind speed (that is, high-speed streams) can be simply related to the same magnetic spatial structure if the empirical findings that fast wind comes from near-equatorial holes while slow wind occurs near sector boundaries are assumed to hold in three dimensions. That is, low speeds occur everywhere near the neutral sheet, high speeds everywhere sufficiently far from this three-dimensional surface. Under this assumption, the pattern of alternating fast and slow solar wind flow familiar from in-place observations performed in the ecliptic plane (and hence within approximately 7° of the solar equatorial plane) would then be characteristic only of near-equatorial latitudes. The polar latitudes, sites of the large, long-lived polar holes, would be regions of uniform high-speed solar wind flow.

This view of a large-scale, three-dimensional solar wind structure organized about a simple magnetic geometry implies a significant revision of many earlier ideas about the solar wind. On the purely phenomenological side, the dominance of the polar influence on the three-dimensional structure and hence the pattern of solar wind variations even in the near-equatorial wind is a radical departure from the previous orthodox model. This revision extends to the view of the basic physical characteristics and origins of the wind. The average solar wind properties determined in the late 1960s and used as the touchstone of theoretical models may deviate markedly from the global averages wherein the high-speed flows from the polar regions would be more heavily weighted. The much higher speeds and energies in the high-speed wind have become the most important, and difficult, characteristics of the solar wind, which will require explanations by future models.

Time-dependent phenomena. It is also recognized that some of the observed variations in solar wind properties are true short-term temporal changes produced by solar activity. The most spectacular time-dependent phenomenon identified in the solar wind is the interplanetary shock wave—a cloud of fast-moving plasma, ejected by a large solar flare, that produces a shock front in sweeping up the slower-moving background solar wind in its path. Direct observations of these shock waves have been made over a large range of distances from the Sun, 0.3 to more than 10 AU, and radio observations imply existence of the frontal shocks as close to the Sun as 2 solar radii. Other, less spectacular time-dependent solar wind phenomena, such as clouds with very high-density plasma, unusual chemical composition, and looplike (rather than open) magnetic fields, have been tentatively identified.

Since the mid-1970s, there has been much discussion of the relationship between these interplanetary phenomena and coronal mass ejections—regions of bright material, often with a looplike form, seen to move outward through the solar corona. **Figure 4** shows an example of a coronal mass ejection observed with an instrument on the *Solar Maximum Mission* spacecraft. The sharp, inner bright loop seen in this event is plasma from a solar prominence that erupted from the surface of the Sun an hour before

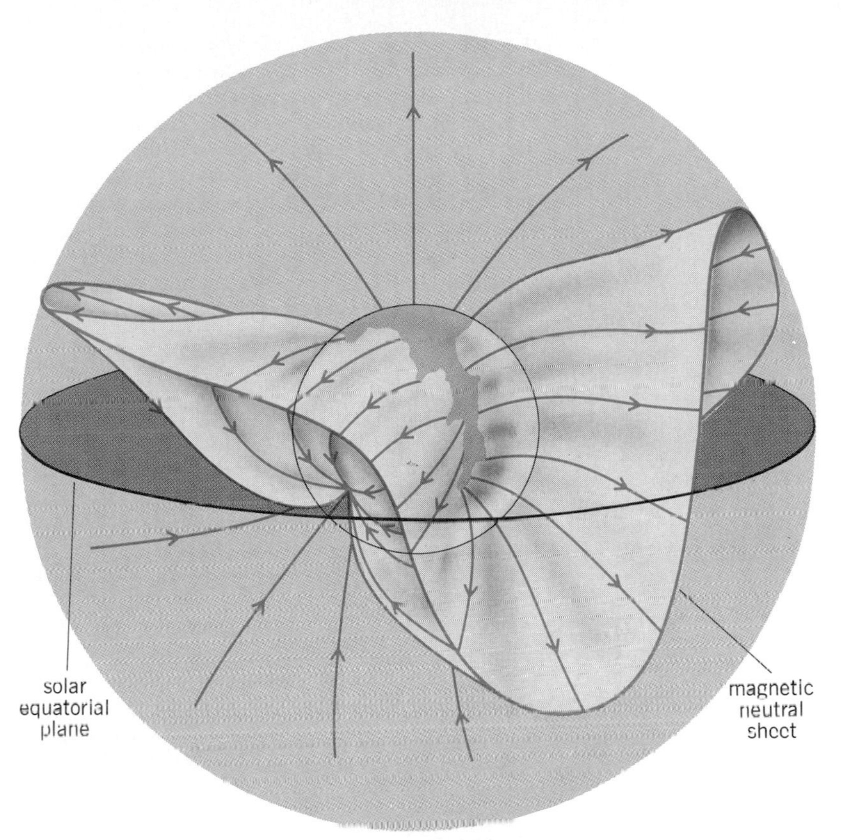

Fig. 3. **Magnetic neutral sheet separating outward-pointing magnetic field lines (positive magnetic polarity) from inward-pointing magnetic field lines (negative magnetic polarity). In this case the positive polarities originate in the north polar hole and its equatorward extensions. The latter warp the neutral sheet from its idealized equatorial location to produce the magnetic polarity or sector structure observed in the near-equatorial solar wind.**

Fig. 4. **Coronal mass ejection observed with a coronagraph on the *Solar Maximum Mission* spacecraft. The dark disk blocks the bright light from the solar photosphere and the inner corona within 1.6 solar radii of the center of the Sun. The two bright looplike features extending above this disk toward the upper left are seen on successive images to move outward at speeds of approximately 120 mi s^{-1} (200 km s^{-1}). (*From A. J. Hundhausen et al., Coronal mass ejections observed during the Solar Maximum Mission: Latitude distribution and rate of occurrence, J. Geophys. Res., 89:2639–2640, 1984*)**

the image in the coronagraph was obtained. The more diffuse, outer loop is thought to be coronal material displaced by the outward motion of the prominence. Detailed examination of coronal mass ejections and interplanetary shock waves shows clear examples of physical relationships; the shock wave is the interplanetary manifestation of some mass ejections. It has also been suggested that some (perhaps the majority of) mass ejections do not produce shock waves and may instead be related to the other, less spectacular time-dependent phenomena seen in the solar wind. SEE SOLAR MAGNETIC FIELD; SUN.

A. J. Hundhausen

Bibliography. S. I. Akasofu and Y. Kamide (eds.), *The Solar Wind and the Earth,* 1987; D. M. Butler and K. Papadopoulos (eds.), *Solar Terrestrial Physics: Present and Future—A Report Based on the Solar Terrestrial Physics Workshop, December, 1982–November 1983*, NASA Ref. Pub. 1120, 1984; A. J. Hundhausen, Solar activity and the solar wind, *Rev. Geophys. Space Phys.*, 17:2034–2048, 1979; R. G. Marsden (ed.), *The Sun and the Heliosphere in Three Dimensions,* 1986; E. J. Smith and J. H. Wolfe, Fields and plasmas in the outer solar system, *Space Sci. Rev.*, 23:217–252, 1979; B. T. Tsurutani and R. G. Stone (eds.), *Collisionless Shock in the Heliosphere: Reviews of Current Research*, 1985; J. B. Zirker (ed.), *Coronal Holes and High-Speed Wind Streams*, 1977.

Solstice

The two days during the year when the Earth is so located in its orbit that the inclination (about 23½°, or 23°45) of the polar axis is toward the Sun. This occurs on June 21, called the summer solstice, when the North Pole is tilted toward the Sun; and on December 22, called the winter solstice, when the South Pole is tilted toward the Sun (see **illus.**). The adjectives summer and winter, used above, refer to the Northern Hemisphere; seasons are reversed in the Southern Hemisphere.

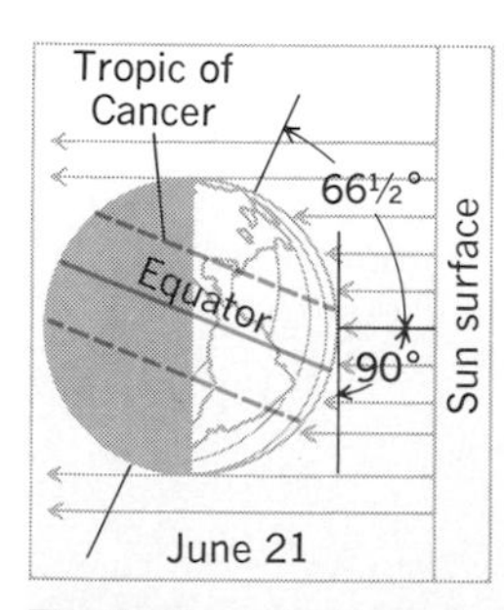

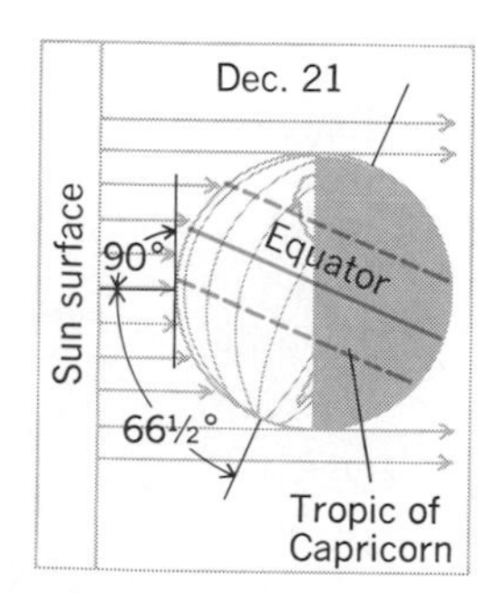

The Earth at the time of the summer and winter solstices. The dates may vary because of the extra one-fourth day in the year.

At the time of the summer solstice the Sun's rays are vertical overhead at the Tropic of Cancer, 23½° north. At the North Pole the Sun will then circle 23½° above the horizon, and at the Arctic Circle, 66½° north, the noon Sun will be 47° above the horizon and the setting Sun will touch the horizon to the north. Thus, on this day every place north of the Arctic Circle will have 24 h of sunlight and the length of day at all places north of the Equator will be more than 12 h, increasing in length with increasing latitude.

Identical conditions are found in the Southern Hemisphere at the time of the Northern Hemisphere's winter solstice when the Sun is vertical above the Tropic of Capricorn, 23½° south, and the South Pole is tilted toward the Sun. SEE MATHEMATICAL GEOGRAPHY.

Van H. English

Speckle

The generation of a random intensity distribution, called a speckle pattern, when light from a highly coherent source, such as a laser, is scattered by a rough surface or inhomogeneous medium. Although the speckle phenomenon has been known since the time of Isaac Newton, the development of the laser is responsible for the present-day interest in speckle. Speckle has proved to be a universal nuisance as far as most laser applications are concerned, and only in the mid-1970s did investigators turn from the unwanted aspects of speckle toward the uses of speckle patterns, in a wide variety of applications.

Basic phenomenon. Objects viewed in coherent light acquire a granular appearance. The detailed irradiance distribution of this granularity appears to have no obvious relationship to the microscopic properties of the illuminated object, but rather it is an irregular pattern that is best described by the methods of probability theory and statistics. Although the mathematical description of the observed granularity is rather complex, the physical origin of the observed speckle pattern is easily described. The surfaces of most materials are extremely rough on the scale of an optical wavelength (approximately 5×10^{-7} m). When nearly monochromatic light is reflected from such a surface, the optical wave resulting at any moderately distant point consists of many coherent wavelets, each arising from a different microscopic element of the surface. Since the distances traveled by these various wavelets may differ by several wavelengths if the surface is truly rough, the interference of the wavelets of various phases results in the granular pattern of intensity called speckle. If a surface is imaged with a perfectly corrected optical system as shown in **Fig. 1**, diffraction causes a spread of the light at an image point, so that the intensity at a given image point results from the coherent addition of contributions from many independent surface areas. As long as the diffraction-limited point-spread function of the imaging system is broad by comparison with the microscopic surface variations, many dephased coherent contributions add at each image point to give a speckle pattern.

The basic random interference phenomenon underlying laser speckle exists for sources other than lasers. For example, it explains radar "clutter," results for scattering of x-rays by liquids, and electron scattering by amorphous carbon films. Speckle theory also explains why twinkling may be observed for stars, but not for planets. SEE TWINKLING STARS.

Applications. The principal applications for speckle patterns fall into two areas: metrology and stellar speckle interferometry.

Metrology. In the metrology area, the most obvious application of speckle is to the measurement of surface roughness. If a speckle pattern is produced by

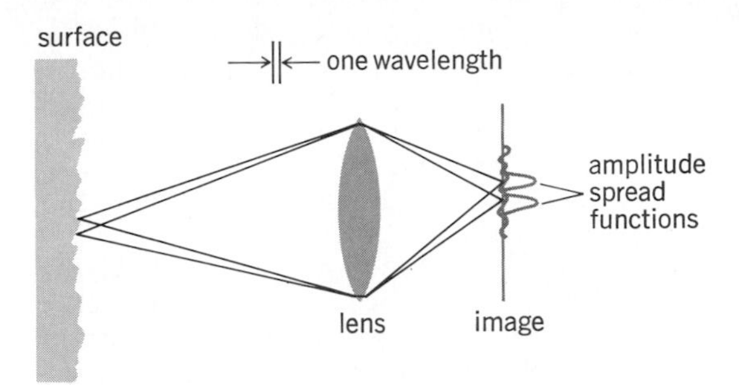

Fig. 1. Physical origin of speckle for an imaging system.

coherent light incident on a rough surface, then surely the speckle pattern, or at least the statistics of the speckle pattern, must depend upon the detailed surface properties. If the surface root-mean-square roughness is small compared to one wavelength of the fully coherent radiation used to illuminate the surface, the roughness can be determined by measuring the speckle contrast. If the root-mean-square roughness is large compared to one wavelength, the radiation should be spatially coherent polychromatic, instead of monochromatic; the roughness is again determined by measuring the speckle contrast.

An application of growing importance in engineering is the use of speckle patterns in the study of object displacements, vibration, and distortion that arise in nondestructive testing of mechanical components. The key advantage of speckle methods in this case is that the speckle size can be adjusted to suit the resolution of the most convenient detector, whether it be film or a television camera, while still retaining information about displacements with an accuracy of a small fraction of a micrometer. Although several techniques for using laser speckle in metrology are available, the basic idea can be explained by the following example for measuring vibration. A vibrating surface illuminated with a laser beam is imaged as shown in Fig. 1. A portion of the laser beam is also superimposed on the image of the vibrating surface. If the object surface is moving in and out with a travel of one-quarter of a wavelength or more, the speckle pattern will become blurred out. However, for areas of the surface that are not vibrating (that is, the nodal regions) the eye will be able to distinguish the fully developed high-contrast speckle pattern.

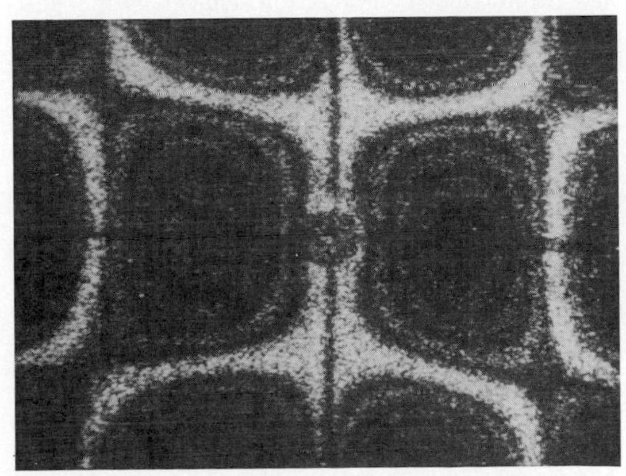

Fig. 2. Electronic speckle interferogram of vibrating object.

Electronic speckle interferometry (ESPI) is finding much use in industrial applications involving measurement of the vibrational properties of structures. In ESPI a television camera is used to view a speckle pattern. The output of the television camera is processed electronically to emphasize the high spatial frequency structure. The result is viewed on a television monitor (**Fig. 2**). The bright fringes correspond to the region of the object which are at the nodes of the vibration pattern. By knowing the angles at which the object is illuminated and viewed, it is possible to calculate the amplitude of the vibration from the number of dark fringes present. Thus, simply by looking at the television monitor it is possible to determine the vibrational pattern over the entire object. The amplitude and frequency of the drive source can be changed and instantly the new vibrational pattern can be observed.

Speckle patterns can also be used to examine the state of refraction of the eye. If a diffusing surface illuminated by laser light moves perpendicular to the line of sight of an observer, the speckles may appear to move with respect to the surface. For a normal eye, movement in a direction opposite to that of the surface indicates underaccommodation, and movement with the surface indicates overaccommodation. If the speckles do not move, but just seem to "boil," the observed surface is imaged on the retina.

Stellar speckle interferometry. Stellar speckle interferometry, which has many similarities with the laser speckle methods used in metrology, is a technique for obtaining diffraction-limited resolution of stellar objects despite the presence of the turbulent atmosphere which limits the resolution of conventional pictures to approximately 1 arc-second. If a short-exposure photograph is taken of a magnified image of an unresolved star and a narrow-bandwidth spectral filter is used, the picture has a specklelike structure. The size of the speckles is equal to the diffraction-limited resolution limit of the telescope, regardless of the resolution limit determined by the turbulent atmosphere. This means that the short-exposure photograph of a resolvable object, for example a binary star, contains information about the object down to the diffraction limit of the telescope, which is approximately 0.02 arc-second for the 200-in. (5-m) Palomar Mountain telescope, one-fiftieth the resolution limit set by the atmosphere. Hence, by extracting correctly the information in short-exposure pictures of objects with more than one resolvable element, detail down to the diffraction limit of the telescope can be observed. The technique has proved to be of enormous value in the study of binary stars and centrosymmetric resolvable stars, and research is directed at making the technique useful for observing objects having a more general shape. Image intensifiers and modern electronics have made it possible to look at much dimmer astronomical objects. The use of solid-state detector arrays and computers utilizing special processors such as array processors makes it possible to eliminate much of the optical analog processing and to do complete digital processing and obtain higher accuracy in the measurements. SEE ASTRONOMICAL PHOTOGRAPHY; OPTICAL TELESCOPE.

James C. Wyant

Bibliography. J. C. Dainty (ed.), *Laser Speckle and Related Phenomena*, 1975; R. K. Erf (ed.), *Speckle Metrology*, 1978; M. Francon, *Laser Speckle and Applications in Optics*, 1979; G. A. Slettemoen, Electronic speckle pattern interferometric system based on a speckle reference beam, *Appl. Opt.*, 19:616–623, 1980.

Spectral type

A label used to indicate the physical and chemical characteristics of a star, as indicated by study of the star's spectrum. Stars possess a remarkable variety of spectra, some simple, others extraordinarily complex. To understand the natures of the stars, it was first necessary to bring order to the subject and to classify the spectra.

Standard sequence. Based in part on earlier work by A. Secchi, who had developed a system that correlated with stellar color, W. C. Pickering initiated the modern system starting about 1890. He ordered the spectra by letter, A through O, largely on the basis of the strengths of the hydrogen lines. He and Mrs. W. P. Fleming then found that several of the letters were unneeded or redundant. On the basis of continuity of lines other than hydrogen, A. Maury and A. J. Cannon found that B preceded A and O preceded B. The result is the classical spectral sequence, OBAFGKM (**Fig. 1**). Cannon also decimalized the classes, setting up the sequence O5, . . ., O9, B0, . . ., B9, A0, and so forth. (Not all the numbers are used, however.) The modern standard sequence, called the Harvard sequence after the observatory where it was formulated, runs from O3 to M8. Over 350,000 stars, classified by Cannon, were included in the final result, the *Henry Draper Catalog*. SEE ASTRONOMICAL CATALOGS.

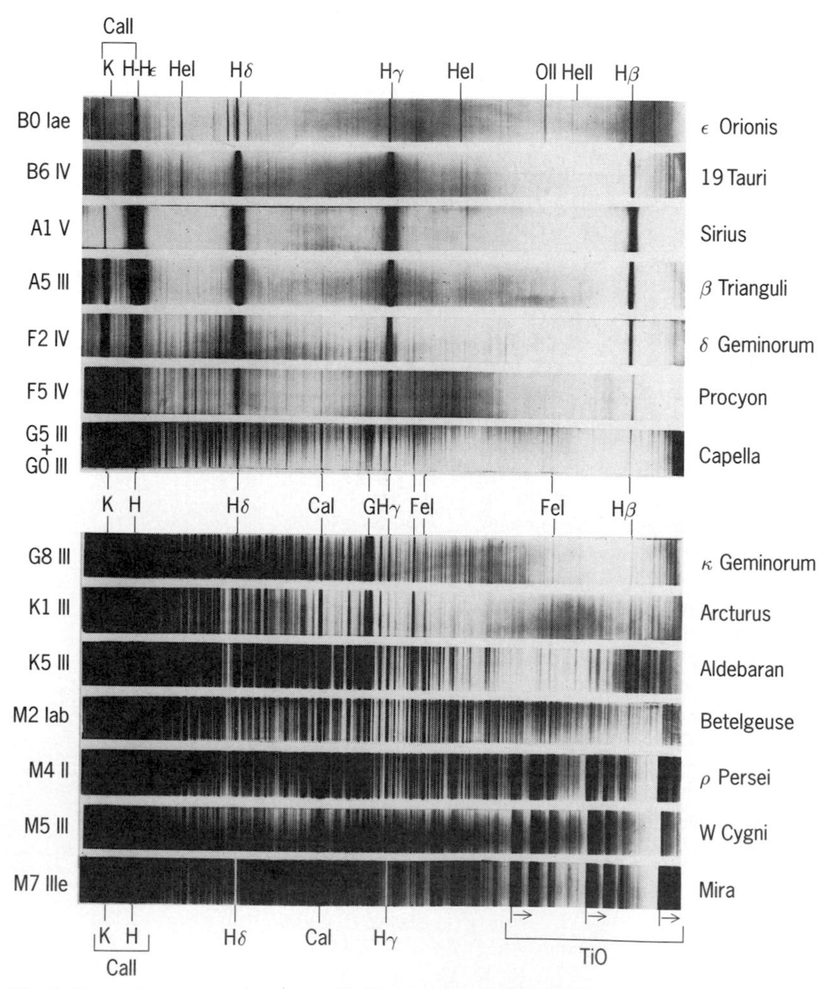

Fig. 1. Spectral sequence from B to M. (*University of Michigan*)

Class A has the strongest hydrogen lines, B is characterized principally by neutral helium (with weaker hydrogen), and O by ionized helium. Hydrogen weakens notably through F and G, but the metal lines, particularly those of ionized calcium, strengthen. In K, hydrogen becomes quite weak, while the neutral metals grow stronger. The M stars effectively exhibit no hydrogen lines at all but instead are dominated by molecules, particularly titanium oxide (TiO). At G, the sequence branches downward into R and N, whose stars are rich in carbon molecules. Class S, in which the titantium oxide molecular bands of class M are replaced by zirconium oxide (ZrO), was introduced in 1923.

Spectral variation with temperature. At first appearance, the different spectral types seem to reflect differences in stellar composition. However, within the standard sequence, OBAFGKM, the elemental abundances are roughly similar. The dramatic variations seen in Fig. 1 are strictly the result of changes in temperature (**Fig. 2**). The strengths of an ion's absorption lines depend on the number of electrons capable of absorbing at those wavelengths, which in turn depends on the product of the atomic absorption probability, the number of electrons excited to the appropriate level of the ion, and the number of ions. In class M, the temperature is too low to allow a significant number of electrons in the second level of hydrogen, from which the optical Balmer series arises. As a result, there are no hydrogen lines. As temperature climbs and collisional activity increases, more electrons are knocked into level 2 and the Balmer lines strengthen from K through A. Above class A, however, the temperature becomes so high that the hydrogen becomes ionized. That ionization and changes in the opacity of the stellar atmospheres cause the hydrogen lines to weaken.

The other ions are subject to the same kind of variation. The second level of neutral helium lies at a very high energy. As a result, the helium lines do not become visible until class B. It also takes a great deal of energy to ionize helium, so the ionized helium lines are not observed until nearly class O. Ionized metal lines are produced at intermediate temperatures and neutral metal lines at low temperatures. Molecules, which are relatively fragile, cannot exist unless the temperature is relatively low.

Spectral variation with composition. The different spectra of the R, N, and S stars, however, are caused by true and dramatic variations in the chemical composition. The M stars have more oxygen than carbon. In the N stars, which have similar temperatures, the ratio is reversed and carbon dominates oxygen. The R stars are carbon-rich versions of classes G and K. In the cool S stars, carbon and oxygen have about the same abundance. That condition and the increased abundance of zirconium (which has a greater affinity for oxygen than titanium) yield strong zirconium oxide lines. These composition variations are the result of internal thermonuclear processing and convection. SEE STELLAR EVOLUTION.

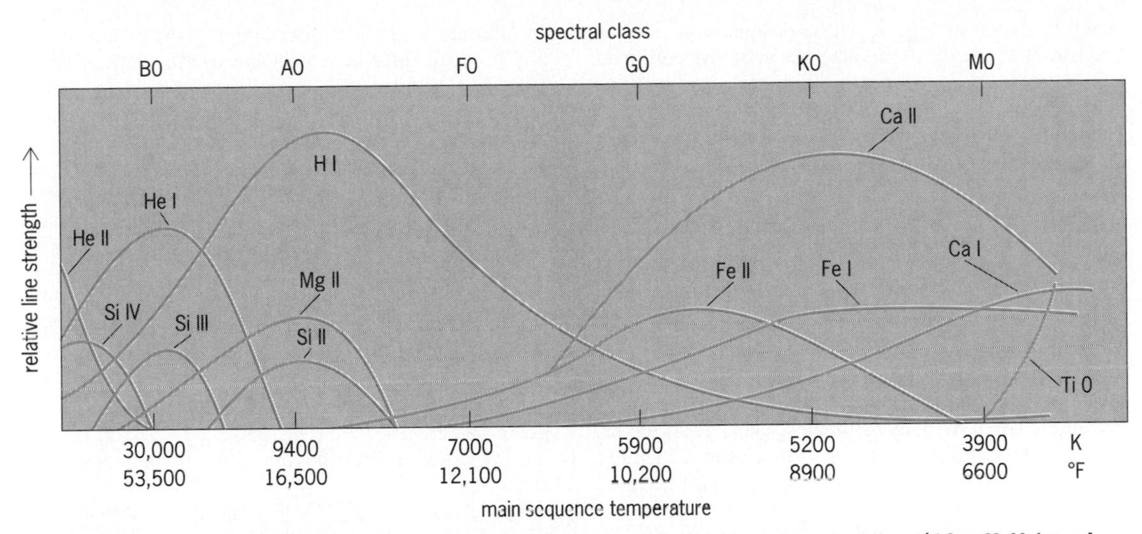

Fig. 2. Variation in line strength for different ions as a function of temperature and spectral class. *(After H. Voigt, ed., Landolt-Börnstein Numerical Data and Functional Relationships in Science and Technology, Group VI, vol. 1: Astronomy and Astrophysics, Springer-Verlag, 1965)*

Morgan-Keenan-Kellman system. In the 1940s, W. W. Morgan, P. C. Keenan, and E. Kellman expanded the Harvard sequence to include luminosity. A system of roman numerals is appended to the Harvard class to indicate position on the Hertzsprung-Russell diagram: I for supergiant, II for bright giant, III for giant, IV for subgiant, and V for dwarf or main sequence. The Morgan-Keenan-Kellman class, which includes the refined Harvard type, is defined by a set of standard stars in the MKK atlas, to which a program star must be compared. On this system, the Sun is a G2 V star. The R, N, and S stars are all giants. Additional refinements have been added to indicate a variety of spectral peculiarities. SEE ASTRONOMICAL SPECTROSCOPY; HERTZSPRUNG-RUSSELL DIAGRAM; STAR.

James B. Kaler

Bibliography. C. Jaschek and M. Jaschek, *The Classification of Stars*, 1990; J. B. Kaler, *Stars and Their Spectra: An Introduction to the Spectral Sequence*, 1989; W. W. Morgan, P. C. Keenan, and E. Kellman, *An Atlas of Stellar Spectra*, 1943; K. Aa. Strand (ed.), *Basic Astronomical Data*, 1963; Y. Yamashita, K. Nariai, and Y. Norimoto, *An Atlas of Representative Spectra*, 1978.

Spectrograph

An optical instrument that consists of an entrance slit, collimator, disperser, camera, and detector and that produces and records a spectrum. A spectrograph is used to extract a variety of information about the conditions that exist where light originates and along its paths. It does this by revealing the details that are stored in the light's spectral distribution, whether this light is from a source in the laboratory or a quasistellar object a billion light-years away.

Proper spectrograph design takes into account the type of light sources to be measured, and the circumstances under which these measurements will be made. Since observational astronomy presents unusual problems in these areas, the design of astronomical spectrographs may also be unique.

Astronomical versus laboratory design. Astronomical spectrographs have the same general features as laboratory spectrographs (**Fig. 1**). The width of the entrance slit influences both spectral resolution and the amount of light entering the spectrograph, two of the most important variables in spectroscopy. The collimator makes this light parallel so that the disperser (a grating or prism) may properly disperse it. The camera then focuses the dispersed spectrum onto a detector, which records it for further study.

Laboratory spectrographs usually function properly only in a fixed orientation under controlled environmental conditions. By contrast, most astronomical spectrographs are used on a moving telescope operating at local temperature. Thus, their structures must be mechanically and optically insensitive to orientation and temperature.

The brightness, spectral characteristics, and geometry of laboratory sources may be tailored to experimental requirements and to the capabilities of a spectrograph. Astronomical sources, in the form of images at the focus of a telescope, cannot be manipulated, and their faintness and spectral diversity make unusual and difficult demands on spectrograph performance.

Typical laboratory spectrographs use either concave gratings, which effectively combine the functions of collimator, grating, and camera in one optical element; or plane reflection gratings with spherical reflectors for collimators and cameras. An example of

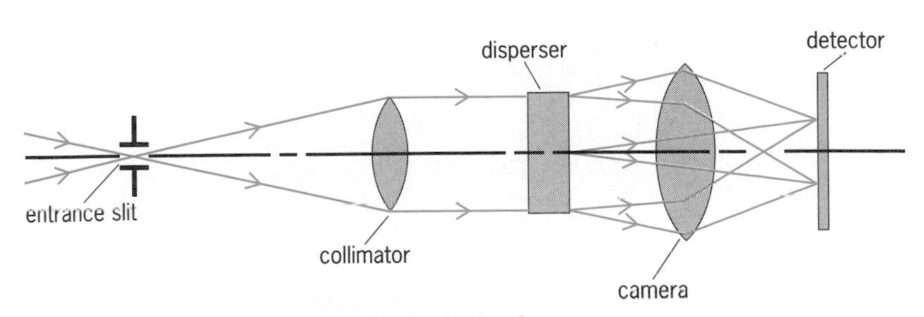

Fig. 1. Basic optical components of a spectrograph.

each is shown in **Fig. 2**. These designs may severely constrain the optical characteristics of the collimator and camera, and sometimes the grating, to obtain acceptable imaging at the detector. As a result, they are generally too limited in the choice of important spectrograph parameters for use in astronomy. In particular, they tend to preclude the use of very short cameras, an essential feature of spectrographs intended for study of faint astronomical objects. The Czerny-Turner design (Fig. 2*b*) is used in some solar spectrographs, where it offers some advantages and ample light is available. *SEE SPECTROHELIOSCOPE; SUN.*

To provide maximum flexibility in its application, the collimator, gratings, and cameras of an astronomical spectrograph must each be relatively free of aberrations when used in collimated light. Then the optical subassemblies can be interchanged without disturbing the spectrograph's optical performance. Grating interchange changes spectral resolution and wavelength coverage with minimal effect on efficiency. Camera and detector interchange achieves operating characteristics that may be crucial for certain applications. Examples of typical astronomical spectrographs are shown in **Fig. 3**.

Details of astronomical spectrographs. Astronomical spectrographs typically use a pair of adjustable polished slit jaws, which are tilted to the telescope beam so that unused light is reflected to guide optics for accurate image positioning. The collimator may be either an off-axis paraboloid or a two-mirror Cassegrain-type optical system. It has a relatively long focal length and an *f*/ratio matching that of the telescope's for optimum slit transmission.

The disperser is usually a plane reflection grating. In laboratory applications, larger gratings are generally used for their increased spectral resolution. However, in astronomy the resolution of the grating (number of grooves times interference order) is seldom realized, and larger gratings are used to obtain longer collimator focal lengths in order to preserve or increase slit demagnification at the camera's focus.

The spectral resolution of many laboratory spectrographs can be increased (grating permitting) only by decreasing the slit width. However, this decrease reduces available light and taxes the imaging capabilities of the optical system and the resolution capabilities of the detector. A better approach, usually used in astronomical spectrographs, is to keep the slit width within a relatively small range of values, for a given camera focal length, and vary spectral resolution by using gratings having different angular dispersions.

The cameras used in astronomy are often quite short (fast) to allow the entrance slit to be as wide as possible for maximum light transmission and to maximize spectral coverage for a given detector size. The camera is often a variation of the classical Schmidt telescope, using a combination of mirrors and ultraviolet transmitting optics. Some spectrographs may have a selection of cameras to permit higher resolutions (with narrower slit openings) when there is sufficient light, or to improve imaging in some wavelength ranges. In particular, very large spectrographs, used at the coudé or Nasmyth foci, may have an array of camera focal lengths (Fig. 3*b*). When this arrangement is combined with interchangeable gratings, it permits a large selection of spectral resolutions and wavelength coverages. *SEE SCHMIDT CAMERA.*

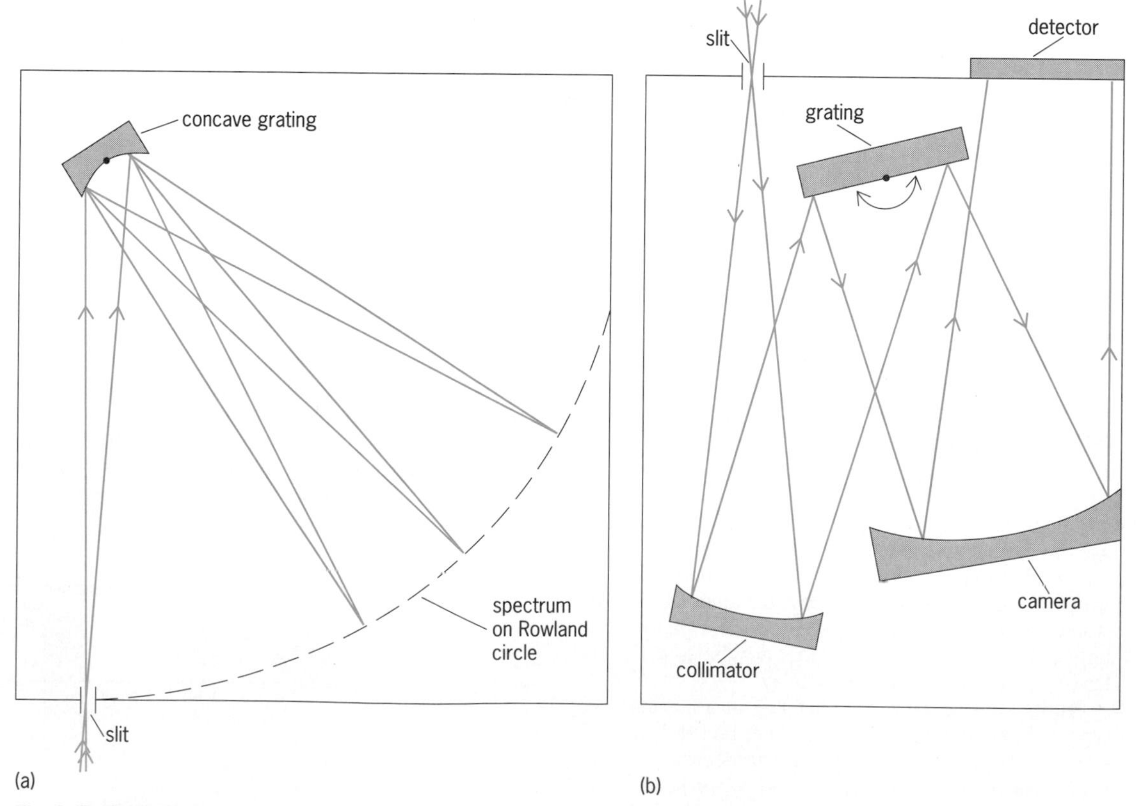

Fig. 2. Typical laboratory spectrographs. (*a*) Classical concave grating spectrograph. (*b*) Czerny-Turner plane grating spectrograph.

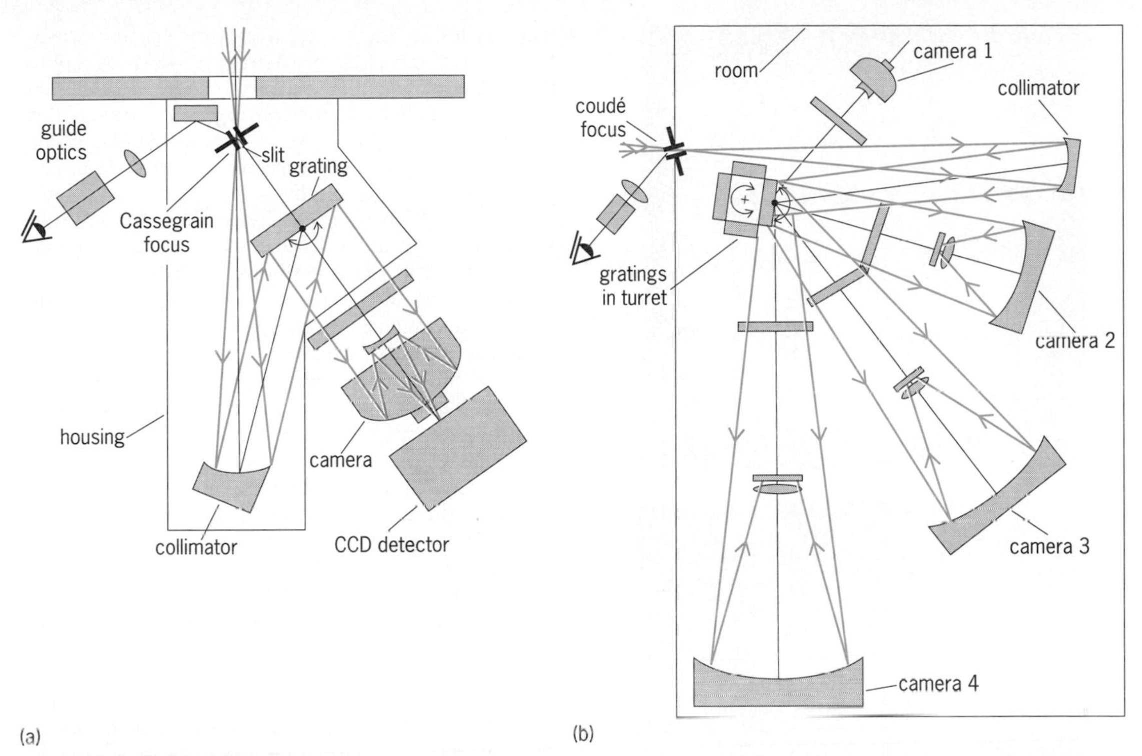

Fig. 3. Typical astronomical spectrographs. (*a*) Spectrograph for use at the Cassegrain focus (often called a Cassegrain spectrograph) using a parabolic collimator and semisolid Schmidt camera. (*b*) Plan view of a coudé spectrograph with four Schmidt cameras and four interchangeable gratings. Camera 1 is the shortest camera and camera 4 is the longest. The spectrograph is shown using camera 2.

In the past, the detector in astronomical spectroscopy was almost always the photographic emulsion. However, image intensifiers, and more recently, the very efficient and accurate cooled charge-coupled devices (CCDs) have become the preferred detectors. SEE CHARGE-COUPLED DEVICES; IMAGE TUBE.

Observing sites and telescopes. Spectrographs are particularly appropriate for use at poor observing sites because, although the atmosphere might reduce available light, it has little effect on its spectrum. Light pollution is also less of a problem, since dispersing the light both dilutes the intensity of the unwanted background and identifies the spectrum of this background so that it can be subtracted later. SEE LIGHT POLLUTION.

Astronomical spectrographs can be used effectively on smaller telescopes. The reason is that star images are typically larger than the entrance-slit width, and the amount of light passing through the slit is thus proportional to the diameter, not area, of a telescope's aperture. Thus, the magnitude limit of a 50-in. (1.25-m) telescope will be only about 1 stellar magnitude fainter than that of a far less costly 20-in. (0.5-m) telescope using the same spectrograph. SEE ASTRONOMICAL SPECTROSCOPY; OPTICAL TELESCOPE.

Ron Hilliard

Bibliography. A. Beer (ed.), *Vistas of Astronomy*, 1956; W. A. Hiltner (ed.), *Astronomical Techniques*, 2d ed., 1962; J. F. James and R. S. Sternberg, *The Design of Optical Spectrometers*, 1969; C. R. Kitchin, *Astrophysical Techniques*, 2d ed., 1991; J. Meaburn, *Detection and Spectrometry of Faint Light*, 1976; R. A. Sawyer, *Experimental Spectroscopy*, 1963.

Spectroheliograph

An instrument for the monochromatic visual observation of the Sun. A telescope projects an image of the Sun on the first slit of a powerful spectroscope (**Fig. 1**). The resulting spectrum is imaged in the plane of a second slit which permits only a single line element of the spectrum to emerge from the instrument. The emergent line element is a monochromatic image of that part of the Sun that falls on the first slit.

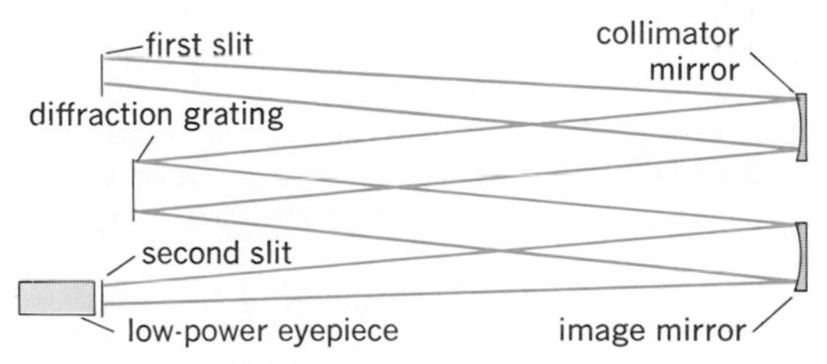

Fig. 1. Hale spectrohelioscope.

The widths of the slits are generally chosen to isolate a spectral interval 0.05 nanometer (0.5 angstrom) or less in width. When the two slits are vibrated synchronously at high frequency, persistence of vision permits monochromatic observation of an area of the solar surface. The slits may also be moved at a slow rate and the image recorded by photographic means

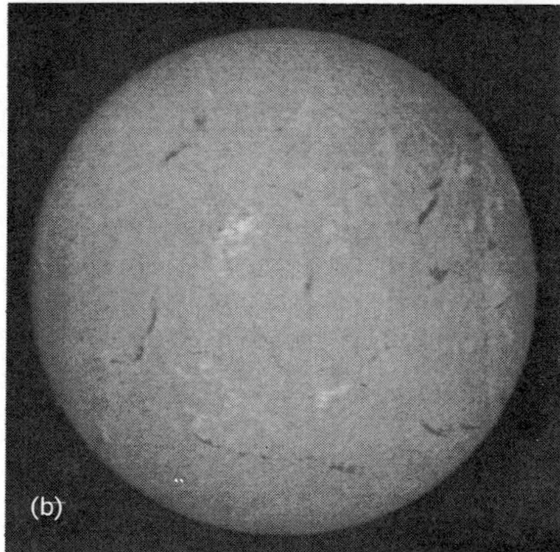

Fig. 2. Spectroheliograms of the Sun (1958 June $20^{d}12^{h}10^{m}$ UT) formed of line elements 0.03 nm wide. (*a*) At 660.0 nm. (*b*) At 656.3 nm (Hα).

(**Fig. 2**). This modification of the spectrohelioscope is a simple form of the spectroheliograph.

Robert R. McMath/John W. Evans

SS 433

A remarkable stellar object with unique properties: it shows evidence of ejection of two narrow streams of cool gas traveling in oppositely directed beams from a central object at a velocity of almost one-quarter the speed of light—the beams executing a repeating, rotating pattern about the central object once every 164 days.

Discovery. SS 433 is an example of an astronomical object discovered, forgotten, and rediscovered several times over a period of two decades. The initial observation of the object was made in 1959 by C. B. Stephenson and N. Sanduleak during a survey of the Milky Way for peculiar stars. The Stephenson-Sanduleak (SS) published survey lists this star as the 433d entry, thus the nomenclature SS 433. The SS survey searched for objects with a specific anomaly in their spectrum—namely, bright emission lines. Ordinarily, when starlight is dispersed via a prism or grating into its component colors, the resulting spectrum is relatively smooth, with comparable amounts of light reaching Earth at every wavelength (color). However, occasionally stars are found which have intense peaks in their spectra at very specific wavelengths. When seen on a spectrum recorded on a photographic plate, this phenomenon is called an emission line, because the convergence of light at this particular wavelength causes a linear thickening of the developed emulsion at a specific location. Emission lines in stars are caused by light emitted from atoms in the star which have become excited due to collisions with other atoms or with electromagnetic radiation (light). The relaxation from this excitation is normally accompanied by emission of light at certain specific, unchanging wavelengths characteristic of the chemical element involved. Thus the detection of emission lines is a valuable insight into the nature of the star, since they provide a ''fingerprint'' enabling identification of chemical elements present in the object. Perhaps 5–10% of all stars show emission lines in their spectra, so this characteristic of SS 433 was not in itself sufficiently unusual to provoke more detailed observations of the object. *See Astronomical spectroscopy.*

During the 1960s and 1970s a variety of astronomers mapping the skies at x-ray and radio wavelengths unknowingly rediscovered SS 433, as they found that intense x-ray and radio emissions were emanating from this region of sky. However, the precision of these observations was insufficient to permit these workers to associate the source of the emissions specifically with SS 433, as opposed to numerous other nearby stars (**Fig. 1**). Finally in 1978 three independent groups of English and Canadian astronomers recognized that the visible object SS 433 and the previously cataloged sources of radio and x-ray emission were in fact all the same object. This is extraordinary, as only the tiniest fraction of all stars emit detectable amounts of either radio or x radiation. The first modern spectroscopy of SS 433 confirmed the results of the SS catalog by showing the object to have extraordinarily intense emission lines of hydrogen and helium. These workers also pointed out that SS 433 is surrounded in the sky by a large, diffuse glow of radio emission, itself a previously cataloged object, termed W50. The structure of W50 has led most astronomers to conclude that it is the remnant of an ancient exploded star, or supernova, in this case probably occurring more than 100,000 years ago. The central location of SS 433 within W50 leads to the speculation, appealing but unproved, that the two objects are in fact associated. *See Radio astronomy; Supernova; X-ray astronomy.*

Jets of matter. The most peculiar characteristics of SS 433 have been revealed by an intensive series of spectroscopic observations. These observations show that the spectrum possesses not only a set of emission lines due to hydrogen and helium, but two further sets of lines, one displaced to longer (redder) wavelengths from the familiar lines, and the second displaced to shorter (bluer) wavelengths. These displacements can be understood in terms of the Doppler effect, a familiar mechanism which lengthens the apparent wavelength of any wave phenomenon (including light or sound) when there is a recessional motion between the source and the observer, and shortens the wavelength if there is approach. Thus the observations imply that in addition to a stationary object, SS 433 possesses some gas (a mixture of hydrogen and helium) approaching the Earth, while some presumably different patch of gas recedes. The remarkable property is the velocity of approach and recession, calculated simply from the magnitude of the observed spectral Doppler shifts. The velocity of this gas in the initial observations was found to be up to 30,000 mi/s (50,000 km/s), that is, about 16% of the speed of light. Because the escape velocity from the Milky Way Galaxy is only a few hundred kilometers per second, one never observes stellar objects with velocities in excess of this, since they would rapidly leave the Galaxy. *See Doppler effect.*

Further spectral monitoring of SS 433 has revealed spectacular changes in these two sets of Doppler-shifted lines. The wavelengths of the lines change every night in a smoothly progressing pattern, indicating that the velocity of the emitting regions is also changing. Each set of lines proves to cycle in a regular pattern between a recessional velocity of 30,000 mi/s (50,000 km/s) and an approach velocity of 20,000 mi/s (30,000 km/s), with the cycle lasting approximately 164 days (**Fig. 2**). The pattern then begins again. The currently accepted interpretation of this periodic behavior is that the "moving" emission lines are due to light from two narrow streams or jets of matter ejected from a central object in opposite directions. A slow rotation of the axis of these jets, once every 164 days, is then responsible for the changing velocities observed at the Earth. This rotation is probably caused by precession, a wobbling motion of the star. Different velocities are seen on different days because the moving axis of the jets may be more or less directly pointed toward Earth at a given time. Interpretation of the observations shown in Fig. 2, using this concept, shows that the true velocity of the ejected beam remains constant throughout the 164-day cycle at a value of about 50,000 mi/s (80,000 km/s), about one-quarter of the velocity of light. A velocity this high is never directly observed, since the Earth would have to be fortuitously located exactly in the conical surface of the rotation pattern to have the jets point exactly toward and away from it. The tremendous beam velocity inferred implies a huge energy source to accelerate a substantial amount of gas to this speed: the kinetic energy in the beams is approximately a million times as large as the total amount of light energy radiated by the Sun.

Time dilation. Especially intriguing is the observation (Fig. 2) that on a given night the average velocity of the approaching and receding beams is not zero, but rather a large positive value, about 7500 mi/s (12,000 km/s), despite the fact that SS 433 is approximately stationary with respect to the Earth. This proves to be a direct consequence of Einstein's special theory of relativity. An outside observer perceives a change in measured times and lengths of a system moving at very large velocity. The beam velocity in SS 433 is large enough that special relativity is important, and this "time dilation" effect causes a permanent redshift of the spectral lines of 7500 mi/s (12,000 km/s); the 164-day rotational component is then superposed on top of this underlying effect.

Radio maps. Elegant confirmation of these indirect inferences concerning the nature of SS 433 has been provided by detailed maps of the radio structure of the central object. These observations, which utilized the Very Large Array, an ensemble of 27 radio astronomy antennas in New Mexico, indicate that this radio structure also changes cyclically every 164 days, as radio-emitting blobs of gas coast outward after ejection from the jets. The radio maps directly reveal the spirallike trail created by the jet rotation, thus

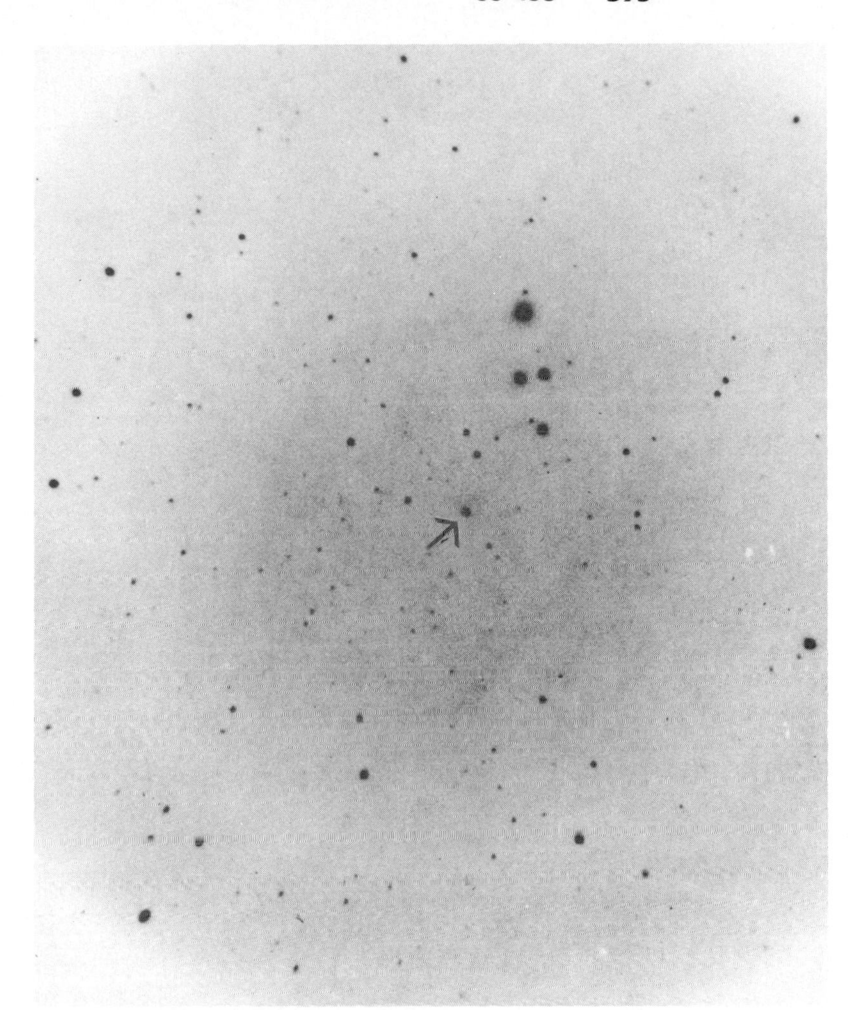

Fig. 1. Photograph of SS 433 made at Lick Observatory, using the 36 in. (0.9 m) refractor telescope. The obvious lack of any features which distinguish the object from its numerous neighbors in this dense portion of the Milky Way helps explain why the star was overlooked for so long.

confirming the concept of SS 433 developed from the optical spectroscopy. The radio observations also provide clues enabling an estimate of the distance from the Earth to SS 433 as 15,000 light-years (9×10^{16} mi or 1.4×10^{17} km), about one-sixth the diameter of the Milky Way Galaxy. *See Radio telescope.*

Star type and matter source. There has been much speculation as to the type of star present in SS 433, with many astronomers now agreeing that the enormous velocities in the beams require a highly collapsed, compact star with a strong gravitational field. A neutron star, the same end point of stellar evolution responsible for pulsars, or possibly a black hole could satisfy this requirement. The source of matter ejected through the beams is also a problem; it seems probable that this may be supplied by a relatively normal, nearby companion star trapped in an orbit about SS 433. The emission lines in SS 433 have also been shown to cycle through a very small-amplitude period every 13 days, implying that the mutual orbital motion of SS 433 and this unseen companion may have been detected. *See Binary star; Black hole; Neutron star; Pulsar.*

Perhaps the most vexing question of all is why there is only one object like SS 433 known in a galaxy of 10^{11} stars. Only the accidental discovery of a

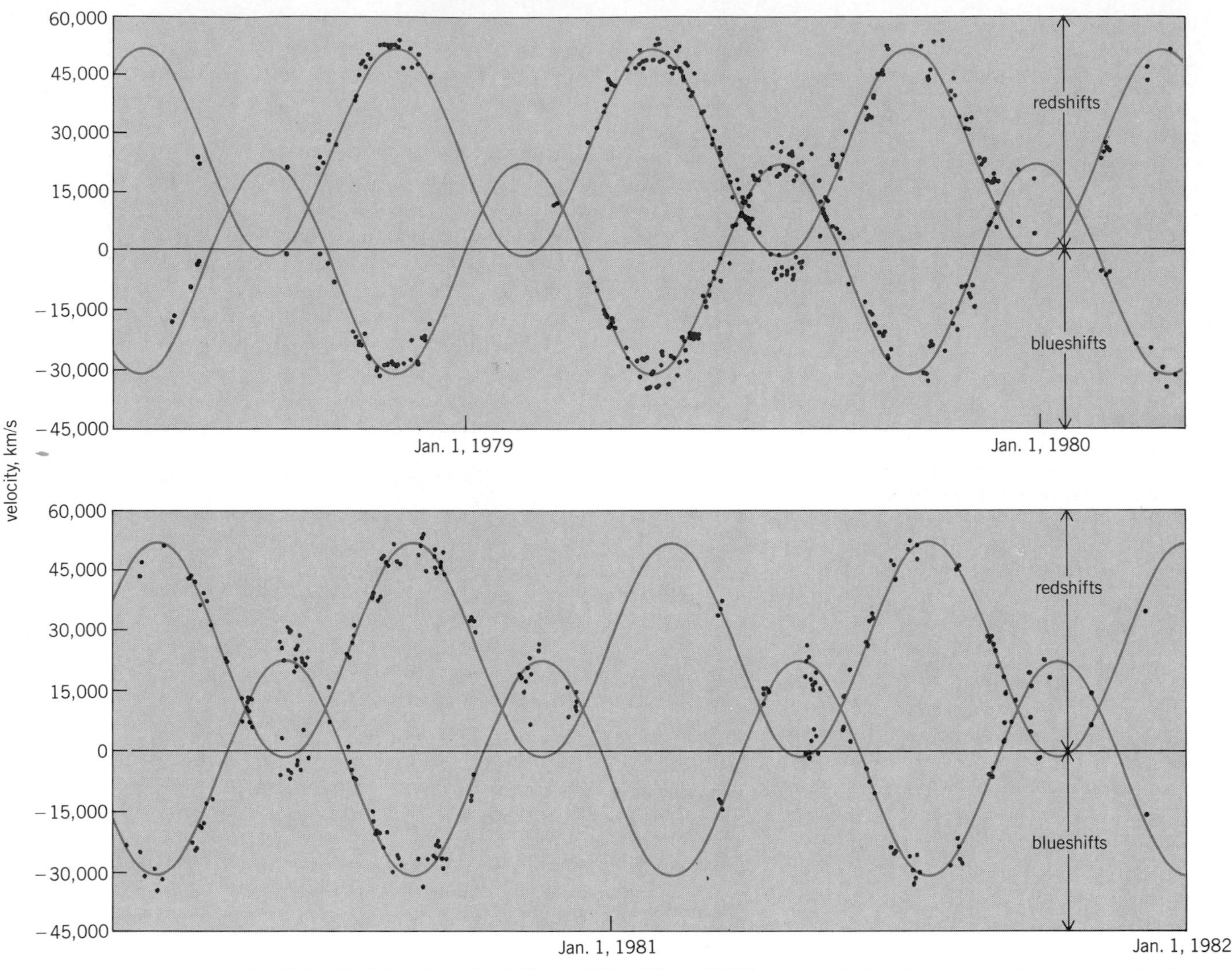

Fig. 2. Graph of the value of redshifts and blueshifts of SS 433 over a period of about 3 years. Large gaps in the data are caused when the object is too close to the Sun for nighttime observation; smaller gaps are due to the proximity of the Moon, making observations difficult. Curves show the predicted behavior of the Doppler-shifted gas is located in two narrow, oppositely directed beams which rotate every 164 days. 1 km/s = 0.6 mi/s.

second such star can determine whether this is a possible end point of stellar evolution for a certain class of binary star, or whether SS 433 is truly a unique and obscure stellar accident. *See Astrophysics, high-energy.*

Bruce Margon

Bibliography. B. Margon, The bizarre spectrum of SS 433, *Sci. Amer.*, 243(4):54–65, October 1980; B. Margon, Relativistic jets in SS 433, *Science*, 215:247–252, 1982; D. Overbye, Does anyone understand SS 433?, *Sky Telesc.*, 58:510–516, 1979.

Star

A celestial body, consisting of a large, self-luminous mass of hot gas held together by its own gravity. The Sun is a typical star; its physical parameters are

$$\text{Radius } R_\odot = 6.9 \times 10^{10}\text{ cm} = 6.9 \times 10^{8}\text{ m} = 4.3 \times 10^{5}\text{ mi}$$
$$\text{Mass } M_\odot = 2 \times 10^{33}\text{ g} = 2 \times 10^{30}\text{ kg} = 4.4 \times 10^{30}\text{ lb}$$
$$\text{Luminosity } L_\odot = 4 \times 10^{33}\text{ ergs/s} = 4 \times 10^{26}\text{ W}$$

The Sun's mean density is 1.45 g/cm^3 (1.45 times that of water), and its central density is about 120 g/cm^3 (120 times that of water). The surface temperature is 5750 K (9900°F), the mean temperature about 5×10^6 K (9×10^6 °F), and the central temperature 1.3×10^7 K (2.3×10^7 °F). In spite of the high density, the gas in the Sun is almost completely ionized from surface to center.

Composition and distribution. The composition by weight of the average star is about 70% hydrogen, 28% helium, 1.5% carbon, nitrogen, oxygen, and neon, and 0.5% iron group and heavier elements.

The stars contain by far the largest known fraction of the mass of the universe. The parameters given above describe their average condition. Stars are born, produce nuclear energy, evolve, and eventually die. Their life-spans range from 10^6 years, for a star of high luminosity, to 10^{10} for the Sun, and up to 10^{13} years for the faintest main-sequence stars. The oldest known in the Milky Way Galaxy are over 10^{10} years old. *See Stellar evolution.*

The nomenclature for the identification of stars gives their general location and brightness. The sky was subdivided into constellations and the brighter stars were named; Greek alphabet letters generally describe them, with α representing the brightest star, visually, in a given constellation. Thus Betelgeuse, or α Orionis, is the visually brightest star in Orion. Fundamentally a star is defined by its coordinates on the celestial sphere, right ascension and declination, and its brightness, or apparent magnitude. Because of the precession of the equinoxes, celestial coordinates must be specified for a given epoch. About 6000 stars are visible to the naked eye, but over 10^{12} exist in the Milky Way Galaxy. Catalogs give positions, brightnesses, motions, parallaxes, spectral types, velocities, and other properties of the stars. Many of these catalogs are now available in magnetic tape format. Several hundred thousand stellar positions are also available, with much of the data on tape suitable for computer search. Some fainter stars are tabulated by the name of the discoverer of any interesting property and are listed by many unrelated and overlapping names. *SEE ASTRONOMICAL CATALOGS; CELESTIAL SPHERE; CONSTELLATION; MAGNITUDE; PRECESSION OF EQUINOXES.*

Near stars. The nearest stars have been so identified by their large angular proper motions in the plane of the sky and by subsequent measurements of parallax. The 26 nearest stars are listed in **Table 1** together with relevant data. A few more, intrinsically faint stars may exist at or within this limit of distance. Table 1 also gives the transverse motions, derived from the proper motion and distance of the star, and the radial velocity, which is from the measured Doppler shift. Many stars have a total space motion of from 60 to 300 mi/s (100 to 500 km/s) with respect to the Sun. Such objects are called high-velocity stars and belong, according to the nomenclature of Walter Baade, to population II, that is, stars found in the spheroidal halo of a galaxy. The more slowly moving stars are younger, and may be members of population I like the Sun; such stars are found in the spiral regions or flattened disk of a galaxy. The absolute visual magnitudes show that 36 of the 40 nearby stars are intrinsically fainter than the Sun. In addition, about half are in multiple systems, doubles or triples. Most of the stars in Table 1 have spectral types which put them on the main sequence or dwarf branch. No red giant or supergiant is included, although there are three white dwarfs in the immediate neighborhood of the Sun. **Figure 1**, a Hertzsprung-Russell (H-R) diagram, is a means for plotting the relation between the luminosity and surface temperature or other temperature-dependent parameter, such as color or spectral type of the stars. Along the main sequence, through which the line is drawn in the diagram, the stars are distinguished chiefly by mass; luminosity is a steep function of the mass. Nearby stars, for which data are most accurate, do not include all interesting types. To include a greater variety of stars, the less accurate data for stars out to 20 parsecs (3.8×10^{14} mi, 6.2×10^{14} m, or 65 light-years) are plotted. The relation between absolute visual magnitude and observed photoelectric color for stars closer than 20 parsecs shows the main sequence in more detail for the somewhat brighter stars and also the red giants and subgiants (**Fig. 2**). The use of photoelectric color, like that of spectral type, is essentially an arrangement by surface temperature and is convenient and accurate. *SEE GIANT STAR; HERTZSPRUNG-RUSSELL DIA-*

Table 1. The 26 nearest stars[a]

Name	Parallax, seconds of arc	Distance, light-years[b]	Annual proper motion, seconds of arc	Radial velocity, km/s (mi/s)	Transverse velocity, km/s (mi/s)	Apparent magnitude and spectrum	Absolute magnitude
Sun						−26.7 G2	+4.8
α Centauri[c]	0.760	4.3	3.68	−25 (−10)	23 (14)	+0.3 G2(1.7 K5)	4.7(6.1)[d]
Barnard's star	.545	6.0	10.30	−108 (−67)	90 (56)	9.5 M5	13.2
Wolf 359	.421	7.7	4.84	+13 (+8)	54 (34)	13.5 M6e, v[e]	16.6
Luyten 726–8	.410	7.9	3.35	+29 (+18)	38 (24)	12.5 M6e (13.0 M6e)v	15.6(16.1)
Lalande 21185	.398	8.2	4.78	−86 (−53)	57 (35)	7.5 M2	10.5
Sirius[f]	.375	8.7	1.32	−8 (−5)	16 (10)	−1.5 A0(8.7 DA)	1.4(11.6)
Ross 154	.351	9.3	0.67	−4 (−2)	9 (6)	10.6 M5e	13.3
Ross 248	.316	10.3	1.58	−81 (−50)	23 (14)	12.2 M6e	14.7
ϵ Eri	.303	10.8	0.97	+15 (+9)	15 (9)	3.8 K2	6.2
Ross 128	.298	10.9	1.40	−13 (−8)	22 (14)	11.1 M5	13.5
61 Cyg	.293	11.1	5.22	−64 (−40)	84 (52)	5.6 K6(6.3 M0)	7.9(8.6)
Luyten 789–6	.292	11.2	3.27	−60 (−37)	53 (33)	12.2 M6	14.5
Procyon	.288	11.3	1.25	−3 (−2)	20 (12)	0.5 F5(10.8 DA?)	2.8(13.1)
ϵ Indi	.285	11.4	4.67	−40 (−25)	77 (48)	4.7 K5	7.0
Σ 2398	.280	11.6	2.29	+1 (+1)	38 (24)	8.9 M4(9.7 M4)	11.1(11.9)
Groombr. 34	.278	11.7	2.91	+14 (+9)	49 (30)	8.1 M2e (10.9 M4e), v	10.3(13.1)
τ Cet	.275	11.8	1.92	−16 (−10)	33 (21)	3.6 G4	5.8
Lacaille 9352	.273	11.9	6.87	+10 (+6)	118 (73)	7.2 M2	9.4
+5° 1668	.263	12.4	3.73	+26 (+16)	67 (42)	9.8 M4	12.0
Lacaille 8760	.255	12.8	3.46	+23 (+14)	64 (40)	6.6 M1	8.6
Kapetyn's star	.251	13.0	8.79	+242 (+150)	166 (103)	9.2 M0	11.2
Ross 614	.251	13.1	0.97	+24 (+15)	18 (11)	11.1 M5e(14.8)	13.1(16.8)
Kruger 60	.249	13.1	0.87	−24 (−15)	16 (10)	9.9 M4 (11.4 M5e)	11.9(13.4)
−12° 4523	.244	13.4	1.24	−13 (−8)	24 (15)	10.0 M5	11.9
vMa 2[g]	.236	13.8	2.98	+70 (+43)	59 (37)	12.3 DG	14.2

[a]From P. van de Kamp. In 26 listings there are 36 individual stars, and some have still-undiscovered faint companions.
[b]1 light-year = 5.9×10^{12} mi = 9.5×10^{12} km.
[c]α Centauri has a second companion (dM5e) of absolute magnitude +15.4.
[d]Parentheses indicate uncertain value.
[e]Here v = variable flare star which may emit bursts of light and even radio noise.
[f]Sirius and Procyon each have a white dwarf companion.
[g]v Ma 2 is a white dwarf.

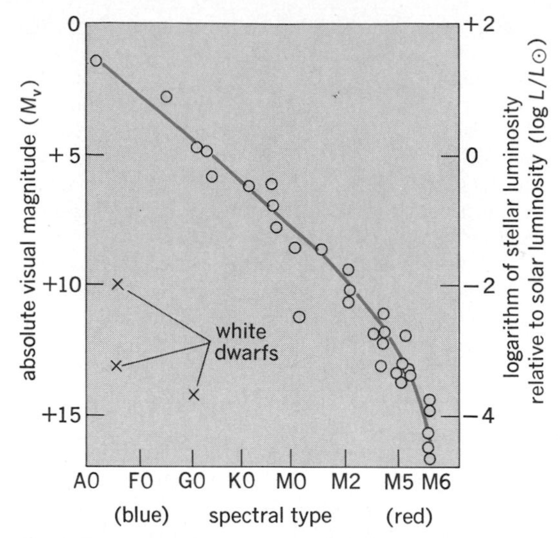

Fig. 1. Hertzsprung-Russell diagram for nearby stars showing main sequence and white dwarfs.

GRAM; PARALLAX; PARSEC; SUPERGIANT STAR; WHITE DWARF STAR.

Bright stars. The apparently brighter stars are listed in **Table 2**. Because stars of high intrinsic luminosity can be seen at great distances, Table 2 includes many such stars, including giants and supergiants. For many of these the parallax is too small to be measured directly and the luminosities are only approximate. Similarly, physically close visual doubles would be missed. The H-R diagram for the brighter stars (**Fig. 3**) illustrates the existence of the other branches in addition to the main-sequence and red-

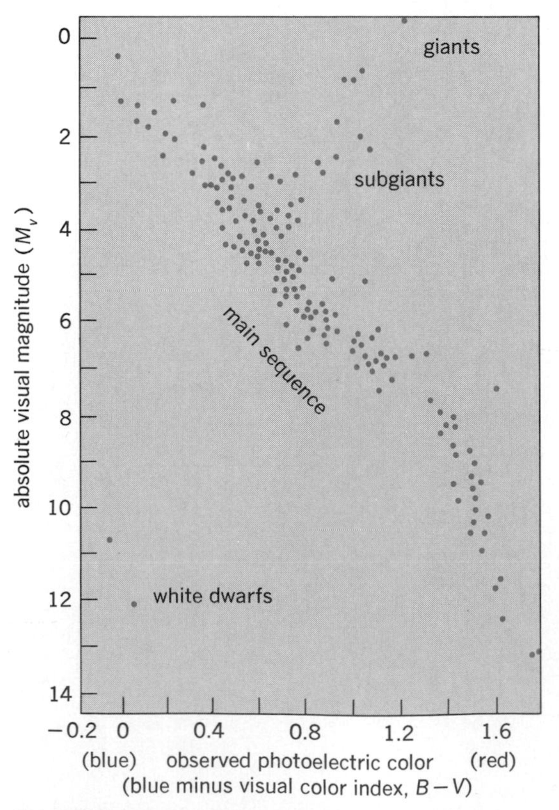

Fig. 2. Main-sequence stars, subgiants, and giants.

giant branch; white dwarfs are missing. A high frequency of binaries is evident in Table 2.

Stellar spectra. The spectrum of a star in a large majority of cases shows absorption lines superposed on a continuous background. The interior of the star is at high temperature and pressure; its spectrum is nearly that of a blackbody. The star shades off into space through a reversing layer or stellar atmosphere in which the continuous spectrum and absorption lines are formed. In the Sun this reversing layer is about 150 mi (250 km) thick; its base is about 5750 K (9900°F), and its outer layer is almost 4200 K (7100°F). The continuum is formed at a depth equal to the mean free path for an average quantum. The emergent continuum resembles, but is not identical with, a blackbody. The atoms in the reversing layer produce absorption lines because of the existence of this temperature gradient. Outside the normal reversing layer in many stars there may be a temperature inversion in the low-density chromosphere and corona. The temperature of the corona eventually reaches 1,500,000 K (2,700,000°F). The ultraviolet rays and x-rays from these outer layers affect only slightly the spectrum of the integrated light of a star, and carry about one-millionth of its total energy. The x-ray radiation of the Sun is highly variable. Some stars have been found by their strong (and variable) x-ray emission. *SEE SUN; X-RAY ASTRONOMY; X-RAY STAR.*

Until the 1970s, stellar spectra were obtained with a slit spectrograph by black and white photography. Astronomical spectrographs largely employ plane gratings, operate in the wavelength region from 300 to 700 mm, and provide dispersions ranging from 1 to about 400 A/mm (0.1 to about 40 nm/mm). Wavelength standards are provided by a comparison spectrum impressed during the exposure of a plate, usually by a laboratory source imaged at both ends of the spectrograph slit. Standards for photometry of the lines are provided by plate calibration devices. Because a point source, like a star, gives only a narrow streak of spectrum with the stigmatic spectrographs used, the spectra are suitably widened by allowing the star image to trail up and down the slit. Low-resolution spectra and some high-resolution spectra of bright stars and of the Sun can be obtained by photoelectric scanning at the telescope.

The development of two-dimensional, solid-state detectors produced a revolution in astronomical spectral measurements. The earliest devices focused and accelerated electrons from a thin photocathode, as in a television camera tube, to intensify the original image and project it onto a phosphor or a film. But the advantages of photon-counting on an array of diode detectors were soon realized; linear arrays of diodes are suited to the format of stellar spectra. Each diode is read out through its own solid-state amplifier, giving the numbers of photons detected at each wavelength. The advantages are even greater when faint or extended objects are observed with a two-dimensional array of diodes. This gives the fluxes at each position on the spectrum (say, the x direction) and along the slit (the y direction). Since the slit is larger than the object, the average brightness of the blank sky at each wavelength is measured and can be subtracted from the brightness of the object plus sky. The spectrum of various parts of a faint extended object can then be obtained. These diode arrays are small, about 1 in. (20–30 mm) on a side, but the diode size is about 20 micrometers. The choice detector is the charge-

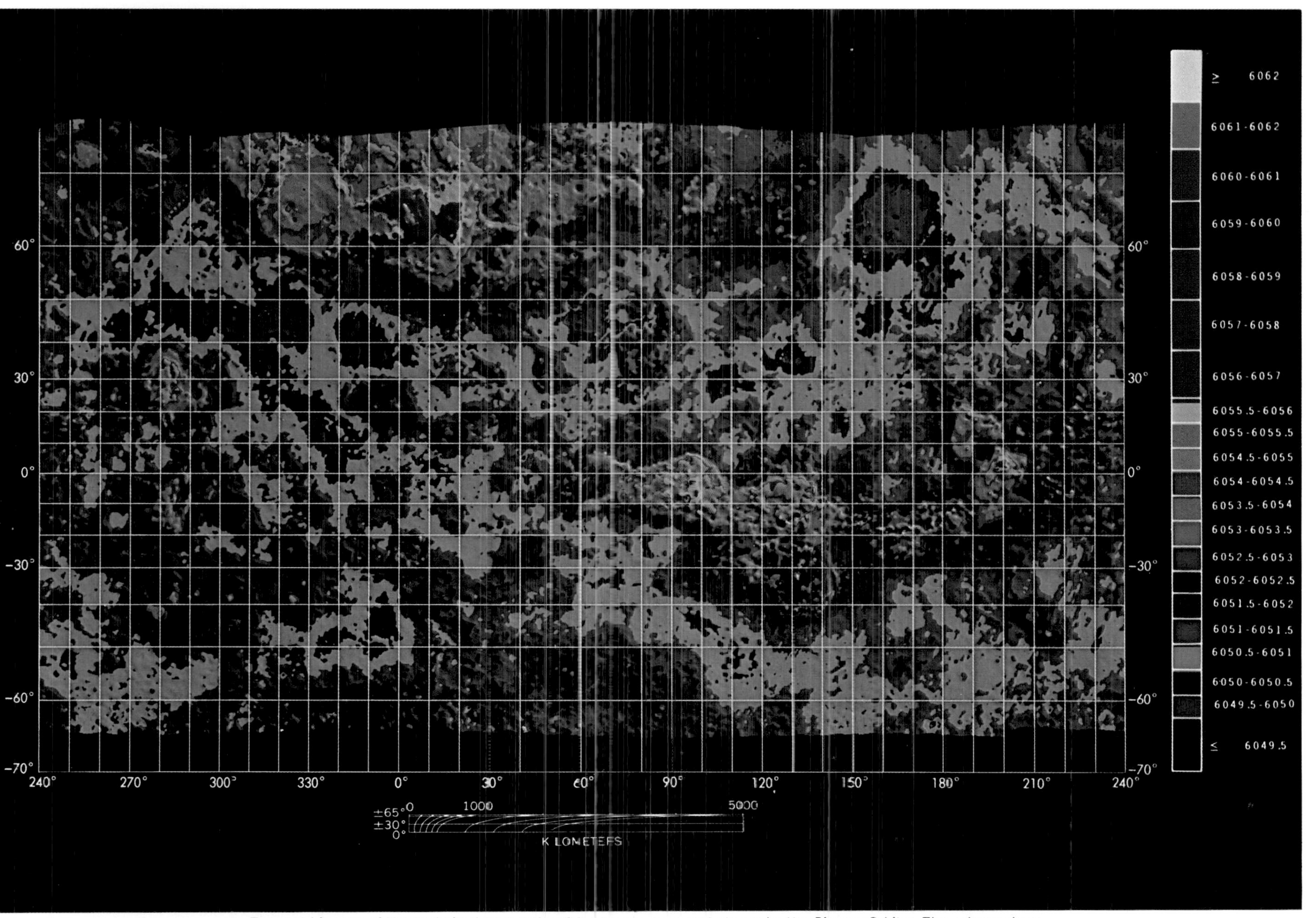

Topographic map of Venus derived from radar altimeter measurements taken by the *Pioneer Orbiter*. The color scale on the right shows the altitude range from the center of Venus in kilometers. High places are represented toward the red end of the spectrum, and low places toward the blue end. (*Courtesy of Dr. M. Masursky, U.S. Geological Survey*)

Table 2. The 25 brightest stars*

Star	Name	Spectrum	Absolute visual magnitude, M_v	Visual brightness, V	Color index, B–V	Remarks
α CMa	Sirius	A1 V	+1.4	−1.43	0.00	
α Car	Canopus	F0 Ia	−4.5	−0.73	+0.15	
α Cen		G2 V	+4.7	−0.27	+0.66	Double
α Boo	Arcturus	K2 IIIp	−0.1	−0.06	+1.23	
α Lyr	Vega	A0 V	+0.5	+0.04	0.00	
α Aur	Capella	G0 IIIp	−0.6	+0.09	+0.80	Spectroscopic binary, double
β Ori	Rigel	B8 Ia	−7	+0.15	−0.04	Double
α CMi	Procyon	F5 IV-V	+2.7	+0.37	+0.41	
α Eri	Achernar	B3 V	−2	+0.53	−0.16	
β Cen		B0.5 V	−4	+0.66	−0.21	Double
α Ori	Betelgeuse	M2 Iab	−5	+0.7	+1.87	Variable
α Aql	Altair	A7 IV-V	+2.2	+0.80	+0.22	
α Tau	Aldebaran	K5 III	−0.7	+0.85	+1.52	Variable, double
α Cru		B0.5 V	−4	+0.87	−0.24	Double
α Sco	Antares	M1 Ib	−4	+0.98	+1.00	Double, variable
α Vir	Spica	B1 V	−3	+1.00	−0.23	Spectroscopic binary
α PsA	Fomalhaut	A3 V	+1.9	+1.16	+0.09	
β Gem	Pollux	K0 III	+1.0	+1.16	+1.01	
α Cyg	Deneb	A2 Ia	−7	+1.26	+0.09	
β Cru		B0.5 IV	−4	+1.31	−0.23	
α Leo	Regulus	B7 V	−0.7	+1.36	−0.11	Double
ϵ CMa	Adhara	B2 II	−5	+1.49	−0.17	
α Gem	Castor	A0	+0.9	+1.59	+0.05	Double, spectroscopic binary
λ Sco	Shaula	B2 IV	−3	+1.62	−0.23	
γ Ori	Bellatrix	B2 III	−4	+1.64	0.23	

*The spectra are from H. L. Johnson and W. W. Morgan; colors and magnitudes are photoelectric, V being the equivalent of a visual brightness and B–V a blue minus visual color index. The absolute visual magnitudes M_V are based on measured parallaxes; when only one significant figure is given, however, they are only estimates.

coupled-diode (CCD) array. They are available with up to 800 × 800 pixels, have high quantum efficiency from the violet to the near infrared, and have a large dynamic range. By exposing them to a uniform source, the pixel-by-pixel sensitivity variation may be corrected for. While much smaller than photographic plates, they are much more nearly linear, and more sensitive. The image exists in the form of countable charges in each pixel; these are read out by a two-dimensional, charge-shifting scheme onto a large disk, and eventually onto magnetic tape, using a fairly large computer at the telescope. With about 5×10^5 picture elements, the sky-subtraction and the image processing requires even larger computers at the astronomer's home laboratory. The effective gain in speed is by about a factor of 5 to 20, depending on project and wavelength region; the greatest advantage is for faint objects which would otherwise be inaccessible. *See* CHARGE-COUPLED DEVICES; IMAGE TUBE.

Another technological gain has opened the ultraviolet region from 900 to 3000 angstroms (90 to 300 nanometers) to observation from orbiting telescopes. The image is recorded electronically and transmitted to ground stations for processing. Relatively bright stars and the interstellar gas were studied with the Copernicus satellite (now extinct). Much fainter stars and nebulae are studied with the *International Ultraviolet Explorer* (*IUE*), launched into geosynchronous orbit in 1978. Its detector is a silicon-intensified vidicon that is used for both high- and low-resolution stellar spectroscopy. These were preludes to the Hubble Space Telescope, launched into orbit in 1990. *See* SATELLITE ASTRONOMY; ULTRAVIOLET ASTRONOMY.

The infrared is best observed with cryogenic solid-state bolometers. Great advances in sensor technology have permitted moderate-resolution spectrometry of many cooler stars, down to 1500 K (2200°F) temperatures. Even cooler stars have been detected at low spectral resolution. *See* ASTRONOMICAL SPECTROSCOPY; INFRARED ASTRONOMY.

Spectral classification. The spectral classification of a star gives in a simple symbolic form the essential features of its complex spectrum. By inspection, at dispersions ranging from 100 to 400 A/mm (10 to 40 nm/mm), many features are found to vary in a smooth way from one star to another. This variation is correlated with the colors of the stars. As a result, spectral classification includes a vast majority of the stars and represents a sequence of decreasing reversing-layer temperature, from the left to the right (**Fig. 4**). There are several side sequences whose temperatures

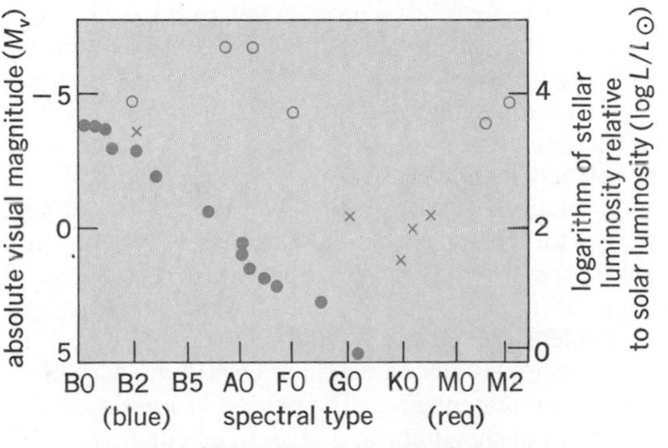

Fig. 3. Diagram of brightest stars showing main sequence (solid circles) as in Fig. 1 and location of other types of stars: giants (crosses) and supergiants (open circles).

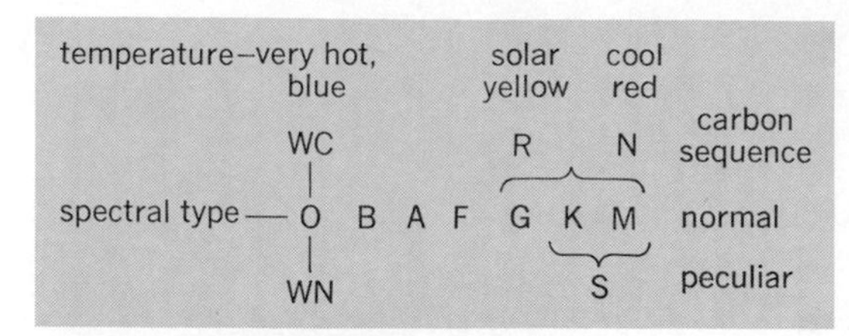

Fig. 4. Spectral classification from Draper catalog.

approximately correspond with those of normal stars, in which apparent abundance differences exist. Both dwarf and giant stars exist over a wide range of spectral type. Decimal subdivisions are used; in more refined analyses prefixes and suffixes are added, such as e for emission lines, n for broad lines because of rotation, q for novalike, p for peculiar, d for dwarf, D for white dwarf, sg for subgiant, g for giant, and c for supergiant. Almost 300,000 stars have been classified on this system.

Peculiar spectra exist, such as the carbon stars (old name R and N, new type C) and S stars, which have about the same temperature as K or M stars. Another very important type of peculiarity has been found associated with stars with magnetic fields. *SEE STELLAR MAGNETIC FIELD*.

The percentage of stars of different spectral types in the Henry Draper catalog is roughly as follows: B, 3%; A, 27%; F, 10%; G, 16%; K, 37%; and M, 7%; other types are rare. This is a selection of stars by apparent brightness and does not represent their true distribution in space, which from the data in Table 1 heavily favors the late-type M dwarfs.

A more refined system of spectral classification developed by W. W. Morgan, with P. C. Keenan and other collaborators, has been widely used. In this system, employing spectra at 120 A/mm (12 nm/mm), a two-parameter set of criteria provides by inspection both spectral type and estimated luminosity. The luminosity is indicated by a suffix ranging from Ia, extremely bright supergiant, to III, normal giant, to V, main-sequence or dwarf star. Thus the designation G2 V represents a star like the Sun, G2 III a giant of nearly the same temperature, with certain luminosity-sensitive features enhanced. This two-dimensional classification is being carried out for all the Henry Draper catalog stars of the southern hemisphere. Quantitative applications of such classifications are numerous. They give distances to stars, groups of stars, or features in the structure of the Milky Way Galaxy. The fact that a finite number of spectral lines or bands indicate temperature, while others indicate luminosity, opens the possibility of quantitative, objective classification systems. Some use photoelectric cells, measuring through filters a number of narrow wavelength bands. Others use photoelectric scans of the spectrum. Special-purpose multicolor photometry, for example, on the Strömgren system, surpasses spectroscopy in precision for certain types of stars.

Temperature and luminosity. Most differences in appearance of stellar spectra are caused by changes in the surface temperature. The degree of ionization and excitation of the atoms at a given temperature dictates whether an element will have an appreciable concentration of atoms in the lower atomic level that produces the absorption line. For example, at very high temperature, helium (He) is all He^{2+} and thus has no lines; at about 35,000 K (63,000°F), He^{+} dominates, and below 20,000 K (36,000°F), He exists. Below about 12,000 K (21,000°F) insufficient atoms of He are excited to the states at 19 eV which produce lines in the normal spectral region; no helium lines will be seen (except in the normally inaccessible ultraviolet) in cooler stars. Stars with He^{+} lines exist and are of type O; He lines occur in type B stars. Stellar temperatures can be accurately determined by quantitative application of these methods, involving the Saha ionization and the Boltzmann excitation equations (**Fig. 5**).

Absolute magnitude effects occur because the ionization equation depends on pressure, the electron concentration setting the recombination rate. Consequently, stars of low surface gravity, which have lower pressures, will show a given percentage of ionization at a temperature about 500 K (900°F) lower than those of the main sequence. A red giant of the same temperature as the Sun, 100 times brighter, has 10 times the radius and about 0.04 times the surface gravity (allowing for the larger mass of the giant). The lower gravity and pressure result in an increased level of ionization of sensitive elements. Thus the luminosity classification of a star, a second parameter, is possible after a temperature classification has been made. These effects are calibrated by stars of otherwise known luminosity.

The luminosity of a star is its energy output, either in ergs per second, or in watts, or in units of solar luminosity, or in absolute magnitude. First, apparent magnitude and distance in parsecs or parallax must be known. Let a star have luminosity L, radius R, and effective temperature T. Then from Stefan's law and the areas of the star Eq. (1) is obtained. If the temperature corresponding to a given spectral type or color is known, one of a variety of types of H-R diagrams can be plotted, connecting luminosity, or absolute magnitude (or apparent magnitudes if all stars of a group are located at the same distance) as ordinates and temperature, or spectral type or color as abscissas. In a diagram in which log L and log T are

$$\frac{L_{\text{star}}}{L_{\odot}} = \left(\frac{R_{\text{star}}}{R_{\odot}}\right)^2 \left(\frac{T_{\text{star}}}{T_{\odot}}\right)^4 \qquad (1)$$

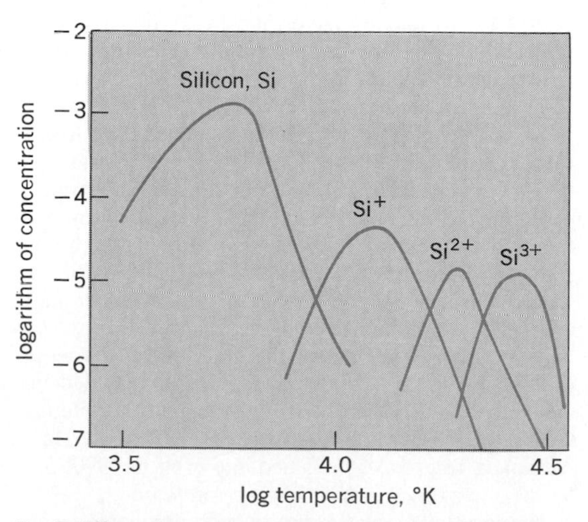

Fig. 5. Effect of temperature on spectral lines of silicon in various states of ionization. °F = (K × 1.8) − 459.67.

used, loci of constant radius are straight lines (**Fig. 6**). Such a diagram, with main-sequence and giant branch, makes clear the location of many of the various sequences of stars. *SEE HEAT RADIATION*.

Age, evolution, and mass. A group of color-magnitude curves for clusters, open (population I) and globular (very old population II), shows that, although the fainter end of the main sequence is essentially the same for all groups of stars, the brighter ends vary from one group to another in accordance with differences in age and composition (**Fig. 7**). Stellar evolution causes such variations. *SEE STAR CLUSTERS*.

These H-R diagrams are most significant for the study of the ages, nuclear energy sources, and evolution of the stars. The location of a star in an H-R diagram is completely determined by its mass and chemical composition, if the latter is specified in detail throughout the star. The effects of rotation and moderate magnetic field are observable but small. Large rotation, near instability, causes hot stars to change luminosity, color, and spectrum.

A point in an H-R diagram, specified by L and R, does not uniquely determine the mass, however, because of the possible variation of composition with depth. (Such variation is caused by the consumption of hydrogen and its conversion to helium by thermonuclear processes.)

The masses of the stars are determined for stars which are members of a double or multiple system, either visual or spectroscopic binaries. The application of Newton's laws provides the mass, although often with considerable uncertainty. If a pair of stars rotates in an orbit whose plane includes the line of sight, so that the star is also an eclipsing binary, and if the lines of both stars are visible in the composite spectrum and show measurable Doppler shifts, it is a two-line spectroscopic binary. Then the size of the orbit, its inclination to the line of sight, the relative masses, and the actual masses are determinable. Such completely observed cases are rare but provide what quantitative information there is about masses, radii, and surface temperatures. Visual binaries with well-observed absolute orbits also give masses. The number of favorable circumstances required makes it difficult to measure directly the luminosities, masses, and radii of all the interesting types of stars. Thus many of the desirable relationships between parameters of a star are statistical in nature. The mass-luminosity relation is of this nature. However, a few classic, well-observed cases provide accurate and complete data for theoretical analysis. *SEE BINARY STAR*.

The stars of the main sequence, for which stellar evolution has not yet been a substantial factor in displacement from their normal positions in an H-R diagram, obey a fairly well-established mass luminosity relationship. However, because of stellar evolution, stars move off the main sequence. If they belong to population I, they essentially move horizontally in an H-R diagram, so that they still obey the mass-luminosity relationship. Population II stars with low metal abundances brighten by factors up to a hundred in the subgiant and red-giant stage. A few such cases of clear violation of the mass-luminosity relation are known: for example, the population II visual binary, ζ Herculis A, for which the brighter component is 4 times brighter than the Sun, although it weighs only 1.07 times as much as the Sun. Other types of stars

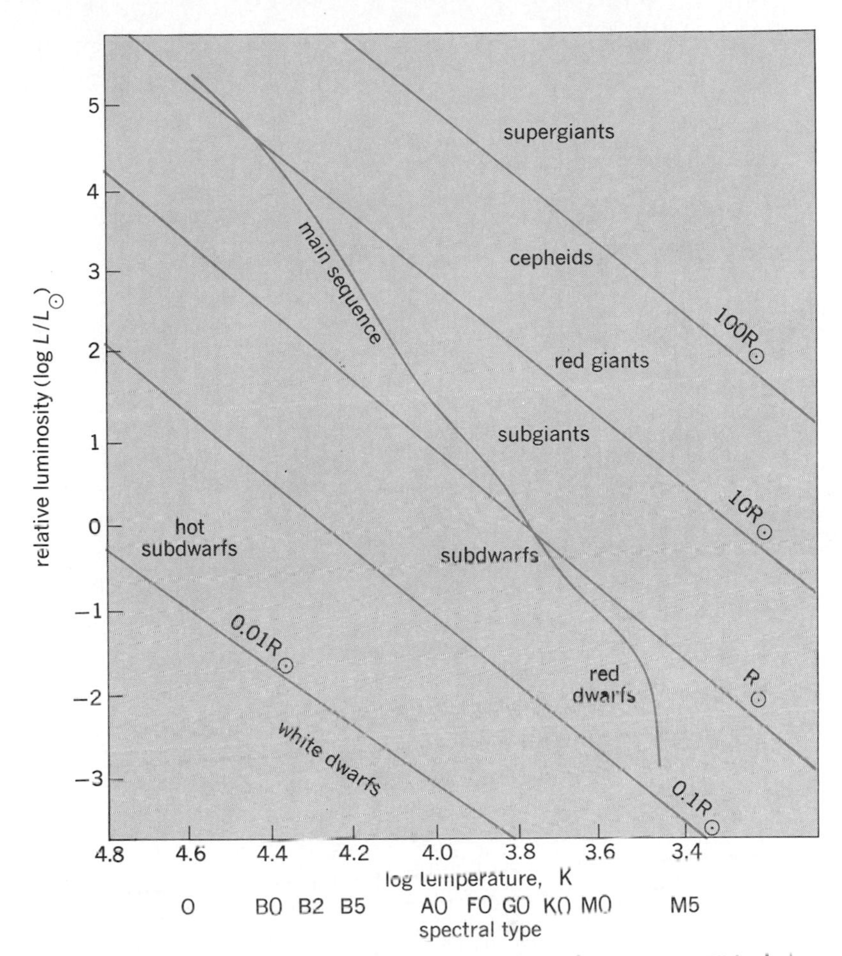

Fig. 6. Main sequence and branch sequences as functions of temperature and relative luminosity; approximate spectral types are listed under temperature scale. °F = (K × 1.8) − 459.67.

which deviate from the mass-luminosity relationship are the white dwarfs, the fainter components of Algol-type eclipsing variables, and some close O-type binaries. *SEE MASS-LUMINOSITY RELATION*.

Table 3 summarizes surface temperature T, luminosity L, and radius R, and mass M of typical stars on the main sequence as a function of spectral type and gives, with a good deal less certainty, similar data for giants and supergiants. From the data surface gravity $g = GM/R^2$ (where G is the gravitational constant) and mean density $\rho = 3M/4\pi R^3$ can be obtained. A few direct measures of the angular diameters of red supergiants were made by stellar interferometers at the Mount Wilson Observatory and agree approximately with the estimates in Table 3. An interferometer has been devised by R. H. Brown and used at Narrabri, Australia, for a very accurate measurement of angular diameters of about 20 nearby stars of high temperature. This correlation-type interferometer, while of limited applicability, yields radii if the stars have accurate parallaxes. Direct effective temperatures are obtained from their luminosities. The radius of Sirius is 1.76 times that of the Sun, slightly smaller than that given in Table 3. The radius of Vega, however, which also is of spectral type A0, is $3.03R_\odot$. The calibration of the mean effective temperature of A0 stars by Brown gives 10,250 K (18,000°F). Other methods have given a slightly lower temperature. The bright F5 star α CMi has

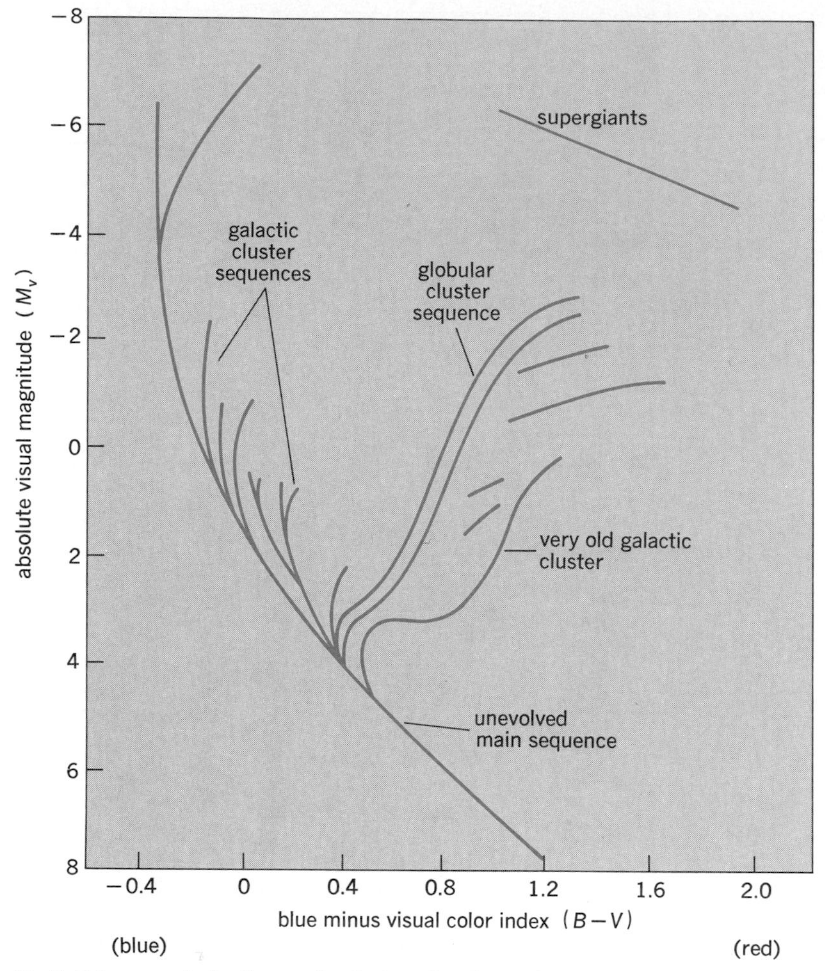

Fig. 7. Color-magnitude diagram for clusters showing variations in brightness for groups of different ages and compositions. Clusters are arranged chronologically, the oldest farthest above, and to the right, clusters of the unevolved main sequence.

2.17 $R_\odot$, appreciably larger than given in Table 3; on the other hand, α CMi has, in fact, evolved above the main sequence; its temperature is about 6450 K (11,150°F).

Increased resolving power has been made possible by technical developments. Fast recording of the signal as a star is occulted by the limb of the Moon shows the diffraction pattern; the nulls in the diffraction are partly filled in, if the star has appreciable angular diameter. Still another technique, speckle interferometry, records electronically the image blurred by the Earth's atmosphere, in a very short time interval. Computer analysis restores the image to its diffraction-limited state. Major questions concerning radius and temperature still exist for the hotter stars. SEE ASTRONOMICAL PHOTOGRAPHY; SPECKLE.

Surface temperature. The determination of surface temperature of stars is carried out by various techniques. Measurements of color, because the star's radiation is approximately that of a blackbody, provide a color temperature. But when combined with the presently highly developed theory of stellar atmospheres, the knowledge of the opacity of stellar material and detailed measures of the energy distribution at many wavelengths yield good measurements of the effective temperatures of stars. Late-type stars with strong molecular bands, however, have highly distorted energy curves, and temperatures can be deduced best from infrared magnitudes, measured with solid-state bolometers out to wavelengths of 20 micrometers.

Because all blackbody curves have essentially the same shape on the long-wavelength side of their energy maxima, all hot stars have essentially the same blue color, because their energy maxima are in the vacuum ultraviolet. In addition, hot stars are usually of high luminosity so that interstellar absorption and reddening impedes the analysis of their colors and luminosities.

Rockets and the first successful orbiting observatory provide the energy distribution in the far-ultraviolet and some information on the line spectra of the hottest and apparently brightest stars. Interstellar reddening does not increase as rapidly toward the violet as from the red to blue. Nevertheless, the flux is less than that predicted from model atmospheres and the standard temperature scale. Absorption lines are of highly ionized elements. In addition, mass loss from hot B stars has been established by displaced absorption lines, formed in an expanding cloud near the star.

Satellite observations of the ultraviolet flux have greatly increased the reliability of the temperature scale for hot stars. These include the Netherlands *ANS* satellite, the United Kingdom *TD1*, and the *International Ultraviolet Explorer*. As a result, the ground-based scale has been largely substantiated. Theoretical model atmospheres fit down to 110 nm. In binary stars, however, interaction may drive mass exchange, producing an optically thick accretion disk and excess ultraviolet radiation.

Strengths of absorption lines of different stages of ionization and excitation may be used to determine the temperature in hot stars. Certain of the hottest O stars, the Wolf-Rayet stars, and the nuclei of planetary nebulae show emission lines excited by a fluorescent conversion of their far-ultraviolet into visible radiations. The process is one of photoelectric ionization followed by recombination and emission of subordinate lines. Temperatures for such objects can be determined by the Zanstra method. The range of stellar temperatures is large. The hottest stars have effective temperatures near 50,000–100,000 K (90,000–180,000°F); the cooler stars are near 1500–2000 K (2200–3100°F). Although these extremes are somewhat unreliable, the temperatures given in Table 3 serve as a guide to the physical conditions in average stars. The x-ray stars observed optically give evidence for envelopes as hot as 50,000,000 K (90,000,000°F). SEE PLANETARY NEBULA.

Stellar rotation. Another important property of stars is their rotation on their axes. The prevalence of wide double stars and of close spectroscopic binaries indicates that a large amount of angular momentum is often contained in the material that condensed to form the stars. In close double systems, revolution and rotation are often synchronous. In single stars, especially those of early spectral type, rotation is rapid.

The Sun has an equatorial velocity v of only 2 km/s (1.2 mi/s). Measurements of the rotational broadening of spectral lines give results contained in **Table 4**, based largely on the work of A. Slettebak. The large rotation of the early type B stars often results in instability, in the form of the ejection of matter from the rapidly rotating equatorial regions. As a result, such stars are often surrounded by disklike rings of low-density material about 10 times the radius of the star. These are detected by the presence of emission lines

Table 3. Approximate physical parameters of the stars

Properties of the Sun*:
$L_\odot$ = 3.9 × 10^{33} ergs/s = 3.9 × 10^{26} W, $M_\odot$ = 2 × 10^{33} g = 2 × 10^{30} kg = 4.4 × 10^{30} lb, T = 5750 K = 9900°F,
M_v = +4.64, $R_\odot$ = 6.96 × 10^{10} cm = 6.96 × 10^{8} m = 4.32 × 10^{5} mi, log g = +4.44, log ρ = +0.16

Type	Color (B − V)	Absolute visual magnitude (M_v)	log $L/L_\odot$	T/1000 K	log $R/R_\odot$	log $M/M_\odot$	Remarks
Main sequence							
O8	−0.3 :†	− 5	+5.05	35	+0.96	+1.25	Uncertain, wide range
B0	−0.32:	− 4.3	+4.66	25	+1.05	+1.15	Uncertain
B1	−0.28	− 3.5	+4.06	22	+0.85	+1.05	
B2	− .24	− 2.8	+3.72	20	+0.76	+0.95	
B5	− .16	− 1.3	+2.96	15	+ .62	+ .78	
A0	0.00	+ 0.8	+1.73	11	+ .31	+ .45	
A5	+ .19	+ 1.9	+1.21	8.7	+ .24	+ .25	
F0	+ .37	+ 2.5	+0.85	7.6	+ .18	+ .14	
F5	+ .47	+ 3.5	+ .49	6.6	+ .12	+ .08	
G0	+ .60	+ 4.2	+ .21	6.0	+ .06	+ .04	
G5	+ .70	+ 5.2	− .19	5.5	− .06	− .17	
K0	+ .86	+ 6.1	− .55	5.1	− .18	− .22	
K5	+1.24	+ 7.5	− .84	4.4	− .18	− .25	
M0	+1.45:	+ 9.0	−1.23	3.6	− .22	− .30	
M2	+1.5 :	+10.0	−1.48	3.2	− .24	− .40	
M4	+1.6 :	+12.0	−1.91	3.1	− .42	− .55	Uncertain, because of steepness of main sequence
M6	+1.8 :	+15.0	−2.71	2.9	− .76	−0.90	
Giant stars							
G0		+ 0.7	+1.65	5.3	+0.90		Masses, to 4 $M_\odot$
K0		+ 0.2	+2.04	4.2	+1.30		Masses, to 4 $M_\odot$
M0		− 0.4	+2.72	3.3	+1.84		Masses, to 4 $M_\odot$
Supergiant stars							
B0		− 7	+5.66	25	+1.55		All data have wide range
A0		− 7	+4.77	10	+1.90		All data have wide range
K0		− 7	+5.26	3.6	+3.03		All data have wide range
M0		− 7	+5.66	3.0	+3.39		All data have wide range

*g = surface gravity in cm/s^2; ρ = mean density in g/cm^3.
†Colons indicate discordant determinations.

and by absorption lines if the disk contains the line of sight. The rotation drops rapidly down the main sequence and usually is too small to be detected in single stars of types later than G0. However, a method has been developed for the direct measurement of the rotation period of some stars with slower rotation. *SEE STELLAR ROTATION.*

Composition. The chemical composition of a star can be deduced from the spectrum of its atmosphere or from the theory of its internal structure. There is evidence that stars need not be chemically homogeneous. Products of nuclear reactions may concentrate in the center of the star so that the ratio of helium to hydrogen increases inward. Very slow mixing, however, counteracts the tendency toward complete diffusive separation by gravity of heavy from light elements.

Detailed studies of the composition of stars can be made only in their atmospheres. The ratio of hydrogen to heavy elements is about 8000–12,000 to 1, by number of atoms, for the Sun and young population I stars. For extreme high-velocity stars, old objects of population II, the abundance of the metals may be up to 400 times lower. Certain elements are unobservable in the spectra of stars, and the abundances of heavy and generally rare elements, as well as the isotopic abundances, are best obtained from the crust of Earth or meteorites. **Table 5** is a composite résumé of current determinations of the abundances. *SEE ELEMENTS, COSMIC ABUNDANCE OF.*

Stellar or solar abundances are determined by a set of measurements and interpreted by theory. Wavelengths of lines yield identification of the elements, and line intensities can be interpreted in terms of the number of atoms in the atomic levels producing the lines. However, to do this, the transition probability in the line must be known, either from quantum mechanics or laboratory measurements. Laboratory data on atomic transition probabilities are still in flux. In addition, corrections for the state of ionization and excitation must be applied to permit computation of the total concentration of the element, and these corrections depend strongly on the temperature. This is especially true for atoms whose visible lines arise from levels of high excitation potential; for example, in the Sun only one atom of helium in 10^{20} is likely

Table 4. Mean rotational surface velocities, in km/s (mi/s)

Type	Dwarfs	Giants
B1–B3	200 (124)	127 (79)
B5–B7	257 (160)	163 (101)
B8–A2	177 (110)	93 (58)
A3–A7	173 (107)	202 (126)
A9–F2	87 (54)	125 (78)
F3–F6	31 (19)	67 (42)
F7–G0	<25 (<16)	34 (21)

Table 5. Abundances of elements in normal stars and Sun*

Atomic number	Symbol	log N†
1	H	12.0
2	He	11.0
3	Li	1.0
4	Be	1.1
5	B	2.1
6	C	8.7
7	N	8.0
8	O	8.9
9	F	5.4
10	Ne	8.5
11	Na	6.3
12	Mg	7.5
13	Al	6.2
14	Si	7.5
15	P	(5.5)
16	S	(7.3)
17	Cl	(5.4)
18	Ar	(6.8)
19	K	4.7
20	Ca	6.2
21	Sc	3.2
22	Ti	4.3
23	V	3.7
24	Cr	5.4
25	Mn	4.9
26	Fe	7.5
27	Co	4.3
28	Ni	5.1
29	Cu	3.9
30	Zn	4.0
31	Ga	2.4
32	Ge	3.3
37	Rb	2.5
38	Sr	3.1
39	Y	2.8
40	Zr	2.3
41	Nb	2.0
42	Mo	2.1
44	Ru	1.8
45	Rh	1.7
46	Pd	1.8
47	Ag	0.8
48	Cd	2.0
49	In	1.7
56	Ba	2.9

*Parentheses indicate poor values. The rare earths have been omitted; they average about 1.0 in log *N*.
†Logarithm of number of atoms.

to be excited into a state capable of producing a visible absorption line. Astrophysical abundance determinations had been severely limited by Earth's atmosphere, which prevents observation of the far-ultraviolet spectra, where the resonance lines ofmany important elements are located. Satellite observations now permit such determinations. The subject of stellar atmospheres is an important branch of modern astrophysics. It relies heavily on the mathematical solution of problems of the diffusion of light outward through an absorbing and emitting medium and on physical information about the shape of atomic absorption lines, the amount of continuous absorption of light (which limits the depth to which we can see), and the atomic theories of line broadening. *See* ATOMIC STRUCTURE AND SPECTRA.

Motions of stars. Galactic dynamics and kinematics are an important part of the subject of stellar statistics and are intimately connected with the distribution of the stars in space. The inner 10,000 parsecs of the flattened part of the Milky Way Galaxy has a mass of about 6×10^{11} Suns, of which only a small fraction is visible from Earth. Stars move in orbits around their common center of gravitation, located about 8500 parsecs (1.6×10^{17} mi or 2.6×10^{17} km) from Earth in Sagittarius. The galaxy is flattened by its systematic rotation, the linear rotational velocity being about 140 mi/s (220 km/s) at the Sun's distance from the galactic center; this corresponds to a rotation period of 2×10^8 years. Work on the possible existence of a large spherical halo surrounding the flattened disk suggests the total mass is four times as large. *See* GALAXY, EXTERNAL; MILKY WAY GALAXY.

Differential rotation. Within the Milky Way Galaxy the rotation is described with respect to an external stationary frame of reference, from which it would also appear as a differential rotation, with angular velocity varying with distance from the galactic center (**Fig. 8**). This rotation curve is established by the study of the relative velocities of distant stars, and of clouds of interstellar hydrogen seen by their 21-cm radio-frequency radiation, both measured with respect to the Sun. Were the galaxy to rotate as a rigid body, there would be no relative velocities of approach or recession, that is, no radial velocity of a systematic character. A differential rotation, however, is detectable in the transverse motions of the stars, with respect to an outside frame of reference, even in the rigid body case. In nonrigid body rotation, the distant stars show systematic velocities in certain preferred directions. The relative radial velocity $V(r,l)$ is given by Eq. (2), where l is the azimuthal coordinate (in the

$$V(r, l) = R_0[\omega(R) - \omega(R_0)] \sin l \qquad (2)$$

galactic plane) of the star, measured from the galactic center; R_0 is the Sun's distance to the galactic center, and R that of the star; the motion is assumed to be in the galactic plane. To a first approximation, for small distances r measured from the Sun to the star, the above formula can be written as Eq. (3), where A is

$$V(r, l) = 2rA \sin 2l \qquad (3)$$

the first-order galactic rotation constant, which lies between about 8 and 11 mi/s (13 and 17 km/s) per 1000 parsecs (1 parsec equals 1.9×10^{13} mi or 3.1×10^{13} km). This double wave is visible in the mean and radial velocities of the stars in the range of distance 300 to 3000 parsecs; at larger distances higher-order terms must be included. The cosmological expansion, seen on a large scale in the redshifts of distant galaxies, does not operate within the Milky Way or within clusters of galaxies.

Peculiar velocity. Superposed on the systematic rotation of the galaxy are individual motions of the stars. Each star moves in a somewhat elliptical orbit and therefore shows a peculiar velocity with respect to the local standard of rest, the standard moving in a circular orbit around the galactic center. The Sun has an orbit of small ellipticity and inclination, so that solar motion with respect to the mean of neighboring stars can be detected by analysis of either radial velocities or proper motions of nearby stars.

Stars can be considered to be particles in a gas which has no collisions, with gravitational distant encounters between pairs of stars only slightly altering their orbits. The velocity dispersion of the stars with respect to their local standard of rest is essentially maxwellian, except that they display greater mobility in the directions toward and away from the galactic center than in other directions.

The velocity distribution of stars in Earth's neighborhood is not a single maxwellian one, but is characterized by at least two different dispersions. One young group (population I) is characterized by a small dispersion of space motion with respect to the local standard of rest, for example, about ±5 mi/s (±8 km/s) for B stars, increasing to ±12 mi/s (±20 km/s) for M stars. Most stars have a higher dispersion, about ±28 mi/s (±45 km/s), and belong to the so-called old-disk population. Such stars are typically 10^{10} years old, but have a chemical composition resembling that of young population I. Very low metal abundances characterize another group, the oldest population II stars, which have a velocity dispersions ranging from ±19 to ±93 mi/s (±30 to ±150 km/s). The high-velocity population II stars are moving in galactocentric orbits of high eccentricity and inclination to the galactic plane. If they were moving with a velocity vector measured with respect to the local standard of rest in the same direction as the galactic rotation, their kinetic energy would be so great compared to the gravitational force of the galaxy as to exceed escape velocity. Thus stars with large velocity vectors, greater than 40 mi/s (65 km/s) with respect to the Sun, are absent in about one quadrant of the sky, in the forward direction of galactic rotation.

Halo. The so-called massive spheroidal halo enveloping the Milky Way Galaxy and extending far outside the galactocentric distance of the Sun, $R_{\odot}$, may have an enormous mass, but it has not yet been detected by any of its individual stars passing the Sun with high velocity. Search for faint extensions of other galaxies, by the integrated light of possible faint, red stars, has also failed. One justification for the existence of the halo is theoretical: in its absence the flattened disk of the Galaxy would be unstable. More direct evidence is that the Milky Way, as well as other galaxies, has rotational velocity curves flatter than the one sketched in Fig. 8. A decrease in rotational velocity would be expected because, at large distances, outside the body of the Milky Way, the gravitational potential should begin to decrease; however, this decrease has not yet been observed. Most other galaxies also show essentially flat rotation curves; that is, the mass within a test point continues to increase outward, canceling the $1/r^2$ part of the gravitational equation. There has been much speculation on the nature of the "invisible mass," and on what this gravitating matter could be, other than luminous stars or interstellar gas. An enveloping cloud of massive neutrinos was a popular but short-lived explanation. The outer, massive halo is supposed to contain about five times the mass of the visible Milky Way, and extend out to at least 100,000 light-years (6×10^{17} mi or 10^{18} km), but it remains a somewhat mysterious component of the Galaxy. The following discussion will disregard the effects of its existence.

Solar motion. The solar motion is defined by a vector giving the direction of A, the apex of solar motion, and U, the velocity with respect to the local standard of rest. The radial velocity of the Sun with respect to a direction located at angular distance λ from the apex is a differential motion, $\delta V = -U \sin \lambda$. By averaging over the observed radial velocities of stars of a given type, at moderate distances from the Sun (or correcting separately for galactic rotation if necessary), both U and λ can be determined.

The apex of solar motion with respect to the nearby population I stars is at right ascension 18^h, declination +29°, with U about 12 mi/s (20 km/s). With respect to the population II stars, the direction of the apex is strongly dependent on the velocity dispersion.

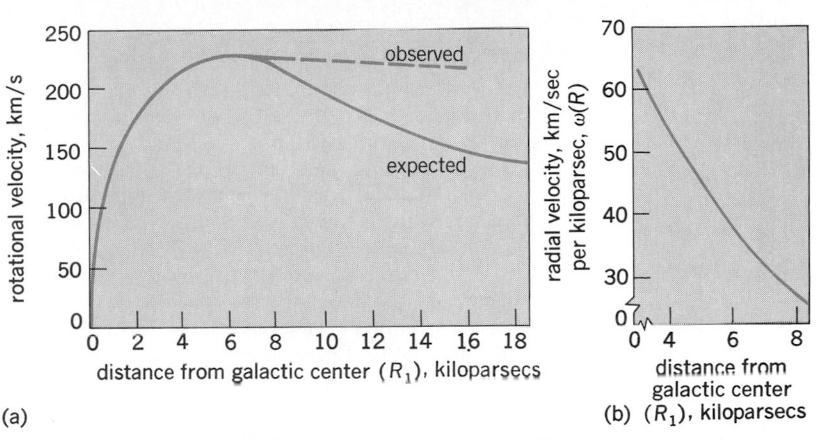

Fig. 8. Velocity curves. (*a*) Rotational velocity varies with distance from galactic center. The broken curve is the known part of the rotation curve; the solid curve is what would be expected if most of the mass were interior to 10 kiloparsecs. (*b*) If the galaxy were a solid, the angular velocity would be constant instead of decreasing with radius as it does. 1 km/s = 0.6 mi/s.

Proper motion. The transverse motions of the stars with respect to a fixed coordinate system on the plane of the sky are seen as proper motions, which are usually given in seconds of arc per year. The total proper motion μ is derived from the component μ_α, in right ascension, in seconds of time, and μ_δ, in declination, in seconds of arc, by Eq. (4). Proper motion of a star

$$\mu^2 = (15\ \mu_\alpha \cos \delta)^2 + \mu_\delta{}^2 \qquad (4)$$

is measured by displacements on photographic plates taken at a sufficiently long interval of time. The standard of rest is usually established by stars of accurately known absolute positions and motions. In principle, an ideal standard of rest would be provided by the faint extragalactic nebulae, and quasars; and work is in progress with this technique. Normally, however, absolute positions of reference stars are determined by visual observations, accurately timed, of the transits of stars, using meridian circles. About 200 stars have motions exceeding 1″ per year; motions down to 0.″005 per year are moderately dependable.

If the Sun is considered at rest, components of the pecular motion of a star as observed from the Sun form a vector diagram (**Fig. 9**). If, in a year the star moves from S to S' at velocity v, the velocity in the line of sight is radial velocity V, and the velocity across the line of sight is proper motion μ, in seconds of arc per year. Parallax p is also expressed in seconds of arc; then tangential velocity T is $T = 4.74\ \mu/p$, the space velocity v is $v^2 = V^2 + T^2$, and the radial and tangential velocities are $V = v \cos \theta$ and $T = v \sin \theta$. All velocities are expressed in kilometers per second (1 km/s = 0.62 mi/s).

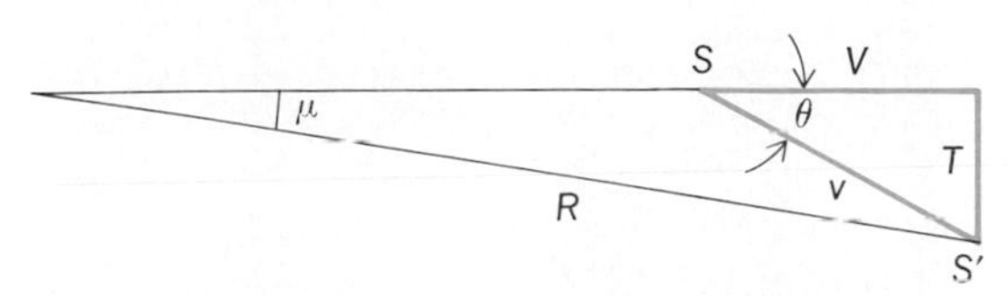

Fig. 9. Vector diagram of components of proper and radial motions of a star as observed from the Sun.

Conversely, if the stars were all standing still and the Sun alone were moving, or if the stars' motions are averaged out, the formula above for the tangential velocity indicates that a star would show a proper motion due to the projection of the solar motion on the plane of the sky. This permits the determination of the solar motion from the proper motions of nearby stars for which p is not too small. Similarly, if the observed motions of a group of stars, after correction for solar motion, are used, an estimate is obtained of the average parallax. Special catalogs give accurate positions and proper motions.

Radial velocity. Spectroscopic measures of Doppler shifts give radial velocity. A general catalog of stellar radial velocities of 15,000 stars is available. Stellar radial velocities are determined with accuracies up to ± 0.06 mi/s (± 0.1 km/s); the largest velocities are about 250 mi/s (400 km/s), but only 4% are greater than 37 mi/s (60 km/s).

Galactic dynamics has as its main result the interpretation of the space motions of stars in terms of galactic rotation, orbits, and the distribution of mass in the Milky Way Galaxy. From the rotational-velocity curve for circular orbits, the mass is determined; this method has been applied to a few extragalactic nebulae, with the general result that masses of galaxies range from 1×10^9 to 2×10^{12} Suns. The dynamics of such features as the spiral arms are not understood without reference to the motion of the interstellar gas, galactic magnetic fields, and the effects of local perturbations of the gravitational potential field. Some features of the spiral arms have been simulated by computer experiments; however, the arms do not survive more than a few rotations.

Spatial distribution. The distribution of stars in space is the subject of studies of stellar statistics and galactic structure. The Milky Way is the dominant feature of the galaxy, even to the eye, and represents the mean plane in which the stars are concentrated. It is nearly a great circle, indicating that the Sun lies near the galactic plane.

The number of stars at a given apparent magnitude is a function of galactic latitude and longitude; in the galactic plane the complex structure of interstellar clouds of dust that absorb light and the irregularities of spiral structure produce an irregular, patchy appearance, with a maximum of faint stars in the direction of the galactic center in Sagittarius, with subsidiary maxima in Cygnus and Carina. These maxima are probably due to spiral arms seen lengthwise.

For an average over galactic longitude, the latitude dependence is shown in **Table 6**, which gives log $N_{m,b}$, where $N_{m,b}$ is the number of stars brighter than apparent magnitude m, per square degree, at latitude b. The light of the entire sky is equivalent to that of one star of magnitude -6.6. The number of stars increases very rapidly with m, about threefold per magnitude, in the galactic plane; the rate of increase is slower for fainter stars and at higher galactic latitudes. The galactic concentration of stars increases for fainter stars, a natural result because the faint stars are on the average more distant.

Table 6. Apparent distribution log $N_{m,b}$ of stars as a function of galactic latitude

m	b				
	0°	10°	25°	50°	90°
6	−0.89	−0.97	−1.16	−1.35	−1.43
8	0.00	−0.08	−0.26	−0.45	−0.56
10	+0.89	+0.79	+0.59	+0.40	+0.26
12	+1.74	+1.63	+1.41	+1.18	+1.00
14	+2.57	+2.43	+2.17	+1.88	+1.65
16	+3.33	+3.19	+2.84	+2.48	+2.21
18	+4.01	+3.87	+3.42	+2.98	+2.68
20	+4.60	+4.46	+3.90	+3.38	+3.07

Table 7. Frequency log $\phi(M_v)$ of stars in the neighborhood of the Sun, as a function of absolute visual magnitude

M_v	log $\phi(M_v)$	M_v	log $\phi(M_v)$	M_v	log $\phi(M_v)$
−6	−8.4	0	−4.0	+6	−2.4
−5	−7.4	+1	−3.5	+8	−2.3
−4	−6.6	+2	−3.2	+10	−2.1
−3	−6.0	+3	−3.0	+12	−1.9
−2	−5.4	+4	−2.7	+14	−1.8
−1	−4.7	+5	−2.5	+16	−2.0

Certain types of objects are highly concentrated toward the galactic plane, particularly cepheid variables, luminous O and B stars, and galactic clusters. Others, like M dwarfs and long-period variables, are less concentrated, while globular clusters and RR Lyrae variable stars show almost no concentration. SEE CEPHEIDS; VARIABLE STAR.

Part of this effect is caused by luminosity differences; for example, M dwarfs are so intrinsically faint that they cannot be seen to great distances. On the other hand, O and B stars are intrinsically highly concentrated to the galactic plane, while the globular clusters and other extreme population II high-velocity stars are found at great heights, up to 15,000 parsecs from the plane.

In the galactic plane the frequency distribution of the types of stars varies from point to point. Interstellar gas and dust are highly concentrated in the spiral arms, together with O and B stars and other population I objects of high luminosity. In the neighborhood of the Sun, the luminosity function given in **Table 7** is a useful average value. It gives the density of stars per cubic parsec as a function of absolute visual magnitude $\phi(M_v)$ within range $M_v + 1/2$ to $M_v - 1/2$. However, outside the spiral arms, and at heights greater than 100 parsecs from the galactic plane, the high-luminosity end of $\phi(M)$ is cut off.

At great heights, stars with M_v less than $+3$ are rare. The space density of all types of stars together is about 0.1 solar masses per cubic parsec in the galactic plane and decreases rapidly with height above the plane, following an approximately exponential law, with a scale height h_0, which varies with the type of star. The B stars and the interstellar gas clouds have h_0 about 100 parsecs; RR Lyrae variables have h_0 in the order of 2500 parsecs. The difference reflects the different kinetic energies of slow- and fast-moving stars. The space density of stars increases greatly toward the center of the Milky Way Galaxy, possibly by a factor of 100. The thickness of the galaxy also increases; the bulge surrounding the galactic center is composed mainly of stars of population II, which show a very large h_0. The center, obscured in visual light, has a point source of infrared and radio waves, presumably coincidental with a density singularity. Other galaxies show starlike objects at their centers. Such a center, or nucleus, may be a cluster of 10^8 stars within a few light-years. The nucleus has a density a billion times higher than does the Galaxy near the Sun. These nuclei are often the seat of violent explosions resembling quasars. Considerable evidence suggests that a massive black hole is present in most galaxies, including the Milky Way Galaxy. SEE ASTROPHYSICS, HIGH-ENERGY; BLACK HOLE; QUASAR.

Jesse L. Greenstein

Bibliography. E. H. Avrett (ed.), *Frontiers of Astrophysics*, 1976; E. Bohm-Vitense, *Introduction to Stellar Astrophysics,* 2 vols., 1989; R. Burnham, Jr., *Burnham's Celestial Handbook*, 1978; G. W. Collins, *The Fundamentals of Stellar Astrophysics,* 1989; D. Hoffleit, *The Bright Star Catalogue*, 4th ed., 1982; L. Jenkins, *General Catalogue of Stellar Parallaxes*, 1963; D. M. Popper, Stellar masses, *Annu. Rev. Astron. Astrophys.*, 18:115–164, 1980; F. H. Shu, *The Physical Universe*, 1982; R. E. Wilson, *General Catalogue of Stellar Radial Velocities*, 1963.

Star clouds

Aggregations of thousands or millions of stars spread over hundreds or thousands of light-years in space. The Milky Way is composed of such star clouds, the heaviest clouds being in the richest parts, such as Cygnus, Sagittarius, Carina, and Scutum (see **illus.**).

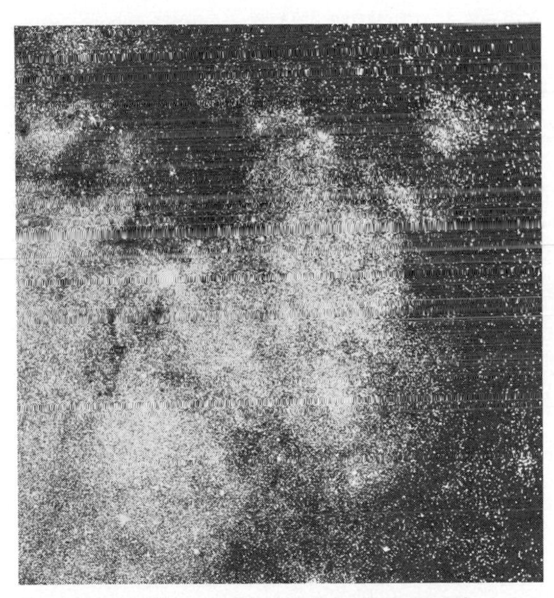

Great star cloud in Scutum. (*Yerkes Observatory*)

The stars in such clouds may appear unevenly distributed because of the presence of obscuring interstellar dust and gas. SEE INTERSTELLAR MATTER; MILKY WAY GALAXY; STAR.

Helen S. Hogg

Star clusters

Groups of stars held together by gravitational attraction. The two chief types are open clusters, containing from a dozen up to many hundreds of stars, and globular clusters, composed of thousands to hundreds of thousands of stars. A relative of the star cluster is the stellar association, a group of dozens or hundreds of relatively young stars spread loosely over a large volume of space. Star clusters are important in outlining the shape and extent of the Milky Way Galaxy and in deriving theories of stellar evolution on the assumption that stars of a given cluster were formed at the same time.

Open clusters. Open clusters, formerly called galactic clusters, lie along the backbone of the Milky Way Galaxy, strongly concentrated to the central plane of the Milky Way. A dozen are visible to the unaided eye, over 1000 are cataloged, and many more must exist. Most open clusters have an asymmetrical appearance (**Fig. 1**).

Distances and dimensions. The distances to open clusters range from 25 parsecs (5×10^{14} mi or 8×10^{14} km) for the sparse Ursa Major cluster up to 5000 parsecs (1.0×10^{17} mi or 1.6×10^{17} km) for the faintest detectable against a rich stellar background. Distances may be determined by geometric methods, including trigonometric parallaxes or stellar motions, and photometric methods. SEE PARALLAX (ASTRONOMY).

Angular diameters of open clusters range from several degrees down to several minutes of arc, and linear diameters range from 15 parsecs (2.9×10^{14} mi or 4.6×10^{14} km) to 2 parsecs (4×10^{13} mi or 6×10^{13} km). From a study of the way in which linear diameters appeared to increase with increasing distance, the absorption of light in space was deduced by R. J. Trumpler in 1930.

Spectral characteristics. The brightest stars in some clusters, like the Pleiades, are blue, of spectral type B; in others, like the Hyades or Praesepe, they are yellow or red. Some stars with luminosities brighter than absolute magnitude −3 are found, and sometimes supergiants up to −7. SEE HYADES; PLEIADES.

Most stars fall along the highly populated branch of the spectrum-luminosity diagram known as the main sequence. The point where this sequence starts furnishes a criterion of the age of the cluster. Open clusters show great diversity in their spectrum-luminosity diagrams. They may be classified on this basis, as well as by richness and central concentration. They may contain such types of stars as bright O stars, visual and spectroscopic binaries, certain kinds of variables, and white dwarfs. Some clusters, like the Pleiades, contain amounts of nebulosity equivalent to many solar masses. SEE HERTZSPRUNG-RUSSELL DIAGRAM; STELLAR EVOLUTION.

Motions. Measures of proper motion and radial velocity show the cluster stars to be sharing a common motion in space, with random velocities up to tens of miles per second. Nearby open clusters whose most prominent characteristic is a large common proper motion of the stars are called moving clusters. The diagram of proper motions in a cluster shows a conspicuous convergent point, where their parallel motions appear to meet in space. This category includes the Taurus and Ursa Major moving clusters. SEE CONSTELLATION.

Age and dissolution. By comparing the spectrum-luminosity diagram with a standard main sequence, the age of a cluster may be determined. At one extreme are young clusters, formed in recent geologic times, like the Orion Nebula cluster with an age of about 1×10^6 years. At the other extreme is NGC 188, with an age of 5×10^9 years, comparable with that of old systems like globular clusters. The lifetime

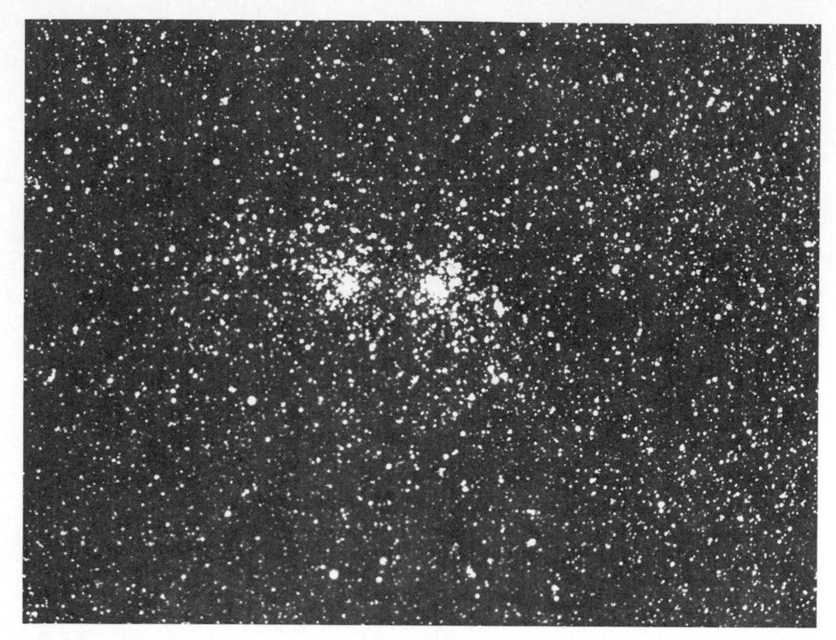

Fig. 1. Double cluster in Perseus. (*Yerkes Observatory*)

of a cluster depends on a balance between its mass and its radius. High-velocity stars may "evaporate" from dense clusters; encounters with interstellar clouds disrupt loose clusters.

Stellar associations. Systems in which early-type (O to B2) stars are more numerous than in the surrounding field are known as stellar associations. About 70 are cataloged. The radii range up to 200 parsecs (4×10^{15} mi or 6×10^{15} km). Stellar associations are perishable, lasting perhaps 10^7 years.

Fig. 2. The great globular cluster in Hercules, Messier 13. Photographed with 200-in. (500-cm) telescope. (*California Institute of Technology, Palomar Observatory*)

Globular clusters. Groups of thousands to hundreds of thousands of stars in globular symmetry constitute globular clusters. Though they are scattered widely in galactic latitude, their strong concentration toward the region of Sagittarius-Scorpio led Harlow Shapley in 1917 to postulate this as the center of the Milky Way Galaxy. Several are visible to the unaided eye, like Messier 13, the great cluster in Hercules (**Fig. 2**). A total of 137 have been cataloged in the Milky Way Galaxy, including several so far distant that they are really intergalactic. A few dozen more may be undetected.

Distances and dimensions. The distances to globular clusters in the Milky Way Galaxy range from 2 kiloparsecs (4×10^{16} mi or 6×10^{16} km) for the nearest to more than 100 kiloparsecs (2×10^{18} mi or 3×10^{18} km). They are too great for geometric methods of distance determination, but photometric methods involving color-magnitude diagrams and RR Lyrae stars can be used. The apparent diameters range from 36′ to 1′, and the linear diameters from 100 parsecs (2×10^{15} mi or 3×10^{15} km) to 10 parsecs (2×10^{14} or 3×10^{14} km).

Structure. Globular clusters differ markedly in their degree of central concentration. A few are noticeably elliptical. Three observable characteristics are the radius of the central core, its brightness, and the tidal radius (from tidal forces imposed by the Galaxy). In some globulars the frequency of stars falls off as the cube of the distance from the center. Star counts in Messier 3, which has 2.45×10^5 solar masses, show that 95% of the visual light comes from stars intrinsically brighter than the Sun, while 90% of the mass is contributed by fainter stars. Density near the center is high, 50 stars per cubic parsec compared with 1 star per 10 cubic parsecs near the Sun.

Color-magnitude diagrams differ appreciably from those of open clusters. Stars of absolute magnitude brighter than -3 are absent. The brightest stars are yellow-red; the main sequence is represented by large numbers of stars from type F down, with a horizontal branch near absolute magnitude 0. The spectra show low metal content (ratio of iron to hydrogen), but the ratio varies from cluster to cluster.

More than 2000 variable stars have been found in globular clusters with Messier 3 the richest, containing more than 200. Short-period RR Lyrae stars make up nearly 90% of the variables. *See* VARIABLE STAR.

In the 1970s the discovery of x-ray sources in eight globular clusters, some of them burst sources, produced a new interest in these clusters. A survey with the *Einstein X-ray Observatory* has increased to 18 the number of galactic globular clusters known to have such sources, of which about half are burst sources. A derived luminosity function indicates two classes of source; both are binary systems, one type probably with an accreting neutron star, the other with a white dwarf as the compact member. *See* ASTROPHYSICS, HIGH-ENERGY; BINARY STAR; NEUTRON STAR; WHITE DWARF STAR; X-RAY ASTRONOMY.

The nature of apparent wisps of nebulosity in globular clusters is disputed. Radio sources have been detected in some globulars, but the radio emission does not indicate large masses of gaseous material.

Motions. The individual stars in a globular cluster are describing orbits about the cluster center, which was first proved observationally for Omega Centauri. The clusters themselves are describing large orbits about the center of the Galaxy. Radial velocities for

70 clusters with respect to the Sun range from 306 to −224 mi/s (493 to −360 km/s).

Age, formation, and dissolution. Many clusters probably formed in various parts of the Galaxy during its first 10^9 years. Some globular cluster stars may be the oldest stars in the Galaxy. Their ages, based on low-metallicity evolutionary models, can be as great as 1.41×10^{11} years. The clusters are so stable dynamically that their individual stars will burn out before the clusters disintegrate. *SEE MILKY WAY GALAXY; STELLAR EVOLUTION.*

Clusters in extragalactic systems. In the 1970s globular clusters were discovered by the thousands in distant external galaxies. Open clusters totaling hundreds have been detected only in the nearest galaxies. Besides normal globular clusters, the Magellanic Clouds have a second type whose bright stars are blue. In the Local Group in 12 low-mass irregular and spiral galaxies, some 500 globulars are known with x-ray sources detected in five in Messier 31. As more distant galaxies are studied, cluster counts become less definitive because of corrections for other image types such as very distant galaxies. Counts in 25 elliptical and S0 galaxies total around 15,000 with marked variation between galaxies, while estimates run around 60,000 total. Two giant ellipticals are outstandingly rich, Messier 87 in Virgo with 6000 counted and 15,000 estimated, and NGC 3311 in Hydra with comparable numbers. Central galaxies in rich clusters of galaxies appear to have unusually large numbers of globular clusters. From their average brightness, globular clusters can be used as distance indicators for very remote galaxies. *SEE GALAXY, EXTERNAL; LOCAL GROUP; MAGELLANIC CLOUDS; STAR.*

Helen S. Hogg

Bibliography. Catalogues of Open and Gobular Clusters, *The Astronomical Almanac*, 1985; C. Payne Gaposchkin, *Stars and Clusters*, 1979; W. E. Harris and R. Racine, Globular clusters in galaxies, *Annu. Rev. Astron. Astrophys.*, 17:241–274, 1979.

Starburst galaxy

A galaxy that is observed to be undergoing an unusually high rate of formation of stars. It is often defined as a galaxy that, if it continues to form stars at the currently observed rate, will exhaust its entire supply of star-forming material, the interstellar gas and dust, in a time period that is very short compared to the age of the universe. For a typical starburst galaxy, this gas exhaustion time scale is less than 10^8 years, that is, less than 1% of the age of the universe. Since such a galaxy must shortly run out of star-forming material, the high star formation rate currently observed not only must end soon but also must have started relatively recently or the gas supply would have run out long ago. It follows that such galaxies must be undergoing a passing burst of star formation. The term starburst usually also implies that the burst of star formation is occurring in the nuclear regions of the galaxy, because the term was coined to describe a sample of luminous spiral galaxies with bright, pointlike nuclei. The prototype starburst nucleus is NGC 7714. However, there exist related objects that meet the definition of short gas-exhaustion time scale but exhibit more widespread star formation. These include extragalactic H II regions, clumpy irregular galaxies, blue compact dwarf galaxies, and the nearest, best-studied starburst galaxy, M82.

Diagnostics. The existence of a starburst in a galaxy is usually deduced from the presence of strong, narrow emission lines in the optical spectrum. These lines originate in hot gaseous nebulae, called H II regions, which form around the hot young stars. Stronger proof of the existence of a large population of hot young stars in a galaxy comes from the detection in the optical spectrum of broad emission lines of helium, characteristic of very massive and short-lived Wolf-Rayet stars. Few such Wolf-Rayet galaxies are known because making the helium-line measurements is too difficult. In a small number of galaxies, the ultraviolet absorption features caused by hot young stars have been detected by the *International Ultraviolet Explorer* (*IUE*) satellite. Additional direct evidence for young massive stars with short lifetimes is provided by the detection of numerous supernova remnants, which have been observed to fade in radio brightness, in the three nearby starburst galaxies M82, NGC 253, and NGC 4736. *SEE INTERSTELLAR MATTER; RADIO ASTRONOMY; SATELLITE ASTRONOMY; STELLAR EVOLUTION; SUPERNOVA; ULTRAVIOLET ASTRONOMY; WOLF-RAYET STAR.*

Radiation processes. Starburst galaxies are often blue because of the dominance of their energy output by hot massive stars. Many blue starburst galaxies were discovered in the extensive ultraviolet-sensitive objective prism surveys of B. E. Markarian (**Fig. 1**). Most starburst galaxies are also strong emitters of far-infrared radiation (**Fig. 2**), because the newly formed stars illuminate their natal clouds of dust and gas with ultraviolet and optical radiation, and as the dust warms, it reradiates this energy at far-infrared wavelengths. The *Infrared Astronomical Satellite* (*IRAS*), which surveyed the entire sky at far-infrared wavelengths in 1983, discovered over 100,000 galaxies that emit strongly at 60 micrometers, the majority of which are likely to be starburst galaxies. Strong millimeter-wavelength emission lines of carbon monoxide, often highly concentrated to the nuclear regions, have also shown that starburst galaxies are rich in molecular gas, the dense, relatively cool material in which star formation takes place. *SEE INFRARED ASTRONOMY; MOLECULAR CLOUD; PROTOSTAR.*

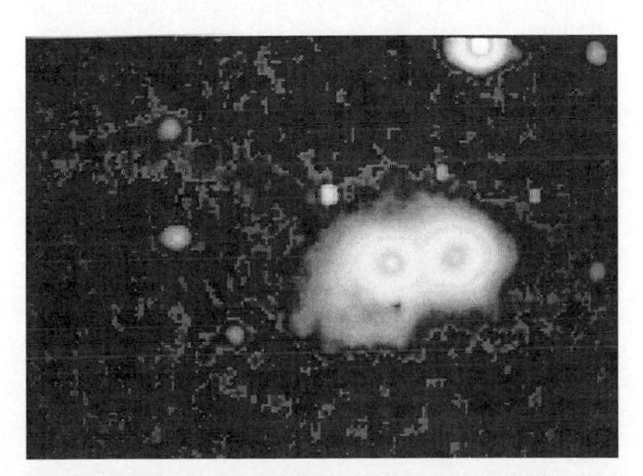

Fig. 1. Blue-light image of the double-nucleus starburst object Markarian 788. The faint tail to the top left is believed to be tidal debris resulting from the collision of two galaxies, the two bright spots being the nuclei of the two galaxies. Optical spectra show that each nucleus is a site of intensive star formation triggered by the collision. (*Photograph by J. Mazzarella*)

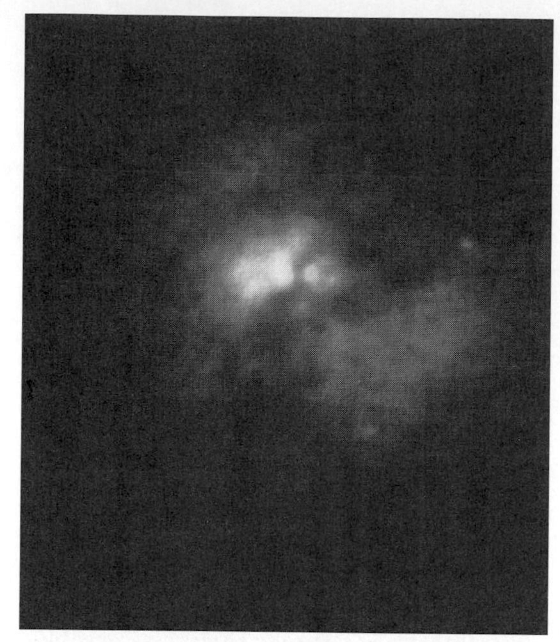

Fig. 2. Image of the central part of the ultraluminous infrared galaxy Arp 220, taken with the Wide Field and Planetary Camera (WF/PC) on the Hubble Space Telescope. It reveals a complex structure of star formation close to the radio nucleus. (*E. Shaya and D. Dowling, University of Maryland; WF/PC Team; NASA*)

Superwinds. Energetic outflows of hot material have been discovered along the minor axes of some nearby starburst galaxies. Revealed in x-rays and emission lines of hydrogen, these so-called superwinds are believed to be ejected by the combined energy of many supernova explosions in the starburst.

Causes of starburst episodes. There are several theories as to why starbursts occur. One likely cause is the interaction between two galaxies as they pass close to, or collide with, one another. The tidal forces generated result in shock-wave compression of the interstellar material, loss of angular momentum and infall of material into the central regions of the galaxy, and star formation in the compressed clouds. Many starburst galaxies show evidence of interactions, including distorted appearance and long, wispy tails of material (Fig. 1). Not all starbursts can be due to interactions, however, since some display no evidence for any recent disturbance. Other mechanisms that are thought responsible for high star-formation rates in galaxies are very strong spiral density waves and central bar instabilities.

Relationship to other objects. Nuclei of starburst galaxies resemble active galactic nuclei in being bright and pointlike. However, active galactic nuclei are believed to differ from starbursts in harboring a central massive black hole, since they display, among other phenomena, an emission-line spectrum that is excited by a more powerful source of ultraviolet energy than can be provided by young stars. Some galaxies display the characteristics of both starbursts and active galactic nuclei, with a ring of star formation surrounding an active galactic nucleus. It is not known whether there are direct physical relationships between starbursts and active galactic nuclei. SEE ASTROPHYSICS, HIGH-ENERGY; BLACK HOLE; GALAXY, EXTERNAL.

Carol J. Lonsdale

Bibliography. D. A. Allen, Star formation and IRAS galaxies, *Sky Telesc.*, 73:372–374, April 1987; W. C. Keel, Crashing galaxies, cosmic fireworks, *Sky Telesc.*, 77:18–25, January 1989; C. Lacey, Galactic evolution: Starbursts, quasars and all that, *Nature*, 340:675–676, 1989; A. Prestwich, Starburst galaxies: More heat than light, *New Scient.*, 114:46–49, 1987; D. W. Weedman et al., NGC 7714: The prototype starburst galactic nucleus, *Astrophys. J.*, 248:105–112, 1981.

Stellar evolution

The large-scale, systematic, and irreversible changes with time of the structure and composition of a star. Stars are born, age, and die much like living beings, but the time scales involved are incomparably longer: for stars like the Sun, the lifetimes are on the order of 10^{10} years, and even for the most massive stars, whose life expectancy is much shorter, the life-span is still on the order of 10^7 or 10^6 years. As a consequence, stellar evolution cannot be observed directly, except for rare cases of objects "caught" at stages of unusually fast structural changes.

Hertzsprung-Russell diagram. Evidence on stellar evolution is therefore mainly indirect. Many stars are observed at various stages of evolution; the task is to identify these stages. It is useful to plot two of the most important stellar characteristics: luminosity (the total amount of energy emitted per second) and effective temperature (which represents the stars' surface conditions) against each other in a Hertzsprung-Russell diagram, as is done schematically in **Fig. 1**. Most stars will fall on the main sequence, which runs diagonally across the diagram from hot and luminous stars in the upper left corner to cool and faint stars in

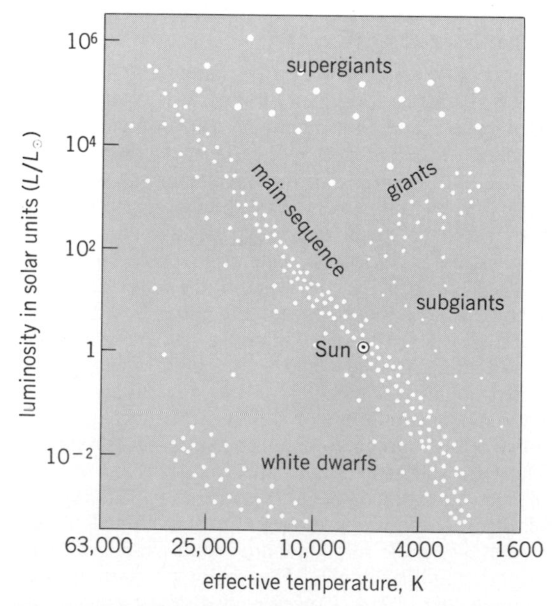

Fig. 1. Hertzsprung-Russell diagram plotted schematically for comparison with evolutionary tracks in Figs. 2–4. °F = (K × 1.8) − 459.67.

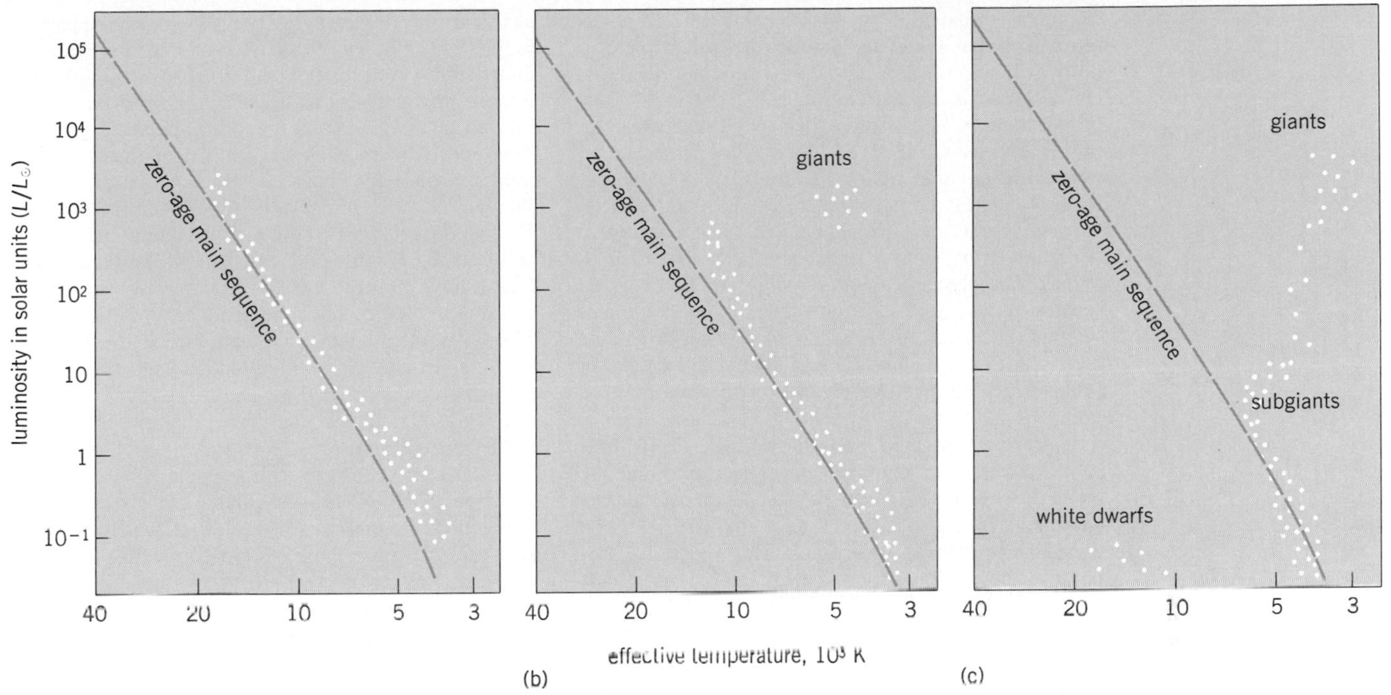

Fig. 2. Schematic Hertzsprung-Russell diagrams for open star clusters based on model calculations. (*a*) A young cluster (age about 10^7 years). Almost all stars lie along the main sequence; only those of very low mass are still contracting to it, while the most massive ones have already begun to move toward the giant region. (*b*) A middle-age cluster (age about 10^8 years). Low-mass stars are now on the main sequence, but most massive ones may have already exploded as supernovas or may have shed most of their mass and become white dwarfs. (*c*) An old open cluster (age about 4×10^9 years). All stars more massive than the Sun have evolved into subgiants or giants, there are some white dwarfs, and a few stars have disappeared in supernova explosions. °F = (K × 1.8) − 459.67.

the lower right corner. (In precise diagrams, based only on directly observed quantities, visual or photographic absolute magnitudes are usually given, and then the main sequence is not a straight line.) In addition, there are two fairly distinct groups: the rather cool but very luminous giants (and even more luminous supergiants) and the fairly hot but quite faint white dwarfs.

A star's luminosity depends on its effective temperature T (which characterizes the temperature at its surface) and on the total surface area as $L = 4\pi\sigma R^2T^4$. The multiplicative constant $4\pi\sigma$ disappears if the star's luminosity L is expressed in units of solar luminosity (which is 4×10^{33} ergs or 4×10^{26} W), and the stellar radius R in terms of the solar radius (7×10^5 km): then $L = R^2\ (T/T_\odot)^4$, where the solar effective temperature is $T_\odot$ = 5780 K. The effective temperatures of most giants are lower than that of the Sun, yet they are much more luminous; therefore their radii must be several orders of magnitude (10^2–10^4) larger than the solar radius. On the other hand, by a similar argument, white dwarfs must be much smaller, and indeed are of about the size of the Earth, that is, 0.01 solar radius. *SEE HERTZSPRUNG-RUSSELL DIAGRAM*.

Observational evidence. Although theoretical understanding of how stars are built and how they generate energy is necessary to determine what makes one star a giant and another one a white dwarf, observations alone contribute important information. Of particular value are studies of star clusters. These stellar systems consist of stars that were born at approximately the same time. **Figure 2** shows schematic Hertzsprung-Russell diagrams of three star clusters. In Fig. 2*a* there are no giants; all stars lie along the main sequence band, and only the hottest and most luminous ones tend to be displaced to the right of it. In Fig. 2*b* this trend is much more pronounced; in fact, the top of the main sequence is missing, and a few cool giants (of reddish color) are observed instead. In Fig. 2*c* all the more luminous stars are now red giants, while the main sequence is populated only at its faint red end. From this evidence it can be inferred, and theoretical studies of stellar evolution confirm it, that stars evolve from the main-sequence to the giant stage. Thus the sequence of the three clusters in Fig. 2 is essentially a time sequence, from the youngest to the oldest cluster. *SEE STAR CLUSTERS*.

Stars are essentially thermal engines: they shine because they produce energy. The main-sequence stage is clearly one in which a typical star remains for a very long time, since so many stars are observed at this particular stage. Obviously, then, hot and luminous stars consume energy so fast that the main-sequence stage is shorter for them than for stars like the Sun. The mass M of certain stars (members of suitable binary systems) can be determined and it is found that along the main sequence there exists a clear mass-luminosity relation: more massive stars are more luminous. The dependence is very steep and can be written approximately as $L \propto M^4$. If the mass of the star is doubled, its total energy output increases 16 times; a star of 10 solar masses radiates as much energy as about 10,000 Suns. As discussed below, the source of stellar energy is mainly thermonuclear conversion of hydrogen into helium. Since the initial chemical composition of all stars in a star cluster is most likely the same, the amount of nuclear fuel initially available for any star is proportional to its mass. However, since the star consumes this energy at a rate proportional to M^4, its life expectancy t decreases rapidly with mass: $t \propto M^{-3}$. SEE MASS-LUMINOSITY RELATION.

Stars do not start their lives on the main sequence. In the youngest clusters (Fig. 2*a*), the faint (that is, low-mass) stars tend to lie somewhat above the main sequence. Very young stars have been identified as rather cool objects, shining mainly in the infrared spectral region, and still surrounded by agglomerates of gas and dust, from which they obviously were born. As Fig. 2*a* shows, even the earliest stages of stellar life proceed more slowly for the less massive stars.

The Hertzsprung-Russell diagram in Fig. 1 has gaps between the main-sequence hot stars and the giants, and again between the main sequence and the white dwarfs: almost no stars have characteristics that would place them in these gaps. This situation arises because the evolution across these gaps is relatively very fast. There is no doubt that the white dwarfs represent the final evolutionary stage for stars of about the solar mass. Since stars evolve first from the main sequence toward the region of the red giants, the subsequent evolution must bring them across the Hertzsprung-Russell diagram from the giants to the white-dwarf region. Such evolution must be rapid, since few transition objects are observed. One such kind of object is the planetary nebulae, of which the Ring Nebula is a well-known example. In contradiction to their traditional but quite misleading name, planetary nebulae are actually old stars that have blown away their outer layers, which are still visible as rather symmetric nebulous shells surrounding the central star. The central star, a core of the parental star, is a small object evolving rapidly into the white-dwarf stage. SEE NEBULA; PLANETARY NEBULA.

Thus observations of various types of stars provide many clues about stellar evolution, but these clues are very incomplete, and at times ambiguous or confusing. The theory of stellar evolution must rely heavily on good understanding of the physical laws and principles on which the stars are built, and according to which they behave. Fortunately, these laws are in many respects simpler than the laws governing biological evolution, and there is confidence that many aspects of stellar evolution are reasonably well understood.

Role of gravitation. Stars are believed to be formed from condensations in the interstellar medium. Once a reasonably massive initial condensation forms (perhaps in a random process, or in local condensations due to traveling shock waves), it attracts surrounding particles by its force of gravitation. Although other forces may also assist, gravitation appears to be the force that builds stars. However, once they have formed, gravitation tends to destroy them. This paradox is inherent in the very nature of gravitation. According to Newton's law of universal gravitation, any two massive particles are mutually attracted by a force inversely proportional to the square of the distance between them. A particle located in the force field of many other particles, as inside or in the vicinity of a star, will be subject to a resultant net force due to the whole assembly. If this assembly is spherically symmetric, it will have a definite center. Any particle at a distance r from this center will be attracted only by particles lying inside the sphere of radius r, that is, closer to the center than the particle under study. And the net combined attraction of those interior particles acts as if their combined mass were located at the center of the star. Specifically, a gas molecule of mass m at the surface of the star is attracted toward the star's center by the total mass of the star M, and the resulting gravitational force is $F = GMm/r^2$, where r now is the radius of the star and G is the gravitational constant. If the particle can fall or move closer to the center, it does so, thereby increasing the force, since the distance r has decreased. This goes on all the time, and for all the particles forming the star. Gravitation tends to squeeze the star to smaller and smaller dimensions, but every contraction only strengthens the force, thereby compelling further contraction. SEE GRAVITATION.

A star's life is a perpetual struggle against self-destruction, namely, collapse under the effect of self-gravitation. At the outset, the object (called a protostar) is little more than a huge contracting cloud of cold material. Its contraction accelerates all the time for the reasons just explained, and would collapse outright into a black hole if forces were not generated to counteract the gravitational contraction. Such a force is the thermal pressure of the gas. As the particles fall closer together, their potential energy is converted into kinetic energy of motion, and redistributed in numerous collisions; the resulting random motions are perceived as rising temperature of the gas, and associated with it is the pressure which eventually begins to balance gravitation. The thermal pressure is proportional to the density of the gas, and to its temperature. As the star contracts, its density increases, but its temperature is more efficient for producing a sufficiently high gas pressure, which can rise to millions of degrees if the star has contracted sufficiently. If a star has more mass, its self-gravitation is stronger; therefore a higher gas temperature is needed to halt the contraction. This results in higher luminosity. The deeper layers in the star must support a greater weight than the layers lying above them; therefore the gas pressure, as well as the temperature within the star must rise with depth and the temperature peaks at the center, where it must be very much higher than at the surface. SEE GAS.

A temperature gradient is established in the star in the sense that temperature steadily decreases from the center to the surface. According to the second law of thermodynamics, energy must flow down the temperature gradient. A star must therefore shine simply because it is hotter inside, even if it does not possess adequate energy sources. Higher central temperature implies a steeper temperature gradient, hence also larger flow of heat. This is why more massive stars (which have a higher central temperature, as explained above) are more luminous. But loss of thermal energy means that the star would cool off, and hence the gas pressure would decrease because of decreasing temperature; gravitation would prevail, and the star would continue to contract. Indeed, this is what happens when the star does not have, or cannot tap its, nuclear energy resources. It is bound to contract, albeit rather slowly; each step in contraction releases a large amount of potential energy, of which about half is converted into internal thermal energy (temperature rises), while the other half escapes as radiation. Because of the stars' enormous masses, gravitational potential energy is a large source of radiative energy, and the Sun could shine at its present luminosity for about 10^8 years without nuclear sources. However, 10^8 years is much shorter than the known geological history of the Earth and of life on it, during which history the Sun must have been shining at about the same luminosity as it is now.

Thermonuclear reactions. When the central temperature of a contracting star reaches about 10^7 K, thermonuclear reactions are ignited in which hydrogen is converted into helium. For every gram of hydrogen involved in this reaction, 0.0071 g of material is converted into energy, yielding 6.4×10^{11} joules of energy. Conversion of 6×10^8 metric tons of hydrogen per second is needed to cover the present energy output of the Sun; yet the Sun has such a large store of hydrogen that it can shine at this rate for about 10^{10} years, and has lived for only about half that time. The speed at which the conversion of hydrogen into helium proceeds is very sensitive to temperature. Practically no reactions occur at temperatures below 10^7 K; but with increasing temperature above this threshold, the rate increases so rapidly that a region at a temperature of 2×10^7 K generates about 130,000 times more energy than a region at a temperature of 10^7 K. Since more massive stars must be hotter at their centers, they also produce much more energy, and this explains the mass-luminosity relation mentioned above. *SEE CARBON-NITROGEN-OXYGEN CYCLES; PROTON-PROTON CHAIN.*

From main sequence to red giant. The onset of thermonuclear reactions provides sufficient heat to keep the pressure sufficiently high to halt the star's contraction. The star enters a long and quiet period of its life, and is now a main-sequence star. Eventually, however, all the available hydrogen in its interior will be consumed. This is only about 15% of all the hydrogen in the star, but the rest cannot be used since it lies in regions where the temperature is not sufficiently high. As the energy source at the star's center vanishes, gravitation prevails again, and contraction resumes. Contraction heats up the star everywhere, but most important is the rise in temperature in the layers immediately surrounding the helium-rich core in which hydrogen has been exhausted. In a relatively thin shell, which still contains hydrogen, the temperature will rise again high enough to ignite the thermonuclear conversion of hydrogen into helium. Instead of a nuclear-burning core, the star now has a nuclear-burning shell. This shell-burning stage is short, since the energetically inactive core cannot maintain a sufficiently high temperature to support the weight of the outer layers by gas pressure, and begins to contract again. However, this time the star is no longer a homogeneous body: internal motions are, roughly speaking, reflected on the nuclear burning shell, and do not reach the outer layers. Thus, while the core is contracting, the envelope outside the nuclear-burning shell begins to expand. The size of the star begins to increase, but at the same time the outer layers cool off as they expand, there is a drop in the surface temperature, and the color changes from white to yellow to red: the star evolves rapidly into a red giant.

From red giant to supernova. If massive enough (above about 0.7 solar mass), the star reaches, in its contracting core, a temperature of or above 10^8 K. Then a new thermonuclear reaction starts in the core, namely, a transformation of three helium nuclei into one nucleus of carbon. This new energy source again temporarily stabilizes the star and halts core contraction and envelope expansion. A second quiet stage ensues, which explains why a fairly distinct and relatively populous category of red giants is observed. The high central temperature demands high energy production, which makes the giants very luminous. However, they are considerably less common objects than the main-sequence stars. This is because the quiet red giant stage is much shorter. As nuclear fuel, helium is much inferior to hydrogen, and is therefore consumed much faster. Another contraction of the core follows, accompanied eventually by an enormous expansion of the outer layers, and the star becomes a red supergiant, with a radius of 500 solar radii or so. If the star is sufficiently massive to become sufficiently hot at its center, further reactions may occur, building up heavier and heavier elements, all the way up to iron, which has the most compact nucleus. Elements heavier than iron cannot be built in thermonuclear reactions that release energy; on the contrary, energy must be supplied. In fact, the energy-releasing interactions of elements heavier than oxygen already yield so little that all the later stages of stellar evolution must be quite short.

The first nuclear reaction, conversion of hydrogen into helium, consumes about 81% of the total available nuclear energy, fusion of helium into carbon and oxygen raises the percentage to 90%; and by the time silicon is dominant in the innermost core, only about 3% of the nuclear energy remains available. This declining fuel efficiency is one important reason for the quickly decreasing durations of the successive nuclear burning stages, but there are others. The star also gradually has less fuel available, since each following reaction occurs always within a smaller core mass, as the required high temperature is reached only there. With less fuel and worse quality of it, the star faces an ever larger demand on energy output, since the increasingly higher central temperature (reaching 3×10^9 K before the collapse) means a greater temperature gradient and therefore higher luminosity. Also, with increasing temperature, a larger and larger fraction of the energy generated eventually turns into neutrinos, which escape without contributing to the required thermal gas pressure. Ultimately, the photons produced become so energetic that they break up the heavier atomic nuclei, usually all the way back to the very stable nucleus of helium, and the energy spent in this photodisintegration is again lost from the much-needed support of the star against gravitational collapse. All these factors make the final nuclear burning stage (silicon into iron) last probably only days, rather than years.

In the early, sufficiently long evolutionary stages, the envelope of the star has sufficient time to adjust to the interior conditions. It expands greatly when the helium core contracts before helium ignition, but contracts when helium is ignited explosively. Thus, in its outward appearance, the star at first crosses the Hertzsprung-Russell diagram from left to right, changing from a blue star into a red supergiant; but during the envelope contraction, it makes a loop backward into the blue region (**Fig. 3**). Depending on mass and chemical composition, several such loops may occur, followed by a return to the red supergiant region. However, at the advanced stages of nuclear evolution of the core, the envelope cannot follow the rapid succession of core expansions and contractions. This is probably why, to astronomers' initial surprise, the

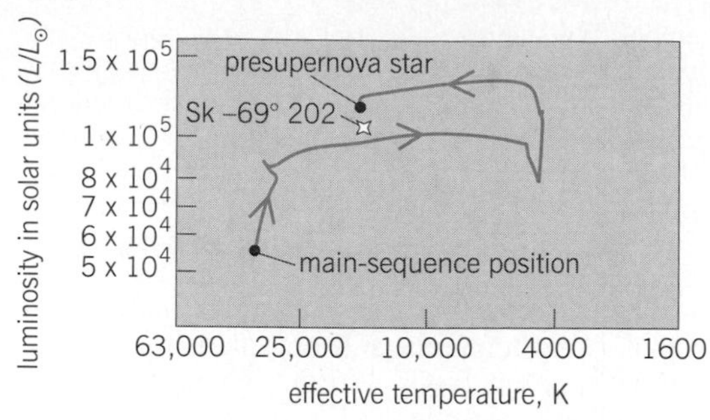

Fig. 3. Evolutionary track on the Hertzsprung-Russell diagram for a star of 20 solar masses and composition appropriate to the Large Magellanic Cloud. This is the approximate mass and composition of the star Sk −69°202, which exploded as supernova 1987A. The star spends 90% of its life near a position on the main sequence before evolving into a red giant on the right-hand side of the figure, and then looping backward into the blue region on the left. Location of the presupernova star agrees well with observed properties of Sk −69°202. (*After S. E. Woosley and M. M. Phillips, Supernova 1987A!, Science, 240:750–759, American Association for the Advancement of Science, 1988*)

star that exploded as supernova 1987A in the Large Magellanic Cloud was a blue supergiant. *See Giant star; Nucleosynthesis; Supergiant star.*

Supernova outburst. When all nuclear sources in the core are exhausted and the gas pressure in the core becomes much smaller than the weight of the outer layers, the innermost core collapses within a fraction of a second from about the size of the Earth to a sphere of at most a few tens of kilometers across. There it stops and a neutron star is formed, provided the collapsing mass is not too large: if it exceeds 3–4 solar masses, not even the pressure of degenerate neutron gas will stop the collapse, and it should continue into a black hole stage. The collapse involves squeezing free electrons into atomic nuclei, where they combine with protons to form neutrons, with a subsequent disintegration of the nuclei into a sea of individual neutrons. This, and associated processes, releases an enormous outpouring of neutrinos, which thus carry away the bulk of the energy released by the exploding supernova. *See Gravitational collapse.*

Shortly after the collapse of the inner core, the outer parts of the star follow, and hit the rigid wall of the neutron star. An extremely powerful shock wave forms, which travels outward and changes the implosion into an explosion, throwing the bulk of the star's body into space. The photons generated by sudden nuclear reactions of the abruptly heated material announce the supernova explosion to optical observers; however, as was actually observed in the case of supernova 1987A, the photon signal is preceded by a neutrino signal.

Rarity. A supernova explosion is a spectacular but very rare event. Only three supernova explosions have been sufficiently well documented and directly observed in the Milky Way Galaxy during historical times: (1) Chinese and Japanese astronomers observed in 1054 the apparition of a "guest star" at the place where the Crab Nebula in Taurus is now observed; (2) Tycho Brahe observed a supernova in Cassiopeia in 1572; (3) Johannes Kepler observed a supernova in Ophiuchus in 1604. But of these three galactic supernovae, only the first one probably was due to a collapse of a massive star. The other two were supernovae of type I, in which the cause of explosion is rather obscure; they seem to be associated with low-mass stars and may be due to an accretion process in which a degenerate white dwarf receives so much mass from a binary star companion that it exceeds about 1.4 solar masses (a value known as the Chandrasekhar limit) and collapses into a neutron star. On the other hand, large parts of the Milky Way Galaxy are hidden from the Earth by clouds of interstellar dust, and even objects intrinsically as luminous as supernovae become invisible. Therefore, when supernova 1987A flared up in the nearest stellar system outside the Milky Way Galaxy, the Large Magellanic Cloud, at a distance of about 160,000 light-years, it was the next best event to a nearby galactic supernova for astronomers.

Supernova 1987A. This supernova flared up on February 23, 1987, and was observed optically right from the onset of the explosion. The neutrinos emitted by the collapsing inner core were recorded by the Kamiokande II detector in Japan and by the IMB detector near Cleveland, Ohio, 3 h before the outburst was noticed optically. At its peak visual luminosity, the supernova was of the third magnitude, easily visible to the unaided eye from southern latitudes. Yet this peak brightness was lower than expected from a typical supernova. The explanation probably is that the star exploded when it was a B3 supergiant, hot even on the surface (Fig. 3) but relatively small (about 50 solar radii). Normally it is expected (although it has not been possible to check it) that just before the explosion the star should be a red supergiant with a radius about 500 solar radii. The parental star of the supernova 1987A probably had been a red supergiant (Fig. 3), perhaps 10,000 years before the outburst. Indeed, as the circumstellar material is gradually illuminated by the advancing flash of the supernova, it appears that there is definite evidence of the material ejected by the ancestor star less violently when it was a red supergiant.

The story of supernova 1987A is understood as follows (Fig. 3). Initially, it was probably a star of about 20 solar masses, with a radius of about 6 solar radii and a luminosity near 60,000 solar luminosities. The entire life-span of the star may have been about 10^7 years. Of that time, 90% was spent near the main sequence in the core hydrogen-burning stage; after a rapid expansion, core helium burning in the red giant stage ensued, lasting about 10^6 years. During all of this time, the star was losing mass from its surface by a powerful stellar wind, which may have carried away some 5 solar masses. *See Supernova.*

Final stages of typical stars. The paucity of supernovae in the Milky Way Galaxy, and the statistics based on more than 600 distant supernovae so far observed in the entire observable universe, show that a supernova explosion is a rare event. Perhaps there are two or three supernova per galaxy per century. The conclusion is unescapable that only a small fraction of stars end their lives in a supernova explosion, probably only stars more massive than about 8 solar masses. The less massive stars, which represent a vast majority, must have a different fate.

An important property of stars of smaller mass, such as the Sun, is that the densities in their central cores are rather high. Near the center of the Sun, the density of material is on the order of 10^2 g/cm^3. In more massive stars, with higher central temperatures, the pressure of radiation becomes important and helps to support the weight of the outer layers; in Sun-like stars the radiation pressure is practically negligible, and the cores must become rather dense in order to develop sufficiently high gas pressure to balance the gravitational contraction. As discussed above, with the exhaustion of hydrogen in the core, further contraction must set in, raising the material density even higher. At a certain stage (at a lower density for less massive, cooler stars), the density becomes so high that electron degeneracy occurs. This phenomenon is a consequence of the uncertainty principle, one of the basic laws of quantum mechanics, which specifies, among other things, that the momentum of a particle known to be within a certain volume cannot be reduced below a value that depends on this volume, and that it increases as the volume decreases. Thus, when more electrons are forced to share the same very small volume of space, they will acquire high velocities; and high velocities imply more violent collisions and therefore larger gas pressure. Thus an increase in the density of the stellar material, diminishing as it does the volume per particle, introduces a new type of pressure, the pressure of the degenerate electrons. This pressure is independent of temperature and eventually stabilizes the star.

When degeneracy sets in at a certain stage of evolution (this happens during core contraction between two nuclear-burning stages), it halts the contraction. As a consequence, the core temperature no longer increases, and the next nuclear-burning stage is not reached. For example, a star with a mass of less than 0.7 solar mass, after it has exhausted all its core hydrogen, will never reach the temperature of 10^8 K needed to ignite the helium conversion into carbon. Such a core then produces no thermal energy through nuclear reactions, and no energy through contraction. However, the temperature gradient persists and the core continues to lose heat; therefore it begins to cool off.

It is important to realize the difference between thermal pressure of ordinary gas, and the pressure of degenerate electron gas. Both can, if sufficiently strong, halt gravitational collapse. But thermal pressure implies a thermal gradient and therefore energy losses. Thus, in the absence of nuclear energy production, the star has to shrink in order to replenish the energy losses; thermal gas pressure only slows down gravitational collapse into slow contraction. The pressure of degenerate electrons is independent of temperature and the size of a degenerate star does not decrease with time.

The final evolutionary stage of a star of small mass is one of complete electron degeneracy, which stabilizes the mechanical structure of the star. The star does not generate energy, and shines only by gradually releasing the heat energy accumulated at earlier stages in the random motions of the particles. A thin nondegenerate surface layer slows down the escape of the heat, so that this evolutionary stage is unusually long (10^9–10^{10} years). The star does not change in size; it only cools off gradually. At the early stages it is still hot enough to maintain a fairly high surface temperature, and it is observed as a white dwarf. As it cools off, it fades, and changes into something that ought to be called a dark dwarf. Because of their long lifetimes, many white dwarfs are known in the vicinity of the Sun, and only their extremely low luminosity (10^{-2} solar luminosity) makes it impossible to observe directly how numerous these objects are. *See White dwarf star.*

The low luminosity, combined with high effective temperature, implies that white dwarfs are very small stars, about the size of the planets (a few percent of the solar radius). Yet their masses are comparable to that of the Sun; only when a solar mass is squeezed into the size of the Earth does the complete electron degeneracy set in. Again, greater weight of the outer layers demands higher counterpressure, and therefore greater density. Therefore the radius of a degenerate dwarf is smaller for a larger mass of the dwarf, and theoretical calculations show that the radius tends to zero when the mass approaches the Chandrasekhar limit. Beyond that limit, no stable configuration is possible for an electron-degenerate dwarf, since the pressure of the degenerate electrons is not strong enough to counteract self-gravitation. By the sheer weight of the material, electrons are squeezed into atomic nuclei and combine with protons to form neutrons. This permits further drastic contraction of the star, which continues until the neutrons are squeezed to such a small volume per particle that the uncertainty principle begins to operate again, this time on the degenerate neutron gas, and generates a new strong pressure opposing further contraction. A star can stabilize again, and exist for a long time as a neutron star. Neutron stars are incomparably smaller than even the white dwarf; a neutron star of 1 solar mass would have a radius of about 6 mi (9 km). Objects so tiny are not easy to detect by usual optical observations. Fortunately, at least some neutron stars are observed as pulsars, since they produce flashes of radiation by spinning rapidly in a strong magnetic field. *See Neutron star; Pulsar.*

One way of producing neutron stars is the supernova explosion. The supernova observed in 1054 by Chinese astronomers in Taurus left behind a nebulosity (Crab Nebula) surrounding a pulsar. It is not clear if all supernovae leave a stellar remnant. Theoretical calculations show that if the mass of the remnant is larger than about 3 solar masses, not even the pressure of the degenerate neutrons can stop its collapse, and no other physical force is known that could counterbalance gravitation under such circumstances. Therefore it has been concluded that such an object would collapse without a limit and become a black hole. Actually, if a star of 3 solar masses were to contract to within a radius of 5.6 mi (9 km), the object would appear as a black hole even if its further collapse were arrested by some unknown force inside that sphere, since the velocity necessary for escape from the surface would already surpass the speed of light, and therefore light would not be able to escape, a characteristic property of a black hole. Systematic searches for stellar black holes concentrate mainly on binary stars emitting strong x-rays. Only one candidate, the x-ray source Cygnus X-1, is universally accepted as a fairly strong candidate for a black hole. A few other x-ray binaries are regarded as possible candidates, the two leading ones being the x-ray source LMC X-3 in the Large Magellanic Cloud, and a transient x-ray source called A0620-00. *See Black hole.*

As a rule, however, the x-rays are generated by a neutron star, not by a black hole. Black holes, although of immense interest, do not seem to play an important role in stellar evolution. The vast majority of stars apparently do not collapse into black holes; some of them end as neutron stars, but for most the degenerate dwarf stage appears to be the end stage.

Stars less massive than about 0.07 solar mass never ignite hydrogen in their cores. For a certain period of time, they generate luminous energy by contraction, and are visible as red dwarfs. When electron degeneracy halts the contraction, the star cools off and becomes a brown dwarf. Since it is not known how many stars fall into this mass range, an intensive search for brown dwarfs is being conducted. The best chance to detect such a faint object is to search for it in binary systems containing a degenerate white dwarf.

Mass loss from stars. The paucity of neutron stars and black holes, and the abundance of white dwarfs, pose an interesting problem considering that no star more massive than the Chandrasekhar limit (1.4 solar masses) can become a degenerate dwarf. It must therefore be assumed that many stars lose so much mass from their surfaces that their total masses eventually drop below the Chandrasekhar limit even before they reach the final decisive stages of their lives; thus they rather peacefully settle down to the degenerate white dwarf stage. Several ways are known by which a star can efficiently shed mass. A good percentage of all stars (30–50%) actually live in binary systems where the components are so close to each other that they cannot expand freely into the giant or supergiant size. At a certain distance from the center of the expanding star (depending on the mass and distance of the companion), the gravitational attraction of the companion prevails, and matter begins to flow either to that star or away from the system. A star can lose a large fraction of its mass in this type of evolution, and many interesting phenomena are associated with mass transfer in binary stars, for example, novae and dwarf novae. *SEE BINARY STAR; NOVA.*

Another way of shedding mass is the phenomenon of planetary nebula, described above, in which the star blows away its envelope and only the core is left; the core of a giant or supergiant is an almost ready-made white dwarf. Possibly most of the stars with masses up to 4 solar masses pass through the planetary nebula stage and thus evolve into white dwarfs. Shedding of a planetary nebula envelope is a mild, rather slow process. An even less conspicuous way of losing mass is through stellar wind. Many stars exhibit continuous streaming of gas from their surfaces. Even the Sun is losing some mass in this way, but the mass loss rate is negligibly small, about 2×10^{-14} solar mass per year. But for stars in the giant and supergiant stages, the mass loss rate in such a stellar wind can be very much higher, between 10^{-10} and 10^{-5} solar mass per year. If such a condition prevails for 10^6 years or more, it can severely reduce the mass of the star, and thus change completely its future evolution. Strong stellar winds are observed in particular in very luminous stars, both hot blue supergiants and cool red giants and supergiants. It appears that the main driving mechanism is radiation pressure: upon gas atoms in hot stars and upon dust grains around cool stars. *SEE SOLAR WIND.*

Evolutionary models of stars. Knowledge of certain evolutionary stages of stars, in particular of the end stages and of the very earliest stages, is rather incomplete. On the contrary, evolution from the main-sequence to the red giant stage can be modeled successfully, and models of stellar structure and evolution match the observations very well. High-speed computers generate model sequences for different masses, such as those shown in **Fig. 4**. The models start on the main sequence at the moment when the star is supposed to start significant thermonuclear reactions (hydrogen burning). At that time, the star is assumed to be chemically homogeneous.

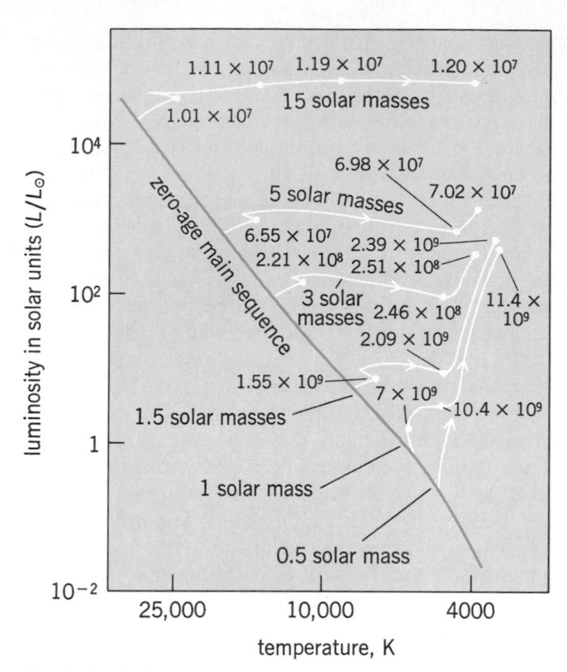

Fig. 4. Evolutionary tracks of stars from the main sequence to red giants based on model calculations. Ages, in years, measured from the time when the star was chemically homogeneous (zero-age main sequence) are indicated along the track. °F = (K × 1.8) − 459.67. *(After G. O. Abell, Realm of the Universe, 2d ed., Saunders College, 1980)*

It is assumed that the stars in the Sun's vicinity and in the spiral arms of the Milky Way Galaxy (population I stars) contain about 70% hydrogen and 27% helium (by weight), the remaining 3% being heavier elements. The multielectron atoms of the heavier elements play an important role, since (contrary to hydrogen and helium) they are not completely stripped of all their electrons even deep inside the star, and contribute significantly to the opacity of the stellar material, that is, to the fact that photons originating deep in the stellar interior typically travel only a fraction of 1 cm before they are absorbed. Successive absorptions and reemissions slow down the progress of the photons to the surface, which then typically takes 10^6–10^8 years. This slow process significantly contributes to the stability of the star, and also provides important shielding for life on Earth from the pernicious radiation generated at the center of the Sun.

Thermonuclear reactions take place only in the innermost part of the Sun and stars; 94% of the entire solar energy output is generated in the innermost 33% of the mass occupying 0.8% of the volume of the Sun, according to one model. The radiation generated there consists of gamma rays and hard x-rays that would be completely lethal if actually emitted by the Sun. However, as these photons filter through the inert and cooler layers, they get degraded to much less energetic photons, until they eventually emerge as mostly visible light or infrared radiation, with some ultraviolet rays, corresponding to the effective surface temperature of 5780 K.

Since the evolution on the main sequence is so slow, conditions of equilibrium may be imposed, in computing models, which must be very nearly exactly fulfilled: (1) hydrostatic equilibrium, requiring that, at each level, gravitation must be exactly balanced by gas pressure; and (2) thermal equilibrium, meaning that exactly all the energy produced per second must be radiated away in a second. An additional simplification is that the gas behaves as a perfect gas, even under conditions prevailing at the center of the Sun. Although the density there is 148 g/cm^3, compressing the atoms close together, the high temperature of 1.5×10^7 K significantly reduces the effective size of the atoms by stripping them of almost all their electrons (ionization is nearly complete), and enhanced thermal motions of particles more than compensate for the increased interactions due to overcrowding.

Energy generated in the core is transported through the star (and eventually to the surface) mainly by radiation; that is, the carriers are the photons. Only when the diffusion of photons becomes too slow because of high opacity does an additional mode of heat transport appear, namely, convection. Convective currents develop as hot masses of material rise upward, lose their surplus heat, and return back in a cooler descending current. Convection prevails in the cores of the hot and luminous stars, while the cores of the Sun-like and smaller stars are radiative. In cooler stars convection appears in the outer layers, where hydrogen and helium are only partially ionized. These extended convective envelopes are responsible for the outer appearance of red giants and supergiants, as well as for the pulsation instability of some of them, such as the cepheids. The third mode of heat transfer, conduction, is efficient only in degenerate dwarfs, where it is favored by the high density of material (10^5–10^8 g/cm^3) and by the large mobility of the free electrons. *See* *Cepheids*.

Although convection may in some cases mix stellar material over large regions of the star, it does not reach far enough to provide effective mixing between the core and the surface. Thus the star (except in the case of very small and cool stars) cannot bring a fresh supply of hydrogen to its core when it becomes depleted there. The gradual differentiation of chemical composition in the deep interior, caused by successive nuclear reactions occurring always in smaller and smaller central regions (where the required high temperature exists), remains largely unmixed, so that an old massive star should have an onionlike chemical stratification, starting with an iron core and proceeding outward through layers of successively lighter elements, until the surface layers should preserve their initial, hydrogen-rich composition.

The figure of 70% hydrogen and 27% helium mentioned above for the initial composition is only approximate in the sense that the ratio of hydrogen to helium is not easy to determine. The relation between mass and luminosity, and mass and radius, is more sensitive to the abundance of the heavier elements, since they determine the opacity in the stellar interior. Population I stars have about 3% of their mass in heavier elements, while the halo-type stars (population II), observed, for example, in globular clusters, contain less than 1% of their mass in the form of the heavier elements.

Peculiar and variable stars. A small number of stars deviate from the norm in various respects. Some have anomalous chemical composition; others oscillate (regularly or less regularly) in size and light; some brighten or fade abruptly; others have unusually extended atmospheres and lose mass at an unusually high rate. In many cases these phenomena occur in interacting binary systems. At other times they signal violation of the hydrostatic or thermal equilibrium, or local events which may be associated with magnetic fields, rapid rotation, and so forth. *See* *Stellar magnetic field*; *Stellar rotation*; *Variable star*.

Star formation. Formation of new stars is going on even now in the Milky Way Galaxy as well as in other spiral and irregular galaxies. Stars tend to be formed in groups called associations. The most likely sites of star formation appear to be dense, dark dust-and-gas clouds scattered along the spiral arms. It is believed that such a cloud gradually contracts and fragments; and each massive fragment condenses to form a protostar, in which internal gravitational attraction already prevails and leads to further shrinking of the body. It would eventually lead to a rapid and complete collapse into a black hole if the increasing gravitational force were not counterbalanced by gas pressure due to its compressional heating.

Theoretically the most difficult problem is to understand why and how the initial contraction starts in a cloud which has low density, large volume, and a considerable amount of random motion of the individual particles. It appears that the contraction is triggered by outside forces. Shock waves are traveling through the Galaxy. Some are created at random by supernova explosions. Others have a more organized character of a traveling spiral wave, which is believed to maintain the spiral structure of the stellar system. Such a spiral wave is a locus of stronger local gravitational potential, hence it tends to cause condensations of interstellar clouds. As other clouds, traveling at a different velocity, slam into this agglomeration, cloud collapse begins along a whole spiral arm and leads to an enhanced rate of star formation.

At the earliest stages, the newly born stars are shrouded by envelopes of in-falling dust, and can be observed only at infrared or radio wavelengths, to which the dust is more transparent. Theoretical models of early evolution agree that the collapsing cloud quickly develops a fairly dense core, which attracts more material from the surrounding cloud. When the star establishes radiative transport of heat in its interior and begins to shine, its radiation pressure pushes away the nonaccreted remnant of the dust "cocoon." The contraction of the star slows down as it progresses toward the stable configuration of a main-sequence star, along a path shown in **Fig. 5** according to model calculations. Stars lying above a line indicated on the figure are still completely surrounded by in-falling material, and are either completely hidden or shine only partially and to a variable degree. Stars of small mass descend at first vertically through the Hertzsprung-Russell diagram along the so-called Hayashi track, where convection prevails over almost the whole star and thoroughly mixes the material. When radiative energy transport prevails, the evolutionary tracks become predominantly horizontal. By that time the star becomes optically visible, but interaction with the remnants of the surrounding gas and dust cocoon, as well as possible internal instabilities, make it variable in luminosity and render its spectrum very peculiar. Such stars are known as the Herbig emission stars (when their masses are about 5 solar masses) and as T Tauri stars (with masses 0.5 to 2.5 solar masses). These objects typically appear in groups (associations) associated with nebulosities. Obviously stars are usually formed in groups, perhaps by fragmentation of a large protostar, and later on they disperse, although at times a strong gravitational bond

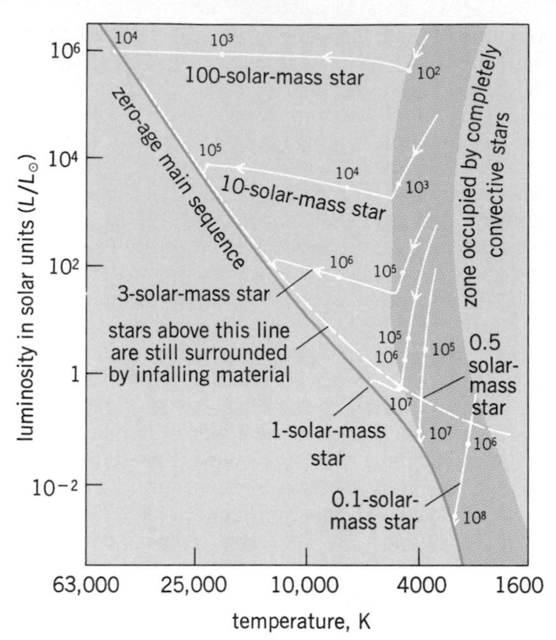

Fig. 5. Evolutionary tracks of young stars contracting toward the main sequence, according to model calculations. Numbers along tracks indicate ages in years. °F = (K × 1.8) − 459.67. *(After G. O. Abell, Realm of the Universe, 2d ed., Saunders College, 1980)*

holds them together in a star cluster. SEE GALAXY, EXTERNAL; MILKY WAY GALAXY.

Solar neutrino puzzle. Although progress in understanding of stellar structure and evolution has been very impressive, there is one major puzzle that affects the understanding of the structure of the nearest star, the Sun. In the course of the thermonuclear reactions converting hydrogen into helium, a certain amount of neutrinos are generated, which should pass through the Sun without interaction. A tiny fraction of these may be captured in an ingeniously devised experiment conducted by R. Davis in a deep mineshaft. However, the observed flux of neutrinos is significantly smaller than the solar models predict. It is not clear where the fault is: in understanding of stellar structure, or in the nature of the neutrinos. SEE SOLAR NEUTRINOS; STAR; SUN.

Mirek J. Plavec

Bibliography. G. T. Bath (ed.), *The State of the Universe*, 1980; I. Iben, The life and times of an intermediate mass star, *Quart. J. Roy. Astron. Soc.*, 26:1–39, 1985; R. Kippenhahn and A. Weigert, *Stellar Structure and Evolution*, 1990; R. P. Kirshner, Supernova: Death of a star, *Nat. Geog. Mag.*, 173(5):618–647, May 1988; C. Payne-Gaposchkin, *Stars and Clusters*, 1979; V. C. Reddish, *Stellar Formation*, 1978; R. A. Schorn, Happy birthday, supernova!, *Sky Telesc.*, 75:134–137, 1988; I. S. Shklovskii, *The Stars, Their Birth, Life, and Death*, 1978; S. E. Woosley and M. M. Phillips, Supernova 1987A!, *Science*, 240:750–759, 1988.

Stellar magnetic field

A magnetic field, far stronger than the Earth's magnetic field, which is possessed by many stars. Magnetic fields have been directly measured in only a few hundred stars. Nevertheless, magnetic fields play a fundamental role in determining the physics and structure of the atmospheres of stars.

The measurement of stellar magnetic field strengths in gauss and determination of their polarity as positive or negative (or as north or south) depends on studying the Zeeman effect in spectral lines. If a magnetic field is present in a stellar atmosphere, then spectral lines, which appear single in the absence of a magnetic field, split into multiple components that are also polarized. The separation of these components is proportional to the strength of the field. Measurement of the polarization yields information about the polarity of the magnetic field and about how the field is distributed over the stellar surface. SEE ZEEMAN EFFECT.

Unfortunately, other effects such as stellar rotation and turbulence—random motions of the gas in the stellar atmosphere—may broaden spectral lines by more than the splitting produced by the magnetic field. In this case, the individual Zeeman components overlap one another and cannot be seen or measured as distinct lines. For this reason, magnetic fields can be measured only in those stars with unusually large magnetic fields or small rotation and turbulence.

Detecting stellar magnetic fields can be extremely difficult. For example, Earth has a dipolar magnetic field (that is, a field with one north pole and one south pole) of about 6×10^{-5} tesla (0.6 gauss); the Sun has fields as large as 0.3 tesla (3000 gauss) or even more in transitory, localized regions, but these regions may have either positive or negative polarity. However, summed over the entire surface of the Sun, the net magnetic field is only two or three times larger than that of the Earth. SEE SOLAR MAGNETIC FIELD.

With modern techniques, the smallest magnetic field that can be measured in stars is 20 to 50 times larger than the magnetic field that the Sun would appear to have if it could be viewed only in the same way that stars are observed. Given this limitation, it is perhaps surprising that any stellar magnetic fields have been found at all, yet observations show that approximately 10% of the stars with temperatures in the range 8000–25,000 K have magnetic fields of a few hundred to several thousand gauss. The fields are predominantly dipolar, and the field is "frozen" into the surface of the star. As the star rotates, the north and south magnetic poles are observed alternately, and so the magnetic field varies in a period equal to the rotation period of the star.

Cooler stars apparently do not have simple dipolar fields but, like the Sun, have many small regions of opposite polarity. The net field is close to zero, and direct measurements are very difficult. There is, however, indirect evidence for magnetic fields in cool stars. In the Sun, localized "active regions" of unusually high temperature are invariably associated with intense magnetic fields. Spectral emission lines that are signatures of these active regions are seen in many stars. Therefore it is probable that these stars also have magnetic fields.

There are two hypotheses concerning the origin of magnetic fields. The simple dipole fields of the hot stars are thought to be "fossil" fields, remnants of the magnetic field originally embedded in the interstellar matter from which stars form. This field is compressed and magnified as the interstellar matter contracts to form stars. The chaotic magnetic fields of cool stars like the Sun are thought to be produced by a dynamo. Rotation, turbulence, and convection twist and concentrate magnetic lines of force within the

Sun. After reaching a critical magnetic intensity, the lines of force become buoyant, are carried upward by convection, burst through the solar surface, and are observed in active regions and sunspots. *SEE MAGNETOHYDRODYNAMICS; STAR; SUN.*

Sidney C. Wolff

Bibliography. R. M. Bonnet and A. K. Dupree, *Solar Phenomena in Stars and Stellar Systems,* 1981; E. F. Borra, J. D. Landstreet, and L. Mestel, *Annual Reviews of Astronomy and Astrophysics,* pp. 191–220, 1982; J. A. Eddy, *The New Solar Physics,* 1978; R. O. Pepin, J. A. Eddy, and R. B. Merrill, *The Ancient Sun,* 1979; A. M. Soward, *Stellar and Planetary Magnetism,* 1983; S. C. Wolff, *The A-Type Stars: Problems and Prospectives,* 1983; Ya.B. Zeldovich, *Magnetic Fields in Astrophysics,* 1984.

Stellar population

One of the categories into which stars may be classified, based on their place in the evolution of the galaxy that they occupy. The stellar component of the Milky Way Galaxy consists of three populations: the thin disk, the thick disk, and the halo. The thin disk, originally referred to as population I, is the youngest, located amid most of the molecular and cold atomic gas, and confined to a height of order 1 kiloparsec (3.3×10^3 light-years or 3.1×10^{16} km or 1.9×10^{16} mi) from the plane. The halo, roughly corresponding to the original population II, is a much older, far more extended structure, having an approximately spherical distribution with a scale length of order 3.5 kpc.

Development of concept. The existence of different ages in the stellar component of external galaxies was announced by W. Baade in 1944. During the wartime blackouts of Los Angeles, he was able to take the first deep multiwavelength photographs of the Andromeda Galaxy, M31. He found that the blue stars are confined to the disk of that galaxy, while the redder stars whose colors are more like the Sun or the globular clusters are more concentrated toward the galactic center and nuclear bulge. Baade referred to these stellar components as populations and labeled the younger blue disk stars population I. Those in the halo and nuclear spheroid, identified with the RR Lyrae variables (also known as cluster variables because of their preponderance in globular clusters), were called population II. A hypothetical population III has also been proposed to represent the first stars formed during the collapse of a galaxy, although there is no evidence that such stars have been discovered so far. *SEE ANDROMEDA GALAXY; VARIABLE STAR.*

The population concept is central to the understanding of chemical evolution of the Milky Way Galaxy. The globular clusters, the prototypical population II objects, have main-sequence turn-off ages of at least 10^{10} years and are among the most metal-deficient stars in the Milky Way Galaxy. These are widely distributed throughout the system and serve as markers in the halo to the mass distribution. On the other hand, the OB associations contain many massive stars with ages not greater than 10^7 years and have the highest metal abundances, often as much as a factor of 2 higher than the Sun. (At an age of about 5×10^9 years, the Sun was formed as a population I star as well.) The subdwarfs, the metal-deficient main-sequence stars, are a systematically older group of stars than those found confined to the galactic disk (see **illus.**). *SEE STAR CLUSTERS.*

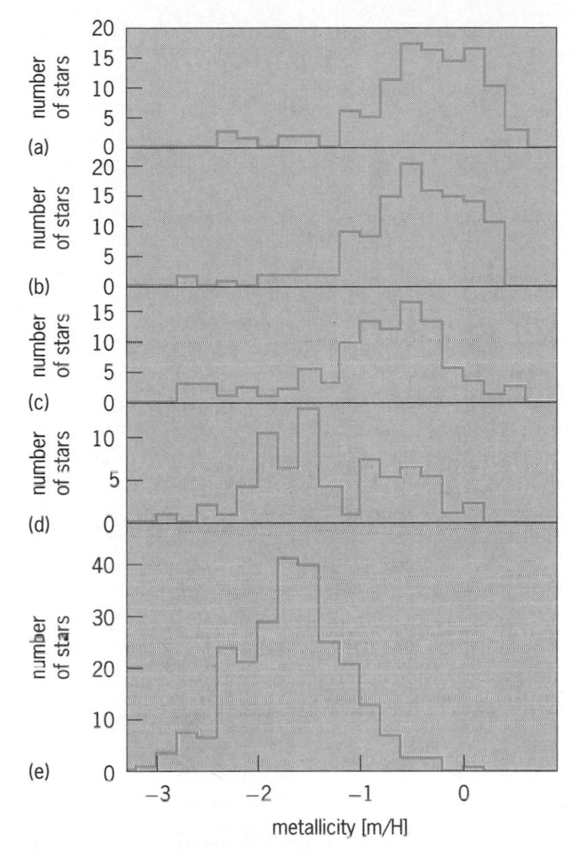

The metallicity distribution of stars as a function of velocity *V* perpendicular to the galactic plane. (Negative values of *V* indicate motion toward the plane.) The metallicity [m/H] is the logarithmic difference in the overall abundance of heavy elements (elements other than hydrogen and helium) with respect to the Sun. For example, in a star with [m/H] = 0 this abundance is equal to that of the Sun, while in a star with [m/H] = −3 it is 10^{-3} that of the Sun. (*a*) −30 km/s ≦ *V* ≦ −10 km/s. (*b*) −60 km/s ≦ *V* ≦ −30 km/s. (*c*) −100 km/s ≦ *V* ≦ −60 km/s. (*d*) −150 km/s ≦ *V* ≦ −100 km/s. (*e*) *V* ≦ −150 km/s. (*After B. W. Carney, D. W. Latham, and J. R. Laird, A survey of proper-motion stars, VIII. On the galaxy's third population, Astron. J., 97:423–430, 1989*)

Populations in galaxies. Fundamentally, stellar populations indicate that the Milky Way Galaxy has undergone an extended period of active star formation. The spheroid population is quite similar to elliptical galaxies, which are assumed to have ended active star formation many billions of years ago. Starburst galaxies give a counterexample of the fixity of the population distinctions, however, since they may have completed their primary star-forming phase long ago, but the injection of a new mass of gas through collisions or accretion in a large-scale flow can produce a dramatic turn-on of the star formation. *SEE GALAXY, EXTERNAL; STARBURST GALAXY.*

Characteristics of populations. The thin disk is the youngest component of the galactic stellar population. Still actively forming massive stars from molecular clouds, it is confined to within about 0.35 kpc of the plane. All of the stars have metallicities within, lying between about one-fifth and twice the solar value (illus. *a*), and star formation appears to have remained constant in this population for about the past 8×10^9

years. One reason for the relatively small thickness of the disk is the low velocity dispersion of the component stars, about 22 km/s (14 mi/s); their motion is completely dominated by the differential rotation of the disk. These stars are found associated with H II regions and OB associations as well as open clusters. Massive stars are uniquely formed in the thin disk, supporting the attribution of youth to the population. Finally, type Ia and type II supernovae, the death throes of massive stars, show a thin-disk distribution throughout the Milky Way Galaxy. *SEE INTERSTELLAR MATTER; MOLECULAR CLOUD; SUPERNOVA.*

The thick disk is an older population, approximately 8–10 × 10^9 years, roughly corresponding to the range between what was once called population II and population I. Its metallicity lies between about one-tenth and one-third of the solar value. The stars in this population are distributed over greater distances from plane, up to 1.5 kpc, and have correspondingly larger velocity dispersion, about 45 km/s (28 mi/s). This population also includes globular clusters and subdwarfs that overlap at the lowest end of the abundances with the properties of the halo globulars, although the system of old disk globulars is distributed differently than those of the halo. Type I supernovae are associated with this population. In the Milky Way Galaxy the spatial structure of this population roughly resembles that of an E7 elliptical galaxy.

Finally, lying around the disk and the nuclear spheroidal bulge, there is a halo that extends to considerable distances from the plane, some as distant as 30 kpc. This population has an age of order 10–15 × 10^9 years and a scale height of order 3.5 kpc or greater. The stars in this region have very large velocity dispersions, about 130 km/s (80 mi/s), and do not appear to participate in the differential rotation as much as other stars. Their metallicities are all lower than about one-twentieth that of the Sun and may extend down to 10^{-3} of the solar value (illus. *e*). The most metal-poor globular cluster belong to this population. This stellar halo is not the same as the so-called dark matter halo, but is probably embedded within it.

Delineation of populations. The population paradigm has become progressively more blurred with time, as the complex character of galactic star formation has become clearer, and there has been considerable debate over whether any of these populations are truly distinct. Indeed, the separation based on elemental abundances renders this question somewhat circular since these abundances are already embedded in the definition of the membership in the different populations. Nonetheless, there are some distinct differences in these populations that permit their delineation. For instance, the globular clusters, although separated into halo and disk systems, are distinctly older than any of the clusters found in the disk. The oldest disk open clusters, NGC 188 and M69, show abundances nearly as low as the most metal-rich globulars like 47 Tucanae, but are younger by nearly 3 × 10^9 years than the youngest of the disk globulars. The most metal-deficient globulars are considerably older than any of the stars in the disk and much lower in abundance. *SEE MILKY WAY GALAXY; STAR.*

Steven N. Shore

Bibliography. J. Bahcall and R. Soniera, Predicted star counts in selected fields and photometric bands: Applications to galactic structure, the disk luminosity, and the detection of a massive halo, *Astrophs. J. Suppl.*, 47:357–403, 1981; B. W. Carney, D. W. Latham, and J. R. Laird, A survey of proper-motion stars, VIII. On the galaxy's third population, *Astron. J.*, 97:423–430, 1930; G. Gilmore, I. King, and P. van der Kruit, *The Milky Way as a Galaxy*, 1990; D. Mihalas and J. Binney, *Galactic Astronomy*, 1981; A. Sandage, The population concept, globular clusters, subdwarfs, and the collapse of the Galaxy, *Annu. Rev. Astron. Astrophys.*, 24:421–458, 1986.

Stellar rotation

Surface rotational equatorial velocities, ranging from a few miles (or kilometers) per second up to 300 mi s^{-1} (500 km s^{-1}), exhibited by stars. At 300 mi s^{-1} (500 km s^{-1}) the stars become unstable. Observable directly in the Sun and in eclipsing stars, stellar rotation is detected through its broadening of the absorption lines by the Doppler effect. If the equatorial velocity is v_R and i is the inclination of the equator to the plane of the sky, the observed velocity at the limb is $\pm v_R \sin i$. A line is broadened, approximately, by $\Delta\lambda/\lambda = (v_R/c) \sin i$, where c is the speed of light. The strength of the line is unchanged, so that the profiles vary with increasing velocity (**Fig. 1**). The line in a nonrotating star is relatively narrow, absorbing about 45% of the light in the center and being about 1.0 nanometer wide. For higher rotations the line is broadened more and more, becoming shallower and wider. At the highest velocity shown, the star is near instability and the line is only 15% deep and 2.0 nm wide. A more direct way of measuring stellar rotation has also been developed, applicable to the more slowly rotating, cooler main-sequence stars. *SEE DOPPLER EFFECT.*

Since rotational angular momentum tends to be conserved, stellar rotation is inherited from the gases out of which the star is formed. Galactic rotation ensures that the gas has a systematic angular velocity large enough to cause rapid stellar rotation. Since the interstellar gas is turbulent, individual eddies, as well as the stars born out of them, need not have parallel angular momentum vectors. Many stars exist in binary or multiple systems with orbital angular momentum of similar origin. If the rotation and orbital revolution are synchronous, the momentum is largely in the orbital motion; probably prestellar condensations with excessive momentum form binaries by fission. *SEE ANGULAR MOMENTUM; ORBITAL MOTION.*

Distribution of velocities. Studies of absorption line broadening by G. A. Shajin and O. Struve in 1929 shows that $\langle v_R \sin i \rangle$ was a steep function of spectral type, with a maximum in the hot stars of type B, and with largest velocities found in stars of type Be. The latter have emission lines indicating mass loss and surrounding gaseous shells. If the axes are randomly distributed, $\langle v_R \rangle = (4/\pi) \langle v_R \sin i \rangle$. **Figure 2** shows schematically the distributions of $v_R \sin i$ for main-sequence stars of different spectral type, that is, different surface temperature, and for the B and Be stars. The F stars are slightly hotter than the Sun, and a few of them rotate with moderately high equatorial velocity. Most of them have very sharp lines near zero rotation. The hotter B and A stars have a relatively flat distribution of rotational velocities, reaching values of the order of 200 mi s^{-1} (300 km s^{-1}).

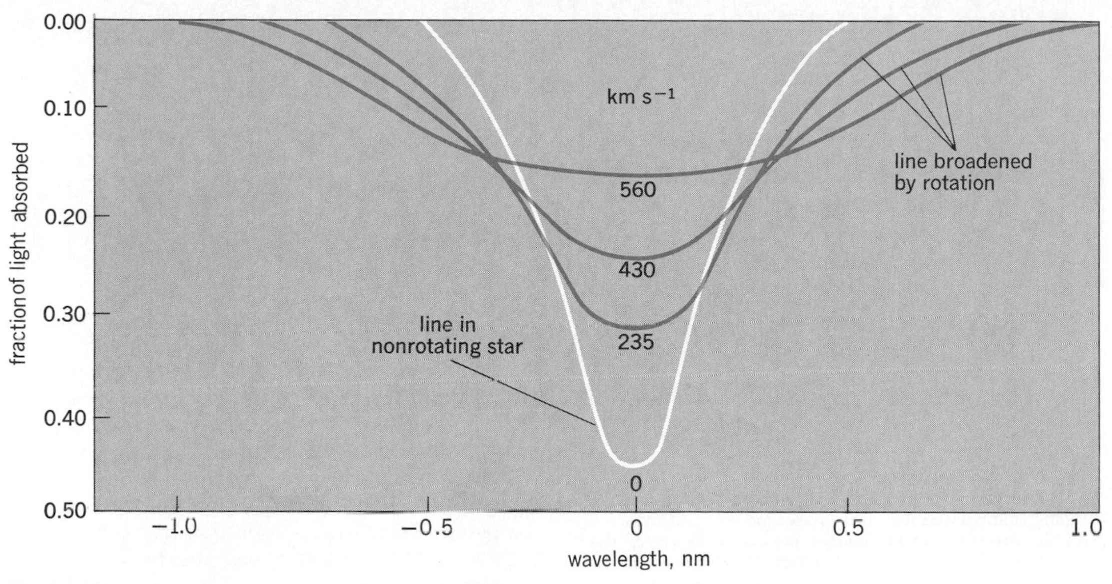

Fig. 1. Effect of rotation on the shape of a spectral absorption line. 1 km s^{-1} = 0.6 mi s^{-1}.

The Be stars are peaked at over 200 mi s^{-1} (300 km s^{-1}). They are objects in which the rotation is so rapid that instability has occurred and gas envelopes surround the star, producing emission lines. The observed rapid decrease of v_R with decreasing temperature is startling, and fundamental to much speculation on convection, stellar winds, and mass loss. *SEE HERTZSPRUNG-RUSSELL DIAGRAM.*

Observation of giants shows their $\langle v_R \rangle$ to be smaller at a given spectral type (with only a few exceptions) than for main-sequence stars. Clearly, their instability would begin at lower v_R, but the reduction of v_R has another cause. From stellar evolution theory, the envelope expands, resulting in a larger radius of gyration. The outcome is a reduction of the linear velocity of the surface layers. Depending on whether expanding stars rotate as rigid bodies, or whether angular momentum is conserved in shells, or is not conserved at all, one can expect different ratios of v_R for giants and main sequence stars. Superglants have turbulently broadened absorption lines, but in general show little or no rotation. Very old stars, which if unevolved must be cooler than the Sun, show little rotation. A few young stars in the process of formation show excessive rotation for their surface temperatures, and the v_R in young clusters is greater than in old. *SEE GIANT STAR; STELLAR EVOLUTION; SUPERGIANT STAR.*

Direct measurement. An improved technique of monitoring the strength of the Ca II emission cores of the broad Ca II absorption lines (known as the H and K lines) in cool stars by a photoelectric device has permitted the direct detection of rotation. On the surface of the Sun, brighter hotter regions of enhanced magnetic field (called plages) are clearly visible as sources of these cores (K at 393.3 and H at 396.8 nm). Even in the integrated light of the Sun, they are weakly visible. Young, main-sequence stars of types F–M have much stronger H and K emission, and apparently the plages in many stars are larger, brighter, and relatively longer-lasting features of their surface. Like a lighthouse beam, the calcium emission varies with the period of rotation (**Fig. 3**); HD 149661 has a 21.3-day period, while HD 152391 an 11.0-day period. The corresponding equatorial velocities of rotation are 1.2 and 2.5 mi s^{-1} (1.9 and 4 km s^{-1}), far less than can be measured for the B, A, and F stars illustrated in Fig. 2. *SEE SUN.*

This enormous advance is based on work begun in 1966 at Mount Wilson Observatory (California) by O. Wilson, who used a photometer that measured H and K lines on a spectrograph to search for long-term stellar emission-line variability resembling the solar cycle

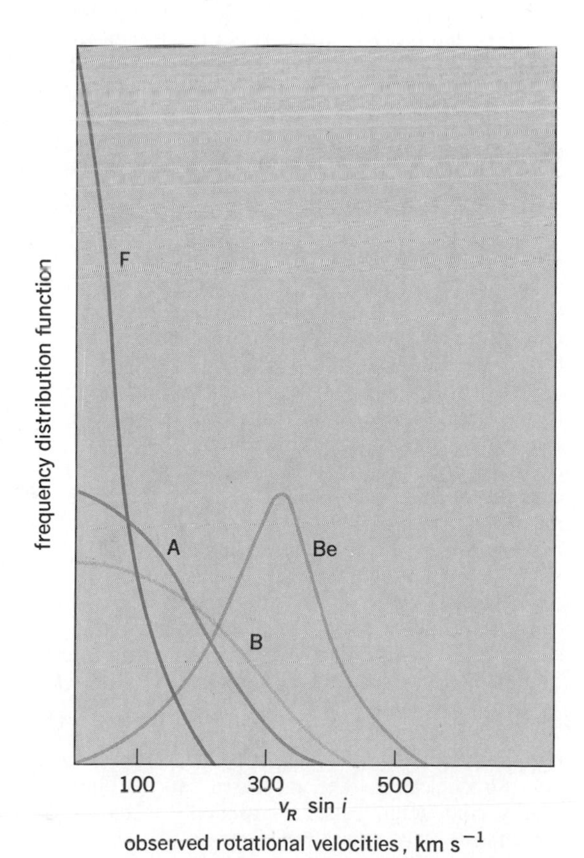

Fig. 2. Frequency distribution of observed rotational velocities for types of stars F, A, B, and Be. 1 km s^{-1} = 0.6 mi s^{-1}.

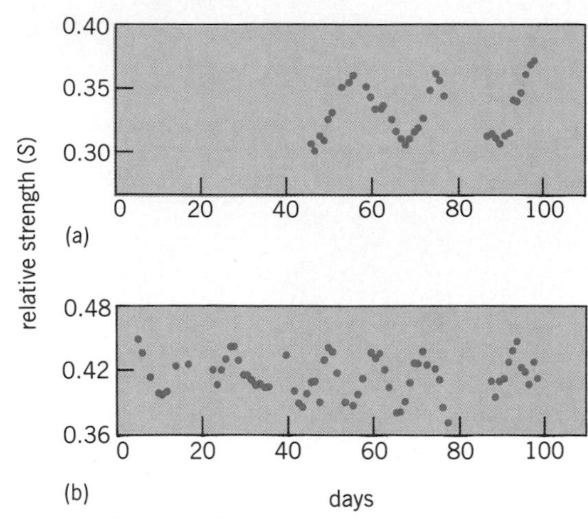

Fig. 3. Relative strength *S* of Ca II emission from two rotating main-sequence stars, observed over a few months. The rotation modulates the line strength as the brighter hemisphere passes. Horizontal scale is number of days after June 29, 1980. (*a*) HD 149661. (*b*) HD 152391. (*After S. L. Baliunas et al., Stellar rotation in lower main-sequence stars measured from time variations in H and K emission-line fluxes, Astrophys. J., 275:752–772, 1983*)

of 11.7 years, visible in magnetic fields, sunspots, and plages. After 11 years of observation, he established the existence of stellar activity cycles by the long-period behavior of such lines. But many of his stars showed rapid variations, appearing as noise in the long cyclic changes. Improved photoelectric apparatus was built for a spectrometer on the 60-in. (1.5-m) telescope on Mount Wilson, which was devoted full time starting in 1980 to the study of the rapid cycles (2.5–54 days).

As long as the magnetically active regions remain at the same longitude, the cycle is accurately observed; if new bright regions appear, there is a phase shift, but the period P is maintained. The observed velocity $V_R \sin i$ in B stars is replaced by $v_R = 2\pi R/P$, and periods can be measured directly and with accuracy improving with elapsed time. The value of $\sin i$ can then be determined if the star's radius is known in other ways, or statistical values of R can be determined from v_R and P.

The periods found (some conditioned by the limitations on observing time) range from a few days to 2 months (the Sun's period is 26 days). Rotation periods are largest for old stars. The distribution of frequencies of rotational velocity, v_R, for old stars has a median near 2.5 mi s^{-1} (4 km s^{-1}); few late-type main-sequence stars rotate as rapidly as 12 mi s^{-1} (20 km s^{-1}), quite unlike the early-type stars in Fig. 2.

The actual luminosity in the H and K emission, called F, decreases strongly with increasing period for stars of a given spectral type (or photoelectric color). The relation has been found to be given by Eq. (1),

$$F_H + F_K = 3.65 \log v_R + 0.55 \qquad (1)$$

indicating that F has a steep dependence on v_R with only small scatter. The measured ratio of line to nearby continuum, called S, generally is larger for cooler stars, with considerable scatter. But when stars of a given small range of color are isolated, slow rotation correlates with weak lines. This is shown in **Fig. 4**, where the curves are average relations of relative strength to rotation period, isolating stars with photoelectric colors (as defined by color index B − V) within a narrow range. SEE MAGNITUDE.

A physical theory connecting rotational velocity and line strengths uses the Rossby number of hydrodynamic theory. If the convective turnover time is long compared to the rotation period, more mechanical energy is dissipated, the stellar surface activity is increased, and presumably, stronger stellar winds are generated, blowing matter, angular momentum, and magnetic field into space. Thus, rotation will slow down and magnetic activity decrease with time.

Binary systems. Complex phenomena occur in close binary systems, where tidal drag ensures strong interaction between the stars. The theory of particle orbits in a rotating frame of reference was developed by J. L. Lagrange and has been extensively applied to the stellar case. Emission-line B stars often occur in double systems and, if an eclipse occurs, can be shown to be surrounded by a thin equatorial ring of hot gases. These gases, when traveling between the hot star and the observer, can also be seen by the absorption lines they produce. Complex, nonuniform streams exist and transfer matter from the larger to the smaller star. If two stars are separated by not more than their radii, a common gaseous envelope is formed (the so-called W Ursae Majoris stars). The brightness is continuously variable, in periods of a few hours, in these and other contact variables. This phenomenon has assumed importance with the discovery of binaries containing massive stars (or black holes), in which matter falls through an enormous potential gradient, emitting hard x-rays. This occurs either as the streams encounter an accretion ring or the collapsed star. SEE BINARY STAR; BLACK HOLE; ECLIPSING VARIABLE STARS.

Slowing of rotation. The Sun's surface rotates with a shorter period at the equator than at the poles. Its internal rotation could be at a greater angular velocity, and meridional currents are required. The outer-convective zone, however, is the most important feature of the entire phenomenon of rotation. All stars with surface temperature below 7500 K (13,000°F), type F2, have their hydrogen partially ionized just below the surface layers; convection develops and trans-

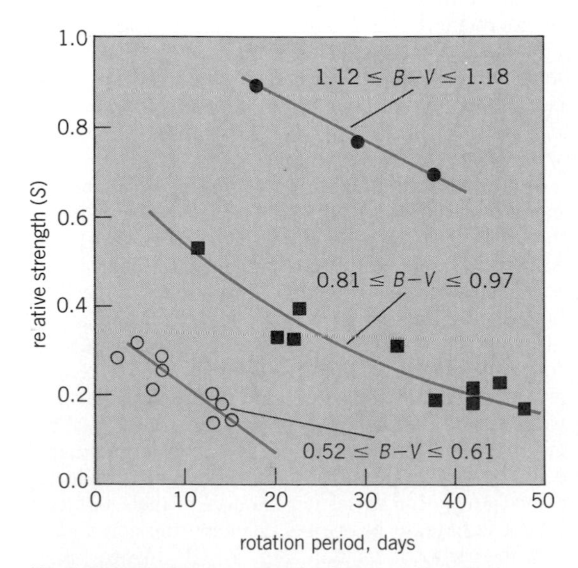

Fig. 4. Relation between relative strength *S* of Ca II emission lines and rotation period for groups of stars within narrow ranges of color, as defined by the color index B − V.

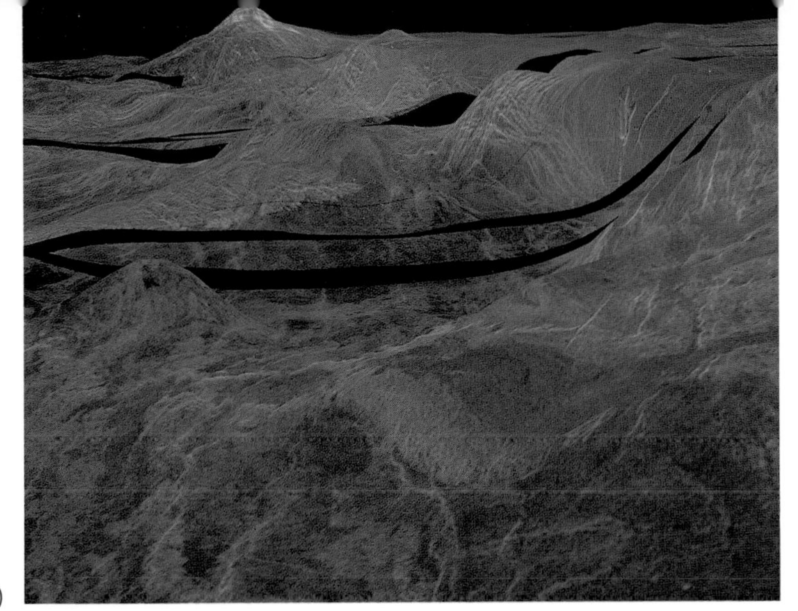

(a)

(b)

Images of the surface of Venus from the *Magellan* radar-mapping mission. Color has been added to simulate the appearance of the Venus surface. (*a*) Perspective view of the southern boundary of Lakshmi Planum, a plateau 1.5–2.5 mi (2.5–4.0 km) high. (*b*) Perspective view of a portion of Western Eistla Regio, with a rift valley in the foreground and the volcanoes Gula Mons (right) and Sif Mons (left) on the horizon. (*c*) Sapas Mons, one of several large volcanic edifices that compose the broad equatorial rise known as Atla Regio.

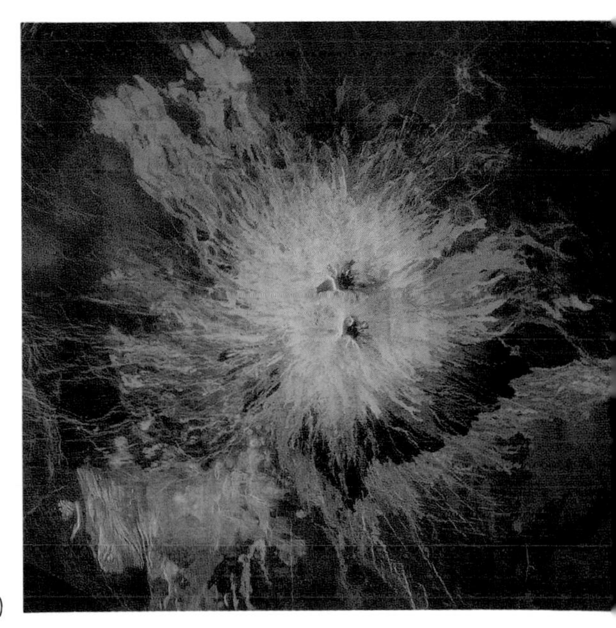

(c)

ports much of the flux outward. In addition, above the region where hydrogen has become neutral, some energy is still transported to and through the surface. This energy is about 10^{-5} the radiated energy in the Sun; in more active stars the ratio may be 10^{-2}. The rotating, magnetized, solar wind can be observed by space probes. The ejection of matter, small in the Sun, may be as high as 10^{-6} $M_\odot$ per year in young stars. Since the matter is also magnetized, a convective star ejects rotating plasma which interacts with the star's environment. One obvious consequence is a mechanism which slows the rotation of the star by a drag, linking it to the stationary interstellar medium. Thus, one expects and finds that young stars with convective envelopes rotate more rapidly than old ones. In addition, the cutoff of rapid rotation for young stars, from observations in Fig. 2, lies near the temperature at which deep convection begins. The Sun cannot stop an initially rapid rotation with its present mass loss, but given an early, more active phase which young stars have, it may have done so. Correlation of Ca II emission with shorter periods of rotation shows that enhanced stellar activity slows the star's rotation. *SEE SOLAR WIND.*

Effect of rotation. The effect of rotation on the properties of stars is complex. The maximum possible angular velocity for a uniformly rotating star is given by Eq. (2), with ω_c the angular velocity, G the gravi-

$$\frac{\omega_c^2}{2\pi G\langle\rho\rangle} = \varphi_c \qquad (2)$$

tational constant, and $\langle\rho\rangle$ the mean density, where $\varphi_c = 0.36075$. A normal hot star of 8 $M_\odot$, 4 $R_\odot$ gives $\omega_c = 1.6 \times 10^{-4}$ radian s^{-1}, about 348 mi s^{-1} (560 km s^{-1}). At this breakup point, which corresponds to the observed Be stars in Fig. 2, the star is nonuniform in brightness, highly distorted in shape, and composite in spectrum and color. Heat must flow through the star from the equator to the hot poles. Thus, rotating stars differ from nonrotating stars, in an amount dependent on the inclination of the poles to the line of sight. Smaller-scale variations of luminosity, temperature, and spectrum are observed in stars with angular velocities less than ω_c, as deviations from the main sequence of nonrotating stars. *SEE STAR*

Jesse L. Greenstein

Bibliography. L. H. Aller and D. McLaughlin (eds.), *Stellar Structure*, 1965, reprint 1981; S. L. Baliunas et al., Stellar rotation in lower main-sequence stars measured from time variations in H and K emission-line fluxes, *Astrophys. J.*, 275:752–772, 1983; K. J. Fricke and R. Kippenhahn, Evolution of rotating stars, *Annu. Rev. Astron. Astrophys.*, 10:45–72, 1972; F. Middlekoop, Magnetic structure in cool stars, IV. Rotation and Ca II H and K emission of main-sequence stars, *Astron. Astrophys.*, 107:31–35, 1982; A. H. Vaughan, Comparison of activity in old and young main-sequence stars, *Publ. Astron. Soc. Pac.*, 92:392–396, 1980; A. H. Vaughan and G. Preston, A survey of chromospheric Ca II H and K emission in field stars of the solar neighborhood, *Publ. Astron. Soc. Pac.*, 92:385–391, 1980.

Subgiant star

A member of the family of stars intermediate between giants and the main sequence in the Hertzsprung-Russell diagram. The mean luminosity of a subgiant is about 10 times the Sun; the surface temperature lies between 6700 and 12,100°F (4000 and 7000 K). The masses are about 1.4 times that of the Sun. The subgiants often violate the mass-luminosity relation; that is, ζ Herculis A, a G subgiant, is four times as bright as its mass would predict. *SEE HERTZSPRUNG-RUSSELL DIAGRAM; MASS-LUMINOSITY RELATION.*

The subgiants are of particular importance in theories of stellar evolution. If a main-sequence star has exhausted about 12% of its mass of hydrogen, the star begins to evolve, expanding, cooling at the surface, and brightening. Old stars of population II, of masses about 1.35 times the Sun, are now evolving into the subgiant region of the Hertzsprung-Russell diagram. The age of the oldest known stellar systems can be determined from the luminosities of the subgiants to be between 5×10^9 and 1.2×10^{10} years. Similar stars, not in clusters, suggest an upper limit to the age of the galaxy, which, subject to some theoretical uncertainties, is of the same order. A different type of subgiant occurs also in close binary systems of the younger population I. Some have been found, unexpectedly, to emit bursts of radio-frequency energy, indicating mass transfer and the presence of some high-energy electrons. *SEE STAR; STELLAR EVOLUTION.*

Jesse L. Greenstein

Submillimeter astronomy

Astronomical observations carried out in the region of the electromagnetic spectrum with wavelengths from approximately 0.3 to 1.0 millimeter. Objects in the universe radiate light of different energies depending on how hot they are. This emission is called thermal radiation and is governed by Planck's radiation law. Stars like the Sun radiate most of their energy at optical wavelengths (350–700 nanometers). Cooler stars and planets radiate most of their energy at infrared wavelengths (0.7–20 micrometers). *SEE HEAT RADIATION.*

Interstellar dust and gas radiate at many wavelengths depending on the proximity of hot stars. Cold molecular and atomic gas in interstellar space is best observed at far-infrared (20–300 mm), submillimeter (0.3–1.0 mm), millimeter, and radio (>1.0 cm) wavelengths. Important nonthermal processes include shock waves, synchrotron emission (from electrons spiraling in magnetic fields), and hyperfine transitions (such as electrons changing their spin states in atomic hydrogen, giving rise to 21-cm radio waves). *SEE INFRARED ASTRONOMY; RADIO ASTRONOMY.*

Observatories and telescopes. Submillimeter astronomy is one of the last windows of the electromagnetic spectrum to be investigated, because of the effect of the Earth's atmosphere. Molecules in the atmosphere, primarily water vapor, absorb (dim) infrared and submillimeter light. This can be circumvented by placing telescopes on high dry mountaintops, flying them on airplanes, or orbiting them above the Earth's atmosphere. The *Infrared Astronomical Satellite* (*IRAS*), which operated in 1983, was such an experiment. Since 1974, observations have been carried out with a 36-in. (0.9-m) telescope on the Kuiper Airborne Observatory, an airplane that flies as high as 45,000 ft (13.7 km). However, in observational astronomy it is desirable to achieve great sensitivity, and this requires a telescope with maximum collecting area, hence of large diameter, which is difficult to mount on a satellite or airplane. Also, it is desirable

Some telescopes for submillimeter astronomy

Name	Location	Elevation, ft (m)	Diameter, ft (m)
Existing installations			
Kuiper Airborne Observatory (KAO)	Suborbital	45,000 (13,700)	3.0 (0.9)
Cologne 3-m Telescope	Gornergrat, near Zermatt, Switzerland	10,285 (3,135)	9.8 (3.0)
Caltech Submillimeter Observatory (CSO)	Mauna Kea, Hawaii	13,360 (4,072)	34.1 (10.4)
Swedish-ESO Submillimeter Telescope (SEST)	La Silla, Chile	7,850 (2,400)	49.2 (15.0)
James Clerk Maxwell Telescope (JCMT)	Mauna Kea, Hawaii	13,425 (4,092)	49.2 (15.0)
Planned installations			
Max-Planck-Institut für Radioastronomie/University of Arizona 10-m Submillimeter Telescope*	Mount Graham, Arizona	10,717 (3,267)	32.8 (10.0)
Stratospheric Observatory for Infrared Astronomy (SOFIA)†	Suborbital	50,000 (15,200)	9.8 (3.0)
Far-Infrared and Submillimeter Telescope (FIRST)‡	Orbital	>200 mi (>321 km)	23.0 (7.0)
Large Deployable Reflector (LDR)§	Orbital	>200 mi (>321 km)	32.8 (10.0)

*Expected to be operational in 1992.
†Possible completion in 1990s (United States). Optical, infrared, and submillimeter.
‡Possible deployment in 1990s (European).
§Possible deployment about 2005 (United States). Operating wavelength 30 μm to 1 mm.

to make maps with the greatest amount of resolved detail. This can be done with a single large telescope or with one or more telescopes at different geographic positions (up to thousands of miles apart). The signals are combined to synthesize a telescope whose size is equal to the maximum separation of the individual dishes. This technique, called very long baseline interferometry (VLBI), was pioneered with radio telescopes but is also planned for submillimeter astronomy. *See Radio telescope.*

Fig. 1. James Clerk Maxwell Telescope, 49-ft-diameter (15-m) submillimeter telescope situated at Mauna Kea, Hawaii.

To achieve these objectives, large reflectors at dry mountain sites are now used for submillimeter astronomy (see **table**). The James Clerk Maxwell Telescope (**Fig. 1**), the Swedish-ESO Submillimeter Telescope, and the Caltech Submillimeter Telescope are not constructed like conventional optical telescopes, with a single aluminized glass mirror. Instead, their parabolic surfaces are formed from a large number of carefully contoured aluminum panels. *See Telescope.*

From the Earth's surface it is not possible to observe all submillimeter wavelengths. At wavelengths outside three atmospheric windows at 0.35, 0.45, and 0.8 mm, it is necessary to fly or orbit telescopes to make submillimeter observations. The atmosphere lets light of wavelength 0.8 mm through to about 1.2 mi (2 km) above sea level. Light of wavelengths 0.35 and 0.45 mm can be observed at altitudes higher than 1.9 mi (3 km), but even on a good night only about one-half of the photons at these wavelengths gets through the atmosphere to the summit of Mauna Kea, Hawaii, with an elevation of 13,800 ft (4205 m).

Areas of study. Submillimeter observations carried out so far and planned for the future involve interstellar chemistry, protostars, interstellar dust, and cosmology.

Interstellar chemistry. Below is a partial list of species observed in cold molecular clouds. Uncertain detections are designated by a question mark.

Simple hydrides, oxides, sulfides, and related molecules:

H_2	CO	CS
NH_3	SO_2	CC
		OCS

Nitriles, acetylene derivatives, and related molecules:

HCN	$H(C\equiv C)_5-CN$
H_3C-CN	$H_3C-C\equiv C-CN$
$C\equiv C-CO$	$H_3C-C\equiv CH$
$C\equiv C-CS$	$H_3C-(C\equiv C)_2-H$
$HC\equiv CCHO$?	$H_3C-(C\equiv C)_2-CN$?
$HC\equiv C-CN$	$H_2C=CH-CN$
$H(C\equiv C)_2-CN$	HNC
$H(C\equiv C)_3-CN$	$HN=C=O$
$H(C\equiv C)_4-CN$	

Aldehydes, alcohols, ethers, ketones, and related molecules:

$H_2C=O$	H_3COH	$HO-CH=O$?
$H_2C=S$		$H_2C=C=O$?
$H_3C-CH=O$		

Cyclic molecules:

C_3H_2
C_3H

Ions:

CH^+	HCS^+
HN_2^+	$HCNH^+$?
HCO^+	SO^+ ?

Radicals:

CH	C_2H	CN	SO
OH	C_3H	C_3N	C_2S
	C_4H		
	C_5H		
	C_6H		

Many of these species were first observed at millimeter and radio wavelengths, but submillimeter astronomy will allow many new discoveries. Such components of interstellar gas are detected by means of spectrometers, which are tuned to the specific frequencies emitted by rotating and vibrating molecules. (Rotational transitions, in particular, result in submillimeter lines.) The observed spectral lines are like astronomical fingerprints, and theoretical calculations and laboratory measurements of known substances stipulate the wavelengths to look for in the interstellar medium. Differences are usually found between laboratory wavelengths and observed wavelengths because of the velocities of the gas in the astronomical sources (according to Doppler's principle). Thus, spectral line observations can give accurate information on the mass motions within clouds of gas. SEE COSMOCHEMISTRY; DOPPLER EFFECT; INTERSTELLAR MATTER.

Protostars. Stellar nurseries such as the Orion Molecular Cloud are prime areas of study at infrared and submillimeter wavelengths. Not only is there interest in what chemical species are to be found in molecular clouds and what processes are involved in star formation, but also observations at long wavelengths serve as a probe of the densely shrouded cloud cores that are invisible at optical and near-infrared wavelengths. An individual protostar may be surrounded by a dense doughnut-shaped disk of dust, which collimates a strong wind flowing from the newborn star (**Fig. 2**). High-velocity gas, shock fronts, and even radiation from water masers are observed. SEE ORION NEBULA; STELLAR EVOLUTION.

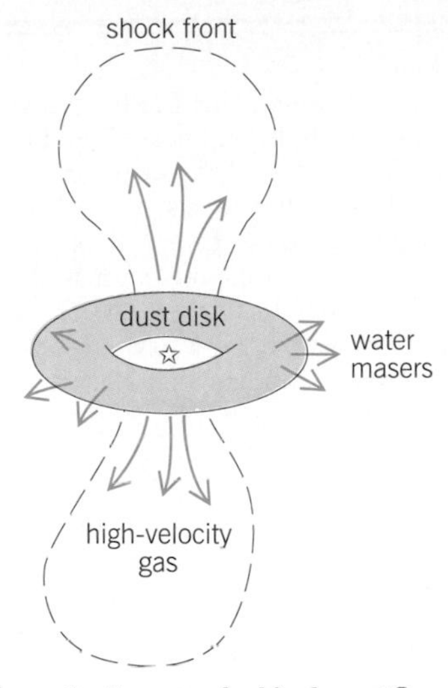

Fig. 2. Schematic diagram of a bipolar outflow source, consisting of a central protostar surrounded by a dense toroid of dust, which in turn collimates the strong wind from the protostar.

Interstellar dust. Planck's radiation law stipulates the distribution of energy of photons given off by a body at a given temperature. Since dust and cold dust radiate principally at infrared and submillimeter wavelengths, the most accurate way to measure the dust content of interstellar clouds and distant galaxies is to observe at such wavelengths. Also, a very large fraction of the energy liberated in the universe ends up being absorbed by dust and reradiated at infrared and submillimeter wavelengths, so a study of these bands is necessary to understand the energy budget of the universe.

Cosmology. It is generally believed that the universe formed some 10–20 × 10^9 years ago in a fiery big bang. Over the course of time the universe has expanded and cooled, so that the remnant of the primeval fireball now corresponds to a source of background radiation at 2.75 kelvins. Copious x-rays have been discovered in the directions of clusters of galaxies, attributable to very hot gases. In 1970 R. A. Sunyaev and Ya. B. Zel'dovich predicted that such hot gas would distort the observed strength of the cosmic background radiation passing through galaxy clusters. At millimeter wavelengths the intensity should be decreased, while at wavelengths shorter than 1.4 mm there should be an observed increase of intensity. In 1984 the diminishing effect was confirmed at a wavelength of 15 mm for at least one galaxy cluster. The corresponding increase of intensity has not yet been observed unambiguously at shorter wavelengths. Such observations with submillimeter telescopes would make it possible to investigate the initial conditions in very distant clusters of galaxies, at their stages of development billions of years in the past. SEE COSMIC BACKGROUND RADIATION.

Kevin Krisciunas

Bibliography. G. B. Field and E. J. Chaisson, *The Invisible Universe: Probing the Frontiers of Astrophysics*, 1985; J. D. Kraus, *Radio Astronomy*, 2d ed., 1986; T. G. Phillips and D. B. Rutledge, Superconducting tunnel detectors in radio astronomy, *Sci. Amer.*, 254(5):96–102, May 1986; W. Sweet, New radiotelescopes open era of submillimeter astronomy, *Phys. Today*, 40(8):65–67, August 1987; G. L. Verschuur, *The Invisible Universe Revealed: The Story of Radio Astronomy*, 1987.

Sun

The star around which the Earth revolves, and the planet's source of light and heat (**Fig. 1**). The Sun is a globe of gas, 1.4×10^6 km (8.65×10^5 mi) in diameter, held together by its own gravity. Because of the weight of the outer layers, the density and temperature increase inward, until a central temperature of over 1.5×10^7 K (2.7×10^7 °F) and density more than 90 times that of water is reached. At these great temperatures and densities, thermonuclear reactions converting hydrogen into helium take place, releasing the energy which streams outward.

The surface temperature of the Sun is about 6000 K (10,000°F); since solids and liquids do not exist at these temperatures, the Sun is entirely gaseous. Almost all the gas is in atomic form, although a few molecules exist in the coolest regions at the surface.

The Sun is a typical member of a numerous class of stars, the spectral type dG2 (d indicates dwarf). Other characteristics are given in **Table 1**. *See Star.*

Besides its great importance to human life, the Sun is of interest to all astronomers because it is the only star near enough for detailed study of its surface structure. Various surface and atmospheric phenomena, such as sunspot activity, and other behavior may be studied, and astronomers try to extrapolate these to the other stars which may be observed only as points of light.

The light and heat of the Sun make the Earth habitable. The Sun is, in fact, the ultimate source of nearly all the energy utilized by industrial civilizations in the form of water, power, fuels, and wind. Only atomic energy, radioactivity, and the lunar tides are examples of nonsolar energy. *See Earth.*

Table 1. Principal physical characteristics of the Sun

Characteristic	Value
Mean distance from Earth (the astronomical unit)	1.4960×10^8 km = 9.2956×10^7 mi
Radius	$(6.960 \pm .001) \times 10^5$ km = $(4.325 \pm .001) \times 10^5$ mi
Mass	$(1.991 \pm .002) \times 10^{33}$ g = $(4.390 \pm .004) \times 10^{30}$ lb
Mean density	$1.410 \pm .002$ g/cm³
Surface gravity	$(2.738 \pm .003) \times 10^4$ cm/s² = $89.8 \pm .1$ ft/s² = 28 × terrestrial gravity
Total energy output	$(3.86 \pm 0.03) \times 10^{33}$ erg/s = $(3.86 \pm .03) \times 10^{26}$ W
Energy flux at surface	$(6.34 \pm .07) \times 10^{10}$ erg/(cm²)(s) = $(6.34 \pm .07) \times 10^7$ W/m²
Effective surface temperature	5780 ± 50 K = 9940 ± 90°F
Stellar magnitude (photovisual)	$-26.73 \pm .03$
Absolute magnitude (photovisual)	$+4.84 + .03$
Inclination of axis of rotation to ecliptic	7°
Period of rotation	About 27 days; the Sun does not rotate as a solid body; it exhibits a systematic increase in period from 25 days at the equator to 36 days at the poles

Solar Structure

The interior of the Sun can be studied only by inference from the observed properties of the entire star. The mass, radius, surface temperature, and luminosity are known. Using the known properties of gases, it is possible to calculate that structure of the Sun which will produce the observed parameters at the surface (**Fig. 2**). The solution is complicated by uncertainties in the behavior of matter and radiation under the high temperature and density that are present in the solar interior. This is particularly true of the nature of the nuclear reactions. However, the general properties of the solution are quite reliable. A number of theoretical models using different assumptions have led to more or less similar results. A central density of near 90 g/cm³ has been found, decreasing to 10^{-7} g/cm³ at the surface. The central temperature is about 1.5×10^7 K (2.7×10^7 °F), decreasing to 6000 K (10,000°F) at the surface. Since this takes place over 700,000 km (430,000 mi), the temperature gradient is only 20 K per kilometer (60°F per mile). The radiation produced at the center by nuclear interactions flows outward rapidly.

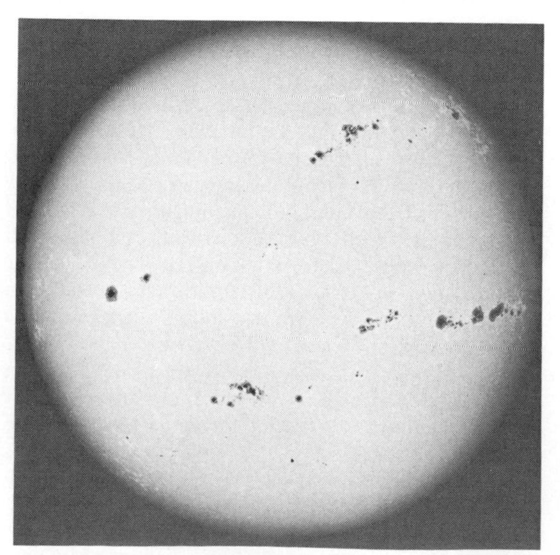

Fig. 1. The Sun, photographed in white light during the 1957 maximum of the sunspot cycle. (*California Institute of Technology/Palomar Observatory*)

The energy of the Sun is produced by the conversion of hydrogen into helium; because each hydrogen atom weighs 1.0078 atomic units and each helium atom is made from four hydrogen atoms, only 4.003 units, it follows that 0.0282 unit, or 0.7% of the mass m, is converted into energy E according to the Einstein formula $E = mc^2$, where c is the speed of light. Since the solar mass is 2×10^{33} g, conversion of 0.7% into helium yields 1.2×10^{52} ergs (1.2×10^{45} joules), enough to maintain the Sun for 10^{19} s, or over 10^{11} years. The rate of conversion required to produce the observed flux is 4×10^{38} atoms/s. For each hydrogen atom converted, one neutrino is produced, giving a flux of 1.3×10^{11} neutrinos/(cm²) (s) at the Earth. These neutrinos cannot be detected; only higher-energy neutrinos produced by subordinate processes can be observed. These have been detected at the Earth, but in smaller quantities than expected. However, the neutrino emission theory has sufficient uncertainties that the theory of nuclear burning is still generally accepted. *See Solar neutrinos.*

Although the material at the center of the Sun is so dense that a few millimeters are opaque, the photons created by nuclear reactions are continually absorbed and reemitted and thus make their way to the surface. The atoms in the center of the Sun are entirely stripped of their electrons by the high temperatures, and most of the absorption is by continuous processes, such as scattering of light by electrons.

In the outer regions of the solar interior, the temperature is low enough for ions and even neutral atoms to form and, as a result, atomic absorption becomes very important. The high opacity makes it very difficult for the radiation to continue outward; steep temperature gradients are established which result in convective currents. Most of the outer envelope of the Sun is in such convective equilibrium. These large-scale mass motions produce many interesting phenomena at the surface, including sunspots and solar activity.

Rotation. Because the Sun is made of a gaseous plasma, it need not rotate as a solid object like the Earth. Indeed, observations of sunspots and other features on the Sun's surface show that the equator rotates once each 25 days, while the polar regions complete a revolution only every 36 days. Helioseismic observations show that this rotation pattern as a function of latitude persists through the convection zone (from relative radius $r/R = 0.7$ to the surface; Fig. 2), and below this the Sun rotates nearly like a rigid body, with a period of 27 days (although to date the rotation rate below $r/R = 0.4$ is undetermined). This so-called differential rotation is known to be caused by convection in the Sun's interior, in combination with the Coriolis force, but the details of this process remain unknown.

Since young stars are observed to rotate more rapidly than the Sun, astronomers have sought evidence that the central part of the Sun ($r/R < 0.4$, which contains most of the Sun's mass) might be rotating rapidly, with only the outer surface rotating at the observed fairly slow rate. To date, there is no convincing evidence for such a rapidly rotating solar core, but neither can it be ruled out. *See* STELLAR ROTATION.

Radiation. Electromagnetic energy is produced by the Sun in essentially all wavelengths. Important radiation has been measured from long radio waves of 300 m down to x-rays of less than 0.1 nanometer (from rockets). In addition, considerable energy is emitted in the form of high-energy particles (cosmic rays). However, more than 95% of the energy is concentrated in the relatively narrow band between 290 and 2500 nm and is accessible to routine observation from ground stations on Earth. The maximum radiation is in the green region, and the eyes of human beings have naturally evolved to be sensitive to this range of the spectrum. The total radiation and its distribution in the spectrum are parameters of fundamental significance, because they measure the total energy output of the Sun and its effective surface temperature. The total radiation received from the Sun is termed the solar constant and has, in fact, been found within the limits of observation to be constant to about $\pm 1\%$. The presently accepted value is 1.97 cal/(cm^2)(min). This is equivalent to 1.374×10^6 ergs/(cm^2)(s) or 1.374×10^3 W/m^2. *See* SOLAR CONSTANT.

The measurement of the solar constant has been greatly complicated by the absorption of solar radiation in all wavelengths by the Earth's atmosphere.

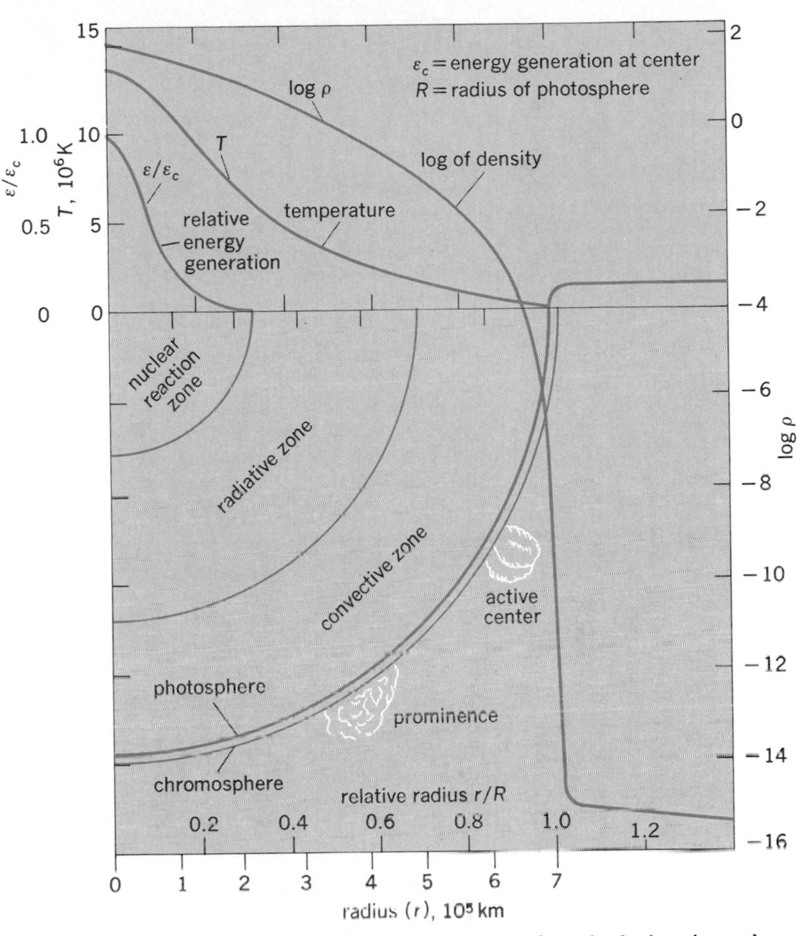

Fig. 2. Solar temperature *T*, density *P*, and energy generation ε in the interior and atmosphere of Sun. Curves in upper section show these as functions of radial distance from the center *r*. Lower section shows the principal zones in the interior of Sun. 1 km = 0.6 mi; °F = (K × 1.8) − 459.67.

Determinations have been made by observations at stations atop high mountains where the atmospheric perturbation is minimized, and by observing the variation during the day as the radiation passes through successively smaller distances in the terrestrial atmosphere. A great bulk of the studies was carried out by the Smithsonian Institution in a continuous program from the 1880s until 1955. The Smithsonian results showed that the solar constant was indeed constant to 1%, but there might be variation at a smaller scale. Spacecraft measurements suggest variations of 1 part in 2000, but cannot be confirmed because the constant is not measured from the ground.

The measurement of this immensely important quantity has been neglected, partly because many feel it is constant, and because the work is difficult, trying, and unexciting. A reduction of 7% in the solar constant would produce a completely frozen Earth, which only a 50% increase would thaw (because of the reflectivity of the ice). However, the fossil record shows that the solar constant has not changed much in the last billion years.

Atmosphere. Because the Sun is visible as a two-dimensional surface rather than as a point, the study of its atmosphere and surface phenomena is most interesting.

Although the Sun is gaseous, it is seen as a discrete surface from which practically all the heat and light are radiated. One sees through the gas to the point at

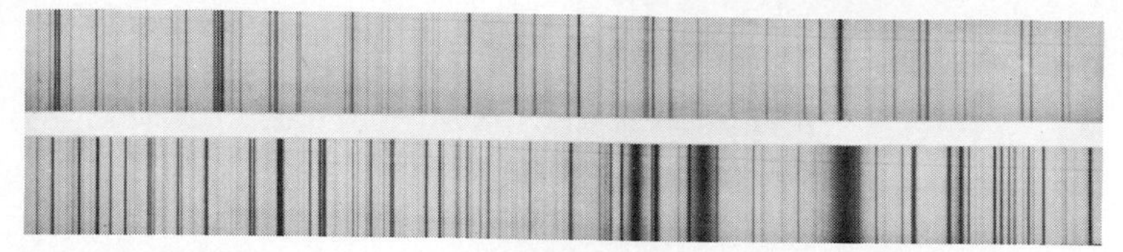

Fig. 3. Two sections of the Fraunhofer spectrum, showing bright continuum and dark absorption lines. The wavelength range covered by each strip is approximately 8.5 nm. The three strongest lines are produced by magnesium; the others, mostly iron. (*Sacramento Peak Observatory, operated by the Association of Universities for Research in Astronomy, Inc.*)

which the density is so high that the material becomes opaque. This layer, the visible surface of the Sun, is termed the photosphere. Light from farther down reaches the Earth by repeated absorption and emission by the atoms, but the deepest layers cannot be seen directly. The surface is actually not sharp; the density drops off by about a factor of 3 each 150 km (90 mi). However, the Sun is so far away that the smallest distance that can be resolved with the best telescope is about 300 km (200 mi), and thus the edge appears sharp. *See Photosphere*.

When the Sun is observed in isolated wavelengths absorbed by its atmospheric gases, it is no longer possible to see down to the photosphere, but instead the higher levels of the atmosphere, known as the chromosphere, are seen. This name results from the rosy color seen in this region at a solar eclipse. The chromosphere is a rapidly fluctuating region of jets and waves coming up from the surface. When all the convected energy coming up from below reaches the surface, it is concentrated in the thin material and produces considerable activity. As a result, this outer region is considerably hotter than the photosphere, with temperatures up to 30,000 K (54,000°F). *See Chromosphere; Eclipse*.

When the Moon obscures the Sun at a total solar eclipse, the vast extended atmosphere of the Sun called the corona can be seen. The corona is transparent and is visible only when seen against the dark sky of an eclipse. Its density is low, but its temperature is high (more than 10^6 K or 1.8×10^6 °F). The hot gas evaporating out from the corona flows steadily to the Earth and farther in what is called the solar wind. *See Solar wind*.

Within the solar atmosphere many transient phenomena occur which may all be grouped under the heading of solar activity. They include sunspots and faculae in the photosphere, flares and plages in the chromosphere, and prominences and a variety of changing coronal structures in the corona. The existence and behavior of all these phenomena are connected with magnetic fields, and their frequency waxes and wanes in a great 22-year cycle called the sunspot cycle. *See Solar magnetic field*.

The sunspots and flares are sources of x-rays, cosmic rays, and radio emission which often have profound influence on interplanetary space and the upper atmosphere of the Earth.

The photospheric features of the Sun (sunspots and faculae) are readily observed with a small telescope by projection of the solar image through the eyepiece onto a shaded white card. This is the only safe method for observing the Sun without a specially designed solar filter. Observations of chromospheric phenomena require special filters to exclude all light except that emitted by the atmospheric gases. Observations of the corona require either a total eclipse or a coronagraph.

Solar physics. Understanding of the Sun is derived from observations of the morphology of various phenomena and from physical analysis with the spectrograph. The former permits determination of what is happening, and the latter determination of the detailed physical parameters of the gas under observation. *See Astronomical spectroscopy*.

When the light of the Sun is broken up into the different frequencies by the spectrograph, a bright continuum interrupted by thousands of dark absorption lines known as the Fraunhofer lines can be seen. Each line represents some discrete transition in a particular atom which occurs when a photon of light is absorbed and one of the atomic electrons jumps from one level to another. Of the 92 natural elements, 64 are represented in the Fraunhofer spectrum (**Fig. 3**), first observed by the German physicist J. Fraunhofer. The remaining atoms are undoubtedly present but remain undetected because they are rare or their lines are produced in spectral regions not accessible to present spectrographs. The relative abundances of the most numerous atoms have been estimated from the line intensities (**Table 2**). Many of these abundances have been confirmed by measurement of the relative abundances of different elements in the streams of particles coming from the Sun at the time of solar flares. The Sun can be described as a globe of chemically (but not spectroscopically) pure hydrogen and helium with traces of the other elements.

The nature of the formation of the spectrum lines can be understood in terms of a few simple rules which are an elaboration of those first put forth by G. R. Kirchhoff. A small amount of a hot gas will be transparent in all wavelengths except those characteristic of the atoms of the gas. In those wavelengths radiation is emitted by the jumping electrons, and an emission line spectrum is produced. If the volume of the gas is large enough and its density high enough, it will be opaque at all frequencies because of the existence of continuous processes which will absorb any wavelength of light. In this case the gas emits a continuous spectrum at all frequencies. However, at the outer edges the atoms of the gas absorb light coming from one direction (the inside) and reemit it in all directions, with a consequent reduction of intensity. This results in dark lines superposed against the bright background. Although the ultimate source of all energy is at the center of the Sun, at any one point the

Table 2. Relative numbers of the most abundant atoms in Sun

Element	Number
Hydrogen, H	1,000,000
Helium, He	50,000–200,000
Oxygen, O	500
Nitrogen, N	400
Carbon, C	200
Magnesium, Mg	33
Silicon, Si	20
Iron, Fe	15
Sulfur, S	8
Aluminum, Al	2
Sodium, Na	2
Calcium, Ca	1.5

radiation is a result of either a scattering of light coming from within or the emission of radiation by an atom excited by collisions with the electrons of the gas.

By analyzing the nature of the spectrum lines observed, it can be determined which atoms are present, as well as their stages of ionization. This is so because, as electrons are successively removed from an atom by ionization (as a result of photoelectric effect or collisions with electrons), the spectrum changes completely. Obviously, the hotter the gas, the higher the stage of ionization. Thus when lines of iron ionized 14 times are observed in the solar corona, it may be assumed that the gas is very hot. On the other hand, by observing the spectra of molecules in the upper photosphere, it may be concluded that the temperature is near the lowest possible in the Sun.

In addition, the shape of the spectrum lines is a key to the physical conditions under which they are radiated. If the temperature is high, the atoms have a high velocity and the lines are broadened by the Doppler effect. If there are strong magnetic fields, the spectrum lines are split by the Zeeman effect. If there are strong electric fields, the lines are broadened by the Stark effect. The spectroscopist must disentangle these various effects. This task is aided by the fact that the spectrum lines can be observed at both the center of the Sun and near the limb, as well as from atoms which behave differently. For example, hydrogen atoms are very light and move extremely fast; hence the hydrogen lines show a strong broadening due to the Doppler effect. If the lines emitted by heavy atoms in the same region do not show this broadening, it may be concluded that this is purely a temperature effect, because the heavy atoms would not be expected to move so fast. *See Doppler effect; Stark effect; Zeeman effect.*

Helium, on the other hand, is inert and radiates only at high temperatures. Thus, when helium lines are seen, the gas is highly excited. In these ways the astrophysicist uses the spectrum lines to interpret the physical conditions in the atmosphere of the Sun. *See Atomic structure and spectra.*

Photosphere. As mentioned above, the photosphere is the visible surface of the Sun. In the visible wavelengths, its brightness decreases smoothly from the center of the solar disk to the limb. This limb darkening results from the fact that the line of sight to the observer passes through the atmosphere at an increasing angle to the normal as the point of observation approaches the limb. Hence the line of sight penetrates to a lower depth at the center than at the limb, where the path through the overlying material is much longer. Thus what is seen at the limb is a higher point in the atmosphere. The fact that the photosphere is darker at the limb indicates that the temperature decreases outward. Once the range of temperature is known, the variation of limb darkening with wavelength can be used to establish the absorbing properties of the materials at different wavelengths.

By combination of all the information on limb darkening at different wavelengths, it has been possible to construct reasonably reliable models of the temperature variation through the photosphere (**Fig. 4**).

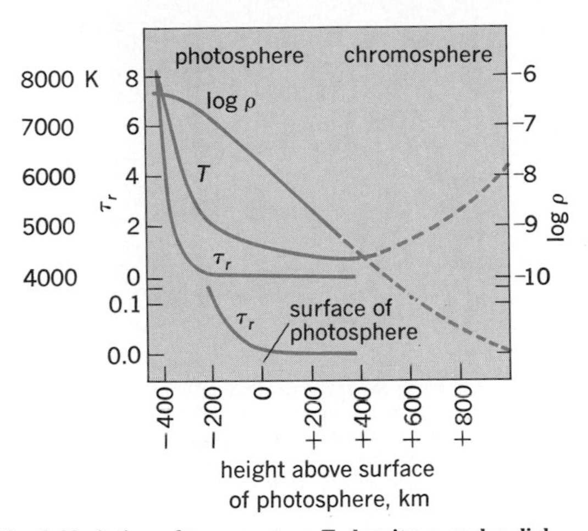

Fig. 4. Variation of temperature T, density ρ, and radial optical depth τ_r with height in the photosphere and (less certainly) in the chromosphere. Lower curve of optical depth shows variation with an expanded ordinate scale. 1 km = 0.6 mi; °F = (K × 1.8) −459.67.

Granulation. Except for sunspots and accompanying activity, the photosphere is quite uniform over the Sun. The only structure visible is the granulation, an irregular distribution having the shape of bright corn kernels with dark lanes in between (**Fig. 5**). The grains are quite small, of the order of 1000 km (600 mi) in diameter (about 1.3 seconds of arc as seen from the Earth), and have a life-span of about 8 min. The dark lanes between the granules are about 200 km (125 mi) across. Since the highest optical resolving power employed is of the order of 200–300 km (100–200 mi), there may be many more fine lanes crossing the granules and dividing them into even smaller elements.

The granulation is visible evidence of convective activity below the surface. The bright grains are presumably the tops of hot, rising columns which bring energy up from the interior, while the dark intergranular may be the cool downward-moving material. High-resolution spectrograms (**Fig. 6**) show that each bright granule is marked by a violet displacement of

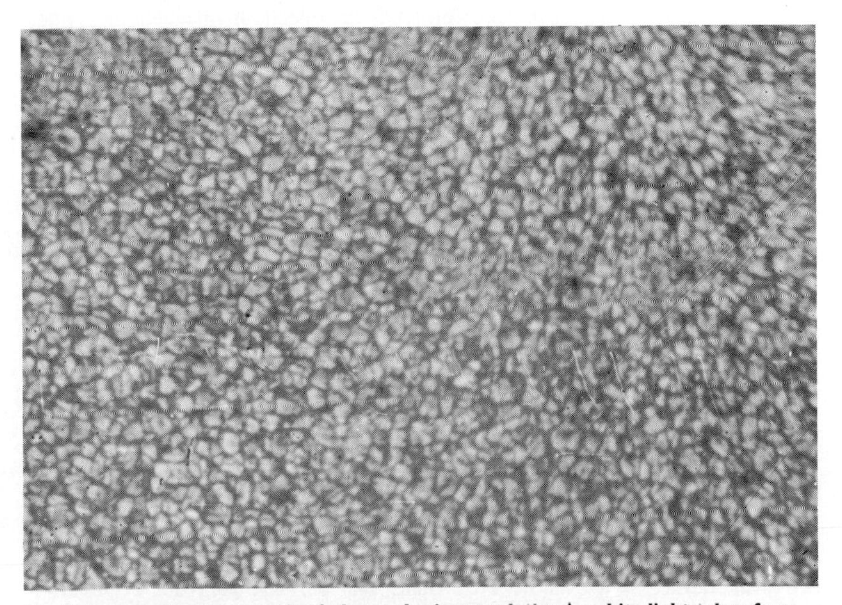

Fig. 5. Large-scale photograph of photospheric granulation in white light taken from an altitude of 80,000 ft (24 km) above sea level. The length of this section is about 55,000 km (34,000 mi) on the Sun. (*Princeton University Observatory*)

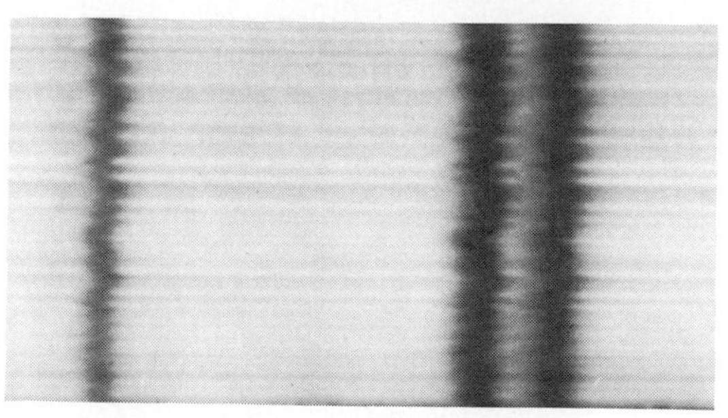

Fig. 6. High-resolution spectrogram at 518.85 nm made by J. W. Evans at Sacramento Peak Observatory, showing lines (left to right): Fe I 518.7922, Ti I 518.8700, and Ca I 518.8848. The total slit length is 120,000 km (75,000 mi) on the Sun. (*Sacramento Peak Observatory, Air Force Cambridge Research Laboratories*)

the spectrum lines, indicating an upward velocity. The measured difference in brightness between a granule and an intergranule area is about 15%, indicating an effective temperature difference of the order of 200 K (360°F). Photographs of the granulation often show what appear to be distinct chains. However, no one has proved that these are anything other than a random association by the eye of the observer.

In Fig. 6 the brightness variations in the continuous spectrum are presumably the white-light granulation. The complex structure of the absorption lines is easy to see; there are some elements where the absorption lines are entirely missing, and others where tilted features indicate rotation. The brightest granules in the continuum correspond to arrow-shaped shifts to the blue.

Supergranulation. In addition to the granulation, which can be seen in broadband pictures of the Sun's surface, another larger scale of convection, called the supergranulation, can be seen in Doppler images of the surface. The supergranular cells are about 40 times larger than the granulation cells, and their lifespan is about 1–2 days. The supergranular motions play an important role in the structure of the upper atmosphere. The systematic outward flow of material in each cell tends to concentrate magnetic fields near the cell edges. The resulting chromospheric network shows strong magnetic fields distributed around the outside of the cells, resulting in enhanced activity at those points.

Helioseismology. In the late 1950s, R. Leighton and his collaborators discovered, by measuring the Doppler shifts of solar spectral lines, that the surface of the Sun is constantly oscillating up and down with a period of about 5 min. It was later found that these motions are the surface manifestation of some 10^7 trapped acoustic oscillation modes inside the Sun. Many of the modes are global modes, coherent over the entire Sun; the lifetimes of the observed modes range from a day to several months. The surface amplitudes of the modes are as high as 20 cm/s (8 in./s), making the oscillation amplitude of any point on the surface (which comes from the incoherent superposition of the individual modes) about 0.4 km/s (0.2 mi/s). By observing the properties of these oscillation modes, a number of interesting things about the Sun have been learned, such as the rotation rate as a function of depth and latitude mentioned above. SEE HELIOSEISMOLOGY.

Chromosphere. The chromosphere was first detected and named by early solar-eclipse observers. They saw it as a beautiful rosy arc which remained visible for a few seconds above the limb of the Moon when the photosphere had been covered. The red color is due to the dominating brightness of the H-α line of hydrogen at 656.28 nm in the chromospheric spectrum, which, except for a bare trace of continuum, is a pure emission spectrum of bright lines (**Fig. 7**). Because eclipses are rare and the chromosphere can be seen only edge-on at that time, astronomers have developed a special method of photographing the chromosphere in its characteristic lines, particularly H-α. In these spectrum lines the chromosphere is no longer transparent, and one looks at it instead of looking through it to the photosphere. The two principal devices for this observation are the birefringent filter, which utilizes combinations of calcite and quartz plates, and polarizers, which isolate all but a narrow band as small as 0.025 nm; and the spectroheliograph, which builds up a picture by moving the spectrograph across the image of the Sun. With these devices the overpowering background of photospheric light is removed, and only the chromosphere is seen. SEE INTERFEROMETRY; SPECTROHELIOSCOPE.

The study of the morphology of the chromosphere is best carried out by time-lapse motion pictures. Photographs are made every 10 s or so, and then run through a projector at the normal rate of 16 frames per second. This gives a remarkable picture of the dynamic variations in the chromosphere. Such films of the center and limb of the Sun have provided remarkable pictures of this very complex zone just above the surface of the photosphere.

While the magnetic-field energy in the photosphere is considerably less than the energy of the material, the rapid falloff in density with height results in a situation where the magnetic field dominates the material motions, and the gas is ordered into large-scale patterns. This is not strikingly evident in the low chromosphere (0–1500 km or 0–900 mi), which seems an irregular extension of the underlying photosphere. In fact, the temperature continues to drop above the surface of the photosphere (as evidenced by limb darkening) to an apparent minimum of about 4000 K (7000°F) at a height of 1500 km (900 mi). Even in the low chromosphere, however, the effect of the chromospheric network in the small bright regions or faculae that mark the edge of each network cell is seen. These faculae may be observed in white light near the limb of the Sun. In them the chromosphere shows a very fine structure of bright points, as though one were looking down at each line of force in the magnetic field (**Fig. 8*a***).

If the edge of the Sun is examined in hydrogen light (H-α), the chromosphere is seen as an irregular band 3000–4000 km (1900–2500 mi) high, from which small jets, called spicules, protrude up to 7000

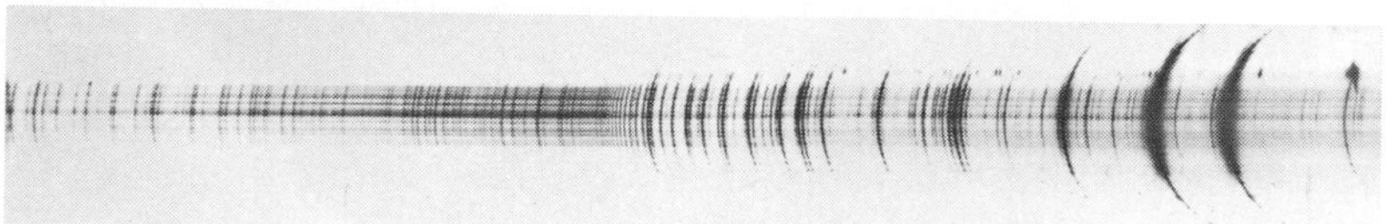

Fig. 7. Flash spectrum of the ultraviolet light from the chromosphere, photographed during an eclipse. This is a negative to enhance details. The strong H and K lines are at the right, the convergence of the Balmer series of hydrogen is near the center, and at the left the lines merge into the Balmer continuum toward shorter wavelengths. (*High Altitude Observatory*)

km (4400 mi). If the wavelength is tuned slightly off the line center, the uniform band disappears because it has only a narrow range of wavelength, and a forest of spicules down to the very surface replaces it. Since the limb is confusing, with many objects in the line of sight, the distribution is better seen on the disk. There the spicules are seen to be clumped and to come only from the small bright faculae that mark the chromospheric network. This is even more marked in the wing of the line (off-band), where the chromosphere is transparent, and only the spicules are seen (Fig. 8*b*). The spicules are visible offband because they are moving rapidly and their spectrum lines are Doppler-broadened. Magnetic measurements show the faculae to have fairly strong magnetic fields, and velocity measurements show a gentle flow from the center to the edge of the chromospheric network cells and a downflow at the edges. Thus the chromosphere can be regarded as consisting of two components, the general chromosphere, which is evenly distributed (hence independent of magnetic effects) and dominates at lower levels, and the spicule component, which is connected with magnetic fields and dominates at greater heights.

Visibility through the Earth's atmosphere and instrumental inadequacies limit the ability to resolve the structure of the spicules (**Fig. 9**). They shoot up to a height of about 6000 km (3700 mi) above the photosphere with a velocity of about 20–30 km/s (10–20 mi/s) and then fade out. They may be very narrow, certainly less than 1000 km (600 mi) across. At the top of their trajectory some spicules fade out and others drop back into the chromosphere. It is believed that the spicule jets occur as a result of the focusing of mass motions in the low chromosphere by the strong magnetic fields at the edges of the chromospheric network. They form a channel by which energy travels from below into the solar corona.

Although the spicules were described above as jets, this structure is not certain. The spicules are difficult to observe, and it is possible that they are small flare-like eruptions or something more complicated. In pictures taken in the ultraviolet from *Skylab*, larger features called macrospicules were seen; H-α pictures from the ground showed small eruptions at their base.

When the solar surface is photographed in the ultraviolet, the hotter regions formed in the transition from chromosphere to corona are observed. These observations show the ultraviolet emission coming primarily from the edges of the network. This may be because the spicules are connected with the heating of the corona, or simply because the density there is higher, because of the spicule eruptions, so that the higher-temperature region is more apparent. It is not known why the chromosphere is hotter than the photosphere, and it may be that the temperature rises only in these network regions. The photosphere shows no variation at all with the chromospheric network, except for the velocity field discussed above.

Physical conditions in the chromosphere can be studied in the ultraviolet and radio ranges. If the emission in microwaves, say 3 cm, is measured, the temperature is found to steadily increase at longer wavelengths, which do not penetrate so deeply. This means the temperature increases outward, and the rate of increase can be estimated. At the edge of the Sun the observed emission should absorb greater (limb brightening) because higher layers should be observed, but the amount of limb brightening is, in fact, fairly small. This probably is because the temperature increase is irregular. Ultraviolet observations from *Skylab* have imaged the Sun and shown that the high-temperature areas are distributed with the chromospheric network, as well as the active regions around sunspots. Since all these places have strong magnetic fields, it can be inferred that the heating of the upper parts of the solar atmosphere depends on phenomena

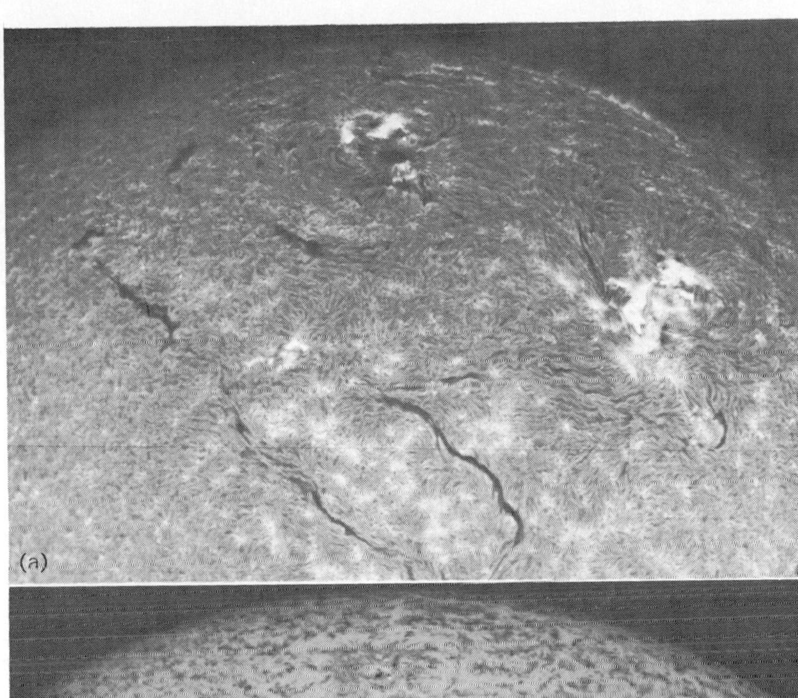

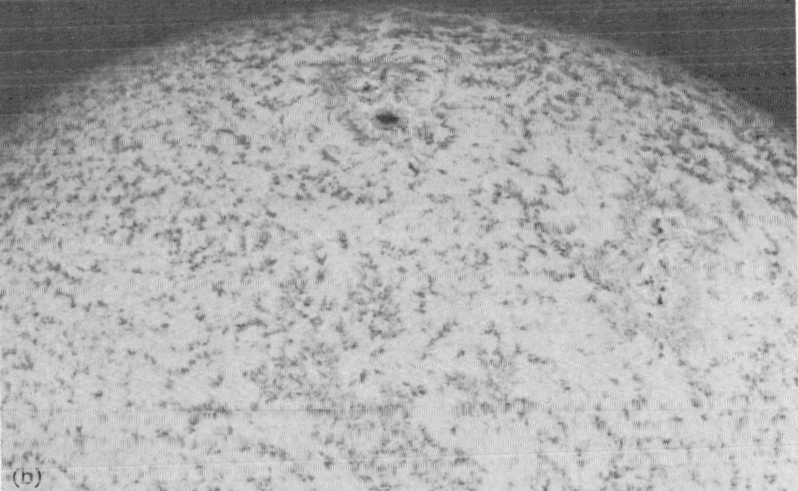

Fig. 8. Spectroheliograms. (*a*) Photocenter of H-α. Dark filaments form the boundary between regions of bright flocculi of different magnetic polarities. The flocculi themselves outline the cells of the chromospheric network and are brighter where the magnetic field is stronger. The fine dark features coming from each bright flocculus are spicules. At the bottom center, two filaments converge in a region of weak magnetic activity. Two active regions around sunspots appear, surrounded by bright plages and "whirlpool structure." The dark filaments that curve into the sunspot regions also mark the boundary between fields of opposite polarity. These fields run horizontally along the filaments, as can be judged from the disk structure. The filament near the limb shows a typical bright rim structure underlying it. (*b*) At 0.07 nm, showing structures with wide profiles. Only the spicules at the edges of the cells remain. These features are the spicules seen at the limb, and are seen to move with appropriate velocity on disk pictures. They must stand above the surface because they appear invariably as dark areas on the limbward side of flocculi near the limb. Spicule profiles at the limb were found to be broader than the photospheric H-α, so that they must be prominent in pictures in the wings. Although light from deeper layers in the Sun may reach the Earth off-band (because the absorption coefficient is lower), the structures one sees are the highest, which have the broadest profiles. The sunspot and penumbra appear clearly, because one sees through overlying material. A bright rim is evident around the edges of the penumbrae. (*R. B. Leighton and coworkers, California Institute of Technology/Palomar Observatory*)

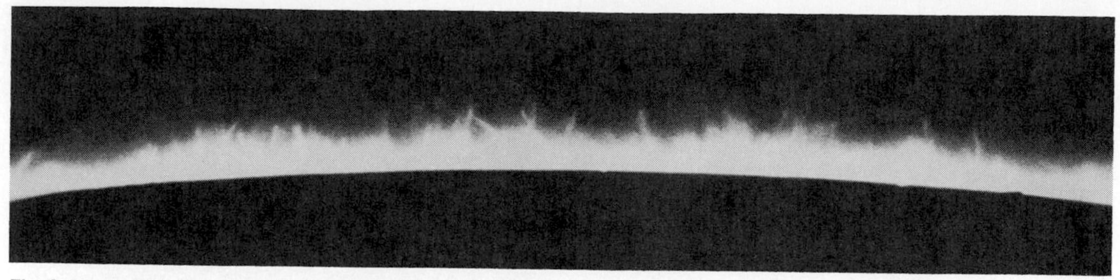

Fig. 9. Large-scale photograph of the chromosphere in H-α light, with the disk of the Sun artificially eclipsed and the hairy spicules projecting above the continuous chromosphere. The length of this section is about 140,000 km (90,000 mi). (*Photograph by R. B. Dunn, through 15-in. telescope, Sacramento Peak Observatory, operated by the Association of Universities for Research in Astronomy, Inc.*)

in regions of strong magnetic fields. Detailed observations of spicules show them to be flarelike, heating the atmosphere by input of magnetic and mechanical energy, and high-resolution observations in spectrum lines of C IV, which can exist only at temperatures above 60,000 K (108,000°F), show continual high-speed (up to 400 km/s or 250 mi/s) ejections.

Another unexpected discovery from the ultraviolet and radio observations is the extreme sharpness of the transition to temperatures of 10^6 K (1.8×10^6 °F) in the corona above; the whole process takes place in a few hundred kilometers, a small distance on the scale of the Sun. *SEE RADIO ASTRONOMY; ULTRAVIOLET ASTRONOMY.*

Inside the chromospheric network cells a confusing and interesting mass of small elements exists; the elements oscillate back and forth, almost like water in a bathtub, with velocities approaching 10 km/s (6 mi/s) and a period of about 180 s. This horizontal oscillation is very easy to see in motion pictures of the chromosphere. It is presumably coupled with the 250-s oscillation in the photosphere, and plays an important role in the transport of energy upward.

An important tool for the understanding of the transition region is the extreme ultraviolet spectrum as recorded by rockets and satellites. Although the chromosphere is invisible against the disk in the visible region, in the ultraviolet region its high temperature makes it the dominant contributor, eventually superseded by the still hotter corona at even shorter wavelengths. In the extreme ultraviolet (XUV) spectrum, the lines of such ions as O II, O III, O IV, O V, O VI, C II, C III, C IV, and so forth, are seen. The roman numeral gives the number of electrons that have been removed, less one. From the relative intensity of these lines one can get some idea of the conditions at the successively greater heights in the atmosphere at which they are radiated. Furthermore, from the distribution of these lines on the disk it is possible to determine the temperature distribution. *SEE ROCKET ASTRONOMY; SATELLITE ASTRONOMY.*

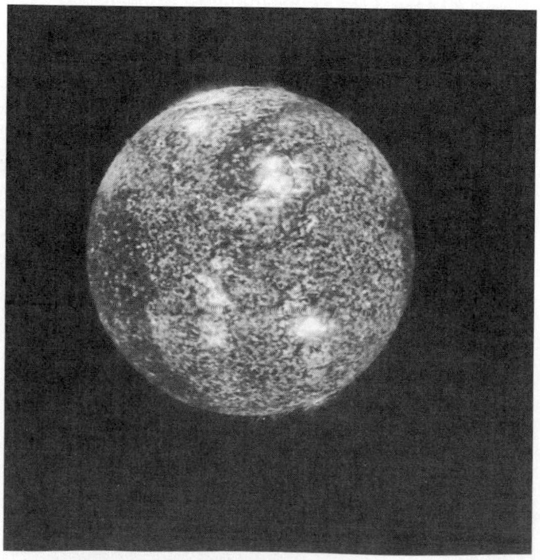

Fig. 10. Photograph of the Sun in ionized helium (He II) at 30.4 nm made by *Skylab*. (*U.S. Navy photo*)

In **Fig. 10**, the He II image shows the bright chromospheric network. This network disappears in regions at the left and right of the He II image; these regions are coronal holes at the poles of the Sun. Prominences are also visible in the photograph. The brightest regions are active regions; since the coronal lines are active ones, they show up only in active regions. The brightest area is a flare which is also marked by a string of other emission lines of ions up to Fe XXIV.

Corona. The chromosphere can be seen only by blocking the bright photosphere, as in an eclipse, or observing in special wavelengths where the photosphere does not emit and the corona, which is even weaker and fainter, requires even more extreme measures. The halo of pearly light seen at solar eclipses (**Fig. 11**) is so weak that it is difficult to photograph even if the sky is clear. And while the chromosphere can be observed in the light of strong spectrum lines, the lines of the corona in the visible spectrum are so weak they can be detected only at the edge of the Sun. In the extreme ultraviolet and x-ray ranges this situation is different; only hot gases can emit at these wavelengths, so a picture of the Sun in x-rays is a picture of the corona. The x-ray pictures from *Skylab* (**Fig. 12**) have thus revealed the inner structure of the corona; unfortunately the radiation from outer regions is too weak to be detected that way. *SEE X-RAY ASTRONOMY.*

The light seen from the corona originates in three distinct processes which distinguish the F (Fraunhofer), K (Kontinuierlich, that is, continuous), and E (emission) components.

The F corona is not directly associated with the Sun. It is a halo produced by the scattering of sunlight by interplanetary dust between the Sun and the Earth, and is properly regarded as the inner zone of the zodiacal light. Because the scattering is by solid particles, the Fraunhofer spectrum is seen in this light, and hence the name. The F component is negligible compared with the K component in the inner corona, but because of the rapid falloff of density in the corona, it dominates beyond 2.5 solar radii. *SEE ZODIACAL LIGHT*.

The K corona is photospheric light scattered by the free electrons in the solar corona. Although this light is not truly emitted by the corona, it is an important tracer of the coronal material, particularly because its intensity is proportional to the electron density and does not fall off so greatly with height as the emission lines and x-rays, which fall off as density squared. Because it has no preferred wavelength, however, the K corona must be observed either during eclipse, or, if the time behavior is desired, from spacecraft, using the dark skies of space. The coronal electrons are moving so fast that the Fraunhofer spectrum lines are blurred and a continuous spectrum is seen. The light is also polarized by the scattering process, and this polarization is used to detect the K corona against the sky background. Because the K corona is produced by the simple process of electron scattering, it is possible to measure directly the electron density (and hence the total density) of the material in the corona. This is done by careful photometric measurements of the coronal brightness at eclipse. Since the material is predominately hydrogen and is entirely ionized, there is essentially one proton for each electron measured, and the density is equal to the proton mass times the number of electrons observed. The density at the base of the corona is found to be about 4×10^8 atoms/cm^3, falling off exponentially with a scale height of 50,000 km (30,000 mi). The structure of the K corona varies markedly with the sunspot cycle. At sunspot maximum it presents a fairly symmetrical globular appearance with many domes and streamers in all directions. At sunspot minimum the globular shell shrinks toward the solar surface and individual dominant streamers are seen. At the poles there are brushes of small, symmetrically diverging streamers reminiscent of an iron-filings pattern. It is this appearance which suggested the presence of a general solar magnetic field.

The electron and temperature distribution in the corona may also be studied with radio telescopes. The propagation of radio waves through the corona is such that at each frequency one can see only down to a point characterized by the plasma frequency, which increases with density. At successively higher frequencies one sees deeper and deeper into the corona. Therefore the brightness at each frequency indicates the temperature at that height. It is possible, therefore, to map the temperature as a function of height. Further information may be obtained from the distribution of brightness across the solar disk in radio radiation. At high frequencies it is possible to see completely through the corona to the chromosphere, which appears as a uniform disk of about 6000 K (10,000°F). At longer wavelengths the corona is no longer transparent in the slanting direction of the limb, and therefore a peak of brightness near the edge of the Sun is seen. At the lower frequencies, where the radiation does not penetrate the corona, the Sun appears much larger than it does in visual wavelengths. *SEE RADIO TELESCOPE*.

The E, or emission-line, corona (**Fig. 13**) is the true emission of the ions in the corona. It is therefore the best source of information on the physical state of the corona. Originally the term E corona referred only to the emission lines in the visible, but it is now taken to include all the intrinsic radiation coming from pro-

Fig. 11. The solar corona observed about 2 s after second contact during the eclipse of November 12, 1966, at Pulacayo, Bolivia (altitude 13,000 ft or 4000 m). This photograph, made with a radially symmetric, neutral density filter in the focal plane to compensate for the steep decline of coronal radiance with increasing distance, allows structural features to be traced from the chromosphere out to 4.5R_0. Typical "helmet" streamers overlie prominences at the SE and SW limbs, while another streamer at high latitude in the NW develops into a narrow ray at large distance. Arches in the corona and the absence of coronal material in domes immediately above prominences are particularly striking at the bases of the NW and SW streamers. A coronal condensation appears on the NW limb at a latitude of about 25°. (*G. A. Newkirk, National Center for Atmospheric Research*)

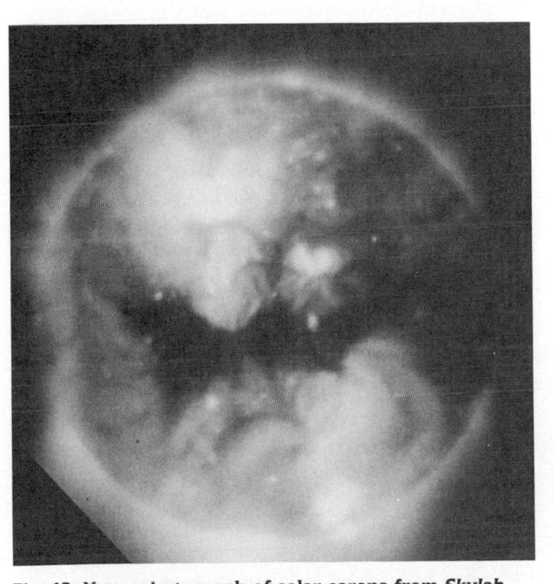

Fig. 12. X-ray photograph of solar corona from *Skylab* with a filter of passband 0.3–3.2 and 4.4–5.4 nm. Large coronal hole extends south from north polar cap. (*Solar Physics Group, American Science and Engineering, Inc.*)

Fig. 13. Typical stable form of structure of the E corona in the light of the green line of Fe XIV photographed without an eclipse through a birefringent filter. (*Sacramento Peak Observatory, operated by the Association of Universities for Research in Astronomy, Inc.*)

cesses in the corona: emission lines, continuous emission produced by interaction of free electrons and ions, and other minor sources. The emission lines are best observed from spacecraft in the extreme ultraviolet, where the photosphere is weak and the lines can be seen against the disk.

In spite of rather formidable technical difficulties, the coronagraph makes it possible to record the spectrum of the E corona in the visible, and even to photograph the structure directly through a birefringent filter whenever the sky is sufficiently clear. Since observations of the K corona outside of eclipse are limited to rather low resolution, the emission-line corona is an important tool in the observation of the day-to-day variation of the corona. The spectrograph dilutes the sky light by spreading it into a long, continuous spectrum. The coronal emission lines, on the other hand, are merely separated by the dispersion without dilution and stand out conspicuously against the sky continuum. **Figure 14** shows a short length of spectrum with a green 530.3-nm line of the corona and lines of a bright prominence superposed on the Fraunhofer spectrum of a sky scatter. The K corona is so weak that it cannot be seen. The emission lines are proportional to the square of the density and fall off sharply with increasing height. *See Coronagraph.*

The history of the identification of the coronal emission lines is a remarkable chapter in astronomy. The strongest, the Fe XIV green line, was independently discovered by W. Harkness and C. A. Young during the 1869 eclipse. This, and weaker coronal emission lines subsequently discovered, corresponded to no known spectrum lines, and were collectively attributed to the unknown element, coronium. More than 70 years later, they were identified by W. Grotrian and, more extensively, by B. Edlen as forbidden lines from highly ionized atoms which could be excited only under conditions of high temperature, low density, and enormous volumes quite beyond any conceivable laboratory resources.

Fig. 14. Green line of Fe XIV in E corona (long arc at left) with bright metallic lines of a short prominence, and yellow chromospheric helium line D_3 (arc at right), photographed through a small coronagraph. (*Sacramento Peak Observatory, operated by the Association of Universities for Research in Astronomy, Inc.*)

The lines of the E corona in the visible are due to highly ionized iron, nickel, calcium, or argon. These complex ions have levels between which magnetic dipole transitions occur which have a relatively long wavelength for such highly ionized matter. In the ultraviolet the normal spectral lines of these and other ions may be observed. Typically the ions of a given element will be ionized down to the member of heliumlike sequence (C V, O VII, Ne IX) which requires great energy to ionize it further. The heliumlike ions are hard to excite; their lines are seen in the soft x-ray range. When the heliumlike ion captures an electron, even briefly, the lithiumlike ion is formed; the odd electron is very active spectroscopically, and the strongest lines in the ultraviolet are *2s-2p* transitions in lithiumlike ions such as O VI and Ne VIII. The berylliumlike sequence is also strong. In solar flares, very highly ionized species up to Fe XXVI have been seen.

If the spectrum of the E corona is analyzed, the ions present are found to have ionization potentials of 250 to 350 eV, and are common at temperatures from 1.5 to 2.5 $\times$ 10^6 K (2.7 to 4.5 $\times$ 10^6 °F). Measurement of the radio emission and the width of the spectrum lines confirms this.

The high temperature of the corona is a fascinating problem. Obviously the temperature of a star is highest at the center and decreases outward to the edge. The steep increase in temperature in the corona has been conjectured to result from boundary effects connected with the steeply decreasing density at the edge of the Sun and the convective currents beneath it. These, together with magnetic fields, give a ''crack of the whip'' effect in which the same energy is concentrated in progressively smaller numbers of atoms. The result is the production of the high temperature of the corona. It is now known by various indirect means that coronas are common in G and K stars like the Sun, and that many have enormously extensive coronas compared to the Sun. Many have been detected to be x-ray sources, similar to but much stronger than the solar corona. *See Star.*

Solar wind. One of the most remarkable surprises of the age of rockets and satellites has been the discovery of the existence of the solar wind. This continual outflow of matter from the Sun was predicted by Eugene Parker on the basis that the high temperature in the corona must lead to a rapid outflow at great distances from the Sun. He predicted a velocity near 1000 km/s (600 mi/s). Spacecraft such as *Mariner 2* and the IMP series have detected a continual flow of plasma from the Sun with a velocity ranging from 300 to 500 km/s (200 to 300 mi/s) and density about 1 atom/cm^3 near the Earth. The flow occurs because the conductivity in the corona is so high that high temperatures exist at some distance from the Sun, where gravity can no longer hang on to the material. The particles flow along a spiral path dictated by magnetic fields from the Sun carried out into the interplanetary medium. The rotation of the Sun produces the spiral pattern. The magnetic field near the Earth is measured to be rather uniform over large sectors of the Sun, corresponding to one dominant polarity or another. The solar-wind flow has a continual effect on the upper atmosphere of the Earth.

Coronal holes. Early coronal observations showed that the corona was occasionally not visible over certain regions. In particular, most of the time it was quite weak over the poles. The x-ray pictures, particularly the long sequences from *Skylab*, revealed great bands of the solar surface essentially devoid of corona for many months. These proved to be regions where the local magnetic fields were connected to quite distant places, so the fields actually reached out to heights from which the solar wind could sweep the gas outward. The poles, of course, were the extreme case because the field lines reached out to the other pole of the Sun, but of course were swept away by the solar wind. Analysis of solar wind data showed that equatorial coronal holes were associated with high-velocity streams in the solar wind, and recurrent geomagnetic storms were associated with the return of these holes. Thus the great intensity of the corona over sunspot regions is partly due to their strong, closed magnetic fields which trap the coronal gas. Coronal holes are particularly prominent in the late stages of the sunspot cycle.

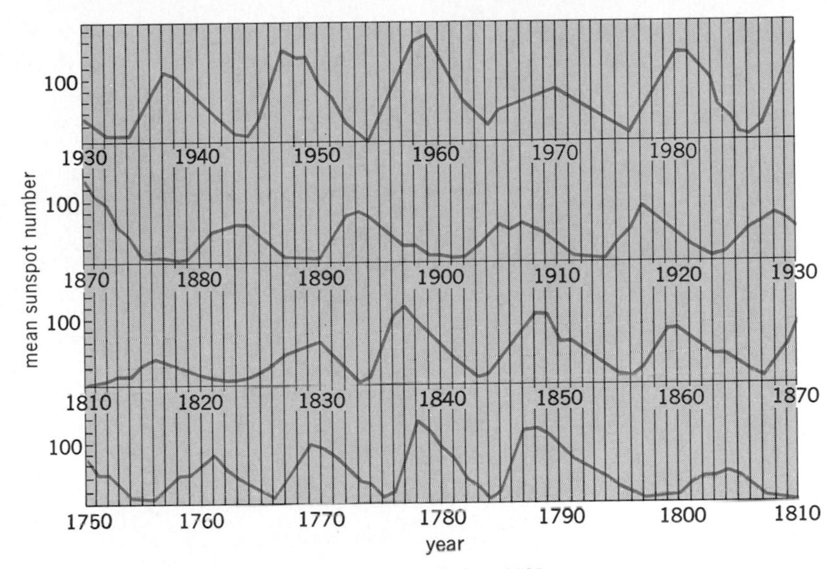

Fig. 15. Annual mean sunspot number from 1750 to 1989.

Solar Activity

The foregoing section has been concerned with the structure of the normal steady state of the undisturbed Sun. There are, in addition, a number of transient phenomena known collectively as solar activity. These are all connected with sunspots or their remnants, and wax and wane in a remarkable cycle of activity. The sunspot cycle consists of variations in the sizes, numbers, and positions of the sunspots, expressed quantitatively as the sunspot number (**Fig. 15**). The number of sunspots peaks soon after the beginning of each cycle and decays to a minimum in 11 years. The magnetic polarity of the sunspot groups reverses in each successive cycle so that the complete cycle lasts 22 years. The first spots of a cycle always occur at higher latitudes, between 20 and 35°, and as the spots increase in size they occur closer to the equator. Almost no spots are observed outside the latitude range of 5–35°. The average duration of the cycle from the first appearance of high-latitude spots to the disappearance of the last low-latitude spots is nearly 14 years. The difference between these two figures is the result of overlap of consecutive cycles. Thus there is hardly ever a time when there are no spots on the Sun at all. In the cycles peaking in 1958 and 1969, there was a great dominance of sunspots in the northern hemisphere, an asymmetry which has since disappeared. For periods of time there were few or no spots in the southern hemisphere. In addition, the two cycles with maximum activity in 1946 and 1957 were the two largest in history. Whether these facts are connected is not known.

Maunder minimum. Sunspots were discovered by C. Scheiner and by Galileo around 1610, and were extensively observed afterward. They can easily be seen with a small telescope or with the camera obscura, a dark room with a pinhole. But between 1650 and 1715 very few spots were seen, although experienced observers searched for them. After 1715 the spots returned. This period, first noted by E. W. Maunder, was associated with a long cold spell in Europe, known as the Little Ice Age. A number of secondary effects, particularly in carbon-14 in tree rings, confirm the low solar activity level at that time. Whether or not the Maunder minimum caused the Little Ice Age cannot be established, but there is little doubt that the sunspots dropped out during this period (although a weak 11-year period remained in the few spots present). So it is quite possible that the present era is a period of relatively high activity. Compared to stars of the Sun's class, it is about average.

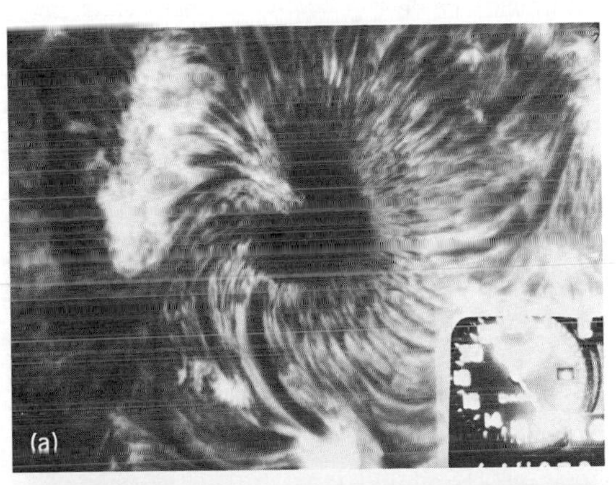

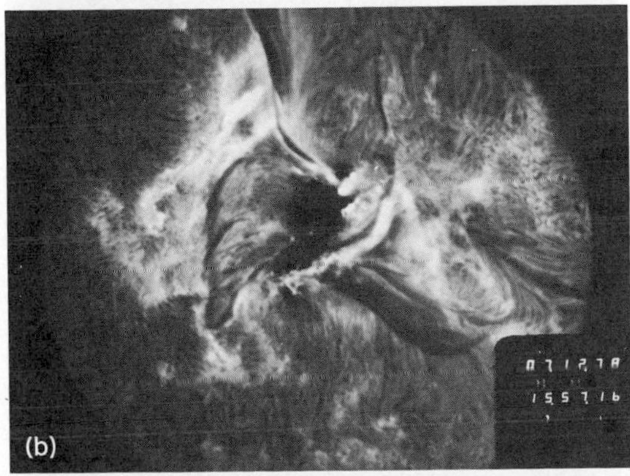

Fig. 16. Sunspots photographed in H-α light. In these photographs it is not possible to see down to the granulation. At this higher level the matter is dominated by field lines that delineate the magnetic configuration. (*a*) Large symmetric sunspot. Some lines of force terminate in the following plage (near bottom of photograph) and some radiate outward. (*b*) Highly active sunspot with twisted field lines connected to satellite polarity all around it. This spot group produced huge flares. (*Big Bear Solar Observatory*)

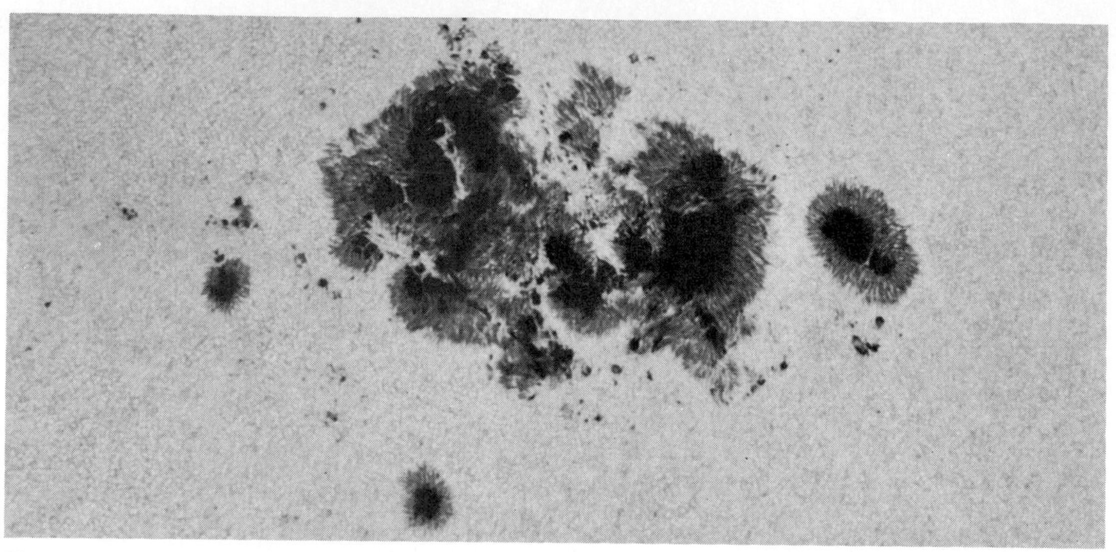

Fig. 17. Large sunspot group of May 17, 1951, photographed in white light showing the filamentary structure of the penumbra and the granulation of the surrounding surface. (*California Institute of Technology/Palomar Observatory*)

Sunspots. Sunspots are the most conspicuous features of solar activity and are easily seen through a small telescope. They are dark areas on the Sun produced by the most intense magnetic fields. The magnetic fields produce a cooling of the surface, most likely by suppression of the normal convection which transports energy from the lower levels. Deprived of this supply of heat, the area cools by radiation and becomes a dark sunspot. *See Sunspot*.

Curiously enough, the region close to the cool sunspot is the scene of the hottest and most intense activity because of the great magnetic energy present in the sunspot.

Fig. 18. Sunspot photographed in white light with extremely high resolution. Granulation at top is separated by fine dark lanes; the granules are about 1000 km (620 mi) across. At lower right is a sunspot umbra and in between is the penumbra, where fine fibrils mark the radial force lines. (*Big Bear Solar Observatory*)

The birth of sunspots is marked by the appearance of pairs of small spots about 20,000 km (12,000 mi) apart. In the light of the H-α line the chromosphere is seen to be particularly bright, with dark arches connecting the members of a pair. The arches mark lines of force connecting the spot fields. These fields are almost always in a particular orientation; in any given cycle, almost all the spot groups in a hemisphere will have the same magnetic polarity leading as the Sun rotates east to west in 27 days; in the other hemisphere the spot polarities will be exactly opposite. The fields are therefore designated p if they are the normal preceding polarity for that hemisphere, and f if they are the normal following magnetic polarity. In the next spot cycle the polarities reverse. If an emerging spot group has the wrong polarity, it usually dies out quickly but occasionally grows into a very great spot group.

As the sunspot group grows, remarkable things happen. The p spot rapidly moves westward while the f spot remains fixed. In most cases the f spot dies out, leaving a bright plage, while the p spot grows into a mature, round spot (**Fig. 16**). If there are little spots around, they merge into the main p spot. After a few days the p spot stops separating, and the sunspots no longer move. Often if the region is born with a tilt to the east-west line, the p spot will drag it into conformity with the rules. Doppler shifts in the dark arches show material moving upward at the center and flowing downward at the ends; this must result from magnetic tubes pushing up from below. When the sunspots form, the dark arches disappear, but they may remain at the center of the group to mark flux still emerging.

The typical spot group grows in a few days to the configuration of p spot and f plage, lasts a week or two, and disappears, leaving two puddles of p and f magnetic fields marked by weak plages. But occasionally a large spot group appears (**Fig. 17**), usually marked from the start by rapid growth of many small spot pairs, sometimes by new eruption of field in an older spot group. The new flux always emerges in pairs, because all magnetic fields must occur in pairs. As new flux pushes its way into old, great stresses and shears occur, which are relieved by flares.

A typical mature sunspot is seen in white light to contain a central dark area, the umbra, where the magnetic field is strong and vertical, surrounded by a less dark band called the penumbra, where the magnetic field spreads out radially, forming an aura of dark fibrils across the granulation (**Fig. 18**). The umbra appears dark because it is quite cool, only about 3000 K (5000°F) compared to 6000 K (10,000°F) in the photosphere. The spot pressure, consisting of magnetic and gas pressure, must balance the outside pressure; hence the spot must somehow cool by magnetic effects until the inside gas pressure is considerably lower than the outside.

Observations of the chromosphere above the sunspot reveal the presence of strong oscillations, called umbral flashes, with a 150-s period. In the penumbra, running waves spread radially outward with a period of 300 s—roughly twice that of the umbral flashes—and a velocity of about 10 km/s (6 mi/s).

There is a strong tendency for sunspot activity to break out all over the Sun at the same time. This enhances the idea that there is a general mechanism for their growth. There also may be preferred meridians for their occurrence, but the evidence is weak.

By means of the Babcock magnetograph, which permits the measurement of weak magnetic fields on the Sun, it has been possible to study the connection between the intense magnetic fields in the sunspot groups and the weak magnetic fields distributed over the surface (**Fig. 19**). The weak fields are concentrated on the edges of the chromospheric network, which consists of cells 30,000 km (20,000 mi) across. The fields vary considerably in strength from cell to cell. The origin of these magnetic fields appears to be the big sunspot groups. As the spots decay or are torn apart by motions in the solar atmosphere, their magnetic fields spread out and drift toward the poles. Because the poles rotate more slowly than the equator, the fields lag behind each sunspot group in a large region of one magnetic polarity shaped like the wing of a butterfly. Sometimes these regions may extend 90–120° in longitude. Their polarity is normally that of the following sunspot polarity. As time goes by, the weak fields reach the pole and establish there a dominant magnetic polarity which, because it is the same as the following spots, is opposite to the polarity of the preceding sunspots which dominate most spot groups.

Thus, a few years after the outbreak of the sunspot cycle, the fields generated by the broken-up spots reach the poles, producing at each pole magnetic fields of polarity opposite to those of the preceding sunspots in each hemisphere. Some models of solar activity propose that, as the sunspots of the parent cycle die away, this large-scale dipole field is amplified by the differential rotation to produce a new cycle of sunspots, which naturally will be opposite in polarity to the preceding cycle.

The large unipolar magnetic regions spreading out from centers of activity are easily recognized on H-α or calcium monochromatic pictures of the Sun. This is because the boundaries between these regions, where the magnetic field becomes horizontal in order to change direction, are usually marked by large

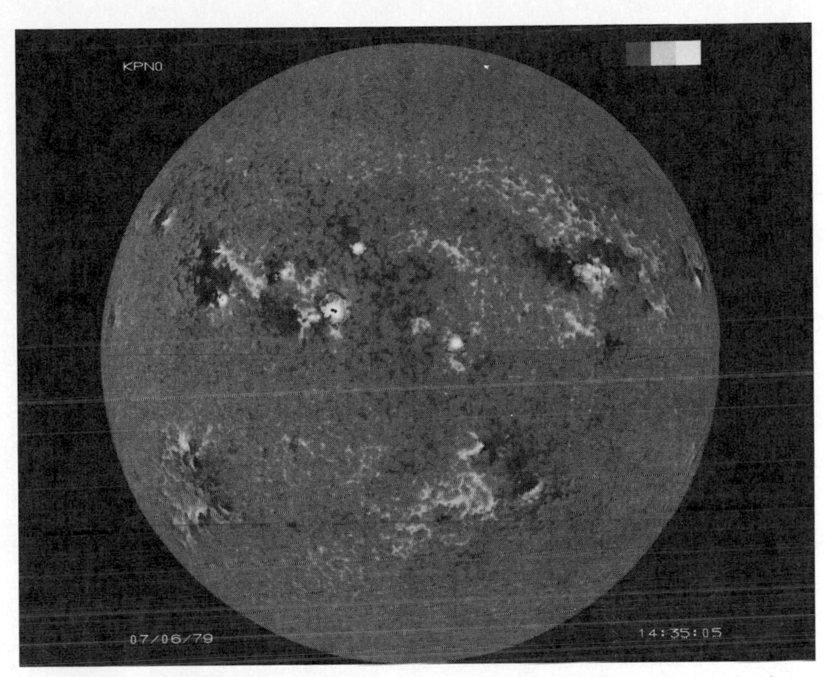

Fig. 19. Distribution of magnetic fields on the Sun at a time of high solar activity (July 6, 1979). Black areas are regions where the magnetic field points away from the Earth; white areas, toward the Earth. The strength of the field is indicated by the degree of blackness or whiteness. The north rotation pole of the Sun is at the top. (*Kitt Peak National Observatory*)

prominences, accumulations of material in the atmosphere supported by the horizontal field. These are seen as dark filaments, such as in Fig. 8. Because the unipolar regions are areas of enhanced magnetic field, they also correspond to enhanced emission in the chromospheric network, and thus can be distinguished in calcium spectroheliograms.

There are models for the evolution of solar magnetic fields and, in turn, the fields produce the sunspots and all their effects. However, these explanations mostly rest on differential rotation, as there is very little understanding of why the equator of the Sun should rotate so much more rapidly than the poles.

Ephemeral active regions. The picture of the sunspot cycle has been confused by the discovery of ephemeral active regions. These are tiny active regions which last only a day on the average and occur all over the Sun. The distribution of active regions in lifetime continues over to these small regions, until most of the flux erupting on the Sun at any one time is in the form of ephemeral regions; but they are so small and short-lived that they have no lasting effect on the field configuration. Ephemeral regions are best visible in x-ray pictures, where they appear as x-ray bright points (Fig. 18). At sunspot minimum they are quite prominent, but at maximum they disappear in x-ray photographs, although they still can be found in magnetic data. So it is possible that the sunspot cycle simply governs the maximum size to which ephemeral regions may grow.

Prominences. Although prominences appear dark against the disk, they appear bright against the dark sky. They occur only in regions of horizontal magnetic fields, because these fields support them against the solar gravity. Thus filaments on the disk are good markers of the transition from one magnetic polarity to the opposite. As can be seen from **Figs. 20** and **21**,

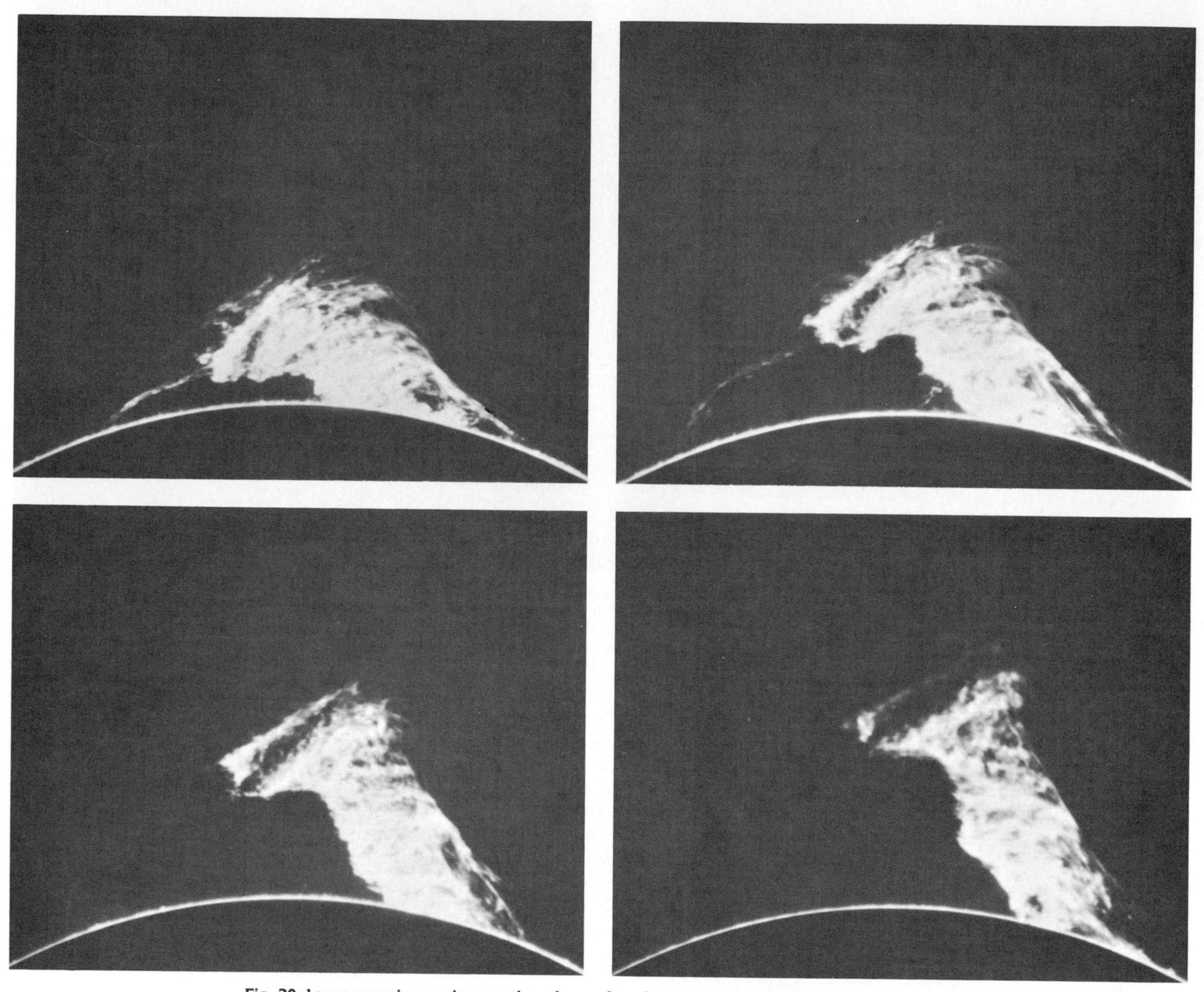

Fig. 20. Large prominence in eruptive phase after days or weeks of static inactivity. Four frames cover interval of 29 min. Horizontal width of single frame is about 670,000 km (420,000 mi). H-α light. (*Sacramento Peak Observatory, operated by the Association of Universities for Research in Astronomy, Inc.*)

prominences are among the most beautiful of solar phenomena. The spectra of prominences show a number of bright emission lines of various elements, mostly singly ionized (**Fig. 22**). Analysis of these lines shows that long-lived, stable prominences have a temperature of about 6000 K (10,000°F), while transient prominences connected with flares show many fewer lines and are over 30,000 K (54,000°F).

Plages. Just as prominences occur when the magnetic field changes from one sign to the other, plages occur whenever the magnetic field is vertical and relatively strong. They are bright regions visible in any strong spectrum line (**Fig. 23**). It is now believed that the field is, in fact, stable only when it exceeds 1000 gauss (0.1 tesla), and the difference between apparently weaker field regions, such as the chromospheric network, and plages is only the density of clumps of strong field. Plages are normally associated with sunspots; a typical active region will have the preceding magnetic field clumped in a sunspot and the following field spread out in a plage. In H-α light, the plage is seen to be connected to the sunspot by dark fibrils outlining the lines of force (**Fig. 24**); since the magnetic fluxes must be equal, the field in the plage is related to that of the spot by the ratio of their areas. The plage is bright in almost all wavelengths—x-ray, radio, ultraviolet—although in x-rays the peak brightness is in the connecting arches. The elements of the chromospheric network resemble small plages. If the field is just as strong as a plage, but horizontal, there is no plage brightening; therefore the vertical field plays a most important role in the heating of the plage and corona.

The appearance of plages in different wavelengths (**Fig. 25**) shows clearly the growing dominance of the magnetic fields with increasing height. If the energy of the magnetic field is compared with the kinetic energy of the gas in the low photosphere, only the magnetic fields inside sunspots are strong enough to dominate material motions. But in the chromosphere, the density drops so rapidly that the magnetic fields spreading out from the sunspot easily dominate. As a result little or no whirlpool structure is seen in white-light photospheric pictures, but it is easily seen in

Fig. 21. Characteristic small prominence, showing fibrous structure. The generally vertical filaments are paths of downward moving material. Horizontal width of a single frame is 125,000 km (78,000 mi). H-α light. (*Sacramento Peak Observatory, operated by the Association of Universities for Research in Astronomy, Inc.*)

H-α, which reveals the chromospheric structure higher up. In white light the plages are visible only near the limb, where one sees higher in the photosphere. They are easily seen in H-α, but are even more widespread in calcium K-line spectroheliograms because of enhanced contrast in the violet. With rockets, spectroheliograms can be made in the high-ionization lines showing emission mainly from active regions and plages. These are the only regions in the atmosphere hot enough to produce considerable ultraviolet radiation.

Although sunspots are always accompanied by plages, the reverse is not true. Occasional fields of small plages develop in the sunspot zone and fade away without the appearance of any spots. In these cases the magnetic fields never reach sufficient strength to generate sunspots. Similarly, the plages usually remain for a few weeks to mark the location where a sunspot has died. Eventually they probably break up and spread out into the chromospheric network, but this has not been observed directly.

Flares. The most spectacular activity associated with sunspots is the solar flare (**Fig. 26**). A flare is defined as an abrupt increase in the H-α emission from the sunspot region. The brightness of the flare may be five times that of the associated plage; the rise time is seldom longer than a few minutes, sometimes only 10 s. Of course, the hydrogen brightening is only a symptom; observations by other means show that there is a tremendous energy release, of as much as 10^{33} ergs (10^{26} W) in a large flare. An active sunspot group will show many small flares the size of a large sunspot. Once or twice in the lifetime of a large group, a great flare will occur covering the entire region. The energy released in the three or four great flares each year equals that in all the small flares put together. The flares are ranked by area in three classes, with class 3 the greatest; there is also a class of subflares for smaller brightenings which are only marginally flares.

Solar flares occur only as the result of sunspot ac-

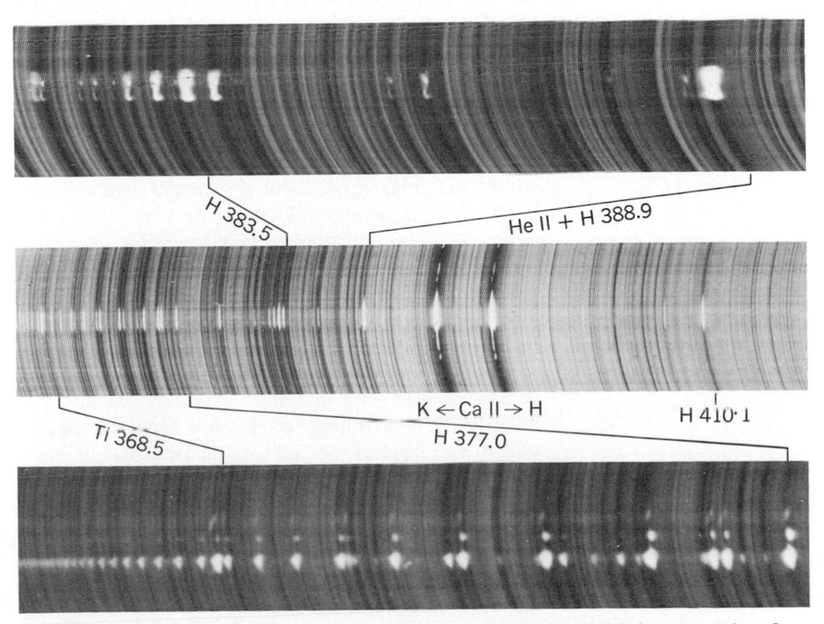

Fig. 22. Spectra of bright prominences at different dispersions with wavelengths of corresponding lines. From top to bottom the wavelength range covered by each spectrum is 8.8, 54.5, and 12.9 nm. The converging Balmer series merges into the Balmer continuum in bottom spectrum. (*Sacramento Peak Observatory, operated by the Association of Universities for Research in Astronomy, Inc.*)

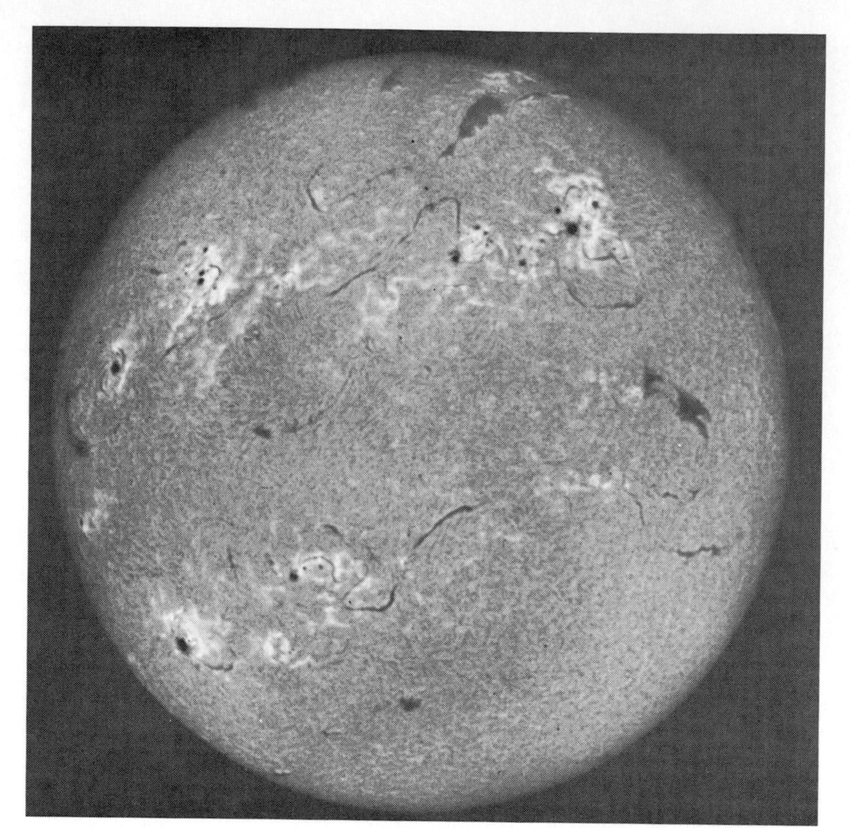

Fig. 23. The disk of the Sun photographed in H-α light, showing bright plages in centers of activity and dark filaments (prominences). (*Sacramento Peak Observatory, operated by the Association of Universities for Research in Astronomy, Inc.*)

tivity. Practically all sunspots produce some flares, but certain spots are prodigiously active. The source of the flare energy is the magnetic fields surrounding the sunspots. Every magnetic field has a certain minimum energy form, normally that where no currents flow, a so-called potential field. But if the fields have been strongly twisted, currents are set up which stabilize the twisted fields until they become unstable and jump to lower energy states, producing flares.

Complex magnetic fields occur only as the result of sunspot motion or emergence. But sunspots move only when they are emerging; hence all of the complexity results from flux emergence. Because the density of the solar atmosphere is low and its conductivity is high, magnetic field lines are very sticky. If a new spot emerges in old magnetic fields, it cannot connect to them immediately, but steep gradients are set up which eventually are released in the flare cataclysm. As mentioned above, when a new dipole emerges, the preceding sunspot runs forward, pulled by unseen forces below the surface. If the emergence is near an old spot and the p spot moves into like polarity, little will happen. If it hits a p spot, the two will merge. But if the field it runs into has opposite polarity, there will be a strong tendency for the new p spot to exchange its previous ties to connect to the local f spot. The boundary between these fields never takes a simple form, but usually is marked by a filament (a prominence seen against the disk), which marks a boundary with field lines running at right angles to the line between the spots. Of course, a lower-energy configuration would have the spots directly connected, but this never occurs. Eventually the filament becomes unstable and a flare occurs. All solar flares involve filaments, and one of the best flare warnings is a sudden lifting of a filament, marking a strong magnetic gradient.

Another source of flare energy is the decay of old sunspot fields. If the fields are not simply connected but are marked by large filaments and sheared magnetic fields, a filament will usually erupt in a large, slow flare as the fields decay.

Flares are rarely seen in white light because they are an atmospheric phenomenon and their density is so low that they are transparent. On the other hand, their temperature is so high that, in the ultraviolet region, they may equal the intensity of the entire Sun. However, because the flares are most easily observed in H-α and because the H-α brightening is an extremely accurate indicator, they are mostly studied in this way.

It has become possible to observe the flare phenomenon over a great range of wavelengths, in which many diverse phenomena have been observed. In x-rays a sharp pulse of radiation is observed, with photon energies of 100,000 eV and higher. At lower x-ray energies, the time behavior of the pulse is not as steep. The x-ray pulse always occurs during the first sharp brightening. Simultaneously, a broad-band

Fig. 24. Two photographs of the same active region matched to scale. The dominant p sunspot is on the left, with following plage on the right. (*a*) Photograph in H-α light. All areas of strong field are bright (except the sunspot), and all boundaries between opposite polarity are marked by dark arched fibrils. (*b*) Photograph of the magnetic field. White polarity is preceding, dark polarity following (the spot itself is saturated). The small inclusions of dark polarity near white are typically the loci of solar flares. (Large flares only occur with much more complex fields and sharper gradients.) (*Big Bear Solar Observatory*)

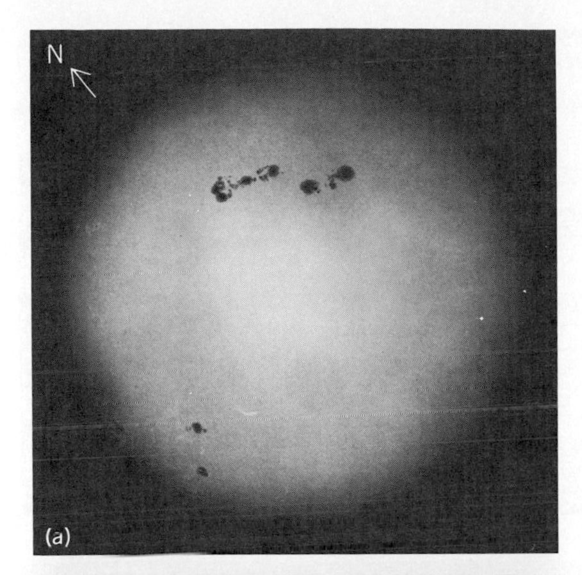

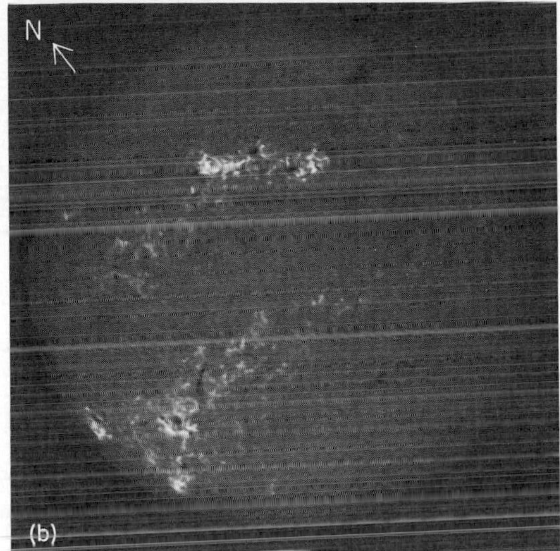

Fig. 25. The Sun photographed on April 8, 1980, in (*a*) white light and (*b*) red H-α light, showing the appearance of plages in different wavelengths. Plages can be seen in white light near the limb. The northern (upper) hemisphere shows two large bipolar groups which had limited flare activity because the magnetic configuration was simple. New groups are rotating onto the disk at lower left. (*Big Bear Solar Observatory*)

burst of radio waves is emitted in the microwave region, peaking around 3000 MHz. This is synchrotron emission from the same energetic electrons which cause the x-ray burst. At longer wavelengths, more complex phenomena (**Fig. 27**) are observed associated with the outward movement of disturbances from the flare through the corona. The corona at each point radiates a certain frequency depending on the density; thus as disturbances pass upward, radiation is excited at successively lower frequencies. The most intense of these is the type II burst, with a downward drift in frequency during a period of 2 to 5 min; this corresponds to an outward velocity of about 1000 km/s (600 mi/s) and is seen only in the largest flares. The type III bursts are much more frequent and show a very rapid frequency drift corresponding to disturbances propagating in the corona at 100,000 km/s (60,000 mi/s) and more. Some type III bursts are observed to turn downward, as though reflected by magnetic fields in the atmosphere. These are called U bursts. Following a large type II burst, one occasionally observes a type IV radio burst or storm which is a long-lasting wide-band emission from a great cloud in the corona. By analogy, the intense broad-band microwave burst accompanying the x-ray emission is called microwave type IV. The outward-moving type III bursts also produce an afteremission called type V emission.

Any large sunspot group is accompanied by a radio noise storm, and the detailed bursts of this storm are normally called type I bursts. They are strongly polarized by the coronal magnetic fields.

In addition to the x-ray and radio emissions from flares, which are all electromagnetic radiation, great numbers of energetic particles, both nuclei and electrons, are observed. Since energetic particles must follow the magnetic fields, the corpuscular radiation is not observed from all big flares, but only from those favorably situated in the western hemisphere of the Sun. Because of the solar rotation, the lines of force from the western side of the Sun (as seen from the Earth) lead back to the Earth and guide the flare particles to the Earth. The nuclei are mostly protons, because hydrogen is the dominant constituent of the Sun.

The flare-produced cosmic rays are most numerous at 3–5 MeV, but range upward in energy to hundreds of megaelectronvolts. Almost all flares produce cosmic rays, but the cosmic-ray production from the biggest flare of any year will probably equal that of all the other flares put together. The high-energy cosmic rays appear moments after the beginning of the H-α flare, but because of time of flight and the guiding by the interplanetary magnetic field, the lower-energy cosmic rays appear much later. A cosmic ray storm may last for days. The flux of low-energy particles in big flares is so intense that it endangers the lives of astronauts outside the terrestrial magnetic field. *See* Cosmic rays.

Relativistic electrons with energies of megaelectronvolts are also observed shortly after the beginning

Fig. 26. The great "sea horse" flare of August 7, 1972, late in the flare, photographed in the blue wing of the H-α line. The neutral line between two bright strands is crossed by an arcade of bright loop prominences raining down from the corona. (*Big Bear Solar Observatory*)

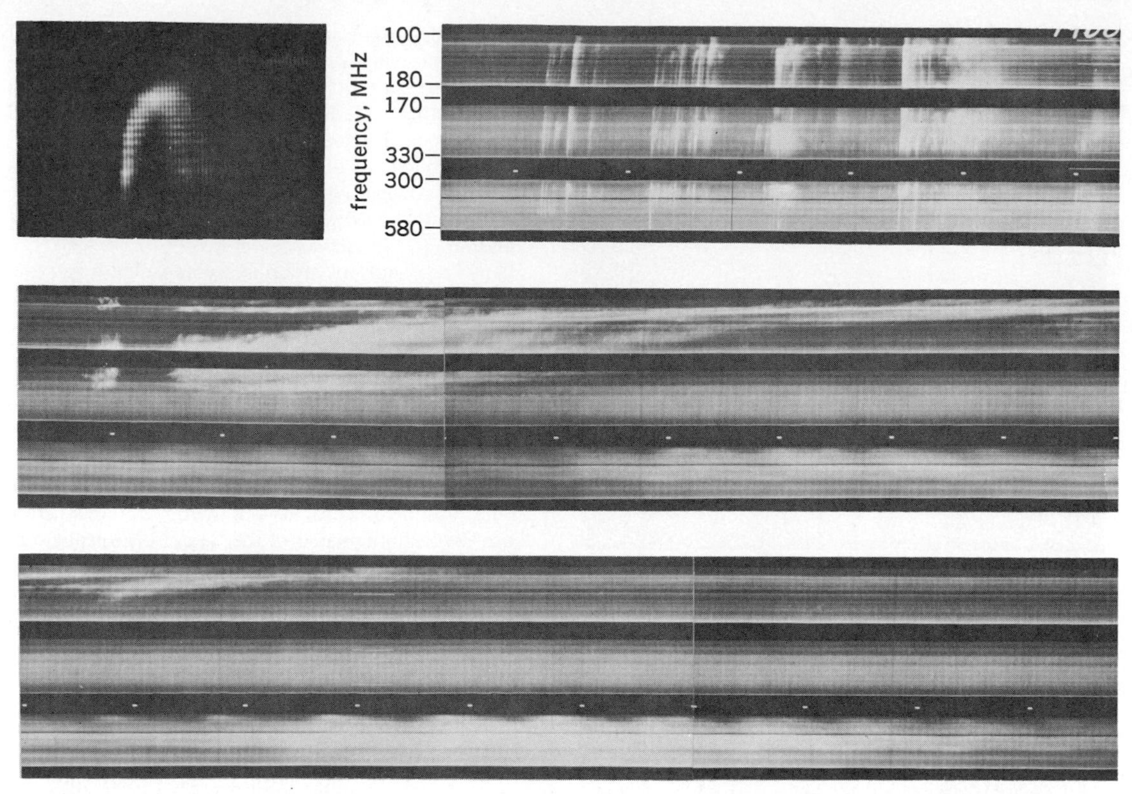

Fig. 27. Time variations in the solar radio spectrum. Frequency range from 100 to 580 MHz (the vertical coordinate) is divided into three overlapping bands with frequencies indicated in top row. Variations with time are shown in horizontal direction. The interval between the white pips between the middle and lower frequency bands represents 1 min. Top left, a highly magnified U burst in the 140-MHz band. The interval between successive vertical scans is 0.3 s. Top right, a series of type III bursts. The middle and bottom rows are continuous, showing the development of a type II burst followed by a strong continuum in the 450-MHz band. (*Recordings from radio spectrometer, Fort Davis Station, Harvard College Observatory*)

of optical flares. They are presumably accelerated by the same phenomenon that produced the cosmic rays.

A day or two after a very large flare, the pulse of arrival of a new group of particles of low energy (1–5 MeV) but of very large numbers may be observed. These particles, which produce what is called a geomagnetic storm, are a cloud ejected directly from the flare with a velocity of 1500 km/s (900 mi/s) or more.

It is known that any large sunspot group is continually producing large numbers of low-energy cosmic rays, presumably as the result of many small flares. The energy spectrum is very steep, about the inverse fifth power; hence, if the magnetic changes in a large flare increase the energy of all cosmic rays by a factor of 10, the number in some high-energy range will be increased by 10^5.

Until 1972 the MeV-energy cosmic rays could be studied only upon their arrival at the Earth, distorted by time of flight and magnetic field effects. Then studies of the high-energy gamma-ray spectrum revealed a series of gamma-ray lines due to nuclear reactions on the surface of the Sun produced by the energetic particles. The following reactions were observed:

1. Positrons created by collisions of protons with surface nuclei annihilate with ambient electrons, producing a line at 0.5 MeV.
2. Neutrons split off from ambient nuclei by protons recombine with ambient protons, forming deuterons, giving a strong line at 2.2 MeV.
3. Protons colliding with carbon and oxygen nuclei excite them to a higher state, giving lines at 4.4 and 6 MeV.

Many other lines are possible, but higher-resolution detectors adequate to detect them have not been orbited.

High-energy phenomena play a fundamental role in flare physics. All evidence suggests that the primary energy input from the magnetic field to flare gases is

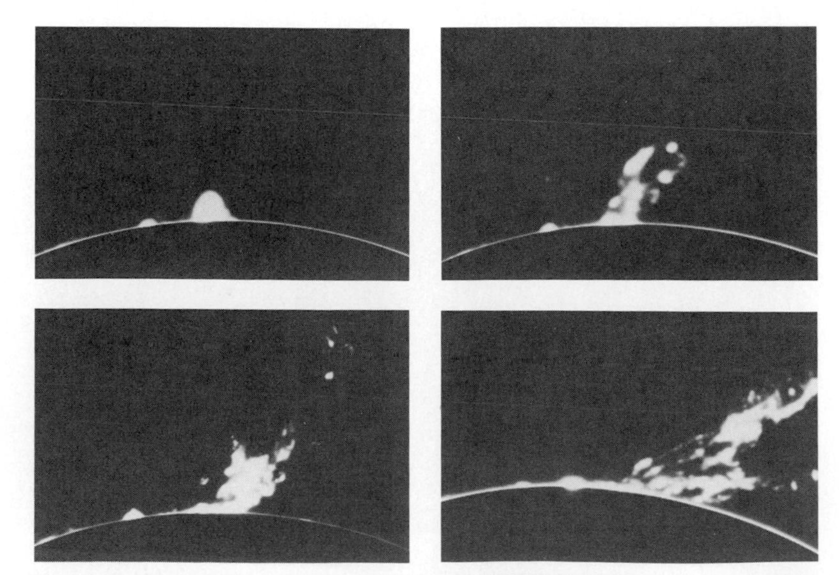

Fig. 28. Flare prominence, or spray, of February 10, 1956. Universal time for successive frames: 211600, 211845, 212115, 213345 cover a total interval of 17 min 45 s. Some fragments exceeded a velocity of 1100 km/s. (700 mi/s). Each frame shows 670,000 km (420,000 mi) on the Sun. (*Sacramento Peak Observatory, operated by the Association of Universities for Research in Astronomy, Inc.*)

due to high-energy nonthermal particles—electrons or protons—that transmit their energy to the ambient atoms by collisions. A few minutes later, thermal energy in the flare is released by compressing material in the corona; temperatures above 2×10^7 K (3.6×10^7°F) are produced, and the resulting material condenses in the form of elegant loop prominences.

Flare emissions. Since the flare introduces a very energetic plasma into a relatively cool region, a whole range of particles and waves are produced. Each produces a characteristic emission.

As noted above, the energetic protons produce gamma-ray lines when they penetrate to the photosphere. If they are sufficiently numerous, they heat the photosphere faster than it can reemit and a white light flare is observed, usually in the form of bright transient flashes at the foot points of the flare loops.

Electrons radiate more efficiently. When they collide with protons, they produce hard x-rays by bremsstrahlung. For the first, impulsive burst of electrons the electron spectrum is hard and so are the x-rays. Later a thermal, softer spectrum is produced. The electrons also produce microwave emission as they spiral in the magnetic fields by a process known as synchrotron emission. From the spectrum of the microwaves and x-rays the number and spectrum of the electrons can be deduced. With huge radio antennas such as the Very Large Array in New Mexico, the source of microwaves can be mapped. It turns out to be a small kernel right at the top of the flare loops. *SEE BREMSSTRAHLUNG.*

When the filament at the site of the flare heats up, a coronal cloud is formed with a temperature above 10^7 K or 18×10^6 °F (usually 30×10^6 K or 54×10^6 °F) and density 10^{10} atoms/cm^3, 10–100 times greater than the normal corona. This cloud is compressed by magnetic effects and produces strong, soft x-rays and ultraviolet emissions. This emission is full of spectrum lines, and these have been extensively studied by x-ray spectroscopists. The lines observed are due to highly ionized atoms all the way up to Fe XXVI (iron with 25 of 26 electrons stripped away). In the first stages of the thermal event the Fe XXVI and XXV lines are seen; then as the plasma cools down, lines of elements in lower ionization dominate, along with iron in lower ionization states. Permitted and forbidden transitions are observed, and the ratio of these is used to deduce the local density, the temperature being given by the ionization. *SEE MAGNETOHYDRODYNAMICS.*

Flare ejecta. Almost every flare is accompanied by the ejection of material; the magnetic field acts like a rifle barrel to collimate the flare energy. Neutral hydrogen in the flare filament is usually ejected in a poorly collimated spray (**Fig. 28**) at speeds up to a few hundred kilometers per second (although 1500 km/s or 900 mi/s has been recorded). The sprays give rise to a magnetohydrodynamic shock wave in the corona which may be recognized in meter-wave radio emission by a so-called type II or slow-drift burst. The frequency of the emission drifts downward as the wave moves to lower densities and the radio emission excited decreases in frequency. The wave velocity is 1000–2000 km/s (600–1200 mi/s), and the waves have been observed all the way out to the Earth. The waves carry energetic trapped particles along, and a large increase in low-energy cosmic rays (1–5 MeV) is often observed when the wave reaches the Earth.

Flare waves are often observed on the solar surface in H-α light when their path is such as to include the surface. They usually appear as a bright front; sometimes filaments wink (that is, they drop out of the H-α bandpass by a down-and-up Doppler shift) as the front goes by.

There is a whole range of jetlike ejecta called surges which may be quite large. They occur when the flare is not too big and the magnetic field is strong enough to contain and collimate the material. Surge velocities are usually less than 200 km/s (120 mi/s), and they often fall back along the same route.

Flares produce intense streams of electrons. These travel at about one-third the speed of light and produce type III or fast-drift radio bursts. In this case the frequency of the radio emission excited by the passage of the electron stream through the corona drifts downward rapidly as the stream reaches lower densities. The streams have been tracked all the way to the Earth, where pulses of electrons have been measured, typically at energies of 40 keV.

The development of spacecraft coronagraphs has permitted the observation of flare ejecta in the corona. Most of the eruptions produce large shock waves traveling outward through the corona.

Flare and corona. The normal processes of heating the corona cannot sustain all the material injected by

Fig. 29. Quiescent prominences in H-α light, seen (*a*) against the sky and (*b*) against the disk. (*Big Bear Solar Observatory*)

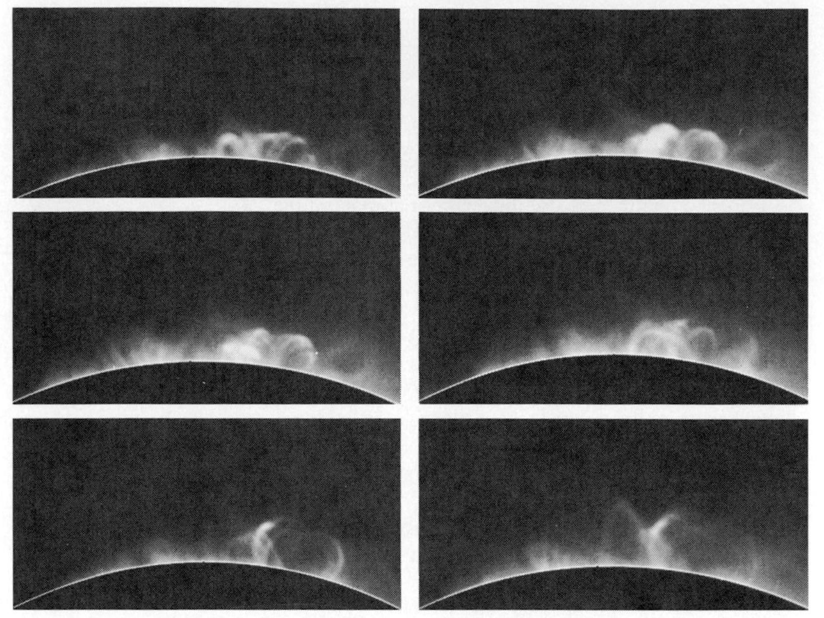

Fig. 30. Six exposures of coronal activity over an interval of 4 h taken in the green line (530.3 nm). Concentric arch structure in the first frame is a characteristic structure. (*Sacramento Peak Observatory, operated by the Association of Universities for Research in Astronomy, Inc.*)

the flare, which therefore cools down by radiation. When this material cools and condenses, it rains down on the surface. Because of the great strength of magnetic fields overlying the sunspots, the gas follows the magnetic lines of force in curved trajectories, forming loop prominences (**Fig. 29**). The cooler the material gets, the lower its degree of ionization and the faster it radiates and cools. In fact, a large fraction of the H-α radiation in flares is thought to be due to the falling material from the great loop prominences coming from the coronal cloud. Since the flare may last only minutes and the loop prominences will go on for many hours, they are often seen as a signal that a great center of activity is at the limb.

The birefringent filter makes it possible to see the appearance of the corona by isolating the coronal emission lines. The coronal material, as well as the prominence material, follows the magnetic lines of force, and many curved arches are seen. However, these do not change as rapidly as the condensing material in the prominences (**Fig. 30**). Often, series of five or six concentric arches are seen over an active center; they expand slowly and occasionally have been observed to break open at the top and whip into vertical streamers with apparent velocities of up to 600 km/s (370 mi/s). This activity coincides with the eruption of material through the magnetic field at the time of a flare. The coronal cloud produced by the flare is responsible for low-energy x-rays in the 0.2–1.2-nm range, as well as for gradual increases of radiation in the microwave region. However, the hard x-rays and impulsive microwave bursts are associated with more transient groups of hard electrons which are accelerated by flares.

Solar terrestrial effects. Aside from heat, light, and the solar tides, the direct influences of the Sun on the Earth are generally too delicate for direct detection by the human senses. One exception is the aurora, which may be seen on occasions in temperate latitudes and more often in Arctic and Antarctic regions. The agents responsible for other solar influences on the Earth are ultraviolet and x-radiation and streams of charged particles emitted by the Sun. These are strongly variable with the level of solar activity, and the effects on the Earth are very large indeed when they are observed with sensitive radio equipment and magnetic compasses, as well as with detectors on artificial Earth satellites.

The terrestrial ionosphere is a result of the steady flux of solar ultraviolet and x-radiation. Even the quiet corona and low-level chromospheric activity is sufficient to make possible an ionosphere with a low electron density. As the solar activity increases, the electron density in the ionosphere increases and higher frequencies are reflected. Most of the ionization is caused by solar ultraviolet rays between 20 and 100 nm, which ionize oxygen and nitrogen in the upper atmosphere. Only a small fraction of the molecules above 100 km (60 mi) are ionized.

The burst of hard x-rays emitted at the onset of a solar flare causes ionization much deeper in the atmosphere, around 60 km (37 mi). At this level the density of the air is sufficiently high so that the electrons cannot oscillate freely but give up their energy by collisions with neutral atoms; thus, when they are excited by radio waves (which normally would penetrate and be reflected from the higher ionospheric layers), they absorb the radio waves and produce a shortwave fade-out (SWF). Long-distance radio communication, which depends on reflection from the ionosphere, deteriorates or is blacked out altogether for a few hours. The effect is almost instantaneous with the beginning of the flare, and serves as an indicator of its occurrence. It may also be observed by recording the radio emission from galactic sources which normally penetrates the ionosphere but is suddenly absorbed during an SWF. This phenomenon is called sudden cosmic noise absorption (SCNA). Temporary ionospheric currents in the beginning of the fade-out produce changes in the geomagnetic field strength, and if they are severe they may induce currents in long land lines sufficient to stop telephone communications. SWF is always associated with a flare, although occasionally flares will not produce SWF.

The ionizing radiation in SWF is hard x-rays in the 0.1–0.2-nm region. By a coincidence, there is also a small amount of ionization in this region (called the D layer) produced by the photoionization of nitric oxide by the Lyman-α line of hydrogen. However, this line does not change much during flares and does not produce a fade-out.

A whole series of other phenomena is associated with the creation of the D layer at the bottom of the ionosphere and its changing height. Among these phenomena are the sudden phase anomalies (SPA), produced by the lowering of the effective reflecting layer; and the sudden enhancement of atmospherics (SEA), produced by increased reflection of distant radio noise.

The particle effects on the Earth are extensive. The cosmic-ray storms produced by big flares do not reach the equatorial regions easily because of the Earth's magnetic field, but they spiral into the polar caps and produce what is called polar-cap absorption (PCA). This is the direct ionization of the ionosphere above the poles. Its effect is a polar blackout of radio communications across the polar regions. The very large number of particles associated with the low-energy pulse manages to penetrate the geomagnetic field to produce the geomagnetic storm. This is most intense near the poles, but may reach down into temperate and even tropic latitudes. The aurora is but one trace of the energetic particles precipitating in the upper atmosphere. The currents induced by these great numbers of particles produce sharp changes in the Earth's magnetic field, and the ionization produces considerable changes in the radio propagation. The magnetic field changes have also been known to produce severe effects in long power lines, creating surges and pulses which trip circuit breakers and produce power outages.

The relation between geomagnetic storms and flares is not completely determinate. Large flares on the western side of the Sun usually produce the most geomagnetic disturbances, but great magnetic storms often occur without a flare. The storms not associated with flares show a strong recurrence and, since for a long time their origin could not be identified, their source was termed the M region.

The M regions have now been identified with coronal holes. Spacecraft measurements have shown a close connection between coronal holes and high-velocity streams. The coronal holes are long-lived and produce a geomagnetic storm at the Earth each time they go by. The biggest geomagnetic storms, however, are produced by the shock waves from big flares.

Measurements of the plasma velocity in the solar wind show that there is a strong correlation between the velocity of the plasma and the geomagnetic activity. The passage of the M region and the beginning of the geomagnetic storm are usually connected with a shock wave in the interplanetary magnetic field which impinges on the Earth's field and produces a sharp lowering of the field, known as a sudden commencement, the signal for the beginning of a geomagnetic storm. The particles which closely follow produce the storm.

Observations of the aurora have identified a smaller phase of the geomagnetic field called the substorm. A substorm is a violent rearrangement in the outer fringes of the Earth's field, where it interacts with the solar wind and is drawn out into a long tail on the antisolar side. It has been found that substorms tend to occur when the north-south component of the magnetic field in the solar wind changes from parallel to antiparallel with the Earth's field.

Many astronomers believe there is a link between the sunspot cycle and long-term weather trends, but it has proved very difficult to detect. The evidence for a Little Ice Age during the Maunder minimum is persuasive, but there was no effect in Asia. There has been an apparent coincidence between great droughts in the western United States in 1910, 1932, 1954, and 1976 and alternate minima of the 11-year cycle; that is, the droughts appeared to agree with the 22-year magnetic cycle. Droughts are so sporadic, however, that direct statistical connections are not very obvious. For example, 1980 was a year of heat and drought too, but at the maximum of the sunspot cycle. However, some analyses show that the likelihood of severe local droughts does peak at the 22-year minima, but the locality of the drought will move about. Measurements of isotope ratios in fossil water, which indicate the temperature at which the water fell as rain, also indicate 22-year periods. So there are bits of evidence but little understanding.

Solar Instruments

Although the observational instruments of the solar astronomer are the same in principle as those used by other astronomers, there are two points which determine the differences. First, the Sun is extremely bright, so that high concentration of light is not so important. Second, because of the disturbance of the

Fig. 31. The 150-ft (45-m) solar-tower telescope of the Mount Wilson Observatory has a vertical underground spectrograph.

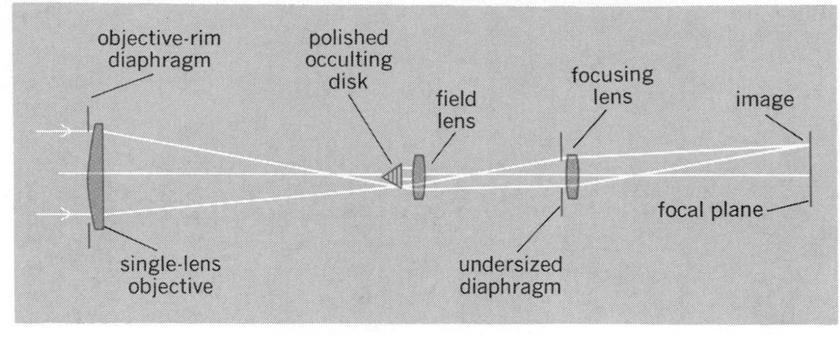

Fig. 32. Optical system of a coronagraph.

Earth's atmosphere by solar heating during the day, the ''seeing'' is considerably worse in the daytime, and solar telescopes cannot hope to utilize as sharp an image as stellar telescopes. Furthermore, the telescopes must be designed so that the great heating produced by the Sun does not distort the images. This is usually done by liberal use of white paint and careful control of the atmospheric conditions inside the telescope.

Solar telescopes fall into two general classes: those designed for observations of the brilliant solar disk, and coronagraphs designed for the study of the much fainter prominences and the still fainter corona through the relatively bright, scattered light of the sky.

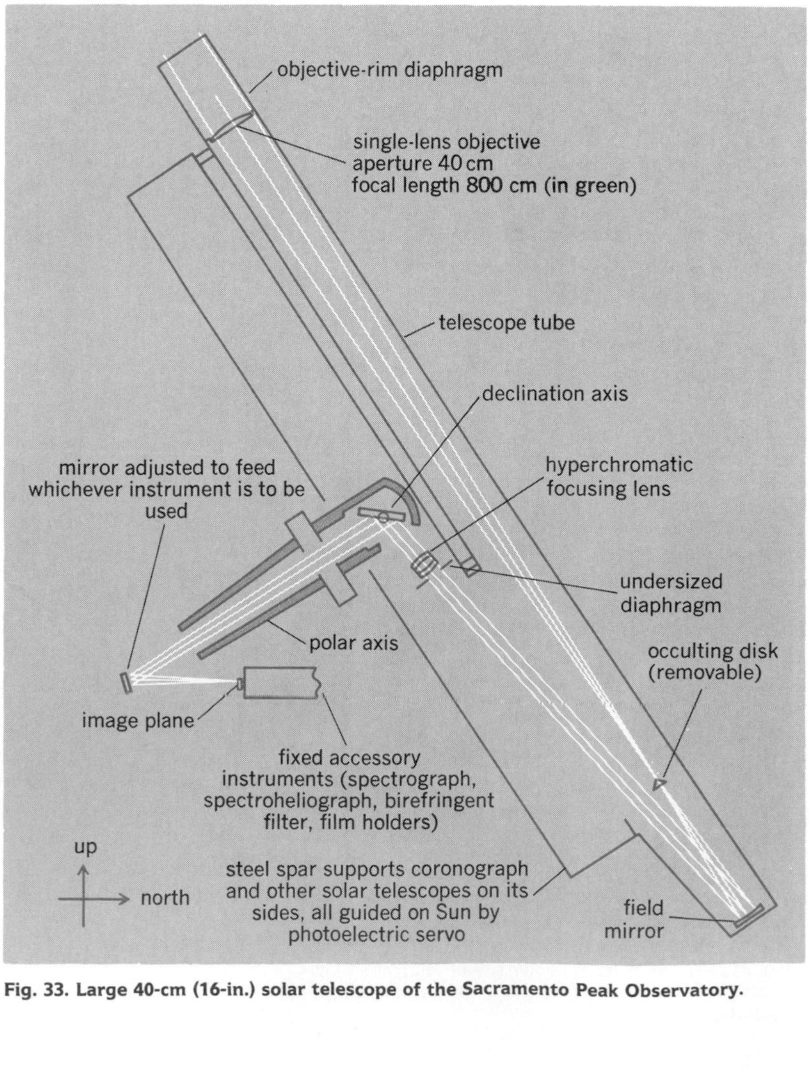

Disk telescope. Excellent definition in a solar image of reasonable size is usually the first requirement in a disk telescope. Definition is limited by the quality of the optical system, by the fundamental diffraction limit of resolution inherent in a given aperture, and by the quality of the seeing. Seeing is the term used to describe the blurring effect of the changing density gradients in the terrestrial atmosphere due to thermal convection. Seeing is never perfect, and there are very few locations where excellent definition is possible during perhaps 100 h in a year. These places are usually in lowland and maritime regions where stabilizing temperature inversions exist. Unfortunately, the transparency is usually poor in such locations. Most solar telescopes have been built on top of mountains, because astronomers were used to building telescopes there and because the air is clear. Unfortunately, the air above such sites is quite turbulent and the seeing is very poor. In addition, clouds tend to form over mountain peaks. A few telescopes have been built in lakes, where daytime convection is reduced and much better results are obtained.

To maintain the definition of the telescope optical system, it must be guarded against the severe effects of solar heating. Solar heating produces air currents in the telescope and distorts the optics. This is not serious in refractors because little heat is absorbed in the lens and the lens acts as a cap against thermal convection. But reflectors, which are preferred for their achromatic properties, have severe difficulties, because the mirror absorbs heat and because the light travels up and down the tube. All modern reflecting solar telescopes are vacuum telescopes with large entrance windows. Special low-expansion materials are used to reduce changes in the mirror, and the gregorian design is preferred because heat load on secondary optics is reduced.

The most convenient type of disk telescope is the solar tower (**Fig. 31**). Two flat mirrors at the top of a tower, one of which is equatorially mounted to follow the diurnal motion of the Sun, reflect sunlight into a long-focus fixed vertical telescope. The telescope may be either a refractor or a compound reflector of 30–50 cm (12–20 in.) aperture. It produces a large image of the Sun near ground level, where the light can be conveniently reflected into any one of a number of large fixed accessory instruments.

A less expensive variant of the tower telescope is the horizontal fixed telescope, with the flat mirrors at ground level. This arrangement, when carefully designed, works well for small apertures, but convective disturbances in the long horizontal air path degrade the definition of large instruments.

For many purposes an equatorial telescope is preferred; in particular, it is easier to evacuate an equatorial system than a large tower, and for some purposes, such as polarimetry, a straight-through system is desirable. However, it is difficult to servo-guide an equatorial, because the whole telescope must be moved instead of a single mirror. *See* *Optical telescope*; *Telescope*.

Coronagraph. The coronagraph, invented by B. Lyot in 1931, is designed with one overriding consideration in mind: the elimination of instrumental scattered light. Its most delicate task is the observation of the corona immediately adjacent to the disk of the Sun, which is about a million times brighter. In an

Fig. 33. Large 40-cm (16-in.) solar telescope of the Sacramento Peak Observatory.

ordinary telescope, a little dust on the objective, diffraction at the edge of the aperture, and otherwise insignificant defects in the glass of the objective are all sources of diffuse scattered light which is usually some hundreds of times brighter than the corona.

The coronagraph is deceptively simple (**Fig. 32**). The critical component is the single-lens objective; it must be made of flawless glass, which is polished and cleaned to a perfection far beyond average standards. The rest of the system is more ordinary. The disk of the Sun is eclipsed by a polished conical disk, and photospheric light diffracted by the rim of the objective is intercepted in its image formed by the field lens, on an undersized diaphragm. The final lens then images the corona on the focal plane. The spectrum of the corona can be observed by placing the slit of the spectrograph at the focal plane, or direct photographs in the green line may be taken by inserting the appropriate birefringent filter behind the focusing lens and letting the image fall on a photographic film or plate.

The usual coronagraph has an aperture of 15 cm (6 in.) or less. The largest is a 40-cm (16-in.) instrument (**Fig. 33**). An internal mirror system reflects the light through the polar axis into an observing laboratory, where the image can be directed to large analyzing instruments, as in a tower telescope. The focusing lens corrects the color aberration of the objective, and the telescope can be used equally well for disk and limb observations. Figure 21 was taken with it, in combination with a birefringent filter.

The coronagraph became outmoded when coronal observations from spacecraft became possible. The coronagraph can record the corona only at the limb, whereas soft x-ray telescopes reveal the whole corona on the disk. Also, a much greater range of spectral lines is available in the ultraviolet. Thus, only a few coronagraphs remain. However, these instruments still have capabilities that cannot be matched in space, in particular high spatial resolution and flexibility. They are unexcelled for prominence observation. Since the coronagraphs must be located on mountains, they have limited good seeing.

Spectrograph. Modern solar spectrographs for use with disk telescopes utilize the great brightness of the solar image to achieve dispersion of the order of 100 mm/nm and spectroscopic resolution in excess of 500,000. These desirable characteristics require long focal lengths (10–25 m or 33–82 ft) and superlative diffraction gratings of the largest sizes that remain consistent with accuracy of ruling.

Large solar spectrographs are nearly always either of the Littrow autocollimating type or the reflecting type in which the collimator and camera element are long-focus concave mirrors (**Fig. 34**). They may be mounted either vertically in a well or horizontally on solid concrete piers. The Fourier transform spectrometer has proved extraordinarily powerful.

The spectrographs used with coronagraphs for observations of solar limb phenomena are generally much smaller and simpler, with medium dispersion of 5 mm/nm or less. They are usually simple Littrow grating spectrographs of about 200 cm focal length. The most important requirements are high light efficiency and stigmatic image. Most of them are small enough to be carried on the same mounting with the coronagraphs that feed them.

Spectroheliograph. The spectroheliograph and birefringent filter present an extended area of the solar image at one sharply defined wavelength. The wavelength chosen is usually at the center of the H-α line of hydrogen or at the H or K line of ionized calcium. The chromosphere is opaque at these particular wavelengths, and the picture obtained is, therefore, that of the chromosphere. It is evident from Figs. 8, 23, 24*a*, and 25*b* that the structure is quite different from that of the photosphere, shown in Figs. 1, 5, and 17.

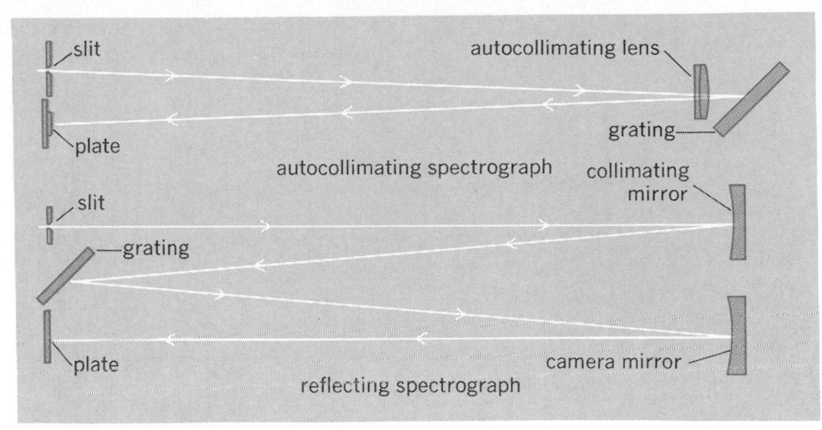

Fig. 34. Two forms of solar spectrograph. Horizontal scale is greatly compressed.

The spectroheliograph is a stigmatic scanning monochromator made by inserting a slit in the focal plane of a spectrograph, accurately centered on the H-α line (or any other wavelength desired), so that only the light of this line is transmitted. Variations in image intensity along the first slit are faithfully reproduced in the exit slit. Several different scanning arrangements have been successfully used; one uses fixed optics. The monochromator is mounted on ways which permit smooth motion in the direction of dispersion. The solar image from a fixed telescope falls on the entrance slit of the monochromator. The light of the H-α line emerging from the second slit falls on a stationary photographic plate. As the monochromator moves along its ways, the first slit scans across the solar image, and the exit slit correspondingly scans the photographic plate. Thus an image of the Sun in the light of the H-α line is built up on the photographic plate continuously. Varying the second slit changes the output wavelength.

Birefringent filter. The birefringent filter consists of a multiple sandwich of alternate layers of polarizing films and plates cut from a birefringent crystal (usually quartz or calcite). The assembly transmits the light in a series of sharp, widely spaced wavelength bands. One or another of the polarizers absorbs the light of all intervening wavelengths. A multiplier filter is used to isolate the desired band and exclude the others. Filters made for observations on the disk of the Sun generally have transmission bands 0.025–0.075 nm wide, centered on the H-α line. For observations at the limb, bandwidths up to 1 nm are used for prominences in H-α, and bandwidths of about 0.2 nm for the green and red coronal lines. The birefringent filter is compact enough to be used with a conventional small telescope but is far less flexible than the bulkier spectroheliograph in choice of bandwidth and wavelength. Fabry-Perot interferometers have also been utilized. *See* Astronomical observatory; Astronomical photography.

Harold Zirin

Bibliography. C. De Jager and Z. Svetska (eds.), *Progress in Solar Physics*, 1986; P. Foukal, *Solar As-*

trophysics, 1990; E. Gibson, *The Quiet Sun*, NASA SP-303, 1972; A. Krüger (ed.), *Introduction to Solar Radio Astronomy and Radio Physics*, 1979; R. W. Noyes, *The Sun, Our Star*, 1982; M. Stix, *The Sun: An Introduction*, 1988; P. A. Sturrock (ed.), *Physics of the Sun*, 3 vols., 1985; P. A. Sturrock (ed.), *Solar Flares*, 1980; Z. Svestka, *Solar Flares*, 1976; E. Tandberg-Hanssen and A. G. Emslie, *Physics of Solar Flares*, 1988; H. Zirin, *Astrophysics of the Sun*, 1988; J. R. Zirker (ed.), *Coronal Holes and High Speed Windstreams*, 1977.

Sundial

An instrument for telling time by the Sun. It is composed of a style that casts a shadow and a dial plate, which is the surface upon which hour lines are marked and upon which the shadow falls. The style lies parallel to Earth's axis. The construction of the hour lines is based on the assumption that the apparent motion of the Sun is always on the celestial equator.

Sundials can be made in any form and on any surface. They may be large and stationary, or small and portable. They may be made for use in a particular place or anywhere. The most widely used form is the horizontal dial that indicates local apparent time (Sun time). Other forms of the sundial indicate local mean time and standard time.

The highest form of sundial construction is found in the heliochronometer, which tells standard time with great accuracy. Incorporated in its construction is the equation of time and the time difference in longitude between the place where it is to be used and the standard time meridian for that locality. This makes possible a sundial that can be read as a clock. The sundial is said to be the oldest scientific instrument to come down to us unchanged. The underlying scientific principle of its construction makes it a useful device for educational purposes as well as for timekeeping. SEE TIME.

Robert N. Mayall

Bibliography. R. N. Mayall and M. L. Mayall, *Skyshooting: Photography for Amateur Astronomers*, 1968; R. N. Mayall and M. L. Mayall, *Sundials: How to Know, Use, and Make Them*, 2d ed., 1973; M. Stoneman, *Easy to Make Wooden Sundials*, 1982.

Sunspot

A dark area in the photosphere of the Sun caused by a lowered surface temperature. The temperature at the center of a spot is about 6700°F (4000 K), and the surface brightness is one-fifth that of the normal photosphere. The sizes and numbers of sunspots vary in the celebrated 11-year sunspot cycle, which is shared by all other forms of solar activity. SEE PHOTOSPHERE; SUN.

John W. Evans

Supergiant star

A member of the family containing the intrinsically brightest stars, populating the top of the Hertzsprung-Russell diagram. Supergiant stars occur at all temperatures, from 54,000 to 5000°F (30,000 to 3000 K), and have luminosities ranging from 10^4 to 10^6 times that of the Sun. The hot supergiants have radii 20 times that of the Sun. The cool supergiants are the largest known stars, reaching several thousand solar radii. Among the bright supergiants are Deneb (α Cygni), Rigel (β Orionis), and Betelgeuse (α Orionis). Because few have masses exceeding 20–40 Suns, the ratio of luminosity to mass is high. Thermonuclear energy sources will therefore be rapidly exhausted, hydrogen being consumed in a time scale of only 10^6 to 10^7 years. The supergiants are thus young but rapidly evolving massive stars. Nearly all are slightly variable in light and radial velocity, have highly turbulent and extended atmospheres, and show broadened spectral lines. SEE HERTZSPRUNG-RUSSELL DIAGRAM; STELLAR EVOLUTION.

The cool supergiants have such low mean density and surface gravity that they are clearly unstable and lose matter into space in an intense version of the solar wind. These stellar winds carry as much as 1 solar mass away from the supergiant in only 10^6 years. Rapid evolution by neutrino emission also shortens the life of supergiants.

An upper limit to the luminosity of the supergiants seems to be fairly established at well above 10^6 times that of the Sun. There is also a theoretical upper limit, set by A. S. Eddington, when the radiative flux carries momentum beyond that value which gravity can balance. Consequently, supergiants can be used to determine distances to a galaxy, with moderate accuracy, by assuming that they have the same limit in that galaxy as in the Milky Way. When observable, the spectra of supergiants in the Magellanic Clouds and nearby galaxies closely resemble those in the Milky Way. SEE GALAXY, EXTERNAL; STAR.

Jesse L. Greenstein

Supergranulation

A system of convective cells, invisible in ordinary photographs, with typical diameters of 30,000 km (19,000 mi), covering the Sun's surface.

Relation to granulation. High-resolution photographs of the Sun's visible surface reveal the granulation, a closely packed cellular grid with bright (hot) centers surrounded by dark (cool) lanes. The granules have lifetimes of 10–30 min and average diameters of 1000 km (600 mi). In the 1930s L. Biermann suggested that, in the outer 25–30% of the Sun, heat from the Sun's interior is transported to the surface by convection, similar to a pot of boiling soup heated from below. The granules are the surface manifestation of this process, corresponding to the boiling bubbles (plumes). A so-called mixing-length theory proposed that the bubble sizes are approximately equal to the local scale height H (the distance in which density or pressure changes by a factor $e \approx 2.7$). Since H varies from about 1000 km (600 mi) at the surface to 100,000 km (60,000 mi) at the base of the convection zone, a very large range of plume sizes was hypothesized.

Discovery. In 1959, R. Leighton modified the spectroheliograph to image the Sun by using the Doppler and Zeeman effects. Images of line-of-sight (approaching and receding) gas motions and of magnetic fields could be obtained. Immediately obvious on the Doppler photographs was a new cell structure with an area 900 times that of the granulation, having

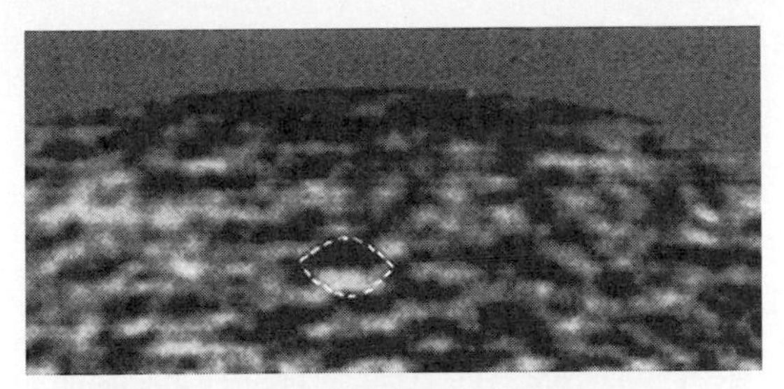

Fig. 1. Dopplergram showing supergranules near the solar limb. The slant view makes horizontal velocities appear alternately bright (approaching) and dark (receding). Supergranules appear elongated because of foreshortening. One supergranule is encircled by a broken line. (*Courtesy of L. Cram, National Solar Observatory, Sunspot, New Mexico*)

a mainly horizontal flow pattern (**Fig. 1**). Such a supergranule typically has a small upward velocity of 50–100 m/s (170–330 ft/s) at its center, horizontally outward flows with peak speeds of about 500 m/s (1700 ft/s), and descending 100–200 m/s (330–660 ft/s) funnels at the cell boundary. Lifetimes of supergranules are 1–2 days. The cells form an irregular polygonal structure, with most of the downflow occurring at the polygon vertices. The supergranules fit neatly within an essentially identical network (grid) structure seen in the magnetic photographs. The kinetic energy of the supergranular gas flow at the Sun's surface exceeds the magnetic field's ability to resist such motions, and thus the magnetic field is dragged forcibly to the supergranule boundaries until the two patterns coincide. This magnetic field, in turn, causes local heating of the upper solar atmosphere (chromosphere), producing a similar chromospheric network pattern seen in high-temperature spectral lines (**Fig. 2**). SEE DOPPLER EFFECT; FRAUNHOFER LINES; SOLAR MAGNETIC FIELD; SPECTROGRAPH; SPECTROHELIOSCOPE; ZEEMAN EFFECT.

Fig. 2. Hydrogen-alpha (Hα) filtergram showing details of the chromospheric network, which is coincident with the magnetic network marking the boundaries of supergranules. The pattern of fibrillar dark vertical structures (spicules) looks like hedgerows separating farmers' fields. (*Courtesy of R. Dunn, National Solar Observatory, Sunspot, New Mexico*)

Invisibility. Convection cells in nature typically are one-third to one-half as deep as their diameters. Thus supergranules were thought to extend perhaps 10,000–15,000 km (6000–9000 mi) below the surface. By the time these hot plumes reach the surface, they have given up essentially all their excess heat to the surrounding gas and hence show little temperature difference (intensity contrast) between their centers and boundaries; this accounts for their invisibility in normal-intensity photographs. Only cellular motion of the gas remains at the Sun's surface, as revealed by the Doppler technique. Since 1985 supergranulation has been studied more precisely through a technique called local correlation tracking. This method uses extremely high-resolution photographs of the normal granulation to measure granule motions. Since the granules are carried on the supergranule currents, they serve to define the horizontal supergranule flow field at and near the center of the solar disk, where the Doppler (line-of-sight) measurements are ineffective.

Structure of convection zone. Discovery of the supergranulation effectively destroyed the mixing-length concept, which had predicted that there were no preferred size scales on the Sun, while only two were seen. G. Simon and N. Weiss argued in 1968 that at most three or four distinct scales might be expected: the two that had been observed, a so-called giant cell, and perhaps an intermediate scale. However, attempts to find giant cells proved futile. Theoretical advances in understanding stellar and laboratory convection led Simon and Weiss to theorize in 1991 that there are no scales larger than supergranules and that supergranule plumes can penetrate to the base of the convection zone. An intermediate convection scale, discovered in 1981, was named mesogranulation, since its size (4000–6000 km or 3000–4000 mi) and lifetime (3–10 h) lie between those of granulation and supergranulation. Mesogranules, observed with the same techniques as supergranules, may be secondary eddies produced within supergranules. Like the granules, they are carried outward by the supergranular flows, and disappear upon reaching the network boundary. SEE SUN.

George W. Simon

Bibliography. L. November et al., The detection of mesogranulation on the Sun, *Astrophys. J.*, 245:L123–L126, 1981; G. Simon and R. Leighton, Velocity fields in the solar atmosphere, III. Large scale motions, the chromospheric network and magnetic fields, *Astrophys. J.*, 140:1120–1147, 1964; G. Simon and N. Weiss, Convective structures in the Sun, *Mon. Not. Roy. Astron. Soc.*, 252:1p–5p, 1991; G. Simon and N. Weiss, Supergranules and the hydrogen convection zone, *Z. Astrophys.*, 69:435–450, 1968.

Superluminal motion

Apparent proper motion exceeding the velocity of light c in an astronomical object. Many quasars exhibit systematic changes in images of the radio emission from their nuclei over periods of months to years. In some cases, features in the image appear to separate at more than 10 times the speed of light, given the great distance of the quasars from Earth.

Superluminal motion was one of the most exciting discoveries to emerge from a technique in radio astronomy first developed in the 1960s called very long baseline interferometry (VLBI). This method involves the tape recording of radio signals from large antennas at up to 10–15 locations across the Earth, and the combination of these signals in a computer to form a radio image of the quasar at extremely high resolution (less than 0.001 second of arc). A sequence of images of quasar 3C 345, one of the best-studied examples, illustrates the phenomenon of superluminal motion (**Fig. 1**). *See Quasar; Radio astronomy; Radio telescope.*

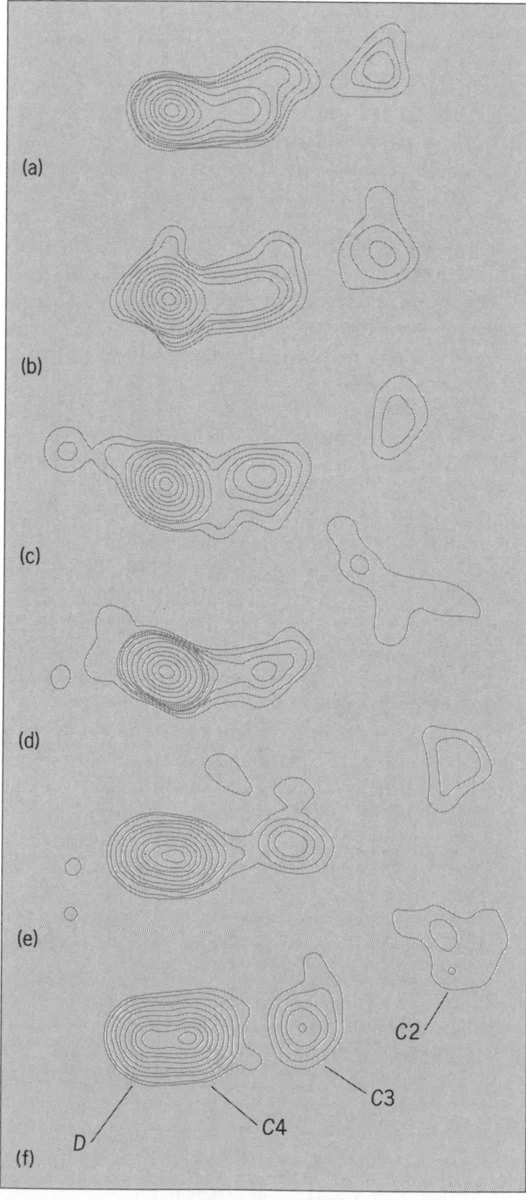

Fig. 1. Series of six very long baseline interferometry (VLBI) images of quasar 3C 345 (redshift $z = 0.6$, distance = 800 megaparsecs) observed over a period of 4.7 years, showing superluminal motion in its jet. Three features, labeled *C2, C3,* and *C4,* move away from the nucleus *D* at approximately six times the speed of light. Overall length of the jet, *D–C2,* is 0.006 arc-second or 23 parsecs. (*a*) June 1979. (*b*) July 1980. (*c*) February 1981. (*d*) February 1982. (*e*) February 1983. (*f*) February 1984. (*After J. A. Zensus and T. A. Pearson, eds., Superluminal Radio Sources, Cambridge University Press, 1987*)

Objects showing superluminal motion. Superluminal motion is seen mostly in quasars but also in some other active galactic nuclei. This rapid motion is confined to within a few tens of parsecs of the nucleus, whose power source is believed to be a massive black hole. Over 30 examples of superluminal motion are now known, with apparent proper motions up to $10c$, and the list continues to grow. A handful of quasars and galaxies exhibit motion that is subluminal (that is, with speeds less than c). The phenomenon is common in objects observed by using VLBI; these are mostly quasars whose appearance in the radio-frequency part of the spectrum is dominated by emission from the nucleus. The phenomenon is rare in quasars as a whole, because most are weak radio sources, and those from which radio emission is detected usually show such emission on scales comparable to, or larger than, the galaxy in which the quasar nucleus is situated (the host galaxy). *See Galaxy, external.*

Certain objects in the Milky Way Galaxy can show superluminal motion also, namely novae and supernovae, the results of violent outbursts in normally quiet stars. An example is the elliptical ring of light centered on, and illuminated by, supernova SN 1987A in the Large Magellanic Cloud, as photographed by the Hubble Space Telescope. *See Nova; Satellite astronomy; Supernova.*

Explanations. Announcement of the discovery of superluminal motion caused widespread concern because of the apparent violation of A. Einstein's special theory of relativity, even though the basic explanation still favored now was in fact predicted some years earlier. Many explanations were proposed (besides Einstein's theory being incorrect), but only the relativistic jet model has stood the test of time and is fully integrated into an overall picture of how many diverse phenomena associated with quasars arise. Among the rejected alternatives are:

1. Overestimation of the distance to quasars. If quasars are actually much closer than is generally believed, the apparent speeds are correspondingly reduced.

2. The Christmas tree model, in which randomly flashing lights represent the superluminal features. Any apparent speed is possible because the events are noncausal.

3. A gravitational lens, a massive object along the line of sight to the quasar, could cause gravitational focusing and consequent changes in image shape.

4. The lighthouse model. A rotating beam from the quasar illuminates stationary clouds that reflect the radio waves to Earth, analogous to a lighthouse beacon.

5. The light-echo model. An outburst from the center illuminates regions at successively greater radii, which reflect the radiation into the line of sight, yielding a minimum speed of $2c$. The ring around SN 1987A and the varying intensity of its emission are explained well by this model. *See Gravitational lens.*

Relativistic jet model. Superluminal motion is explained in the relativistic jet model as mainly a geometric effect (**Fig. 2**). A feature moves away from the nucleus of the quasar at high (relativistic) speed βc (but less than c) at a small angle to the line of sight to the Earth. Radio waves from the moving feature arrive only slightly later than waves from the nucleus,

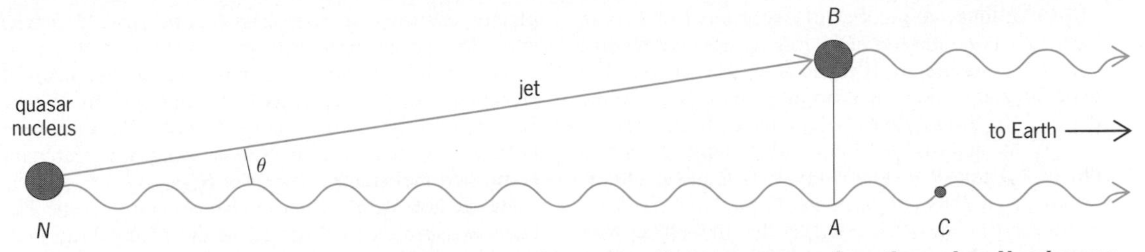

Fig. 2. Geometry of the relativistic jet model. A feature that emits radio waves emerges from the nucleus *N* and moves along the jet at speed βc (with $\beta < 1$) at angle θ to the line of sight to Earth, reaching *B* after time *t*. Meanwhile, a radio wave emitted from the nucleus has reached *C*. Waves from *A* and *B* reach Earth at time δt later, when the quasar is seen as a double source. Since $NB = \beta ct$, $NA = \beta ct \cos \theta$, and $NC = ct$, the time delay $\delta t = t(1 - \beta \cos \theta)$. In this time, the feature appears to have moved a distance across the sky $AB = \beta ct \sin \theta$, so the apparent speed is $AB/\delta t = \beta c \sin \theta/(1 - \beta \cos \theta)$, which can be much larger than *c* if θ is small and β approaches 1.

whereas the feature took a much longer time to reach its current position; the motion appears superluminal because the former measure of time is used rather than the latter. As the speed approaches *c* and the angle θ decreases, the apparent speed can be arbitrarily large.

The relativistic jet model forms the basis for an overall picture that explains many diverse phenomena in quasars. A rotating black hole at the very center generates a pair of powerful jets that emerge in opposite directions along the rotation axis. Radio emission is observed both from the immediate vicinity of the black hole (the nucleus) and from features in the jet that is approaching the Earth. The approaching jet appears brighter because of relativistic beaming, and the receding jet is dimmed, usually to a level at which it is undetectable. This phenomenon explains why a series of features is observed moving away from one side of an (apparently) stationary nucleus, as in Fig. 1, and why the properties of the different components are not the same. On larger scales, the jets expand and eventually yield their kinetic energy to ionized plasma either inside the host galaxy or in the intergalactic medium. This plasma is seen as a pair of large radio lobes, one on each side of the galaxy. *See* *Astrophysics, high-energy; Black hole*.

This model has a number of observational consequences, many of which are still being actively studied. Two of the key issues center on the distribution of jet angles and the statistics of measured jet speeds. Only very few jets can point almost at the Earth, and it is not known what the others look like. Relativistic beaming is a major complication to consider: most of the quasars well studied with VLBI are precisely those that are beamed almost at the Earth, since these are the brightest and easiest to observe. It is not known whether all quasars possess jets, but only those pointed toward the Earth are visible as superluminal sources. If a characteristic (actual) speed βc for quasars exists, superluminal motion may be used to study the cosmological evolution of the universe. *See* *Cosmology*.

Stephen C. Unwin

***Bibliography*.** M. C. Cohen et al., Expanding quasars and the expansion of the universe, *Astrophys. J.*, 329:11–7, 1988; A. P. Marscher and J. S. Scott, Superluminal motion in compact radio sources, *Publ. Astron. Soc. Pacific*, 92:127–133, 1980; J. A. Zensus and T. J. Pearson (eds.), *Superluminal Radio Sources*, 1987; J. A. Zensus and T. J. Pearson (eds.), *Parsec-scale Radio Jets*, 1990.

Supermassive stars

Hypothetical objects with masses exceeding 60 solar masses, the mass of the largest known ordinary stars (1 solar mass equals 4.4×10^{30} lbm or 2×10^{30} kg). The term is most often used in connection with objects larger than 10^4 solar masses that might be the energy source in quasars and active galaxies. Since these objects are hypothetical at present, very little is known with certainty about their nature and behavior. The description of their properties presented here is based entirely on theoretical calculations.

Dimensions. Just prior to the onset of instability, models indicate that a 10^4-solar-mass supermassive star would have a radius of 2×10^6 mi (3×10^6 km). The central temperature of such a nonrotating object would be roughly 10^9 K (1.8×10^9 °F), and nuclear reactions would have occurred in a considerable part of the stellar interior. The corresponding numbers for a 10^8-solar-mass object are a radius of 2×10^{12} mi (3×10^{12} km or 500 times the size of Pluto's orbit) and a central temperature of 300,000 K (540,000°F), too low for fusion reactions to start.

Dynamics and evolution. About the only thing known for certain about supermassive stars is that they are unstable. Arthur Eddington, describing the properties of stars in general in the 1920s, showed that objects larger than ordinary stars would be supported mainly by radiation pressure rather than gas pressure. A star completely supported by radiation pressure, like a supermassive star, is unstable (because the ratio of specific heats γ equals 4/3). However, supermassive stars might last a significant length of time before instability occurs. The lower-mass objects, in which nuclear reactions can provide some energy, could live 10^6 years. In the higher-mass objects, larger than 10^6 solar masses, for example, the lifetime is determined by the time taken for gravitational binding energy to dissipate, and is on the order of decades for 10^8 solar masses. Rotation affects these lifetimes. A rapidly rotating supermassive star, also called a relativistic disk, might have a much longer lifetime (and also a lower temperature). *See* *Star*.

When instability occurs, a supermassive star can collapse catastrophically or explode. Models producing explosions have so far been confined to the lower-mass objects. The most reasonable scenario is that an unstable supermassive star will collapse extremely rapidly, possibly emit gravitational radiation, and become a giant black hole. *See* *Black hole; Gravitation; Gravitational collapse*.

Observations. A number of exceptional objects in nearby galaxies may contain low-mass supermassive stars, with masses of 1000 solar masses or so. The best analyzed of these is a bright patch of light at the center of the Tarantula Nebula, a huge cloud of glowing gas located in the Large Magellanic Cloud, a companion galaxy to the Milky Way Galaxy. Observations from the *International Ultraviolet Explorer* satellite in the late 1970s suggest that this object may be a single star with a mass of 2000 solar masses. However, other astronomers have argued that the central source (also known as R, or Radcliffe, 136 a) is a very dense star cluster, containing about 40 very massive stars. *See* MAGELLANIC CLOUDS; SATELLITE ASTRONOMY.

Another place where supermassive stars could be found is in the nuclei of galaxies, particularly active galaxies. William Fowler and Fred Hoyle suggested in 1963 that they could provide the energy for the quasars, distant objects with luminosities as high as 10^{48} ergs/s (10^{41} W), 10^5 times the power emitted by the stars in the Milky Way Galaxy. This suggestion could be verified by the detection of periodicities in the light emission from quasars, dimly suggested but not confirmed by the observations of some objects. The final catastrophic collapse might generate long-wavelength gravitational waves which could be detected by satellite facilities. Exploding supermassive stars may also be responsible for the rapid motions which occur in many galaxies, including the Milky Way Galaxy (as shown by the expanding spiral arms in the inner regions), and they may be the sites for some nucleosynthesis. *See* GALAXY, EXTERNAL; MILKY WAY GALAXY; QUASAR.

Harry L. Shipman

Bibliography. M. Kafatos (ed.), *Supermassive Black Holes,* 1988; A. MacRobert, The supermassive star debate, *Sky Telesc.*, 67(2):134–135, February 1984; F. Pacini and M. Rees, Rotation in high-energy astrophysics, *Sci. Amer.*, 228(2):98–105, February 1973; S. Shapiro and S. Teukolsky, *Black Holes, White Dwarfs, and Neutron Stars*, 1983; H. L. Shipman, *Black Holes, Quasars, and the Universe*, 2d ed., 1980.

Supernova

The sudden, temporary brightening of a star to a luminosity comparable with that of an entire galaxy. Brightness increases to billions of times that of the Sun in days and decreases again in months to years.

A typical supernova puts out more than 10^{42} joules of electromagnetic radiation, much of it in visible light, but significant amounts in infrared, ultraviolet, x-ray, and radio radiation as well. In addition, the star blows off one to ten solar masses of material (2×10^{30} to 2×10^{31} kg or 4×10^{30} to 4×10^{33} lb) at speeds of 3000 to 10,000 km (1800 to 6000 mi/s), and so contributes about 10^{44} J of kinetic energy to the heating and stirring of interstellar gas. Supernovae are significant in galactic evolution as the major contributors of elements heavier than hydrogen and helium. Shock waves caused by their expanding remnants may also accelerate cosmic rays and trigger formation of new stars by colliding with dense clouds of molecular gas. Some supernova explosions leave behind pulsars or other neutron stars. Others completely destroy the star. *See* COSMIC RAYS; NEUTRON STAR; NUCLEOSYNTHESIS; PULSAR.

The basic connection among supernovae, neutron stars, and cosmic rays was first suggested by Walter Baade and Fritz Zwicky in 1934. The 1968 discovery of a pulsar in the Crab Nebula supernova remnant confirmed their ideas. *See* CRAB NEBULA.

Rates and types. Supernovae are rare. None has been unequivocally observed in the Milky Way Galaxy since 1604, but this is partly because, with the limitations of current receivers, it is not possible at present to survey this dusty galaxy very thoroughly. Studies of other galaxies reveal that a typical large spiral can expect to host two to six supernovae per century, while a large elliptical galaxy might have one or two. Pretelescopic astronomers in China, Japan, the Arab world, and Europe recorded probable supernovae in the years 185, 393, 1006, 1054, 1181, 1408, 1572, 1604, and (possibly) 1679. Since all occurred in the one-seventh of the galaxy nearest to the Earth, eight or nine events in 2000 years is consistent with the expected rate of a few per century. *See* GALAXY, EXTERNAL; MILKY WAY GALAXY.

Most supernovae fall into one of two classes, called type I and type II, and defined by the absence or presence of lines of hydrogen (the most abundant element in the universe) in their spectra. Type II events occur only in galaxies and parts of galaxies where massive, young stars have recently formed. They characteristically leave neutron stars (or perhaps sometimes black holes) behind, and spectra taken long past the time of maximum brightness show that, while the surface of the exploding star consisted mostly of hydrogen, the interior contained a range of heavy elements (carbon to iron) built up by nuclear reactions in the core of the star. *See* ASTRONOMICAL SPECTROSCOPY; BLACK HOLE.

Type I events are further subdivided into Ia's (associated with the full range of galaxy types, including old stellar populations) and Ib's (associated with either young or intermediate-age stars). Both show spectroscopic evidence for the production of significant amounts of new heavy elements, and the Ia's, at least, are believed to disrupt the parent stars completely.

Progenitors and mechanisms. Astrophysical objects have two major sources of energy, nuclear reactions and gravitational contraction. One of these must occur very rapidly to produce a supernova. Calculations show that nuclear reactions will explode when some fuel gets hot enough to burn while it is very dense; gravitational collapse sets in when the pressure at the center of a star drops below a critical value that depends again on density and temperature. Thus the occurrence of supernovae must be associated with stars evolving to particular values of central temperature and density.

Computer models of stellar evolution demonstrate that stars that begin their lives with more than about eight times the mass of the Sun will eventually experience core collapse, after several million years of nuclear reactions have given them layered interiors of heavy elements. These become type II supernovae, and the collapsed core is revealed as a neutron star or pulsar after the debris clears away. Calculations of what the brightness and spectrum of such a star should be as a function of time agree with observations of type II supernovae. The chief remaining problem is in understanding how about 1% of the

10^{46} J available from core collapse is deposited in the outer layers of the star to blow it off. The other 99% of the energy is expected to come out as neutrinos and gravitational radiation, and was regarded as undetectable until the advent of the supernova SN 1987A. *See* GRAVITATIONAL COLLAPSE; STELLAR EVOLUTION.

Explosive nuclear reactions, on the other hand, produce type I supernovae. Models of spectra and brightness versus time based on this scenario agree well with the observations. The chief problem here is in understanding precisely which stars will evolve to explosive conditions. They must be of relatively low mass (since type I supernovae are observed among old stars), and most current ideas involve interaction between a close pair of stars, at least one of which is a white dwarf, and so already at very high density. *See* WHITE DWARF STAR.

Supernova remnants. In addition to the expanding gas clouds associated with the supernovae of 1054 (the Crab Nebula), 1572 (Tycho's supernova), 1604 (Kepler's supernova), and so forth, about 200 other similar sources of radio, visible, and radiation are observed in the Milky Way Galaxy. These are the remnants of other, mostly earlier, unrecorded explosions. Some, but by no means all, have known pulsars at their centers. A typical remnant remains detectable for about 10,000 years; hence the few hundred observed confirm a birth rate of a few per century. Polarization of the radio (and occasionally optical) emission indicates that it is synchrotron radiation, the kind produced by very high-speed electrons moving through a magnetic field. This shows that supernovae are capable of accelerating particles to high energy, and strengthens the evidence that they are responsible for cosmic rays. *See* RADIO ASTRONOMY; X-RAY ASTRONOMY.

Supernova remnants eventually blend into the general interstellar gas, after heating and stirring it. In the process, they will often overlap and merge (the more so as the young stars that produce type II supernovae normally form in clusters). The products look like large, hollow shells in maps of cool interstellar gas, but they are really filled with hotter, lower-density gas. The solar system appears to be inside one of these superbubbles. *See* INTERSTELLAR MATTER.

SN 1987A. The first supernova visible to the unaided eye in nearly three centuries brightened suddenly on February 23, 1987 (and received the name SN 1987A only because it was the first event cataloged that year, not because of its importance). It occurred not in the Milky Way Galaxy but in a nearby companion galaxy called the Large Magellanic Cloud (LMC). Although visible only from the Southern Hemisphere, SN 1987A has become the most intensively studied supernova ever. The many details seen for the first time have essentially confirmed the basic understanding of core collapse events, but have not been able to solve the major outstanding problem of how energy is deposited in the stellar envelope to blow it off. *See* MAGELLANIC CLOUDS.

At least as important as the visible supernova was a burst of neutrinos (chargeless, nearly massless particles, capable nevertheless of transporting large quantities of energy) that reached the Earth about 7 h before optical brightening. Detectors built in Japan and the United States for other purposes (and perhaps ones in the Soviet Union and Europe as well) recorded the passage of about 10 neutrinos each. In combination with properties of neutrinos measured at accelerators and reactors, these sufficed to demonstrate that $2–5 \times 10^{46}$ J had been liberated in a few seconds and at a characteristic temperature near 4×10^{10} K. These are just the parameters that theorists had long predicted for a core-collapse supernova in a massive star.

The spectrum showed lines of hydrogen as expected for a type II event. In addition, the Large Magellanic Cloud had been extensively studied, and the exploding star had been photographed and had its color and spectrum measured in the 1970s. As a result, the progenitor is known to have been a star of about 20 solar masses, with a chemical composition like that of other stars in the Large Magellanic Cloud (deficient in heavy elements compared to the Sun and other Milky Way stars). The preexplosion lifetime must have been about 10^7 years.

The nonsolar composition explains an initially puzzling feature of SN 1987A, the fact that core collapse occurred when the star was a blue rather than a red supergiant. Stellar evolution models should have led to the expectation of blue progenitors among stars in the Large Magellanic Cloud, and the different progenitor structure in turn explains why SN 1987A brightened faster than most type II supernovae and did not get quite as bright at maximum light as most type II's. The ultraviolet spectrum through 1988 and 1989 showed lines from material around the star, indicating that it had been a red supergiant in the past, and that part of its envelope had blown off in a presupernova wind, carrying some helium, nitrogen, and other elements made by nuclear reactions early in the star's life.

Postmaximum spectra at other wavelengths show that other elements were produced by the progenitor and in the explosion. Of particular importance are gamma ray lines resulting from the decay of nickel-56 to cobalt-56, and cobalt-56 to iron-56. Detection of these means that nucleosynthesis was occurring right up to the instant of core collapse (since nickel-56 and cobalt-56 have half-lives of only 7 and 66 days), and that there is a source of energy from the nuclear decays that keeps type II events shining longer than they would if only the energy from core collapse were available. In addition to these properties of SN 1987A about which there is general agreement, there have been incompletely confirmed reports of several different kinds of companion objects. *See* GAMMA-RAY ASTRONOMY.

SN 1987A has made astronomers supernova-conscious. Several groups have made preparations to study the next galactic core collapse, at least its neutrinos, even if the light should be blocked by dust. Still more detailed data should be available, and the time history of the neutrino burst can be expected to clarify how energy is transferred from the collapsing core to the expanding stellar envelope. *See* ASTROPHYSICS, HIGH-ENERGY; VARIABLE STAR.

Virginia Trimble

Bibliography. M. Kafatos and A. Michalitsianos (eds.), *Supernova 1987A in the Large Magellanic Cloud*, 1988; A. Petschek (ed.), *Supernovae*, 1990; V. Trimble, 1987A; The greatest supernova since Kepler, *Rev. Mod. Phys.*, 60:859–871, 1988; V. Trimble, Supernovae, *Rev. Mod. Phys.*, 54:1183–1224, 1982, and 55:511–563, 1983; K. W. Weiler and R. A. Sramek, Supernovae and supernova remnants, *Annu. Rev. Astron. Astrophys.*, 26:295–341, 1988; S. E. Woosley and T. A. Weaver, The physics of supernova explosions, *Annu. Rev. Astron. Astrophys.*, 24:204–253, 1986.

Symbiotic star

A stellar object whose optical spectrum displays features indicative of two very different thermal regimes: a stellar spectrum whose flux distribution and absorption lines suggest the presence of a cool star, and emission lines which can be formed only in a much hotter medium. In 1941, P. W. Merrill labeled this kind of composite spectrum a combination spectrum, and the object emitting it a symbiotic star.

This definition is rather vague and, in order to isolate a nearly homogeneous class of objects, ''cool star'' and ''much hotter medium'' must be defined more clearly. There is no unanimity in this respect, and as a consequence the number of symbiotic stars listed by various astronomers differs. The most comprehensive lists contain somewhat more than 100 objects. Symbiotic stars are thus quite rare, but they are very important for understanding stellar evolution.

Spectra. **Figure 1** shows a fairly typical spectrum of a symbiotic star. The optical region (wavelengths longer than 320 nanometers) displays continuous radiation typical for a cool star, declining at shorter wavelengths and characterized by wide absorption bands due to molecules, in particular titanium oxide (TiO). Starting with the Balmer limit (at 365 nm) and going to the ultraviolet, a rising continuum appears, typical for a much hotter star. Observation of such ultraviolet spectra, which was a decisive step toward understanding the nature of symbiotic stars, was made possible by the *International Ultraviolet Explorer (IUE)* satellite. All over the spectrum, emission lines are seen in which the flux considerably exceeds the continuum level. These emission lines are due to neutral helium (He I), to ionized helium (He II), and to various ions of iron from singly ionized (Fe II) to six-times ionized (Fe VII), or, exceptionally, thirteen-times ionized (Fe XIV). More often, doubly ionized oxygen (O III) and neon (Ne III) are observed. SEE SATELLITE ASTRONOMY; ULTRAVIOLET ASTRONOMY.

In many cases, the parental ions are placed in square brackets, for example [Ne III], [Fe VII], and [O III], in order to indicate that the corresponding spectral lines are forbidden; that is, the probability is very low in this case that an electron will jump from a higher-energy level to a lower-energy level with emission of a radiation quantum (a photon). In that case, it is far more likely that the higher level will be deexcited by a collision with another electron which will carry away the excess energy, and the electron in question will return to the lower level without any emission of a photon. However, if the density of the observed plasma is low, collisions are sufficiently rare that a radiative transition can be realized. Thus the presence of the forbidden lines in symbiotic spectra indicates that large regions in the object are occupied

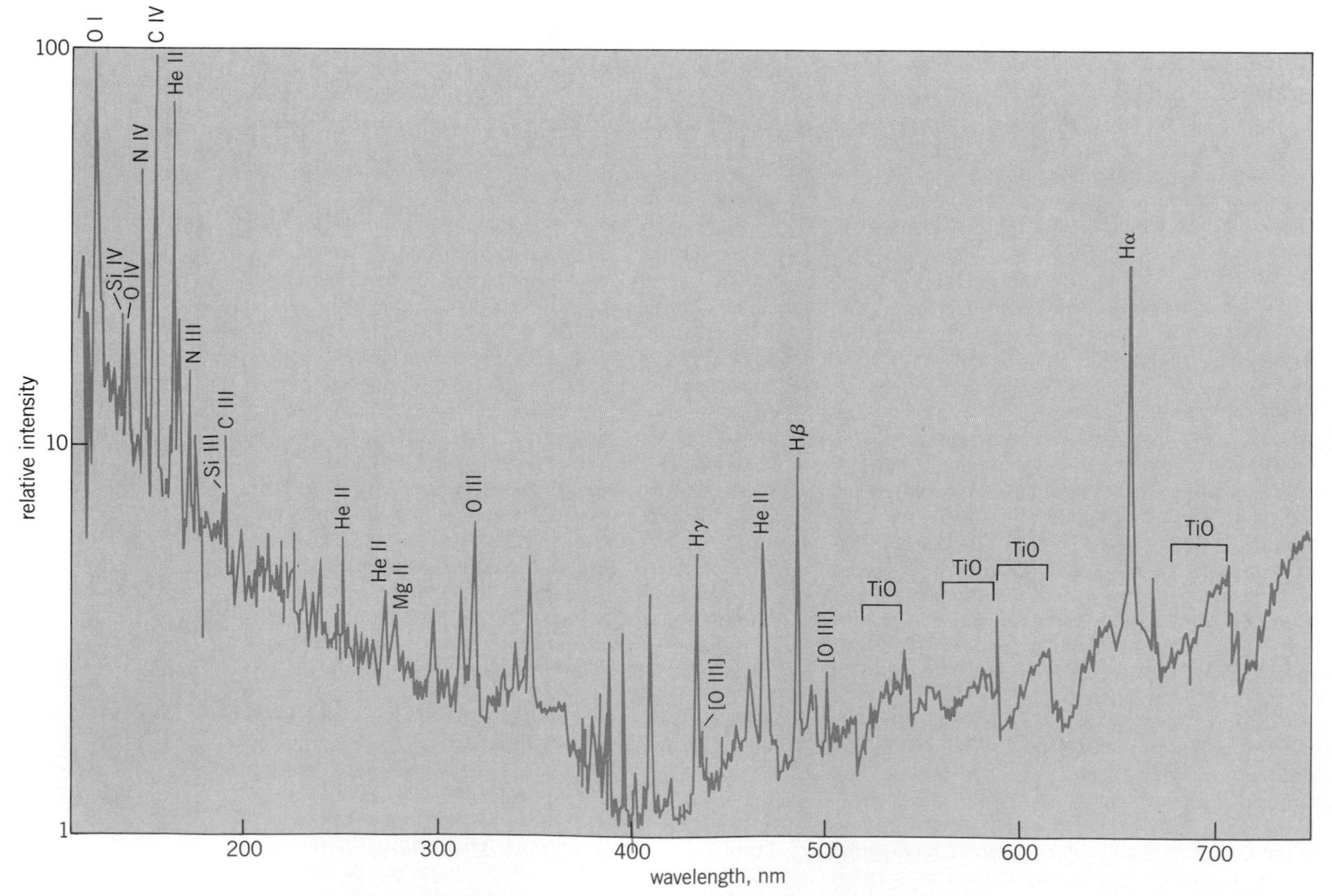

Fig. 1. Spectrum of AG Pegasi from 110 to 750 nm, obtained by combining terrestrial and satellite data. (*After M. J. Plavec, D. M. Popper, and R. K. Ulrich, eds., Close Binary Stars: Observations and Interpretation, Reidel, 1980*)

by a low-density plasma; at the same time, the high degree of ionization shows that this plasma must be at a fairly high temperature, say in the range of 10^4–10^6 K. *SEE ASTRONOMICAL SPECTROSCOPY; ATOMIC STRUCTURE AND SPECTRA; SELECTION RULES (PHYSICS).*

Interacting binary models. The consensus is that most (if not all) symbiotic stars are actually binary stars. The "cool" continuum, observed in the optical and infrared spectral regions, is due to a cool star. The "hot" continuum observed in the ultraviolet is due to a much hotter star. The emission lines, observed both in the optical and in the ultraviolet, originate in a plasma which surrounds either component or the whole system. The energy necessary to ionize the plasma and excite the observed emission lines comes from the hot star. But at least one of the components must be postulated to be the source of this radiating plasma. Therefore the system is assumed to be an interacting binary star, in which one of the components is losing mass, which then interacts either with the other component or with its radiation.

At any given wavelength a hotter star will always radiate more energy per unit area of its surface than a cooler star. Yet in the optical spectrum of the symbiotics, the cool star always dominates, and the continuous radiation of the hot star is practically imperceptible. Thus, the proper combination of objects must pair a large cool star (a red giant or even a supergiant) with a much smaller hot star. According to the nature of the hot object, three types can be distinguished: subdwarf symbiotic, algol symbiotic, and novalike symbiotic. *SEE GIANT STAR; SUPERGIANT STAR.*

Subdwarf symbiotic. A subdwarf symbiotic or a PN (planetary nebula) symbiotic would be a combination of a cool red giant with a small hot subdwarf star. These subdwarfs lie to the left of, and below, the main sequence in the Hertzsprung-Russell diagram. These stars are probably the inner cores of former giants and supergiants, which have shed the cool outer envelope and are now contracting to become genuine white degenerate dwarfs. In other words, subdwarfs are akin to the central stars of planetary nebulae. They are very hot, but their total radiation (luminosity) is small because they have small radii, of the order of 10 earth radii or so. Hot gases may still be streaming out of their surfaces. Thus in this type of a symbiotic the hot star would be both the source of the circumstellar plasma and the source of its luminous energy. The red giant need not be active. Since the subdwarf stage is relatively very short, these "pure natural" symbiotics should be very rare. *SEE HERTZSPRUNG-RUSSELL DIAGRAM; PLANETARY NEBULA.*

Algol symbiotic. This consists of a red giant combined with a main-sequence star. A main-sequence star cannot be at the same time sufficiently hot to ionize the circumstellar plasma and sufficiently small not to overshadow the red giant in the optical region. However, if the red giant loses mass which flows toward the main-sequence star, an accretion disk can form around the latter star. If the rate of mass transfer is sufficiently high (on the order of 10^{-4} solar mass per year or so), the inner parts of the disk become hot enough to provide the ionizing photons, and yet are small. This type of interaction is known to exist, mostly with lower mass-loss rates, in a large class of interacting binaries called the Algol-type semidetached binaries. **Figure 2** shows a pole-on view of an Algol-type interacting binary with an accretion disk. The stars Z Andromedae, CI Cygni, AR Pavonis, and several other symbiotics may be of this type.

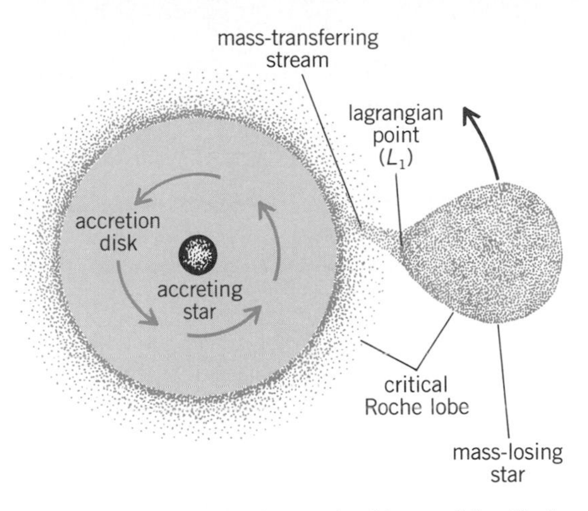

Fig. 2. Pole-on view of an interacting binary of the Algol type. (*After M. J. Plavec, IUE looks at the Algol paradox, Sky Telesc., 65:413–416, May 1983*)

Novalike symbiotic. This consists of a red giant combined with a white dwarf. An isolated degenerate white dwarf is too small (hardly larger than the Earth) to produce enough photons to ionize the plasma, and it may even be not hot enough. However, if again the cool giant loses mass, its accretion onto the white dwarf may ignite thermonuclear reactions of hydrogen and helium in the surface layers of the white dwarf. Low rates of accretion suffice to generate large output of energy in this case, and this energy may be coming in outbursts. This model of a symbiotic is in principle similar to contemporary models of novae. Since many symbiotic stars show a tendency toward irregular or semiregular outbursts in which their brightness increases significantly, this is an important model. *SEE NOVA; THERMONUCLEAR REACTION; WHITE DWARF STAR.*

Mass transfer. Both the Algol and the novalike models postulate that the cool giant be unstable and lose mass from its surface. In the Algol-like model, this happens via Roche lobe overflow. More frequent is the case of mass loss via stellar wind, which may be strengthened if the cool giant or supergiant is pulsating as a Mira variable. *SEE BINARY STAR; MIRA; STAR; STELLAR EVOLUTION; VARIABLE STAR.*

Mirek J. Plavec

Bibliography. M. Friedjung and R. Viotti (eds.), *The Nature of the Symbiotic Stars*, 1982; S. J. Kenyon, *The Symbiotic Stars*, 1986; S. J. Kenyon and R. F. Webbink, The nature of symbiotic stars, *Astrophys. J.*, 279:252–283, 1984.

Syzygy

The alignment of three celestial objects within a solar system. Syzygy is most often used to refer to the alignment of the Sun, Earth, and Moon at the time of new or full moon. Although syzygy is strongly associated with these two lunar phases in many minds, it must be emphasized that the alignment of any three celestial objects within the solar system (or within any other system of objects in orbit about a star) constitutes syzygy. Alignments need not be perfect in order for syzygy to occur: because the orbital planes for any three bodies in the solar system rarely coincide, the

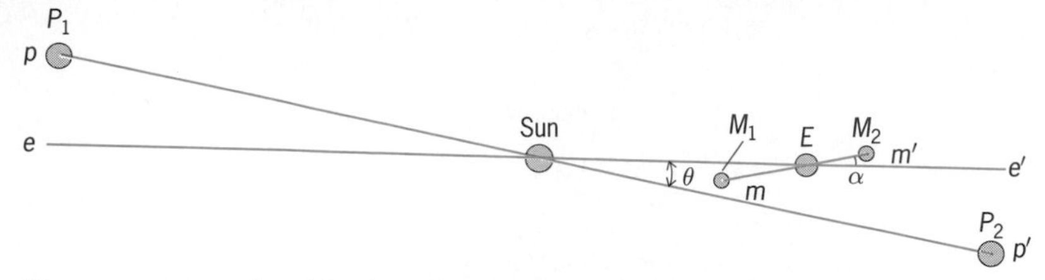

Diagram, not to scale, of the Sun, the Earth, the Moon, and a planet, illustrating syzygy. Symbols are explained in text.

geometric centers of three objects that are in syzygy almost never lie along the same line. *SEE PHASE; SOLAR SYSTEM.*

In general, syzygy occurs whenever an observer on one of the three objects would see the other two objects either in opposition or in conjunction. Opposition occurs when two objects appear 180° apart in the sky as viewed from a third object. Conjunction occurs when two objects appear near one another in the sky as seen from a third object.

In the **illustration,** *e-e′* is the orbital plane of the Earth (*E*), *p-p′* is the orbital plane of an outer planet (*P*) that is tilted at an angle θ to the orbital plane of the Earth, and *m-m′* is the orbital plane of the Moon (*M*), which is tilted at an angle α to the Earth's orbital plane. Syzygy occurs for the Earth, Sun, and the planet when the planet is either at P_1 (conjunction) or P_2 (opposition). The Sun, Earth, and Moon are in syzygy when the Moon is either at M_1 (new moon) or M_2 (full moon). If the Moon is at M_2 when the planet is at P_2, then the Earth, Moon, and the planet are in syzygy. (As seen from the Earth, the Moon and the planet would be seen near one another in the sky; that is, the Moon and the planet would be in conjunction.) The Earth, Moon, and a planet will be in syzygy any time that the Moon and the planet are seen either in the same direction from the Earth or in opposite directions: the Moon need not be new or full and the planet need not be in opposition or conjunction. The maximum angle for θ is about 17° (Pluto); the value for α is slightly more than 5°. There is no accepted upper limit for θ; if a new planet were discovered with an orbital plane tilted 25° from that of the Earth, syzygy would still occur at P_1 and P_2. *SEE MOON; PLANET.*

Solar and lunar eclipses are dramatic results of syzygy. During a solar eclipse, when the Moon is in its new phase, the alignment of the Sun, Earth, and Moon is so nearly perfect that the Moon's shadow falls on the Earth; during a lunar eclipse, which occurs at the time of the full moon, the Moon passes through the Earth's shadow. *SEE ECLIPSE.*

An occultation is another type of eclipse that can occur during syzygy. For an Earth-based observer, an occultation occurs when the Moon is seen to pass in front of a planet or other member of the solar system. The occultation of a star by the Moon does not qualify as syzygy, since the star is far beyond the limits of the solar system. *SEE OCCULTATION.*

Harold P. Coyle

Bibliography. V. Illingworth, *The Facts on File Dictionary of Astronomy*, 2d ed., 1985; J. Mitton, *A Concise Dictionary of Astronomy*, 1991; J. M. Pasachoff, *Contemporary Astronomy*, 4th ed., 1989; C. Payne-Gaposchkin and K. Haramundaris, *Introduction to Astronomy*, 2d ed., 1970.

T Tauri star

A member of a class of very young, optically visible, solar-mass stars with a variety of peculiarities, including variability and evidence for mass loss. T Tauri stars were discovered through their unusually strong emission lines by K. Himpel and independently by A. Joy. Many radiate unexpectedly intensely at infrared and ultraviolet wavelengths. Their name derives from Joy's brightest example, a variable star designated T in the constellation Taurus. Two subclasses have been defined, the classical T Tauri stars, identified from hydrogen-emission-line surveys, and the weak-line, or naked, T Tauri stars, discovered through their x-ray emission. In the 1980s it became clear that most of the apparently anomalous properties of T Tauri stars can be attributed to the fact that many are still surrounded by disks of gas and dust. *SEE VARIABLE STAR.*

Youth. The youth of the T Tauri stars was originally suspected because of their association with star-forming clouds. Their erratic brightness variations also indicated that they had not yet become stable. Ages between 10^5 and 10^7 years are confirmed by their effective temperatures and luminosities. These ages place them, as a group, well above the zero-age main sequence in the Hertzsprung-Russell diagram, in a location where, on theoretical grounds, forming stars are expected. The large amounts of lithium implied by their spectra also support the contention that they are extremely young, since the nuclear reactions that fuel stars use up lithium at a very early stage of evolution. *SEE HERTZSPRUNG-RUSSELL DIAGRAM.*

Emission lines. The unusually strong emission lines observed in the spectra of classical T Tauri stars often have the extended wings characteristic of significant mass outflow. Outflow velocities are high, typically 100 km/s (60 mi/s). In optical images, high-velocity, oppositely directed jets can often be seen emanating from the poles of the stars themselves. Although they are already visible, T Tauri stars are evidently still in the process of shedding the dust and gas from which they formed. From theories about how stars form, the remnant material is expected to be distributed in disks. Such disks are the intrinsic source of the excess infrared radiation. The excess ultraviolet emission, as well as the very strong emission lines, may derive from the so-called boundary layer between the rapidly rotating disk and the more slowly rotating star. The frequently noted absence of redshifted high-velocity wings from the emission-line profiles can be attributed to a disk obscuring from view the gas that is flowing away from the direction of the Earth, while allowing the blueshifted, approaching gas to be observed. Many of the major differences between classical and weak-line T Tauri stars lie in the fact that there is little evidence that massive disks are around the weak-line stars. *SEE DOPPLER EFFECT.*

Properties. Properties of the T Tauri disks have been determined from observations at far-infrared and millimeter radio wavelengths. Masses range from 1 to 10% of the mass of the Sun, and sizes are about 100 astronomical units (1.5×10^{10} km or 1×10^{10} mi), the diameter of Pluto's orbit. In a few cases, the gas appears to rotate about the star in keplerian orbits, like the planets in the solar system. These disks are therefore similar to the primitive solar nebula, before the planets formed. If, as seems likely, the Sun experienced a T Tauri phase in its early history, T Tauri stars may be the birth sites of other planetary systems. *See* PROTOSTAR; SOLAR SYSTEM; STELLAR EVOLUTION.

Anneila I. Sargent

Bibliography. C. Bertout, T Tauri stars: Wild as dust, *Annu. Rev. Astron. Astrophys.*, 27:351–395, 1989; G. H. Herbig, The properties and problems of T Tauri stars and related objects, *Adv. Astron. Astrophys.*, 1:47–103, 1962; E. H. Levy, J. I. Lunine, and M. S. Matthews (eds.), *Protostars and Planets III*, 1992.

Taurus

The Bull, in astronomy, a winter constellation. Taurus is the second sign of the zodiac. The group contains two notable star clusters, the Hyades and the Pleiades. The Hyades is a V-shaped cluster, the V forming the head of the charging bull, with the fiery bright star Aldebaran in the right eye (see **illus.**). This star has long been used in navigation. The long horns of the bull extend northeast to the constellation Auriga. Farther west lies the compact, beautiful cluster of six stars, the famous Pleiades, sometimes called the seven sisters, suggesting thereby that one of the stars has faded from naked-eye view. This group in the Bull's shoulder has the shape of a tiny dipper. *See* CONSTELLATION; HYADES; PLEIADES; ZODIAC.

Ching-Sung Yu

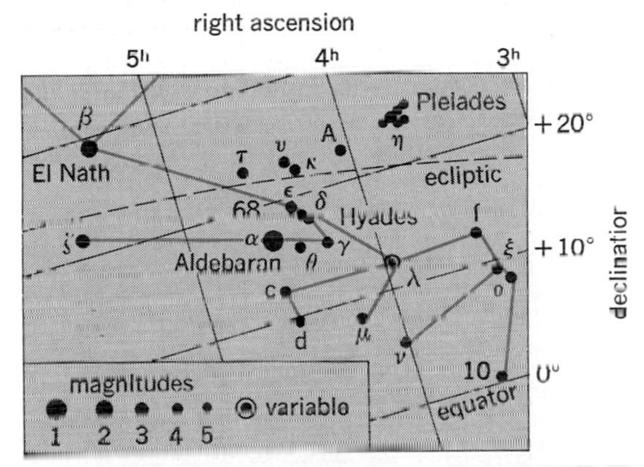

Line pattern of the constellation Taurus. The grid lines represent the coordinates of the sky. The apparent brightness, or magnitude, of the stars is shown by the sizes of the dots, which are graded by appropriate numbers as indicated.

Tektite

A member of one of several groups of objects that are composed almost entirely of natural glass formed from the melting and rapid cooling of terrestrial rocks by the energy accompanying impacts of large extraterrestrial bodies. Tektites are dark brown to green, show laminar to highly contorted flow structure on weathered surfaces and in thin slices, are brittle with excellent conchoidal fracture, and occur in masses ranging to as much as tens of kilograms but are mostly much smaller to microscopic in size. The shapes of tektites are those of common fluid splash and rotational forms including drops, spheres, and dumbbells, unless they have been abraded together with surface gravels. A few tektites have shapes that are caused by two different heating events: the impact that melted the parent rock to form the glass, and a second event apparently due to reentry aerodynamic heating.

Occurrences, groups, and ages. With the discovery of tektites in the Soviet Union, there are now five major groups known: (1) North American, 3.4×10^7 years old, found in Texas (bediasites) and the Georgia Coastal Plains, with a single specimen reported from Martha's Vineyard, Massachussetts; (2) Czechoslovakian (moldavites), 1.5×10^7 years old, found both in Bohemia (green and transparent) and in Moravia (brown and turbid); (3) Ivory Coast, 1.3×10^7 years old; (4) Russian (irgizites; see **illus.**), 1.1×10^7 years old, found in the Northern Aral Region; and (5) Australasian, 700,000 years old, occurring notably in Australia, the Philippines, Belitung, Thailand, and numerous other localities. The North American, Ivory Coast, and Australasian tektites also occur as microtektites in oceanic sediment cores near the areas of their land occurrences. In the land occurrences, virtually all of the tektites are found mixed with surface gravels and recent sediments that are younger than their formation ages.

Tektites, in general, are not rare objects. Millions of tektites have been recovered from the Australasian occurrence. However, some of the more pleasingly colored specimens from Bohemia and Georgia are prized by collectors and command substantial prices, either as specimens or potential gem material.

Three of the tektite groups are associated with known large impact craters. The irgizites occur in and immediately around the Zhamanshin meteoritic impact crater [10–15 km (6–9 mi) in diameter] in the Northern Aral Region of the Commonwealth of Independent States. Both the Ivory Coast tektites and the moldavites occur close to probable source impact craters of identical age, the Bosumtwi Crater, Ghana [10.5 km (6.5 mi) in diameter] and the Ries Crater, Germany [24 km (15 mi) in diameter], respectively. The source craters for the North American and Australasian tektites have not yet been identified positively.

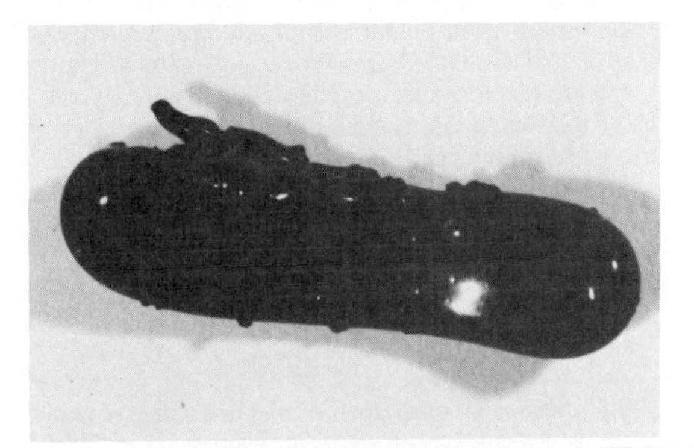

Dumbbell-shaped tektite (irgizite) from the Zhamanshin meteoritic impact crater, Northern Aral Region, Commonwealth of Independent States. The glass is so dark brown that it appears black in reflected light.

Composition. The chemical compositions of tektites differ from those of ordinary terrestrial rocks principally in that they contain less water and have a greater ratio of ferrous to ferric iron, both of which are almost certainly a result of their very high-temperature history. The extreme ranges of major element compositions, expressed as oxides in weight percent, are as follows: SiO_2, 48–85 wt %; Al_2O_3, 8–18 wt %; FeO, 1.4–11 wt %; MgO, 0.4–28 wt %; CaO, 0.3–10 wt %; Na_2O, 0.3–3.9 wt %; K_2O, 1.3–3.8 wt %; TiO_2, 0.3–1.1 wt %. However, portions of the foregoing ranges are based on analyses of glass particles from oceanic sediment cores that may or may not be tektites, as this identification is difficult for these very small particles. If other glass occurrences from impact craters were included as tektites, such as splash form glass from the Wabar Craters in Saudi Arabia, and the Lonar Crater in India, the ranges would expand significantly. As indicated by the wide range of chemical compositions, there are corresponding wide ranges of properties such as specific gravity and refractive index.

Inclusions. Spherical vesicles ranging in size from microscopic to as much as several centimeters are common in most tektites, as are small lechatelierite (silica glass) particles. The presence of coesite, a high-pressure polymorph of silica, in some tektites from Southeast Asia is additional evidence of the impact genesis of tektites. Coesite is known to form at the Earth's surface only from the very high pressures of transient shock waves caused by large hypervelocity meteorite impacts and cratering events. Baddeleyite, monoclinic ZrO_2, is present in some tektites as a high-temperature decomposition product of the mineral zircon. Meteoritic nickel-iron also has been observed in a few tektites, which is further evidence of the origin of tektites as meteoritic impact melts. Such inclusions of coesite, baddeleyite, and nickel-iron are common in the glassy fusion products in and around a number of terrestrial impact craters. SEE METEORITE.

Historical perspective. Tektites have been recognized as unusual objects for more than a millennium, and they were the objects of intensive scientific research from the turn of the century until a few years before the return of lunar samples by the Apollo missions. This activity was generated by the possibility that tektites originated from the Moon as secondary ejecta from impact craters on the lunar surface. This point of view was effectively advocated by a number of scientists until the recognition of the associated impact craters with some tektite groups on the Earth. Later analyses of lunar samples demonstrated that they are not suitable parent rocks for tektites. Unfortunately, the discovery of the Russian tektites, in a clear relation to an impact crater on the Earth, came too late to influence these arguments significantly.

Unanswered questions. The source crater of the Australasian tektites has not yet been identified, although this is the youngest and most widespread tektite group. Also, present understanding of the mechanics and sequence of events in very large impact crater formation does not permit specifying unambiguously the mode by which some tektites escape the Earth's atmosphere to reenter and form the remelted layer as a result of aerodynamic heating.

Elbert A. King

Bibliography. P. W. Florenski, The meteoritic crater Zhamanshin (Northern Aral Region, USSR) and its tektites and impactites [in German], *Chem. Erde*, 36:83–95, 1977; E. A. King, The origin of tektites: A brief review, *Amer. Sci.*, 65:212–218, 1977; E. A. King, *Space Geology: An Introduction*, pp. 69–80, 1976; E. A. King and J. Arndt, Water content of Russian tektites, *Nature*, 269:48–49, 1977; J. A. O'Keefe, *Tektites and Their Origin*, 1976; S. R. Taylor, Tektites: A post-Apollo view, *Earth-Sci. Rev.*, 9:101–123, 1973.

Telescope

An instrument used to collect, measure, or analyze radiation from a distant object. Most commonly, telescope refers to an assemblage of lenses or mirrors, or both, that enhances the ability of the eye either to see objects more distinctly or to see fainter objects. In its most general meaning, telescope refers to a device that collects radiation, which may be in the form of electromagnetic radiation or particle radiation, from a limited direction in space.

Optical telescopes. Optical telescopes may be classified as refracting telescopes, which use only lenses; reflecting telescopes, which use mirrors to focus light; and catadioptric telescopes, which use both lenses and mirrors. Special types of reflecting telescopes which depart radically from conventional ground-based systems are the multimirror telescope and the space telescope.

Refracting telescopes. Small refracting telescopes are used in binoculars, cameras, gunsights, galvanometers, periscopes, surveying instruments, rangefinders, and a great variety of other devices. Parallel or nearly parallel light from the distant object enters from the left, and the objective lens forms an inverted image of it (**illus**. *a*). The inverted image is viewed with the aid of a second lens, called the eyepiece. The eyepiece is adjusted (focused) to form a parallel bundle of rays so that the image of the object may be viewed by the eye without strain. The angular separation A of two points on the object is magnified to the angle A' as seen by the eye. The magnification (A'/A) is numerically equal to the ratio of the focal lengths of the lenses (f_1/f_2). Both the objective lens and the eyepiece usually consist of several lenses combined in order to minimize deleterious effects (aberrations) that are caused by the wavelength dependence of the refraction of glass and by the use of lenses with spherical surfaces. Additional lenses are included in telescopes that are intended for terrestrial viewing to provide an upright image. SEE ABERRATION.

The telescope has been used in astronomy since its introduction by Galileo in the seventeenth century. Improvements in the ability to see faint objects have resulted largely from an increase in the diameter of the objective. The largest refracting telescope is the 1-m (40-in.) telescope at Yerkes Observatory. This size is about the limit for optical glass lenses.

Reflecting telescopes. Much larger telescopes have been made using mirrors instead of lenses (illus. *b*). The reflecting surface of the main mirror is a thin deposited layer of aluminum. Since the bulk of the mirror serves only to support the aluminum, the bulk does not have to be clear optical glass. However, it must be able to accept a high polish by an optician and also be mechanically stable. In the reflecting telescope shown in illus. *b*, auxiliary mirrors may be used near the prime focus at A to provide additional focus positions behind the main mirror at B or at a

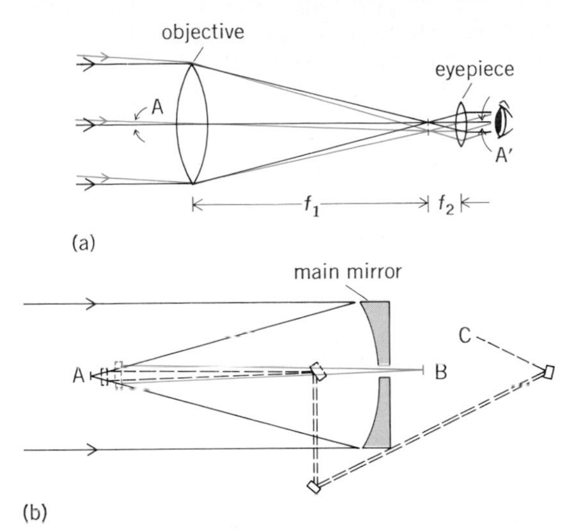

Simplified optical diagrams of (*a*) refracting telescope and (*b*) large, versatile astronomical reflecting telescope.

fixed position *C* where the collected light from a star may be analyzed in a laboratory. Although the eye is frequently used to verify the identification of the object to be studied, most scientific investigations are made by using photographic plates or electronic light detectors. The largest conventional (single-mirror) reflecting telescope is the 6-m (236-in.) instrument near Zelenchukskaya in the Crimea. *See* ASTRONOMICAL PHOTOGRAPHY; IMAGE TUBE.

Catadioptric telescopes. Catadioptric telescopes combine both mirrors and lenses. This combination is generally used to image a wide field of the sky. A notable example is the 1.2-m (48-in.) Schmidt telescope on Palomar Mountain in California.

Multimirror and segmental mirror telescopes. The practical and economical limit to the size of mirrors has been nearly reached in the Crimean telescope. Therefore, newer designs and constructions have been directed toward utilizing a number of mirrors which are mounted so that the light collection by them may be brought to a common focus.

Two possible configurations have been studied: a multimirror telescope (MMT) and a segmented mirror telescope (SMT). In the multimirror plan, all mirrors are of similar shape, but additional optical elements are introduced to bring the light to a common focus. An example is on Mount Hopkins in Arizona, in which six 1.8-m (72-in.) mirrors are combined to collect as much light as a single 4.5-m (177-in.) mirror.

A segmental mirror telescope, the Keck telescope, has been constructed on Mauna Kea, Hawaii. It has 36 hexagonal mirrors, each about 2 m (80 in.) across. Each segment has its surface shaped appropriately for its position in the array, but only six different shapes are required. The light collected by the array is equivalent to that collected by a single mirror of 10-m (394-in.) aperture, making it the largest reflecting telescope.

Space telescope. The ability of large telescopes to resolve fine detail is limited by a number of factors. The ultimate limitation set by diffraction of light is rarely or never achieved by large telescopes. Instead, the finest resolved detail is limited by atmospheric inhomogeneities which set a limit of about 0.2 arc-second. Distortion due to the mirror's own weight causes additional problems in astronomical telescopes. The Earth-orbiting Hubble Space Telescope, with an aperture of 2.4 m (94 in.), was designed to eliminate these problems. The telescope is designed to be used in ultraviolet as well as visible light, resulting in a great improvement in resolution of small angles not only by the elimination of the aforementioned terrestrial effects but by the reduced blurring by diffraction in the ultraviolet.

Launched in April 1990, the telescope was soon discovered to produce blurred images because incorrect assembly of a lens that was to test the shape of the main mirror resulted in an incorrectly shaped mirror. The problem will most likely be corrected by specially designed lenses that will be included in new cameras that will replace the present cameras when the telescope is serviced by a space shuttle. In the meantime, some of the high-resolution imaging science can be accomplished by extensive computer processing of the blurred images. *See* OPTICAL TELESCOPE; SATELLITE ASTRONOMY.

Collectors of radiant energy. As collectors of radiation from a specific direction, telescopes may be classified as focusing and nonfocusing. Nonfocusing telescopes are used for radiation with energies of x-rays and above. Focusing telescopes, intended for nonvisible wavelengths, are similar to optical ones, but they differ in the details of construction.

Radio telescopes. Radio telescopes utilize mirrors of very large size which, because of the long wavelength for which they are used, may consist of only an open wire mesh. Because of the limitation by diffraction, arrays of large reflectors have been constructed. The output of each mirror of the array is combined in a process called aperture synthesis to yield a resolution roughly equivalent to that provided by a telescope the size of the array. An example of this type is the Very Large Array (VLA) near Socorro, New Mexico. *See* RADIO TELESCOPE.

Infrared telescopes. In infrared telescopes the secondary mirror (that near *A* in illus. *b*) is caused to oscillate rotationally about an axis through a diameter. This motion causes an infrared detector at *B* to see alternately the sky and the sky plus the desired object. The signals received at these two mirror positions are subtracted and, as a consequence, the large background radiation received both from the atmosphere and from the telescope is canceled. However, because of the random nature of thermal radiation, the fluctuations of the background emission are not canceled. Thus infrared telescopes are additionally designed to reduce the telescope background radiation and its fluctuations. Two large infrared telescopes are the 3.8-m (150-in.) United Kingdom Infrared Telescope and the 3-m (120-in.) NASA Infrared Telescope Facility on Mauna Kea. *See* INFRARED ASTRONOMY.

Ultraviolet telescopes. Ultraviolet telescopes have special mirror coatings with high ultraviolet reflectivity. Since the atmosphere is not transparent below 300 nanometers, ultraviolet telescopes are usually flown above the atmosphere either in rockets or in orbiting spacecraft. An example is the *International Ultraviolet Explorer*. *See* ULTRAVIOLET ASTRONOMY.

X-ray telescopes. X-ray telescopes must be used above the atmosphere and are flown in rockets or satellites. The focusing type uses an unusual optical design in which the reflection from the surfaces occurs at nearly grazing incidence. This is the only way of achieving reflective optics for x-rays.

Nonfocusing x-ray telescopes use opaque heavy-metal (lead) channels or tubes in front of an x-ray detector to confine the directional sensitivity of the detector. The detectors may be proportional counters or scintillation detectors. SEE X-RAY TELESCOPE.

Gamma-ray telescopes. Gamma-ray telescopes use coincidence and anticoincidence circuits with scintillation or semiconductor detectors to obtain directional discrimination. With coincidence counting, two or more detectors in a line must give a simultaneous detection for a gamma ray to be counted. Other detectors are often used to surround the telescope to reduce the unwanted background arising from undesired particles. A simultaneous count received in one or more of these shielding detectors nullifies (by anticoincidence) the detection otherwise registered in the coincidence detectors. Thus gamma rays that trigger only the coincidence circuits are detected. Since many gamma rays are produced within the atmosphere by other particles, the telescopes are usually flown in balloons, rockets, or satellites. However, at very high energies (more than 100 GeV), ground-based techniques have been used either to detect the Cerenkov light from the shower of electrons produced when a gamma ray hits the atmosphere or to detect directly the particles that penetrate to the ground. SEE GAMMA-RAY ASTRONOMY.

Cosmic-ray telescopes. Cosmic-ray telescopes are used to detect primary protons or heavier-element nuclei or to detect the products produced when these particles interact with the atmosphere. In its simplest form a cosmic-ray telescope may consist of nuclear track emulsions borne aloft in balloons or spacecraft. A very large cosmic-ray telescope deep in a mine in Utah detects penetrating mesons in the 1–100-teraelectronvolt range. This telescope uses Cerenkov detectors in combination with plane parallel arrays of cylindrical spark-tube counters. The Cerenkov counter detects the presence of a high-energy particle and triggers the acoustic sensing of sparks in the counters. A computer is used to analyze the spark data and determine the direction of the incoming meson. The arrival of high-energy cosmic rays appears to be isotropic, most likely because of scattering by magnetic fields of the galaxy. SEE COSMIC RAYS.

Neutrino telescope. Planning and preliminary experimentation for a neutrino telescope has been undertaken. Called the deep underwater muon and neutrino detector (DUMAND), it will consist of 1.8×10^6 m^3 (6×10^7 ft^3) of ocean water 4.5 km (2.8 mi) beneath the ocean surface. An array of 216 Cerenkov counters, suspended in the water, will sense the muon which results from the interaction of an extraterrestrial neutrino with the ocean water around the detector. Since the muon preserves the direction of the original neutrino, computer analysis of which detectors were excited will yield the neutrino's direction to within 1°.

Neutrinos from several astrophysical sources have already been recorded by omnidirectional detectors operating in deep mines. Neutrinos from the Sun have been observed since the 1970s by a detector in a mine in South Dakota. Several underground detectors registered neutrinos from supernova 1987A in February 1987. SEE SOLAR NEUTRINOS; SUPERNOVA.

William M. Sinton

Bibliography. P.-Y. Bely, C. J. Burrows, and G. D. Illingworth (eds.), *The Next Generation Space Telescope*, 1989; D. S. Hayes, R. M. Genet, and D. R. Genet, *New Generation Small Telescopes*, 1987; D. J. Schroeder (ed.), *Astronomical Optics*, 1987; R. N. Sinnott, The Keck telescope's giant eye, *Sky Telesc.*, 80:15–22, 1990; B. J. Thompson and R. R. Shannon (eds.), *Space Optics*, 1974.

Tide

Stresses exerted in a body by the gravitational action of another, and related phenomena resulting from these stresses. Every body in the universe raises tides, to some extent, on every other. This article deals only with tides on the Earth, since these are fundamentally the same as tides on all bodies. Sometimes variations of sea level, whatever their origin, are referred to as tides.

Introduction. The tide-generating forces arise from the gravitational action of Sun and Moon, the effect of the Moon being about twice as effective as that of the Sun in producing tides. The tidal effects of all other bodies on the Earth are negligible. The tidal forces act to generate stresses in all parts of the Earth and give rise to relative movements of the matter of the solid Earth, ocean, and atmosphere. The Earth's rotation gives these movements an alternating character having principal periodicities of 12.42 and 12.00 h, corresponding to half the mean lunar and solar day, respectively.

In the ocean the tidal forces act to generate alternating tidal currents and displacements of the sea surface. These phenomena are important to shipping and have been studied extensively. The main object of tidal studies has been to predict the tidal elevation or current at a given seaport or other place in the ocean at any given time.

The prediction problem may be attacked in two ways. Since the relative motions of Earth, Moon, and Sun are known precisely, it is possible to specify the tidal forces over the Earth at any past or future time with great precision. It should be possible to relate tidal elevations and currents at any point in the oceans to these forces, making use of classical mechanics and hydrodynamics. Such a theoretical approach to tidal prediction has not yet yielded any great success, owing in great part to the complicated shape of the ocean basins. However, use of numerical-hydrodynamical models (such as the work of K. T. Bogdanov, N. Grijalva, W. Hansen, M. C. Henderschott, and C. L. Pekeris) has yielded some satisfactory results and undoubtedly will have practical importance.

The other approach, which consists of making use of past observations of the tide at a certain place to predict the tide for the same place, has yielded practical results. The method cannot be used for a location where there have been no previous observations. In the harmonic method the frequencies of the many tidal constituents are derived from knowledge of the movements of Earth, Moon, and Sun. The amplitude and epoch of each constituent are determined from the tidal observations. The actual tide can then be synthesized by summing up an adequate number of harmonic constituents. The method might loosely be thought of as extrapolation.

A ''convolution'' method of tidal analysis and prediction has been proposed by W. H. Munk and D. E. Cartwright. In this method past observations at a

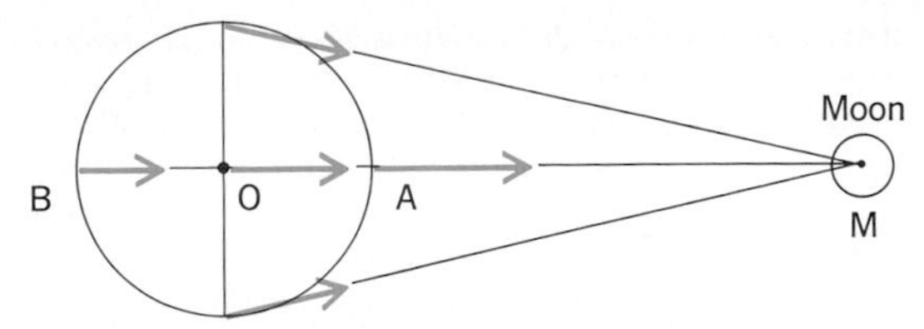

Fig. 1. Schematic diagram of the lunar gravitational force on different points in the Earth.

place are used to determine a numerical operator which, when applied to the known tide-producing forces, will calculate the resulting tide.

In the following discussion only the lunar effect is considered, and it is understood that analogous statements apply to the solar effect.

Tide-generating force. If the Moon attracted every point within the Earth with equal force, there would be no tide. It is the small difference in direction and magnitude of the lunar attractive force, from one point of the Earth's mass to another, which gives rise to the tidal stresses.

According to Newton's laws, the Moon attracts every particle of the Earth with a force directed toward the center of the Moon, with magnitude proportional to the inverse square of the distance between the Moon's center and the particle. At point A in **Fig. 1**, the Moon is in the zenith and at point B the Moon is at nadir. It is evident that the upward force of the Moon's attraction at A is greater than the downward force at B because of its closer proximity to the Moon. Such differential forces are responsible for stresses in all parts of the Earth. The Moon's gravitational pull on the Earth can be expressed as the vector sum of a constant force, equal to the Moon's attraction on the Earth's center, and a small deviation which varies from point to point in the Earth (**Fig. 2**). This small deviation is referred to as the tide-generating force. The larger constant force is balanced completely by acceleration (centrifugal force) of the Earth in its orbital motion around the center of mass of the Earth-Moon system, and plays no part in tidal phenomena. *SEE GRAVITATION*.

The tide-generating force is proportional to the mass of the disturbing body (Moon) and to the inverse cube of its distance. This inverse cube law accounts for the fact that the Moon is 2.17 times as important, insofar as tides are concerned, as the Sun, although the latter's direct gravitational pull on the Earth, which is governed by an inverse-square law, is about 180 times the Moon's pull.

The tide-generating force, as illustrated in Fig. 2, can be expressed as the gradient of the tide-generating potential, Eq. (1), where λ is the zenith distance of

$$\psi = \frac{3}{2}\frac{\gamma M r^2}{c^3}\left(\frac{1}{3} - \cos^2\lambda\right) \quad (1)$$

the Moon, r is distance from the Earth's center, c is distance between the centers of Earth and Moon, γ is the gravitational constant, and M is the mass of the Moon. In this expression, terms containing higher powers of the smaller number r/c have been neglected. As ψ depends only on the space variables r and λ, it is symmetrical about the Earth-Moon axis.

It helps one visualize the form of the tide-generating potential to consider how a hypothetical "inertialess" ocean covering the whole Earth would respond to the tidal forces. In order to be in equilibrium with the tidal forces, the surface must assume the shape of an equipotential surface as determined by both the Earth's own gravity and the tide-generating force. The elevation of the surface is given approximately by Eq. (2), where ψ is evaluated at the Earth's surface

$$\bar{\zeta} = -\frac{\psi}{g} + \text{const} \quad (2)$$

and g is the acceleration of the Earth's gravity. The elevation $\bar{\zeta}$ of this hypothetical ocean is known as the equilibrium tide. Knowledge of the equilibrium tide over the entire Earth determines completely the tide-generating potential (and hence the tidal forces) at all points within the Earth as well as on its surface. Therefore, when the equilibrium tide is mentioned, it shall be understood that reference to the tide-generating force is also being made.

Harmonic development of the tide. The equilibrium tide as determined from relations (1) and (2) has the form of a prolate spheroid (football-shaped) whose major axis coincides with the Earth-Moon axis. The Earth rotates relative to this equilibrium tidal form so that the nature of the (equilibrium) tidal variation with time at a particular point on the Earth's surface is not immediately obvious. To analyze the character of this variation, it is convenient to express the zenith angle of the Moon in terms of the geographical coordinates θ, ϕ of a point on the Earth's surface (θ is colatitude, ϕ is east longitude) and the declination D and west hour angle reckoned from Greenwich α of the Moon. When this is done, the equilibrium tide can be expressed as the sum of the three terms in Eq. (3), where a is the Earth's radius.

$$\bar{\zeta} = \frac{3}{4}\frac{\gamma M}{g}\frac{a^2}{c^3}\Big[(3\sin^2 D - 1)(\cos^2\theta - 1/3) + \sin 2D \sin 2\theta \cos(\alpha + \phi) + \cos^2 D \sin^2\theta \cos 2(\alpha + \phi)\Big] \quad (3)$$

The first term represents a partial tide which is symmetrical about the Earth's axis, as it is independent of longitude. The only time variation results from the slowly varying lunar declination and distance from Earth. This tide is called the long-period tide. Its actual geographical shape is that of a spheroid whose axis coincides with the Earth's axis and whose oblateness slowly but continuously varies.

The second term of Eq. (3) represents a partial tide having, at any instant, maximum elevations at 45°N

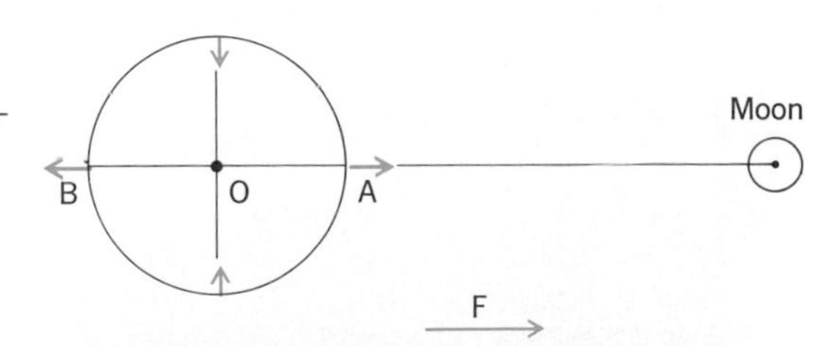

Fig. 2. Schematic diagram of the tide-generating force on different points in the Earth. The vector sum of this tide-generating force and the constant force *F* (which does not vary from point to point) produce the force field indicated in Fig. 1. Force *F* is compensated by the centrifugal force of the Earth in its orbital motion.

and 45°S on opposite sides of the Earth, and two minimum elevations lying at similar, alternate positions on the same great circle passing through the poles. Because of the factor cos $(\alpha + \phi)$ the tide rotates in a westerly direction relative to the Earth, and any geographical position experiences a complete oscillation in a lunar day, the time taken for α to increase by the amount 2π. Consequently, this partial tide is called the diurnal tide. Because of the factor sin $2D$, the diurnal equilibrium tide is zero at the instant the Moon crosses the Equator; because of the factor sin 2θ, there is no diurnal equilibrium tidal fluctuation at the Equator or at the poles.

The third term of Eq. (3) is a partial tide having, at any instant, two maximum elevations on the Equator at opposite ends of the Earth, separated alternately by two minima also on the Equator. This whole form also rotates westward relative to the Earth, making a complete revolution in a lunar day. But any geographic position on the Earth will experience two cycles during this time because of the factor cos $2(\alpha + \phi)$. Consequently, this tide is called the semidiurnal tide. Because of the factor $\sin^2 \theta$, there is no semidiurnal equilibrium tidal fluctuation at the poles, while the fluctuation is strongest at the Equator.

It has been found very convenient to consider the equilibrium tide as the sum of a number of terms, called constituents, which have a simple geographical shape and vary harmonically in time. This is the basis of the harmonic development of the tide. A great number of tidal phenomena can be adequately described by a linear law; that is, the effect of each harmonic constituent can be superimposed on the effects of the others. Herein is the great advantage of the harmonic method in dealing with tidal problems. The three terms of Eq. (3) do not vary with time in a purely harmonic manner. The parameters c and D themselves vary, and the rapidly increasing α does not do so at a constant rate owing to ellipticity and other irregularities of the Moon's orbit. Actually, each of the three partial tides can be separated into an entire species of harmonic constituents. The constituents of any one of the three species have the same geographical shape, but different periods, amplitudes, and epochs.

The solar tide is developed in the same way. As before, the three species of constituents arise: long-period, diurnal, and semidiurnal. The equilibrium tide at any place is the sum of both the lunar and solar tides. When the Sun and Moon are nearly in the same apparent position in the sky (new Moon) or are nearly at opposite positions (full Moon), the lunar and solar effects reinforce each other. This condition is called the spring tide. During the spring tide the principal lunar and solar constituents are in phase. At quadrature the solar effect somewhat cancels the lunar effect, the principal lunar and solar constituents being out of phase. This condition is known as the neap tide.

The entire equilibrium tide can now be expressed by Eq. (4), where $H = 3\gamma Ma^2/g\bar{c}^3 = 54$ cm, and $1/\bar{c}$

$$\bar{\zeta} = H\Big[{}^1\!/\!_2(1 - 3\cos^2\theta)\sum_L f_i C_i \cos A_i + \sin 2\theta \sum_D f_i C_i \cos(A_i + \phi) + \sin^2\theta \sum_S f_i C_i \cos(A_i + 2\phi)\Big] \quad (4)$$

represents the mean (in time) value of $1/c$. Each term in the above series represents a constituent. Terms of higher powers of the Moon's parallax (a/c) are not included in Eq. (4) because of their different latitude dependence, but they are of relatively small importance. The subscripts L, D, and S indicate summation over the long-period, diurnal, and simidiurnal constituents, respectively. The C's are the constituent coefficients and are constant for each constituent. They account for the relative strength of all lunar and solar constituents. In a purely harmonic development, such as carried out by A. T. Doodson in 1921, the A parts of the arguments increase linearly with time, and the node factors f are all unity. In George Darwin's "almost harmonic" development of 1882, the constituents undergo a slow change in amplitude and epoch with the 19-year nodal cycle of the Moon. The node factors f take this slow variation into account. The A's increase almost linearly with time. Tables in U.S. Coast and Geodetic Survey Spec. Publ. 98 enable one to compute the phase of the argument of any of Darwin's constituents at any time, and values of the node factors for each year are given.

In spite of the many advantages of the purely harmonic development, Darwin's method is still used by most agencies engaged in tidal work. In Darwin's classification, each constituent is represented by a symbol with a numerical subscript, 0, 1, or 2, which designates whether the constituent is long-period, diurnal, or semidiurnal. Some of the most important of Darwin's constituents are listed in the **table**.

The periods of all the semidiurnal constituents are grouped about 12 h, and the diurnal periods about 24 h. This results from the fact that the Earth rotates much faster than the revolution of the Moon about the Earth or of the Earth about the Sun. The principal lunar semidiurnal constituent M_2 beats against the others giving rise to a modulated semidiurnal waveform whose amplitude varies with the Moon's phase (the spring-neap effect), distance, and so on. Similarly, the amplitude of the modulated diurnal wave varies with the varying lunar declination, solar declination, and lunar phase. For example, the spring tide at full Moon or new Moon is manifested by constituents M_2 and S_2 being in phase, thus reinforcing each other. During the neap tide when the Moon is at quadrature, the constituents M_2 and S_2 are out of phase, and tend to cancel each other. The other variations in the intensity of the tide are similarly reflected in the "beating" of other groups of constituents.

Darwin's constituents

Constituent	Speed, deg/h	Coefficient
Long-period		
Mf, lunar fortnightly	1.098	0.157
Ssa, solar semiannual	0.082	0.073
Diurnal		
K_1, lunisolar	15.041	0.530
O_1, larger lunar	13.943	0.377
P_1, larger solar	14.959	0.176
Semidiurnal		
M_2, principal lunar	28.984	0.908
S_2, principal solar	30.000	0.423
N_2, larger lunar elliptic	28.440	0.176
K_2, lunisolar	30.082	0.115

Tides in the ocean. The tide in the ocean deviates markedly from the equilibrium tide, which is not surprising if one recalls that the equilibrium tide is based on neglect of the inertial forces. These forces are appreciable unless the periods of all free oscillations in the ocean are small compared with those of the tidal forces. Actually, there are free oscillations in the ocean (ordinary gravity seiches) having periods of the order of a large fraction of a day, and there may be others (planetary modes) having periods of the order of several days. For the long-period constituents the observed tide should behave like the equilibrium tide, but this is difficult to show because of their small amplitude in the presence of relatively large meteorological effects.

At most places in the ocean and along the coasts, sea level rises and falls in a regular manner. The highest level usually occurs twice in any lunar day, the times bearing a constant relationship with the Moon's meridional passage. The time between the Moon's meridional passage and the next high tide is called the lunitidal interval. The difference in level between successive high and low tides, called the range of the tide, is generally greatest near the time of full or new Moon, and smallest near the times of quadrature. This results from the spring-neap variation in the equilibrium tide. Tide range usually exhibits a secondary variation, being greater near the time of perigee (when the Moon is closest to the Earth) and smaller at apogee (when it is farthest away).

The above situation is observed at places where the tide is predominantly semidiurnal. At many other places, it is observed that one of the two maxima in any lunar day is higher than the other. This effect is known as the diurnal inequality and represents the presence of an appreciable diurnal variation. At these places, the tide is said to be of the "mixed" type. At a few places, the diurnal tide actually predominates, there generally being only one high and low tide during the lunar day.

Both observation and theory indicate that the ocean tide can generally be considered linear. As a result of this fact, the effect in the ocean of each constituent of the series in Eq. (4) can be considered by itself. Each equilibrium constituent causes a reaction in the ocean. The tide in the ocean is the sum total of all the reactions of the individual constituents. Furthermore, each constituent of the ocean tide is harmonic (sinusoidal) in time. If the amplitude of an equilibrium constituent varies with the nodal cycle of the Moon, the amplitude of the oceanic constituent varies proportionately.

As a consequence of the above, the tidal elevation in the ocean can be expressed by Eq. (5), where

$$\zeta = \sum f_i h_i \cos (A_i - G_i) \qquad (5)$$

$h_i(\theta,\phi)$ is called the amplitude and $G_i(\theta,\phi)$ the Greenwich epoch of each constituent. The summation in Eq. (5) extends over all constituents of all species. The f's and the A's have the same meaning as in Eq. (4) for the equilibrium tide and are determined from astronomic data.

To specify completely the tidal elevation over the entire surface of the ocean for all time, one would need ocean-wide charts of $h(\theta,\phi)$, called corange charts, and of $G(\theta,\phi)$, called cotidal charts, for each important constituent. Construction of these charts would solve the ultimate problem in tidal prediction. Many attempts have been made to construct cotidal charts, the most notable being those of W. Whewell, 1833; R. A. Harris, 1904; R. Sterneck, 1920; and G. Dietrich, 1944. These attempts have been based on a little theory and far too few observations.

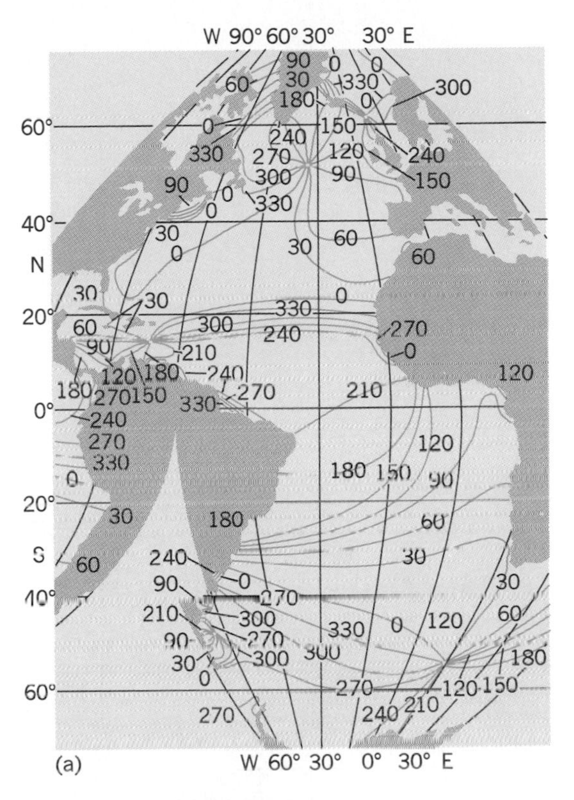

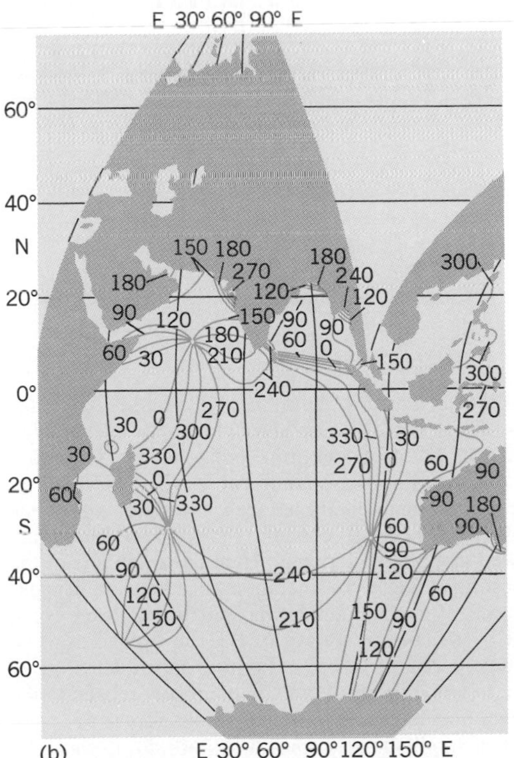

Fig. 3. Cotidal chart for M_2. (*a*) Atlantic Ocean. (*b*) Indian Ocean. (*After G. Dietrich, Veroeff. Inst. Meeresk., n.s. A, Geogr.-naturwiss. Reihe, no. 41, 1944*)

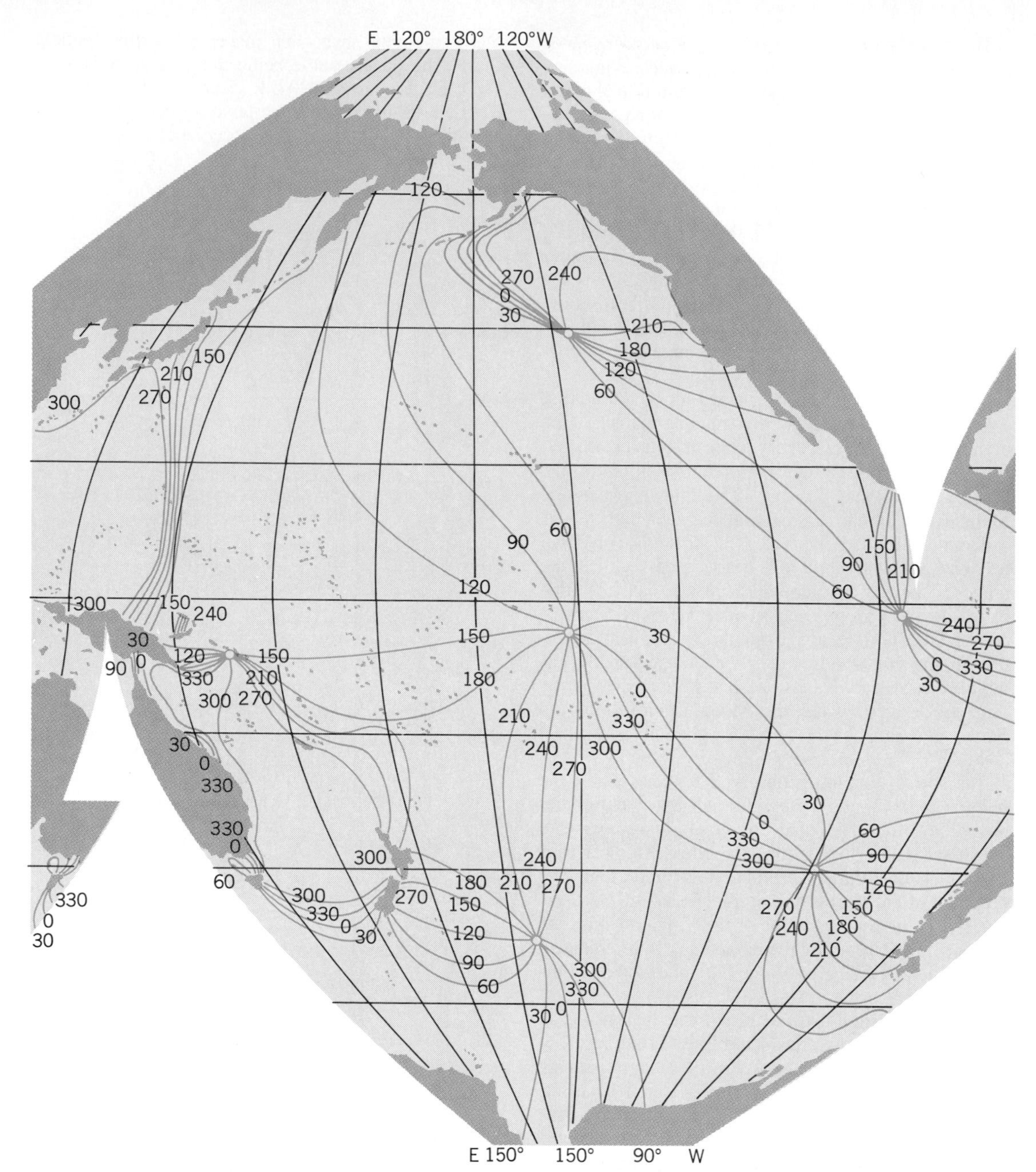

Fig. 4. Pacific Ocean cotidal chart for M_2. ***(After G. Dietrich, Veroeff. Inst. Meeresk., n.s. A, Geogr.-naturwiss. Reihe, no. 41, 1944)***

Figures 3 and **4** show Dietrich's cotidal chart for M_2. Each curve passes through points having high water at the same time, time being indicated as phase of the M_2 equilibrium argument. A characteristic feature of cotidal charts is the occurrence of points through which all cotidal curves pass. These are called amphidromic points. Here the amplitude of the constituent under consideration must be zero. The existence of such amphidromic points has been borne out by theoretical studies of tides in ocean basins of simple geometric shape. The mechanism which gives rise to amphidromic points is intimately related to the rotation of the Earth and the Coriolis force.

The amplitude of a constituent, $h(\theta,\phi)$, is generally high in some large regions of the oceans and low in others, but in addition there are small-scale erratic variations, at least along the coastline. Perhaps this is partly an illusion caused by the placement of some tide gages near the open coast and the placement of others up rivers and estuaries. It is well known that the phase and amplitude of the tide change rapidly as the tidal wave progresses up a river. *See River tides.*

The range of the ocean tide varies between wide limits. The highest range is encountered in the Bay of Fundy, where values exceeding 50 ft (15 m) have been observed. In some places in the Mediterranean, South Pacific, and Arctic, the tidal range never exceeds 2 ft (0.6 m).

The tide may be considerably different in small adjacent seas than in the nearby ocean, and here resonance phenomena frequently occur. The periods of free oscillation of a body of water are determined by their boundary and depth configurations. If one of these free periods is near that of a large tidal constituent, the latter may be amplified considerably in the small sea. The large tidal range in the Bay of Fundy is an example of this effect. Here the resonance period is nearly 12 h, and it is the semidiurnal constituents that are large. The diurnal constituents are not extremely greater in the Bay of Fundy than in the nearby ocean.

In lakes and other completely enclosed bodies of water the periods of free oscillation are usually much smaller than those of the tidal constituents. Therefore the tide in these places obeys the principles of statics. Since there is no tidal variation in the total volume of water in lakes the mean surface elevation does not change with the tide. The surface slope is determined by the slope of the equilibrium tide, and the related changes in elevation are usually very small, of the order of a fraction of a millimeter for small lakes.

Tidal currents. The south and east components of the tidal current can be developed in the same way as the tidal elevation since they also depend linearly on the tidal forces. Consequently, the same analysis and prediction methods can be used. Expressions similar to Eq. (5) represent the current components, each constituent having its own amplitude and phase at each geographic point. It should be emphasized that the current speed or direction cannot be developed in this way since these are not linearly related to the tidal forces.

Only in special cases are the two tidal current components exactly in or out of phase, and so the tidal current in the ocean is generally rotatory. A drogue or other floating object describes a trajectory similar in form to a Lissajous figure. In a narrow channel only the component along its axis is of interest. Where shipping is important through such a channel or port entrance, current predictions, as well as tidal height predictions, are sometimes prepared.

Owing to the rotation of the Earth, there is a gyroscopic, or Coriolis, force acting perpendicularly to the motion of any water particle in motion. In the Northern Hemisphere this force is to the right of the current vector. The horizontal, or tractive, component of the tidal force generally rotates in the clockwise sense in the Northern Hemisphere. As a result of both these influences the tidal currents in the open ocean generally rotate in the clockwise sense in the Northern Hemisphere, and in the counterclockwise sense in the Southern Hemisphere. There are exceptions, however, and the complete dynamics should be taken into account.

The variation of the tidal current with depth is not well known. It is generally agreed that the current would be constant from top to bottom were it not for stratification of the water and bottom friction. The variation of velocity with depth due to the stratification of the water is associated with internal wave motion. Serial observations made from anchored or drifting ships have disclosed prominent tidal periodicities in the vertical thermal structure of the water.

Dynamics of ocean tide. The theoretical methods for studying tidal dynamics in the oceans were put forth by Laplace in the eighteenth century. The following assumptions are introduced: (1) The water is homogeneous; (2) vertical displacements and velocities of the water particles are small in comparison to the horizontal displacements and velocities; (3) the water pressure at any point in the water is given adequately by the hydrostatic law, that is, it is equal to the head of water above the given point; (4) all dissipative forces are neglected; (5) the ocean basins are assumed rigid (as if there were no bodily tide), and the gravitational potential of the tidally displaced masses is neglected; and (6) the tidal elevation is small compared with the water depth.

If assumptions (1) and (3) are valid, it can readily be shown that the tidal currents are uniform with depth. This is a conclusion which is not in complete harmony with observations, and there are internal wave modes thus left out of Laplace's theory. Nevertheless the main features of the tide are probably contained in the equations.

The water motion in the oceans is, in theory, determined by knowledge of the shape of the ocean basins and the tide-generating force (or equilibrium tide) at every point in the oceans for all time. The theory makes use of two relations: (1) the equation of continuity, which states that the rate of change of water mass in any vertical column in the ocean is equal to the rate at which water is flowing into the column; and (2) the equations of motion, which state that the total acceleration of a water "particle" (relative to an inertial system, thus taking into account the rotation of the Earth) is equal to the total force per unit mass acting on that particle. Under the above assumptions, the equation of continuity takes the form of Eq. (6),

$$\frac{\partial \zeta}{\partial t} = -\frac{1}{a \sin \theta}\left[\frac{\partial}{\partial \theta}(ud \sin \theta) + \frac{\partial}{\partial \phi}(vd)\right] \tag{6}$$

where $d(\theta,\phi)$ is the water depth. The equations of motion in the southward and eastward directions, respectively, are given by Eqs. (7), where ω designates

$$\begin{aligned} \frac{\partial u}{\partial t} - 2\omega v \cos \theta &= \frac{g}{a}\frac{\partial}{\partial \theta}(\zeta - \bar{\zeta}) \\ \frac{\partial v}{\partial t} + 2\omega u \cos \theta &= \frac{g}{a}\csc\theta \frac{\partial}{\partial \phi}(\zeta - \bar{\zeta}) \end{aligned} \tag{7}$$

the angular rate of rotation of the Earth, and u and v the south and east components of the tidal current. All other quantities are as previously defined.

It is probable that exact mathematical solutions to Eqs. (6) and (7), taking even approximately into account the complicated shape of the ocean basins, will never be obtained. However, the equations have certain features which serve to give us some insight into the nature of ocean tides. For instance, it is evident that if many equilibrium tides are acting simultaneously on the ocean, then the ocean tide will be the sum of the individual reactions. This linearity results directly from assumption (6). In certain shallow regions of the ocean the tides are noticeably distorted, as would be expected if assumption (6) were violated. This distortion is usually considered as resulting from the presence of so-called shallow-water constituents having frequencies equal to harmonics and to beat frequencies of the equilibrium constituents. These must be considered, at some places, or there will be large discrepancies between prediction and observation. Certain mathematical solutions to Eqs. (6) and (7) have been obtained for hypothetical ocean basins of simple geometric shape. Laplace solved them for an ocean of constant depth covering the entire Earth. Several solutions have been obtained for an ocean of constant depth bounded by two meridians. The result of one of the solutions obtained by J. Proudman and A. Doodson is shown in **Fig. 5**, which represents a cotidal chart of the K_2 tide in an ocean of depth

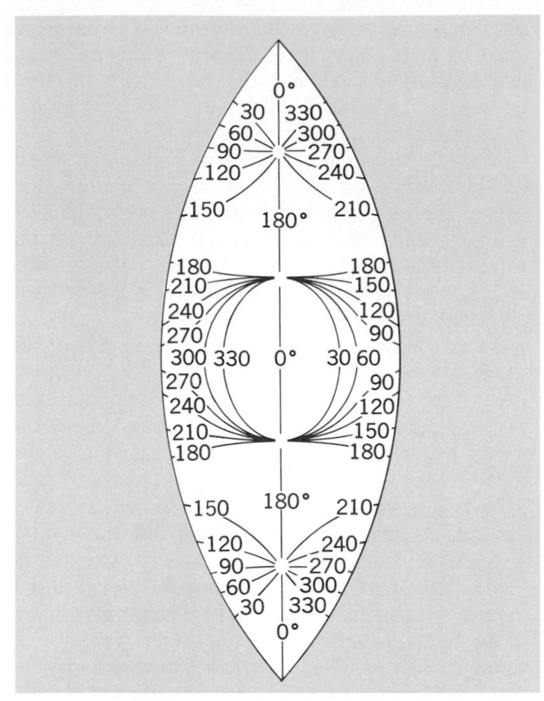

Fig. 5. Cotidal chart for K_2 in a hypothetical ocean of constant depth bounded by meridians 70° apart. (*After A. T. Doodson and H. D. Warburg, Admiralty Manual of Tides, London, 1941*)

14,520 ft (4426 m) bounded by meridians 70° apart. The K_2 tide was calculated because of mathematical simplifications, but the M_2 tide should be quite similar. Comparison of Fig. 5 with the Atlantic Ocean in Fig. 3 discloses no striking similarities except for the general occurrence of amphidromic systems.

The bodily tide. The solid part of the Earth suffers periodic deformation resulting from the tide-generating forces just as the oceans do.

The gravest known modes of free oscillation of the solid Earth have periods of the order of an hour, much shorter than those of the principal tidal constituents. Therefore, the principles of statics can be used to describe the bodily tide, in contrast to tides in the oceans and atmosphere, where the inertial effect is important.

Associated with the bodily tide are periodic changes in gravity, manifesting themselves as (1) a variation of the vertical, or plumb line, with respect to any solid structure embedded in the Earth's crust; and (2) a variation in the magnitude of the acceleration of gravity at any point. These effects arise from the gravitational attraction of the tidally displaced matter of the Earth (solid, ocean, and atmosphere) as well as directly from the tide-generating forces. The magnitude of the former factor is of the order of several tens of microgals (1 gal = 1 cm/s^2).

Atmospheric tides. Since air, as other matter, is subject to gravitational influence, there are tides in the atmosphere possessing many features of similarity with those in the ocean. One of the characteristics of these tides is a small oscillatory variation in the atmospheric pressure at any place. This fluctuation of pressure, as in the case of the ocean tide, may be considered as the sum of the usual tidal constituents, and standard tidal analysis and prediction methods may be used. The principal lunar semidiurnal constituent M_2 of the pressure variation has been determined for a number of places, and found to have an amplitude of the order of 0.03 millibar (3 pascals). The dynamical theory of these tides has been the subject of considerable study. The equations which have been considered have the same general form as those for ocean tides. The S_2 constituent shows a much larger oscillation with an amplitude of the order of 1 millibar (10^2 Pa), but here diurnal heating dominates the gravitational effects. If diurnal heating were the whole story one would expect an even larger S_1 effect, and the fact that S_2 is larger is attributed to an atmospheric resonance near 12 h.

Tidal analysis and prediction. The distribution in space and time of the tidal forces within the Earth is precisely known from astronomic data. The effects of these forces on the oceans cannot, by present methods, be described in detail on a worldwide basis because of the difficult nature of the dynamical relationships and the complicated shape of the ocean basins. Practical prediction methods make use of past observations at the place under consideration.

The procedure is the same for prediction of any tidal variable—such as the atmospheric pressure, component displacements of the solid Earth, components of the tidal current, and so on—which depends linearly on the tidal forces. In the harmonic method the frequencies, or periods, of the tidal constituents are determined by the astronomic data, and the harmonic constants (amplitudes and epochs) are obtained from the observations. Equation (5) then represents the tide at all past and future times for the place under consideration, where the values of h are the amplitudes of whatever tidal variable is being predicted. In this discussion the sea-level elevation will be used as an example, since it is the variable for which predictions are most commonly made. The procedure is basically the same for each constituent, but is most easily described for the series of constituents, S_1, S_2, S_3, . . . , whose periods are submultiples of 24 h. Suppose that the tidal elevation at 1:00 A.M. is averaged for all the days of the tide record, and similarly for 2:00 A.M., 3:00 A.M., and for each hour of the day. The 24 values thus obtained represent the average diurnal variation during the entire record. Any constituent whose period is not a submultiple of 24 h will contribute very little to the average of all the 1:00 A.M. values since its phase will be different from one day to the next, and its average value at 1:00 A.M. will be very close to zero for a long record. The same is true for each hour of the day, and so its average diurnal variation is small. The longer the record the freer will be the average diurnal oscillation from the effects of the other constituents. The diurnal oscillation is then analyzed by the well-known methods of harmonic analysis to determine the amplitudes and phases of all the harmonics of the 24-h oscillation.

The same procedure is used for each other constituent; that is, the tide record is divided into consecutive constituent days, each equal to the period (or double the period in the case of the semidiurnal constituents) of the constituent. If the tide record is tabulated each solar hour, there is a slight complication due to the fact that the constituent hours do not coincide with the solar hours. This difficulty is overcome by substituting the tabulated value nearest the required time and later compensating the consistent error introduced by an augmenting factor.

Since the record length is always finite, the harmonic constants of a constituent determined by this method are somewhat contaminated by the effects of other constituents. A first-order correction of these effects can be made by an elimination procedure. In general it is more efficient to take the record length equal to the synodic (beat) period of two or more of the principal constituents. Of course, the longer the record the better. Standard analyses consist of 29 days, 58 days, 369 days, and so on.

It is not practical to determine the harmonic constants of the lesser constituents in this way if errors or uncertainties of the data are of the same order of magnitude as their amplitudes. If tidal oscillations in the oceans were far from resonance then the amplitude H of each constituent should be expected to be approximately proportional to its theoretical coefficient C, and the local epochs G all to be near the same value. In other words, for the semidiurnal constituent X, Eqs. (8) should hold. Here X is referred to

$$\frac{H(X)}{C(X)} = \frac{H(M_2)}{C(M_2)} \qquad G(X) = G(M_2) \qquad (8)$$

M_2 for the reason that the latter is one of the principal constituents whose harmonic constants can be determined with best accuracy. Any other important constituent could be used. Inferring the harmonic constants of the lesser constituents by means of Eqs. (8) is sometimes preferable to direct means. It should be borne in mind that a constituent of one species cannot be inferred from one of another species because their equilibrium counterparts have different geographic shapes and no general relationship such as Eqs. (8) exists.

Once the harmonic constants are determined, the tide is synthesized according to Eq. (5), usually with the help of a special tide-predicting machine, although any means of computation could be used. Usually only the times and heights of high and low water are published in the predictions.

Tidal friction. The dissipation of energy by the tide is important in the study of planetary motion because it is a mechanism whereby angular momentum can be transferred from one type of motion to another. An appreciable amount of tidal dissipation takes place in the ocean, and possibly also in the solid Earth. In 1952 Sir Harold Jeffreys estimated that about half the tidal energy present in the ocean at any time is dissipated each day. A large part of this dissipation takes place by friction of tidal currents along the bottom of shallow seas and shelves and along the coasts. The rate of dissipation is so large that there should be a noticeable effect on the tide in the oceans.

If the planet's speed of rotation is greater than its satellite's speed of revolution about it, as is the case in the Earth-Moon system, then tidal dissipation always tends to decelerate the planet's rotation, with the satellite's speed of revolution changing to conserve angular momentum of the entire system. The Moon's attraction on the irregularly shaped tidal bulge on the Earth exerts on it a decelerating torque. Thus tidal friction tends to increase the length of day, to increase the distance between Earth and Moon, and to increase the lunar month, but these increases are infinitesimal. The day may have lengthened by 1 s during the last 120,000 years because of tidal friction and other factors.

Gordon W. Groves

Bibliography. P. Crean, T. S. Murty, and J. A. Stronach, *Mathematical Modeling of Tides and Estuarine Circulation*, 1988; G. Godin, *Analysis of Tides*, 1972; H. Lamb, *Hydrodynamics*, 6th ed., 1945; G. I. Marchuk and B. A. Kagan, *Dynamics of Ocean Tides*, 1989; P. Melchior, *The Tides of the Planet Earth*, 2d ed, 1983.

Time

The dimension of the physical universe which orders the sequence of events at a given place; also, a designated instant in this sequence, such as the time of day, technically known as an epoch.

Measurement. Time measurement consists of counting the repetitions of any recurring phenomenon and possibly subdividing the interval between repetitions. Two aspects to be considered in the measurement of time are frequency, or the rate at which the recurring phenomena occur, and epoch, or the designation to be applied to each instant.

A determination of time is equivalent to the establishment of an epoch or the correction that should be applied to the reading of a clock at a specified epoch. A time interval may be measured as the duration between two known epochs or by counting from an arbitrary starting point, as is done with a stopwatch. Time units are the intervals between successive recurrences of phenomena, such as the period of rotation of the Earth or a specified number of periods of radiation derived from an atomic energy-level transition. Other units are arbitrary multiples and subdivisions of these intervals, such as the hour being 1/24 of a day, and the minute being 1/60 of an hour. SEE DAY; MONTH; YEAR.

Time bases. Several phenomena are used as bases with which to determine time. The phenomenon traditionally used has been the rotation of the Earth, where the counting is by days. Days are measured by observing the meridian passages of stars and are subdivided with the aid of precision clocks. The day, however, is subject to variations in duration. Thus, when a more uniform time scale is required, other bases for time must be used.

Sidereal time. The hour angle of the vernal equinox is the measure of sidereal time. It is reckoned from 0 to 24 hours, each hour being subdivided into 60 sidereal minutes and the minutes into 60 sidereal seconds. Sidereal clocks are used for convenience in most astronomical observatories because a star or other object outside the solar system comes to the same place in the sky at virtually the same sidereal time.

Solar time. The hour angle of the Sun is the apparent solar time. The only true indicator of apparent solar time is a sundial. Mean solar time has been devised to eliminate the irregularities in apparent solar time that arise from the obliquity of the ecliptic and the varying speed of the Earth in its orbit around the Sun. It is the hour angle of a fictitious point moving uniformly along the celestial equator at the same rate as the average rate of the Sun along the ecliptic. In practice, it is intervals of sidereal time that are directly observed and afterward converted into intervals of mean solar time by division by 1.00273790935. Both sidereal and solar time depend on the rotation of the Earth for their time base.

Universal Time. The mean solar time determined from the rotation of the Earth by using astronomical observations is referred to as UT1. Observations are made at a number of observatories around the world. The raw observations of time are referred to as UT0 and must be corrected for the polar motion of the

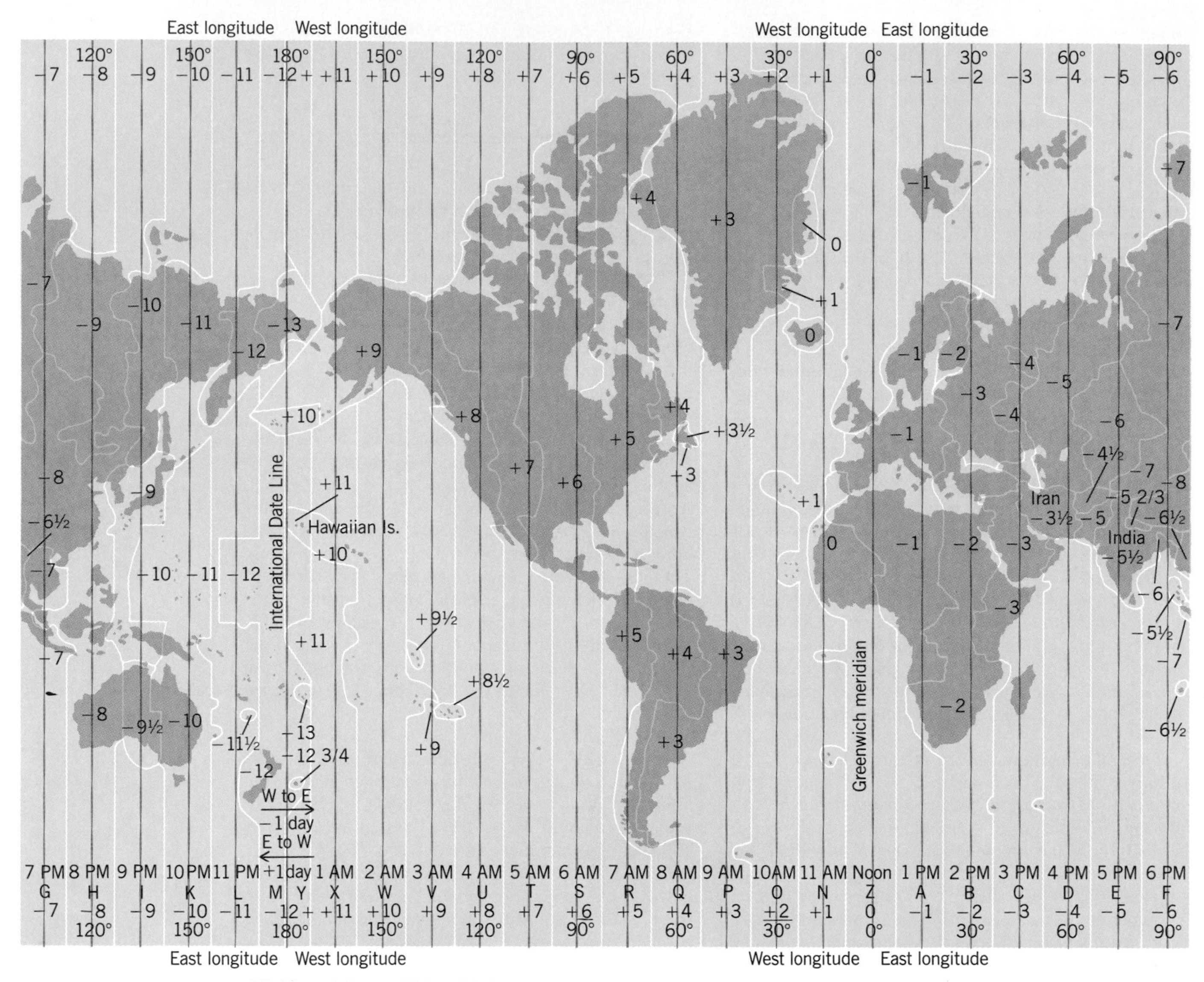

Division of the world into 24 time zones, progressively differing from Greenwich by 1 h. Some countries use half-hour intervals or fractional hours. Numerical designations indicate number of hours by which zone time must be increased or decreased to obtain Coordinated Universal Time. Longitudes of standard meridians, letter designations, and the times in the zones when it is noon at Greenwich are also shown. (*After J. M. Pasachoff, Contemporary Astronomy, 3d ed., 1985*)

Earth in order to obtain UT1. The effect of polar motion may amount to several hundredths of a second. The Bureau International de l'Heure (BIH) in Paris, France, receives these data and maintains a UT1 time scale. *See* EARTH ROTATION AND ORBITAL MOTION.

Because the Earth has a nonuniform rate of rotation and since a uniform time scale is required for many timing applications, a different definition of a second was adopted in 1967. The international agreement calls for the second to be defined as 9,192,631,770 periods of the radiation derived from an energy-level transition in the cesium atom. This second is referred to as the international or SI (International System) second and is independent of ongoing astronomical observations. International Atomic Time (TAI) is maintained by the BIH from data contributed by time-keeping laboratories around the world. *See* ATOMIC TIME.

Coordinated Universal Time (UTC) uses the atomic second as its time base. However, the designation of the epoch may be changed at certain times so that UTC does not differ from UT1 by more than 0.9 s. UTC forms the basis for civil time in most countries and may sometimes be referred to as Greenwich mean time. The adjustments to UTC to bring this time scale into closer accord with UT1 consist of the insertion or deletion of integral seconds. These "leap seconds" may be applied at 23 h 59 m 59 s of June 30 or December 31 of each year according to decisions made by the BIH. UTC differs from TAI by an integral number of atomic seconds.

Dynamical time. Dynamical time is based on the orbital motion of the Sun, Moon, and planets. It is the time inferred in the ephemerides of the positions of these objects, and from its inception in 1952 until 1984 was referred to as ephemeris time. Barycentric Dynamical Time (TDB) refers to ephemerides which have been computed by using the barycenter of the solar system as a reference. Terrestrial Dynamical Time (TDT) is the practical realization of dynamical time and is defined as being equal to TAI + 32.184 seconds. *SEE DYNAMICAL TIME.*

Civil and standard times. Because rotational time scales are defined as hour angles, at any instant they vary from place to place on the Earth. When the fictitious point is directly over the meridian of Greenwich, the mean solar time is 12 noon at Greenwich. At that instant the mean solar time for all places west of Greenwich is earlier than noon and for all places east of Greenwich later than noon, the difference being 1 hour for each 15° of longitude. Thus, at the same instant at short distances east of the 180th meridian the mean solar time is 12:01 A.M. and at a short distance west of the same meridian it is 11:59 P.M. of the same day. Thus persons traveling westward around the Earth must advance their time 1 day, and those traveling eastward must retard their time 1 day in order to be in agreement with their neighbors when they return home. The International Date Line is the name given to a line where the change of date is made. It follows approximately the 180th meridian but avoids inhabited land. To avoid the inconvenience of the continuous change of mean solar time with longitude, zone time or civil time is generally used. The Earth is divided into 24 time zones, each approximately 15° wide and centered on standard longitudes of 0°, 15°, 30°, and so on (see **illus.**). Within each of these zones the time kept is the mean solar time of the standard meridian. *SEE INTERNATIONAL DATE LINE.*

Zone time is reckoned from 0 to 24 hours for most official purposes, the time in hours and minutes being expressed by a four-figure group followed by the zone designation. For example, "1009 zone plus five" refers to the zone 75° west of Greenwich, where zone time must be increased by 5 hours to obtain UTC. The various zones are sometimes designated by letters, especially the Greenwich zone which is Z, "1509 Z" meaning 1509 UTC. The zone centered on the 180th meridian is divided into two parts, the one east of the date line being designated plus 12 and the other minus 12. The time July 2, 2400 is identical with July 3, 0000.

In civil life the designations A.M. and P.M. are often used, usually with punctuation between hours and minutes. Thus 1009 may be written as 10:09 A.M. and 1509 as 3:09 P.M. The designations for noon and midnight, however, are often confused, and it is better to write 12:00 noon and July 2–3, 12:00 midnight, in order to avoid ambiguity. In some occupations where time is of special importance, there is a rule against using 12:00 at all, 11:59 or 12:01 being substituted. The time 1 minute after midnight is 12:01 A.M. and 1 minute after noon is 12:01 P.M.

The illustration shows the designations of the various time zones, the longitudes of the standard meridians, and the letter designations and the times in the various zones when it is noon at Greenwich. In the United States the boundaries of the time zones are fixed by the Department of Transportation. Frequently the actual boundaries depart considerably from the meridians exactly midway between the standard meridians. Ships at sea and transoceanic planes use UTC for navigation and communication, but for regulating daily activities on board they use any convenient approximation to zone time, avoiding frequent changes during daylight hours.

Many countries, including the United States, advance their time 1 hour, particularly during the summer months, into "daylight saving time." For example, 6 A.M. is redesignated as 7 A.M. Such a practice effectively transfers an hour of little-used early morning light to the evening.

Time scales are coordinated internationally by the BIH. Most countries maintain local time standards to provide accurate time within their borders by radio, telephone, and TV services. These national time scales are often intercompared by using radio time signals, portable clocks, or time signals transferred by artificial Earth satellites or radio interferometry.

Dennis D. McCarthy

Transit

The apparent passage of a planet across the surface of the Sun, of a satellite across the surface of the parent planet, or of a star, planet, or reference point across an adopted line of reference.

Only Mercury and Venus are seen in transit across the surface of the Sun, because they are the only planets that orbit inside the path of the Earth. *SEE PLANET.*

Transits of Venus. These are very rare. The Earth must be essentially in a straight line with Venus and the Sun, therefore Venus must be in inferior conjunction at the same time that it passes one of the nodes of its orbit. Conjunction must occur within 2 days of June 7 and December 9 to fulfill the conditions. The last transit of Venus took place in 1882. There will not be another one until 2004. *SEE VENUS.*

Transits of Mercury. These are relatively more frequent and occur at the rate of about 13 per century. The same geometric conditions are required as for transits of Venus, but the limits are not as narrow. Conjunction must occur within 3 days of May 8 or within 5 days of November 10 (see **illus.**). Because of this, November transits are about twice as frequent as May transits. Observations of November transits yield more accurate results than those of May transits, because in November the motion of the planet is more rapid, thus permitting better timing of the contacts.

Four contacts are observed: exterior ingress, inte-

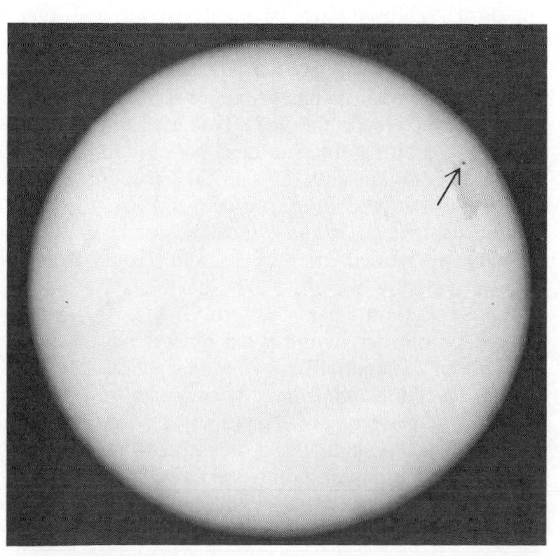

Photograph of transit of Mercury (arrow) on November 14, 1953, taken at Washington, D.C. (*Official U.S. Navy Photograph*)

rior ingress, interior egress, and exterior egress, designating respectively the exterior and interior points of tangency between the planet and the Sun at the beginning of the transit (ingress) and the interior and exterior points of tangency at the end (egress).

Transits of Mercury are observed for the purpose of determining the exact position of the planet and to improve data on the elements of the orbit. *See Mercury.*

Transits of Jupiter's satellites. Transits of the galilean satellites of Jupiter occur at each of their inferior conjunctions with the exception of satellite IV, which occasionally passes clear of the planet's disk. They are difficult to observe and are used mainly to estimate the albedo (reflectivity) of the satellites relative to that of Jupiter. As each satellite passes in front of the planet, it casts its shadow on the planet's disk and causes the phenomenon of shadow-transit. *See Jupiter.*

Transits of stars. Passages of stars across the local meridian are observed extensively for solving problems of fundamental positional astronomy, timekeeping, and navigation. At the precise instant of transit of a star across the local meridian, the local sidereal time is exactly equal to the star's right ascension, and the latitude of the observer is equal to the sum of the star's declination and its zenith distance.

Furthermore, the difference between the local sidereal time and the Greenwich sidereal time gives a measure of the longitude of the observer. Any of those quantities may be treated as the unknown to be determined by observations of meridian transits. *See Astronomical coordinate systems; Eclipse.*

Simone D. Gossner

Trojan asteroids

Asteroids located near the equilateral lagrangian stability points of the Sun-Jupiter system (see **illus.**). As shown by J. L. Lagrange in 1772, these are two of the five stable points in the circular, restricted, three-body system, the other three points being located along a line through the two most massive bodies in the system. In 1906 Max Wolf discovered an asteroid located near the lagrangian point following Jupiter in its orbit. Within a year, two more were found, one of which was located near the preceding lagrangian point. It was quickly decided to name these asteroids after participants in the Trojan War as given in Homer's *Iliad.* Hence the term Trojan asteroid refers to asteroids orbiting the Sun near one of Jupiter's equilateral lagrangian points. With the exception of (624) Hektor in the preceding "cloud" and (617) Patroclus in the following "cloud," asteroids in each of these clouds are named after Greek and Trojan warriors, respectively.

One sometimes uses the term Trojans in a generic sense to refer to hypothetical objects occupying the equilateral lagrangian points of other pairs of bodies. Unsuccessful searches have been made for Trojans of the Earth, Saturn, and Neptune, as well as for the Earth-Moon system. It is doubtful whether bodies near these lagrangian points would be in stable orbits, because of perturbations by the major planets. Indeed most of Jupiter's Trojans do not move in the plane of its orbit, but in orbits inclined by as much as 25° and at longitudes differing by up to 40° from the longitudes of the theoretical lagrangian points.

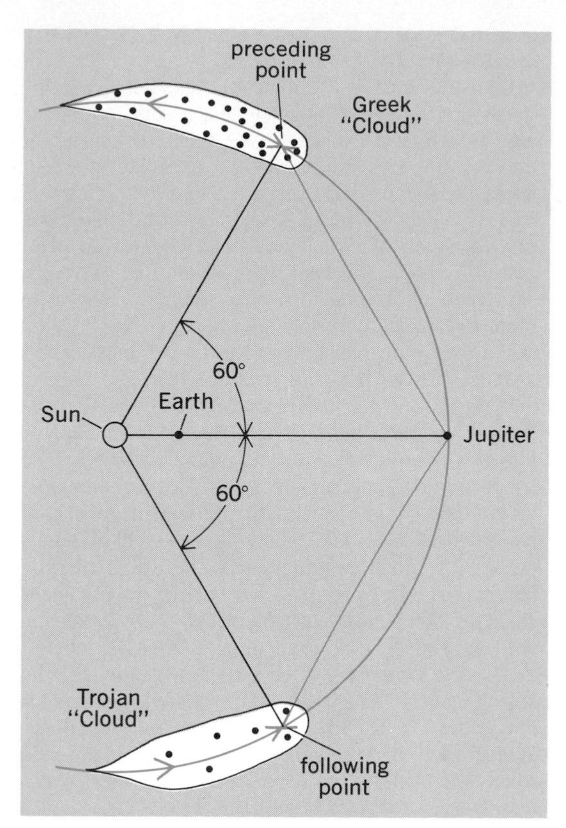

Lagrangian points and Trojan asteroids.

There are at least 25 numbered Trojan asteroids, about a dozen of which have diameters greater than 60 mi (100 km). However, photographic surveys have shown that there are about 900 with diameters exceeding 9 mi (15 km). Approximately 700 of these are located around the preceding point, while only about 200 are found near the following point.

Trojans are dark objects reflecting only between 2 to 5% of the visual light they receive. The majority are compositionally similar to the most common type of outer main-belt asteroid, but some, perhaps as many as one-third, have no known analog among the asteroids or meteorites. *See Asteroid; Meteorite.*

Edward F. Tedesco

Bibliography. R. P. Binzel, T. Gehrels, and M. S. Matthews (eds.), *Asteroids II,* 1990; T. Gehrels (ed.), *Asteroids,* 1979; C. T. Kowal, *Asteroids: Their Nature and Utilization,* 1988.

Twilight

The period between sunset and darkness in the evening and darkness and sunrise in the morning. The following statements apply to evening twilight; the reverse would apply to morning twilight.

The characteristic light is caused by atmospheric scattering, which transmits sunlight to the observer for some time after the Sun has set. It depends geometrically on latitude, longitude, and elevation of the observer, and on the time of the year. Physically it depends also on local conditions, particularly the weather.

Three degrees of twilight are conventionally distinguished. Civil twilight ends when the center of the

Sun is 6° below the horizon; if the sky is clear, it is usually practicable to carry on ordinary outdoor occupations without artificial light during civil twilight. Nautical twilight ends when the depression of the Sun is 12°; at this time both the horizon and the brighter stars are visible. Astronomical twilight ends when the depression of the Sun is 18°; at this time no trace of illumination by the Sun appears in the sky. As thus defined, the times of ending of the three sorts of twilight can be precisely calculated.

Gerald M. Clemence

Bibliography. G. V. Rozenberg, *Twilight*, 1966; U.S. Nautical Almanac Office and Gale Research Co., *Sunrise and Sunset Table for Key Cities and Weather Stations in the United States*, 1977; U.S. Naval Observatory, *Tables of Sunrise, Sunset, and Twilight*, supplement to *American Ephemeris*, 1946, reprint 1962.

Twinkling stars

A phenomenon by which light from the stars, as it passes through fluctuations in the Earth's atmosphere, is rapidly modulated and redirected to make the starlight appear to flicker. Although it is familiar to those who have looked with the unaided eye at the night sky, the twinkling phenomenon affects all wavelengths that manage to penetrate the Earth's atmosphere, from the visible to the radio wavelengths. At visible wavelengths, atmospheric fluctuations are caused predominantly by temperature irregularities along the line of sight. Minor contributions are made by irregularities in atmospheric density and in water vapor content. All such irregularities introduce slight changes in the index of refraction of air, and these changes affect light waves in two ways: they modulate the intensity of the light, and they deflect the light waves in one direction and then another. An analogous phenomenon is often observed when light grazes across the surface of a hot highway: light is bent and distorted by pockets of hot air rising over the pavement's surface. At radio wavelengths, electron density irregularities in the ionosphere modulate and redirect radio waves.

Unaided-eye observations. High-altitude winds and local air currents carry atmospheric irregularities across the line of sight. Because the entrance pupil of the human eye is much smaller than the characteristic size of the atmospheric irregularities, any twinkling that is visible to the unaided eye is caused primarily by the modulation of the light intensity and not by the deflection of the light waves. Modulation is best understood as an interference phenomenon acting on adjacent light waves. Like random waves in water, sometimes the waves add constructively, causing the star to brighten, and other times the waves add destructively, causing the star to dim.

While stars twinkle, planets generally do not. This difference occurs because stars are essentially infinitesimally small points of light. All light from a star travels along the same path through the atmosphere, and all the light from a star is modulated simultaneously. Planets have a definite angular extension. Light from one side of the planet traverses one path through the atmosphere, while light from the other side of the planet traverses another path through the atmosphere. Along each path the modulation of light is different, so that when all the light is added together by the human eye the total modulation or twinkling is averaged out. SEE PLANET; STAR.

Telescope observations. The twinkling phenomenon is of utmost interest to astronomers who view the skies from ground-based telescopes. While modulation variations are present, it is the deflection of light that causes the most serious problems. The entrance pupil of an optical telescope is often much larger than the characteristic size of atmospheric irregularities. The composite star image produced by a large telescope is a blurry circle that results when the randomly deflected light waves are added together in an extended time exposure. To diminish atmospheric effects, telescopes are built on high mountains, and are placed at least 30–45 m (100–150 ft) above the ground. The best observatories are those where the twinkling phenomenon is minimized. Examples include the Mauna Kea Observatory in Hawaii and observatories in the foothills of the Andes Mountains in northern Chile. SEE ASTRONOMICAL OBSERVATORY; OPTICAL TELESCOPE.

To completely remove the twinkling effects of the atmosphere, there are two alternatives. The first is to place a telescope in orbit above the atmosphere, as with the Hubble Space Telescope. The second alternative is to monitor the random deflections of the atmosphere and, within the telescope, to bend the deflected light back onto its original path. This optical technique is given the name adaptive optics. SEE ADAPTIVE OPTICS; SATELLITE ASTRONOMY.

Radio sources. Small radio sources have been found to vary rapidly in brightness because of the flow of density fluctuations in the solar wind across the path between the radio source and the Earth. The method of interplanetary scintillations has proved a useful means of probing the interplanetary medium near the Sun and also of establishing the very small size of some of the radio sources. Radio sources whose angular diameters are large do not show interplanetary scintillation at all, just as planets do not share in the stellar twinkling. Pulsars show erratic changes of amplitude in their radio emission, caused by radio scintillation both near the neutron star and in interstellar space. SEE PULSAR; RADIO ASTRONOMY; SOLAR WIND.

Laird A. Thompson

Bibliography. H. G. Booker and W. E. Gordon, A theory of radio scattering in the troposphere, *Proc. Inst. Radio Eng.*, 38:401–412, 1950; B. J. Rickett, Interstellar scattering and scintillation of radio waves, *Annu. Rev. Astron. Astrophys.*, 15:479–504, 1977; N. J. Woolf, High resolution imaging from the ground, *Annu. Rev. Astron. Astrophys.*, 19:367–398, 1981; A. T. Young, Seeing: Its cause and cure, *Astrophys. J.*, 189:587–604, 1974.

Ultraviolet astronomy

Astronomical observations carried out in the region of the electromagnetic spectrum with wavelengths from approximately 10 to 350 nanometers. The ultraviolet spectrum is divided into the extreme-ultraviolet (EUV; 10–90 nm), far-ultraviolet (FUV; 90–200 nm), and near-ultraviolet (near-UV; 200–350 nm). Ultraviolet radiation from astronomical sources contains important diagnostic information about the composition and physical conditions of these objects. This information includes atomic absorption and emission lines of all the most abundant elements in many states of ionization. The hydrogen molecule (H_2), the most abundant molecule in the universe, has its absorption and emission lines in the far-ultraviolet. Thus, ultraviolet observations make it possible to probe a very wide range of physical conditions of matter in the universe, from the very cold gas in dense interstellar regions with temperatures of perhaps 30 K (−406°F) to the hot gas found in supernovae remnants and in the coronas of stars with temperatures approaching 10^7 K. *SEE ASTRONOMICAL SPECTROSCOPY; ULTRAVIOLET RADIATION.*

Observations. Ultraviolet radiation with wavelengths less than 310 nm is strongly absorbed by molecules in the atmosphere of the Earth. Therefore, ultraviolet observations must be carried out by using instrumentation situated above the atmosphere. Ultraviolet astronomy began with instrumentation at high altitudes aboard sounding rockets for brief glimpses of the Sun and stars. The first major ultraviolet satellite observatories to be placed in space were the United States *Orbiting Astronomical Observatories* (*OAO*s). *OAO 2* operated from 1968 to 1972 and provided the first full survey of the many kinds of ultraviolet sources in the sky, while *OAO 3* (*Copernicus*) operated from 1972 to 1980 and obtained high-resolution spectra of bright ultraviolet-emitting stars in order to probe the composition and physical state of intervening interstellar gas and to study the stellar winds of hot stars. Also, a number of smaller satellites, including the European *TD 1* and the Dutch *ANS,* provided very important survey measurements on the ultraviolet brightnesses of astronomical sources. *SEE ROCKET ASTRONOMY.*

With the launch of the *International Ultraviolet Explorer* (*IUE*) into a geosynchronous orbit on January 26, 1978, the full potential of ultraviolet astronomy to probe a wide range of scientific problems became a reality. The *IUE* satellite is a collaborative project of the U.S. National Aeronautics and Space Administration (NASA), the European Space Agency, and the United Kingdom Science and Engineering Research Council. It consists of a reflecting telescope of modest size (18 in. or 45 cm in diameter) followed by several spectrographs with ultraviolet-sensitive television cameras that produce ultraviolet spectra over the wavelength region from 120 to 320 nm. Between 1978 and 1990 the *IUE* obtained approximately 50,000 ultraviolet spectra of a wide range of astronomical objects, including comets and planets, cool and hot stars in the Milky Way Galaxy, and external galaxies and quasars. Research with the *IUE* is generally conducted at either of two ground-control centers, and observations are obtained through real-time communication with the remotely controlled satellite.

Discoveries. The important discoveries of ultraviolet astronomy span all areas of modern astronomy and astrophysics. Some of the notable discoveries in the area of solar system astronomy include new information on the upper atmospheres of the planets, including planetary aurorae and the discovery of the enormous hydrogen halos surrounding comets. In studies of the interstellar medium, ultraviolet astronomy has provided fundamental information about the molecular hydrogen content of cold interstellar clouds along with the discovery of the hot phase of the interstellar medium, which is created by the supernova explosions of stars. In stellar astronomy, ultraviolet measurements led to important insights about the processes of mass loss through stellar winds and have permitted comprehensive studies of the conditions in the outer chromospheric and coronal layers of cool stars. In galactic and extragalactic astronomy, two significant results are the discovery of the hot gaseous halo of the Milky Way Galaxy and new insights about the mysterious energetic sources at the centers of active galaxies and quasars. *SEE COMET; GALAXY, EXTERNAL; INTERSTELLAR MATTER; MILKY WAY GALAXY; PLANETARY PHYSICS; QUASAR; SUPERNOVA.*

Prospects. Ultraviolet astronomy appears to have a very bright future with the advent of a number of powerful new space observatories, including the Hubble Space Telescope, the *Extreme Ultraviolet Explorer (EUVE)* satellite, and the *Lyman/Far-Ultraviolet Explorer* (*FUSE*) satellite (scheduled for launch in 1997). In addition, a number of smaller telescopes are being placed above the atmosphere for short missions aboard the NASA space shuttle as part of the Astro Program, which had its first flight in 1990, and as part of various international collaborative ventures.

Although it had an inauspicious beginning, the Hubble Space Telescope, an observatory designed for visible and ultraviolet measurements (see **illus.**), is expected to become the centerpiece of ultraviolet astronomy for the 1990s. Unfortunately, the 94-in. (2.4-m) primary mirror of the Hubble Space Telescope was ground incorrectly; and the images produced by the observatory exhibit spherical aberration. This causes the pictures of celestial objects taken by the observatory to be blurry. It is expected that the imaging problem will be fixed in 1993 by replacing one of the scientific instruments with an instrument containing corrective optical elements. The observatory should then be capable of obtaining an angular resolution of better than 0.1 second of arc at visible and, hopefully, ultraviolet wavelengths. The combination of large collecting area and high angular resolution will allow the Hubble Space Telescope to image objects about 40 times fainter than is possible from the ground. The angular resolution of the corrected Hubble Space Telescope will be 10–20 times better than can typically be recorded from the ground. This will permit the resolution of fine details in such extended sources as planets, star clusters, nebulae, and galaxies. The telescope is expected to operate for more than 10 years. Its initial complement of instruments includes two imaging cameras, two spectrographs, and a high-speed photometer. Each instrument has substantial capability at ultraviolet wave-

Hubble Space Telescope during orbital deployment by the space shuttle *Discovery* in 1989. *(NASA)*

lengths down to approximately 120 nm. Before its imaging capabilities are improved, it is expected that many of the scientific studies to be pursued with the Hubble Space Telescope will involve ultraviolet spectroscopy. In addition, image computer-processing techniques have proven valuable in improving the angular resolution in the degraded images of those objects containing high-contrast small-scale structures. SEE ABERRATION; OPTICAL TELESCOPE.

The *EUVE* and *FUSE* satellites will exploit the information found in the region of the electromagnetic spectrum from 8 to 120 nm. Both missions are part of the NASA Explorer program, which provides modest-sized satellites for special research initiatives. The *EUVE* mission will produce ultraviolet brightness measurements and spectra of sources in the wavelength range from 8 to 80 nm, and the *FUSE* satellite will, in addition, have a major capability in the 90- to 120-nm region. The *EUVE* and *FUSE* telescopes have diameters of 20 and 28 in. (0.5 and 0.7 m), respectively. Both facilities will be used to study a wide range of astronomical problems that cannot be pursued with the Hubble Space Telescope. SEE SATELLITE ASTRONOMY.

Blair D. Savage

Bibliography. J. N. Bahcall and L. Spitzer, Jr., The space telescope, *Sci. Amer.*, 247(1):40–51, July 1982; Y. Kondo, A. Boggess, and S. P. Maran, Astrophysical contribution of the IUE, *Annu. Rev. Astron. Astrophys.*, 27:397–420, 1989; L. A. Shore, IUE: Nine years of astronomy, *Astronomy*, 15(4):14–22, April 1987.

Universe

The totality of matter and energy that exists and the space that contains it. The modern concept of the universe and its scale is quite different from that held a few hundred years ago, when it was believed that the Earth was in the center with the Sun, Moon, stars, and planets revolving around it. In the sixteenth century, when Copernicus theorized that the Sun, not the Earth, was at the center of planetary orbits, it was still believed that this solar system was at the center of the universe and that the solar system, in fact, constituted the universe. When it was recognized in the nineteenth century that the Sun is just one of an enormous number of stars in a great, disk-shaped system called the Milky Way Galaxy, it was still believed that the Sun was near the center of that system. But early in the twentieth century, H. Shapley showed that the Sun and Earth are located not at the center, but more than two-thirds of the way out toward the edge, of the visible Milky Way Galaxy. In 1924 E. Hubble proved that the Milky Way is only one of millions of galaxies in the sea of space; later observations showed that the universe has no center.

Distance and time. The two quantities distance and time merge in their meanings when the universe is discussed. A critical concept is that light has a finite speed. Though much faster than speeds ordinarily encountered, light, traveling at approximately 186,000 mi/s (300,000 km/s), still requires billions of years to traverse the universe. This light travel time must be taken into account when interpreting light received from distant objects. For example, the Moon is one-half a light-second from Earth, the Sun 8 light-minutes away. A light-year is defined as the distance that light travels in 1 year, the equivalent of 6×10^{12} mi (9.5×10^{12} km). Most celestial objects in the universe change exceedingly slowly, and millions to billions of years must pass before any difference can be noticed. For a relatively nearby galaxy, therefore, the light received now can be interpreted as representing the current state of that object. However, for a very distant galaxy billions of light-years away, the light detected now has taken so long to reach the Earth that the galaxy has had time to change appreciably, and might look very different if its light could be seen immediately. Thus, merely observing light from very distant objects makes it possible to look into the past, to see conditions as they were long ago, not as they are now. The more distant the object, the farther back in time its light originated, and the earlier it is possible to look into the universe's history. SEE LIGHT-YEAR.

Structure. All discussions about the properties of the universe are based on a fundamental assumption without which astronomers are powerless to come to any conclusions about the nature of the universe. This assumption, the cosmological principle, states that the same laws of physics operate throughout the universe, and that the Earth, Sun, and Milky Way Galaxy are not in any privileged place in the universe. It is assumed that, on a sufficiently large scale, all observations are independent of location, so that an observer in a remote galaxy would conclude the same things about the universe as a whole that astronomers on Earth do.

The hierarchy of structure in the universe proceeds from stars (around which there may be planets), to galaxies, to clusters of galaxies, to clusters of clusters (called superclusters), and perhaps beyond, to clusters of superclusters. Each galaxy contains from 10^6 to 10^{13} stars, each cluster anywhere from 10 to 10,000 galaxies. The Milky Way Galaxy contains roughly 10^{11} stars, and is a member of a rather small cluster called the Local Group. The Local Group numbers only about two dozen galaxies, including the great

Fig. 1. The spiral galaxy nearest to Earth, M31 (NGC 224), in Andromeda. Also shown are two of its dwarf elliptical companions, NGC 205 (upper left) and 221 (middle right). *(California Institute of Technology/Palomar Observatory)*

spiral galaxy in Andromeda, some 2×10^6 light-years away (**Fig. 1**). *See Galaxy, external; Local Group; Milky Way Galaxy.*

The universe is composed overwhelmingly of hydrogen, the simplest atom, containing a single proton and an electron. Ninety of every hundred atoms are hydrogen. Nine are helium. The last 1% represents all the rest of the elements, including carbon, nitrogen, oxygen, and iron, essential to life on Earth.

Planets. The Sun is the only star known to possess a planetary system; however, there are tantalizing hints of orbiting matter or solid bodies around other stars. For example, there are several stars relatively near the Sun whose motions through space may indicate gravitationally the presence of planetary-size companions. Observations of the star Beta Pictoris indicate a cloud of material in orbit around it, perhaps a solar system in formation (**Fig. 2**). There may be millions of planetary systems throughout the Milky Way Galaxy.

The solar system contains a great variety of objects. In addition to the nine known planets orbiting the Sun, there are a multitude of planetary satellites, including Phobos and Deimos, the tiny, irregularly shaped moons of Mars; Jupiter's moon Io, on which active volcanoes have been discovered; Saturn's Titan, large enough to retain its own atmosphere; and the surprising Miranda (**Fig. 3**), orbiting the planet Uranus. Between the orbits of Mars and Jupiter are found the asteroids, or minor planets. Beyond Pluto is the realm of the comets, chunks of rock and frozen gas whose occasional excursions into the inner solar system provide some spectacular sights. *See Asteroid; Comet; Jupiter; Mars; Satellite; Saturn; Uranus.*

The solar system is immense: the distance from the Sun to Pluto is roughly 4×10^9 mi (6×10^9 km), corresponding to 6 light-hours. The radius of the comet cloud is nearly a light-year; yet this vast extent is only about one-quarter of the distance to even the nearest star. *See Planet; Solar system.*

Stars. Stars are gaseous spheres shining because of the energy produced deep in their interiors. This energy comes from nuclear fusion. Most stars are fusing hydrogen to helium; the Sun is a typical example. Stars range in mass from about one-tenth to more than 100 times the mass of the Sun (4×10^{30} lb or 2×10^{30} kg). The length of time that a star can maintain hydrogen fusion depends on its mass. The Sun is about 5×10^9 years old, and can last for another 5×10^9 years. Paradoxically, more massive stars last for a shorter time, since they are also brighter and consume their fuel more quickly, some in as little as 10^7 years. Less massive stars can survive for over 10^{12} years, longer than the age of the universe. *See Stellar evolution.*

After a star's supply of hydrogen is exhausted, drastic changes take place. A typical star, having less than a few times the mass of the Sun, will become a cool (3000–4000 K or 5000–7000°F), swollen giant with a radius hundreds of times larger than the current Sun (4×10^5 mi or 7×10^5 km). When the Sun reaches such a stage, it will engulf the four inner planets in its bloated atmosphere. The next step in evolution is the loss of the star's outer layers and exposure of the small, hot, dense core. A star in this phase is called a white dwarf and is only about the size of the Earth, or 1% of the Sun's current radius. A white dwarf glows with stored heat and has no further energy sources. It is doomed to fade slowly into oblivion unless it is in a binary system and can gain enough mass from a companion to explode as a supernova. *See Giant star; Supernova; White dwarf star.*

For a star more than several times as massive as the Sun, the evolution is more dramatic. After the hydrogen in the core is exhausted, the star swells not to giant but to supergiant proportions, thousands of times the Sun's present size. Then, instead of a relatively calm loss of the outer layers, they are violently ejected in another type of cataclysmic supernova explosion. The energy produced in such an explosion can, for a few weeks, rival that produced by an entire galaxy. The remnant core left behind by such a supernova is even smaller, hotter, and denser than a white dwarf. The radius is only a few tens of miles, and the density is so high that all of the protons and electrons in the atoms of the core have merged to form neutrons. This object is called a neutron star, and is likely to be rotating very rapidly, many times per second. These rotating neutron stars can be de-

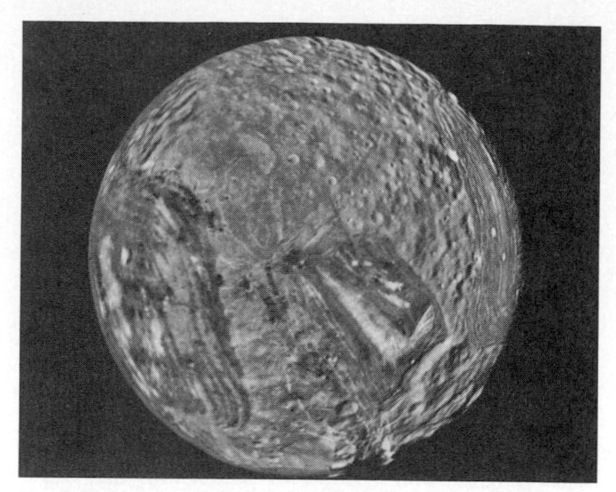

Fig. 3. Uranus's satellite Miranda, as photographed by *Voyager 2* in 1986. Along with the expected craters on its surface are surprising grooves and angular features. (*NASA/Jet Propulsion Laboratory*)

Fig. 2. Photograph of the region around the star Beta Pictoris, taken with a ground-based telescope, and using an occulting disk to block the light from the star. A surrounding disk of material can be seen, perhaps a forming solar system. (*Photograph by J. Gradie, Hawaii Institute of Geophysics; courtesy of B. Zuckerman*)

tected as radio pulsars. *See Neutron star; Pulsar; Supergiant star; Supernova.*

Most stars are not alone in space; binary and multiple star systems are extremely common, and many may have planets associated with them. Stars are separated by distances that average several light-years. The nearest star to the Sun, Proxima Centauri, is a member of a triple system located 4.3 light-years away. Within 17 light-years of the Sun there are fewer than 50 stars. *See Binary star; Star.*

Galaxies. Galaxies, composed of stars and interstellar gas and dust, are the fundamental large-scale building blocks of the universe. Galaxies come in several types whose present structures reflect the initial conditions under which they were formed. One important quantity is the amount of angular momentum per unit mass possessed by the protogalactic cloud; the more angular momentum, the greater the current rotation and flattening of the galaxy.

Spirals. Many galaxies are spirals, like the Milky Way, whose flattened disk does indeed display conspicuous rotation about the galactic center (Fig. 1). This rotation is differential, with material farther from the center revolving more slowly than matter closer in. Embedded within the disk are the spiral arms that contain young, hot stars and large amounts of gas and dust and that are sites of ongoing star formation. Surrounding and pervading this disk is a spheroidal component, composed of older, less massive stars, stars almost as old as the galaxy itself. Some spiral galaxies have arms that emanate not directly from the center but from a bar-shaped structure of stars and gas that passes through the center; these are called barred spiral galaxies. Both kinds of spirals typically contain between 10^9 and 10^{12} times the mass of the Sun. They range in radius from 10,000 to several times 100,000 light years.

The Milky Way Galaxy itself had been thought to contain roughly 2×10^{11} times the Sun's mass. Studies of interstellar molecular clouds have led to an upward revision of the amount of material in the Galaxy's halo; the total mass may be in excess of 10^{12} times the Sun's mass. The Sun is situated on the inside edge of a spiral arm approximately 30,000 light-years from the center of the Galaxy; the entire disk has a diameter of at least 100,000 light-years. At the Sun's distance from the galactic center, it takes 2×10^8 years for a complete revolution. In its lifetime of 5×10^9 years, therefore, the Sun has made 25 trips around the center of the Milky Way.

Ellipticals. There are two other main classes of galaxies: ellipticals and irregulars. Elliptical galaxies are ellipsoidal, lacking any well-defined plane or rotation axis. They contain only a spheroidal (or nearly so) distribution of stars. Interstellar gas and dust are generally not found in abundance in elliptical galaxies, an indication that early star formation was quite efficient. Ellipticals constitute the majority of galaxies, and they exhibit the broadest range in size and mass. The tiny, dwarf ellipticals can contain as little as 10^6 times the Sun's mass, and have radii as small as 1000 light-years. They are quite common; for example, more than half the members of the Local Group are dwarf ellipticals. At the other extreme are the giant elliptical galaxies, with masses up to 10^{13} times the Sun's mass, and radii of 10^6 light-years or more.

Irregulars. Irregular galaxies have no symmetry to their shapes. There is speculation that irregular galaxies were originally formed not as they are now seen but as normal spiral or elliptical galaxies. They may have been acted on by some force (perhaps a tidal force from a massive neighboring galaxy, or a colossal explosion from within) that has distorted their original shapes. In fact, both the Large and Small Magellanic Clouds, small companion galaxies to the Milky Way, were first classified as irregulars. Subsequent studies of the stellar populations and motions within these galaxies have led to their reclassification as barred spirals. Their shapes have been affected by their proximity to the Milky Way.

Active galaxies. The most distant galaxies are also the youngest in the sense that their light began its journey early in the history of the universe. At these early times, galaxies were more active than now. Many of the most distant galaxies possess central regions that are sites of violent, energetic events; the phenomena called quasars are thought to be manifestations of this activity in the nuclei of adolescent galaxies. *See Quasars.*

Expansion. In A. Einstein's general theory of relativity (1915), gravitation produces a universe that is in motion, either expanding or contracting with time. The data then available bore no indication of this, so Einstein reluctantly included in the theory an artificial repulsive force, embodied in a parameter called the cosmological constant, which would support the universe against its own weight. Then, in 1929, following up on earlier work by V. Slipher and others, Hubble studied Doppler shifts of the absorption lines found in the spectra of galaxies. The Doppler shift can be used to calculate radial velocity, the relative movement of one object toward or away from another. For speeds much less than that of light, the fractional change in the wavelength of a spectral line (the so-called redshift) is equal to the radial velocity of the object divided by the speed of light. Based on his observed Doppler shifts, Hubble concluded that the redshifts are proportional to distance. This observation was subsequently interpreted to show that the distant galaxies are all flying away from the Milky Way at speeds proportional to their distances, with the most distant galaxies receding the fastest. This motion was just the kind that Einstein's theory had predicted: the universe was expanding. *See Doppler effect; Redshift; Relativity.*

The relationship is described by the Hubble law,

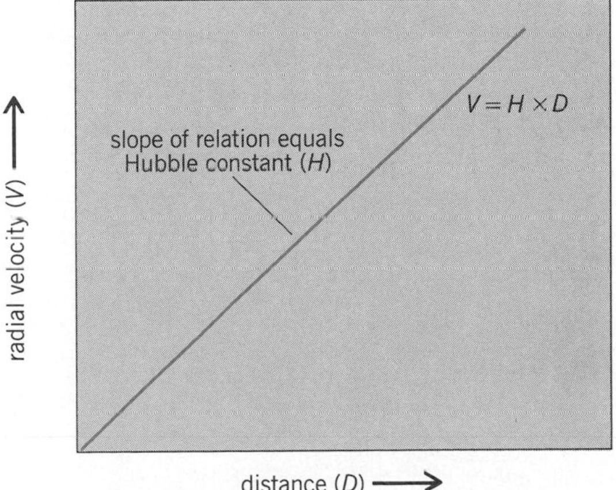

Fig. 4. Graph of the Hubble law. For a distant galaxy, the observed radial velocity *V* is equal to the distance *D* multiplied by the Hubble constant *H*.

$V = H \times D$, where V is the observed radial velocity, H is a constant of proportionality called the Hubble constant, and D is the distance to the object in question (**Fig. 4**). Determination of the Hubble constant is one of the most active areas in observational astronomy; if the value of the Hubble constant is known, it is straightforward to invert the equation and derive distances to faraway galaxies. *SEE HUBBLE CONSTANT.*

The expansion of the universe does not involve motion relative to a center, like the orbit of a planet around the Sun, or of the Sun around the center of the Milky Way Galaxy, or even of the Milky Way around the center of mass of the Local Group. These are all examples of objects moving through space. The expansion of the universe means that space itself is expanding, carrying the galaxies and clusters of galaxies along with it. The observation that distant galaxies are receding would be the same no matter where in the universe it was made; if there are astronomers on a distant planet in a distant galaxy, they will observe all galaxies (including the Milky Way) moving away from them. The proportionality between distance and recession speed means that the universe was smaller in the past, and that objects in it were closer together. Carrying this reasoning further, it can be concluded that all the space in the universe must once have been coincident: the universe must have had a beginning.

The age of the universe can be estimated from the Hubble constant. Extrapolating back in time, all of space would have coincided approximately 2×10^{10} years ago. Actually, since the matter in the universe has been exerting its gravity and slowing the expansion, the universe is somewhat younger than this value. The oldest stars in the Galaxy are close to this age, roughly $13–15 \times 10^9$ years old.

Big bang theory. The theory currently best able to explain the observed properties of the universe is the big bang theory, which suggests that the universe began in a primeval fireball of almost unimaginable temperature and density. Values of 10^{12} K are predicted for the earliest moments of the universe; under these conditions, matter as it is now known could not exist. Yet, a mere 10 s after creation, the temperature was only a few billion kelvins, and ordinary particles such as protons, neutrons, and electrons could survive. Radiation, primarily in the form of gamma rays, was in equilibrium with matter, constantly being absorbed and reradiated; the universe was opaque. Within the first few minutes of the universe's existence, some 20% of the protons (hydrogen nuclei) were converted into helium nuclei. This primordial synthesis of helium is now evident when chemical abundances in stars, nebulae, and galaxies are assessed. *SEE BIG BANG THEORY.*

For the next million years or so, the universe cooled and continued to expand. When the temperature reached a few thousand kelvins, hydrogen atoms could form, and it became too cool for them to remain in equilibrium with the radiation. This decoupling of matter and radiation meant that the universe became transparent, and the radiation could travel freely. The remnant of this radiation is now seen as a whisper of radio emission, characteristic of a temperature of only 3 K. This radiation comes from all directions because it fills the universe, having been produced long ago when the universe was more compact. Further cooling permitted clumps of material to fall together under the influence of gravity, and galaxies could form. *SEE COSMIC BACKGROUND RADIATION.*

Future. The future of the universe depends upon the total amount of matter it contains, or, equivalently, its average density. The initial explosion imparted to the universe a certain amount of kinetic energy; if the gravitational potential energy of all the matter in the universe exceeds this kinetic energy, then the universe has not achieved escape velocity and is said to be closed. The expansion will slow and ultimately reverse itself. Conditions will once again be as they were in the initial explosion as the universe collapses in on itself, perhaps to begin another cycle of expansion. Such a repeating universe is said to be oscillating. If, on the other hand, the universe's gravity is not sufficient to overcome its expansion, then the expansion will go on forever, slowing, to be sure, but continuing. The universe can then be described as open. If the expansion is exactly balanced by gravity, then the universe is on the borderline between open and closed, and is just barely open (a possibility now of interest as a prediction of the inflationary universe theory, discussed below); it will expand forever but will decelerate to zero velocity.

The behavior of the various possible universes can be shown on a graph (**Fig. 5**) in which the ordinate is a scale factor $R(t)$ that can be thought of as indicating relative separations between galaxies, and the abscissa is time: different universes exhibit different behavior of the scale factor as a function of time. If the universe is closed, galaxy separations will reach a maximum and then decrease, while if it is open, galaxy separations will always increase. The figure shows that for a given value of the Hubble constant the age of an open universe is greater than that of a closed universe. The straight line is the extrapolation back in time assuming no deceleration, and hence represents the maximum possible age of the universe.

Current observations indicate that the universe is open and does not possess sufficient mass to overcome its expansion. For the universe to be closed, it would have to contain about 100 times more matter

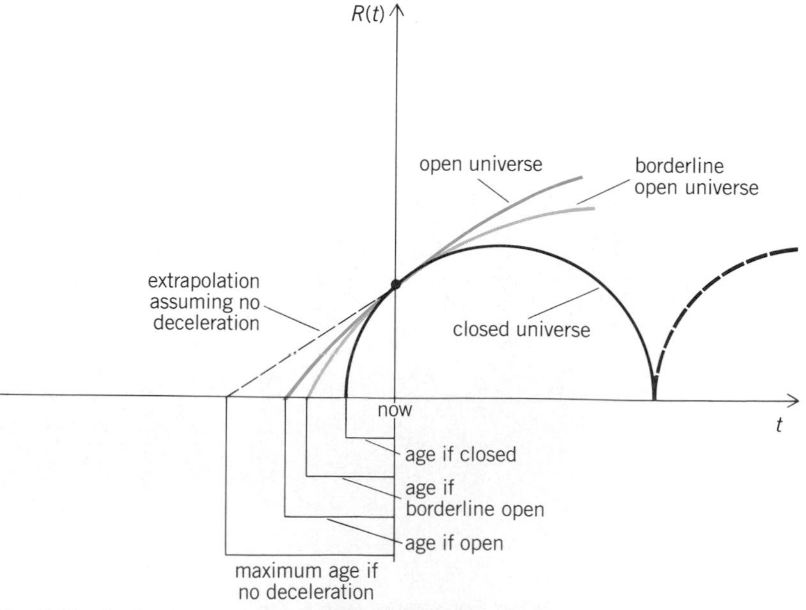

Fig. 5. Graph of scale factor $R(t)$, indicating relative separations between galaxies, versus time t, for various possible universes.

than can be accounted for. There have been suggestions that dark, nonluminous matter in and between galaxies may represent enough mass to close the universe, but this has not been proven. The abundance of deuterium (heavy hydrogen, whose nucleus contains a neutron in addition to a proton) is a sensitive indicator of the density of matter early in the universe's history. Since it is assumed that all of the deuterium currently observed is primordial (that is, created in the big bang), the amount seen now reflects the density of the early universe during the era of nucleosynthesis. The more observed now, the less dense the universe was when deuterium was formed, and the less dense it is now. Current studies show that the deuterium abundance is relatively high, indicating an open universe. Finally, the change in the expansion rate itself can give some clue as to the destiny of the universe: it appears that the actual deceleration has not been as great as it would have been in a closed universe. Therefore, the bulk of the current evidence indicates that the universe is open and will expand forever.

Karen B. Kwitter

Large-scale structure. The large-scale structure of the universe can be visualized as spongelike: galaxies are arrayed in filaments and sheets around astonishingly empty cells. If the galaxies can be used as tracers of the overall distribution of mass in the universe (that is, if large amounts of undetected matter do not exist at significant distances from galaxies), then perhaps as much as 95–98% of the universe is empty. The irregular cells (often called voids) have diameters on the order of 25–50 megaparsecs (1 Mpc equals 3.26×10^6 light-years or 1.9×10^{19} mi or 3.1×10^{19} km), and the interconnected filaments of galaxies have been traced to lengths of up to a few hundred megaparsecs (using a distance scale consistent with a value of the Hubble constant H of 100 km s^{-1} Mpc^{-1}).

Areal distribution of nearby galaxies. Until the mid-1970s, observational studies of the distribution of matter in the universe were primarily confined to the examination of the areal distribution of galaxies over the sky since even relative distances for the vast majority of the objects were lacking. Nonetheless, the filamentary structure is so striking that the nearest such filament was found by W. Herschel in 1784, when he noted that most of the 500 "nebulae" that he had discovered the previous year seemed to be concentrated in a wide belt across the northern sky, oriented roughly perpendicularly to the Milky Way. The Virgo Cluster, found a few years earlier by C. Messier and P. Méchain, formed a dense concentration near the pole of the Milky Way inside Herschel's "stratum" of nebulae. This great band of nebulae was further explored and commented on, but only in the 1950s did G. de Vaucouleurs realize that observers on Earth are seeing from the inside a flattened system of groups and clusters of galaxies, with a few individual objects scattered between. Vaucouleurs called this the Local Supercluster in 1958, though his work of 1953 had called attention to neglected previous studies. SEE VIRGO CLUSTER.

A view of nearby galaxies, namely, the 1200 or so galaxies brighter than apparent photographic magnitude 13.0 cataloged in 1932 by H. Shapley and A. Ames (**Fig. 6**), clearly shows Herschel's "stratum" (the Local Supercluster) as well as several other clumps and filaments. The Milky Way runs around the outside of the figures.

Areal distribution of more distant galaxies. During the 1830s, visual surveys of nebulae were extended to the southern hemisphere by J. Herschel. The two Herschels' surveys are reasonably complete to about the 14th photographic magnitude over the entire sky and

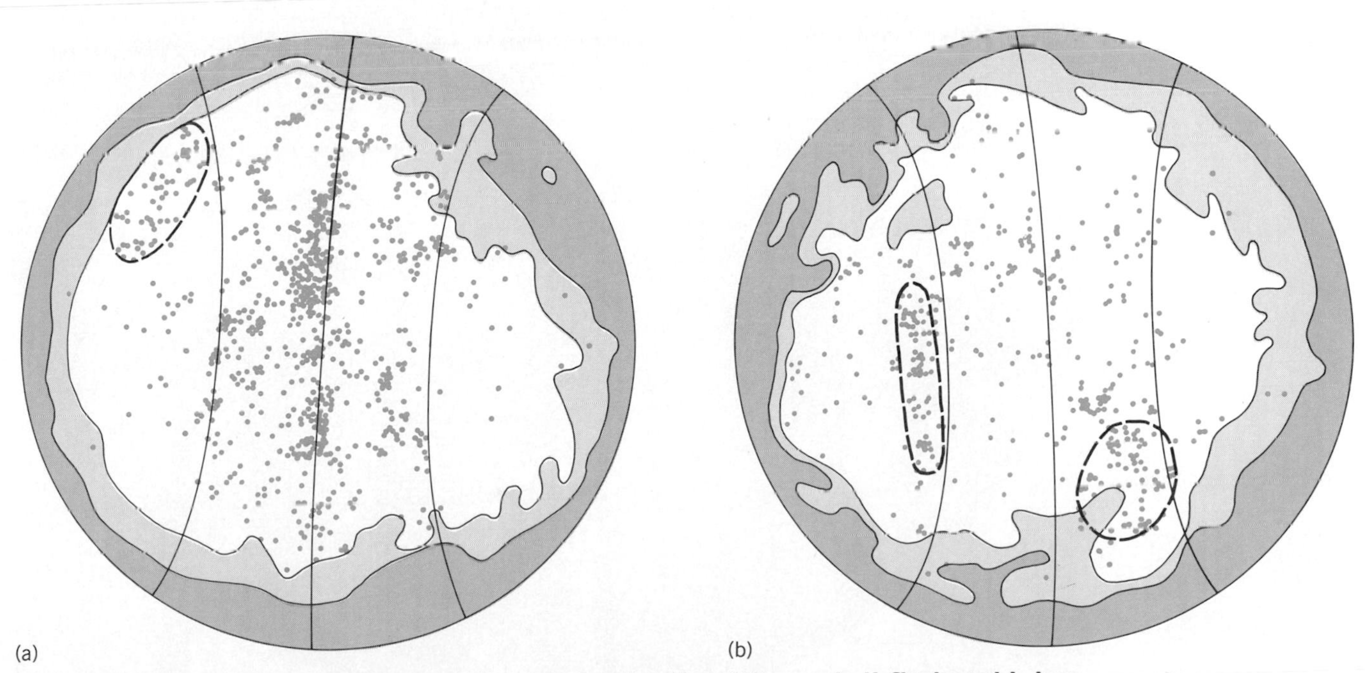

Fig. 6. Galaxies brighter than the 13th photographic magnitude included in a 1932 survey by H. Shapley and A. Ames. (*a*) Northern galactic hemisphere. The Local Supercluster is the broad vertical belt of galaxies. (*b*) Southern galactic hemisphere. The solid lines show the supergalactic equator and the parallels at 30° latitude, and broken lines enclose three other clouds of galaxies probably not associated with the Local Supercluster. The Milky Way runs around the outside of the figures. (*After G. de Vaucouleurs, The Local Supercluster of galaxies, Bull. Astron. Soc. India, 9:1–23, 1981*)

show features in the distribution of galaxies fainter than those in the Shapley-Ames catalog. In particular, a strong filament containing several groups and clusters of galaxies runs from Pisces to the Perseus Cluster at the edge of the Milky Way. Other streams of galaxies are visible, too, particularly in the southern hemisphere.

Just before 1900, it became generally possible to photograph large areas of the sky, and E. Pickering began a photographic survey at the several observing stations of Harvard Observatory. This survey continued for half a century and eventually covered about two-thirds of the sky. Summarizing the main results in 1957, Shapley noted (1) the general smoothness of galactic absorption (as shown by analysis of star counts to faint limits), (2) the tendency of galaxies to cluster, and (3) the tendency of the clusters themselves to be found in "metagalactic clouds." Statistical analysis by B. Bok in 1934 of the distributions of stars and galaxies seen on the Harvard photographs, and of Hubble's galaxy counts made at Mount Wilson, showed that none of these samples of galaxies were distributed at random. Other statistical studies of the distribution of galaxies from about 1930 to about 1960 gave the same result: the distribution of galaxies is nonrandom.

In the 1950s and 1960s, other important surveys of galaxies and clusters of galaxies became available. Particularly valuable have been the galaxy counts to the 19th photographic magnitude made by C. D. Shane and C. Wirtanen at Lick Observatory, covering the northern two-thirds of the sky (**Fig. 7**); and the catalogs of rich galaxy clusters prepared by G. Abell and by F. Zwicky from the Palomar Sky Survey plates. Abell, working with H. G. Corwin and R. P. Olowin, extended his survey to the southern hemisphere, and Zwicky's work also includes an important survey of galaxies in the northern sky that is essentially complete to the 15th photographic magnitude. Shane and Wirtanen counted so many galaxies (over a million) that the numbers had to be represented by successively lighter shades of gray, from black for no galaxies to white for 10 or more galaxies in squares of 10 arc-minutes on a side. The filamentary, sponge-like structure in the galaxy distribution is clearly shown in Fig. 7, as are many dense clusters (and even

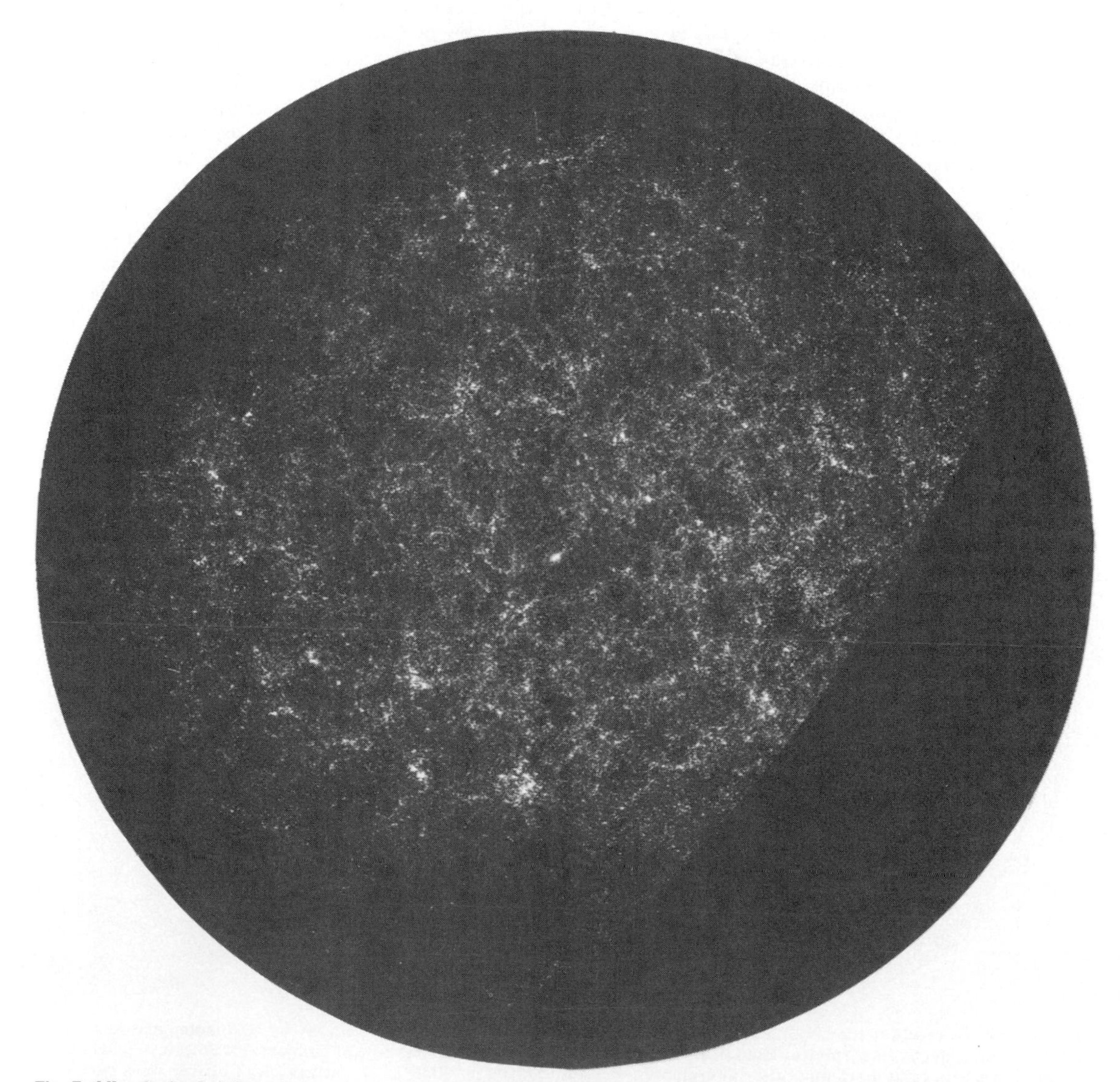

Fig. 7. All galaxies brighter than the 9th photographic magnitude as counted on photographic plates taken at Lick Observatory. Many clusters and filaments of galaxies are visible, as well as the great voids between them. (*After M. Seldner et al., New reduction of the Lick catalogue of galaxies, Astron. J., 82:249–256, 1977*)

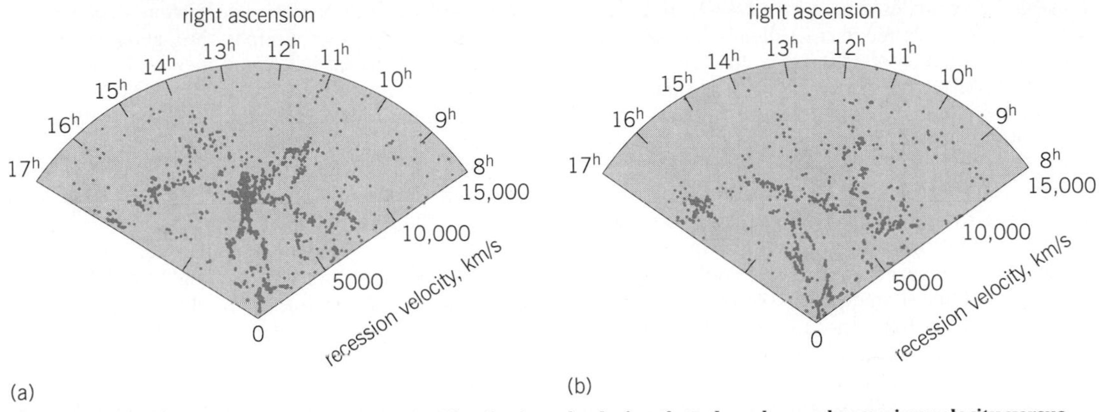

Fig. 8. Two slices of the universe, showing the distribution of galaxies plotted as observed recession velocity versus right ascension for two declination zones 6° wide centered at (*a*) +29.5° and (*b*) +35.5°. The spongelike structure in the distribution is apparent. In particular, the structures, voids as well as filaments, are clearly correlated in these adjacent slices. 1 km/s = 0.62 mi/s. (*After M. Geller and J. Huchra, in V. C. Rubin and G. V. Coyne, eds., Large-Scale Motions in the Universe, Princeton University Press, 1988*)

some clumps of clusters).

Statistical studies of these catalogs have been made by many investigators using many different techniques, including the *n*-point correlation functions of which the 2-point, or covariance, function has been the most widely used; chi-squared tests; nearest-neighbor tests; dispersion-subdivision tests; and cluster, or percolation, analysis. All have given much the same answer: clumping in Zwicky's galaxy survey and in the Lick data occurs on scales up to about 10 Mpc. When these tests are applied to the cluster catalogs, the scale of clumping extends to perhaps 50 Mpc. Other, less complete galaxy and cluster catalogs give the same answers, though suggestions (not generally accepted) have been made of clumping on scales of hundreds of megaparsecs in all the catalogs with sufficient data that have so far been studied.

Therefore, it would seem that on scales small compared with the radius of the universe, matter is indeed distributed nonuniformly, and that the usual assumption of homogeneity and isotropy is unwarranted. This leaves open the question of the distribution of matter at larger-scale lengths approaching the horizon size. Galaxy counts to increasingly fainter limiting magnitudes in small areas but in different directions of the sky suggest that the portion of the universe sampled by the Lick Survey and by the cluster catalogs is a "fair sample" of the universe as a whole. However, there is a difficulty in interpreting this data. From about the 17th or 18th magnitude to the limiting magnitude of the photograph or other detector (charge-coupled devices allow counts to be made to beyond the 25th magnitude), the counts—in whatever direction they are made—are identical to within the errors of the data. This could result merely from the fact that galaxies at all distances are counted indiscriminately in the areal surveys: bright galaxies at great distance appear at first sight to be little different from nearby faint galaxies (which dominate the counts), and all are simply projected onto the plane of the sky. This results in an apparent—and misleading—uniformity of the universe. Other data are needed to find answers to questions about the large-scale distribution beyond the local region.

Spatial distribution of galaxies. The missing set of data comprises accurate distances to extragalactic objects. That galaxies are nonrandomly distributed, at least on small scales up to 50 or perhaps even 100 Mpc, became generally accepted during the mid-1970s. However, the statistical tests giving these scale lengths say little or nothing about the nature of the nonrandom distribution: whether it is filametary and spongy, as suggested in the mid-1970s by J. Einasto and colleagues, or whether it is more clumpy, perhaps in a hierarchical manner (galaxies clumping to form clusters, clusters clumping to form superclusters, and so forth) as advocated by J. Peebles and others. It is obvious that a knowledge of the distances of the galaxies would go a long way toward resolving this question.

The redshift can be used as a fairly good distance indicator beyond the disturbing gravitational influence of the Local Supercluster. As discussed above, once the Hubble constant is known, it is a simple matter to find a galaxy's distance by measuring the redshift of its spectrum. Since the mid-1960s, this process has been enormously speeded up by the use of image intensifier technology. It is now possible to obtain—within a reasonable amount of time—the spectrum of virtually any object that can be detected. Bright galaxies, such as the Shapley-Ames objects, can have their spectra obtained in less than a minute. SEE IMAGE TUBE.

The resultant virtual explosion of known galaxy redshifts has made it possible to map the spatial distribution of moderately distant galaxies as well as their areal distribution. The frothy spongelike structure turns out to be real and not the result of patchy galactic absorption (distant voids can be seen through nearby ones just as easily as they can be seen through the nearby filaments). The filaments have typical lengths on the order of 40–60 Mpc, though one, the Pisces-Perseus Supercluster and its extensions, may stretch through 600 Mpc or more of space. The voids are typically 25–50 Mpc across, and the rich clusters are generally found at the intersections of two or more filaments (**Fig. 8**). This is the origin of the typical scale length of about 50 Mpc found in analyses of the areal data.

A tremendous sheet of galaxies, called the Great Wall by its discoverers, M. Geller and J. Huchra, can be traced across the entire 200-Mpc survey area. Its full extent is unknown, being limited by the area surveyed. The Great Wall is shown in Fig. 8; it is the thin band of galaxies running from about right ascension (RA) 8 h and recession velocity (V) 7000 km/s, to RA 17 h and V 10,000 km/s. It is easily seen in both slices of the universe shown in the figure, and is just as easily seen in several other slices north and south of these two.

B. Tully and his colleagues analyzed the whole-sky Abell catalog of rich galaxy clusters. They found not only the Great Wall but several other, even larger structures with sizes of up to 400 Mpc. They and other groups have also made still-controversial suggestions that these tremendous sheets and filaments show geometric alignments and patterns, with nearby superclusters being aligned both parallel to and perpendicular to the plane of the Local Supercluster. Similarly, T. Broadhurst and his colleagues have found a remarkable 130-Mpc periodicity in deep so-called pencil-beam redshift surveys at both north and south galactic poles. This finding has been interpreted to mean that the observed galaxies are found preferentially in at least eight evenly spaced sheets 130 Mpc apart. Alternatively, the galaxies may be defining the surfaces of gigantic voids three to five times larger than those found in less deep surveys.

Further analysis by Tully of the spatial positions of galaxies within the Local Supercluster shows that their distribution mimics that of the 50-Mpc distribution in the universe but on a scale five to ten times smaller. The suggestion then is that the finding via correlation analysis, cluster analysis, and so forth of scale lengths of up to about 10 Mpc in the galaxy counts is also not accidental.

All of this evidence, preliminary as some of it is, suggests that the large-scale distribution of galaxies in space is fractal; that is, the structure is similar on all scale lengths. This suggestion apparently confirms a finding from theoretical work on the big bang that there are no preferred scale lengths for the growth of inhomogeneities within the primordial fireball. This work is further described below.

Other important observational results have emerged from these studies. The filaments themselves are composed primarily of spiral galaxies clumped into groups and clouds, while most of the known elliptical and lenticular galaxies are located in the clusters at the intersections of the filaments. The clusters include relatively few spirals. The clouds and clusters also tend to have their major axes aligned along the axes of the filaments.

Observations at nonoptical wavelengths. Nearly all of the evidence set out above for the large-scale structure comes from optical observations of galaxies (though radio and infrared work on nearby galaxies has played an important role in finding redshift-independent distances). These optical data are known to be biased by selection effects, though these can be corrected to some extent. The more objectively sampled selection of far-infrared sources from the *Infrared Astronomical Satellite* (*IRAS*) survey of the sky has been used to construct an all-sky catalog of galaxies in which the selection effects are well understood. Studies of this sample have confirmed the surprisingly large (velocities of approximately 400 mi/s or 600 km/s) coherent peculiar motions of galaxies over very large volumes of space (on the order of 50 Mpc across) that were found through optical studies of the distances and motions of elliptical and lenticular galaxies. Since the *IRAS* sample is dominated by spiral galaxies, the similar conclusions from the two studies provide impressive evidence, from two independent data sets, for large-scale streaming motions throughout the nearby universe. *See* Infrared astronomy.

These streaming motions are most easily explained as retardation of the Hubble flow by large concentrations of mass. Indeed, maps of the velocity field of nearby galaxies are fit quite well by models with just two mass concentrations: the Virgo Cluster at a distance of 12–14 Mpc, and the Hydra-Centaurus Supercluster, often called the Great Attractor, at about 35–40 Mpc. More distant clusters and superclusters are also suspected of having some influence on local galaxy motions, but the extensive observational data required to confirm or refute these suggestions are not yet available.

For more distant objects, the radio studies have not given as clear a picture as the optical and infrared data. Distant radio sources are apparently not significantly clustered, whatever their nature (radio galaxies, quasars, and so forth), though most nearby sources are associated with galaxies and therefore must share their distribution in space. Most rich clusters of galaxies emit x-rays, apparently thermal radiation from hot gas distributed throughout the clusters. However, no extended x-ray sources have been unambiguously identified outside of clusters that would indicate the presence of similar gas clouds in intercluster space. Similarly, radio observations of extragalactic neutral hydrogen clouds have shown that these too are apparently closely associated with galaxies or with groups of galaxies. Thus, it would seem that the galaxies themselves highlight the distribution of most matter in the universe and that the distribution is indeed nonuniform as far as it can be determined. *See* Radio astronomy; X-ray astronomy.

Even the microwave background radiation shows fluctuations. If the usual interpretation of the origin of this 2.7-K radiation is valid—that is, if the radiation is the cooled relic of the hot big bang that created the universe some $1.0–1.5 \times 10^{10}$ years ago—then the fluctuations recorded by the *Cosmic Background Explorer* (*COBE*) satellite and announced in April 1992 show a nonuniform distribution of matter and radiation in the universe when the universe was only a few hundred thousand years old. The fluctuations seen in the *COBE* data are scale-invariant; that is, they show no preferred angular sizes down to the smallest fluctuations presently detectable. This finding is in accord with a theoretical prediction made independently in 1970 by Y. Zel'dovitch and E. Harrison. Interestingly, the smallest of the *COBE* fluctuations correspond to about the same size as the largest structures so far observed in the galaxy distribution, a few hundred megaparsecs. This result confirms the suggestion that at least the large-scale galaxy distribution perhaps reflects primordial conditions. Smaller fluctuations that might have led to individual galaxies or clusters of galaxies may be seen in observations from high-altitude balloon flights and a microwave observatory at the South Pole. The very small amplitude of the fluctuations in the *COBE* data, only 1.1×10^{-5} times the smoothed average intensity of the radiation itself, also matches the predictions of several models of the formation of the large-scale structure.

Superimposed on these smaller fluctuations is an all-sky bipolar anisotropy caused by the Earth's motion through the background radiation. Its amplitude is much greater (1.2×10^{-3} times the average intensity), so this anisotropy was discovered in the 1970s using high-altitude aircraft, rocket, and balloon flights. This amplitude also indicates a total velocity of about 360 km/s for the Earth through the radiation. This velocity is similar in magnitude to the large peculiar velocities discussed above and is thought to arise from the same cause: the solar system is being carried along through space with the Local Group of galaxies under the gravitational influence of nearby superclusters of galaxies. The velocity through the background radiation is well enough established that it has become common to correct the redshifts of extragalactic objects for the motion, in essence reducing the observed velocities to a reference frame that is nonmoving (or at least comoving) with respect to the background radiation.

Missing mass problem. Another set of observations that are relevant to the origin and evolution of large-scale structure suggests that there may be large amounts of ''dark'' matter in the universe, which so far has been detected only indirectly. This is the so-called missing mass.

Many galaxy clusters appear to be reasonably spherical, and the distribution of galaxies in them can be reasonably modeled by a gravitationally bound and relaxed sphere of particles. Yet when the redshifts of the galaxies in these clusters are measured, it is found that the individual galaxies are moving too fast to be gravitationally bound to the clusters if all the cluster mass is in the galaxies. (This assumes, of course, that the redshifts are entirely Doppler shifts due to the galaxies' motions.) Conversely, there must be more mass in the cluster to hold it together than can be accounted for simply by adding up the masses of the individual galaxies.

Similarly, spectroscopic studies of individual galaxies have shown that their speeds of rotation do not fall off with increasing radius as expected if Kepler's and Newton's laws operate for galaxies as they do for the solar system or for double-star systems, which have one or a few concentrated objects. The fact that the rotation speed stays constant or even (in some cases) increases with radius leads to the inescapable conclusion that there is more mass in the outer parts of galaxies than can be detected by their radiation.

Thus, there may be much more mass in the universe than can be observed. (There might even be enough mass to eventually stop the expansion of the universe.) Both observational and theoretical work on the form this mass might take appears to rule out ordinary baryonic matter in the form of stars, planets, black holes, and so forth. Physicists have instead suggested that the missing mass may be in the form of massive neutrinos, gravitinos, photinos, axions, and other exotic weakly interacting massive particles (WIMPs) that could have been produced in prodigious numbers early in the big bang. Though several experiments since 1980 have suggested a mass for the neutrino of anywhere from about 15 to 17,000 eV, there is still no evidence for the existence of other such particles except that they are required in theories of supersymmetric gravity. *See* WEAKLY INTERACTING MASSIVE PARTICLE (WIMP).

Origin and evolution. Most of the work in explaining the observations of the large-scale structure in the universe has been done within the confines of a hot big bang model of the origin of the universe. Tremendous progress has been made toward a theory consistent with the general outlines of the big bang model that will give rise not only to the clumpy filamentary structure but to the smooth microwave background as well.

So far, the most successful theories explaining the origin of large-scale structure are the cold dark-matter models. These generally rely on a hot big bang leading to an initial power-law expansion of the universe. During a phase transition at about 10^{-35} s after the big bang, the strong nuclear force separated from the electroweak force. (Analogous phase transitions occur when water cools to form ice or heats to form steam.) This step led to a short period of exponential expansion triggered by energy released during the phase transition, lasting until about 10^{-32} s after the big bang. This period of exponential expansion has been graphically termed inflation by its originator, A. Guth. Because the diameter of the universe grew from about 10^{-23} cm to about 10 cm during the inflationary period, the cosmic horizon retreated much faster than the speed of light. This phenomenon explains why the present curvature of the universe is too small to measure accurately (the universe appears and acts flat in a true euclidean sense), and also why the universe looks so very much the same in any direction at any distance. *See* INFLATIONARY UNIVERSE COSMOLOGY.

Inflation also smoothed out quantum fluctuations in the initial radiation field, leading to the generally uniform background radiation that is now observed. However, inflationary big bang theories predict that the quantum fluctuations would not be completely washed away but would be inflated into random, scale-invariant inhomogeneities. As the universe continued to expand and cool, matter fell toward the slightly increased gravitational fields of the subtle fluctuations, forming gravity wells into which more matter fell, until the density in the gravity wells was great enough to build the galaxies, the clusters, and the filamentary large-scale structure that is now observed.

Because the baryons were still coupled to the radiation field when this process started (decoupling of matter and radiation took place roughly 700,000 years after the big bang), they should have dragged some of the radiation along with them as they fell toward the gravity wells. The theory predicts that this action would leave ripples in the microwave background radiation that should be observable now, and that these ripples should have the same scale-invariant properties as the initial quantum fluctuations. This prediction is exactly what the *COBE* observations have shown. The amplitude of the *COBE* fluctuations is also just that predicted by the theory.

Because the ''dark'' (that is, nonbaryonic) matter was ''cold'' (that is, slow moving), it would not have had time to fall so far into the gravity wells and would therefore be more widely dispersed than the normal baryonic matter. This explanation leads directly to the explanation of the coronae (halos) of dark matter around galaxies and the dispersion of dark matter through clusters. A related suggestion is that the voids are not as empty as they look; they, too, should have a considerable amount of this dark matter in them. The dark matter needs to be nonbaryonic to satisfy the constraints imposed by the observed abundances of the light elements. Nucleosynthesis within the big bang leads to the observed abundances only if the density of baryonic matter is less than 10% of the critical value needed to close, or at least flatten, the universe.

Since galaxies formed where the quantum fluctuations were, the large-scale structure may well be a reflection of the conditions in the universe when it was very young indeed. Therefore, if this basic cold dark-matter picture is essentially correct, an observational window opens up on the big bang that allows investigation back much further into its early days than even the microwave background radiation can. Even if this interesting family of theories turns out to be incorrect, the very large structures that are now observed in the distribution of the galaxies are far too large to be anything but primordial: at the average random speeds of galaxies moving with respect to each other, virtually all galaxies must still be close to their birthplaces; they are simply moving too slowly to have traveled far during the lifetime of the universe. Thus, through better understanding of the forms and sizes of the large-scale structures, more will be learned about the origin of the universe itself.

Unfortunately, it has so far proven impossible to meet all the observational constraints to build the large-scale structure out of this conceptually simple picture. Whatever success the cold dark-matter models have had has been confined to building galaxies and clusters (and even here, much adjustment of the theory is still needed). The observations of structure on scales of hundreds of megaparsecs has forced unwelcomed modification of the cold dark-matter theories, leading some scientists to investigate alternative explanations.

This outlook does not mean, however, that the basic big bang model itself has to be abandoned. The big bang explains too well too many observed features of the universe: its apparent expansion, the microwave background, the relative abundances of the light elements, and the evolving population of quasars, among others. Thus, work to explain the large-scale structure problem will almost certainly take place within the accepted framework provided by the hot big bang. SEE COSMOLOGY.

Harold G. Corwin, Jr.

Bibliography. J. Audouze, M.-C. Pelletan, and A. Szalay (eds.), *Large Scale Structures in the Universe*, Int. Astron. Union Symp. 130, 1988; M. Davis et al., The end of dark matter?, *Nature*, 356:489–494, 1992; V. de Lapparent, M. J. Geller, and J. P. Huchra, A slice of the universe, *Astrophys. J.*, 302: L1–L5, 1986; J. H. Oort, Superclusters, *Annu. Rev. Astron. Astrophys.*, 21:373–428, 1983; J. M. Pasachoff, *Contemporary Astronomy*, 4th ed., 1989; V. C. Rubin and G. V. Coyne (eds.), *Large-Scale Motions in the Universe*, 1988; J. S. Silk, *The Big Bang*, rev. ed., 1989; M. Zeilik and J. Gaustad, *Astronomy: The Cosmic Perspective*, 1983; M. Zeilik, S. A. Gregory, and E. V. P. Smith, *Introductory Astronomy and Astrophysics*, 3d ed., 1992.

Uranus

The first planet to be discovered with the telescope and the seventh in the order of distance from the Sun. It was found accidentally by W. Herschel in England on March 13, 1781. Herschel's telescope was good enough to show that this object was not starlike; the object appeared as a fuzzy patch of light, not a point. At first he thought that it was a comet, but subsequent calculations of the orbit demonstrated that Uranus was indeed a planet, about twice as far from the Sun as Saturn and therefore in an orbit that agreed almost exactly with the prediction of the Titius-Bode relation for planetary distances. SEE PLANET.

The planet and its orbit. The main orbital elements are the semimajor axis (mean distance to the Sun) of 1.79×10^9 mi (2.87×10^9 km); the eccentricity of 0.046; the sidereal period of 84.01 years; the orbital velocity of 4.25 mi/s (6.8 km/s); and the inclination of orbital plane to ecliptic of 0.8°.

These characteristics indicate a normal, well-behaved orbital motion. Therefore it is surprising to discover that the obliquity (inclination of rotational axis to orbit plane) of Uranus is 98°, exceeded only by the obliquities of Pluto and Venus. This means that the axis is almost in the plane of the orbit, and thus the seasons on Uranus are very unusual. During summer in one hemisphere, the pole points almost directly toward the Sun while the other hemisphere is in total darkness. Forty-two years later, the situation is reversed. Thus, "day" and "night" for an observer at the north or south pole of Uranus each last more than 40 years. Recent years of solstice (pole toward Sun) are 1944 and 1985, and recent and future years of equinox (equator toward Sun) are 1966 and 2007. The south pole was almost directly facing the Sun when the *Voyager 2* spacecraft sped past the planet in January 1986. The actual period of rotation of Uranus is 17.24 h as determined by the *Voyager 2* observations of radio emissions from the planet. SEE PLUTO; VENUS.

The mean apparent equatorial diameter of the disk of this distant planet is about 3″.6 (the Moon has an apparent diameter of 31′). The corresponding linear equatorial diameter is 31,770 mi (51,120 km). The mass is 14.53 times the mass of the Earth, a value that can be well determined from the observed motions of the satellites. The corresponding mean density is 1.29 g/cm^3, which is greater than that of Saturn even though Uranus is smaller than Saturn. This means that Uranus is richer than Saturn (or Jupiter) in elements heavier than hydrogen and helium but not nearly so rich in these elements as Earth (mean density 5.5). The same is true for Neptune; thus these two planets constitute a distinct subgroup in the outer solar system when compared with Jupiter and Saturn, which are composed of a nearly solar distribution of the elements. SEE JUPITER; NEPTUNE; SATURN.

The apparent visual magnitude of Uranus at mean opposition, that is, when closest to Earth, is +5.5. Therefore, this planet is just visible to the naked eye in a dark sky when its position among the stars is known. The corresponding value of the albedo, or reflectivity, is 0.4, which is a typical value for the outer planets. The temperature Uranus would assume in simple equilibrium with the incident solar radiation is about 55 K (−360°F), according to calculations. This is essentially identical with the value that is obtained by direct measurement. There is no evidence for the existence of an internal energy source as observed for Jupiter, Saturn, and Neptune. The temperature measured by *Voyager 2* was the same at both the sunlit and dark poles. Observations at radio frequencies have detected thermal radiation from the lower atmosphere of Uranus, indicating that the temperature increases with depth but much more slowly than in the case of the other major planets. At a wavelength of 21 cm, a temperature of 250 ± 60 K (9 ± 90°F) has been measured.

Through the telescope, Uranus appears as a small, slightly elliptical blue-green disk. This appearance is confirmed in the nearly featureless pictures of the planet obtained by *Voyager 2*. Uranus owes its characteristic aquamarine color to the relatively high proportion of methane in its atmosphere. This gas absorbs the orange and red wavelengths from incident sunlight, scattering back blue-green light to an observer. The methane abundance is 20 to 30 times the amount corresponding to a solar distribution of the elements, in agreement with the enrichment of heavy elements in the planet deduced from its average density. In striking contrast, *Voyager 2* found the proportion of helium to hydrogen, the two most abundant gases in the planet's atmosphere, to be essentially equal to that observed in the Sun.

Like Jupiter and Saturn, Uranus exhibits zonal winds that are parallel to the equator, despite the unusual orientation of the planet's rotational axis. These winds appear to increase in speed with increasing latitude, unlike the winds on the other two planets, which are strongest at the equator. They move in the same direction as the rotation of Uranus, producing a circulation pattern like that on Neptune. The causes of this circulation and the way it maintains the two poles at the same temperature have not yet been determined.

Voyager 2 discovered that Uranus has a magnetic field which is inclined at an angle of 60° to the axis of rotation. This is similar to the magnetic obliquity on Neptune, five times that found on Earth. This field traps belts of electrons and protons (like Earth's Van Allen belts) in a magnetosphere whose unusual configuration is determined by the orientations of the planet's magnetic and rotational axes. *SEE MAGNETOSPHERE; PLANETARY PHYSICS.*

Satellites. Before the *Voyager 2* encounter, only five satellites of Uranus had been discovered. They form a remarkably regular system with low orbital eccentricities and inclinations close to the plane of the planet's equator. The *Voyager* cameras found 10 additional small satellites, including 2 that may serve as gravitational "shepherds" for the outermost ring (see **table**).

The spacecraft was able to obtain pictures of the surfaces of all five of the larger satellites and resolved features on Puck as well. For reasons that are not yet understood, there is increasing evidence of internal geological activity on the larger satellites with decreasing distance from the planet. Miranda, the smallest but also the closest, revealed the most spectacular diversity of surface features (**Fig. 1**). Impact craters are present on all of these objects as well.

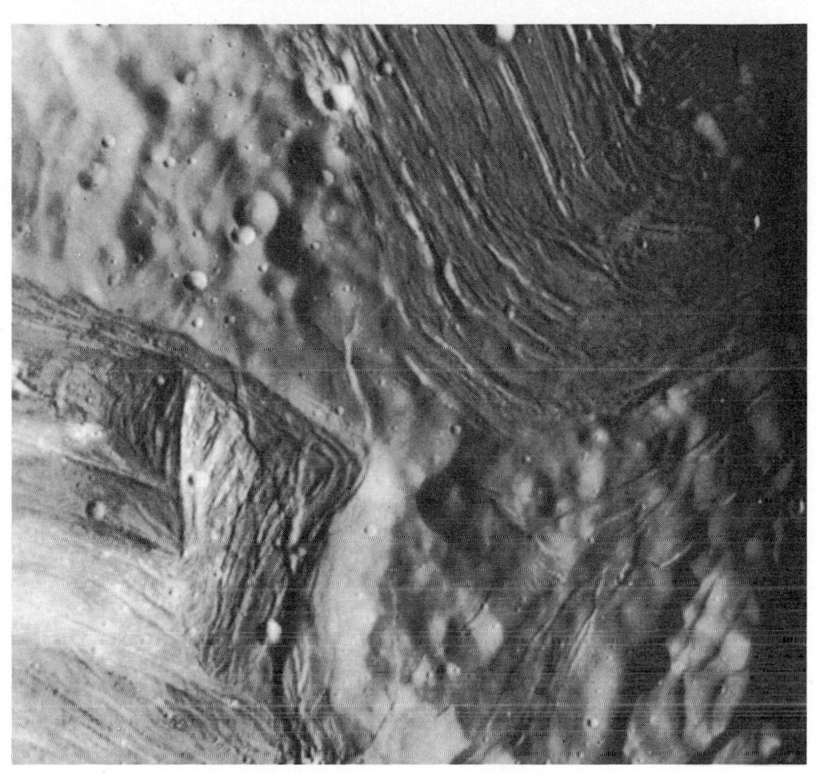

Fig. 1. Miranda viewed from *Voyager 2*. Features as small as 2000 ft (600 m) across are visible. The two regions of grooved terrain are remarkable, showing evidence of large-scale geological activity on this tiny satellite.

With densities of approximately 1.5 g/cm^3, these satellites are presumably composed of a mixture of abundant ices (dominated by water ice) and some rocky material. The generally low reflectivities and the deposits of dark material observed in some locations on the satellites' surfaces are ascribed to the presence of carbon-rich compounds. These com-

Satellites of Uranus

Satellite	Mean distance from Uranus, 10^3 mi	Mean distance from Uranus, 10^3 km	Sidereal period, days	Radius, mi	Radius, km	Albedo (reflectivity)	Density, g/cm^3
Cordelia*	30.9	49.7	0.34	12	20		
Ophelia*	33.4	53.8	0.38	15	25		
Bianca	36.8	59.2	0.44	15	25		
Cressida	38.4	61.8	0.46	19	30		
Desdemona	39.0	62.7	0.48	19	30		
Juliet	40.1	64.6	0.50	25	40		
Portia	41.1	66.1	0.51	25	40		
Rosalind	43.4	69.9	0.56	19	30		
Belinda	46.8	75.3	0.63	19	30		
Puck	53.5	86.0	0.76	53	85	0.07	
Miranda	80.8	130	1.41	151	243	0.3	1.3
Ariel	119	191	2.52	342	580	0.4	1.6
Umbriel	165	266	4.14	370	595	0.2	1.4
Titania	271	436	8.71	500	805	0.3	1.6
Oberon	362	583	13.5	482	775	0.2	1.5

*Shepherding satellites for ϵ ring.

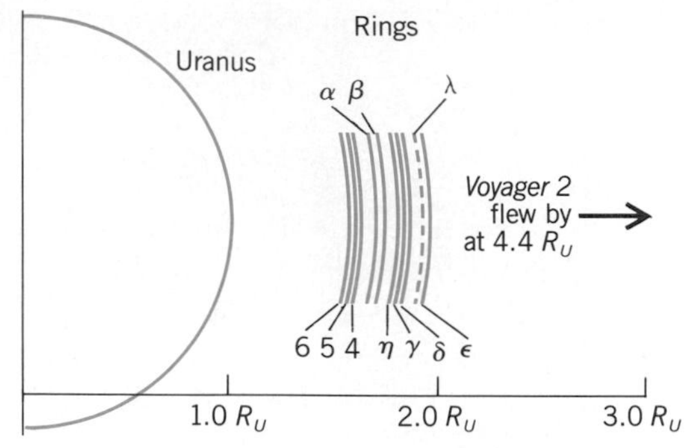

Fig. 2. Diagram of Uranus ring system, showing one hemisphere of Uranus at the left and segments of the 10 narrow rings. The rings all lie between 1.5 and 2 times the radius of Uranus (R_U) from the planet's center.

pounds are more neutral in reflectivity than the dark material found on Saturn's satellite Iapetus, however, suggesting that more than one type of organic material is present in the outer solar system.

Rings. In 1977 a system of nine rings (**Fig. 2**) was discovered around Uranus during observations of a stellar occultation. They are designated by a mixture of Greek letters and Arabic numerals (Fig. 2). Since the rotational axis of Uranus is almost in the plane of the planet's orbit around the Sun, the rings are nearly perpendicular to this plane, which made it possible to discover them when they passed in front of a star.

The *Voyager* cameras added a tenth ring between δ and ε (Fig. 2). All of these rings are remarkably narrow. The widest is ε, whose mean width of only 36 mi (58 km) is still six to seven times larger than that of the next widest rings. Like the small satellites, the rings are as dark as coal, with reflectivities of only 3–5%. (The Earth's Moon has a reflectivity of 11%.) Looking back toward the Sun from within the shadow of Uranus, the *Voyager* cameras found that the spaces between these narrow rings are often thinly populated with dust. Measurements of the diminution of starlight by the rings using a photometer on the spacecraft as well as measurements of the effect of the rings on radio transmission from the spacecraft itself suggest an absence of small particles within the ε ring. The origin and means of confinement of these rings are topics of ongoing research.

Cosmogony. It appears that all four giant planets have cores of rock and ice that are approximately the same mass: between 10 and 20 times the mass of the Earth. The great difference between Jupiter and Saturn on the one hand and Uranus and Neptune on the other can then be understood in terms of differences in their atmospheres. The outer two planets acquired less gas from the original solar nebula when they formed, perhaps because the nebula was simply thinner at this large distance from its center. Since the nebular gases were predominantly hydrogen and helium, Uranus and Neptune began with a smaller endowment of these two light elements and thus ended up with mean densities larger than Jupiter and Saturn. The excess methane remarked on above was produced by the core, which was unable to trap the two lightest gases. The hydrogen and helium are therefore present in a solarlike ratio.

The large obliquity of Uranus and its lack of an internal heat source remain unsolved problems. The obliquity is commonly attributed to the impact by a large body during the late phases of core formation, but a detailed theory has not yet been developed.

Tobias C. Owen

Bibliography. J. K. Beatty, B. O'Leary, and A. Chaikin (eds.), *The New Solar System,* 3d ed., 1989; W. K. Hartmann, *Moons and Planets*, 2d ed., 1983; G. Hunt, *Uranus and the Outer Planets,* 1982; D. Morrison and T. Owen, *The Planetary System,* 1988; Reports of *Voyager 2* encounter with Uranus, *Science,* 233:39–109, 1986.

Ursa Major

The most widely known and oldest of the astronomical constellations. Ursa Major, or the Great Bear, is a circumpolar group as viewed from the middle latitudes of the Northern Hemisphere. One part of the configuration, a group of seven bright stars, which is pictured as the tail of the Great Bear, is commonly known in the United States as the Big Dipper which it resembles (see **illus.**). This group of stars is also known in various lands as Charles' Wain (wagon) and the Plough. The Chinese call it the Northern Bushel. The two stars α and β at the front of the bowl of the dipper are called pointers, because a line joining them points to Polaris, the North Star. One of the pointers, the northern one, called Dubhe, is a navigational star.

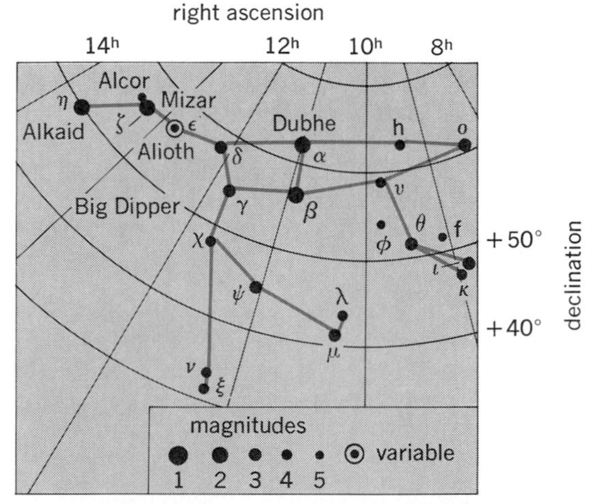

Line pattern of the constellation Ursa Major. The grid lines represent the coordinates of the sky. The apparent brightness, or magnitude, of the stars is shown by the sizes of the dots, which are graded by appropriate numbers as indicated.

The star next to the end of the handle is Mizar, another navigational star, with its close companion Alcor. Next to Mizar is Alioth, the third navigational star in this group. SEE CONSTELLATION; URSA MINOR.

Ching-Sung Yu

Ursa Minor

The astronomical constellation Little Bear. Ursa Minor is a circumpolar constellation whose brightest star, Polaris, is almost at the north celestial pole. Seven of the eight stars appear to form a dipper, hence the constellation is alternately known as the

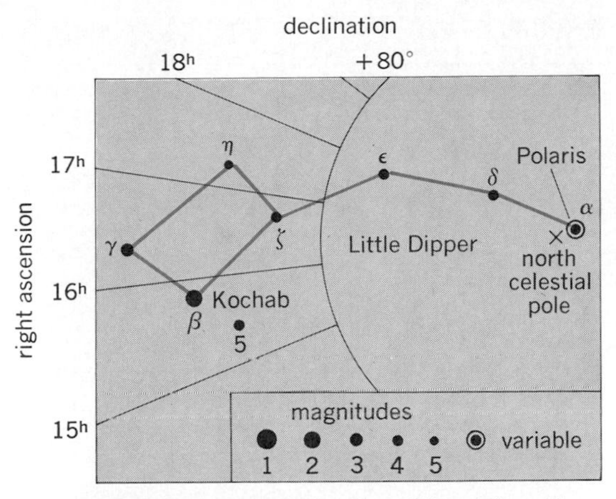

Line pattern of the constellation Ursa Minor. The grid lines represent the coordinates of the sky. The apparent brightness, or magnitude, of the stars is shown by the sizes of the dots, which are graded by appropriate numbers as indicated.

Little Dipper (see **illus.**). Polaris is at the end of the handle. Situated about 1° from the true celestial pole, Polaris is a variable star, pulsating in brightness periodically. The two bright stars β and γ at the front of the bowl are often called Guardian of the Pole, because they circle about Polaris closer than other conspicuous stars. Kochab, the brighter of the two, was at one time closer to the true pole. It is an Arab word meaning polestar. The Little Bear is also known as Smaller Chariot by the Danes. *See Constellation; Polaris; Ursa Major.*

Ching-Sung Yu

Variable star

A star that has a detectable change in brightness which is often accompanied by other physical changes. During the life of a star, changes in brightness occur in the early stages, while it is forming, or in the late stages, close to its death. Therefore, variability provides important clues about the evolution and nature of stars. Depending upon the type of star, variability in brightness can provide information about its size, radius, mass, temperature, luminosity, internal and external structure, and distance from the Earth.

Over 25,000 stars are known to vary in brightness. Some variable stars are very bright and can be observed with the naked eye, while others are observable only with telescopes. The variation from maximum to minimum brightness (the range) can be anywhere from thousandths of a magnitude to many full magnitudes, and the period can be anywhere from a fraction of a second to several years. The change in brightness may be caused by intrinsic physical changes in the star or stellar system such as pulsations or eruptions, or extrinsic changes, such as the eclipse of the star by one or more stars. *See Magnitude.*

Observation. Most variable stars need to be monitored in a systematic way over years and decades in order to determine their long-term behavior, period, and peculiarities and changes in brightness variation. The simplest method of determining the brightness of variable stars is visually through a telescope or binoculars. Groups of observers worldwide, such as the American Association of Variable Star Observers (AAVSO), carry out systematic visual brightness determinations for thousands of variable stars. Group members, mostly amateur astronomers, observe with their own binoculars and telescopes. They use special star maps (finding charts) which identify the variable star and stars of constant and known brightness (comparison stars) used in estimating the brightness of the variable. Observational data provided by these dedicated observers have been used extensively by professional astronomers to analyze the behavior of the changes in brightness of variable stars, to schedule and correlate observations made with large telescopes and special instruments (both on the Earth and aboard satellites orbiting the Earth), and to make theoretical variable-star models with the aid of powerful computers in order to understand better the physical structure of these stars.

Other methods for systematically monitoring variable stars, used by both professionals and increasing numbers of amateurs, include photographic and photoelectric photometry techniques. Photographic plates provide a lasting record of the brightness of stars. Photoelectric photometry uses photomultipliers to amplify and measure the intensity of the light from a star. Photometry is useful for observing stars that have a small range of variation, from thousandths to tenths of a magnitude. *See Astronomical photography.*

Certification and nomenclature. Once a star is discovered to be changing in brightness, and its position in the sky and the kind (type) and the amount (range) of its variability are determined, it is ready to be certified and named as a variable star. The naming of a variable star is based on a system developed in the mid-1800s by F. W. A. Argelander. The first variable star discovered in a constellation is given the letter R, followed by the genitive form of the Latin name of the constellation. For example, R Leonis is the variable star R in the constellation of Leo. When other variables are discovered in that constellation, they receive the letters S, T, and on to Z. Subsequent vari-

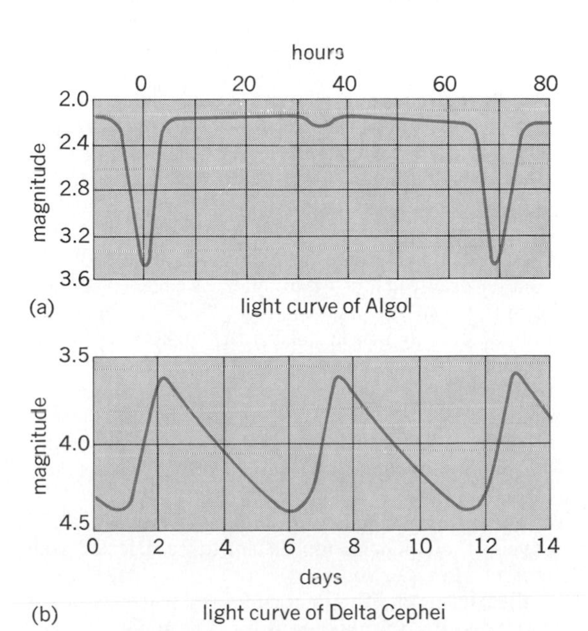

Fig. 1. Comparison of eclipsing variable stars and intrinsic variable stars. (*a*) An eclipsing variable star, with a period of 2.87 days. (*b*) A typical intrinsic variable star, with a period of 5.37 days.

Representative types of intrinsic variables

Type	Range of period, days	Amplitude, magnitude	Spectra and luminosity	Example	Number known
Pulsating					
RR Lyrae	<1	<2	A to F (blue), giants	RR Lyrae	5983
Cepheids (types I and II)	1–70	0.1–2	F to G (yellow), supergiants	δ Cephei	780
Long-period					
Mira	80–1000	2.5–6+	M (red), giants	o Ceti (Mira)	5200
Semiregular	30–1000	1–2	M (red), giants and supergiants	Z Ursae Majoris	3044
RV Tauri	30–150	Up to 3	G to K (yellow and red), supergiants	RV Tauri	108
Irregular	Irregular	Up to several magnitudes	All types		3335
Others					135
Eruptive					
Supernovae	?	15 or more		CM Tauri (Crab Nebula)	7
Novae	Centuries?	7–16	O to A, subdwarfs	GK Persei	224
Recurrent novae	20 years?	7–9		RS Ophiuchi	10
Dwarf novae (U Geminorum and Z Camelopardalis)	10–600	2–6	A to F, subdwarfs	SS Cygni, Z Camelopardalis	289
Flare	?	Up to 6	M, dwarfs	UV Ceti	890
Nebular	Rapid and erratic	Up to a few magnitudes	B to M, main sequence and subgiants	T Tauri	1211
R Coronae Borealis		1–9	F to K, supergiants	R Coronae Borealis	38
Others					52

ables are named RR, RS to RZ, then SS to SZ, and so on to ZZ, after which follow AA, AB to AZ, and so on to QZ. (The letter J is never used in this system.) The next variable after QZ, the 334th variable in a constellation, is named V335. (For example, the variable V335 Sagittarii is the 335th variable in Sagittarius.) Stars that have been assigned Greek letters, such as o (omicron) Ceti, or small roman letters, such as g Herculis, prior to the start of this system continue to keep those names.

Until World War II, the certification and naming of a variable was done in Germany. The International Astronomical Union continued the certification and naming until 1952. Since then it has been carried out by the compilers of the *General Catalogue of Variable Stars* in the former Soviet Union. The fourth edition of this catalog and name lists that supplement it list 30,199 known variables designated through April 1989. An additional 14,810 stars listed in the *New Catalogue of Suspected Variables* (1982) are in need of confirmation of their variability.

Not all constellations have the same number of variables, because the distribution of variables is not uniform in the Milky Way Galaxy. While the concentration is very high in the disk, particularly toward the center of the Galaxy, as in the case of the constellation Sagittarius, with 3891 variable stars listed, the corona of the Milky Way Galaxy has the lowest number, as in the case of the constellation Caelum, which has only nine variables listed. *See Milky Way Galaxy.*

There may be thousands of variable stars yet undiscovered, too faint or too distant to be detected with present instrumention.

Classification. The light curve and the spectrum of a variable star form the basis for determining its classification. A light curve (**Fig. 1**) is the graph of the observed brightness (usually in magnitudes) of the star plotted against time (usually in Julian days). The important parameters are the shape of the light curve, the brightness range, and the regularity and the period of the brightness variation. *See Light curves.*

The spectrum of a star helps to measure its temperature, luminosity, chemical composition, age (population), and mass. The spectra of variable stars range from very hot, blue, O or B, to cool, red, M spectral classes. Some lines in the spectra help to identify the nature of the variable. For example, metallic oxides and carbon lines, accompanied by bright lines of hydrogen, are usually seen in cool, red, long-period variables. *See Astronomical spectroscopy.*

Variable stars can be divided into two major types: extrinsic and intrinsic variables (Fig. 1). The extrinsic variables are those stars in which the variability in brightness occurs because of the occultation of one star by another. In this eclipsing binary type of star, the change in the brightness of the system is the result of a geometric phenomenon. The primary eclipse occurs when the brighter star is eclipsed by the fainter, and the secondary eclipse occurs when the fainter star is eclipsed by the brighter. Information on the size, shape, and the distance to the stars from the Earth may be obtained from the eclipse phenomenon of these stars. The prototype of eclipsing binaries is Algol (β Persei), a bright, naked-eye star that changes its brightness from 2.1 to 3.4 magnitudes as it is eclipsed every 69 h (Fig. 1*a*). *See Binary star; Eclipsing variable stars.*

Intrinsic variables are those stars in which the variability of brightness occurs because of physical change in or on the star itself. These stars are divided into two classes: pulsating and eruptive (see **table**).

Pulsating variables. Periodic pulsation (contraction and expansion) of the star and its outer layer results in the variation in brightness, as well as variations in the star's temperature, spectrum, and radius.

Cepheids. Cepheids are rare, highly luminous (supergiant), yellow variable stars. They vary with periods from 1 to 100 days, and have a range of variation from 0.1 to 2 magnitudes. The prototype of the class is δ Cephei, a naked-eye star which varies in brightness from magnitude 4.1 to 5.2 every 5.4 days (Fig. 1*b*). Polaris, the North Star, is also a cepheid variable and varies between magnitudes 2.5 and 2.6 every 4 days.

Cepheids show an important correlation between their period of variation and their relative brightness, in that those stars with longer periods are also brighter. This period-luminosity relationship was first noticed in 1912 by H. Leavitt while studying the cepheid variables in the Small Magellanic Cloud (**Fig. 2**). H. Shapley extended this relationship to cepheids in the Galaxy, and converted the relationship to period and intrinsic luminosity of these stars. In 1950, W. Baade determined the zero point of this relationship, with his work on these stars in the Andromeda Galaxy. SEE ANDROMEDA GALAXY; MAGELLANIC CLOUDS.

Thus, by observing the apparent brightness change and determining the period of this variation, use of the period-luminosity relationship leads to the absolute magnitude of the star. Once the absolute magnitude of a star is known, the distance to it can be determined from the equation below, where M is the

$$M = m + 5 - 5 \log D$$

absolute magnitude, m is the apparent magnitude, and D is the distance in parsecs. Due to this important correlation, cepheids have been used to measure distances to nearby galaxies and in the determination of the distance scale of the universe.

A similar group of variable stars, W Virginis stars, are found in globular star clusters and the corona of the Milky Way Galaxy. These stars are bluer (and thus hotter) and older (population III) than the cepheids. They have periods from 10 to 30 days, and obey a similar period–luminosity relationship. They are sometimes called type II cepheids to distinguish them from the classical type I cepheids. SEE CEPHEIDS.

RR Lyrae variables. RR Lyrae variables are the second most common type of variable in the Galaxy. They have periods of less than 1 day, and have a small range of variation, from 0.5 to 1.5 magnitudes. They are particularly numerous in globular clusters, and thus are sometimes referred to as cluster variables. Their spectral classes range from A to F. The prototype of the class is RR Lyrae, which within 13.5 h varies from magnitude 6.9 to 8.0 and back. All RR Lyrae variables have the same intrinsic luminosity, of magnitude 0.5. Therefore these variables do not follow the period-luminosity relationship. Thus, if an RR Lyrae star can be identified in a star cluster and its apparent magnitude determined, the above equation and the star's known absolute magnitude can be used to obtain the distance to the cluster.

Long-period variables. Long-period variable stars are the most abundant of all variables in the Milky Way Galaxy. They are red, cool, giant or supergiant stars with spectral class M or R, S, or C carbon types. Long-period variables are old stars which have evolved from the main sequence and are in the late stages of their evolution. Their size (diameter) is several hundred times that of the Sun; their mass, however is about the same as the Sun. Their mass is concentrated in the core, which is surrounded by an extended gaseous atmosphere. SEE STELLAR EVOLUTION.

The light variability of long-period variables is due to the pulsation (contraction and expansion) of the

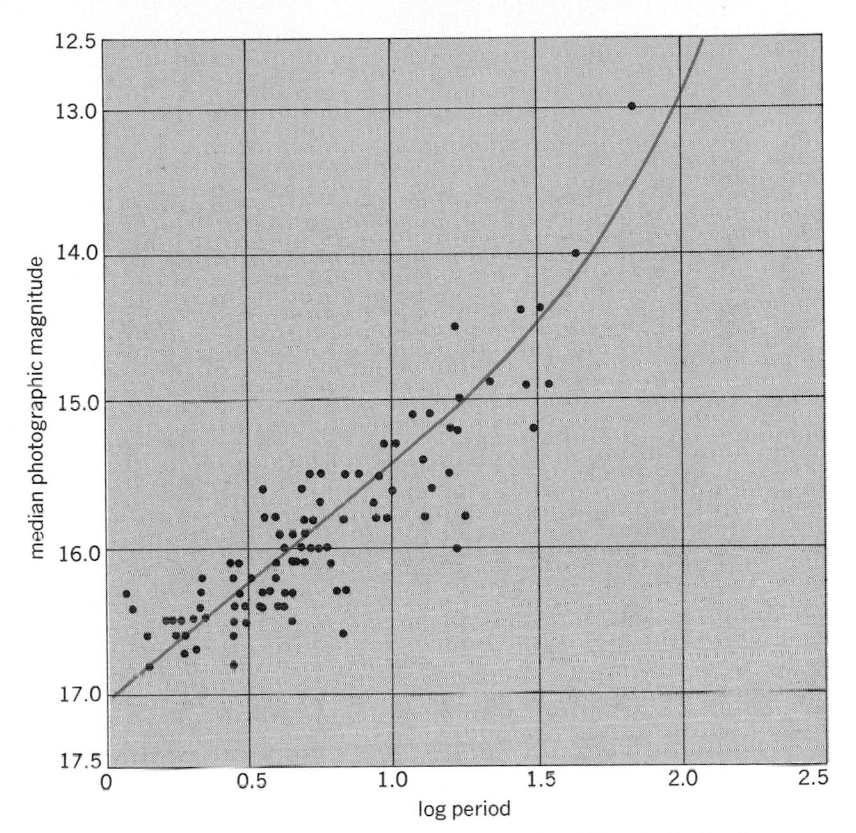

Fig. 2. Period-luminosity curve for cepheid variables in Small Magellanic Cloud.

star and its radiating surface. The prototype of long-period variables is Ceti (Mira), which has given its name to those long-period variables that have a range of variation of 2.5 magnitudes and more. Mira variables vary with time intervals from 100 to 1000 days. Although the change in brightness is periodic, the brightness of individual cycles may not be the same, some cycles being much brighter or fainter than others. Generally the rise to maximum brightness is faster than the decline to minimum brightness. SEE MIRA.

The spectra of long-period variables are abundant with lines of metallic oxides such as titanium oxide (TiO) and bright emission lines of hydrogen, seen particularly near maximum brightness. These bright lines are thought to be generated by propagating shock waves in the atmosphere of these stars.

Red giant and supergiant stars with less regularity of variation, shorter periods, and smaller ranges of variation, less than 2.5 magnitudes, are called semi-regular variables. Other red variable stars that do not exhibit any regularity in their brightness change are called irregular stars.

Rare, very luminous, yellow supergiant stars that generally show alternating shallow and deep fadings are called RV Tauri stars. These pulsating variables vary by 2 to 3 magnitudes within 30 to 150 days.

Eruptive variables. Eruptive variables are those stars that have one or more eruptions—the ejection of matter into space—in their lifetime. There are many types of eruptive variables, ranging from supernovae and novae, in which spectacular explosions take place which brighten the system by many magnitudes, to nebular variables, which are young stars in the early stages of stellar formation, to R Coronae Borealis variables, which have sudden unpredictable decreases in their brightness. Only a few of the eruptive variable types will be discussed; a more complete listing of types is given in the table.

Supernovae. The most spectacular type of eruption is that of a supernova, wherein either a hot and very massive star undergoes an explosion or a white dwarf in a binary system is incinerated. The luminosity increases within a few days to a few thousands to tens of thousands times the original brightness. Bright emission lines in the spectrum indicate that matter is ejected at the time of the spectacular eruption. The star slowly fades within a few months and disappears from telescopic view. At maximum light a supernova outshines the total brightness of its galaxy. In the Milky Way Galaxy the most recent supernova observed was seen in 1604 in the constellation Serpens. However, Supernova 1987A, which exploded on February 23, 1987, in the Large Magellanic Cloud, a nearby companion galaxy to the Milky Way Galaxy, was visible to the unaided eye and was studied intensely. *SEE SUPERNOVA.*

Novae, recurrent novae, and dwarf novae. The best known of the eruptive variables are stars that brighten by 7 to 16 magnitudes within about a day. They stay at maximum brightness for a few days or weeks and then slowly fade. The word nova, meaning new, was used for these stars. Actually a nova is not a new star, but an already existing star which due to the eruption has become very luminous and thus visible.

Recurrent novae are stars that about evey 20 years or more have eruptions during which the system brightens by 7 to 9 magnitudes.

Dwarf novae are divided into two categories: U Geminorum and Z Camelopardalis stars. Dwarf novae have smaller-scale eruptions, in which the star brightens by 2 to 6 magnitudes within a day, stays bright 1 to 2 weeks, and then fades to the original brightness. The quasiperiodic eruptions may occur from ten to several hundred days apart. Z Camelopardalis stars are differentiated from U Geminorum variables by their periods of nonvariability midway between maximum and minimum brightness. These standstills can last days, months, or sometimes years.

The above classes of variables are sometimes referred to as cataclysmic variables because of the cataclysmic explosions they undergo. Cataclysmic variables are very close binary systems made up of an evolved, hot, dense white dwarf and a less evolved cool star which transfers mass onto the white dwarf. *SEE NOVA.*

R Coronae Borealis stars. R Coronae Borealis stars, instead of brightening through eruptions, irregularly decrease in brightness every 2 to 3 years by 1 to 9 magnitudes. These bright, highly luminous supergiant stars are rich in carbon and poor in hydrogen in composition. The decrease in brightness is caused by the veiling of the star by thick carbon clouds expelled to the star's atmosphere. *SEE STAR.*

Janet Akyüz Mattei

Bibliography. C. Hoffmeister, G. Richter, and W. Wentzel, *Variable Stars,* 1985; D. Levy, *Variable Star Observing,* 1989; J. A. Mattei, E. H. Mayer, and M. E. Baldwin, Observing variable stars, *Sky Telesc.*, 60:285–289, October 1980; J. A. Mattei, E. H. Mayer, and M. E. Baldwin, Variable stars and the AAVSO, *Sky Telesc.*, 60:180–184, September 1980; J. R. Percy (ed.), *The Study of Variable Stars Using Small Telescopes,* 1987; M. Petit, *Variable Stars,* 1987.

Venus

The second planet in distance from the Sun. This neighbor of the Earth is very similar to it in such gross characteristics as mass, radius, and density (see **table**). In other ways Venus is apparently different. Its atmospheric mass is almost a hundred times that of the Earth; its atmosphere is mostly carbon dioxide instead of nitrogen and oxygen; an extensive cloud layer of concentrated sulfuric acid is present; its surface temperature is an unbearable 900°F (750 K); and it rotates with a period of 243 days, and from east to west, in the opposite sense of most other planets. Some of these differences are due more to alternate evolutionary paths of the two planets than to totally different initial conditions.

Appearance. To the naked eye, Venus is the brightest starlike object in the sky. It is usually visible during the night either soon after sunset or close to sunrise. It can sometimes be seen during the daytime. As observed through a pair of binoculars or a telescope, Venus exhibits a crescentlike or gibbous appearance. Similar to the Moon, Venus ranges over a full set of phases from a "new moon" to a "full moon." This phase variation is caused by a changing fraction of the Sun-illuminated hemisphere facing toward the Earth. At new moon, only the dark nighttime side is seen, while at full moon all of the daytime hemisphere is seen. Venus undergoes a complete set of phase changes over its synodic period of 584 days. *SEE PHASE.*

Clouds. The light seen coming from Venus is entirely due to sunlight that is reflected from a dense cloud layer whose top is located about 45 mi (70 km) above the surface and whose bottom lies within 30 mi (50 km) of the surface. In contrast to the Earth's approximately 50% cloud cover, the clouds of Venus are present over the entire planet. In yellow or red light they present a uniform appearance. However, the clouds show a banded and spotted pattern when viewed in ultraviolet light (**Fig. 1**). These ultraviolet markings provide information about atmospheric motions.

The clouds of Venus consist of a large number of tiny particles, about 1 micrometer in size, that are made of a water solution of concentrated sulfuric acid. Such a composition may at first seem very sur-

Characteristics of Venus

Characteristics	Values
Mass	0.82 Earth's mass
Radius	0.95 Earth's radius or 3760 mi (6052 km)
Mean density	0.94 Earth's value or 5.2 g/cm^3
Orbital distance from the Sun	0.72 Earth's distance or 6.7×10^7 mi (1.1×10^8 km)
Orbital period*	0.62 of an Earth year or 225 Earth days
Orbital eccentricity	0.4 Earth's value or 0.007
Orbital inclination to Earth's orbital plane	3.4°
Rotational period†	243 Earth days with respect to the stars, 117 Earth days with respect to the Sun

*Length of a Venus year.
†Length of a Venus day.

prising when compared with the water clouds of the Earth's lower atmosphere. However, sulfuric acid particles are the dominant type of particles in the Earth's upper atmosphere, although the amount there is much less than the amount present in Venus's atmosphere. In the case of both planets, sulfuric acid is produced primarily from sulfur-containing gases that combine with water vapor and oxygen-containing gases. Compositional measurements made from the United States *Pioneer Venus* Sounder probe, which descended through Venus's atmosphere on December 9, 1978, show that sulfur dioxide is the principal sulfur-containing gas in Venus's atmosphere. Sulfur dioxide is also the major gas species injected into the Earth's stratosphere by volcanic explosions. Such injections cause a large, but temporary, increase in the amount of sulfuric acid there.

Atmospheric composition. Measurements conducted within Venus's atmosphere from United States and Soviet (now C.I.S.) probes and remotely from spacecraft and the Earth have provided a good definition of the gases that constitute Venus's atmosphere. By far, the chief gas species is carbon dioxide, which makes up 96% of the atmospheric molecules, while nitrogen accounts for almost all the remainder. Trace amounts of sulfur dioxide (~150 parts per million), water vapor (~50 ppm), carbon monoxide (~30 ppm), argon (~70 ppm), helium (~10 ppm), neon (~7 ppm), hydrogen chloride (~0.4 ppm), and hydrogen fluoride (~0.005 ppm) are present in the lower atmosphere, with the concentration of the first two of these declining dramatically near the cloud tops due to the formation of new sulfuric acid there. Chemical tranformations also occur in the deeper portions of the atmosphere, aided by the high temperatures there. For example, at altitudes below about 25 mi (40 km), carbon monoxide is gradually converted into carbonyl sulfide, a gas containing carbon, oxygen, and sulfur atoms.

Carbon dioxide. In contrast to the dominance of carbon dioxide in Venus's atmosphere, the Earth's atmosphere consists mostly of nitrogen and oxygen, with carbon dioxide being present at a level of only 340 ppm. In part, this difference may stem more from temperature differences than from intrinsic differences. Over the lifetime of the Earth, an amount of carbon dioxide comparable to that in Venus's atmosphere was vented out of the Earth's hot interior. The outgassed carbon dioxide remained in the atmosphere for only a short time. Almost all of it participated with rain in dissolving land rocks. Rivers carried the dissolved rock and carbon dioxide into the oceans, where they subsequently precipitated to form carbonate rocks, such as limestone. Venus's surface is much too hot for oceans of water to be present, and hence its atmosphere has been able to retain essentially all of the carbon dioxide vented from its interior.

Rare gases. There are two varieties of rare gases found in planetary atmospheres: those that were derived from the gas cloud (solar nebula) from which the planets formed (primitive rare gases), and those that were produced from the radioactive decay of certain elements, such as potassium, in the interior of the planets. A fundamental finding about the composition of Venus's atmosphere by spacecraft observations is the detection of much more primitive argon and neon than in Earth's atmosphere. Furthermore, Mars's atmosphere has even less of these rare gases than does the Earth's. These differences in the abundances of

Fig. 1. Clouds of Venus photographed in ultraviolet light from the *Pioneer Venus* spacecraft. The south pole is near the bottom of the image.

rare gases among the atmospheres of Venus, Earth, and Mars may, in part, be due to sizable differences in the rates at which their earliest atmospheres were lost. Such losses may have been caused by the blowoff of portions of these atmospheres by massive, high-velocity stray bodies that were particularly abundant in the first several hundred million years of these planets' lifetimes. The blowoff was powered by hot rock vapor generated when these bodies collided with the planets' surfaces. Mars was especially vulnerable to atmospheric blowoff because of its low mass and hence its low surface gravity. Conceivably, the Earth lost a good fraction of its earliest atmosphere when a body with a mass comparable to that of Mars hit it. Such a giant impact is considered by many as the most likely way by which the Earth's Moon formed. Since Venus and Earth have comparable amounts of carbon dioxide and nitrogen in their atmospheres and the rocks of their interiors, these atmospheric components may have been added at a later stage, perhaps from volatile-rich bodies that came from the outer solar system. SEE EARTH; MARS; MOON; SOLAR SYSTEM.

Water vapor. The amount of water vapor in Venus's atmosphere is much less (about 100,000 times) than the amount of water in the Earth's oceans. Since it is unlikely that Venus was initially endowed with so much less water than the Earth, Venus probably lost almost all of its original water over its lifetime. The loss of water from a planet may be determined primarily by the amount that is in the upper atmosphere, the stratosphere, where solar ultraviolet radiation decomposes water vapor molecules into hydrogen and oxygen. The light gas hydrogen can eventually escape the planet's gravity and be lost to space, while the leftover oxygen can combine with other gases, such as carbon monoxide, or with iron at the planet's surface. Because Venus is closer to the Sun than the Earth, its lower atmosphere was hotter, water vapor was more abundant at early times there, and hence much more of the water vapor was able to penetrate into its stratosphere than was the case for the Earth. Consequently, Venus could have lost much more water over its lifetime than did the Earth. Some confirmation of this viewpoint has been given by the finding of one of the *Pioneer Venus* experiments that there is about 100 times as much heavy water, or water containing deuterium, on Venus as on Earth. Such an enrichment of deuterium is expected since light hydrogen can more easily escape to space than heavy hydrogen.

Temperature. By detecting long-wavelength heat radiation produced at the surface, radio telescopes first showed that the surface temperature was 850°F (730 K), that is, over 600°F (350 K) higher than the boiling point of water. This result has been confirmed by direct temperature measurements of the atmosphere made from Soviet and United States spacecraft that have descended through Venus's atmosphere. The atmospheric temperature has a relatively cool value of −10°F (250 K), about 45°F (25 K) below the freezing point of water, near the top of the cloud layer, which is at a pressure of about 1/20 that at the Earth's surface. The temperature gradually increases with decreasing altitude until it reaches 850°F (730 K) at the surface, where the pressure is 90 times that at the Earth's surface.

The high value of Venus's surface temperature is not due to its being closer to the Sun than the Earth. Because its cloud layer reflects to space about 75% of the incident sunlight, Venus actually absorbs less solar energy than does the Earth. Rather, the high temperature is the result of a very efficient greenhouse effect that allows a small but significant fraction of the incident sunlight to penetrate to the surface (about 2.5% according to Soviet and United States spacecraft measurements), but prevents all except a negligible fraction of the heat generated by the surface from escaping directly to space. The thermal energy produced by the surface and hot lower atmosphere, which occurs at infrared wavelengths, is very effectively absorbed by the carbon dioxide, water vapor, sulfur dioxide, and sulfuric acid particles of the atmosphere. However, these materials are poor absorbers at visible wavelengths, where most of the solar energy lies. The greenhouse effect raises Venus's surface temperature by almost 900°F (500 K), but causes only a modest 60°F (35 K) rise in the surface temperature of the Earth. This difference is due to the Earth's atmosphere being partially transparent at some infrared wavelengths, thus permitting some surface heat to escape to space. SEE GREENHOUSE EFFECT.

Meteorology. Studies of the motion of the ultraviolet markings, the horizontal displacement experienced by atmospheric probes, and other data indicate that the atmosphere near the cloud tops is moving with a jet-stream-like velocity of about 200 mi/h (100 m/s) from east to west in the direction of Venus's rotation and only about 10 mi/h (5 m/s) from the equator toward the pole. In contrast to the situation for the Earth where only a small portion of the atmosphere moves at jet-stream speeds, the entire cloud top region on Venus moves at these large speeds. The wind speed for the most part gradually, but in a few places sharply, decreases with declining altitude and achieves values of a modest few meters per second within 6 mi (10 km) of the surface. Like the winds on the Earth, the winds on Venus are produced ultimately by differences in the amount of solar energy absorbed by different areas of the planet, such

Fig. 2. Panoramic view, in two parts, of the surface of Venus obtained by the Soviet *Venera 13* spacecraft shortly after it landed. Parts of the spacecraft and apparatus extending from it can be seen at the bottoms of the frames.

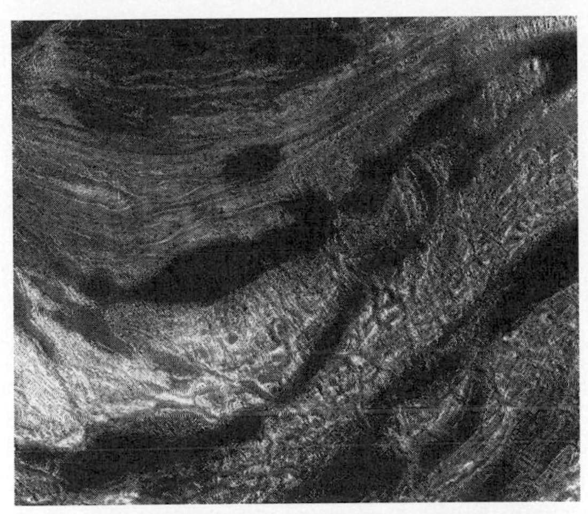

Fig. 3. *Magellan* image of the central and northern parts of the highland region Ovda Regio (north at top). Ridges running approximately east-west (horizontally) have been interpreted as resulting from crustal shortening (compression). Ridges are cut by through-going fractures and graben (fault-bounded valleys), indicating that an episode of northeast-southwest extension followed the compressional event. The youngest event was flooding of low-lying areas by smooth (radar-dark) lava flows. (*Jet Propulsion Laboratory; NASA*)

as areas at different latitudes. The large wind speeds near the cloud tops may be the result of Venus's atmosphere being very deep. The transport of momentum by the mean circulation and the eddies from the deep dense portions near the surface to the much less dense regions near the cloud tops translates sluggish motions into fast ones.

At midlatitudes and in the polar regions of the Earth, the east-west wind speeds are determined primarily by a balance between variations of pressure with latitude and the Coriolis force due to the rotation of the Earth's surface, which is shared by its atmosphere. But Venus's surface rotates too slowly for the Coriolis force to be important in its atmosphere. Rather, centrifugal forces (analogous to those on a merry-go-round) owing to the rotating motions of the winds themselves around the planet may provide the balance to the pressure gradient forces.

Because Venus's atmosphere is massive, atmospheric motions are very effective in reducing the horizontal variations of temperature in its lower atmosphere. In the region of the clouds and at lower altitudes, temperature variations occur mostly in the north-south direction, with the equator being only a few degrees warmer than the poles close to the surface and some tens of degrees warmer within the cloud region. Above the clouds, the atmosphere is actually somewhat warmer at the poles than near the equator because of the effects of heat transported by atmospheric motions. As Venus's axis of rotation is almost exactly perpendicular to its orbital plane, its climate has little seasonal variability.

Interior. The very similar mean densities of Venus and the Earth imply that Venus is made of rocks similar to those that make up the Earth. However, because Venus formed closer to the Sun, in perhaps a somewhat warmer environment, it may have initially contained a smaller amount of sulfur and water-bearing compounds. Such an environmental difference would probably not have significantly affected Venus's content of the long-lived radioactive elements uranium, potassium, and thorium. Over the lifetime of the Earth, and presumably Venus, the decay of these elements may have generated enough heat to cause these planets to become chemically differentiated, as free iron melted and sank toward their centers. Also, both planets probably formed "hot" because of gravitational energy released in bringing small chunks of rock together to form them. In this case, they may have undergone substantial differentiation in their early histories, with an accompanying release of much of their atmospheric gases. Thus, Venus's interior may be qualitatively similar to that of the Earth in having a central iron core, a middle mantle made of rocks rich in silicon, oxygen, iron, and magnesium, and a thin outer crust containing rocks enriched in silicon in comparison with the rocks of the mantle. However, in contrast to the situation for the Earth, Venus's core may now be either entirely solid or entirely liquid, which could account for the absence of a detectable magnetic field.

Rotation. In contrast to the Earth and almost all other planets, Venus rotates in the opposite direction with respect to its orbital motion about the Sun. It rotates so slowly that there are only two sunrises and sunsets per Venus year. Tides raised in the body of the planet by the Sun may have greatly reduced the rate of rotation from an initially large value, similar to the Earth's, to its present low value. Because Venus's atmosphere is so massive, tides raised by the Sun in the atmosphere may have also been important, with its current rate of rotation being determined

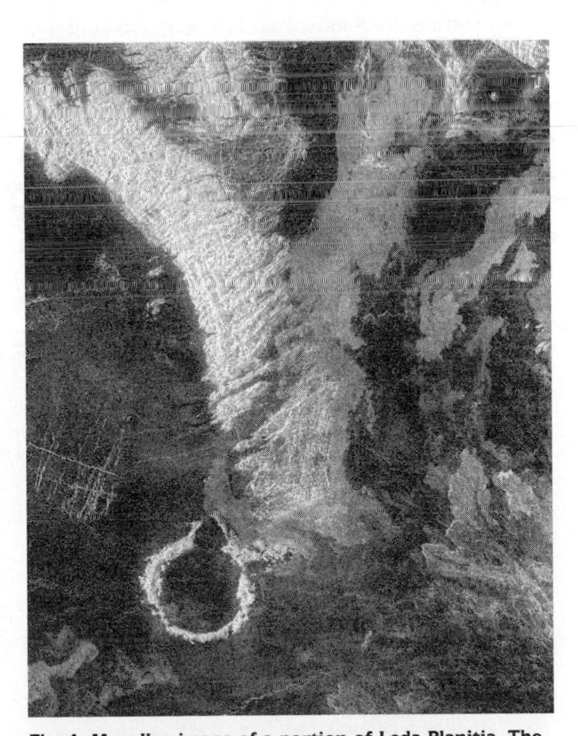

Fig. 4. *Magellan* image of a portion of Leda Planitia. The smooth (radar-dark) plains were formed by volcanic lava flows that covered the region. These flows embayed the older, radar-bright, highly fractured or chaotic highlands, sometimes called tessera, that rise out of the plains in the upper-left or northwest part of the image, as well as the circular ring structure in the lower left, which is probably an impact crater. Subsequent volcanism produced radar-bright flows in the upper right. (*Jet Propulsion Laboratory; NASA*)

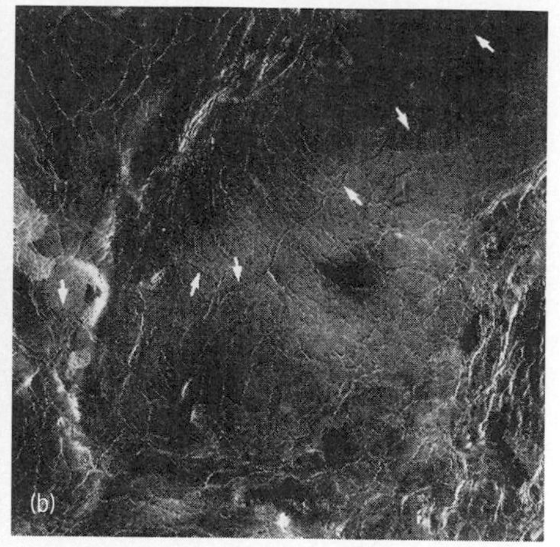

Fig. 5. ***Magellan*** **images of channels on Venus. (*a*) Part of a channel in the Lada Terra region, 750 mi (1200 km) long and 12 mi (20 km) wide, displaying numerous streamlined structures that attest to the very fluid lavas responsible for carving the channel. (*b*) A 360-mi (600-km) segment of the Hildr Channel (indicated by arrows), whose length of 4200 mi (6800 km) makes it the longest known channel in the solar system (*Jet Propulsion Laboratory; NASA*)**

chiefly by a balance between the oppositely directed torques arising from these two types of solar tides. Venus apparently presents the same face to the Earth at the times of closest approach. This suggests that the tidal forces that were exerted by the Earth also played a role and helped lock Venus in its present rotational state.

Life. The current high surface temperature and acid constitution of the clouds preclude the existence of living organisms like those that inhabit the Earth. But Venus may have had a hot ocean of water in its early history when the Sun put out less energy than it does now. If so, it is unclear whether life could have arisen then. Any early life would have been destroyed when Venus lost its oceans and achieved its current high surface temperature.

James B. Pollack

Spacecraft. Venus has been more intensely explored by spacecraft than any other planet. Some 31 (9 of them unsuccessful) United States and Soviet spacecraft have been sent to Venus. The Soviet Union *Venera* and *Vega* series of spacecraft and the United States *Mariner* and *Pioneer* series provided data on the composition of the surface and atmosphere, and the dynamics of the upper atmosphere. *Venera* landers transmitted back the first images from the surface of another planet (**Fig. 2**). The surface of Venus was seen in different places to be rocky with little granular material. The United States *Magellan* mapping mission began mapping operations on September 15, 1990, and by the end of 1992 had mapped nearly 98% of the surface. Venus is the last of the inner planets to be mapped in detail but is the most important planet for understanding how the Earth evolved as a life-sustaining habitat.

Magellan has one instrument, a 12.5-cm-wavelength radar capable of imaging the surface and mapping surface topography as an altimeter. The radar also measures radio emissions from the surface, revealing the electrical properties of the surface mate-

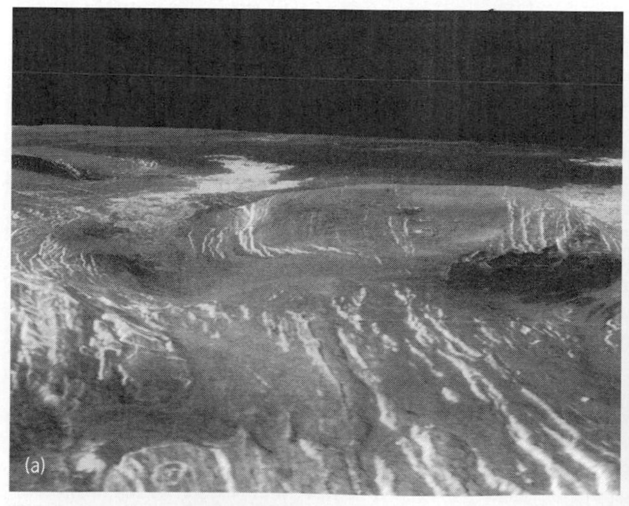

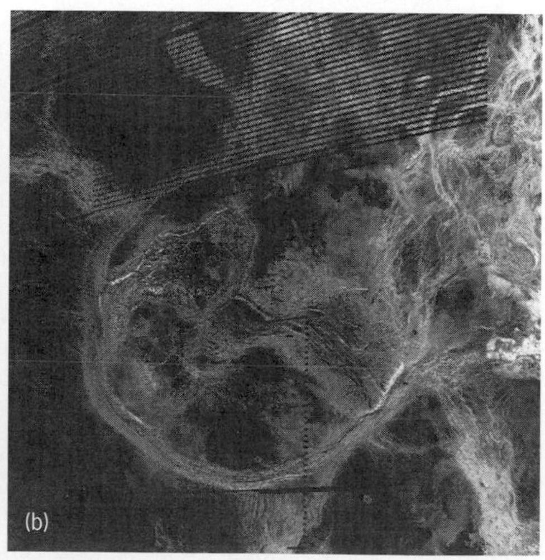

Fig. 6. ***Magellan*** **images of coronae on Venus. (*a*) Computer-simulated view of the corona Idem-Kuva, 60 mi (97 km) in diameter. Lava flows extend for hundreds of miles across the fractured plains in the background. (*b*) Artemis Chasma, whose diameter of 1300 mi (2100 km) makes it the largest corona identified. The interior contains complex systems of fractures, numerous flows, and small volcanoes, and at least two impact craters. The margin forms a steep trough with raised rims. (*Jet Propulsion Laboratory; NASA*)**

rial. The synthetic aperture radar (SAR) resolves features measuring about 400 ft (120 m) through the thick clouds that perpetually hide the planet. The altimeter measures surface elevations accurate to about 100 ft (30 m), and the main antenna, when used as a radiometer, records the natural thermal emissions from the surface to help determine the surface composition.

Surface. *Magellan*'s mapping mission has revealed a unique global volcanic and tectonic style on Venus. Broad volcanic plains make up about 85% of the surface of Venus. The rest is tectonically deformed, higher-standing terrains with complex systems of folds and faults. Regional tectonism is evident in the widespread compressional and extensional deformation of much of the surface material (**Fig. 3**). Venus apparently has a dynamic mantle that drives ongoing crustal warping. However, while various regions of the planet show evidence of motion, no evidence of Earth-like plate tectonics has been found.

Long, narrow troughs are seen in many areas where the crust has ruptured; these linear rift zones are associated with extensive broad, domical rises and shield-volcano complexes. Examples are the Sif-Gula region, Beta Regio, and Alta Regio. Large areas of the planet are covered by lava that flowed from volcanic vents (**Fig. 4**). Volcanism on Venus occurs globally, unlike Earth where volcanism is mostly restricted to linear zones that define plate boundaries.

The planet also has some unexplained surface features, including long channels meandering across the plains (**Fig. 5***a*). One channel is 4200 mi (6800 km) long (Fig. 5*b*). Its origin remains unknown. Many of the channels are clearly formed by lava, but even under the high temperature (850°F or 730 K) of the surface, most known volcanic lava compositions should solidify before they could flow far in open channels. A more exotic lava such as sulfur or a carbonate may have formed some of the longer channels. Another unique landform is the scattered flat, circular volcanoes that resemble giant pancakes. These range in diameter from about 9 to 50 mi (15 to 80 km) and are a few hundred meters thick.

Coronae are classified as volcanic-tectonic (**Fig. 6**). These features are oval to circular and range in diameter from 60 to 1300 mi (100 to 2100 km). Most are around 190 mi (300 km) in diameter. They have low relief, up to 1–2 mi (2–3 km). The margins are generally irregular concentric ridges and troughs. The interiors may be quite irregular. Many coronae have associated volcanic flows. A mantle plume model has been postulated for the formation of coronae, in which the rising lower-density plume causes doming, with contemporaneous or subsequent relaxation to form the bounding ridges and troughs.

The higher elevations of Venus, Maxwell Mons for example, have distinctly higher-than-average radar reflectivity and low emissivity. One explanation for this behavior is that, at slightly lower temperature and pressure in the more elevated regions, a metallic mineral is stable and distributed throughout the surface material to create higher reflectivity. Possible minerals could be pyrrhotite, pyrite, or magnetite.

Impact craters. Impact craters are much less abundant on Venus than on the Moon or Mars. Approximately 940 impact craters have been identified in *Magellan* images (**Fig. 7**). Most appear to be unmodified by erosion. Based on the abundance of impact craters, the average age of the surface appears to be about 5×10^8 years old. This age would imply a low rate of surface modification by erosional or even by volcanic and tectonic processes. There are small areas devoid of craters, suggesting that parts of the planet have undergone relatively recent volcanic or tectonic activity. On a global scale, impact craters are randomly distributed, allowing the determination of the mean surface age of 5×10^8 years but providing little information at a regional scale.

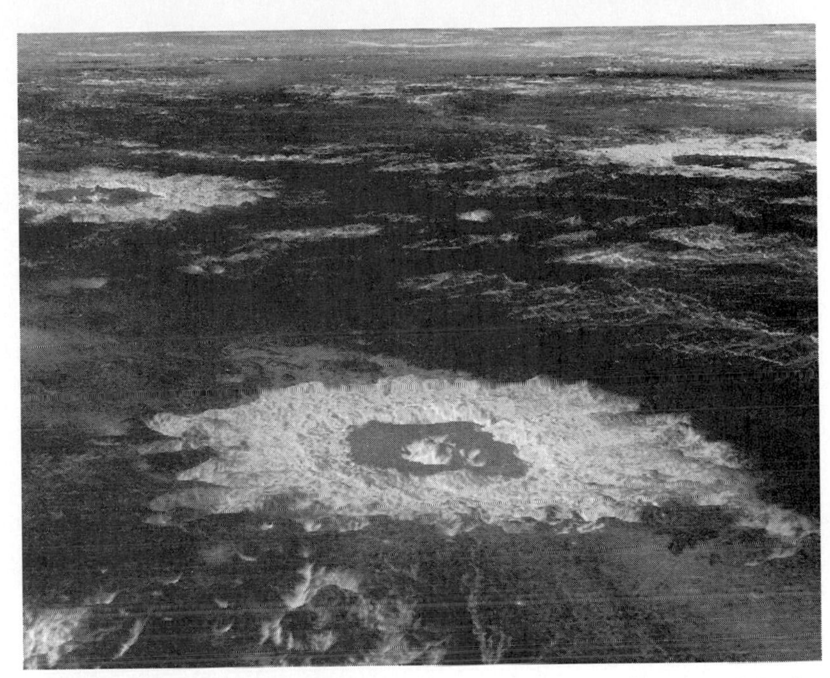

Fig. 7. Computer-simulated perspective view, based on data from *Magellan* spacecraft, showing three impact craters in the northwestern portion of Lavinia Planitia with diameters ranging from 23 to 39 mi (37 to 63 km). (*Jet Propulsion Laboratory; NASA*)

Asteroids and comets that collide with Venus should have typical velocities of about 12 mi/s (20 km/s). Venus's thick atmosphere plays a significant role in meteorite impacts. Because the atmosphere is so dense, only craters larger than 2 mi (3 km) in diameter can form, except in crater clusters where large projectiles apparently broke up before impact, peppering an area with smaller fragments. Ordinarily, a projectile that would produce a crater smaller than about 2 mi (3 km) in diameter would vaporize or break up in its transit to the surface.

Wind features. It was of major scientific interest to discover wind activity on Venus. *Magellan*'s radar saw in many areas abundant bright and dark wind streaks near topographic barriers such as small volcanic ridges. Wind streaks have been found to be most frequent near large impact craters. The craters may have provided the material that is moved by the wind. Also, the impact process itself may have produced some streaks. Further study of the orientations of some wind streaks may provide a better understanding of Venus's global wind patterns. Some of the larger impact craters have large parabolic features hundreds of miles long. Many of the wind streaks are in the vicinity of these parabolas. The parabolas may be bright or dark, and all are open to the west. This orientation is consistent with some interaction with the east-to-west winds tens of miles above the surface of Venus, which reach speeds of 120 mi/h (200 km/h). Particulate crater ejecta may have been distributed downwind. Alternatively, the turbulent atmospheric disturbance created by the impact may have been carried to the west by the general atmospheric circulation, creating deposition or erosion patterns in the lee of the event.

History of volcansim. Currently active volcanism has not been detected on Venus. One theory postulates cessation of a violent episode of global volcanism 5×10^8 years ago. Subsequently there would have been less volcanic activity, forming the large shield volcanoes such as Maat Mons that are scattered over the planet. It is estimated that the activity represented by the younger volcanoes is less than 5% of the older plains flows. An alternate hypothesis holds that there is an equilibrium between the formation of impact craters and the volcanic outpourings that erase them, so that the surface always appears to be about 5×10^8 years old. *See Planet; Planetary physics.*

R. Stephen Saunders

Bibliography. J. K. Beatty, B. O'Leary, and A. Chaikin (eds.), *The New Solar System*, 3d ed., 1989; D. J. Eicher, *Magellan* scores at Venus, *Astronomy*, 19(1):34–42, January 1991; D. M. Hunten et al., *Venus,* 1983; *Magellan* reports, *Science*, 252:247–312, April 12, 1991; *Pioneer Venus* results, *Science*, 203:743–808, February 23, 1979, and 205:41–121, July 6, 1979; A. Young and L. Young, Venus, *Sci. Amer.*, 223(3):70–78, September 1975.

Virgo

In astronomy, a constellation handed down from antiquity, visible throughout the summer months. This sixth sign of the zodiac represents a maiden in a half reclining position (see **illus**.). It is identified with the

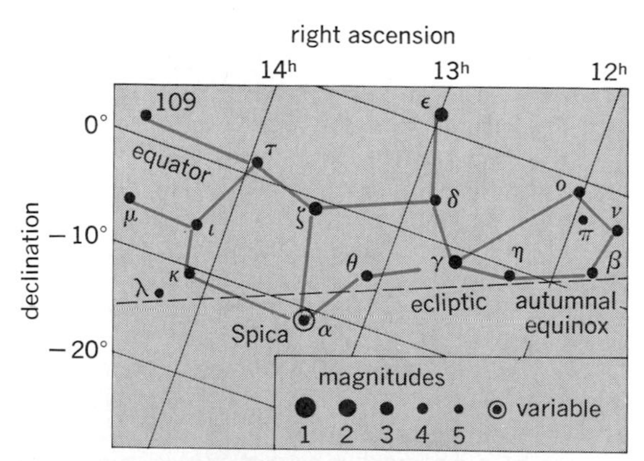

Line pattern of the constellation Virgo. The grid lines represent the coordinates of the sky. The apparent brightness, or magnitude, of the stars is shown by the sizes of the dots, which are graded by appropriate numbers as indicated.

goddess of justice, Astraea. The balance is seen by her side in the sky. The brightest star in the constellation, Spica, is a spectroscopic binary with a massive dark companion. It is a fine first-magnitude star of the purest white tint, and also a navigational star. *See Constellation: Libra; Zodiac.*

Ching-Sung Yu

Virgo Cluster

The nearest large cluster of galaxies, and one of the best studied. It dominates the distribution of bright galaxies in the sky, forming a conspicuous clump in the distribution of easily visible elliptical and spiral galaxies. The clump lies mostly in the constellation of Virgo but also extends into neighboring constellations, especially Coma Berenices. The well-known spirals M58, M61, and M100, for example, are members of the Virgo Cluster. *See Constellation; Virgo.*

Investigation. The Virgo Cluster was not recognized for what it is until the twentieth century. Although C. Messier noted that several of the nebulous objects in his catalog seemed to concentrate in the area of Virgo, the significance of this concentration was not realized for nearly 150 years. Early in the twentieth century, astronomers speculated about the "nebulae" there, but until it was established by E. Hubble and his contemporaries that these objects were, in fact, galaxies, the nature of the clustering remained obscure. *See Messier catalog.*

H. Shapley and A. Ames were the first astronomers to map out the Virgo Cluster in detail and to explore its characteristics. Their studies, published in 1930, showed that the 100 or so bright galaxies in Virgo were only the tip of the iceberg, with more than 2000 fainter galaxies showing up in deeper searches of the cluster. They found, furthermore, that the cluster extends south into what is now called the Southern Extension, where several dozen additional galaxies lie.

More modern exploration of the cluster was carried out by G. de Vaucouleurs, who promoted the idea, now universally recognized, that the Virgo Cluster is so extensive in size that it includes a gaint halo of many outlying groups, including the Local Group, to which the Milky Way Galaxy belongs. He called this structure the Local Supercluster. *See Local Group; Universe.*

In the 1980s, the realization that the Virgo Cluster held the key to many extragalactic problems led to a great upsurge in the amount of research being done on the cluster and its members. One particularly important study was the thorough catalog and analysis carried out with the Carnegie Institution's Du Pont Telescope by A. Sandage, B. Binggeli, and G. Tammann.

Structure. A realistic understanding of the structure of the Virgo Cluster can be obtained only when the distances of all of its members are reliably known. In the meantime, it is necessary to attempt to understand the structure from a study of the projected distribution of galaxies, the membership of which is not always certain. Some bright galaxies in Virgo are probably foreground objects, and most of the faintest galaxies seen there are certainly background objects at much greater distances.

In the Du Pont survey, the cluster has the appearance of two superimposed clusters (**Fig. 1**). The two concentrations are approximately 5° apart; one is centered near the very luminous elliptical galaxy M87 and the other near M49, another giant elliptical galaxy. They lie along a nearly north-south line. The spiral galaxies are much less concentrated toward these central peaks than are either the giant ellipticals or the dwarf ellipticals. Furthermore, the spirals are more common, relative to other types, in the southern concentration than in the northern. It is also found

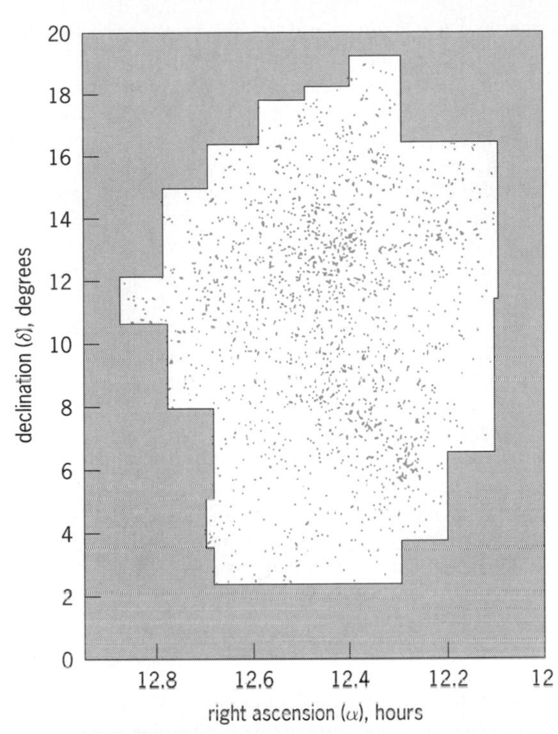

Fig. 1. Distribution of members of the Virgo Cluster on the sky, as determined in the Du Pont Telescope Survey by A. Sandage and his collaborators. (*After O.-G. Richter and B. Binggeli, eds., The Virgo Cluster, European Southern Observatory, 1985*)

that the dwarf elliptical galaxies are less concentrated toward the center of the entire complex than are any of the other types, possibly the result of mass segregation, with the more massive galaxies, under the effects of gravity and encounters, having fallen toward the center of mass more than the low-mass ones.

Populations. The Virgo Cluster is an example of a mixed-population cluster, with both early-type (elliptical and S0) and late-type (spiral and irregular) galaxies present. For members brighter than apparent blue magnitudes of approximately 15.7, about 75% are early types; most of these are dwarf galaxies of low intrinsic luminosity. The remainder are mostly spiral and dwarf irregulars. The most common type of spiral is Sc, of which there are 32 members out of the whole sample of 1343 probable members brighter than this limit (**Fig. 2**).

The total population of all types and of all intrinsic luminosities is not known, but it can be estimated from the surveys and by extrapolating to the faint end of the galaxy luminosity function (the relationship between number and luminosity of galaxies). For galaxies brighter than an absolute magnitude of about −10, a total population of 1900 galaxies is calculated. This excludes the several outlying groups and clusters, such as the Local Group (which includes 30 galaxies).

Peculiarities of members. As the nearest large cluster, the Virgo Cluster provides a good testing ground for theories of how the dense cluster environment affects the properties of galaxies. The most obvious peculiarity of the Virgo galaxies is the rather limited extent of the disks of the spirals. Especially evident in their distribution of neutral hydrogen, as measured at radio wavelengths, the spiral galaxies have possibly been stripped of their outer gas by encounters with other galaxies in the cluster. A related phenomenon is the presence of what are called anemic galaxies, spirals that have unusually low surface brightness and inconspicuous spiral arms. There are several such galaxies in the Virgo Cluster, possibly the result of direct collisions or other tidal disturbances that have removed gas from the disks, inhibiting star formation there and leaving unusually pale galaxies in place of once-robust ones.

Dynamics. The motions of the galaxies of the Virgo Cluster give information about the past and the future of the entity. How it formed in the first place may be deduced from the present state of the motions, which have been gradually altered by gravitational interactions but which still show some of the characteristics of the initial conditions. The present state of the cluster's interactions can tell about the prospects for the smoothing out of its motions or for eventual collapse or evaporation. In the 1980s a considerable amount of effort was expended in obtaining many velocities of Virgo members so as to refine knowledge of the cluster's dynamical state.

One of the conclusions of work on the velocities is that the cluster is more complex than was once thought. It is not a simple, single group, acting under gravity in a symmetrical way. The velocities in the core are complicated. The elliptical galaxies form a relatively regular, concentrated subcluster that may constitute a dynamically smoothed system, in which gravitational interactions have erased any irregularity of shape. However, the spirals have a wider distribution in space and an accompanying larger spread in velocities; they seem to be still in the process of falling into the cluster core.

Fig. 2. Two members of the Virgo Cluster, an Sc galaxy (top) and an edge-on Sb galaxy (bottom). (*Lick Observatory photograph by P. Hodge*)

The velocities can give value for the entire cluster mass and for the ratio of mass to light, a useful ratio for gauging the amount of dark matter in a cluster. For the Virgo Cluster, the mass-to-light ratio is found to be in the range of 500 in solar units, depending on how well smoothed the velocities have become, which is still uncertain. This value indicates that the cluster contains mostly unseen matter, as the average mass-to-light ratio for a population of normal stars is about 1. The total mass of the cluster is calculated to be 4×10^{14} times the mass of the Sun. *See Cosmology*.

Distance. Determining the distance to the Virgo Cluster has been one of the more important tasks of twentieth-century astronomy. The Virgo Cluster forms an essential stepping-stone for the cosmic distance scale, as it provides a connection between local distance criteria and the use of the Hubble law of gauging distances. The best measurements of the distance to the Virgo Cluster are based on the study of its globular star clusters, its planetary nebulae, its novae, and its supernovae. These give a mean distance of 16 megaparsecs (5×10^7 light-years). When the infall velocity of the Milky Way Galaxy toward the center of the Virgo Cluster is taken into account, this distance indicates that the local cosmic value of the Hubble constant, which relates expansion velocity to distance, is about 70 km per second per megaparsec. *See Galaxy, external; Hubble constant*.

Paul Hodge

Bibliography. O.-G. Richter and B. Binggeli (eds.), *The Virgo Cluster*, 1985; A. Sandage and G. Tammann, *A Revised Shapley Ames Catalog of Galaxies*, 1981; S. van den Bergh and C. Pritchet (eds.), *The Extragalactic Distance Scale*, 1988; F. Zwicky, *Morphological Astronomy*, 1957.

Weakly interacting massive particle (WIMP)

A hypothetical elementary particle that could provide simultaneous solutions to at least four long-standing astrophysical puzzles: the solar neutrino and pulsation problems, and the nature of the mysterious galactic dark-matter background and the so-called missing mass of the universe. The solar neutrino problem sets the most stringent requirements on WIMP characteristics. Consequently, detection schemes have been proposed and are under way. Only WIMPs that meet these conditions will be considered here.

The concept dates from 1977, when it was suggested that a suitably massive particle, interacting only via gravity and the weak interaction, would survive the big bang in sufficient numbers to dominate conventional matter and possibly to eventually reverse the expansion of the universe. If present in stars like the Sun, such particles would have remarkable consequences. Almost paradoxically, the great strength of WIMPs results from the weakness of their interactions: the resulting long mean-free-path ensures communication over great distances, the most important factor in efficient energy transport.

Effects in Sun. WIMPs with masses and scattering cross sections in the respective ranges 4–10 GeV/c^2 (where c is the speed of light) and 10^{-42} to 10^{-40} m^2 soon reach a quasithermodynamic equilibrium near the solar center, occupying a mass-dependent scale height (the radius at which the density drops to $1/e$ = 0.37 of its density at the center) of approximately 0.05–0.08 of the solar radius. Collisions take place on gravitationally confined but otherwise free (nonkeplerian) orbits (period of approximately 15 min) at a frequency from approximately four times per orbit to once every 25 orbits. Energy removed by these collisions from hotter, conventional, central particles is delivered to cooler regions farther out. Energy transport in the Sun is enhanced over conventional photon diffusion by the ratio of the photon diffusion time scale (approximately 10^5 years) to WIMP collision times (4 min to 6 h). Because of this enormous factor, WIMP abundances of only about 1 in 10^{11} solar particles suffice to dominate central energy transport. *See Sun*.

Solar neutrino problem. The efficient, gravitationally confined, central energy transport reduces the temperature gradient and thus lowers central temperatures (see **illus.**) in precisely those regions of the Sun responsible for the highly temperature-dependent flux of higher-energy neutrinos detectable by the chlorine-37 experiment, and could therefore explain why less than the expected number of solar neutrinos are detected in this experiment. Since a minor part of the photon luminosity is produced in this part of the Sun, modest, self-consistent adjustments elsewhere easily maintain the total luminosity constraint. *See Solar neutrinos*.

Solar pulsation problem. The frequency separations of successive low-degree, p-mode solar oscillations are well observed and sensitive to the sound speed near the solar center. Separations in the standard solar model are too large by approximately 10%; other attempts to solve the solar neutrino problem make them too large by up to 60%. Only the WIMP model satisfies both the neutrino and the p-mode separation constraints without any arbitrary adjustment of the initially published model. It also makes distinctly different predictions for the properties of the controver-

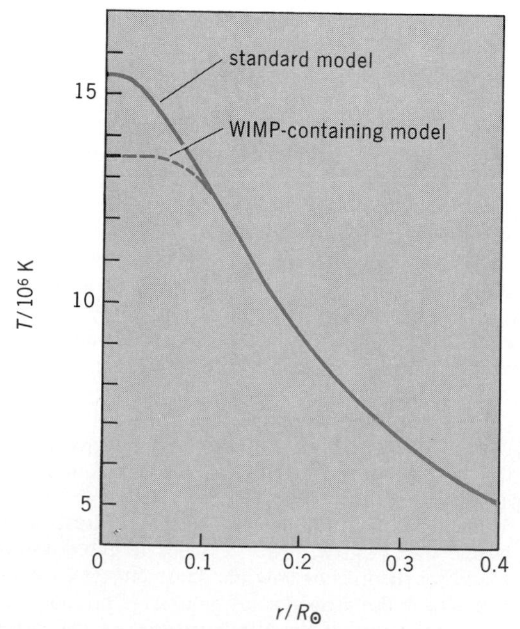

Plot of absolute temperature T versus radius r (as fraction of solar radius $R_\odot$ in central regions of the current Sun for the standard model and a WIMP-containing model.

sial g-mode oscillations. Improved data from space-based observations may help test these predictions. *See* HELIOSEISMOLOGY.

Galactic dark-matter background. Dynamical arguments indicate the presence, but not the nature, of this background. If it is composed of WIMPs with the very properties needed in the solar neutrino problem, the Sun would indeed accrete the desired number while orbiting in the Milky Way Galaxy for its known lifetime. This is either deeply significant or a remarkable coincidence.

The Milky Way Galaxy's dark-matter background would be a concentrated fossil of that universal background apparently required to produce galaxies with the broad statistical properties now observed. The solar neutrino problem may thus be the last link in a chain that stretches, via the WIMP connection, from an origin in the big bang at the formation of the universe to the present time and locality in the heart of the Sun. *See* COSMOLOGY; GALAXY, EXTERNAL; UNIVERSE.

Detection. Long development times are expected for cryogenic detectors of a broader class of dark-matter candidates. However, solar-neutrino-satisfying WIMPs may be detectable with conventional, noncryogenic, silicon diode ionization detectors, if their interaction is nonaxial. If the WIMP interaction is axial, hydrogen or deuterium in a modified low-pressure time-projection chamber could work. With silicon, anticipated event rates of up to several thousand per day may well lead to the discovery of WIMP drift, a seasonal modulation in the detected galactic background due to the Earth's changing direction of motion around the Sun.

John Faulkner

Bibliography. J. Christensen-Dalsgaard, D. O. Gough, and J. Toomre, Seismology of the Sun, *Science*, 229:923–931, 1985; J. Faulkner, D. O. Gough, and M. N. Vahia, Weakly interacting massive particles and solar oscillations, *Nature*, 321:226–229, 1986; R. L. Gilliland et al., Solar models with energy transport by weakly interacting particles, *Astrophys. J.*, 306:703–709, 1986; W. H. Press and D. N. Spergel, Capture by the Sun of a galactic population of weakly interacting, massive particles, *Astrophys. J.*, 296:679–684, 1985; J. Primack, D. Seckel, and B. Sadoulet, Detection of cosmic dark matter, *Annu. Rev. Nucl. Part. Sci.*, 38:751–807, 1988.

White dwarf star

An intrinsically faint star of very small radius and high density. The mass of a typical white dwarf is about 0.7 that of the Sun, and the average radius is 5000 mi (8000 km). The mean density is 600,000 g/cm^3 (or about 10 tons/in.^{3}). Surface temperatures range from 4000 K (7000°F) to over 100,000 K (180,000°F), and brightness is from 10^{-4} to 10 times that of the Sun. Such stars are composed largely of helium, carbon, or heavier elements.

Observations of the hotter white dwarfs of normal spectrum have confirmed the Einstein gravitational redshift. The average of 50 white dwarfs with hydrogen lines shows an apparent velocity or recession from the Earth between 25 and 30 mi/s (40 and 50 km/s). The most probable mass deduced from these observations is about 0.7 times that of the Sun, and the radius 1% that of the Sun. *See* GRAVITATIONAL REDSHIFT.

The theoretical upper limit of a nonrotating white dwarf mass is about 1.4 times greater than the Sun. Given the high density, at any plausible temperature the electronics can become degenerate. All low-electron-energy states are occupied, with only electrons of exceptionally high energy being able to change their positions and momenta freely.

Atmosphere. Spectroscopic observations of white dwarfs demonstrate atmospheric compositions reflecting various stages of nuclear evolution of their red giant parents. Their cores must contain helium, carbon, and heavier elements. Their atmospheres, however, show the strangest dichotomy in composition of any stars. Either they are essentially pure hydrogen with no metals (hydrogen to helium ratio greater than 10^3), or they are pure helium, sometimes with some metals (helium to hydrogen ratio greater than 10^4). White dwarfs with hot hydrogen or helium atmospheres have been found from space observations of *Apollo-Soyuz*, *Skylab*, and the *International Ultraviolet Explorer* satellites. Results from space have, in general, confirmed those from the ground, and have given confidence in the higher surface temperatures, which surpass 100,000 K (180,000°F) in a few stars.

Fundamental data. The number of white dwarfs for which substantial information has been collected has tripled since the mid-1970s. Some 1500 stars have been observed photoelectrically and spectroscopically. A new quantitative classification scheme gives both temperature and dominant composition parameters for these stars. White dwarfs have a small spread in radius, so that their absolute luminosities can be derived from their surface temperatures; the latter are estimated from their colors. More elaborate temperature and composition determinations are made from the energy distribution observed over a wide range of wavelengths (infrared to ultraviolet) fitted to theoretical models of the atmospheres. The properties of such models, computed for atmospheres dominated by hydrogen, have been published for temperatures between 5000 and 100,000 K (8500 and 180,000°F), and fit the observed fluxes quite well. Atmospheres which are dominantly helium (with carbon) have been analyzed in somewhat less detail.

White dwarfs had usually been recognized by their blue color (that is, high temperature) and small distance indicated by large angular proper-motion; numerous blue stars, selected only by color, have now been found by multicolor photography supplemented by photoelectric colors and spectra. The total number of hot white dwarfs in space is thus well known, about 10^{-3} pc^{-3} (1 parsec = 1.9×10^{13} mi = 3.1×10^{13} km); since the search for cool degenerate stars is difficult and incomplete, their total number can only be estimated below 10^{-2} pc^{-3}, including stars that are as cool as the Sun. Another useful generalization concerns the frequency of white dwarfs with atmospheres containing helium, carbon, or metals, which is about 20% of those containing dominantly hydrogen.

Thermal energy. A white dwarf is one final stage of stellar evolution, with thermonuclear energy sources nearly extinct. The heavy nuclei are still mobile, and their thermal energy maintains the star's radiation. The star cools and becomes fainter and redder. The thermal energy, which is normally $\frac{3}{2}kT$ (where T is the thermodynamic temperature and k is Boltzmann's constant) for a gas, per ion, becomes 3 kT for a solid. The transition between the state of free mobility and a lattice state occurs at about 2×10^7 K (3.6×10^7°F) at these high densities. The total energy available is proportional to the total number of nucleons and is therefore inversely proportional to the atomic number; the lifetime at a given brightness depends on these considerations. About 3% of the stars are now white dwarfs, but in about 5×10^9 years all present stars of higher luminosity than the Sun will have burned out to the white dwarf stage. *See Stellar evolution*.

An anomaly has been found in the number and relative frequency of cool, red white dwarfs. It had been expected that these would be very common, but, in fact, objects that are more than 10,000 times fainter than the Sun are rare. A further peculiarity of the available thermal energy may account for this. At surface temperature below 4000 K (7000°F), the solid lattice of the core is below its Debye temperature. The specific heat is therefore very small, and the surface radiation cools the star at an unexpectedly rapid rate.

Interpretation of composition. The increased knowledge of the surface composition of white dwarf atmospheres has introduced more problems than solutions. About 50 white dwarfs have had their compositions determined, while many others are obviously hydrogen-dominated. Since normal stellar composition is dominantly hydrogen (hydrogen-to-helium ratio of about 10), those white dwarfs which have high helium or carbon content are explained as the result of thermonuclear processes which have destroyed hydrogen, yielding helium, or have burnt some helium, yielding carbon. The carbon/helium ratio is found to be widely variable. Many of the stars with helium atmospheres also show lines of metals, such as calcium, magnesium, and iron, but almost always with much lower metal/helium ratios than in normal stars.

Such complications must be understood as the result of competition between gravitational accretion of matter (of normal composition) from interstellar space, and the rapid, downward diffusion of heavy elements in the intense gravitational field. This field is 10^4 times greater than in the Sun, that is, about 10^6 m/s^2 (3×10^6 ft/s^2); the differential diffusion velocities can separate iron from hydrogen at the surface of a white dwarf in a few hundred years. Thus the hydrogen-dominated atmospheres may arise from an episode of accretion of interstellar matter, followed by an interval of settling-out. Stars which have had extensive mixing between their atmospheres and dense cores, without an accretion episode, might be dominated by helium and carbon. Another force which may lift and support heavy elements is the pressure of radiation, which is appreciable at high temperatures for elements not completely stripped of their electrons.

The composition of the interior of white dwarfs must be one with a negligible hydrogen/helium ratio, and appreciable helium and carbon, as a result of thermonuclear exhaustion of hydrogen. The observable atmosphere and envelope contain at most a few percent of the total mass. *See Nucleosynthesis*.

Intrinsic variability. An unexpected discovery, made possible by rapid photoelectric photometry, was that of a new class of variable white dwarfs, called the ZZ Ceti stars. About 20 have been fully studied, all of them showing multiperiodic, small-amplitude variations; the star can switch from one period to another in a few days. Several periods may coexist with appreciable amplitude. These are interpreted as surface waves (the so-called g-mode) with many radial nodes, in an atmosphere whose composition varies with depth. The luminosity variation reflects chiefly a temperature variation of parts of the visible atmosphere and, in contrast to pulsating variables such as the cepheids, negligible radial motion occurs. In white dwarfs with dominantly hydrogen atmospheres, it appears that nearly all stars with effective temperatures from 11,000 to 12,000 K (19,000 to 21,000°F) are ZZ Ceti variables; that is, every hydrogen-atmosphere white dwarf as it cools through this temperature range develops high-mode, nonradial oscillations. The periodicities observed lie in the range 100 to 1200 s, much too long to be radial pulsations (or overtones); the amplitudes are from 0.005 to 0.30 magnitude, and the periods observed, P, are remarkably stable ($dP/P < 10^{-13}$, where dP is the variation in the period).

A triumph of the interpretation of the ZZ Ceti phenomenon was a prediction of the existence of the same type of instability in much hotter stars with helium atmospheres. In hydrogen-atmosphere white dwarfs, hydrogen must be confined to a thin layer (about 10^{-10} the mass of the star) floating on top of a helium envelope. The force that drives the oscillations is found in the partially ionized hydrogen, which becomes convectionally unstable. This model develops nonradial oscillations near the surface temperature range given by observations. If instead theorists assumed a thin helium-dominated atmosphere, they found that at temperatures between 20,000 and 30,000 K (36,000 and 54,000°F) partial ionization of helium could drive nonradial pulsations. Search among helium-atmosphere degenerates resulted in the discovery of one hot variable of period near 600 s, in the correct temperature range. Finally, two even hotter degenerates near 100,000 K (180,000°F) were found, in 1983, to be rapid pulsators. *See Variable star*.

Binary systems. White dwarfs have been found to be one member of certain close, interacting binary stars of the type that may become classical novae and supernovae (undergoing a giant explosion), recurrent novae, or dwarf novae. This class of stars, now called cataclysmic variables, provides fascinating insights into what must be happening on a larger scale near neutron stars and black holes. A main-sequence star near a white dwarf evolves and increases in size until the matter overflows its Roche lobe and finds itself in the gravitational grasp of the white dwarf. Since the velocity of infall at the surface of a white dwarf is 3700 mi/s (6000 km/s), a hydrogen atom would have an energy of 25 keV on impact. Furthermore, excessive hydrogen in a white dwarf would result eventually in a thermonuclear runaway. In fact, a dense accretion disk, like the rings of Saturn, is formed around the white dwarf. This provides far-ultraviolet and x-ray flux, rapid flickering in intensity, and occasionally violent outbursts. The rate of mass transfer affects the thickness, and even the existence, of the accretion disk. The magnitude of the outburst and the interval between outbursts are correlated, and depend on the mass-transfer rate. *See Nova*.

It is possible for both stars in an initially close binary to become white dwarfs with summed mass exceeding the Chandrasekhar limit of 1.4 solar masses. An extraordinary situation may arise if tidal interactions bring the stars close enough together for substantial orbital energy to be lost as gravity waves. Gravitational energy loss increases as $(v/c)^5$, where v is the stars' velocity and c is the speed of light; that is, this loss becomes catastrophically rapid as the pair narrows. The relative kinetic energy at the beginning of a merger is of the order of 10^{44} J (10^{51} ergs), much of which may supply energetic photons or particles. But since no stable structural configuration exists for a degenerate star of mass exceeding the Chandrasekhar limit, the merged star implodes to the radius typical of a neutron star. The difference of gravitational potentials at the white dwarf radius (4000 mi or 7000 km) and the ultimate neutron star radius (6 mi or 10 km) is of the order of 10^{46} J (10^{53} ergs), only slightly less than the rest energy of the masses involved. This dramatic scenario has been suggested as a possible origin of luminous supernovae of type I, which occur in elliptical galaxies lacking massive young stars. In contrast, supernovae of type II occur in spiral galaxies, by the collapse of young stars of 100 or more solar masses to a black hole. Until now, no binary white dwarfs have been found which approach the ultrashort periods of a few seconds required for rapid orbital evolution by gravitational-wave radiation. SEE BINARY STAR; GRAVITATION; GRAVITATIONAL COLLAPSE; NEUTRON STAR; SUPERNOVA.

Magnetic fields. Magnetic fields have been found to exist in the atmospheres of about 15 white dwarfs, with well-established fields commonly in the range of 3 to 30 megagauss (0.3 to 3 kiloteslas). In this range of fields, spectral lines are split by the Zeeman effect, the classical Zeeman triplet of hydrogen being commonly observed. At stronger fields, circular and linear polarization is found in the continuous spectrum. At a very strong field of 300 MG (30 kT), the gyrofrequency of an electron corresponds to the wavelength of ultraviolet light; thus, a strong interaction between visible light and an ionized gas may be expected with so strong a field. At least one white dwarf exhibits such cyclotron absorption, with possible overtones. In another strong-field white dwarf, molecular bands of molecular carbon (C_2) are strengthened and shifted. SEE ZEEMAN EFFECT.

A field in the range 300 to 700 MG (30 to 70 kT) has been found. The two lowest states of the hydrogen atom produce the Lyα line at 121.6 nanometers, but at such enormous fields both hydrogen states are shifted so that the wavelength of the one observable Zeeman component reaches 134.5 nm, where it becomes stationary as the field varies from 300 to 800 MG (30 to 80 kT). Thus, an absorption line resembling a molecular bandhead at that wavelength is produced by atomic hydrogen on a star whose surface is covered by such enormous fields. At still larger fields, the Lyα line would be blurred into invisibility. It was in this same white dwarf that magnetic fields were first detected by polarization of the continuous spectrum. A few spectral features still remain unidentified in other magnetic white dwarfs.

The volume force represented by strong magnetic fields is enormous, about 10^{16} ergs/cm^3 (10^{15} joules/m^3), which is the pressure in a gas at 10,000 K (18,000°F), and density 10,000 g/cm^3 (150 lb/in.^{3}). Therefore, gas dynamics at the surface of such a strongly magnetized gas is dominated by the magnetic forces. In some cataclysmic binaries, much weaker magnetic fields control the motion of infalling gases focusing matter into an accretion funnel or spot. SEE MAGNETOHYDRODYNAMICS; STAR.

Jesse L. Greenstein

Bibliography. J. R. P. Angel, Magnetic white dwarfs, *Annu. Rev. Astron. Ap.*, 16:487–519, 1978; J. L. Greenstein, Spectrophotometry of the white dwarfs, *Astrophys. J.*, 276:602–620, 1984; J. Liebert, White dwarf stars, *Annu. Rev. Astron. Ap.*, 18:363–398, 1980; G. P. McCook and E. M. Sion, *Catalogue of Spectroscopically Observed White Dwarfs*, 2d ed., 1984; H. M. Van Horn and V. Weidemann (eds.), *White Dwarfs and Variable Degenerate Stars*, 1979; G. Wegner (ed.), *White Dwarfs*, 1989; F. Wesemael et al., Atmospheres for hot, high-gravity stars, *Astrophys. J. Suppl.*, 43:159–303, 1980, and 45:177–257, 1981.

Wolf-Rayet star

A member of a class of very hot stars (180,000–63,000°F or 100,000–35,000 K) which characteristically show broad, bright emission lines in their spectra. Wolf-Rayet stars are classified into main groups, the WN (nitrogen) and the WC (carbon) stars. The emission lines show very high excitation (lines of He^+ and even up to O^{5+} are strong) and are broadened by an unexplained mechanism. The Australian stellar interferometer has shown that the angular size of the emission-line region is five times that of the region producing the continuous spectrum. Displaced absorption lines caused by fast-moving gas shells are sometimes seen, especially in the ultraviolet. The simple explanation of expansion of an outer layer with velocities up to 1250 mi/s (2000 km/s) is found insufficient from studies of Wolf-Rayet stars which happen to be members of a spectroscopic binary system. Scattering of light by fast-moving electrons has also been suggested. Luminosities must be very high, in the range of 10^4–10^5 times that of the Sun. These stars are probably very young and represent an unstable and short-lived early stage in stellar evolution. Many are close binaries with two stars nearly in contact. SEE STAR; STELLAR EVOLUTION.

Jesse L. Greenstein

X-ray astronomy

The study of x-ray emission from extrasolar sources. It includes the study of virtually all types of astronomical objects, from stars to galaxies and quasars. X-ray astronomy is a recent addition to the ancient science of astronomy. The first nonsolar x-ray source was detected in 1962 during a rocket flight from White Sands, New Mexico. Riccardo Giacconi and coworkers designed the experiment for this flight and observed the source, later identified with a star in the constellation Scorpius (Sco X-1), and a uniform background of x-rays which comes from beyond the Milky Way Galaxy.

X-ray astronomy is a space science that requires rockets or space satellites to carry experiments above the Earth's atmosphere, which would otherwise absorb the radiation. The x-ray region of the electromagnetic spectrum extends from wavelengths of about 10 picometers to a few tens of nanometers, with shorter wavelengths corresponding to higher-energy photons (1 nm corresponds to about 1000 eV). X-ray astronomy is subdivided into broad bands—soft and hard—depending upon the energy of the radiation being studied. Observations in the soft band (below about 10 keV) must be carried out above the atmosphere, while hard x-ray observations can be made at high altitudes achieved with balloons. There is a limit to the observable spectrum at low energies due to absorption by neutral interstellar hydrogen gas. The low-energy cutoff depends on direction, but can be as low as 250 eV. In the direction of the galactic center the cutoff is as high as 2000 eV. At high energies (short wavelength) space remains transparent through the entire x-ray spectrum. *See* ROCKET ASTRONOMY; SATELLITE ASTRONOMY.

Historical development. The short history of x-ray astronomy has seen rapid advances in the number of objects detected and their variety, and in understanding the nature of these objects. This can be compared to the surge in radio astronomy, following its early history, and is typical of the era of discovery in most fields. The first exploratory phase in x-ray astronomy was carried out with sounding rockets and balloon flights which spanned the late 1960s. In 1970 the first satellite devoted to x-ray astronomy was launched (*SAS-A* or *Uhuru*).

During the early 1970s this, and other satellites which followed, conducted the first full sky surveys for x-ray emissions. While rather insensitive compared to the current level of development, these early instruments provided rich returns, including the discovery of new types of objects previously unsuspected. Examples are x-ray emission from binary systems containing a collapsed star (white dwarf, neutron star, and possible black holes) and, in the extragalactic realm, the discovery of intergalactic high-temperature gas associated with clusters of galaxies which was previously unobservable.

Einstein Observatory. In 1978 a major advance came to the field with the launch of the *Einstein Observatory* (*HEAO 2*). This satellite introduced the use of focusing high-resolution optics to x-ray astronomy. The ensuing increase in sensitivity and the new ability to obtain images have resulted in a qualitative change in the scope of the field. *See* X-RAY TELESCOPE.

Advanced observatories. A joint United States–British–West German satellite devoted to x-ray astronomy, *ROSAT,* was launched in June 1990. It first surveyed the entire sky with higher sensitivity than was previously achieved. Following this 6-month activity, *ROSAT* was used in a pointed mode to observe specific sources or regions of the sky for detailed studies and searches for extremely faint objects. The observatory uses two imaging detectors. One is a high-resolution device similar to the *Einstein Observatory* high-resolution imager, which can be moved into the focus of the x-ray telescope to obtain pictures with the best sharpness. A second detector, the position-sensitive proportional counter, gives observers a larger-sized image with less resolution but more information on the energy content of the x-rays from a source.

In the late 1990s NASA expects to launch the *Advanced X-Ray Astrophysics Facility* (*AXAF*), which will be one of the four Great Observatories planned by NASA as cornerstones of the space astronomy program. *AXAF* will continue the investigations started with the *Einstein Observatory* and *ROSAT*, and because the x-ray telescope for *AXAF* will have about 10 times better resolution than previous observatories, it will allow detection of sources much fainter than before, as well as more detailed studies of these sources. *AXAF* is a long-lived observatory, designed to last at least 15 years in orbit, with occasional servicing from the space shuttle or the space station *Freedom*.

Other x-ray astronomy missions planned for the 1990s include international projects such as *ASTRO-D* with Japan, an x-ray telescope designed to function at higher energies than the *ROSAT* telescope, providing a first use of charge-coupled-device-type x-ray detectors in space to detect emission from heavy elements in the hot gases associated with x-ray sources. Another international project is the Soviet *SPECTRUM X* mission, which will include a United States experiment to measure polarization from x-ray celestial sources for the first time. Both of these missions are scheduled for launch in 1993 or 1994. In the same time frame, the National Aeronautics and Space Administration (NASA) is planning to launch the *X-Ray Timing Explorer* (*XTE*), which will be able to measure the time variations in strong x-ray sources in order to help in understanding the fundamental sources of energy for these objects. *XTE* will be able to observe sources up to fairly high energies, which will be important in resolving the different physical mechanisms that are producing the observed emission.

Identification of sources. The first nonsolar x-ray source to be identified with a specific celestial object was the Crab Nebula. This was done in 1964 with a rocket-borne experiment. Taking advantage of the lunar occulation of the Crab Nebula, the experiment detected the decrease in x-ray counting rate as the source was obscured by the Moon. This allowed a positive identification with the nebula, and also demonstrated that this particular x-ray source was itself extended in size (about 1 minute of arc in diameter). This first successful identification was quickly followed by several more as the experimental techniques were refined and better rocket control systems became available. The identification of Sco X-1 (first detected in the 1962 rocket flight) came in 1966 as a result of an accurate x-ray position and optical studies of objects within the x-ray position. The object is a faint (thirteenth-magnitude) ultraviolet star which is emitting about 1000 times as much power in x-rays as in visible light. This makes Sco X-1 a type of object which is entirely different from previously known objects, and its existence could not have been foreseen on the basis of either optical or radio observations alone. *See* CRAB NEBULA.

Types of sources. X-ray astronomy is traditionally subdivided into subspecialties depending upon the origin of sources—galactic or extragalactic—and the energy range of observations—soft (about 0.25–10 keV) and hard (greater than 10 keV). Presently, most observations have been in the soft band and have been carried out with a series of scientific satellites. These satellite observatories have progressed through the initial discovery phase of the field to systematic surveys of the entire sky, and now to detailed studies of specific objects and classes of objects. Sky surveys have detected and located hundreds of sources at a sensitivity of about 1/10,000 the strength of the brightest source Sco X-1. This is comparable to a survey of the sky in visible light which extends from the brightest stars to about ninth-magnitude stars. There are about 250,000 stars in this range of brightness and only 500 x-ray sources. **Figure 1** shows how these x-ray objects are distributed about the sky. The concentration of sources along the equator of this plot cor-

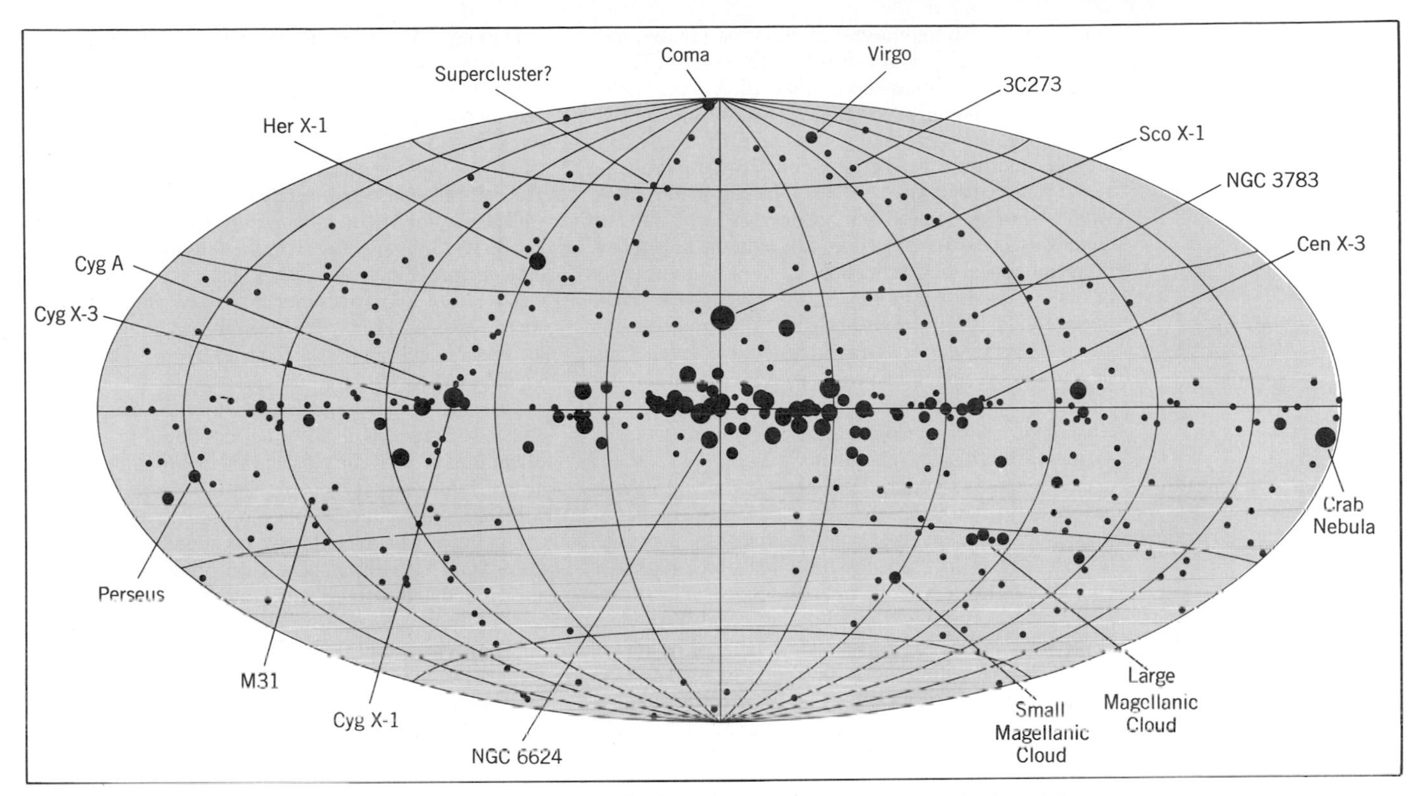

Fig. 1. Locations of about 400 x-ray sources in the Fourth Uhuru Catalog, plotted on an equal-area projection of the celestial sphere. The plane of the Milky Way Galaxy is the equator of this plot. ***(After W. Forman et al., The Fourth Uhuru Catalog of x-ray sources, Astrophys. J. Suppl., vol. 38, 1978)***

responds to objects which lie in the Milky Way Galaxy, particularly in the disk, which contains most of the galactic stars and the spiral arms. The other x ray sources which are plotted in Fig. 1 are associated mainly with extragalactic objects such as individual galaxies, clusters of galaxies, and quasars. *SEE MILKY WAY GALAXY.*

Galactic sources. The first two identified nonsolar x-ray sources (Sco X-1 and the Crab Nebula) are galactic objects which are unusual and distinct from one another.

Supernova remnants. The Crab is a supernova remnant which is left over from the explosive death of a star. This particular supernova remnant is about 900 years old and contains a rapidly rotating neutron star at its center as well as a nebula consisting of hot gas and energetic particles. The neutron star is the remaining core of the star which exploded, and its rotation causes it to radiate pulses of electromagnetic energy in radio waves, visible light, and even x-rays. Other supernova remnants have been observed and found to emit x-rays. The youngest known supernova remnant in the Milky Way Galaxy is Cas-A, in the constellation Cassiopeia, and is about 350 years old. Its appearance in x-rays is shown in **Fig. 2**, where the expanding shell of the explosion can easily be seen. More detailed analysis from the spectrum of x-ray emission shows that heavy elements, formed by nucleosynthesis within the preeruptive star, have been blown off during the explosion and are mixing into the interstellar medium. *SEE NEUTRON STAR; NUCLEOSYNTHESIS; PULSAR; SUPERNOVA.*

Compact sources. Studies of galactic objects in x-ray astronomy have led to new insights concerning the later stage of a star's life—old age, death, and in some cases metamorphosis into collapsed objects, such as a white dwarf, neutron star, or black hole. Several types of galactic x-ray sources have been identified. In addition to the supernova remnants described above, there are compact sources which form the majority of galactic emitters. These sources are further classified by their variability, which ranges from time scales as long as years to as short as milliseconds. There are transient sources which suddenly

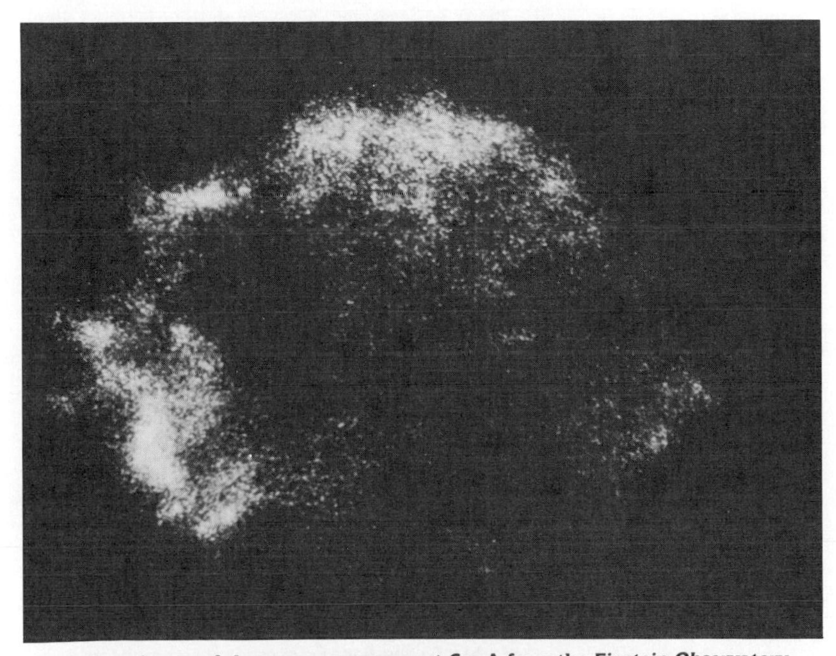

Fig. 2. X-ray image of the supernova remnant Cas-A from the *Einstein Observatory.* *(S. S. Murray, Harvard-Smithsonian Center for Astrophysics)*

appear and reach high levels of emission rapidly, then slowly fade over periods of weeks to months. This is similar to the behavior of optical novae. Other sources have periodic variations which result from binary systems where an x-ray emitting star orbits around a normal star and is periodically blocked from view. The x-ray star is often a collapsed object—a white dwarf, a neutron star, or even a black hole. In some instances the x-rays are not emitted as a steady flux, but appear pulsed with a period of a few seconds or less. This is similar to the observations of radio pulsars, and is believed to be due to the rotation of a neutron star beaming x-rays toward the Earth. Another type of time behavior is represented by the "bursters." These sources usually emit at some constant level, with occasional short bursts or flares of increased brightness. The bursts are minutes long and occur at random times. Sometimes groups of bursts occur, and often there are long quiescent periods followed by active intervals. These outbursts may well be instabilities in the x-ray emission processes associated with this type of source. *SEE BINARY STAR; BLACK HOLE; NOVA; STELLAR EVOLUTION; WHITE DWARF STAR.*

Galactic sources are also classified according to their locations within the Milky Way. Several sources are associated with globular clusters—collections of many thousands of stars—and are also burster-type sources. Other associations are those clustered toward the galactic center and those mainly associated with the spiral arms. It is widely accepted that many of the galactic x-ray sources are different types of collapsed objects which are accreting material from their surroundings, and that the in-falling matter heats up and radiates in x-rays as it comes close to the surface of the compact stars. The variety in behavior is due to differing local conditions, so that the amount of in-falling material and the path taken vary from class to class. *SEE STAR CLUSTERS; X-RAY STAR.*

Normal stars. With the increase in sensitivity obtained with the x-ray telescope of the *Einstein Observatory*, low levels of x-ray emission have been detected from all types of normal stars. This surprising result has revived interest in models of stellar structure and evolution, which must now be reevaluated. In these sources the x-ray emission is most likely coming from the corona of the stars, with energy being transported from the stellar interior, through the surface, and into the extended atmosphere. *SEE SOLAR CORONA; STAR; SUN.*

Extragalactic sources. As with galactic x-ray astronomy, virtually all types of objects have been detected as x-ray sources as a result of the increased sensitivity available. These distant objects are radiating enormous quantities of energy in the x-ray band, in some cases more than visible light. Among the more interesting types of objects detected have been apparently normal galaxies, galaxies with active nuclei, radio galaxies, clusters of galaxies, and quasars.

Galaxies and quasars. The nearest spiral galaxy is M31, the Andromeda Galaxy. This was first detected as an x-ray source with the survey satellites of the early 1970s. It has now been observed with the *Einstein Observatory*, which makes it possible to see a normal galaxy like the Milky Way resolved into individual stellar sources. The distribution and association with galactic features is similar to the Milky Way. There appears to be no emission associated with M31 as a whole, but just the summed emission of the stars. *SEE ANDROMEDA GALAXY.*

Observations of seemingly normal galaxies have led to the discovery of a new class of x-ray source, which might be labeled optically dull. These objects emit unusually large fluxes of x-rays, much more than the summed emission of their stars, but appear to be otherwise normal in their optical and radio properties. The mechanisms that produce these x-rays are not well understood, and will require further observations and study. In some cases, more detailed studies of the optical emission from such galaxies have led to the discovery of signs of nuclear activity in the form of weak emission lines. This may be an important clue in understanding the presence of a strong x-ray source.

In active nuclei galaxies, those with strong optical emission lines and nonthermal continuum spectra, x-ray emission is usually orders of magnitude in excess of that from normal galaxies, and is clearly associated with the galaxy itself. In most cases the emission comes from the galactic nucleus, and must be confined to a relatively small region on the basis of variability and lack of structure at the current observational limit of a few seconds of arc. Many astrophysicists believe that galactic nuclei are the sites of massive black holes, at least 10^6 to 10^9 times the mass of the Sun, and that the radiation from these objects is due to gravitational energy released by in-falling material. Some scientists speculate that this is a common feature of all galaxies, and that the broad range of properties such as x-ray luminosity reflects the size of the black hole and the availability of in-falling matter. Thus, normal galaxies have dormant nuclei where there is little or no material available, while Seyfert galaxies are quite active, and the quasars represent the extreme case of this mechanism. X-ray observations are particularly relevant in understanding the nature of galactic nuclei, since the high-energy emission is a more direct probe of the basic energy source than either optical or radio. *SEE QUASAR.*

Clusters of galaxies. Clusters of galaxies are another class of extragalactic x-ray sources. These are collections of hundreds to thousands of individual galaxies which form gravitationally bound systems. They are among the largest aggregates of matter in the universe and can be detected at very large distances. This allows them to be studied at early epochs in the development of the universe, and they may yield information on the order of formation of galaxies and their evolution. The space between galaxies in such clusters has been found to contain hot (approximately 10^8 K), tenuous gas which glows in x-rays. The mass of the matter is equal to the mass of the visible galaxies; it contains heavy elements which were not present in the primordial mixture from which stars and galaxies are believed to have formed.

Studies of clusters can be used to help determine the order of formation in the early universe. There are two competing scenarios for the relationship between the time that objects formed and their sizes. In one, the largest structures form first and then these evolve to form clusters of galaxies and galaxies. In the other, galaxies form first and then clump together to form the clusters and superclusters. X-ray observations of clusters of galaxies that have subclustering structure have been interpreted as examples where cluster formation is in progress, giving support to the hierarchy of small to large formation processes. At the same time, optical observations of superclusters and large voids in the space between these superclusters have been taken as evidence of the initial formation of large structures followed by the smaller objects. Attempts to detect x-rays from hot gas filling the spaces within superclusters have been inconclusive thus far and will require future observations with more sensitive instruments. *See* Cosmology; Galaxy, external; Universe.

X-ray background. In addition to observing a bright source which later was identified as Sco X-1, the 1962 rocket experiment discovered a uniform background of x-ray emission around the sky. This was confirmed by subsequent experiments and found to contain at least two components. At low energies the background is dominated by galactic emission, while above about 1 keV the radiation comes from beyond the Milky Way Galaxy. For this harder component, there are two possible sources. Either there is a truly diffuse source of emission such as hot gas or very high-energy particles, or the background is the superposition of many discrete sources associated with various extragalactic objects. Measurements of the spectrum of the background indicate a smooth shape. This can be well approximated by the combined spectra of several types of extragalactic sources or by the spectrum expected from a very hot gas. Studies of source counts have been carried out since the first all-sky survey found a large number of extragalactic objects. Depending upon the extrapolation models used, observed classes of individual sources account for about 30% and perhaps all of the background observed. This result limits the amount of hot gas that can be present in the universe to less than the critical amount necessary for closure. That is, there is not enough matter presently known to exist in the universe to allow gravitational forces to overcome the initial expansion of the big bang and cause a later contraction. If such matter exists, it must be in some unobservable form such as dark stars or isolated black holes.

Source count data indicate that the mix of different classes of extragalactic sources varies with their distance. That is, the strongest, and closest, extragalactic sources consist mainly of clusters of galaxies and active galaxies such as Seyferts. At fainter fluxes, and thus greater distances, there are mainly active galaxies such as quasars. This fact provides evidence for rapid evolution of quasars in the sense that they were more common in the earlier age of the universe than in the present epoch. An unexpected result of source-count studies has been the discovery of very few BL Lacerta–type objects at faint x-ray fluxes. These are active galaxies that have no emission lines in their spectra, but are otherwise similar to quasars. There is no widely accepted explanation for this observation. Future observations with more sensitivity and greater sky coverage will be needed to determine the total contribution of discrete sources to the x-ray background, and to probe the early universe for the composition of objects that formed at such times. *See* Astrophysics, high-energy.

Stephen S. Murray

Bibliography. J. A. Bleeker and W. Hermsen (eds.), *X-ray and Gamma Ray Astronomy,* 1989; M. Elvis (ed.), *X-ray Astronomy: 10 Years from Einstein to AXAF,* 1989; R. Giacconi, *X-ray Astronomy with the Einstein Satellite,* 1981; P. Joss (ed.), *High-energy Astrophysics in the Twenty-first Century,* 1990; C. Sarazin, *X-ray Emission from Clusters of Galaxies,* 1988; W. Tucker and R. Giacconi, *The X-ray Universe,* 1985.

X-ray star

A source of x-rays from outside the solar system. Although the Sun emits some x-radiation, it cannot be called an x-ray star because its energy output in the optical region of the spectrum vastly exceeds that in the x-ray region. Observations of the 100 or so known galactic x-ray stars show that these objects have just the reverse characteristic: the overwhelming amount of their energy is given off in the x-ray band, while in visible light they are inconspicuous. *See* Star; Sun; X-ray astronomy.

Location. The majority of galactic x-ray stars are located in a narrow layer coincident with the plane of the Milky Way Galaxy. Because this coincides with the spatial distribution of bright young stars in the Galaxy, there is at least circumstantial evidence that many x-ray stars may be associated with such young stars, or at least may have had similar stages of early evolution. However, a few x-ray stars have also been observed in the spherical halo of the Galaxy, far from the plane, and in globular clusters; both of these regions are known to contain the oldest stars in the Galaxy. It is uncertain whether the x-ray stars in the plane and the halo are closely related or are a different phenomenon. *See* Milky Way Galaxy; Stellar evolution.

Binary star systems. The study of both x-ray and optical data from the region of x-ray stars has led to the conclusion that most, or perhaps even all, such stars are members of binary star systems, consisting of one relatively normal visible star and one subluminous compact star, in a gravitationally bound orbit about each other. In the case of the x-ray data, in several sources eclipses are observed; that is, the x-ray intensity regularly vanishes or is drastically reduced, later to reappear, in a strictly periodic cycle. These eclipses are interpreted as the passage of the normal star through the line of sight between the Earth and the x-ray star.

In cases where the position of the x-ray source has been very accurately measured, it is sometimes possible to locate a visible star associated with the system. About 10 such optical counterparts are known, although all are too faint to be seen with the unaided eye. In most cases, studies of the visible component show evidence independent of the x-ray data for the binary nature of these systems. The spectral lines of the visible star indicate a periodic shift in wavelength from the Doppler effect, as the two stars move about their common center of mass. Often the x-ray star heats or gravitationally distorts the normal star sufficiently that the side facing the x-rays and the side facing away differ substantially in brightness. The effect observed at Earth is then a cyclical change in the visible star's brightness, as the x-ray star moves in its orbit, with the period of the brightness variation equal to the orbital period of the star. *See* Binary star.

Nature of x-ray stars. Although the geometry of the orbits of several x-ray binary systems is now quite accurately measured, and the visible components studied in some detail, the exact nature of the x-ray stars themselves is a matter of controversy. Most theories assume these objects to be highly evolved and quite compact. The compact star accretes gaseous matter from its nearby visible companion. This infalling matter reaches a very high temperature before it is stopped at or near the surface of the x-ray star, and the very hot gas then emits x-rays. In effect, a machine is at work to convert gravitational energy into x-radiation.

Support is lent to this theoretical picture by much of the observational data. In particular, many x-ray stars are seen to rapidly fluctuate in intensity on a time scale of a fraction of a second, implying that the emitting region is quite compact. However, these data cannot help to distinguish the precise evolutionary state of the compact star. One known type of highly evolved compact object which may be responsible for the x-ray emission is the white dwarf. Such stars have radii about equal to that of the Earth, and are numerous and well studied in the solar neighborhood. However, no x-ray system is close enough to Earth that the faint visible radiation from a white dwarf would be directly detectable. SEE WHITE DWARF STAR.

Some x-ray stars are known to emit extremely regular and rapid bursts of x-rays, in an analogous fashion to the radio pulsars. The pulsars are generally believed to be neutron stars: stellar remnants collapsed even more than white dwarfs, until they have radii of only several miles, but still possess a mass equivalent to that of the Sun. It seems likely that most of the x-ray pulsars are also neutron stars. SEE NEUTRON STAR; PULSAR.

Finally, the possibility exists that some of the x-ray stars may be black holes, the ultimate stage of stellar collapse. In these as yet hypothetical stars, matter is compressed into such a dense configuration that the intense gravitational field prevents even light from escaping. They could evidence themselves only indirectly, for example, in a binary system where the hot accreting matter could radiate x-rays just prior to falling into the black hole. SEE BLACK HOLE; GRAVITATIONAL COLLAPSE.

It is difficult to devise observational tests that will make it possible to choose unambiguously among these possible candidates for x-ray stars. One possibility may lie in measuring the mass of the compact object, by carefully noting its effects on the visible star. There are theoretical limits on the maximum masses of stable white dwarfs and neutron stars, of about 1.2 and 3 times the mass of the Sun, respectively. X-ray stars which can definitely be proven more massive than either of these limits may well provide the first observational evidence for the elusive black holes. It may emerge that all three types of compact stars—white dwarfs, neutron stars, and black holes—are candidates for different x-ray stars. SEE ASTROPHYSICS, HIGH-ENERGY.

Joseph Silk; Bruce Margon

X-ray telescope

An instrument designed to collect and detect x-rays emitted from a source outside the Earth's atmosphere and to resolve the x-rays into an image. Absorption by the atmosphere requires that x-ray telescopes be carried to high altitudes. Balloons are used for detection systems designed for higher-energy (harder) x-ray observations, whereas rockets and satellites are required for softer x-ray detectors. SEE X-RAY ASTRONOMY.

Image formation. An image-forming telescopic lens for x-ray wavelengths can be based either on the phenomenon of total external reflection at a surface where the index of refraction changes (grazing-incidence telescope) or on the principles of constructive interference (multilayer telescope).

Grazing-incidence telescope. In the case of x-rays, the index of refraction in matter is slightly less than unity. By application of Snell's laws, the condition for total external reflection is that the radiation be incident at small grazing angles, less than a critical angle of about 1°, to the reflecting surface. The value of the critical angle depends on the wavelength of the radiation and the material used. As the wavelength decreases (higher energy), the grazing angle required is smaller; as the atomic number (Z) of the material increases, the grazing angle for a given wavelength increases. The detailed reflectivity as a function of energy is complicated by x-ray absorption edges in the material.

Based on these properties, x-ray mirrors have been constructed which focus an image in two dimensions. Various configurations of surfaces are possible; **Fig. 1** shows one which has been built and used successfully for x-ray astronomy on the *Einstein Observatory* (*HEAO 2*). Two reflective surfaces of revolution are used to produce a high-quality image. These mirrors are manufactured from fused quartz, which is coated with nickel after being shaped and polished. The surfaces are extremely smooth in order to focus well, and the shape of the mirror must also be within narrow tolerances. The *Einstein Observatory* mirrors are the largest ever built; they are about 2 ft (0.6 m) in diameter and 4 ft (1.2 m) long, and the total area of polished surface is about the same as that of the 200-in. (5-m) telescope at Palomar Mountain; yet the effective area of the telescope is only about 150 in.^{2} (1000 cm^2), because of the grazing-angle geometry.

Multilayer telescope. A second type of x-ray telescope is based on the principles of constructive interference in extremely thin layers of material deposited on a mirror surface. Unlike the grazing-incidence telescope, multilayer telescopes do not require the x-rays to strike at shallow angles in order to be reflected. Instead these mirrors are similar to normal optical telescope mirrors where the incoming radiation strikes the mirror at nearly normal incidence to be reflected and focused.

The multilayers are coatings of specially selected materials that have crystal structures of regularly spaced atoms. They are evaporated onto a mirror surface that has been highly polished so that it is the

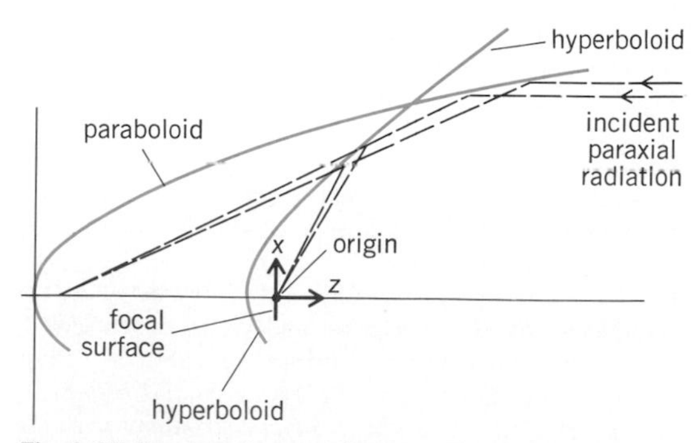

Fig. 1. Wolter type I x-ray telescope. Two surfaces of revolution (paraboloid and hyperboloid) reflect the incident x-rays to a common focus. The horizontal scale is greatly compressed. The *y* axis is perpendicular to the plane of the figure.

correct shape to focus x-rays and is so smooth that the average bumpiness on the mirror surface is comparable to the wavelength of the x-rays being reflected. This is the same quality of surface that is required for grazing-incidence mirrors, but is somewhat easier to achieve when the surface being polished is nearly flat. Several layers of coatings are used in order to achieve high reflectivity, and each layer must be of a precise thickness and composition to work properly. These coatings are deposited in vacuum to achieve the required high degree of cleanliness and purity of the multilayers if they are to work efficiently.

The advantage of a multilayer mirror is that all of the mirror area is used in collecting the x-ray radiation, whereas the grazing-incidence telescopes have only a small projected area of the actual mirror surface collecting radiation. There is an offsetting disadvantage, as the multilayer mirror reflects x-rays only within a very narrow range of energy while the grazing-incidence mirror reflects over a broad range of energies. The effect is similar to using a narrow-band filter with an optical telescope, and in some cases this can be very useful.

The first astronomical use of multilayer, normal-incidence x-ray telescopes was in photographing the Sun. Images of the solar surface of very high angular resolution were made in a selected wavelength interval near 6.7 nanometers. They showed that the filamentary structure of solar flares extends down to as small a size as can be resolved in the pictures, a result that has direct consequences for the theoretical work being done on understanding these flares.

As the technology for making multilayer mirrors improves, their use can be expected not only for astronomical purposes but as components of x-ray microscopes useful for other research applications.

Image detection. The telescope mirrors focus x rays, producing an image in two dimensions in the same manner as the lenses of optical telescopes result in images of the sky. Suitable devices are required to detect and record these images, completing the functional requirements of an astronomical telescope. Various types of x-ray detectors have been developed for this purpose. These are position sensitive devices which in effect are electronic cameras suitable for x-ray wavelengths.

The high angular resolution of the grazing-incidence telescope requires a camera that has correspondingly good spatial resolution. One type of detector uses microchannel plates (MCP; **Fig. 2**) and yields about 20 micrometers resolution for x-rays in the soft energy band (about 200–4000 eV). The microchannel plate is an array of small hollow tubes (about 15 μm in diameter) or channels which are processed to have high secondary electron yield from their inner walls. A single x-ray photon which strikes the surface of a channel produces a free electron. The electric field produced by placing a high voltage across the microchannel plate accelerates this electron, which collides with the wall of the tube to produce more electrons. This results in a cascade of electrons through the channel, multiplying in number until a sufficient signal is produced to be recorded electronically, giving the location of the event. *SEE IMAGE TUBE.*

Another type of detector being developed for x-ray imaging applications is the charge-coupled device (CCD). This is a solid-state detector that consists of microscopic silicon picture elements (pixels) in which

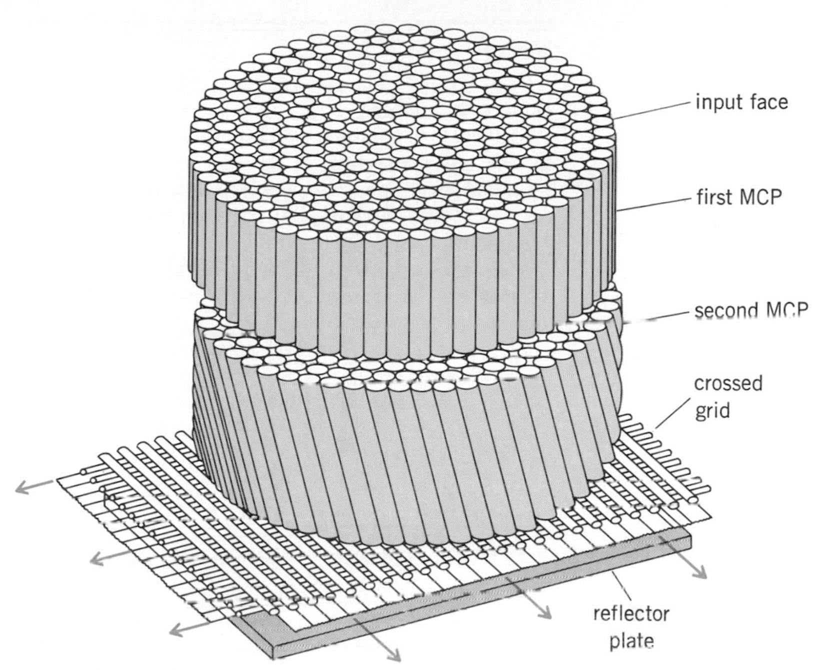

Fig. 2. Blow-up (not to scale) of the major components of a high-resolution imaging detector for an x-ray telescope. Microchannel plates (MCP) detect incident radiation and amplify the signal for position determination by the grid of crossed wires below.

electronic charges produced by the passage of an x-ray photon are collected. The charge-coupled device is periodically read out by shifting the collected charges from one pixel to the next, as in a bucket brigade. The charge is detected in a sensitive amplifier, and the corresponding position is determined by keeping count of the number of charge transfers that were made. Typical charge-coupled devices have pixels which are 15–25 μm in size, and there are up to 800 × 800 pixels in one such device. The charge-coupled device not only records the position of an event but can also yield information on the energy content of the photon. Thus this type of detector is useful as an imaging spectrometer, providing an image of an x-ray source which can be analyzed for the energy distribution of x-rays at different parts of the image. Such devices are needed to study supernova remnants and other extended sources in order to understand the conditions within such objects and how they vary. *SEE CHARGE-COUPLED DEVICES.*

Yet another type of detector used with x-ray telescopes is a form of gas counter in which x-rays are photoelectrically absorbed, yielding an electron which is detected by the ionization it produces in the gas. By operation of the counter in the proportional mode and use of planes of wires to localize the electrical signals, the position and amplitude of each event are recorded. These detectors generally have lower spatial resolution (about 1 mm) than do microchannel plate detectors, but they can be made larger in size and provide some energy measurement which is not possible in the microchannel plate device. *SEE TELESCOPE.*

Stephen S. Murray

Bibliography. G. Burbidge and A. Hewitt, *Telescopes for the '80s*, 1981; G. W. Fraser, *X-ray Detectors in Astronomy*, 1989; L. Golub (ed.), *X-ray Instrumentation in Astronomy II*, 1989; P. Joss (ed.), *High-Energy Astrophysics in the Twenty-First Century*, 1990.

Year

Any of several units of time based on the revolution of Earth around the Sun. The tropical year, to which the calendar is adjusted, is the period that is required for the mean longitude of the Sun to increase 360°. Its duration is approximately 365.24220 mean solar days. It is also the period after which the seasons repeat themselves.

The sidereal year, 365.25636 mean solar days in duration, is the average period of revolution of Earth with respect to a fixed direction in space.

The anomalistic year, 365.25964 mean solar days in duration, is the average interval between successive closest approaches of Earth to the Sun. SEE TIME.

Gerald M. Clemence

Zeeman effect

The splitting of a spectrum line formed in the presence of a magnetic field into two or more components. The effect was discovered by P. Zeeman in 1897. The details of the splitting pattern depend on the quantum states of the atom or molecule involved. For a given spectrum line, the amount of splitting is proportional to the magnetic field strength, and the intensity and polarization of the components depend on the field direction. Historically, the Zeeman effect has played a crucial role in the classification of spectrum lines. In astronomy, the effect now permits the measurement of magnetic fields in the Sun, stars, and other celestial objects.

The explanation of the Zeeman effect depends on the interaction between the internal electron magnetic moments and the external ambient magnetic field. When internal electron spin and orbital fields are large compared with the external field, they couple in a quantized fashion to produce the Zeeman effect. When the fields are comparable or the external field is larger, splitting patterns no longer follow the same rules, and the Paschen-Back effect is observed. The transition takes place for field strengths of around 1 tesla (10,000 gauss) or more.

Normal Zeeman triplet. The simplest splitting case is for a normal Zeeman triplet. If the field H is along the line of sight, two so-called sigma (σ) components are produced that are separated by a wavelength difference $\Delta\lambda$ and are oppositely circularly polarized (**Fig. 1***a*). If the field is transverse, there are three components, one undisplaced and plane-polarized in a direction perpendicular to the field, plus two displaced components polarized parallel to field direction (Fig. 1*b*).

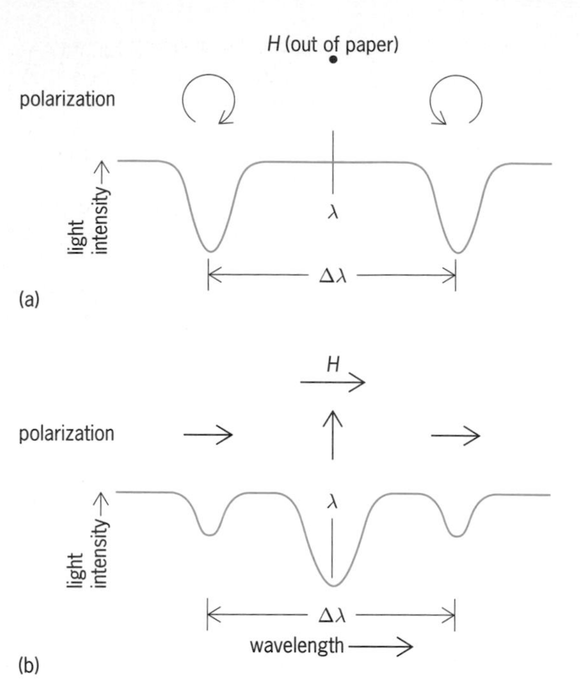

Fig. 1. Splitting patterns for magnetic fields. (*a*) Line-of-sight (longitudinal) magnetic field. (*b*) Transverse magnetic field.

The magnitude of the splitting is given by the equation below, where $\Delta\lambda$ is the splitting in nanometers,

$$\Delta\lambda = 9.34\ (10^{-8})\ g\lambda^2 H$$

g is the Landé factor and depends on the quantum levels, λ is the wavelength in nanometers, and H is the field in teslas (1 gauss = 10^{-4} tesla). The Landé factor g ranges from 0 to over 3, with an average value around unity.

In the real world there are two complicating factors: spectrum lines have finite widths, and the magnetic field direction is arbitrary, not just longitudinal or transverse. Line broadening may cause the Zeeman components to overlie each other, and the splitting is then incomplete. For a longitudinal field, however, the sigma components are oppositely polarized. By introducing alternately right-circular and left-circular polaroids, the blending can be suppressed. This is the principle of a so-called longitudinal magnetograph. However, this scheme fails for blended transverse-field components, since the displaced components are of like polarization. Transverse fields are much more difficult to measure unless the components are completely separated.

In general, spectrum lines broaden linearly with λ, but Zeeman splitting goes as λ^2. This implies that, for a given Landé factor g, separation of the sigma components increases with λ. At optical wavelengths there is an infrared advantage, and at radio wavelengths splitting conditions are even better. In the visible, the $g = 3$ iron (Fe) line at 525.0 nm is split for 0.15-T (1500-G) fields; in the infrared, the $g = 3$ Fe line at 1564.8 nm is split for 0.07 T (700 G); in radio wavelengths at the 21-cm hydrogen line, a few tenths of a nanotesla (a few microgauss) can be measured.

Anomalous Zeeman effect. A simple triplet is the exception, and most spectroscopic transitions lead to a multiplicity of components and the anomalous Zeeman effect. Twenty sigma components are not unusual. Almost always the intensity pattern is symmetric, typically rising or falling, about the undisplaced ($H = 0$) line.

Exceptions to the rules occur, however. For example, molecular cyanogen (CN) at 1163 nm produces a one-sided pattern without symmetry; Fe at 547.0 nm has a negative g; and scattered throughout the visible are a number of $g = 0$ lines.

Solar magnetographs. In 1908, G. E. Hale discovered Zeeman splittings in the spectra of sunspots. Using a favorable line, such as the $g = 3$ Fe line at 525.0 nm, he visually measured fields in sunspot umbrae up to 0.35 T (3500 G). He was not able to do much outside of spots because of inadequate splitting there. No real progress was made until after World

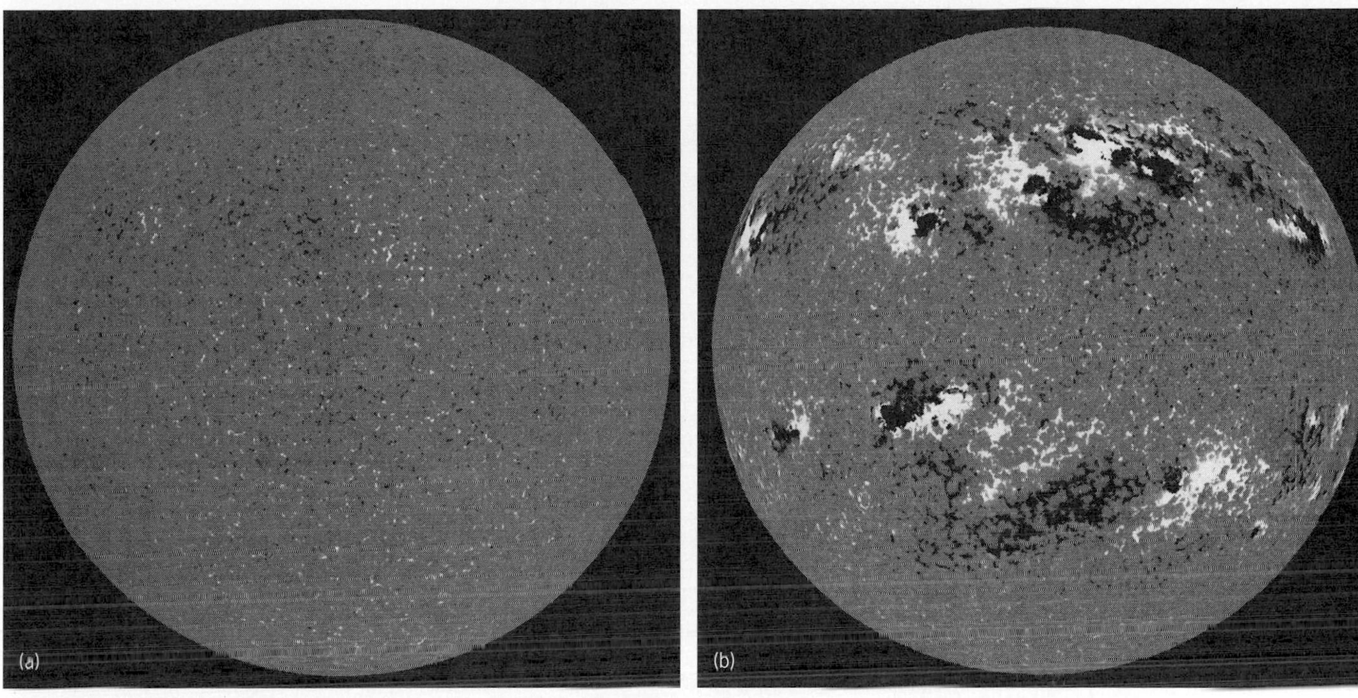

Fig. 2. Longitudinal Zeeman maps of the Sun. (*a*) Quiet disk at solar minimum. (*b*) Active disk at solar maximum. (*National Optical Astronomy Observatories*)

War II, when K. O. Kiepenheuer substituted a photomultiplier for the eye. By placing an exit slit on the wing of the Fe line at 525.0 nm while rotating a quarter-wave plate in front of the spectrograph entrance aperture, he was able to observe a periodically varying signal resulting from the alternate extinguishing of the sigma components. His results were not reproducible, however, because of false modulation from imperfections in the waveplate, because of telescope polarization, and because of line Doppler shifts. Finally in 1954, H. D. Babcock and H. W. Babcock solved the problem by introducing two photomultipliers, one in each line wing, and taking the difference signal. They replaced the rotating wave plate with an electrooptic circular analyzer. False modulation was reduced to such an extremely low level that signals of a few hundred microteslas (a few gauss) were measurable. By slowly making a raster scan of the solar image, a full-disk magnetogram was produced. With only minor alterations for computer control, their instrument continues to chart daily the Sun's magnetism from the 150-ft (45-m) tower telescope on Mount Wilson, in California. *SEE DOPPLER EFFECT; FRAUNHOFER LINES; SPECTROGRAPH.*

Practically all modern solar longitudinal magnetographs follow the Babcock recipe but with the improved spatial resolution afforded by solid-state diode arrays and charge-coupled devices. For example, while the resolution of the Mount Wilson magnetograph is about 12 arc-seconds, the system at Kitt Peak, Arizona, has a resolution of 1 arc-second, limited only by seeing (**Fig. 2**). The lower resolution of the Mount Wilson magnetograph has been retained in order to provide continuity in the data record, which goes back to 1960. *SEE CHARGE-COUPLED DEVICES.*

A vector magnetograph senses orthogonal states of plane polarization and thus measures the transverse component of magnetic field. A number of such magnetographs have been built, but because the blended plane-polarized components cannot be separated, they are not very useful for observations outside of sunspots. Observing lines in the infrared should improve this situation in the future. *SEE SOLAR MAGNETIC FIELD; SUN.*

Stellar magnetographs. Compared with the Sun, stellar magnetism is difficult to observe because not only is there less light but also the star's rotation widens the lines at the expense of the Zeeman effect. A differential analyzer, invented by H. W. Babcock, produces simultaneous spectra in right- and left-circular polarized light. Recorded photographically or with a charge-coupled device, the line shift is plotted against the effective Landé factor g, corrected for the λ^2 term. This method has been successful for Ap-type (peculiar-A-type) stars where fields from 0.1–0.2 up to 1 T (1–2 up to 10 kilogauss) are observed. Another method involves no polarization but makes use of a Fourier analysis of differences between the profiles of lines of differing Landé factor g. In this way, magnetic fields are deduced in the range 0.1–0.5 T (1000–5000 G) for a variety of late-type (spectral types G, K, and M) dwarfs. *SEE STELLAR MAGNETIC FIELD.*

Other Zeeman observations. At radio wavelengths the ability of the Zeeman effect to detect weak fields is greatly enhanced. For example, the longitudinal magnetic field in interstellar clouds may be deduced from Zeeman splitting of the 21-cm line in emission of neutral atomic hydrogen (H I). Fields of 0.5–2.0 nanoteslas (5–20 microgauss) are observed. In regions of star formation, hydroxyl (OH) maser lines at 1612 MHz display a Zeeman splitting to yield field strengths of 0.1–1.0 microtesla (1–10 milligauss). *SEE ASTRONOMICAL SPECTROSCOPY; RADIO ASTRONOMY.*

William C. Livingston

Bibliography. R. J. Bray and R. E. Loughhead, *Sunspots*, 1979; G. Herzberg, *Atomic Spectra and Atomic Structure*, 2d ed., 1944; S. Svanberg, *Atomic* *and Molecular Spectroscopy*, 1990; G. L. Verschur and K. I. Kellermann (eds.), *Galactic and Extragalactic Radio Astronomy*, 1988.

Zenith

The point directly overhead in the sky. The astronomical zenith, which is that usually meant, is the upper intersection of a plumb line with the celestial sphere. The zenith distance of a celestial object is its angular distance from the astronomical zenith and is identical with the complement of its altitude. The geocentric zenith is the upper intersection with the celestial sphere of an imagined line through the center of Earth and the observer. The point diametrically opposite the zenith is called the nadir. *See Astronomical coordinate systems; Celestial sphere.*

Gerald M. Clemence

Zodiac

A band of the sky extending 9° on each side of the ecliptic, within which the Moon and principal planets remain. It is divided into 12 conventional signs, each containing 30° of celestial longitude. These signs are named Aries, Taurus, Gemini, Cancer, Leo, Virgo, Libra, Scorpio, Sagittarius, Capricornus, Aquarius, and Pisces. The names are identical with the names of the constellations through which the Sun moves, but the actual constellations are not each 30° in extent, nor does the Sun actually enter the constellation of Aries at the vernal equinox, although the vernal equinox is sometimes conventionally called the first point of Aries. The signs of the zodiac are without practical astronomical significance. *See Ecliptic.*

Gerald M. Clemence

Zodiacal light

A diffuse, night-sky luminosity easily seen at low to middle geographic latitudes in the absence of moonlight. It is caused by sunlight scattered and absorbed by interplanetary (solar system) dust particles. Zodiacal light extends over the entire sky, but it is brightest toward the Sun and in the zodiacal band. (In the ecliptic at 30° from the Sun its visual brightness is three times that of the brightest part of the Milky Way.) It is best seen in the west after evening twilight and in the east before morning twilight (**Fig. 1**), when the ecliptic is close to the vertical. In the Northern Hemisphere this corresponds to spring evenings and autumn mornings. *See Ecliptic.*

Fig. 1. Morning zodiacal light as it typically appears at a low, northern latitude site, Mount Haleakala, Hawaii, at an altitude of 10,000 ft (3300 m). The outcropping of rock at the base of the zodiacal cone is approximately 20° from the Sun. (*P. B. Hutchison*)

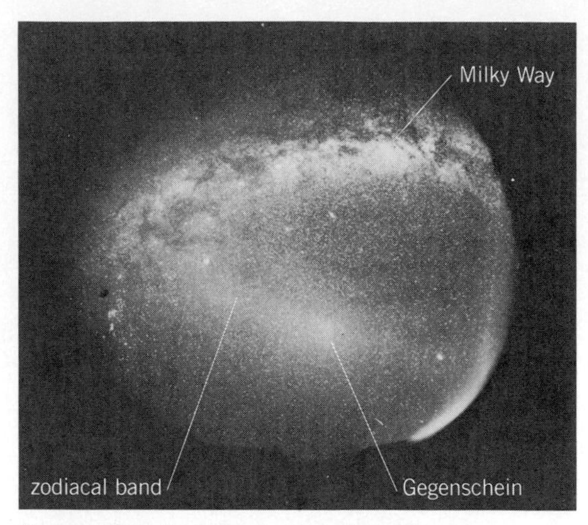

Fig. 2. All-sky photograph, taken from Mount Komagatake, Japan.

Zodiacal light brightness decreases monotonically with elongation (angular distance from the Sun) to a relatively flat minimum in the ecliptic at 120 to 140°, after which it gradually increases out to the antisolar point at 180°. The corresponding polarization reaches a maximum of approximately 20% at 70° elongation, falling to zero at 160°. The polarization changes direction between 160 and 180°, where it is again zero. Such characteristics of brightness and polarization, when combined with information on color and angular dependence, can be used to provide information on the optical properties and spatial distribution of the dust. Ground and space observations have shown the zodiacal light to have the same color as the Sun and to be remarkably constant over periods as long as a solar cycle (11 years). The density of particles responsible for the zodiacal light falls off somewhat faster than $1/R$, where R is heliocentric distance. The particles are primarily in the size range of tens to hundreds of micrometers in diameter, and are believed to originate primarily from comets. Band structure seen in infrared measurements of thermal emission from the dust may be due to particles arising from collisions in and near the asteroid belt.

Gegenschein. The enhanced brightness near the antisolar point is called the Gegenschein or counterglow (**Fig. 2**). It is barely above the visible threshold and is described by visual observers as being oval in appearance, 6 by 10° or larger, with the long axis in the ecliptic. Modern observations have shown the Gegenschein to be an intrinsic part of the zodiacal light, rather than a separate phenomenon due to the Earth's atmosphere or to a concentration of dust opposite the Earth from the Sun; that is, the zodiacal dust particles have an increased scattering efficiency near backscattering.

Fraunhofer corona. The solar corona has two principal parts: a K-corona, which is due to Thomson scattering by free electrons located very close to the Sun, and an F-corona (also referred to as Fraunhofer or false corona or inner zodiacal light), which arises from scattering by zodiacal dust seen at small elongations. Although the Earth's atmosphere complicates attempts to observe zodiacal light closer than approximately 30° to the Sun, eclipse and space observations have shown the brightness to increase smoothly all the way in to the F-corona. The F-corona or inner zodiacal light and the ''primary'' zodiacal light seen at larger elongations come primarily from dust particles located relatively close to the Sun. *See Solar corona; Sun.*

J. L. Weinberg

Contributors

Contributors

A

Akasofu, Prof. S. I. *Department of Geophysics, University of Alaska.*

Allen, Dr. Richard G. *Steward Observatory, University of Arizona.*

Aller, Prof. Lawrence H. *Department of Astronomy, University of California, Los Angeles.*

Altrock, Dr. Richard C. *National Optical Astronomy Observatories, National Solar Observatory/Sacramento Peak, Sunspot, New Mexico.*

Anderson, Dr. John D. *Jet Propulsion Laboratory, California Institute of Technology.*

B

Backer, Dr. Donald C. *Department of Astronomy, University of California.*

Bahcall, Dr. John N. *School of Natural Sciences, Institute for Advanced Studies, Princeton, New Jersey.*

Bally, Dr. John. *AT&T Bell Laboratories, Holmdel, New Jersey.*

Barger, Prof. Vernon D. *Department of Physics, University of Wisconsin, Madison.*

Beatty, J. Kelly. *Consultant, Chelmsford, Massachusetts.*

Blitz, Dr. Leo. *Laboratory for Millimeter-Wave Astronomy, University of Maryland.*

Bocko, Dr. Mark F. *Department of Physics and Astronomy, College of Arts and Science, University of Rochester.*

Bolz, Dr. Ray E. *Leonard S. Case Professor and Dean of Engineering, Case Western Reserve University.*

Boss, Dr. Alan P. *Department of Terrestrial Magnetism, Carnegie Institution, Washington, D.C.*

Brandt, Dr. John C. *Laboratory for Atmospheric and Space Physics, University of Colorado.*

Brown, Dr. Robert L. *National Radio Astronomy Observatory, Charlottesville, Virginia.*

Brownlee, Dr. Donald E., II. *Department of Astronomy, University of Washington, Seattle.*

Burke, Dr. James D. *Jet Propulsion Laboratory, California Institute of Technology.*

C

Cameron, Dr. A. G. W. *Associate Director for Planetary Sciences, Center for Astrophysics, Harvard College Observatory.*

Caughlan, Prof. Georgeanne R. *Department of Physics, Montana State University.*

Chapman, Dr. Clark R. *Planetary Science Institute, Tucson, Arizona.*

Chapman, Dr. Robert D. *Section Head, Laboratory for Solar Physics and Astrophysics, NASA Goddard Space Flight Center, Beltsville, Maryland.*

Clemence, Dr. Gerald M. *Deceased; formerly, Observatory, Yale University.*

Cleminshaw, Dr. Clarence H. *Deceased; formerly, Director Emeritus, Griffith Observatory, Los Angeles, California.*

Corwin, Dr. Harold G., Jr. *Infrared Processing and Analysis Center, California Institute of Technology.*

Cox, Dr. Arthur N. *Los Alamos Scientific Laboratory, University of California, Los Alamos, New Mexico.*

Coyle, Harold P. *Harvard-Smithsonian Center for Astrophysics, Cambridge, Massachusetts.*

Crawford, Dr. David L. *Kitt Peak National Observatory, Tucson, Arizona.*

D

Danby, Dr. J. M. A. *Department of Mathematics, North Carolina State University.*

Dent, Dr. William. *Department of Physics and Astronomy, University of Massachusetts.*

de Vaucouleurs, Prof. Gerard. *Department of Astronomy, University of Texas.*

di Cicco, Dennis. *Associate Editor, Sky & Telescope, Sky Publishing Corporation, Cambridge, Massachusetts.*

Dick, Dr. Steven. *U.S. Naval Observatory, Washington, D.C.*

Douglass, Prof. David H. *Department of Physics and Astronomy, College of Art and Science, University of Rochester.*

Duncombe, Dr. Raynor L. *Department of Aerospace Engineering, University of Texas, Austin.*

E

Elliot, Dr. James L. *Department of Earth, Atmospheric, and Planetary Sciences, Massachusetts Institute of Technology.*

Elliot, Lyn E. *Department of Earth, Atmospheric, and Planetary Sciences, Massachusetts Institute of Technology.*

English, Prof. Van H. *Department of Geography, Dartmouth College.*

Evans, Dr. John W. *National Solar Observatory, Sunspot, New Mexico.*

Evenson, Dr. Paul. *Bartol Research Institute, University of Delaware.*

F

Faulkner, Prof. John. *Board of Studies in Astronomy and Astrophysics, Lick Observatory, University of California, Santa Cruz.*

Fiala, Dr. Alan D. *U.S. Naval Observatory, Washington, D.C.*

Foster, Prof. Robert J. *Department of Geology and Geophysics, San Jose State University.*

Fretter, Prof. William B. *Department of Physics, University of California, Berkeley.*

G

Golub, Dr. Leon. *Harvard Smithsonian Center for Astrophysics, Smithsonian Astrophysical Observatory, Cambridge, Massachusetts.*

Gooding, Dr. James L. *Space Scientist, Planetary Materials Branch, NASA Lyndon B. Johnson Space Center, Houston, Texas.*

Gossner, Simone D. *U.S. Naval Observatory, Washington, D.C.*

Green, Dr. Louis C. *Department of Astronomy, Haverford College.*

Green, Dr. Richard F. *National Optical Astronomy Observatory, Tuscon, Arizona.*

Greenstein, Prof. Jesse L. *Department of Astronomy, California Institute of Technology.*

Grindlay, Dr. Jonathan E. *Center for Astrophysics, Harvard University.*

Groth, Dr. Edward J. *Department of Physics, Princeton University.*

Groves, Dr. Gordon W. *Instituto de Geofisica, Torre de Ciencias, Ciudad Universitaria, Mexico.*

H

Hagar, Charles F. *Department of Physics and Astronomy, San Francisco State University.*

Halbedel, Dr. Elaine M. *Corralitos Observatory, Las Cruces, New Mexico.*

Hardy, Dr. John. *Litton Optical Systems, Lexington, Massachusetts.*

Harrington, Dr. Robert S. *U.S. Naval Observatory, Washington, D.C.*

Harrison, Prof. Edward R. *Department of Physics and Astronomy, University of Massachusetts.*

Hewins, Dr. Roger H. *Department of Geological Sciences, Faculty of Arts and Sciences, Wright Geological Laboratory, Rutgers University.*

Hilliard, Dr. Ron. *Optomechanics Research, Inc., Vail, Arizona.*

Hodge, Dr. Paul. *Department of Astronomy, University of Washington.*

Hogg, Helen S. *David Dunlap Observatory, University of Toronto, Richmond Hill, Ontario, Canada.*

Hollweg, Prof. Joseph V. *Space Science Center, University of New Hampshire.*

Hudson, Dr. Hugh. *Center for Astrophysics and Space Sciences, University of California, San Diego.*

Hundhausen, Dr. A. J. *High Altitude Observatory, National Center for Atmospheric Research, Boulder, Colorado.*

I

Itano, Dr. Wayne M. *U.S. Department of Commerce, Time and Frequency Division, National Institute of Standards and Technology, Boulder, Colorado.*

K

Kaler, Dr. James B. *Department of Astronomy, College of Liberal Arts and Sciences, University of Illinois, Urbana-Champaign.*

Keil, Dr. Stephen L. *Chief, Solar Research Branch, National Optical Astronomy Observatories, National Solar Observatory, Air Force Geophysics Laboratory, Sunspot, New Mexico.*

King, Dr. Elbert A. *Department of Geology, University of Houston.*

Kleinmann, Dr. Douglas E. *Astronomy Program, Department of Physics and Astronomy, University of Massachusetts.*

Kleinmann, Dr. Susan G. *Astronomy Program, Department of Physics and Astronomy, University of Massachusetts.*

Klock, Dr. Benny L. *Plans and Requirements, Advanced Weapons Technology Division, Defense Mapping Agency, Washington, D.C.*

Kohman, Dr. Truman P. *Department of Chemistry, Carnegie-Mellon Institute.*

Kowal, Dr. Charles T. *Department of Astrophysics, California Institute of Technology.*

Krauss, Dr. Lawrence M. *Sloan Physics Laboratory, Center for Theoretical Physics, Yale University.*

Krisciunas, Dr. Kevin. *Joint Astronomy Centre, Hilo, Hawaii.*

Kuiper, Prof. Gerard P. *Deceased; formerly, Lunar and Planetary Laboratory, University of Arizona.*

Kulsrud, Prof. Russell M. *Department of Astrophysical Sciences, Princeton University.*

Kwitter, Dr. Karen B. *Department of Physics and Astronomy, Williams College.*

L

Lamb, Prof. Richard C. *Department of Physics, Iowa State Unversity.*

Landshoff, Dr. Rolf K. M. *Consulting Scientist, Lockheed Missiles and Space Company, Palo Alto, California.*

Latham, Dr. David W. *Center for Astrophysics, Cambridge, Massachusetts.*

Layzer, Prof. David. *Department of Astronomy, Harvard University.*

Liller, Dr. William. *Department of Astronomy, Harvard University.*

Livingston, Dr. William C. *National Solar Observatory, National Optical Astronomy Observatories, Tucson, Arizona.*

Lonsdale, Dr. Carol J. *Science and Science Support Supervisor, Infrared Processing and Analysis Center, California Institute of Technology.*

Lynds, Dr. Beverly T. *Center for Astrophysics and Space Astronomy, University of Colorado.*

M

McCarthy, Dr. Dennis D. *Time Service Division, U.S. Naval Observatory, Washington, D.C.*

McMath, Dr. Robert R. *Deceased; formerly, Director, McMath-Hulbert Observatory, University of Michigan.*

Malin, Dr. David F. *Anglo-Australian Observatory, Epping, Australia.*

Maran, Dr. Stephen P. *American Astronomical Society, Washington, D.C.*

Margon, Dr. Bruce. *Department of Astronomy, University of Washington, Seattle.*

Markowitz, Dr. William. *Department of Physics, Nova University; Editor, "Geophysical Surveys."*

Marschall, Prof. Laurence A. *Department of Physics, Gettysburg College, Gettysburg, Pennsylvania.*

Martin, Dr. Peter G. *Department of Astronomy, University of Toronto, Ontario, Canada.*

Mashhoon, Prof. Bahram. *Department of Physics and Astronomy, College of Arts and Sciences, University of Missouri.*

Mattei, Janet Akyüz. *Director, American Association of Variable Star Observers, Cambridge, Massachusetts.*
Mayall, Robert N. *Director, Planning and Research Associates, Boston, Massachusetts.*
Meeus, Dr. Jean. *Consultant, Heuvestraat, Belgium.*
Meisel, Dr. David D. *Department of Physics and Astronomy, State University of New York, Genesee.*
Moody, Capt. Alton B. *Navigation Consultant, La Jolla, California.*
Muller, Dr. Richard A. *Lawrence Berkeley Laboratory, University of California, Berkeley.*

N

Noyes, Dr. Robert W. *Harvard-Smithsonian Center for Astrophysics, Smithsonian Astrophysical Observatory, Harvard College Observatory, Cambridge, Massachusetts.*

O

Olsen, Dr. Edward J. *Department of Geology, Field Museum of Natural History, Chicago, Illinois.*
Ostro, Dr. Steven J. *Jet Propulsion Laboratory, California Institute of Technology.*
Otten, Prof. Ernst Wilhelm. *Institut für Physik, Johannes Gutenberg-Universität Mainz, Germany.*
Owen, Dr. Tobias C. *Institute for Astronomy, Honolulu, Hawaii.*

P

Page, Dr. Thornton. *National Aeronautics and Space Administration, Houston, Texas.*
Papagiannis, Prof. Michael D. *President, IAU Commission 51, Department of Astronomy, Boston University.*
Paresce, Dr. Francesco. *Space Telescope Science Institute, Baltimore, Maryland.*
Parker, Dr. Eugene. *Laboratory for Astrophysics and Space Research, Enrico Fermi Institute, University of Chicago.*
Pasachoff, Prof. Jay M. *Director, Hopkins Observatory, Williams College, Williamstown, Massachusetts.*
Paul, Dr. Jacques. *Comissariat à l'Energie Atomique, Division des Sciences de la Matière, DAPNIA/SAp, CE Suclay, Gif-sur-Yvette, France.*
Peters, Prof. Philip C. *Department of Physics, University of Washington.*
Plavec, Prof. Mirek J. *Department of Astronomy, University of California, Los Angeles.*
Pollack, Dr. James B. *Ames Research Center, National Aeronautics and Space Administration, Moffett Field, California.*
Popper, Dr. Daniel M. *Department of Astronomy, University of California, Los Angeles.*

R

Rockett, Frank H. *Engineering Consultant, Charlottesville, Virginia.*
Rust, Dr. David M. *Applied Physics Laboratory, Johns Hopkins University, Laurel, Maryland.*

S

Sandage, Dr. Allan. *Mount Wilson and Palomar Observatories, Pasadena, California.*
Sargent Dr. Anneila I. *Owens Valley Radio Observatory, California Institute of Technology.*
Saunders, Dr. R. Stephen. *Jet Propulsion Laboratory, California Institute of Technology.*
Savage, Prof. Blair D. *Washburn Observatory, Department of Astronomy, University of Wisconsin.*
Schild, Dr. Rudolph E. *Harvard-Smithsonian Center for Astrophysics, Cambridge, Massachusetts.*
Scott, Dr. Edward R. D. *Institute of Meteoritics, Department of Geology, University of New Mexico.*
Seidelmann, Dr. P. K. *Director, U.S. Naval Observatory, Washington, D.C.*
Shao, Dr. Michael. *Jet Propulsion Laboratory, California Institute of Technology.*
Shipman, Dr. Harry L. *Department of Physics, University of Delaware.*
Shore, Dr. Steven N. *Goddard Space Flight Center, NASA, Greenbelt, Maryland.*
Silk, Dr. Joseph. *Department of Astronomy, University of California, Berkeley.*
Simon, Dr. George W. *Senior Scientist, Space Physics Division, Phillips Laboratory, National Solar Observatory, Sunspot, New Mexico.*
Sinton, Dr. William M. *Department of Astronomy, University of Hawaii, Manoa.*
Smartt, Dr. Raymond. *National Solar Observatory, Sunspot, New Mexico.*
Smith, Prof. Harlan J. *Deceased; formerly, Department of Astronomy, University of Texas, Austin.*
Spergel, Dr. David N. *Princeton University Observatory.*
Stolper, Dr. Edward M. *Division of Geological and Planetary Science, California Institute of Technology.*
Strand, Dr. Kaj Aa. *Retired; formerly, Director, U.S. Naval Observatory.*
Strom, Dr. Robert G. *Department of Planetary Sciences, Lunar and Planetary Laboratory, University of Arizona.*
Suess, Dr. Hans E. *Department of Chemistry, University of California, San Diego.*

T

Tedesco, Dr. Edward F. *Jet Propulsion Laboratory, California Institute of Technology.*
Terrell, Dr. James. *Earth and Space Science Division, Los Alamos National Laboratory, Los Alamos, New Mexico.*
Thompson, Dr. Laird A. *Department of Astronomy, College of Liberal Arts and Sciences, University of Illinois.*
Trimble, Dr. Virginia. *Department of Physics, University of California, Irvine.*
Tully, Dr. R. Brent. *Institute for Astronomy, University of Hawaii.*
Turner, Dr. Edwin L. *Peyton Hall, Princeton University Observatory.*
Turner, Joyce B. *Peyton Hall, Princeton University Observatory.*
Turok, Dr. Neil. *Department of Physics, Joseph Henry Laboratories, Princeton University.*

U

Unwin, Dr. Stephen C. *Owens Valley Radio Observatory, California Institute of Technology.*

V

Veverka, Prof. Joseph. *Laboratory for Planetary Studies, Cornell University.*

W

Warner, Deborah Jean. *Curator, History of Astronomy, National Museum of American History, Washington, D.C.*
Weinberg, Dr. J. L. *Space Astronomy Laboratory, University of Florida.*
Westfall, Dr. Richard S. *Department of History and Philosophy of Science, University of Indiana.*
Wing, Dr. Robert F. *Department of Astronomy, Ohio State University.*
Wolff, Dr. Sidney C. *Institute for Astronomy, University of Hawaii, Manoa.*
Wood, Dr. Frank Bradshaw. *Department of Physics and Astronomy, University of Florida.*

Wright, Prof. Edward L. *Department of Astronomy, University of California, Los Angeles.*
Wyant, Prof. James C. *Optical Sciences Center, University of Arizona.*

Y

Yeomans, Dr. Donald K. *Jet Propulsion Laboratory, California Institute of Technology.*
Yu, Dr. Ching-Sung. *Deceased; formerly, Professor Emeritus of Astronomy, Hood College, Frederick, Maryland.*

Z

Zeilik, Dr. Michael. *AstroNet, Santa Fe, New Mexico.*
Zinner, Dr. Ernst. *McDonnell Center for the Space Sciences, Washington University, St. Louis, Missouri.*
Zirin, Dr. Harold. *Department of Physics, California Institute of Technology.*

Index

Index

Asterisks indicate page references to article titles.

A

B

D

E

F

Q

R

T

U

V

W

X

Z